Conversion Factors

Length

1 m = 39.37 in. = 3.281 ft
1 in. = 2.54 cm
1 km = 0.621 mi
1 mi = 5280 ft = 1.609 km
1 lightyear = 9.461×10^{15} m
1 angstrom (Å) = 10^{-10} m

Mass

1 kg = 10^3 g = 6.85×10^{-2} slug
1 slug = 14.59 kg
1 u = 1.66×10^{-27} kg

Time

1 min = 60 s
1 h = 3600 s
1 day = 8.64×10^4 s
1 year = 365.242 days = 3.156×10^7 s

Volume

1 liter = 1000 cm^3 = 3.531×10^{-2} ft^3
1 ft^3 = 2.832×10^{-2} m^3
1 gallon = 3.786 liter = 231 in.^{3}

Angle

$180° = \pi$ rad
1 rad = 57.30°
1° = 60 min = 1.745×10^{-2} rad

Speed

1 km/h = 0.278 m/s = 0.621 mi/h
1 m/s = 2.237 mi/h = 3.281 ft/s
1 mi/h = 1.61 km/h = 0.447 m/s = 1.47 ft/s

Force

1 N = 0.2248 lb = 10^5 dynes
1 lb = 4.448 N
1 dyne = 10^{-5} N = 2.248×10^{-6} lb

Work and energy

1 J = 10^7 erg = 0.738 ft·lb = 0.239 cal
1 cal = 4.186 J
1 ft·lb = 1.356 J
1 Btu = 1.054 10^3 J = 252 cal
1 J = 6.24×10^{18} eV
1 eV = 1.602×10^{-19} J
1 kWh = 3.60×10^6 J

Pressure

1 atm = 1.013×10^5 N/m^2 (or Pa) = 14.70 lb/in.^{2}
1 Pa = 1 N/m^2 = 1.45×10^{-4} lb/in.^{2}
1 lb/in.^{2} = 6.895×10^3 N/m^2

Power

1 hp = 550 ft·lb/s = 0.746 kW
1 W = 1 J/s = 0.738 ft·lb/s
1 Btu/h = 0.293 W

FOURTH EDITION

COLLEGE PHYSICS

Raymond A. Serway

James Madison University

Jerry S. Faughn

Eastern Kentucky University

SAUNDERS GOLDEN SUNBURST SERIES

SAUNDERS COLLEGE PUBLISHING

Harcourt Brace College Publishers

Fort Worth Philadelphia San Diego New York Orlando Austin San Antonio
Toronto Montreal London Sydney Tokyo

Text Typeface: New Baskerville
Compositor: Progressive Information Technologies
Publisher: John Vondeling
Development Editor: Laura Maier
Senior Project Editor: Sally Kusch
Copy Editors: Mary Patton, Judy Patton
Managing Editor: Carol Field
Manager of Art and Design: Carol Bleistine
Art and Design Coordinator: Sue Kinney
Text Designer: Tracy Baldwin
Cover Designer: Lawrence R. Didona
Text Artwork: Rolin Graphics
Layout Artists: Dorothy Chattin, Rebecca Lemna
Photo Researcher: Sue Howard
Director of EDP: Tim Frelick
Production Manager: Charlene Squibb

Cover: Aerial view of windsurfer riding on the crest of a wave. This dramatic photograph illustrates many physical principles that are described in the text. For example, the water wave carries energy and momentum as it travels from one location to another. The surfer and the surfboard move in a complex path under the action of several forces, including gravity, wind resistance, and the force of water on the surfboard. *(Darrell Wong/Tony Stone Images, Inc.)*

Printed in the United States of America

COLLEGE PHYSICS, 4th edition

ISBN 0-03-003562-7

Library of Congress Catalog Card Number: 94-067168

67890123   032   10 987654

PREFACE

T his textbook, now in its fourth edition, is appropriate for a one-year course in introductory physics commonly taken by students majoring in biology, the health professions, and other disciplines including environmental, earth, and social sciences, and technical fields such as architecture. The mathematical techniques used in the book include algebra, geometry, and trigonometry, but not calculus.

CHANGES IN THE FOURTH EDITION

A number of changes and improvements have been made in preparing the fourth edition of this textbook. Many of these changes are in response to comments and suggestions offered by users of the third edition and reviewers of the manuscript.

Major changes and new features in the fourth edition are as follows:

1. The fourth edition is accompanied by computer disks containing a set of about 100 physics simulations for the highly acclaimed program Interactive Physics II from Knowledge Revolution. Most of these simulations are keyed to selected worked-out example problems in the text and to selected end-of-chapter homework problems. The remainder are demonstrations that complement concepts or applications discussed in the text. Simulations can be used by the instructor in the classroom or laboratory to help students understand physics concepts by developing better visualization skills. Experiments are easily created by first drawing objects on the screen, defining parameters such as mass, charge, elasticity, and spring constants. The simulation is started by simply clicking the RUN button. The simulation engine calculates the motion of the defined system and displays it in smooth animation. The results can be measured and analyzed in graphical, digital, tabular, and bar graph formats. The acquired data can also be exported to a spreadsheet of your choice for other types of analyses. The Interactive Physics

   icon identifies the examples, problems, and figures for which a simulation exists. A complete list follows.

   List of Interactive Physics Simulations

CHAPTER 2	Problem 2.47	Example 3.6
Example 2.5	Problem 2.56	Example 3.8
Example 2.6		Problem 3.15
Problem 2.36	CHAPTER 3	Problem 3.19
Problem 2.42	Example 3.3	Problem 3.22
Problem 2.43	Example 3.4	Problem 3.24

2. A number of sections in the book have been deleted, shortened, or combined with other sections to produce a textbook that is somewhat shorter and with a more balanced presentation than the third edition. The fourth edition contains 30 chapters with 14 of them containing material normally considered the core of the first semester and 16 that are customarily covered in the second half of the course. The purpose of this reorganization was to make it easier to cover all the significant topics in both classical and quantum physics in a one-year course while still retaining most of the optional sections so that a professor can tailor the course to his or her own interests and needs.

3. A large number of student users of this textbook are majoring in biology or pre-professional programs such as pre-medicine, pre-dentistry, and so forth. For this reason, several reviewers and users of the textbook have suggested that we specifically point out those topics in the book that are of special interest to them.

   A DNA icon 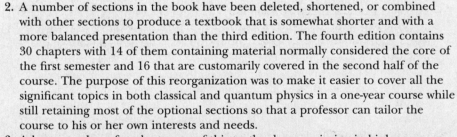 is used to identify the applications of physics to biology and the health professions.

4. Some material, including relevant problems, has been moved out of the main body of the text to "For Further Study" mini-chapters at the end of the book. The derivation of the expression for gravitational potential energy valid for large distances above the surface of the Earth and applications of this concept to escape speed and astrophysics is an example of one topic that has been moved. Likewise, discussions of surface tension, the surface of liquids, capillary action, viscosity, Poiseuille's law, transport phenomena, and motion through a viscous medium have been moved from the chapter on fluids to the end material. Finally, a discus-

sion of Gauss' law in electrostatics with applications has been moved. The authors feel that moving these optional topics improves the continuity of the flow of the material in the text while keeping these important subjects available for those who want to discuss and use them in their classes.

5. Many sections have been rewritten to improve the clarity of the presentations and the precision of language. We hope that the result is a book that is both accurate and appealing to read.

6. One feature of the previous editions that has received much critical acclaim from users and reviewers is the application of physics to the real world. Virtually every topic in the book is related to real-life situations through optional sections, examples within the body of the text, or through special **Physics in Action** focus boxes. A common question heard in many physics classes is, "Why do I have to know this stuff? I don't need it in my major or in the things I do in my life." The authors hope that these optional features will help to answer this question and demonstrate that physics is a science that students can always apply to the world around them.

7. Other features of previous editions that have received much praise are the worked-out example problems in the text and the end-of-chapter problems and questions. All of these have been enhanced in this edition. A substantial revision of the problem sets was made in an effort to provide a greater variety and to reduce repetition. Also, about 25% more problems at the intermediate level have been added. In addition, the conceptual questions at the end of the chapter have been revised and increased by about 25%. The conceptual questions section features three to six worked-out examples, followed by a good selection of questions for the student to work through.

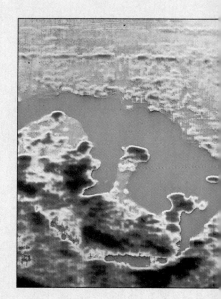

8. Much of the line art and many of the color photographs have been replaced or modified to improve the clarity of presentation and to improve the visual appeal of the text. However, color in this textbook is used primarily for pedagogical purposes. A chart explaining the pedagogical use of color is included on the inside front cover for easy reference.

9. The guest essays have been revised and updated. Three new essays have been added on the topics of scaling, how the body uses energy, and recent developments in fusion research.

10. Significant figures in both worked examples and end-of-chapter problems have been handled with care. Most numerical examples and problems are worked out to either two or three significant figures, depending on the accuracy of the data provided.

OBJECTIVES

The main objectives of this introductory physics textbook are twofold: to provide the student with a clear and logical presentation of the basic concepts and principles of physics, and to strengthen an understanding of the concepts and principles through a broad range of interesting applications to the real world. In order to meet these objectives, emphasis is placed on sound physical arguments and discussions of everyday experiences. At the same time, we have attempted to motivate the student through practical examples that demonstrate the role of physics in other disciplines.

COVERAGE

This book is concerned with standard topics in classical physics and 20th century physics. The book is divided into six parts: Part I (Chapters 1–9) deals with the fundamentals of Newtonian mechanics and the physics of fluids; Part II (Chapters 10–12) is concerned with heat and thermodynamics; Part III (Chapters 13–14) covers wave motion and sound; Part IV (Chapters 15–21) is concerned with electricity and magnetism; Part V (Chapters 22–25) treats the properties of light and the field of geometric and wave optics; Part VI (Chapters 26–30) represents an introduction to special relativity, quantum physics, atomic, and nuclear physics.

FEATURES

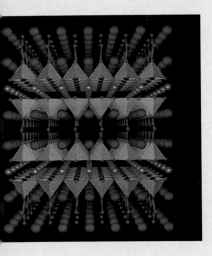

Most instructors would agree that the textbook assigned in a course should be the student's major guide for understanding and learning the subject matter. With this in mind, we have included many pedagogic features in the textbook that should enhance its usefulness to both the student and instructor. These are as follows:

Style

We have attempted to write the book in a style that is clear, relaxed, and pleasing to the reader. New terms are carefully defined, and we have tried to avoid jargon. At the same time, we have attempted to keep the presentation accurate and precise.

Organization

The book is divided into six parts: mechanics, thermodynamics, vibrations and wave motion, electricity and magnetism, light and optics, and modern physics. Each part includes an overview of the subject matter to be covered in that part and some historical perspectives.

Introductory Chapter

The introductory chapter, which "sets the stage" for the text, discusses the building blocks of matter, the units of physical quantities, order-of-magnitude calculations, dimensional analysis, significant figures, and mathematical notation.

Units

The international system of units (SI) is used throughout the book. The British engineering system of units is used only to a limited extent in the problem sets of the early chapters on mechanics.

Previews

Most chapters begin with a chapter preview, which includes a brief discussion of the chapter objectives and content.

Marginal Notes and Equations

Important equations are highlighted with a light gold screen, and marginal notes are often used to describe their meaning. Marginal notes are also used to locate specific definitions and important statements in the text.

Biographical Sketches

Throughout the text short biographies of important scientists are included to add historical and human perspective.

Problem-solving Strategies

General strategies and suggestions are included for solving the types of problems featured in both the worked examples and end-of-chapter problems. This feature should help students identify the essential steps in solving problems and increase their skills as problem solvers. This feature is highlighted by a light tan screen for emphasis, which helps the student locate the strategies quickly.

Physics in Action

This boxed material focuses on photographs of interesting demonstrations and phenomena in physics, accompanied by detailed explanations. The material can also serve as a source of information for initiating classroom discussions.

Worked Examples

A large number of worked examples, including many new ones, are presented as an aid in understanding concepts. In many cases, these examples serve as models for solving end-of-chapter problems. The examples are set off from the text for ease of location, and most examples are given titles to describe their content. Many examples include a Reasoning section to illustrate the underlying concepts and methodology used in arriving at a correct solution. This will help students understand the logic behind the solution and the advantage of using a particular approach to solve the problem. The solution answer is highlighted with a light blue screen.

Special Topics

Many chapters include special topic sections which are intended to expose the student to various practical and interesting applications of physical principles. Those topics dealing with applications of physics to the life sciences are identified by the DNA icon.

Guest Essays

In this fourth edition, we have included 14 essays, 12 written by guest authors. Three of these essays are new, while the remaining 11 are revisions of material that appeared in the third edition. Seven of the essays examine areas of physics research; the remaining essays are entitled "Biological Perspective" and explore life science applications of physics. The essays discuss the concepts of scaling, the Big Bang, arch structures, the circulatory system, how the human body consumes energy, alternative sources of energy, superconductivity, voltage measurements in medicine, the nervous system, general relativity, vacuum tunneling spectroscopy, applications of lasers, medical applications of radiation, and recent advances in fusion research. Although the essays are intended as supplemental reading for the student, most include questions and problems that could be assigned when these topics are covered in class. The essays have been placed at the ends of the chapters to avoid interrupting the main textual material.

Important Statements

Many important statements and definitions are highlighted with a tan screen for added emphasis and ease of review.

Illustrations and Photographs

The readability and effectiveness of the text material and worked examples are enhanced by the large number of figures, diagrams, photographs, and tables. Full color is used to add clarity to the artwork and to provide realistic renditions of physical situations wherever possible. Vectors are color-coded to avoid confusion between different vector quantities, and curves in xy plots are drawn in color. Three-dimensional effects are produced using color air-brushed areas, where appropriate, and an object in motion is often drawn with "ghost" areas representing previous locations of the object. Approximately 300 color photographs have been carefully selected, and their accompanying captions have been written to serve as an added instructional and learn-

ing tool. Vector overlays on many photographs enable students to visualize physics at work in real situations. A complete description of the pedagogical use of color appears on the inside front cover.

Topics for Further Study

In an effort to improve the continuity in the flow of material, and reduce the overall length of the main text, a few topics have been moved from the main body of the text to special mini-chapters preceding the Appendices. This material, which is cross-referenced in the text, includes the following topics: the discussion of gravitational potential energy for large distances above the Earth's surface and escape speed (a supplement to Chapter 7); topics from fluids, including surface tension, the surface of liquids, capillary action, viscosity, Poiseuille's law, transport phenomena, and motion through a viscous medium (a supplement to Chapter 9); and Gauss' law in electrostatics (a supplement to Chapter 15).

Summaries

Each chapter contains a summary that reviews the important concepts and equations discussed in that chapter.

Conceptual Questions

A list of conceptual questions is provided at the end of each chapter. The questions, which require verbal responses, provide the student with a means of self-testing the concepts presented in the chapter. Many of these are intended to test the students' ability to perform order-of-magnitude calculations and to apply the concepts they have learned to real-life situations without performing detailed calculations. The section of conceptual questions begins with a sampling of three to six worked examples, which provide reasoning and answer statements for the student.

Problems

An extensive set of problems is included at the end of each chapter. In this fourth edition, many new problems have been included in the text, with particular emphasis placed on more intermediate-level problems. Answers to odd-numbered problems are given at the end of the book; these pages have colored edges for ease of location. Problems that have complete solutions available in the Student Solutions Manual are indicated by a boxed problem number for the benefit of both student and instructor. For the convenience of assigning problems, most are keyed to specific sections of the chapter. The remaining problems, labeled ''Additional Problems,'' are not keyed to specific sections. Problem numbers are color-coded to indicate one of three levels of difficulty. Straightforward problems are numbered in black, intermediate-level problems are numbered in blue, and a small number of highly challenging problems are in magenta. In our opinion, assigned problems should consist mainly of those from the first two categories to help build self-confidence. Those problems that are accompanied by Interactive Physics II computer simulations are labeled with the Interactive Physics icon, and those with a focus on the life sciences are identified by the DNA icon.

Appendices

Several appendices are provided at the end of the text. Most of the appendix material represents a review of mathematical techniques used in the text, such as scientific notation, algebra, geometry, and trigonometry. Reference to these appendices is made as needed throughout the text. Most of the mathematical review sections include worked examples and exercises with answers. Some appendices contain useful tables that supplement textual information. For easy reference, the front endpapers contain a chart explaining the pedagogical use of color throughout the text and a list of often used conversion factors. The back endpapers include a table of physical constants and other useful data.

ANCILLARIES

STUDENT ANCILLARIES

New Options to Help Students Learn Physics   Saunders College Publishing is now offering several items to supplement and enhance the classroom experience. These ancillaries will allow instructors to customize the textbook to their student's needs and their own style of instruction. One or more of these ancillaries may be shrink-wrapped with the text at a reduced price:

Pocket Guide to Accompany College Physics by V. Gordon Lind, Utah State University   This $5'' \times 7''$ handbook provides formulas and helpful hints at a glance and gives students a convenient reference booklet—an invaluable companion when it comes to reviewing concepts and solving problems just before exams.

Student Solutions Manual and Study Guide by John R. Gordon, James Madison University and Raymond A. Serway   The manual features detailed solutions to selected end-of-chapter problems from the text, a multiple-choice chapter quiz, a skills section that reviews mathematical concepts, and suggested approaches to problem-solving methodology. Each chapter also includes a review checklist, important notes from key sections of the text, and a list of important equations and concepts.

Interactive Physics Simulations by Raymond Serway and Knowledge Revolution   Approximately 150 simulations are available on computer disk in either Macintosh or IBM format to be used in conjunction with the program Interactive Physics II from Knowledge Revolution. About 100 of these simulations are visual representations of selected worked examples and end-of-chapter problems from the text. The remaining simulations are demonstrations that complement concepts or applications discussed in the text. A student version of the Interactive Physics II program will be available at a reduced cost.

Discovery Exercises for Interactive Physics by Jon Staib, James Madison University   This is a workbook that was developed to be used in conjunction with the *Interactive Physics Simulations* described above. The workbook features directed student exercises to accompany the simulations provided on disk that will be packaged with the workbook. The workbook can be used as a tutorial for reviewing physical concepts or as the basis for a number of computer-generated laboratory experiences.

Physics Laboratory Manual by David Loyd   Supplements the learning of basic physical principles while introducing laboratory procedures and equipment. Each chapter of the manual includes a pre-laboratory assignment, objectives, an equipment list, the theory behind the experiment, experimental procedures, graphs, and questions. A laboratory report is provided for each experiment so the student can record data, calculations, and experimental results.

So You Want to Take Physics A Preparatory Course by Rodney Cole, University of California—Davis: Introductory level coverage of numerous physical principles ideal for strengthening essential physics-mathematical skills.

INSTRUCTOR ANCILLARIES

Instructor's Manual with Complete Solutions by Jerry Faughn and Charles Teague This manual consists of complete solutions to *all* the problems in the text, as well as answers to the even-numbered problems. This manual, which was prepared using the word processing software Microsoft Word, is also available on disk in Macintosh format.

Transparency Masters for Selected Solutions Transparency masters provide full solutions in bold print for all the selected solutions available in the Student Solutions Manual Study Guide.

Printed Test Bank by Steve Van Wyk The printed test bank contains approximately 2300 problems and questions (both open-ended and multiple choice), generated from the Computerized Test Bank. The format allows instructors to duplicate pages for distribution to students.

Computerized Test Bank Available for the IBM and Macintosh computers, the test bank contains approximately 2300 open-ended and multiple choice problems and questions, representing every chapter of the text. The test bank allows instructors to customize tests by rearranging, editing, or adding questions. The questions are graded in level of difficulty for the instructor's convenience; the program places the answers to all questions on a separate grading key.

Overhead Transparency Acetates The collection of transparencies consists of approximately 200 full-color figures and photographs from the text to enhance lectures; they feature large print for easy viewing in the classroom.

Interactive Physics Demonstrations by Raymond A. Serway Many physical situations discussed in the textbook involving the motion of objects are brought to life with computer simulations developed using the highly acclaimed Interactive Physics II program. Most simulations are keyed to specific figures and/or examples in the text. Others represent material that is difficult to visualize or present in a textbook format, such as the tracking of a specific point on a rotating object. Instructors who use computer simulations as a teaching tool could use the templates as part of classroom presentations.

Practice Problems with Solutions About 400 Level 1 problems that do not appear in the text are available with full solutions in a printed version and on Macintosh disk. These can be used for homework assignments, student practice and drill exercises, or included on examinations.

Instructor's Manual for Physics Laboratory Manual by David Loyd Each chapter contains a discussion of the experiment, teaching hints, answers to selected questions from the student laboratory manual, and a post-laboratory quiz with short answers and essay questions. The author has also included a list of the suppliers of scientific equipment and a summary of the equipment needed for all the experiments in the manual.

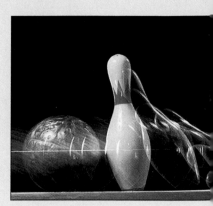

Physics Demonstration Videotapes   A unique collection of 70 physics demonstrations is provided on videotape to supplement classroom presentations and to help motivate students.

Saunders Multimedia Presentation Package   Includes the unique Saunders Physics Videodisc, version 2, with all the physics demonstrations and more than 1200 still images from Saunders physics texts, including *College Physics* 4/e. Unique Lecture-Active presentation software and Directory of Images enables the instructor to customize a presentation swiftly and easily using the multimedia package. A barcode manual is also available.

    Videodisc also operates without the software by simply using a remote control or barcode scanner.

Videotapes from NOVA and Infinite Voyage   Several physics-related videotapes are available to adopters in accordance with Saunders' video policy. Instructional materials accompany each tape to provide support for classroom use.

Infinite Voyage Videodisc   The video images from the series are also available in a videodisc format.

TEACHING OPTIONS

This book contains more than enough material for a one-year course in introductory physics. This serves two purposes. First, it gives the instructor more flexibility in choosing topics for a specific course. Second, the book becomes more useful as a resource for students. On the average, it should be possible to cover about one chapter each week for a class that meets three hours per week. Many sections containing interesting applications are considered optional and are therefore marked with an asterisk (*). Those optional sections dealing with applications of physics to the life sciences are identified with the DNA icon . Instructors are encouraged to cover those optional sections which best match their students' interests. We would also like to offer the following suggestions for shorter courses or for those instructors who choose to move at a slower pace through the year:

    Option A: If you wish to place more emphasis on contemporary topics in physics, you should consider omitting all or parts of Chapter 8 (Rotational Equilibrium and Rotational Dynamics), Chapter 21 (Alternating Current Circuits and Electromagnetic Waves), and Chapter 25 (Optical Instruments).

    Option B: If you wish to place more emphasis on classical physics, you could omit all or parts of Part VI of the textbook, which deals with special relativity and other topics in 20th century physics.

ACKNOWLEDGMENTS

In preparing the fourth edition of this textbook, we have been guided by the expertise of many people who reviewed part or all of the manuscript. We wish to acknowledge the following scholars and express our sincere appreciation for their helpful suggestions, criticisms, and encouragement:

Subhash Antani, Edgewood College
Charles R. Bacon, Ferris State University
Dilip Balamore, Nassau Community College
Michael Bretz, University of Michigan, Ann Arbor

Michael E. Browne, University of Idaho
Ronald W. Canterna, University of Wyoming
Clinton M. Case, Western Nevada Community College
Lawrence B. Coleman, University of California, Davis
Robert J. Endorf, University of Cincinnati
Lothar Frommhold, University of Texas at Austin
Simon George, California State University, Long Beach
George W. Greenlees, University of Minnesota
Grant W. Hart, Brigham Young University
Robert C. Hudson, Roanoke College
Bo Lou, Ferris State University
Steven McCauley, California State Polytechnic University, Pomona
John Morack, University of Alaska, Fairbanks
Lawrence S. Pinsky, University of Houston
Michael Ram, State University of New York at Buffalo

We wish to thank the following scholars for their comments and suggestions during prior revisions of the text:

Albert Altman, University of Lowell
John Anderson, University of Pittsburgh
Neil W. Ashcroft, Cornell University
Gordon Aubrecht, Ohio State University
Charles R. Bacon, Ferris State University
Ralph Barnett, Florissant Valley Community College
Louis Barrett, Western Washington University
Paul Bender, Washington State University
Jeffrey Braun, University of Evansville
John Brennan, University of Central Florida
Joseph Catanzarite, Cypress College
Roger W. Clapp, University of South Florida
Giuseppe Colaccico, University of South Florida
Jorge Cossio, Miami-Dade Community College
Terry T. Crow, Mississippi State College
Paul Feldker, Florissant Valley Community College
Albert Thomas Fromhold, Jr., Auburn University
Teymoor Gedayloo, California Polytechnic State University
John R. Gordon, James Madison University
Wlodzimierz Guryn, Brookhaven National Laboratory
James Harmon, Oklahoma State University
Robert Hudson, Roanoke College
Fred Inman, Mankato State University
Ronald E. Jodoin, Rochester Institute of Technology
Drasko Jovanovic, Fermilab
Frank Kolp, Trenton State College
Joan P. S. Kowalski, George Mason University
Sol Krasner, University of Chicago
Karl F. Kuhn, Eastern Kentucky University
David Lamp, Texas Tech University
Harvey S. Leff, California State Polytechnic University
Joel Levine, Orange Coast College
Michael Lieber, University of Arkansas
James Linblad, Saddleback Community College

Bill Lochslet, Pennsylvania State University
Jeffrey V. Mallow, Loyola University of Chicago
David Markowitz, University of Connecticut
Joe McCauley, Jr., University of Houston
Ralph V. McGrew, Broome Community College
Bill F. Melton, University of North Carolina at Charlotte
H. Kent Moore, James Madison University
Carl R. Nave, Georgia State University
Blaine Norum, University of Virginia
M. E. Oakes, University of Texas at Austin
Lewis J. Oakland, University of Minnesota
Lewis O'Kelly, Memphis State University
T. A. K. Pillai, University of Wisconsin, La Crosse
William D. Ploughe, Ohio State University
Brooke M. Pridmore, Clayton State College
Joseph Priest, Miami University
James Purcell, Georgia State University
Kurt Reibel, Ohio State University
Virginia Roundy, California State University, Fullerton
William R. Savage, The University of Iowa
Reinhard A. Schumacher, Carnegie Mellon University
Donald D. Snyder, Indiana University at South Bend
Carey E. Stronach, Virginia State University
Thomas W. Taylor, Cleveland State University
L. L. Van Zandt, Purdue University
Howard G. Voss, Arizona State University
Larry Weaver, Kansas State University
Donald H. White, Western Oregon State College
George A. Williams, The University of Utah
Jerry H. Wilson, Metropolitan State College
Robert M. Wood, University of Georgia
Clyde A. Zaidins, University of Colorado at Denver
Peter D. Zimmerman, Louisiana State University

We thank the following individuals for writing the interesting essays that appear throughout the text: Isaac D. Abella, University of Chicago; Gordon Batson, Clarkson University; William G. Buckman, Western Kentucky University; Paul Davidovitz, Boston College; Ronald C. Davidson, Princeton University; Roger A. Freedman, University of California, Santa Barbara; David Griffing, Miami University; Paul K. Hansma, University of California, Santa Barbara; Laurent Hodges, Iowa State University; David Markowitz, University of Connecticut; Philip Morrison, Massachusetts Institute of Technology; Clifford M. Will, Washington University; and Sidney C. Wolff, National Optical Astronomy Observatory.

We are especially grateful to John R. Gordon for his fine work on the Student Solutions Manual and Study Guide, to Joan Fu at Los Angeles Harbor College for her suggestions on improvements to the Study Guide, and to Linda Miller for typing the Study Guide. Our gratitude to Charles Teague for checking solutions to all the problems and for his assistance in preparing the Instructor's Manual, to Michael Carchidi for coordinating and contributing to the new problems, and to Karen Andsager for accuracy-checking solutions to all new problems. Robert Hudson and Michael Ram were instrumental in the reviewing process, checking the entire text for accuracy and helping to ensure the consistent use of significant figures throughout the text. Special thanks to Sue Howard for locating many excellent photographs and to Henry Leap and Jim Lehman for providing numerous photographs of physics demonstrations. Our gratitude to Irene Nunes for skillfully editing the material in the most heavily revised chapters, and to Mary Patton and Judy Patton for their excellent copyediting. We wish to thank the professional staff at Saunders College Publishing for their expertise and perseverance in the project. In particular we thank Laura Maier, Sally Kusch, Carol Bleistine, Charlene Squibb, Tim Frelick, and Marjorie Waldron. Our special thanks to our publisher John Vondeling for his expert advice in this revision.

Finally, we dedicate this book to our wives and children for their love, support, and long-term sacrifices.

Raymond A. Serway
James Madison University
Harrisonburg, Virginia

Jerry S. Faughn
Eastern Kentucky University
Richmond, Kentucky

July 1994

TO THE STUDENT

We feel it is appropriate to offer some words of advice that should be of benefit to you, the student. Before doing so, we shall assume that you have read the preface, which describes the various features of the text that will help you through the course.

HOW TO STUDY

Very often we are asked "How should I study physics and prepare for examinations?" There is no simple answer to this question, but we would like to offer some suggestions based on our own experiences in learning and teaching over the years.

First and foremost, maintain a positive attitude towards the subject matter, keeping in mind that physics is the most fundamental of all natural sciences. Other science courses that follow will use the same physical principles, so it is important that you understand and be able to apply the various concepts and theories discussed in the text.

CONCEPTS AND PRINCIPLES

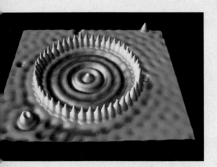

It is essential that you understand the basic concepts and principles *before* attempting to solve assigned problems. This is best accomplished through a careful reading of the textbook before attending your lecture on that material. In the process, it is useful to jot down certain points that are not clear to you. Take careful notes in class, and then ask questions pertaining to those ideas that require clarification. Keep in mind that few people are able to absorb the full meaning of scientific material after one reading. Several readings of the text and notes may be necessary. Your lectures and laboratory work should supplement the text and clarify some of the more difficult material. You should reduce memorization of material to a minimum. Memorizing passages from a text, equations, and derivations does not necessarily mean you understand the material. Your understanding of the material will be enhanced through a combination of efficient study habits, discussions with other students and instructors, and your ability to solve the problems in the text. Ask questions whenever you feel it is necessary. If you are reluctant to ask questions in class, seek private consultation or initiate discussions with your classmates. Many individuals are able to speed up the learning process when the subject is discussed on a one-to-one basis.

STUDY SCHEDULE

It is important to set up a regular study schedule, preferably on a daily basis. Be sure to read the syllabus for the course and adhere to the schedule set by your instructor. The lectures will be much more meaningful if you have read the corresponding textual material *before* attending the lecture. As a general rule, you should devote about two hours of study time for every hour in class. If you are having trouble with the course, seek the advice of the instructor or students who have already taken the course. You may find it necessary to seek further instruction from experienced students. Very often, instructors offer review sessions in addition to regular class periods. It is important that you avoid the practice of delaying study until a day or two before an exam. More often than not, this will lead to disastrous results. Rather than an all-night study session before an exam, it is better to briefly review the basic concepts and equations, followed by a good night's rest. The Pocket Guide that accompanies the text provides formulas and helpful hints and can be a useful review tool before exams. If you feel you need additional help in understanding the concepts, preparing for exams, or in problem solving, we suggest that you acquire a copy of the student study guide that accompanies the text; it should be available at your college bookstore.

USE THE FEATURES

You should make *full* use of the various features of the text discussed in the preface. For example, marginal notes are useful for locating and describing important equations, while important statements and definitions are highlighted with a tan screen. Many useful tables are contained in the appendices, but most are incorporated into the text where they are used most often. Appendix A is a convenient review of mathematical techniques. Answers to odd-numbered problems are given at the end of the text, and answers to many end-of-chapter conceptual questions are provided in the study guide. Exercises (with answers), which follow some worked examples, represent extensions of those examples, and in most cases you are expected to perform a simple calculation. Their purpose is to test your problem-solving skills as you read through the text. Problem-solving strategies are included in selected chapters throughout the text to give you additional information to help you solve the problems. An overview of the entire text is given in the table of contents, while the index will enable you to locate specific material quickly. Footnotes are sometimes used to supplement the discussion or to cite other references on the subject. A list of suggested additional readings is given at the end of each chapter.

After reading a chapter, you should be able to define any new quantities introduced in that chapter and to discuss the principles and assumptions that were used to arrive at certain key relations. The chapter summaries and the review sections of the study guide should help you in this regard. In some cases, it will be necessary to refer to the index of the text to locate certain topics. You should be able to correctly associate with each physical quantity a symbol used to represent that quantity and the unit in which the quantity is specified. Furthermore, you should be able to express each important relation in a concise and accurate prose statement.

PROBLEM SOLVING

R. P. Feynman, Nobel laureate in physics, once said, "You do not know anything until you have practiced." In keeping with this statement, we strongly advise that you develop the skills necessary to solve a wide range of problems. Your ability to solve problems will be one of the main tests of your knowledge of physics, and therefore you should try to solve as many problems as possible. It is essential that you understand basic concepts and principles before attempting to solve problems. It is good practice to try to find alternative solutions to the same problem. For example, problems in mechanics can be solved using Newton's laws, but very often an alternative method using energy considerations is more direct. You should not deceive yourself into thinking you understand the problem after seeing its solution in class. You must be able to solve the problem and similar problems on your own.

The method of solving problems should be carefully planned. A systematic plan is especially important when a problem involves several concepts. First, read the problem several times until you are confident you understand what is being asked. Look for any key words that will help you interpret the problem, and perhaps allow you to make certain assumptions. Your ability to interpret the question properly is an integral part of problem solving. You should acquire the habit of writing down the information given in a problem, and decide what quantities need to be found. You might

want to construct a table listing quantities given, and quantities to be found. This procedures is sometimes used in the worked examples of the text. After you have decided on the method you feel is appropriate for the situation, proceed with your solution. General problem-solving strategies of this type are included in the text and are highlighted by a light tan screen.

We often find that students fail to recognize the limitations of certain formulas or physical laws in a particular situation. It is very important that you understand and remember the assumptions underlying a particular theory or formalism. For example, we shall find that certain equations in kinematics apply only to a object moving with constant acceleration. They are not valid for situations in which the acceleration is not constant, as in the cases of the motion of an object connected to a spring and the motion of an object through a fluid.

EXPERIMENTS

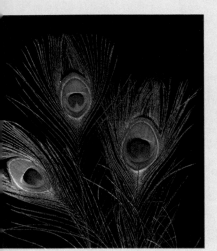

Physics is a science based upon experimental observations. In view of this fact, we recommend that you make every effort to supplement the text through various types of "hands-on" experiments, either in the laboratory or at home. Many simple experiments can be used to test ideas and models discussed in class or in the text. For example, an object swinging on the end of a long string together with a wristwatch can be used to investigate pendulum motion; an object attached to the end of a vertical spring or a rubber band can be used to determine the nature of restoring forces and to investigate periodic motion; collisions between equal masses can be observed while playing billiards; an approximate value for the acceleration due to gravity can be obtained by dropping an object from a known height and simply measuring the time of its fall with a stopwatch; traveling waves can be investigated with the aid of a stretched rope or the common "Slinky" toy (a stretched spring); a pair of Polaroid sunglasses and some discarded lenses or a magnifying glass can be used to perform various experiments in optics. The list is endless. When physical models are not available, be imaginative and try to develop models of your own.

We strongly encourage you to make use of the computer disk that is available to accompany this book. It contains a set of simulations keyed to selected worked-out examples in the text and to certain end-of-chapter problems. It is far easier to understand physics if you see it in action, and these simulations will enable you to be a part of that action.

CLOSING COMMENTS

Someone once said that there are only two professions in which people truly enjoy what they are doing: professional sports and physics. Although this statement is most likely an exaggeration, both professions are truly exciting and stretch your skills to the limit. It is our sincere hope that you too will find physics exciting and that you will benefit from this experience, regardless of your chosen profession.

Welcome to the exciting world of physics.

> *To see a World in a Grain of Sand*
> *And a Heaven in a Wild Flower,*
> *Hold infinity in the palm of your hand*
> *And Eternity in an hour.*
>
> W. Blake, "Auguries of Innocence"

CONTENTS OVERVIEW

CONTENTS

xix

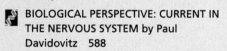

PART

5   LIGHT AND OPTICS   699

Mechanics

Physics, the most fundamental science, is concerned with the basic principles of the Universe. It is one of the foundations upon which the other physical sciences—astronomy, chemistry, and geology—are based. The beauty of physics lies in the simplicity of its fundamental theories and in the way just a small number of basic concepts, equations, and assumptions can alter and expand our view of the world around us.

The myriad physical phenomena in our world are parts of one or more of the following five areas of physics:

1. Mechanics, which is concerned with the effects of forces on material objects
2. Thermodynamics, which deals with heat, temperature, and the behavior of large numbers of particles
3. Electromagnetism, which deals with charges, currents, and electromagnetic fields
4. Relativity, a theory that describes particles moving at any speed, and connects space and time
5. Quantum mechanics, a theory dealing with the behavior of particles at the submicroscopic level as well as the macroscopic world

The first part of this textbook addresses mechanics, sometimes referred to as classical mechanics or Newtonian mechanics. This is an appropriate subject with which to begin an introductory text because many of the basic principles of mechanical systems can later be used to describe such natural phenomena as waves and heat transfer. Furthermore, the laws of conservation of energy and momentum to be introduced in our study of mechanics retain their importance in the fundamental theories that follow, including the theories of modern physics.

The first serious attempts to develop a theory of motion were made by the Greek astronomers and philosophers. Although they devised a complex model to describe the motions of heavenly bodies, their model lacked correlation between such motions and the motions of objects on Earth. The study of mechanics was enhanced by a number of careful astronomical investigations by Copernicus, Brahe, and Kepler in the 16th century. In that and the following century, Galileo attempted to relate the motions of falling bodies and projectiles to the motions of planetary bodies, and Sevin and Hooke studied forces and their relationship to motion. A major development in the theory of mechanics was provided by Newton in 1687 when he published his *Principia*. Newton's elegant theory, which remained unchallenged for more

(C.O. Rentmeester/The Image Bank)

than 200 years, was based on his hypothesis of universal gravitation together with contributions made by Galileo and others.

Today, mechanics is of vital importance to students from all disciplines. It is highly successful in describing the motions of material bodies, such as planets, rockets, and baseballs. In the first nine chapters of the text, we shall describe the laws of mechanics and examine a wide range of phenomena that can be understood through these laws.

1

INTRODUCTION

This painting by the Englishman Joseph Wright (1734–1797) is entitled "A Philosopher Lecturing on the Orrery." An orrery is a mechanical model of the solar system. Through his paintings, Wright explored the impact of new scientific and industrial developments on our view of the world. (The Bridgeman Art Library)

The goal of physics is to provide an understanding of nature by developing theories based on experiments. The theories are usually expressed in mathematical form. Fortunately, it is possible to explain the behavior of a variety of physical systems with a limited number of fundamental laws.

Scientists continually work at improving our comprehension of fundamental laws, and new discoveries are made every day. In many research areas, a great deal of overlap occurs among physics, chemistry, and biology. The numerous recent technological advances are results of the efforts of many scientists, engineers, and technicians. Some of the most notable of these advances are unmanned space missions and manned Moon landings, microcircuitry and high-speed computers, and sophisticated imaging techniques used in scientific research and medicine. The impacts of such developments on our society have indeed been great, and it is very likely that future discoveries will be exciting, challenging, and of great benefit to humanity.

Since following chapters will be concerned with the laws of physics, we must begin by clearly defining the basic quantities involved in these laws. For example, such physical quantities as force, velocity, volume, and acceleration can be described in terms of more fundamental quantities. In the next several chapters we shall encounter three basic quantities: **length** (L), **mass** (M), and **time** (T). In later chapters we will need to add two other standard units to our list, for temperature (the kelvin) and for electric current (the ampere). In our study of mechanics, however, we shall be concerned only with the units of length, mass, and time.

1.1
STANDARDS OF LENGTH, MASS, AND TIME

If we are to report the result of a measurement of a certain quantity to someone who wishes to reproduce this measurement, a unit for the quantity must be defined. For example, if someone familiar with our system of measurement and weights reports that a wall is 2.0 meters high and our fundamental unit of length is defined to be 1.0 meter, we know that the height of the wall is twice the fundamental unit of length. Likewise, if we are told that a person has a mass of 75 kilograms and our fundamental unit of mass is defined as 1.0 kilogram, then that person has a mass 75 times as great as the fundamental unit of mass. In 1960, an international committee agreed on a system of standards and designations for these fundamental quantities, called the **SI system** (Système International) of units. Its units of length, mass, and time are the meter, kilogram, and second, respectively.

LENGTH

In 1799, the legal standard of length in France became the meter, defined as one ten-millionth of the distance from the equator to the North Pole. As recently as 1960, the official length of the meter was the distance between two lines on a specific bar of platinum-iridium alloy stored under controlled conditions. This standard was abandoned for several reasons, a principal one being that the limited accuracy with which the separation between the lines can be determined does not meet the current requirements of science and technology. Then the meter was defined as 1 650 763.73 wavelengths of orange-red light emitted from a krypton-86 lamp. In October 1983, this definition too was abandoned and the meter was redefined as *the distance traveled by light in vacuum during an interval of 1/299 792 458 second.* This latest definition arose because the speed of light is now defined as a universal constant having a value of 299 792 458 meters per second.

MASS

The SI unit of mass, the **kilogram,** is defined as *the mass of a specific platinum-iridium alloy cylinder kept at the International Bureau of Weights and Measures at Sèvres, France.* As we shall see in Chapter 4, mass is a quantity used to measure the resistance to a change in state of motion of an object. It is more difficult to cause a change in the state of motion of an object with a large mass than an object with a small mass.

TIME

Before 1960, the time standard was defined in terms of the average length of a solar day in the year 1900. (A solar day is the time interval between successive appearances of the Sun at the highest point it reaches in the sky each day.) The

(Top) The National Standard Kilogram No. 20, an accurate copy of the International Standard Kilogram kept at Sèvres, France, is housed under a double bell jar in a vault at the National Institute of Standards and Technology. *(Bottom)* The primary frequency standard (an atomic clock) at the National Institute of Standards and Technology. This device keeps time with an accuracy of about one millionth of a second per year. *(Courtesy of National Institute of Standards and Technology, U.S. Dept. of Commerce)*

Timepieces are commonly used to measure time intervals in sporting events and in laboratory and industrial environments. *(Don Allen Sparks, The Image Bank)*

TABLE 1.1
Approximate Values of Some Measured Lengths

	Length (m)
Distance from Earth to most remote known quasar	1×10^{26}
Distance from Earth to most remote known normal galaxies	4×10^{25}
Distance from Earth to nearest large galaxy (M31 in Andromeda)	2×10^{22}
Distance from Earth to nearest star (Proxima Centauri)	4×10^{16}
One lightyear	9×10^{15}
Mean orbit radius of the Earth about the Sun	2×10^{11}
Mean distance from Earth to Moon	4×10^{8}
Mean radius of the Earth	6×10^{6}
Typical altitude of a satellite orbiting Earth	2×10^{5}
Length of a football field	9×10^{1}
Length of a housefly	5×10^{-3}
Size of smallest dust particles	1×10^{-4}
Size of cells of most living organisms	1×10^{-5}
Diameter of a hydrogen atom	1×10^{-10}
Diameter of an atomic nucleus	1×10^{-14}
Diameter of a proton	1×10^{-15}

basic unit of time, the second, was defined to be $(1/60)(1/60)(1/24) = 1/86\ 400$ of the average solar day. In 1967, the second was redefined to take advantage of the high precision attainable with a device known as an atomic clock, which uses the characteristic frequency of the light emitted from the cesium-133 atom as its "reference clock." The **second** is now defined as *9 192 631 700 times the reciprocal of this frequency.*

APPROXIMATE VALUES FOR LENGTH, MASS, AND TIME INTERVALS

Approximate values of some lengths, masses, and time intervals are presented in Tables 1.1, 1.2, and 1.3, respectively. Note the wide ranges of values. Study these tables and get a feel for what is meant by a kilogram of mass (this book has a mass of about 2 kilograms) or a time interval of 10^{10} seconds (one year is about 3×10^{7} seconds). Study Appendix A if you need to learn or review the notation for powers of 10, such as the expression of the number 50 000 in the form 5×10^{4}.

Systems of units commonly used are the *SI system,* in which the units of length, mass, and time are the meter (m), kilogram (kg), and second (s), respectively; the *cgs* or *gaussian system,* in which the units of length, mass, and time are the centimeter (cm), gram (g), and second, respectively; and the *British engineering system* (sometimes called the conventional system), in which the units of length, mass, and time are the foot (ft), slug, and second, respectively. Throughout most of this book we shall use SI units, since they are almost universally accepted in science and industry. We shall make limited use of British engineering units in the study of mechanics.

Some of the most frequently used metric prefixes representing powers of 10 are listed in Table 1.4 along with their abbreviations. For example, 10^{-3} m is equivalent to 1 millimeter (mm), and 10^{3} m is 1 kilometer (km). Likewise, 1 kg is equal to 10^{3} g, and 1 megavolt (MV) is equivalent to 10^{6} volts (V).

TABLE 1.2
Approximate Values of Some Masses

	Mass (kg)
Universe	10^{52}
Milky Way galaxy	7×10^{41}
Sun	2×10^{30}
Earth	6×10^{24}
Moon	7×10^{22}
Shark	1×10^{2}
Human	7×10^{1}
Frog	1×10^{-1}
Mosquito	1×10^{-5}
Bacterium	1×10^{-15}
Hydrogen atom	2×10^{-27}
Electron	9.1×10^{-31}

TABLE 1.3
Approximate Values of Some Time Intervals

	Time Interval (s)
Age of the Universe	5×10^{17}
Age of the Earth	1×10^{17}
Average age of a college student	6×10^{8}
One year	3×10^{7}
One day (time required for one revolution of Earth about its axis)	9×10^{4}
Time between normal heartbeats	8×10^{-1}
Period[a] of audible sound waves	1×10^{-3}
Period of typical radio waves	1×10^{-6}
Period of vibration of an atom in a solid	1×10^{-13}
Period of visible light waves	2×10^{-15}
Duration of a nuclear collision	1×10^{-22}
Time required for light to cross a proton	3×10^{-24}

[a] A *period* is defined as the time required for one complete vibration.

A triple beam balance is used to make accurate mass measurements. This particular balanced system contains 215 g of potassium ferrocyanide. *(Michael Dalton, Fundamental Photographs)*

TABLE 1.4
Some Prefixes for Powers of Ten Used with Metric Units

Power	Prefix	Abbreviation
10^{-18}	atto-	a
10^{-15}	femto-	f
10^{-12}	pico-	p
10^{-9}	nano-	n
10^{-6}	micro-	μ
10^{-3}	milli-	m
10^{-2}	centi-	c
10^{-1}	deci-	d
10^{1}	deka-	da
10^{3}	kilo-	k
10^{6}	mega-	M
10^{9}	giga-	G
10^{12}	tera-	T
10^{15}	peta-	P
10^{18}	exa-	E

1.2

THE BUILDING BLOCKS OF MATTER

A 1-kg cube of solid gold has a length of about 3.73 cm on a side. Is this cube nothing but wall-to-wall gold, with no empty space? If the cube is cut in half, the two resulting pieces still retain their chemical identity as solid gold. But what if the pieces of cube are cut again and again, indefinitely? Will the smaller and smaller pieces always be the same substance, gold? Questions such as these can be traced back to early Greek philosophers. Two of them—Leucippus and Democritus—could not accept the idea that such cutting could go on forever. They speculated that the process ultimately must end when it produces a particle

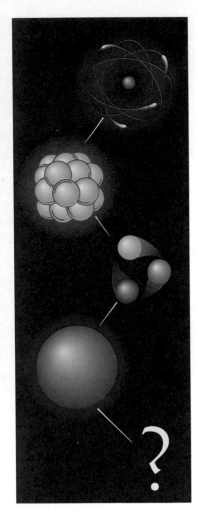

FIGURE 1.1
Distances at the frontier of nuclear physics are astonishingly short. An atom is so small that a single-file line of 250 000 of them would fit within the thickness of aluminum foil. The nucleus at the atom's center is a cluster of nucleons, each 100 000 times smaller than the atom itself. The three quarks inside each nucleon are smaller still. *(Courtesy of SURA, Inc.)*

that can no longer be cut. In Greek, *atomos* means "not sliceable." From this comes our English word *atom* for the smallest, ultimate particle of matter. Elementary-particle physicists still engage in speculation and experimentation concerning the ultimate building blocks of matter.

Let us review briefly what is known about the ultimate structure of the world around us. It is useful to view the atom as a miniature Solar System with a dense, positively charged nucleus occupying the position of the Sun and negatively charged electrons orbiting like the planets. This model of the atom enables us to understand certain properties of the simpler atoms, such as hydrogen, but fails to explain many fine details of atomic structure.

Following the discovery of the nucleus in the early 1900s, the question arose: Does it have structure? That is, is the nucleus a single particle or a collection of particles? The exact composition of the nucleus has not been defined completely even today, but by the early 1930s a model evolved that helps us understand how the nucleus behaves. Specifically, scientists determined that occupying the nucleus are two basic entities, protons and neutrons. The *proton* is nature's fundamental carrier of positive charge, and the number of protons in a nucleus determines what element the material is. For instance, one proton in the nucleus means that the atom is an atom of hydrogen, regardless of how many neutrons may be present; two protons mean an atom of helium.

The existence of *neutrons* was verified conclusively in 1932. A neutron has no charge and a mass about equal to that of a proton. One of its primary purposes is to act as a "glue" to hold the nucleus together. If neutrons were not present, the repulsive force between the positively charged particles would cause the nucleus to fly apart.

But is this where the breaking down stops? As we shall explore more carefully in Chapter 30, even more elementary building blocks than protons and neutrons exist. Protons, neutrons, and a zoo of other exotic particles are now thought to be composed of six particles called **quarks,** which have been given the names *up, down, strange, charmed, bottom,* and *top* (Fig. 1.1). The up, charmed, and top quarks have charges of $+2/3$ that of the proton, whereas the down, strange, and bottom quarks have charges of $-1/3$ that of the proton. The proton consists of two up quarks and one down quark, which you can easily show leads to the correct charge for the proton. Likewise, the neutron is composed of two down quarks and one up quark, giving a net charge of zero.

1.3

DIMENSIONAL ANALYSIS

The word *dimension* has a special meaning in physics. It denotes the qualitative nature of a physical quantity. Whether the separation between two points is measured in units of feet or meters or furlongs, the measured quantity is a distance. We say that the dimension of distance is *length*.

The symbols we will use to specify length, mass, and time are L, M, and T, respectively. We shall often use brackets [ ] to denote the dimensions of a physical quantity. For example, in this notation the dimensions of velocity, v, are written $[v] = $ L/T, and the dimensions of area, A, are $[A] = $ L^2. The dimensions of area, volume, velocity, and acceleration are listed in Table 1.5, along with their units in the three common systems. The dimensions of other quantities, such as force and energy, will be described later as they are introduced.

TABLE 1.5
Dimensions and Some Units of Area, Volume, Velocity, and Acceleration

System	Area (L^2)	Volume (L^3)	Velocity (L/T)	Acceleration (L/T^2)
SI	m^2	m^3	m/s	m/s^2
cgs	cm^2	cm^3	cm/s	cm/s^2
British engineering (conventional)	ft^2	ft^3	ft/s	ft/s^2

In many situations, you may be faced with having to derive or check a specific formula. Although you may have forgotten the details of the derivation, you can use a powerful procedure called **dimensional analysis.** Dimensional analysis makes use of the fact that *dimensions can be treated as algebraic quantities:* Quantities can be added or subtracted only if they have the same dimensions, and the quantities on the two sides of an equation must have the same dimensions. By following these simple rules, you can determine whether or not you have the correct form of an expression. The relationship can be correct only if the dimensions on the two sides of the equation are the same.

To illustrate this procedure, suppose you wish to derive a formula for the distance x traveled by a car in a time t if the car starts from rest and moves with constant acceleration a. In Chapter 2 we shall find that the correct expression for this special case is $x = \frac{1}{2}at^2$. Let us check the validity of this expression with a dimensional analysis approach.

The quantity x on the left side has the dimension length. In order for the equation to be dimensionally correct, the quantity on the right side must also have the dimension length. We can perform a dimensional check by substituting the basic units for acceleration, L/T^2, and time, T, into the equation. That is, the dimensional form of the equation $x = \frac{1}{2}at^2$ can be written

$$L = \frac{L}{T^2} \cdot T^2 = L$$

The units of time cancel as shown, leaving the unit of length. Note that the factor $\frac{1}{2}$ was ignored because it has no dimensions. Dimensional analysis cannot give us numerical factors; this is its characteristic limitation.

EXAMPLE 1.1 Analysis of an Equation

Show that the expression $v = v_0 + at$ is dimensionally correct, where v and v_0 represent velocities, a is acceleration, and t is a time interval.

Solution Since

$$[v] = [v_0] = \frac{L}{T}$$

and the dimensions of acceleration are L/T^2, the dimensions of at are

$$[at] = \frac{L}{T^2}(T) = \frac{L}{T}$$

and the expression is dimensionally correct. On the other hand, if the expression were given as $v = v_0 + at^2$, it would be dimensionally *incorrect.* Try it and see!

A device called a micrometer is being used to measure a flange. A typical micrometer can measure lengths to a precision of about 0.01 mm. (*Jeffrey Coolidge, Stockphotos, Inc.*)

1.4

SIGNIFICANT FIGURES

When measurements are performed on certain quantities, the measured values are known only to within the limits of the experimental uncertainty. The value of the uncertainty can depend on factors such as the quality of the apparatus, the skill of the experimenter, and the number of measurements performed.

Suppose that in a laboratory experiment we are asked to measure the area of a rectangular plate with a meter stick. Let us assume that the accuracy to which we can measure a particular dimension of the plate is ± 0.1 cm. If the length of the plate is measured to be 16.3 cm, we can claim only that its length lies somewhere between 16.2 cm and 16.4 cm. In this case, we say that the measured value has three significant figures. Likewise, if the plate's width is measured to be 4.5 cm, the actual value lies between 4.4 cm and 4.6 cm. (This measured value has only two significant figures.) Note that the significant figures include the first estimated digit. Thus, we could write the measured values as 16.3 ± 0.1 cm and 4.5 ± 0.1 cm.

Suppose now that we would like to find the area of the plate by multiplying the two measured values together. If we were to claim that the area is (16.3 cm)(4.5 cm) = 73.35 cm^2, our answer would be unjustifiable since it contains four significant figures, which is greater than the number of significant figures in either of the measured lengths. A good rule of thumb for determining the number of significant figures that can be claimed is as follows. **When several quantities are multiplied, the number of significant figures in the final answer is the same as the number of significant figures in the *least* accurate of the quantities being multiplied, where "least accurate" means "having the lowest number of significant figures." The same rule applies to division.**

Applying this rule to the preceding multiplication example, we see that the answer for the area can have only two significant figures because the dimension 4.5 cm has only two significant figures. Thus, we can only claim the area to be 73 cm^2, realizing that the value can range between (16.2 cm)(4.4 cm) = 71 cm^2 and (16.4 cm)(4.6 cm) = 75 cm^2.

The presence of zeros in an answer may be misinterpreted. For example, suppose the mass of an object is measured to be 1500 g. This value is ambiguous because it is not known whether the last two zeros are being used to locate the decimal point or represent significant figures in the measurement. In order to remove this ambiguity, it is common to use scientific notation to indicate the number of significant figures. In this case, we would express the mass in scientific notation as 1.5×10^3 g if there were two significant figures in the measured value or as 1.50×10^3 g if there were three significant figures. Likewise, a number such as 0.000 15 should be expressed in scientific notation as 1.5×10^{-4} if it has two significant figures or as 1.50×10^{-4} if it has three significant figures. The three zeros between the decimal point and the digit 1 in the number 0.000 15 are not counted as significant figures, because they are present only to locate the decimal point. In general, a **significant figure** is a reliably known digit, other than a zero used to locate the decimal point.

Decimal places must be considered for addition and subtraction. **The number of decimal places in the sum should match the smallest number of decimal places of all the terms.** For example, if we compute 123 + 5.35, the answer is 128

Could this be the result of poor data analysis? Perhaps the improper handling of significant figures caused this disaster. What do you think? *(Roger-Viollet.* © *ND-Viollet)*

and not 128.35. In this book, *most of the numerical examples and end-of-chapter problems will yield answers having either two or three significant figures.*

EXAMPLE 1.2   Installing a Carpet

A carpet is to be installed in a room whose length is measured to be 12.71 m (four significant figures) and whose width is measured to be 3.46 m (three significant figures). Find the area of the room.

Solution   If you multiply 12.71 m by 3.46 m on your calculator, you will get the answer 43.9766 m². How many of these numbers should you claim? Our rule of thumb for multiplication says that you can claim only the number of significant figures in the least accurate of the quantities being measured. In this example, we have three significant figures in our less accurate measurement, so we should express our final answer as 44.0 m². Note that in arriving at this answer, we used a general rule for rounding off numbers which states that the last digit retained is to be increased by 1 if the first digit dropped was 5 or greater.

1.5
CONVERSION OF UNITS

Sometimes it is necessary to convert units from one system to another. Conversion factors between the SI and conventional systems for units of length are as follows:

This road sign near Raleigh, North Carolina, shows distances in miles and kilometers. How accurate are the conversions? *(Billy E. Barnes/Stock Boston)*

$$1 \text{ mile} = 1609 \text{ m} = 1.609 \text{ km} \qquad 1 \text{ ft} = 0.3048 \text{ m} = 30.48 \text{ cm}$$

$$1 \text{ m} = 39.37 \text{ in.} = 3.281 \text{ ft} \qquad 1 \text{ in.} = 0.0254 \text{ m} = 2.54 \text{ cm}$$

A more extensive list of conversion factors can be found on the inside of the front cover.

Units can be treated as algebraic quantities that can cancel or multiply each other. For example, suppose we wish to convert 15.0 in. to centimeters. Since 1 in. = 2.54 cm, we find that

$$15.0 \text{ in.} = 15.0 \cancel{\text{in.}} \times 2.54 \frac{\text{cm}}{\cancel{\text{in.}}} = 38.1 \text{ cm}$$

EXAMPLE 1.3   Pull Over, Buddy!

If a car is traveling at a speed of 28.0 m/s, is it exceeding the speed limit of 55.0 mi/h?

Solution   We can perform the conversion from meters per second to miles per hour in two steps—meters to miles and then seconds to hours. Converting the length units, we find that

$$28.0 \text{ m/s} = \left(28.0 \frac{\cancel{\text{m}}}{\text{s}}\right)\left(\frac{1 \text{ mi}}{1609 \cancel{\text{m}}}\right) = 1.74 \times 10^{-2} \text{ mi/s}$$

Now we finish the conversion by converting seconds to hours, as follows:

$$1.74 \times 10^{-2} \text{ mi/s} = \left(1.74 \times 10^{-2} \frac{\text{mi}}{\cancel{\text{s}}}\right)\left(60 \frac{\cancel{\text{s}}}{\cancel{\text{min}}}\right)\left(60 \frac{\cancel{\text{min}}}{\text{h}}\right) = 62.6 \text{ mi/h}$$

The car should slow down because it is exceeding the speed limit.

An alternative approach is to use the single conversion relationship 1 m/s = 2.237 mi/h:

$$28.0 \text{ m/s} = \left(28.0 \frac{\cancel{\text{m}}}{\cancel{\text{s}}}\right)\left(\frac{2.237 \text{ mi/h}}{1 \cancel{\text{m/s}}}\right) = 62.6 \text{ mi/h}$$

Exercise   Use the speedometer pictured in Figure 1.2 to see whether the foregoing result is in agreement with the speedometer dial (calibrated in miles per hour and kilometers per hour).

FIGURE 1.2

(Example 1.3)   This modern car speedometer gives speed readings in both miles per hour and kilometers per hour. See if you can confirm the conversions for a few readings on the dial.
(Paul Silverman, Fundamental Photographs)

1.6

ORDER-OF-MAGNITUDE CALCULATIONS

Often it is useful to estimate an answer to a problem in which little information is given. This answer can then be used to determine whether a more precise calculation is necessary. Such approximations are usually based on certain assumptions, which must be modified if greater precision is needed. Sometimes it is necessary to know a quantity only within a factor of 10. In such a case we refer to the **order of magnitude** of the quantity, by which we mean the power of ten that is closest to the actual value of the quantity. For example, the mass of a person might be 75 kg $\approx 10^2$ kg. We would say that the person's mass is *on the order of* 10^2 kg. Usually, when an order-of-magnitude calculation is made, the results are reliable to within a factor of 10. If a quantity increases in value by three orders of magnitude, this means that its value increases by a factor of $10^3 = 1000$.

EXAMPLE 1.4 How Much Gasoline Do We Use?

Estimate the number of gallons of gasoline used annually by all cars in the United States.

Solution Since there are about 200 million people in the United States, an estimate of the number of cars in the country is 40 million (assuming one car and five people per family). We shall also estimate that the average distance traveled per car per year is 10 000 miles. If we assume a gasoline consumption of 20 mi/gal, each car uses about 500 gal/year. Multiplying this by the total number of cars in the United States gives an estimated total consumption of 2×10^{10} gal. This corresponds to a yearly consumer expenditure of more than $20 billion! This is probably a low estimate since we haven't accounted for commercial consumption and for such factors as two-car families.

EXAMPLE 1.5 The Number of Atoms in a Solid

Estimate the number of atoms in 1 cm³ of a solid.

Solution From Table 1.1 we note that the diameter of an atom is about 10^{-10} m. If in our model we assume that the atoms in the solid are solid spheres of this diameter, then the volume of each sphere is about 10^{-30} m³ (more precisely, volume = $4\pi r^3/3 = \pi d^3/6$, where $r = d/2$). Therefore, since 1 cm³ = 10^{-6} m³, the number of atoms in the solid is on the order of $10^{-6}/10^{-30} = 10^{24}$ atoms.

A more precise calculation would require knowledge of the density of the solid and the mass of each atom. However, our estimate agrees with the more precise calculation to within a factor of 10.

1.7

MATHEMATICAL NOTATION

Many mathematical symbols will be used throughout this book. You are no doubt familiar with some, such as the symbol = to denote the equality of two quantities.

The symbol $\propto$ denotes a proportionality. For example, $y \propto x^2$ means that y is proportional to the square of x.

The symbol $<$ means *is less than,* and $>$ means *is greater than.* For example, $x > y$ means x is greater than y.

The symbol $\ll$ means *is much less than,* and $\gg$ means *is much greater than.*

The symbol $\approx$ indicates that two quantities are *approximately equal* to each other.

The symbol $\equiv$ means *is defined as.* This is a stronger statement than a simple $=$.

It is convenient to use the notation Δx (read as "delta x") to indicate the *change in the quantity* x. (Note that Δx does not mean "the product of Δ and x.") For example, suppose that a person out for a morning stroll starts measuring her distance away from home when she is 10 m from her doorway. The person then moves along a straight-line path and stops strolling 50 m from the door. Her change in position during the walk is $\Delta x = 50\ \text{m} - 10\ \text{m} = 40\ \text{m}$ or, in symbolic form,

$$\Delta x = x_f - x_i$$

In this equation x_f is the *final position* and x_i is the *initial position.*

We shall often have occasion to add several quantities. A useful abbreviation for representing such a sum is the Greek letter Σ (capital sigma). Suppose we wish to add a set of five numbers represented by x_1, x_2, x_3, x_4, and x_5. In the abbreviated notation, we would write the sum as

$$x_1 + x_2 + x_3 + x_4 + x_5 = \sum_{i=1}^{5} x_i$$

where the subscript i on x represents any one of the numbers in the set. For example, if there are five masses in a system, m_1, m_2, m_3, m_4, and m_5, the total mass of the system $M = m_1 + m_2 + m_3 + m_4 + m_5$ could be expressed as

$$M = \sum_{i=1}^{5} m_i$$

Finally, the magnitude of a quantity x, written $|x|$, is simply the absolute value of that quantity. The sign of $|x|$ is always positive, regardless of the sign of x. For example, if $x = -5$, $|x| = 5$; if $x = 8$, $|x| = 8$.

1.8

COORDINATE SYSTEMS AND FRAMES OF REFERENCE

Many aspects of physics deal in some way with locations in space. For example, the mathematical description of the motion of an object requires a method for describing the position of the object. Thus, it is fitting that we first discuss how to describe the position of a point in space. This is done by means of coordinates. A point on a line can be located with one coordinate; a point in a plane is located with two coordinates; and three coordinates are required to locate a point in space.

A coordinate system used to specify locations in space consists of

1. A fixed reference point O, called the origin

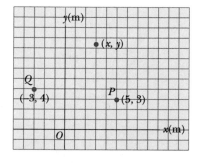

FIGURE 1.3

Designation of points in a two-dimensional cartesian coordinate system. Every point is labeled with coordinates (x, y).

2. A set of specified axes or directions with an appropriate scale and labels on the axes
3. Instructions that tell us how to label a point in space relative to the origin and axes

One convenient coordinate system that we shall use frequently is the *cartesian coordinate system,* sometimes called the rectangular coordinate system. Such a system is illustrated in Figure 1.3. The point P in the figure has coordinates (5, 3). This notation means that if we start at the origin (O), we can reach P by moving 5 units along the positive x axis and then 3 units parallel to the positive y axis. (Positive x is usually selected to the right of the origin and positive y upward from the origin. Negative x is usually to the left of the origin and negative y downward from the origin. This convention is not an absolute necessity, however. There may be instances in which you would like to take positive x to the left of the origin and negative x to the right. Feel free to do so.) The point Q has coordinates $(-3, 4)$, corresponding to 3 units in the negative x direction and 4 units in the positive y direction.

The cartesian coordinate system will be used far more often than any other in this book. However, sometimes it is more convenient to locate a point in space by its plane polar coordinates (r, θ), as shown in Figure 1.4. When this system is used, an origin and a reference line are selected, as shown. A point is then specified by the distance r from the origin to the point and by the angle θ between r and the reference line. (Frequently the reference line is selected to be the positive x axis of a cartesian coordinate system, although this is not necessary.) The angle θ is considered positive when measured counterclockwise from the reference line and negative when measured clockwise. For example, if a point is specified by the polar coordinates 3 m and 60°, one locates this point by moving out 3 m from the origin at an angle of 60° above (or counterclockwise from) the reference line. A point specified by polar coordinates 3 m and $-60°$ is located 3 m out from the origin and 60° below (or clockwise from) the reference line.

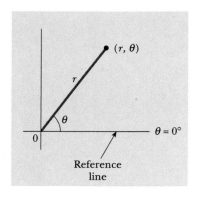

FIGURE 1.4
A polar coordinate system.

1.9

TRIGONOMETRY

The portion of mathematics that is based on the special properties of a right triangle is called trigonometry. Many of the concepts of this branch of mathematics are of utmost importance in the study of physics. We will now review some of the more inherent fundamental definitions and ideas that you will need to know.

Consider the right triangle shown in Figure 1.5, where side a is opposite the angle θ, side b is adjacent to the angle θ, and side c is the hypotenuse of the triangle. The basic trigonometry functions defined by such a triangle are the ratios of the lengths of certain sides of the triangle. Specifically, the most important relationships are called the sine (sin), cosine (cos), and tangent (tan) functions. In terms of the angle θ, these functions are[1]

[1] In order to recall these trigonometric relationships, consider using the following memory device. Recall the word *SOHCAHTOA* whose letters stand for *S*ine = *O*pposite/*H*ypotenuse, *C*osine = *A*djacent/*H*ypotenuse, and *T*angent = *O*pposite/*A*djacent. (Thanks go to Professor Don Chodrow for pointing this out.)

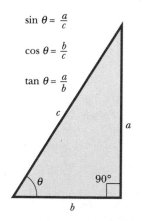

$$\sin \theta = \frac{a}{c}$$

$$\cos \theta = \frac{b}{c}$$

$$\tan \theta = \frac{a}{b}$$

FIGURE 1.5
Certain trigonometric functions for a right triangle.

$$\sin \theta = \frac{\text{side opposite } \theta}{\text{hypotenuse}} = \frac{a}{c}$$

$$\cos \theta = \frac{\text{side adjacent to } \theta}{\text{hypotenuse}} = \frac{b}{c}$$ **[1.1]**

$$\tan \theta = \frac{\text{side opposite } \theta}{\text{side adjacent to } \theta} = \frac{a}{b}$$

For example, if the angle θ is equal to 30°, it is found that the ratio of a to c is always 0.50; that is, sin 30° = 0.50. Note that the sin, cos, and tan are quantities without units, since each represents the ratio of two lengths.

Another important relationship, called the **Pythagorean theorem,** exists between the lengths of the sides of a right triangle; it is

$$c^2 = a^2 + b^2$$ **[1.2]**

Finally, it will often be necessary to find the value of **inverse** relationships. For example, suppose you know the value of the sine of an angle, but you need to know the value of the angle itself. The inverse sine function may be expressed as $\sin^{-1}(0.866)$, which is a shorthand way of asking the question "What angle has a sine of 0.866?" Punching a couple of buttons on your calculator reveals that this angle is 60°. Try it for yourself and show that $\tan^{-1}(0.400) = 21.8°$.

The following examples will give you a little practice in working with these trigonometric principles in practical situations.

EXAMPLE 1.6   How High Is the Building?

A person attempts to measure the height of a building by walking out a distance of 46.0 m from its base and shining a flashlight beam toward its top. He finds that when the beam is elevated at an angle of 39.0° with respect to the horizontal, as shown in Figure 1.6, the beam just strikes the top of the building. Find the height of the building and the distance the flashlight beam has to travel before it strikes the top of the building.

Solution We know the length of the adjacent side of the red triangle shown in the figure, and we know the value of the angle. Thus, we can use the defini-

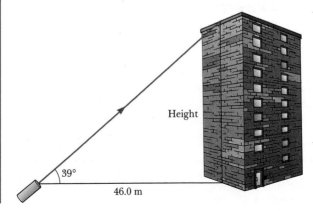

FIGURE 1.6
(Example 1.6)

Height

39°

46.0 m

tion of the tangent function to find the height of the building, which is the opposite side of the triangle.

$$\tan 39.0° = \frac{\text{height of building}}{46.0 \text{ m}}$$

or

Height of building $= (\tan 39.0°)(46.0 \text{ m}) = (0.810)(46.0 \text{ m}) = \boxed{37.3 \text{ m}}$

Now that we know the lengths of both the adjacent side and the opposite side of the triangle, we can find the length of the hypotenuse, which is the distance, c, that the beam travels before it strikes the top of the building.

$$c = \sqrt{a^2 + b^2} = \sqrt{(37.3 \text{ m})^2 + (46.0 \text{ m})^2} = \boxed{59.2 \text{ m}}$$

Exercise Try using a different method to find the distance traveled by the light. Would another trigonometric function, such as the sin or cos, give you the answer? Try it.

EXAMPLE 1.7   Things To Do When the CB Goes Bad

A truck driver moves up a straight mountain highway, as shown in Figure 1.7. Elevation markers at the beginning and ending points of the trip show that he has risen vertically 0.530 km, and the mileage indicator on the truck shows that he has traveled a total distance of 3.00 km during the ascent. Find the angle of incline of the hill.

Solution This is an example in which we must use an inverse trigonometric relationship. From the sine of the angle, defined as the opposite side over the hypotenuse, we find

$$\sin \theta = \frac{0.530}{3.00} = 0.177$$

To find θ, we use the inverse sine relationship:

$$\theta = \sin^{-1}(0.177) = \boxed{10.2°}$$

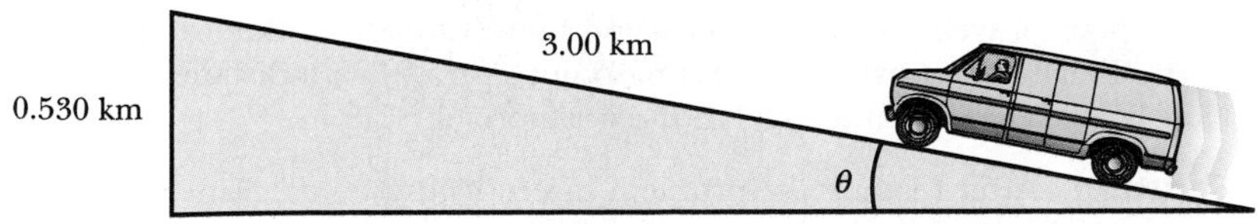

FIGURE 1.7
(Example 1.7)

A GUIDE FOR
PROBLEM SOLVING

| Read Problem |

↓

| Draw Diagram |

↓

| Identify Data |

↓

| Choose Equation(s) |

↓

| Solve Equation(s) |

↓

| Evaluate and Check
Answer |

1.10

PROBLEM-SOLVING STRATEGY

Most courses in general physics require the student to learn the skills of problem solving, and examinations are composed largely of problems that test such skills. This brief section presents some ideas that will enable you to increase your accuracy in solving problems, enhance your understanding of physical concepts, eliminate initial panic or lack of direction in approaching a problem, and organize your work. One help is to adopt a problem-solving strategy. Many chapters in this text will include a section labeled ''Problem-Solving Strategies'' that should help you through the rough spots.

Six basic steps are commonly used to develop a problem-solving strategy:

1. Read the problem carefully at least twice. Be sure you understand the nature of the problem before proceeding further.
2. Draw a suitable diagram with appropriate labels and coordinate axes, if needed.
3. As you examine what is being asked in the problem, identify the basic physical principle (or principles) that are involved, listing the knowns and unknowns.
4. Select a basic relationship or derive an equation that can be used to find the unknown, and symbolically solve the equation for the unknown.
5. Substitute the given values with the appropriate units into the equation.
6. Obtain a numerical value for the unknown. The problem is verified if the following questions can be properly answered: Do the units match? Is the answer reasonable? Is the plus or minus sign proper or meaningful?

One of the purposes of this strategy is to promote accuracy. Properly drawn diagrams can eliminate many sign errors. Diagrams also help to isolate the physical principles of the problem. Symbolic solutions and carefully labeled knowns and unknowns will help eliminate other careless errors. The use of symbolic solutions should help you think in terms of the physics of the problem. A check of units at the end of the problem can indicate a possible algebraic error. The physical layout and organization of your problem will make the final product more understandable and easier to follow. Once you have developed an organized system for examining problems and extracting relevant information, you will become a more confident problem solver.

The following example illustrates this procedure.

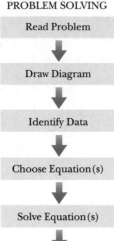

a = 450 km
c = 525 km
b = ?

FIGURE 1.8
(Example 1.8)

EXAMPLE 1.8 A Round Trip by Air

An airplane travels 450 km due east and then travels an unknown distance due north. Finally, it returns to its starting point by traveling a distance of 525 km. How far did the airplane travel in the northerly direction?

Solution After reading the problem carefully, draw a diagram of the situation, as in Figure 1.8. Since the path of the airplane forms a right triangle, you can use the Pythagorean theorem, which relates the three sides of the triangle:

$$c^2 = a^2 + b^2$$

In this problem, the hypotenuse c and the side a are known quantities. We must solve for the unknown quantity, b. Subtracting a^2 from both sides of the preceding equation gives

$$b^2 = c^2 - a^2$$

We now take the square root of both sides to find the solution for b:

$$b = \sqrt{c^2 - a^2}$$

Note that the negative solution has been disregarded because it is not physically meaningful.

Finally, we substitute the given values for the quantities c and a to find a numerical value for b:

$$b = \sqrt{(525 \text{ km})^2 - (450 \text{ km})^2} = \boxed{270 \text{ km}}$$

SUMMARY

The physical quantities we shall encounter in our study of mechanics can be expressed in terms of three fundamental quantities, length, mass, and time—which have the units meters (m), kilograms (kg), and seconds (s), respectively, in the SI system. It is often helpful to use dimensional analysis to check equations and to assist in deriving equations.

When inserting numerical factors into equations, you must be sure that the dimensions of these factors are consistent throughout the equation. In many cases it will be necessary to use the table of conversion factors on the inside front cover to convert from one system of units to another.

Often it is useful to estimate an answer to a problem in which little information is given. Such estimates are called **order-of-magnitude calculations.**

The three most basic trigonometric functions for a right triangle are the sine, cosine, and tangent, defined as

$$\sin \theta = \frac{\text{side opposite } \theta}{\text{hypotenuse}}$$

$$\cos \theta = \frac{\text{side adjacent to } \theta}{\text{hypotenuse}}$$

$$\tan \theta = \frac{\text{side opposite } \theta}{\text{side adjacent to } \theta}$$

The **Pythagorean theorem** is an important relationship between the lengths of the sides of a right triangle:

$$c^2 = a^2 + b^2$$

where c is the hypotenuse of the triangle and a and b are its other two sides.

ADDITIONAL READING

A. V. Astin, "Standards of Measurements," *Sci. American,* June 1968, p. 50.

H. Butterfield, "The Scientific Revolution," *Sci. American,* September 1960, p. 173.

J. Friberg, "Numbers and Measures in the Earliest Written Records," *Sci. American,* February 1984, p. 110.

G. Goth, "Dimensional Analysis by Computer," *The Physics Teacher,* February 1986, p. 75.

L. M. Lederman, "The Value of Fundamental Science," *Sci. American,* November 1984, p. 40.

Philip and Phyllis Morrison and the office of Charles and Ray Eames, *The Powers of Ten,* New York, The Sci. American Library, W. H. Freeman, 1982.

CONCEPTUAL QUESTIONS

Example Estimate the number of breaths taken during an average life span of 70 years.

Reasoning The only estimate that one must make in this exercise is the average number of breaths that a person takes in 1 min. This number varies, depending on whether one is exercising, sleeping, angry, serene, etc. We shall choose 8 breaths per minute as our estimate of the average. The number of minutes in a year is

$$1 \text{ yr} \times 365 \,\frac{\text{days}}{\text{yr}} \times 24 \,\frac{\text{hr}}{\text{day}} \times 60 \,\frac{\text{min}}{\text{hr}} = 5.26 \times 10^5 \text{ min}$$

Thus, in 70 years there will be $(70)(5.26 \times 10^5 \text{ min}) = 3.68 \times 10^7$ min. At a rate of 8 breaths/min the individual would take about 3×10^8 breaths. You may want to check your own rate and repeat the above calculation.

1. Suppose that two quantities, A and B, have different dimensions. Determine which of the following arithmetic operations *could* be physically meaningful: (a) A + B, (b) A/B, (c) B − A, (d) AB.
2. Estimate the order of magnitude of the length, in meters, of each of the following: (a) a ladybug, (b) a femur bone, (c) your campus administration building, (d) a giraffe, (e) a city block, (f) a football field.
3. Estimate the number of times your heart beats in a day.
4. Estimate your age in seconds.
5. The height of a horse is sometimes given in units of "hands." Why is this a poor standard of length?
6. How many of the lengths or time intervals given in Tables 1.2 and 1.3 could you verify using only equipment found in a typical dormitory room?
7. An ancient unit of length called the cubit was equal to six palms, where a palm was the width of the four fingers of an open hand. Noah's ark was 300 cubits long, 50 cubits wide, and 30 cubits high. Estimate the volume of the ark in cubic feet. Also, estimate the volume of a typical home and compare it to the volume of the ark.

8. If an equation is dimensionally correct, does this mean that the equation must be true?
9. An automobile tire is rated to last for 50 000 miles. Estimate the number of revolutions the tire will make in its lifetime.
10. Figure 1.9 is a photograph showing unit conversions on the labels of some grocery-store items. Check the accuracy of these conversions. Are the manufacturers using significant figures correctly?

FIGURE 1.9 (Question 10) *(Courtesy of Henry Leap and Jim Lehman)*

PROBLEMS

Section 1.3   Dimensional Analysis

1. In a desperate attempt to come up with an equation to use during an examination, a student tries $v^2 = ax$. Use dimensional analysis to determine whether this equation might be valid.

2. (a) Suppose that the displacement of an object is related to the time according to the expression $x = Bt^2$. What are the dimensions of B? (b) A displacement is related to the time as $x = A \sin(2\pi ft)$, where A and f are constants. Find the dimensions of A. (*Hint:* A trigonometric function appearing in an equation must be dimensionless.)

3. (a) One of the fundamental laws of motion states that the acceleration of an object is directly proportional to the resultant force on it and inversely proportional to its mass. From this statement, determine the dimensions of force. (b) The newton is the SI unit of force. According to the results for (a), how can you express a force having units of newtons using the fundamental units of mass, length, and time?

4. The radius of a circle inscribed in any triangle whose sides are a, b, and c is given by

$$r = \sqrt{\frac{(s-a)(s-b)(s-c)}{s}}$$

where s is an abbreviation for $(a + b + c)/2$. Check this formula for dimensional consistency.

5. A spherical droplet of blood is drawn into a pipette, as in Figure 1.10. Write a formula for the height, h, of the column of blood in the pipette in terms of the radius, r, of the pipette and the radius, R, of the original droplet. Check the dimensions of your formula.

6. The period of a simple pendulum, defined as the time for one complete oscillation, is measured in time units and is given by

$$T = 2\pi \sqrt{\frac{\ell}{g}}$$

where ℓ is the length of the pendulum and g is the acceleration due to gravity, in units of length divided by time squared. Show that this equation is dimensionally consistent.

Section 1.4   Significant Figures

7. The value of the speed of light is now known to be $2.99\,7924\,58 \times 10^8$ m/s. Express the speed of

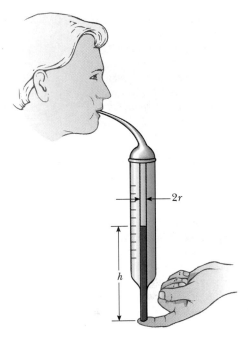

FIGURE 1.10   (Problem 5)

light to (a) three significant figures, (b) five significant figures, and (c) seven significant figures.

8. How many significant figures are there in (a) 78.9 ± 0.2, (b) 3.788×10^9, (c) 2.46×10^{-6}, (d) 0.0032?

9. Carry out the following arithmetic operations: (a) the sum of the numbers 756, 37.2, 0.83, and 2.5; (b) the product 3.2×3.563; (c) the product $5.67 \times \pi$.

10. The fisherman in the photograph on p. 20 (who bears a strong resemblance to Ernest Hemingway) catches two sturgeon. The smaller of the two has a measured length of 93.46 cm (two decimal places, four significant figures), and the larger fish has a measured length of 135.3 cm (one decimal place, four significant figures). What is the total length of fish caught for the day?

11. Calculate (a) the circumference of a circle of radius 3.5 cm and (b) the area of a circle of radius 4.65 cm.

12. A farmer measures the distance around a rectangular field. The length of each long side of the rectangle is found to be 38.44 m, and the length of each short side is found to be 19.5 m. What is the total distance around the field?

☐ indicates problems that have full solutions available in the Student Solutions Manual and Study Guide.

(Problem 10)

Section 1.5   Conversion of Units

13. A billionaire offers to give you $1 billion if you can count it out using only one-dollar bills. Should you accept her offer? Assume that you can count at an average rate of one bill every second, and be sure to allow for the fact that you need about 8 hours a day for sleeping and eating.

14. A quart container of ice cream is to be made in the form of a cube. What should be the length of a side, in centimeters? (Use the conversion 1 gallon = 3.786 liter.)

15. Estimate the age of the Earth in years, using the data in Table 1.3 and the appropriate conversion factors.

16. Estimate the distance to the nearest star, in feet, using the data in Table 1.1 and the appropriate conversion factors.

17. A house is 50 ft long and 26 ft wide, and has 8.0-ft-high ceilings. What is the volume of the house in cubic meters and in cubic centimeters?

18. (a) How many seconds are there in a year? (b) If one micrometeorite (a sphere with a diameter of 1.0×10^{-6} m) struck each square meter of the Moon each second, how many years would it take to cover the Moon to a depth of 1.0 m? (*Hint:* Consider a cubic box, 1.0 m on a side, on the Moon, and find how long it would take to fill the box.)

19. A painter is to cover the walls in a room that is 8.0 ft high and 12 ft along a side. What surface area, in square meters, must he cover?

20. You can obtain a rough estimate of the size of a molecule by the following simple experiment. Let a droplet of oil spread out on a smooth water surface. The resulting "oil slick" will be approximately one molecule thick. Given an oil droplet of mass 9.00×10^{-7} kg and density 918 kg/m^3 that spreads out into a circle of radius 41.8 cm on the water surface, what is the diameter of an oil molecule?

21. One cubic meter (1.00 m^3) of aluminum has a mass of 2.70×10^3 kg, and 1.00 m^3 of iron has a mass of 7.86×10^3 kg. Find the radius of an aluminum sphere whose mass is the same as that of an iron sphere of radius 2.00 cm. (*Note:* Density is defined as the mass of an object divided by its volume.)

22. One cubic centimeter (1.0 cm^3) of water has a mass of 1.0×10^{-3} kg. (a) Determine the mass of 1.0 m^3 of water. (b) Assuming biological substances are 98% water, estimate the masses of a cell with a diameter of 1.0 μm, a human kidney, and a fly. Assume a kidney is roughly a sphere with a radius of 4.0 cm, and a fly is roughly a cylinder 4.0 mm long and 2.0 mm in diameter. (See the hint in Problem 21.)

23. (a) Find a conversion factor to convert from miles per hour to kilometers per hour. (b) For a while, federal law mandated that the maximum highway speed would be 55 mi/h. Use the conversion factor from part (a) to find the speed in kilometers per hour. (c) The maximum highway speed has been raised to 65 mi/h in some places. In kilometers per hour, how much of an increase is this over the 55-mi/h limit?

24. The speed of light is about 3.00×10^8 m/s. Convert this to miles per hour.

25. Use the fact that the speed of light in free space is about 3.00×10^8 m/s to determine how many miles a pulse from a laser beam travels in 1 hour.

26. The base of a pyramid covers an area of 13.0 acres (1 acre = 43 560 ft^2) and has a height of 481 ft (Fig. 1.11). If the volume of a pyramid is given by the expression $V = (1/3) bh$, where b is the area of the base and h is the height, find the volume of this pyramid in cubic meters.

27. The pyramid described in Problem 26 contains approximately 2.0 million stone blocks that average 2.5 tons each. Find the weight of this pyramid in pounds and in newtons.

FIGURE 1.11 (Problems 26 and 27)

Section 1.6 Order-of-Magnitude Calculations

Note: In developing answers to the problems in this section, you should state your important assumptions, including the numerical values assigned to parameters used in the solution. Since only order-of-magnitude results are expected, do not be surprised if your results differ from those given in the answer section.

28. Imagine that you are the equipment manager of a professional baseball team. One of your jobs is to keep baseballs on hand for games. Balls are sometimes lost when players hit them into the stands as either home runs or foul balls. Estimate how many baseballs you have to buy per season in order to make up for such losses. Assume your team plays an 81-game home schedule in a season.

29. A hamburger chain advertises that it has sold more than 50 billion hamburgers. Estimate how many pounds of hamburger meat must have been used by the restaurant chain and how many head of cattle were required to furnish the meat.

30. Estimate the number of piano tuners living in New York City. This question was raised by the physicist Enrico Fermi, who was well known for making order-of-magnitude calculations.

31. Estimate the number of Ping-Pong balls that would fit (without being crushed) into a room 4 m long, 4 m wide, and 3 m high. Assume that the diameter of a Ping-Pong ball is 3.8 cm.

Section 1.8 Coordinate Systems and Frames of Reference

32. A point is located in a polar coordinate system by the coordinates $r = 2.5$ m and $\theta = 35°$. Find the x and y coordinates of this point, assuming the two coordinate systems have the same origin.

33. A certain corner of a room is selected as the origin of a rectangular coordinate system. If a fly is crawling on an adjacent wall at a point having coordinates (2.0, 1.0), where the units are meters, what is the distance of the fly from the corner of the room?

34. Express the location of the fly in Problem 33 in polar coordinates.

35. Two points in a rectangular coordinate system have the coordinates (5.0, 3.0) and (−3.0, 4.0), where the units are centimeters. Determine the distance between these points.

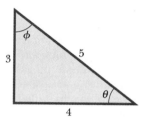

FIGURE 1.12 (Problem 36)

Section 1.9 Trigonometry

36. In Figure 1.12, find (a) the side opposite θ, (b) the side adjacent to ϕ, (c) cos θ, (d) sin ϕ, and (e) tan ϕ.

37. In a certain right triangle, the two sides that are perpendicular to each other are 5.00 m and 7.00 m long. What is the length of the third side of the triangle?

38. In Problem 37, what is the tangent of the angle for which 5 is the opposite side?

39. A right triangle has a hypotenuse of length 3.00 m, and one of its angles is 30.0°. What are the lengths of (a) the side opposite the 30.0° angle and (b) the side adjacent to the 30.0° angle?

40. For the triangle shown in Figure 1.13, what are (a) the length of the unknown side, (b) the tangent of θ, and (c) the sin of ϕ?

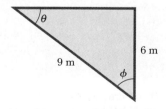

FIGURE 1.13 (Problem 40)

ADDITIONAL PROBLEMS

41. Use the standard prefixes in Table 1.4 and the well-established prefixes for the numbers 1 through 10 to determine appropriate words for the following conversions: (a) 10^{12} microphones, (b) 10^{21} picolos, (c) 10 rations, (d) 10^6 bicycles, (e) 10^{12} pins, (f) $3\frac{1}{3}$ tridents, (g) 2000 mocking-birds, (h) 10^{-6} phone, (i) 10^{-9} goat. (For example, 10^{18} miner = *examiner*. Some of these amusing conversions are from a table published by Solomon W. Golomb of UCLA.)

42. The radius of the planet Saturn is 5.85×10^7 m, and its mass is 5.68×10^{26} kg (Fig. 1.14). (a) Find the density of Saturn (its mass divided by its volume) in grams per cubic centimeter. (The volume of a sphere is given by $(4/3)\pi r^3$.) (b) Find the surface area of Saturn in square feet. (The surface area of a sphere is given by $4\pi r^2$.)

43. The consumption of natural gas by a company satisfies the empirical equation $V = 1.5t + 0.0080t^2$, where V is the volume in millions of cubic feet and t the time in months. Express this equation in

cubic feet and seconds. Put the proper units on the coefficients. Assume a month is 30 days.

44. Soft drinks are commonly sold in aluminum containers. Estimate the number of such containers thrown away each year by U.S. consumers. Approximately how many tons of aluminum does this represent?

FIGURE 1.14 (Problem 42) A view of Saturn from Voyager 2. *(Courtesy of NASA)*

FIGURE 1.15 (Problem 47) *(Dennis Brack/Black Star)*

THE WIZARD OF ID

By permission of John Hart and Field Enterprises, Inc.

45. At the time of this book's printing, the U.S. national debt is about $4 trillion. (a) If this debt were paid at a rate of $1000 per second, how long (in years) would it take to pay off the debt, assuming no interest were charged? (b) A dollar bill is about 15.5 cm long. If 4 trillion dollar bills were laid end to end around the Earth's equator, how many times would they encircle the Earth? Take the radius of the Earth to be about 6378 km. (*Note:* Before doing any of the calculations, try to guess at the answers. You may be very surprised.)

46. The displacement of an object moving under uniform acceleration is some function of time and the acceleration. Suppose we write this displacement as $s = k\, a^m t^n$, where k is a dimensionless constant. Show by dimensional analysis that this expression is satisfied if $m = 1$ and $n = 2$. Can this analysis give the value of k?

47. The hour hand and minute hand of the famous London Parliament clock, whose tower houses the bell "Big Ben," are 2.7 m and 4.5 m long, respectively. How far apart are the tips of these hands when the clock strikes five o'clock (Fig. 1.15)? (*Hint:* Use the law of cosines.)

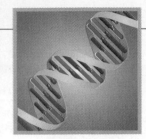

PHILIP MORRISON   Massachusetts Institute of Technology

SCALING—THE PHYSICS OF LILLIPUT*

The fictional traveler Lemuel Gulliver spent a busy time in a kingdom called Lilliput, where all living things—people, cattle, trees, grass—were exactly similar to their counterparts in our world, except that they were all built on the scale of one inch to the foot. Lilliputians were a little under 6 inches high, on the average, and built proportionately just as we are. Gulliver also visited Brobdingnag, the country of the giants, who were exactly like "regular" people but 12 times as tall. As Swift described it, daily life in both kingdoms was about like ours (in the 18th century). His commentary on human behavior is still worth reading, but we shall see that people of such sizes just could not have been as he described them.

Long before Swift lived, Galileo understood why very small or very large models of humans could not be like us, but apparently Dean Swift had never read what Galileo wrote. One character in Galileo's "Two New Sciences" says, "Now since . . . in geometry, . . . mere size cuts no figure, I do not see that the properties of circles, triangles, cylinders, cones, and other solid figures will change with their size. . . ." However, his physicist friend replies, "The common opinion is here absolutely wrong." Let us see why.

We start with the strength of a rope. It is easy to see that if one man who pulls with a certain strength can almost break a certain rope, two such ropes will just withstand the pull of two men. A single large rope with the same cross-section as the two smaller ropes combined will contain just double the number of fibers of one of the small ropes, and it will also do the job. In other words, the breaking strength of a wire or rope is proportional to its area of cross-section or to the square of its diameter. Experience and theory agree in this conclusion. Furthermore, the same proportionality holds, not only for ropes or cables supporting a pull but also for columns or struts supporting a thrust. The thrust that a column will support, comparing only those of a given material, is also proportional to the cross-sectional area of the column.

Now the body of a human or any other animal is held up by a set of columns or struts—the skeleton—supported by various braces and cables, which are muscles and tendons. However, the weight of the body that must be supported is proportional to the amount of flesh and bone present, that is, to the volume.

Let us now compare Gulliver with the Brobdingnagian giant, 12 times his height. Since the giant is exactly like Gulliver in construction, every one of his linear dimensions is 12 times the corresponding one of Gulliver's. Because the strength of his columns and braces is proportional to their cross-sectional area and thus to the square of their linear dimension (strength $\propto L^2$), his bones will be 12^2 or 144 times as strong as Gulliver's. Because his weight is proportional to his volume and thus to L^3, it will be 12^3, or 1728, times as great as Gulliver's. So the giant will have a strength-to-weight ratio a dozen times smaller than ours. Just to support his own weight, he would have as much trouble as we should have in carrying 11 people on our back.

In reality, of course, Lilliput and Brobdingnag do not exist; but we can see real effects of a difference in scale if we compare similar animals of very different sizes. The smaller ones are not scale models of the larger ones. Figure 1 shows the corresponding leg bones of two closely related animals of the deer family: one a tiny gazelle, the

*Adapted from PSSC PHYSICS, 2nd edition, 1965; D.C. Heath and Company with Education Development Center, Inc., Newton, MA.

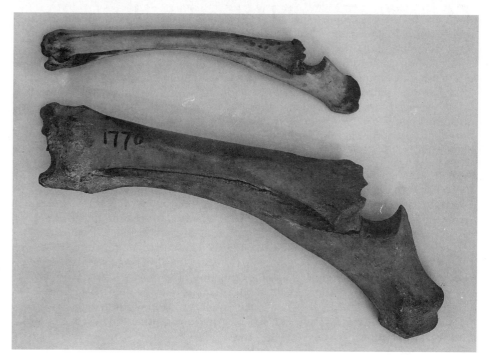

FIGURE 1

The front leg bones of a gazelle and a bison. The animals are related, but the gazelle is much smaller. The photos show the approximate relative sizes of the bones. Note that the bone of the larger animal is much thicker in comparison to its length than that of the gazelle. The small deer is generally more lightly and gracefully built. Can you visualize how much different Lilliputians must have been from men of normal size?

other a bison. Notice that the bone of the large animal is not at all similar geometrically to that of the smaller. The former is much thicker for its length, thus counteracting the scale change, which would make a strictly similar bone too weak.

Galileo wrote very clearly on this point, disproving the possibility of Brobdingnag or of any normal-looking giants: ". . . if one wishes to maintain in a great giant the same proportion of limb as that found in an ordinary man he must either use a harder and stronger material for making the bones, or he must admit a diminution of strength in comparison with men of medium stature; for if his height be increased inordinately he will fall and be crushed under his own weight. Whereas, if the size of a body be diminished, the strength of that body is not diminished in the same proportion; indeed, the smaller the body the greater its relative strength. Thus a small dog could probably carry on his back two or three dogs of his own size; but I believe that a horse could not carry even one of his own size." The sketch of Figure 2 is taken from Galileo, who drew it to illustrate the paragraph just quoted.

An elephant is already so large that its limbs are clumsily thickened. However, a whale, the largest of all animals, may weigh 40 times as much as an elephant; yet the

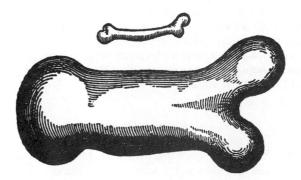

FIGURE 2

Galileo's drawing illustrating scaling. Over 300 years ago, Galileo wrote concerning the fact that a bone of greater length must be increased in thickness in greater proportion in order to be comparably strong.

whale's bones are not proportionately thickened. They are strong enough because the whale is supported by water. What is the fate of a stranded whale? Its ribs break. Some of the dinosaurs of old were animals of whale-like size; how did *they* get along?

Following Galileo, we have investigated the problems of scaling up to giants. Now let's take a look at some of the problems that arise when we scale down.

When you climb dripping wet out of a pool, there is a thin film of water on your skin. Your fingers are no less wet than your forearm; the thickness of the water film is much the same over most of your body. Roughly, at least, the amount of water you bring out is proportional to the surface area of your body. You can express this by the relationship

$$\text{amount of water} \propto L^2$$

where L is your height. The original load on your frame is, as before, proportional to your volume. So, the ratio *extra load/original load* is proportional to L^2/L^3, or to $1/L$. Perhaps you carry out of the pool a glassful or so of water, which amounts to about a 1% increase in what you have to move about. But a Lilliputian will bring out about 12% of his or her weight, which would be equivalent to a heavy winter suit of clothing with an overcoat. Getting out of the pool would be no fun! If a fly gets wet, its body load doubles, and it is all but imprisoned by the drop of water.

There is a still more important effect of the scale of a living body. Your body loses heat mainly through the skin (and to a lesser extent through breathing out warm air). It is easy to believe—and it can be checked by experiment—that the heat loss is proportional to the surface area, so that

$$\text{heat loss} \propto L^2$$

keeping other factors, like the temperature, nature of skin, and so on, constant. The food taken in must supply this heat, as well as the surplus energy we use in moving about. So minimum food needs go as L^2. If a man like Gulliver can live off a leg of lamb and a loaf of bread for a day or two, a Lilliputian with the same body temperature will require a volume of food only $(\frac{1}{12})^2$ as large. But the leg of lamb for the Lilliputian, scaled down to that world, will be smaller in volume by a factor of $(\frac{1}{12})^3$. Therefore, the Lilliputian would need a dozen roasts and loaves to feel as well fed as Gulliver did after one. Lilliputians must be a hungry lot, restless, active, graceful, but easily waterlogged. You can recognize these qualities in many small mammals, like a mouse.

We can see why there are no warm-blooded animals much smaller than the mouse. Fish and frogs and insects can be very much smaller because their temperature is not higher than that of their surroundings. In accord with the scaling laws of area and volume, small, warm-blooded animals need relatively a great deal of food; really small ones could not gather or even digest such an enormous amount. Certainly the agriculture of the Lilliputians could not have supported a kingdom like the one Gulliver described.

Now we see that neither Brobdingnag nor Lilliput can really be a scale model of our world. But what have these conclusions to do with physics?

Let's start again with the very large. As we scale up any system, the load will eventually be greater than the strength of the structure. This effect applies to every physical system, not just to animals, of course. Buildings can be very large because their materials are stronger than bone, their shapes are different, and they do not move. These facts determine the constants like K in the equation

$$\text{strength} = KL^2$$

but the same laws hold. No building can be made that will look like the Empire State but be as high as a mountain, say 10 000 m. Mountains are solid structures, for the most part, without interior cavities. Just as the bones of a giant must be thick, an object of mountainous size on the Earth must be either all but solid or else built of new materials yet unknown.

Our arguments are not restricted to the surface of the Earth. We can imagine building a tremendous structure far out in space away from the gravitational pull of the Earth. The load then is not given by the Earth's gravitational pull, but as the structure is built larger and larger, each part pulls gravitationally on every other, and soon the outside of the structure is pulled in with great force. The inside, built of ordinary materials, is crushed, and large protuberances on the surface break off or sink in. As a result, any large structure like a planet has a simple shape, and if it is large enough, the shape is close to spherical. Any other shape will be unable to support itself. Here is the essential reason why the planets and the Sun tend to be spherical. The pull of gravity is important for us on Earth, but as we extend the range of dimensions we study, it becomes absolutely dominant in the very large. Only motion can change this result. The great masses of gas that are nebulae, for example, are changing in time, and hence the law that large objects must be simple in shape is modified.

When we go from our size to the very small, gravitational effects cease to be important. As we saw in investigating Lilliput, however, surface effects become significant. If we go far enough toward the very small, surfaces no longer appear smooth but are so rough that we have difficulty in defining a surface. Other descriptions must be used. In any case, it will not come as a complete surprise that in the domain of the atom, the very small, scale factors demonstrate that the dominant pull is one that is not easily observed in everyday experience.

Such arguments as these run through all of physics. Like order-of-magnitude measurements, they are extremely valuable when we begin the study of any physical system. How the behavior of a system will change with changes in the scale of its dimensions, its motion, and so on, is often the best guide to a detailed analysis.

Even more, it is by the study of systems built on many unusual scales that physicists have been able to uncover unsuspected physical correlations. When changing scale, one aspect of the physical world may be much emphasized and another one may be minimized. In this way we may discover, or at least get a clearer view of, things that are less obvious on our normal scale of experience. It is largely for this reason that physicists examine, in and out of their laboratories, the very large and the very small, the slow and the rapid, the hot and the cold, and all the other unusual circumstances they can contrive. In examining what happens in these circumstances, we use instruments both to produce the unusual circumstances and to extend our senses in making measurements.

It is hard to resist pointing out how much the scale of our own size affects the way we see the world. It has been largely the task of physics to try to form a picture of the world that does not depend upon the way we happen to be built; but it is hard to get rid of these effects of our own scale. We can build big roads and bridges that are long and thin, but these are essentially not three-dimensional, complex structures. The very biggest things we can make that have some roundness, that are fully three-dimensional, are buildings and great ships. These lack a good deal of being a thousand times larger than humans in their linear dimensions.

Within our present technology, our scaling arguments are important. If we design a new large object on the basis of a small one, we are warned that new effects too small to detect on our scale may enter and even become the most important things to consider. We cannot just scale up and down blindly, geometrically, but by scaling in the light of physical reasoning, we can sometimes foresee what changes will occur. In this way we can employ scaling in intelligent airplane design, for example, and not arrive at a jet transport that looks like a bee—and won't fly.

Questions and Problems

1. The leg bones of one animal are twice as strong as those of another closely related animal of similar shape. (a) What would you expect to be the ratio of these animals' heights? (b) What would you expect to be the ratio of their weights?

2. A hummingbird must eat very frequently and even then must have a highly concentrated form of food, such as sugar. What does the concept of scaling tell you about the size of a hummingbird?

3. About how many Lilliputians would it take to equal the mass of one citizen of Brobdingnag?

4. The total surface area of a rectangular solid is the sum of the areas of the six faces. If each dimension of a given rectangular solid is doubled, what effect does this have on the total surface area?

5. A hollow metal sphere has a wall thickness of 2 cm. If you increase both the diameter and thickness of this sphere so that the overall volume is three times the original overall volume, how thick will the shell of the new sphere be?

6. If your height and all your other dimensions were doubled, by what factor would this change (a) your weight? (b) the ability of your leg bones to support your weight?

7. An elephant of mass 4.0×10^3 kg consumes 3.4×10^2 times as much food as a guinea pig of mass 0.70 kg. They are both warm-blooded, plant-eating, similarly shaped animals. Find the ratio of their surface areas, which is approximately the ratio of their heat losses, and compare it with the ratio of food consumed.

8. A rectangular water tank is supported above the ground by four pillars 5.0 m long whose diameters are 20 cm. If the tank were made 10 times longer, wider, and deeper, what diameter pillars would be needed? How much more water would the tank hold?

9. How many state maps of scale $1:1\,000\,000$ would you need to cover the state with those maps?

MOTION IN ONE DIMENSION

2

An apple and a feather fall from rest in a four-foot vacuum chamber. Regardless of their mass, in the absence of air resistance, all objects fall to the Earth with the same acceleration.

(© 1993 James Sugar, Black Star)

People have always been concerned with some form of motion. For example, in today's world you consider motion when you describe to someone how fast your new car will go or how much pickup it has. Likewise, prehistoric cave dwellers must have, in their own way, pondered motion as they devised methods for capturing a rapidly moving antelope. *The branch of physics concerned with the study of the motion of an object and the relationship of this motion to such physical concepts as force and mass is called* **dynamics.** *The part of dynamics that describes motion without regard to its causes is called* **kinematics.** In this chapter we shall focus on kinematics and on one-dimensional motion, that is, motion along a straight line. We shall start by discussing displacement, velocity, and acceleration. These concepts will enable us to study the motion of objects undergoing constant acceleration. In Chapter 3 we shall discuss the motions of objects in two dimensions.

Mechanics has had a long history. The first recorded evidence of its study can be traced to ancient Sumerian and Egyptian civilizations, whose primary concern was to understand the motions of heavenly bodies. The most systematic and detailed early studies of the heavens were conducted by the Greeks from about 300 B.C. to A.D. 300. Ancient scientists and lay people regarded the Earth as the center of the Universe. This geocentric model was accepted by such notables as Aristotle (384–322 B.C.) and Claudius Ptolemy (about A.D. 140). Largely because of the authority of Aristotle, the geocentric model became the accepted theory of the Universe and stayed in that position until the 17th century.

About 250 B.C. the Greek philosopher Aristarchus worked out the details of a model of the Solar System based on a moving Earth. He proposed that the sky appears to turn westward because the Earth is really turning eastward. This model was not given much consideration because it was believed that if the Earth turned, it would set up a great wind as it moved through the air. The critics could

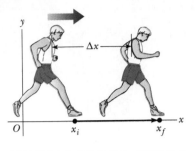

FIGURE 2.1
A sprinter moving along the x axis from x_i to x_f undergoes a displacement of $\Delta x = x_f - x_i$.

not see that the Earth was carrying the air and everything else with it as it rotated.

The Polish astronomer Nicholaus Copernicus (1473–1543) is credited with initiating the revolution that finally replaced the geocentric model. In his system, called the heliocentric model, the Earth and the other planets revolve in circular orbits around the Sun.

This early knowledge formed the foundation for the work of Galileo Galilei (1564–1642). Galileo stands out as perhaps the dominant facilitator of the entrance of physics into the modern era. In 1609 he became one of the first to make astronomical observations with a telescope. He observed mountains on the Moon, the larger satellites of Jupiter, the rings of Saturn, spots on the Sun, and the phases of Venus. His observations convinced him of the correctness of the Copernican theory. Galileo's work with motion is particularly well known, and because of his leadership, experimentation has become an important part of our search for knowledge.

2.1

DISPLACEMENT

In order to describe the motion of an object, one must be able to specify its position at all times using some convenient coordinate system and a specified origin. For example, consider a sprinter moving along the x axis from an initial position, x_i, to some final position, x_f, as in Figure 2.1. The **displacement** of an object, defined as its *change in position*, is given by the difference between its final and initial coordinates, or $x_f - x_i$. As mentioned in Chapter 1, we use the Greek letter delta (Δ) to denote a change in a quantity. Therefore, we write the displacement, or change in position, of the object as

| Definition of displacement |

$$\Delta x \equiv x_f - x_i \qquad \text{[2.1]}$$

From this definition, we see that Δx is positive if x_f is greater than x_i and negative if x_f is less than x_i. If the sprinter moves from an initial position of $x_i = -3$ m to a final position of $x_f = 5$ m, his displacement is 8 m. [That is, the displacement is $\Delta x = 5$ m $-$ (-3 m) $= 5$ m $+$ 3 m $= 8$ m.]

| A vector has both direction and magnitude |

Displacement is an example of a vector quantity. Many physical quantities in this book, including displacement, velocity, and acceleration, are vectors. In general, a **vector** is a physical quantity that requires the specification of both direction and magnitude. By contrast, a **scalar** is a quantity that has magnitude and no direction. Scalar quantities, such as mass and temperature, can be specified by numbers with appropriate units.

From here on we shall designate vector quantities with boldface type. For example, **v** will denote a velocity vector, and **a** will denote an acceleration vector. In this chapter, which deals with one-dimensional motion, it will not be necessary to use this notation. The reason is that in one-dimensional motion there are only two directions in which an object can move, and these two directions are easily specified by plus and minus signs. For example, if a truck moves from an initial position of 10 m to a final position of 80 m, as in Figure 2.2a, its displacement is

$$\Delta x = x_f - x_i = 80 \text{ m} - 10 \text{ m} = +70 \text{ m}$$

In this case, the displacement has a magnitude of 70 m and is directed in the positive x direction, as indicated by the plus sign of the result. (Sometimes the

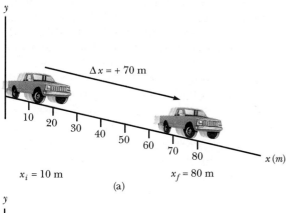

$\Delta x = + 70$ m

$x_i = 10$ m

$x_f = 80$ m

(a)

x (m)

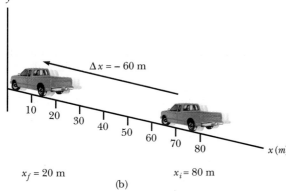

$\Delta x = - 60$ m

$x_f = 20$ m

$x_i = 80$ m

(b)

x (m)

FIGURE 2.2
(a) A truck moving to the right from $x_i = 10$ m to $x_f = 80$ m undergoes a displacement $\Delta \mathbf{x} = + 70$ m. (b) A truck moving to the left from $x_i = 80$ m to $x_f = 20$ m undergoes a displacement $\Delta \mathbf{x} = - 60$ m.

plus sign is omitted, but a result of, say, 70 m for the displacement is understood to be the same as + 70 m.) The displacement vector is usually represented by an arrow, as in Figure 2.2a. The length of the arrow represents the magnitude of the displacement, and the head of the arrow indicates its direction.

Now suppose the truck moves to the left from an initial position of 80 m to a final position of 20 m, as in Figure 2.2b. In this situation, the displacement is

$$\Delta x = x_f - x_i = 20 \text{ m} - 80 \text{ m} = - 60 \text{ m}$$

The minus sign in this result indicates that the displacement is in the negative x direction. Likewise, the arrow representing the displacement vector is to the left, as in Figure 2.2b.

2.2

AVERAGE VELOCITY

Consider a truck moving along a highway (the x axis), as in Figure 2.2. Let the truck's position be x_i at some time t_i, and let its position be x_f at time t_f. (The indices i and f refer to the initial and final locations, respectively.) In the time interval $\Delta t = t_f - t_i$, the displacement of the truck is $\Delta x = x_f - x_i$.

The **average velocity,** $\bar{v}$, is defined as the displacement, Δx, divided by the time interval during which the displacement occurred:

$$\bar{v} \equiv \frac{\Delta x}{\Delta t} = \frac{x_f - x_i}{t_f - t_i}$$

[2.2]   Average velocity

FIGURE 2.3
A drag race viewed from a blimp. One car follows the red straight-line path from *P* to *Q*, and a second car follows the blue curved path.

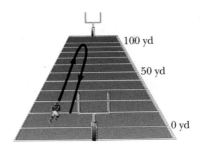

FIGURE 2.4
The path followed by a confused football player.

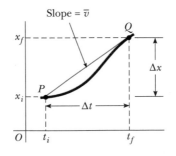

FIGURE 2.5
Position-time graph for an object moving along the *x* axis. The average velocity $\bar{v}$ in the time interval $\Delta t = t_f - t_i$ is the slope of the blue straight line connecting the points *P* and *Q*.

The average velocity of an object can be either positive or negative, depending on the sign of the displacement. (The time interval, Δt, is always positive.) For example, if we select a coordinate system so that a car moves from a position 100 m from the origin to a position 50 m from that origin in a time interval of 2 s, the *x* component of the average velocity is -25 m/s. In this case, the minus sign indicates motion to the left. If an object moves in the negative *x* direction during some time interval, its velocity in the *x* direction during that interval is necessarily negative.

In order to understand the vector nature of velocity, consider the following situation. Suppose a friend tells you that she will be taking a trip in her car and will travel at a constant rate of 55 mi/h in a straight line for 1 h. If she starts from her home, where will she be at the end of the trip? Obviously, you cannot answer this question because the direction in which she will travel is not specified. All you can say is that she will be located 55 mi from her starting point. However, if she tells you that she will be driving at the rate of 55 mi/h directly northward, then her final location will be known exactly. The velocity of an object is known only if its direction and its magnitude (speed) are specified. In general, *any physical quantity that is a vector must be characterized by both a magnitude and a direction.*

As an example of the use of Equation 2.2, suppose that the truck in Figure 2.2a moves 100 m to the right in 5.0 s. The substitution of these values into Equation 2.2 gives the average velocity in this time interval as

$$\bar{v} = \frac{\Delta x}{\Delta t} = \frac{100 \text{ m}}{5.0 \text{ s}} = +20 \text{ m/s}$$

Note that the units of average velocity are units of length divided by units of time. These are meters per second (m/s) in SI. Other units for velocity might be feet per second (ft/s) in the conventional system or centimeters per second (cm/s) if we decided to measure distances in centimeters.

Let us assume we are watching a drag race from the Goodyear blimp. In one run we see a car follow the straight-line path from *P* to *Q* shown in Figure 2.3 during the time interval Δt, and in a second run a car follows the curved path during the same time interval. If you examine the definition of average velocity (Eq. 2.2) carefully, you will see that the two cars had the same average velocity. This is because they had the same displacement $(x_f - x_i)$ during the same time interval (Δt), as indicated in Figure 2.3.

Figure 2.4 shows the unusual path of a confused football player. He receives a kickoff at his own goal, runs downfield to within inches of a touchdown, and then reverses direction to race backward until he is tackled at the exact location where he first caught the ball. What is his average velocity during this run? From the definition of average velocity, we see that the football player's average velocity is zero because his displacement is zero. In other words, since x_i and x_f have the same value, $\bar{v} = \Delta x/\Delta t = 0$. From this we see that displacement should not be confused with distance traveled. The confused football player clearly traveled a great distance, almost 200 yards; yet his displacement is zero.

GRAPHICAL INTERPRETATION OF VELOCITY

Figure 2.5 is a graph of the motion of an object moving along a straight-line path from the position x_i at time t_i to the position x_f at time t_f. (Note that the motion is along a straight line, yet the position-time graph is *not* a straight line. Why?)

The straight line connecting points P and Q provides us with a geometric interpretation of average velocity. As indicated on the graph, the slope of the line is $\Delta x \, (= x_f - x_i)$ divided by the time interval for the motion, $\Delta t \, (= t_f - t_i)$. Therefore,

> the average velocity of an object during the time interval t_i to t_f is equal to the slope of the straight line joining the initial and final points on a graph of the position of the object plotted versus time.

2.3

INSTANTANEOUS VELOCITY

Let's imagine that you take a trip in your car along a perfectly straight highway. At the end of your journey, it is a relatively simple task to calculate your average velocity. The car's odometer gives you the distance traveled, a compass can give you direction, and a watch can supply the time interval. However, such a calculation would omit a great deal of information about what actually occurred on your trip. For example, if your calculation indicated that your average velocity was 55 mi/h, this would not necessarily mean that your velocity at every instant was 55 mi/h. Instead, you might have traveled at 55 mi/h for a short distance, stopped for lunch and then made up some time by traveling at 70 mi/h, paused again as a police officer wrote out a ticket, and then raced forward again when the coast was clear.

More precisely, the instantaneous velocity, v, is defined as the limit of the average velocity as the time interval Δt becomes infinitesimally short. In mathematical language this is written

$$v \equiv \lim_{\Delta t \to 0} \frac{\Delta x}{\Delta t}$$ [2.3]

Definition of instantaneous velocity

The notation $\lim \Delta t \to 0$ means that the ratio $\Delta x/\Delta t$ is to be evaluated as the time interval Δt approaches zero.

To better understand the meaning of instantaneous velocity as expressed by Equation 2.3, consider the data in Table 2.1. Assume you have been observing a runner racing along a track. One second after starting into motion, the runner has moved to a position 1.00 m from the starting point; at $t = 1.50$ s, the runner is 2.25 m from the starting point; and so on. After collecting these data, suppose you wish to determine the velocity of the runner at the time $t = 1.00$ s. Table 2.2 presents some of the calculations you might perform to determine the velocity in question. Let us use $t = 1.00$ s as our initial time, and first consider the top row of the table. For the observed portion of the run (from 1.00 s to 3.00 s), the time interval $\Delta t = 2.00$ s and the displacment $\Delta x = +8.00$ m. Thus, the average velocity in this interval is $\Delta x/\Delta t = +4.00$ m/s. This gives only a rough approximation of the instantaneous velocity at $t = 1.00$ s. According to Equation 2.3, we can find increasingly more nearly correct answers by letting the time interval become smaller and smaller. Therefore, consider the second entry in Table 2.2, from 1.00 s to 2.00 s, corresponding to a time interval of $\Delta t = 1.00$ s. In this interval, $\Delta x = +3.00$ m and so the average velocity is $+3.00$ m/s. This is closer

TABLE 2.1
Positions of a Runner at Specific Instants of Time

t (s)	x (m)
1.00	1.00
1.01	1.02
1.10	1.21
1.20	1.44
1.50	2.25
2.00	4.00
3.00	9.00

TABLE 2.2
Calculated Values of the Time Intervals, Displacements, and Average Velocities for the Runner of Table 2.1

Time Interval (s)	Δt (s)	Δx (m)	$\bar{v}$ (m/s)
1 to 3.00	2.00	8.00	4.00
1 to 2.00	1.00	3.00	3.00
1 to 1.50	0.50	1.25	2.5
1 to 1.20	0.20	0.44	2.2
1 to 1.10	0.10	0.21	2.1
1 to 1.01	0.01	0.02	2

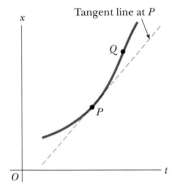

FIGURE 2.6
Geometric construction for obtaining the instantaneous velocity from the x versus t curve. The instantaneous velocity at P is defined as the slope of the line tangent to the curve at P.

to the correct answer because the time interval is smaller. Now let us consider a very short time interval indicated by the last entry in Table 2.2. In this case, $\Delta t = 0.01$ s and the displacement $\Delta x = +0.02$ m. Thus, the average velocity in this interval is $+2$ m/s. We could improve the reliability of our calculation by allowing the time interval to become even smaller, but we can state with some degree of confidence that the instantaneous velocity of the runner was $+2.00$ m/s at the time $t = 1.00$ s.

In day-to-day usage, the terms *speed* and *velocity* are used interchangeably. In physics, however, there is a clear distinction between these two quantities.

> The **instantaneous speed** of an object, which is a scalar quantity, is defined as the magnitude of the instantaneous velocity. Hence, by definition, speed can never be negative.

*GRAPHICAL INTERPRETATION OF INSTANTANEOUS VELOCITY

Figure 2.6 is a repeat of the graph drawn earlier (Fig. 2.5) for the position of an object versus time. To find the instantaneous velocity of the object at point P, we must find the average velocity during an infinitesimally short time interval. This means that point Q on the curve must be brought closer and closer to point P until the two points nearly overlap each other. From this construction we see that the line joining P and Q is approaching the line tangent to the curve at point P.

> The slope of the line tangent to the position-time curve at P is defined to be the instantaneous velocity at that time.

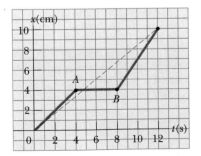

FIGURE 2.7
(Example 2.1)

EXAMPLE 2.1   A Toy Train

A toy train moves slowly along a straight portion of track according to the graph of position versus time in Figure 2.7. Find (a) the average velocity for the total trip, (b) the average velocity during the first 4.0 s of motion, (c) the average velocity during the next 4.0 s of motion, (d) the instantaneous velocity at $t = 2.0$ s, and (e) the instantaneous velocity at $t = 5.0$ s.

Solution   (a) The slope of the line joining the starting point and end point on the graph (the dashed line) provides the average velocity for the total trip. A measurement of the slope gives

$$\bar{v} = \frac{\Delta x}{\Delta t} = \frac{10 \text{ cm}}{12 \text{ s}} = \boxed{+0.83 \text{ cm/s}}$$

(b) The slope of the line joining the starting point to the point at $t = 4.0$ s on the curve gives us the average velocity during the first 4 s:

$$\bar{v} = \frac{\Delta x}{\Delta t} = \frac{4.0 \text{ cm}}{4.0 \text{ s}} = \boxed{+1.0 \text{ cm/s}}$$

(c) Following the same procedure for the next 4.0-s interval, we see that the slope of the line between points A and B is zero. During this time interval, the train has remained at the same location, 4.0 cm from the starting point.

(d) A line drawn tangent to the curve at the point corresponding to $t = 2.0$ s has the same slope as the line in part (b). Thus, the instantaneous velocity at this time is $+1.0$ cm/s. This has to be true because the graph indicates that during the first 4.0 s of motion the train covers equal distances in equal intervals of time. In other words, the train moves at a constant velocity during the first 4.0 s. Under these conditions the average velocity and the instantaneous velocity are identical at all times.

(e) At $t = 5.0$ s, the slope of the position-time curve is zero. Therefore, the instantaneous velocity is zero at this instant. In fact, the train is at rest during the entire time interval between 4.0 s and 8.0 s.

2.4

ACCELERATION

As you travel from place to place in your car, you normally do not travel long distances at a constant velocity. The velocity of the car increases when you step on the gas and decreases when you apply the brakes. Furthermore, the velocity changes when you round a curve, altering your direction of motion. The rate of change of velocity is defined as acceleration. Let us put this definition in mathematical terms.

Suppose a car moves along a straight highway as in Figure 2.8. At time t_i it has a velocity of v_i, and at time t_f its velocity is v_f.

Definition of average acceleration

> The **average acceleration** during this time interval is defined as the change in velocity divided by the time interval during which this change occurs:

$$\bar{a} \equiv \frac{\Delta v}{\Delta t} = \frac{v_f - v_i}{t_f - t_i} \qquad \text{[2.4]}$$

For example, suppose the car shown in Figure 2.8 accelerates from an initial velocity of $v_i = +10$ m/s to a final velocity of $v_f = +30$ m/s in a time interval of 2.0 s. (Note that both velocities are toward the right, the direction selected as the positive direction.) These values can be inserted into Equation 2.4 to give the average acceleration:

$$\bar{a} = \frac{30 \text{ m/s} - 10 \text{ m/s}}{2.0 \text{ s}} = +10 \text{ m/s}^2$$

FIGURE 2.8
A car moving to the right accelerates from a velocity of v_i to a velocity of v_f in the time interval $\Delta t = t_f - t_i$.

A car in a drag race undergoes large positive and negative accelerations during a race. *(© Jeffrey Sylvester, FPG)*

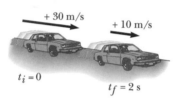

FIGURE 2.9
The velocity of the car decreases from + 30 m/s to + 10 m/s in a time interval of 2 s.

Acceleration is measured in dimensions of length divided by the square of time. Some units of acceleration are meters per second per second [(m/s)/s, which is usually written m/s²] and feet per second per second (ft/s²). The acceleration calculated in the preceding case was + 10 m/s². This notation means that, on the average, the car was moving so that its velocity increased at a rate of 10 m/s every second.

As a second example of the computation of acceleration, consider the car pictured in Figure 2.9. In this case, the velocity of the car has changed from an initial value of + 30 m/s to a final value of + 10 m/s in a time interval of 2.0 s. The average acceleration during this time interval is

$$\bar{a} = \frac{10 \text{ m/s} - 30 \text{ m/s}}{2.0 \text{ s}} = -10 \text{ m/s}^2$$

The minus sign indicates that the acceleration vector is in the negative x direction (to the left). For the case of motion in a straight line, the direction of the velocity of an object and the direction of its acceleration are related as follows. *When the object's velocity and acceleration are in the same direction, the speed of the object increases with time.* (The first case we cited demonstrates this situation.) *When the object's velocity and acceleration are in opposite directions, the speed of the object decreases with time.* (Decreases in speed are sometimes called decelerations.)

To clarify this point, consider the following situation. Suppose the velocity of a car changes from − 10 m/s to − 30 m/s in a time interval of 2.0 s. The minus signs here indicate that the velocity of the car is in the negative x direction. The average acceleration of the car in this time interval is

$$\bar{a} = \frac{-30 \text{ m/s} - (-10 \text{ m/s})}{2.0 \text{ s}} = -10 \text{ m/s}^2$$

The minus sign indicates that the acceleration vector is also in the negative x direction. Since the velocity and acceleration vectors are in the same direction, the speed of the car must increase as the car moves to the left. Note that a negative value for the acceleration does not always indicate a deceleration.

INSTANTANEOUS ACCELERATION

In some situations, the value of the average acceleration differs in different time intervals. It is therefore useful to define instantaneous acceleration. This concept is analogous to the definition of instantaneous velocity discussed in Section 2.3. In mathematical terms, the **instantaneous acceleration,** a, is defined as the limit of the average acceleration as the time interval Δt goes to zero. That is,

Definition of instantaneous acceleration

$$a \equiv \lim_{\Delta t \to 0} \frac{\Delta v}{\Delta t} \qquad \text{[2.5]}$$

Here again, the notation lim $\Delta t \to 0$ means that the ratio $\Delta v / \Delta t$ is to be evaluated as the time interval Δt approaches zero.

Figure 2.10 is useful for understanding the concept of instantaneous acceleration. It plots the velocity of an object versus time. This could represent, for example, the motion of a car along a busy street. The average acceleration of the car between times t_i and t_f can be found by determining the slope of the line joining points P and Q. If we imagine that point Q is brought closer and closer to point P, the value that we find for the average acceleration between these points approaches the value of the acceleration of the car at point P. The instantaneous acceleration at point P, for example, is the slope of the graph at the point. That is,

> the instantaneous acceleration of an object at a certain time equals the slope of the velocity-time graph at that instant of time.

From now on we shall use the term *acceleration* to mean "instantaneous acceleration."

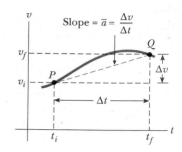

FIGURE 2.10
Velocity-time graph for an object moving in a straight line. The slope of the blue line connecting points P and Q is defined as the average acceleration in the time interval $\Delta t = t_f - t_i$.

EXAMPLE 2.2   A Fly Ball Is Caught

A baseball player moves in a straight-line path in order to catch a fly ball hit to the outfield. His velocity as a function of time is shown in Figure 2.11. Find his instantaneous acceleration at points A, B, and C on the curve.

Solution   The instantaneous acceleration at any time is the slope of the velocity-time curve at that instant. The slope of the curve at point A is

$$a = \frac{\Delta v}{\Delta t} = \frac{4 \text{ m/s}}{2 \text{ s}} = \boxed{+2 \text{ m/s}^2}$$

At point B, the slope of the line is zero and hence the instantaneous acceleration at this time is also zero. Note that even though the instantaneous acceleration has dropped to zero at this instant, the velocity of the player is not zero. (In general, if $a = 0$, the velocity is constant in time but *not necessarily* zero.) Instead, the player is running at a constant velocity of $+4$ m/s. A value of zero for an instantaneous acceleration means that at a particular instant the velocity of the object is not changing.

Finally, at point C we calculate the slope of the velocity-time curve and find that the instantaneous acceleration is

$$a = \frac{\Delta v}{\Delta t} = \frac{-2 \text{ m/s}}{1 \text{ s}} = \boxed{-2 \text{ m/s}^2}$$

For this situation, the minus sign indicates that the speed of the player is decreasing as he approaches the location where he will attempt to catch the baseball.

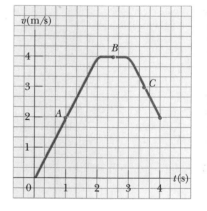

FIGURE 2.11
(Example 2.2)

2.5

MOTION DIAGRAMS

The concepts of velocity and acceleration are often confused with each other, but in fact they are quite different quantities. It is instructive to make use of motion diagrams to describe the velocity and acceleration vectors as time progresses during the motion of an object. In order not to confuse these two vector quantities in Figure 2.12, we use red for velocity vectors and violet for acceleration vectors. The vectors are sketched at several instants during the motion of the object, and the time intervals between adjacent positions are assumed to be equal. Figure 2.12a represents a car moving to the right with a constant positive

FIGURE 2.12
(a) Motion diagram for an object whose constant acceleration is in the direction of its velocity. The velocity vectors at each instant are red, and the constant acceleration vector is violet.
(b) Motion diagram for an object whose constant acceleration points *opposite* the velocity at each instant.

(a)

(b)

acceleration. In this case, note that the velocity vector increases in time. Because the car moves faster as it travels to the right, its displacement between adjacent positions increases as time progresses. If the car moves initially to the right with a constant negative acceleration (or deceleration) as in Figure 2.12b, the velocity vector decreases in time and eventually reaches zero. (This type of motion is exhibited by a car that skids to a stop after its brakes are applied.) In this case, the car moves more slowly as it moves to the right, so its displacement between adjacent positions decreases as time progresses. From this diagram, we see that the acceleration and velocity vectors are *not* in the same direction.

You should be able to construct a motion diagram corresponding to a particle that moves initially to the left with a constant positive or negative acceleration. You should also construct appropriate motion diagrams after completing the mathematical solutions to kinematic problems to see if your answers are consistent with the diagrams.

2.6

ONE-DIMENSIONAL MOTION WITH CONSTANT ACCELERATION

Most of the applications in this book will be concerned with objects moving with *constant acceleration*. This type of motion is important because it applies to many objects in nature. For example, an object in free fall near the Earth's surface moves in the vertical direction with constant acceleration, assuming that air resistance can be neglected. When an object moves with constant acceleration, *the average acceleration equals the instantaneous acceleration.* Consequently, the velocity increases or decreases at the same rate throughout the motion.

Since the average acceleration equals the instantaneous acceleration when a is constant, we can eliminate the bar used to denote average values from our defining equation for acceleration. That is, since $\bar{a} = a$, we can write Equation 2.4 as

$$a = \frac{v_f - v_i}{t_f - t_i}$$

For convenience, let $t_i = 0$ and t_f be any arbitrary time t. Also, we shall let $v_i = v_0$ (the initial velocity at $t = 0$) and $v_f = v$ (the velocity at any arbitrary time t). With this notation, we can express the acceleration as

$$a = \frac{v - v_0}{t}$$

or

$$v = v_0 + at \qquad \text{(for constant } a\text{)} \qquad \textbf{[2.6]}$$

One of the features of one-dimensional motion with constant acceleration is the manner in which the initial, final, and average velocities are related. Because the velocity is increasing or decreasing *uniformly* with time, we can express the **average velocity** in any time interval as the arithmetic average of the initial velocity, v_0, and the final velocity, v:

$$\overline{v} = \frac{v_0 + v}{2} \qquad \text{(for constant } a\text{)} \qquad \textbf{[2.7]}$$

Note that *this expression is valid only when the acceleration is constant, that is, when the velocity changes uniformly with time.* The graphical interpretation of $\overline{v}$ is shown in Figure 2.13. As you can see, the velocity varies linearly with time according to Equation 2.6.

We can now use this result along with the defining equation for average velocity, Equation 2.2, to obtain an expression for the displacement of an object as a function of time. Again we choose $t_i = 0$, and for convenience we shall also assume that we have selected our coordinate system so that the initial position of the object under consideration is at the origin; that is, $x_i = 0$. This gives

$$x = \overline{v}t = \left(\frac{v_0 + v}{2}\right)t$$

$$x = \tfrac{1}{2}(v + v_0)t \qquad \textbf{[2.8]}$$

We can obtain another useful expression for displacement by substituting the equation for v (Eq. 2.6) into Equation 2.8:

$$x = \tfrac{1}{2}(v_0 + at + v_0)t$$

$$x = v_0 t + \tfrac{1}{2}at^2 \qquad \text{(for constant } a\text{)} \qquad \textbf{[2.9]}$$

Finally, we can obtain an expression that does not contain time by substituting the value of t from Equation 2.6 into Equation 2.8. This gives

$$x = \tfrac{1}{2}(v + v_0)\left(\frac{v - v_0}{a}\right) = \frac{v^2 - v_0^2}{2a}$$

$$v^2 = v_0^2 + 2ax \qquad \text{(for constant } a\text{)} \qquad \textbf{[2.10]}$$

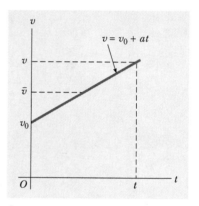

FIGURE 2.13
The velocity varies linearly with time for a particle moving with constant acceleration. The average velocity is just the mean value of the initial and final velocities.

TABLE 2.3
Equations for Motion in a Straight Line Under Constant Acceleration

Equation	Information Given by Equation
$v = v_0 + at$	Velocity as a function of time
$x = v_0 t + \frac{1}{2}at^2$	Displacement as a function of time
$v^2 = v_0^2 + 2ax$	Velocity as a function of displacement

Note: Motion is along the x axis. At $t = 0$, the particle is at the origin ($x_0 = 0$) and its velocity is v_0.

Equations 2.6 through 2.10 may be used to solve any problem in one-dimensional motion with constant acceleration. The three equations that are used most often are listed in Table 2.3 for convenience.

The best way to gain confidence in the use of these equations is to work a number of problems. Many times you will discover that there is more than one method for solving a given problem.

PROBLEM-SOLVING STRATEGY
Accelerated Motion

The following procedure is recommended for solving problems involving accelerated motion:

1. **Make sure all the units in the problem are consistent. That is, if distances are measured in meters, be sure that velocities have units of meters per second and accelerations have units of meters per second per second.**
2. **Choose a coordinate system.**
3. **Make a list of all the quantities given in the problem and a separate list of those to be determined.**
4. **Select from the list of kinematic equations the one or ones that will enable you to determine the unknowns.**
5. **Construct an appropriate motion diagram, and check to see if your answers are consistent with the diagram.**

EXAMPLE 2.3   The Indianapolis 500

A race car starting from rest accelerates at a rate of 5.00 m/s². What is the velocity of the car after it has traveled 100 ft?

Reasoning and Solution   Refer to the Problem-Solving Strategy section to see how the steps indicated there are applied to this example. First, you must be sure that the units are consistent. The units in this problem are *not* consistent. If we choose to leave the distance in feet, we must change the length dimension of the units of acceleration from meters to feet. Alternatively, we can leave the units of acceleration as they are and convert the distance traveled to meters. Let's do the latter. The table of conversion factors in the inside front cover gives 1 ft = 0.305 m; thus, 100 ft = 30.5 m.

Second, you must choose a coordinate system. A convenient one is shown in Figure 2.14. The origin of the coordinate system is at the initial location of the car, and the positive direction is to the right. Using this convention, we require that velocities, accelerations, and displacements to the right are positive, and vice versa.

Next, you will find it convenient to make a list of the quantities given in the problem and a separate list of those to be determined:

Given	To Be Determined
$v_0 = 0$	v
$a = +5.00$ m/s^2	
$x = +30.5$ m	

The final step is to select from kinematic equations (Table 2.3) those that will allow you to determine the unknowns. In the present case, the equation

$$v^2 = v_0{}^2 + 2ax$$

is our best choice because it will give us a value for v directly:

$$v^2 = (0)^2 + 2(5.00 \text{ m/s}^2)(30.5 \text{ m}) = 305 \text{ m}^2/\text{s}^2$$

from which

$$v = \sqrt{305 \text{ m}^2/\text{s}^2} = \pm 17.5 \text{ m/s}$$

Since the car is moving to the right, we choose $+17.5$ m/s as the correct solution for v. Alternatively, the problem can be solved by using $x = v_0 t + \frac{1}{2}at^2$ to find t and then using the expression $v = v_0 + at$ to find v. Try it!

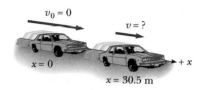

FIGURE 2.14
(Example 2.3)

EXAMPLE 2.4   The Supercharged Sports Car

A certain automobile manufacturer claims that its super-deluxe sports car will accelerate uniformly from rest to a speed of 87.0 mi/h in 8.00 s (Fig. 2.15).

(a) Determine the acceleration of the car.

Solution   First note that $v_0 = 0$ and the speed after 8.00 s is 87.0 mi/h = 38.9 m/s, where the conversion 1 mi/h = 0.447 m/s has been used. Using $v = v_0 + at$, we can find the acceleration:

$$a = \frac{v - v_0}{t} = \frac{38.9 \text{ m/s}}{8.00 \text{ s}} = \boxed{+4.86 \text{ m/s}^2}$$

(b) Find the displacement of the car in the first 8.00 s.

FIGURE 2.15
(Example 2.4)

Solution   The distance traveled by the car can be found from Equation 2.8:

$$x = v_0 t + \frac{1}{2}at^2 = 0 + \frac{1}{2}(4.86 \text{ m/s}^2)(8.00 \text{ s})^2 = \boxed{156 \text{ m}}$$

Frequently in this book, an example will be followed by an exercise whose purpose is to test your understanding of the example. The answer will be provided immediately after the exercise, when appropriate. Here is your first exercise, which relates to Example 2.4.

Exercise   What is the speed of the car 10.0 s after it begins its motion, assuming it continues to accelerate at the rate of + 4.86 m/s²?

Answer   48.6 m/s, or 109 mi/h.

2.7

FREELY FALLING BODIES

It is now well known that all objects dropped near the surface of the Earth in the absence of air resistance fall toward the Earth with the same constant acceleration. It was not until about 1600 that this conclusion was accepted. Prior to that time, the teachings of the great philosopher Aristotle (384–322 B.C.) had held that heavier objects fell faster than lighter ones.

It was Galileo who originated our present-day ideas concerning falling bodies. There is a legend that he discovered the law of falling bodies by observing that two different weights dropped simultaneously from the Leaning Tower of Pisa hit the ground at approximately the same time. Although it is doubtful that this particular experiment was carried out, it is well established that Galileo performed many systematic experiments on objects moving on inclined planes. In his experiments he rolled balls down a slight incline and measured the distances they covered in successive time intervals. The purpose of the incline was to reduce the acceleration and enable Galileo to make accurate measurements of the time intervals. (Some people refer to this experiment as "diluting gravity.") By gradually increasing the slope of the incline, he was finally able to draw conclusions about freely falling objects, because a falling ball is equivalent to a ball falling down a vertical incline. Galileo's achievements in the science of mechanics paved the way for Newton in his development of the laws of motion, which we shall study in Chapter 4.

You might want to try the following experiment. Simultaneously drop a coin and a crumpled-up piece of paper from the same height. If the effects of air friction are negligible, both will have the same motion and hit the floor at the same time. In the idealized case, where air resistance is absent, such motion is referred to as free fall. If this same experiment could be conducted in a vacuum, where air friction is zero, the paper and coin would fall with the same acceleration, regardless of the shape of the paper. (See the photographs in Physics in Action.) On August 2, 1971, such a demonstration was conducted on the Moon by astronaut David Scott. He simultaneously released a hammer and a feather, and they fell with the same acceleration to the lunar surface. This demonstration surely would have pleased Galileo!

PHYSICS IN *ACTION*

On the left is a multiflash photograph of a falling ball. The time interval between successive images is (1/60) s. As the ball falls, the spacing between successive images increases, indicating that the ball accelerates downward. Can you estimate **g** from this photograph?

The middle photograph shows a multiflash image of two freely falling balls released simultaneously. The ball on the left side is solid, and the ball on the right is a hollow Ping-Pong ball. The time interval between flashes is (1/30) s, and the scale is in centimeters. Can you determine **g** from these data? Note that the effect of air resistance is greater for the smaller, hollow ball on the right, as indicated by the smaller separation between successive images.

In the photograph on the right, a feather and an apple are released from rest in a 4-ft vacuum chamber. The trap door that released the two objects was opened with an electronic switch at the same instant the camera shutter was opened. The two objects fell at approximately the same rate, as indicated by the horizontal alignment of the multiple images. When air resistance is negligible, all objects fall with the same acceleration regardless of their masses.

FREELY FALLING BODIES

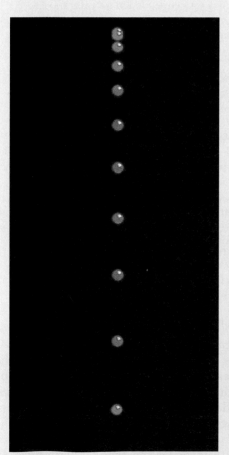

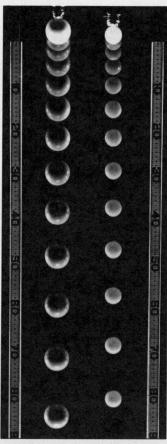

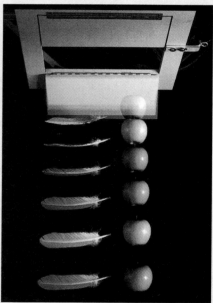

(© James Sugar, Black Star)

(Ken Edward/Science Source/Photo Researchers)   *(Education Development Center, Newton, MA)*

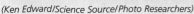

**Galileo Galilei
(1564–1642)**

(North Wind)

Galileo Galilei, an Italian physicist and astronomer, was born into the family of a Florentine cloth merchant in Pisa. As a young boy, Galileo was sent to school at the local monastery until his father, who was interested in many things (including mathematics and music), decided to teach him. The young Galileo entered medical school, but he soon withdrew because he could not accept many beliefs of the profession. Because of his mathematical skills, and through the influence of his father's friends, he received the chair in mathematics at the University of Pisa at age 25.

An interesting legend asserts that Galileo climbed the Leaning Tower of Pisa and simultaneously dropped a bullet and a cannonball to prove that they would fall with the same acceleration. Although there is no real evidence to support this story, Galileo formulated the laws that govern the motion of objects in free fall (including projectiles). He also investigated the motion of an object on an inclined plane, established the concept of relative motion, invented the thermometer, and discovered that the motion of a swinging pendulum could be used to measure time intervals. Some claim that he discovered the properties of a pendulum by using his pulse to time the motion of a swinging lamp in the Pisa Cathedral.

In 1609, Galileo learned of the invention of the telescope in Holland and immediately designed and constructed his own telescope, which he then used to make several major discoveries in astronomy. He discovered four of Jupiter's moons, found that the Moon's surface is rough (contrary to Aristotle's view), discovered sunspots and the phases of Venus, and showed that the Milky Way consists of an enormous number of stars.

As a result of his observations in astronomy, Galileo publicly defended Nicholaus Copernicus's assertion that the Sun is at the center of the Universe (the heliocentric system). The Church meanwhile had accepted Aristotle's viewpoint that the Earth was fixed and the heavens revolved around it, and declared the Copernican view to be heretical. Following many skirmishes with the Church, Galileo published *Dialogue Concerning Two New World Systems* to support the Copernican model. He was taken to Rome in 1633 on a charge of heresy and was sentenced to life imprisonment. The sentence was commuted to house arrest, and he was confined to his villa at Arcetri, near Florence, for the rest of his life. In his last years, Galileo summed up his life's work in a manuscript entitled *Discourses and Mathematical Discoveries Concerning Two New Sciences*. It was smuggled out of Italy and published in Holland in 1638. After completing this work, he went blind and died in Arcetri in 1642.

We shall denote the free-fall acceleration with the symbol **g**. The magnitude of **g** decreases with increasing altitude. Furthermore, slight variations in the magnitude of **g** occur with latitude. However, at the surface of the Earth the magnitude of **g** is approximately 9.8 m/s^2, or 980 cm/s^2, or 32 ft/s^2. Unless stated otherwise, we shall use the value 9.80 m/s^2 when doing calculations. Furthermore, we shall assume that the vector **g** is directed downward toward the center of the Earth.

When we use the expression *freely falling body,* we do not necessarily mean an object dropped from rest.

A freely falling body is an object moving freely under the influence of gravity only, regardless of its initial motion. Objects thrown upward or downward and those released from rest are all falling freely once they are released!

It is important to emphasize that any freely falling object experiences an acceleration directed downward. This is true regardless of the initial motion of the object. An object thrown upward (or downward) experiences the same acceleration as an object released from rest.

Once they are in free fall, all objects have an acceleration downward, which is the free-fall acceleration g.

If we neglect air resistance and assume that the free-fall acceleration does not vary with altitude over short vertical distances, then the motion of a freely falling body is equivalent to motion in one dimension under constant acceleration. Therefore, the equations developed in Section 2.6 for objects moving with constant acceleration can be applied. The only modification that we need to make in these equations for freely falling bodies is to note that the motion is in the vertical direction (the y direction) rather than the horizontal (x) direction, and that the acceleration is downward and has a magnitude of 9.80 m/s^2. Thus, we always take $a = -g = -9.80 \text{ m/s}^2$, where the minus sign means that the acceleration of a freely falling body is downward. In Chapter 7 we shall study how to deal with variations in g with altitude.

EXAMPLE 2.5   Look Out Below! [1]

A golf ball is released from rest at the top of a very tall building. Neglecting air resistance, calculate the position and velocity of the ball after 1.00, 2.00, and 3.00 s.

Solution   We choose our coordinates so that the starting point of the ball is at the origin ($y_0 = 0$ at $t = 0$) and remember that we have defined y to be positive upward. Since $v_0 = 0$, and $a = -g = -9.80 \text{ m/s}^2$, Equations 2.6 and 2.9 become

$$v = at = (-9.80 \text{ m/s}^2)t$$

$$y = \tfrac{1}{2}at^2 = \tfrac{1}{2}(-9.80 \text{ m/s}^2)t^2$$

where t is in seconds, v is in meters per second, and y is in meters. These expressions give the velocity and displacement at any time t after the ball is released. Therefore, at $t = 1.00$ s,

$$v = (-9.80 \text{ m/s}^2)(1.00 \text{ s}) = \quad -9.80 \text{ m/s}$$

$$y = \tfrac{1}{2}(-9.80 \text{ m/s}^2)(1.00 \text{ s})^2 = \quad -4.90 \text{ m}$$

[1] Worked examples and end-of-chapter problems marked with icon are accompanied by Interactive Physics II simulations on disk.

Likewise, at $t = 2.00$ s, we find that $v = -19.6$ m/s and $y = -19.6$ m. Finally, at $t = 3.00$ s, $v = -29.4$ m/s and $y = -44.1$ m. The minus sign for v indicates that the velocity vector is directed downward, and the minus sign for y indicates displacement in the negative y direction.

Exercise   Calculate the position and velocity of the ball after 4 s.

Answer -78.4 m, -39.2 m/s.

EXAMPLE 2.6   Not a Bad Throw for a Rookie!

A stone is thrown from the top of a building with an initial velocity of 20.0 m/s straight upward. The building is 50.0 m high, and the stone just misses the edge of the roof on its way down, as in Figure 2.16. Determine (a) the time needed for the stone to reach its maximum height, (b) the maximum height, (c) the time needed for the stone to return to the level of the thrower, (d) the velocity of the stone at this instant, and (e) the velocity and position of the stone at $t = 5.00$ s.

Solution   (a) To find the time necessary for the stone to reach the maximum height, use Equation 2.6, $v = v_0 + at$, noting that $v = 0$ at maximum height:

$$20.0 \text{ m/s} + (-9.80 \text{ m/s}^2)t = 0$$

$$t = \frac{20.0 \text{ m/s}}{9.80 \text{ m/s}^2} = \boxed{2.04 \text{ s}}$$

(b) This value of time can be substituted into Equation 2.9, $y = v_0 t + \frac{1}{2}at^2$, to give the maximum height as measured from the position of the thrower:

$$y_{max} = (20.0 \text{ m/s})(2.04 \text{ s}) + \frac{1}{2}(-9.80 \text{ m/s}^2)(2.04 \text{ s})^2 = \boxed{20.4 \text{ m}}$$

(c) When the stone is back at the height of the thrower, the y coordinate is zero. From the expression $y = v_0 t + \frac{1}{2}at^2$ (Eq. 2.9), with $y = 0$, we obtain the expression

$$0 = 20.0t - 4.90t^2$$

This is a quadratic equation and has two solutions for t. The equation can be factored to give

$$t(20.0 - 4.90t) = 0$$

One solution is $t = 0$, corresponding to the time at which the stone starts its motion. The other solution is $t = 4.08$ s, which is the solution we are after.

(d) The value for t found in (c) can be inserted into $v = v_0 + at$ (Eq. 2.6) to give

$$v = 20.0 \text{ m/s} + (-9.80 \text{ m/s}^2)(4.08 \text{ s}) = \boxed{-20.0 \text{ m/s}}$$

Note that the velocity of the stone when it arrives back at its original height is equal in magnitude to the stone's initial velocity but opposite in direction. This indicates that the motion is symmetric.

(e) From $v = v_0 + at$ (Eq. 2.6), the velocity after 5.00 s is

$$v = 20.0 \text{ m/s} + (-9.80 \text{ m/s}^2)(5.00 \text{ s}) = \boxed{-29.0 \text{ m/s}}$$

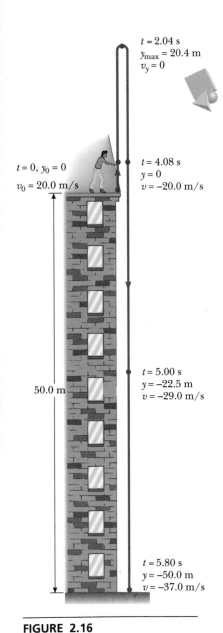

$t = 2.04$ s
$y_{max} = 20.4$ m
$v_y = 0$

$t = 0, y_0 = 0$
$v_0 = 20.0$ m/s

$t = 4.08$ s
$y = 0$
$v = -20.0$ m/s

$t = 5.00$ s
$y = -22.5$ m
$v = -29.0$ m/s

50.0 m

$t = 5.80$ s
$y = -50.0$ m
$v = -37.0$ m/s

FIGURE 2.16

(Example 2.6)   Position and velocity versus time for a freely falling object thrown initially upward with a velocity of $v_0 = +20.0$ m/s.

We can use $y = v_0 t + \frac{1}{2}at^2$ (Eq. 2.9) to find the position of the particle at $t = 5.00$ s:

$$y = (20.0 \text{ m/s})(5.00 \text{ s}) + \tfrac{1}{2}(-9.80 \text{ m/s}^2)(5.00 \text{ s})^2 = \boxed{-22.5 \text{ m}}$$

Exercise   Find the velocity of the stone just before it hits the ground.

Answer -37 m/s.

SUMMARY

The **displacement** of an object moving along the x axis is defined as the change in position of the object, and is given by

$$\Delta x \equiv x_f - x_i \qquad \text{[2.1]}$$

Displacement

where x_i is the initial position of the object and x_f is its final position.

The **average velocity** of an object moving along the x axis during some time interval is equal to the displacement of the object, Δx, divided by the time interval, Δt, during which the displacement occurred:

$$\bar{v} \equiv \frac{\Delta x}{\Delta t} = \frac{x_f - x_i}{t_f - t_i} \qquad \text{[2.2]}$$

Average velocity

The average velocity is equal to the slope of the straight line joining the initial and final points on a graph of the displacement of the object versus time.

The slope of the line tangent to the position-time curve at some point is equal to the instantaneous velocity at that time. The **instantaneous speed** of an object is defined as the magnitude of the instantaneous velocity.

The **average acceleration** of an object during some time interval is defined as the change in velocity, Δv, divided by the time interval, Δt, during which the change occurred:

$$\bar{a} \equiv \frac{\Delta v}{\Delta t} = \frac{v_f - v_i}{t_f - t_i} \qquad \text{[2.4]}$$

Average acceleration

The **instantaneous acceleration** of an object at a certain time equals the slope of a velocity-time graph at that instant.

A **vector** quantity is any quantity that is characterized by both a magnitude and a direction. In contrast, a **scalar** quantity such as mass has a magnitude only. Displacement, velocity, and acceleration are all vector quantities. In the case of motion along a straight line, we use algebraic signs to describe the directions of the vectors. For example, an object moving to the right with a speed of 8 m/s would have a velocity of $+8$ m/s, assuming the positive direction has been chosen to point to the right. If the object were moving to the left, its velocity would be -8 m/s.

Vectors and scalars

The equations that describe the motion of an object moving with constant acceleration along the x axis are

$$v = v_0 + at \qquad \text{[2.6]}$$

$$x = v_0 t + \tfrac{1}{2}at^2 \qquad \text{[2.9]}$$

$$v^2 = v_0{}^2 + 2ax \qquad \text{[2.10]}$$

Equations of kinematics

An object falling in the presence of the Earth's gravity experiences a free-fall acceleration directed toward the center of the Earth. If air friction is neglected and if the altitude of the falling object is small compared with the Earth's radius, then one can assume that the free-fall acceleration, **g**, is constant over the range of motion, where g is equal to 9.80 m/s², or 32.0 ft/s². Assuming that the positive direction for y is chosen to be upward, the acceleration is $-g$ (downward) and the equations describing the motion of the falling object are the same as the foregoing equations, with the substitutions $x \rightarrow y$ and $a \rightarrow -g$.

ADDITIONAL READING

I. B. Cohen, "Galileo," *Sci. American,* August 1949, p. 40.
S. Drake, "Galileo's Discovery of the Law of Free Fall," *Sci. American,* May 1973, p. 84.
Galileo Galilei, "Dialogues Concerning Two New Sci-ences," translated by H. Crew and A. de Salvio, Evanston, IL, Northwestern University Press, 1939.
O. Gingerich, "The Galileo Affair," *Sci. American,* August 1982, p. 132.

CONCEPTUAL QUESTIONS

Example   Can the equations of kinematics be used in a situation where the acceleration varies in time? Can they be used when the acceleration is zero?
Reasoning   The equations of kinematics cannot be used if the acceleration varies continuously. However, if the acceleration changes in steps, having one constant value for a while and then another constant value for some later time interval, the equations for constant-acceleration motion can be used to follow each section of the motion separately. If the acceleration is zero during some time interval, the velocity is constant, and the kinematics equations can be used; in this case, since $a = 0$, the equations become $v = v_0$, and $x - x_0 = vt$.

Example   A ball is thrown upward. While the ball is in the air, (a) what happens to its velocity? (b) does its acceleration increase, decrease, or remain constant?
Reasoning   (a) The velocity of the ball changes continuously. As it travels upward, its speed decreases by 9.80 m/s during each second of its motion, and its velocity vector is directed upward. When it reaches the peak of its motion, its speed becomes zero. As the ball moves downward, its speed increases by 9.80 m/s each second, and its velocity vector is directed downward. (b) The acceleration of the ball is constant in magnitude and direction throughout the free flight of the ball, from the instant it leaves the hand until the instant before it strikes the ground. Its magnitude is the free-fall acceleration, $g = 9.80$ m/s². (If the acceleration were zero at the peak when the velocity is zero, this would say that there would be no change in velocity thereafter, so the ball would stop at the peak and remain there, which is not the case.)

Example   A pebble is dropped into a water well, and the splash is heard 16.0 s later. Determine the *approximate* distance from the rim of the well to the water's surface.
Reasoning   If air resistance is neglected, then we can treat the pebble as an object undergoing free-fall acceleration with an initial velocity of zero. Under these conditions, the displacement versus time is equal to $\tfrac{1}{2}gt^2$. Since $g = 9.80$ m/s², and $t = 16.0$ s, we find a displacement of 1250 m.

Example If a car is traveling eastward, can its acceleration be westward? Explain.

Reasoning Yes. This occurs when a car is slowing down, so that the direction of its acceleration is opposite to its direction of motion.

Example A child throws a marble into the air with an initial velocity v_0. Another child drops a ball at the same instant. Compare the accelerations of the two objects while they are in flight.

Reasoning Once the objects leave the hand, both are freely-falling objects, and hence both experience the same acceleration equal to the free-fall acceleration, $-g$.

1. The speed of sound in air is 331 m/s. During the next thunderstorm, try to estimate your distance from a lightning bolt by measuring the time lag between the flash and the thunderclap. You can ignore the time it takes for the light flash to reach you. Why?

2. If the average velocity of an object is zero in some time interval, what can you say about the displacement of the object for that interval?

3. Average velocity and instantaneous velocity are generally different quantities. Can they ever be equal for a specific type of motion?

4. Can the instantaneous velocity of an object ever be greater in magnitude than the average velocity? Can it ever be less?

5. The following strobe photographs were taken of a disk moving from left to right under different conditions, and the time interval between images is constant. Taking the direction to the right as being positive, describe the motion of the disk in each case. For

(a)

(b)

(c)

(Question 5) *(Courtesy David Rogers)*

which case is (a) the acceleration positive? (b) the acceleration negative? (c) the velocity constant?

6. If the average velocity is nonzero for some time interval, does this mean that the instantaneous velocity is never zero during this interval? Explain.

7. If an object has zero velocity at some instant, is its acceleration necessarily zero at that time?

8. At the end of its arc, the velocity of a pendulum is zero. Is its acceleration also zero at this point?

9. If the velocity of a particle is not zero, can its acceleration ever be zero?

10. A ball is thrown vertically upward. What are its velocity and acceleration when it reaches its maximum altitude? What is its acceleration just before it hits the ground?

11. Can the sign of the acceleration ever be positive for an object that is slowing down? Explain.

12. An object is thrown upward. Discuss the sign of the free-fall acceleration relative to the velocity while the object is in the air.

13. Car A, traveling from New York to Miami, has a speed of 25 m/s. Car B, traveling from New York to Chicago, also has a speed of 25 m/s. Are their velocities equal? Explain.

14. Two cars are moving in the same direction in parallel lanes along a highway. At some instant, the velocity of car A exceeds the velocity of car B. Does this mean that A's acceleration is greater than B's? Explain.

15. Suppose that you are caught speeding and are presented with a traffic ticket. Will your fine be based on your average speed or on your instantaneous speed?

16. A rule of thumb for driving is that a separation of one car length for each 10 mi/h of speed should be maintained between moving vehicles. Assuming a constant reaction time, discuss the relevance of this rule for (a) motion with constant velocity and (b) motion with constant acceleration.

17. A stone is thrown upward from the top of a building. Does the stone's displacement depend on the location of the origin of the coordinate system? Does the stone's velocity depend on the origin?

18. A baseball is thrown vertically upward into the air and returns to its starting point. Sketch graphs of (a) its velocity versus time and (b) its acceleration versus time.

19. Two children stand atop a tall building. One drops a rock over the edge, while simultaneously the second throws a rock downward so that it has an initial speed of 10 m/s. Compare the accelerations of the two objects while in flight.

20. A ball is thrown vertically upward with an initial speed of 8.0 m/s. What will its speed be when it returns to its starting point?

21. A student at the top of a building of height h throws

one ball upward with a speed of v_0 and then throws a second ball downward with the same initial speed, v_0. How do the final velocities of the balls compare when they reach the ground?

22. A book is moved once around the perimeter of a tabletop with dimensions 1.00 m $\times$ 2.00 m. If the book ends up at its initial position, what is its displacement?

23. A space shuttle takes off from Florida and circles the globe several times, finally landing in California. While the shuttle is in flight, the wife of an astronaut flies from Florida to California so she can greet her husband when he steps off the shuttle. Who undergoes the greater displacement, the wife or the astronaut?

24. Does the displacement between two objects locate the objects? Discuss this in terms of the information carried by a vector.

PROBLEMS

Section 2.2   Average Velocity

Section 2.3   Instantaneous Velocity

1. The Olympic record for the marathon is 2 h, 9 min, 21 s. The marathon distance is 26 mi, 385 yd. Determine the average speed (in miles per hour) of this record.

2. If the average speed of an orbiting space shuttle is 19 800 mi/h, determine the time required for it to circle the Earth. Make sure you consider the fact that the shuttle is orbiting about 200 mi above the Earth's surface, and assume that the Earth's radius is 3963 miles.

3. A certain bacterium swims with a speed of 3.5 μm/s. How long would it take this bacterium to swim across a petri dish having a diameter of 8.4 cm?

4. A person travels by car from one city to another. She drives for 30.0 min at 80.0 km/h, 12.0 min at 100 km/h, and 45.0 min at 40.0 km/h, and spends 15.0 min eating lunch and buying gas. (a) Determine the average speed for the trip. (b) Determine the distance between the cities along this route.

5. An athlete swims the length of a 50.0-m pool in 20.0 s and makes the return trip to the starting position in 22.0 s. Determine her average velocities in (a) the first half of the swim, (b) the second half of the swim, and (c) the round trip.

6. A person takes a trip, driving with a constant speed of 89.5 km/h except for a 22.0-min rest stop. If the person's average speed is 77.8 km/h, how long is the trip?

7. A speedy tortoise can run with a speed of 10.0 cm/s, and a hare can run 20 times as fast. In a race, they both start at the same time, but the hare stops to rest for 2.0 minutes. The tortoise wins by a shell (20 cm). (a) How long does the race take? (b) What is the length of the race?

8. Runner A is initially 4.0 mi west of a flagpole and is running with a constant velocity of 6.0 mi/h due east. Runner B is initially 3.0 mi east of the flagpole and is running with a constant velocity of 5.0 mi/h due west. How far are the runners from the flagpole when their paths cross?

9. The position-time graph for a bug crawling along the x axis is shown in Figure 2.17. Determine whether the velocity is positive, negative, or zero for the times (a) t_1, (b) t_2, (c) t_3, (d) t_4.

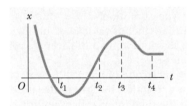

FIGURE 2.17   (Problem 9)

10. A runner moves so that his positions at certain times are given by the data in the following table. Use these data to construct a table like Table 2.2 in the text. From your table, find (a) the average velocity during the complete interval and (b) the instantaneous velocity at $t = 2.00$ s.

t (s)	x (m)	t (s)	x (m)
2.00	5.66	2.50	6.32
2.01	5.674	3.00	6.92
2.20	5.93	4.00	8.00

11. A race car moves such that its position fits the relationship

$$x = (5.0 \text{ m/s})t + (0.75 \text{ m/s}^3)t^3$$

where x is measured in meters and t in seconds. (a) Plot a graph of position versus time. (b) Determine the instantaneous velocity at $t = 4.0$ s, using time intervals of 0.40 s, 0.20 s, and 0.10 s. (c) Compare the average velocity during the first 4.0 s with the results of (b).

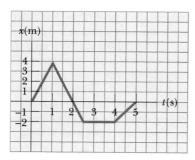

FIGURE 2.18   (Problems 13 and 14)

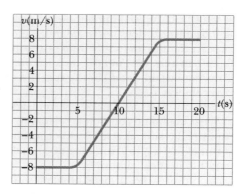

FIGURE 2.19   (Problem 20)

12. In order to qualify for the finals in a racing event, a race car must achieve an average speed of 250 km/h on a track with a total length of 1600 m. If a particular car covers the first half of the track at an average speed of 230 km/h, what minimum average speed must it have in the second half of the event in order to qualify?

13. A tennis player moves in a straight-line path as shown in Figure 2.18. Find her average velocities in the time intervals (a) 0 to 1.0 s, (b) 0 to 4.0 s, (c) 1.0 s to 5.0 s, (d) 0 to 5.0 s.

14. Find the instantaneous velocities of the tennis player of Figure 2.18 at (a) 0.50 s, (b) 2.0 s, (c) 3.0 s, (d) 4.5 s.

15. Two cars travel in the same direction along a straight highway, one at a constant speed of 55 mi/h and the other at 70 mi/h. (a) Assuming that they start at the same point, how much sooner does the faster car arrive at a destination 10 mi away? (b) How far must the faster car travel before it has a 15-min lead on the slower car?

Section 2.4   Acceleration

16. A car traveling in a straight line has a velocity of + 5.0 m/s at some instant. After 4.0 s, its velocity is + 8.0 m/s. What is its average acceleration in this time interval?

17. Jules Verne in 1865 proposed sending men to the Moon by firing a space capsule from a 220-m-long cannon with final velocity of 10.97 km/s. What would have been the unrealistically large acceleration experienced by the space travelers during launch? Compare your answer with the free-fall acceleration, 9.80 m/s².

18. A tennis ball with a speed of 10.0 m/s is thrown perpendicularly at a wall. After striking the wall, the ball rebounds in the opposite direction with a speed of 8.0 m/s. If the ball is in contact with the wall for 0.012 s, what is the average acceleration of the ball while it is in contact with the wall?

19. A car traveling initially at + 7.0 m/s accelerates at the rate of + 0.80 m/s² for an interval of 2.0 s. What is its velocity at the end of the acceleration?

20. The velocity-versus-time graph for an object moving along a straight path is shown in Figure 2.19. (a) Find the average accelerations of this object during the time intervals 0 to 5.0 s, 5.0 s to 15 s, and 0 to 20 s. (b) Find the instantaneous accelerations at 2.0 s, 10 s, and 18 s.

21. A record of travel along a straight path is as follows:
    1. Start from rest with constant acceleration of 2.77 m/s² for 15.0 s
    2. Constant velocity for the next 2.05 min
    3. Constant negative acceleration − 9.47 m/s² for 4.39 s

    (a) What was the total displacement for the complete trip? (b) What were the average speeds for legs 1, 2, and 3 of the trip as well as for the complete trip?

22. A certain car is capable of accelerating at a rate of + 0.60 m/s². How long does it take for this car to go from a speed of 55 mi/h to a speed of 60 mi/h?

23. The engine of a model rocket accelerates the rocket vertically upward for 2.0 s as follows: at $t = 0$, its speed is zero; at $t = 1.0$ s, its speed is 5.0 m/s; at $t = 2.0$ s, its speed is 16 m/s. Plot a velocity-time graph for this motion, and from it determine (a) the average acceleration during the 2.0-s interval and (b) the instantaneous acceleration at $t = 1.5$ s.

Section 2.6   One-Dimensional Motion with Constant Acceleration

24. A speedboat increases in speed from 20 m/s to 30 m/s in a distance of 200 m. Find (a) the magnitude of its acceleration and (b) the time it takes the boat to travel this distance.

25. A driver in a car traveling at a speed of 60 mi/h sees a deer 100 m away on the road. Calculate the minimum constant acceleration that is necessary

for the car to stop without hitting the deer (assuming that the deer does not move in the meantime).

26. A car accelerates uniformly from rest to a speed of 40 mi/h in 12 s. (a) Find the distance the car travels during this time and (b) the constant acceleration of the car.

27. A Cessna aircraft has a lift-off speed of 120 km/h. (a) What minimum constant acceleration does this require if the aircraft is to be airborne after a take-off run of 240 m? (b) How long does it take the aircraft to become airborne?

28. A racing car reaches a speed of 40 m/s. At this instant, it begins a uniform negative acceleration, using a parachute and a braking system, and comes to rest 5.0 s later. (a) Determine the acceleration of the car. (b) How far does the car travel after acceleration starts?

29. Two cars are traveling along a straight line in the same direction, the lead car at 25 m/s and the other car at 30 m/s. At the moment the cars are 40 m apart, the lead driver applies the brakes, causing her car to have an acceleration of -2.0 m/s^2. (a) How long does it take for the lead car to stop? (b) Assuming that the chasing car brakes at the same time as the lead car, what must be the chasing car's minimum negative acceleration so as not to hit the lead car? (c) How long does it take for the chasing car to stop?

30. An electron moving in a straight line has an initial speed of 3.0×10^5 m/s. It undergoes an acceleration of 8.0×10^{14} m/s^2. (a) How long will it take to reach a speed of 5.4×10^5 m/s? (b) How far will it have traveled in this time?

31. A jet plane lands with a velocity of $+100$ m/s and can accelerate at a maximum rate of -5.0 m/s^2 as it comes to rest. (a) From the instant it touches the runway, what is the minimum time needed before it can come to rest? (b) Can this plane land on a small island airport where the runway is 0.80 km long?

32. A train 400 m long is moving on a straight track with a speed of 82.4 km/h. The engineer applies the brakes at a crossing, and later the last car passes the crossing with a speed of 16.4 km/h. Assuming constant acceleration, determine how long the train blocked the crossing. Disregard the width of the crossing.

33. A car starts from rest and travels for 5.0 s with a uniform acceleration of $+1.5$ m/s^2. The driver then applies the brakes, causing a uniform negative acceleration of -2.0 m/s^2. If the brakes are applied for 3.0 s, how fast is the car going at the end of the braking period, and how far has it gone?

34. In order to pass a physical education class at a university, a student must run 1.0 mi in 12 min. After running for 10 min, she still has 500 yd to go. If her maximum acceleration is 0.15 m/s^2, can she make it? If the answer is no, determine what acceleration she would need to be successful.

35. A hockey player is standing on his skates on a frozen pond when an opposing player, moving with a uniform speed of 12 m/s, skates by with the puck. After 3.0 s, the first player makes up his mind to chase his opponent. If he accelerates uniformly at 4.0 m/s^2, (a) how long does it take him to catch his opponent, and (b) how far has he traveled in this time? (Assume the player with the puck remains in motion at constant speed.)

Section 2.7   Freely Falling Bodies

36. A ball is thrown vertically upward with a speed of 25.0 m/s. (a) How high does it rise? (b) How long does it take to reach its highest point? (c) How long does it take to hit the ground after it reaches its highest point? (d) What is its speed when it returns to the level from which it started?

37. A pebble is dropped into a deep well, and 3.0 s later the sound of a splash is heard as the pebble reaches the bottom of the well. The speed of sound in air is about 340 m/s. (a) How long does it take for the pebble to hit the water? (b) How long does it take for the sound to reach the observer? (c) What is the depth of the well?

38. In Mostar, Bosnia, the ultimate test of a young man's courage once was to jump off a 400-year-old bridge (now destroyed) into the River Neretva, 23 m below the bridge. (a) How long did the jump last? (b) How fast was the diver traveling upon impact with the river? (c) If the speed of sound in air is 340 m/s, how long after the diver took off did a spectator on the bridge hear the splash?

39. A model rocket is launched straight upward with an initial speed of 50.0 m/s. It accelerates with a constant upward acceleration of 2.00 m/s^2 until its engines stop at an altitude of 150 m. (a) What is the maximum height reached by the rocket? (b) How long after lift-off does the rocket reach its maximum height? (c) How long is the rocket in the air?

40. A ball thrown vertically upward is caught by the thrower after 20.0 s. Find (a) the initial velocity of the ball and (b) the maximum height it reaches.

41. A peregrine falcon dives at a pigeon. The falcon starts downward from rest and falls with free-fall acceleration. If the pigeon is 76.0 m below the initial position of the falcon, how long does it take the falcon to reach the pigeon? Assume that the pigeon remains at rest.

42. A parachutist descending at a speed of 10 m/s drops a camera from an altitude of 50 m. (a) How long does it take the camera to reach the ground? (b) What is the velocity of the camera just before it hits the ground?

43. A rocket moves upward, starting from rest with an acceleration of $+ 29.4$ m/s² for 4.00 s. It runs out of fuel at the end of this 4.00 s and continues to move upward. How high does it rise?

44. The tallest volcano in the Solar System is the 24-km-tall Martian volcano, Olympus Mons. Assume an astronaut drops a ball off the rim of the crater and that the free-fall acceleration remains constant throughout the ball's 24-km fall at a value of 3.7 m/s². (We assume the crater is as deep as the volcano is tall, which is not the case in nature.) Find (a) the time for the ball to reach the crater floor and (b) the velocity with which it hits. (In light of your answer for the velocity, does it seem reasonable that air resistance, even in Mars' thin atmosphere, can really be neglected in this problem?)

45. A small mailbag is released from a helicopter that is descending steadily at 1.5 m/s. After 2 s, (a) what is the speed of the mailbag, and (b) how far is it below the helicopter? (c) What are your answers to parts (a) and (b) if the helicopter is rising steadily at 1.5 m/s?

ADDITIONAL PROBLEMS

46. A bullet is fired through a board 10.0 cm thick in such a way that the bullet's line of motion is perpendicular to the face of the board. If the initial speed of the bullet is 400 m/s and it emerges from the other side of the board with a speed of 300 m/s, find (a) the acceleration of the bullet as it passes through the board and (b) the total time the bullet is in contact with the board.

47. Two students are on a balcony 19.6 m above the street. One student throws a ball vertically downward at 14.7 m/s; at the same instant the other

student throws a ball vertically upward at the same speed. The second ball just misses the balcony on the way down. (a) What is the difference in their time in air? (b) What is the velocity of each ball as it strikes the ground? (c) How far apart are the balls 0.800 s after they are thrown?

48. A train travels along a straight track between stations 1 and 2 as shown in Figure 2.20. The engineer of the train is instructed to start from rest at station 1 and accelerate uniformly between points A and B, then coast with a uniform velocity between points B and C, and finally decelerate uniformly between points C and D (at the same rate as between points A and B) until the train stops at station 2. If the distances AB, BC, and CD are all equal, and if it takes 5.00 min to travel between the two stations, determine how much of this 5.0-min period the train spends between points (i) A and B, (ii) B and C, and (iii) C and D.

49. One swimmer in a relay race has a 0.50-s lead and is swimming at a constant speed of 4.0 m/s. He has 50 m to swim before reaching the end of the pool. A second swimmer moves in the same direction as the leader. What constant speed must the second swimmer have in order to catch up to the leader at the end of the pool?

50. A certain cable car in San Francisco can stop in 10 s when traveling at maximum speed. On one occasion, the driver sees a dog a distance of d m in front of the car and slams on the brakes instantly. The car reaches the dog 8.0 s later, and the dog jumps off the track just in time. If the car travels 4.0 m beyond the position of the dog before coming to a stop, how far was the car from the dog? (*Hint:* You will need three equations.)

51. A ball is thrown upward from the ground with an initial speed of 25 m/s; at the same instant, a ball is dropped from a building 15 m high. After how long will the balls be at the same height?

52. A ranger in a National Park is driving at 35 mi/h when a deer jumps into the road 200 ft ahead of the vehicle. After a reaction time of t s, the ranger applies the brakes to produce an acceleration at

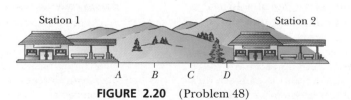

FIGURE 2.20 (Problem 48)

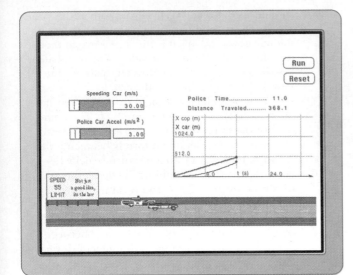

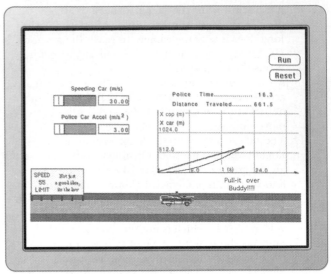

An example of the Interactive Physics simulation to accompany Problem 56.

$a = -9.0$ ft/s². What is the maximum reaction time allowed if she is to avoid hitting the deer?

53. A mountain climber stands at the top of a 50-m cliff that overhangs a calm pool of water. He throws two stones vertically downward 1.0 s apart and observes that they cause a single splash. The first stone has an initial velocity of $+2.0$ m/s. (a) How long after release of the first stone will the two stones hit the water? (b) What initial velocity must the second stone have if they are to hit simultaneously? (c) What will the velocity of each stone be at the instant they hit the water?

54. A person sees a lightning bolt pass close to an airplane that is flying in the distance. The person hears thunder 5.0 s after seeing the bolt, and sees the airplane overhead 10 s after hearing the thunder. The speed of sound in air is 1100 ft/s. (a) Find the distance of the airplane from the person at the instant of the bolt. (Neglect the time it takes the light to travel from the bolt to the eye.) (b) Assuming that the plane travels with a constant speed toward the person, find the velocity of the airplane. (c) Look up the speed of light in air, and defend the approximation used in (a).

55. A sports car enthusiast buys a super-deluxe machine that can accelerate at the rate of $+16$ ft/s². She decides to test her car in a drag race with an-

other speedster in a souped-up stock car. Both start from rest, but the experienced stock car driver leaves 1.0 s before the driver of the sports car. The stock car moves with a constant acceleration of $+12$ ft/s². Find (a) the time it takes the sports car to overtake it, (b) the distance the two travel before they are side by side, and (c) the velocities of both cars at the instant they are side by side.

 56. A speeder passes a parked police car at 30.0 m/s. The police car starts from rest with a uniform acceleration of 2.44 m/s². (a) How much time passes before the speeder is overtaken by the police car? (b) How far does the speeder get before being overtaken by the police car?

57. An ice sled powered by a rocket engine starts from rest on a large frozen lake and accelerates at $+40$ ft/s². After some time t_1 the rocket engine is shut down and the sled moves with constant velocity v for a time of t_2. If the total distance traveled by the sled is 17 500 ft and the total time is 90 s, find (a) the times t_1 and t_2 and (b) the velocity v. At the 17 500-ft mark, the sled begins to accelerate at -20 ft/s². (c) What is the final position of the sled when it comes to rest? (d) How long does it take to come to rest?

VECTORS AND TWO-DIMENSIONAL MOTION

3

This interesting set of directional arrows was photographed in Nome, Alaska. A particular location relative to Nome is specified by the direction of an arrow and the distance to that location. This is one type of vector quantity. We shall examine it and other vector quantities in this chapter. (David Bartruff, FPG)

In our discussion of one-dimensional motion in Chapter 2, we used the concept of vectors only to a very limited extent. As we progress in our study of motion, the ability to manipulate vector quantities will become increasingly important. As a result, much of this chapter will be devoted to techniques for adding vectors, subtracting them, and so forth. We will then apply our newfound skills to a special case of two-dimensional motion—projectiles. We shall also see that an understanding of vector manipulation is necessary in order to work with and understand relative motion.

3.1

VECTORS AND SCALARS REVISITED

Each of the physical quantities we shall encounter in this book can be categorized as either a scalar or a vector quantity. A scalar is a quantity that can be completely specified by its magnitude with appropriate units; that is, a **scalar** has

55

Jennifer pointing in the right direction. *(Raymond A. Serway)*

only magnitude and no direction. A **vector** is a physical quantity that requires the specification of both direction and magnitude.

Temperature is an example of a scalar quantity. If someone tells you that the temperature of an object is $-5°C$, that information completely specifies the temperature of the object; no direction is required. Other examples of scalars are masses, time intervals, and the number of pages in this textbook. Scalar quantities can be manipulated using the rules of ordinary arithmetic. For example, if you have 2 liters of water in a container (where 1 liter is defined to be 1000 cm³) and you add 3 more liters, by ordinary arithmetic the amount of water in the container is then 5 liters.

An example of a vector quantity is force. If you are told that someone is going to exert a force of 10 lb on an object, that is not enough information to let you know what will happen to the object. The effect of a force of 10 lb exerted horizontally is different from the effect of a force of 10 lb exerted vertically upward or downward. In other words, you need to know the direction of the force as well as its magnitude.

Velocity is another example of a vector quantity. If we wish to describe the velocity of a moving vehicle, we must specify both its speed (say, 30 m/s) and the direction in which the vehicle is moving (say, northeast). Other examples of vector quantities include displacement and acceleration, which were defined in Chapter 2.

3.2

SOME PROPERTIES OF VECTORS

Equality of Two Vectors.   Two vectors, **A** and **B**, are defined as equal if they have the same magnitude and the same direction. This property allows us to translate a vector parallel to itself in a diagram without affecting the vector. In fact, for most purposes, any vector can be moved parallel to itself without being affected.

Adding Vectors.   When two or more vectors are added together, they must all have the same units. For example, it would be meaningless to add a velocity vector to a displacement vector, because they are different physical quantities. Scalars also obey this rule. For example, it would be meaningless to add temperatures and areas.

When a vector quantity is handwritten, it is often represented with an arrow over the letter ($\vec{A}$). In this book, a vector quantity will be represented by boldface type (for example, **A**). The magnitude of a vector such as **A** will be represented by italic type such as *A*. Likewise, italic type will be used to represent scalars.

The procedures for adding vectors rely on geometric methods. (Later we shall develop an algebraic technique for adding vectors that is much more convenient and will be used throughout the remainder of the text.) To add vector **B** to vector **A**, first draw **A** on a piece of graph paper to some scale such as 1 cm = 1 m. Vector **A** must be drawn so that its direction is specified relative to a coordinate system. Then draw vector **B** to the same scale and with the tail of **B** starting at the tip of **A**, as in Figure 3.1a. Vector **B** must be drawn along the direction that makes the proper angle relative to vector **A**. The resultant vector, **R**, given by **R** = **A** + **B**, is the vector drawn from the tail of **A** to the tip of **B**. This is known as the *triangle method of addition*.

Weather vanes can be used to determine the direction of the wind velocity vector at any instant. *(© Steven Simpson/FPG)*

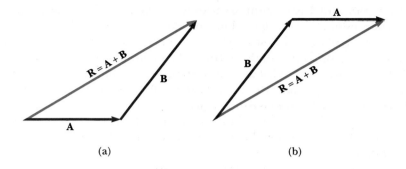

FIGURE 3.1
(a) When vector **B** is added to vector **A**, the vector sum, **R**, is the vector that runs from the tail of **A** to the tip of **B**. (b) Here the resultant runs from the tail of **B** to the tip of **A**. These constructions prove that **A** + **B** = **B** + **A**.

When two vectors are added, the sum is independent of the order of the addition. That is, **A** + **B** = **B** + **A**. This can be seen from the geometric construction in Figure 3.1b.

An alternative graphical procedure for adding two vectors, known as the *parallelogram rule of addition,* is shown in Figure 3.2. In this construction, the tails of vectors **A** and **B** are joined together, and the resultant vector, **R**, is the diagonal of the parallelogram formed with **A** and **B** as its sides.

This same general approach can also be used to add more than two vectors, as is done in Figure 3.3 for four vectors. The resultant vector sum, **R** = **A** + **B** + **C** + **D**, is the vector drawn from the tail of the first vector to the tip of the last vector. Again, the order in which vectors are added is unimportant.

Negative of a Vector.   The negative of the vector **A** is defined as the vector that when added to **A** gives zero for the vector sum. This means that **A** and −**A** have the same magnitude but opposite directions.

Subtraction of Vectors.   Vector subtraction makes use of the definition of the negative of a vector. We define the operation **A** − **B** as vector −**B** added to vector **A**:

$$\mathbf{A} - \mathbf{B} = \mathbf{A} + (-\mathbf{B}) \tag{3.1}$$

The geometric construction for subtracting two vectors is shown in Figure 3.4.

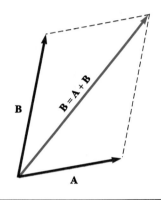

FIGURE 3.2
In this construction, the resultant **R** is the diagonal of a parallelogram with sides **A** and **B**.

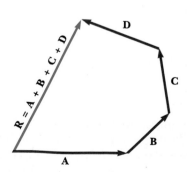

FIGURE 3.3
A geometric construction for summing four vectors. The resultant vector, **R**, is the vector that completes the polygon.

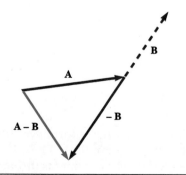

FIGURE 3.4
This construction shows how to subtract vector **B** from vector **A**. The vector −**B** has the same magnitude as the vector **B** but points in the opposite direction.

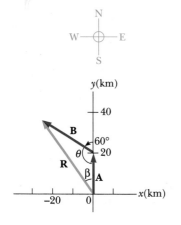

FIGURE 3.5
(Example 3.1) A graphical method for finding the resultant displacement vector **R** = **A** + **B**.

Multiplication and Division of Vectors by Scalars. The multiplication or division of a vector by a scalar gives a vector. For example, if a vector, **A**, is multiplied by the scalar number 3, the result, written 3**A**, is a vector with a magnitude three times that of the original vector **A**, pointing in the same direction as **A**. On the other hand, if we multiply vector **A** by the scalar −3, the result is a vector with a magnitude three times that of **A**, pointing in the direction opposite **A** (because of the negative sign).

EXAMPLE 3.1   Taking a Trip

A car travels 20.0 km due north and then 35.0 km in a direction 60.0° west of north, as in Figure 3.5. Find the magnitude and direction of the car's resultant displacement.

Solution   The problem can be solved geometrically using graph paper and a protractor, as shown in Figure 3.5. The resultant displacement, **R**, is the sum of the two individual displacements **A** and **B**.

The length of **R**, drawn to the same scale as **A** and **B**, indicates that the displacement of the car is 48.2 km, and a measurement of the angle β shows that the displacement is approximately 38.9° west of north.

3.3

COMPONENTS OF A VECTOR

One method of adding vectors makes use of the projections of a vector along the axes of a rectangular coordinate system. These projections are called **components**. Any vector can be completely described by its components.

Consider a vector, **A**, in a rectangular coordinate system, as shown in Figure 3.6. Note that **A** can be expressed as the sum of two vectors, **A**$_x$ parallel to the x axis, and **A**$_y$ parallel to the y axis. That is,

$$\mathbf{A} = \mathbf{A}_x + \mathbf{A}_y$$

A$_x$ and **A**$_y$ are the component vectors of **A**. The projection of **A** along the x axis, A_x, is called the x component of **A**, and the projection of **A** along the y axis, A_y, is called the y component of **A**. *These components can be either positive or negative numbers with units.* From the definitions of sine and cosine of an angle, we see that $\cos \theta = A_x/A$ and $\sin \theta = A_y/A$. Hence, the magnitudes of the components of **A** are

$$A_x = A \cos \theta$$
$$A_y = A \sin \theta \qquad \text{[3.2]}$$

These components form two sides of a right triangle, the hypotenuse of which has the magnitude A. Thus, it follows that **A**'s magnitude and direction are related to its components through the Pythagorean theorem and the definition of the tangent:

$$A = \sqrt{A_x^2 + A_y^2} \qquad \text{[3.3]}$$

$$\tan \theta = \frac{A_y}{A_x} \qquad \text{[3.4]}$$

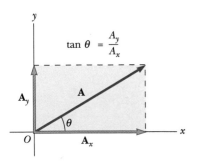

FIGURE 3.6
Any vector **A** lying in the xy plane can be represented by its rectangular components, **A**$_x$ and **A**$_y$.

To solve for the angle θ, we can write Equation 3.4 in the form

$$\theta = \tan^{-1}\left(\frac{A_y}{A_x}\right)$$

If a coordinate system other than the one shown in Figure 3.6 is chosen, the components of the vector must be modified accordingly. In many applications it is more convenient to express the components of a vector in a coordinate system having axes that are not horizontal and vertical but are still perpendicular to each other. Suppose a vector, **B**, makes an angle of θ' with the x' axis defined in Figure 3.7. The rectangular components of **B** along the axes of Figure 3.7 are given by $B_{x'} = B\cos\theta'$ and $B_{y'} = B\sin\theta'$, as in Equation 3.2. The magnitude and direction of **B** are obtained from expressions equivalent to Equations 3.3 and 3.4. Thus, we can express the components of a vector in any coordinate system that is convenient for the situation.

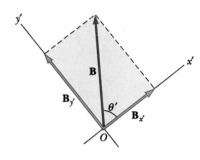

FIGURE 3.7
The components of vector **B** in a tilted coordinate system.

ADDING VECTORS

In order to add vectors algebraically rather than graphically, the following procedure is used. First, find the components of all the vectors in some coordinate system that is appropriate for the problem. Next, add all the x components to find the resultant component in the x direction. Similarly, add all the y components to find the resultant component in the y direction. Because the resultant x and y components are at right angles to each other, you can use the Pythagorean theorem to determine the magnitude of the resultant vector. Finally, use a suitable trigonometric function and the components of the resultant vector to find the angle that the resultant vector makes with the x axis. These steps are summarized as a Problem-Solving Strategy and are illustrated in the worked examples that follow.

PROBLEM-SOLVING STRATEGY
Adding Vectors

When two or more vectors are to be added, the following steps are used:

1. **Select a coordinate system.**
2. **Draw a sketch of the vectors to be added (or subtracted), with a label on each vector.**
3. **Find the x and y components of all vectors.**
4. **Find the resultant components (the algebraic sum of the components) in both the x and y directions.**
5. **Use the Pythagorean theorem to find the magnitude of the resultant vector.**
6. **Use a suitable trigonometric function to find the angle the resultant vector makes with the x axis.**

EXAMPLE 3.2   Help Is on the Way!

Find the horizontal and vertical components of the 100-m displacement of a superhero who flies from the top of a tall building along the path shown in Figure 3.8a.

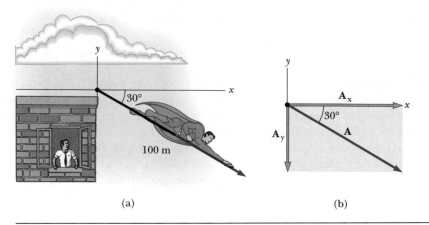

FIGURE 3.8
(Example 3.2)

Solution The triangle formed by the displacement and its components is shown in Figure 3.8b. Since $A = 100$ m and $\theta = -30.0°$ (θ is negative because it is measured clockwise from the x axis), we have

$$A_y = A \sin \theta = (100 \text{ m}) \sin(-30.0°) = \boxed{-50.0 \text{ m}}$$

Note that $\sin(-\theta) = -\sin \theta$. The negative sign for A_y reflects the fact that displacement in the y direction is *downward* from the origin.

The x component of displacement is

$$A_x = A \cos \theta = (100 \text{ m}) \cos(-30.0°) = \boxed{+86.6 \text{ m}}$$

Note that $\cos(-\theta) = \cos \theta$. Also, from an inspection of the figure, you should be able to see that A_x is positive in this case.

EXAMPLE 3.3 **Taking a Hike**

A hiker begins a trip by first walking 25.0 km due southeast from her base camp. On the second day she walks 40.0 km in a direction 60.0° north of east, at which point she discovers a forest ranger's tower.

(a) Determine the components of the hiker's displacements in the first and second days.

Solution If we denote the displacement vectors on the first and second days by **A** and **B**, respectively, and use the camp as the origin of coordinates, we get the vectors shown in Figure 3.9. Displacement **A** has a magnitude of 25.0 km and is 45.0° south of east. Its components are

$$A_x = A \cos(-45.0°) = (25.0 \text{ km})(0.707) = \boxed{17.7 \text{ km}}$$

$$A_y = A \sin(-45.0°) = -(25.0 \text{ km})(0.707) = \boxed{-17.7 \text{ km}}$$

The negative value of A_y indicates that the y coordinate decreased in this displacement. The signs of A_x and A_y are also evident from Figure 3.9.

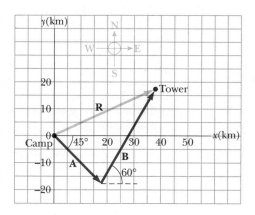

FIGURE 3.9
(Example 3.3)

The second displacement, **B**, has a magnitude of 40.0 km and is 60.0° north of east. Its components are

$$B_x = B \cos 60.0° = (40.0 \text{ km})(0.500) = \boxed{20.0 \text{ km}}$$

$$B_y = B \sin 60.0° = (40.0 \text{ km})(0.866) = \boxed{34.6 \text{ km}}$$

(b) Determine the components of the hiker's total displacement for the trip.

Solution The resultant displacement for the trip, **R** = **A** + **B**, has components given by

$$R_x = A_x + B_x = 17.7 \text{ km} + 20.0 \text{ km} = \boxed{37.7 \text{ km}}$$

$$R_y = A_y + B_y = -17.7 \text{ km} + 34.6 \text{ km} = \boxed{16.9 \text{ km}}$$

Exercise Determine the magnitude and direction of the total displacement.

Answer 41.3 km, 24.1° north of east from the base camp.

3.4

VELOCITY AND ACCELERATION IN TWO DIMENSIONS

In our discussion of one-dimensional motion in Chapter 2, the vector nature of displacement, velocity, and acceleration was taken into account through the use of positive and negative signs. To completely describe the motion of an object in two or three dimensions, we must make use of vectors.

Consider an object moving along some curve in space as shown in Figure 3.10. When the object is at some point P at time t_i, its position is described by the position vector $\mathbf{r}_i$, drawn from the origin to P. Likewise, when the object has moved to some other point, Q, at time t_f, its position vector is $\mathbf{r}_f$. As you can see from the vector diagram in Figure 3.10, the final position vector is the sum of the initial position vector and $\Delta\mathbf{r}$. Since $\mathbf{r}_f = \mathbf{r}_i + \Delta\mathbf{r}$, the displacement of the object is defined as the change in the position vector:

$$\Delta\mathbf{r} \equiv \mathbf{r}_f - \mathbf{r}_i \qquad \text{[3.5]}$$

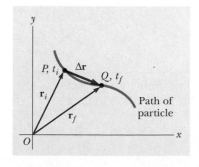

FIGURE 3.10
An object moving along some curved path between points P and Q. The displacement vector, $\Delta\mathbf{r}$, is the difference in the position vectors. That is, $\Delta\mathbf{r} = \mathbf{r}_f - \mathbf{r}_i$.

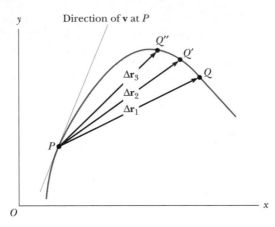

FIGURE 3.11

As the time interval between the final and initial positions approaches zero, the direction of the displacement vector **Δr** approaches that of the line tangent to the curve at *P*. By definition, the instantaneous velocity at *P* is in the direction of this tangent line.

Note that the magnitude of the displacement vector is *not* equal to the distance traveled measured along the curved path. In fact, the displacement is less than this distance when the path is curved.

The **average velocity** of a particle during the time interval Δt is the ratio of the displacement to the time interval for this displacement.

Average velocity

$$\bar{\mathbf{v}} \equiv \frac{\Delta \mathbf{r}}{\Delta t} \qquad [3.6]$$

Since the displacement is a vector and the time interval is a scalar, we conclude that the average velocity is a *vector* quantity directed along **Δr**. Note that the average velocity between points *P* and *Q* is *independent of the path* between the two points. This is because the average velocity is proportional to the displacement, which in turn depends only on the initial and final position vectors. As in the case of one-dimensional motion, we conclude that if a particle starts its motion at some point and returns to this point via any path, its average velocity for the trip is zero since its displacement is zero.

Consider again the motion of a particle between two points in the *xy* plane, as in Figure 3.11. As the time intervals become smaller and smaller, the displacements, $\Delta \mathbf{r}_1$, $\Delta \mathbf{r}_2$, $\Delta \mathbf{r}_3$, . . ., get progressively smaller and the direction of the displacement approaches that of the line tangent to the path at the point *P*. The **instantaneous velocity, v,** is defined as the limit of the average velocity, $\Delta \mathbf{r}/\Delta t$, as Δt goes to zero:

Instantaneous velocity

$$\mathbf{v} \equiv \lim_{\Delta t \to 0} \frac{\Delta \mathbf{r}}{\Delta t} \qquad [3.7]$$

The direction of the velocity vector is along a line that is tangent to the path of the particle and in the direction of motion.

The average acceleration of an object whose velocity changes by **Δv** in the time interval Δt is a vector defined as the ratio $\Delta \mathbf{v}/\Delta t$.

Average acceleration

$$\bar{\mathbf{a}} \equiv \frac{\Delta \mathbf{v}}{\Delta t} \qquad [3.8]$$

Instantaneous acceleration

The instantaneous acceleration vector, **a**, is defined as the limit of the average acceleration vector as Δt goes to zero.

It is important to recognize that a particle can accelerate in several ways. First, the magnitude of the velocity vector (the speed) may change with time. Second, a particle accelerates when the direction of the velocity vector changes with time (makes a curved path) even though the speed is constant. Finally, acceleration may be due to changes in both the magnitude and the direction of the velocity vector.

3.5

PROJECTILE MOTION

In the situations we considered in Chapter 2, objects moved along straight-line paths, such as the x axis. Now let us look at some cases in which an object moves in a plane. By this we mean that the object may move in both the x and y directions simultaneously, or move in two dimensions. The particular form of two-dimensional motion we shall concentrate on is called **projectile motion**. Anyone who has observed a baseball in motion (or, for that matter, any object thrown into the air) has observed projectile motion. It is surprisingly simple to analyze if the following three assumptions are made:

1. The free-fall acceleration, g, has a magnitude of 9.80 m/s², is constant over the range of motion, and is directed downward.
2. The effect of air resistance is negligible.
3. The rotation of the Earth does not affect the motion.

With these assumptions, we shall find that the path of a projectile is curved as shown in Figure 3.12. Such a curve is called a parabola.

Let us choose our coordinate system so that the y direction is vertical and positive upward. In this case,

the acceleration in the y direction is $-g$, just as in free fall, and the acceleration in the x direction is 0 (because air friction is neglected).

In a volcanic eruption in Mount Etna, Sicily, the lava particles from the eruption follow parabolic paths, as one would expect, since they are projectiles near the surface of the Earth. (© *Otto Hahn, Peter Arnold, Inc.*)

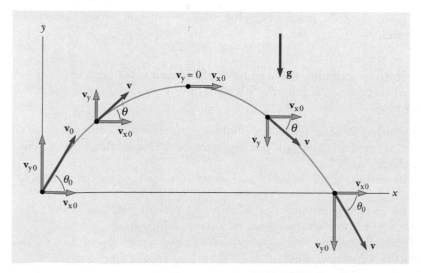

FIGURE 3.12
The parabolic trajectory of a particle that leaves the origin with a velocity of $\mathbf{v}_0$. Note that the velocity, $\mathbf{v}$, changes with time. However, the x component of the velocity, v_x, remains constant in time. Also, $v_y = 0$ at the peak, but the acceleration is always equal to the free-fall acceleration and acts vertically downward.

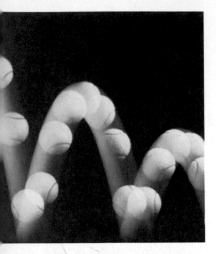

A multiflash photograph of a tennis ball undergoing several bounces off a hard surface. Note the parabolic path of the ball following each bounce. *(© Richard Megna 1992, Fundamental Photographs)*

Furthermore, let us assume that at $t = 0$, the projectile leaves the origin with a velocity of $\mathbf{v}_0$, as shown in Figure 3.12. If the velocity vector makes an angle of θ_0 with the horizontal where θ_0 is called the projection angle, then from the definitions of cosine and sine functions and Figure 3.12, we have

$$v_{x0} = v_0 \cos \theta_0 \quad \text{and} \quad v_{y0} = v_0 \sin \theta_0$$

In order to analyze projectile motion, we shall separate the motion into two parts, the x (or horizontal) motion and the y (or vertical) motion, and solve each part separately. We shall look first at the x motion. As noted before, motion in the x direction occurs with $a_x = 0$. This means that *the velocity component along the x direction remains constant.* Thus, if the initial value of the velocity component in the x direction is $v_{x0} = v_0 \cos \theta_0$, this is also the value of the velocity at any later time. That is,

$$v_x = v_{x0} = v_0 \cos \theta_0 = \text{constant} \tag{3.9}$$

Equation 3.9 can be substituted into the defining equation for velocity (Eq. 2.2) to give us an expression for the horizontal position of the projectile as a function of time:

$$x = v_{x0}t = (v_0 \cos \theta_0)t \tag{3.10}$$

These equations tell us all we need to know about the motion in the x direction. Let us now consider the y motion. Since it has constant acceleration, the equations developed in Section 2.6 can be used. In these equations, v_{y0} shall denote the initial velocity in the y direction and $-g$, the free-fall acceleration. The negative sign for g indicates that the positive direction for the vertical motion is assumed to be upward. With this choice of signs, we have

$$v_y = v_{y0} - gt \tag{3.11}$$

$$y = v_{y0}t - \tfrac{1}{2}gt^2 \tag{3.12}$$

$$v_y^2 = v_{y0}^2 - 2gy \tag{3.13}$$

where $v_{y0} = v_0 \sin \theta_0$. The speed, v, of the projectile at any instant can be calculated from the components of velocity at that instant, using the Pythagorean theorem:

$$v = \sqrt{v_x^2 + v_y^2}$$

Before we examine some numerical examples dealing with projectile motion, let us pause to summarize what we have learned so far about this kind of motion:

1. Provided air resistance is negligible, the horizontal component of velocity, v_x, remains constant because there is no horizontal component of acceleration.
2. The vertical component of acceleration is equal to the free-fall acceleration, g.
3. The vertical component of velocity, v_y, and the displacement in the y direction are identical to those of a freely falling body.
4. Projectile motion can be described as a superposition of the two motions in the x and y directions.

PROBLEM-SOLVING STRATEGY
Projectile Motion

We suggest that you use the following approach to solving projectile motion problems:

1. **Select a coordinate system.**
2. **Resolve the initial velocity vector into x and y components.**
3. **Treat the horizontal motion and the vertical motion independently.**
4. **Follow the techniques for solving problems with constant velocity to analyze the horizontal motion of the projectile.**
5. **Follow the techniques for solving problems with constant acceleration to analyze the vertical motion of the projectile.**

EXAMPLE 3.4   The Stranded Explorers

An Alaskan rescue plane drops a package of emergency rations to a stranded party of explorers, as shown in Figure 3.13. The plane is traveling horizontally at 40.0 m/s at a height of 100 m above the ground.

(a) Where does the package strike the ground relative to the point at which it was released?

Solution   The coordinate system for this problem is selected as shown in Figure 3.13, with the positive x direction to the right and the positive y direction upward.

Consider first the horizontal motion of the package. The only equation available to us is

$$x = v_{x0} t$$

The initial x component of the package velocity is the same as the velocity of the plane when the package was released, 40.0 m/s. Thus, we have

$$x = (40.0 \text{ m/s}) t$$

If we know t, the length of time the package is in the air, we can determine x, the distance traveled by the package along the horizontal. To find t, we move to the equations for the vertical motion of the package. We know that at the in-

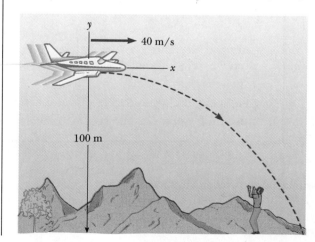

FIGURE 3.13
(Example 3.4)   From the point of view of an observer on the ground, a package released from the rescue plane travels along the path shown.

stant the package hits the ground its y coordinate is -100 m. We also know that the initial velocity of the package in the vertical direction, v_{y0}, is zero because the package was released with only a horizontal component of velocity. From Equation 3.12 we have

$$y = -\tfrac{1}{2}gt^2$$

$$-100 \text{ m} = -\tfrac{1}{2}(9.80 \text{ m/s}^2)\,t^2$$

$$t^2 = 20.4 \text{ s}^2$$

$$t = 4.51 \text{ s}$$

This value for the time of flight substituted into the equation for the x coordinate gives

$$x = (40.0 \text{ m/s})(4.51 \text{ s}) = \boxed{180 \text{ m}}$$

(b) What are the horizontal and vertical components of the velocity of the package just before it hits the ground?

Solution   We already know the horizontal component of the velocity of the package just before it hits, because the velocity in the horizontal direction remains constant at 40.0 m/s throughout the flight.

The vertical component of the velocity just before the package hits the ground may be found by using Equation 3.11, with $v_{y0} = 0$:

$$v_y = v_{y0} - gt = 0 - (9.80 \text{ m/s}^2)(4.51 \text{ s}) = \boxed{-44.1 \text{ m/s}}$$

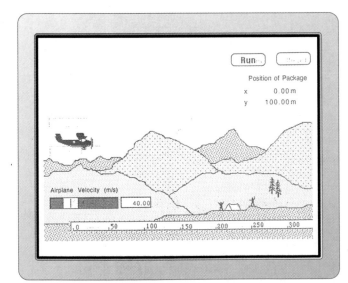
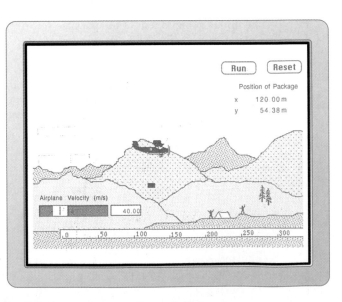

An example of the Interactive Physics simulation to accompany Example 3.4.

EXAMPLE 3.5 The Long Jump

A long jumper leaves the ground at an angle of 20.0° to the horizontal and at a speed of 11.0 m/s, as in the photograph.

(a) How far does he jump? (Assume that the motion of the long jumper is equivalent to that of a particle.)

Solution His horizontal motion is described by using Equation 3.10:

$$x = (v_0 \cos \theta_0)t = (11.0 \text{ m/s})(\cos 20.0°)t$$

The value of x can be found if t, the total duration of the jump, is known. We can find t with $v_y = v_0 \sin \theta_0 - gt$ (Eq. 3.11) by noting that at the top of the jump the vertical component of velocity goes to zero:

$$v_y = v_0 \sin \theta_0 - gt$$
$$0 = (11.0 \text{ m/s}) \sin 20.0° - (9.80 \text{ m/s}^2)t_1$$
$$t_1 = 0.384 \text{ s}$$

Note that t_1 is the time interval to the *top* of the jump. Because of the symmetry of the vertical motion, an identical time interval passes before the jumper returns to the ground. Therefore, the *total time* in the air is $t = 2t_1 = 0.768$ s, and the distance jumped is

$$x = (11.0 \text{ m/s})(\cos 20.0°)(0.768 \text{ s}) = \boxed{7.94 \text{ m}}$$

(b) What is the maximum height reached?

Solution The maximum height reached is found using $y = (v_0 \sin \theta_0)t - \frac{1}{2}gt^2$ (Eq. 3.12) with $t = t_1 = 0.384$ s.

$$y_{max} = (11.0 \text{ m/s})(\sin 20.0°)(0.384 \text{ s}) - \tfrac{1}{2}(9.80 \text{ m/s}^2)(0.384 \text{ s})^2 = \boxed{0.722 \text{ m}}$$

Although the assumption that the motion of the long jumper is that of a projectile is an oversimplification of the situation, the values obtained are reasonable.

In a long-jump event, Willie Banks can leap horizontal distances of at least 8 meters. *(© R. Mackson/FPG)*

EXAMPLE 3.6   That's Quite an Arm

A stone is thrown upward from the top of a building at an angle of 30.0° to the horizontal and with an initial speed of 20.0 m/s, as in Figure 3.14. The height of the building is 45.0 m.

(a) How long is the stone "in flight"?

Solution   The initial x and y components of the velocity are

$$v_{x0} = v_0 \cos \theta_0 = (20.0 \text{ m/s})(\cos 30.0°) = +17.3 \text{ m/s}$$

$$v_{y0} = v_0 \sin \theta_0 = (20.0 \text{ m/s})(\sin 30.0°) = +10.0 \text{ m/s}$$

To find t, we can use $y = v_{y0}t - \frac{1}{2}gt^2$ (Eq. 3.12) with $y = -45.0$ m and $v_{y0} = 10.0$ m/s (we have chosen the point of release as the origin, as shown in Fig. 3.14):

$$-45.0 \text{ m} = (10.0 \text{ m/s})t - \tfrac{1}{2}(9.80 \text{ m/s}^2)t^2$$

Solving the quadratic equation for t (see Appendix A) gives, for the positive root, the value $t = 4.22$ s. Does the negative root have any physical meaning? (Why not think of another way of finding t from the information given?)

(b) What is the speed of the stone just before it strikes the ground?

Solution   The y component of the velocity just before the stone strikes the ground can be obtained using Equation 3.11 with $t = 4.22$ s:

$$v_y = v_{y0} - gt = 10.0 \text{ m/s} - (9.80 \text{ m/s}^2)(4.22 \text{ s}) = -31.4 \text{ m/s}$$

Since $v_x = v_{x0} = 17.3$ m/s, the required speed is given by

$$v = \sqrt{v_x^2 + v_y^2} = \sqrt{(17.3)^2 + (-31.4)^2} \text{ m/s} = \boxed{35.9 \text{ m/s}}$$

Exercise   Where does the stone strike the ground?

Answer   73.0 m from the base of the building.

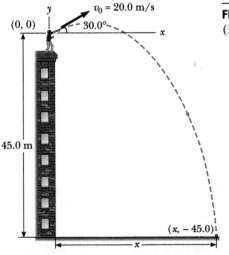

FIGURE 3.14
(Example 3.6)

PHYSICS IN *ACTION*

PARABOLIC PATHS

On the left is a multiflash photograph of a popular laboratory demonstration in which a projectile is fired at a falling target. The conditions of the experiment are that the gun is aimed at the target and that the projectile leaves the gun at the same instant that the target is released from rest. Under these conditions the projectile will hit the target, as shown, independent of the initial velocity of the projectile. The reason is that both objects move toward the Earth with the same acceleration.

The middle photograph shows an evening view of an illuminated water fountain in La Place de la Concorde, Paris. The individual water streams follow parabolic trajectories. The photograph shows that the horizontal range and maximum height of a given stream depend on the elevation angle of its initial velocity.

On the right, a welder cuts holes through a heavy metal construction beam with a hot torch. The sparks generated in the process follow parabolic paths.

(Courtesy of Central Scientific Company)

(© The Telegraph Colour Library/FPG)

(© The Telegraph Colour Library/FPG)

*3.6

RELATIVE VELOCITY

Observers in different frames of reference may measure different displacements or velocities for an object in motion. That is, two observers moving with respect to each other would generally not agree on the outcome of a measurement.

For example, if two cars were moving in the same direction with speeds of 50 mi/h and 60 mi/h, a passenger in the slower car would measure the speed of the faster car relative to the slower car as 10 mi/h. Of course, a stationary observer would measure the speed of the faster car as 60 mi/h. This simple example demonstrates that velocity measurements differ in different frames of reference.

Another simple example is a package dropped from an airplane flying parallel to the Earth with a constant velocity. An observer from the airplane would describe the motion of the package as a straight line toward the Earth. On the other hand, an observer on the ground would view the trajectory of the package as that of a projectile. Relative to the ground, the package would have a vertical component of velocity (resulting from the acceleration due to gravity and equal to the velocity measured by the observer in the airplane) *and* a horizontal component of velocity (given to it by the airplane's motion). If the airplane continued to move horizontally with the same velocity, the package would hit the ground directly beneath the airplane (assuming negligible air resistance).

In order to develop a general approach to solving relative-motion problems, let us return to the case of the fast car and the slower car. The situation was a fast car moving at 60 mi/h and a slower car moving at 50 mi/h, and we wanted to know the speed of the fast car with respect to the slower car. We solved this problem with a minimum of thought and effort, but you will encounter many situations in which a more systematic method for attacking such problems is beneficial. To develop this method, let us write down all the information we are given and that which we want to know in the form of velocities with subscripts appended. We have

$v_{se} = +50$ mi/h   (Here the subscript se means the velocity of the *slower* car with respect to the *Earth.*)

$v_{fe} = +60$ mi/h   (The subscript fe means the velocity of the *fast* car with respect to the *Earth.*)

We want to know v_{fs}, which is the velocity of the *fast* car with respect to the slower car. To find this, we write an equation for v_{fs} in terms of the other velocities, so on the right side of the equation the subscripts start with f and eventually end with s. Also, each velocity subscript starts with the letter that ended the preceding velocity subscript. For this case, we have

$$v_{fs} = v_{fe} + v_{es}$$

The boldface notation is, of course, indicative of the fact that velocity is a vector quantity. As we shall see in the following examples, this vector nature of the velocity is of paramount importance in certain instances.

We know that $v_{es} = -v_{se}$, so

$$v_{fs} = +60 \text{ mi/h} - 50 \text{ mi/h} = +10 \text{ mi/h}$$

There is no general equation for you to memorize in order to work relative velocity problems; instead, you should develop the necessary equations on your own by following the technique already demonstrated for writing subscripts. We suggest that you practice this technique to work the following examples.

EXAMPLE 3.7   Where Does the Ball Land?

A passenger at the rear of a train, traveling at 15 m/s relative to the Earth, throws a baseball with a speed of 15 m/s in the direction opposite the motion of the train. What is the velocity of the baseball relative to the Earth?

Solution We first write down our knowns and unknowns with appropriate subscripts:

$$\mathbf{v}_{te} = +15 \text{ m/s} \quad (\text{velocity of the } train, \text{ t, relative to the } Earth, \text{ e})$$

$$\mathbf{v}_{bt} = -15 \text{ m/s} \quad (\text{velocity of the } baseball, \text{ b, relative to the } train, \text{ t})$$

We need $\mathbf{v}_{be}$, the velocity of the *baseball* relative to the *Earth*. Following the strategy for watching the subscripts while we write down the equation for our unknown, we have

$$\mathbf{v}_{be} = \mathbf{v}_{bt} + \mathbf{v}_{te}$$

or

$$\mathbf{v}_{be} = +15 \text{ m/s} - 15 \text{ m/s} = 0$$

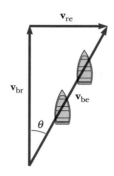

EXAMPLE 3.8 Crossing a River

A boat heading due north crosses a wide river with a velocity of 10.0 km/h relative to the water. The river has a uniform velocity of 5.00 km/h due east. Determine the velocity of the boat with respect to an observer on the riverbank.

FIGURE 3.15
(Example 3.8)

Solution We have

$\mathbf{v}_{br} = 10.0$ km/h due north (velocity of the *boat*, b, with respect to the *river*, r)

$\mathbf{v}_{re} = 5.00$ km/h due east (velocity of the *river*, r, with respect to the *Earth*, e)

and we want, $\mathbf{v}_{be}$, the velocity of the *boat* with respect to the *Earth*. Our equation becomes

$$\mathbf{v}_{be} = \mathbf{v}_{br} + \mathbf{v}_{re}$$

Because the velocities are not along the same direction as in preceding examples, the terms in the equation must be manipulated as vector quantities, which are shown in Figure 3.15. The quantity $\mathbf{v}_{br}$ is due north, $\mathbf{v}_{re}$ is due east, and the vector sum of the two, $\mathbf{v}_{be}$, is at the angle θ, as defined in Figure 3.15. Thus, we see that the velocity of the boat with respect to the Earth can be found from the Pythagorean theorem as

$$v_{be} = \sqrt{(v_{br})^2 + (v_{re})^2} = \sqrt{(10.0 \text{ km/h})^2 + (5.00 \text{ km/h})^2} = 11.2 \text{ km/h}$$

and the direction of $\mathbf{v}_{be}$ is

$$\theta = \tan^{-1}\left(\frac{v_{re}}{v_{br}}\right) = \tan^{-1}\frac{5.00}{10.0} = 26.6°$$

Therefore, the boat travels at a speed of 11.2 km/h in the direction 63.4° north of east with respect to the Earth.

EXAMPLE 3.9 Which Way Should the Boat Head?

If the boat of the preceding example travels with the same speed of 10.0 km/h relative to the water and is to travel due north, as in Figure 3.16, in what direction should it travel?

Solution We know

$$\mathbf{v}_{br} = \text{velocity of the } boat, \text{ b, with respect to the } river, \text{ r}$$

$$\mathbf{v}_{re} = \text{velocity of the } river, \text{ r, with respect to the } Earth, \text{ e}$$

and we need $\mathbf{v}_{be}$, the velocity of the *boat*, b, with respect to the *Earth*, e. We have

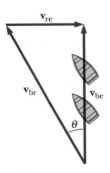

FIGURE 3.16
(Example 3.9)

$$\mathbf{v}_{be} = \mathbf{v}_{br} + \mathbf{v}_{re}$$

The relationship among these three vectors is shown in Figure 3.16; it agrees with our intuitive guess that the boat must head upstream in order to be pushed directly northward across the water. The speed v_{be} can be found from the Pythagorean theorem:

$$v_{be} = \sqrt{(v_{br})^2 - (v_{re})^2} = \sqrt{(10.0 \text{ km/h})^2 - (5.00 \text{ km/h})^2} = \boxed{8.66 \text{ km/h}}$$

and the direction of v_{be} is

$$\theta = \tan^{-1} \frac{v_{re}}{v_{be}} = \tan^{-1}\left(\frac{5.00}{8.66}\right) = 30°$$

where θ is west of north.

SPECIAL TOPIC

RELATIVE MOTION AT HIGH SPEEDS

As mentioned in the preceding section, we have techniques to transform the velocity of a particle as measured in the Earth's reference frame to velocities measured in a frame moving with uniform motion with respect to the Earth. However, these techniques are valid *only* at particle speeds (relative to both observers) that are small compared with the speed of light, c (where $c \approx 3 \times 10^8$ m/s). When the particle speed according to either observer approaches the speed of light, it is necessary to use equations developed by Einstein in his special theory of relativity. Although we shall discuss the theory of relativity in Chapter 26, a preview of some of its predictions is in order.

You may wonder how one can test the validity of the transformation equations. The velocity transformation equations given in the preceding section are used in Newtonian mechanics, whereas Einstein's theory uses relativistic transformations. From experiments on high-speed particles such as electrons and protons in particle accelerators, one finds that Newtonian mechanics *fails* at particle speeds approaching the speed of light. On the other hand, Einstein's theory of special relativity is in agreement with experiment at *all* speeds. (Thus, both types of transformation equations give the results found in the last section when the particle speed is small compared with the speed of light.) Finally, Newtonian mechanics places no upper limit on speed of a particle. In contrast, the relativistic velocity transformation equations predict particle speeds that *can never exceed the speed of light*. Electrons and protons accelerated through very high voltages can attain speeds close to the speed of light but can never reach that value. Hence, experimental results are in complete agreement with the theory of relativity.

SUMMARY

Two vectors, **A** and **B**, can be added geometrically by either the triangle method or the parallelogram rule. In the **triangle method,** the two vectors are drawn to scale, on graph paper, so that the tail of the second vector starts at the tip of the first. The resultant vector is the vector drawn from the tail of the first to the tip of the second. In the **parallelogram method,** the vectors are drawn to scale on graph paper with the tails of the two vectors joined. The resultant vector is the diagonal of the parallelogram formed with the two vectors as its sides.

The negative of a vector **A** is a vector with the same magnitude as **A** but pointing in the opposite direction.

The x component of a vector is equivalent to its projection along the x axis of a coordinate system. Likewise, the y component is the projection of the vector along the y axis of this coordinate system. The resultant of two or more vectors can be found mathematically by resolving all vectors into their x and y components, finding the resultant x and y components, and then using the Pythagorean theorem to find the resultant vector. The angle of the resultant vector with respect to the x axis can be found by use of a suitable trigonometric function.

An object moving above the surface of the Earth in both the x and y directions simultaneously is said to be undergoing **projectile motion**. The object moves along the horizontal (x) direction so that its velocity in this direction, v_x, is a constant, and the object also moves in the vertical (y) direction with a constant downward free-fall acceleration of magnitude $g = 9.80$ m/s^2. The equations describing the motion of a projectile are

$$x = v_{x0} t \qquad \text{[3.10]}$$

$$v_y = v_{y0} - gt \qquad \text{[3.11]}$$

$$y = v_{y0} t - \tfrac{1}{2} g t^2 \qquad \text{[3.12]}$$

$$v_y{}^2 = v_{y0}{}^2 - 2gy \qquad \text{[3.13]}$$

Projectile motion

where $v_{y0} = v_0 \sin \theta_0$ is the initial vertical component of the velocity and $v_{x0} = v_0 \cos \theta_0$ is the initial horizontal component of the velocity.

Observations made in different frames of reference can be related to one another through the techniques of the transformation of relative velocities.

ADDITIONAL READING

P. Brancazio, "The Trajectory of a Fly Ball," *The Physics Teacher,* January 1985, p. 20.

S. Drake and J. MacLachlan, "Galileo's Discovery of the Parabolic Trajectory," *Sci. American,* March 1975, p. 102.

W. F. Magie, *Source Book in Physics,* Cambridge, Mass., Harvard University Press, 1963. Contains many excerpts from Galileo on projectile motion.

CONCEPTUAL QUESTIONS

Example   Can a force directed vertically on an object ever cancel a force directed horizontally?
Reasoning   No. Pushes or pulls can only cancel when the two forces are in opposite directions along the same line. A vertical force can never cancel a horizontal force.

Example   Does a ball dropped out the window of a moving car take longer to reach the ground than one dropped from a car at rest at the same height?
Reasoning   They both take the same time to reach the ground. Neglecting air resistance, once released, they

both experience the same downward free-fall acceleration, $-\mathbf{g}$.

Example   (a) Can an object accelerate if its speed is constant? (b) Can an object accelerate if its velocity is constant?
Reasoning   (a) Yes. Although its speed may be constant, the *direction* of its motion (that is, the direction of $\mathbf{v}$) may change, causing an acceleration. For example, an object moving in a circle with constant speed has an acceleration directed toward the center of the circle (a centripetal acceleration). (b) No. An object that moves with constant velocity has zero acceleration. Note that constant velocity means that both the direction and magnitude of $\mathbf{v}$ remain constant.

Example   As a projectile moves in its parabolic path, is there any point along its path where the velocity and acceleration are (a) perpendicular to each other? (b) parallel to each other?
Reasoning   (a) At the top of its flight, $\mathbf{v}$ is horizontal and $\mathbf{a}$ is vertical. This is the only point where the velocity and acceleration vectors are perpendicular. (b) If the object is thrown straight up or down, then $\mathbf{v}$ and $\mathbf{a}$ will be parallel throughout the downward motion. Otherwise, the velocity and acceleration vectors are never parallel.

Example   A ball is thrown upward in the air by a passenger on a train that is moving with constant velocity.
(a) Describe the path of the ball as seen by the passenger. Describe the path as seen by a stationary observer outside the train. (b) How would these observations change if the train were accelerating along the track?
Reasoning   (a) The passenger sees the ball moving vertically up and then down with a velocity that changes with time. The external observer sees the ball moving in a parabola, with constant horizontal velocity equal to that of the train, as well as a continuously changing vertical velocity. (b) If the train were accelerating horizontally forward, the passenger would see the ball accelerating horizontally backward as well as vertically downward. Relative to the passenger, the ball would move in a parabola with its axis inclined to the vertical. The outside observer would see pure parabolic motion for the ball, with a horizontal acceleration of zero, a vertical acceleration of $-\mathbf{g}$, and a horizontal velocity equal to that of the train at the moment the ball was released.

Example   A student accurately uses the method for combining vectors. The two vectors she combines have magnitudes of 55 and 25 units. The answer that she gets is either 85, 20, or 55. Pick the correct answer and state why it is the only one of the three that can be correct.
Reasoning   The maximum vector one can obtain by adding vectors of 50 and 25 units occurs when the vectors are in the same direction. In this case, the maximum vector is 80 units in length, so we have to exclude the vector of length 85 as a possible answer. Likewise, the minimum vector is obtained when the vectors being added are in opposite directions. For our case, the minimum vector is 30 units in length, so we must exclude the vector of length 20 units as a possible answer. The vectors being added must be at some angle with respect to each other to give a resultant vector of length 55 units.

1. If $\mathbf{B}$ is added to $\mathbf{A}$, under what conditions does the resultant vector have a magnitude equal to $A + B$? Under what conditions is the resultant vector equal to zero?
2. Two vectors have unequal magnitudes. Can their sum be zero? Explain.
3. Vector $\mathbf{A}$ lies in the xy plane. For what orientations will both of its rectangular components be negative? For what orientations will its components have opposite signs?
4. Can a vector have a component greater than its magnitude?
5. Under what circumstances would a vector have components that are equal in magnitude?
6. If the component of vector $\mathbf{A}$ along the direction of vector $\mathbf{B}$ is zero, what can you conclude about these two vectors?
7. Can a vector have a component equal to zero and still have a nonzero magnitude? Explain.
8. If $\mathbf{A} + \mathbf{B} = 0$, what can you say about the components of the two vectors?
9. List the steps in a procedure for reconstructing a vector from a set of components. Sketch any appropriate diagrams to accompany the list.
10. "If a set of vectors laid tail to head forms a closed polygon, the resultant is zero." Is this statement true? Discuss.
11. "The magnitude of a vector is a scalar." Explain this statement.
12. Can the magnitude of a vector have a negative value?
13. State which of the following are vectors and which are not: force, temperature, the amount of water in a can, the weight of a book, the height of a building, the velocity of a sports car, the age of the Universe.
14. Is it possible to add a vector quantity to a scalar quantity? Explain.
15. If a rock is dropped from the top of a sailboat's mast, will it hit the deck at the same point whether the boat is at rest or in motion at constant velocity?
16. A rock is dropped at the same instant that a ball, at the same initial elevation, is thrown horizontally. Which will have the greater velocity when it reaches ground level?
17. In one scene of a Bugs Bunny cartoon, we find Bugs standing on a large flat rock as both he and the rock fall through the air under gravity. Just before impact with the ground, Bugs Bunny steps off the rock and walks away, completely unharmed; the rock is totally destroyed. Why is this scene unrealistic? What is a more realistic fate for Bugs?

PROBLEMS

Section 3.2   Some Properties of Vectors

1. A person walks 25.0° north of east for 3.10 km. How far would a person walk due north and due east to arrive at the same location?

2. A golfer takes two putts to get his ball into the hole once he is on the green. The first putt displaces the ball 6.00 m east, and the second, 5.40 m south. What displacement would have been needed to get the ball into the hole on the first putt?

3. Vector **A** is 3.00 units in length and points along the positive x axis. Vector **B** is 4.00 units in length and points along the negative y axis. Use graphical methods to find the magnitude and direction of the vectors (a) **A** + **B** and (b) **A** − **B**.

4. Each of the displacement vectors **A** and **B** shown in Figure 3.17 has a magnitude of 3.00 m. Graphically find (a) **A** + **B**, (b) **A** − **B**, (c) **B** − **A**, (d) **A** − 2**B**

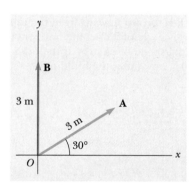

FIGURE 3.17   (Problem 4)

5. A roller coaster moves 200 ft horizontally, then rises 135 ft at an angle of 30.0° above the horizontal. It then travels 135 ft at an angle of 40.0° downward. What is its displacement from its starting point at the end of this movement? Use graphical techniques.

6. A dog searching for a bone walks 3.50 m south, then 8.20 m at an angle 30.0° north of east, and finally 15.0 m west. Find the dog's resultant displacement vector, using graphical techniques.

7. A jogger runs 100 m due west, then changes direction for the second leg of the run. At the end of the run, she is 175 m away from the starting point at an angle of 15.0° north of west. What were the direction and length of her second displacement? Use graphical techniques.

8. A man lost in a maze makes three consecutive displacements so that at the end of the walks he is right back where he started. The first displacement is 8.00 m westward, and the second is 13.0 m northward. Find the magnitude and direction of the third displacement, using the graphical method.

Section 3.3   Components of a Vector

9. A roller coaster travels 135 ft at an angle of 40.0° above the horizontal. How far does it move horizontally and vertically?

10. The eye of a hurricane passes over Grand Bahama Island. It is moving in a direction 60.0° north of west with a speed of 41.0 km/h. Three hours later, the course of the hurricane suddenly shifts due north, and its speed slows to 25.0 km/h. How far from Grand Bahama is the hurricane 4.50 h after it passes over the island?

11. A quarterback takes the ball from the line of scrimmage, runs backward for 10.0 yards, then runs sideways parallel to the line of scrimmage for 15.0 yards. At this point, he throws a 50.0-yard forward pass straight downfield, perpendicular to the line of scrimmage. What is the magnitude of the football's resultant displacement?

12. A person going for a walk follows the path shown in Figure 3.18. The total trip consists of four straight-line paths. At the end of the walk, what is the person's resultant displacement measured from the starting point?

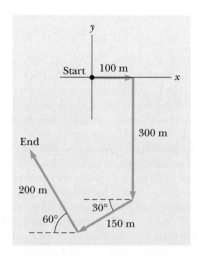

FIGURE 3.18   (Problem 12)

□ indicates problems that have full solutions available in the Student Solutions Manual and Study Guide.

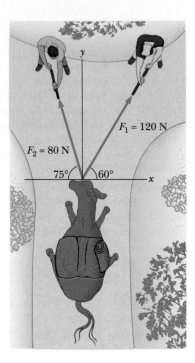

FIGURE 3.19 (Problem 15)

13. A shopper pushing a cart through a store moves 40.0 m down one aisle, then makes a 90.0° turn and moves 15.0 m. He then makes another 90.0° turn and moves 20.0 m. Where is the shopper relative to his original position, in magnitude and direction? Note that you are not given the direction moved after any of the 90.0° turns. As a result, there could be more than one answer.

14. A girl delivering newspapers covers her route by traveling 3.00 blocks west, 4.00 blocks north, then 6.00 blocks east. (a) What is her resultant displacement? (b) What is the total distance she travels?

15. Two people pull on a stubborn mule, as seen from a helicopter in Figure 3.19. Find (a) the single force that is equivalent to the two forces shown, and (b) the force that a third person would have to exert on the mule to make the net force equal to zero.

16. A man pushing a mop across a floor causes it to undergo two displacements. The first has a magnitude of 150 cm and makes an angle of 120° with the positive *x* axis. The resultant displacement has a magnitude of 140 cm and is directed at an angle of 35.0° to the positive *x* axis. Find the magnitude and direction of the second displacement.

Section 3.4 Velocity and Acceleration in Two Dimensions

Section 3.5 Projectile Motion

17. A golfball with an initial speed of 50 m/s lands exactly 240 m downrange on a level course. (a) Neglecting air friction, what *two* projection angles would achieve this result? (b) What is the maximum height reached by the ball, using the two angles determined in part (a)?

18. Tom the cat is chasing Jerry the mouse across a table surface 1.5 m off the floor. Jerry steps out of the way at the last second, and Tom slides off the edge of the table at a speed of 5.0 m/s. Where will Tom strike the floor, and what velocity components will he have just before he hits?

19. A place kicker must kick a football from a point 36.0 m (about 40.0 yd) from the goal, and the ball must clear the crossbar, which is 3.05 m high. When kicked, the ball leaves the ground with a speed of 20.0 m/s at an angle of 53° to the horizontal. (a) By how much does the ball clear or fall short of clearing the crossbar? (b) Does the ball approach the crossbar while still rising or while falling?

20. A brick is thrown upward from the top of a building at an angle of 25° to the horizontal and with an initial speed of 15 m/s. If the brick is in flight for 3.0 s, how tall is the building?

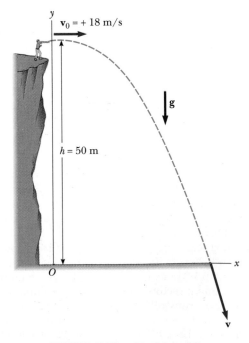

FIGURE 3.20 (Problem 22)

21. A ball is thrown straight upward and returns to the thrower's hand after 3.00 s in the air. A second ball is thrown at an angle of 30.0° with the horizontal. At what speed must the second ball be thrown so that it reaches the same height as the one thrown vertically?

22. A student stands at the edge of a cliff and throws a stone horizontally over the edge with a speed of 18 m/s. The cliff is 50 m above a flat horizontal beach, as shown in Figure 3.20. How long after being released does the stone strike the beach below the cliff? With what speed and angle of impact does it land?

23. The fastest recorded pitch in major-league baseball, thrown by Nolan Ryan in 1974, was clocked at 100.8 mi/h. If a pitch were thrown horizontally with this velocity, how far would the ball fall vertically by the time it reached home plate, 60.0 ft away?

24. A car is parked on a cliff overlooking the ocean on an incline that makes an angle of 24.0° below the horizontal. The negligent driver leaves the car in neutral, and the emergency brakes are defective. The car rolls from rest down the incline with a constant acceleration of 4.00 m/s² and travels 50.0 m to the edge of the cliff. The cliff is 30.0 m above the ocean. Find (a) the car's position relative to the base of the cliff when the car lands in the ocean, and (b) the length of time the car is in the air.

Section 3.6   Relative Velocity

25. A hunter wishes to cross a river that is 1.5 km wide and flows with a speed of 5.0 km/h parallel to its

banks. The hunter uses a small powerboat that moves at a maximum speed of 12 km/h with respect to the water. What is the minimum time necessary for crossing?

26. How long does it take an automobile traveling in the left lane at 60.0 km/h to overtake (become even with) another car that is traveling in the right lane at 40.0 km/h, when the cars' front bumpers are initially 100 m apart?

27. An escalator is 20.0 m long. If a person stands on the "up" side, it takes 50.0 s to ride to the top. (a) If a person walks up the moving escalator with a speed of 0.500 m/s relative to the escalator, how long does it take to get to the top? (b) If a person walks down the "up" escalator with the same relative speed as in (a), how long does it take to reach the bottom?

28. A boat moves through the water of a river at 10 m/s relative to the water, regardless of the boat's direction. If the water in the river is flowing at 1.5 m/s, how long does it take the boat to make a round trip consisting of a 300-m displacement downstream followed by a 300-m displacement upstream?

29. A skater (water spider) maintains an average position on the surface of a stream by darting upstream (against the current), then drifting downstream (with the current) to its original position. The current in the stream is 0.500 m/s relative to the shore, and the skater darts upstream 0.560 m (relative to a spot on shore) in 0.800 s during the first part of its motion. Take upstream as the positive direction. (a) Determine the velocity of the skater relative to the water (i) during its dash upstream and (ii) during its drift downstream. (b) How far upstream relative to the water does the skater move during one cycle of this motion? (c) What is the average velocity of the skater relative to the water?

30. Two canoeists in identical canoes exert the same effort paddling and hence maintain the same speed relative to the water. One paddles directly upstream (and moves upstream), whereas the other paddles directly downstream. If downstream is the positive direction, an observer on shore determines the velocities of the two canoes to be −1.2 m/s and +2.9 m/s, respectively. (a) What is the speed of the water relative to shore? (b) What is the velocity of each canoe relative to the water?

31. A river flows due east at 1.50 m/s. A boat crosses the river from the south shore to the north shore by maintaining a constant velocity of 10.0 m/s due north relative to the water. (a) What is the velocity of the boat relative to shore? (b) If the river is

(Problem 23) Nolan Ryan throwing his fast ball.
(Michael Layton, Duomo)

300 m wide, how far downstream has the boat moved by the time it reaches the north shore?

32. A rowboat crosses a river with a velocity of 3.30 mi/h at an angle 62.5° north of west relative to the water. The river is 0.505 mi wide and carries an eastward current of 1.25 mi/h. How far upstream is the boat when it reaches the opposite shore?

33. The pilot of an aircraft wishes to fly due west in a 50.0-km/h wind blowing toward the south. If the speed of the aircraft in the absence of a wind is 200 km/h, (a) in what direction should the aircraft head, and (b) what should its speed be relative to the ground?

34. A car travels due east with a speed of 50.0 km/h. Rain is falling vertically with respect to the Earth. The traces of the rain on the side windows of the car make an angle of 60.0° with the vertical. Find the velocity of the rain with respect to (a) the car and (b) the Earth.

35. A science student is riding on a flatcar of a train traveling along a straight horizontal track at a constant speed of 10.0 m/s. The student throws a ball along a path that she judges to make an initial angle of 60.0° with the horizontal and to be in line with the track. The student's professor, who is standing on the ground nearby, observes the ball to rise vertically. How high does the ball rise?

ADDITIONAL PROBLEMS

36. An artillery shell is fired with an initial speed of 1.70×10^3 m/s (approximately five times the speed of sound) at an initial angle of 55.0° to the horizontal. Neglecting air resistance, find (a) its horizontal range and (b) the time it is in motion.

37. Two swimmers start at the same point in a stream that flows with a speed of v. Each swimmer is capable of swimming with the same speed, c ($c > v$), relative to the stream. One swimmer swims downstream a distance of L and then upstream the same distance, while the other swimmer swims directly perpendicular to the stream's flow a distance of L and then back the same distance, so that both swimmers return to the same point. Which swimmer returns first? (*Note:* First guess at the answer.)

38. A shopper in a department store can walk up a stationary (stalled) escalator in 30 s. If the normally functioning escalator can carry the standing shopper to the next floor in 20 s, how long would it take the shopper to walk up the moving escalator? Assume the same walking effort for the shopper whether the escalator is stalled or moving.

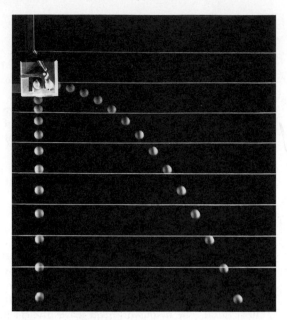

FIGURE 3.21   (Problem 40) *(Educational Development Center)*

39. A ball is projected horizontally from the edge of a table that is 1.00 m high, and it strikes the floor at a point 1.20 m from the base of the table. (a) What is the initial speed of the ball? (b) How high is the ball above the floor when its velocity vector makes a 45.0° angle with the horizontal?

40. Figure 3.21 is a multiflash photograph of two golf balls released simultaneously. The time interval between flashes is (0.033) s, and the white parallel lines were placed 15 cm apart. (a) Find the speed at which the right ball was projected, and (b) show that both balls should be expected to reach the floor simultaneously.

41. If a person can jump a horizontal distance of 3.0 m on the Earth, how far could he jump on the Moon, where the free-fall acceleration is $g/6$ and $g = 9.80$ m/s²? Repeat for Mars, where the acceleration due to gravity is $0.38g$.

42. A home run is hit in such a way that the baseball just clears a wall 21 m high, located 130 m from home plate. The ball is hit at an angle of 35° to the horizontal, and air resistance is negligible. Find (a) the initial speed of the ball, (b) the time it takes the ball to reach the wall, and (c) the velocity components and the speed of the ball when it reaches the wall. (Assume the ball is hit at a height of 1.0 m above the ground.)

43. Towns A and B in Figure 3.22 are 80.0 km apart. A couple arranges to drive from town A and meet

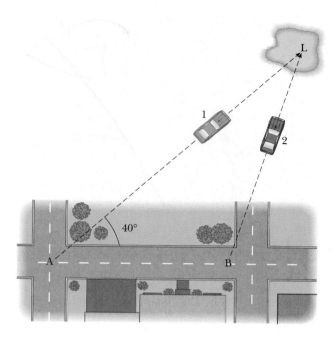

FIGURE 3.22 (Problem 43)

a couple driving from town B at the lake, L. The two couples leave simultaneously and drive for 2.50 h in the directions shown. Car 1 has a speed of 90.0 km/h. If the cars arrive simultaneously at the lake, what is the speed of car 2?

44. A daredevil is shot out of a cannon at 45.0° to the horizontal with an initial speed of 25.0 m/s. A net is positioned a horizontal distance of 50.0 m from the cannon. At what height above the cannon should the net be placed in order to catch the daredevil?

45. A daredevil decides to jump a canyon of width 10 m. To do so, he drives a motorcycle up an incline sloped at an angle of 15°. What minimum speed must he have in order to clear the canyon?

46. A rocket is launched at an angle of 53° above the horizontal with an initial speed of 100 m/s. It moves for 3.0 s along its initial line of motion with an acceleration of 30 m/s². At this time its engines fail and the rocket proceeds to move as a free body. Find (a) the maximum altitude reached by the rocket, (b) its total time of flight, and (c) its horizontal range.

47. A dart gun is fired while being held horizontally at a height of 1.00 m above ground level, and at rest relative to the ground. The dart from the gun travels a horizontal distance of 5.00 m. A child holds the same gun in a horizontal position while sliding down a 45.0° incline at a constant speed of 2.00 m/s. How far will the dart travel if the child fires the gun when it is 1.00 m above the ground?

48. The determined coyote is out once more to try to capture the elusive roadrunner. The coyote wears a new pair of Acme power roller skates, which provide a constant horizontal acceleration of 15 m/s², as shown in Figure 3.23. The coyote starts off at rest 70 m from the edge of a cliff at the instant the roadrunner zips by in the direction of the cliff.

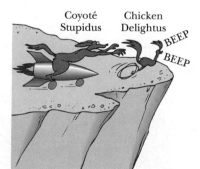

FIGURE 3.23 (Problem 48)

(a) If the roadrunner moves with constant speed, determine the minimum speed he must have in order to reach the cliff before the coyote. (b) If the cliff is 100 m above the base of a canyon, determine where the coyote lands in the canyon. (Assume that his skates are still in operation when he is in "flight" and that his horizontal component of acceleration remains constant at 15 m/s².)

49. A projectile is fired with an initial speed of v_0 at an angle of θ_0 to the horizontal, as in Figure 3.12. When it reaches its peak, it has (x, y) coordinates given by $(R/2, h)$, and when it strikes the ground, its coordinates are $(R, 0)$, where R is called the horizontal range. (a) Show that it reaches a maximum height, h, given by

$$h = \frac{v_0{}^2 \sin^2 \theta_0}{2g}$$

(b) Show that its horizontal range is given by

$$R = \frac{v_0{}^2 \sin 2\theta_0}{g}$$

50. In a very popular lecture demonstration, a projectile is fired at a falling target as in Figure 3.24. The projectile leaves the gun at the same instant that the target is dropped from rest. Assuming that the gun is initially aimed at the target, show that the projectile will hit the target. (One restriction of this experiment is that the projectile must reach the target before the target strikes the floor.)

51. An enemy ship is on the east side of a mountainous island, as shown in Figure 3.25. The enemy ship can maneuver to within 2500 m of the 1800-m-high mountain peak and can shoot projectiles with an initial speed of 250 m/s. If the western shoreline is horizontally 300 m from the peak, what are the distances from the western shore at which a ship can be safe from the bombardment of the enemy ship?

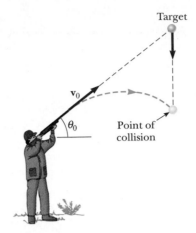

FIGURE 3.24   (Problem 50)

52. A boy can throw a ball a maximum horizontal distance of R on a level field. How far can he throw the same ball vertically upward? Assume that his muscles give the ball the same speed in each case. (Is this assumption valid?)

53. A projectile is fired up an incline (having an angle of ϕ) with an initial velocity of v_0 at an angle of θ_0 with respect to the horizontal ($\theta_0 > \phi$), as shown in Figure 3.26. Show that the projectile will travel a distance of d up the incline, where

$$d = \frac{2v^2{}_0 \cos(\theta_0) \sin(\theta_0 - \phi)}{g \cos^2(\phi)}.$$

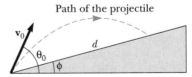

FIGURE 3.26   (Problem 53)

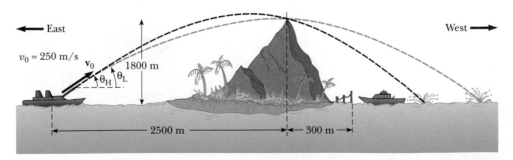

FIGURE 3.25   (Problem 51)

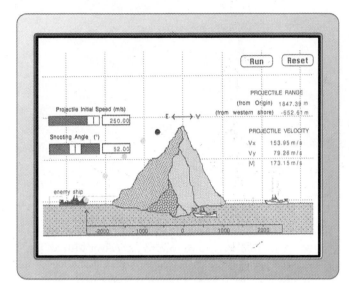 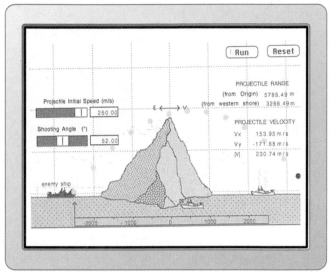

An example of the Interactive Physics simulation to accompany Problem 51.

54. Figure 3.27 illustrates the difference in proportions between the male and female anatomies. The displacements $\mathbf{d}_{1m}$ and $\mathbf{d}_{1f}$ from the bottom of the feet to the navel have magnitudes of 104 cm and 84.0 cm, respectively. The displacements $\mathbf{d}_{2m}$ and $\mathbf{d}_{2f}$ have magnitudes of 50.0 cm and 43.0 cm, respectively. (a) Find the vector sum of the displacements $\mathbf{d}_1$ and $\mathbf{d}_2$ in each case. (b) The male figure is 180 cm tall, the female 168 cm. Normalize the displacements of each figure to a common height of 200 cm, and re-form the vector sums as in part (a). Then find the vector difference between the two sums.

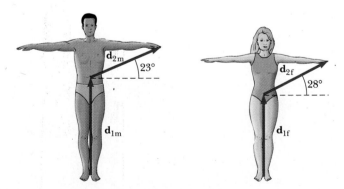

FIGURE 3.27   (Problem 54)

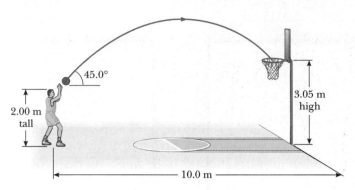

45.0°

2.00 m
tall

3.05 m
high

10.0 m

FIGURE 3.28   (Problem 55)

 55. A 2.00-m tall basketball player wants to make a goal from 10.0 m from the basket, as in Figure 3.28. If he shoots the ball at a 45.0° angle, at what initial speed must he throw the basketball so that it goes through the hoop without striking the backboard?

56. A football is thrown toward a receiver with an initial speed of 20 m/s, at an angle of 30° above the horizontal. At that instant, the receiver is 20 m from the quarterback. In what direction and with what constant speed should the receiver run in order to catch the football at the level at which it was thrown?

THE LAWS OF MOTION

4

A multiflash exposure of a tennis player executing a backhand stroke. The parabolic path of the ball before it is struck is visible, as is the follow-through of the swing. The force exerted by the racket on the ball is equal to and opposite the force exerted by the ball on the racket. (© Zimmerman/FPG)

In the foregoing chapters on kinematics, we used the definitions of displacement, velocity, and acceleration to describe motion, without concerning ourselves with the causes of that motion. Now, however, we would like to be able to answer such questions as "What mechanism causes motion?" and "Why do some objects accelerate at higher rates than others?" In this chapter, therefore, we shall investigate motion in terms of the forces that cause it. We shall then discuss the three fundamental laws of motion, which are based on experimental observations and were formulated by Sir Isaac Newton three centuries ago.

4.1

INTRODUCTION TO CLASSICAL MECHANICS

Classical mechanics deals with objects that (a) are large compared with the dimensions of atoms ($\approx 10^{-10}$ m) and (b) move at speeds that are much less than the speed of light (3×10^8 m/s). If either criterion is violated, the equations and results of this chapter do not apply. Our study of relativity and quantum mechanics in later chapters will enable us to work with the complete range of speeds and sizes.

In this chapter we shall see that it is possible to describe the acceleration of an object in terms of its mass and the external force acting on it. This force represents the interaction of the object with its environment. The mass of an object is a measure of the object's inertia, that is, the tendency of the object to resist a change in its state of motion.

83

4.2

THE CONCEPT OF FORCE

When we think of a force, we usually imagine a push or a pull exerted on some object. For instance, you exert a force on a ball when you throw it or kick it, and you exert a force on a chair when you sit down on it. What happens to an object when it is acted on by a force depends on the magnitude and the direction of the force. Force is a vector quantity; thus, we denote it with a directed arrow, just as we do velocity and acceleration.

If you pull on a spring, as in Figure 4.1a, the spring stretches. If a child pulls hard enough on a wagon, as in Figure 4.1b, the wagon moves. When a football is kicked, as in Figure 4.1c, it is deformed and set in motion. These are all examples of **contact forces,** so named because they result from physical contact between two objects.

Another class of forces does not involve physical contact between two objects. Early scientists, including Newton, were uneasy with the concept of forces that act between two disconnected objects. To overcome this conceptual difficulty, Michael Faraday (1791–1867) introduced the concept of a *field*. The corresponding forces are called *field forces*. According to this approach, when a mass, *m*, is placed at some point *P* near a second mass, *M*, we say that *m* interacts

FIGURE 4.1
Examples of forces applied to various objects. In each case a force acts on the object surrounded by the dashed lines. Something in the environment external to the boxed area exerts this force.

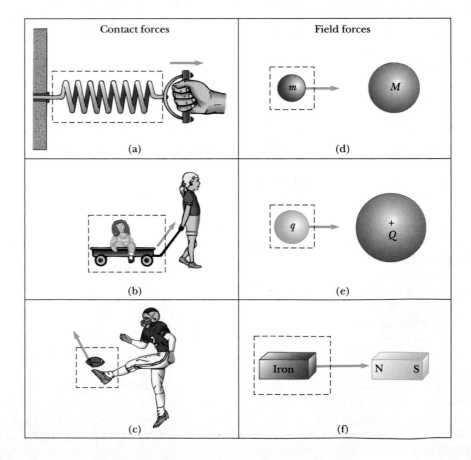

A football is set in motion by the contact force, **F**, on it due to the kicker's foot. The ball is deformed during the short time in contact with the foot. *(Ralph Cowan, Tony Stone Worldwide)*

with M by virtue of the gravitational field that exists at P. Thus, the force of gravitational attraction between two objects, illustrated in Figure 4.1d, is an example of a field force. This force keeps objects bound to the Earth and gives rise to what we commonly call the *weight* of the object. The planets of our Solar System are under the actions of gravitational forces.

Another common example of a field force is the electric force that one electric charge exerts on another, as in Figure 4.1e. A third example is the force exerted by a bar magnet on a piece of iron, as in Figure 4.1f. The known fundamental forces in nature are all field forces. These are, in order of decreasing strength, (1) strong nuclear forces between subatomic particles, (2) electromagnetic forces between electric charges at rest or in motion, (3) gravitational attractions between objects, and (4) weak nuclear forces, which arise in certain radioactive decay processes. In classical physics we are concerned only with gravitational and electromagnetic forces.

Whenever a force is exerted on an object, its shape can change. For example, when you squeeze a rubber ball or strike a punching bag with your fist, the object deforms to some extent. Even more rigid objects, such as automobiles, are deformed under the action of external forces. Often the deformations are permanent, as in the case of a collision between vehicles. If you pull on a coiled spring, as in Figure 4.1a, the spring stretches.

4.3

NEWTON'S FIRST LAW

Consider the following simple experiment. Suppose a book is lying on a table. Obviously, the book remains at rest if left alone. Now imagine that you push the book with a horizontal force great enough to overcome the force of friction between book and table, so that the book is set in motion. If the magnitude of

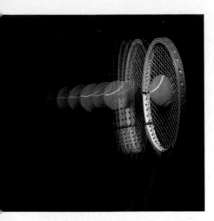

A multiflash exposure of a tennis ball being struck by a racket. The ball experiences a large contact force as it is struck, and a correspondingly large acceleration. *(Ben Rose/The Image Bank)*

your applied force is equal to the magnitude of the friction force, the book moves with constant velocity. If the magnitude of your applied force exceeds the magnitude of the friction force, the book accelerates. If you stop applying the force, the book stops sliding after traveling a short distance because the force of friction retards its motion. Now imagine pushing the book across a smooth, waxed floor. The book again comes to rest once the force is no longer applied, but not as soon as before. Finally, imagine that the book is moving on a horizontal, frictionless surface. In this situation, the book continues to move in a straight line with constant velocity until it hits a wall or some other object. We call such motion **uniform.**

Before about 1600, scientists felt that the natural state of matter was the state of rest. Galileo was the first to take a different approach. He devised thought experiments—such as an object moving on a frictionless surface, as just described—and concluded that it is not the nature of an object to *stop*, once set in motion; rather, it is an object's nature to *resist acceleration*. In his words, "Any velocity, once imparted to a moving body, will be rigidly maintained as long as the external causes of retardation are removed." This approach to motion was later formalized by Newton in a form that has come to be known as **Newton's first law of motion:**

> **An object at rest remains at rest, and an object in motion continues in motion with constant velocity (that is, constant speed in a straight line), unless it experiences a net external force.**

By external force, we mean any force that results from the interaction between the object and its environment, such as the force exerted on an object when it is lifted. In simpler terms, Newton's first law says that *when the net external force on an object is zero, its acceleration is zero.* That is, when $\Sigma\mathbf{F} = 0$, then $\mathbf{a} = 0$. From the first law, we conclude that an isolated body (a body that does not interact with its environment) is either at rest or moving with constant velocity.

Another example of uniform motion on a nearly frictionless surface is the motion of a lightweight disk on a column of air, as in Figure 4.2. If the disk is given an initial velocity, it will coast a great distance before coming to rest because it is moving on a layer of air rather than a rough surface. This fact is central to the game of air hockey, in which the disk makes many collisions with the sides of the table before coming to rest.

Finally, consider a spaceship traveling in space, far from any planets or other matter. The spaceship requires a propulsion system to change its velocity. However, if the propulsion system is turned off when a velocity, **v**, is reached, the spaceship coasts in space with a constant velocity and the astronauts get a "free ride."

v = constant

Air flow

Electric blower

FIGURE 4.2
A disk moving on a column of air is an example of uniform motion, that is, motion in which the acceleration is zero and the velocity remains constant.

MASS AND INERTIA

Imagine a bowling ball and a golf ball sitting side by side on the ground. Newton's first law tells us that both remain at rest as long as no net external force acts on them. Now imagine supplying a net force by striking each ball with a golf club. Both balls resist your attempt to change their states of motion. But you know from everyday experience that if the two are struck with equal force, the golf ball will travel much farther than the bowling ball. That is, the bowling ball is more

Isaac Newton, a British physicist and mathematician, is regarded as one of the greatest scientists in history. Before the age of 30 he formulated the basic concepts and laws of motion, discovered the universal law of gravitation, and invented the calculus. Newton was able to explain the motions of the planets, the ebb and flow of the tides, and many special features of the motion of the Moon and the Earth. He also made many important discoveries in optics, showing, for example, that white light is composed of a spectrum of colors. His contributions to physical theories dominated scientific thought for two centuries and remain important today.

Newton was born prematurely on Christmas Day in 1642, shortly after his father's death. When he was three, his mother remarried and he was left in his grandmother's care. Because he was small in stature as a child, he was bullied by other children and took refuge in such solitary activities as the building of water clocks, kites carrying fiery lanterns, sundials, and model windmills powered by mice. His mother withdrew him from school at the age of 12 with the intention of turning him into a farmer. Fortunately for later generations, his uncle recognized his scientific and mathematical abilities and helped send him to Trinity College in Cambridge.

In 1665, the year Newton completed his Bachelor of Arts degree, the university was closed because of the bubonic plague that was raging through England. Newton returned to the family farm at Woolsthorpe to study. During this especially creative period, he laid the foundations of his work in mathematics, optics, motion, celestial mechanics, and gravity.

Newton was a very private person who studied alone and labored day and night in his laboratory, conducting experiments, performing calculations, and immersing himself in theological studies. His greatest single work, *Mathematical Principles of Natural Philosophy*, was published in 1687. In his later years he spent much of his time quarreling with other eminent minds, including the mathematician Gottfried Leibnitz, who worked independently on the development of calculus; Christian Huygens, who developed the wave theory of light; and Robert Hooke, who supported Huygens' theory. These disputes, the strain of his studies, and his work in alchemy, which involved mercury (a poison), caused him in 1692 to suffer severe depression. He was elected president of the Royal Society in 1703, and he retained that office until his death in 1727.

Isaac Newton (1642–1727)

(Giraudon/Art Resource)

successful in resisting your attempt to change the state of motion. The tendency of an object to resist any attempt to change its motion is called the **inertia** of the object.

Mass is a measurement of inertia, and the SI unit of mass is the kilogram. The greater the mass of a body, the less it accelerates under the action of an applied force. For example, if a given force acting on a 3-kg mass produces an acceleration of 4 m/s^2, the same force applied to a 6-kg mass will produce an acceleration of only 2 m/s^2. Shortly we will apply this idea to a quantitative description of the concept of mass. Mass is a scalar quantity that obeys the rules of ordinary arithmetic. Furthermore, the mass of an object is independent of the coordinate system used in describing the motion of that object.

Inertia is the principle that underlies the operation of seat belts. In the event of an accident, the purpose of the seat belt is to hold the passenger firmly in place

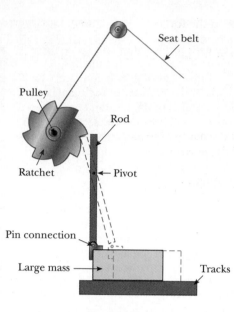

Seat belt

Pulley

Rod

Ratchet

Pivot

Pin connection

Large mass

Tracks

FIGURE 4.3
A mechanical arrangement for a safety belt.

Astronaut Edwin E. Aldrin, Jr., walking on the Moon after the Apollo 11 lunar landing. The weight of this astronaut on the Moon is less than it is on Earth, but his mass remains the same.
(Courtesy of NASA/Black Star)

relative to the car, to prevent serious injury. Figure 4.3 illustrates how one type of shoulder harness operates. Under normal conditions, the ratchet turns freely to allow the harness to wind on or unwind from the pulley as the passenger moves. If an accident occurs, the car undergoes a large negative acceleration and rapidly comes to rest. The large mass under the seat, because of its inertia, continues to slide forward along the tracks. The pin connection between the mass and the rod causes the rod to pivot about its center and engage the ratchet wheel. At this point the ratchet wheel locks in place, and the harness no longer unwinds.

A slight modification enables this device to also activate an air bag in the car. In this case, movement of the mass and pivoting of the rod open a valve on a cylinder that contains nitrogen under pressure. The nitrogen rushes into the air bag, causing it to expand rapidly so that it serves as a protective cushion for the passenger.

4.4

NEWTON'S SECOND LAW

Newton's first law explains what happens to an object when the resultant force acting on it is zero: the object either stays at rest or keeps moving with constant velocity. Newton's second law answers the question of what happens to an object that has a nonzero resultant force acting on it.

Imagine pushing a block of ice across a frictionless horizontal surface. When you exert a certain force on the block, it moves with an acceleration of, say, 2 m/s^2. If you push twice as hard, you find that the acceleration also doubles. Pushing three times as hard triples the acceleration, and so on. From observations such as these has come the conclusion that *the acceleration of an object is directly proportional to the resultant force acting on it.*

Common experience with pushing objects tells you that mass also affects acceleration. Imagine that you stack identical blocks of ice on top of each other while pushing the stack with constant force. If the force when you push one block produces an acceleration of 2 m/s², the acceleration will drop to half that value when two blocks are pushed, one-third that initial value when three blocks are pushed, and so on. From this we see that *the acceleration of an object is inversely proportional to its mass*. **Newton's second law** summarizes these observations:

> **The acceleration of an object is directly proportional to the resultant force acting on it and inversely proportional to its mass. The direction of the acceleration is the direction of the resultant force.**

In equation form, we can state Newton's second law as

$$\sum \mathbf{F} = m\mathbf{a} \qquad \text{[4.1]}$$

Newton's second law

where **a** is the acceleration of the object, *m* is its mass, and Σ**F** represents the *vector sum of all external forces acting on the object*. (An external force is one that results from interaction between the object and its environment.) You should note that, because this is a vector equation, it is equivalent to the following three scalar equations:

$$\sum F_x = ma_x \qquad \sum F_y = ma_y \qquad \sum F_z = ma_z \qquad \text{[4.2]}$$

Note that if the resultant force is zero, then **a** = 0, which corresponds to the equilibrium situation where **v** is either constant or zero. Hence, *the first law of motion is a special case of the second law.*

UNITS OF FORCE AND MASS

> **The SI unit of force is the newton, defined as the force that, when acting on a 1-kg mass, produces an acceleration of 1 m/s².**

From this definition and Newton's second law, we see that the newton can be expressed in terms of the fundamental units of mass, length, and time:

$$1 \text{ N} \equiv 1 \text{ kg} \cdot \text{m/s}^2 \qquad \text{[4.3]}$$

Definition of newton

The unit of force in the cgs system is called the **dyne** and is defined as the force that, when acting on a 1-g mass, produces an acceleration equal to 1 cm/s²:

$$1 \text{ dyne} \equiv 1 \text{ g} \cdot \text{cm/s}^2 \qquad \text{[4.4]}$$

Definition of dyne

In the British engineering system the unit of force is the **pound**. The following conversion from pounds to newtons will be useful to you in many problems:

$$1 \text{ lb} \equiv 4.448 \text{ N} \qquad \text{[4.5]}$$

Definition of pound

When we speak of going on a diet to lose a few pounds, we really mean that we want to lose a few kilograms; that is, we want to reduce our mass. When we lose those few kilograms, the force of gravity (pounds) on our reduced mass decreases, and that is how we "lose a few pounds."

TABLE 4.1
Units of Mass, Acceleration, and Force

System	Mass	Acceleration	Force
SI	kg	m/s^2	$N = kg \cdot m/s^2$
cgs	g	cm/s^2	$dyne = g \cdot cm/s^2$

Note: $1 \text{ N} = 10^5 \text{ dyne} = 0.225 \text{ lb}.$

Since $1 \text{ kg} = 1 \times 10^3 \text{ g}$ and $1 \text{ m} = 1 \times 10^2 \text{ cm}$, it follows that $1 \text{ N} = 1 \times 10^5$ dynes. (It is left to an exercise to show that $1 \text{ N} = 0.225 \text{ lb}$.) The units of mass, acceleration, and force in the SI and cgs system are summarized in Table 4.1.

WEIGHT

We are well aware that objects are attracted to the Earth. The force exerted by the Earth on an object is called the **weight** of the object, **w**. This force, a vector quantity, is directed approximately toward the center of the Earth, and its magnitude varies with location. In contrast, the mass of an object is a scalar quantity whose value is unchanging regardless of location. An object having a mass of 100 kg on Earth also has a mass of 100 kg on the Moon—or anywhere else, for that matter.

We have seen that a freely falling object experiences an acceleration, **g**, acting toward the center of the Earth. Application of Newton's second law to the freely falling object shown in Figure 4.4, with **a** = **g** and **F** = **w**, gives

$$\mathbf{w} = m\mathbf{g} \qquad [4.6]$$

Weight varies with geographic location because it depends on **g**. For example, a person who weighs 180 lb on Earth weighs only about 30 lb on the Moon. Furthermore, objects weigh less at higher altitudes than at sea level because g decreases with increasing distance from the center of the Earth. Hence, weight, unlike mass, is not an inherent property of an object. For example, if an object has a mass of 70.0 kg, then the magnitude of its weight at a location where $g = 9.80 \text{ m/s}^2$ is $mg = 686 \text{ N}$. At the top of a mountain, where g might be 9.76 m/s^2, the object's weight would be 683 N. Therefore, if you want to lose weight without going on a diet, move to the top of a mountain or weigh yourself at an altitude of 30 000 ft during a flight on a jet airplane.

4.5

NEWTON'S THIRD LAW

In Section 4.2 we found that a force is exerted on an object when that object comes into contact with some other object. For example, consider the task of driving a nail into a block of wood, as illustrated in Figure 4.5. To accelerate the nail and drive it into the block, a net force must be supplied to the nail by the hammer. However, Newton recognized that a single isolated force (such as the force on the nail by the hammer) cannot exist. Instead, *forces in nature always exist in pairs.* According to Newton, the hammer exerts a force on the nail, and the nail

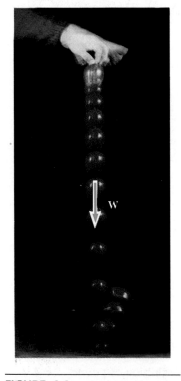

FIGURE 4.4
The only force acting on an object in free fall is its weight, **w**. Newton's second law applied in the vertical direction gives $w = mg$.

PHYSICS IN *ACTION*

(Left) An athlete running at sunset. The external forces acting on the athlete, as described by the blue vectors, are (a) the force exerted by the Earth, **w**, (b) the force exerted by the ground, **F**, and (c) the force of air resistance, **R**. *(Right)* The golfer is able to drive the ball large distances by swinging the club at high speeds and following through on the swing to increase the contact time between club and ball. Why does the longer contact time increase the ball's momentum?

FORCES AND MOTION

(Mitchell Funk/The Image Bank)

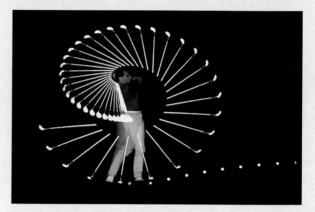

(M. Hans Vandystadt/Photo Researchers)

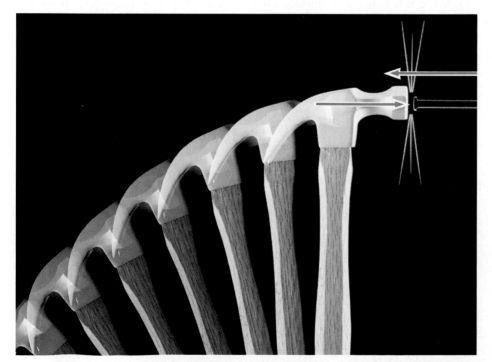

FIGURE 4.5
The force exerted by the hammer on the nail is equal in magnitude and opposite in direction to the force exerted by the nail on the hammer. *(John Gillmoure/ The Stock Market)*

exerts a force on the hammer. There is clearly a net force on the hammer, because it rapidly slows down after coming into contact with the nail.

Newton described this type of situation in terms of his third law of motion:

> **If two bodies interact, the magnitude of the force exerted on body 1 by body 2 is equal to the magnitude of the force exerted on body 2 by body 1, and these two forces are opposite in direction.**

An alternative statement of this law is that *for every action there is an equal and opposite reaction.*

Action-reaction pairs are always exerted on *different* objects. In our hammer-and-nail example, we could call the force the hammer exerts on the nail the action and the force the nail exerts on the hammer the reaction. However, there is nothing special about this assignment of names. If you choose to call the force the hammer exerts on the nail the reaction and the force the nail exerts on the hammer the action, it is perfectly satisfactory to do so.

Let us discuss some other examples of action-reaction pairs. The force acting on a freely falling object is its weight, **w**. Let's call this force the action. What is the reaction? Since the weight is the force exerted on the falling object by the Earth, the reaction is the force exerted on the Earth by the falling object. Thus, as the falling object accelerates toward the Earth, the Earth accelerates toward the object. However, since the mass of the Earth is so much greater than that of the falling object, the acceleration of the Earth due to this reaction force is negligibly small.

Figure 4.6a shows a television (TV) set at rest on a table. Two forces, indicated by **n** and **w** in the figure, are acting on the TV. The force **w** is the weight of the TV, and the force **n** is the upward force exerted on the TV by the table. The force **n** is called the **normal** force. The word *normal* is used because the

FIGURE 4.6
When a TV set is sitting on a table, the forces acting on the set are the normal force, **n**, and the force of gravity, **w**, as illustrated in (b). The reaction to **n** is the force exerted by the TV set on the table, **n**′. The reaction to **w** is the force exerted by the TV set on the Earth, **w**′.

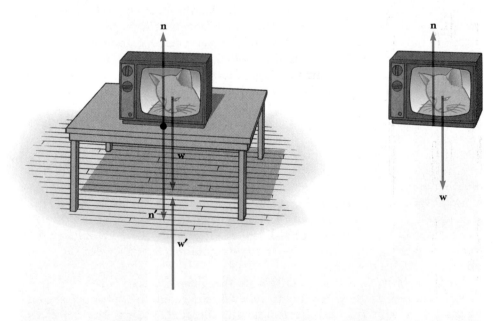

(a)   (b)

direction of **n** is perpendicular to the table surface and *normal* is a synonym for perpendicular. Because the TV is not accelerating, we know from the first law that **w** and **n** must be equal in magnitude and opposite in direction. However, these two forces do not constitute an action-reaction pair, because such forces always act on different objects. In this case both **w** and **n** are exerted on the same object, the TV. If we choose to call **w** the action, the reaction is the force, **w**′, exerted on the Earth by the TV. The reaction to **n**, denoted by **n**′ in the figure, is the force exerted on the table by the TV. Note that the only forces acting on the TV are **w** and **n**, as indicated in Figure 4.6b. From the first law, we see that since the TV is in equilibrium (**a** = 0), it follows that $n = w = mg$.

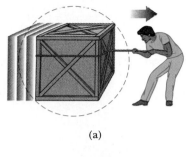

(a)

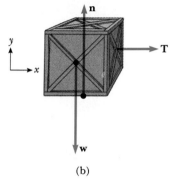

(b)

FIGURE 4.7
(a) A block being pulled to the right on a smooth surface.
(b) The free-body diagram that represents the external forces on the block.

4.6

SOME APPLICATIONS OF NEWTON'S LAWS

This section applies Newton's laws to objects moving under the actions of constant external forces. We assume that objects behave as particles, and so we need not worry about rotational motion. We also neglect any friction effects. Finally, we neglect the masses of any ropes or strings involved; in these approximations, the magnitude of the force exerted along a rope (the tension) is the same at all points in the rope.

When we apply Newton's law to an object, we are interested only in those *external* forces that act *on the body*. For example, in Figure 4.6b, the only external forces acting on the TV are **n** and **w**. The reactions to these forces, **n**′ and **w**′, act on the table and on the Earth, respectively, and do not appear in Newton's second law as applied to the TV.

Consider a block being pulled to the right on the horizontal surface of a table, as in Figure 4.7a. Suppose you want to find the acceleration of the block and the force exerted by the table on it. The horizontal force applied to the block acts through the rope that is attached to it. The force that the rope exerts on the block is denoted by **T**. The magnitude of **T** is the tension in the rope.

A dashed circle is drawn around the block in Figure 4.7a to remind you to isolate the block from its surroundings. Since we are interested only in the motion of the block, we must be able to identify all external forces acting on it. These are illustrated in Figure 4.7b. The force diagram for the block includes, in addition to the force **T**, the block's weight, **w**, and the normal force, **n**. As before, **w** is the force of gravity pulling down on the block and **n** is the upward force exerted on the block by the table.

A force diagram like Figure 4.7b is referred to as a **free-body diagram.** The construction of a free-body diagram is a crucial step in applying Newton's laws; its importance cannot be overemphasized. The reactions to the forces we have listed—namely, the force exerted by the rope on the hand doing the pulling, the force exerted by the block on the Earth, and the force exerted by the block on the table—are not included in the free-body diagram because they act on other bodies and not on the block.

We now apply Newton's second law to the block. First we must choose an appropriate coordinate system. In this case it is convenient to use the one shown in Figure 4.7b, with the x axis horizontal and the y axis vertical. We can apply Newton's second law in the x direction, y direction, or both, depending on what we are asked to find in a problem. In addition, we may be able to use the kinematic equations of motion found in Chapter 2 if the acceleration in the problem is constant. For example, if the force $\mathbf{T}$ in Figure 4.7 is constant, then the acceleration in the x direction is also constant because $\mathbf{a}_x = \mathbf{T}_x / m$. Hence, if we need to find the displacement or the velocity of the object at some instant, we can use the equations of motion with constant acceleration.

OBJECTS IN EQUILIBRIUM AND NEWTON'S FIRST LAW

Objects that are either at rest or moving with constant velocity are said to be in equilibrium, and Newton's first law is a statement of one condition that must be true for equilibrium conditions to prevail. In equation form, the first law implies that if a body is in equilibrium, then

$$\sum \mathbf{F} = 0 \qquad \text{[4.7]}$$

This statement signifies that the *vector* sum of all the forces (that is, the net force) acting on an object in equilibrium is zero. Usually, equilibrium problems are more easily solved if we work with Equation 4.7 expressed in terms of the components of the external forces acting on an object. By this we mean that, in a two-dimensional problem, the sum of all the external forces in the x direction and the sum of those in the y direction must separately equal zero; that is,

$$\sum F_x = 0$$
$$\sum F_y = 0 \qquad \text{[4.8]}$$

if the object is to remain in equilibrium.

This set of equations is often referred to as the **first condition for equilibrium.** We shall not consider three-dimensional problems in this text, but the extension of Equation 4.8 to a three-dimensional situation can be made by adding a third equation, $\Sigma F_z = 0$.

PROBLEM-SOLVING STRATEGY
Objects in Equilibrium

The following procedure is recommended for problems involving objects in equilibrium:

1. Make a sketch of the situation described in the problem statement.
2. Draw a free-body diagram for the *isolated* object under consideration, and label all external forces acting on it.
3. Resolve all forces into x and y components, choosing a convenient coordinate system.
4. Use the equations $\Sigma F_x = 0$ and $\Sigma F_y = 0$. Keep track of the signs of the various force components.
5. Application of Step 4 leads to a set of equations with several unknowns. Solve the simultaneous equations for the unknowns in terms of the known quantities.

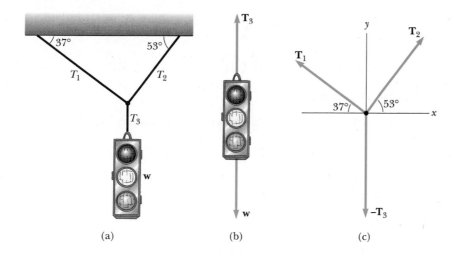

(a)   (b)   (c)

FIGURE 4.8
(Example 4.1)   (a) A traffic light suspended by cables. (b) A free-body diagram for the traffic light. (c) A free-body diagram for the knot joining the cables.

EXAMPLE 4.1   A Traffic Light at Rest

A traffic light weighing 100 N hangs from a vertical cable tied to two other cables that are fastened to a support, as in Figure 4.8a. The upper cables make angles of 37.0° and 53.0° with the horizontal. Find the tension in each of the three cables.

Reasoning   We must construct two free-body diagrams in order to work this problem. The first of these is for the traffic light, shown in Figure 4.8b; the second is for the knot that holds the three cables together, as in Figure 4.8c. The knot is a convenient point to choose because all forces in question act at this point.

Solution   From the free-body diagram in Figure 4.8b, we see that $T_3 = w = 100$ N. Next, we choose the coordinate axes shown in Figure 4.8c and resolve the forces into their x and y components:

Force	x Component	y Component
$\mathbf{T}_1$	$-T_1 \cos 37.0°$	$T_1 \sin 37.0°$
$\mathbf{T}_2$	$T_2 \cos 53.0°$	$T_2 \sin 53.0°$
$\mathbf{T}_3$	0	-100 N

The first condition for equilibrium gives us the equations

$$(1) \qquad \sum F_x = T_2 \cos 53.0° - T_1 \cos 37.0° = 0$$

$$(2) \qquad \sum F_y = T_1 \sin 37.0° + T_2 \sin 53.0° - 100 \text{ N} = 0$$

From (1) we see that the horizontal components of $\mathbf{T}_1$ and $\mathbf{T}_2$ must be equal in magnitude, and from (2) we see that the sum of the vertical components of $\mathbf{T}_1$ and $\mathbf{T}_2$ must balance the weight of the light. We can solve (1) for T_2 in terms of T_1 to give

$$T_2 = T_1 \left(\frac{\cos 37.0°}{\cos 53.0°} \right) = T_1 \left(\frac{0.799}{0.602} \right) = 1.33 T_1$$

This value for T_2 can be substituted into (2) to give

$$T_1 \sin 37.0° + (1.33T_1)(\sin 53.0°) - 100 \text{ N} = 0$$

$$T_1 = \boxed{60.1 \text{ N}}$$

$$T_2 = 1.33T_1 = 1.33(60.0 \text{ N}) = \boxed{79.9 \text{ N}}$$

Exercise   When will $T_1 = T_2$?

Answer   When the supporting cables make equal angles with the horizontal support.

EXAMPLE 4.2   Sled on a Frictionless Hill

A child holds a sled at rest on a frictionless, snow-covered hill, as shown in Figure 4.9a. If the sled weighs 77.0 N, find the force exerted on the rope by the child and the force exerted on the sled by the hill.

Reasoning   Figure 4.9b shows the forces acting on the sled and a convenient coordinate system to use for this type of problem. Note that **n**, the force exerted on the sled by the ground is perpendicular to the hill. The ground can exert a component of force along the incline only if there is friction between the sled and the hill. Because the sled is at rest, we are able to apply the first condition for equilibrium as $\Sigma F_x = 0$ and $\Sigma F_y = 0$.

Solution   Applying the first condition for equilibrium to the sled, we find that

$$\sum F_x = T - (77.0 \text{ N})(\sin 30.0°) = 0$$

$$T = \boxed{38.5 \text{ N}}$$

$$\sum F_y = n - (77.0 \text{ N})(\cos 30.0°) = 0$$

$$n = \boxed{66.7 \text{ N}}$$

Note that n is *less* than the weight of the sled. This is so because the sled is on an incline and **n** is equal to and opposite the component of weight perpendicular to the incline.

Exercise   What happens to the normal force as the angle of incline increases?

Answer   It decreases.

Exercise   When is the normal force equal to the weight of the sled?

Answer   When the sled is on a horizontal surface and the applied force is either zero or along the horizontal.

ACCELERATING OBJECTS AND NEWTON'S SECOND LAW

Whenever a net force acts on an object, the object is accelerated, and we must use Newton's second law in order to describe the object's motion. The representative problems and suggestions that follow should help you to solve problems of this kind.

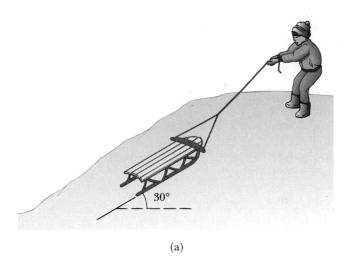

(a)

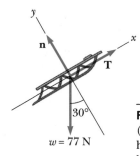

(b)

FIGURE 4.9
(Example 4.2) (a) A child holding a sled on a frictionless hill. (b) A free-body diagram for the sled.

PROBLEM-SOLVING STRATEGY
Newton's Second Law

The following procedure is recommended for working problems that involve the application of Newton's second law:

1. Draw a simple, neat diagram of the system.
2. Isolate the object of interest whose motion is being analyzed. Draw a free-body diagram for this object. When there are multiple objects of interest, draw a *separate* diagram for each object.
3. Establish convenient coordinate axes for each object, and find the components of the forces along these axes.
4. Apply Newton's second law in the x and y directions for each object.
5. Solve the component equations for the unknowns. Remember that, in order to obtain a complete solution, you must have as many independent equations as you have unknowns.
6. If necessary, use the equations of kinematics (motion with constant acceleration) from Chapter 2 to find all the unknowns.

EXAMPLE 4.3   Moving a Crate

The combined weight of the crate and dolly in Figure 4.10 is 300 N. If the person pulls on the rope with a constant force of 20.0 N, what is the acceleration of the system (crate plus dolly), and how far will it move in 2.00 s? Assume that the system starts from rest and that there are no frictional forces opposing the motion of the system.

Reasoning We can find the acceleration of the system from Newton's second law. Because the force is constant, its acceleration is constant. Therefore, we can apply the equations of motion with constant acceleration to find the distance traveled.

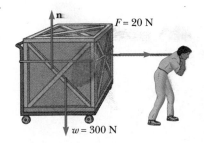

FIGURE 4.10
(Example 4.3)

Solution   In order to apply Newton's second law to the system, we must first know the system's mass:

$$m = \frac{w}{g} = \frac{300 \text{ N}}{9.80 \text{ m/s}^2} = 30.6 \text{ kg}$$

Now we can find the acceleration from the second law:

$$a_x = \frac{F_x}{m} = \frac{20.0 \text{ N}}{30.6 \text{ kg}} = 0.654 \text{ m/s}^2$$

Since the acceleration is constant, we can find the distance the system moves in 2.00 s using the relation $x = v_0 t + \frac{1}{2}at^2$ with $v_0 = 0$:

$$x = \tfrac{1}{2}at^2 = \tfrac{1}{2}(0.654 \text{ m/s}^2)(2.00 \text{ s})^2 = \boxed{1.31 \text{ m}}$$

It is important to note that the constant applied force of 20.0 N is assumed to act on the system all during its motion. If the force were removed at some instant, the system would continue to move with constant velocity and hence zero acceleration.

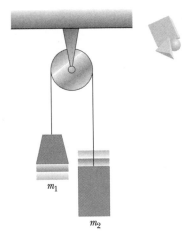

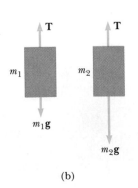

(a)

(b)

FIGURE 4.11
(Example 4.4)   Atwood's machine.

EXAMPLE 4.4   Atwood's Machine

When two unequal masses are hung vertically over a light frictionless pulley as in Figure 4.11a, the arrangement is called *Atwood's machine*. If $m_1 = 2.00$ kg and $m_2 = 4.00$ kg, calculate the acceleration of the two masses and the tension in the string.

Solution   The free-body diagrams for the two masses are shown in Figure 4.11b. When we apply Newton's second law to m_1, with **a** upward for this mass (since $m_2 > m_1$), we find that

$$(1) \qquad \sum F_y = T - m_1 g = m_1 a$$

Similarly, for m_2 we find that

$$(2) \qquad \sum F_y = T - m_2 g = -m_2 a$$

The negative sign on the right-hand side of (2) indicates that m_2 accelerates downward.

When (2) is subtracted from (1), we get

$$-m_1 g + m_2 g = m_1 a + m_2 a$$

$$(3) \qquad a = \left(\frac{m_2 - m_1}{m_1 + m_2}\right) g$$

Substituting (3) into (1), we get

$$(4) \qquad T = \left(\frac{2 m_1 m_2}{m_1 + m_2}\right) g$$

Substituting the given values of m_1 and m_2 into (3) and (4) gives

$$a = \boxed{3.27 \text{ m/s}^2} \qquad T = \boxed{26.1 \text{ N}}$$

Exercise What are the values of a and T in the special cases when (a) $m_1 = m_2$ and (b) $m_2 \gg m_1$?

Answer (a) $a = 0$, $T = m_1 g = m_2 g$; (b) $a = g$, $T \approx 2 m_1 g$.

EXAMPLE 4.5 The Run-Away Car

A car of mass m is on an icy driveway inclined at an angle of $\theta = 20.0°$, as in Figure 4.12a. Determine the acceleration of the car, assuming the incline is frictionless.

Reasoning The free-body diagram for the car is shown in Figure 4.12b. The only forces on the car are the normal force, **n**, acting perpendicular to the driveway surface, and the weight, **w**, acting vertically downward. It is convenient to choose the coordinate axes with the x axis along the incline and the y axis perpendicular to it. Then we replace the weight vector with a component of magnitude $mg \sin \theta$ along the positive x axis and a component of magnitude $mg \cos \theta$ in the negative y direction.

Solution Applying Newton's second law in component form, with $a_y = 0$, gives

$$(1) \qquad \sum F_x = mg \sin \theta = ma_x$$
$$(2) \qquad \sum F_y = n - mg \cos \theta = 0$$

From (1) we see that the acceleration along the driveway is provided by the component of weight directed down the incline:

$$(3) \qquad a_x = g \sin \theta$$

Note that the acceleration given by (3) is *constant* and *independent of the mass* of the car—it depends only on the angle of inclination and on g. In our example, $\theta = 20.0°$ and so, we find that

$$a_x = \boxed{3.35 \text{ m/s}^2}$$

Exercise A compact car and a large luxury sedan are at rest at the top of an ice-covered (frictionless) driveway. If they both slide down the driveway, which reaches the bottom first?

Answer Because the acceleration is independent of the mass, and thus is the same for both vehicles, they arrive at the bottom simultaneously.

(a)

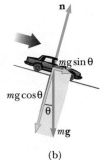

(b)

FIGURE 4.12
(Example 4.5)

Exercise   If the length of the driveway is 25.0 m and a car starts from rest at the top, how long does it take to travel to the bottom? What is the car's speed at the bottom?

Answer   3.86 s; 12.9 m/s.

4.7

FORCE OF FRICTION

When a body is moving on a rough surface or falling through a fluid medium such as air or water, the interaction between the body and its surroundings causes resistance to the motion. We call such resistance a force of friction. Forces of friction are very important in our everyday lives. They allow us to walk and run and are necessary for the motions of wheeled vehicles.

Consider a block on a table as in Figure 4.13a. If we apply to the block an external horizontal force, **F**, acting to the right, the block will remain stationary if **F** is not too large. The force that keeps the block from moving acts to the left and is called the force of static friction, f_s. As long as the block is not moving, $f_s = F$. If

FIGURE 4.13
The direction of the force of friction, **f**, between a block and a horizontal surface is opposite the direction of an applied force, **F**. (a) The force of static friction equals the applied force. (b) When the magnitude of the applied force exceeds the maximum static friction force, $f_{s,\text{max}}$, the block accelerates to the right. (c) A graph of the magnitude of the frictional force versus the applied force. Note that $f_{s,\text{max}} > f_k$.

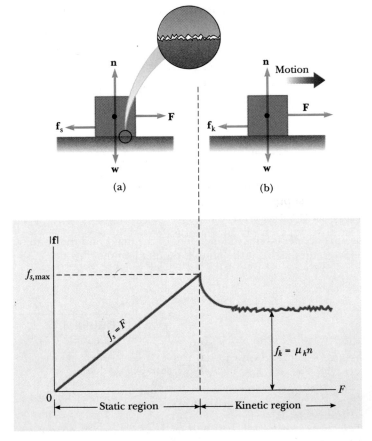

(a)   (b)

(c)

F increases, f_s also increases; if **F** decreases, f_s also decreases. Experiments show that the frictional force arises from the nature of the two surfaces that are touching; because of their roughness, contact is made only at a few points, as shown in the magnified view of the surface in Figure 4.13a. Actually, the frictional force is much more complicated than it appears here, since it ultimately involves forces between atoms or molecules where the surfaces are in contact.

If we increase the magnitude of **F** enough, as in Figure 4.13b, the block will eventually move. When the block is on the verge of slipping, f_s is a maximum. When F exceeds $f_{s,\mathrm{max}}$, the block moves and accelerates to the right. When the block is in motion, the retarding frictional force becomes less than $f_{s,\mathrm{max}}$ (Fig. 4.13c). We call the retarding force on an object in motion the **force of kinetic friction, f_k**. The unbalanced force in the x direction, $\mathbf{F} - \mathbf{f_k}$, produces an acceleration in that direction. If after the block is set in motion, $F = f_k$, the block moves to the right with constant speed. If the applied force is removed, then the frictional force acting to the left decelerates the block and eventually brings it to rest.

The force of kinetic friction is less than $f_{s,\mathrm{max}}$ for the following reason. When an object is stationary, the contact points between it and the surface on which it rests are said to be cold-welded. While the object is in motion, these small welds can no longer form and the frictional force decreases.

Both f_s and f_k are proportional to the normal force acting on an object and depend on the natures of the two surfaces in contact. The experimental observations can be summarized as follows:

1. The direction of the force of static friction between any two surfaces in contact is opposite the direction in which there is a tendency for an object to move. This force can have the values

$$f_s \leq \mu_s n \qquad \text{[4.9]}$$

where the dimensionless constant μ_s is called the **coefficient of the static friction** and n is the magnitude of the normal force. The equality in Equation 4.9 holds when an object is on the verge of slipping. Thus, when an object is on the verge of slipping, $f_s = f_{s,\mathrm{max}} \equiv \mu_s n$. The inequality holds when the applied force is less than this value. In general, when an object is at rest relative to a surface, the frictional force always acts in such a way as to maintain a velocity of zero relative to the surface.

2. The direction of the force of kinetic friction is opposite the direction of motion and its magnitude is

$$f_k = \mu_k n \qquad \text{[4.10]}$$

where μ_k is the **coefficient of kinetic friction.**

3. The values of μ_k and μ_s depend on the natures of the surfaces, but μ_k is generally less than μ_s. Typical values of μ range from around 0.01 to 1.5. Table 4.2 lists some reported values.

Finally, the coefficients of friction are nearly independent of the area of contact between the surfaces. Although the coefficient of kinetic friction varies with speed, we shall ignore any such variations.

TABLE 4.2
Coefficients of Friction[a]

	μ_s	μ_k
Steel on steel	0.74	0.57
Aluminum on steel	0.61	0.47
Copper on steel	0.53	0.36
Rubber on concrete	1.0	0.8
Wood on wood	0.25–0.5	0.2
Glass on glass	0.94	0.4
Waxed wood on wet snow	0.14	0.1
Waxed wood on dry snow	—	0.04
Metal on metal (lubricated)	0.15	0.06
Ice on ice	0.1	0.03
Teflon on Teflon	0.04	0.04
Synovial joints in humans	0.01	0.003

[a] All values are approximate.

EXAMPLE 4.6   Moving into the Dormitory

At the beginning of a new school term, a student moves a box of books by attaching a rope to the box and pulling with a force of 90.0 N at an angle of 30.0°, as shown in Figure 4.14. The box of books has a mass of 20.0 kg, and the coefficient of kinetic friction between the bottom of the box and the floor is 0.500. Find the acceleration of the box.

Reasoning   There are basically three steps required for the solution to this problem. (1) First find the normal force **n** by applying the first condition of equilibrium, $\Sigma F_y = 0$, in the vertical direction. (2) Calculate the force of kinetic friction on the box from $f_k = \mu_k n$. (3) Apply Newton's second law along the horizontal direction to find the acceleration of the box.

Solution   The box is not accelerating in the vertical direction, and so we find the normal force from $\Sigma F_y = 0$. The forces in the y direction are the weight, **w**, the normal force, **n**, and the vertical component of the applied 90.0-N force. This vertical component has a magnitude equal to $(90.0 \text{ N})(\sin 30.0°)$. We find that

$$\Sigma F_y = n + (90.0 \text{ N})(\sin 30.0°) - (20.0 \text{ kg})(9.80 \text{ m/s}^2) = 0$$
$$n = 151 \text{ N}$$

The normal force is *not* equal to the weight of the box, because the vertical component of the 90.0-N force is helping to support some of that weight.

Knowing the normal force, we can find the force of kinetic friction:

$$f_k = \mu_k n = (0.500)(151 \text{ N}) = 75.5 \text{ N} \qquad \text{(to the left)}$$

Finally, we determine the horizontal acceleration using Newton's second law:

$$\Sigma F_x = (90.0 \text{ N})(\cos 30.0°) - 75.5 \text{ N} = (20.0 \text{ kg})(a_x)$$

$$a_x = \boxed{+0.122 \text{ m/s}^2}$$

Exercise   If the initial speed of the box is zero, what is its speed after it has traveled 2.00 m? How long does it take to pull it this distance?

Answer   0.699 m/s; 5.73 s.

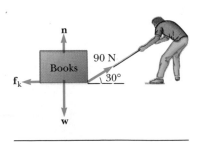

FIGURE 4.14
(Example 4.6)   A box of books being pulled to the right at an angle of 30.0°.

EXAMPLE 4.7   The Sliding Hockey Puck

A hockey puck is given an initial speed of 20.0 m/s on a frozen pond, as in Figure 4.15. The puck remains on the ice and slides 120 m before coming to rest. Determine the coefficient of friction between puck and ice.

Reasoning   The puck slides to rest with a constant acceleration along the horizontal. Thus, we can use the kinematic equation $v^2 = v_0^2 + 2ax$ to find a. Newton's second law applied in the horizontal direction is $-f_k = -\mu_k n = ma$. To find μ_k, we first find the normal force $\mathbf{n}$ by applying $\Sigma F_y = 0$ in the vertical direction.

Solution   With the final speed, v, equal to zero; the initial speed, $v_0 = 20.0$ m/s; and the displacement, $x = 120$ m:

$$v^2 = v_0^2 + 2ax$$

$$0 = (20.0 \text{ m/s})^2 + 2a(120 \text{ m})$$

$$a = -1.67 \text{ m/s}^2$$

The negative sign means that the acceleration is to the left, *opposite* the direction of the velocity.

The magnitude of the force of kinetic friction is found from $f_k = \mu_k n$, and n is found from $\Sigma F_y = 0$ as follows:

$$\sum F_y = n - w = 0$$

$$n = w = mg$$

Thus,

$$f_k = \mu_k n = \mu_k mg$$

Now we apply Newton's second law along the horizontal direction, taking the positive direction toward the right:

$$\sum F_x = -f_k = ma$$

$$-\mu_k mg = m(-1.67 \text{ m/s}^2)$$

$$\mu_k = \frac{1.67 \text{ m/s}^2}{9.80 \text{ m/s}^2} = \boxed{0.170}$$

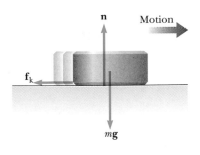

FIGURE 4.15
(Example 4.7)   *After* the puck is given an initial velocity to the right, the external forces acting on it are its weight, $m\mathbf{g}$, the normal force, $\mathbf{n}$, and the force of kinetic friction, $\mathbf{f_k}$.

EXAMPLE 4.8   Connected Objects

Two objects are connected by a light string that passes over a frictionless pulley, as in Figure 4.16a. The coefficient of sliding friction between the 4.00-kg object and the surface is 0.300. Find the acceleration of the two objects and the tension in the string.

Reasoning   Connected objects are handled by applying Newton's second law separately to each. The free-body diagrams for the block on the table and for the hanging block are shown in Figure 4.16b. We shall obtain two equations involving the unknowns T and a that can be solved simultaneously.

Solution   With the positive x direction to the right and the positive y direction upward, Newton's second law applied to the 4.00-kg object gives

$$\sum F_x = T - f_k = (4.00 \text{ kg})(a)$$

$$\sum F_y = n - (4.00 \text{ kg})(g) = 0$$

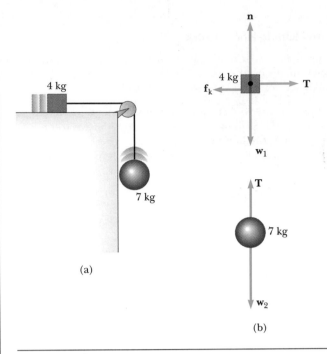

FIGURE 4.16
(Example 4.8)   (a) Two objects connected by a light string that passes over a frictionless pulley. (b) Free-body diagrams for the objects.

Since $f_k = \mu_k n$ and $n = 4g = 39.2$ N, we have $f_k = (0.300)(39.2$ N$) = 11.8$ N. Therefore,

$$(1) \qquad T = f_k + (4.00 \text{ kg})(a) = 11.8 \text{ N} + (4.00 \text{ kg})(a)$$

Now we apply Newton's second law to the 7.00-kg object moving in the vertical direction, this time selecting downward as the positive direction:

$$\sum F_y = (7.00 \text{ kg})(g) - T = (7.00 \text{ kg})(a)$$

or

$$(2) \qquad T = 68.6 \text{ N} - (7.00 \text{ kg})(a)$$

Subtracting (1) from (2) eliminates T:

$$56.8 \text{ N} - (11.0 \text{ kg})(a) = 0$$

$$a = \boxed{5.16 \text{ m/s}^2}$$

When this value for the acceleration is substituted into (1), we get

$$T = \boxed{32.4 \text{ N}}$$

FRICTION AND THE MOTION OF A CAR

Forces of friction are important in the analysis of the motion of cars and other wheeled vehicles. There are several types of friction forces to consider, the main ones being the force of friction between tires and road surface and the retarding force produced by air resistance.

FIGURE 4.17
The horizontal forces on the car are the *forward* forces, **f**, exerted on each tire by the road and the force of air resistance, **R**, which acts *opposite* the car's velocity. (The wheels of the car exert a backward force on the road not shown in the diagram.) (© *Williams/Edwards Concepts/The Image Bank*)

As each of four wheels turns to propel a car forward, it exerts a backward force on the road through its tire. The reaction to this backward force is a forward force, **f**, exerted by the road on the tire (Fig. 4.17). If we assume that the same forward force **f** is exerted on each tire, the net forward force on the car is 4**f**, and the car's acceleration is therefore $\mathbf{a} = 4\mathbf{f}/m$.

When the car is in motion, we must also consider the force of air resistance, **R**, which acts in the direction opposite its velocity. The net force on the car is therefore $4\mathbf{f} - \mathbf{R}$, and so the car's acceleration is $\mathbf{a} = (4\mathbf{f} - \mathbf{R})/m$. At normal driving speeds, the magnitude of **R** is proportional to the first power of the speed. That is, $R = bv$, where b is a constant. Thus, the force of air resistance increases with increasing speed. When R is equal to $4f$ the acceleration is zero, and the car moves at a constant speed.

A similar situation occurs when an object falls through air. When the upward force of air resistance balances the downward force of gravity, the net force on the object is zero and hence the object's acceleration is zero. Once this condition is reached, the object continues to move downward with some constant maximum speed called the **terminal speed.**

SUMMARY

Newton's first law states that an object at rest remains at rest, and an object in motion continues in motion with a constant velocity, unless it experiences a net force.

The resistance of an object to a change in its state of motion is called **inertia.** Mass is the physical quantity used to measure inertia.

Newton's second law states that the resultant force acting on an object is equal to the product of the mass of the object and its acceleration:

$$\sum \mathbf{F} = m\mathbf{a} \tag{4.1}$$

The **weight** of an object is equal to the product of its mass and the acceleration due to gravity:

$$\mathbf{w} = m\mathbf{g} \tag{4.6}$$

Newton's third law states that, if two bodies exert force on each other, the force exerted by body 1 on body 2 is equal in magnitude and opposite in direction to the force exerted by body 2 on body 1. Thus, an isolated force can never occur in nature.

An **object in equilibrium** has no net force acting on it, and the first law, in component form, implies that $\Sigma F_x = 0$ and $\Sigma F_y = 0$.

The maximum force of static friction, $\mathbf{f}_{s,\text{max}}$, between an object and a surface is proportional to the normal force acting on the object. This maximum force occurs when the object is on the verge of slipping. In general,

$$f_s \leqslant \mu_s n \qquad\qquad \textbf{[4.9]}$$

where μ_s is the **coefficient of static friction.** When a body slides over a surface, the direction of the force of kinetic friction, $\mathbf{f}_k$, is opposite the direction of the motion, and the magnitude is proportional to that of the normal force. The magnitude of $\mathbf{f}_k$ is

$$f_k = \mu_k n \qquad\qquad \textbf{[4.10]}$$

where μ_k is the **coefficient of kinetic friction.** In general, $\mu_k < \mu_s$.

ADDITIONAL READING

P. Brancazio, "The Physics of Kicking a Football," *The Physics Teacher,* October 1985, p. 403.

B. I. Cohen, "Isaac Newton," *Sci. American,* December 1955, p. 73.

W. F. Magie, *Source Book in Physics,* Cambridge, Mass., Harvard University Press, 1963. Excerpts from Newton on the laws of motion.

M. McCloskey, "Intuitive Physics," *Sci. American,* April 1983, p. 122.

M. McCloskey, A. Caramaza, and B. Gross, "Curvilinear Motion in the Absence of External Forces: Naive Beliefs About the Motion of Objects," *Science,* December 1980, p. 1139.

I. Newton, *Mathematical Principles of Natural Philosophy (Principia),* translated by A. Motte, revised by F. Cajori, Berkeley, University of California Press, 1947.

F. Palmer, "Friction," *Sci. American,* February 1951, p. 54.

R. Zimmerer. "The Measurement of Mass," *The Physics Teacher,* September 1983, p. 354.

CONCEPTUAL QUESTIONS

Example   Is it possible to have motion in the absence of a force? Explain.

Reasoning   Yes. Motion requires no force. Newton's first law says that motion needs no cause, hence an object in motion continues to move by itself in the absence of external forces. The simplest motion to think of is that of a meteoroid in outer space (similar to a glider moving on an air track).

Example   Is there any relation between the total force acting on an object and the direction in which it moves? Explain.

Reasoning   There is no relation between the total force on an object and the direction of its current motion. Force describes what the rest of the Universe does to the object, and the environment can push it forward, backward, sideways, or not at all.

Example   A 0.150-kg baseball is thrown upward with an initial speed of 20.0 m/s. If air resistance is neglected, what is the force on the ball (a) when it reaches half its maximum height? (b) when it reaches its peak?

Reasoning   The only force on the ball at *all* points in its trajectory is the force of gravity acting downward. The magnitude of this force is $w = mg = 1.47$ N.

Example If a small sports car collides head-on with a massive truck, which vehicle experiences the greater impact force? Which vehicle experiences the greater acceleration? Explain.

Reasoning The car and truck experience equal but oppositely directed forces. A calibrated spring scale placed between the colliding vehicles reads the same whichever way it faces. Since the car has the smaller mass, it stops with much greater acceleration.

Example Consider a sky diver falling through air *before* reaching her maximum speed, or terminal speed. As the speed of the sky diver increases, what happens to her acceleration? What is her acceleration when she reaches terminal speed?

Reasoning The forces on the sky diver are the constant downward force of gravity (her weight) and an upward force of air resistance, which is less than her weight before she reaches terminal speed. As her downward speed increases, the force of air resistance increases. The vector sum of her weight and the force of air resistance gives a total force which decreases with time, so her acceleration decreases. Once she reaches terminal speed, the two forces balance each other, the total force is zero, and her acceleration is zero.

Example A child pulls a wagon with a horizontal force, causing it to accelerate. Newton's third law says that the wagon exerts an equal and opposite force on the child. How can the wagon accelerate?

Reasoning The motion of any object is determined by the forces that act on it. In this situation, the horizontal forces exerted on the wagon are the forward force exerted by the child and the backward force of friction between the wagon and the surface. The resultant of these two forces causes the wagon to accelerate. Note that the horizontal forces that act on the child are the forward force of friction between the child's feet and surface and the backward force of the wagon.

Example A passenger sitting in the rear of a bus claims he was injured when the driver slammed on the brakes, causing a suitcase to come flying toward the passenger from the front of the bus. If you were the judge in this case, what disposition would you make? Why?

Reasoning The inertia of the suitcase would keep it moving forward as the bus stops. There would be no tendency for the suitcase to be thrown backward toward the passenger. Throw the case out of court!

Example Identify the action-reaction pairs in the following situations: (1) a girl takes a step; (2) a snowball hits a girl on the back; (3) a baseball player catches a ball; (4) a gust of wind strikes a window.

Reasoning (1) The action is the force exerted on the Earth by her foot; the reaction is the force exerted on her foot by the Earth. (2) The action is the force exerted on the girl's back by the snowball; the reaction is the force exerted on the snowball by the girl's back. (3) The action is the force exerted on the ball by the glove; the reaction is the force exerted on the glove by the ball. (4) The action is the force exerted on the window by the air molecules; the reaction is the force exerted on the air molecules by the window.

1. If an object is at rest, can we conclude that no external forces are acting on it?

2. If gold were sold by weight, would you rather buy it in Denver or in Death Valley? If it were sold by mass, in which of the two locations would you prefer to buy it? Why?

3. A space explorer is moving through space far from any planet or star. She notices a large rock, taken as a specimen from an alien planet, floating around the cabin of the ship. Should she push it gently or kick it toward the storage compartment? Why?

4. How much does an astronaut weigh out in space, far from any planets?

5. Although the frictional force between two surfaces may decrease as the surfaces are smoothed, it will increase if the surfaces are made extremely smooth and flat. Explain.

6. Why is the frictional force involved in the rolling of one body over another less than that involved in sliding motion?

7. In discussing friction, we have treated forces exerted by a surface as though there are two separate forces, a normal force and a friction force. Is this really necessary? Explain.

8. A massive metal object on a rough metal surface may undergo contact welding to that surface. Discuss how this affects the frictional forces that arise between object and surface.

9. Analyze the motion of a rock dropped in water in terms of its speed and acceleration as it falls. Assume that a resistive force is acting on the rock that increases as the velocity increases.

10. The force of air resistance acting on a falling object is approximated by the statement that it is proportional to the object's velocity and is directed upward. If the object falls fast enough, it should eventually start to move upward, since the force of air resistance must eventually exceed the weight of the object. Does this statement make sense? Why?

11. A ball is held in a person's hand. (a) Identify all the external forces acting on the ball and the reaction to each. (b) If the ball is dropped, what force is exerted on it while it is falling? Identify the reaction force in this case. (Neglect air resistance.)

12. Identify all the action-reaction pairs that exist for a horse pulling on a cart. Include the Earth in your examination.

13. If a car is traveling westward at a constant speed of 20 m/s, what is the resultant force acting on it?

14. A large crate is placed on the bed of a truck but not tied down. (a) As the truck accelerates forward, the crate remains at rest relative to the truck. What force causes the crate to accelerate forward? (b) If the driver slammed on the brakes, what could happen to the crate?

15. Explain why a rope climber must pull downward on the rope in order to move upward. Discuss the force exerted by the arms in relation to the weight of the person during the various stages of each "step" up the rope.

16. In a tug-of-war between two athletes, each pulls on the rope with a force of 200 N. What is the tension in the rope? If the rope does not move, what horizontal force does each athlete exert against the ground?

17. A rubber ball is dropped onto a floor. What force causes the ball to bounce back into the air?

18. Can the force of air resistance ever lift an object?

19. Suppose you are driving a car at a high speed. Why should you avoid "slamming on" your brakes when you want to stop in the shortest possible distance?

20. The driver of a speeding empty truck slams on the brakes and skids to a stop through a distance *d*. (a) If the truck carried a load that doubled its mass, what would be the truck's "skidding distance"? (b) If the initial speed of the truck were halved, what would be the truck's "skidding distance"?

21. A child in a car holds onto a string attached to a helium-filled balloon. What happens to the balloon when the car accelerates forward?

22. A janitor finds it quite easy to sweep a floor using a long-handled broom when the angle between handle and floor is small. However, if the angle is large, it becomes very difficult to push the broom. Explain.

23. The force of gravity is twice as great on a 20-N rock as it is on a 10-N rock. Why doesn't the 20-N rock have a greater free-fall acceleration?

24. The Earth is attracted to an object with a force equal to and opposite the force the Earth exerts on the object. Explain why the Earth's acceleration is not equal to and opposite that of the object.

25. The ropes and strings used in the text examples are all assumed to be massless. In reality, however, they must have mass. Discuss the effect of the mass of the rope on a typical acceleration problem involving a rope. Draw diagrams of the forces.

26. Can an object be in equilibrium if only one force acts on it?

27. An object thrown into the air stops at the highest point in its path. Is it in equilibrium at this point? Explain.

28. The mayor of a city decides to fire some city employees because they refuse to remove the sag from the cables that support the city's traffic lights. If you were a lawyer, what defense would you give on behalf of the employees? Who do you think would win the case in court?

29. Suppose the head of a hammer is loose, and you wish to tighten it. In terms of inertia, explain how you can accomplish this by banging the bottom of the handle (rather than the hammerhead) against a hard surface.

30. Suppose a truck loaded with sand accelerates at 0.5 m/s² on a highway. If the driving force on the truck remains constant, what happens to the truck's acceleration if its trailer leaks sand at a constant rate through a hole in its bottom?

31. As a rocket is fired from a launching pad, its speed and acceleration increase with time as its engines continue to operate. Explain why this occurs even though the thrust of the engines remains constant.

32. Draw a free-body diagram for each of the following objects: (a) a projectile in motion in the presence of air resistance, (b) a rocket leaving the launch pad with its engines operating, (c) an athlete running along a horizontal track.

PROBLEMS

Sections 4.2 through 4.5

1. A freight train has a mass of 1.5×10^7 kg. If the locomotive can exert a constant pull of 7.5×10^5 N, how long does it take to increase the speed of the train from rest to 80 km/h?

2. (a) An 850-kg car is moving to the right at 1.44 m/s. What is the net force on the car? (b) What would be the net force on the car if it were moving to the left?

3. After falling from rest at a height of 30 m, a 0.50-kg ball rebounds upward, reaching a height of

☐ indicates problems that have full solutions available in the Student Solutions Manual and Study Guide.

20 m. If the contact between ball and ground lasted 2.0 ms, what average force was exerted on the ball?

4. A bag of sugar weighs 5.00 lb on Earth. What should it weigh in newtons on the Moon, where the acceleration due to gravity is 1/6 that on Earth? Repeat for Jupiter, where g is 2.64 times Earth gravity. Find the mass in kilograms at each of the three locations.

5. In an avalanche, a mass of snow and ice high on a mountain breaks loose and starts an essentially frictionless "ride" down the mountain on a cushion of compressed air. If you were on a 30° slope and an avalanche started 400 m up the slope, how much time would you have to get out of the way?

6. (a) Draw a free-body diagram that indicates all the forces acting on a freely falling baseball. For each force, specify the reaction force. (b) Repeat for the baseball moving as a projectile toward an outfielder.

7. A 6.0-kg object undergoes an acceleration of 2.0 m/s². (a) What is the magnitude of the resultant force acting on it? (b) If this same force is applied to a 4.0-kg object, what acceleration is produced?

8. A 3.00-kg ball is dropped from the roof of a building 176.4 m high. While the ball is falling to Earth, a horizontal wind exerts a constant force of 12.0 N on it. (a) How far from the building does the ball hit the ground? (b) How long does it take to hit the ground? (c) What is its speed when it hits the ground?

9. A football punter accelerates a football from rest to a speed of 10 m/s during the time in which his toe is in contact with the ball (about 0.20 s). If the football has a mass of 0.50 kg, what average force does the punter exert on the ball?

10. The air exerts a forward force of 10 N on the propeller of a 0.20-kg model airplane. If the plane accelerates forward at 2.0 m/s², what is the magnitude of the resistive force exerted by the air on the airplane?

11. A 5.0-g bullet leaves the muzzle of a rifle with a speed of 320 m/s. What force (assumed constant) is exerted on the bullet while it is traveling down the 0.82-m-long barrel of the rifle?

12. The force of the wind on the sails of a sailboat is 390 N north. The water exerts a force of 180 N east. If the boat (including crew) has a mass of 270 kg, what are the magnitude and direction of its acceleration?

13. Two forces are applied to a car in an effort to move it, as shown in Figure 4.18. (a) What is the resultant of these two forces? (b) If the car has a mass of 3000 kg, what acceleration does it have?

14. A boat moves through the water with two forces

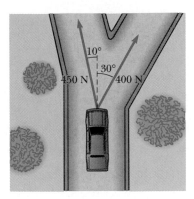

FIGURE 4.18 (Problem 13)

acting on it. One is a 2000-N forward push by the motor, and the other is an 1800-N resistive force due to the water. (a) What is the acceleration of the 1000-kg boat? (b) If it starts from rest, how far will it move in 10.0 s? (c) What will its velocity be at the end of this time?

Section 4.6 Some Applications of Newton's Laws
Newton's First Law and the First Condition of Equilibrium

15. Two persons are pulling a boat through the water as in Figure 4.19. Each exerts a force of 600 N directed at a 30.0° angle relative to the forward motion of the boat. If the boat moves with constant velocity, find the resistive force, F, exerted on the boat by the water.

16. A block of mass $m = 2.0$ kg is held in equilibrium on an incline of angle $\theta = 60°$ by the horizontal force F, as shown in Figure 4.20. (a) Determine

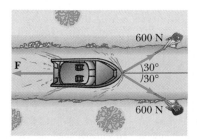

FIGURE 4.19 (Problem 15) Two people pulling a boat.

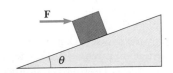

FIGURE 4.20 (Problems 16 and 58)

FIGURE 4.21 (Problem 17)

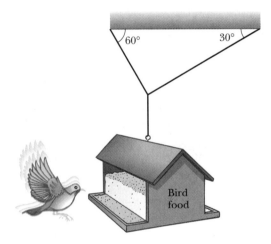

FIGURE 4.22 (Problem 18)

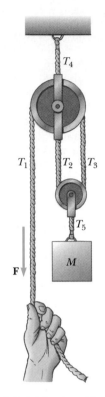

FIGURE 4.23 (Problem 19)

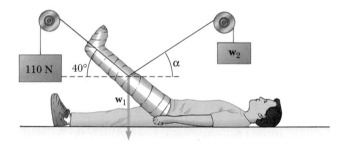

FIGURE 4.24 (Problem 20)

the value of *F*. (b) Determine the normal force exerted by the incline on the block (ignore friction).

17. Find the tension in the two wires that support the 100-N light fixture in Figure 4.21.

18. A 150-N bird feeder is supported by three cables as shown in Figure 4.22. Find the tension in each cable.

19. A mass, *M*, is held in place by the applied force **F** and a pulley system, as shown in Figure 4.23. The pulleys are massless and frictionless. Determine (a) the tension in each section of rope and (b) the magnitude of the applied force.

20. The leg and cast in Figure 4.24 weigh 220 N (w_1). Determine the weight w_2 and the angle α needed in order that there be no force exerted on the hip joint by the leg plus cast.

Newton's Second Law

21. A shopper in a supermarket pushes a loaded cart with a horizontal force of 10 N. The cart has a mass of 30 kg. (a) How far will it move in 3.0 s, starting from rest? (Ignore friction.) (b) How far

will it move in 3.0 s if the shopper places her 30-N child in the cart before she begins to push it?

22. An 80-kg person escapes from a burning building by jumping from a window situated 30 m above a catching net. Assuming that air resistance exerts a 100-N force on the person as he falls, determine his velocity just before he hits the net.

23. A 2000-kg sailboat experiences an eastward force of 3000 N from the ocean tide and a wind force against its sails of magnitude 6000 N, directed 45° north of west. What are the magnitude and direction of the resultant acceleration?

24. A train has a mass of 5.22×10^6 kg and is moving at 90.0 km/h. The engineer applies the brakes,

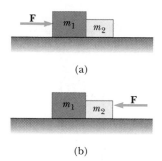

(a)

(b)

FIGURE 4.25   (Problem 25)

which results in a net backward force of 1.87×10^6 N on the train. The brakes are held on for 30.0 s. (a) What is the new speed of the train? (b) How far does it travel during this period?

25. A small bug is placed between two blocks of masses m_1 and m_2 ($m_1 > m_2$) on a frictionless table. A horizontal force, **F**, can be applied to either m_1, as in Figure 4.25a, or m_2, as in Figure 4.25b. Show that the bug has a greater chance of surviving when the force is applied to m_1. (*Hint:* Determine the contact force in each case.)

26. A 2.0-kg mass starts from rest and slides down an inclined plane 80 cm long in 0.50 s. What *net force* is acting on the mass along the incline?

27. A 40.0-kg wagon is towed up a hill inclined at 18.5° with respect to the horizontal. The tow rope is parallel to the incline and has a tension of 140 N in it. Assume that the wagon starts from rest at the bottom of the hill, and neglect friction. How fast is the wagon going after moving 80 m up the hill?

28. A mass, $m_1 = 5.00$ kg, resting on a frictionless horizontal table is connected to a cable that passes over a pulley and then is fastened to a hanging mass, $m_2 = 10.0$ kg, as in Figure 4.26. Find the acceleration of each mass and the tension in the cable.

FIGURE 4.26   (Problems 28, 40, and 84)

(Problem 23) *(Superstock)*

29. The parachute on a race car of weight 8820 N opens at the end of a quarter-mile run when the car is traveling at 35 m/s. What total retarding force must be supplied by the parachute to stop the car in a distance of 1000 m?

30. A 5.0-kg bucket of water is raised from a well by a rope. If the upward acceleration of the bucket is 3.0 m/s², find the force exerted by the rope on the bucket.

31. Two people pull as hard as they can on ropes attached to a 200-kg boat. If they pull in the same direction, the boat has an acceleration of 1.52 m/s² to the right. If they pull in opposite directions, the boat has an acceleration of 0.518 m/s² to the left. What is the force exerted by each person on the boat? (Disregard any other forces on the boat.)

32. On takeoff, the combined action of the engines and wings of an airplane exerts an 8000-N force on the plane, directed upward at an angle of 65.0° above the horizontal. The plane rises with constant velocity in the vertical direction while continuing to accelerate in the horizontal direction. (a) What is the weight of the plane? (b) What is its horizontal acceleration?

33. Two objects of masses 10.0 kg and 5.00 kg are connected by a light string that passes over a frictionless pulley as in Figure 4.27. The 5.00-kg object

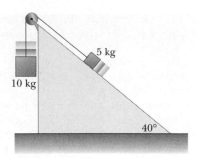

FIGURE 4.27 (Problem 33)

lies on a smooth incline of angle 40.0°. Find the accelerations of each object and the tension in the string.

34. A 1000-kg car is pulling a 300-kg trailer. Together the car and trailer have an acceleration of 2.15 m/s² in the forward direction. Neglecting frictional forces on the trailer, determine (a) the net force on the car; (b) the net force on the trailer; (c) the force exerted on the car by the trailer; (d) the force exerted on the road by the car.

35. Two blocks are fastened to the ceiling of an elevator as in Figure 4.28. The elevator accelerates upward at 2.00 m/s². Find the tension in each rope.

36. Two blocks are fastened to the top of an elevator as in Figure 4.28. The mass of each rope is 1.00 kg. The elevator accelerates upward at 4.00 m/s². Find the tensions in the ropes at points A, B, C, and D.

37. Two masses of 3.00 kg and 5.00 kg are connected by a light string that passes over a frictionless pulley as in Figure 4.29. Determine (a) the tension in the string, (b) the acceleration of each mass, and (c) the distance each mass will move in the first second of motion if both masses start from rest.

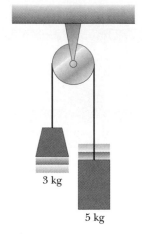

FIGURE 4.29 (Problem 37)

Section 4.7 Force of Friction

38. A dockworker loading crates on a ship finds that a 20-kg crate, initially at rest on a horizontal surface, requires a 75-N horizontal force to set it in motion. However, after the crate is in motion, a horizontal force of 60 N is required to keep it moving with a constant speed. Find the coefficients of static and kinetic friction between crate and floor.

39. A copper block of mass $m = 2.00$ kg rests on a steel table and is connected by a light cord to a mass, M, and a wall as in Figure 4.30. Given that $\theta = 30.0°$, (a) what is the maximum value M can have before it and m begin to move? (b) Would your answer be different if the apparatus were placed on the Moon? Explain.

40. Masses $m_1 = 10.0$ kg and $m_2 = 5.00$ kg are connected by a light string that passes over a frictionless pulley as in Figure 4.26. If m_1, initially held at rest on the table, falls 1.0 m in 1.2 s, determine the coefficient of kinetic friction between it and the table.

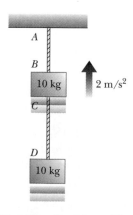

FIGURE 4.28 (Problems 35 and 36)

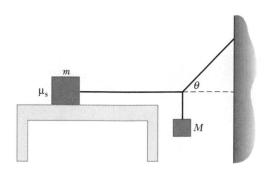

FIGURE 4.30 (Problem 39)

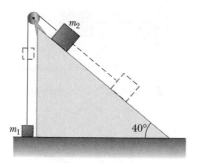

FIGURE 4.31 (Problem 42)

41. A box of books weighing 300 N is shoved across the floor of an apartment by a force of 400 N exerted downward at an angle of 35.2° below the horizontal. If the coefficient of kinetic friction between box and floor is 0.570, how long does it take to move the box 4.00 m, starting from rest?

42. Masses $m_1 = 4.00$ kg and $m_2 = 9.00$ kg are connected by a light string that passes over a frictionless pulley. As shown in Figure 4.31, m_1 is held at rest on the floor and m_2 rests on a fixed incline of $\theta = 40.0°$. The masses are released from rest, and m_2 slides 1.00 m down the incline in 4.00 s. Determine (a) the coefficient of kinetic friction between m_2 and the incline, (b) the acceleration of each mass, and (c) the tension in the string.

43. Two blocks of masses 40.0 kg and 20.0 kg are stacked on a table with the heavier block on top. The coefficient of static friction is 0.600 between the two blocks and 0.300 between the bottom block and the table. A horizontal force is slowly applied to the top block until one of the blocks moves. (a) Where does slippage occur first: between the two blocks or between the bottom block and the table? Explain. (b) What value of the coefficient of static friction between the bottom block and the table would change the answer to part (a)?

 44. An object falling under the pull of gravity experiences a frictional force of air resistance. The magnitude of this force is approximately proportional to the speed of the object, $f = bv$. Let us say $b = 15$ kg/s and $m = 50$ kg. (a) What is the maximum speed the object can reach while falling? (b) Does your answer to part (a) depend on the initial speed of the object? Explain.

45. The board sandwiched between two other boards in Figure 4.32 weighs 95.5 N. If the coefficient of friction between the boards is 0.663, what must be the magnitude of the compression forces (assume horizontal) acting on both sides of the center board to keep it from slipping?

46. The person in Figure 4.33 weighs 170 lb. The crutches each make an angle of 22.0° with the vertical (as seen from the front). Half of the person's weight is supported by the crutches. The other half is supported by the vertical forces of the ground on his feet. Assuming the person is at rest and the force of the ground on the crutches acts along the crutches, determine (a) the smallest possible coefficient of friction between crutches and ground and (b) the magnitude of the compression force supported by each crutch.

47. A student decides to move a box of books into her dormitory room by pulling on a rope attached to the box. She pulls with a force of 80.0 N at an angle of 25.0° with the horizontal. The box has a mass of 25.0 kg, and the coefficient of kinetic friction between box and floor is 0.300. (a) Find the acceleration of the box. (b) Along the way, she must move the box up a ramp inclined at 10.0° with the horizontal. If the box starts from rest at the bottom of the ramp and is pulled at an angle of 25.0° with respect to the incline and with the same 80.0-N force, can it be moved up the ramp? If so, what is the acceleration up the ramp?

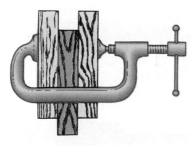

FIGURE 4.32 (Problem 45)

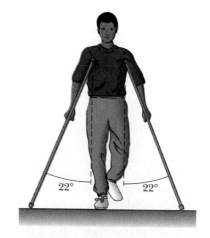

FIGURE 4.33 (Problem 46)

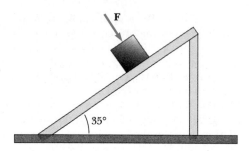

FIGURE 4.34 (Problem 50)

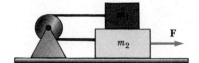

FIGURE 4.35 (Problem 53)

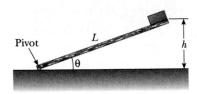

FIGURE 4.36 (Problems 54 and 85)

48. A box slides down a 30.0° ramp with an acceleration of 1.20 m/s². Determine the coefficient of kinetic friction between the box and the ramp.

49. A car is traveling at 50.0 km/h on a flat highway. (a) If the coefficient of friction between road and tires on a rainy day is 0.100, what is the minimum distance in which the car will stop? (b) What is the stopping distance when the surface is dry and the coefficient of friction is 0.600?

50. The coefficient of static friction between the 3.00-kg crate and the 35.0° incline of Figure 4.34 is 0.300. What minimum force **F** must be applied to the crate perpendicular to the incline to prevent the crate from sliding down the incline?

51. A car with a speed of 40.0 km/h approaches the bottom of an icy hill. The hill has an angle of inclination of 10.5°. The driver applies the brakes, which makes the car skid up the hill. If the coefficient of kinetic friction between ice and tires is 0.153, how far, measured along the incline, is the car on the hill when it stops?

52. A 35.0-kg child tries to climb up an icy slope, to no avail. The angle of the incline is 35.0° with respect to the horizontal, and the coefficient of static friction between the child's boot and the incline is 0.200. (a) Compare the force of gravity (along the incline) that is exerted on the child with the force needed to make her slide. Why can't she avoid sliding? (b) Does the mass of the child have anything to do with the fact that she can't help sliding? Explain. (c) Does the angle of the incline have anything to do with the fact that she can't help sliding? Explain. (d) To what value must the angle be decreased in order for the child not to slide down?

53. A light cord is passed around a frictionless pulley and connected to the two masses shown in Figure 4.35, with m_2 pulled by a constant horizontal force, **F**. The coefficient of kinetic friction between m_1 and m_2 is the same as that between m_2 and the horizontal surface. What is the value of this coefficient if the blocks are to move at a constant speed?

54. As a part of a laboratory investigation, a student wishes to measure the coefficients of friction between a block of metal and a wooden board. The board has a length of L, and the block is placed at one end of it. The board is raised, and the block begins to slide when it is a distance of h above the lower end of the board, as in Figure 4.36. At this angle, the block slides down the length of the board in the time t. Determine (a) the coefficient of static friction between block and board, (b) the acceleration of the block, (c) the smallest angle that causes the block to move, and (d) the coefficient of kinetic friction between block and board.

55. A block of mass $m = 2.00$ kg rests on the left edge of a block of length $L = 3.00$ m and mass $M = 8.00$ kg. The coefficient of kinetic friction between the two blocks is $\mu_k = 0.300$, and the surface on which the 8.00-kg block rests is frictionless. A constant horizontal force of magnitude $F = 10.0$ N is applied to the 2.00-kg block, setting it in motion as shown in Figure 4.37a. (a) How long will it take before this block makes it to the right side of the

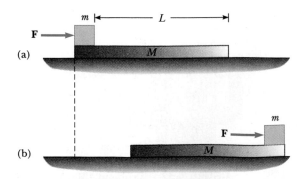

FIGURE 4.37 (Problem 55)

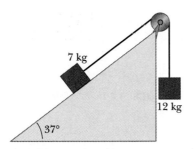

FIGURE 4.38   (Problem 56)

FIGURE 4.40   (Problem 60)

8.00-kg block, as shown in Figure 4.37b? (*Note:* Both blocks are set in motion when **F** is applied.) (b) How far does the 8.00-kg block move in the process?

56. Find the acceleration experienced by each of the two masses shown in Figure 4.38 if the coefficient of friction between the 7.00-kg mass and the plane is 0.250.

57. A hockey puck is hit on a frozen lake and starts moving with a speed of 12.0 m/s. Five seconds later, its speed is 6.00 m/s. (a) What is its average acceleration? (b) What is the average value of the coefficient of kinetic friction between puck and ice? (c) How far does the puck travel during this 5.00-s interval?

58. A 2.00-kg block is held in equilibrium on an incline of angle $\theta = 60.0°$ by a horizontal force, **F**, applied in the direction shown in Figure 4.20. If the coefficient of static friction between block and incline is $\mu_s = 0.300$, determine (a) the minimum value of **F** and (b) the normal force of the incline on the block. (c) Answer part (a) for $\theta = 30.0°$.

59. A sled weighing 60.0 N is pulled horizontally across snow so that the coefficient of kinetic friction between sled and snow is 0.100. A penguin weighing 70.0 N rides on the sled. (See Fig. 4.39.) If the coefficient of static friction between penguin and sled is 0.700, find the maximum horizontal force that can be exerted on the sled before the penguin begins to slide off.

ADDITIONAL PROBLEMS

60. A man attempting to train his dog is shown in Figure 4.40 pulling on the dog with a force of 70.0 N at an angle of 30.0° to the horizontal. Find the *x* and *y* components of this force.

61. Four forces act on a boat, shown from the top in Figure 4.41. Find the magnitude and direction of the resultant force on the boat.

62. (a) What is the resultant force exerted by the two cables supporting the traffic light in Figure 4.42? (b) What is the weight of the light?

63. Figure 4.43 shows the speed of a person's body, during a chin-up. Assuming the motion is vertical

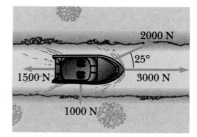

FIGURE 4.41   (Problem 61)

FIGURE 4.39   (Problem 59)

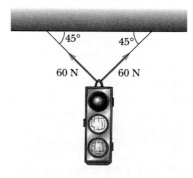

FIGURE 4.42   (Problem 62)

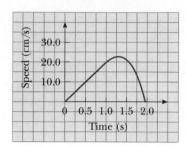

FIGURE 4.43   (Problem 63)

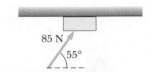

FIGURE 4.45   (Problem 69)

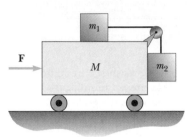

FIGURE 4.46   (Problem 70)

and the mass of the person (excluding the arms) is 64.0 kg, determine the magnitude of the force exerted on the body by the arms at various stages of the motion.

64. Some baseball pitchers are capable of throwing a fast ball at 100 mi/h. The pitcher achieves this speed by moving his arm through a distance of about 1.50 m. What average force must he exert on the 0.150-kg ball during this time?

65. As a protest against the umpire's calls, a baseball pitcher throws a ball straight up into the air at a speed of 20.0 m/s. In the process, he moves his hand through a distance of 1.50 m. If the ball has a mass of 0.150 kg, find the force he exerts on the ball to give it this upward speed.

66. A 2.00-kg aluminum block and a 6.00-kg copper block are connected by a light string over a frictionless pulley. They are allowed to move on a fixed steel block-wedge (of angle $\theta = 30.0°$) as shown in Figure 4.44. Making use of Table 4.2, determine (a) the acceleration of the two blocks and (b) the tension in the string.

67. A 2000-kg car is slowed down uniformly from 20.0 m/s to 5.00 m/s in 4.00 s. What average force acted on the car during this time, and how far did the car travel during the deceleration?

68. A girl coasts down a hill on a sled, reaching the bottom with a speed of 7.0 m/s. If the coefficient of friction between runners and snow is 0.050 and the girl and sled together weigh 600 N, how far

does the sled travel on the level surface before coming to rest?

69. A 4.00-kg block is pushed along the ceiling with a constant applied force of 85.0 N that acts at an angle of 55.0° with the horizontal, as in Figure 4.45. The block accelerates to the right at 6.00 m/s². Determine the coefficient of kinetic friction between block and ceiling.

70. (a) What horizontal force **F** must be applied to the cart in Figure 4.46 in order that the blocks remain stationary relative to the cart? Assume that all surfaces, wheels, and pulleys are frictionless. (*Hint:* Note that the tension force in the string ac-

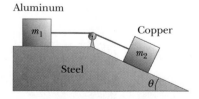

FIGURE 4.44   (Problem 66)

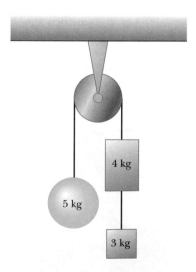

FIGURE 4.47   (Problem 71)

FIGURE 4.48   (Problem 72)

celerates m_1.) (b) What is the acceleration of the system?

71. Three masses are connected by light strings as shown in Figure 4.47. The string connecting the 4.00-kg mass and the 5.00-kg mass passes over a light frictionless pulley. Determine (a) the acceleration of each mass and (b) the tension in the two strings.

72. Find the tension in each cable supporting the 600-N cat burglar in Figure 4.48.

73. (a) What is the minimum force of friction required to hold the system of Figure 4.49 in equilibrium? (b) What coefficient of static friction between the 100-N block and the table ensures equilibrium? (c) If the coefficient of kinetic friction between the 100-N block and the table is 0.250, what hanging weight should replace the 50.0-N weight to allow the system to move at a constant speed once it is set in motion?

74. A box rests on the back of a truck. The coefficient

of friction between box and truck bed is 0.300. (a) When the truck accelerates forward, what force accelerates the box? (b) Find the maximum acceleration the truck can have before the box slides.

75. A 3.00-kg block starts from rest at the top of a 30.0° incline and slides 2.00 m down the incline in 1.50 s. Find (a) the acceleration of the block, (b) the coefficient of kinetic friction between it and the incline, (c) the frictional force acting on the block, and (d) the speed of the block after it has slid 2.00 m.

76. A 72-kg man stands on a spring scale in an elevator. Starting from rest, the elevator ascends, attaining its maximum speed of 1.2 m/s in 0.80 s. It travels with this constant speed for 5.0 s, undergoes a uniform *negative* acceleration for 1.5 s, and comes to rest. What does the spring scale register (a) before the elevator starts to move? (b) during the first 0.80 s? (c) while the elevator is traveling at constant speed? (d) during the negative acceleration?

77. A 3.0-kg mass hangs at one end of a rope that is attached to a support on a railroad car. When the car accelerates to the right, the rope makes an angle of 4.0° with the vertical, as shown in Figure 4.50. Find the acceleration of the car.

78. Two blocks on a frictionless horizontal surface are connected by a light string as in Figure 4.51, where $m_1 = 10$ kg and $m_2 = 20$ kg. A force of 50 N is applied to the 20-kg block. (a) Determine the acceleration of each block and the tension in the string. (b) Repeat the problem for the case where the coefficient of kinetic friction between each block and the surface is 0.10.

79. A 5.0-kg penguin sits on a 10-kg sled, as in Figure 4.52. A horizontal force of 45 N is applied to the sled, but the penguin attempts to impede the motion by holding onto a cord attached to a wall. The coefficient of kinetic friction between the moving surfaces is 0.20. (a) Draw a free-body diagram for the penguin and one for the sled, and

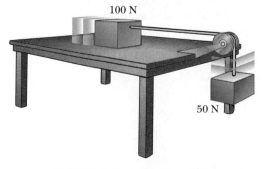

FIGURE 4.49   (Problem 73)

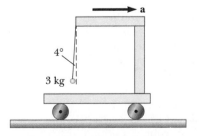

FIGURE 4.50   (Problem 77)

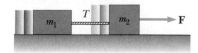

FIGURE 4.51   (Problem 78)

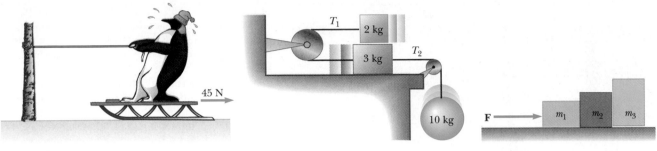

FIGURE 4.52   (Problem 79)   **FIGURE 4.53**   (Problem 81)   **FIGURE 4.54**   (Problem 83)

identify the reaction force for each force you include. (b) Determine the tension in the cord and the acceleration of the sled.

80. A rope of mass m and length L hangs from a ceiling. If a mass, M, is connected to the rope at a point that is a distance of x from the ceiling $(0 \leq x \leq L)$, (a) determine an expression for the tension in the rope at that point in terms of m, L, M, and x. (b) For $m = 5.00$ kg, $L = 5.00$ m, and $M = 20.0$ kg, and a rope that can support a maximum tension of $T = 215$ N, how far from the ceiling can the 20.0-kg mass be hung without the rope breaking?

81. In Figure 4.53, the coefficient of kinetic friction between the two blocks is 0.30. The table surface and the pulleys are frictionless. (a) Draw a free-body diagram for each block. (b) Determine the acceleration of each block. (c) Find the tension in the strings.

82. Bob and Kathy, two construction workers on the roof of a building, are about to raise a keg of nails from the ground by means of a light rope passing over a light frictionless pulley 10.0 m above the ground. Bob weighs 900 N, Kathy 600 N, the keg 300 N, and the nails 600 N. Both workers slip off the roof, and the following unfortunate sequence of events takes place. Hanging together on the rope, Bob and Kathy strike the ground just as the keg hits the pulley. Unnerved by his fall, Bob lets go of the rope, and the keg pulls Kathy up to the roof, where she cracks her head against the pulley but gamely hangs on. However, the nails spill out of the keg when it strikes the ground, and the empty keg rises as Kathy returns to the ground. Finally, she has had enough, lets go of the rope, and remains on the ground, only to be hit by the empty keg again. Ignoring the possible mid-air collisions that merely added insult to injury, how long did it take this industrial accident to run its course? Assume that all collisions with the ground and pulley serve to start each subsequent motion from rest, so that each trip up and down begins with zero velocity.

83. Three blocks are in contact with each other on a frictionless horizontal surface as in Figure 4.54. A horizontal force, **F**, is applied to m_1. If $m_1 = 2.00$ kg, $m_2 = 3.00$ kg, $m_3 = 4.00$ kg, and **F** = 180 N, find (a) the acceleration of the blocks, (b) the resultant force on each block, and (c) the magnitude of the contact forces between the blocks.

84. A steel block of mass $m_1 = 2.0$ kg rests on a steel table and is connected to a hanging block by a light cord over a frictionless pulley, as shown in Figure 4.26. (a) Weights are added to the hanging block so that its mass, m_2, is increased slowly. What value of m_2 causes m_1 to begin moving? (b) For the value of m_2 determined in (a), how long does m_1 take to move 0.50 m along the table? Does this result depend on the values of m_1 and m_2? Explain.

85. Figure 4.55 represents a simple experimental arrangement for measuring coefficients of friction. The angle of inclination is increased until the block slips at the critical angle, θ_c. (a) Show that $\mu_s = \tan \theta_c$. (b) Once the block is in motion, the angle of inclination can be reduced to a value, θ_c', such that the block moves down the incline at constant speed. Show that $\mu_k = \tan \theta_c'$.

86. The three blocks in Figure 4.55 are connected by light strings that pass over frictionless pulleys. The acceleration of the 5.00-kg block is 2.00 m/s² to the left, and the surfaces are rough. Find (a) the tension in each string and (b) the coefficient of kinetic friction between blocks and surface. (Assume the same μ_k for both blocks.)

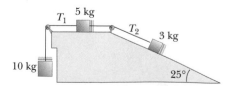

FIGURE 4.55   (Problems 85 and 86)

WORK AND ENERGY

5

Cyclists work hard and expend energy racing at top speed.
(© Index/Stock).

The concept of energy is one of the most important in the world of science. In everyday usage, the term *energy* has to do with the cost of fuel for transportation and heating, electricity for lights and appliances, and the foods we consume. However, these ideas do not really define energy. They tell us only that fuels are needed to do a job and that those fuels provide us with something we call energy.

Energy is present in the Universe in a variety of forms, including mechanical energy, chemical energy, electromagnetic energy, heat energy, and nuclear energy. Although energy can be transformed from one form to another, the total amount of energy in the Universe remains the same. If an isolated system loses energy in some form, then, by the principle of conservation of energy, the system must gain an equal amount of energy in other forms. For example, when an electric motor is connected to a battery, chemical energy is converted to electrical energy, which in turn is converted to mechanical energy. The transformation of energy from one form into another is an essential part of the study of physics, chemistry, biology, geology, and astronomy.

In this chapter we are concerned only with mechanical energy. We introduce the concept of *kinetic energy*, which is defined as the energy associated with

119

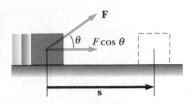

FIGURE 5.1
If an object undergoes a displacement of **s**, the work done by the force **F** is $(F \cos \theta)\, s$.

motion, and the concept of *potential energy,* the energy associated with position. We shall see that the ideas of work and energy can be used in place of Newton's laws to solve certain problems.

We begin by defining *work,* a concept that provides a link between force and energy. With this as a foundation, we can then discuss the principle of conservation of energy and apply it to problems.

5.1

WORK

Almost all the terms we have used thus far have conveyed the same meaning in physics as in everyday life. Now, however, we encounter a term whose meaning in physics is distinctly different from its meaning in our day-to-day affairs. This new term is **work**. It can be defined with the help of Figure 5.1. Here we see an object that undergoes a displacement of **s** along a straight line while acted on by a constant force, **F**, that makes an angle of θ with **s**.

> The work W done by an agent exerting a constant force is defined as the product of the component of the force along the direction of displacement and the magnitude of the displacement:

$$W \equiv (F \cos \theta)\, s \qquad\qquad \textbf{[5.1]}$$

As an example of the distinction between this definition of *work* and our everyday meaning of the word, consider holding a heavy chair at arm's length for 10 min. At the end of this time interval, your tired arms may lead you to think that you have done a considerable amount of work. According to our definition, you have done no work on the chair.[1] You have exerted a force in order to support the chair, but you have not moved it. A force does no work on an object if the object does not move. This can be seen by noting that if $s = 0$, Equation 5.1 gives $W = 0$.

Also note from Equation 5.1 that the work done by a force is zero when the force is perpendicular to the displacement. That is, if $\theta = 90°$, then $W = 0$ because $\cos 90° = 0$. For example, no work is done when a bucket of water is carried horizontally at constant velocity, because the upward force exerted to support the bucket is perpendicular to the displacement of the bucket, as shown in Figure 5.2. Likewise, the work done by the force of gravity during the horizontal displacement is also zero, for the same reason.

The sign of the work depends on the direction of **F** relative to **s**. Work is positive when the component $F \cos \theta$ is in the same direction as the displacement. For example, when you lift a box as in Figure 5.3, the work done by the force you exert on the box is positive because that force is upward, in the same direction as the displacement. Work is negative when **F** $\cos \theta$ is in the direction opposite the displacement. The negative sign comes from the fact that $\theta = 180°$

FIGURE 5.2
No work is done when a bucket is moved horizontally, because the applied force, **F**, is perpendicular to the displacement.

[1] Actually, you do burn calories while holding a chair at arm's length, since your muscles are continuously contracting and relaxing while the chair is being supported. Thus, work is being done on your body but not on the chair.

TABLE 5.1
Units of Work in the Three Common Systems of Measurement

System	Unit of Work	Name of Combined Unit
SI	newton-meter (N·m)	joule (J)
cgs	dyne-centimeter (dyne·cm)	erg
British engineering (conventional)	foot-pound (ft·lb)	foot-pound

Although this athlete is pushing against the wall, no work is done on the wall since it does not move. *(Fourby-Five)*

and cos 180° = −1, which, from Equation 5.1, gives a negative value for W. Finally, if an applied force acts along the direction of the displacement, then $\theta = 0°$. Since cos 0° = 1, in this case Equation 5.1 becomes

$$W = Fs \qquad \text{[5.2]}$$

Work is a scalar quantity, and its units are force times length. Therefore, the SI unit of work is the **newton-meter** (N·m). Another name for the newton-meter is the **joule** (J). The unit of work in the cgs system is the **dyne-centimeter** (dyne·cm), which is also called the **erg,** and the unit in the conventional (British engineering) system is the **foot-pound** (ft·lb). These are summarized in Table 5.1. Note that 1 J = 10^7 ergs.

EXAMPLE 5.1   Mr. Clean

A man cleaning his apartment pulls a vacuum cleaner with a force of magnitude $F = 50.0$ N at an angle of 30.0°, as shown in Figure 5.4. He moves the vacuum cleaner a distance of 3.0 m. Calculate the work done by the 50.0-N force.

Solution   We can use $W = (F \cos \theta)s$, with $F = 50.0$ N, $\theta = 30.0°$, and $s = 3.00$ m, to get

$$W = (50.0 \text{ N})(\cos 30.0°)(3.00 \text{ m}) = \boxed{130 \text{ J}}$$

Note that the normal force, **n**, the weight, $m\mathbf{g}$, and the upward component of the applied force (50.0 N) cos 30.0° do *no* work because they are perpendicular to the displacement.

Exercise   Find the work done on the vacuum cleaner by the man if he pulls it 3.00 m with a horizontal force of $F = 50.0$ N.

Answer   150 J.

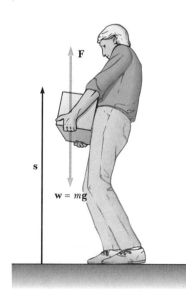

FIGURE 5.3
Positive work is done by this person when the box is lifted, because the applied force, **F**, is in the same direction as the displacement. When the box is lowered to the floor, the work done by the person is negative.

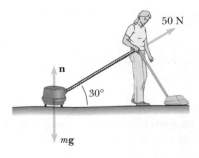

FIGURE 5.4
(Example 5.1)   A vacuum cleaner being pulled at an angle of 30° with the horizontal.

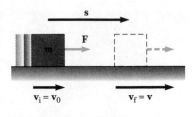

FIGURE 5.5
An object undergoing a displacement and a change in velocity under the action of a constant net force, **F**.

5.2

KINETIC ENERGY AND THE WORK-ENERGY THEOREM

Solving problems with Newton's second law can be difficult if the forces involved are complex. An alternative approach to such problems is to relate the speed of an object to its displacement under the influence of some net force. If the work done by the net force on the object can be calculated for a given displacement, the change in the object's speed is easy to evaluate.

Figure 5.5 shows a particle of mass m moving to the right under the action of a constant net force, **F**. Because the force is constant, we know from Newton's second law that the particle moves with a constant acceleration, **a**. If the particle is displaced a distance of s, the work done by **F** is

$$W_{net} = Fs = (ma)s \qquad [5.3]$$

However, in Chapter 2 we found that the following relationship holds when an object undergoes constant acceleration:

$$v^2 = v_0^2 + 2as \qquad \text{or} \qquad as = \frac{v^2 - v_0^2}{2}$$

This expression is now substituted into Equation 5.3 to give

$$W_{net} = m\left(\frac{v^2 - v_0^2}{2}\right)$$

$$W_{net} = \tfrac{1}{2}mv^2 - \tfrac{1}{2}mv_0^2 \qquad [5.4]$$

The quantity $mv^2/2$ has a special name in physics: **kinetic energy.** Any object of mass m and speed v is defined to have a kinetic energy, KE, of

Kinetic energy

$$KE \equiv \tfrac{1}{2}mv^2 \qquad [5.5]$$

Kinetic energy is a scalar quantity and has the same units as work. For example, a 1.0-kg mass moving with a speed of 4.0 m/s has a kinetic energy of 8.0 J. We can think of kinetic energy as the energy associated with the motion of an object.

It is often convenient to write Equation 5.4 as

Work-energy theorem

$$W_{net} = KE_f - KE_i \qquad [5.6]$$

This equation says that the effect of the work done by a *net* force acting on an object is to change the kinetic energy of the object from some initial value, KE_i, to some final value, KE_f. Equation 5.6 is an important result known as the **work-energy theorem.** Thus, we conclude that

> **The net work done on an object by a net force acting on it is equal to the change in the kinetic energy of the object.**

The work-energy theorem also says that the speed of the object increases if the net work done on it is positive, because the final kinetic energy is greater than the initial kinetic energy. The object's speed decreases if the net work is negative, because the final kinetic energy is less than the initial kinetic energy. Notice that the speed and kinetic energy of an object will change only if work is done on the object by some external force.

Consider the relationship between the work done on an object and the change in its kinetic energy as expressed by Equation 5.4. Because of this connection, we can also think of kinetic energy as the work the object can do in coming to rest. For example, suppose a hammer is on the verge of striking a nail, as in Figure 5.6. The moving hammer has kinetic energy and can do work on the nail. The work done on the nail is Fs, where F is the average force exerted on the nail by the hammer and s is the distance the nail is driven into the wall.

For convenience, Equation 5.6 was derived under the assumption that the net force acting on the object was constant. A more general derivation would show that this equation is valid under all circumstances, including that of a variable force.

FIGURE 5.6
The moving hammer has kinetic energy and thus can do work on the nail, driving it into the wall.

EXAMPLE 5.2   Towing a Car

A 1400-kg car has a net forward force of 4500 N applied to it. The car starts from rest and travels down a horizontal highway. What are its kinetic energy and speed after it has traveled 100 m? (Ignore losses in kinetic energy due to friction and air resistance.)

Solution   With the initial velocity given as zero, Equation 5.4 reduces to

$$(1) \qquad W_{net} = \tfrac{1}{2}mv^2$$

The work done by the net force on the car is

$$W_{net} = Fs = (4500 \text{ N})(100 \text{ m}) = 4.50 \times 10^5 \text{ J}$$

This work all goes into changing the kinetic energy of the car; thus the final value of the kinetic energy, from (1), is also 4.50×10^5 J.

The speed of the car can be found from (1) as follows:

$$\tfrac{1}{2}mv^2 = 4.50 \times 10^5 \text{ J}$$

$$v^2 = \frac{2(4.50 \times 10^5 \text{ J})}{1400 \text{ kg}} = 643 \text{ m}^2/\text{s}^2$$

$$v = \boxed{25.4 \text{ m/s}}$$

5.3

POTENTIAL ENERGY

In the preceding section we saw that an object that has kinetic energy can do work on another object, as illustrated by the moving hammer driving a nail into the wall. Now we introduce another form of energy associated with the position of an object.

GRAVITATIONAL POTENTIAL ENERGY

As an object falls freely in a gravitational field, the field exerts a force on it, doing positive work on it and thereby increasing its kinetic energy. Consider Figure 5.7, which shows a brick of mass m held at a height of y_i above a nail in a board lying

Brick

FIGURE 5.7
Because of its position in space, the brick can do work on the nail.

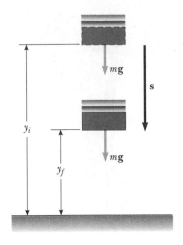

FIGURE 5.8
The work done by the gravitational force as the block falls from y_i to y_f is equal to $mgy_i - mgy_f$.

on the ground. When the brick is released, it falls toward the ground, gaining speed and thereby gaining kinetic energy. As a result of its position in space, the brick has potential energy (it has the *potential* to do work), and this potential energy is converted to kinetic energy as the brick falls. When the brick reaches the ground, it does work on the nail, driving it into the board. The energy that an object has as a result of its position in space near the surface of the Earth is called **gravitational potential energy.**

Let us now derive an expression for the gravitational potential energy of an object at a given location. Consider a block of mass m at an initial height of y_i above the ground, as in Figure 5.8, and let us neglect air resistance. As the block falls, the only force that does work on it is the gravitational force, mg. The work done by the gravitational force as the block undergoes a downward displacement of $s = y_i - y_f$ is

$$W_g = mgs = mgy_i - mgy_f$$

We define the quantity mgy to be the gravitational potential energy, PE:

$$PE \equiv mgy \qquad \text{[5.7]}$$

Thus, the gravitational potential energy associated with any object at a point in space is the product of the object's weight and its vertical coordinate. It is important to note that this relationship is valid only for objects near the Earth's surface, where g is approximately constant. The origin of the coordinate system can be located at any convenient point.

If we substitute this expression for PE into the expression for W_g, we have

$$W_g = PE_i - PE_f \qquad \text{[5.8]}$$

The work done on any object by the force of gravity is equal to the object's initial potential energy minus its final potential energy.

The units of gravitational potential energy are the same as those of work: joules, ergs, or foot-pounds. Potential energy, like work and kinetic energy, is a scalar quantity.

Note that the gravitational potential energy associated with an object depends only on the vertical height of the object above the surface of the Earth. From this fact we see that the same amount of work is done on an object if it falls vertically to the Earth as if it starts at the same point and slides down a frictionless incline to the Earth.

PROBLEM-SOLVING STRATEGY
Choosing a Zero Level

In working problems involving gravitational potential energy, it is always necessary to choose a location at which the gravitational potential energy is zero. This choice is completely arbitrary because the important quantity is the *difference* in potential energy, and that difference is independent of the location of zero. It is often convenient, but not essential, to choose the surface of the Earth as the reference position for zero potential energy. In most cases, the statement of the problem suggests a convenient level to use. The following example illustrates this important point.

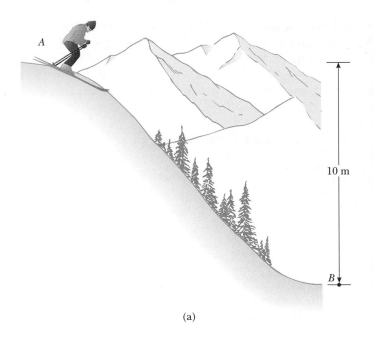

(a)

(b)

FIGURE 5.9
(Example 5.3)
(Telegraph Colour Library/FPG)

EXAMPLE 5.3 Wax Your Skis

A 60.0-kg skier is at the top of a slope, as shown in Figure 5.9a. At the initial point A, the skier is 10.0 m vertically above point B.

(a) Setting the zero level for gravitational potential energy at B, find the gravitational potential energy of the skier at A and B, and then find the difference in potential energy between these two points.

Solution The gravitational potential energy at B is zero by choice. Hence, the potential energy at A is

$$PE_i = mgy_i = (60.0 \text{ kg})(9.80 \text{ m/s}^2)(10.0 \text{ m}) = \boxed{5880 \text{ J}}$$

Since $PE_f = 0$, the difference in potential energy is

$$PE_i - PE_f = 5880 \text{ J} - 0 \text{ J} = \boxed{5880 \text{ J}}$$

(b) Repeat this problem with the zero level at point A.

Solution In this case, the initial potential energy is zero because of the chosen reference level. The final potential energy is

$$PE_f = mgy_f = (60.0 \text{ kg})(9.80 \text{ m/s}^2)(-10.0 \text{ m}) = \boxed{-5880 \text{ J}}$$

The distance y_f is negative because the final point is 10.0 m *below* the zero reference level. The difference in potential energy is now

$$PE_i - PE_f = 0 \text{ J} - (-5880 \text{ J}) = \boxed{5880 \text{ J}}$$

These calculations show that the potential energy of the skier at the top of the slope is greater than the potential energy at the bottom by 5880 J, *regardless of the zero level selected.*

> **Exercise**   If the zero level for gravitational potential energy is selected to be midway down the slope, at a height of 5.00 m, find the initial potential energy, the final potential energy, and the difference in potential energy between points *A* and *B*.
>
> **Answer**   2940 J, −2940 J, 5880 J.

5.4
CONSERVATIVE AND NONCONSERVATIVE FORCES

Forces in nature can be divided into two categories, conservative and nonconservative. Let us examine their properties.

CONSERVATIVE FORCES

Definition of a conservative force

> A force is conservative if the work it does on an object moving between two points is independent of the path the object takes between the points. In other words, the work done on an object by a conservative force depends only on the initial and final positions of the object. Furthermore, a force is conservative if the work it does on an object moving through any closed path is zero.

The force of gravity is conservative. As we learned in the preceding section, the work done by the gravitational force on an object moving between any two points near the Earth's surface is

$$W_g = mgy_i - mgy_f$$

From this we see that W_g depends only on the initial and final coordinates of the object and hence is independent of path. Furthermore, note that W_g is zero when the object moves over any closed path (where $y_i = y_f$).

We can associate a potential energy function with any conservative force. In the preceding section, the potential energy function associated with the gravitational force was found to be

$$PE = mgy$$

Potential energy functions can be defined only for conservative forces. In general, the work, W_c, done on an object by a conservative force is equal to the initial potential energy of the object minus the final value:

$$W_c = PE_i - PE_f \qquad \textbf{[5.9]}$$

NONCONSERVATIVE FORCES

Definition of a nonconservative force

> A force is *nonconservative* if it leads to a dissipation of mechanical energy. If you moved an object on a horizontal surface, returning it to the same location and same state of motion, but found it necessary to do net work on the object, then something must have dissipated the energy transferred to the object. That dissipative force is recognized as friction between object and surface. Friction is a dissipative, or nonconservative, force.

PHYSICS IN *ACTION*

The multiflash photo of a pole vaulter on the left illustrates various forms of energy, including energy of motion (kinetic energy) and energy associated with position in space (potential energy). The skillful rock climber in the photograph on the right moves cautiously up the mountain. As she ascends the mountain, her internal (biochemical) energy is transformed into gravitational potential energy. Can you identify other forms of energy transformations if she descends the mountain with the help of supporting ropes?

WORK AND ENERGY IN SPORTS

(Estate of Harold Edgerton. Courtesy of Palm Press, Inc.)

(Telegraph Colour Library/FPG)

Suppose you displace a book between two points on a table. If you displace it in a straight line along the blue path in Figure 5.10, the change in mechanical energy due to friction is $-fd$, where d is the distance between the two points. However, if you move the book along any other path between the two points, the change in mechanical energy due to friction is greater (in absolute magnitude) than $-fd$. For example, the change in mechanical energy due to friction along the red semicircular path in Figure 5.10 is equal to $-f(\pi d/2)$, where d is the diameter of the circle.

5.5

CONSERVATION OF MECHANICAL ENERGY

Conservation principles play a very important role in physics, and you will encounter a number of them as you continue in this course. The first, conservation of energy, is one of the most important. Before we describe its mathematical details, let us pause briefly to examine what is meant by conservation. When we say that a physical quantity is *conserved*, we simply mean that the value of the

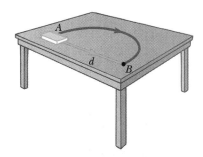

FIGURE 5.10

The loss in mechanical energy due to the force of friction depends on the path taken as the book is moved from A to B.

Twin Falls on the Island of Kauai, Hawaii. The potential energy of the water at the top of the falls is converted to kinetic energy at the bottom. In many locations, this mechanical energy is used to produce electrical energy. *(Bruce Byers, FPG)*

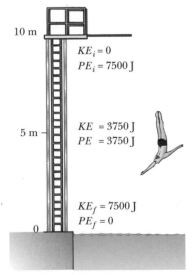

FIGURE 5.11
(Example 5.4)   The kinetic energy and potential energy of a diver at various heights. The zero of potential energy is taken to be at the surface of the pool.

quantity remains constant. Although the form of the quantity may change in some manner, its final value is the same as its initial value. For example, the energy in an isolated system may change from gravitational potential energy to kinetic energy or to one of a variety of other forms, but energy is never lost from the system.

In order to develop the principle of conservation of energy, let us return to the work-energy theorem, Equation 5.6, which says that the net work done on a system equals the change in the system's kinetic energy. Let us assume that the only force doing work on the system is conservative. In this case the net work on the system is equal to W_c, and from Equation 5.9 we have

$$W_{net} = W_c = PE_i - PE_f$$

We now substitute this expression into Equation 5.6 for W_{net} to find

$$PE_i - PE_f = KE_f - KE_i$$

or

$$KE_i + PE_i = KE_f + PE_f \qquad \text{[5.10]}$$

More formally, the principle of conservation of mechanical energy states that **the total mechanical energy in any isolated system of objects remains constant if the objects interact only through conservative forces.** It is important to note that Equation 5.10 is valid *provided* no energy is added to or removed from the system. Furthermore, there must be no nonconservative forces within the system. Therefore, if a conservative system (one subject only to conservative forces) has some initial combined amount of kinetic and potential energy, some of that energy may later transform from one kind into the other, but the total mechanical energy remains constant. This is equivalent to saying that, if the kinetic energy of a conservative system increases (or decreases) by some amount, the potential energy of the system must decrease (or increase) by the same amount.

If the force of gravity is the *only* force doing work on an object, then the total mechanical energy of the object remains constant, and the principle of conservation of mechanical energy takes the form

$$\tfrac{1}{2}mv_i^2 + mgy_i = \tfrac{1}{2}mv_f^2 + mgy_f \qquad \text{[5.11]}$$

Because there are other forms of energy besides kinetic and gravitational potential energy, we shall modify Equation 5.11 to include these new forms as we learn about them.

EXAMPLE 5.4   The Daring Diver

A 755-N diver drops from a board 10.0 m above the water surface, as in Figure 5.11.

(a) Use conservation of mechanical energy to find his speed 5.00 m above the water surface.

Reasoning   As the diver falls toward the water, only one force acts on him: the force of gravity. (This assumes that air resistance can be neglected.) Therefore, we can be assured that only the force of gravity does any work on him and that mechanical energy is conserved. In order to find the diver's speed at the 5-m mark, we will choose the water surface as the zero level for potential energy. Also, note that the diver drops from the board; that is, he leaves with zero velocity and therefore zero kinetic energy.

Solution   Conservation of mechanical energy, Equation 5.11, gives

$$\tfrac{1}{2}mv_i^2 + mgy_i = \tfrac{1}{2}mv_f^2 + mgy_f$$

$$0 + (755 \text{ N})(10.0 \text{ m}) = \tfrac{1}{2}(77.0 \text{ kg})v_f^2 + (755 \text{ N})(5.00 \text{ m})$$

$$v_f = \boxed{9.90 \text{ m/s}}$$

(b) Find the speed of the diver just before he strikes the water.

Solution   With the final position of the diver at the surface of the pool, Equation 5.11 gives

$$0 + (755 \text{ N})(10.0 \text{ m}) = \tfrac{1}{2}(77.0 \text{ kg})v_f^2 + 0$$

$$v_f = \boxed{14.0 \text{ m/s}}$$

Exercise   If the diver pushes off so that he leaves the board with an initial speed of 2.00 m/s, find his speed when he strikes the water. Use conservation of energy.

Answer   14.1 m/s.

EXAMPLE 5.5   Sliding Down a Frictionless Hill

A sled and its rider together weigh 800 N. They move down a frictionless hill through a vertical distance of 10.0 m, as shown in Figure 5.12a. Use conservation of mechanical energy to find the speed of the sled-rider system at the bottom of the hill, assuming the rider pushes off with an initial speed of 5.00 m/s.

Reasoning   The forces acting on the sled and rider as they move down the hill are shown in Figure 5.12b. In the absence of a frictional force, the only forces acting are the normal force, **n**, and the gravitational force. At all points along the path, **n** is perpendicular to the direction of travel and hence does no work. Likewise, the component of weight perpendicular to the incline ($w \cos \theta$) does no work. The only force that does any work is the component of the gravitational force, $w \sin \theta$, along the slope of the hill. As a result, we are justified in using Equation 5.11.

Solution   In this case the initial energy includes kinetic energy because of the initial speed:

$$\tfrac{1}{2}mv_i^2 + mgy_i = \tfrac{1}{2}mv_f^2 + mgy_f$$

or, after canceling m throughout the equation,

$$\tfrac{1}{2}v_i^2 + gy_i = \tfrac{1}{2}v_f^2 + gy_f$$

If we set the origin at the bottom of the incline, we see that the initial and final y coordinates are $y_i = 10.0$ m and $y_f = 0$. Thus we get

$$\tfrac{1}{2}(5.00 \text{ m/s})^2 + (9.80 \text{ m/s}^2)(10.0 \text{ m}) = \tfrac{1}{2}v_f^2 + (9.80 \text{ m/s}^2)(0)$$

$$v_f = \boxed{14.9 \text{ m/s}}$$

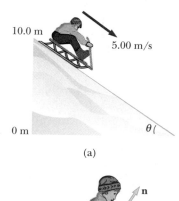

(a)

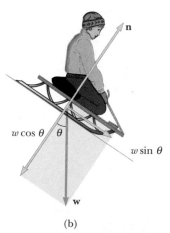

(b)

FIGURE 5.12
(Example 5.5)   (a) A sled and rider start from the top of a frictionless hill with a speed of 5.00 m/s. (b) A free-body diagram for the system (sled plus rider).

> **Exercise**   If the sled and rider start at the bottom of the incline and are given
> an initial speed of 5.00 m/s up the incline, how high will they rise vertically, and
> what will their speed be when they return to the bottom of the hill?
>
> **Answer**   1.28 m; 5.00 m/s.

POTENTIAL ENERGY STORED IN A SPRING

The concept of gravitational potential energy is of tremendous value in descriptions of certain types of mechanical motion. One of these, the motion of a mass
attached to a stretched or compressed spring, is discussed more completely in
Chapter 13, but it is instructive to examine it here.

Consider Figure 5.13a, which shows a spring in its equilibrium position—
that is, the spring is neither compressed or stretched. If we push a block against
the spring as in Figure 5.13b, the spring is compressed a distance of x. In order to
compress the spring, we must exert on the block a force of magnitude $F = kx$,
where k is a constant for a particular spring called the **spring constant.** For a
flexible spring, k is a small number, whereas for a stiff spring, k is large. The
equation $F = kx$ says that the more we compress the spring, the larger the force
we must exert. Since the block is in equilibrium in Figure 5.13b, the spring must
be exerting on it a force of $F_s = -kx$. This equation for F_s, describing the force
exerted by a stretched or compressed spring, is often called **Hooke's law** after Sir
Robert Hooke, who discovered the relationship between F_s and x.

To find an expression for the potential energy associated with the spring
force, called the **elastic potential energy,** let us determine the work required to
compress a spring from its equilibrium position to some final arbitrary position

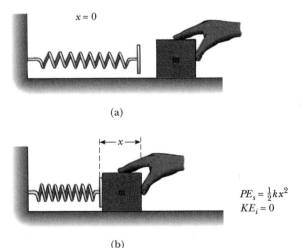

$x = 0$

(a)

$\leftarrow x \rightarrow$

$PE_s = \frac{1}{2}kx^2$
$KE_i = 0$

(b)

$x = 0$

$PE_s = 0$
$KE_f = \frac{1}{2}mv^2$

(c)

FIGURE 5.13
A block of mass m on a frictionless surface is pushed
against a spring and released
from rest. If x is the compression in the spring, as in (b),
the potential energy stored in
the spring is $\frac{1}{2}kx^2$. This energy
is transferred to the block in
the form of kinetic energy, as
in (c).

x. The force that must be applied to compress the spring varies from $F = 0$ to $F = kx$ at maximum compression. Because the force increases linearly with position (that is, $F \propto x$), the average force that must be applied is

$$\bar{F} = \frac{F_0 + F_x}{2} = \frac{0 + kx}{2} = \tfrac{1}{2} kx$$

Therefore, the work *done by the applied force* is

$$W = \bar{F}x = \tfrac{1}{2} kx^2$$

This work is stored in the compressed spring as elastic potential energy. Thus, we define the elastic potential energy, PE_s, as

$$PE_s \equiv \tfrac{1}{2} kx^2 \qquad\qquad \textbf{[5.12]}$$

Figure 5.13 shows how this stored elastic potential energy can be recovered. When the block is released, the spring snaps back to its original length and the stored elastic potential energy is converted to kinetic energy of the block. The elastic potential energy stored in the spring is zero when the spring is in equilibrium $(x = 0)$. Note that energy is also stored in the spring when it is stretched, and is given by Equation 5.12. Furthermore, the elastic potential energy is a maximum when the spring has reached its maximum compression or extension. Finally, the potential energy in the spring is always positive because it is proportional to x^2.

This new form of energy is included as another term in our equation for conservation of mechanical energy:

$$(KE + PE_g + PE_s)_i = (KE + PE_g + PE_s)_f$$

where PE_g is the gravitational potential energy.

EXAMPLE 5.6   A Block Projected on a Frictionless Surface

A 0.50-kg block rests on a horizontal, frictionless surface as in Figure 5.13. The block is pressed against a light spring having a spring constant of $k = 80.0$ N/m. The spring is compressed a distance of 2.0 cm and released. Consider the system to be the spring and block, and find the speed of the block when the block is at the $x = 0$ position.

Reasoning   Our expression for the conservation of mechanical energy is

$$(KE + PE_g + PE_s)_i = (KE + PE_g + PE_s)_f$$

The initial kinetic energy of the block is zero, and since the block remains at the same level throughout its motion, the gravitational potential energy terms on both sides of the equation are equal. Thus,

$$PE_{si} = KE_f$$

Solution   From the last expression we have

$$\tfrac{1}{2} kx_i^2 = \tfrac{1}{2} mv_f^2$$

$$\tfrac{1}{2}(80.0 \text{ N/m})(2.0 \times 10^{-2} \text{ m})^2 = \tfrac{1}{2}(0.50 \text{ kg})v_f^2$$

$$v_f = \boxed{0.25 \text{ m/s}}$$

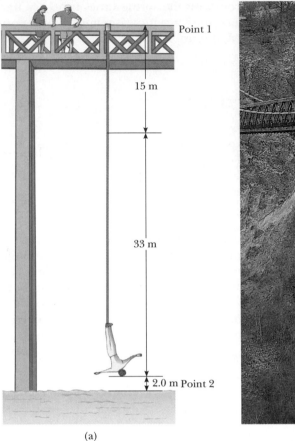

(a)

(b)

FIGURE 5.14
(Example 5.7)   A bungee-cord
jumper. (*John Cancalosi/Peter
Arnold, Inc.*)

EXAMPLE 5.7   The Bungee-Cord Daredevil

A 70-kg daredevil, attached by his ankles to a bungee cord with an unstretched
length of 15 m, drops from the top of a 50-m-high bridge. The cord stops him
at water level, as shown in Figure 5.14. Assuming he is 2.0 m tall, find the spring
constant, k, of the bungee cord.

Reasoning   Since mechanical energy is conserved in this situation, the sum of
the kinetic energy, the elastic potential energy, and the gravitational potential
energy of the system (jumper plus bungee cord) remains constant. That is,

$$\tfrac{1}{2}mv^2 + \tfrac{1}{2}kx_1^2 + mgy_1 = \tfrac{1}{2}mv^2 + \tfrac{1}{2}kx_2^2 + mgy_2$$

Because the daredevil drops from rest and stops when the cord is fully extended,
the initial and final values of the kinetic energy are zero. Hence, in this case the
preceding expression simplifies to

$$mgy_1 = \tfrac{1}{2}kx_2^2$$

Solution   With zero gravitational potential energy set at water level, and noting that the cord stretches 33 m (see Fig. 5.14), the preceding expression gives

$$(70 \text{ kg})(9.8 \text{ m/s}^2)(50 \text{ m}) = \tfrac{1}{2}k(33 \text{ m})^2$$

$$k = 63 \text{ N/m}$$

Exercise What would happen to the daredevil if the spring constant were 80 N/m?

Answer He would stop 3.7 m above water level.

5.6

NONCONSERVATIVE FORCES AND THE WORK-ENERGY THEOREM

In realistic situations, nonconservative forces such as friction are usually present. In such situations, the total mechanical energy of the system is not constant, and one cannot apply Equation 5.10. In order to account for nonconservative forces, let us return to the work-energy theorem:

$$W_{\text{net}} = \tfrac{1}{2}mv_f^2 - \tfrac{1}{2}mv_i^2$$

Let us separate the net work into two parts, that done by the nonconservative forces, W_{nc}, and that done by the conservative forces, W_c. The work-energy relationship then becomes

$$W_{nc} + W_c = \tfrac{1}{2}mv_f^2 - \tfrac{1}{2}mv_i^2 \qquad \textbf{[5.13]}$$

Solving for W_{nc} and substituting the expression for W_c provided by Equation 5.9, we have

$$W_{nc} = \tfrac{1}{2}mv_f^2 - \tfrac{1}{2}mv_i^2 - (PE_i - PE_f)$$

$$W_{nc} = \underset{\text{Change in kinetic energy}}{(KE_f - KE_i)} + \underset{\text{Change in potential energy}}{(PE_f - PE_i)} \qquad \textbf{[5.14]}$$

The work done by all nonconservative forces equals the change in kinetic energy plus the change in potential energy.

PROBLEM-SOLVING STRATEGY
Conservation of Energy

Take the following steps in applying the principle of conservation of energy:

1. Define your system, which may consist of more than one object.
2. Select a reference position for the zero point of gravitational potential energy.
3. Determine whether or not nonconservative forces are present.
4. If mechanical energy is conserved (that is, if only conservative forces are present), you can write the total initial energy at some point as the sum of the kinetic and potential energies at that point. Then, write an expression for the total final energy, $KE_f + PE_f$, at the final point of interest. Since mechanical energy is conserved, you can equate the two total energies and solve for the unknown.

5. **If nonconservative forces such as friction are present (and thus mechanical energy is not conserved), first write expressions for the total initial and total final energies. In this case, the difference between the two total energies is equal to the work done by the nonconservative force(s). That is, you should apply Equation 5.14.**

EXAMPLE 5.8   A Crate Sliding Down a Ramp

A 3.00-kg crate slides down a ramp at a loading dock. The ramp is 1.00 m long and inclined at an angle of 30.0°, as shown in Figure 5.15. The crate starts from rest at the top, experiences a constant frictional force of magnitude 5.00 N, and continues to move a short distance on the flat floor. Use energy methods to determine the speed of the crate when it reaches the bottom of the ramp.

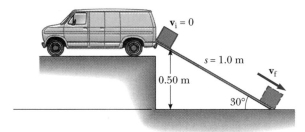

FIGURE 5.15
(Example 5.8)   A crate slides down an incline under the influence of gravity. The potential energy of the crate decreases while its kinetic energy increases.

Reasoning   The crate's initial energy is all in the form of gravitational potential energy, and its final energy is all kinetic. However, we cannot say that $PE_i = KE_f$, because there is an external nonconservative force that removes mechanical energy from the crate: the force of friction. Thus, we must use Equation 5.13 to find the final speed.

Solution   Because $v_i = 0$, the initial kinetic energy is zero. If the y coordinate is measured from the bottom of the incline, then $y_i = 0.500$ m. Therefore, the total mechanical energy of the crate at the top is all potential energy, given by

$$PE_i = mgy_i = (3.00 \text{ kg})(9.80 \text{ m/s}^2)(0.500 \text{ m}) = 14.7 \text{ J}$$

When the crate reaches the bottom, its potential energy is zero because its elevation is $y_f = 0$. Therefore, the total mechanical energy at the bottom is all kinetic energy:

$$KE_f = \tfrac{1}{2}mv_f^2$$

In this case, $W_{nc} = -fs$, where s is the displacement along the ramp. (Recall that the forces normal to the ramp do no work on the crate because they are perpendicular to the displacement.) With $f = 5.00$ N and $s = 1.00$ m, we have

$$W_{nc} = -fs = (-5.00 \text{ N})(1.00 \text{ m}) = -5.00 \text{ J}$$

This says that some mechanical energy is lost because of the presence of the retarding frictional force. Applying Equation 5.13 gives

$$-fs = \tfrac{1}{2}mv_f^2 - mgy_i$$

$$\tfrac{1}{2}mv_f^2 = 14.7\,\text{J} - 5.00\,\text{J} = 9.7\,\text{J}$$

$$v_f^2 = \frac{19.4\,\text{J}}{3.00\,\text{kg}} = 6.47\,\text{m}^2/\text{s}^2$$

$$v_f = \boxed{2.54\,\text{m/s}}$$

Exercise   Find the speed at the bottom of the ramp if the ramp surface is frictionless.

Answer $v_f = 3.13$ m/s.

EXAMPLE 5.9   Fun on the Slide

A child of mass 20.0 kg takes a ride on an irregularly curved slide of height 6.00 m, as in Figure 5.16. The child starts from rest at the top. Determine her speed at the bottom, assuming no friction is present.

Reasoning   The normal force does no work on the child because this force is always perpendicular to the displacement. Furthermore, because there is no friction, $W_{nc} = 0$ and we can apply the principle of conservation of mechanical energy.

Solution   If we measure the y coordinate from the bottom of the slide, then $y_i = 6.00$ m, $y_f = 0$, and we get

$$KE_i + PE_i = KE_f + PE_f$$

$$0 + (20.0\,\text{kg})(9.80\,\text{m/s}^2)(6.00\,\text{m}) = \frac{1}{2}(20.0\,\text{kg})(v_f^2) + 0$$

$$v_f = \boxed{10.8\,\text{m/s}}$$

Note that this speed is the same as if the child had fallen vertically a distance of 6.00 m! From the free-fall equation, $v_f^2 = v_i^2 - 2gy$, we get the same result:

$$v_f = \sqrt{2(9.80\,\text{m/s}^2)(6.00\,\text{m})} = 10.8\,\text{m/s}$$

The effect of the slide is to direct the motion at the bottom horizontally rather than vertically.

Exercise   If the child's speed at the bottom is 8.00 m/s rather than 10.8 m/s, how much mechanical energy is removed from the system?

Answer   536 J.

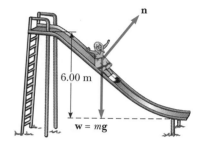

FIGURE 5.16
(Example 5.9)   If the slide is frictionless, the speed of the child at the bottom depends on the height of the slide and is independent of its shape.

5.7

CONSERVATION OF ENERGY IN GENERAL

We can generalize the energy conservation principle to include all forces, both conservative and nonconservative, acting on a system. In the study of thermodynamics we shall find that mechanical energy can be transformed into internal energy of the system. For example, when a block slides over a rough surface, the mechanical energy lost is transformed into heat stored in the block and the surface, as evidenced by measurable increases in the temperatures of both. We shall see that, on an atomic scale, this internal energy is associated with the vibration of atoms about their equilibrium positions. Such internal atomic motion has kinetic and potential energy, and so one can say that frictional forces arise fundamentally from forces that are conservative at the atomic level. Therefore, if we include this increase in internal energy in our energy expression, total energy is conserved.

This is just one example of how you can analyze an isolated system and always find that its total energy does not change, as long as you account for all forms of energy. That is, *energy can never be created or destroyed. Energy may be transformed from one form into another, but the total energy of an isolated system is always constant.* From a universal point of view, we can say that *the total energy of the Universe is constant:* if one part of the Universe gains energy in some form, another part must lose an equal amount of energy. No violation of this principle has been found.

Total energy is always conserved

5.8

POWER

From a practical viewpoint, it is interesting to know not only the work done on an object but also the rate at which that work is done. Power is defined as the *time rate of doing work.*

If an external force is applied to an object and if the work done by this force is W in the time interval Δt, then the average power, $\bar{P}$, during this interval is defined as the ratio of the work done to the time interval:

Average power

$$\bar{P} = \frac{W}{\Delta t} \qquad [5.15]$$

It is sometimes useful to rewrite Equation 5.15 by substituting $W = F\Delta s$ and noting that $\Delta s/\Delta t$ is the average speed of the object during the time Δt:

$$\bar{P} = \frac{W}{\Delta t} = \frac{F\Delta s}{\Delta t} = F\bar{v} \qquad [5.16]$$

This result says that the average power delivered either to or by an object is equal to the product of the force acting on the object during some time interval and the object's average speed during this time interval. In Equation 5.16, F is the component of force in the direction of the average velocity.

A result of the same form as Equation 5.16 is obtained for instantaneous values. That is, the instantaneous power delivered to an object is equal to the product of the force on the object at that instant and the instantaneous speed, or $P = Fv$.

The units of power in the SI system are joules per second, which are also called **watts** (W) after James Watt:

$$1 \text{ W} = 1 \text{ J/s} = 1 \text{ kg} \cdot \text{m}^2/\text{s}^3 \qquad [5.17]$$

The unit of power in the British engineering system is the horsepower (hp), where

$$1 \text{ hp} \equiv 550 \frac{\text{ft} \cdot \text{lb}}{\text{s}} = 746 \text{ W} \qquad [5.18]$$

The watt is commonly used in electrical applications, but it can be used in other scientific areas as well. For example, European sports-car engines are now rated in kilowatts.

We now define units of energy (or work) in terms of the unit of power, the watt. One kilowatt-hour (kWh) is the energy converted or consumed in 1 h at the constant rate of 1 kW = 1000 J/s. Therefore,

$$1 \text{ kWh} = (10^3 \text{ W})(3600 \text{ s}) = (10^3 \text{ J/s})(3600 \text{ s}) = 3.60 \times 10^6 \text{ J}$$

It is important to realize that a kilowatt-hour is a unit of energy, not power. When you pay your electric bill, you are buying energy, and the amount of electricity used by an appliance is usually expressed in multiples of kilowatt-hours. For example, an electric bulb rated at 100 W would "consume" 3.60×10^5 J of energy in 1 h.

EXAMPLE 5.10   Power Delivered by an Elevator Motor

A 1000-kg elevator carries a maximum load of 800 kg. A constant frictional force of 4000 N retards its motion upward, as in Figure 5.17. What minimum power, in kilowatts and in horsepower, must the motor deliver to lift the fully loaded elevator at a constant speed of 3.00 m/s?

Solution   The motor must supply the force, **T**, that pulls the elevator upward. From Newton's second law and from the fact that **a** = 0 because **v** is constant, we get

$$T - f - Mg = 0$$

where M is the total mass (elevator plus load), equal to 1800 kg. Therefore,

$$T = f + Mg$$
$$= 4.00 \times 10^3 \text{ N} + (1.80 \times 10^3 \text{ kg})(9.80 \text{ m/s}^2)$$
$$= 2.16 \times 10^4 \text{ N}$$

From $P = Fv$ and the fact that **T** is in the same direction as **v**, we have

$$P = Tv$$
$$= (2.16 \times 10^4 \text{ N})(3.00 \text{ m/s}) = 6.48 \times 10^4 \text{ W}$$
$$= 64.8 \text{ kW} = \boxed{86.9 \text{ hp}}$$

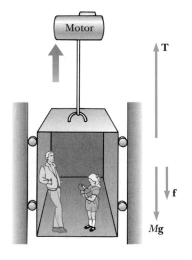

FIGURE 5.17
(Example 5.10)   The motor provides an upward force, T, on the elevator. A frictional force, f, and the total weight, Mg, act downward.

---

*5.9

WORK DONE BY A VARYING FORCE

Consider an object being displaced along the x axis under the action of a varying force. Such a situation occurs, for instance, when a person pulls an object at a constant speed along a surface that varies in roughness from point to point. The magnitude of the required force changes as the roughness increases or decreases. In such a situation, we cannot use $W = Fs \cos \theta$ to calculate the work done by the force, because *this relationship applies only when the force is constant in magnitude and direction.* However, we can calculate the work done as the object

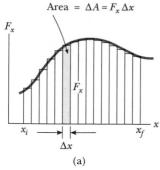

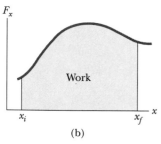

FIGURE 5.18
(a) The work done by the force F_x for the small displacement Δx is $F_x \Delta x$, which equals the area of the blue rectangle. The total work done for the displacement x_i to x_f is approximately equal to the sum of the areas of all the rectangles. (b) The work done by the variable force F_x as the object moves from x_i to x_f is equal to the area under the curve.

undergoes a very small displacement, Δx, as shown in Figure 5.18a. During this small displacement the force is approximately constant, and so we can express the work done by the force as

$$\Delta W = F_x \Delta x$$

Note that this is just the area of the blue rectangle in Figure 5.18a. Now, if we imagine that the force-displacement curve is divided into a large number of such intervals, as in Figure 5.18a, then the total work done for the displacement from x_i to x_f is approximately equal to the sum of a large number of such terms:

$$W \approx \sum F_x \Delta x \qquad \textbf{[5.19]}$$

The right side of this equation is the sum of the areas of all the rectangles pictured. If we decrease Δx so that the rectangles are made very narrow, however, the right side of the equation turns out to be just the total area under the curve, which is shown in Figure 5.18b. Thus, we can say that

> **The total work done by a varying force is equal to the area under the force-displacement curve.**

EXAMPLE 5.11   Work Done by a Changing Force

A force acting on an object varies with x as shown in Figure 5.19.

(a) Calculate the work done by the force as the object moves from $x = 0$ to $x = 6.0$ m.

Solution   The work is equal to the area under the curve from $x = 0$ to $x = 6.0$ m. This area is equal to the area of the rectangular section between $x = 0$ and $x = 4.0$ m plus the area of the triangular section between $x = 4.0$ m and $x = 6.0$ m. The area of the rectangle is (4.0×5.0) N·m = 20 J, and the area of the triangle is $\frac{1}{2}(2.0 \times 5.0)$ N·m = 5.0 J. Therefore, the total work done is 25 J.

Following the motion described in part (a), the force immediately drops to a constant value of 3.0 N in the negative direction for the next 2.0 m, as shown. Find (b) the work done by this force during this interval and (c) the net work done for the entire 8.0 m of motion.

Solution   (b) The work done is the area enclosed by the rectangular curve between 6.0 and 8.0 m, which is

$$W = Fs = (-3.0 \text{ N})(2.0 \text{ m}) = -6.0 \text{ J}$$

Why is W negative in this case?

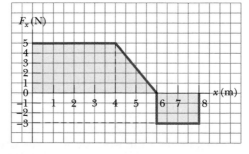

FIGURE 5.19
(Example 5.11)   The force is constant for the first 4.0 m of motion and then decreases linearly with x from $x = 4.0$ m to $x = 6.0$ m. Finally, the force becomes -3.0 N from $x = 6.0$ m to $x = 8.0$ m. The net work done by this force is the total area under this curve, shaded in tan.

(c) The net work done is

$$W_{net} = 25\,J - 6.0\,J = \boxed{19\,J}$$

SUMMARY

The **work** done by a *constant* force, **F**, acting on an object is defined as the product of the component of the force in the direction of the object's displacement and the magnitude of the displacement. If the force makes an angle of θ with the displacement **s**, the work done by **F** is

$$W \equiv (F \cos \theta)s \qquad\qquad \text{[5.1]}$$

The **kinetic energy** of an object of mass m moving with speed v is defined as

$$KE \equiv \tfrac{1}{2}mv^2 \qquad\qquad \text{[5.5]}$$

The **work-energy theorem** states that the net work done on an object by external forces equals the change in kinetic energy of the object:

$$W_{net} = KE_f - KE_i \qquad\qquad \text{[5.6]}$$

The **gravitational potential energy** of an object of mass m that is elevated a distance of y above the Earth's surface is given by

$$PE \equiv mgy \qquad\qquad \text{[5.7]}$$

A force is **conservative** if the work it does on an object depends only on the initial and final positions of the object and not on the path taken between those positions. A force for which this is not true is said to be **nonconservative**.

The **principle of conservation of mechanical energy** states that if the only force acting on a system is conservative, total mechanical energy remains constant.

$$KE_i + PE_i = KE_f + PE_f \qquad\qquad \text{[5.10]}$$

The **elastic potential energy** stored in a spring is given by

$$PE_s \equiv \tfrac{1}{2}kx^2 \qquad\qquad \text{[5.12]}$$

where k is the spring constant and x is the distance the spring is compressed or extended from its unstretched (equilibrium) position.

Energy can never be created or destroyed. Energy may be transformed from one form into another, but the total energy of an isolated system is always constant. The **work-energy theorem** states that the work done by all nonconservative forces acting on a system equals the change in the total mechanical energy of the system:

$$W_{nc} = (KE_f - KE_i) + (PE_f - PE_i) \qquad\qquad \text{[5.14]}$$

Average power is the ratio of work done to the time interval during which the work is done:

$$\bar{P} = \frac{W}{\Delta t} \qquad\qquad \text{[5.15]}$$

If a force, F, acts on an object moving with an average speed of $\bar{v}$, the average power delivered to the object is

$$\bar{P} = F\bar{v} \qquad\qquad [5.16]$$

The total work done by a varying force is equal to the area under the force-displacement curve.

ADDITIONAL READING

G. R. Davis, "Energy for Planet Earth," *Sci. American,* September 1990, p. 54.

Freeman J. Dyson, "Energy in the Universe," *Sci. American,* September 1971, p. 50.

A. Einstein and L. Infeld, *Evolution of Physics,* New York, Simon and Schuster, 1938.

"Energy and Power," *Sci. American,* September 1971. This entire issue is devoted to energy-related topics.

H. M. Hubbard, "The Real Cost of Energy," *Sci. American,* April 1991, p. 36.

G. Waring, "Energy and the Automobile," *The Physics Teacher, 18,* 1980, p. 494. An informative article on fuel consumption by automobiles.

R. Wilson and W. J. Jones, *Energy, Ecology, and the Environment,* New York, Academic Press, 1974.

CONCEPTUAL QUESTIONS

Example   An Earth satellite is in a circular orbit at an altitude of 500 km. Explain why the work done by the gravitational force acting on the satellite is zero. Using the work-energy theorem, what can you say about the speed of the satellite?

Reasoning   As the satellite moves in a circular orbit about the Earth, its displacement along the circular path during any small time interval is always perpendicular to the gravitational force, which always acts toward the center of the Earth. Therefore, the work done by the gravitational force during any displacement is zero. (Recall that the work done by a force is defined to be $Fs \cos \theta$, where θ is the angle between the force and the displacement. In this case, the angle is 90°, so the work done is zero.) Since the work-energy theorem says that the net work done on an object during any displacement is equal to the change in its kinetic energy, and the work done in this case is zero, the change in the satellite's kinetic energy is zero, hence its speed remains constant.

Example   A car traveling at 50 km/h skids a distance of 35 m after its brakes lock. Estimate how far it will skid if its brakes lock when its initial speed is 100 km/h. What happens to the car's kinetic energy as it comes to rest?

Reasoning   Let us assume that the friction force between the car and road surface is constant and the same in both cases. The work required to stop the car is equal to its initial kinetic energy. If the speed is doubled as in this example, the work required is quadrupled. Since the work done is equal to force times the distance traveled, the car will travel four times as far when its speed is doubled, so

the estimated distance it skids is 140 m. The kinetic energy of the car is changed into internal energy associated with the tires, brake pads, and road as they heat up.

Example   In most situations we have encountered in this chapter, frictional forces tend to reduce the kinetic energy of an object. However, frictional forces can sometimes increase an object's kinetic energy. Describe a few situations in which friction causes an increase in kinetic energy.

Reasoning   If a crate is located on the bed of a truck, and the truck accelerates, the friction force acting on the crate causes it to undergo the same acceleration as the truck, assuming the crate doesn't slip. Another example is a car that accelerates because of the frictional forces between the road surface and its tires. This force is in the direction of motion of the car and produces an increase in the car's kinetic energy.

Example   A weight is connected to a spring that is suspended vertically from the ceiling. If the weight is displaced downward from its equilibrium position and released, it will oscillate up and down. If air resistance is neglected, will the total energy of the system (weight plus spring) be conserved? How many form of potential energy are there for this situation?

Reasoning   Yes, the total mechanical energy of the system is conserved since the only forces acting are conservative in nature: the force of gravity and the spring force. There are two forms of potential energy in this case, gravitational potential energy and elastic potential energy stored in the spring.

(Question 1)   *(© Andy Hayt/Focus on Sports)*

1. When a punter kicks a football, is he doing work on the ball while his toe is in contact with it? Is he doing work on the ball after it loses contact with his toe? Are any forces doing work on the ball while it is in flight?

2. Discuss whether any work is being done by each of the following agents and, if so, whether the work is positive or negative: (a) a chicken scratching the ground, (b) a person studying, (c) a crane lifting a bucket of concrete, (d) the force of gravity on the bucket in part (c), (e) the leg muscles of a person in the act of sitting down.

3. As a pendulum swings back and forth, does the tension force in the string do any work on the pendulum bob? Does the force of gravity do any work on the bob?

4. Can the kinetic energy of an object be negative?

5. Can the speed of an object change if the net work done on it is zero?

6. In a collision between two cars, one car left skid marks twice the length of those left by the other car. A bystander who viewed the collision claims that the cars applied their brakes at the same time. What conclusions can you draw?

7. Two identical objects move with speeds of 5 m/s and 20 m/s. What is the ratio of their kinetic energies?

8. A ball is thrown straight up. At what position is its kinetic energy a maximum? At what position is its gravitational potential energy a maximum?

9. Can the gravitational potential energy of an object ever be negative? Explain.

10. A pile driver is a device used to drive objects into the Earth by dropping a heavy weight on them. By how much does the energy of a pile driver increase when the weight it drops is doubled? (Assume the weight is dropped from the same height each time.)

11. A bowling ball is suspended from the ceiling of a lecture hall by a strong cord. The ball is drawn away from its equilibrium position and released from rest at the tip of the demonstrator's nose. If the demonstrator remains stationary, explain why she is not struck by the ball on its return swing. Would this demonstrator be safe if the ball were given a slight push from its starting position at her nose?

12. A person drops a ball from the top of a building while another person at the bottom observes the ball's motion. Will these two people always agree on the value of the ball's potential energy? on its change in potential energy? on its kinetic energy?

13. How can the work-energy theorem explain why the force of sliding friction usually reduces the kinetic energy of a particle?

14. Discuss the production and dissipation of mechanical energy as an athlete (a) lifts a weight, (b) holds it up, (c) lowers it slowly. Include the muscles in your discussion.

15. Advertisements for the Superball once stated that it would rebound to a height greater than the height from which it was dropped. Is this possible?

16. Discuss the work done by a pitcher throwing a baseball. What is the approximate distance through which the force acts as the ball is thrown?

17. Discuss the kinetic and potential energies of various parts of the body when a person is jumping off the ground.

18. Discuss the energy transformations that occur during a pole-vaulting event.

19. Discuss the role of kinetic and potential energies in (a) baseball, (b) football, (c) tennis, (d) basketball, (e) track.

20. During a stress test of the cardiovascular system, a patient walks and runs on a treadmill. (a) Is the energy expended by the patient equivalent to the energy of walking and running on the ground? Explain. (b) If the treadmill is tilted upward, what effect does this have? Discuss.

21. Many mountain roads are built so that they spiral around the mountain rather than go straight up toward the peak. Discuss such a spiral design from the viewpoint of energy and power.

22. Estimate the time it takes you to climb a flight of stairs. (For a ballpark estimate, assume you can travel upward a distance of 8 m in 6 s.) Then approximate the horsepower required to perform this task.

23. One bullet has twice the mass of a second bullet. If both are fired so that they have the same speed, which has more kinetic energy? What is the ratio of the two kinetic energies?

24. Two sharpshooters fire .30-caliber rifles, using identical shells. The barrel of rifle A is 2 cm longer than that of rifle B. Which rifle has the higher muzzle speed? (*Hint:* The force of the expanding gases in the barrel accelerates the bullets.)

25. A team of furniture movers wishes to load a truck using a ramp from the ground to the rear of the truck. One of the movers claims that less work would be required if the ramp's length were increased, reducing its angle with the horizontal. Is this claim valid? Explain.

26. Estimate the work done by a pitcher throwing a baseball at 40 m/s (90 mi/h). The mass of a baseball is approximately 0.15 kg.

(Question 26)   Nolan Ryan throwing his fast ball. *(Michael Layton, Duomo)*

27. A catcher "gives" with the ball when she catches a 0.15-kg baseball moving at 40.0 m/s. If she moves her catcher's mitt a distance of 2.0 cm, what is the average force acting on her hand?

28. Three identical balls are thrown from the top of a building, all with the same initial speed. One ball is thrown horizontally, the second at some angle above the horizontal, and the third at some angle below the horizontal. Neglecting air resistance, compare the speeds of the balls as they reach the ground.

29. An Olympic high-jumper whose height is 2.0 m makes a record leap of 2.3 m over a horizontal bar. Estimate the speed with which he must leave the ground to perform this feat.

30. Estimate the kinetic energy of an 80 000-kg airliner flying at 600 km/h.

(Question 29)   *(FPG)*

PROBLEMS

Section 5.1   Work

1. A tugboat exerts a constant force of 5.00×10^3 N on a ship moving at constant speed through a harbor. How much work does the tugboat do on the ship if each moves a distance of 3.00 km?

2. A weight lifter lifts a 350-N set of weights from ground level to a position over his head, a vertical distance of 2.00 m. How much work does the weight lifter do, assuming he moves the weights at constant speed?

3. If a man lifts a 20.0-kg bucket from a well and does 6.00 kJ of work, how deep is the well? Assume that the speed of the bucket remains constant as it is lifted.

4. Starting from rest, a 5.0-kg block slides 2.5 m down a rough 30.0° incline in 2.0 s. Determine (a) the work done by the force of gravity, (b) the mechanical energy lost due to friction, and (c) the work done by the normal force between block and incline.

(Problem 1)   *(© David Plowden/Photo Researchers, Inc.)*

□ indicates problems that have full solutions available in the Student Solutions Manual and Study Guide.

(Problem 2) *(Gerard Vandystadt/Photo Researchers)*

5. A shopper in a supermarket pushes a cart with a force of 35 N directed at an angle of 25° downward from the horizontal. Find the work done by the shopper as she moves down a 50-m length of aisle.

6. A driver of a 7.50×10^3 N car passes a sign stating "Bridge Out 30 Meters Ahead." She slams on the brakes, coming to a stop in 10.0 s. How much work must be done by the brakes on the car if it is to stop just in time? Neglect the weight of the driver, and assume that the negative acceleration of the car caused by the braking is constant.

7. Batman, whose mass is 80.0 kg, holds onto the free end of a 12.0-m rope, the other end of which is fixed to a tree limb above. He puts the rope in motion as only Batman knows how, eventually getting it to swing enough that he can reach a ledge when the rope makes a 60.0° angle with the downward vertical. How much work is done against the force of gravity in this maneuver?

8. A 70.0-kg base runner begins his slide into second base when moving at a speed of 4.0 m/s. The coefficient of friction between his clothes and Earth is 0.70. He slides so that his speed is zero just as he reaches the base. (a) How much mechanical energy is lost due to friction acting on the runner? (b) How far does he slide?

9. A horizontal force of 150 N is used to push a 40.0-kg packing crate a distance of 6.00 m on a rough horizontal surface. If the crate moves at constant speed, find (a) the work done by the force and (b) the coefficient of kinetic friction.

10. A stewardess pulls her 70-N flight bag a distance of 200 m along an airport floor at constant speed. The force she exerts is 40 N at an angle of 50° above the horizontal. Find (a) the work she does, (b) the work done by the force of friction, and

(c) the coefficient of kinetic friction between her flight bag and the floor.

Section 5.2 Kinetic Energy and the Work-Energy Theorem

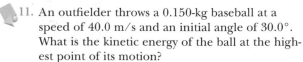

11. An outfielder throws a 0.150-kg baseball at a speed of 40.0 m/s and an initial angle of 30.0°. What is the kinetic energy of the ball at the highest point of its motion?

12. A 5.00-g bullet moving at 600 m/s penetrates a tree trunk to a depth of 4.00 cm. (a) Use work and energy considerations to find the average frictional force that stops the bullet. (b) Assuming that the frictional force is constant, determine how much time elapses between the moment the bullet enters the tree and the moment it stops moving.

13. A 0.60-kg particle has a speed of 2.0 m/s at point A and kinetic energy of 7.5 J at point B. What is (a) its kinetic energy at A? (b) its speed at point B? (c) the total work done on the particle as it moves from A to B?

14. A person doing a chin-up weighs 700 N exclusive of the arms. During the first 25.0 cm of the lift, each arm exerts an upward force of 355 N on the torso. If the upward movement starts from rest, what is the person's speed at this point?

15. In the special theory of relativity, the kinetic energy of a particle of mass m and speed v is given by the expression

$$KE = mc^2 \left(\frac{1}{\sqrt{1 - v^2/c^2}} - 1 \right)$$

where $c = 3.00 \times 10^8$ m/s is the speed of light in a vacuum. Compare this expression with the classical result $KE = mv^2/2$ for a 10.0-kg particle moving at (a) 1.00×10^5 m/s, (b) 1.00×10^6 m/s, (c) 1.00×10^7 m/s, (d) 1.00×10^8 m/s, (e) 1.00×10^9 m/s. What happens to your comparison in part (e)?

16. A 2.0-g bullet leaves the barrel of a gun at a speed of 300 m/s. (a) Find its kinetic energy. (b) Find the average force exerted on the bullet by the expanding gases as the bullet moves the length of the 50-cm-long barrel.

17. A 7.00-kg bowling ball moves at 3.00 m/s. How fast must a 2.45-g Ping-Pong ball move so that the two balls have the same kinetic energy?

18. A mechanic pushes a 2.50×10^3-kg car from rest to a speed of v, doing 5000 J of work in the process. During this time, the car moves 25.0 m. Neglecting friction between car and road, find (a) v and (b) the horizontal force exerted on the car.

19. A 10-kg crate is pulled up a rough incline with an initial speed of 1.5 m/s. The pulling force is 100 N

parallel to the incline, which makes an angle of 20° with the horizontal. If the coefficient of kinetic friction is 0.40 and the crate is pulled a distance of 5.0 m, (a) how much work is done by the gravitational force? (b) How much work is done by the 100-N force? (c) What is the change in kinetic energy of the crate? (d) What is the speed of the crate after it is pulled 5.0 m?

 20. A 2000-kg car moves down a level highway under the actions of two forces. One is a 1000-N forward force exerted on the drive wheels by the road; the other is a 950-N resistive force. Use the work-energy theorem to find the speed of the car after it has moved a distance of 20 m, assuming it starts from rest.

 21. On a frozen pond, a 10-kg sled is given a kick that imparts to it an initial speed of $v_0 = 2.0$ m/s. The coefficient of kinetic friction between sled and ice is $\mu_k = 0.10$. Use the work-energy theorem to find the distance the sled moves before coming to rest.

Section 5.3 Potential Energy

Section 5.4 Conservative and Nonconservative Forces

 22. A 2.0-m-long pendulum is released from rest when the support string is at an angle of 25° with the vertical. What is the speed of the bob at the bottom of the swing?

23. A 2.00-kg ball is attached to a ceiling by a 1.00-m-long string. The height of the room is 3 m. What is the gravitational potential energy of the ball relative to (a) the ceiling? (b) the floor? (c) a point at the same elevation as the ball?

24. A 1.00×10^3-kg roller-coaster car is initially at the top of a rise, at point A. It then moves 50.0 m at an angle of 40.0° below the horizontal to a lower point, B. (a) Choosing point B as the zero level for gravitational potential energy, find the potential energy of the car at A and B, and the difference in potential energy between these points. (b) Repeat part (a), choosing point A as the zero reference level.

25. A softball pitcher hurls a 0.250-kg ball around a vertical circular path of radius 0.6 m before releasing it. The pitcher maintains a 30.0-N component of force in the direction of motion around the complete circular path. The speed of the ball at the top of the circle is 15.0 m/s. If the ball is released at the bottom of the circle, what is its speed upon release?

26. A 40-N child is in a swing that is attached to ropes 2.0 m long. Find the gravitational potential energy of the child relative to her lowest position (a) when the ropes are horizontal, (b) when the

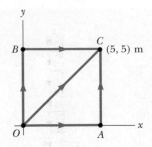

FIGURE 5.20 (Problem 27)

ropes make a 30° angle with the vertical, and (c) at the bottom of the circular arc.

27. A 3.00-kg particle moves from the origin to the position $x = 5.00$ m, $y = 5.00$ m under the influence of gravity acting in the negative y direction (Fig. 5.20.) Using Equation 5.1, calculate the work done by gravity in movements from O to C along the paths (a) OAC, (b) OBC, and (c) OC. Your results should all be identical. Why?

28. A 5.0-kg block is pushed 3.0 m up a vertical wall with constant speed by a constant force of magnitude F applied at an angle of $\theta = 30°$ with the horizontal, as shown in Figure 5.21. If the coefficient of kinetic friction between block and wall is 0.30, determine the work done by (a) **F**, (b) gravity, and (c) the normal force between block and wall. (d) By how much does the gravitational potential energy of the block increase?

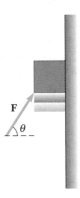

FIGURE 5.21 (Problem 28)

Section 5.5 Conservation of Mechanical Energy

 29. The launching mechanism of a toy gun consists of a spring of unknown spring constant, as shown in Figure 5.22a. If its spring is compressed a distance of 0.120 m, the gun can launch a 20.0-g projectile to a maximum height of 20.0 m when fired vertically from rest. Neglecting all resistive forces, de-

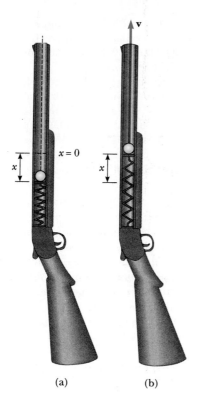

(a)   (b)

FIGURE 5.22 (Problem 29)

termine (a) the spring constant and (b) the speed of the projectile as it moves through the equilibrium position of the spring (where $x = 0$), as shown in Figure 5.22b.

30. A child and sled with a combined mass of 50.0 kg slide down a frictionless hill. If the sled starts from rest and has a speed of 3.00 m/s at the bottom, what is the height of the hill?

31. A 50-kg pole-vaulter running at 10 m/s jumps over the bar. Assuming that her horizontal component of velocity over the bar is 1.0 m/s, and neglecting air resistance, how high did she jump?

32. A surprising demonstration involves dropping an egg from a third-floor window so that the egg lands on a foam-rubber pad without breaking. If a 56.0-g egg falls (from rest) 12.0 m, and the 5.00-cm-thick foam pad stops it in 6.25 ms, by how much is the pad compressed? (*Note:* Assume constant upward acceleration as the egg compresses the foam-rubber pad.)

33. Tarzan swings on a 30.0-m-long vine initially inclined at an angle of 37.0° with the vertical. What is his speed at the bottom of the swing (a) if he starts from rest? (b) if he pushes off with a speed of 4.00 m/s?

34. The masses of the javelin, discus, and shot are 0.80 kg, 2.0 kg, and 7.2 kg, respectively, and record throws in the corresponding track events are about 89 m, 69 m, and 21 m, respectively. Neglecting air resistance, (a) calculate the minimum initial kinetic energies that would produce these throws, and (b) estimate the average force exerted on each object during the throw, assuming the force acts over a distance of 2.0 m. (c) Do your results suggest that air resistance is an important factor?

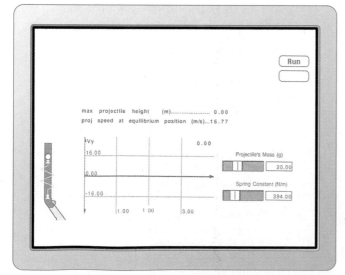

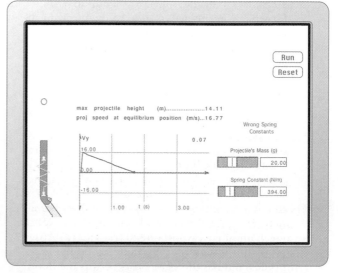

An example of the Interactive Physics simulation to accompany Problem 29.

Section 5.6   Nonconservative Forces and the Work-Energy Theorem

35. A child starts from rest at the top of a 4.00-m-high slide. (a) What is her speed at the bottom if the slide is frictionless? (b) If she reaches the bottom with a speed of 6.00 m/s, what percentage of her total energy at the top is lost as a result of friction?

36. A 2.10×10^3-kg car starts from rest at the top of a 5.0-m-long driveway that is sloped at 20° with the horizontal. If an average friction force of 4.0×10^3 N impedes the motion, find the speed of the car at the bottom of the driveway.

37. An airplane of mass 1.5×10^4 kg is moving at 60 m/s. The pilot then revs up the engine so that the forward thrust of the propeller becomes 7.5×10^4 N. If the force of air resistance has a magnitude of 4.0×10^4 N, find the speed of the airplane after it has traveled 500 m. Assume that the airplane is in level flight throughout this motion.

38. A 70-kg diver steps off a 10-m tower and drops straight down into the water. If he comes to rest 5.0 m beneath the surface, determine the average resistance force exerted on him by the water.

39. A 25.0-kg child on a 2.00-m-long swing is released from rest when the swing supports make an angle of 30.0° with the vertical. (a) Neglecting friction, find the child's speed at the lowest position. (b) If the speed of the child at the lowest position is 2.00 m/s, what is the mechanical energy lost due to friction?

40. A 5.00-g feather is held 50.0 cm above a table. It is dropped from rest and subjected to the nonconservative, constant force of air friction **f**. When the feather reaches the table, it is falling at a speed of 10.0 cm/s. Calculate (a) the magnitude of the force of air resistance, *f*, and (b) the mechanical energy lost by the feather due to air friction.

41. In a circus performance, a monkey on a sled is given an initial speed of 4.0 m/s up a 20° incline. The combined mass of monkey and sled is 20 kg, and the coefficient of kinetic friction between sled and incline is 0.20. How far up the incline does the sled move?

42. Starting from rest, a 10.0-kg block slides 3.00 m down a frictionless ramp (inclined at 30.0° from the floor) to the bottom. The block then slides an additional 5.00 m along the floor before coming to a stop. Determine (a) the speed of the block at the bottom of the ramp, (b) the coefficient of kinetic friction between block and floor, and (c) the mechanical energy lost due to friction. (d) Answer parts (a) through (c) for a ramp that is not frictionless but has the same properties as the floor.

43. An 80-N box is pulled 20 m up a 30°-incline by an applied force of 100 N that points along the incline. If the coefficient of kinetic friction between box and incline is 0.22, calculate the change in the kinetic energy of the box.

44. A skier starts from rest at the top of a hill that is inclined at 10.5° with the horizontal. The hillside is 200 m long, and the coefficient of friction between snow and skis is 0.0750. At the bottom of the hill, the snow is level and the coefficient of friction is unchanged. How far does the skier move along the horizontal portion of the snow before coming to rest?

Section 5.8   Power

45. A 1.50×10^3-kg car accelerates uniformly from rest to 10.0 m/s in 3.00 s. Find (a) the work done on the car in this time interval, (b) the average power delivered by the engine in this time interval, and (c) the instantaneous power delivered by the engine at $t = 2.00$ s.

46. A 1.50×10^3-kg car starts from rest and accelerates to 18.0 m/s in 12.0 s. Assume that air resistance remains constant at 400 N during this time. Find (a) the average power developed by the engine and (b) the instantaneous power output of the engine at $t = 12.0$ s.

47. A 2.0×10^3-kg car starts from rest and accelerates along a horizontal roadway to +20 m/s in 15 s. Assume that air resistance remains constant at −500 N during this time. Find (a) the average power developed by the engine and (b) the instantaneous power developed at $t = 15$ s.

48. While running, a person dissipates about 0.60 J of mechanical energy per step per kilogram of body mass. If a 60-kg person develops a power of 70 W during a race, how fast is the person running? Assume that a running step is 1.5 m long.

49. When an automobile moves with constant velocity, the power developed is used to overcome the frictional forces exerted by air and road. If the power developed in an engine is 50.0 hp, what total frictional force acts on the car at 55.0 mi/h (24.6 m/s)? (Note that one horsepower equals 746 W.)

50. A skier of mass 70 kg is pulled up a slope by a motor-driven cable. (a) How much work is required to pull him 60 m up a 30° slope (assumed frictionless) at a constant speed of 2.0 m/s? (b) How many horsepower must a motor have to perform this task?

51. A 50.0-kg student climbs a 5.00-m-long rope and stops at the top. (a) What must her average speed be in order to match the power output of a 200-W lightbulb? (b) How much work does she do?

(Problem 54)   *(Ron Dorman/FPG)*

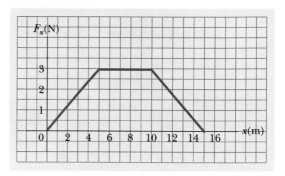

FIGURE 5.24   (Problem 56)

52. A rain cloud contains 2.66×10^7 kg of water vapor. How long would it take for a 2.00-kW pump to raise the same amount of water to the cloud's altitude, 2.00 km?

53. A 650-kg elevator starts from rest. It moves upward for 3 s with constant acceleration until it reaches its cruising speed, 1.75 m/s. (a) What is the average power of the elevator motor during this period? (b) How does this compare with its power during an upward cruise with constant speed?

54. Water flows over a section of Niagara Falls at the rate of 1.2×10^6 kg/s and falls 50 m. How much power is generated by the falling water?

Section 5.9 Work Done by a Varying Force

55. A 1.00-kg object initially at rest is acted upon by a nonconstant force that causes it to move 3.00 m. The force varies with position as shown in Figure 5.23. (a) How much work is done on the object by this force? (b) What is the speed of the object at $x = 3.00$ m?

56. An object is subjected to a force that varies with position as shown in Figure 5.24. Find the work done by the force on the object as it moves (a) from $x = 0$ to $x = 5.00$ m, (b) from $x = 5.00$ m to $x = 10.00$ m, and (c) from $x = 10.00$ m to $x = 15.00$ m. (d) What is the total work done by the force over the distance $x = 0$ to $x = 15.0$ m?

57. If the object of Problem 56 has a mass of 3.00 kg and a speed of 0.500 m/s at $x = 0$, find its speed at (a) $x = 5.00$ m, (b) $x = 10.0$ m, and (c) $x = 15.00$ m.

ADDITIONAL PROBLEMS

58. A student wants to measure the spring constant of a massless spring by connecting the spring to the bottom of a fixed steel incline of angle $\theta = 45°$, then slowly placing a uniform copper cube (5 cm on an edge) on the incline against the spring until the spring exhibits a maximum compression, as shown in Figure 5.25. Determine the spring constant, k, given that the maximum compression of the spring is 0.40 m. Take the density of copper to be 8920 kg/m^3 and refer to Table 4.2.

59. Two masses are connected by a light string passing over a light, frictionless pulley as in Figure 5.26. The 5.00-kg mass is released from rest. (a) Determine the speed of each mass when the two pass each other. (b) Determine the speed of each mass at the moment the 5.00-kg mass hits the floor.

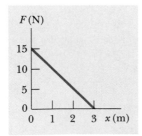

F (N)

15
10
5
0
    0   1   2   3 x (m)

FIGURE 5.23   (Problem 55)

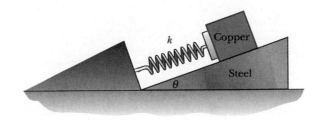

FIGURE 5.25   (Problem 58)

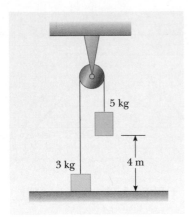

FIGURE 5.26   (Problem 59)

(c) How much higher does the 3.00-kg mass travel after the 5.00-kg mass hits the floor?

60. The plane shown in Figure 5.27 is designed for vertical takeoff and landing. If it has a mass of 8.0×10^3 kg when fueled, find the net work done on the plane as it accelerates upward at 1.0 m/s² for a distance of 30 m after starting from rest.

61. A 0.400-kg bead slides on a curved wire, starting from rest at point *A* in Figure 5.28. If the wire is frictionless, find the speed of the bead (a) at *B* and (b) at *C*.

62. A 300-g piece of putty is slowly placed on the top of a massless vertical spring of constant 19.6 N/m. (a) Determine the maximum compression of the spring. (b) What is the answer to part (a) if the putty is dropped onto the spring from a height of 0.500 m from the top of the spring?

63. A 2.0-kg block is released with an initial speed of 5.0 m/s along a horizontal tabletop toward a light

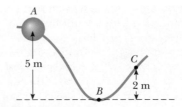

FIGURE 5.28   (Problems 61 and 65)

spring of force constant 10 N/m. The coefficient of kinetic friction between block and table is 0.25, and the block is initially 2.0 m from the spring. (a) What is the maximum compression of the spring? (b) How far from the spring is the block when it comes to rest?

64. A spring of length 0.80 m rests along a frictionless 30° incline, as in Figure 5.29. A 2.0-kg mass compresses the spring by 0.10 m. (a) Determine the spring constant *k*. (b) The mass is pushed down, compressing the spring an additional 0.60 m, and then released. If the incline is 2.0 m long, determine the maximum height attained by the mass and how far from the base of the incline it lands.

65. If the wire in Problem 61 (Fig. 5.28) is frictionless between points *A* and *B* and rough between *B* and *C*, and if the bead starts from rest at *A*, (a) find its speed at *B*. (b) If the bead comes to rest at *C*, find the loss in mechanical energy as it goes from *B* to *C*.

66. A catcher "gives" with the ball when he catches a 0.15-kg baseball moving at 25 m/s. (a) If he moves his glove a distance of 2.0 cm, what is the average force acting on his hand? (b) Repeat for the case in which his glove and hand move 10 cm.

67. A 98.0-N grocery cart is pushed 12.0 m by a shopper who exerts a constant horizontal force of 40.0 N. If all frictional forces are neglected and the cart starts from rest, what is its final speed?

68. Three men, each of whom develops 200 W, lift a 4.0×10^3-N piano, at a constant speed, to a penthouse 30 m above the street. If the pulley system is 70% efficient (so that 30% of the work is wasted

FIGURE 5.27   (Problem 60) *(Courtesy of NASA)*

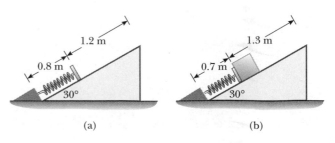

FIGURE 5.29   (Problem 64)

as heat, countering the friction in the pulley), how much time is required to lift the piano? Assume that the pulley is massless.

69. (a) A 75-kg man jumps from a window 1.0 m above a sidewalk. What is his speed just before his feet strike the pavement? (b) If the man jumps with his knees and ankles locked, the only cushion for his fall is an approximately 0.50-cm give in the pads of his feet. Calculate the average force exerted on him by the ground in this situation. This average force is sufficient to cause cartilage damage in the joints or to break bones.

70. Tarzan and Jane, whose total mass is 130 kg, start their swing on a 5.0-m-long vine when the vine is at an angle of 30° with the horizontal. At the bottom of the arc, Jane, whose mass is 50 kg, steps off. What is the maximum height at which Tarzan can land on a branch after his swing continues?

71. A 50.0-kg parachutist jumps out of an airplane at a height of 1.00 km. The parachute deploys, and she lands on the ground with a speed of 5.00 m/s. How much mechanical energy was lost to air resistance during this jump?

72. A ski jumper starts from rest 50.0 m above the ground on a frictionless track, and flies off the track at an angle of 45.0° above the horizontal and at a height of 10.0 m from the ground. Neglect air resistance. (a) What is his speed when he leaves the track? (b) What is the maximum height, y_{max}, he attains? (c) Where does he land relative to the end of the track?

73. Two identical massless springs of constant $k = 200$ N/m are fixed at opposite ends of a level track, as shown in Figure 5.30. A 5.00-kg block is pressed against the left spring, compressing it by 0.150 m. The block (initially at rest) is then released, as shown in Figure 5.30a. The entire track is frictionless *except* for the section between A and B. Given that the coefficient of kinetic friction be-

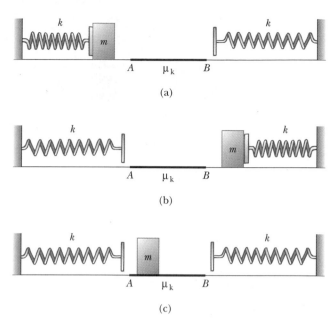

FIGURE 5.30   (Problem 73)

tween block and track along AB is $\mu_k = 0.0800$, and given that the length AB is 0.250 m, (a) determine the maximum compression of the spring on the right (see Fig. 5.30b). (b) Determine where the block eventually comes to rest, as measured from A (see Fig. 5.30c).

74. A 3.00-kg mass is moving so that its component of velocity along the x direction is 5.00 m/s and its component of velocity along the y direction is -3.00 m/s. (a) What is the kinetic energy at this time? (b) Find the change in kinetic energy if the velocity changes so that its new x component is 8.00 m/s and its new y component is 4.00 m/s.

75. A projectile of mass m is shot horizontally with an initial speed of v_0 from a height of h above a flat desert surface. At the instant before the projectile hits the surface, find (a) the work done on the projectile by gravity, (b) the change in kinetic energy since the projectile was fired, and (c) the final kinetic energy of the projectile.

76. Two blocks, A and B (with mass 50 kg and 100 kg, respectively), are connected by a string as shown in Figure 5.31. The pulley is frictionless and of negligible mass. The coefficient of kinetic friction between block A and the incline is $\mu_k = 0.25$. Determine the change in the kinetic energy of block A as it moves from C to D, a distance of 20 m up the incline.

77. A particle of mass m starts at rest and slides down a frictionless track as in Figure 5.32. It leaves the

(Problem 72)   *(T. Zimmerman/FPG)*

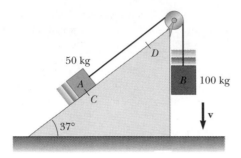

FIGURE 5.31   (Problem 76)

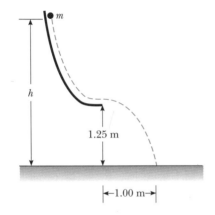

FIGURE 5.32   (Problem 77)

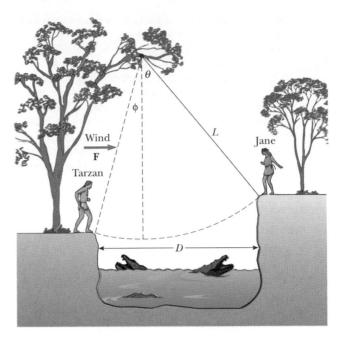

FIGURE 5.33   (Problem 79)

track horizontally, striking the ground. At what height above the ground did it start?

78. A toy gun uses a spring to horizontally project a 5.3-g soft rubber sphere. The spring constant is 8.0 N/m, the barrel of the gun is 15 cm long, and a constant frictional force of 0.032 N exists between barrel and projectile. With what speed is the projectile launched if the spring is compressed 5.0 cm?

79. Jane, whose mass is 50.0 kg, needs to swing across a river filled with man-eating crocodiles in order to save Tarzan from danger. However, she must swing into a *constant* horizontal wind force, **F**, on a vine that is initially at an angle of θ with the vertical (see Fig. 5.33). $D = 50.0$ m, $F = 110$ N, $L = 40.0$ m, and $\theta = 50.0°$. (a) With what minimum speed must Jane begin her swing in order to just make it to the other side? (*Hint:* First determine the potential energy associated with the wind force. Since the wind force is constant, use an analogy with the constant gravitational force.) (b) Once the rescue is complete, Tarzan and Jane must swing back across the river. With what mini-

mum speed must they begin their swing? Assume that Tarzan has a mass of 80.0 kg.

80. A 200-g particle is released from rest at point A on the inside of a smooth hemispherical bowl of radius $R = 30.0$ cm (Fig. 5.34). Calculate (a) its gravitational potential energy at A relative to B, (b) its kinetic energy at B, (c) its speed at B, and (d) its kinetic energy and potential energy at C.

81. The particle described in Problem 80 (Fig. 5.34) is released from point A at rest. Its speed at B is 1.50 m/s. (a) What is its kinetic energy at B? (b) How much mechanical energy is lost as a result of friction as the particle goes from A to B?

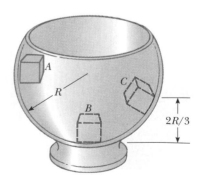

FIGURE 5.34   (Problems 80 and 81)

(c) Is it possible to determine μ from these results in a simple manner? Explain.

82. A spring has a force constant of 500 N/m. Show that the energy stored in the spring is (a) 0.400 J when the spring is stretched 4.00 cm from equilibrium, (b) 0.225 J when the spring is compressed 3.00 cm from equilibrium, and (c) zero when the spring is unstretched.

83. A 0.250-kg block is placed on a vertical spring ($k = 5.00 \times 10^3$ N/m) and pushed downward, compressing the spring 0.100 m. As the block is released, it leaves the spring and continues to travel upward. Show that it rises to a height of 10.2 m above the point of release.

84. The light horizontal spring in Figure 5.13 has a force constant of $k = 100$ N/m. A 2.00-kg block is pressed against one end of the spring, compressing it 0.100 m. After the block is released, it moves 0.250 m to the right before coming to rest. Show that the coefficient of kinetic friction between the horizontal surface and the block is 0.102.

MOMENTUM AND COLLISIONS

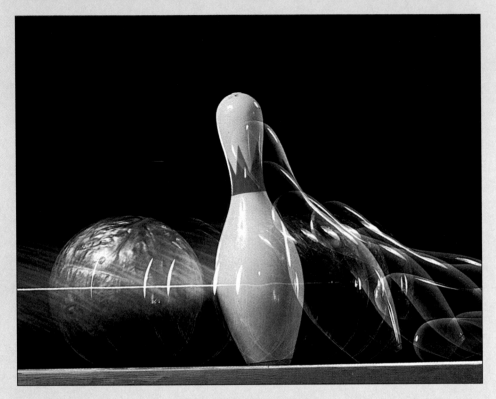

As a result of the collision between the bowling ball and pin, part of the ball's momentum is transferred to the pin. Consequently, the pin acquires momentum and kinetic energy, while the ball loses momentum and kinetic energy. However, the total momentum of the system (ball and pin) remains constant. *(Ben Rose/The Image Bank)*

Consider what happens when a golf ball is struck by a club. The ball is given a very large initial velocity by the collision; consequently, it is able to travel more than 100 meters through the air. It experiences a great change in velocity and a correspondingly great acceleration. Furthermore, because the ball experiences this acceleration over a very short time interval, the average force on the ball during collision with the club is very large. By Newton's third law, the club experiences a reaction force that is equal to and opposite the force on the ball. This reaction force produces a change in the velocity of the club. Because the club is much more massive than the ball, however, its change in velocity is much less than the change in velocity of the ball.

One of the main objectives of this chapter is to help you understand and analyze such events. As a first step, we shall introduce the term *momentum*. We often use the concept of momentum in describing objects in motion. For example, a very massive football player is often said to have a great deal of momentum as he runs down the field. A much less massive player, such as a halfback, can have equal or greater momentum if he moves with a higher velocity. This follows from the definition of momentum as the product of mass and velocity.

The concept of momentum leads us to a second conservation law: conservation of momentum. This law is especially useful for treating problems that involve collisions between objects.

6.1

MOMENTUM AND IMPULSE

The **linear momentum** of an object of mass m moving with a velocity $\mathbf{v}$ is defined as the product of the mass and the velocity:

$$\mathbf{p} \equiv m\mathbf{v} \qquad \text{[6.1]}$$

As its definition shows, momentum is a vector quantity, with its direction matching that of the velocity. Momentum has dimensions ML / T, and its SI units are kilogram-meters per second (kg · m/s).

Often we shall find it advantageous to work with the components of momentum. For two-dimensional motion, these are

$$p_x = mv_x \qquad p_y = mv_y$$

where p_x represents the momentum of an object in the x direction and p_y its momentum in the y direction.

The definition of momentum in Equation 6.1 coincides with our everyday usage of the word. When we think of a massive object moving with a high velocity, we often say that the object has a large momentum, in accordance with Equation 6.1. Likewise, a small object moving slowly is said to have a small momentum. On the other hand, a small object moving with a high velocity can have a large momentum.

When Newton first expressed the second law in mathematical form, he wrote it not as $\mathbf{F} = m\mathbf{a}$ but as

$$\mathbf{F} = \frac{\text{change in momentum}}{\text{time interval}} = \frac{\Delta \mathbf{p}}{\Delta t} \qquad \text{[6.2]}$$

where Δt is the time interval during which the momentum changes by $\Delta \mathbf{p}$. This equation says that the time rate of change of momentum of an object is equal to the resultant force acting on the object. To see that this is equivalent to $\mathbf{F} = m\mathbf{a}$ for an object of constant mass, consider a constant force, $\mathbf{F}$, acting on a particle and producing a constant acceleration. We can write Equation 6.2 as

$$\mathbf{F} = \frac{\Delta \mathbf{p}}{\Delta t} = \frac{m\mathbf{v}_f - m\mathbf{v}_i}{\Delta t} = \frac{m(\mathbf{v}_f - \mathbf{v}_i)}{\Delta t} \qquad \text{[6.3]}$$

Now recall that the velocity of an object moving with constant acceleration varies with time as

$$\mathbf{v}_f = \mathbf{v}_i + \mathbf{a}t$$

If we take $\Delta t = t$ and substitute for v_f in Equation 6.3, we see that $\mathbf{F}$ reduces to the familiar equation

$$\mathbf{F} = m\mathbf{a}$$

Note from Equation 6.2 that if the resultant force $\mathbf{F}$ is zero, the momentum of the object does not change. In other words, the linear momentum and velocity

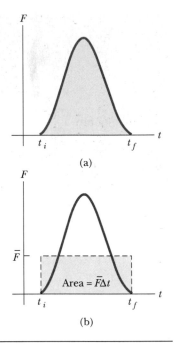

(a)

(b)

FIGURE 6.1

(a) A force acting on an object may vary in time. The impulse is the area under the force-time curve. (b) The average force (horizontal dashed line) gives the same impulse to the object in the time interval Δt as the real time-varying force described in (a).

of a particle are conserved when $\mathbf{F} = 0$. This property of momentum is important in the analysis of collisions, a subject we shall take up in a later section.

Equation 6.2 can be written as $\mathbf{F}\,\Delta t = \Delta\mathbf{p}$ or

$$\mathbf{F}\,\Delta t = \Delta\mathbf{p} = m\mathbf{v}_f - m\mathbf{v}_i \qquad [6.4]$$

This result is often called the **impulse-momentum theorem**. The term on the left side of the equation, $\mathbf{F}\,\Delta t$, is called the **impulse** of the force $\mathbf{F}$ for the time interval Δt. According to this result, *the impulse of the force acting on an object equals the change in momentum of that object.*

This equation tells us that if we exert a force on an object for time interval Δt, the effect of this force is to change the momentum of the object from some initial value, $m\mathbf{v}_i$, to some final value, $m\mathbf{v}_f$. For example, suppose a pitcher throws a baseball with a velocity of $\mathbf{v}_i$, and a batter hits the ball head on so as to reverse the direction of its velocity. The force, $\mathbf{F}$, that the bat exerts on the ball can change both the direction and the magnitude of the initial velocity to a higher value, $\mathbf{v}_f$.

A word of caution. If you tried to solve a problem such as this using Newton's second law, you would encounter some difficulty in measuring the value to be used for $\mathbf{F}$, because the force exerted on the ball is large, of short duration, and not constant. Instead, it might be represented by a curve like that in Figure 6.1a. The force starts out small as the bat comes in contact with the ball, rises to a maximum value when they are firmly in contact, and then drops off as the ball leaves the bat. In such instances, it is necessary to define an **average force**, $\overline{F}$, shown as a dashed line in Figure 6.1b. This average force can be thought of as the constant force that would give the same impulse to the object in the time interval Δt as the actual time-varying force gives in this interval. The brief collision between a bullet and an ace of clubs is illustrated in Figure 6.2.

> The impulse imparted by a force during the time interval Δt is equal to the area under the force-time graph from the beginning to the end of the time interval.

FIGURE 6.2

Cutting the card quickly: A bullet traveling at about 900 m/s collides with the ace of clubs. This photograph was taken using a microflash stroboscope at an exposure time of less than 1 μs. You should be able to show that the bullet moved less than 1 mm during this time interval (which explains why the action is "frozen"). *(Dr. Gary S. Settles/Photo Researchers)*

EXAMPLE 6.1   Teeing Off

A 50-g golf ball is struck with a club, as in Figure 6.3. The force on the ball varies from zero when contact is made up to some maximum value (when the ball is deformed) and then back to zero when the ball leaves the club. Thus, the force-time graph is somewhat like that in Figure 6.1. Assume that the ball leaves the club face with a velocity of +44 m/s.

(a) Estimate the impulse due to the collision.

Solution   Before the club hits the ball, the ball is at rest on the tee, and as a result its initial momentum is zero. The magnitude of the momentum immediately after the collision is

$$p_f = mv_f = (50 \times 10^{-3} \text{ kg})(44 \text{ m/s}) = +2.2 \text{ kg·m/s}$$

and thus the impulse imparted to the ball, which equals its change in momentum is

$$\Delta p = mv_f - mv_i = \quad +2.2 \text{ kg·m/s}$$

(b) Estimate the duration of the collision and the average force on the ball.

Solution   From Figure 6.3, it appears that a reasonable estimate of the distance the ball travels while in contact with the club is the radius of the ball, about 2.0 cm. The time it takes the club to move this distance (the contact time) is then

$$\Delta t = \frac{\Delta x}{v_i} = \frac{2.0 \times 10^{-2} \text{ m}}{44 \text{ m/s}} = 4.5 \times 10^{-4} \text{ s}$$

Finally, the magnitude of the average force is estimated to be

$$\bar{F} = \frac{\Delta p}{\Delta t} = \frac{2.2 \text{ kg·m/s}}{4.5 \times 10^{-4} \text{ s}} = \quad 4.9 \times 10^3 \text{ N}$$

FIGURE 6.3
(Example 6.1)   A golf ball being struck by a club. *(Courtesy of Dr. Harold E. Edgerton, MIT)*

EXAMPLE 6.2   Follow the Bouncing Ball

A 100-g ball is dropped from a height of $h = 2.00$ m above the floor (Fig. 6.4). It rebounds vertically to a height of $h' = 1.50$ m after colliding with the floor.

(a) Find the momentum of the ball immediately before it collides with the floor and immediately after it rebounds.

Solution   We can find the velocity of the ball just before it strikes the floor by using the principle of conservation of mechanical energy. Equating the initial potential energy to the final kinetic energy, with the floor as the reference level, gives

$$mgh = \tfrac{1}{2}mv_i^2$$
$$v_i = -\sqrt{2gh} = -\sqrt{2(9.80 \text{ m/s}^2)(2.00 \text{ m})} = -6.26 \text{ m/s}$$

where the minus sign is chosen to indicate that the initial velocity is in the negative y direction, as in Figure 6.4a.

Likewise, v_f, the ball's velocity after colliding with the floor, is obtained from the energy expression

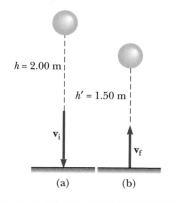

FIGURE 6.4
(Example 6.2)   (a) A ball is dropped from height h and reaches the floor with a velocity of $\mathbf{v_i}$. (b) The ball rebounds from the floor with a velocity of $\mathbf{v_f}$ and reaches height h'.

$$\tfrac{1}{2}mv_f^2 = mgh'$$

$$v_f = \sqrt{2gh'} = \sqrt{2(9.80 \text{ m/s}^2)(1.50 \text{ m})} = +5.42 \text{ m/s}$$

Because $m = 0.100$ kg, the initial and final momenta are

$$p_i = mv_i = (0.100 \text{ kg})(-6.26 \text{ m/s}) = \boxed{-0.626 \text{ kg} \cdot \text{m/s}}$$

and

$$p_f = mv_f = (0.100 \text{ kg})(5.42 \text{ m/s}) = \boxed{+0.542 \text{ kg} \cdot \text{m/s}}$$

The initial momentum is negative because the velocity, and hence the momentum, are directed downward, in the negative direction.

(b) Determine the average force exerted by the floor on the ball. Assume that the time interval of the collision is 1.00×10^{-2} s (a typical value).

Solution   From the impulse-momentum theorem, we find

$$\bar{F}\Delta t = mv_f - mv_i$$

$$\bar{F} = \frac{[0.542 - (-0.626)]\text{kg} \cdot \text{m/s}}{1.00 \times 10^{-2} \text{ s}} = \boxed{+1.17 \times 10^2 \text{ N}}$$

Before

−15.0 m/s

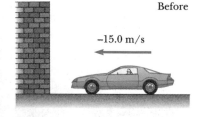

After

2.6 m/s

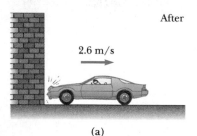

(a)

(b)

FIGURE 6.5
(Example 6.3)   (a) This car's momentum changes as a result of its collision with the wall. (b) In a crash test (an inelastic collision), much of the car's initial kinetic energy is transformed into the energy it took to damage the vehicle. *(b, Courtesy of General Motors)*

EXAMPLE 6.3   How Good Are the Bumpers?

In a particular crash test, a 1.50×10^3 kg automobile collides with a wall as in Figure 6.5a. The initial and final velocities of the automobile are $v_i = -15.0$ m/s and $v_f = +2.60$ m/s, respectively. If the collision lasts for 0.150 s, find the impulse due to the collision and the average force exerted on the automobile.

Solution   The initial and final momenta of the automobile are

$$p_i = mv_i$$

$$= (1.50 \times 10^3 \text{ kg})(-15.0 \text{ m/s}) = -2.25 \times 10^4 \text{ kg} \cdot \text{m/s}$$

$$p_f = mv_f = (1.50 \times 10^3 \text{ kg})(2.60 \text{ m/s}) = +0.390 \times 10^4 \text{ kg} \cdot \text{m/s}$$

Hence, the impulse, which equals the change in momentum, is

$$\bar{F}\Delta t = \Delta p = p_f - p_i$$

$$= 0.390 \times 10^4 \text{ kg} \cdot \text{m/s} - (-2.25 \times 10^4 \text{ kg} \cdot \text{m/s})$$

$$\bar{F}\Delta t = \boxed{+2.64 \times 10^4 \text{ kg} \cdot \text{m/s}}$$

The average force exerted on the automobile is

$$\bar{F} = \frac{\Delta p}{\Delta t} = \frac{2.64 \times 10^4 \text{ kg} \cdot \text{m/s}}{0.150 \text{ s}} = \boxed{+1.76 \times 10^5 \text{ N}}$$

6.2
CONSERVATION OF MOMENTUM

In this section we shall use the law of conservation of momentum to describe what happens when two particles collide with each other. *The force due to the collision is assumed to be much greater than any external forces present.*

A collision may be the result of physical contact between two objects, as illustrated in Figure 6.6a. This is a common observation when two macroscopic objects, such as two billiard balls or a baseball and a bat, strike each other. But "contact" on a submicroscopic scale is ill defined and hence meaningless, and so the notion of *collision* must be generalized. More accurately, forces between two bodies arise from the electrostatic interaction of the electrons in the surface atoms of the bodies.

To understand the distinction between macroscopic and microscopic collisions, consider the collision of a proton with an alpha particle (the nucleus of the helium atom), such as occurs in Figure 6.6b. Because the two particles are positively charged, they repel each other.

Figure 6.7 shows two isolated particles before and after they collide. By *isolated,* we mean that no external forces, such as the gravitational force or friction, are present. Before the collision, the velocities of the two particles are v_{1i} and v_{2i}; after the collision, the velocities are v_{1f} and v_{2f}. The impulse-momentum theorem applied to m_1 becomes

$$\overline{F}_1 \, \Delta t = m_1 v_{1f} - m_1 v_{1i}$$

Likewise, for m_2 we have

$$\overline{F}_2 \, \Delta t = m_2 v_{2f} - m_2 v_{2i}$$

where $\overline{F}_1$ is the force on m_1 due to m_2 during the collision and $\overline{F}_2$ is the force on m_2 due to m_1 during the collision (Fig. 6.8).

We are using average values for $\overline{F}_1$ and $\overline{F}_2$ even though the actual forces may vary in time in a complicated way, as is the case in Figure 6.9. Newton's third law states that at all times these two forces are equal in magnitude and opposite in direction ($F_1 = -F_2$). Additionally, the two forces act for the same time interval. Thus,

$$\overline{F}_1 \, \Delta t = -\overline{F}_2 \, \Delta t$$

or

$$m_1 v_{1f} - m_1 v_{1i} = -(m_2 v_{2f} - m_2 v_{2i})$$

from which we find

$$m_1 v_{1i} + m_2 v_{2i} = m_1 v_{1f} + m_2 v_{2f} \qquad \text{[6.5]}$$

This result is known as **conservation of momentum.**

> The principle of conservation of momentum states that, when no external forces act on a system consisting of two objects, the total momentum of the system before the collision is equal to the total momentum of the system after the collision.

Note that momentum is conserved for a *system* of objects. In the example used to derive Equation 6.5, the system was taken to be two colliding objects. More generally, a system includes all the objects that are interacting with one another. Additionally, we have assumed that the only forces acting during the collision are internal forces, meaning the forces that arise between the interacting objects of the system. For example, in the collision of two objects shown in Figure 6.8, the internal forces are F_1 and F_2. If a third object outside the system

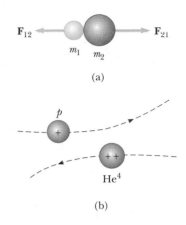

FIGURE 6.6
(a) A collision between two objects resulting from direct contact. (b) A collision between two charged objects.

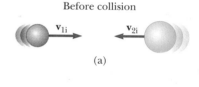

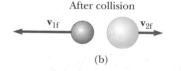

FIGURE 6.7
Before and after a head-on collision between two objects. The momentum of each object changes as a result of the collision, but the total momentum of the system remains constant.

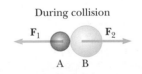

FIGURE 6.8
When two objects collide, the force F_1 exerted on object A is equal in magnitude and opposite in direction to the force F_2 exerted on object B.

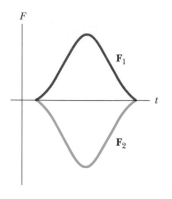

FIGURE 6.9

Force as a function of time for the two colliding particles in Figure 6.8. Note that $\mathbf{F}_1 = -\mathbf{F}_2$.

(consisting of m_1 and m_2) were to exert a force on either m_1 or m_2 (or both objects) during the collision, momentum would not be conserved for the system. In all the example problems that we shall consider, the system will be assumed to be isolated.

Our derivation has assumed that only two objects interact, but the result remains valid regardless of the number involved. In its most general form, conservation of momentum can be stated as follows:

> **The total momentum of an isolated system of objects is conserved regardless of the nature of the forces between the objects.**

To understand what this statement means, imagine that you initially stand at rest and then jump upward, leaving the ground with a velocity of v. Obviously, your momentum is not conserved, because before the jump it was zero and it became mv as you began to rise. However, the total momentum of the system is conserved if the system selected includes all objects that exert forces on one another. You must include the Earth as part of this system because you exert a downward force on the Earth when you jump. The Earth in turn exerts on you an upward force of the same magnitude, as required by Newton's third law. Momentum is conserved for the system consisting of you and the Earth. Thus, as you move upward with some momentum mv, the Earth moves in the opposite direction with momentum of the same magnitude. The recoil velocity of the Earth due to this event is imperceptibly small, of course, because the Earth is so massive, but its momentum is finite.

EXAMPLE 6.4   **The Recoiling Pitching Machine**

A baseball player attempts to use a pitching machine to help him improve his batting average. He places the 50-kg machine on a frozen pond as in Figure 6.10. The machine fires a 0.15-kg baseball horizontally with a speed of 36 m/s. What is the recoil velocity of the machine?

Solution   We take the system to consist of the baseball and the pitching machine. Because of the force of gravity and the normal force, the system is not really isolated. However, both forces are directed perpendicularly to the motion of the system. Therefore, momentum is conserved in the x direction since there are no external forces in this direction (assuming the surface is frictionless).

Because the baseball and pitching machine are at rest before firing, the total momentum of the system is zero. Therefore, the total momentum after firing must also be zero; that is,

$$m_1\mathbf{v}_1 + m_2\mathbf{v}_2 = 0$$

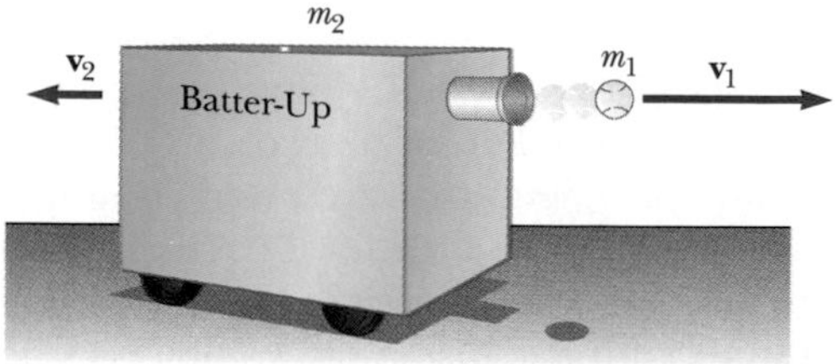

FIGURE 6.10

(Example 6.4)   When the baseball is fired to the right, the pitching machine recoils to the left.

With $m_1 = 0.15$ kg, $\mathbf{v}_1 = +36$ m/s, and $m_2 = 50$ kg, we find the recoil velocity of the pitching machine to be

$$\mathbf{v}_2 = -\frac{m_1}{m_2}\mathbf{v}_1 = -\left(\frac{0.15 \text{ kg}}{50 \text{ kg}}\right)(36 \text{ m/s}) = \boxed{-0.11 \text{ m/s}}$$

The negative sign for $\mathbf{v}_2$ indicates that the pitching machine is moving to the left after firing, in the direction *opposite* the motion of the baseball.

6.3

COLLISIONS

We have seen that, for any type of collision, the total momentum of the system just before collision equals the total momentum just after collision. We can say that the total momentum is always conserved for any type of collision. However, the total kinetic energy is generally not conserved in a collision, because some of the kinetic energy is converted to thermal energy and internal elastic potential energy when the objects deform.

We define an **inelastic collision** as a collision in which momentum is conserved but kinetic energy is not. The collision of a rubber ball with a hard surface is inelastic, because some of the kinetic energy is lost when the ball is deformed during contact with the surface. When two objects collide and stick together, the collision is called **perfectly inelastic.** For example, if two pieces of putty collide, they stick together and move with some common velocity after the collision. If a meteorite collides head on with the Earth, it becomes buried in the Earth and the collision is considered perfectly inelastic. Not all of the initial kinetic energy is necessarily lost in a perfectly inelastic collision.

An **elastic collision** is defined as one in which *both momentum and kinetic energy are conserved*. Billiard-ball collisions and the collisions of air molecules with the walls of a container at ordinary temperatures are highly elastic. Macroscopic collisions such as those between billiard balls are only approximately elastic, because some permanent deformation, and hence some loss of kinetic energy, takes place. Perfectly elastic collisions do occur, however, between atomic and subatomic particles. Elastic and perfectly inelastic collisions are *limiting* cases; most actual collisions fall into a category between them.

Inelastic collision

Elastic collision

The force from a nitrogen-propelled, hand-controlled device allows an astronaut to move about freely in space without restrictive tethers. *(Courtesy of NASA)*

PHYSICS IN *ACTION*

COLLISIONS

On the left is a multiflash photograph of an apparatus called the ballistic pendulum, commonly used in introductory laboratories (see Example 6.7). A metal ball is fired from a spring-loaded gun into a pendulum that contains a cup, which "catches" the ball. The ball makes a totally inelastic collision with the pendulum. If one measures the mass of the ball, the mass of the pendulum, and the vertical distance traveled by the system (ball + pendulum), one can calculate the velocity of the ball before the collision. It is important to recognize that only momentum is conserved in this collision. That is, the initial momentum of the ball equals the momentum of the ball and pendulum just after the collision. Although mechanical energy is *not* conserved, we can equate the kinetic energy of the system just *after* the collision to the gravitational potential energy of the system at its highest position. Why is it incorrect to equate the initial kinetic energy of the ball to the final potential energy of the system?

In the photograph on the right, a strawberry is being pierced by a bullet traveling at a supersonic speed of 450 m/s. (What would William Tell think?) This collision was photographed with a microflash stroboscope using an exposure time of about 0.33 μs. The velocity of the bullet decreases from v_{1i} to v_{1f} because of the collision, so the bullet loses kinetic energy. This loss appears in the strawberry, which disintegrates completely after the collision. Note that the points of both entry and exit of the bullet are visually explosive.

(Courtesy of Central Scientific Company)

(Dr. Gary S. Settles and Stephen S. McIntyre/SPL/Photo Researchers)

We can summarize the types of collisions as follows:

1. An elastic collision is one in which both momentum and kinetic energy are conserved.
2. An inelastic collision is one in which momentum is conserved but kinetic energy is not.
3. A perfectly inelastic collision is an inelastic collision in which the two objects stick together after the collision, so that their final velocities are the same and the momentum of the system is conserved.

In the remainder of this section, we shall treat collisions in one dimension and of two extreme types: perfectly inelastic and elastic.

PERFECTLY INELASTIC COLLISIONS

Consider two objects of masses m_1 and m_2 moving with initial velocities v_{1i} and v_{2i} along a straight line, as in Figure 6.11. We shall assume that the objects collide head on so that they move along the same line of motion after the collision. If the two objects stick together and move with some common velocity, v_f, after the collision, then only the momentum of the system is conserved, and we can say that the total momenta before and after the collision are equal:

$$m_1 v_{1i} + m_2 v_{2i} = (m_1 + m_2) v_f \qquad [6.6]$$

It is important to note that v_{1i}, v_{2i}, and so on represent the x components of the vectors $\mathbf{v}_{1i}$, $\mathbf{v}_{2i}$, and so on, so one must be careful with signs. For example, in Figure 6.11, v_{1i} would have a positive value (m_1 moving to the right), whereas v_{2i} would have a negative value (m_2 moving to the left).

In a typical inelastic collision problem, only one quantity in the preceding equation is unknown, and so conservation of momentum is sufficient to tell us what we need to know.

EXAMPLE 6.5   The Cadillac Versus the Compact Car

An 1800-kg luxury sedan stopped at a traffic light is struck from the rear by a compact car with a mass of 900 kg. The two cars become entangled as a result of the collision. If the compact car was moving at a velocity of +20.0 m/s before the collision, what is the velocity of the entangled mass after the collision?

Solution   The momentum before the collision is that of the compact car alone, because the large car is initially at rest. Thus, for the momentum before the collision we have

$$p_i = m_1 v_i = (900 \text{ kg})(20.0 \text{ m/s}) = +1.80 \times 10^4 \text{ kg} \cdot \text{m/s}$$

After the collision, the mass that moves is the sum of the masses of the large car and the compact car, and the momentum of the combination is

$$p_f = (m_1 + m_2) v_f = (2700 \text{ kg})(v_f)$$

Equating the momentum before the collision to the momentum after the collision and solving for v_f, the velocity of the wreckage, we get

$$v_f = \frac{p_i}{m_1 + m_2} = \frac{1.80 \times 10^4 \text{ kg} \cdot \text{m/s}}{2700 \text{ kg}} = \boxed{+6.67 \text{ m/s}}$$

Before collision

(a)

After collision

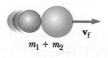

(b)

FIGURE 6.11
(a) Before and (b) after a perfectly inelastic head-on collision between two objects.

EXAMPLE 6.6   Here's Mud in Your Eye

Two balls of mud collide head on in a perfectly inelastic collision, as in Figure 6.11. Suppose $m_1 = 0.500$ kg, $m_2 = 0.250$ kg, $v_{1i} = +4.00$ m/s, and $v_{2i} = -3.00$ m/s.

(a) Find the velocity of the composite ball of mud after the collision.

Solution   Writing Equation 6.6 for conservation of momentum with the positive direction for velocity to the right, we can find the velocity of the combined mass after the collision:

$$m_1 v_{1i} + m_2 v_{2i} = (m_1 + m_2) v_f$$

$$(0.500 \text{ kg})(4.00 \text{ m/s}) + (0.250 \text{ kg})(-3.00 \text{ m/s}) = (0.750 \text{ kg})(v_f)$$

$$v_f = \boxed{+1.67 \text{ m/s}}$$

(b) How much kinetic energy is lost in the collision?

Solution   The kinetic energy before the collision is

$$KE_i = KE_1 + KE_2 = \tfrac{1}{2} m_1 v_{1i}{}^2 + \tfrac{1}{2} m_2 v_{2i}{}^2$$

$$= \tfrac{1}{2}(0.500 \text{ kg})(4 \text{ m/s})^2 + \tfrac{1}{2}(0.250 \text{ kg})(-3.00 \text{ m/s})^2 = 5.13 \text{ J}$$

The kinetic energy after the collision is

$$KE_f = \tfrac{1}{2}(m_1 + m_2) v_f^2 = \tfrac{1}{2}(0.750 \text{ kg})(1.67 \text{ m/s})^2 = 1.05 \text{ J}$$

Hence, the loss in kinetic energy is

$$KE_i - KE_f = \boxed{4.08 \text{ J}}$$

Most of this lost energy is converted to thermal energy and internal elastic potential energy as the objects collide and deform.

EXAMPLE 6.7   The Ballistic Pendulum

The ballistic pendulum (Fig. 6.12; also see the photograph in "Physics in Action," p. 160) is a device used to measure the velocity of a fast-moving projectile such as a bullet. The bullet is fired into a large block of wood suspended from some light wires. The bullet is stopped by the block, and the entire system swings through the vertical distance h. It is possible to obtain the initial velocity of the bullet by measuring h and the two masses. As an example of the technique, assume that the mass of the bullet, m_1, is 5.00 g, the mass of the pendulum, m_2, is 1.00 kg, and h is 5.00 cm. Find the initial velocity of the bullet, v_{1i}.

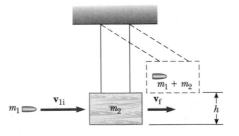

FIGURE 6.12
(Example 6.7)   Diagram of a ballistic pendulum. Note that $\mathbf{v}_f$ is the velocity of the system just after the perfectly inelastic collision.

Solution The collision between the bullet and the block is perfectly inelastic. Writing the conservation of momentum for the collision in the form of Equation 6.6, we have

$$m_1 v_{1i} = (m_1 + m_2) v_f$$

(1) $\quad (5.00 \times 10^{-3} \text{ kg})(v_{1i}) = (1.0050 \text{ kg})(v_f)$

There are two unknowns in this equation, v_{1i} and v_f; the latter is the velocity of the block plus embedded bullet *immediately after the collision*. We must look for additional information if we are to complete the problem. Kinetic energy is not conserved during an inelastic collision. However, mechanical energy is conserved after the collision, and so the kinetic energy of the system at the bottom is transformed into the potential energy of the bullet plus block at height h:

$$\tfrac{1}{2}(m_1 + m_2) v_f^2 = (m_1 + m_2) gh$$

$$\tfrac{1}{2}(1.0050 \text{ kg})(v_f^2) = (1.0050 \text{ kg})(9.80 \text{ m/s}^2)(5.00 \times 10^{-2} \text{ m})$$

giving

$$v_f = 0.990 \text{ m/s}$$

With v_f now known, (1) yields v_{1i}:

$$v_{1i} = \frac{(1.0050 \text{ kg})(0.990 \text{ m/s})}{5.00 \times 10^{-3} \text{ kg}} = \boxed{199 \text{ m/s}}$$

Exercise Explain why it would be incorrect to equate the initial kinetic energy of the incoming bullet to the final gravitational potential energy of the bullet-block combination.

ELASTIC COLLISIONS

Now consider two objects that undergo an elastic head-on collision (Fig. 6.13). In this situation, *both momentum and kinetic energy are conserved;* we can write these conditions as

$$m_1 v_{1i} + m_2 v_{2i} = m_1 v_{1f} + m_2 v_{2f} \qquad \textbf{[6.7]}$$

$$\tfrac{1}{2} m_1 v_{1i}^2 + \tfrac{1}{2} m_2 v_{2i}^2 = \tfrac{1}{2} m_1 v_{1f}^2 + \tfrac{1}{2} m_2 v_{2f}^2 \qquad \textbf{[6.8]}$$

where v is positive if an object moves to the right and negative if it moves to the left.

In a typical problem involving elastic collisions, there are two unknown quantities, and Equations 6.7 and 6.8 can be solved simultaneously to find them. An alternative approach, employing a little mathematical manipulation of Equation 6.8, often simplifies this process. To see this, let's cancel the factor $\tfrac{1}{2}$ in Equation 6.8 and rewrite it as

$$m_1 (v_{1i}^2 - v_{1f}^2) = m_2 (v_{2f}^2 - v_{2i}^2)$$

Here we have moved the terms containing m_1 to one side of the equation and those containing m_2 to the other. Next, let us factor both sides of the equation:

$$m_1 (v_{1i} - v_{1f})(v_{1i} + v_{1f}) = m_2 (v_{2f} - v_{2i})(v_{2f} + v_{2i}) \qquad \textbf{[6.9]}$$

We now separate the terms containing m_1 and m_2 in the equation for the conservation of momentum (Eq. 6.7) to get

Before collision

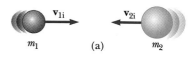

After collision

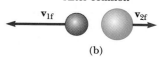

FIGURE 6.13
(a) Before and (b) after an elastic head-on collision between two hard spheres.

$$m_1(v_{1i} - v_{1f}) = m_2(v_{2f} - v_{2i}) \qquad \text{[6.10]}$$

To obtain our final result, we divide Equation 6.9 by Equation 6.10 and get

$$v_{1i} + v_{1f} = v_{2f} + v_{2i}$$

$$v_{1i} - v_{2i} = -(v_{1f} - v_{2f}) \qquad \text{[6.11]}$$

This equation, in combination with the condition for conservation of momentum, will be used to solve problems dealing with perfectly elastic, head-on collisions. According to Equation 6.11, the relative velocity of the two objects before the collision, $v_{1i} - v_{2i}$, equals the negative of the relative velocity of the two objects after the collision, $-(v_{1f} - v_{2f})$.

PROBLEM-SOLVING STRATEGY
Conservation of Momentum

The following procedure is recommended for problems involving collisions between two objects:

1. Set up a coordinate system and define your velocities with respect to that system. That is, objects moving in the direction selected as the positive direction of the x axis are considered to have a positive velocity, and those moving in the negative x direction, a negative velocity. It is convenient to make the x axis coincide with one of the initial velocities.
2. In your sketch of the coordinate system, draw all velocity vectors with labels and include all the given information.
3. Write expressions for the momenta of each object before and after the collision. (In two-dimensional collision problems, write expressions for the x and y components of momentum before and after the collision.) Remember to include the appropriate signs for the velocity vectors.
4. Now write expressions for the *total* momentum *before* and *after* the collision and equate the two. (For two-dimensional collisions, this expression should be written for the momentum in both the x and y directions.) Remember, it is the momentum of the *system* (the two colliding objects) that is conserved, *not* the momenta of the individual objects.
5. If the collision is *inelastic*, kinetic energy is not conserved. Proceed to solve the momentum equations for the unknown quantities.
6. If the collision is *elastic*, kinetic energy is conserved, so you can equate the total kinetic energies before and after the collision. This gives an additional relationship between the velocities.

EXAMPLE 6.8   Let's Play Pool

Two billiard balls move toward one another as in Figure 6.13. The balls have identical masses, and the collision between them is perfectly elastic. If the initial velocities of the balls are +30 cm/s and −20 cm/s, what is the velocity of each ball after the collision?

Solution We turn first to Equation 6.7. The equal masses cancel on each side, and after substituting the appropriate values for the initial velocities, we have

$$30 \text{ cm/s} + (-20 \text{ cm/s}) = v_{1f} + v_{2f}$$

$$(1) \qquad 10 \text{ cm/s} = v_{1f} + v_{2f}$$

Since kinetic energy is also conserved, we can apply Equation 6.11, which gives

$$30 \text{ cm/s} - (-20 \text{ cm/s}) = v_{2f} - v_{1f}$$

$$(2) \qquad 50 \text{ cm/s} = v_{2f} - v_{1f}$$

Solving (1) and (2) simultaneously, we find

$$v_{1f} = \boxed{-20 \text{ cm/s}} \qquad v_{2f} = \boxed{+30 \text{ cm/s}}$$

That is, the balls *exchange velocities!* This is always the case when two objects of equal mass collide elastically, head on.

Exercise Find the final velocity of the two balls if the ball with initial velocity −20 cm/s has a mass equal to half that of the ball with initial velocity +30 cm/s.

Answer $v_{1f} = -3.0$ cm/s; $v_{2f} = +47$ cm/s.

EXAMPLE 6.9 A Two-Body Collision with Spring

A block of mass $m_1 = 1.60$ kg, moving to the right with a speed of 4.00 m/s on a frictionless, horizontal track, collides with a spring attached to a second block, of mass $m_2 = 2.10$ kg, that is moving to the left with a speed of 2.50 m/s (Fig. 6.14a). The spring constant is 600 N/m. For the instant when m_1 is moving to the right with a speed of 3.00 m/s, determine (a) the velocity of m_2 and (b) the distance, x, that the spring is compressed.

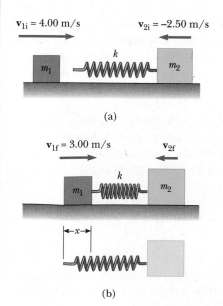

$\mathbf{v}_{1i} = 4.00$ m/s $\mathbf{v}_{2i} = -2.50$ m/s

(a)

$\mathbf{v}_{1f} = 3.00$ m/s $\mathbf{v}_{2f}$

(b)

FIGURE 6.14
(Example 6.9)

Solution   (a) First, note that the initial velocity of m_2 is -2.50 m/s because its direction is to the left. Since the total momentum of the system is constant, we have

$$m_1 v_{1i} + m_2 v_{2i} = m_1 v_{1f} + m_2 v_{2f}$$

$$(1.60 \text{ kg})(4.00 \text{ m/s}) + (2.10 \text{ kg})(-2.50 \text{ m/s}) = (1.60 \text{ kg})(3.00 \text{ m/s})$$
$$+ (2.10 \text{ kg}) v_{2f}$$

$$v_{2f} = \boxed{-1.74 \text{ m/s}}$$

This result shows that m_2 is still moving to the left at that instant.

(b) To determine the compression, x, in the spring in Figure 6.14b, we can make use of conservation of energy since no friction forces are acting on the system. Thus, we have

$$\tfrac{1}{2} m_1 v_{1i}^2 + \tfrac{1}{2} m_2 v_{2i}^2 = \tfrac{1}{2} m_1 v_{1f}^2 + \tfrac{1}{2} m_2 v_{2f}^2 + \tfrac{1}{2} k x^2$$

Substitution of the given values and the result of part (a) into this expression gives

$$x = \boxed{0.173 \text{ m}}$$

Exercise   Find the velocity of m_1 and the compression in the spring at the instant m_2 is at rest.

Answer   0.719 m/s; 0.251 m.

6.4

GLANCING COLLISIONS

The collisions we have considered until now have been head-on collisions, in which the incident mass strikes a second mass head on and both rebound along a straight-line path that coincides with the line of motion of the incident mass. Anyone who has every played billiards knows that such collisions are the exception rather than the rule. A more common type of collision is a *glancing collision*, in which the colliding masses rebound at some angle relative to the line of motion of the incident mass. Figure 6.14a shows a blue ball that travels with an initial speed of v_{1i} and strikes a red ball obliquely (off center). After the collision, the blue ball caroms off at an angle of θ relative to its incident line of motion, and the red ball rebounds at an angle of ϕ.

As we emphasized earlier, *momentum is conserved in all collisions when no external forces are acting*, and glancing collisions are no exception. Since momentum is a vector quantity, the conservation of momentum principle must be written $\mathbf{p}_i = \mathbf{p}_f$. That is, the total initial momentum of the system (the two balls) must equal the total final momentum of the system. For a collision in two dimensions, such as that in Figure 6.15, this implies that the total momentum is conserved along the x direction *and* along the y direction. We can state this in equation form as

$$\sum p_{ix} = \sum p_{fx} \qquad\qquad \textbf{[6.12]}$$

and

$$\sum p_{iy} = \sum p_{fy} \qquad\qquad \textbf{[6.13]}$$

Billiard balls undergo collisions with each other that are nearly elastic. *(Henry Groskinsky/Peter Arnold, Inc.)*

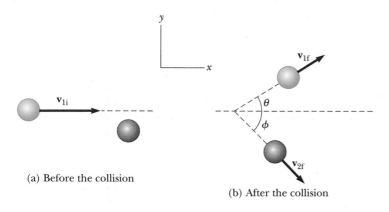

(a) Before the collision

(b) After the collision

FIGURE 6.15
(a) Before and (b) after a glancing collision between two balls.

The following example illustrates how to use the principle of conservation of momentum to treat glancing collisions.

EXAMPLE 6.10   Collision at an Intersection

At an intersection, a 1500-kg car traveling east at 25 m/s collides with a 2500-kg van traveling north at 20 m/s, as shown in Figure 6.16. Find the direction and magnitude of the velocity of the wreckage immediately after the collision, assuming that the vehicles undergo a perfectly inelastic collision (that is, they stick together).

Solution   Let us choose the positive x direction to be east and the positive y direction to be north, as in Figure 6.16. Before the collision, the only object having momentum in the x direction is the car. Thus, the total initial momentum of the system (car plus van) in the x direction is

$$\sum p_{ix} = (1500 \text{ kg})(25 \text{ m/s}) = 37\ 500 \text{ kg} \cdot \text{m/s}$$

Now let us assume that the wreckage moves at an angle of θ and a speed of v after the collision, as in Figure 6.16. The total momentum in the x direction after the collision is

$$\sum p_{fx} = (4000 \text{ kg})(v \cos \theta)$$

Because momentum is conserved in the x direction, we have

$$\sum p_{ix} = \sum p_{fx}$$

(1) $37\ 500 \text{ kg} \cdot \text{m/s} = (4000 \text{ kg})(v \cos \theta)$

Similarly, the total initial momentum of the system in the y direction is that of the van, which equals $(2500 \text{ kg})(20 \text{ m/s})$. Applying conservation of momentum to the y direction, we have

$$\sum p_{iy} = \sum p_{fy}$$

$$(2500 \text{ kg})(20 \text{ m/s}) = (4000 \text{ kg})(v \sin \theta)$$

(2) $50\ 000 \text{ kg} \cdot \text{m/s} = (4000 \text{ kg})(v \sin \theta)$

If we divide (2) by (1), we get

$$\tan \theta = \frac{50\ 000}{37\ 500} = 1.33$$

$$\theta = \boxed{53°}$$

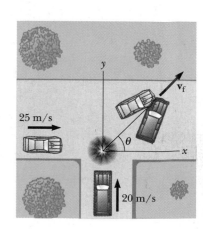

FIGURE 6.16
(Example 6.10)   A top view of a perfectly inelastic collision between a car and a van.

When this angle is substituted into (2)—or, alternatively, into (1)—the value of v is

$$v = \frac{50\ 000\ \text{kg}\cdot\text{m/s}}{(4000\ \text{kg})(\sin 53°)} = \boxed{16\ \text{m/s}}$$

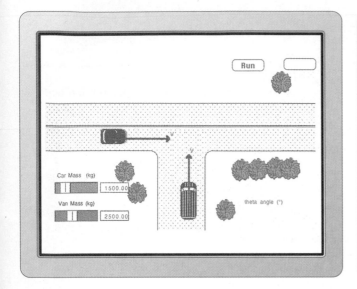 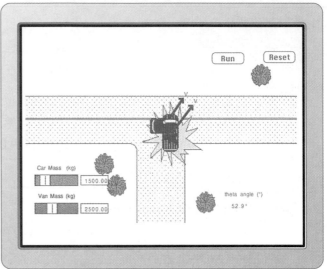

An example of the Interactive Physics simulation to accompany Example 6.10.

SUMMARY

The **linear momentum** of an object of mass m moving with a velocity of **v** is defined to be

$$\mathbf{p} \equiv m\mathbf{v} \qquad\qquad \textbf{[6.1]}$$

The **impulse** of a force, **F**, acting on an object is equal to the product of the force and the time interval during which the force acts:

$$\text{Impulse} = \mathbf{F}\,\Delta t$$

The **impulse-momentum theorem** states that the impulse of a force on an object is equal to the change in momentum of the object:

$$\mathbf{F}\,\Delta t = \Delta\mathbf{p} = m\mathbf{v}_\text{f} - m\mathbf{v}_\text{i} \qquad\qquad \textbf{[6.4]}$$

Conservation of momentum of two interacting objects means that, when no external forces act on a system consisting of two objects, the total momentum of the system before the collision is equal to the total momentum of the system after the collision.

$$m_1\mathbf{v}_{1\text{i}} + m_2\mathbf{v}_{2\text{i}} = m_1\mathbf{v}_{1\text{f}} + m_2\mathbf{v}_{2\text{f}} \qquad\qquad \textbf{[6.5]}$$

An **inelastic collision** is one in which momentum is conserved but kinetic energy is not. A **perfectly inelastic collision** is one in which the colliding objects stick together after the collision. An **elastic collision** is one in which both momentum and kinetic energy are conserved.

In glancing collisions, conservation of momentum can be applied along two perpendicular directions, that is, along an x axis and a y axis.

ADDITIONAL READING

A. Einstein and L. Infeld, *Evolution of Physics,* New York, Simon and Schuster, 1938.

H. W. Lewis, "Ballistocardiography," *Sci. American,* February 1958, p. 89.

F. Ordway, "Principles of Rocket Engines," *Sky and Telescope, 14,* 1954, p. 48. An introduction to the principles of rocket propulsion.

CONCEPTUAL QUESTIONS

Example   A boxer wisely moves his head backward just before receiving a punch. How does this maneuver help reduce the force of impact?

Reasoning   As the boxer moves away from the moving fist, the time his head is in contact with the fist is increased, so his body has more time to absorb the momentum of his opponent's fist. Since the impulse of the impact force is the average force multiplied by the time of impact, the impact force decreases as the impact time increases.

Example   A karate expert is able to break a stack of boards with a swift blow with the side of his bare hand. How is this possible?

Reasoning   The arm and hand have a large momentum before the collision with the boards. This momentum is reduced quickly as the hand is in contact with the boards for a short time. Since the contact time is very small, the impact force (the force of the hand on the boards) is very large. Thus, a karate expert obtains the best result by delivering the blow in a short time.

Example   A piece of clay is thrown against a brick wall and sticks to the wall. What happens to the momentum of the clay? Is momentum conserved? Explain.

Reasoning   Initially the clay has momentum directed toward the wall. When the clay collides and sticks to the wall, nothing appears to have any momentum, and you may (wrongfully) conclude that momentum is not conserved. However, the "lost" momentum is actually imparted to the wall and Earth, causing both to move. Because of the enormous mass of the Earth, its recoil motion is too small to detect.

Example   As a ball falls toward the Earth, its momentum increases. How would you reconcile this fact with the law of conservation of momentum?

Reasoning   As the momentum of the ball increases in the downward direction, some other object must be gaining momentum in the opposite direction because of momentum conservation. The second object is the Earth.

Again, because the Earth's mass is so large, its motion toward the ball is negligibly small.

Example   If the speed of a particle is doubled, by what factor is its momentum changed? What happens to it kinetic energy?

Reasoning   Since momentum is defined as the product mv, doubling the speed v will double the momentum. On the other hand, since kinetic energy is defined as $\frac{1}{2}mv^2$, doubling v will quadruple the kinetic energy.

Example   An open box slides across the icy (frictionless) surface of a frozen lake. What happens to the speed of the box as water from a rain shower collects in it, assuming that the rain falls vertically downward into the box? Explain.

Reasoning   There are no external horizontal forces acting on the box, so its momentum cannot change as it moves along the horizontal surface. As the box slowly fills with water, its mass increases with time. Since the product mv must be a constant, and m is increasing, the speed of the box must decrease.

Example   A boy stands at one end of a floating raft that is stationary relative to the shore. He then walks to the opposite end of the raft, away from the shore. Does the raft move? Explain.

Reasoning   Yes, the raft moves toward the shore. Neglecting friction between the raft and water, there are no horizontal forces acting on the system consisting of the boy and raft. Therefore, the center of mass of the system remains fixed relative to the shore (or any stationary point). As the boy moves away from the shore, the boat must move toward the shore such that the center of mass of the system remains constant. An alternative explanation is that the momentum of the system remains constant if friction is neglected. As the boy acquires a momentum away from the shore, the boat must acquire an equal momentum toward the shore such that the total momentum of the system is always zero.

1. If a particle's kinetic energy is zero, what is its momentum? If a particle's total energy is zero, is its momentum necessarily zero? Explain.

2. Consider a field of insects, all of essentially equal mass. If the total momentum of the insects is zero, what does this imply about their motion? If the total kinetic energy of the insects is zero, what does this imply?

3. Does a large force always produce a larger impulse on a body than a smaller force does? Explain.

4. If two objects collide and one is initially at rest, is it possible for both to be at rest after the collision? Is it possible for one to be at rest after the collision? Explain.

5. Is it possible for a collision to occur in which all of the kinetic energy is lost? If so, cite an example.

6. If the forward momentum of a bullet is the same as the backward momentum of the gun that shot it, why isn't it as dangerous to be hit by the gun as by the bullet?

7. A skater is standing still on a frictionless ice rink. Her friend throws a Frisbee straight at her. In which of the following cases is the largest momentum transferred to the skater? (a) The skater catches the Frisbee and holds onto it. (b) The skater catches the Frisbee momentarily, but then drops it vertically downward. (c) The skater catches the Frisbee, holds it momentarily, and throws it back to her friend.

8. If two particles have equal kinetic energies, are their momenta necessarily equal? Explain.

9. In a perfectly elastic collision between two objects, do both objects have the same kinetic energy after the collision? Explain.

10. If two automobiles collide, they usually do not stick together. Does this mean the collision is perfectly elastic? Explain why a head-on collision is likely to be more dangerous than other types of collisions.

11. Early in this century, Robert Goddard proposed sending a rocket to the Moon. Critics took the position that in a vacuum such as exists between the Earth and the Moon, the gases emitted by the rocket would have nothing to push against to propel the rocket. According to *Scientific American* (January 1975), Goddard placed a gun in a vacuum and fired a blank cartridge from it. (A blank cartridge fires only the hot gases of the burning gunpowder.) What happened when the gun was fired?

12. An astronaut walking in space accidentally severs the safety cord attaching him to the spacecraft. If he happens to have a can of spray deodorant, how can he use it to return safely to his ship?

13. Explain why momentum is conserved when a ball bounces from a floor.

14. Consider a perfectly inelastic collision between a car and a large truck. Which vehicle loses more kinetic energy as a result of the collision?

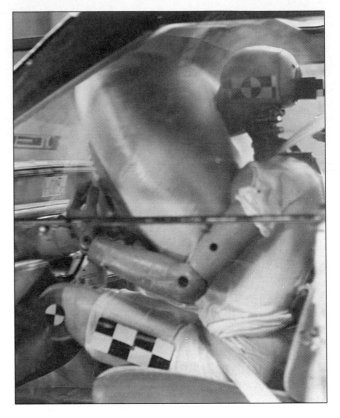

(Question 21)   *(Courtesy of General Motors)*

15. Gymnasts always perform on padded mats. Use the impulse-momentum theorem to discuss how these mats protect the athletes.

16. An open bed sheet is held loosely at its sides by two students to form a "catching net." The instructor asks a third student to throw a raw egg into the middle of the sheet as hard as possible. Why doesn't the egg break?

17. How do car bumpers that collapse on impact help to protect the driver?

18. A pole-vaulter falls from a height of 5 m onto a foam-rubber pad. Could you calculate her velocity just before she reaches the pad? Would you be able to calculate the force exerted on the pole-vaulter due to the collision? Explain.

19. A toy gun shoots rubber bullets at a target. Compare the impulse delivered to the target when the bullets embed in it with the case in which they strike the target and bounce off.

20. A magician places some dishes and silverware on a table covered by a tablecloth. He then rapidly removes the tablecloth without disturbing the dishes

and utensils. On the basis of what you have learned in this chapter, explain how this trick is possible.

21. When a collision occurs, an air bag is inflated, which protects the passenger (the dummy, in this case) from serious injury. Why does the air bag soften the blow? Discuss the physics involved in this dramatic picture.

PROBLEMS

Section 6.1   Momentum and Impulse

1. Calculate the magnitude of the linear momentum for the following cases: (a) a proton with mass 1.67×10^{-27} kg, moving with a speed of 5.00×10^{6} m/s; (b) a 15.0-g bullet moving with a speed of 300 m/s; (c) a 75.0-kg sprinter running with a speed of 10.0 m/s; (d) the Earth (mass = 5.98×10^{24} kg) moving with an orbital speed equal to 2.98×10^{4} m/s.

2. An object is moving so that its kinetic energy is 150 J and the magnitude of its momentum is 30.0 kg·m/s. Determine the mass and speed of the object.

3. A pitcher claims he can throw a 0.145-kg baseball with as much momentum as a 3.00-g bullet moving with a speed of 1.50×10^{3} m/s. (a) What must the baseball's speed be if the pitcher's claim is valid? (b) Which has greater kinetic energy, the ball or the bullet?

4. A 0.10-kg ball is thrown straight up into the air with an initial speed of 15 m/s. Find the momentum of the ball (a) at its maximum height and (b) halfway to its maximum height.

5. A garden hose is held parallel to the ground as in Figure 6.17. What force is necessary to hold the nozzle stationary if the discharge rate of the water is 0.600 kg/s with a speed of 25.0 m/s?

FIGURE 6.17   (Problem 5)

6. A 1500-kg car moving with a speed of 15 m/s collides with a utility pole and is brought to rest in 0.30 s. Find the average force exerted on the car during the collision.

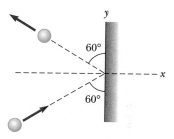

FIGURE 6.18   (Problem 7)

7. A 3.0-kg steel ball strikes a massive wall at 10 m/s at an angle of 60° with the plane of the wall. It bounces off with the same speed and angle (Fig. 6.18). If the ball is in contact with the wall for 0.20 s, what is the average force exerted on the ball by the wall?

8. A 0.50-kg football is thrown with a speed of 15 m/s. A stationary receiver catches the ball and brings it to rest in 0.020 s. (a) What is the impulse delivered to the ball? (b) What is the average force exerted on the receiver?

9. An estimated force-time curve for a baseball struck by a bat is shown in Figure 6.19. From this curve, determine (a) the impulse delivered to the ball and (b) the average force exerted on the ball.

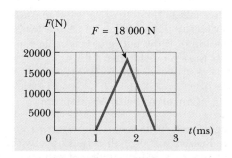

FIGURE 6.19   (Problem 9)

□ indicates problems that have full solutions available in the Student Solutions Manual and Study Guide.

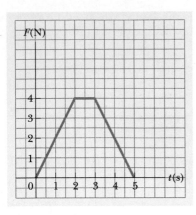

FIGURE 6.20 (Problems 10 and 11)

FIGURE 6.21 (Problem 17)

10. The force, $\mathbf{F_x}$, acting on a 2.00-kg particle varies in time as shown in Figure 6.20. Find (a) the impulse of the force, (b) the final velocity of the particle if it is initially at rest, and (c) the final velocity of the particle if it is initially moving along the x axis with a velocity of -2.00 m/s.

11. Find the average force exerted on the particle graphed in Figure 16.20 for the time interval $t_i = 0$ to $t_f = 5.0$ s.

12. A 0.15-kg baseball is thrown with a speed of 20 m/s. It is hit straight back at the pitcher with a final speed of 22 m/s. (a) What is the impulse delivered to the ball? (b) Find the average force exerted by the bat on the ball if the two are in contact for 2.0×10^{-3} s.

13. A 0.50-kg object is at rest at the origin of a coordinate system. A 3.0-N force in the $+x$ direction acts on the object for 1.50 s. (a) What is the velocity at the end of this interval? (b) At the end of this interval, a constant force of 4.0 N is applied in the $-x$ direction for 3.0 s. What is the velocity at the end of the 3.0 s?

14. Water falls at the rate of 250 g/s from a height of 60 m into a 750-g bucket on a scale (without splashing). If the bucket is originally empty, what does the scale read after 3.0 s?

Section 6.2 Conservation of Momentum

15. High-speed stroboscopic photographs show that the head of a 200-g golf club is traveling at 55 m/s just before it strikes a 46-g golf ball at rest on a tee. After the collision, the club head travels (in

the same direction) at 40 m/s. Find the speed of the golf ball just after impact.

16. A rifle with a weight of 30 N fires a 5.0-g bullet with a speed of 300 m/s. (a) Find the recoil speed of the rifle. (b) If a 700-N man holds the rifle firmly against his shoulder, find the recoil speed of man and rifle.

17. A 60.0-kg astronaut is on a space walk away from the shuttle when her tether line breaks! She is able to throw her 10.0-kg oxygen tank away from the shuttle with a velocity of 12.0 m/s to propel herself back to the shuttle (Fig. 6.21). Assuming that she starts from rest (relative to the shuttle), determine the maximum distance she can be from the craft when the line breaks and still return within 60.0 s (the amount of time she can hold her breath).

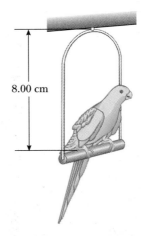

FIGURE 6.22 (Problem 18)

18. The bird perched on the swing in Figure 6.22 has a mass of 52.0 g, and the base of the swing has a mass of 153 g. Assume that the swing and bird are originally at rest and that the bird then takes off

horizontally at 2.00 m/s. If the base can swing freely (i.e., without friction) around the pivot, how high will the base of the swing rise above its original level?

19. A 45.0-kg girl is standing on a 150-kg plank. The plank, originally at rest, is free to slide on a frozen lake, which is a flat, frictionless surface. The girl begins to walk along the plank at a constant velocity of 1.50 m/s relative to the plank. (a) What is her velocity relative to the ice surface? (b) What is the velocity of the plank relative to the ice surface?

20. A 730-N man stands in the middle of a frozen pond of radius 5.0 m. He is unable to get to the other side because of a lack of friction between his shoes and the ice. To overcome this difficulty, he throws his 1.2-kg physics textbook horizontally toward the north shore, at a speed of 5.0 m/s. How long does it take him to reach the south shore?

21. A 7.00-kg bowling ball is dropped from rest at an initial height of 3.00 m. (a) What is the speed of the Earth coming up to meet the ball just before the ball hits the ground? Use 5.98×10^{24} kg as the mass of the Earth. (b) Use your answer to part (a) to justify ignoring the motion of the Earth when dealing with the motions of terrestrial objects.

22. A 65.0-kg person throws a 0.0450-kg snowball forward with a ground speed of 30.0 m/s. A second person, with a mass of 60.0 kg, catches the snowball. Both people are on skates. The first person is initially moving forward with a speed of 2.50 m/s, and the second person is initially at rest. What are the velocities of the two people after the snowball is exchanged? Disregard the friction between the skates and the ice.

Section 6.3 Collisions

Section 6.4 Glancing Collisions

23. A 2000-kg car traveling at 10.0 m/s collides with a 3000-kg car that is initially at rest at a stoplight. The cars stick together and move 2.00 m before friction causes them to stop. Determine the coefficient of kinetic friction between the cars and the road, assuming that the negative acceleration is constant and all wheels on both cars lock at the time of impact.

24. (a) Three carts of masses 4.0 kg, 10 kg, and 3.0 kg move on a frictionless horizontal track with speeds of 5.0 m/s, 3.0 m/s, and 4.0 m/s, as shown in Figure 6.23. The carts stick together after colliding. Find the final velocity of the three carts. (b) Does your answer require that all carts collide and stick together at the same time?

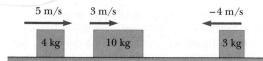

FIGURE 6.23 (Problem 24)

 25. A 1.20-kg skateboard is coasting along the pavement at a speed of 5.00 m/s when a 0.800-kg cat drops from a tree vertically downward onto the skateboard. What is the speed of the skateboard-cat combination?

 26. A railroad car of mass 2.00×10^4 kg moving at 3.00 m/s collides and couples with two coupled railroad cars, each of the same mass as the single car and moving in the same direction at 1.20 m/s. (a) What is the speed of the three coupled cars after the collision? (b) How much kinetic energy is lost in the collision?

27. A 3.00-kg sphere makes a perfectly inelastic collision with a second sphere that is initially at rest. The composite system moves with a speed equal to one-third the original speed of the 3.00-kg sphere. What is the mass of the second sphere?

 28. A 5.0-g object moving to the right at 20 cm/s makes an elastic head-on collision with a 10-g object that is initially at rest. Find (a) the velocity of each object after the collision and (b) the fraction of the energy transferred to the 10-g object.

29. A 10.0-g object moving to the right at 20.0 cm/s makes an elastic head-on collision with a 15.0-g object moving in the opposite direction at 30.0 cm/s. Find the velocity of each object after the collision.

 30. A 25.0-g object moving to the right at 20.0 cm/s overtakes and collides elastically with a 10.0-g object moving in the same direction at 15.0 cm/s. Find the velocity of each object after the collision.

31. An alpha particle of mass 4.00 u, moving to the right at 1.00×10^6 m/s, collides with a proton of mass 1.00 u that is at rest before the collision. Find (a) the speed of each particle after the collision, assuming a perfectly elastic collision, and (b) the kinetic energy of each particle before and after the collision (1 u = 1 atomic mass unit = 1.67×10^{-27} kg).

32. An 8.0-g bullet is fired into a 2.5-kg ballistic pendulum and becomes embedded in it. If the pendulum rises a vertical distance of 6.0 cm, calculate the initial speed of the bullet.

33. An 8.00-kg mass moving east at 15.0 m/s on a frictionless horizontal surface collides with a 10.0-kg mass that is initially at rest. After the collision, the 8.00-kg mass moves south at 4.00 m/s. (a) What is the velocity of the 10.0-kg mass after the collision? (b) What percentage of the initial kinetic energy is lost in the collision?

34. A 90-kg fullback moving east with a speed of 5.0 m/s is tackled by a 95-kg opponent running north at 3.0 m/s. If the collision is perfectly inelastic, calculate (a) the velocity of the players just after the tackle and (b) the kinetic energy lost as a result of the collision. Can you account for the missing energy?

35. Two blocks of masses $m_1 = 2.00$ kg and $m_2 = 4.00$ kg are released from a height of 5.00 m on a frictionless track, as shown in Figure 6.24, and undergo an elastic head-on collision. (a) Determine the velocity of each block just before the collision. (b) Determine the velocity of each block immediately after the collision. (c) Determine the maximum heights to which m_1 and m_2 rise after the first collision. (d) What happens to the masses

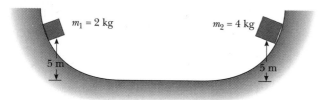

FIGURE 6.24   (Problem 35)

after a second collision, and what is the ultimate behavior of this motion? (e) If, after the first collision, the masses stuck together, how high would the combined system rise on the track?

36. The mass of the blue puck in Figure 6.25 is 20%

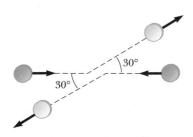

FIGURE 6.25   (Problem 36)

greater than the mass of the green one. Before colliding, the pucks approach each other with equal and opposite momentum, and the green puck has an initial speed of 10.0 m/s. Find the speeds of the pucks after the collision if half the kinetic energy is lost during the collision.

37. Two shuffleboard disks of equal mass, one orange and the other yellow, are involved in a perfectly elastic glancing collision. The yellow disk is initially at rest and is struck by the orange disk moving initially to the right at 5.00 m/s. After the collision, the orange disk moves in a direction that makes an angle of 37.0° with its initial direction, and the velocity of the yellow disk is perpendicular to that of the orange disk (after the collision). Determine the speed of each disk after the collision.

38. A neutron in a reactor makes an elastic head-on collision with a carbon atom that is initially at rest. (The mass of the carbon nucleus is about 12 times that of the neutron.) (a) What fraction of the neutron's kinetic energy is transferred to the carbon nucleus? (b) If the neutron's initial kinetic energy is 1.6×10^{-13} J, find its final kinetic energy and the kinetic energy of the carbon nucleus after the collision.

39. Consider the ballistic pendulum described in Example 6.7 and shown in Figure 6.12. If the mass of the bullet is 8.00 g and the mass of the pendulum is 2.00 kg, find the ratio of the kinetic energy after the collision to the kinetic energy before the collision. What accounts for the missing energy?

40. A cue ball traveling at 4.0 m/s makes a glancing, elastic collision with a target ball of equal mass that is initially at rest. The cue ball is deflected so that it makes an angle of 30° with its original direction of travel. Find (a) the angle between the velocity vectors of the two balls after the collision and (b) the speed of each ball after the collision.

ADDITIONAL PROBLEMS

41. A 0.40-kg soccer ball approaches a player horizontally with a speed of 15 m/s. The player illegally strikes the ball with her hand and causes it to move in the opposite direction with a speed of 22 m/s. What impulse was delivered to the ball by the player?

42. A student observes a completely elastic collision between two objects on a frictionless horizontal table and measures the following velocity components:

	Before the Collision		After the Collision	
	v_x (m/s)	v_y (m/s)	v_x (m/s)	v_y (m/s)
Object 1	1.0	2.0	3.0	0.0
Object 2	3.0	−1.0	1.0	1.0

The student neglects to record the mass of each object. The instructor, however, can see that there is a problem with these measurements. What is it?

43. A block of mass m_1 moves *east* on a tabletop, with a speed of v_0 toward a second block of mass m_2, which is at rest. After the collision, the first block is observed to move *south* with a speed v. (a) Show that, in general,

$$v \leqq \sqrt{\frac{m_2 - m_1}{m_2 + m_1}} \, v_0$$

(*Hint:* You may assume that $KE_{after} \leqq KE_{before}$.)
(b) What does this tell you about m_1 and m_2?

44. A 2.0-g particle moving at 8.0 m/s makes a perfectly elastic head-on collision with a resting 1.0-g object. (a) Find the speed of each after the collision. (b) If the stationary particle has a mass of 10 g, find the speed of each particle after the collision. (c) Find the final kinetic energy of the incident 2.0-g particle in the situations described in (a) and (b). In which case does the incident particle lose more kinetic energy?

45. An 80-kg man drops from a 3.0-m diving board. Two (2.0) seconds after reaching the water, the man comes to rest. What average force did the water exert on him?

46. A 79.5-kg man holding a 0.500-kg ball stands on a frozen pond next to a wall. He throws the ball at the wall with a speed of 10.0 m/s (relative to the ground), and then catches the ball after it rebounds from the wall. (a) How fast is he moving after he catches the ball? (Ignore the projectile motion of the ball, and assume that the ball loses no energy in its collision with the wall.) (b) How many times does the man have to go through this process before his speed reaches 1.00 m/s relative to the ground?

47. A billiard ball rolling across a table at 1.50 m/s makes a head-on elastic collision with an identical ball. Find the speed of each ball after the collision (a) when the second ball is initially at rest, (b) when the second ball is moving toward the first at a speed of 1.00 m/s, and (c) when the second ball is moving away from the first at a speed of 1.00 m/s.

48. A 0.03-kg bullet is fired vertically at 200 m/s into a 0.15-kg baseball that is initially at rest. How high does the combination rise after the collision, assuming the bullet embeds in the ball?

49. An 80.0-kg astronaut is working on the engines of his ship, which is drifting through space with a constant velocity. The astronaut, wishing to get a better view of the Universe, pushes against the ship and later finds himself 30.0 m behind the ship and moving so slowly that he can be considered to be at rest. Without a thruster, the only way to return to the ship is to throw his 0.500-kg wrench directly away from the ship. If he throws the wrench with a speed of 20.0 m/s, how long does it take him to reach the ship?

50. An unstable nucleus of mass 17×10^{-27} kg, initially at rest, disintegrates into three particles. One of the particles, of mass 5.0×10^{-27} kg, moves along the positive y axis with a speed of 6.0×10^6 m/s. Another particle, of mass 8.4×10^{-27} kg, moves along the positive x axis with a speed of 4.0×10^6 m/s. Determine the third particle's speed and direction of motion. For now you may assume that mass is also conserved in the disintegration process.

51. A 2000-kg car moving east at 10.0 m/s collides with a 3000-kg car moving north. The cars stick together and move as a unit after the collision, at an angle of 40.0° north of east and at a speed of 5.22 m/s. Find the velocity of the 3000-kg car before the collision.

52. A 0.400-kg bead slides on a curved frictionless wire, starting from rest at point A in Figure 6.26. At point B the bead collides elastically with a 0.600-kg ball at rest. Find the distance the ball moves up the wire.

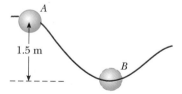

FIGURE 6.26   (Problem 52)

53. Tarzan, whose mass is 80.0 kg, swings from a 3.00-m vine that is horizontal when he starts. At the bottom of his arc, he picks up 60.0-kg Jane in an inelastic collision. What maximum height do they reach, after their upward swing?

54. (a) A 12-g bullet is fired horizontally into a 100-g wooden block that is initially at rest on a friction-

less horizontal surface and connected to a spring of constant 150 N/m. If the bullet-block system compresses the spring by a maximum of 0.80 m, what was the velocity of the bullet at impact with the block? (b) Solve part (a) if the coefficient of kinetic friction between the block and the surface is 0.60.

55. The force shown in the force-time diagram in Figure 6.27 acts on a 1.5-kg mass. Find (a) the impulse of the force, (b) the final velocity of the mass if it is initially at rest, and (c) the final velocity of the mass if it is initially moving along the x axis with a velocity of -2.0 m/s.

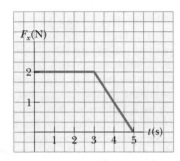

FIGURE 6.27   (Problem 55)

56. Two blocks of masses M and m approach each other on a horizontal table with the same constant speed, v_0, as measured by a laboratory observer. The blocks undergo a perfectly elastic collision, and it is observed that M stops while m moves opposite its original motion with some constant speed, v. (a) Determine the ratio of the two masses, M/m. (b) What is the ratio of their speeds, v/v_0?

57. A 0.30-kg puck, initially at rest on a frictionless horizontal surface, is struck by a 0.20-kg puck that is initially moving along the x axis with a velocity of 2.0 m/s. After the collision, the 0.20-kg puck has a speed of 1.0 m/s at an angle of $\theta = 53°$ to the positive x axis. (a) Determine the velocity of the 0.30-kg puck after the collision. (b) Find the fraction of kinetic energy lost in the collision.

58. The forces shown in the force-time diagram in Figure 6.28 act on a 1.5-kg mass. Find (a) the impulse for the interval $t = 0$ to $t = 3.0$ s and (b) the impulse for the interval $t = 0$ to $t = 5.0$ s. (c) If the forces act on a 1.5-kg particle that is initially at rest, find the particle's speed at $t = 3.0$ s and at $t = 5.0$ s.

59. A 0.11-kg tin can is resting on top of a 1.7-m-high fencepost. A 0.0020-kg bullet is fired horizontally at the can. It strikes the can with a speed of

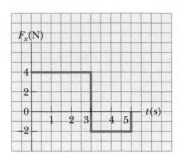

FIGURE 6.28   (Problem 58)

900 m/s, passes through it, and emerges with a speed of 720 m/s. When the can hits the ground, how far is it from the fencepost? Disregard friction while the can is in contact with the post.

60. An 8.00-g bullet is fired into a 250 g block that is initially at rest at the edge of a 1-m-high table (Fig. 6.29). The bullet remains in the block, and after the impact the block lands 2.00 m from the bottom of the table. Determine the initial speed of the bullet.

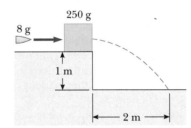

FIGURE 6.29   (Problem 60)

61. A 7.0-g bullet is fired into a 1.5-kg ballistic pendulum. The bullet emerges from the block with a speed of 200 m/s, and the block rises to a maximum height of 12 cm. Find the initial speed of the bullet.

62. An unstable nucleus of mass 1.7×10^{-26} kg, initially at rest at the origin of a coordinate system, disintegrates into three particles. One particle, having a mass of $m_1 = 5.0 \times 10^{-27}$ kg, moves along the y axis with a speed $v_1 = 6.0 \times 10^6$ m/s. Another particle, of mass $m_2 = 8.4 \times 10^{-27}$ kg, moves along the x axis with a speed $v_2 = 4.0 \times 10^6$ m/s. Find the magnitude and direction of the velocity of the third particle.

63. A cannon of mass $m_1 = 800$ kg (when unloaded) is loaded with a "shot" of mass $m_2 = 10.0$ kg. The cannon is aimed at mass $m_3 = 7990$ kg, which is connected to a massless spring of spring constant $k = 4500$ N/m, as in Figure 6.30a. The cannon is

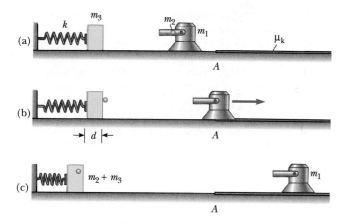

FIGURE 6.30 (Problem 63)

then fired, and the shot inelastically collides with mass m_3 and sticks in it, as shown in Figure 6.30b. The combined system then compresses the spring a maximum distance of $d = 0.5$ m, as in Figure 6.30c. (a) Determine the speed of m_2 just before it collides with m_3. (You may assume that m_2 travels in a straight line.) (b) Determine the recoil speed of the cannon. (c) The cannon recoils towards the right, and when it passes point A there is friction (with $\mu_k = 0.600$) between the cannon and the ground. How far to the right of point A does the cannon slide before coming to rest?

64. A 0.500-kg block is released from rest at the top of a frictionless track 2.50 m above the top of a table. It then collides elastically with a 1.00-kg mass that is initially at rest on the table, as shown in Figure 6.31. (a) Determine the speeds of the two masses just after the collision. (b) How high up the track does the 0.500-kg mass travel back after the collision? (c) How far away from the bottom of the table does the 1.00-kg mass land, given that the table is 2.00 m high? (d) How far away from the bottom of the table does the 0.500-kg mass eventually land?

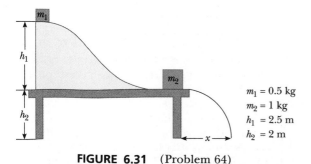

$m_1 = 0.5$ kg
$m_2 = 1$ kg
$h_1 = 2.5$ m
$h_2 = 2$ m

FIGURE 6.31 (Problem 64)

65. A student performs a ballistic pendulum experiment, using an apparatus similar to that shown in the multiflash photo (Fig. 6.32a). She obtains the following data (averaged over a series of five measurements): $h = 8.68$ cm, $m_1 = 68.8$ g, $m_2 = 263$ g. (a) Determine the initial speed v_{1i}, of the projectile. (b) The second part of the student's experiment is to obtain v_{1i} by firing the same projectile horizontally (with the pendulum removed from the path of the projectile) and measuring its horizontal displacement, X, and vertical displacement, Y, before it strikes the floor (Fig. 6.32b). Show that the initial speed of the projectile is related to X and Y through the relation

$$v_{1i} = \frac{X}{\sqrt{2Y/g}}$$

What numerical value does the student obtain for v_{1i} based on her measured values of $X = 257$ cm and $Y = 85.3$ cm? What factors might account for the difference in this value compared to that obtained in part (a)?

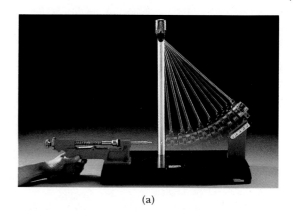

(a)

(b)

FIGURE 6.32 (Problem 65). *(Photo courtesy of Central Scientific Company)*

66. A block of mass M lying on a rough horizontal surface is given an initial velocity of $\mathbf{v}_0$. After traveling a distance d, it makes a head-on elastic collision with a block of mass $2M$. How far does the second block move before coming to rest? (Assume the coefficient of friction is the same for both blocks.)

67. Two carts of equal mass, $m = 0.250$ kg, are placed on a frictionless track that has a light spring of force constant $k = 50.0$ N/m attached to one end of it, as in Figure 6.33. The blue cart is given an initial velocity of $\mathbf{v}_0 = 3.00$ m/s to the right, and the red cart is initially at rest. If the carts collide elastically, find (a) the velocity of the carts just after the first collision and (b) the maximum compression in the spring.

68. A small block of mass $m_1 = 0.500$ kg is released from rest at the top of a curved wedge of mass $m_2 = 3.00$ kg, which sits on a frictionless horizontal surface as in Figure 6.34a. When the block leaves the wedge, its velocity is measured to be 4.00 m/s to the right, as in Figure 6.34b. (a) What is the velocity of the wedge after the block reaches the horizontal surface? (b) What is the height, h, of the wedge?

FIGURE 6.33   (Problem 67)

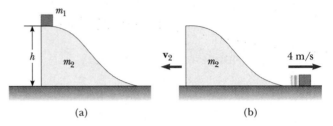

(a)   (b)

FIGURE 6.34   (Problem 68)

CIRCULAR MOTION AND THE LAW OF GRAVITY

7

This long exposure photograph shows the Delicate Arch located at the Arches National Park in Utah. The arch is silhouetted against a background of circular star trails produced as the Earth rotates about its axis and orbits the Sun. The brightest track inside the arch represents the star Polaris. (David Nunuk/SPL/Photo Researchers)

In this chapter we shall investigate circular motion, a specific type of two-dimensional motion. We shall encounter such terms as *angular speed, angular acceleration, centripetal acceleration,* and *centripetal force*. The results derived here will help you understand the motions of a diverse range of objects in our environment, from a car moving around a circular race track to clusters of galaxies orbiting a common center.

We shall also introduce Newton's universal law of gravitation, one of the fundamental laws in nature, and show how this law, together with Newton's laws of motion, enables us to understand a variety of familiar phenomena.

Circular motion and the universal law of gravitation are related historically in that Newton discovered the law of gravity as a result of attempting to explain the circular motion of the Moon about the Earth and the motions of the planets about the Sun. Thus, it is appropriate to consider these two important physical topics together. A more general treatment of gravitational potential energy, the concept of escape velocity, and a discussion of black holes are provided in a section at the back of the textbook entitled "For Further Study in Chapter 7."

7.1

ANGULAR SPEED AND ANGULAR ACCELERATION

We began our study of linear motion by defining the terms *displacement, velocity,* and *acceleration*. We will take the same basic approach now as we turn to a study of rotational motion. Consider Figure 7.1a, a top view of a phonograph record

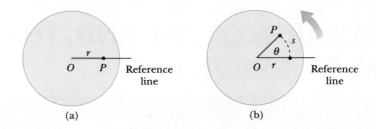

FIGURE 7.1
(a) The point *P* on a rotating record at *t* = 0. (b) As the record rotates, the point *P* moves through an arc length of *s*.

rotating on a turntable. The axis of rotation is at the center of the record, at *O*. A point *P* on the record is at the distance *r* from the origin and moves about *O* in a circle of radius *r*. In fact, every point on the record undergoes circular motion about *O*. To analyze such motion, it is convenient to set up a *fixed* reference line, as shown in Figure 7.1a. Let us assume that at time *t* = 0, the point *P* is on the reference line as in Figure 7.1a and a line is drawn on the record from the origin out to *P*. After an interval of Δt has elapsed, *P* has advanced to a new position (Fig. 7.1b). In this time interval, the line *OP* has moved through the angle θ with respect to the reference line. Likewise, *P* has moved a distance of *s* measured along the circumference of the circle; *s* is called an *arc length*.

In situations we have encountered thus far, angles have been measured in degrees. However, in scientific work angles are often measured in *radians* (rad) rather than degrees, for the effect of simplifying certain equations. In fact, almost all of the equations derived in this chapter and the next require that angles be measured in radians. With reference to Figure 7.1b, when the arc length *s* is equal to the radius *r*, the angle θ swept out by *r* is equal to one radian. In general, any angle θ, measured in radians, is defined by the relation

$$\theta \equiv \frac{s}{r} \tag{7.1}$$

The radian is a pure number with no dimensions. This can be seen from Equation 7.1, since θ is the ratio of an arc length (a distance) to the radius of the circle (also a distance).

To convert degrees to radians, note that when point *P* in Figure 7.1 moves through an angle of 360° (one revolution), the arc length *s* is equal to the circumference of the circle, $2\pi r$. From Equation 7.1 we see that the corresponding angle in radians is $2\pi r/r = 2\pi$ rad. Hence,

$$1 \text{ rad} \equiv \frac{360°}{2\pi} \approx 57.3°$$

From this definition, it follows that any angle in degrees can be converted to an angle in radians with the expression

$$\theta(\text{rad}) = \frac{\pi}{180°} \, \theta(\text{deg})$$

For example, 60° equals $\pi/3$ rad and 45° equals $\pi/4$ rad.

Returning to the phonograph record, we see from Figure 7.2 that, as the record rotates and a point on it moves from *P* to *Q* in a time of Δt, the angle through which the record rotates is $\Delta\theta = \theta_2 - \theta_1$. We define $\Delta\theta$ as the **angular displacement**. The *average angular speed*, $\overline{\omega}$ (ω is the Greek letter omega), of a

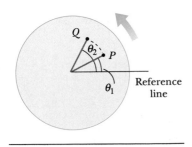

FIGURE 7.2
As a point on the record moves from *P* to *Q*, the record rotates through the angle $\Delta\theta = \theta_2 - \theta_1$.

rotating rigid object is the ratio of the angular displacement, $\Delta\theta$, to the time interval Δt:

$$\bar{\omega} \equiv \frac{\theta_2 - \theta_1}{t_2 - t_1} = \frac{\Delta\theta}{\Delta t}$$ [7.2]   Average angular speed

By analogy with linear speed, the **instantaneous angular speed,** ω, is defined as the limit of the average speed, $\Delta\theta/\Delta t$, as the time interval Δt approaches zero:

$$\omega \equiv \lim_{\Delta t \to 0} \frac{\Delta\theta}{\Delta t}$$ [7.3]   Instantaneous angular speed

Angular speed has the units radians per second (rad/s). We shall take ω to be positive when θ is increasing (counterclockwise motion) and negative when θ is decreasing (clockwise motion). When the angular speed is constant, the instantaneous angular speed is equal to the average angular speed.

EXAMPLE 7.1   Whirlybirds

The rotor on a helicopter turns at an angular speed of 320 revolutions per minute (in this book, we shall sometimes use the abbreviation rpm, but in most cases we shall use rev/min). Express this in radians per second.

Solution   We shall use the conversion factors 1 rev = 2π rad and 60 s = 1 min, to give

$$320 \frac{\text{rev}}{\text{min}} = 320 \frac{\cancel{\text{rev}}}{\cancel{\text{min}}} \left(\frac{2\pi \text{ rad}}{\cancel{\text{rev}}} \right) \left(\frac{1 \cancel{\text{min}}}{60 \text{ s}} \right) = 10.7\pi \frac{\text{rad}}{\text{s}}$$

Figure 7.3 shows a bicycle turned upside down so that a repair person can work on the rear wheel. The bicycle pedals are turned so that at time t_1 the wheel

FIGURE 7.3
An accelerating bicycle wheel rotates with (a) angular speed ω_1 at time t_1 and (b) angular speed ω_2 at time t_2.

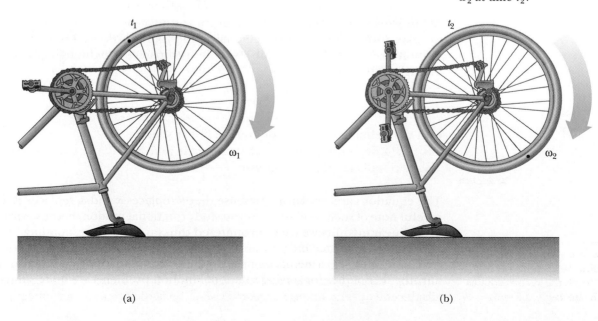

(a)   (b)

has angular speed ω_1 (Fig. 7.3a), and at a later time, t_2, it has angular speed ω_2 (Fig. 7.3b).

> The **average angular acceleration,** $\overline{\alpha}$ (α is the Greek letter alpha), of an object is defined as the ratio of the change in the angular speed to the time, Δt, it takes the object to undergo the change:

Average angular acceleration

$$\overline{\alpha} \equiv \frac{\text{change in angular speed}}{\text{time interval}} = \frac{\omega_2 - \omega_1}{t_2 - t_1} = \frac{\Delta \omega}{\Delta t} \qquad \text{[7.4]}$$

Instantaneous angular acceleration

The **instantaneous angular acceleration** is defined as the limit of the ratio $\Delta \omega / \Delta t$ as Δt approaches zero. Angular acceleration has the units radians per second per second (rad/s^2). Note that

> When a rigid object rotates about a fixed axis, as does the bicycle wheel, every portion of the object has the same angular speed and the same angular acceleration.

This, in fact, is precisely what makes these variables so useful for describing rotational motion.

The following argument should convince you that ω and α are the same for every point on the wheel. If a point on the rim of the wheel had a greater angular speed than a point nearer the center, the shape of the wheel would be changing. The wheel remains circular (symmetrically distributed about the axle) only if all points have the same angular speed and the same angular acceleration.

7.2

ROTATIONAL MOTION UNDER CONSTANT ANGULAR ACCELERATION

Let us pause for a moment to consider some similarities between the equations we have found thus far for rotational motion and those we found for linear motion in earlier chapters. For example, compare the defining equation for average angular speed,

$$\overline{\omega} \equiv \frac{\theta_f - \theta_i}{t_f - t_i} = \frac{\Delta \theta}{\Delta t}$$

with the defining equation for average linear speed,

$$\overline{v} \equiv \frac{x_f - x_i}{t_f - t_i} = \frac{\Delta x}{\Delta t}$$

The equations are similar in the sense that θ replaces x and ω replaces v. Take careful note of such similarities as you study rotational motion, because virtually every linear quantity we have encountered thus far has a corresponding "twin" in rotational motion. Once you are adept at recognizing such analogies, you will find it unnecessary to memorize many of the equations in this chapter. Additionally, the techniques for solving rotational motion problems are quite similar to those you have already learned for linear motion. For example, problems concerned with objects that rotate with constant angular acceleration can be solved in much the same manner as those dealing with linear motion under constant acceleration. If you understand problems involving objects that move with constant linear acceleration, these rotational motion problems should be little more than a review for you.

An additional analogy between linear motion and rotational motion is revealed when we compare the defining equation for average angular acceleration,

$$\bar{\alpha} \equiv \frac{\omega_f - \omega_i}{t_f - t_i} = \frac{\Delta\omega}{\Delta t}$$

with the defining equation for average linear acceleration,

$$\bar{a} \equiv \frac{v_f - v_i}{t_f - t_i} = \frac{\Delta v}{\Delta t}$$

In light of the analogies between variables in linear motion and those in rotational motion, it should not surprise you that the equations of rotational motion involve the variables θ, ω, and α. In Chapter 2, Section 2.6, we developed a set of kinematic equations for linear motion under constant acceleration. The same procedure can be used to derive a similar set of equations for rotational motion under constant angular acceleration. The resulting equations of rotational kinematics, along with the corresponding equations for linear motion under constant acceleration, are as follows:

Rotational Motion About a Fixed Axis with α Constant (Variables: θ and ω)	Linear Motion with a Constant (Variables: x and v)	
$\omega = \omega_0 + \alpha t$	$v = v_0 + at$	(7.5)
$\theta = \omega_0 t + \frac{1}{2}\alpha t^2$	$x = v_0 t + \frac{1}{2}at^2$	(7.6)
$\omega^2 = \omega_0^2 + 2\alpha\theta$	$v^2 = v_0^2 + 2ax$	(7.7)

Again, note the one-to-one correspondence between the rotational equations involving the angular variables θ, ω, and α and the equations of linear motion involving the variables x, v, and a.

EXAMPLE 7.2   The Rotating Wheel

The bicycle wheel of Figure 7.3 rotates with a constant angular acceleration of 3.5 rad/s². If the initial angular speed of the wheel is 2.0 rad/s at $t_0 = 0$, (a) through what angle does the wheel rotate in 2.0 s?

Solution   Since we are given $\omega_0 = 2.0$ rad/s and $\alpha = 3.5$ rad/s², we use

$$\theta = \omega_0 t + \tfrac{1}{2}\alpha t^2$$

$$\theta = (2.0 \text{ rad/s})(2.0 \text{ s}) + \tfrac{1}{2}(3.5 \text{ rad/s}^2)(2.0 \text{ s})^2$$

$$= 11 \text{ rad} = \boxed{630°}$$

(b) What is the angular speed at $t = 2.0$ s?

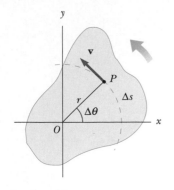

FIGURE 7.4
Rotation of an object about an axis through O that is perpendicular to the plane of the figure (the z axis). Note that a point, P, on the object rotates in a circle of radius r centered at O.

Solution   Making use of Equation 7.5, we find that

$$\omega = \omega_0 + \alpha t = 2.0 \text{ rad/s} + (3.5 \text{ rad/s}^2)(2.0 \text{ s}) = \boxed{9.0 \text{ rad/s}}$$

Exercise   Find the angular speed of the wheel at $t = 2.0$ s by making use of Equation 7.7 and the results of part (a).

7.3

RELATIONS BETWEEN ANGULAR AND LINEAR QUANTITIES

In this section we shall derive some useful relations between the angular speed and acceleration of a rotating object and the linear speed and acceleration of an arbitrary point in the object. Bear in mind that, when a rigid object rotates about a fixed axis, every point in the object moves in a circle whose center is along the axis of rotation.

Consider the arbitrarily shaped object in Figure 7.4, rotating about the z axis through the point O. Assume that the object rotates through the angle $\Delta\theta$, and hence point P moves through the arc length Δs in the interval Δt. We know from the defining equation for radian measure that

$$\Delta\theta = \frac{\Delta s}{r}$$

Let us now divide both sides of this equation by Δt, the interval during which the rotation occurred:

$$\frac{\Delta\theta}{\Delta t} = \frac{1}{r}\frac{\Delta s}{\Delta t}$$

If Δt is very small, then the angle $\Delta\theta$ through which the object rotates is small and the ratio $\Delta\theta/\Delta t$ is the instantaneous angular speed, ω. Also, Δs is very small when Δt is very small, and the ratio $\Delta s/\Delta t$ equals the instantaneous linear speed, v. Hence, the preceding equation is equivalent to

$$\omega = \frac{v}{r}$$

Figure 7.4 allows us to interpret this equation. The distance Δs is traversed along an arc of the circular path followed by the point P as it moves during the time Δt. Thus, v must be the linear speed of a point lying along this arc, where the direction of v is *tangent to the circular path*. As a result, we often refer to this linear speed as the *tangential speed* of a particle moving in a circular path, and write

Tangential speed

$$v_t = r\omega \tag{7.8}$$

That is,

> The **tangential speed** of a point on a rotating object equals the distance of that point from the axis of rotation multiplied by the angular speed.

Note that, although every point on the rotating object has the same angular speed, not every point has the same linear, or tangential, speed. In fact, Equation 7.8 shows that the linear speed of a point on the rotating object increases with movement outward from the center of rotation toward the rim, as one would intuitively expect.

Exercise caution when using Equation 7.8. It has been derived using the defining equation for radian measure and hence is valid only when ω is measured in radians per unit time. Other measures of angular speed, such as degrees per second and revolutions per second, are not to be used in Equation 7.8.

To find a second equation relating linear and angular quantities, imagine that an object rotating about a fixed axis (Fig. 7.4) changes its angular speed by $\Delta\omega$ in the interval Δt. At the end of this time, the speed of a point on the object, such as P, has changed by the amount Δv_t. From Equation 7.8 we have

$$\Delta v_t = r\,\Delta\omega$$

Dividing by Δt gives

$$\frac{\Delta v_t}{\Delta t} = r\frac{\Delta\omega}{\Delta t}$$

If the time interval Δt is very small, then the ratio $\Delta v_t/\Delta t$ is the tangential acceleration of that point and $\Delta\omega/\Delta t$ is the angular acceleration. Therefore, we see that

$$a_t = r\alpha \qquad\qquad \text{[7.9]}$$

Tangential acceleration

That is,

> The **tangential acceleration** of a point on a rotating object equals the distance of that point from the axis of rotation multiplied by the angular acceleration.

Again, radian measure must be used for the angular acceleration term in this equation.

There is one more equation that relates linear quantities to angular quantities, but we shall defer its derivation to the next section.

EXAMPLE 7.3   Computer Disks

A floppy disk in a computer rotates from rest up to an angular speed of 31.4 rad/s in a time of 0.892 s.

(a) What is the angular acceleration of the disk, assuming the angular acceleration is uniform?

Solution   If we use $\omega = \omega_0 + \alpha t$ and the fact that $\omega_0 = 0$ at $t = 0$, we get

$$\alpha = \frac{\omega}{t} = \frac{31.4 \text{ rad/s}}{0.892 \text{ s}} = \boxed{35.2 \text{ rad/s}^2}$$

(b) How many rotations does the disk make while coming up to speed?

Solution   Equation 7.6 enables us to find the angular displacement during the 0.892-s time interval:

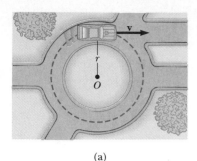

(a)

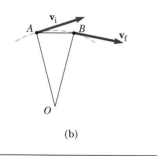

(b)

FIGURE 7.5
(a) Circular motion of a car moving with constant speed. (b) As the car moves along the circular path from *A* to *B,* the direction of its velocity vector changes, so the car undergoes a centripetal acceleration.

$$\theta = \omega_0 t + \tfrac{1}{2}\alpha t^2 = \tfrac{1}{2}(35.2 \text{ rad/s}^2)(0.892 \text{ s})^2 = \boxed{14.0 \text{ rad}}$$

Since 2π rad = 1 rev, this corresponds to 2.23 rev.

(c) If the radius of the disk is 4.45 cm, find the final linear speed of a microbe riding on the rim of the disk.

Solution   The relation $v_t = r\omega$ and the given values lead to

$$v_t = r\omega = (0.0445 \text{ m})(31.4 \text{ rad/s}) = \boxed{1.40 \text{ m/s}}$$

(d) What is the magnitude of the tangential acceleration of the microbe at this time?

Solution   We use $a_t = r\alpha$, which gives

$$a_t = r\alpha = (0.0445 \text{ m})(35.2 \text{ rad/s}^2) = \boxed{1.57 \text{ m/s}^2}$$

Exercise   What is the angular speed and angular displacement of the disk 0.300 s after it begins to rotate?

Answer   10.6 rad/s; 1.58 rad.

7.4
CENTRIPETAL ACCELERATION

Figure 7.5a shows a car moving in a circular path with *constant linear speed v.* Students are often surprised to find that *even though the car moves at a constant speed, it still has an acceleration.* To see why this occurs, consider the defining equation for acceleration to be

$$\mathbf{a} = \frac{\mathbf{v}_f - \mathbf{v}_i}{t_f - t_i} \qquad [7.10]$$

Note that the acceleration depends on the *change in the velocity vector.* Because velocity is a vector, there are two ways in which an acceleration can be produced: by a change in the *magnitude* of the velocity and by a change in the *direction* of the velocity. It is the latter situation that occurs for the car moving in a circular path with constant speed (see Fig. 7.5b). We shall show that the acceleration vector in this case is perpendicular to the path and always points toward the center of the circle. An acceleration of this nature is called a **centripetal** (center-seeking) **acceleration.** Its magnitude is given by

$$a_c = \frac{v^2}{r} \qquad [7.11]$$

To derive Equation 7.11, consider Figure 7.6a. Here an object is seen first at point *A* with velocity $\mathbf{v}_i$ at time t_i and then at point *B* with velocity $\mathbf{v}_f$ at a later time, $t_f.$ Let us assume that here $\mathbf{v}_i$ and $\mathbf{v}_f$ differ only in direction; their magnitudes are the same (that is, $v_i = v_f = v$). To calculate the acceleration, we begin with Equation 7.10, which indicates that we must vectorially subtract $\mathbf{v}_i$ from $\mathbf{v}_f$:

$$\mathbf{a} = \frac{\mathbf{v}_f - \mathbf{v}_i}{t_f - t_i} = \frac{\Delta\mathbf{v}}{\Delta t} \qquad [7.12]$$

where $\Delta\mathbf{v} = \mathbf{v}_f - \mathbf{v}_i$ is the change in velocity. That is, $\Delta\mathbf{v}$ is obtained by adding to $\mathbf{v}_f$ the vector $-\mathbf{v}_i$. This can be accomplished graphically, as shown by the vector triangle in Figure 7.6b. Note that when Δt is very small, Δs and $\Delta\theta$ will also be very small. In this case, $\mathbf{v}_f$ will almost parallel $\mathbf{v}_i$, and the vector $\Delta\mathbf{v}$ will be approximately perpendicular to them, pointing toward the center of the circle. In the limiting case where Δt becomes vanishingly small, $\Delta\mathbf{v}$ will point exactly toward the center of the circle. Furthermore, in this limiting case the acceleration will also be directed toward the center of the circle because it is in the direction of $\Delta\mathbf{v}$.

Now consider the triangle in Figure 7.6a, which has sides Δs and r. This triangle and the one formed by the vectors in Figure 7.6b are similar. (Two triangles are *similar* if the angle between any two sides is the same for both triangles and if the ratio of the lengths of these sides is the same.) This enables us to write a relationship between the lengths of the sides:

$$\frac{\Delta v}{v} = \frac{\Delta s}{r}$$

or

$$\Delta v = \frac{v}{r}\Delta s$$

This can be substituted into Equation 7.12 for Δv to give

$$a = \frac{v}{r}\frac{\Delta s}{\Delta t} \qquad \textbf{[7.13]}$$

In this situation, Δs is a small distance measured along the arc of the circle (a tangential distance), so $v = \Delta s/\Delta t$, where v is the tangential speed. Therefore, Equation 7.13 reduces to Equation 7.11:

$$a_c = \frac{v^2}{r}$$

Since the tangential speed is related to the angular speed through the relation $v_t = r\omega$ (Eq. 7.8), an alternative form of Equation 7.11 is

$$a_c = \frac{r^2\omega^2}{r} = r\omega^2 \qquad \textbf{[7.14]}$$

Thus, we conclude that

> In circular motion the *centripetal acceleration* is directed inward toward the center of the circle and has a magnitude given by either v^2/r or $r\omega^2$.

You should show that the dimensions of a_c are L/T^2, as required.

In order to clear up any misconceptions that might exist concerning centripetal and tangential acceleration, let us consider a car moving around a circular race track. Since the car is moving in a circular path, it always has a centripetal component of acceleration because its direction of travel, and hence the direction of its velocity, is continuously changing. If the car's speed is increasing or decreasing, it also has a tangential component of acceleration. To summarize, the tangential component of acceleration is due to changing speed; the centripetal component of acceleration is due to changing direction.

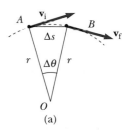

FIGURE 7.6
(a) As the particle moves from A to B, the direction of its velocity vector changes from $\mathbf{v}_i$ to $\mathbf{v}_f$.
(b) The construction for determining the direction of the change in velocity, $\Delta\mathbf{v}$, which is toward the center of the circle.

Centripetal acceleration

When both components of acceleration exist simultaneously, the tangential acceleration is tangent to the circular path and the centripetal acceleration points toward the center of the circular path. Because these components of acceleration are perpendicular to each other, we can find the magnitude of the **total acceleration** using the Pythagorean theorem:

Total acceleration

$$a = \sqrt{a_t^2 + a_c^2} \qquad\qquad [7.15]$$

EXAMPLE 7.4   Let's Go for a Spin

A test car moves at a constant speed of 10 m/s around a circular road of radius 50 m. Find the car's (a) centripetal acceleration and (b) angular speed.

Solution   (a) From Equation 7.11, the magnitude of the centripetal acceleration of the car is found to be

$$a_c = \frac{v^2}{r} = \frac{(10 \text{ m/s})^2}{50 \text{ m}} = \boxed{2.0 \text{ m/s}^2}$$

By definition, the direction of $\mathbf{a}_c$ is toward the center of curvature of the road.

(b) The angular speed of the car can be found using the expression $v_t = r\omega$, which gives

$$\omega = \frac{v_t}{r} = \frac{10 \text{ m/s}}{50 \text{ m}} = \boxed{0.20 \text{ rad/s}}$$

Note that we can also find the centripetal acceleration by using this value of ω, in an alternative method to part (a). From Equation 7.12 we have

$$a_c = r\omega^2 = (50 \text{ m})(0.20 \text{ rad/s})^2 = 2.0 \text{ m/s}^2$$

As expected, the two methods for finding the centripetal acceleration give the same answer.

Exercise   Find the car's tangential acceleration and total acceleration.

Answer   The tangential acceleration is zero because the speed of the car remains constant. Since $a_t = 0$, the total acceleration equals the centripetal acceleration, 2.0 m/s², found in part (a).

A car rounding the curve at the Grand Prix. (*Jo Anne Kalish/Joe Di-Maggio/Peter Arnold, Inc.*)

7.5

CENTRIPETAL FORCE

Consider a ball of mass m tied to a string of length r and being whirled in a horizontal circular path, as in Figure 7.7. Let us assume that the ball moves with constant speed. Because the velocity vector, **v**, changes its direction continuously during the motion, the ball experiences a centripetal acceleration directed toward the center of motion, as described in Section 7.4, with magnitude

$$a_c = \frac{v_t^2}{r}$$

The inertia of the ball tends to maintain motion in a straight-line path; however, the string overcomes this by exerting a force on the ball that makes it

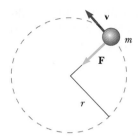

FIGURE 7.7
A ball attached to a string of length r, rotating in a circular path at constant speed.

instead follow a circular path. This force (equal to the force of tension) is directed along the length of the string toward the center of the circle, as shown in Figure 7.7, and is called a **centripetal force.** Thus, the equation for Newton's second law along the radial direction is

$$F_c = ma_c = m\frac{v_t^2}{r}$$ [7.16] Centripetal force

A centripetal force is necessary to produce the acceleration of a mass toward the center of its circular path.

Because it acts at right angles to the motion, a centripetal force causes a change in the direction of the velocity. Beyond this, it is no different from any of the other forces we have studied. For example, friction between tires and road provides the centripetal force that enables a race car to travel in a circular path on a flat road, and the gravitational force exerted on the Moon by the Earth provides the centripetal force necessary to keep the Moon in its orbit. (We shall discuss the gravitational force in more detail in Section 7.7.)

Regardless of the example used, if the centripetal force acting on an object moving initially in a circle vanishes, the object does not continue to move in its circular path; instead, it moves along a straight-line path tangent to the circle. To illustrate this point, consider a ball that is attached to a string and is being whirled in a vertical circle as in Figure 7.8a. If the string happened to break when the ball was at position A, the centripetal force (the tension in the string) would vanish and the ball would move vertically upward. Its subsequent motion would be that of a freely falling body. If the string happened to break when the ball was

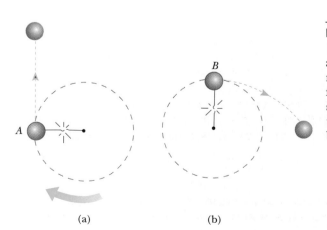

(a) (b)

FIGURE 7.8
(a) When the string breaks at position A, the ball moves vertically upward in free fall. (b) When the string breaks at position B, the ball moves along a **parabolic path.**

at the top of its circular path, shown as point B in Figure 7.8b, the ball would initially fly off horizontally in a direction tangent to the path, and would then move in a parabolic path according to the equations of a projectile.

PROBLEM-SOLVING STRATEGY
Centripetal Forces

Use the following steps when dealing with centripetal forces and centripetal accelerations:

1. **Draw a free-body diagram of the object(s) under consideration, showing all forces that act on it.**
2. **Choose a coordinate system with one axis perpendicular to the circular path followed by the object, and one axis tangent to the circular path.**
3. **Find the net force toward the center of the circular path. This is the centripetal force.**
4. **From here on the steps are virtually identical to those encountered when solving Newton's second law problems with $\Sigma F = ma$. Also, note that the magnitude of the centripetal acceleration can always be written $a_c = v_t{}^2/r$.**

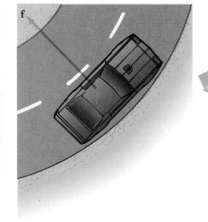

(a)

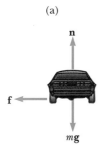

(b)

FIGURE 7.9
(Example 7.5)   (a) Top view of a car on a curved path. (b) A free-body diagram for the car, showing an end view.

EXAMPLE 7.5   Buckle Up for Safety

A car travels at a constant speed of 30.0 mi/h (13.4 m/s) on a level circular turn of radius 50.0 m, as shown in the bird's-eye view in Figure 7.9a. What is the minimum coefficient of static friction between the tires and roadway in order that the car makes the circular turn without sliding?

Reasoning   The centripetal force acting on the car is the force of static friction directed toward the center of the circular path as in Figure 7.9a. Thus, the objective in this example will be to find an expression for the frictional force $f = \mu_s n$ and use this in Equation 7.16.

Solution   Thus, Equation 7.16 becomes

$$(1) \qquad f = m\frac{v_t{}^2}{r}$$

All the forces acting on the car are shown in Figure 7.9b. Because equilibrium is in the vertical direction, the normal force upward is balanced by the force of gravity downward, so that

$$n = mg$$

From this expression we can find the minimum force of friction, as follows:

$$(2) \qquad f = \mu_s n = \mu_s mg$$

By setting the right-hand sides of (1) and (2) equal to each other, we find that

$$\mu_s mg = m\frac{v_t{}^2}{r}$$

or

$$(3) \qquad \mu_s = \frac{v_t^2}{rg}$$

Therefore, the minimum coefficient of static friction that is required for the car to make the turn without sliding outward is

$$\mu_s = \frac{v_t^2}{rg} = \frac{(13.4 \text{ m/s})^2}{(50.0 \text{ m})(9.80 \text{ m/s}^2)} = \boxed{0.366}$$

The value of μ_s for rubber on dry concrete is very close to 1; thus, the car can negotiate the curve with ease. If the road were wet or icy, however, the value for μ_s could be 0.2 or lower. Under such conditions, the centripetal force provided by static friction would not be great enough to enable the car to follow the circular path, and it would slide off the roadway.

EXAMPLE 7.6   Having Fun with a Yo-Yo

A child swings a yo-yo of weight mg in a horizontal circle so that the cord makes an angle of 30.0° with the vertical, as in Figure 7.10a. Find the centripetal acceleration of the yo-yo.

Reasoning   Two forces act on the yo-yo: its weight and the tension in the cord, **T**, as in Figure 7.10b. The first condition for equilibrium applied in the y direction, $\Sigma F_y = 0$, will enable us to find T. The component of **T** toward the center of the circular path will be the centripetal force, and we can finally find a_c from $F_c = ma_c$.

Solution   Applying ΣF_y gives

$$T \cos 30.0° - mg = 0$$

$$T = \frac{mg}{\cos 30.0°}$$

The net force acting toward the center of the circular path is the centripetal force, which in this case is the *horizontal component* of **T**.

$$F_c = T \sin 30.0° = \frac{mg \sin 30.0°}{\cos 30.0°} = mg \tan 30.0°$$

The centripetal acceleration can now be found by applying Newton's second law along the horizontal direction:

$$F_c = ma_c$$

$$a_c = \frac{F_c}{m} = \frac{mg \tan 30.0°}{m} = g \tan 30.0° = (9.80 \text{ m/s}^2)(\tan 30.0°) = \boxed{5.66 \text{ m/s}^2}$$

7.6

DESCRIBING MOTION OF A ROTATING SYSTEM

We have seen that an object moving in a circle of radius r with constant speed v has a centripetal acceleration whose magnitude is v_t^2/r and whose direction is toward the center of rotation. The force necessary to maintain this centripetal acceleration—that is, the centripetal force—must also act toward the center of rotation. In the case of a ball rotating at the end of a string, the force exerted on the ball by the string (equal to the tension in the string) is the centripetal force;

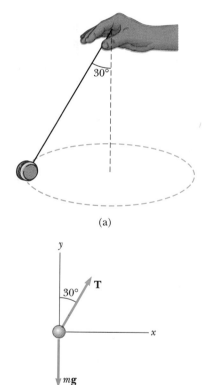

(a)

(b)

FIGURE 7.10
(Example 7.6)   (a) A yo-yo swinging in a horizontal circle so that the cord makes a constant angle of 30° with the vertical. (b) A free-body diagram for the yo-yo.

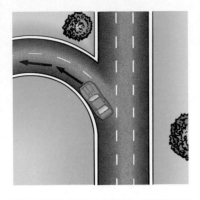

FIGURE 7.11
A car approaching a curved exit ramp.

for a satellite in a circular orbit around the Earth, the force of gravity is the centripetal force; the centripetal force acting on a car rounding a curve on a level road is the force of friction between the tires and the pavement; and so forth.

In order to better understand the motion of a rotating system, consider a car approaching a curved exit ramp at high speed, as in Figure 7.11. As the car takes the sharp left turn onto the ramp, a person sitting in the passenger seat slides to the right across the seat and hits the door. At that point, the force of the door keeps him from being ejected from the car. What causes the passenger to move toward the door? A popular but *erroneous* explanation is that some mysterious force pushes him outward. (This is sometimes called the "centrifugal" force, but we shall not use that term since it always creates confusion.)

The phenomenon is correctly explained as follows. Before the car enters the ramp, the passenger is moving in a straight-line path. As the car enters the ramp and travels a curved path, the passenger, because of inertia, tends to move along the original straight-line path. This is in accordance with Newton's first law: the natural tendency of a body is to continue moving in a straight line. However, if a sufficiently large centripetal force (toward the center of curvature) acts on the passenger, he moves in a curved path along with the car. The origin of this centripetal force is the force of friction between the passenger and the car seat. If this frictional force is not great enough, the passenger slides across the seat as the car turns under him. Because of the passenger's inertia, he continues to move in a straight-line path. Eventually the passenger encounters the door, which provides a large enough centripetal force to enable him to follow the same curved path as the car. The passenger slides toward the door not because of some mysterious outward force but because *there is no centripetal force great enough to enable him to travel along the circular path followed by the car.*

As a second example, consider what happens when you run clothes through the rinse cycle of a washing machine. In the last phase of this cycle, the drum spins rapidly to remove water from the clothes. Why is the water thrown off? An *erroneous* explanation is that the rotating system creates some mysterious outward force on each drop of water, and this force causes the water to be hurled to the outer drum of the machine. The correct explanation goes as follows. When the clothes are at rest in the machine, water is held to them by molecular forces

Passengers on a roller coaster at Knott's Berry Farm experience the thrill of circular motion. Can you identify the origin of the centripetal force(s) acting on a passenger during a loop-the-loop maneuver? *(Superstock)*

between the water and the fabric. During the spin cycle, the clothes rotate and the molecular forces are not great enough to provide the necessary *centripetal* force to keep the water molecules moving in a circular path along with the clothes. Hence, the drops of water, because of their inertia, move in straight-line paths until they encounter the sides of the spinning drum.

In summary, in describing motion in an accelerating frame, one must be very careful to distinguish real forces from fictitious ones. An observer in a car rounding a curve is in an accelerating frame and invents a fictitious outward force to explain why he or she is thrown outward. A stationary observer outside the car, however, considers only real forces on the passenger. To the observer, the mysterious outward force *does not exist!* The only real external force on the passenger is the centripetal (inward) force due to friction or the contact force of the door.

7.7

NEWTON'S UNIVERSAL LAW OF GRAVITATION

Prior to 1686 a great mass of data had been collected on the motions of the Moon and the planets, but a clear understanding of the forces that cause these celestial bodies to move the way they do was not available. In that year, Isaac Newton provided the key that unlocked the secrets of the heavens. He knew, from the first law, that a net force had to be acting on the Moon. If it were not, the Moon would move in a straight-line path rather than in its almost circular orbit. Newton reasoned that this force arises as a result of an attractive field force between Moon and Earth, which we call the force of gravity. He also concluded that there could be nothing special about the Earth-Moon system or the Sun and its planets that would cause gravitational forces to act on them alone. In other words, he saw that the same force of attraction that causes the Moon to follow its path also causes an apple to fall to Earth from a tree. He wrote, "I deduced that the forces which keep the planets in their orbs must be reciprocally as the squares of their distances from the centers about which they revolve; and thereby compared the force requisite to keep the Moon in her orb with force of gravity at the surface of the Earth; and found them answer pretty nearly."

In 1687 Newton published his work on the **universal law of gravitation**, which states that

> Every particle in the Universe attracts every other particle with a force that is directly proportional to the product of their masses and inversely proportional to the square of the distance between them.

If the particles have masses m_1 and m_2 and are separated by the distance r, the magnitude of the gravitational force is

$$F = G \frac{m_1 m_2}{r^2} \qquad \text{[7.17]}$$

where G is a universal constant called the **constant of universal gravitation**, which has been measured experimentally. Its value in SI units is

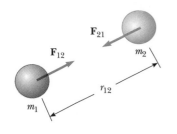

FIGURE 7.12
The gravitational force between two particles is attractive. Note that, according to Newton's third law, $\mathbf{F}_{12} = -\mathbf{F}_{21}$.

$$G = 6.673 \times 10^{-11} \, \frac{\text{N} \cdot \text{m}^2}{\text{kg}^2} \qquad \textbf{[7.18]}$$

This force law is an example of an **inverse-square law** in that it varies as the inverse square of the separation. The force acts so that the objects are always attracted to one another. From Newton's third law, we also know that the force on m_2 due to m_1, designated $\mathbf{F}_{21}$ in Figure 7.12, is equal in magnitude to the force on m_1 due to m_2, $\mathbf{F}_{12}$, and in the opposite direction. That is, these forces form an action-reaction pair.

Several features of the universal law of gravitation deserve some attention:

1. The gravitational force is a field force that always exists between two particles regardless of the medium that separates them.
2. The force varies as the inverse square of the distance between the particles and therefore decreases rapidly with increasing separation.
3. The force is proportional to the product of the particles' masses.

Another important fact is that *the gravitational force exerted by a spherical mass on a particle outside the sphere is the same as if the entire mass of the sphere were concentrated at its center.* For example, the force on a particle of mass m at the Earth's surface has the magnitude

$$F = G \frac{M_E m}{R_E{}^2}$$

where M_E is the Earth's mass and R_E is its radius. This force is directed toward the center of the Earth.

MEASUREMENT OF THE GRAVITATIONAL CONSTANT

The gravitational constant was first measured in an important experiment by Henry Cavendish in 1798. His apparatus consisted of two small spheres, each of mass m, fixed to the ends of a light horizontal rod suspended by a thin metal wire, as in Figure 7.13. Two large spheres, each of mass M, were placed near the smaller spheres. The attractive force between the smaller and larger spheres caused the rod to rotate and the wire to twist. The angle through which the

FIGURE 7.13
(a) A schematic diagram of the Cavendish apparatus for measuring G. The smaller spheres of mass m are attracted to the large spheres of mass M, and the rod rotates through a small angle. A light beam reflected from a mirror on the rotating apparatus measures the angle of rotation. (b) A student Cavendish apparatus. *(Courtesy of PASCO Scientific)*

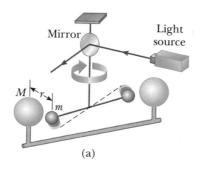

(a)

(b)

suspended rod rotated was measured with a light beam reflected from a mirror attached to the vertical suspension. (Such a moving spot of light is an effective technique for amplifying the motion.) The experiment was carefully repeated with different masses at various separations. In addition to providing a value for G, the results showed that the force is attractive, proportional to the product mM, and inversely proportional to the square of the distance r.

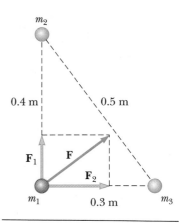

EXAMPLE 7.7   Billiards, Anyone?

Three 0.300-kg billiard balls are placed on a table at the corners of a right triangle, as shown from overhead in Figure 7.14. Find the net gravitational force on the ball designated as m_1 due to the forces exerted on it by the other two balls.

Solution   To find the net gravitational force on m_1, we first calculate the force exerted on m_1 due to m_2. Then we find the force on m_1 due to m_3. Finally, we add these two forces *vectorially* to obtain the net force on m_1.

The force exerted on m_1 due to m_2, denoted by $\mathbf{F}_1$ in Figure 7.14, is upward. Its magnitude is calculated using Equation 7.17:

$$F_1 = G\frac{m_1 m_2}{r^2} = (6.67 \times 10^{-11}\ \text{N} \cdot \text{m}^2/\text{kg}^2)\,\frac{(0.300\ \text{kg})(0.300\ \text{kg})}{(0.400\ \text{m})^2}$$

$$= 3.75 \times 10^{-11}\ \text{N}$$

This result shows that gravitational forces between the common objects that surround us have extremely small magnitudes.

Now let us calculate the gravitational force exerted on m_1 due to m_3, denoted by $\mathbf{F}_2$ in Figure 7.14. It is directed to the right, and its magnitude is

$$F_2 = G\frac{m_1 m_3}{r^2} = (6.67 \times 10^{-11}\ \text{N} \cdot \text{m}^2/\text{kg}^2)\,\frac{(0.300\ \text{kg})(0.300\ \text{kg})}{(0.300\ \text{m})^2}$$

$$= 6.67 \times 10^{-11}\ \text{N}$$

The net gravitational force exerted on m_1 is found by adding $\mathbf{F}_1$ and $\mathbf{F}_2$ as *vectors*. The magnitude of this net force is given by

$$F = \sqrt{F_1{}^2 + F_2{}^2} = \sqrt{(3.75)^2 + (6.67)^2} \times 10^{-11}\ \text{N} = \boxed{7.65 \times 10^{-11}\ \text{N}}$$

Exercise   Find the direction of the resultant force on m_1.

Answer   The vector $\mathbf{F}$ makes an angle of 29.3° with the line joining m_1 and m_3.

FIGURE 7.14
(Example 7.7)

EXAMPLE 7.8   The Mass of the Earth

Use the gravitational force law to find an approximate value for the mass of the Earth.

Solution   Figure 7.15 (obviously not to scale) shows a baseball falling toward the Earth at a location where the free-fall acceleration is g. We know that the gravitational force exerted on the baseball by the Earth is the same as the weight of the ball and that this is given by $w = m_b g$. Since the force in the gravitational law is the weight of the ball, we find that

$$m_b g = G\frac{M_E m_b}{R_E{}^2}$$

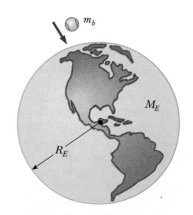

FIGURE 7.15
(Example 7.8)   A baseball falling toward the Earth (not drawn to scale).

TABLE 7.1
Free-Fall Acceleration, g, at Various Altitudes

Altitude (km)[a]	g (m/s²)
1000	7.33
2000	5.68
3000	4.53
4000	3.70
5000	3.08
6000	2.60
7000	2.23
8000	1.93
9000	1.69
10 000	1.49
50 000	0.13

[a] All values are distances above the Earth's surface.

Johannes Kepler (1571–1630), a German astronomer, is best known for developing the laws of planetary motion based on the careful observations of Tycho Brahe. Throughout his life, Kepler was sidetracked by mystic ideas dating back to the ancient Greeks. For example, he believed in the notion of the "music of the spheres" proposed by Pythagoras, in which each planet in its motion sounds out an exact musical note. After spending several years trying to work out a "regular-solid theory" of the planets, he concluded that the Copernican view of circular planetary orbits had to be abandoned for the view that the planetary orbits are ellipses with the Sun always at one of the foci. *(Art Resource)*

We can divide each side of this equation by m_b and solve for the mass of the Earth, M_E:

$$M_E = \frac{gR_E{}^2}{G}$$

The falling baseball is close enough to the Earth so that the distance of separation between the center of the ball and the center of the Earth can be taken as the radius of the Earth, 6.38×10^6 m. Thus, the mass of the Earth is

$$M_E = \frac{(9.80 \text{ m/s}^2)(6.38 \times 10^6 \text{ m})^2}{6.67 \times 10^{-11} \text{ N} \cdot \text{m}^2/\text{kg}^2} = \boxed{5.98 \times 10^{24} \text{ kg}}$$

EXAMPLE 7.9   Gravity and Altitude

Derive an expression that shows how the acceleration due to gravity varies with distance from the center of the Earth at an exterior point.

Solution   The falling baseball of Example 7.8 can be used here also. Now, however, assume that the ball is located at some arbitrary distance r from the Earth's center. The first equation in Example 7.8, with r replacing R_E and m_b removed from both sides, becomes

$$g = G\frac{M_E}{r^2}$$

This indicates that the free-fall acceleration at an exterior point decreases as the inverse square of the distance from the center of the Earth. Our assumption of Chapter 2 that objects fall with constant acceleration is obviously incorrect in light of the present example. For short falls, however, this change in g is so small that neglecting the variation does not introduce a significant error in the result.

Because the true weight of an object is mg, we see that a change in the value of g produces a change in the weight of the object. For example, if you weigh 800 N at the surface of the Earth, you will weigh only 200 N at a height above the Earth equal to the radius of the Earth. Also, we see that if the distance of an object from the Earth becomes infinitely large, the true weight approaches zero. Values of g at various altitudes are listed in Table 7.1.

Exercise   If an object weighs 270 N at the Earth's surface, what will it weigh at an altitude equal to twice the radius of the Earth?

Answer   30.0 N.

*7.8

KEPLER'S LAWS

The movements of the planets, stars, and other celestial bodies have been observed for thousands of years. In early history, scientists regarded the Earth as the center of the Universe. This *geocentric model* was developed extensively by the Greek astronomer Claudius Ptolemy in the second century A.D. and was accepted for the next 1400 years. In 1543 the Polish astronomer Nicolas Copernicus (1473–1543) showed that the Earth and the other planets revolve in circular orbits about the Sun (the *heliocentric hypothesis*).

The Danish astronomer Tycho Brahe (pronounced Brah or BRAH-huh; 1546–1601) made accurate astronomical measurements over a period of 20 years and provided the data for the currently accepted model of the Solar System. Brahe's precise observations of the planets and 777 stars were carried out with nothing more elaborate than a large sextant and compass; the telescope had not yet been invented.

The German astronomer Johannes Kepler, who was Brahe's assistant, acquired Brahe's astronomical data and spent about 16 years trying to deduce a mathematical model for the motions of the planets. After many laborious calculations, he found that Brahe's precise data on the revolution of Mars about the Sun provided the answer. Kepler's analysis first showed that the concept of circular orbits about the Sun had to be abandoned. He eventually discovered that the orbit of Mars could be accurately described by an ellipse with the Sun at one focus. He then generalized this analysis to include the motions of all planets. The complete analysis is summarized in three statements known as **Kepler's laws:**

1. **All planets move in elliptical orbits with the Sun at one of the focal points.**
2. **A line drawn from the Sun to any planet sweeps out equal areas in equal time intervals.**
3. **The square of the orbital period of any planet is proportional to the cube of the average distance from the planet to the Sun.**

Newton later demonstrated that these laws are consequences of a simple force that exists between any two masses. Newton's universal law of gravity, together with his laws of motion, provides the basis for a full mathematical solution to the motions of planets and satellites. More important, Newton's universal law of gravity correctly describes the gravitational attractive force between *any* two masses.

KEPLER'S FIRST LAW

We shall not attempt to derive Kepler's first law here. Suffice it to say that it can be shown that the first law arises as a natural consequence of the inverse-square nature of Newton's law of gravitation. That is, any object bound to another by a force that varies as $1/r^2$ will move in an elliptical orbit. As shown in Figure 7.16a, an ellipse is a curve drawn so that the sum of the distances from any point on the curve to two internal points called focal points or foci (singular, focus) is always the same. For the Sun-planet configuration (Fig. 7.16b), the Sun is at one focus and the other focus is empty. Because the orbit is an ellipse, the distance from the Sun to the planet continuously changes.

KEPLER'S SECOND LAW

Kepler's second law states that a line drawn from the Sun to any planet sweeps out equal areas in equal time intervals. Consider a planet in an elliptical orbit about the Sun, as in Figure 7.17. Imagine that at some instant of time we draw a line from the Sun to a planet, tracing out the line *AS* where *S* represents the position of the Sun. Exactly 30 days later, we repeat the process and draw the line

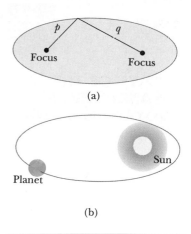

FIGURE 7.16
(a) The sum $p + q$ is the same for every point on the ellipse. (b) In the Solar System, the Sun is at one focus of the elliptical orbit of each planet and the other focus is empty.

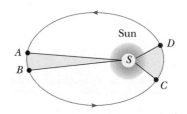

FIGURE 7.17
The two areas swept out by the planet in its elliptical orbit about the Sun are equal if the time interval between points *A* and *B* is equal to the time interval between points *C* and *D*.

PHYSICS IN *ACTION*

VIEWS OF THE PLANETS FROM VOYAGERS 1 AND 2

A View of Saturn from Voyager 2.   In the photograph on the left, Saturn's rings are bright and its northern hemisphere defined by bright features as NASA's Voyager 2 approaches the planet, which it encountered on August 25, 1981. Three images, taken through ultraviolet, violet, and green filters on July 12, 1981, were combined to make this photograph. Voyager 2 was 43 million km (27 million miles) from Saturn when it took this photograph.

A View of Neptune's Largest Moon.   In the middle photograph, Neptune's largest moon, Triton, was recorded by the Voyager 2 spacecraft on August 25, 1989, from a distance of 40 000 km (25 000 miles). The dark smudges on Triton's surface are thought to be material blown upward by active ice volcanoes and downward by high-altitude winds. The slight pink color might be the result of an evaporating layer of nitrogen ice.

A View of Jupiter from Voyager 1.   Voyager 1 took the photograph on the right, of Jupiter and two of its satellites (Io, left, and Europa), on February 13, 1979. Io is about 350 000 km (220 000 miles) above Jupiter's Great Red Spot; Europa is about 600 000 km (375 000 miles) above Jupiter's clouds. Although the two satellites have about the same brightness, Io's color is very different from Europa's. Io's equatorial region shows two types of material—dark orange, broken by several bright spots—producing a mottled appearance. The color variations within and between the polar regions suggest that Io's surface is a composite. Io's surface composition is a mixture of sulfur and sulfur compounds. Other photographs show that Io (like the Earth) has many active volcanoes. Jupiter is about 20 million km (12 million miles) from the spacecraft. At this resolution (about 400 km, or 250 miles) there is evidence of circular motion in Jupiter's atmosphere. While the dominant large-scale motions are west to east, small-scale movement includes eddy-like circulation within and between the bands.

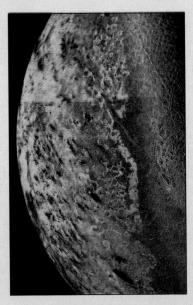

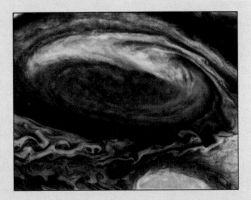

(Left, center, courtesy of NASA; right, courtesy of NASA/SB/FPG)

BS. The area swept out is shown in color in the diagram. When the planet is at point *C*, we repeat the process, drawing the line *CS* from the planet to the Sun. We now wait 30 days, exactly the same amount of time we waited before, and draw the line *DS*. The area swept out in this interval is also shown in color in the figure. The two colored areas are equal.

KEPLER'S THIRD LAW

The derivation of Kepler's third law is simple enough to carry out here. Consider a planet of mass M_p moving about the Sun, which has a mass of M_S, in a circular orbit (Fig. 7.18). (The assumption of a circular rather than an elliptical orbit will not introduce serious error into our approach, because the orbits of all planets except Mercury and Pluto are very nearly circular.) Because the gravitational force on the planet is the centripetal force needed to keep it moving in a circle,

$$\frac{GM_SM_p}{r^2} = \frac{M_pv^2}{r}$$

The speed, *v*, of the planet in its orbit is equal to the circumference of the orbit divided by the time required for one revolution, *T*, called the **period** of the planet. That is, $v = 2\pi r/T$, and the preceding expression becomes

$$\frac{GM_S}{r^2} = \frac{(2\pi r/T)^2}{r}$$

$$T^2 = \left(\frac{4\pi^2}{GM_S}\right)r^3 = K_Sr^3 \qquad \textbf{[7.19]}$$

where K_S is a constant given by

$$K_S = \frac{4\pi^2}{GM_S} = 2.97 \times 10^{-19}\ \text{s}^2/\text{m}^3$$

Equation 7.19 is Kepler's third law. It is also valid for elliptical orbits if *r* is replaced by a length equal to the semimajor axis of the ellipse. Note that K_S is independent of the mass of the planet. Therefore, Equation 7.19 is valid for any planet. If we consider the orbit of a satellite, such as the Moon, about the Earth,

The Earth as seen from the surface of the Moon. The prominent colors are the blues of oceans and the white swirls of clouds. *(© Index Stock Photography, Inc.)*

Kepler's third law

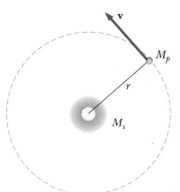

FIGURE 7.18
A planet of mass M_p moving in a circular orbit about the Sun. The orbits of all planets except Mercury and Pluto are nearly circular.

TABLE 7.2
Useful Planetary Data

Body	Mass (kg)	Mean Radius (m)	Period (s)	Mean Distance from Sun (m)	$\frac{T^2}{r^3} \times 10^{-19} \left(\frac{s^2}{m^3}\right)$
Mercury	3.18×10^{23}	2.43×10^6	7.60×10^6	5.79×10^{10}	2.97
Venus	4.88×10^{24}	6.06×10^6	1.94×10^7	1.08×10^{11}	2.99
Earth	5.98×10^{24}	6.38×10^6	3.156×10^7	1.496×10^{11}	2.97
Mars	6.42×10^{23}	3.37×10^6	5.94×10^7	2.28×10^{11}	2.98
Jupiter	1.90×10^{27}	6.99×10^7	3.74×10^8	7.78×10^{11}	2.97
Saturn	5.68×10^{26}	5.85×10^7	9.35×10^8	1.43×10^{12}	2.99
Uranus	8.68×10^{25}	2.33×10^7	2.64×10^9	2.87×10^{12}	2.95
Neptune	1.03×10^{26}	2.21×10^7	5.22×10^9	4.50×10^{12}	2.99
Pluto	$\approx 1 \times 10^{23}$	$\approx 3 \times 10^6$	7.82×10^9	5.91×10^{12}	2.96
Moon	7.36×10^{22}	1.74×10^6	—	—	—
Sun	1.991×10^{30}	6.96×10^8	—	—	—

then the constant has a different value, with the mass of the Sun replaced by the mass of the Earth. In this case, K_E equals $4\pi^2/GM_E$.

A collection of useful planetary data is presented in Table 7.2, where the last column verifies that T^2/r^3 is a constant.

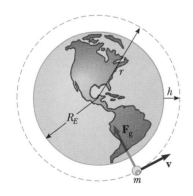

FIGURE 7.19

(Example 7.10)   A satellite of mass m moving in a circular orbit of radius r and with constant speed v around the Earth (not drawn to scale). The centripetal force is provided by the gravitational force acting on the satellite.

EXAMPLE 7.10   An Earth Satellite

A satellite of mass m moves in a circular orbit about the Earth with a constant speed of v and a height of $h = 1000$ km above the Earth's surface, as in Figure 7.19. (For clarity, this figure is not drawn to scale.) Find the orbital speed of the satellite. The radius of the Earth is 6.38×10^6 m, and its mass is 5.98×10^{24} kg.

Reasoning   The only force acting on the satellite is the gravitational force directed toward the center of the circular orbit. This can be equated to mv^2/r to find v. (The distance r is the distance from the center of the Earth to the satellite.)

Solution   The only external force on the satellite is that of the gravitational attraction exerted by the Earth. This force is directed toward the center of the satellite's circular path and is the centripetal force acting on the satellite. Since the force of gravity is $GM_E m/r^2$, we find that

$$F_c = G\frac{M_E m}{r^2} = m\frac{v^2}{r}$$

$$v^2 = \frac{GM_E}{r}$$

In this expression, the distance r is the Earth's radius plus the height of the satellite. That is, $r = R_E + h = 7.38 \times 10^6$ m, and so

$$v^2 = \frac{(6.67 \times 10^{-11}\ \text{N} \cdot \text{m}^2/\text{kg}^2)(5.98 \times 10^{24}\ \text{kg})}{7.38 \times 10^6\ \text{m}} = 5.40 \times 10^7\ \text{m}^2/\text{s}^2$$

Therefore,

$$v = \boxed{7.35 \times 10^3\ \text{m/s} \approx 16\ 400\ \text{mi/h}}$$

Note that v is independent of the mass of the satellite!

Exercise Calculate the period of revolution (the time required for one revolution about the Earth), T, of the satellite.

Answer 105 min.

SUMMARY

The **average angular speed,** $\bar{\omega}$, of a rigid object is defined as the ratio of the angular displacement, $\Delta\theta$, to the time interval, Δt:

$$\bar{\omega} \equiv \frac{\theta_2 - \theta_1}{t_2 - t_1} = \frac{\Delta\theta}{\Delta t} \qquad \textbf{[7.2]}$$

where $\bar{\omega}$ is in radians per second (rad/s).

The **average angular acceleration,** $\bar{\alpha}$, of a rotating object is defined as the ratio of the change in angular speed, $\Delta\omega$, to the time interval, Δt.

$$\bar{\alpha} \equiv \frac{\omega_2 - \omega_1}{t_2 - t_1} = \frac{\Delta\omega}{\Delta t} \qquad \textbf{[7.4]}$$

where $\bar{\alpha}$ is in radians per second per second (rad/s^2).

If an object undergoes rotational motion about a fixed axis under constant angular acceleration α, one can describe its motion with the following set of equations:

$$\omega = \omega_0 + \alpha t \qquad \textbf{[7.5]}$$

$$\theta = \omega_0 t + \tfrac{1}{2}\alpha t^2 \qquad \textbf{[7.6]}$$

$$\omega^2 = \omega_0{}^2 + 2\alpha\theta \qquad \textbf{[7.7]}$$

When an object rotates about a fixed axis, the angular speed and angular acceleration are related to the tangential speed and tangential acceleration through the relationships

$$v_t = r\omega \qquad \textbf{[7.8]}$$

$$a_t = r\alpha \qquad \textbf{[7.9]}$$

Any object moving in a circular path has an acceleration directed toward the center of the circular path, called a **centripetal acceleration.** Its magnitude is given by

$$a_c = \frac{v_t{}^2}{r} = r\omega^2 \qquad \textbf{[7.11, 7.14]}$$

Any object moving in a circular path must have a net force exerted on it that is directed toward the center of the circular path. This net force is called a **centripetal force.** From Newton's second law, the centripetal force and centripetal acceleration are related by

$$F_c = m\frac{v_t{}^2}{r} \qquad \textbf{[7.16]}$$

Some examples of centripetal forces are the force of gravity (as in the motion of a satellite) and the force of tension in a string.

Newton's law of universal gravitation states that every particle in the Universe attracts every other particle with a force that is directly proportional to the product of their masses and inversely proportional to the square of the distance, r, between them:

$$F = G\frac{m_1 m_2}{r^2} \qquad \textbf{[7.17]}$$

where $G = 6.673 \times 10^{-11}$ N·m²/kg² is the **constant of universal gravitation**.

Kepler's laws of planetary motion state that

1. All planets move in elliptical orbits with the Sun at one of the focal points.
2. A line drawn from the Sun to any planet sweeps out equal areas in equal time intervals.
3. The square of the orbital period of a planet is proportional to the cube of the mean distance from the planet to the Sun:

$$T^2 = \left(\frac{4\pi^2}{GM_S}\right)r^3 \qquad \textbf{[7.19]}$$

ADDITIONAL READING

J. Beams, "Ultra-High-Speed Rotation," *Sci. American,* April 1961, p. 134.

I. B. Cohen, "Newton's Discovery of Gravity," *Sci. American,* March 1981, p. 166.

R. Feynman, *The Character of Physical Law,* Cambridge, Mass., MIT Press, 1965, Chaps. 1, 2, and 3.

G. Gamow, *Gravity,* Science Study Series, Garden City, N.Y., Doubleday, 1962.

D. Layzer, *Constructing the Universe,* Scientific American Library, New York, W. H. Freeman and Co., 1984.

F. Pipkin, "Gravity Up in the Air," *The Sciences,* July/August 1984, p. 24.

R. Ruthen, "Catching the Wave," *Sci. American,* March 1992, p. 90.

G. Spetz, "Detection of Gravity Waves," *The Physics Teacher,* May 1984, p. 282.

E. L. Turner, "Gravitational Lenses," *Sci. American,* July 1988, p. 54.

CONCEPTUAL QUESTIONS

Example   It has been suggested that rotating cylinders about 10 miles long and 5.0 miles in diameter be placed in space for colonies. The purpose of their rotations is to simulate gravity for the inhabitants. Explain the concept behind this proposal.

Reasoning   Consider an individual standing against the inside wall of the cylinder with her head pointed toward the axis of the cylinder. As the cylinder rotates, the person tends to move in a straight line path tangent to the circular path followed by the cylinder wall. As a result, the person is forced against the wall, and the normal force exerted on her provides the centripetal force required to keep her moving in a circular path. If the rotational speed is adjusted such that this normal force is equal to her weight on Earth, she would not be able to distinguish between the artificial gravity of the colony and ordinary gravity.

Example Why does an astronaut in a space capsule orbiting the Earth experience a feeling of weightlessness?
Reasoning The astronaut and space capsule both have the same acceleration equal to the free-fall acceleration directed toward the Earth. In the rotating frame of reference of the astronaut and space capsule, the astronaut sees everything at rest and believes there is no force on him. If he tried to weigh himself on a spring scale, the scale would read zero. The scale and the astronaut would float freely around in the capsule.

Example Why does a pilot sometimes black out when pulling out of a steep dive?
Reasoning When a pilot is pulling out of a dive, blood leaves the head because there is not a great enough centripetal force to cause it to follow the circular path of the airplane. The loss of blood from the brain can cause the pilot to black out.

Example An object moves in a circular path with constant speed v. (a) Is the object's velocity constant? (b) Is its acceleration constant? Explain.
Reasoning (a) As the object moves in its circular path with constant speed, the *direction* of the velocity vector changes. Thus, the velocity of the object is not constant. (b) The magnitude of its acceleration remains constant, and is equal to v_t^2/r. The acceleration vector is always directed toward the center of the circular path.

Example Is it possible for a car to move in a circular path in such a way that it has a tangential acceleration but no centripetal acceleration?
Reasoning Any object that moves such that the *direction* of its velocity changes has an acceleration. A car moving in a circular path will always have a centripetal acceleration.

Example A pail of water can be whirled in a vertical path such that no water is spilled. Why does the water remain in the pail, even when the pail is above your head?
Reasoning The tendency of the water is to move in a straight line path tangent to the circular path followed by the container. As a result, at the top of the circular path, the water is forced against the bottom of the pail, and the normal force exerted on the water by the pail provides the centripetal force required to keep the water moving in its circular path.

1. What is the magnitude of the angular speed, ω, of the second hand of a clock? What is the angular acceleration, α, of the second hand?
2. When a wheel of radius R rotates about a fixed axis, do all points on the wheel have the same angular speed? Do they all have the same linear speed? If the angular speed is constant and equal to ω_0, describe the linear speeds and linear accelerations of the points at $r = 0$, $r = R/2$, and $r = R$.
3. If a car's wheels are replaced with wheels of greater diameter, will the reading of the speedometer change? Explain.
4. Are the kinematic expressions for θ, ω, and α valid when the angular displacement is measured in degrees instead of radians?
5. Cite an example of a situation in which an automobile driver can have a centripetal acceleration but no tangential acceleration.
6. Cite a situation in which you could use Equation 7.6 to describe the motion of a record accelerating on a turntable and another situation in which this equation would not be valid.
7. Correct the following statement: "The racing car rounds the turn at a constant velocity of 90 miles per hour."
8. Describe the path of a moving body in the event that the acceleration is constant in magnitude at all times and (a) perpendicular to the velocity; (b) parallel to the velocity.
9. Imagine that you attach a heavy object to one end of a spring and then, holding the spring's other end, whirl the spring and object in a horizontal circle. Does the spring stretch? If so, why? Discuss in terms of centripetal force.
10. Why does mud fly off of a rapidly turning wheel?
11. Because of the Earth's rotation about its axis, you would weigh slightly less at the equator than at the poles. Why?
12. Centrifuges are often used in dairies to separate the cream from the milk. Which remains on the bottom?
13. Use Kepler's second law to convince yourself that the Earth must orbit faster during the winter, when it is closest to the Sun, than during the summer, when it is farthest from the Sun.
14. Estimate the gravitational force between you and a person 2 m away from you.
15. Consider the yo-yo in Figure 7.10. Is it possible to whirl it fast enough so that the cord makes an angle of 90° with the vertical? Explain.
16. Explain why the Earth is not spherical in shape and bulges at the equator.
17. We often think of the brakes and the gas pedal on a car as the mechanisms that accelerate and decelerate the car. Could a steering wheel also fall into either of these categories? Explain.
18. Explain the difference between centripetal and angular acceleration.
19. If someone told you that astronauts are weightless in orbit because they are beyond the pull of gravity, would you accept this statement? Explain.

PROBLEMS

Section 7.1   Angular Speed and Angular Acceleration

1. Convert the following angles in degrees to radians: 30°, 45°, 60°, 90°, 180°, 270°, and 360°.

2. Astronomers commonly use the parsec as the unit of measurement for very great distances. The parsec is defined as the distance at which 1 AU (astronomical unit) subtends an angle of 1 second of arc, where 1 AU equals the distance from the Earth to the Sun, and 1 second of arc equals 1/3600 degree (see Fig. 7.20). (a) Determine the length of one parsec in meters. (b) A lightyear is the distance traveled by light (in vacuum) in one year. Express one parsec in lightyears.

3. A wheel has a radius of 4.1 m. How far (path length) does a point on the circumference travel if the wheel is rotated through angles of 30°, 30 rad, and 30 rev, respectively?

4. (a) The hour and minute hands on a clock face coincide at 12 o'clock. Determine all the other times (up to the second) when these hands coincide. (b) If the clock also has a second hand, determine all the times when all three hands coincide.

5. Find the angular speed of the Earth about the Sun in radians per second and degrees per day.

6. A record has an angular speed of 33 rev/min. (a) What is its angular speed in rad/s? (b) Through what angle, in radians, does it rotate in 1.5 s?

7. A potter's wheel moves from rest to an angular speed of 0.20 rev/s in 30 s. Find its angular acceleration in radians per second per second.

8. (a) A full Moon is observed to subtend an angle of 0.52° when viewed from Earth. Verify this observation by using the diameter of the Moon and its distance from the Earth. (b) When the Sun is viewed from Earth, what angle does it subtend in the sky? (c) Why does the Sun appear to be the same size as the full Moon?

Section 7.2   Rotational Motion Under Constant Angular Acceleration

9. A dentist's drill starts from rest. After 3.20 s of constant angular acceleration it turns at a rate of 2.51 × 10⁴ rev/min. (a) Find the drill's angular acceleration. (b) Determine the angle (in radians) through which the drill rotates during this period.

10. An electric motor rotating a workshop grinding wheel at a rate of 100 rev/min is switched off. Assume constant negative angular acceleration of magnitude 2.00 rad/s². (a) How long does it take for the grinding wheel to stop? (b) Through how many radians has the wheel turned during the interval found in (a)?

11. A mass attached to a 50.0-cm-long string starts from rest and is rotated 40 times in one minute before reaching a final angular speed. (a) Determine the angular acceleration of the mass, assuming that it is constant. (b) What is the angular speed of the mass after 1.00 min?

12. A tire placed on a balancing machine in a service station starts from rest and turns through 4.7 revolutions in 1.2 s before reaching its final angular speed. Calculate its angular acceleration.

13. The tub of a washer goes into its spin-dry cycle, starting from rest and reaching an angular speed of 5.0 rev/s in 8.0 s. At this point the person doing the laundry opens the lid, and a safety switch turns off the washer. The tub slows to rest in 12.0 s. Through how many revolutions does the tub turn? Assume constant angular acceleration while it is starting and stopping.

14. A grinding wheel, initially at rest, is rotated with constant angular acceleration α = 5.0 rad/s² for 8.0 s. The wheel is then brought to rest, with uniform negative acceleration, in 10 rev. Determine the negative angular acceleration required and the time needed to bring the wheel to rest.

15. The driver of a car traveling at 30.0 m/s applies the brakes and undergoes a constant negative acceleration of 2.00 m/s². How many revolutions does each tire make before the car comes to a stop, assuming that the car does not skid and that the tires have radii of 0.300 m?

16. A machine part rotates at an angular speed of 0.60 rad/s; its speed is then increased to 2.2 rad/s at an angular acceleration of 0.70 rad/s². Find the angle through which the part rotates before reaching this final speed.

Section 7.3   Relations Between Angular and Linear Quantities

Section 7.4   Centripetal Acceleration

17. A race car starts from rest on a circular track of radius 400 m. The car's speed increases at the

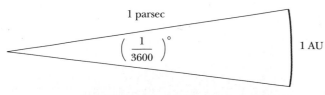

FIGURE 7.20   (Problem 2)

□ indicates problems that have full solutions available in the Student Solutions Manual and Study Guide.

constant rate of 0.500 m/s². At the point where the magnitudes of the centripetal and tangential accelerations are equal, determine (a) the speed of the race car, (b) the distance traveled, and (c) the elapsed time.

18. A coin with a diameter of 2.40 cm is dropped onto a horizontal surface. The coin starts out with an initial angular speed of 18.0 rad/s and rolls in a straight line without slipping. If the rotation slows with an angular acceleration of magnitude 1.90 rad/s², how far does the coin roll before coming to rest?

19. An air puck of mass 0.25 kg is tied to a string and allowed to revolve in a circle of radius 1.0 m on a frictionless horizontal table. The other end of the string passes through a hole in the center of the table, and a mass of 1.0 kg is tied to it (Fig. 7.21). The suspended mass remains in equilibrium while the puck on the tabletop revolves. (a) What is the tension in the string? (b) What is the centripetal force acting on the puck? (c) What is the speed of the puck?

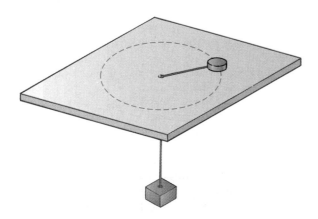

FIGURE 7.21 (Problem 19)

20. The turntable of a record player rotates initially at 33 rev/min and takes 20 s to come to rest. (a) What is the angular acceleration of the turntable, assuming it is uniform? (b) How many rotations does the turntable make before coming to rest? (c) If the radius of the turntable is 0.14 m, what is the initial linear speed of a bug riding on the rim?

21. A car is traveling at 17.0 m/s on a straight horizontal highway, which we assume extends from our left to our right. The wheels of the car have radii of 48.0 cm. If the car speeds up with an acceleration of 2.00 m/s² for 5.00 s, find the number of revolutions of the wheels during this period.

22. A bicycle is turned upside down while its owner repairs a flat tire. A friend spins the other wheel and observes that drops of water fly off tangentially. She measures the heights reached by drops moving vertically (Fig. 7.22). A drop that breaks loose from the tire on one turn rises 54.0 cm above the tangent point. A drop that breaks loose on the next turn rises 51.0 cm above the tangent point. The radius of the wheel is 0.381 m. She correctly infers that the height to which the drops rise decreases because the angular speed of the wheel decreases. Using only the observed heights and the radius of the wheel, estimate the wheel's angular acceleration (assuming it to be constant).

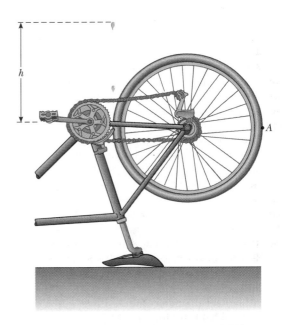

FIGURE 7.22 (Problems 22 and 63)

23. Young David experimented with slings before tackling Goliath. He found that he could develop an angular speed of 8.0 rev/s in a sling 0.60 m long. If he increased the length to 0.90 m, he could revolve the sling only six times per second. (a) Which angular speed gives the greater linear speed? (b) What is the centripetal acceleration at 8.0 rev/s? (c) What is the centripetal acceleration at 6.0 rev/s?

24. The speed of a moving bullet can be determined by allowing the bullet to pass through two rotating paper disks that are mounted a distance of *d* apart

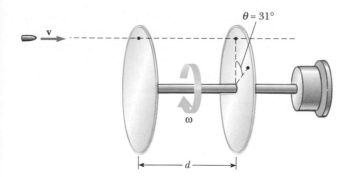

$\theta = 31°$

FIGURE 7.23   (Problem 24)

(Problem 29)   *(Color Box/FPG)*

on the same axle (Fig. 7.23). Find the bullet speed for the following data: $d = 80$ cm, ω (angular speed of the disks) = 900 rev/min, and $\Delta\theta$ (angular displacement of the two bullet holes in the disks) = 31°.

25. Find the centripetal accelerations of (a) a point on the equator of the Earth and (b) a point at the north pole of the Earth.

26. It has been suggested that rotating cylinders about 10 mi long and 5.0 mi in diameter be placed in space and used as colonies. What angular speed must such a cylinder have so that the centripetal acceleration at its surface equals Earth's gravity?

27. (a) What is the tangential acceleration of a bug on the rim of a 78-rpm, 10.0-in. diameter record if the record moves from rest to its final angular speed in 3.0 s? (b) When the record is at its final speed, what is the tangential velocity of the bug? (c) One second after the bug starts from rest, what are its tangential acceleration, radial acceleration, and total acceleration?

28. A 60.0-cm diameter wheel rotates with a constant angular acceleration of 4.00 rad/s². It starts from rest at $t = 0$, and a chalk line drawn to a point, P, on the rim of the wheel makes an angle of 57.3° with the horizontal at this time. At $t = 2.00$ s, find (a) the angular speed of the wheel, (b) the linear velocity and tangential acceleration of P, and (c) the position of P.

Section 7.5   Centripetal Force

29. A 40.0-kg child takes a ride on a Ferris wheel that rotates four times each minute and has a diameter of 18.0 m. (a) What is the centripetal acceleration of the child? (b) What force (magnitude and direction) does the seat exert on the child at the lowest point of the ride? (c) What force does the seat exert on the child at the highest point of the ride? (d) What force does the seat exert on the

child when the child is halfway between the top and bottom?

30. Tarzan ($m = 85$ kg) tries to cross a river by swinging from a 10-m-long vine. His speed at the bottom of the swing (as he just clears the water) is 8.0 m/s. Tarzan doesn't know that the vine has a breaking strength of 1000 N. Does he make it safely across the river? Justify your answer.

31. A 2000-kg car rounds a circular turn of radius 20 m. If the road is flat and the coefficient of friction between tires and road is 0.70, how fast can the car go without skidding?

32. A rotating wheel at an amusement park (Fig. 7.24) is 90 ft in diameter and makes one revolution in 6.0 s. Each 300-kg car is held in place by two bolts. Find the force each bolt exerts on the car when the wheel is rotating in a horizontal plane.

33. A roller-coaster vehicle has a mass of 500 kg when fully loaded with passengers (Fig. 7.25). (a) If the vehicle has a speed of 20.0 m/s at point A, what is the force of the track on the vehicle at this point?

FIGURE 7.24   (Problem 32) *(© David J. Cross/Peter Arnold, Inc.)*

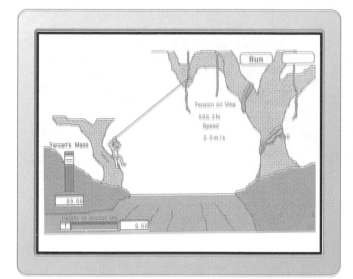

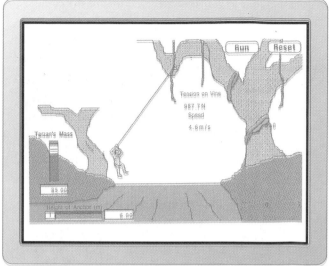

An example of the Interactive Physics simulation to accompany Problem 30.

(b) What is the maximum speed the vehicle can have at *B* in order for gravity to hold it on the track?

34. A sample of blood is placed in a centrifuge of radius 15.0 cm. The mass of a red corpuscle is 3.0×10^{-16} kg, and the magnitude of the centripetal force required to make it settle out of the plasma is 4.0×10^{-11} N. At how many revolutions per second should the centrifuge be operated?

35. An airplane is flying in a horizontal circle at a speed of 100 m/s. The 80.0-kg pilot does not want his radial acceleration to exceed 7 g. (a) What is the minimum radius of the circular path? (b) At this radius, what is the *net* centripetal force exerted on the pilot by the seat belts, the friction between him and the seat, and so forth?

36. A 50.0-kg child stands at the rim of a merry-go-round of radius 2.00 m, rotating with an angular speed of 3.00 rad/s. (a) What is the child's centripetal acceleration? (b) What is the minimum

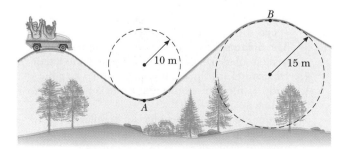

FIGURE 7.25 (Problem 33)

force between her feet and the floor of the carousel that is required to keep her in the circular path? (c) What minimum coefficient of static friction is required? Is the answer you found reasonable? In other words, is she likely to be able to stay on the merry-go-round?

37. An aerobatic airplane pilot experiences "weightlessness" at the top of a loop-the-loop maneuver. If her speed is 200 m/s at the time, find the radius of the loop.

38. An engineer wishes to design a curved exit ramp for a toll road in such a way that a car will not have to rely on friction to round the curve without skidding. He does so by banking the road in such a way that the necessary centripetal force will be supplied by the component of the normal force toward the center of the circular path. (a) Show that for a given speed of v and a radius of r, the curve must be banked at the angle θ such that $\tan \theta = v^2/rg$. (b) Find the angle at which the curve should be banked if a typical car rounds it at a 50.0-m radius and a speed of 13.4 m/s.

39. A pail of water is rotated in a vertical circle of radius 1.00 m (the approximate length of a person's arm). What must be the minimum speed of the pail at the top of the circle if no water is to spill out?

40. A stuntman whose mass is 70 kg swings from the end of a 4.0-m-long rope along the arc of a vertical circle. Assuming he starts from rest when the rope is horizontal, find the tensions in the rope that are required to make him follow his circular path, (a) at the beginning of his motion, (b) at a

height of 1.5 m above the bottom of the circular arc, and (c) at the bottom of his arc.

Section 7.7 Newton's Universal Law of Gravitation

Section 7.8 Kepler's Laws

41. A satellite moves in a circular orbit around the Earth at a speed of 5000 m/s. Determine (a) the satellite's altitude above the surface of the Earth and (b) the period of the satellite's orbit.

42. A 3.00×10^4-kg spaceship is halfway between the Earth and the Moon. Use the data in Table 7.2 to find the net gravitational force exerted on the ship by the Earth and the Moon.

43. A satellite whose orbital period is synchronous with the rotational period of the Earth is called a *geostationary* satellite because it has a fixed geographic longitude. Almost all communications satellites, including those used for cable TV, are geostationary. (a) Determine the orbital radius of a geostationary satellite. (b) How high (in miles) above the surface of the Earth is this?

(Problem 43) *(NASA/Peter Arnold, Inc.)*

 44. A coordinate system (in meters) is constructed on the surface of a pool table, and three masses are placed on the coordinate system as follows: a 2.0-kg mass at the origin, a 3.0-kg mass at (0, 2.0 m), and a 4.0-kg mass at (4.0, 0 m). Find the resultant gravitational force exerted on the mass at the origin by the other two masses.

45. During a solar eclipse, the Moon, Earth, and Sun all lie on the same line, with the Moon between the Earth and the Sun. (a) What force is exerted

on the Moon by the Sun? (b) What force is exerted on the Moon by the Earth? (c) What force is exerted on the Earth by the Sun?

46. Given that the Moon's period about the Earth is 27.32 days and the distance from the Earth to the Moon is 3.84×10^8 m, estimate the mass of the Earth. Assume the orbit is circular. Why do you suppose your estimate is high?

47. A satellite is in a circular orbit just above the surface of the Moon. (The radius of the Moon is 1738 km.) What are the satellite's (a) acceleration and (b) speed? (c) What is the period of the satellite orbit?

48. Show that Kepler's third law for the planets revolving around the Sun can be written $T^2 = r^3$, if T is expressed in years and r is expressed in astronomical units (AU), where 1 AU equals the distance from the Earth to the Sun.

49. An artificial satellite circling the Earth completes each orbit in 90.0 minutes. What is the value of g at the location of this satellite?

50. A 600-kg satellite is in a circular orbit about the Earth at a height above the Earth equal to the Earth's mean radius. Find (a) the satellite's orbital speed, (b) the period of its revolution, and (c) the gravitational force acting on it.

51. Use the data of Table 7.2 to find the point between the Earth and the Sun at which an object can be placed so that the net gravitational force exerted on it by these two objects is zero.

52. Io, a small moon of the giant planet Jupiter, has an orbital period of 1.77 days and an orbital radius of 4.22×10^5 km. From these data, determine the mass of Jupiter.

ADDITIONAL PROBLEMS

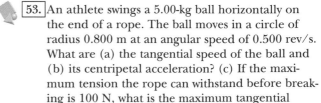

53. An athlete swings a 5.00-kg ball horizontally on the end of a rope. The ball moves in a circle of radius 0.800 m at an angular speed of 0.500 rev/s. What are (a) the tangential speed of the ball and (b) its centripetal acceleration? (c) If the maximum tension the rope can withstand before breaking is 100 N, what is the maximum tangential speed the ball can have?

54. The Solar Maximum Mission Satellite was placed in a circular orbit about 150 mi above the Earth. Determine (a) the orbital speed of the satellite and (b) the time required for one complete revolution.

55. A rotating bicycle wheel has an angular speed of 3.00 rad/s at some instant of time. It is then given an angular acceleration of 1.50 rad/s². A chalk line drawn on the wheel is horizontal at $t = 0$.

(a) What angle does this line make with its original direction at $t = 2.00$ s? (b) What is the angular speed of the wheel at $t = 2.00$ s?

56. A 0.50-kg ball that is tied to the end of a 1.5-m light cord is revolved in a horizontal plane with the cord making a 30° angle with the vertical (see Fig. 7.10). (a) Determine the ball's speed. (b) If the ball is revolved so that its speed is 4.0 m/s, what angle does the cord make with the vertical? (c) If the cord can withstand a maximum tension of 9.8 N, what is the highest speed at which the ball can move?

57. A high-speed sander has a disk 6.00 cm in radius that rotates about its axis at a constant rate of 1200 rev/min. Determine (a) the angular speed of the disk, in radians per second, (b) the linear speed of a point 2.00 cm from the disk's center, (c) the centripetal acceleration of a point on the rim, and (d) the total distance traveled by a point on the rim in 2.00 s.

58. Three masses are aligned along the x axis of a rectangular coordinate system so that a 2.0-kg mass is at the origin, a 3.0-kg mass is at (2.0, 0) m, and a 4.0-kg mass is at (4.0, 0) m. Find (a) the gravitational force exerted on the 4.0-kg mass by the other two masses and (b) the magnitude and direction of the gravitational force exerted on the 3.0-kg mass by the other two.

59. A copper block rests 30.0 cm from the center of a steel turntable. The coefficient of static friction between block and surface is 0.53. The turntable starts from rest at $t = 0$ and rotates with a constant angular acceleration of 0.50 rad/s². After what interval will the block start to slip on the turntable?

60. A massless spring of constant $k = 78.4$ N/m is fixed on the left side of a level track. A block of mass $m = 0.50$ kg is pressed against the spring and compresses it a distance of d, as in Figure 7.26. The block (initially at rest) is then released and travels toward a circular loop-the-loop of radius $R = 1.5$ m. The entire track and the loop-the-loop are frictionless except for the section of track between points A and B. Given that the coefficient of

kinetic friction between the block and the track along AB is $\mu_k = 0.30$, and that the length of AB is 2.5 m, determine the minimum compression, d, of the spring that enables the block to just make it through the loop-the-loop at point C. (*Hint:* The force of the track on the block will be zero if the block barely makes it through the loop-the-loop.)

61. A 0.400-kg pendulum bob passes through the lowest part of its path at a speed of 3.00 m/s. (a) What is the tension in the pendulum cable at this point if the pendulum is 80.0 cm long? (b) When the pendulum reaches its highest point, what angle does the cable make with the vertical? (c) What is the tension in the pendulum cable when the pendulum reaches its highest point?

62. A small block of mass $m = 0.50$ kg is fired with an initial speed of $v_0 = 4.0$ m/s along a horizontal section of frictionless track, as shown in the top portion of Figure 7.27. The mass then moves along the frictionless semicircular, *vertical* tracks of radius $R = 1.5$ m. (a) Determine the force of the track on the block at points A and B. (b) The bottom of the track consists of a section ($L = 0.40$ m) with friction. Determine the coefficient of kinetic friction between the block and that portion of the bottom track if the mass just makes it to point C on the first trip. (*Hint:* If the block just makes it to point C, the force of contact exerted on the block by the track at that point should be zero.)

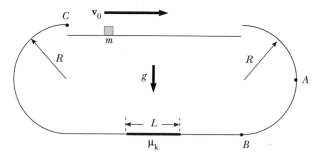

FIGURE 7.27 (Problem 62)

63. A piece of clay is initially at point A on the rim of a bicycle wheel (of radius 0.381 m) rotating counterclockwise about a horizontal axis at a constant angular speed (Fig. 7.22). The clay is dislodged from point A when the wheel diameter through A is horizontal. The clay then rises vertically and returns to point A the instant the wheel completes one revolution. Find the angular speed of the bicycle wheel.

64. A car moves at 10.0 m/s across a bridge made in the shape of a circular arc of radius 30.0 m.

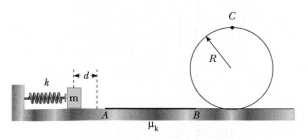

FIGURE 7.26 (Problem 60)

(a) Find the normal force acting on the car when it is at the top of the arc. (b) At what speed does the normal force go to zero? (*Hint:* The normal force becomes zero when the car loses contact with the road.)

65. Because of the Earth's rotation about its axis, a point on the equator experiences a centripetal acceleration of 0.0340 m/s² while a point at the poles experiences no centripetal acceleration. (a) Show that at the equator the gravitational force on an object (the true weight) must exceed the object's apparent weight. (b) What are the apparent weights at the equator and at the poles of a 75.0-kg person? (Assume the Earth is a uniform sphere, and take $g = 9.800$ m/s².)

66. A skier starts at rest at the top of a large hemispherical hill (Fig. 7.28). Neglecting friction, show that the skier will leave the hill and become airborne at a distance of $h = R/3$ below the top of the hill. (*Hint:* At this point, the normal force goes to zero.)

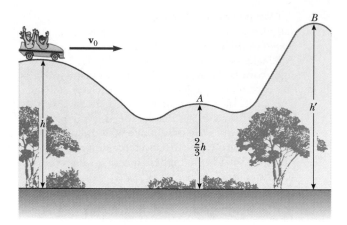

FIGURE 7.29   (Problem 67)

FIGURE 7.28   (Problem 66)

67. A frictionless roller coaster is given an initial velocity of v_0 at height h, as in Figure 7.29. The radius of curvature of the track at point A is R. (a) Find the maximum value of v_0 so that the roller coaster stays on the track at A solely because of gravity. (b) Using the value of v_0 calculated in (a), determine the value of h' that is necessary if the roller coaster is to just make it to point B.

68. A car rounds a banked curve where the radius of curvature of the road is R, the banking angle is θ, and the coefficient of static friction is μ. (a) Determine the range of speeds the car can have without slipping up or down the road. (b) What is the range of speeds possible if $R = 100$ m, $\theta = 10°$, and $\mu = 0.10$ (slippery conditions)?

69. In a popular amusement park ride, a rotating cylinder of radius 3.00 m is set in rotation at an an-

gular speed of 5.00 rad/s, as in Figure 7.30. The floor then drops away, leaving the riders suspended against the wall in a vertical position. What minimum coefficient of friction between a rider's clothing and the wall is needed to keep the rider from slipping? (*Hint:* Recall that the magnitude of the maximum force of static friction is equal to μn, where n is the normal force—in this case, the centripetal force.)

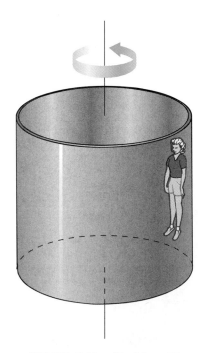

FIGURE 7.30   (Problem 69)

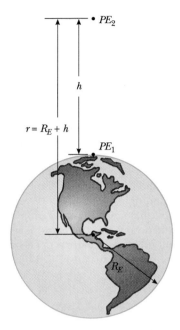

FIGURE 7.31   (Problem 70)

"I'll be working on the largest and smallest objects in the Universe—superclusters and neutrinos. I'd like you to handle everything in between."

70. The general expression for the gravitational potential energy for an object of mass m at a distance of r from the center of the Earth is

$$PE = -G\frac{M_E m}{r}$$

where M_E is the mass of the Earth. An object is moved from the surface of the Earth to a point a distance of h above the Earth's surface, as in Figure 7.31. (a) Show that the difference in potential energy between these two points is

$$PE_2 - PE_1 = G\frac{M_E mh}{R_E(R_E + h)}$$

(b) Show that the difference in potential energy given by the preceding expression reduces to the familiar relation $PE = mgh$ when h is small compared to the Earth's radius.

71. An object of mass m is projected vertically upward from the Earth's surface. Neglect air resistance.

(a) Show that the *escape speed*—the minimum initial speed that will enable the object to reach infinity with a speed of zero—is

$$v_{esc} = \sqrt{\frac{2GM_E}{R_E}}$$

where M_E is the mass of the Earth and R_E is the radius of the Earth. (*Hint:* Make use of the general expression for potential energy given in Problem 70, and note that the total energy of the system remains constant.) (b) Calculate the escape speed from the Earth for a 5000-kg spacecraft.

E
S
S
A
Y

THE BIG BANG

The Universe is expanding. This fundamental observation underlies all of modern cosmological thought. In a classic paper published in 1931, Edwin Hubble and Milton Humason compared distances and velocities of remote galaxies, which are gigantic, gravitationally bound systems of billions of stars. They established that these galaxies are moving away from us with velocities that are proportional to their distances from us. This result, which has been substantiated and extended to still more distant galaxies, demonstrates that the Universe is expanding.

This concept can be easily understood by analogy. Suppose a cook is making raisin bread. After yeast has been mixed into the dough and the bread is set aside to rise, it doubles in each dimension during the next hour, as in Figure 1. The distance between each pair of raisins also doubles. Since each distance doubled during the hour,

FIGURE 1

Expanding raisin bread. (From George O. Abell, David Morrison, and Sydney C. Wolff, *Exploration of the Universe*, 5th ed., Philadelphia, Saunders College Publishing, 1987; p. 659.)

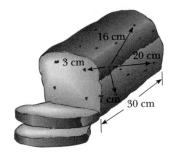

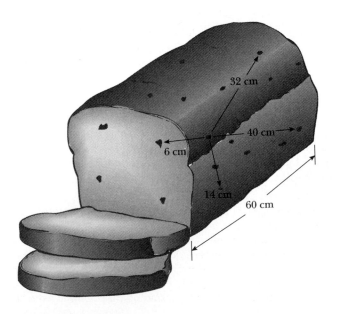

if one raisin is selected as the origin of a coordinate system, every other raisin must move away from this origin at a speed proportional to its distance from the origin. The same is true no matter which raisin you select.

From this analogy, it is clear that, if the Universe is uniformly expanding, all observers, regardless of location, must see all other galaxies moving away from them at speeds that are proportional to their distances. As Hubble and Humason showed, that is precisely what we observe on Earth.

Knowing that our Universe is expanding, we can imagine what must have happened in the past. If we extrapolate backward in time, we find all the galaxies coming together, until some time in the distant past when all the matter in the Universe was crowded to an extreme density. This is a condition that marks the beginning of the Universe, or at least of that Universe we can know about. At that beginning, the Universe suddenly began its expansion with a phenomenon called the "Big Bang."

With the data now available, we can even estimate when our Universe began. The total amount of matter and energy of the Universe creates gravitation. This mutual attraction must slow the expansion, which means that in the past the expansion rate must have been higher than it is today. At the extreme, if the total mass-energy density is low enough that gravitation is ineffective (an essentially "empty" Universe), the deceleration would be zero and the Universe would always have been expanding at the present rate.

That extreme of an empty Universe corresponds to the greatest age of the Universe. Consequently, we can estimate the upper limit to the age of the Universe by asking how long it took for distant galaxies, always moving away from us at their present rates, to have reached those distances. Call that maximum possible age T_0. According to the work of Hubble and Humason, the speed v of a galaxy is proportional to its distance from us, a result that is expressed by the equation $v = Hr$, where H is the constant of proportionality between speed and distance (called the Hubble constant) and r is the present distance between that galaxy and Earth. Since speed is distance divided by time, we can express v as $v = r/T_0$. Hence,

$$\frac{r}{T_0} = Hr$$

$$T_0 = \frac{1}{H}$$

In other words, the maximum age of the Universe is just the reciprocal of the Hubble constant. The value of the Hubble constant is uncertain, primarily because of the uncertainties in deriving the distances to remote galaxies. The best estimate is that the Big Bang occurred 10 to 15 billion years ago.

If our Universe had a beginning, will it also have an end? Will the expansion go on forever or will gravity stop the expansion and force the galaxies to fall together again in a "big crunch"? A major focus of astronomical research at the present time is the effort to answer these questions. One approach combines the law of gravitation and the cosmological principle, which asserts that, apart from local irregularities, the Universe at any given time is the same everywhere. If this principle is valid, we can reach conclusions about the Universe as a whole from measurements of the small portion close enough to study in detail.

Let us consider a spherical region of the Universe that contains a large number of galaxies. The rest of the Universe can be regarded as a hollow spherical shell since, according to the cosmological principle, the matter in this shell is distributed uniformly in all directions. Newton showed that the gravitational forces exerted by a spherical shell on an object in its interior all cancel, so that the net force is zero. We can then picture our sphere as being isolated and uninfluenced by any gravitational forces from the rest of the Universe. The cosmological principle assures us that any conclusion we reach about this sphere must apply to the rest of the Universe. If the sphere chosen is a small one so that the speeds of the galaxies within it are small relative to the speed of light, then Newton's law of gravitation can be applied.

Consider a galaxy located at the surface of our imaginary sphere. It is subject to the gravitational attraction of all the galaxies in the sphere and is decelerated. The situation is analogous to an object launched upward from the surface of the Earth. If the speed of the object exceeds the escape speed, the object will never return to the Earth's surface. If the speed is less than the escape speed, the object will fall back to the ground. Similarly, if the speed of a given galaxy is high enough to overcome the attractive force of all the other galaxies in the sphere, that galaxy will continue to move away from them, and the expansion of the Universe will continue. If the speed is too low, however, gravity will decelerate the galaxy until its speed becomes zero, and the galaxy will then reverse direction and begin to move closer to the other galaxies in the sphere.

For an object launched from the Earth, the escape speed is determined by the mass of the Earth. The escape speed of a galaxy at the surface of our sphere of galaxies is determined by the total mass within the sphere or by the average density of the Universe (since by the cosmological principle, the density of the Universe is everywhere the same). If the density of the Universe is larger than some critical value, gravity will be strong enough to halt the expansion. If the density is smaller than this critical value, the expansion will go on forever.

This density can be estimated from the mass associated with the galaxies in our imaginary sphere. Most galaxies occur in clusters that can be "weighed" by measuring the gravitational influence exerted by one galaxy on another. For example, if two galaxies are in orbit around each other, and if their separation and speeds are known, their combined mass can be estimated from Kepler's third law. There are mathematical procedures that are not much more complex for estimating the total mass of a cluster containing not just two but many galaxies.

The best estimates of the average density of the Universe suggest that the amount of matter it contains is no more than 20 percent of the value required to halt expansion. There are theoretical arguments that the density may equal the critical density required to slow the expansion to a halt, but the deceleration is so gradual in this case that an infinite amount of time is required to reduce the expansion speed to zero. Therefore, both observation and theory suggest that galaxies will never reverse direction.

If this picture is right, there can be no rejuvenation of the Universe. At the time of the Big Bang, the Universe was very dense and hot. As it expanded, it cooled, and hydrogen and helium nuclei and then atoms formed. Somehow—theory is not yet very good at explaining this point—hydrogen and helium atoms came together to form concentrations of matter that later became galaxies. Stars formed within these protogalaxies and in many galaxies star formation has continued to the present. Each star eventually exhausts its sources of nuclear energy. Some stars die gently, other explosively, hurling a portion of their mass into space to be incorporated into a new generation of stars. As they die, most stars leave behind a portion of their mass in the form of a dense core that becomes a black dwarf, a neutron star, or a black hole. This material is forever lost to the Universe; it radiates no energy and cannot, under any conditions we now believe likely to occur, be broken apart into raw material for new stars. Ultimately, therefore, after all the available gas and dust are used up, star formation must cease, a final generation of stars will die, and the galaxies will fade into blackness.

Astronomy and physics as we understand them today offer no alternative to this outcome. However, today's knowledge almost certainly is not complete. With time, the picture presented here of the origin, evolution, and ultimate fate of the material Universe may undergo a radical transformation. Indeed, many astronomers are attracted to this area of research precisely because the subject is not fully understood and seems likely to yield its secrets to a systematic scientific attack. To learn something new, something never known before, is the ultimate challenge and satisfaction of scientific research.

Questions and Problems

1. When Hubble and Humason first reported their results for the relationship between distance and speed for galaxies, they thought the constant of proportionality H was equal to about 200 km/s per million lightyears. How did their estimate for the age of the Universe differ from the modern value based on $H = 20$ km/s per million lightyears? The oldest stars in the Universe are about 15 billion years old. Is this consistent with Hubble and Humason's value for H?
2. Explain why the observation that all galaxies are moving away from us does not mean that we are at the center of the Universe.
3. Derive an expression for escape speed in terms of mean density. Now apply this result to two galaxies, separated by a distance R and moving away from each other at the escape speed, and derive an expression for the critical density at which the escape and expansion speeds are equal. Assume that $H = 20$ km/s per million lightyears. How does your value for the critical density compare with the observed density, currently estimated to be less than about 10^{-30} g/cm^3? What does your calculation indicate about the ultimate fate of the Universe?

8 ROTATIONAL EQUILIBRIUM AND ROTATIONAL DYNAMICS

Derek Swinson, professor of physics at the University of New Mexico, demonstrating the "gyro-ski" technique. The skier initiates a turn by lifting the axle of the rotating bicycle wheel. The direction of the turn depends on whether the left or right hand is used to lift the axle from the horizontal. Ignoring friction and gravity, the angular momentum of the system (the skier and the bicycle wheel) is conserved. (Courtesy of Derek Swinson)

We shall begin this chapter by studying objects that are at rest or moving with constant velocity; such objects are said to be in equilibrium. Knowledge of the conditions that prevail when an object is in equilibrium is important in a variety of fields. For example, students of architecture or industrial technology benefit from an understanding of the forces that act on buildings or large machines, and biology students should understand the forces at work in muscles and bones. Newton's first law, as discussed in Chapter 4, is the basis for much of our work in this chapter. In addition, in order to fully understand objects in equilibrium, we must consider torque. This concept will also play a key role in our discussion of rotational motion.

In this chapter we will complete our study of rotational motion. We shall build on the definitions of angular speed and angular acceleration encountered in Chapter 7 by examining the relationship between these concepts and the forces that produce rotational motion. Specifically, we shall find the rotational analog of Newton's second law and define a term that needs to be added to our equation for conservation of mechanical energy: rotational kinetic energy. One of the central purposes of this chapter is to develop the concept of angular momentum, a quantity that plays a key role in rotational motion. Finally, just as we found that linear momentum is conserved, we shall also find that the angular momentum of any isolated system is always conserved.

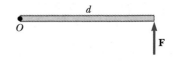

FIGURE 8.1
A bird's-eye view of a door hinged at O, with a force applied perpendicularly to the door.

8.1

TORQUE

Consider Figure 8.1, an overhead view of a door hinged at point O. From this viewpoint, the door is free to rotate about a line perpendicular to the page and passing through O. When the force **F** is applied at the outer edge, as shown, the door can easily be caused to rotate counterclockwise; that is, the rotational effect of the force is quite large. On the other hand, the same force applied at a point nearer the hinges produces a smaller rotational effect on the door.

> The ability of a force to rotate a body about some axis is measured by a quantity called the **torque, τ.** The torque due to a force of **F** has the magnitude
>
> $$\tau = Fd \qquad \text{[8.1]}$$

In this equation, τ (the Greek letter tau) is the torque, and the distance d is the **lever arm** (or moment arm) of the force **F**.

> The lever arm is the perpendicular distance from the axis of rotation to a line drawn along the direction of the force.

As an example, consider the wrench pivoted about the axis O in Figure 8.2a. In this case, the applied force **F** acts at an angle of ϕ with the horizontal. If you examine the definition of lever arm just given, you will see that in this case the lever arm is the distance d shown in the figure and not L, the length of the wrench. That is, d is the perpendicular distance from the axis of rotation to the

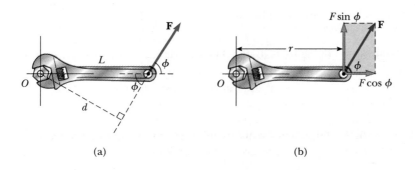

(a)   (b)

FIGURE 8.2
(a) A force, **F**, acting at an angle of ϕ with the horizontal produces a torque of magnitude Fd about the pivot O. (b) The component $F \sin \phi$ tends to rotate the system about O. The component $F \cos \phi$ produces no torque about O.

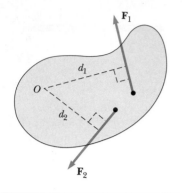

FIGURE 8.3
The force $\mathbf{F}_1$ tends to rotate the body counterclockwise about O, while $\mathbf{F}_2$ tends to rotate the body clockwise.

line along which the applied force acts. In this case, d is related to L by the expression $d = L \sin \phi$. Thus, the net torque produced by the force $\mathbf{F}$ is given by $\tau = FL \sin \phi$. Actually, computing torque in situations such as this is usually best accomplished by resolving the force into components, as shown in Figure 8.2b. The torque about the axis of rotation produced by the component $F \cos \phi$ is zero because the lever arm of this component is zero. That is, the distance from the pivot to the line along which the force acts is zero, because the line along which the force acts passes through the axis of rotation. The component $F \sin \phi$ has a lever arm of L, and the torque produced by this component is, as before, $\tau = FL \sin \phi$.

If two or more forces are acting on an object, as in Figure 8.3, then each has a tendency to produce a rotation about the pivot O. For example, $\mathbf{F}_2$ tends to rotate the object clockwise and $\mathbf{F}_1$ tends to rotate the object counterclockwise. We shall use the convention that the sign of the torque resulting from a force is positive if its turning tendency is counterclockwise and negative if its turning tendency is clockwise. In Figure 8.3, then, the torque resulting from $\mathbf{F}_1$, which has a moment arm of d_1, is positive and equal to $F_1 d_1$; the torque associated with $\mathbf{F}_2$ is negative and equal to $-F_2 d_2$. The *net torque* acting on the object about O is found by summing the torques:

$$\sum \tau = \tau_1 + \tau_2 = F_1 d_1 - F_2 d_2$$

Notice that the units of torque are units of force times length, such as the newton-meter (N·m) or the pound-foot (lb·ft).

EXAMPLE 8.1   The Spinning Safe

Figure 8.4 is a top view of a safe being pushed by two equal and opposite forces acting as shown. Find the net torque exerted on the safe if its width is 1.0 m. Assume an axis of rotation through the center of the safe.

Solution   The torque produced by $\mathbf{F}_1$ is

$$\tau_1 = F_1 d_1 = -(500 \text{ N})(0.50 \text{ m}) = -250 \text{ N·m}$$

and the torque produced by $\mathbf{F}_2$ is

$$\tau_2 = F_2 d_2 = -(500 \text{ N})(0.50 \text{ m}) = -250 \text{ N·m}$$

Because each force produces clockwise rotation, the torques are both negative. Thus, the net torque is -500 N·m.

EXAMPLE 8.2   The Swinging Door

Find the torque produced by the 300-N force applied at an angle of 60° to the door of Figure 8.5a.

Reasoning   In Figure 8.5b, the 300-N force has been replaced by its horizontal and vertical components:

$$F_x = (300 \text{ N}) \cos 60° = 150 \text{ N}$$

$$F_y = (300 \text{ N}) \sin 60° = 260 \text{ N}$$

A convenient and obvious location for the axis of rotation is the hinge of the door.

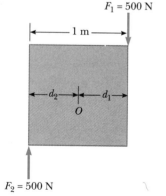

FIGURE 8.4
(Example 8.1)   A top view of a safe being pushed with equal but opposite forces.

Solution   In this case, the 150-N force produces zero torque about the axis of rotation, because the line along which the force acts passes through the axis of rotation and hence the lever arm is zero. The 260-N force has a lever arm of 2.0 m and thus produces a torque of

$$\tau = (260 \text{ N})(2.0 \text{ m}) = \boxed{520 \text{ N}\cdot\text{m}}$$

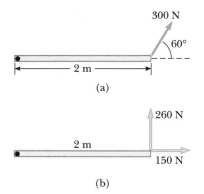

FIGURE 8.5
(Example 8.2)   (a) A top view of a door being pulled by a 300-N force. (b) The components of the 300-N force.

8.2

TORQUE AND THE SECOND CONDITION FOR EQUILIBRIUM

In Chapter 4 we examined some situations, under the heading of the first condition of equilibrium, in which an object has no net external force acting on it; that is, $\Sigma F_x = 0$ and $\Sigma F_y = 0$. We found there that such an object either remains at rest if it is at rest, or moves with a constant velocity if it is in motion. However, the first condition is not sufficient to ensure that an object is in complete equilibrium. This can be understood by considering the situation illustrated in Figure 8.4. In this situation, the two applied forces acting on the safe are equal in magnitude and opposite in direction. The first condition for equilibrium is satisfied because the two external forces balance each other, and yet the safe can still move—it will rotate clockwise because the forces do not act through a common point.

The Chinese acrobats in this difficult formation represent a balanced mechanical system. The external forces acting on the system, as shown by the blue vectors, are the weights of the acrobats, $\mathbf{w}_1$ and $\mathbf{w}_2$, and the upward force of the support, $\mathbf{n}$, on the lower acrobat. The vector sum of these external forces must be zero in such a balanced system. The net external torque acting on such a balanced system must also be zero. (*J. P. Lafont, Sygma*)

This case illustrates that, to fully understand the effect of a force or group of forces on an object, one must know not only the magnitude and direction of the force(s) but the point of application. That is, the net torque acting on an object must be considered. An object in rotational equilibrium has no angular acceleration.

> The **second condition for equilibrium** asserts that if an object is in rotational equilibrium, the net torque acting on it about any axis must be zero. That is,
>
> $$\sum \tau = 0 \qquad \text{[8.2]}$$

You can see that a body in static equilibrium must satisfy two conditions:

The two conditions for static equilibrium

1. The resultant external force must be zero.

$$\sum \mathbf{F} = 0$$

2. The resultant external torque must be zero.

$$\sum \tau = 0$$

The first condition is a statement of translational equilibrium, and the second is a statement of rotational equilibrium. We shall discuss torque and its relation to rotational motion in more detail later in the chapter.

POSITION OF THE AXIS OF ROTATION

In the cases we have been describing so far, the axes of rotation for calculating torques have been selected without explanation. Often the nature of a problem suggests a convenient location for the axis, but just as often no single location stands out as being preferable. You might ask, "What axis should I choose in calculating the net torque?" The answer is:

> If the object is in equilibrium, it does not matter where you put the axis of rotation for calculating the net torque; the location of the axis is completely arbitrary.

Example 8.3 will illustrate this important point.

8.3

THE CENTER OF GRAVITY

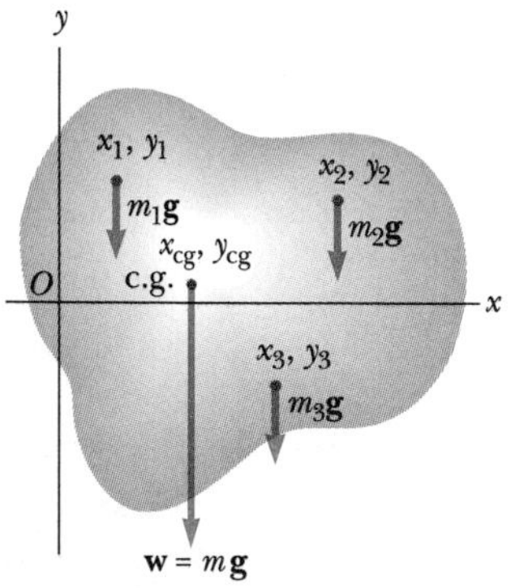

One of the forces that must be considered in dealing with a rigid object is the object's weight—that is, the force of gravity acting on the object. To compute the torque due to the weight force, all of the weight can be thought of as concentrated at a single point called the center of gravity.

Consider an object of arbitrary shape lying in the xy plane, as in Figure 8.6. The object is divided into a large number of very small particles of weight $m_1 g$,

FIGURE 8.6
The center of gravity of an object is the point where all the weight of the object can be considered to be concentrated.

$m_2 g$, $m_3 g$, . . . having coordinates (x_1, y_1), (x_2, y_2), (x_3, y_3), ... Each particle contributes a torque about the origin that is equal to its weight multiplied by its lever arm. For example, the torque due to the weight $m_1 g_1$ is $m_1 g_1 x_1$, and so forth.

We wish to locate the one position of the single force w—the total weight of the object—whose effect on the rotation of the object is the same as that of the individual particles. This point is called the **center of gravity** of the object. Equating the torque exerted by w at the center of gravity to the sum of the torques acting on the individual particles gives

$$(m_1 g + m_2 g + m_3 g + \cdots)x_{cg} = m_1 g x_1 + m_2 g x_2 + m_3 g x_3 + \cdots$$

If we assume that g is uniform over the object (which will be the case in all the situations we will examine), then the g terms in the preceding equation cancel and we get

$$x_{cg} = \frac{m_1 x_1 + m_2 x_2 + m_3 x_3 + \cdots}{m_1 + m_2 + m_3 + \cdots} = \frac{\Sigma m_i x_i}{\Sigma m_i} \qquad \textbf{[8.3]}$$

Similarly, the y coordinate of the center of gravity of the system can be found from

$$y_{cg} = \frac{\Sigma m_i y_i}{\Sigma m_i} \qquad \textbf{[8.4]}$$

The center of gravity of a homogeneous, symmetric body must lie on the axis of symmetry. For example, the center of gravity of a homogeneous rod must lie midway between the ends of the rod. The center of gravity of a homogeneous sphere or a homogeneous cube must lie at the geometric center of the object. One can determine the center of gravity of an irregularly shaped object, such as a wrench, experimentally by suspending the wrench from two different points (Fig. 8.7). The wrench is first hung from point A, and a vertical line, AB (which can be established with a plumb bob), is drawn when the wrench is in equilibrium. The wrench is then hung from point C, and a second vertical line, CD, is drawn. The center of gravity coincides with the intersection of these two lines. In fact, if the wrench is hung freely from any point, the vertical line through that point must pass through the center of gravity.

In several examples in Section 8.4, we shall be concerned with homogeneous, symmetric objects whose centers of gravity coincide with their geometric centers. A rigid object in a uniform gravitational field can be balanced by a single force equal in magnitude to the weight of the object, as long as the force is directed upward through the object's center of gravity.

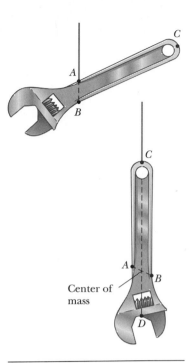

FIGURE 8.7
An experimental technique for determining the center of gravity of a wrench. The wrench is hung freely from two different pivots, A and C. The intersection of the two vertical lines, AB and CD, locates the center of gravity.

EXAMPLE 8.3 **Where Is the Center of Gravity?**

Three particles are located in a coordinate system as shown in Figure 8.8. Find the center of gravity.

Reasoning The y coordinate of the center of gravity is zero because all the particles are on the x axis. To find the x coordinate of the center of gravity, we use Equation 8.3:

$$x_{cg} = \frac{\Sigma m_i x_i}{\Sigma m_i}$$

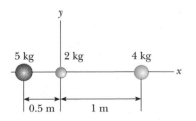

FIGURE 8.8
(Example 8.3) Locating the center of gravity for a system of three particles.

Solution   For the numerator, we find

$$\sum m_i x_i = m_1 x_1 + m_2 x_2 + m_3 x_3$$

$$= (5.00 \text{ kg})(-0.500 \text{ m}) + (2.00 \text{ kg})(0 \text{ m}) + (4.00 \text{ kg})(1.00 \text{ m})$$

$$= 1.50 \text{ kg} \cdot \text{m}$$

The denominator is $\sum m_i = 11.0 \text{ kg}$; therefore,

$$x_{cg} = \frac{1.50 \text{ kg} \cdot \text{m}}{11.0 \text{ kg}} = \boxed{0.136 \text{ m}}$$

Exercise   If a fourth particle, of mass 2.00 kg, is placed at $x = 0$, $y = 0.250$ m, find the x and y coordinates of the center of gravity for this system of four particles.

Answer $x_{cg} = 0.115$ m; $y_{cg} = 0.0380$ m.

8.4
EXAMPLES OF OBJECTS IN EQUILIBRIUM

In Chapter 4 we discussed some techniques for solving problems concerned with objects in equilibrium. Recall that when the objects were treated as geometric points, it was sufficient to simply apply the condition that the net force on the object must be zero. In this chapter we have shown that for objects of finite dimensions, a second condition for equilibrium must be satisfied—namely that the net torque on the object must also be zero. The following general procedure is recommended for solving problems that involve objects in equilibrium.

The Golden Gate Bridge in San Francisco is an example of a structure in equilibrium. The supporting cables are under tension induced by the weight and loads on the bridge. *(Superstock)*

PROBLEM-SOLVING STRATEGY
Objects in Equilibrium

1. Draw a simple, neat diagram of the system.
2. Isolate the object that is being analyzed. Draw a free-body diagram showing all external forces that are acting on this object. For systems that contain more than one object, draw a *separate* diagram for each object. Do not include forces that the object exerts on its surroundings.
3. Establish convenient coordinate axes for each body and find the components of the forces along these axes. Now apply the first condition of equilibrium (the net force on the object in the *x* and *y* direction must be zero) for each object under consideration.
4. Choose a convenient origin for calculating the net torque on the object. Now apply the second condition of equilibrium (the net torque on the object about any origin must be zero). Remember that the choice of the origin for the torque equation is arbitrary; therefore, choose an origin that will simplify your calculation as much as possible. Note that a force that acts along a line passing through the point chosen as the axis of rotation gives zero contribution to the torque.
5. The first and second conditions for equilibrium will give a set of simultaneous equations with several unknowns. All that is left to complete your solution is to solve for the unknowns in terms of the known quantities.

EXAMPLE 8.4   The Seesaw

A uniform 40.0-N board supports two children weighing 500 N and 350 N (Fig. 8.9). The support (often called the *fulcrum*) is under the center of gravity of the board, and the 500-N child is 1.50 m from the center.

(a) Determine the upward force, **n**, exerted on the board by the support.

Reasoning   First note that, in addition to **n**, the external forces acting on the board are the weights of the children and the weight of the board, all of which

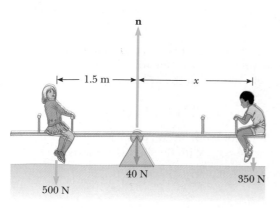

FIGURE 8.9
(Example 8.4)   Two children balanced on a seesaw.

act downward. We can assume that the board's center of gravity is at its geometric center because we were told that the board is uniform. Since the system is in equilibrium, the upward force **n** must balance all the downward forces.

Solution   From $\Sigma F_y = 0$, we have

$$n - 500 \text{ N} - 350 \text{ N} - 40.0 \text{ N} = 0 \quad \text{or} \quad n = \boxed{890 \text{ N}}$$

Although the equation $\Sigma F_x = 0$ also applies to this situation, it is unnecessary to consider it because no forces are acting horizontally on the board.

(b) Determine where the 350-N child should sit to balance the system.

Reasoning   To find this position, we must invoke the second condition for equilibrium. We take the center of gravity of the board as the axis for the torque equation. This choice simplifies the problem, because the torques produced by both **n** and the 40.0-N weight are zero about this axis (because the lever arm of each is zero).

Solution   We apply $\Sigma \tau = 0$ to find

$$(500 \text{ N})(1.50 \text{ m}) - (350 \text{ N})(x) = 0$$

$$x = \boxed{2.14 \text{ m}}$$

(c) Repeat part (b), using another axis for the torque computations.

Reasoning   We stated that when an object is in equilibrium, the choice for the axis about which to compute torques is completely arbitrary. To illustrate this point, let us choose an axis perpendicular to the page and passing through the location of the 500-N child.

Solution   In this case $\Sigma \tau = 0$ yields

$$n(1.50 \text{ m}) - (40.0 \text{ N})(1.50 \text{ m}) - (350 \text{ N})(1.50 + x) = 0$$

From part (a) we know that $n = 890$ N. Thus, we can solve for x to find $x = 2.14$ m, in agreement with the result of part (b).

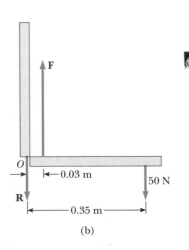

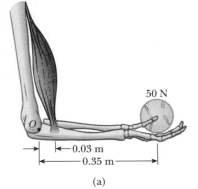

FIGURE 8.10
(Example 8.5)   (a) A weight held with the forearm horizontal. (b) The mechanical model for the system.

EXAMPLE 8.5   A Weighted Forearm

A 50.0-N weight is held in a person's hand with the forearm horizontal, as in Figure 8.10a. The biceps muscle is attached 0.0300 m from the joint, and the weight is 0.350 m from the joint. Find the upward force exerted on the forearm (the ulna) by the biceps and the downward force on the forearm (the humerus) acting at the joint. Neglect the weight of the forearm.

Solution   The forces acting on the forearm are equivalent to those acting on a bar of length 0.350 m, as shown in Figure 8.10b, where **F** is the upward force of the biceps and **R** is the downward force at the joint. From the first condition for equilibrium, we have

$$(1) \quad \sum F_y = F - R - 50.0 \text{ N} = 0$$

From the second condition for equilibrium, we know that the sum of the torques about any point must be zero. With the joint O as the axis, we have

$$F(0.0300 \text{ m}) - (50.0 \text{ N})(0.350 \text{ m}) = 0$$

$$F = \boxed{583 \text{ N}}$$

This value for F can be substituted into (1) to give $R = 533$ N. The two values correspond to $F = 131$ lb and $R = 119$ lb. Clearly, the forces at joints and in muscles can be extremely large.

EXAMPLE 8.6   Walking a Horizontal Beam

A uniform, horizontal 300-N beam, 5.00 m long, is attached to a wall by a pin connection that allows the beam to rotate. Its far end is supported by a cable that makes an angle of 53.0° with the horizontal (Fig. 8.11a). If a 600-N person stands 1.50 m from the wall, find the tension in the cable and the force exerted on the beam by the wall.

Reasoning   First we must identify all the external forces acting on the beam and sketch them on a free-body diagram. This is shown in Figure 8.11b. The forces on the beam consist of the weight of the beam, 300 N, acting downward; the downward force exerted on the beam by the man, which is equal in magnitude to his weight, 600 N; the tension force, **T**, in the cable; and the force of the wall on the beam, **R**. We now resolve the forces **T** and **R** into their horizontal and vertical components, as shown in Figure 8.11c. Note that the *x* component of the tension force ($T \cos 53.0°$) is to the left, whereas the *y* component ($T \sin 53.0°$) is upward. The horizontal and vertical components of **R** are denoted by R_x and R_y, respectively. The first condition for equilibrium can now be applied in the *x* and *y* direction to give us two equations in terms of our unknowns R_x, R_y, and T. The necessary third equation can be found from the second condition of equilibrium.

Solution   From the first condition for equilibrium, we find

(1) $\sum F_x = R_x - T \cos 53.0° = 0$

(2) $\sum F_y = R_y + T \sin 53.0° - 600 \text{ N} - 300 \text{ N} = 0$

The unknowns are R_x, R_y, and T. Because there are three unknowns and only two equations, we cannot find a solution from just the first condition of equilibrium.

Now let us use the second condition of equilibrium. The axis that passes through the pivot at the wall is a convenient one to choose for the torque equation because the forces R_x, R_y, and $T \cos 53.0°$ all have lever arms of zero and hence have zero torque about this pivot. Recalling our sign convention for the torque about an axis and noting that the lever arms of the 600-N, 300-N, and $T \sin 53.0°$ forces are 1.50 m, 2.50 m, and 5.00 m, respectively, we get

(3) $\sum \tau_O = (T \sin 53.0°)(5.00 \text{ m}) - (300 \text{ N})(2.50 \text{ m}) - (600 \text{ N})(1.50 \text{ m}) = 0$

$$T = \boxed{413 \text{ N}}$$

Thus, the torque equation using this axis gives us one of the unknowns immediately! This value for T is then substituted into (1) and (2) to give

$$R_x = \boxed{249 \text{ N}} \qquad R_y = \boxed{570 \text{ N}}$$

If we selected some other axis for the torque equation, the solution would be the same. For example, if the axis were to pass through the center of gravity of the beam, the torque equation would involve both T and R_y; together with (1) and (2), however, it could still be solved for the unknowns. Try it!

Exercise   Repeat this problem, but with the direction of R_x opposite that shown in Figure 8.11c. What answers do you get for T, R_x, and R_y?

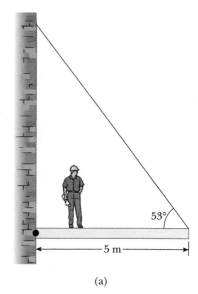

(a)

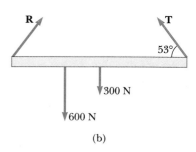

(b)

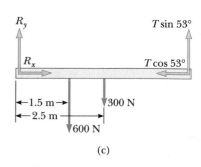

(c)

FIGURE 8.11
(Example 8.6)   (a) A uniform beam attached to a wall and supported by a cable. (b) A free-body diagram for the beam. (c) The component form of the free-body diagram.

Answer $T = 413$ N, $R_x = -249$ N, and $R_y = 570$ N. The negative sign for R_x means its direction was chosen incorrectly. The direction of R_x must be to the right, as shown in Figure 8.11c.

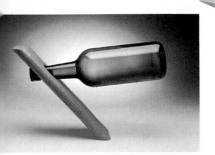

This one-bottle wine holder is an interesting example of a balanced mechanical system, which seems to defy gravity. The system (wine holder + bottle) is balanced when its center of gravity is directly over the lowest support point so that the net torque acting on the system is zero.
(Courtesy of Charles Winters)

EXAMPLE 8.7   Don't Climb the Ladder

A uniform 10-m-long, 50-N ladder rests against a smooth vertical wall as in Figure 8.12a. If the ladder is just on the verge of slipping when the angle it makes with the ground is 50°, find the coefficient of static friction between the ladder and ground.

Reasoning   Figure 8.12b is the free-body diagram for the ladder, showing all external forces acting on it. At the base of the ladder the Earth exerts an upward normal force, **n**, and a force of static friction, **f**, acts to the right. The wall exerts the force **P** to the left. Note that **P** is horizontal because the wall is smooth. (If the wall were rough, an upward frictional force would be exerted on the ladder.) The first condition for equilibrium can now be applied in the *x* and *y* directions to give us two equations in terms of our unknowns *f*, *P*, and *n*. The necessary third equation can be found from the second condition of equilibrium.

Solution   From the first condition for equilibrium applied to the ladder, we have

$$(1) \qquad \sum F_x = f - P = 0$$

$$(2) \qquad \sum F_y = n - 50 \text{ N} = 0$$

From (2) we see that $n = 50$ N. Furthermore, when the ladder is on the verge of slipping, the force of static friction must be maximum and given by the relation $f_{s,\max} = \mu_s n = \mu_s (50 \text{ N})$. Thus, (1) reduces to

$$(3) \qquad \mu_s (50 \text{ N}) = P$$

Let us now apply the second condition of equilibrium and take the torques about the axis O at the bottom of the ladder, as in Figure 8.12c. The force **P** and the weight of the ladder are the only forces that contribute to the torque about this axis, and their lever arms are shown in Figure 8.11c. Note that, since the length of the ladder is 10 m, the lever arm for **P** is $d_1 = 10$ m sin 50°. Likewise, the lever arm for the 50-N weight force is $d_2 = 5.0$ m cos 50°, where the weight vector acts through the center because the ladder is uniform. Thus, we find that

$$\sum \tau_O = P(10 \text{ m sin } 50°) - (50 \text{ N})(5.0 \text{ m cos } 50°) = 0$$

$$P = 21 \text{ N}$$

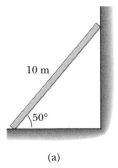

(a)

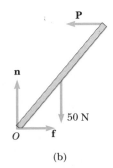

(b)

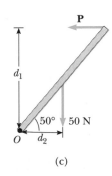

(c)

FIGURE 8.12
(Example 8.7)   (a) A ladder leaning against a frictionless wall. (b) A free-body diagram for the ladder. (c) Lever arms for the forces **w** and **P**.

Now that P is known, we can substitute its value into (3) to find μ_s:

$$\mu_s = \frac{21 \text{ N}}{50 \text{ N}} = \boxed{0.42}$$

8.5

RELATIONSHIP BETWEEN TORQUE AND ANGULAR ACCELERATION

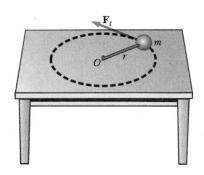

FIGURE 8.13
A mass, m, attached to a light rod of length r moves in a circular path on a frictionless horizontal surface while the tangential force $\mathbf{F}_t$ acts on it.

Earlier in this chapter we considered the situation in which both the net force and the net torque acting on an object are zero. Such objects are said to be in equilibrium. We shall now examine the behavior of an object when the net torque acting on it is not zero. As you shall see, when a rigid object is acted on by a net torque, it undergoes an angular acceleration. Furthermore, the angular acceleration is directly proportional to the net torque. The end result of our investigation will be an expression that is analogous to $\mathbf{F} = m\mathbf{a}$ in translational motion.

Let us begin by considering the system shown in Figure 8.13, which consists of a mass m connected to a very light rod of length r. The rod is pivoted at the point O, and its movement is confined to rotation on a frictionless *horizontal* table. Now let us assume that a force, $\mathbf{F}_t$, perpendicular to the rod and hence tangent to the circular orbit, is acting on m. Since there is no force to oppose this tangential force, the mass undergoes a tangential acceleration according to Newton's second law:

$$F_t = ma_t$$

Multiplying the left and right sides of this equation by r gives

$$F_t r = mra_t$$

In Chapter 7 we found that the tangential acceleration and angular acceleration for a particle rotating in a circular path are related by the expression

$$a_t = r\alpha$$

so now we find that

$$F_t r = mr^2 \alpha \qquad \textbf{[8.5]}$$

The left side of Equation 8.5, which should be familiar to you, is the torque acting on the mass about its axis of rotation. That is, the torque is equal in magnitude to the force on m multiplied by the perpendicular distance from the pivot to the line along which the force acts, or $\tau = F_t r$. Hence, we can write Equation 8.5 as

$$\boxed{\tau = mr^2 \alpha} \qquad \textbf{[8.6]}$$

Equation 8.6 shows that the torque on the system is proportional to angular acceleration, where the constant of proportionality, mr^2, is called the *moment of inertia* of the mass m. (Because the rod is very light, its moment of inertia can be neglected.)

TORQUE ON A ROTATING OBJECT

Now consider a solid disk rotating about its axis as in Figure 8.14a. The disk consists of many particles at various distances from the axis of rotation, as in Figure 8.14b. The torque on each one of these particles is given by Equation 8.6.

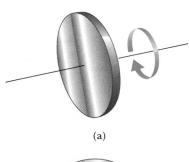

(a)

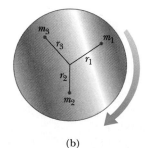

(b)

FIGURE 8.14
(a) A solid disk rotating about its axis. (b) The disk consists of many particles, all with the same angular acceleration.

The *total* torque on the disk is given by the sum of the individual torques on all the particles:

$$\sum \tau = \left(\sum mr^2 \right) \alpha \qquad \text{[8.7]}$$

Note that, because the disk is rigid, all particles have the *same* angular acceleration, so α is not involved in the sum. If the masses and distances of the particles are labeled with subscripts as in Figure 8.14b, then

$$\sum mr^2 = m_1 r_1{}^2 + m_2 r_2{}^2 + m_3 r_3{}^2 + \cdots$$

This quantity is called the **moment of inertia** of the whole body and is given the symbol I:

Moment of inertia

$$I \equiv \sum mr^2 \qquad \text{[8.8]}$$

The moment of inertia has the SI units $kg \cdot m^2$. Using this result in Equation 8.7, we see that the total torque on a rigid body rotating about a fixed axis is given by

Net torque

$$\sum \tau = I\alpha \qquad \text{[8.9]}$$

> **The angular acceleration of an object is proportional to the net torque acting on it. The proportionality constant, I, between the net torque and angular acceleration is the moment of inertia.**

It is important to note that the equation $\Sigma \tau = I\alpha$ (Eq. 8.9) is the rotational counterpart to Newton's second law, $\Sigma F = ma$. Thus, the correspondence between rotational motion and linear motion continues. Recall from Chapter 7 that the linear variables x, v, and a are replaced in rotational motion by the variables θ, ω, and α. Likewise, we now see that *the force and mass in linear motion correspond to torque and moment of inertia in rotational motion.* In this chapter we shall develop other equations for the rotational kinetic energy and the angular momentum of a body rotating about a fixed axis. Based on the analogies already presented, you should be able to predict the form of these equations.

MORE ON THE MOMENT OF INERTIA

Because the moment of inertia of a body—as defined by $I = \Sigma mr^2$ (Eq. 8.8)—will be used throughout the remainder of this chapter, it will be useful to examine it in more detail before discussing other aspects of rotational motion. As seen earlier, a small object (or particle) has a moment of inertia equal to mr^2 about some axis. Now consider a somewhat more complicated system, the baton being twirled by a majorette in Figure 8.15. Let us assume that the baton can be modeled as a very light rod of length 2ℓ with a heavy mass at each end. (The rod of a real baton has significant mass relative to its ends.) Since we are neglecting the mass of the rod, the moment of inertia of the baton about an axis through its center and perpendicular to its length is given by Equation 8.8:

$$I = \sum mr^2$$

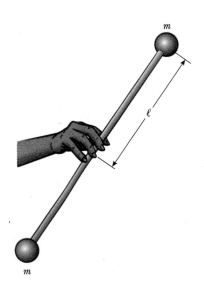

m

ℓ

m

FIGURE 8.15

A baton of length 2ℓ and mass $2m$ (the mass of the connecting rod is neglected). The moment of inertia about the axis through the baton's center and perpendicular to its length is $2m\ell^2$.

Because in this system there are two equal masses equidistant from the axis of rotation, we see that $r = \ell$ for each mass, and the sum is

$$I = \sum mr^2 = m\ell^2 + m\ell^2 = 2m\ell^2$$

We pointed out earlier that I is the rotational counterpart of m. However, there are some important distinctions between the two. For example, mass is an intrinsic property of an object that does not change, whereas *the moment of inertia of a system depends upon the axis of rotation and upon the manner in which the mass is distributed.* Examples 8.8 and 8.9 illustrate this point.

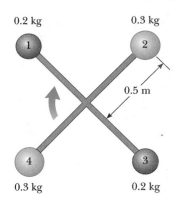

FIGURE 8.16
(Example 8.8)   Four masses connected to light rods rotating in the plane of the page.

EXAMPLE 8.8   The Baton Twirler

In an effort to be the star of the half-time show, a majorette twirls a highly unusual baton made up of four masses fastened to the ends of light rods (Fig. 8.16). Each rod is 1.0 m long. Find the moment of inertia of the system about an axis perpendicular to the page and passing through the point where the rods cross.

Solution   Applying Equation 8.8, we get

$$I = \sum mr^2 = m_1 r_1^2 + m_2 r_2^2 + m_3 r_3^2 + m_4 r_4^2$$

$$= (0.20 \text{ kg})(0.50 \text{ m})^2 + (0.30 \text{ kg})(0.50 \text{ m})^2 + (0.20 \text{ kg})(0.50 \text{ m})^2$$

$$+ (0.30 \text{ kg})(0.50 \text{ m})^2$$

$$= \boxed{0.25 \text{ kg} \cdot \text{m}^2}$$

EXAMPLE 8.9   The Baton Twirler—Second Act

Not satisfied with the crowd reaction from the baton twirling of Example 8.8, the majorette tries spinning her strange baton about the axis OO', as shown in Figure 8.17. Calculate the moment of inertia about this axis.

Solution   Again applying $I = \Sigma mr^2$, we have

$$I = (0.20 \text{ kg})(0)^2 + (0.30 \text{ kg})(0.50 \text{ m})^2 + (0.20 \text{ kg})(0)^2 + (0.30 \text{ kg})(0.50 \text{ m})^2$$

$$= \boxed{0.15 \text{ kg} \cdot \text{m}^2}$$

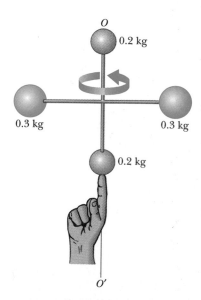

FIGURE 8.17
(Example 8.9)   A double baton rotating about the axis OO'.

CALCULATION OF MOMENTS OF INERTIA FOR EXTENDED OBJECTS

The method used for calculating moments of inertia in Examples 8.8 and 8.9 is simple enough when you have only a few small masses rotating about an axis. The situation becomes much more complex when the object is an extended mass, such as a sphere, a cylinder, or a cone. One type of extended object that is amenable to a simple solution is a hoop rotating about an axis perpendicular to its plane and passing through its center, as shown in Figure 8.18. A bicycle tire, for example, would fit in this category.

To evaluate the moment of inertia of the hoop, we can still use the equation $I = \Sigma mr^2$ (Eq. 8.8) and imagine that the hoop is divided into a number of small segments having masses $m_1, m_2, m_3, \ldots$, as in Figure 8.18. This approach is just

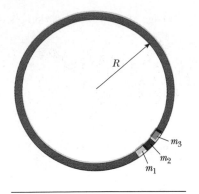

FIGURE 8.18
A uniform hoop can be divided into a large number of small segments that are equidistant from the center of the hoop.

an extension of the baton problem described in foregoing examples, except that now we have a large number of small masses in rotation instead of only four.

We can express the sum for I as

$$I = \sum mr^2 = m_1 r_1^2 + m_2 r_2^2 + m_3 r_3^2 + \cdots$$

All of the segments around the hoop are at the *same distance, R,* from the axis of rotation; thus, we can drop the subscripts on the distances and factor out the common factor R^2:

$$I = (m_1 + m_2 + m_3 + \cdots)R^2$$

We know, however, that the sum of the masses of all the segments must equal the total mass of the hoop, M:

$$M = m_1 + m_2 + m_3 + \cdots$$

and so we can express I as

$$I = Mr^2 \qquad\qquad \textbf{[8.10]}$$

This expression can be used for the moment of inertia of any ring-shaped object rotating about an axis through its center and perpendicular to its plane. Note that the result is strictly valid only if the thickness of the ring is small relative to its inner radius.

The hoop we selected as an example is unique in that we were able to find an expression for its moment of inertia by using only simple algebra. Unfortunately, most extended objects are more difficult to work with, and the methods of

TABLE 8.1
Moments of Inertia for Various Rigid Bodies of Uniform Composition

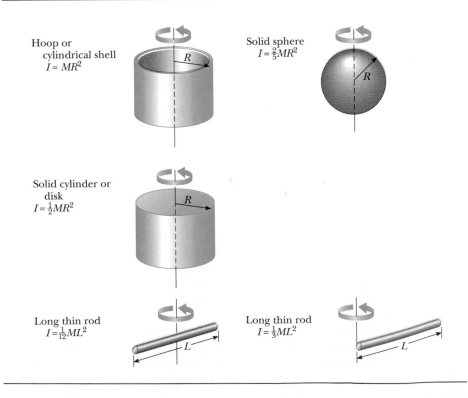

Hoop or cylindrical shell
$I = MR^2$

Solid sphere
$I = \frac{2}{5}MR^2$

Solid cylinder or disk
$I = \frac{1}{2}MR^2$

Long thin rod
$I = \frac{1}{12}ML^2$

Long thin rod
$I = \frac{1}{3}ML^2$

integral calculus are required. Such methods are beyond the scope of this text. The moments of inertia for some common shapes are given without proof in Table 8.1. When the need arises, you can use this table to determine the moment of inertia of a body having any one of the listed shapes.

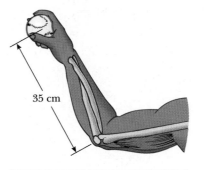

35 cm

FIGURE 8.19
(Example 8.10)   A ball being tossed by a pitcher. The forearm is being used to accelerate the ball.

EXAMPLE 8.10   Warming Up

A baseball player loosening up his arm before a game tosses a 0.150-kg baseball using only the rotation of his forearm to accelerate the ball (Fig. 8.19). The ball starts at rest and is released with a speed of 30.0 m/s in 0.300 s.

(a) Find the constant angular acceleration of the arm and ball.

Solution   During its acceleration, the ball moves through an arc of a circle with a radius of 0.350 m. We can determine the angular acceleration using $\omega = \omega_0 + \alpha t$. Since $\omega_0 = 0$, however, $\omega = \alpha t$, or

$$\alpha = \frac{\omega}{t}$$

We also know that $v = r\omega$, and so we get

$$\alpha = \frac{\omega}{t} = \frac{v}{rt} = \frac{30.0 \text{ m/s}}{(0.350 \text{ m})(0.300 \text{ s})} = \boxed{286 \text{ rad/s}^2}$$

(b) Find the torque exerted on the ball to give it this angular acceleration.

Solution   The moment of inertia of the ball about an axis that passes through the elbow, perpendicularly to the arm, is

$$I = mr^2 = (0.150 \text{ kg})(0.350 \text{ m})^2 = 1.84 \times 10^{-2} \text{ kg·m}^2$$

Thus, the required torque is

$$\tau = I\alpha = (1.84 \times 10^{-2} \text{ kg·m}^2)(286 \text{ rad/s}^2) = \boxed{5.26 \text{ N·m}}$$

EXAMPLE 8.11   The Falling Bucket

A solid, frictionless cylindrical pulley of mass $M = 3.00$ kg and radius $R = 0.400$ m is used to draw water from a well (Fig. 8.20a). A bucket of mass $m = 2.00$ kg is attached to a cord that is wrapped around the cylinder. If the bucket starts from rest at the top of the well and falls for 3.00 s before hitting the water, how far does it fall?

Reasoning   Figure 8.20b shows the two forces on the bucket as it falls: **T** is the tension in the cord, and $m\mathbf{g}$ is the weight of the bucket. We shall choose downward as the positive direction and write Newton's second law for the bucket as

$$mg - T = ma$$

When the given quantities are substituted into this equation, we have

$$(1) \qquad (2.00 \text{ kg})(9.80 \text{ m/s}^2) - T = (2.00 \text{ kg})a$$

With one equation and two unknowns, we must develop an additional equation to complete the problem. To obtain this second equation, let us consider the cylinder's rotational motion. Equation 8.9 applied to the cylinder gives the necessary expression:

$$\tau = I\alpha = \tfrac{1}{2}MR^2\alpha$$

(a)

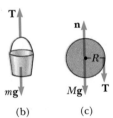

(b)   (c)

FIGURE 8.20
(Example 8.11)   (a) A water bucket attached to a rope passing over a frictionless pulley. (b) A free-body diagram for the bucket. (c) The tension produces a torque on the cylinder about its axis of rotation.

Figure 8.20c shows that the only force producing a torque on the cylinder as it rotates about an axis through its center is **T**, the force due to the tension in the cord. Actually, two other forces act on the cylinder—its weight and the upward force of the axle; but we do not have to consider them here because the lever arm of each about the axis of rotation is zero. Thus, we have

$$T(0.400 \text{ m}) = \tfrac{1}{2}(3.00 \text{ kg})(0.400 \text{ m})^2(\alpha)$$

$$(2) \qquad T = (0.600 \text{ kg} \cdot \text{m})\alpha$$

At this point, it is important to recognize that the downward acceleration of the bucket is equal to the tangential acceleration of a point on the rim of the cylinder. Therefore, the angular acceleration of the cylinder and the linear acceleration of the bucket are related by $a_t = r\alpha$. When this relation is used in (2), we get

$$(3) \qquad T = (1.50 \text{ kg})a_t$$

Solution   Equations (1) and (3) can now be solved simultaneously to find a_t and T. This procedure gives

$$a_t = 5.60 \text{ m/s}^2 \qquad T = 8.40 \text{ N}$$

Finally, we turn to the equations for motion with constant linear acceleration to find the distance, d, that the bucket falls in 3.00 s. Since $v_0 = 0$, we get

$$d = v_0 t + \tfrac{1}{2}at^2 = \tfrac{1}{2}(5.60 \text{ m/s}^2)(3.00 \text{ s})^2 = \boxed{25.2 \text{ m}}$$

8.6

ROTATIONAL KINETIC ENERGY

In Chapter 5 we defined the kinetic energy of a particle moving through space with a speed of v as the quantity $\frac{1}{2}mv^2$. Analogously,

> A body rotating about some axis with an angular speed of ω has rotational kinetic energy given by $\frac{1}{2}I\omega^2$.

To prove this, consider a rigid plane body rotating about some axis perpendicular to its plane, as in Figure 8.21. The body consists of many small particles, each of mass m. All these particles rotate in circular paths about the axis. If r is the distance of one of the particles from the axis of rotation, the speed of this particle is $v = r\omega$. Since the *total* kinetic energy of the body is the sum of all the kinetic energies associated with all the particles making up the body, we have

$$KE_r = \sum \left(\tfrac{1}{2}mv^2\right) = \sum \left(\tfrac{1}{2}mr^2\omega^2\right) = \tfrac{1}{2}\left(\sum mr^2\right)\omega^2$$

$$KE_r \equiv \tfrac{1}{2}I\omega^2 \qquad\qquad \textbf{[8.11]}$$

where $I = \sum mr^2$ is the moment of inertia of the body. Note that the ω^2 term is factored out since it is the same for every particle.

We found the energy concept to be extremely useful for describing the linear motion of a system. It can be equally useful for simplifying the analysis of rotational motion. We have now developed expressions for three types of energy: *gravitational potential energy, PE_g, translational kinetic energy, KE_t,* and *rotational kinetic energy, KE_r.* We must include all these forms of energy in our equation for conservation of mechanical energy.

$$(KE_t + KE_r + PE_g)_i = (KE_t + KE_r + PE_g)_f \qquad\qquad \textbf{[8.12]}$$

where i and f refer to initial and final values, respectively. This relation is *only* true if we ignore dissipative forces such as friction.

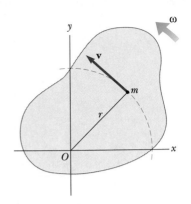

FIGURE 8.21
A rigid plane body rotating about the z axis with angular speed ω. The kinetic energy of a particle of mass m is $\frac{1}{2}mv^2$. The total kinetic energy of the body is $\frac{1}{2}I\omega^2$.

EXAMPLE 8.12   A Ball Rolling Down an Incline

A ball of mass M and radius R starts from rest at a height of 2.00 m and rolls down a 30.0° slope, as shown in Figure 8.22. What is the linear speed of the ball when it leaves the incline? Assume that the ball rolls without slipping.

Reasoning   The initial energy of the ball is gravitational potential energy, and when the ball reaches the bottom of the ramp, this potential energy has been converted to translational and rotational kinetic energy. The conservation of mechanical energy equation becomes

$$(PE_g)_i = (KE_t + KE_r)_f$$

$$Mgh = \tfrac{1}{2}Mv^2 + \tfrac{1}{2}\left(\tfrac{2}{5}MR^2\right)\omega^2$$

where h is the distance through which the ball's center of gravity falls and where we have used $I = \tfrac{2}{5}MR^2$ from Table 8.1 as the moment of inertia of the ball. If the ball rolls without slipping, a point on its surface must have the same instan-

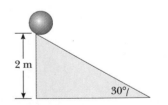

FIGURE 8.22
(Example 8.12)   A ball starts from rest at the top of an incline and rolls to the bottom without slipping.

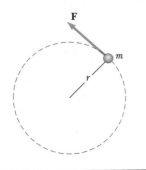

FIGURE 8.23
An object of mass m rotating in a circular path under the action of a constant torque.

taneous speed as its center of gravity has relative to the incline.[1] Thus, we can relate the ball's linear speed to its rotational speed:

$$v = R\omega$$

This expression can be used to eliminate ω from the equation for conservation of mechanical energy:

$$Mgh = \tfrac{1}{2}Mv^2 + \tfrac{1}{5}Mv^2$$

Solution   Solving for v,

$$v = \sqrt{\frac{10\,gh}{7}} = \sqrt{\frac{10(9.80 \text{ m/s}^2)(2.00 \text{ m})}{7}} = \boxed{5.29 \text{ m/s}}$$

Exercise   Repeat this example for a solid cylinder of the same mass and radius as the ball and released from the same height. In a race between the two objects on the incline, which one would win?

Answer $v = \sqrt{4gh/3} = 5.11$ m/s; the ball would win.

8.7
ANGULAR MOMENTUM

In Figure 8.23, an object of mass m is positioned a distance of r away from a center of rotation. Under the action of a constant torque on the object, its angular speed will increase from the value ω_0 to the value ω in a time of Δt. Thus, we can write

$$\tau = I\alpha = I\left(\frac{\omega - \omega_0}{\Delta t}\right) = \frac{I\omega - I\omega_0}{\Delta t}$$

If we define the product

Angular momentum

$$L \equiv I\omega \qquad\qquad \textbf{[8.13]}$$

as the angular momentum of the object, then we can write

$$\tau = \frac{\text{change in angular momentum}}{\text{time interval}} = \frac{\Delta L}{\Delta t} \qquad\qquad \textbf{[8.14]}$$

Equation 8.14 is the rotational analog of Newton's second law, $F = \Delta p / \Delta t$, and states that *the torque acting on an object is equal to the time rate of change of the object's angular momentum.*

When the net external torque ($\Sigma \tau$) acting on the system is zero, we see from Equation 8.14 that $\Delta L / \Delta t = 0$. In this case, the rate of change of the system's angular momentum is zero. Therefore, *the product $I\omega$ remains constant in time.* That is, $L_i = L_f$, or

---

[1] Note that a point on the surface of the ball travels a distance of $2\pi R$ (relative to the center of the ball) during one rotation. If no slippage occurs, this is also the distance traveled by the center of gravity in the same time interval. Thus, the center of gravity has the same speed relative to the incline as a point on the surface of the ball has relative to the center.

$$I_i \omega_i = I_f \omega_f \qquad \text{if} \quad \sum \tau = 0 \qquad \textbf{[8.15]}$$

The angular momentum of a system is conserved when the net external torque acting on the system is zero. That is, **when $\Sigma \tau = 0$, the initial angular momentum equals the final angular momentum.**

In Equation 8.15 we have a third conservation law to add to our list: the **conservation of angular momentum.** We now can state that

The energy, linear momentum, and angular momentum of an isolated system all remain constant.

There are many examples of conservation of angular momentum, some of which should be familiar to you. You may have observed a figure skater spinning in the finale of her act. The skater's angular speed increases when she pulls her hands and feet close to the trunk of her body, as in Figure 8.24. That is, $\omega_2 > \omega_1$. Neglecting friction between skater and ice, we see that there are no external torques on the skater. The moment of inertia of her body decreases as her hands and feet are brought in. The resulting change in angular speed is accounted for as follows. Since angular momentum must be conserved, the product $I\omega$ has to remain constant, and a decrease of the moment of inertia of the skater must be compensated by a corresponding increase in the angular speed.

Similarly, when a diver or acrobat wishes to make several somersaults, he pulls his hands and feet close to the trunk of his body in order to rotate at a greater angular speed. In this case, the external force due to gravity acts through the center of gravity and hence exerts no torque about the axis of rotation, and the angular momentum about the center of gravity is conserved. For example, when a diver wishes to double her angular speed, she must reduce her moment of inertia to half its initial value.

FIGURE 8.24
(a) The angular speed of this skater increases when she pulls her arms in close to her body, demonstrating that angular momentum is conserved. (b) Photograph of Kristi Yamaguchi, winner of a Gold Medal in the 1992 Olympics. *(PIC/Rick Stewart)*

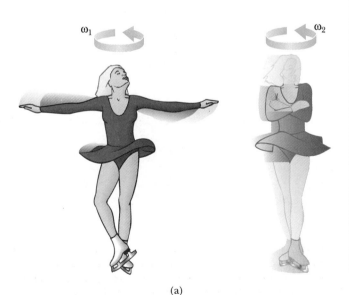

(a)

(b)

Satellite photograph of Hurricane Elena showing the eye of the hurricane. *(NASA)*

PROBLEM-SOLVING STRATEGY
Rotational Motion

Keep the following facts and procedures in mind when solving rotational motion problems.

1. Very few new techniques must be learned in order to solve rotational motion problems. For example, problems involving the equation $\Sigma \tau = I\alpha$ are very similar to those encountered in Newton's second law problems, $\Sigma F = ma$. Note the correspondences between linear and rotational quantities, in that F is replaced by τ, m by I, and a by α.

2. Other analogs between rotational quantities and linear quantities include the replacement of x by θ and v by ω. These are helpful as memory devices for such rotational motion quantities as rotational kinetic energy, $KE_r = \frac{1}{2}I\omega^2$, and angular momentum, $L = I\omega$.

3. With the analogs mentioned in Step 2, techniques for conservation of energy are the same as those examined in Chapter 5, except for the fact that the object's rotational kinetic energy must be included in the expression for the conservation of energy.

4. Likewise, the techniques for solving problems in conservation of angular momentum are essentially the same as those for solving problems in conservation of linear momentum, except that total initial angular momentum is equated to total final angular momentum as $I_i \omega_i = I_f \omega_f$.

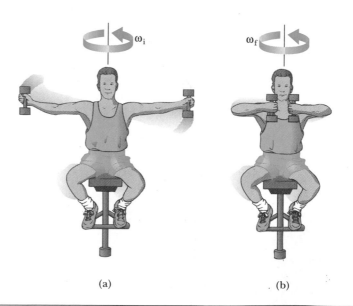

(a)   (b)

FIGURE 8.25
(Example 8.13)   (a) This student is given an initial angular speed while holding two masses as shown. (b) When the masses are pulled in close to the body, the angular speed of the system increases. Why?

EXAMPLE 8.13   The Spinning Stool

A student sits on a pivoted stool while holding a pair of masses (Fig. 8.25). The stool is free to rotate about a vertical axis with negligible friction. The moment of inertia of student, masses, and stool is 2.25 kg·m². The student is set in rotation with an initial angular speed of 5.00 rad/s, with masses outstretched. As he rotates, he pulls the masses inward so that the new moment of inertia of the system (student, masses, and stool) becomes 1.80 kg·m². What is the new angular speed of the system?

Reasoning   We shall apply the principle of conservation of momentum to find the new angular speed. The initial angular momentum, $I_i\omega_i$ will be equated to the final angular momentum $I_f\omega_f$.

Solution   The initial angular momentum of the system is

$$L_i = I_i\omega_i = (2.25 \text{ kg·m}^2)(5.00 \text{ rad/s}) = 11.3 \text{ kg·m}^2/\text{s}$$

When the masses are pulled in, they are closer to the axis of rotation, and as a result the moment of inertia of the system is reduced. The new angular momentum is

$$L_f = I_f\omega_f = (1.80 \text{ kg·m}^2)\omega_f$$

Because the net external torque on the system is zero, angular momentum is conserved. Thus, we find that

$$(11.3 \text{ kg·m}^2/\text{s}) = (1.80 \text{ kg·m}^2)\omega_f$$

$$\omega_f = \boxed{6.28 \text{ rad/s}}$$

EXAMPLE 8.14   The Merry-Go-Round

A student stands at the edge of a circular platform that rotates in a horizontal plane about a frictionless vertical axle (Fig. 8.26). The platform has a mass of $M = 100$ kg and a radius of $R = 2.00$ m. The student, whose mass is $m =$

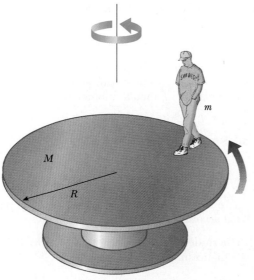

FIGURE 8.26
(Example 8.14)   As this student walks toward the center of the rotating platform, the angular speed of the system increases because the angular momentum of the system (student + platform) must remain constant.

This student is holding the axle of a spinning bicycle wheel while seated on a pivoted stool. The student and stool are initially at rest while the wheel spins in a horizontal plane. When the wheel is inverted about its center by 180°, the student and stool begin to rotate. That is because the total angular momentum of the system (wheel + student + stool) must be conserved. *(Courtesy of Central Scientific Company)*

60.0 kg, walks slowly from the rim of the disk toward the center. If the angular speed of the system is 2.00 rad/s when the student is at the rim, calculate the angular speed when the student reaches a point 0.500 m from the center.

Reasoning   We shall use the principle of conservation of angular momentum. The initial angular momentum of the system is the sum of the angular momentum of the platform plus that of the student when he is at the rim of the merry-go-round. The final angular momentum is the sum of the angular momentum of the platform plus that of the student when he is 0.500 m from the center.

Solution   The moment of inertia of the platform, I_p, is

$$I_p = \tfrac{1}{2}MR^2 = \tfrac{1}{2}(100 \text{ kg})(2.00 \text{ m})^2 = 200 \text{ kg} \cdot \text{m}^2$$

Treating the student as a point mass, his initial moment of inertia is

$$I_s = mR^2 = (60.0 \text{ kg})(2.00 \text{ m})^2 = 240 \text{ kg} \cdot \text{m}^2$$

Thus, the initial angular momentum of the platform plus student is

$$L_i = (I_p + I_s)(\omega_i) = (200 \text{ kg} \cdot \text{m}^2 + 240 \text{ kg} \cdot \text{m}^2)(2.00 \text{ rad/s}) = 880 \text{ kg} \cdot \text{m}^2/\text{s}$$

When the student has walked to the position 0.500 m from the center, his moment of inertia is

$$I_s' = mr_f^2 = (60.0 \text{ kg})(0.500 \text{ m})^2 = 15.0 \text{ kg} \cdot \text{m}^2$$

No change occurs in the moment of inertia of the platform. Because there are no external torques on the *system* (student plus platform) about the axis of rotation, we can apply the law of conservation of angular momentum:

$$L_i = L_f$$

$$880 \text{ kg} \cdot \text{m}^2/\text{s} = 200\omega_f + 15.0\omega_f = 215\omega_f$$

$$\omega_f = \boxed{4.09 \text{ rad/s}}$$

Exercise   Calculate the change in kinetic energy of the system (student plus platform) for this situation. What accounts for this change in energy?

Answer $KE_f - KE_i = 918$ J. The student must perform positive work in order to walk toward the center of the platform.

SPECIAL TOPIC

ANGULAR MOMENTUM AS A FUNDAMENTAL QUANTITY

We have seen that the concept of angular momentum is very useful for describing the motion of macroscopic objects. The concept is also valid on a microscopic scale. It has been used extensively in the development of modern theories of atomic, molecular, and nuclear physics, where it was found that the angular momentum of a system is a *fundamental* quantity. The word *fundamental* in this context implies that angular momentum is an inherent property of atoms, molecules, and their constituents.

In order to explain the results of a variety of experiments on atomic and molecular systems, it is necessary to assign discrete values to their angular momentum. These discrete values are multiples of a fundamental unit of angular momentum, $\hbar = h/2\pi$, where h is Planck's constant.

$$\hbar = 1.054 \times 10^{-34} \frac{\text{kg} \cdot \text{m}^2}{\text{s}^2}$$

Let us accept this postulate for the time being and show how it can be used to estimate the rotational frequency of a diatomic molecule. Consider the O_2 molecule as a rigid rotor—that is, as two atoms separated by a fixed distance, d, and rotating about the center of mass (Fig. 8.27). Equating the rotational angular momentum to the fundamental unit $\hbar$, we can estimate the lowest rotational frequency:

$$I_c \omega \approx \hbar \qquad \text{or} \qquad \omega \approx \frac{\hbar}{I_c}$$

The moment of inertia of the O_2 molecule about this axis of rotation is equal to 2.03×10^{-46} kg·m². Therefore,

$$\omega \approx \frac{\hbar}{I_c} = \frac{1.054 \times 10^{-34} \text{ kg} \cdot \text{m}^2/\text{s}}{2.03 \times 10^{-46} \text{ kg} \cdot \text{m}^2} = 5.19 \times 10^{11} \text{ rad/s}$$

This result is in good agreement with measured rotational frequencies. Furthermore, the rotational frequencies are much lower than the vibrational frequencies of the molecule, which are typically on the order of 10^{13} Hz.

This simple example shows that certain classical concepts and mechanical models may be useful for describing some features of atomic and molecular systems. However, a wide variety of phenomena on the submicroscopic scale can be explained only if one assumes discrete values of the angular momentum associated with a particular type of motion.

The Danish physicist Niels Bohr (1885–1962) was the first to suggest this radical idea in his theory of the hydrogen atom. Strictly classical models had been unsuccessful in describing many properties of the hydrogen atom, such as the fact that the atom absorbs and emits radiation at discrete frequencies. Bohr postulated that the electron could occupy only orbits about the proton for which the orbital angular momentum was equal to $n\hbar$, where n was an integer. From this rather simple model, one can estimate the rotational frequencies of the electron in the various orbits.

Although Bohr's theory provided some insight concerning the behavior of matter at the atomic level, it is basically incorrect. Subsequent developments in quantum mechanics from 1924 to 1930 provided models and interpretations that are now accepted.

Later developments in atomic physics indicated that the electron possesses another kind of angular momentum, called *spin*, which is also an inherent property of the electron. Like angular momentum, spin angular momentum is restricted to discrete values. We shall return to this important property and discuss its great impact on modern physical science in Chapter 28.

Fundamental unit of angular momentum

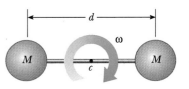

FIGURE 8.27
A rigid-rotor model of the diatomic molecule. The rotation occurs about the center of mass in the plane of the diagram.

SUMMARY

The ability of a force to rotate an object about some axis is measured by a quantity called **torque**, τ. The magnitude of the torque is given by

$$\tau = Fd \tag{8.1}$$

In this equation, d is the **lever arm**—the perpendicular distance from the axis of rotation to a line drawn along the direction of the force. The sign of the torque is negative if the turning tendency of the corresponding force is clockwise and positive if the turning tendency is counterclockwise.

An object is in **equilibrium** when the following conditions are satisfied: (1) the resultant external force must be zero and (2) the resultant external torque must be zero about any origin. That is,

$$\sum \mathbf{F} = 0$$

$$\sum \tau = 0$$

The first equation is called the **first condition for equilibrium**. When this equation holds, an object is either at rest or moving with a constant velocity. The second equation is called the **second condition for equilibrium**. When it holds, an object is said to be in rotational equilibrium.

The **moment of inertia** of a group of particles is

$$I \equiv \sum mr^2 \tag{8.8}$$

If a rigid body free to rotate about a fixed axis has a net external torque acting on it, the body undergoes an angular acceleration, α, where

$$\sum \tau = I\alpha \tag{8.9}$$

If a rigid object rotates about a fixed axis with angular speed ω, its **rotational kinetic energy** is

$$KE_r \equiv \tfrac{1}{2} I\omega^2 \tag{8.11}$$

where I is the moment of inertia about the axis of rotation.

The **angular momentum** of a rotating object is

$$L \equiv I\omega \tag{8.13}$$

If the net external torque acting on a system is zero, the total angular momentum of the system is constant. Applying the law of conservation of angular momentum to an object whose moment of inertia changes with time gives

$$I_i \omega_i = I_f \omega_f \tag{8.15}$$

ADDITIONAL READING

H. Brody, ''The Moment of Inertia of a Tennis Racket,'' *The Physics Teacher,* April 1985, p. 213.

C. Frohlich, ''The Physics of Somersaulting and Twisting,'' *Sci. American,* March 1980, p. 154.

D. F. Griffing, *The Dynamics of Sports: Why That's the Way the Ball Bounces,* Mohican Publishing Co., 1982.

J. G. Kreifeldt and M. Chuang, ''Moment of Inertia: Psychophysical Study of an Overlooked Sensation,'' *Science,* 1979, p. 588.

K. Laws, ''The Physics of Dance,'' *Physics Today,* February 1985, p. 25.

R. F. Post and S. F. Post, ''Flywheels,'' *Sci. American,* December 1973, p. 17.

CONCEPTUAL QUESTIONS

Example Is it possible to change the translational kinetic energy of an object without changing its rotational kinetic energy?

Reasoning Yes. For example, imagine pushing an object across the floor such that it slides (that is, it moves without rotating). Its translational kinetic energy is changing, yet its rotational kinetic energy is zero and does not change.

Example A mouse is initially at rest on a horizontal turntable mounted on a frictionless vertical axle. If the mouse begins to walk clockwise around the perimeter, what happens to the turntable? Explain.

Reasoning The initial angular momentum of the system (mouse plus turntable) is zero. As the mouse begins to walk clockwise, its angular momentum increases, so the turntable must rotate in the counterclockwise direction with an angular momentum whose magnitude equals that of the mouse. This follows from the fact that the final angular momentum of the system must equal the initial angular momentum, which is zero in this case.

Example Space colonies have been proposed that would consist of large cylinders, as described in Problem 7.26. Gravity would be simulated in these cylinders by setting them into rotation. Discuss the difficulties that would be encountered in an attempt to set a cylinder into rotation in empty space.

Reasoning If the cylinder is not rotating initially, it has zero angular momentum. Thus, to make it rotate clockwise, some other object must receive an equal angular momentum counterclockwise in order for the total angular momentum to remain zero. (That is, the total angular momentum of the system must be conserved in the absence of external torques.) One method that would accomplish this to attach two nearby cylinders together by belts that would cause one to rotate clockwise, the other counterclockwise.

Example A particle moves in a straight line, and you are told that the torque acting on it is zero about some unspecified origin. Does this necessarily imply that the total force on the particle is zero? Can you conclude that its velocity is constant? Explain.

Reasoning No, the total force is not necessarily zero. If the line along which the total force acts passes through the origin, the total torque about that origin is zero, yet the total force is not zero. Therefore, you cannot conclude that its velocity is constant.

Example Two cylinders having the same dimensions are set into rotation about their axes with the same angular speed. One is hollow, while the other is filled with water. Which cylinder would be easier to stop rotating?

Reasoning A torque would have to be supplied to decrease the angular momentum of the cylinders to zero.

Thus, the water-filled cylinder would require the larger torque because of its larger initial angular momentum. Furthermore, the freedom of the water to rotate inside the cylinder would make it more difficult to stop.

Example Stars originate as large bodies of slowly rotating gas. Because of gravity, these clumps of gas slowly decrease in size. What happens to the angular speed of a star as it shrinks? Explain.

Reasoning The angular momentum of the gas cloud is conserved. Thus, the product $I\omega$ remains constant. Hence, as the cloud shrinks, its moment of inertia I decreases, so its angular speed ω must increase.

1. Why are the units of torque given as newton-meters and not joules?
2. Is it possible to calculate the torque acting on a rigid body without specifying the origin? Is the torque independent of the location of the origin?
3. Under what conditions would the tension in the string attached to the bucket in Example 8.11 be equal to the weight of the bucket?
4. Explain why changing the axis of rotation of an object should change its moment of inertia.
5. Why does a long pole help a tightrope walker stay balanced?
6. It is more difficult to do a sit-up with your hands behind your head than with your arms stretched out in front of you. Why?
7. In order for a helicopter to be stable as it flies, it must have two propellers. Why?
8. Three homogeneous rigid bodies—a solid sphere, a solid cylinder, and a hollow cylinder—are placed at the top of an incline (Fig. 8.28). If all are released from rest at the same elevation and roll without slipping, which reaches the bottom first? Which reaches last? You should try this at home and note that the result is *independent* of the masses and radii.

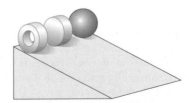

FIGURE 8.28 (Question 8) Which object wins the race?

9. Two spheres look identical and weigh the same. One is hollow, the other solid. Discuss how you might determine which is which.

10. Suppose you remove two eggs from the refrigerator, one hard-boiled and the other uncooked. You wish to determine which is hard-boiled without breaking the eggs. This can be done by spinning the two eggs on the floor and comparing the rotational motions. Which egg spins faster? Which rotates more uniformly? Explain. (This is a neat and helpful trick for homemakers.)

11. As a tetherball winds around a pole, what happens to its angular speed? Explain.

12. Often when a high diver wants to turn a flip in midair, she draws her legs up against her chest. Why does this make her rotate faster? What should she do when she wants to come out of her flip?

13. Why is it easier to keep your balance on a moving bicycle than on a bicycle at rest?

14. Suppose the planets of the Solar System evolved from the condensation of a primordial gas. From the observation that all the planets revolve around the Sun in the same direction, what can you infer about the state of the gas prior to its condensation?

15. A student sits on a stool that is free to rotate about its vertical axis. The student and stool are set into rotation while the student holds a pair of weights with outstretched arms. If she suddenly drops the weights to the floor, what happens to her angular speed? Explain.

16. If angular momentum is conserved for a propeller-driven airplane, it seems that if the propeller is turning clockwise, the airplane should be turning counterclockwise. Why doesn't this occur?

17. A cat usually lands on its feet regardless of the position from which it is dropped. A slow-motion film of a cat falling shows that the upper half of its body twists in one direction while the lower half twists in the opposite direction. Why does this type of rotation occur?

(Question 17)   A falling, twisting cat. *(Photo Researchers, Inc.)*

18. A ladder rests inclined against a wall. Would you feel safer climbing up the ladder if you were told that the floor is frictionless but the wall is rough or that the wall is smooth but the floor is rough? Justify your answer.

19. Give an example in which the net force acting on an object is zero and yet the net torque is nonzero.

20. Give an example in which the net torque acting on an object is zero and yet the net force is nonzero.

21. Imagine that you measure the net torque and the net force on a system to be zero. (a) Can the system still be rotating with respect to you? (b) Can it be translating with respect to you?

22. Why is it easier to balance a basketball on one finger if it is spinning than if it is not spinning?

PROBLEMS

Section 8.1   Torque

1. If the torque required to loosen a nut that is holding a flat tire in place on a car has a magnitude of 40.0 N·m, what *minimum* force must be exerted by the mechanic at the end of a 30.0-cm lug wrench to accomplish the task?

2. A simple pendulum consists of a 3.0-kg point mass hanging at the end of a 2.0-m-long light string that is connected to a pivot point. Calculate the magnitude of the torque (due to the force of gravity) about this pivot point when the string makes a 5.0° angle with the vertical.

3. A fishing pole is inclined to the horizontal at an angle of 20.0° (Fig. 8.29). What is the torque exerted by the fish about an axis perpendicular to the page and passing through the hand of the person holding the pole?

4. Calculate the net torque (magnitude and direction) on the beam in Figure 8.30 about (a) an axis through *O*, perpendicular to the page, and (b) an axis through *C*, perpendicular to the page.

5. A steel band exerts a horizontal force of 80.0 N on a tooth at point *B* in Figure 8.31. What is the torque on the root of the tooth about point *A*?

6. The person in Figure 8.32 weighs 800 N. The

□ indicates problems that have full solutions available in the Student Solutions Manual and Study Guide.

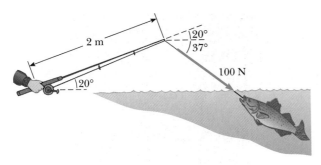

FIGURE 8.29   (Problem 3)

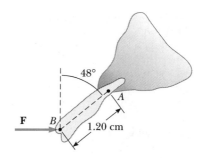

FIGURE 8.31   (Problem 5)

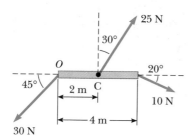

FIGURE 8.30   (Problem 4)

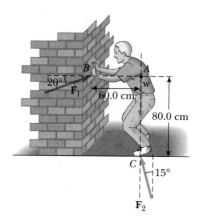

FIGURE 8.32   (Problem 6)

forces $\mathbf{F}_1$ and $\mathbf{F}_2$ have magnitudes of 100 N and 900 N, respectively. Assume that the force of gravity acts downward through point *A*, as shown. Determine the net torque on the person about axes through points *A*, *B*, and *C* perpendicular to the plane of the paper. (*Hint:* Replace $\mathbf{F}_1$ and $\mathbf{F}_2$ with their horizontal and vertical components.)

Section 8.2   **Torque and the Second Condition for Equilibrium**

Section 8.3   **The Center of Gravity**

Section 8.4   **Examples of Objects in Equilibrium**

7. A 45-kg, 5.0-m-long uniform ladder rests against a frictionless wall and makes an angle of 60° with a frictionless floor. Can an 80-kg person stand safely on the ladder, 2.0 m from the top, without causing the ladder to slip if a second person exerts a horizontal force of 500 N toward the wall at a point 3.5 m from the top of the ladder? (*Note:* All distances are measured along the ladder.)

8. A 0.100-kg meter stick is supported at its 40.0-cm mark by a string attached to the ceiling. A 0.700-kg mass hangs vertically from the 5.00-cm mark. A mass, *m*, is attached somewhere on the meter stick to keep it horizontal and in rotational and translational equilibrium. If the tension in the string attached to the ceiling is 19.6 N, determine (a) the value of *m* and (b) its point of attachment on the stick.

 9. A window washer is standing on a scaffold supported by a vertical rope at each end. The scaffold weighs 200 N and is 3.00 m long. What is the tension in each rope when the 700-N worker stands 1.00 m from one end?

 10. A 20.0-kg floodlight in a park is supported at the end of a horizontal beam of negligible mass that is hinged to a pole, as shown in Figure 8.33. A

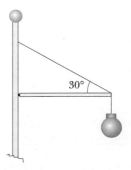

FIGURE 8.33   (Problem 10)

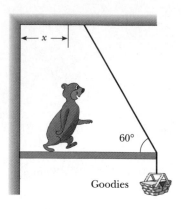

FIGURE 8.34 (Problem 11)

cable at an angle of 30.0° with the beam helps to support the light. Find (a) the tension in the cable and (b) the horizontal and vertical forces exerted on the beam by the pole.

11. A hungry 700-N bear walks out on a beam in an attempt to retrieve some "goodies" hanging at the end (Fig. 8.34). The beam is uniform, weighs 200 N, and is 6.00 m long; the goodies weigh 80.0 N. (a) Draw a free-body diagram for the beam. (b) When the bear is at $x = 1.00$ m, find the tension in the wire and the components of the reaction force at the hinge. (c) If the wire can withstand a maximum tension of 900 N, what is the maximum distance the bear can walk before the wire breaks?

12. A uniform plank of length 2.00 m and mass 30.0 kg is supported by three ropes, as indicated by the blue vectors in Figure 8.35. Find the tension in each rope when a 700-N person is 0.500 m from the left end.

13. The arm in Figure 8.36 weighs 41.5 N. The weight of the arm acts through point A. Determine the magnitudes of the tension force $\mathbf{F}_t$ in the deltoid muscle and the force $\mathbf{F}_s$ of the shoulder on the hu-

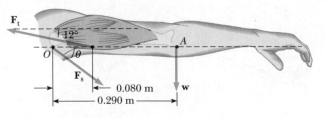

FIGURE 8.36 (Problem 13)

merus (upper-arm bone) to hold the arm in the position shown.

14. Two window washers, Bob and Joe, are on a 3.00-m-long, 345-N scaffold supported by two cables attached to its ends. Bob, who weighs 750 N stands 1.00 m from the left end, as shown in Figure 8.37. Two meters from the left end is the 500-N washing equipment. Joe is 0.500 m from the right end and weighs 1000 N. Given that the scaffold is in rotational and translational equilibrium, what are the forces on each cable?

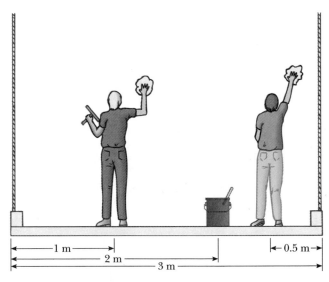

FIGURE 8.37 (Problem 14)

15. The chewing muscle, the masseter, is one of the strongest in the human body. It is attached to the mandible (lower jawbone) as shown in Figure 8.38a. The jawbone is pivoted about a socket just in front of the auditory canal. The forces acting on the jawbone are equivalent to those acting on the curved bar in Figure 8.38b: $\mathbf{C}$ is the force exerted against the jawbone by the food being chewed, $\mathbf{T}$ is the tension in the masseter, and $\mathbf{R}$ is the force exerted on the mandible by the socket. Find $\mathbf{T}$ and $\mathbf{R}$ if you bite down on a piece of steak with a force of 50.0 N.

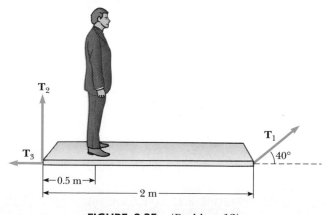

FIGURE 8.35 (Problem 12)

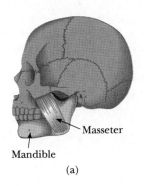

Masseter

Mandible

(a)

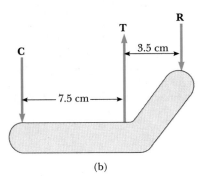

R

T

3.5 cm

C

7.5 cm

(b)

FIGURE 8.38 (Problem 15)

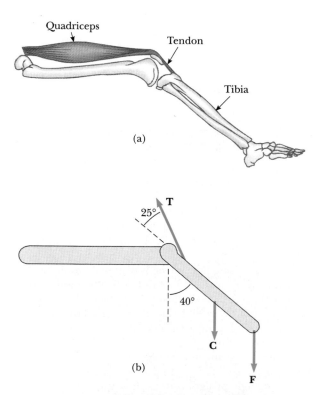

Quadriceps

Tendon

Tibia

(a)

T

25°

40°

C

F

(b)

FIGURE 8.39 (Problem 16)

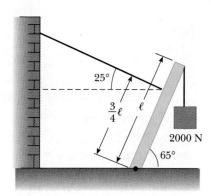

25°

$\frac{3}{4}\ell$ ℓ

2000 N

65°

FIGURE 8.40 (Problem 17)

16. The large quadriceps muscle in the upper leg terminates at its lower end in a tendon attached to the upper end of the tibia (Fig. 8.39a). The forces on the lower leg when the leg is extended are modeled as in Figure 8.39b, where **T** is the tension in the tendon, **C** is the weight of the lower leg, and **F** is the weight of the foot. Find **T** when the tendon is at an angle of 25.0° with the tibia, assuming that $C = 30.0$ N, $F = 12.5$ N, and the leg is extended at an angle of 40.0° with the vertical ($\theta = 40.0°$). Assume that the center of gravity of the lower leg is at its center, and that the tendon attaches to the lower leg at a point one fifth of the way down the leg.

17. A 1200-N uniform boom is supported by a cable as in Figure 8.40. The boom is pivoted at the bottom, and a 2000-N weight hangs from its top. Find the tension in the supporting cable and the components of the reaction force on the boom at the hinge.

18. A hemispherical sign 1.00 m in diameter and of uniform mass density is supported by two wires as shown in Figure 8.41. What fraction of the sign's weight is supported by each wire?

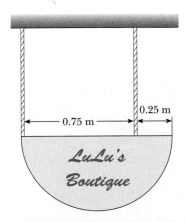

0.25 m

0.75 m

LuLu's Boutique

FIGURE 8.41 (Problem 18)

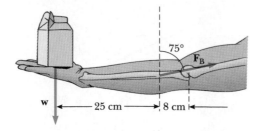

FIGURE 8.42   (Problem 21)

19. A 15.0-m, 500-N uniform ladder rests against a frictionless wall, making an angle of 60.0° with the horizontal. (a) Find the horizontal and vertical forces exerted on the base of the ladder by the Earth when an 800-N fire fighter is 4.00 m from the bottom. (b) If the ladder is just on the verge of slipping when the fire fighter is 9.00 m up, what is the coefficient of static friction between ladder and ground?

20. An 8.0-m, 200-N uniform ladder rests against a smooth wall. The coefficient of static friction between the ladder and the ground is 0.60, and the ladder makes a 50.0° angle with the ground. How far up the ladder can an 800-N person climb before the ladder begins to slip?

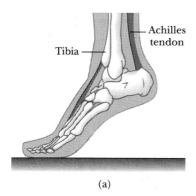

(a)

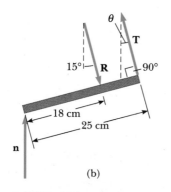

(b)

FIGURE 8.43   (Problem 22)

21. A cook holds a 2.00-kg carton of milk at arm's length (Fig. 8.42). What force $\mathbf{F}_B$ must be exerted by the biceps muscle? (Ignore the weight of the forearm.)

22. When a person stands on tiptoe (a strenuous position), the position of the foot is as shown in Figure 8.43a. The total weight, **w**, is supported by the force, **n**, of the foot on the toe. A mechanical model for the situation is shown in Figure 8.43b, where **T** is the tension force in the Achilles tendon and **R** is the force on the foot due to the tibia. Find the values of T, R, and θ using the model and dimensions given, with **w** = 700 N.

23. A shelf bracket is mounted on a vertical wall by a single screw, as in Figure 8.44. Neglecting the weight of the bracket, find the horizontal force component exerted on the bracket by the screw when an 80.0-N vertical force is applied as shown. (*Hint:* Imagine that the bracket is slightly loose.)

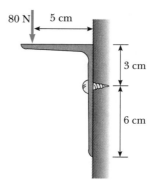

FIGURE 8.44   (Problem 23)

Section 8.5   Relationship Between Torque and Angular Acceleration

24. The length of each of the bonds between the atoms of a molecule of nitrogen, N_2, is 1.10×10^{-10} m, and the mass of each nitrogen atom is 2.32×10^{-26} kg. Determine the moment of inertia of the molecule about an axis passing through the center of mass perpendicular to the line joining the two atoms.

25. A 4.00-m-long light string is wrapped around a solid cylindrical spool of radius 0.500 m and mass 0.500 kg. A 5.00-kg mass is hung from the string, causing it to unwind (Fig. 8.45). How fast will the spool be rotating after all of the string has unwound, assuming that no slippage takes place between the string and the spool?

26. Four masses are held in position at the corners of a rectangle by light rods as shown in Figure 8.46. Find the moment of inertia of the system about

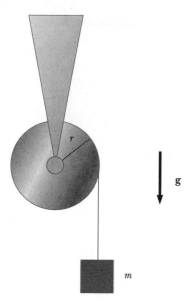

FIGURE 8.45 (Problem 25)

FIGURE 8.47 (Problem 34)

(a) the x axis, (b) the y axis, and (c) an axis through O and perpendicular to the page.

27. If the system shown in Figure 8.46 is set in rotation about each of the axes mentioned in Problem 26, find the torque that will produce an angular acceleration of 1.50 rad/s² in each case.

28. If a constant torque of 20.0 N·m acts on the system shown in Figure 8.46, causing it to rotate about the y axis, find the system's angular speed after 3.00 s, assuming that it starts from rest.

29. A 3.00-cm-diameter coin rolls up a 30.0° inclined plane. The coin starts with an initial angular speed of 60.0 rad/s and rolls in a straight line without slipping. How far does it roll up the inclined plane?

30. A regulation basketball has a 25-cm diameter and may be approximated as a thin spherical shell. Starting from rest, how long will it take a basket-

ball to roll, without slipping, 4.0 m down an incline that makes an angle of 30° with the horizontal? (The moment of inertia of a thin spherical shell of radius R and mass M is $I = \frac{2}{3}MR^2$.)

31. A cylindrical fishing reel has a mass of 0.85 kg and a radius of 4.0 cm. A friction clutch in the reel exerts a restraining torque of 1.3 N·m if a fish pulls on the line. The fisherman gets a bite, and the reel begins to spin with an angular acceleration of 66 rad/s². (a) What is the force of the fish on the line? (b) How much line unwinds in 0.50 s?

32. A potter's wheel, a thick stone disk of radius 0.50 m and mass 100 kg, is freely rotating at

(Problem 32)   The speed of this potter's wheel is controlled by the combined action of craftsman's feet for pumping and the friction between the potter's hands and the clay pot. *(Cary Wolinsky/Stock Boston)*

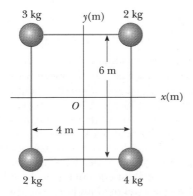

FIGURE 8.46 (Problems 26, 27, and 28)

50 rev/min. The potter can stop the wheel in 6.0 s by pressing a wet rag against the rim and exerting a radially inward force of 70 N. Find the effective coefficient of kinetic friction between the wheel and the wet rag.

33. A 150-kg merry-go-round in the shape of a horizontal disk of radius 1.50 m is set in motion by wrapping a rope about the rim of the disk and pulling on the rope. What constant force would have to be exerted on the rope to bring the merry-go-round from rest to an angular speed of 0.500 rev/s in 2.00 s?

34. A cylindrical 5.00-kg pulley with a radius of 0.600 m is used to lower a 3.00-kg bucket into a well (Fig. 8.47). The bucket starts from rest and falls for 4.00 s. (a) What is the linear acceleration of the falling bucket? (b) How far does it drop? (c) What is the angular acceleration of the cylinder?

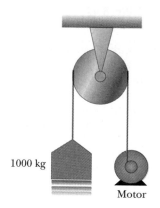

1000 kg

Motor

FIGURE 8.48   (Problem 36)

35. A cable passes over a pulley. Because of friction, the tension in the cable is not the same on opposite sides of the pulley. The force on one side is 120 N, and the force on the other side is 100 N. Assuming that the pulley is a uniform disk of mass 2.1 kg and radius 0.81 m, determine its angular acceleration.

36. When the motor in Figure 8.48 raises the 1000-kg mass, it produces a tension of 1.14×10^4 N in the cable on the right side of the pulley. The pulley has a moment of inertia of 79.8 kg·m² and a radius of 0.762 m. The cable rides over the pulley without slipping. Determine the acceleration of the 1000-kg mass. (*Hint:* Draw free-body diagrams of the mass and the pulley. Do not assume that the tension in the cable is the same on both sides of the pulley.)

37. A string is wrapped around a uniform 2.00-kg cyl-

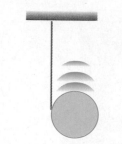

FIGURE 8.49   (Problem 37)

inder of radius 15.0 cm (Fig. 8.49). One end of the string is attached to the ceiling, and the cylinder is allowed to fall from rest. (a) Write down Newton's second law for the cylinder. (b) Find the net torque about the center of the cylinder and equate this to $I\alpha$. (c) Use the equations found in (a) and (b) along with $a = r\alpha$ to show that the linear acceleration of the cylinder is given by $a = (2/3)g$.

38. The combination of an applied force and a frictional force produces a constant torque of 36 N·m on a wheel rotating about a fixed axis. The applied force acts for 6.0 s, during which time the angular speed of the wheel increases from 0 to 10 rad/s. The applied force is then removed, and the wheel comes to rest in 60 s. Find (a) the moment of inertia of the wheel, (b) the magnitude of the frictional torque, and (c) the total number of revolutions of the wheel.

39. An airliner lands with a speed of 50.0 m/s. Each wheel of the plane has a radius of 1.25 m and a moment of inertia of 110 kg·m². At touchdown the wheels begin to spin under the action of friction. Each wheel supports a weight of 1.40×10^4 N, and the wheels attain the angular speed of rolling without slipping in 0.480 s. What is the coefficient of kinetic friction between the wheels and the runway? Assume that the speed of the plane is constant.

Section 8.6   Rotational Kinetic Energy

40. A solid sphere rolls along a horizontal, smooth surface at a constant linear speed without slipping. Show that its rotational kinetic energy about the center of the sphere is 2/7 of its total kinetic energy.

41. The hour hand and the minute hand for the famous Parliament clock tower Big Ben in London are 2.7 m and 4.5 m long and have masses of 60 kg and 100 kg, respectively. Determine the total rotational kinetic energy of these hands. (You may model the hands as long thin rods.)

(Problem 41) *(Ron Watts/Black Star)*

42. Three particles are connected by rigid rods of negligible mass lying along the *y* axis (Fig. 8.50). The system rotates about the *x* axis with an angular speed of 2.00 rad/s. (a) Find the moment of inertia about the *x* axis and the total kinetic energy evaluated from $\frac{1}{2} I\omega^2$. (b) Find the linear speed of each particle and the total kinetic energy evaluated from $\Sigma \frac{1}{2} m_i v_i^2$.

43. The net work done in accelerating a propeller from rest to an angular speed of 200 rad/s is 3000 J. What is the moment of inertia of the propeller?

44. A horizontal 800-N merry-go-round of radius 1.5 m is started from rest by a constant horizontal force of 50 N applied tangentially to the merry-go-round. Find the kinetic energy of the merry-go-round after 3.0 s. (Assume it is a solid cylinder.)

45. A bowling ball is both sliding and spinning on a horizontal surface so that its rotational kinetic energy equals its translational kinetic energy. What is the ratio of the ball's center-of-mass speed to the speed of a point on the ball's surface?

46. A 10.0-kg cylinder rolls without slipping on a rough surface. At the instant its center of mass has a speed of 10.0 m/s, determine (a) the translational kinetic energy of its center of mass, (b) the rotational kinetic energy about its center of mass, and (c) its total kinetic energy.

47. A constant net torque, τ, is applied to an object, causing it to rotate. Show that the work done in rotating the object through an angular displacement of $\theta_2 - \theta_1$ is given by $W = \tau(\theta_2 - \theta_1)$. (*Hint:* Consider the change in kinetic energy of the object during the angular displacement, $\theta_2 - \theta_1$.)

48. A solid 2.0-kg ball of radius 0.50 m starts at a height of 3.0 m above the surface of the Earth and *rolls* down a 20° slope. A solid disk and a ring start at the same time and the same height. The ring and disk each have the same mass and radius as the ball. Which of the three wins the race to the bottom if all roll without slipping?

49. A car is designed to get its energy from a rotating flywheel with a radius of 2.00 m and a mass of 500 kg. Before a trip, the flywheel is attached to an electric motor, which brings the flywheel's rotational speed up to 5000 rev/min. (a) Find the kinetic energy stored in the flywheel. (b) If the flywheel is to supply energy to the car as would a 10.0-hp motor, find the length of time the car could run before the flywheel would have to be brought back up to speed.

50. The top in Figure 8.51 has a moment of inertia of 4.00×10^{-4} kg·m² and is initially at rest. It is free to rotate about a stationary axis, AA'. A string, wrapped around a peg along the axis of the top, is pulled in such a manner as to maintain a constant

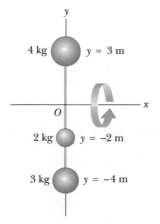

FIGURE 8.50 (Problem 42)

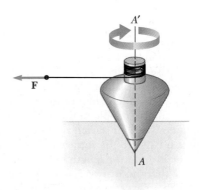

FIGURE 8.51 (Problem 50)

tension of 5.57 N in the string. If the string does not slip while wound around the peg, what is the angular speed of the top after 80.0 cm of string has been pulled off the peg? (*Hint:* Consider the work done.)

51. Assume that the Earth is a homogeneous sphere, and calculate the kinetic energy of rotation about its axis. (Use the table of physical constants inside the front cover.) For how long could this energy be tapped to supply the world's energy needs (assumed constant at 2.0×10^{20} J/yr) if, in the process, the length of the day were not to be increased by more than 1 min?

52. In a circus performance, a large 5.0-kg hoop of radius 3.0 m rolls without slipping. If the hoop is given an angular speed of 3.0 rad/s while rolling on the horizontal and allowed to roll up a ramp inclined at 20° with the horizontal, how far (measured along the incline) does the hoop roll?

Section 8.7   Angular Momentum

53. (a) Calculate the angular momentum of the Earth that arises from its spinning motion on its axis and (b) the angular momentum of the Earth that arises from its orbital motion about the Sun.

54. The hour hand and the minute hand for the famous Parliament Building tower clock Big Ben in London are 2.7 m and 4.5 m long and have masses of 60 kg and 100 kg, respectively. Calculate the total angular momentum of these hands about the center point. (You may model the hands as long thin rods.)

55. A solid, horizontal cylinder of mass 10.0 kg and radius 1.00 m rotates with an angular speed of 7.00 rad/s about a fixed vertical axis through its center. A 0.250-kg piece of putty is dropped vertically onto the cylinder at a point 0.900 m from the center of rotation, and sticks to the cylinder. Determine the final angular speed of the system.

56. A skater spins with an angular speed of 12.0 rad/s with her arms outstretched. She lowers her arms, decreasing her moment of inertia by 10%. Ignoring friction on the skates, determine the percentage change in her kinetic energy. Is the kinetic energy increased or decreased? Explain the result.

57. We have all complained that there aren't enough hours in a day. In an attempt to change that, suppose that all the people in the world lined up at the equator, and all started running east at 2.5 m/s relative to the surface of the Earth. By how much would the length of a day increase? (Assume that there are 5.5×10^9 people in the world with an average mass of 70 kg each, and that the Earth is a solid homogeneous sphere. In addition, you may use the result $1/(1 - x) \approx 1 + x$ for x small.)

58. A student sits on a rotating stool holding two 3.0-kg masses. When his arms are extended horizontally, the masses are 1.0 m from the axis of rotation, and he rotates with an angular speed of 0.75 rad/s. The moment of inertia of the student plus stool is 3 kg·m² and is assumed to be constant. The student then pulls the masses horizontally to 0.30 m from the rotation axis. (a) Find the new angular speed of the student. (b) Find the kinetic energy of the student before and after the masses are pulled in.

59. Halley's comet moves about the Sun in an elliptical orbit, with its closest approach to the Sun being 0.59 AU and its greatest distance being 35 AU (1 AU = the Earth-Sun distance). If the comet's speed at closest approach is 54 km/s, what is its speed when it is farthest from the Sun? You may assume that its angular momentum about the Sun is conserved.

60. A merry-go-round rotates at the rate of 0.20 rev/s with an 80-kg man standing at a point 2.0 m from the axis of rotation. (a) What is the new angular speed when the man walks to a point 1.0 m from the center? Assume that the merry-go-round is a solid 25-kg cylinder of radius 2.0 m. (b) Calculate the change in kinetic energy due to this movement. How do you account for this change in kinetic energy?

61. The puck in Figure 8.52 has a mass of 0.120 kg. Its original distance from the center of rotation is 40.0 cm, and the puck is moving with a speed of 80.0 cm/s. The string is pulled downward 15.0 cm through the hole in the frictionless table. Determine the work done on the puck. (*Hint:* Consider the change of kinetic energy of the puck.)

62. A 60-kg woman stands at the rim of a horizontal turntable having a moment of inertia of 500 kg·m² and a radius of 2.0 m. The system is initially at rest, and the turntable is free to rotate

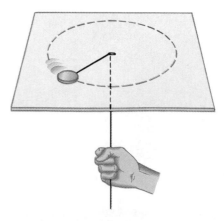

FIGURE 8.52   (Problem 61)

about a frictionless vertical axle through its center. The woman then starts walking clockwise (looking downward) around the rim at a constant speed of 1.5 m/s relative to the Earth. (a) In what direction and with what angular speed does the turntable rotate? (b) How much work does the woman do to set the system in motion?

ADDITIONAL PROBLEMS

63. A cylinder with moment of inertia I_1 rotates with angular velocity ω_0 about a frictionless vertical axle. A second cylinder, with moment of inertia I_2, initially not rotating, drops onto the first cylinder (Fig. 8.53). Since the surfaces are rough, the two eventually reach the same angular velocity, ω. (a) Calculate ω. (b) Show that energy is lost in this situation, and calculate the ratio of the final to the initial kinetic energy.

64. The hour hand and the minute hand for the famous Parliament clock tower Big Ben in London

are 2.7 m and 4.5 m long and have masses of 60 kg and 100 kg, respectively. Calculate the total torque due to the weight of these hands about the axis of rotation when the time reads (a) 3:00, (b) 5:15, (c) 6:00, (d) 8:20, (e) 9:45. (You may model the hands as long thin rods.)

65. A 240-N sphere 0.20 m in radius rolls 6.0 m down a ramp that is inclined at 37° with the horizontal. What is the angular velocity of the sphere at the bottom of the hill if it starts from rest?

66. A person bends over and lifts a 200-N weight as in Figure 8.54a, with his back horizontal. The muscle that attaches two thirds of the way up the spine maintains the position of the back; the angle between the spine and this muscle is 12°. Using the mechanical model in Figure 8.54b and taking the weight of the upper body to be 350 N, find the tension in the back muscle and the compressional force in the spine.

67. A uniform 10.0-N picture frame is supported as shown in Figure 8.55. Find the tension in the cords and the magnitude of the horizontal force at P that are required to hold the frame in the position shown.

68. One end of a uniform 4.0-m-long rod of weight W is supported by a string. The other end rests against the wall, where it is held by friction (see

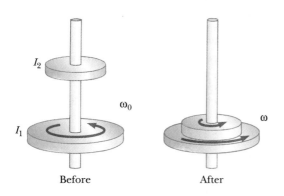

FIGURE 8.53 (Problem 63)

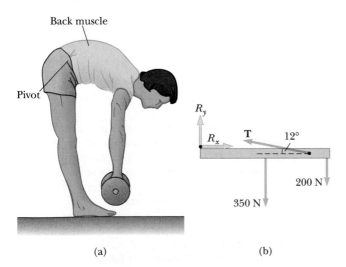

(a)

(b)

FIGURE 8.54 (Problem 66)

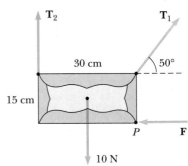

FIGURE 8.55 (Problem 67)

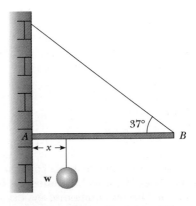

FIGURE 8.56 (Problem 68)

Fig. 8.56). The coefficient of static friction between the wall and the rod is $\mu_s = 0.50$. Determine the distance from A of the nearest point
from which an additional weight w can be hung
without the rod's slipping from A.

69. A 3000-kg crane supports a load of 10 000 kg as in
Figure 8.57. The crane is pivoted with a smooth
pin at A and rests against a smooth support at B.
Find the reaction forces at A and B.

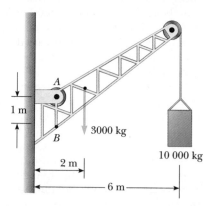

FIGURE 8.57   (Problem 69)

70. The system of point masses shown in Figure 8.58 is
rotating at an angular speed of 2.0 rev/s. The
masses are connected by light, flexible spokes that
can be lengthened or shortened. What is the new
angular speed if the spokes are shortened to
0.50 m? (An effect similar to that illustrated in this
problem occurred in the early stages of the formation of our Galaxy. As the massive cloud of dust
and gas that was the source of the stars and planets contracted, an initially small rotation increased
with time.)

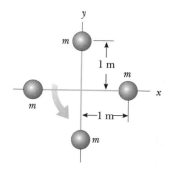

FIGURE 8.58   (Problem 70)

71. A 12.0-kg mass is attached to a cord that is
wrapped around a wheel of radius $r = 10.0$ cm
(Fig. 8.59). The acceleration of the mass down the

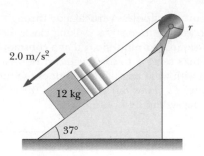

FIGURE 8.59   (Problem 71)

frictionless incline is measured to be 2.00 m/s².
Assuming the axle of the wheel to be frictionless,
determine (a) the tension in the rope, (b) the moment of inertia of the wheel, and (c) the angular
speed of the wheel 2.00 s after it begins rotating,
starting from rest.
72. A block of mass $m = 4.00$ kg is connected (by a
massless string over a massless and frictionless pulley) to a wheel of radius $R = 0.500$ m, mass $M =
8.00$ kg, and moment of inertia $I = 2.00$ kg·m²
(Fig. 8.60a). The block is released from rest at a
height of $H = 2.00$ m above the ground. (a) What
will be its speed just before it strikes the ground?
(b) How much time will pass before the block
strikes the ground? (c) How many revolutions will
the wheel make during this motion?

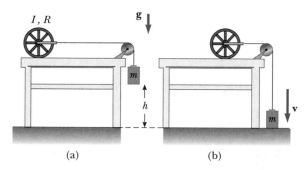

FIGURE 8.60   (Problem 72)

73. A uniform ladder is leaning against a vertical wall.
The ladder slips when it makes a 60.0° angle with
the horizontal floor. Assuming that the coefficient
of static friction between the ladder and the floor
is the same as that between the ladder and the
wall, determine that coefficient.
74. The pulley in Figure 8.61 has a moment of inertia
of 5.0 kg·m² and a radius of 0.50 m. The cord
supporting the masses m_1 and m_2 does not slip,
and the axle is frictionless. (a) Find the accelera-

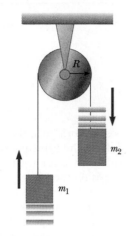

FIGURE 8.61   (Problem 74)

tion of each mass when m_1 = 2.0 kg and m_2 = 5.0 kg. (b) Find the tension in the cable supporting m_1 and the tension in the cable supporting m_2 (note that they are different).

75. A 4.00-kg mass is connected by a light cord to a 3.00-kg mass on a smooth surface (Fig. 8.62). The pulley rotates about a frictionless axle and has a moment of inertia of 0.500 kg·m² and a radius of 0.300 m. Assuming that the cord does not slip on the pulley, find (a) the acceleration of the two masses and (b) the tensions T_1 and T_2.

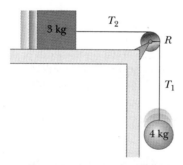

FIGURE 8.62   (Problem 75)

76. A uniform, solid cylinder of mass M and radius R rotates on a frictionless horizontal axle (Fig. 8.63). Two equal masses hang from light cords wrapped around the cylinder. If the system is released from rest, find (a) the tension in each cord and (b) the acceleration of each mass after the masses have descended a distance of h.

FIGURE 8.63   (Problem 76)

77. A 40.0-kg child stands at one end of a 70.0-kg boat that is 4.00 m long (Fig. 8.64). The boat is initially 3.00 m from the pier. The child notices a turtle on a rock beyond the far end of the boat and proceeds to walk to that end to catch the turtle. (a) Neglecting friction between the boat and water, describe the motion of the system (child + boat). (b) Where will the child be relative to the pier when he reaches the far end of the boat? (c) Will he catch the turtle? (Assume that he can reach out 1.00 m from the end of the boat.)

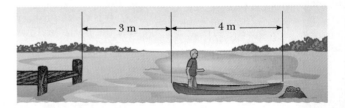

FIGURE 8.64   (Problem 77)

GORDON BATSON   Clarkson University, Potsdam N.Y.

ARCH STRUCTURES

Of all structures built for various utilitarian purposes, a bridge and its structural components are the most visible. The load-carrying tasks of the principal structural components can be comprehended easily; the supporting cables of a suspension bridge are under tension induced by the weight of and loads on the bridge.

The arch is another type of structure whose shape indicates that the loads are carried by compression. The arch can be visualized as an upside-down suspension cable.

The stone arch is one of the oldest existing structures found in buildings, walls, and bridges. Other materials, such as timber, may have been used prior to stone, but nothing of these remains today most likely because of fires, warfare, and the decay processes of nature. Although stone arches were constructed prior to the Roman Empire, the Romans constructed some of the largest and most enduring stone arches.

Before the development of the arch, the principal method of spanning a space was the simple post-and-beam construction (Fig. 1a), in which a horizontal beam is supported by two columns. This type of construction was used to build the great Greek temples. The columns of these temples are closely spaced because of the limited length of available stones. Much larger spans can now be achieved using steel beams, but the spans are limited because the beams tend to sag under heavy loads.

The corbeled arch (or false arch) shown in Figure 1b is another primitive structure; it is only a slight improvement over post-and-beam construction. The stability of this false arch depends upon the horizontal projection of one stone over another and the downward weight of stones from above.

The semicircular arch (Fig. 2a) developed by the Romans was a great technological achievement in architectural design. The stability of this true (or voussoir) arch depends on the compression between its wedge-shaped stones. (That is, the stones are forced to squeeze against each other.) This squeezing results in horizontal outward forces at the springing of the arch (where it starts curving), which must be supported by the foundation (abutments) on the stone wall shown on the sides of the arch (Fig. 2a). It is common to use very heavy walls (buttresses) on either side of the arch to provide the horizontal stability. If the foundation of the arch should move, the compres-

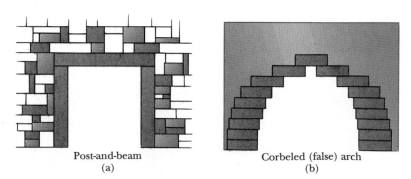

Post-and-beam
(a)

Corbeled (false) arch
(b)

FIGURE 1   Some methods of spanning space: (a) simple post-and-beam structure and (b) corbeled, or false, arch.

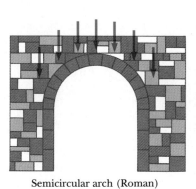

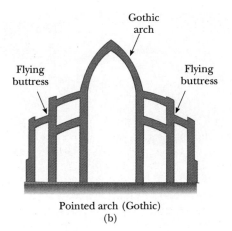

Semicircular arch (Roman)
(a)

Pointed arch (Gothic)
(b)

FIGURE 2 (a) The semicircular arch developed by the Romans. (b) Gothic arch with flying buttresses to provide lateral support. (Typical cross-section of a church or cathedral.) The buttresses transfer the spreading forces of the arch to the foundation of the structure.

sive forces between the wedge-shaped stones may decrease to the extent that the arch collapses. The surfaces of the stones used in the semicircular arches constructed by the Romans were cut, or "dressed," to make a very tight joint; it is interesting to note that mortar was usually not used in these joints. The resistance to slipping between stones was provided by the compression force and the friction between the stone faces.

Another important architectural innovation was the pointed Gothic arch shown in Figure 2b. This type of structure was first used in Europe beginning in the 12th century, followed by the construction of several magnificent Gothic cathedrals in France in the 13th century. One of the most striking features of these cathedrals is their extreme height. For example, the cathedral at Chartres rises to 40 m and the one at Reims has a height of 46 m. It is interesting to note that such magnificent Gothic structures evolved over a very short period of time, without the benefit of any mathematical theory of structures. However, Gothic arches required flying buttresses to prevent the spreading of the arch supported by the tall, narrow columns. The fact that they have been stable for more than 700 years attests to the technical skill of their builders and architects, skill probably acquired through experience and intuition.

Figure 3 shows how the horizontal force at the base of an arch varies with arch height for an arch hinged at the peak. For a given load P, the horizontal force at the base is doubled when the height is reduced by a factor of 2. This explains why the horizontal force required to support a high pointed arch is less than that required for a circular arch. For a given span L, the horizontal force at the base is proportional to the total load P and inversely proportional to the height h. Therefore, in order to minimize the horizontal force at the base, the arch must be made as light and high as possible.

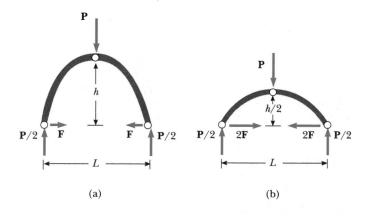

(a)

(b)

FIGURE 3 When the height of an arch is reduced by a factor of 2 and the load force **P** remains the same, the horizontal force at the base is doubled.

FIGURE 4 The St. Louis Gateway Arch seen from the Mississippi River. This beautiful structure has the shape of an inverted freely hanging cable, so that all of its members are under compression.

With the advent of more advanced methods of structural analysis, it has become possible to determine the optimum shape of an arch under given load conditions.

Today, the arch is still the most common structure used to span large distances. One of the most impressive modern arches, the St. Louis Gateway Arch, designed by Eero Saarinen, has a span of 192 m and a height of 192 m (Fig. 4). The largest steel-truss arch bridge, the New River Gorge Bridge in West Virginia, has a span of 520 m. Beautiful concrete arch bridges were designed and built in the 1920s and 1930s by Robert Maillart in Switzerland. The Sando Bridge in Sweden, a single arch of reinforced concrete, spans 264 m.

SOLIDS AND FLUIDS

9

By spreading their arms and legs out from their bodies while keeping the planes of their bodies parallel to the ground, sky divers experience maximum air drag. The forces acting on the sky diver are his or her weight, and the force of air resistance. When these two forces balance each other, the sky diver has zero acceleration and reaches a terminal speed of about 60 m/s. (Heinz Fischer/The Image Bank)

In this chapter we consider some properties of solids and fluids (both liquids and gases). We spend some time looking at properties that are peculiar to solids, but much of our emphasis is on the properties of fluids. We take this approach because an understanding of the behavior of fluids is of fundamental importance to students in the life sciences. We open the fluids part of the chapter with a study of fluids at rest and finish with a discussion of the properties of fluids in motion. Additional topics in fluids including surface tension, viscosity, and transparent phenomena are discussed in a section at the back of the textbook entitled "For Further Study in Chapter 9."

9.1

STATES OF MATTER

Matter is normally classified as being in one of three states—solid, liquid, or gaseous. Often this classification system is extended to include a fourth state, referred to as a plasma. When matter is heated to high temperatures, many of the electrons surrounding each atom are freed from the nucleus. The resulting substance is a collection of free, electrically charged particles—negatively charged electrons and positively charged ions. Such a highly ionized substance containing equal amounts of positive and negative charges is a **plasma**. Plasmas exist inside stars, for example. If we were to take a grand tour of our Universe, we would find that there is far more matter in the plasma state than in the more familiar solid, liquid, and gaseous states because there are far more stars around than any other form of celestial matter. In this chapter, however, we ignore

Crystals of natural quartz (SiO_2), one of the most common minerals on Earth. Quartz crystals are used to make special lenses and prisms and in certain electronic applications. *(Charles Winters)*

plasmas and concentrate on the more familiar solid, liquid, and gaseous forms that make up the environment on our planet.

Everyday experience tells us that a solid has definite volume and shape. A brick, for example, maintains its familiar shape and size day in and day out. We also know that a liquid has a definite volume but no definite shape. For instance, when you fill the tank on a lawn mower, the gasoline changes its shape from that of the original container to that of the tank on the mower. If there is a gallon of gasoline before you pour, however, there still is a gallon after. Finally, a gas has neither definite volume nor definite shape.

All matter consists of some distribution of atoms or molecules. The atoms in a solid are held, by forces that are mainly electrical, at specific positions with respect to one another and vibrate about these equilibrium positions. At low temperatures, however, the vibrating motion is slight and the atoms can be considered to be essentially fixed. As thermal energy (heat) is added to the material, the amplitude of the vibrations increases. A vibrating atom can be viewed as being bound in its equilibrium position by springs attached to neighboring atoms. A collection of such atoms and imaginary springs is shown in Figure 9.1. If a solid is compressed by external forces, we can picture the forces as compressing these tiny internal springs. When the external forces are removed, the solid tends to return to its original shape and size. Consequently, a solid is said to have *elasticity*.

Solids can be classified as being either crystalline or amorphous. A **crystalline solid** is one in which the atoms have an ordered structure. For example, in the sodium chloride crystal (common table salt), sodium and chlorine atoms occupy alternate corners of a cube, as in Figure 9.2a. In an **amorphous solid,** such as glass, the atoms are arranged randomly, as in Figure 9.2b.

For any given substance, the liquid state exists at a higher temperature than the solid state. The intermolecular forces in a liquid are not strong enough to keep the molecules in fixed positions, and they wander through the liquid in a random fashion (Fig. 9.2c). Solids and liquids have the following property in common: When an attempt is made to compress them, strong repulsive atomic forces act internally to resist compression.

In the gaseous state, the molecules are in constant random motion and exert only weak forces on each other. The average separation distances between the molecules of a gas are quite large compared with the size of the molecules.

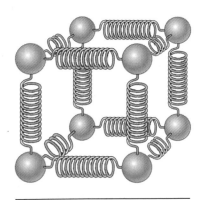

FIGURE 9.1
A model of a solid. The atoms (spheres) are imagined as being attached to each other by springs, which represent the elastic nature of the interatomic forces.

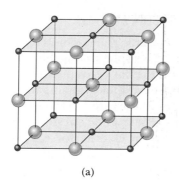

(a)

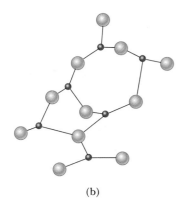

(b)

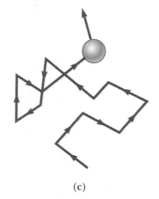

(c)

FIGURE 9.2
(a) The NaCl structure, with the Na^+ (red) and Cl^- (blue) ions at alternate corners of a cube. (b) In an amorphous solid, the atoms are arranged randomly. (c) Erratic motion of a molecule in a liquid.

Occasionally the molecules collide with each other, but most of the time they move as nearly free, noninteracting particles. We shall say more about gases in subsequent chapters.

9.2

THE DEFORMATION OF SOLIDS

We usually think of a solid as an object having definite shape and volume. In our study of mechanics, to keep things simple, we assumed that objects remain undeformed when external forces act on them. In reality, all objects are deformable. That is, it is possible to change the shape or size of an object (or both) through the application of external forces. When the forces are removed, the object tends to return to its original shape and size, which means that the deformation exhibits an elastic behavior.

The elastic properties of solids are discussed in terms of stress and strain. **Stress** is related to the force causing a deformation; **strain** is a measure of the degree of deformation. It is found that, for sufficiently small stresses, stress is proportional to strain, and the constant of proportionality depends on the material being deformed and on the nature of the deformation. We call this proportionality constant the **elastic modulus**:

$$\text{Elastic modulus} \equiv \frac{\text{stress}}{\text{strain}} \qquad \textbf{[9.1]}$$

YOUNG'S MODULUS: ELASTICITY IN LENGTH

Consider a long bar of cross-sectional area A and length L_0, clamped at one end (Fig. 9.3). When an external force, F, is applied along the bar, perpendicularly to the cross section, internal forces in the bar resist the distortion (''stretching'') that F tends to produce, but the bar nevertheless attains an equilibrium in which (1) its length is greater than L_0 and (2) the external force is balanced by internal forces. In such a situation the bar is said to be stressed. We define the **tensile stress** as the ratio of the magnitude of the external force, F, to the cross-sectional area, A. The SI units of stress are newtons per square meter (N/m²). One N/m² is also given a special name, the **pascal** (Pa):

The pascal

$$1 \text{ Pa} \equiv 1 \text{ N/m}^2 \qquad \textbf{[9.2]}$$

The **tensile strain** in this case is defined as the ratio of the change in length, ΔL, to the original length, L_0, and is therefore a dimensionless quantity. Thus, we can use Equation 9.1 to define **Young's modulus**, Y:

$$Y \equiv \frac{\text{tensile stress}}{\text{tensile strain}} = \frac{F/A}{\Delta L/L_0} = \frac{FL_0}{A\,\Delta L} \qquad \textbf{[9.3]}$$

This quantity is typically used to characterize a rod or wire stressed under *either tension or compression*. Note that because the strain is a dimensionless quantity, Y is in pascals. Typical values are given in Table 9.1. Experiments show that (1) the change in length for a fixed external force is proportional to the original length and (2) the force necessary to produce a given strain is proportional to the cross-sectional area.

It is possible to exceed the *elastic limit* of a substance by applying a sufficiently great stress (Fig. 9.4). At the *elastic limit,* the stress-strain curve departs from a straight line. A material subjected to a stress beyond this level ordinarily does not return to its original length when the external force is removed. As the stress is increased further, the material ultimately breaks.

SHEAR MODULUS: ELASTICITY OF SHAPE

Another type of deformation occurs when a body is subjected to a force, **F**, tangential to one of its faces while the opposite face is held fixed (Fig. 9.5a). If the object is originally a rectangular block, such a tangential force results in a

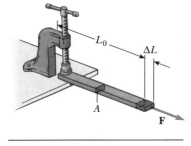

FIGURE 9.3
A long bar clamped at one end is stretched by the amount ΔL under the action of a force, **F**.

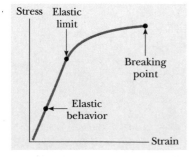

FIGURE 9.4
A stress-strain curve for a solid.

TABLE 9.1
Typical Values for Elastic Modulus

Substance	Young's Modulus (Pa)	Shear Modulus (Pa)	Bulk Modulus (Pa)
Aluminum	7.0×10^{10}	2.5×10^{10}	7.0×10^{10}
Brass	9.1×10^{10}	3.5×10^{10}	6.1×10^{10}
Copper	11×10^{10}	4.2×10^{10}	14×10^{10}
Steel	20×10^{10}	8.4×10^{10}	16×10^{10}
Tungsten	35×10^{10}	14×10^{10}	20×10^{10}
Glass	$6.5{-}7.8 \times 10^{10}$	$2.6{-}3.2 \times 10^{10}$	$5.0{-}5.5 \times 10^{10}$
Quartz	5.6×10^{10}	2.6×10^{10}	2.7×10^{10}
Water	—	—	0.21×10^{10}
Mercury	—	—	2.8×10^{10}

shape whose cross-section is a parallelogram. In this situation, the stress is called a shear stress. A book pushed sideways as in Figure 9.5b is under a shear stress. There is no change in volume with this deformation. We define the **shear stress** as F/A, the ratio of the magnitude of the tangential force to the area, A, of the face being sheared. The **shear strain** is the ratio $\Delta x/h$, where Δx is the horizontal distance the sheared face moves and h is the height of the object. In terms of these quantities, the **shear modulus,** S, is

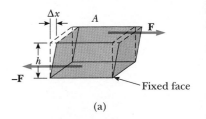

$$S \equiv \frac{\text{shear stress}}{\text{shear strain}} = \frac{F/A}{\Delta x/h} \qquad \textbf{[9.4]}$$

Shear moduli for some representative materials are listed in Table 9.1. Note that the units of shear modulus are force per unit area (Pa).

BULK MODULUS: VOLUME ELASTICITY

Bulk modulus characterizes the response of a substance to uniform squeezing. Suppose that the external forces acting on an object are at right angles to all of its faces (Fig. 9.6) and distributed uniformly over all the faces. As we shall see later, this occurs when an object is immersed in a fluid. A body subject to this type of deformation undergoes a change in volume but no change in shape. The **volume stress,** ΔP, is defined as the ratio of the magnitude of the normal force, F, to the area, A. (When dealing with fluids, we shall refer to the quantity $\Delta P = F/A$ as the **pressure.**) The volume strain is equal to the change in volume, ΔV, divided by the original volume, V. Thus, from Equation 9.1 we can characterize a volume compression in terms of the **bulk modulus,** B, defined as

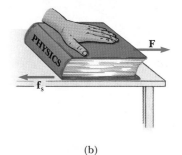

FIGURE 9.5
(a) A shear deformation in which a rectangular block is distorted by a force applied tangent to one of its faces. (b) A book under shear stress.

$$B \equiv \frac{\text{volume stress}}{\text{volume strain}} = -\frac{F/A}{\Delta V/V} = -\frac{\Delta P}{\Delta V/V} \qquad \textbf{[9.5]}$$

Bulk modulus

Note that a negative sign is included in this defining equation so that B is always positive. An increase in pressure (positive ΔP) causes a decrease in volume (negative ΔV), and vice versa.

Table 9.1 lists bulk modulus values for some materials. If you look up such values in a different source, you will often find that the reciprocal of the bulk modulus, called the **compressibility** of the material, is listed. Note from Table 9.1 that both solids and liquids have bulk moduli. However, there is no shear modulus and no Young's modulus for liquids because a liquid will not sustain a shearing stress or a tensile stress (it flows instead).

EXAMPLE 9.1   Built to Last

A vertical steel beam in a building supports a load of 6.0×10^4 N. If the length of the beam is 4.0 m and its cross-sectional area is 8.0×10^{-3} m², find the distance it is compressed along its length.

Solution   Since the beam is under compression, we can use Equation 9.3. Taking Young's modulus for steel, $Y = 20 \times 10^{10}$ Pa, from Table 9.1, we have

$$Y = \frac{FL_0}{A\,\Delta L}$$

or

$$\Delta L = \frac{FL_0}{YA} = \frac{(6.0 \times 10^4 \text{ N})(4.0 \text{ m})}{(20 \times 10^{10} \text{ Pa})(8.0 \times 10^{-3} \text{ m}^2)} = \boxed{1.5 \times 10^{-4} \text{ m}}$$

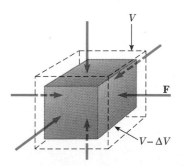

FIGURE 9.6
When a solid is under uniform pressure, it undergoes a change in volume but no change in shape. This cube is compressed on all sides by forces normal to its six faces.

EXAMPLE 9.2 **Squeezing a Lead Sphere**

A solid lead sphere of volume 0.50 m³ is dropped in the ocean to a depth where the water pressure is 2.0×10^7 Pa. Lead has a bulk modulus of 7.7×10^9 Pa. What is the change in volume of the sphere?

Solution   From the definition of bulk modulus (Eq. 9.5), we have

$$B = -\frac{\Delta P}{\Delta V / V}$$

$$\Delta V = -\frac{V \Delta P}{B}$$

In this case, when the sphere is at the surface, where its volume is 0.50 m³, the pressure on it is atmospheric pressure. The increase in pressure, ΔP, when the sphere is submerged is 2.0×10^7 Pa. Therefore, the change in its volume when it is submerged is

$$\Delta V = -\frac{(0.50 \text{ m}^3)(2.0 \times 10^7 \text{ Pa})}{7.7 \times 10^9 \text{ Pa}} = \boxed{-1.3 \times 10^{-3} \text{ m}^3}$$

The negative sign indicates a *decrease* in volume.

FIGURE 9.7
The force exerted by a fluid on a submerged object at any point is perpendicular to the surface of the object. The force exerted by the fluid on the walls of the container is perpendicular to the walls at all points.

Density

9.3

DENSITY AND PRESSURE

The **density** of a substance of uniform composition is defined as its *mass per unit volume.*

In symbolic form, a substance of mass M and volume V has a density, ρ (Greek rho), given by

$$\rho \equiv \frac{M}{V} \qquad\qquad \textbf{[9.6]}$$

The units of density are kilograms per cubic meter in the SI system and grams per cubic centimeter in the cgs system. Table 9.2 lists the densities of some substances. The densities of most liquids and solids vary slightly with changes in temperature and pressure; the densities of gases vary greatly with such changes. Note that under normal conditions the densities of solids and liquids are about 1000 times greater than the densities of gases. This difference implies that the average spacing between molecules in a gas under these conditions is about ten times greater than that in a solid or liquid.

The **specific gravity** of a substance is the ratio of its density to the density of water at 4°C, which is 1.0×10^3 kg/m³. By definition, specific gravity is a dimensionless quantity. For example, if the specific gravity of a substance is 3.0, its density is $3.0(1.0 \times 10^3 \text{ kg/m}^3) = 3.0 \times 10^3$ kg/m³.

We have seen that fluids do not sustain shearing stresses, and thus the only stress that can exist on an object submerged in a fluid is one that tends to compress the object. The force exerted by the fluid on the object is always perpendicular to the surfaces of the object, as shown in Figure 9.7.

TABLE 9.2
Density of Some Common Substances

Substance	ρ (kg/m³)ᵃ	Substance	ρ (kg/m³)ᵃ
Ice	0.917×10^3	Water	1.00×10^3
Aluminum	2.70×10^3	Glycerin	1.26×10^3
Iron	7.86×10^3	Ethyl alcohol	0.806×10^3
Copper	8.92×10^3	Benzene	0.879×10^3
Silver	10.5×10^3	Mercury	13.6×10^3
Lead	11.3×10^3	Air	1.29
Gold	19.3×10^3	Oxygen	1.43
Platinum	21.4×10^3	Hydrogen	8.99×10^{-2}
Uranium	18.7×10^3	Helium	1.79×10^{-1}

ᵃ All values are at standard atmospheric pressure and temperature (STP), defined as
0°C (273 K) and 1 atm (1.01×10^5 Pa). To convert to grams per cubic centimeter,
multiply by 10^{-3}.

The pressure at a specific point in a fluid can be measured with the device pictured in Figure 9.8—an evacuated cylinder enclosing a light piston connected to a spring. As the device is submerged in a fluid, the fluid presses down on the top of the piston and compresses the spring until the inward force of the fluid is balanced by the outward force of the spring. The fluid pressure can be measured directly if the spring is calibrated in advance. This is accomplished by applying a known force to the spring to compress it a given distance.

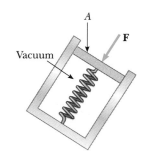

If F is the magnitude of the force exerted by the fluid on the piston and A is the area of the piston, then the **average pressure,** P, of the fluid at the level to which the device has been submerged is defined as the ratio of force to area:

$$P \equiv \frac{F}{A} \qquad \text{[9.7]}$$

FIGURE 9.8
A simple device for measuring pressure in a fluid.

As you can see from this definition, one needs to know the magnitude of the force exerted on a surface *and* the area over which that force is applied. For example, a 700-N man can stand on a vinyl-covered floor in regular street shoes without damaging the surface. If he wears golf shoes with numerous metal cleats protruding from each sole, he does considerable damage to the floor. In both cases the net force applied to the floor is 700 N. However, when the man wears ordinary shoes, the area of his contact with the floor is considerably larger than when he wears golf shoes. (In the latter case, the only area in contact with the floor is the sum of the small cross-sectional areas of the metal cleats.) Hence, the *pressure* on the floor is much smaller when he wears ordinary shoes.

Snowshoes use this principle. The weight of the body—the force—is distributed over the very large areas of the snowshoes so that the pressure at any given point is relatively low and the person does not penetrate very deeply into the snow.

Since pressure is defined as force per unit area, it has units of pascals (newtons per square meter).

Snowshoes prevent the person from sinking into the soft snow because the person's weight is spread over a larger area, which reduces the pressure on the snow's surface. *(Earl Young/FPG)*

EXAMPLE 9.3   The Water Bed

A water bed is 2.00 m on a side and 30.0 cm deep.

(a) Find its weight.

Solution   Since the density of water is 1000 kg/m³ (Table 9.2) and the bed's volume is $(2.00 \times 2.00 \times 0.300)$ m³ = 1.20 m³, the mass of the bed is

$$M = \rho V = (1000 \text{ kg/m}^3)(1.20 \text{ m}^3) = 1.20 \times 10^3 \text{ kg}$$

and its weight is

$$w = Mg = (1.20 \times 10^3 \text{ kg})(9.80 \text{ m/s}^2) = \boxed{1.18 \times 10^4 \text{ N}}$$

This is equivalent to approximately 2640 lb. In order to support such a heavy load, you would be well advised to keep your water bed in the basement or on a sturdy, well-supported floor.

(b) Find the pressure that the water bed exerts on the floor. Assume that the entire lower surface of the bed makes contact with the floor.

Solution   The weight of the water bed is 1.18×10^4 N. Its cross-sectional area is 4.00 m². The pressure exerted on the floor is therefore

$$P = \frac{1.18 \times 10^4 \text{ N}}{4.00 \text{ m}^2} = \boxed{2.95 \times 10^3 \text{ Pa}}$$

Exercise   Calculate the pressure exerted by the bed on the floor if the bed rests on its side.

Answer   Since the area of the bed's side is 0.6 m², the pressure is 1.96×10^4 Pa.

9.4
VARIATION OF PRESSURE WITH DEPTH

If a fluid is at rest in a container, *all* portions of the fluid must be in static equilibrium. Furthermore, *all points at the same depth must be at the same pressure.* If this were not the case, a given portion of the fluid would not be in equilibrium. For example, consider the small block of fluid shown in Figure 9.9a. If the pressure were greater on the left side of the block than on the right, F_1 would be greater than F_2, and the block would accelerate and would not be in equilibrium.

Now let us examine the portion of the fluid contained within the volume indicated by the darker region in Figure 9.9b. This region has a cross-sectional area of A and extends to a depth of h below the surface of the water. Three external forces act on this volume of fluid: its weight, Mg; the upward force, PA, exerted by the fluid below it; and a downward force, P_0A, exerted by the atmosphere, where P_0 is atmospheric pressure. Since this volume of fluid is in equilibrium, these forces must add to zero, and so we get

$$PA - Mg - P_0A = 0 \qquad [9.8]$$

From the relation $M = \rho V = \rho Ah$, the weight of the fluid in the volume is

$$w = Mg = \rho gAh \qquad [9.9]$$

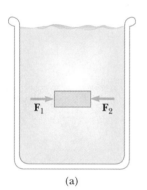

(a)

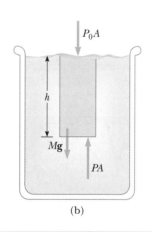

(b)

FIGURE 9.9
(a) If the block of fluid is to be in equilibrium, the force $\mathbf{F}_1$ must balance the force $\mathbf{F}_2$.
(b) The net force on the volume of water within the darker region must be zero.

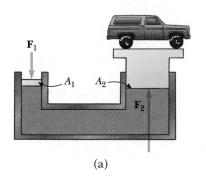

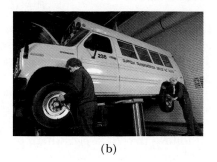

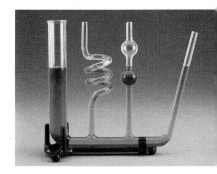

(a)

(b)

FIGURE 9.10
(a) Diagram of a hydraulic press. Since the pressure is the same at the left and right sides, a small force, $\mathbf{F}_1$, at the left produces a much larger force, $\mathbf{F}_2$, at the right. (b) A bus under repair is supported by a hydraulic lift in a garage. *(Superstock)*

When Equation 9.9 is substituted into Equation 9.8, we get

$$P = P_0 + \rho g h \qquad \textbf{[9.10]}$$

where normal atmospheric pressure at sea level is $P_0 = 1.01 \times 10^5$ Pa (equivalent to 14.7 lb/in.²). According to Equation 9.10,

> **The pressure, P, at a depth of h below the surface of a liquid open to the atmosphere is greater than atmospheric pressure by the amount $\rho g h$. Moreover, the pressure is not affected by the shape of the vessel.**

Pressure in a fluid depends only upon depth. Any increase in pressure at the surface is transmitted to every point in the fluid. This was first recognized by the French scientist Blaise Pascal (1623–1662) and is called **Pascal's principle:**

> **Pressure applied to an enclosed fluid is transmitted undiminished to every point of the fluid and to the walls of the containing vessel.**

An important application of Pascal's principle is the hydraulic press (Fig. 9.10a). A downward force, $\mathbf{F}_1$, is applied to a small piston of area A_1. The pressure is transmitted through a fluid to a larger piston of area A_2. Because the pressure is the same on both sides, we see that $P = F_1/A_1 = F_2/A_2$. Therefore, the magnitude of the force $\mathbf{F}_2$ is larger than the magnitude of $\mathbf{F}_1$ by the factor A_2/A_1. That is why a large load, such as a car, can be supported on the large piston by a much smaller force on the smaller piston. Hydraulic brakes, car lifts, hydraulic jacks, forklifts, and other machines make use of this principle.

This photograph illustrates that the pressure in a liquid is the same at all points having the same elevation. Note that the shape of the vessel does not affect the pressure. *(Courtesy of Central Scientific Company)*

9.5

PRESSURE MEASUREMENTS

Another simple device for measuring pressure is the open-tube manometer (Fig. 9.11). One end of a U-shaped tube containing a liquid is open to the atmosphere, and the other end is connected to a system of unknown pressure P. The pressure at point B equals $P_0 + \rho g h$, where ρ is the density of the fluid. The pressure at B, however, equals the pressure at A, which is also the unknown pressure P. Therefore, we conclude that

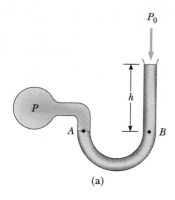

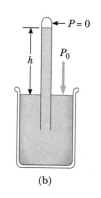

FIGURE 9.11
Two devices for measuring pressure: (a) an open-tube manometer and (b) a mercury barometer.

$$P = P_0 + \rho g h$$

The pressure P is called the *absolute pressure*, and $P - P_0$ is called the *gauge pressure*. Thus, if P in the system is greater than atmospheric pressure, h is positive. If P is less than atmospheric pressure (a partial vacuum), h is negative.

A third instrument that is used to measure pressure is the *barometer* (Fig. 9.11b), invented by Evangelista Torricelli (1608–1647). A long tube that is closed at one end is filled with mercury and then inverted into a dish of mercury. The closed end of the tube is nearly a vacuum, and so its pressure can be taken as zero. It follows that $P_0 = \rho g h$, where ρ is the density of the mercury and h is the height of the mercury column.

One atmosphere of pressure is defined to be the pressure equivalent of a column of mercury that is exactly 0.76 m in height at 0°C with $g = 9.806\ 65$ m/s². At this temperature mercury has a density of 13.595×10^3 kg/m³; therefore,

$$P_0 = \rho g h = (13.595 \times 10^3 \text{ kg/m}^3)(9.806\ 65 \text{ m/s}^2)(0.7600 \text{ m})$$

$$= 1.013 \times 10^5 \text{ Pa}$$

Interestingly, the force of the atmosphere on our bodies (assuming a body area of 2000 in.²) is extremely large, on the order of 30 000 lb! A natural question to raise is: How can we exist under such great forces attempting to collapse our bodies? The answer is that our body cavities and tissues are permeated with fluids and gases that are pushing outward with this same atmospheric pressure. Consequently, our bodies are in equilibrium under the force of the atmosphere pushing in and an equal internal force pushing out.

EXAMPLE 9.4   The Car Lift

In a car lift, compressed air exerts a force on a piston with a radius of 5.00 cm. This pressure is transmitted to a second piston, of radius 15.0 cm. What force must the compressed air exert in order to lift a car weighing 1.33×10^4 N? What air pressure produces this force? Neglect the weights of the pistons.

Solution   Because the pressure exerted by the compressed air is transmitted undiminished throughout the fluid, we have

$$F_1 = \left(\frac{A_1}{A_2}\right) F_2 = \frac{\pi(5.00 \times 10^{-2} \text{ m})^2}{\pi(15.0 \times 10^{-2} \text{ m})^2}(1.33 \times 10^4 \text{ N}) = \boxed{1.48 \times 10^3 \text{ N}}$$

The air pressure that produces this force is

$$P = \frac{F_1}{A_1} = \frac{1.48 \times 10^3 \text{ N}}{\pi(5.00 \times 10^{-2} \text{ m})^2} = \boxed{1.88 \times 10^5 \text{ Pa}}$$

This pressure is approximately twice atmospheric pressure. (Note that pressure here means gauge pressure.)

EXAMPLE 9.5   Pressure in the Ocean

Calculate the absolute pressure at an ocean depth of 1000 m. Assume that the density of water is 1.0×10^3 kg/m³ and that $P_0 = 1.01 \times 10^5$ Pa.

Solution

$$P = P_0 + \rho g h$$

$$= 1.01 \times 10^5 \text{ Pa} + (1.0 \times 10^3 \text{ kg/m}^3)(9.80 \text{ m/s}^2)(1.00 \times 10^3 \text{ m})$$

$$P \approx \boxed{9.9 \times 10^6 \text{ Pa}}$$

This is approximately 100 times greater than atmospheric pressure! Obviously, the design and construction of vessels that can withstand such enormous pressures are not trivial matters.

Exercise   Calculate the total force exerted on the outside of a 30-cm-diameter circular submarine window at this depth.

Answer 7.0×10^5 N.

9.6

BUOYANT FORCES AND ARCHIMEDES' PRINCIPLE

A fundamental principle affecting objects submerged in fluids was discovered by the Greek mathematician Archimedes. **Archimedes' principle** can be stated as follows:

> **Any body completely or partially submerged in a fluid is buoyed up by a force equal to the weight of the fluid displaced by the body.**

Archimedes' principle

Everyone has experienced Archimedes' principle. For example, recall that it is relatively easy to lift someone if you are both standing in a swimming pool, whereas lifting that same individual on dry land would be a difficult task. Evidently, water provides partial support to any object placed in it. We say that an object placed in a fluid is buoyed up by the fluid, and we call this upward force the **buoyant force**. According to Archimedes' principle, *the magnitude of the buoyant force always equals the weight of the fluid displaced by the object.* The buoyant force acts vertically upward through what was the center of gravity of the fluid before the fluid was displaced.

Archimedes' principle can be verified in the following manner. Suppose we focus our attention on the cube of water that is colored red in the container of Figure 9.12. This cube of water is in equilibrium under the action of the forces on it. One of those forces is its weight. What cancels that downward force? Apparently, the water beneath the cube is buoying it up and holding it in equilibrium. Thus, the buoyant force, **B**, on the cube of water must be exactly equal in magnitude to the weight of the water inside the cube: $B = w$.

Now imagine that the cube of water is replaced by a cube of steel of the same dimensions. What is the buoyant force on the steel? The water surrounding a cube behaves in the same way whether the cube is made of water or steel; therefore, *the buoyant force acting on the steel is the same as the buoyant force acting on a cube of water of the same dimensions.* This result applies for a totally submerged object of any shape, size, or density.

Let us show explicitly that the buoyant force is equal in magnitude to the weight of the displaced fluid. The pressure at the bottom of the cube in Figure

FIGURE 9.12
The external forces on a cube of water are its weight, **w**, and the buoyancy force, **B**. Under equilibrium conditions, $B = w$.

Hot-air balloons over Albuquerque, New Mexico. Since hot air is less dense than cold air, there is a net upward buoyant force on the balloons. *(L. Menzies/The Image Works)*

9.12 is greater than the pressure at the top by the amount $\rho_f gh$, where ρ_f is the density of the fluid and h is the height of the cube. Since the pressure difference, ΔP, is equal to the buoyant force per unit area, or, $\Delta P = B/A$, we see that $B = (\Delta P)(A) = (\rho_f gh)(A) = \rho_f gV$, where V is the volume of the cube. Since the mass of the fluid in the cube is $M = \rho_f V$, we see that

$$B = \rho_f Vg = Mg = w_f \qquad \text{[9.11]}$$

where w_f is the weight of the displaced fluid.

The weight of the submerged object is $w_0 = mg = \rho_0 Vg$, where ρ_0 is the density of the object. Since $w_f = \rho_f Vg$ and $w_0 = \rho_0 Vg$, we see that, if the density of the object is greater than the density of the fluid, the unsupported object sinks. If the density of the object is less than that of the fluid, the unsupported submerged object accelerates upward and ultimately floats. When a *floating* object is in equilibrium, only part of it is submerged. In this case, *the buoyant force equals the weight of the object.*

It is instructive to compare the forces on a totally submerged object with those on a floating object.

Case I: A Totally Submerged Object. When an object is *totally* submerged in a fluid of density ρ_f, the upward buoyant force has a magnitude of $B = \rho_f V_0 g$, where V_0 is the volume of the object. If the object has density ρ_0, the downward force of its weight is equal to $w = mg = \rho_0 V_0 g$, and the net force on it is $B - w = (\rho_f - \rho_0)V_0 g$. Hence, if the density of the object is less than the density of the fluid, as in Figure 9.13a, the net force is positive (upward) and the unsupported object accelerates upward. If the density of the object is greater than the density of the fluid, as in Figure 9.13b, the net force is negative and the unsupported object sinks.

(a)

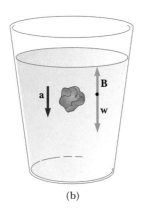

(b)

FIGURE 9.13

(a) A totally submerged object that is less dense than the fluid in which it is submerged experiences a net upward force. (b) A totally submerged object that is denser than the fluid sinks.

Case II: A Floating Object. Now consider an object in static equilibrium floating on a fluid—that is, a partially submerged object. In this case, the upward buoyant force is balanced by the downward weight of the object. If V_f is the volume of the fluid displaced by the object (which corresponds to the volume of the part of the object beneath the fluid level), then the buoyant force has magnitude $B = \rho_f V_f g$. Since the weight of the object is $w = mg = \rho_0 V_0 g$, and $w = B$, we see that $\rho_f V_f g = \rho_0 V_0 g$, or

$$\frac{\rho_0}{\rho_f} = \frac{V_f}{V_0} \qquad \text{[9.12]}$$

Under normal conditions, the average density of a fish is slightly greater than the density of water. This being the case, a fish would sink if it did not have a mechanism for adjusting its density: the internal regulation of the size of the swim bladder. In this manner, fish maintain neutral buoyancy as they swim to various depths.

The human brain is immersed in a fluid of density 1007 kg/m^3, which is slightly less than the average density of the brain, 1040 kg/m^3. Consequently, most of the weight of the brain is supported by the buoyant force of the surrounding fluid. In some clinical procedures, it is necessary to remove a portion of this fluid for diagnostic purposes. During such procedures, the nerves and blood vessels in the brain are placed under great strain, which in turn can

Archimedes, a Greek mathematician, physicist, and engineer, was perhaps the greatest scientist of antiquity. He was the first to accurately compute the ratio of a circle's circumference to its diameter, and he also showed how to calculate the volume and surface area of spheres, cylinders, and other geometric shapes. He is well known for discovering the nature of the buoyant force and was a gifted inventor. One of his practical inventions, still in use today, is Archimedes' screw, an inclined, rotating, coiled tube used originally to lift water from the holds of ships. He also invented the catapult and devised systems of levers, pulleys, and weights for raising heavy loads. Such inventions were successfully used to defend Archimedes' native city, Syracuse, during a two-year siege by the Romans.

According to legend, Archimedes was asked by King Hieron to determine whether the king's crown was made of pure gold or merely a gold alloy. The task was to be performed without damaging the crown. Archimedes allegedly arrived at a solution while taking a bath, noting a partial loss of weight after submerging his arms and legs in the water. As the story goes, he was so excited about his great discovery that he ran naked through the streets of Syracuse shouting, ''Eureka!'' which is Greek for ''I have found it.''

**Archimedes
(287–212 B.C.)**

cause extreme discomfort and pain. Great care must be exercised with such patients until the initial brain fluid volume has been restored by the body.

When service station attendants check the antifreeze in your car or the condition of your battery, they often use devices that apply Archimedes' principle. Figure 9.14 shows a common device that is used to check the antifreeze in a car radiator. The small balls in the enclosed tube vary in density so that all of them float when the tube is filled with pure water, none floats in pure antifreeze, one floats in a 5% mixture, two in a 10% mixture, and so forth. The number of balls that float thus serves as a measure of the percentage of antifreeze in the mixture, which in turn is used to determine the lowest temperature the mixture can withstand without freezing.

Similarly, the charge of some newer car batteries can be determined with a so-called ''magic-dot'' process that is built into the battery (Fig. 9.15). When one

Mercury has a density about 13.6 times that of water. As a result, this steel ball floats in a pool of mercury. (*Courtesy of Henry Leap and Jim Lehman*)

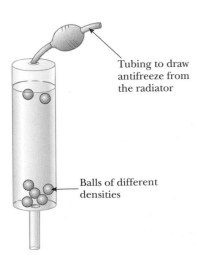

Tubing to draw antifreeze from the radiator

Balls of different densities

FIGURE 9.14
The number of balls that float in this device is a measure of the density of the antifreeze solution in a vehicle's radiator, and consequently a measure of the temperature at which freezing will occur.

FIGURE 9.15
The red ball in the plastic tube inside the battery serves as an indicator of whether the battery is charged or discharged. As the battery loses its charge, the density of the battery fluid decreases, and the ball sinks out of sight.

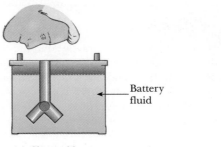

Battery fluid

(a) Charged battery

(b) Discharged battery

looks down into a viewing port in the top of the battery, a red dot indicates that the battery is sufficiently charged; a black dot, that the battery has lost its charge. This is because, if the battery has sufficient charge, the density of the battery fluid is high enough to cause the red ball to float. As the battery loses its charge, the density of the battery fluid decreases and the ball sinks beneath the surface of the fluid, where the dot appears black.

EXAMPLE 9.6   A Red Tag Special on Crowns

A bargain hunter purchases a "gold" crown at a flea market. After she gets home, she hangs it from a scale and finds its weight to be 7.84 N (Fig. 9.16a). She then weighs the crown while it is immersed in water of density 1000 kg/m³, as in Figure 9.16b, and now the scale reads 6.86 N. Is the crown made of pure gold?

Reasoning   When the crown is suspended in air, the scale reads the true weight, w (neglecting the buoyancy of air). When it is immersed in water, the buoyant force, **B**, reduces the scale reading to an apparent weight of $T_2 = w - B$. Hence, the buoyant force on the crown is the difference between its weight in air and its weight in water:

$$B = w - T_2 = 7.84 \text{ N} - 6.86 \text{ N} = 0.98 \text{ N}$$

Because the buoyant force is equal in magnitude to the weight of the displaced fluid, w_f, we have

$$w_f = \rho_f g V_f = 0.98 \text{ N}$$

where V_f is the displaced fluid's volume, and ρ_f is its density (1000 kg/m³). Also, the volume of the crown, V_c, is equal to the volume of the displaced fluid (because the crown is completely submerged).

Solution   We find that

$$V_c = V_f = \frac{0.98 \text{ N}}{g\rho_f} = \frac{0.98 \text{ N}}{(9.8 \text{ m/s}^2)(1000 \text{ kg/m}^3)} = 1.0 \times 10^{-4} \text{ m}^3$$

Finally, the density of the crown is

$$\rho_c = \frac{m_c}{V_c} = \frac{w_c}{gV_c} = \frac{7.84 \text{ N}}{(9.8 \text{ m/s}^2)(1.0 \times 10^{-4} \text{ m}^3)} = 8.0 \times 10^3 \text{ kg/m}^3$$

From Table 9.2 we see that the density of gold is 19.3×10^3 kg/m³. Thus, either the crown is hollow or it is made of an alloy. This was not a good day for the bargain hunter!

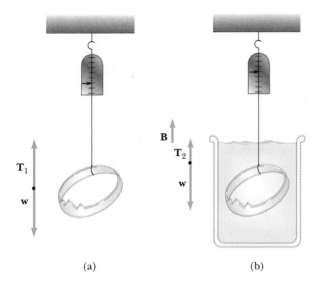

FIGURE 9.16

(Example 9.6) (a) When the crown is suspended in air, the scale reads the true weight, **w**. (b) When the crown is immersed in water, the buoyant force, **B**, reduces the scale reading to the apparent weight, $T_2 = w - B$.

(a)   (b)

EXAMPLE 9.7 Floating Down the River

A raft is constructed of wood having a density of 600 kg/m³. Its surface area is 5.7 m², and its volume is 0.60 m³. When the raft is placed in fresh water of density 1000 kg/m³, as in Figure 9.17, how much of it is below water level?

Reasoning The magnitude of the upward buoyant force acting on the raft must equal (in magnitude) the weight of the raft if the raft is to float. Additionally, from Archimedes' principle the buoyant force is equal to the weight of the displaced water.

Solution The weight of the raft is

$$w_r = \rho_r g V_r = (600 \text{ kg/m}^3)(9.8 \text{ m/s}^2)(0.60 \text{ m}^3) = 3.5 \times 10^3 \text{ N}$$

The upward buoyant force acting on the raft equals the weight of the displaced water, which in turn must equal the weight of the raft:

$$B = \rho_w g V_w = \rho_w g A h = 3.5 \times 10^3 \text{ N}$$

Since the area A and density ρ_w are known, we can find the depth, h, the raft sinks into the water:

$$h = \frac{B}{\rho_w g A} = \frac{3.5 \times 10^3 \text{ N}}{(1000 \text{ kg/m}^3)(9.8 \text{ m/s}^2)(5.7 \text{ m}^2)} = \boxed{0.06 \text{ m}}$$

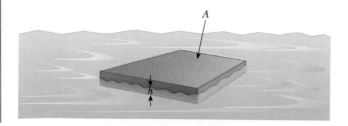

A

FIGURE 9.17

(Example 9.7) A raft partially submerged in water.

9.7

FLUIDS IN MOTION

When a fluid is in motion, its flow can be characterized in one of two ways. The flow is said to be **streamline,** or **laminar,** if every particle that passes a particular point moves exactly along the smooth path followed by particles that passed that point earlier. The path is called a *streamline* (Fig. 9.18a). Different streamlines cannot cross each other under this steady-flow condition, and the streamline at any point coincides with the direction of fluid velocity at that point.

In contrast, the flow of a fluid becomes irregular, or **turbulent,** above a certain velocity or under any conditions that can cause abrupt changes in velocity. Irregular motions of the fluid, called *eddy currents,* are characteristic in turbulent flow, as shown in Figure 9.18b.

In discussions of fluid flow, the term *viscosity* is used for the degree of internal friction in the fluid. This internal friction is associated with the resistance between two adjacent layers of the fluid moving relative to each other. A fluid such as kerosene has a lower viscosity than crude oil or molasses.

Many features of fluid motion can be understood by considering the behavior of an **ideal fluid,** which satisfies the following conditions:

1. *The fluid is nonviscous;* that is, there is no internal friction force between adjacent layers.
2. *The fluid is incompressible,* which means that its density is constant.
3. *The fluid motion is steady,* meaning that the velocity, density, and pressure at each point in the fluid do not change in time.
4. *The fluid moves without turbulence.* This implies that each element of the fluid has zero angular velocity about its center; that is, there can be no eddy currents present in the moving fluid.

EQUATION OF CONTINUITY

Figure 9.19 represents a fluid flowing through a pipe of nonuniform size. The particles in the fluid move along the streamlines in steady-state flow. In a small time interval, Δt, the fluid entering the bottom end of the pipe moves a distance of $\Delta x_1 = v_1 \Delta t$, where v_1 is the speed of the fluid at this location. If A_1 is the cross-sectional area in this region, then the mass contained in the bottom blue region is $\Delta M_1 = \rho A_1 \Delta x_1 = \rho A_1 v_1 \Delta t$, where ρ is the density of the fluid. Similarly, the fluid that moves out of the upper end of the pipe in the same interval, Δt, has a mass of $\Delta M_2 = \rho A_2 v_2 \Delta t$. However, because mass is conserved and because the flow is steady, the mass that flows into the bottom of the pipe through A_1 in the time Δt must equal the mass that flows out through A_2 in the same interval. Therefore, $\Delta M_1 = \Delta M_2$, or

$$\rho A_1 v_1 = \rho A_2 v_2 \qquad \textbf{[9.13]}$$

Equation 9.13 reduces to

$$A_1 v_1 = A_2 v_2 \qquad \textbf{[9.14]}$$

This expression is called the **equation of continuity.**

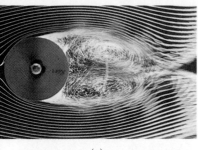

(a)

(b)

FIGURE 9.18
(a) This photograph was taken in a water tunnel using hydrogen bubbles to visualize the flow pattern around a cylinder. The flow was started from rest, and at this instant the pattern shows the development of a complex wake structure on the downstream side of the cylinder. *(Courtesy of Dr. Kenneth W. McAlister, Department of the Army)* (b) Hot gases from a cigarette made visible by smoke particles. The smoke first moves in streamline flow at the bottom and then in turbulent flow above. *(Werner Wolff/Black Star)*

The product of any cross-sectional area of the pipe and the fluid speed at that cross section is a constant.

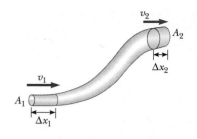

Therefore, the speed is high where the tube is constricted and low where the tube has a larger diameter. The product Av, which has dimensions of volume per unit time, is called the **flow rate**.

The condition Av = constant is equivalent to the fact that the amount of fluid that enters one end of the tube in a given time interval equals the amount of fluid leaving the tube in the same interval, assuming the absence of leaks.

FIGURE 9.19
A fluid moving with streamline flow through a pipe of varying cross-sectional area. The volume of fluid flowing through A_1 in a time interval of Δt must equal the volume flowing through A_2 in the same time interval. Therefore, $A_1 v_1 = A_2 v_2$.

EXAMPLE 9.8   Filling a Water Bucket

A water hose 1.00 cm in radius is used to fill a 20.0-liter bucket. If it takes 1.00 min to fill the bucket, what is the speed, v, at which the water leaves the hose? (1 liter = 10^3 cm³.)

Solution   The cross-sectional area of the hose is

$$A = \pi r^2 = \pi(1.00 \text{ cm})^2 = \pi \text{ cm}^2$$

The flow rate is equal to the product Av. Thus,

$$Av = 20.0 \frac{\text{liters}}{\text{min}} = \frac{20.0 \times 10^3 \text{ cm}^3}{60.0 \text{ s}}$$

$$v = \frac{20.0 \times 10^3 \text{ cm}^3}{(\pi \text{ cm}^2)(60.0 \text{ s})} = \boxed{106 \text{ cm/s}}$$

Exercise   If the radius of the hose is reduced to 0.500 cm, what is the speed of the water as it leaves the hose, assuming the same flow rate?

Answer   424 cm/s.

BERNOULLI'S EQUATION

As a fluid moves through a pipe of varying cross section and elevation, the pressure changes along the pipe. In 1738 the Swiss physicist Daniel Bernoulli (1700–1782) derived a fundamental expression that relates pressure to fluid speed and elevation. Bernoulli's equation is not a freestanding law of physics. It is, instead, a consequence of energy conservation as applied to the ideal fluid.

In deriving Bernoulli's equation, we again assume that the fluid is incompressible and nonviscous and flows in a nonturbulent, steady-state manner. Consider the flow through a nonuniform pipe in the time Δt, as illustrated in Figure 9.20. The force on the lower end of the fluid is $P_1 A_1$, where P_1 is the pressure at the lower end. The work done on the lower end of the fluid by the fluid behind it is

$$W_1 = F_1 \, \Delta x_1 = P_1 A_1 \, \Delta x_1 = P_1 V$$

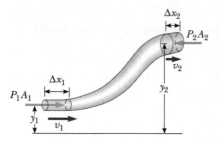

FIGURE 9.20
A fluid flowing through a constricted pipe with streamline flow. The fluid in the section with a length of Δx_1 moves to the section with a length of Δx_2. The volumes of fluid in the two sections are equal.

where V is the volume of the lower green region in Figure 9.20. In a similar manner, the work done on the fluid on the upper portion in the time Δt is

$$W_2 = -P_2 A_2 \, \Delta x_2 = -P_2 V$$

(Remember that the volume of fluid that passes through A_1 in the time Δt equals the volume that passes through A_2 in the same interval.) The work W_2 is negative because the force on the fluid at the top is opposite its displacement. Thus, the net work done by these forces in the time Δt is

$$W = P_1 V - P_2 V$$

Part of this work goes into changing the fluid's kinetic energy, and part goes into changing its gravitational potential energy. If m is the mass of the fluid passing through the pipe in the time interval Δt, then the change in kinetic energy of the volume of fluid is

$$\Delta KE = \tfrac{1}{2} m v_2^2 - \tfrac{1}{2} m v_1^2$$

The change in its potential energy is

$$\Delta PE = mgy_2 - mgy_1$$

We can apply the work-energy theorem in the form $W = \Delta KE + \Delta PE$ (Chapter 5) to this volume of fluid to give

$$P_1 V - P_2 V = \tfrac{1}{2} m v_2^2 - \tfrac{1}{2} m v_1^2 + mgy_2 - mgy_1$$

If we divide each term by V and recall that $\rho = m/V$, this expression reduces to

$$P_1 - P_2 = \tfrac{1}{2}\rho v_2^2 - \tfrac{1}{2}\rho v_1^2 + \rho gy_2 - \rho gy_1$$

Let us move those terms that refer to point 1 to one side of the equation and those that refer to point 2 to the other side:

$$P_1 + \tfrac{1}{2}\rho v_1^2 + \rho gy_1 = P_2 + \tfrac{1}{2}\rho v_2^2 + \rho gy_2 \qquad \text{[9.15]}$$

This is **Bernoulli's equation.** It is often expressed as

Bernoulli's equation

$$P + \tfrac{1}{2}\rho v^2 + \rho gy = \text{constant} \qquad \text{[9.16]}$$

> Bernoulli's equation says that the sum of the pressure (P), the kinetic energy per unit volume ($\tfrac{1}{2}\rho v^2$), and the potential energy per unit volume (ρgy) has the same value at all points along a streamline.

Daniel Bernoulli was a Swiss physicist and mathematician who made important discoveries in hydrodynamics. Born into a family of mathematicians on February 8, 1700, he was the only member of the family to make a mark in physics. He was educated and received his doctorate in Basel, Switzerland.

Bernoulli's most famous work, *Hydrodynamica,* was published in 1738. It is both a theoretical and a practical study of equilibrium, pressure, and velocity of fluids. He showed that, as the velocity of fluid flow increases, its pressure decreases. Chemists working in laboratories use this phenomenon, known as "Bernoulli's principle," when they produce a vacuum by connecting a vessel to a tube through which water is running rapidly. Bernoulli's principle is an early formulation of the idea of conservation of energy.

Bernoulli's *Hydrodynamica* also attempted the first explanation of the behavior of gases with changing pressure and temperature; this was the beginning of kinetic theory of gases.

Daniel Bernoulli (1700–1782)

(North Wind Picture Archives)

An important consequence of Bernoulli's equation can be demonstrated by considering Figure 9.21a, which shows water flowing through a horizontal constricted pipe from a region of large cross-sectional area into a region of smaller cross-sectional area. This device, called a *Venturi tube,* can be used to measure the speed of fluid flow. Let us compare the pressure at point 1 to the pressure at point 2. Because the pipe is horizontal, $y_1 = y_2$ and Equation 9.15 applied to points 1 and 2 gives

$$P_1 + \tfrac{1}{2}\rho v_1{}^2 = P_2 + \tfrac{1}{2}\rho v_2{}^2 \qquad \textbf{[9.17]}$$

Since the water is not backing up in the pipe, its speed in the constriction, v_2, must be greater than the speed v_1. From Equation 9.17, $v_2 > v_1$ means that P_2 must be less than P_1. This result is often expressed by the statement that

Swiftly moving fluids exert less pressure than do slowly moving fluids.

As we shall see in the next section, this important result enables us to understand a wide range of everyday phenomena.

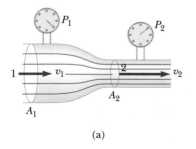

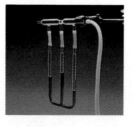

(a)

FIGURE 9.21

(a) The pressure P_1 is greater than the pressure P_2, since $v_1 < v_2$. This device can be used to measure the speed of fluid flow. (b) A Venturi tube. *(Courtesy of Central Scientific Company)*

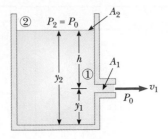

FIGURE 9.22
(Example 9.9)   The water speed, v_1, from the hole in the side of the container is given by $v_1 = \sqrt{2gh}$.

EXAMPLE 9.9   Shoot-Out at the Old Water Tank

A nearsighted sheriff fires at a cattle rustler with his trusty six-shooter. Fortunately for the cattle rustler, the bullet misses him and penetrates the town water tank to cause a leak (Fig. 9.22). If the top of the tank is open to the atmosphere, determine the speed at which the water leaves the hole when the water level is 0.500 m above the hole.

Reasoning   If we assume that the cross-sectional area of the tank is large relative to that of the hole ($A_2 \gg A_1$), then the water level drops very slowly and we can assume $v_2 \approx 0$. Let us apply Bernoulli's equation to points 1 and 2. If we note that $P_1 = P_0$ at the hole, we get

$$P_0 + \tfrac{1}{2}\rho v_1^2 + \rho g y_1 = P_0 + \rho g y_2$$
$$v_1 = \sqrt{2g(y_2 - y_1)} = \sqrt{2gh}$$

Solution   This says that the speed of the water emerging from the hole is equal to the speed acquired by a body falling freely through the vertical distance h. This is known as **Torricelli's law.** If the height h is 0.500 m, for example, the speed of the stream is

$$v = \sqrt{2(9.80 \text{ m/s}^2)(0.500 \text{ m})} = \boxed{3.13 \text{ m/s}}$$

Exercise   If the head of the cattle rustler is 3.00 m below the level of the hole in the tank, where must he stand to get doused with water?

Answer   2.45 m from the base of the tank.

9.8

OTHER APPLICATIONS OF BERNOULLI'S EQUATION

In this section we describe some common phenomena that can be explained, at least in part, by Bernoulli's equation.

First, consider the circulation of air around a thrown baseball. If the ball is not spinning (Fig. 9.23a), the motion of the airstream past the ball is nearly streamline. In this figure the ball is moving from right to left. Hence, from its point of view, the airstream is moving from left to right. A symmetric region of turbulence occurs behind the ball as shown. When the ball is spinning counterclockwise (Fig. 9.23b), layers of air near its surface are carried in the direction of

FIGURE 9.23
(a) The airstream around a nonrotating baseball moving from right to left. The streamlines represent the flow relative to the baseball. Note the symmetric region of turbulence behind the ball. (b) The airstream around a spinning baseball. The ball experiences a deflecting force because of the Bernoulli effect.

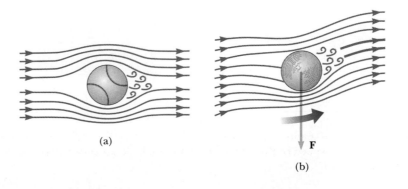

(a)

(b)

PHYSICS IN *ACTION*

THE BERNOULLI EFFECT

In the large-scale Venturi tube shown in the left-hand photograph, air is being passed through a constricted open tube. The balls float above the jets of air from the vertical tubes. The pressure below each ball is greater than the pressure above it as a result of the difference in air speeds. Hence, there is an upward force on each ball which balances its weight.

The different heights of the balls can be explained as follows. The air pressure is lowest at points where the tube is constricted and the air speed is greatest. Thus, the pressure is lowest at the center, and so the center ball is at the lowest elevation. Note that the outer balls are *not* at the same height, even though the tube diameter is the same at these points. This difference indicates that the air has slowed down as it passes from right to left. Furthermore, some air has escaped from the two vertical tubes on the right before reaching the vertical tube on the left.

In the photograph on the right, a ski jumper in "flight" bends his body forward and keeps his arms close to his body to maximize the distance he travels. In this configuration, the shape of the jumper's body is similar to that of an airfoil. The air moving over the top of this shape moves faster than the air moving beneath, causing a pressure difference and a net upward force (lift). If the jumper extends his arms to maintain balance during flight, the drag force increases, resulting in a shorter jump. The length of the jump depends on other factors as well, such as the jumper's speed and the timing of his leap at the end of the ramp before he becomes airborne.

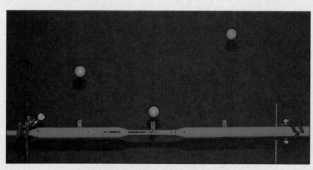

(Courtesy of Henry Leap and Jim Lehman)

(Galen Powell/Peter Arnold, Inc.)

spin because of viscosity. The combined effect of the steady flow of air and the air dragged along due to the spinning motion produces the streamlines shown in 9.23b and a corresponding turbulence pattern. The speed of the air below the ball is greater than the speed above the ball. Thus, from Bernoulli's equation, the air pressure above the ball is greater than the air pressure below the ball, and the ball experiences a downward deflecting force. When a pitcher wishes to throw a curve ball that deflects sideways, the spin axis should be vertical (perpendicular to the page in Fig. 9.23b). In contrast, if he wishes to throw a "sinker," the spin axis should be horizontal. Tennis balls and golf balls with spin also exhibit dynamic lift.

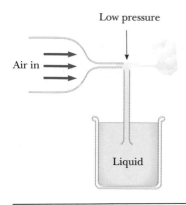

FIGURE 9.24
A stream of air passing over a tube dipped in a liquid causes the liquid to rise in the tube as shown. This effect is used in perfume bottles and paint sprayers.

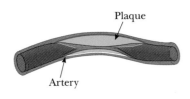

FIGURE 9.25
A flow of blood through a constricted artery.

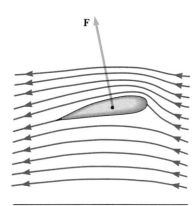

FIGURE 9.26
Streamline flow around an airplane wing. The pressure above is less than the pressure below, and there is a dynamic lift force upward.

Many devices operate in the manner illustrated in Figure 9.24. A stream of air passing over an open tube reduces the pressure above the tube. This causes the liquid to rise into the airstream. The liquid is then dispersed into a fine spray of droplets. You might recognize that this so-called atomizer is used in perfume bottles and paint sprayers. The same principle is used in the carburetor of a gasoline engine. In that case, the low-pressure region in the carburetor is produced by air drawn in by the piston through the air filter. The gasoline vaporizes, mixes with the air, and enters the cylinder of the engine for combustion.

In a person with advanced arteriosclerosis, the Bernoulli effect produces a symptom called vascular flutter. In this situation, the artery is constricted as a result of accumulated plaque on its inner walls, as in Figure 9.25. To maintain a constant flow rate, the blood must travel faster than normal through the constriction. If the blood speed is sufficiently high in the constricted region, the artery may collapse under external pressure, causing a momentary interruption in blood flow. At this moment there is no Bernoulli effect, and the vessel reopens under arterial pressure. As the blood rushes through the constricted artery, the internal pressure drops and again the artery closes. Such variations in blood flow can be heard with a stethoscope. If the plaque becomes dislodged and ends up in a smaller vessel that delivers blood to the heart, the person can suffer a heart attack.

An aneurysm is a weakened spot on an artery where the artery walls have ballooned outward. Blood flows more slowly though this region, as can be seen from the equation of continuity, resulting in an increase in pressure in the vicinity of the aneurysm relative to the pressure in other parts of the artery. This condition is dangerous because the excess pressure can cause the artery to rupture.

The lift on an aircraft wing can also be explained, in part, by the Bernoulli effect. Airplane wings are designed so that the air speed above the wing is greater than that below. As a result, the air pressure above the wing is less than the pressure below, and there is a net upward force on the wing, called the "lift." Another factor influencing the lift on a wing is shown in Figure 9.26. The wing has a slight upward tilt that causes air molecules striking the bottom to be deflected downward. The air molecules bouncing off the wing at the bottom produce an upward force on the wing and a significant lift on the aircraft. Finally, turbulence also has an effect. If the wing is tilted too much, the flow of air across the upper surface becomes turbulent, and the pressure difference across the wing is not as great as that predicted by Bernoulli's equation. In an extreme case, this turbulence may cause the aircraft to stall.

The Bernoulli effect plays an important role in household plumbing systems (Fig. 9.27). The drain pipe beneath a sink has a U-shaped section, called a trap, that is always filled with water to prevent sewer gas from backing up into the home. In a poorly plumbed home (Fig. 9.27a), water rushing through the main pipe reduces the pressure at A while point B, at the sink, remains at atmospheric pressure. This pressure difference causes the water to be pushed out of the trap into the main sewer line. The empty trap then allows sewer gas to enter the home. This chain of events is prevented by installing vents to the outdoors at strategic locations in the plumbing system (Fig. 9.27b). The vents are always open to the atmosphere, thus ensuring that A and B both remain at atmospheric pressure.

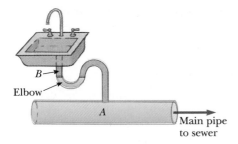

(a) System with no vent

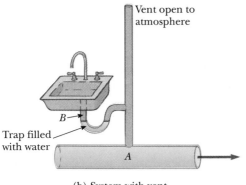

(b) System with vent

FIGURE 9.27
A vent open to the atmosphere is necessary in any plumbing system to prevent sewer gas from entering the home.

SUMMARY

Matter is normally classified as being in one of three states: solid, liquid, or gaseous.

The elastic properties of a solid can be described using the concepts of stress and strain. **Stress** is related to the force producing a deformation; **strain** is a measure of the degree of deformation. Stress is proportional to strain, and the constant of proportionality is the **elastic modulus:**

$$\text{Elastic modulus} \equiv \frac{\text{stress}}{\text{strain}} \qquad \textbf{[9.1]}$$

Three common types of deformation are: (1) the resistance of a solid to elongation or compression, characterized by **Young's modulus,** Y; (2) the resistance to displacement of the faces of a solid sliding past each other, characterized by the **shear modulus,** S; (3) the resistance of a solid or liquid to a volume change, characterized by the **bulk modulus,** B.

In the SI system, pressure is expressed in pascals (Pa), where $1\text{ Pa} \equiv 1\text{ N/m}^2$.

The **density,** ρ, of a substance of uniform composition is its mass per unit volume—kilograms per cubic meter (kg/m^3) in the SI system:

$$\rho \equiv \frac{M}{V} \qquad \textbf{[9.6]}$$

The **pressure**, P, in a fluid is the force per unit area that the fluid exerts on an object immersed in it:

$$P \equiv \frac{F}{A}$$ [9.7]

The pressure in a fluid varies with depth, h, according to the expression

$$P = P_0 + \rho g h$$ [9.10]

where P_0 is atmospheric pressure (1.01×10^5 Pa) and ρ is the density of the fluid.

Pascal's principle states that, when pressure is applied to an enclosed fluid, the pressure is transmitted undiminished to every point of the fluid and to the walls of the containing vessel.

When an object is partially or fully submerged in a fluid, the fluid exerts an upward force, called the **buoyant force,** on the object. According to **Archimedes' principle,** the buoyant force is equal to the weight of the fluid displaced by the object.

Certain aspects of a fluid in motion can be understood by assuming that the fluid is nonviscous and incompressible and that its motion is in a steady state with no turbulence.

1. The flow rate through the pipe is a constant, which is equivalent to stating that the product of the cross-sectional area, A, and the speed, v, at any point is constant.

$$A_1 v_1 = A_2 v_2$$ [9.14]

This relation is referred to as the **equation of continuity.**

2. The sum of the pressure, the kinetic energy per unit volume, and the potential energy per unit volume has the same value at all points along a streamline:

$$P + \tfrac{1}{2}\rho v^2 + \rho g y = \text{constant}$$ [9.16]

This is known as **Bernoulli's equation.**

ADDITIONAL READING

E. Denton, "The Buoyancy of Marine Animals," *Sci American,* July 1960, p. 118.

J. J. Gilman, "Fracture in Solids," *Sci. American,* February 1960, p. 94.

N. Smith, "Bernoulli and Newton in Fluid Mechanics," *The Physics Teacher, 10,* 1972, p. 451.

CONCEPTUAL QUESTIONS

Example   When you drink through a straw, you reduce the pressure in your mouth and let the atmosphere move the liquid. Could you use a straw to drink on the Moon? **Reasoning**   The principle here is similar to that employed in the operation of a barometer. In this case, the pressure inside the mouth is reduced by the sucking action, and the downward force of the atmosphere on the surface of the liquid in the cup forces the liquid up through the straw and into the mouth. Since the Moon has no atmosphere, it cannot exert an atmospheric force

(Example Question)

(Example Question) *(Courtesy of Henry Leap)*

and one would be unable to sip a drink through a straw there. (It would be possible to sip the drink in your space suit, which of course would have a pressure-controlled environment.)

Example The daring physics professor, after a long lecture, stretches out for a nap on a bed of nails as in the photograph above. How is this possible?

Reasoning If you try to support your entire weight on a single nail, the pressure on your body is your weight divided by the very small area of the nail. This pressure is sufficiently large to penetrate the skin. But, if you distribute your weight over several hundred nails, as the professor is doing, the pressure is considerably reduced, since the area that supports your weight is the total area of all the nails in contact with your body. (Note that lying on the bed of nails is much more comfortable than sitting on the bed. Standing on the bed with no shoes on is not recommended.)

Example A woman wearing high-heeled shoes is invited into a home in which the kitchen has vinyl floor covering. Why should the homeowner be concerned?

Reasoning She can exert enough pressure on the floor to puncture or dent the floor covering. The large pressure is caused by the fact that her weight is distributed over the very small cross-sectional area of her high heels. She should be asked to remove her high heels and put on some slippers.

Example Will a ship ride higher in an inland lake or in the ocean? Why?

Reasoning According to Archimedes' principle, the buoyant force on the ship is equal to the weight of the water displaced by the ship. Because the density of salty ocean water is greater than fresh lake water, less ocean water needs to be displaced to enable the ship to float. Thus, the boat floats higher in the ocean than in the inland lake.

Example A typical silo on a farm has many bands wrapped around its perimeter, as shown in the photograph. Why is the spacing between successive bands smaller at the lower portions of the silo?

Reasoning If you think of the grain stored in the silo as a fluid, the pressure the grain exerts on the walls of the silo increases with increasing depth just as water pressure in a lake increases with increasing depth. Thus, the spacing between bands is made smaller at the lower portions to overcome the larger outward forces in these regions.

Example If you hold a sheet of paper and blow across the top surface, the paper rises. Explain.

Reasoning The rapidly moving air over the top of the paper exerts a smaller downward force on the paper than does the slower moving air on the under surface of the paper, which exerts an upward force. Hence, there is a net upward force on the paper because of the Bernoulli effect.

1. What kind of deformation does a cube of Jello exhibit when it jiggles?
2. Why doesn't the roof of a building cave in under the tremendous pressure (14.7 lb/in.^{2}) exerted by the atmosphere?
3. Municipal water supplies are often provided by reservoirs built on high ground. Why is the flow of water from such a reservoir more rapid out of a faucet on the ground floor of a building than out of an identical faucet on a higher floor?
4. How much force does the atmosphere exert on one square kilometer of land?

5. A small amount of water is placed at the bottom of a flexible container and brought to a boil. The container is then capped and placed under cool water. The walls collapse inward. Why?

6. Explain the need for an extra air hole in the top of a can of liquid if the liquid is to be poured smoothly.

7. Two glass tumblers with different shapes and cross-sectional areas are filled to the same level with water. According to the expression $P = P_0 + \rho g h$, the pressures at the bottoms of the two tumblers are the same. In view of this, why does one tumbler weigh more than the other?

8. Why is it impractical to use water for a typical barometer?

9. During inhalation, the pressure in the lungs is slightly less than external pressure and the muscles controlling exhalation are relaxed. Underwater, the body equalizes internal and external pressures. Discuss the condition of the muscles if a person underwater is breathing through a snorkel. Would a snorkel work in deep water?

10. If an inflated beachball is placed beneath the surface of a pool of water and released, it shoots upward, out of the water. Use Archimedes' principle to explain why.

11. Will an ice cube float higher in water or in an alcoholic beverage?

12. Steel is much denser than water. How, then, do steel boats float?

13. A helium-filled balloon will rise until its density matches that of the air. If a sealed submarine begins to sink, will it go all the way to the bottom, or will it stop when its density matches that of the surrounding water?

14. A fish rests on the bottom of a bucket of water while the bucket is being weighed. When it begins to swim around, does the weight change?

15. If 1 000 000 N were placed on the deck of the World War II battleship *North Carolina*, the ship would sink only 2.5 cm in the water. Estimate the cross-sectional area of the ship at water level.

16. An ice cube is placed in a glass of water. What happens to the level of the water as the ice melts?

17. Which dam has to be stronger, one that must hold back 100 000 m³ of 10-m-deep water or one that must hold back 1000 m³ of 20-m-deep water?

18. A small piece of steel is tied to a block of wood. When the wood is placed in a tub of water with the steel on top, half of the block is submerged. If the block is inverted so that the steel is underwater, will the submerged amount of the block increase, decrease, or remain the same? What happens to the water level in the tub when the block is inverted?

19. Why do many trailer trucks use wind deflectors on the tops of their cabs (see photograph)? How do such devices reduce fuel consumption?

20. When a fast-moving train passes a train at rest, the two tend to be drawn together. How does the Bernoulli effect explain this phenomenon?

21. Prairie dogs live in underground burrows with at least two entrances. They ventilate their burrows by building a mound over one entrance, which is open to a stream of air as in the photograph. A second entrance at ground level is open to almost stagnant air. How does this construction create an air flow through the burrow?

22. The device in Figure 9.28 is called a water aspirator. It is used for suction. Explain how it works. What is meant by "suction"?

(Question 19)   *(Courtesy of Henry Leap)*

(Question 21)   *(Pamela Zilly/The Image Bank)*

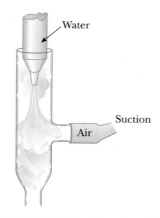

FIGURE 9.28 (Question 22)

FIGURE 9.29 (Question 25)

23. Tornadoes and hurricanes often lift the roofs of houses. Use the Bernoulli effect to explain why. Why should you keep your windows open under these conditions?

24. If air from a hair dryer is blown over the top of a Ping-Pong ball, the ball can be suspended in air. Explain.

25. The device shown in Figure 9.29 is a novice's attempt to design an atomizer. Will this design work? If not, what must be done to fix it? (*Hint:* Consider the Bernoulli effect.)

26. A person claims to be able to make a coin "fly" into a cup without touching the coin. How do you suppose this might be accomplished?

PROBLEMS

Section 9.2 The Deformation of Solids

1. The heels on a pair of women's shoes have radii of 0.50 cm at the bottom. If 30% of the weight of a woman weighing 480 N is supported by each heel, find the stress on each heel.

2. For safety in climbing, a mountaineer uses a nylon rope that is 50 m long and 1.0 cm in diameter. When supporting a 90-kg climber, the rope elongates 1.6 m. Find its Young's modulus.

3. Find the minimum diameter of an 18.0-m-long steel wire that will stretch no more than 9.00 mm when a load of 380 kg is hung on the lower end.

4. A 6.0-m steel beam 1.5 cm in radius supports an overhead walkway. The beam is designed to not stretch more than 5.0×10^{-5} m. What is the maximum load, in newtons, it can withstand?

5. A child slides across a floor in a pair of rubber-soled shoes. The friction force acting on each foot is 20 N, the cross-sectional area of each foot is 14 cm², and the height of the soles is 5.0 mm. Find the horizontal distance traveled by the sheared face of the sole. The shear modulus of the rubber is 3.0×10^6 Pa.

6. If the elastic limit of steel is 5.0×10^8 Pa, determine the minimum diameter a steel wire can have

if it is to support a 70-kg circus performer without its elastic limit being exceeded.

7. The total cross-sectional area of the load-bearing calcified portion of the two forearm bones (radius and ulna) is approximately 2.5 cm². During a car crash, the forearm is slammed against the dashboard. The arm comes to rest from an initial speed of 80 km/h in 5.0 ms. If the arm has an effective mass of 3.0 kg and bone material can normally withstand a maximum compressive stress of 16×10^7 Pa, is the arm likely to withstand the crash?

8. A walkway is supported by pairs of hollow carbon steel pipes, each having an outer diameter of 1.8 cm and walls that are 2.0 mm thick. Each pair of pipes supports a section of walkway weighing 4000 N. If the elastic limit for carbon steel is 29×10^7 Pa, determine the number of people that can safely stand on a single section of the walkway. (Assume an average weight of 670 N/person.)

9. What increase of pressure is required to change the volume of a sample of water by 1.00%?

10. A 2.0-m-long cylindrical steel wire with a cross-sectional diameter of 4.0 mm is placed over a frictionless pulley, with one end of the wire connected to a 500-kg mass and the other end

□ indicates problems that have full solutions available in the Student Solutions Manual and Study Guide.

FIGURE 9.30 (Problem 12)

connected to a 300-kg mass. By how much does the wire stretch while the masses are in motion?

11. If the shear stress in steel exceeds about 4.0×10^8 Pa, the steel ruptures. Determine the shear force necessary to punch a 1-cm-diameter hole in a steel plate that is 0.5 cm thick.

12. A stainless-steel orthodontic wire is applied to a tooth as in Figure 9.30. The wire has an unstretched length of 3.1 cm and a diameter of 0.22 mm. If the wire is stretched 0.10 mm, find the magnitude and direction of the force on the tooth. Disregard the width of the tooth and assume that Young's modulus for stainless steel is 18×10^{10} Pa.

13. Determine the elongation of the rod in Figure 9.31 if it is under a tension of 5.8×10^3 N.

FIGURE 9.31 (Problem 13)

Section 9.3 Density and Pressure

14. If 1.0 m³ of concrete weighs 5.0×10^4 N, what is the height of the tallest cylindrical concrete pillar that will not collapse under its own weight? The compression strength of concrete (the maximum pressure that can be exerted on the base of the structure) is 1.7×10^7 Pa.

15. The four tires of an automobile are inflated to a gauge pressure of 2.0×10^5 Pa. Each tire has an area of 0.024 m² in contact with the ground. Determine the weight of the automobile.

16. A 70-kg man in a 5.0-kg chair tilts back so that all the weight is balanced on two legs of the chair. Assume that each leg makes contact with the floor over a circular area with a radius of 1.0 cm, and find the pressure exerted on the floor by each leg.

17. A pipe contains water at 5.00×10^5 Pa above atmospheric pressure. If the only material you have available to patch a 4.00-mm-diameter hole in the pipe is a piece of bubble gum, how much force must the gum be able to withstand?

Section 9.4 Variation of Pressure with Depth

Section 9.5 Pressure Measurements

18. Water is to be pumped to the top of the Empire State Building, which is 1200 ft high. What gauge pressure is needed in the water line at the base of the building to raise the water to this height?

19. Engineers have developed a bathyscaph that can reach ocean depths of 7.0 mi. (a) What is the absolute pressure at that depth? (b) If the inside of the vessel is maintained at atmospheric pressure, what is the net force on a porthole of diameter 15 cm?

20. The air pressure above the liquid in Figure 9.32 is 1.1 atm. The liquid is ethanol. Determine the pressure in the bulb suspended in the ethanol.

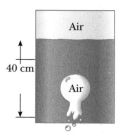

FIGURE 9.32 (Problem 20)

21. A collapsible plastic bag (Fig. 9.33) contains a glucose solution. If the average gauge pressure in the artery is 1.33×10^4 Pa, what must be the minimum height, h, of the bag in order to infuse glu-

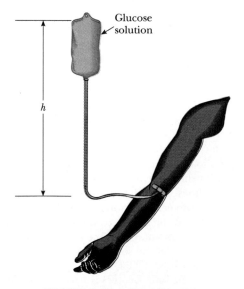

FIGURE 9.33 (Problem 21)

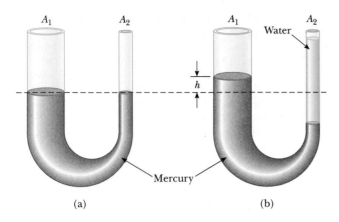

(a)   (b)

FIGURE 9.34   (Problem 22)

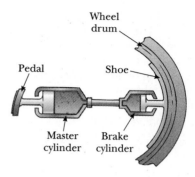

FIGURE 9.36   (Problem 24)

cose into the artery? Assume that the specific gravity of the solution is 1.02.

22. Mercury is poured into a U-tube as in Figure 9.34a. The left arm of the tube has a cross-sectional area of $A_1 = 10.0$ cm², and the right arm has a cross-sectional area of $A_2 = 5.00$ cm². One hundred grams of water are then poured into the right arm (Fig. 9.34b). (a) Determine the length of the water column in the right arm of the U-tube. (b) Given that the density of mercury is 13.6 g/cm³, what distance, h, does the mercury rise in the left arm?

23. A U-tube of constant cross-sectional area, open to the atmosphere, is partially filled with mercury. Water is then poured into both arms. If the equilibrium configuration of the tube is as shown in Figure 9.35, with $h_2 = 1.00$ cm, determine the value of h_1.

24. Figure 9.36 shows the essential parts of a hydraulic brake system. The area of the piston in the master cylinder is 6.4 cm², and that of the piston in the

brake cylinder is 1.75 cm². The coefficient of friction between shoe and wheel drum is 0.50. If the wheel has a radius of 34 cm, determine the frictional torque about the axle when a force of 44 N is exerted on the brake pedal.

25. Piston 1 in Figure 9.37 has a diameter of 0.25 in.; piston 2 has a diameter of 1.5 in. In the absence of friction, determine the force, **F**, necessary to support the 500-lb weight.

26. A tube of uniform cross-sectional area is open to the atmosphere and has the shape shown in Fig-

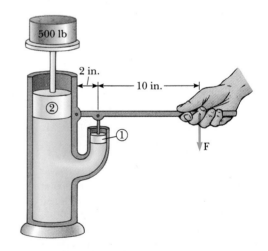

FIGURE 9.37   (Problem 25)

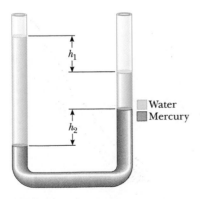

FIGURE 9.35   (Problem 23)

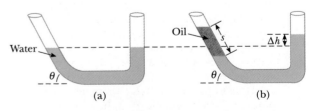

(a)   (b)

FIGURE 9.38   (Problem 26)

ure 9.38, with $\theta = 30°$. It is initially filled with water as in Figure 9.38a, and then oil (density = 750 kg/m³) is poured into the left arm, forming a column 0.80 m long (slant height, s) as in Figure 9.38b. By what distance, h, does the water column rise in the right arm?

27. A container is filled to a depth of 20.0 cm with water. On top of the water floats a 30.0-cm-thick layer of oil with specific gravity 0.700. What is the absolute pressure at the bottom of the container?

Section 9.6   Buoyant Forces and Archimedes' Principle

28. A 1-kg beaker containing 2 kg of oil (density = 916 kg/m³) rests on a scale. A 2-kg block of iron is suspended from a spring scale and completely submerged in the oil (Fig. 9.39). Find the equilibrium readings of both scales.

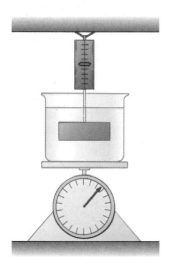

FIGURE 9.39   (Problem 28)

29. A frog in a hemispherical pod finds that he just floats without sinking in a fluid of density 1.35 g/cm³. If the pod has a radius of 6.00 cm and negligible mass, what is the mass of the frog? (See Fig. 9.40.)

30. An empty rubber balloon has a mass of 0.0120 kg.

FIGURE 9.40   (Problem 29)

The balloon is filled with helium at a density of 0.181 kg/m³. At this density the balloon has a radius of 0.500 m. If the filled balloon is fastened to a vertical line, what is the tension in the line?

31. A light spring of constant $k = 90.0$ N/m rests vertically on a table (Fig. 9.41a). A 2.00-g balloon is filled with helium ($\rho = 0.179$ kg/m³) to a volume of 5.00 m³ and connected to the spring, causing it to expand (Fig. 9.41b). Determine the expansion length, L, when the balloon is in equilibrium.

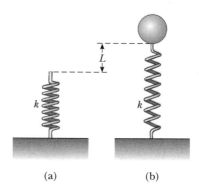

FIGURE 9.41   (Problem 31)

32. A thin spherical shell of mass 4.0 kg and diameter 0.20 m is filled with helium ($\rho = 0.179$ kg/m³). It is then released from rest on the bottom of a pool of water that is 4.0 m deep. (a) Show that the shell rises with constant acceleration, and determine the value of that acceleration. (b) How long will it take the top of the shell to reach the water surface? Neglect frictional effects.

33. A light spring of constant $k = 16.0$ N/m rests vertically on the bottom of a large beaker of water (Fig. 9.42a). A 5.00-kg block of wood (density = 650 kg/m³) is connected to the spring and the

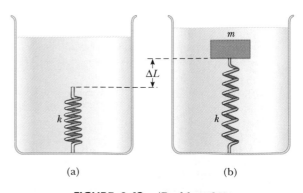

FIGURE 9.42   (Problem 33)

mass-spring system is allowed to come to static equilibrium (Fig. 9.42b). What is the elongation, ΔL, of the spring?

34. A ferry boat is 4.00 m wide and 6.00 m long. When a large loaded truck pulls onto it, it sinks 4.00 cm in the water. What is the weight of the truck?

35. An object weighing 300 N in air is immersed in water after being tied to a string connected to a balance. The scale now reads 265 N. Immersed in oil, the object weighs 275 N. Find (a) the density of the object and (b) the density of the oil.

36. A rectangular air mattress is 2.0 m long, 0.50 m wide, and 0.08 m thick. If it has a mass of 2.3 kg, what additional mass can it support in water?

37. A hollow brass tube (diam. = 4.00 cm) is sealed at one end and loaded with lead shot to give a total mass of 0.2 kg. When the tube is floated in pure water, what is the depth, z, of its bottom end?

38. A 2.0-cm-thick bar of soap is floating on a water surface so that 1.5 cm of the bar is underwater. Bath oil of specific gravity 0.60 is poured into the water and floats on top of the water. What is the depth of the oil layer when the top of the soap is just level with the oil surface?

39. A light balloon is filled with helium at standard temperature and pressure, and is then released from the ground. Determine its initial acceleration in terms of the densities of helium and air.

Section 9.7 Fluids in Motion

Section 9.8 Other Applications of Bernoulli's Equation

40. A natural-gas pipeline with a diameter of 0.250 m delivers 1.55 m³ of gas per second. What is the flow speed of the gas?

41. A cowboy at a dude ranch fills a horse trough that is 1.5 m long, 60 cm wide, and 40 cm deep. He uses a 2.0-cm-diameter hose from which water emerges at 1.5 m/s. How long does it take him to fill the trough?

42. The approximate inside diameter of the aorta is 0.50 cm; that of a capillary is 10 μm. The approximate average blood flow speed is 1.0 m/s in the aorta and 1.0 cm/s in the capillaries. If all the blood in the aorta eventually flows through the capillaries, estimate the number of capillaries in the circulatory system.

43. A fountain sends a stream of water 20 m up into the air. If the base of the stream is 10 cm in diameter, what power is required to send the water to this height?

44. A U-tube open at both ends is partially filled with water (Fig. 9.43a). Oil ($\rho = 750$ kg/m³) is then poured into the right arm and forms a column $L = 5.00$ cm high (Fig. 9.43b). (a) Determine the

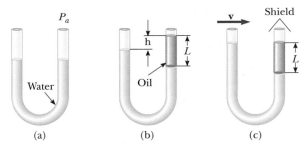

FIGURE 9.43 (Problem 44)

difference, h, in the heights of the two liquid surfaces. Assume that the density of air is 1.29 kg/m³, but be sure to include differences in the atmospheric pressure due to changes in altitude.
(b) The right arm is then shielded from any air motion while air is blown across the top of the left arm until the surfaces of the two liquids are at the same height (Fig. 9.43c). Determine the speed of the air being blown across the left arm.

45. A liquid ($\rho = 1.65$ g/cm³) flows through two horizontal sections of tubing joined end to end. In the first section the cross-sectional area is 10.0 cm², the flow speed is 275 cm/s, and the pressure is 1.20×10^5 Pa. In the second section the cross-sectional area is 2.50 cm². Calculate the smaller section's (a) flow speed and (b) pressure.

46. When a person inhales, air moves down the bronchus (windpipe) at 15 cm/s. The average flow speed of the air doubles through a constriction in the bronchus. Assuming incompressible flow, determine the pressure drop in the constriction.

47. A hypodermic syringe contains a medicine with the density of water (Fig. 9.44). The barrel of the syringe has a cross-sectional area of 2.50×10^{-5} m². The cross-sectional area of the needle is 1.00×10^{-8} m². In the absence of a force on the plunger, the pressure everywhere is 1.00 atm. A force, **F**, of magnitude 2.00 N is exerted on the plunger, making medicine squirt from the needle. Determine the medicine's flow speed. Assume that the pressure in the needle remains equal to 1.00 atm, that the syringe is horizontal, and that the speed of the emerging fluid is the same as the speed of the fluid in the needle.

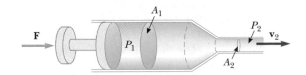

FIGURE 9.44 (Problem 47)

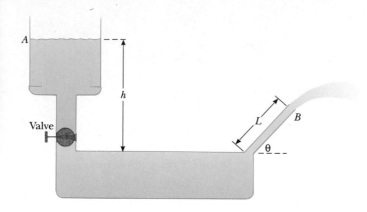

FIGURE 9.45   (Problem 50)

48. (a) If wind blows at 30.0 m/s over the roof of your house, what is the pressure difference at the roof between the inside and outside air? (b) What net force does this pressure difference produce on a roof having an area of 175 m²?

49. What is the net upward force on an airplane wing of area 20 m² if the speed of flow is 300 m/s across the top of the wing and 280 m/s across the bottom?

50. Figure 9.45 shows a water tank with a valve at the bottom. If this valve is opened, what is the maximum height attained by the water stream coming out of the right side of the tank? Assume that $h =$ 10 m, $L =$ 2.0 m, and $\theta =$ 30° and that the cross-sectional area at A is very large compared with that at B.

51. Water flows through a 30-cm-radius pipe at the rate of 0.20 m³/s. The pressure in the pipe is atmospheric. The pipe slants downhill and feeds into a second pipe of radius 15 cm, positioned 60 cm lower. What is the gauge pressure in the lower pipe?

52. (a) Calculate the flow rate (in grams per second) of blood ($\rho =$ 1.0 g/cm³) in an aorta with a cross-sectional area of 2.0 cm² if the flow speed is 40 cm/s. (b) Assume that the aorta branches to form a large number of capillaries with a combined cross-sectional area of 3.0×10^3 cm². What is the flow speed in the capillaries?

53. A jet of water squirts out horizontally from a hole near the bottom of the tank in Figure 9.46. If the hole has a diameter of 3.50 mm, what is the height, h, of the water level in the tank?

54. The inside diameters of the larger portions of the horizontal pipe in Figure 9.47 are 2.5 cm. Water flows to the right at a rate of 1.8×10^{-4} m³/s. Determine the inside diameter of the constriction.

55. The water supply of a building is fed through a main entrance pipe 6.0 cm in diameter. A

2.0-cm-diameter faucet tap positioned 2.0 m above the main pipe fills a 25-liter container in 30 s. (a) What is the speed at which the water leaves the faucet? (b) What is the gauge pressure in the main pipe? (Assume that the faucet is the only outlet in the system.)

56. A large storage tank, open to the atmosphere at the top and filled with water, develops a small hole in its side at a point 16 m below the water level. If the rate of flow from the leak is 2.5×10^{-3} m³/min, determine (a) the speed at which

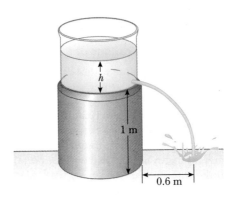

FIGURE 9.46   (Problem 53)

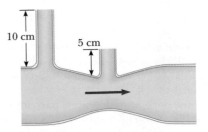

FIGURE 9.47   (Problem 54)

the water leaves the hole and (b) the diameter of the hole.

57. A water tank open to the atmosphere at the top has two holes punched in its side, one above the other. The holes are 5.00 cm and 12.0 cm above the floor. How high does water stand in the tank if the two streams of water hit the floor at the same place?

ADDITIONAL PROBLEMS

58. Bone has a Young's modulus of about 14.5×10^9 Pa. Under compression, it can withstand a stress of about 160×10^6 Pa before breaking. Estimate the length of your femur (thigh bone), and calculate the amount of compression this bone can withstand before breaking.

59. Blood of density 1050 kg/m³ that is to be administered to a patient is raised about 1.00 m higher than the level of the patient's arm. How much greater is the pressure of the blood than it would be if the container were at the same level as the arm?

60. A person rides up a mountain in a lift, but his ears fail to "pop"—that is, the pressure of the inner ear does not equalize with the outside atmosphere. The radius of each eardrum is 0.40 cm. On the way up, the pressure of the atmosphere drops from 1.010×10^5 Pa to 0.998×10^5 Pa. (a) What is the gauge pressure of the inner ear at the top of the mountain? (b) What is the net force on each eardrum?

61. The density of ice is 920 kg/m³, and that of seawater is 1030 kg/m³. What fraction of the total volume of an iceberg is exposed?

62. A sample of an unknown material weighs 300 N in air and 200 N when immersed in alcohol of specific gravity 0.70. What are (a) the volume and (b) the density of the material?

63. The distortion of the Earth's crustal plates is an example of shear on a large scale. A particular crustal rock has a shear modulus of 1.5×10^{10} Pa. What shear stress is involved when a 10-km layer of this rock is sheared through a distance of 5 m?

64. A helium-filled balloon is tied to a 2.0-m-long, 0.050-kg string. The balloon is spherical with a radius of 0.40 m. When released, it lifts a length (h) of the string and then remains in equilibrium as in Figure 9.48. Determine the value of h. When deflated, the balloon has a mass of 0.25 kg. (*Hint:* Only that part of the string above the floor contributes to the load's being held up by the balloon.)

65. A solid copper ball with a diameter of 3.00 m at sea level is placed at the bottom of the ocean, at a depth of 10 000 m. If the density of the seawater

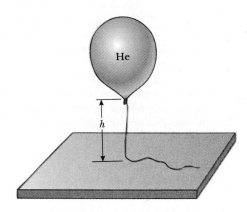

FIGURE 9.48   (Problem 64)

is 1030 kg/m³, how much does the diameter of the ball decrease when it reaches bottom?

66. A particular piece of metal weighs 50 N in air, 36 N in water, and 41 N in oil. Find the densities of (a) the metal and (b) the oil.

67. A 600-kg weather balloon is designed to lift a 4000-kg package. What volume should the balloon have after being inflated with helium at standard temperature and pressure, in order that the total load can be lifted?

68. One method of measuring the density of a liquid is illustrated in Figure 9.49. One side of the U-tube is in the liquid being tested; the other side is in water of density ρ_w. When the air is partially removed at the upper part of the tube, show that the density of the liquid on the left is given by $\rho = (h_w/h)\rho_w$.

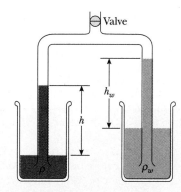

FIGURE 9.49   (Problem 68)

69. A circular swimming pool has a flat bottom and a 6.00-m diameter. It is filled with water to a depth of 1.50 m. There is 1.00 atm of pressure on the top surface. (a) What is the absolute pressure at

the bottom? (b) Two people with a combined mass of 150 kg get into the pool and float there quietly. What is the resulting increase in the absolute pressure at the bottom?

70. A block of wood weighs 50.0 N when weighed in air. A sinker is attached to the block, and the weight of the wood-sinker combination is 200 N when the sinker alone is immersed in water. Finally, the wood-sinker combination is completely immersed and the weight is 140 N. Find the density of the block.

71. A 1.0-kg hollow ball with a radius of 0.10 m, filled with air, is released from rest at the bottom of a 2.0-m-deep pool of water. How high above the water does the ball shoot upward? Neglect all frictional effects, and neglect the ball's motion when it is only partially submerged.

72. A high-speed lifting mechanism supports an 800-kg mass with a steel cable 25 m long and 4.0 cm² in cross-sectional area. (a) Determine the elongation of the cable. (b) By what additional amount does the cable increase in length if the mass is accelerated upward at a rate of 3.0 m/s²? (c) What is the greatest mass that can be accelerated upward at 3.0 m/s² if the stress in the cable is not to exceed the elastic limit of the cable, 2.2×10^8 Pa?

73. A small sphere 0.60 times as dense as water is dropped from a height of 10 m above the surface of a smooth lake. Determine the maximum depth to which the sphere will sink. Neglect any energy transferred to the water during impact and sinking.

74. In 1657 Otto von Guericke, inventor of the air pump, evacuated a sphere made of two brass hemispheres (Fig. 9.50). Two teams of eight horses each could pull the hemispheres apart only on some trials, and then with the greatest difficulty. (a) Show that the force required to pull the evacuated hemispheres apart is $\pi R^2 (P_0 - P)$, where R is the radius of the hemispheres and P is the pressure inside the sphere, which is much less than atmospheric pressure, P_0. (b) Determine the force if $P = 0.10\, P_0$ and $R = 0.30$ m.

75. Oil having a density of 930 kg/m³ floats on water. A rectangular block of wood 4.00 cm high and with a density of 960 kg/m³ floats partly in the oil and partly in the water. The oil completely covers the block. How far below the interface between the two liquids is the bottom of the block?

76. A hollow object with an average density of 900 kg/m³ floats in a pan containing 500 cm³ of water. Ethanol is added to the water until the object is just on the verge of sinking. What volume of ethanol has been added? Disregard the loss of volume caused by mixing.

77. A balloon filled with helium at atmospheric pressure is designed to support a mass of M (payload + empty balloon). Show that the volume of the balloon must be at least $V = M/(\rho_a - \rho_{He})$, where ρ_a is the density of air and ρ_{He} is the density of helium. (Ignore the volume of the payload.)

78. The *spirit-in-glass thermometer,* invented in Florence, Italy, around 1654, consists of a tube of liquid (the spirit) containing a number of submerged glass spheres with slightly different masses (Fig. 9.51). At sufficiently low temperatures all the spheres float, but as the temperature rises, the spheres sink one after the other. The device is a crude but

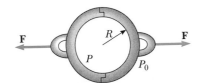

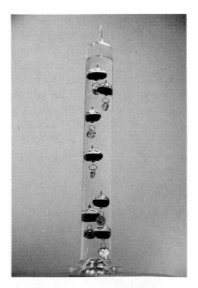

FIGURE 9.50   (Problem 74) *(Courtesy of Henry Leap and Jim Lehman)*

FIGURE 9.51   (Problem 78) *(Photograph and idea courtesy of Prof. Ray Guenther, University of Nebraska at Omaha)*

interesting tool for measuring temperature. Suppose the tube is filled with ethyl alcohol, whose density is 0.789 45 g/cm³ at 20.0°C and decreases to 0.780 97 g/cm³ at 30.0°C. (a) If one of the spheres in the tube has a radius of 1.000 cm and is in equilibrium halfway up the tube at 20.0°C, determine its mass. (b) When the temperature increases to 30.0°C, what mass must a second sphere of radius 1.000 cm have in order to be in equilibrium at the halfway point? (c) At 30.0°C the first sphere has fallen to the bottom of the tube. What upward force does the bottom of the tube exert on this sphere?

79. Water at a pressure of 3.00×10^5 Pa flows through a horizontal pipe at a speed of 1.00 m/s. If the pipe narrows to one-fourth its original diameter, find (a) the flow speed in the narrow section and (b) the pressure in the narrow section.

80. The water jet from an amusement park fountain reaches a height of 10 m. If the fountain nozzle has an opening 10 cm in diameter, what power is required to operate the fountain? (Assume that the length of the nozzle is negligible and that the speed of the water in the supply line to the nozzle is also negligible.)

81. A siphon is a device that allows a fluid to seemingly defy gravity (Fig. 9.52). The flow must be initiated by a partial vacuum in the tube, as in a drinking straw. (a) Show that the speed at which the water emerges from the siphon is given by $v = \sqrt{2gh}$. (b) For what values of y will the siphon work?

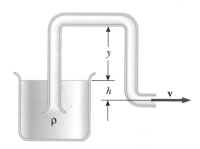

FIGURE 9.52   (Problem 81)

WILLIAM G. BUCKMAN   Western Kentucky University

PHYSICS OF THE HUMAN CIRCULATORY SYSTEM

The circulatory system is an extremely complex and vital part of the human body. The blood supplies food and oxygen to the tissues, carries away waste products from the cells, distributes the heat generated by the cells to equalize body temperature, carries hormones that stimulate and coordinate the activity of organs, distributes antibodies to fight infection, and performs numerous other functions.

William Harvey (1579–1657), an English physician and physiologist, studied blood flow and the action of the heart. He established the essential mechanics of the heart and found that the blood flows from the arterial system through capillary beds and into the veins to be returned to the heart.

The Physical Properties of Blood

Blood is a liquid tissue consisting of two principal parts: the plasma, which is the intercellular fluid, and the cells, which are suspended in the plasma. Plasma is about 90% water, 9% proteins, and 0.9% salts, sugar, and traces of other materials. Blood contains white blood cells and red blood cells. The individual red blood cells are biconcave and have an average diameter of 7.5 μm. There are about 5×10^6 red blood cells per cubic millimeter of blood. The five types of white blood cells found in the blood have an average concentration of 8000 per cubic millimeter, with the concentration normally varying between 4500 and 11 000 per cubic millimeter. The density of blood is about 1.05×10^3 kg/m^3, and its viscosity varies from 2.5 to 4 times that of water.

The Heart as a Pump

The heart can be considered as a double pump, with each side consisting of an atrium and a ventricle (Fig. 1a). Oxygen-poor blood enters the right atrium, flows into the right ventricle, is pumped by the right ventricle to the lungs, and returns through the left atrium to the left ventricle. The left ventricle then pumps the oxygenated blood out through the aorta to the rest of the body. The heart has a system of one-way valves to assure that the blood flows in the proper direction. The heart's pumping cycle has the two ventricles pumping at the same time, as shown in Figure 1b.

The pressure generated by the right ventricle is quite low (about 25 mm Hg), and the lungs offer a low resistance to blood flow. The left ventricle generates a larger pressure, typically greater than 120 mm Hg, at the peak (systole) of the pressure. During the resting stage (diastole) of the heartbeat, the pressure is typically about 80 mm Hg.

We shall now calculate the mechanical work done by the heart. Consider the fluid in the vessel shown in Figure 2. The net force on the fluid is equal to the product of the pressure drop across the fluid, ΔP, and the cross-sectional area, A. The power expended is equal to the net force times the average velocity: $(\Delta PA)(\bar{v})$. Because $A\bar{v} = AL/t =$ volume/time, which is the flow rate, we may now write for the power expended by the heart

$$\text{Power} = (\text{flow rate})(\Delta P)$$

If a normal heart pumps blood at the rate of 97 cm^3/s and the pressure drop from the arterial system to the venous system is 1.17×10^4 Pa, we then have

$$\text{Power} = (97 \text{ cm}^3/\text{s})(10^6 \text{ m}^3/\text{cm}^3)(1.17 \times 10^4 \text{ Pa}) = 1.1 \text{ W}$$

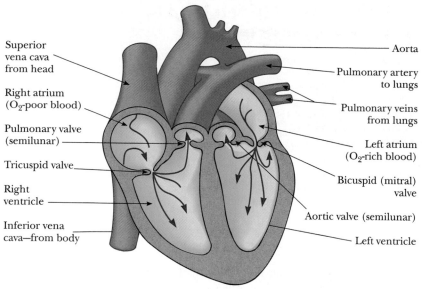

Superior
vena cava
from head

Right atrium
(O_2-poor blood)

Pulmonary valve
(semilunar)

Tricuspid valve

Right
ventricle

Inferior vena
cava—from body

Aorta

Pulmonary artery
to lungs

Pulmonary veins
from lungs

Left atrium
(O_2-rich blood)

Bicuspid (mitral)
valve

Aortic valve (semilunar)

Left ventricle

(a) Anatomy of the heart

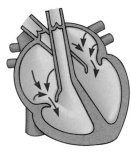
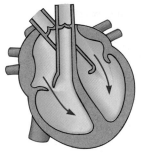
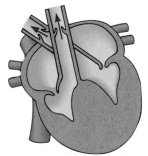

1. Blood fills both atria,
   some blood flows into
   ventricles—diastole
   phase of atria.

2. Atria contract,
   squeezing blood
   into ventricles—
   ventricular diastole.

3. Ventricles contract,
   squeezing blood into
   aorta and pulmonary
   arteries—ventricular
   systole phase.

(b) Pumping cycle of the heart

By measuring oxygen consumption, it is found that the heart of a 70-kg man at rest consumes about 10 W. In the calculation above, it was determined that 1.1 W is required to do the mechanical work of pumping blood; hence, the heart is typically about 10% efficient. During strenuous exercise, the blood pressure may increase by 50% and the blood volume pumped may increase by a factor of 5 to yield an increase of 7.5 times in the power generated by the left ventricle. Because the right ventricle has a systolic pressure about one fifth that of the left ventricle, its power requirement is about one fifth that of the left ventricle.

When we listen to a heart with a stethoscope, we hear two sharp sounds. The first corresponds with the closing of the tricuspid and mitral valves, and the second corresponds with the closing of the aortic and pulmonary valves. Other sounds that are heard are those associated with the flow and turbulence of the blood.

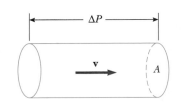

FIGURE 2
The power required to maintain blood flow against viscous forces.

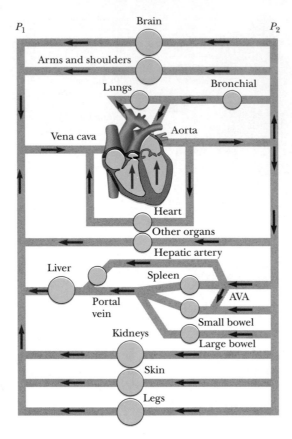

P_1 P_2

FIGURE 3

A diagram of the mammalian circulatory system. Pressure P_2 is that in the arterial system, and P_1 is that in the venous system. Arrows indicate the direction of blood flow.

The Cardiovascular System

The cardiovascular system includes the heart to pump the blood; arteries to carry the blood to the organs, muscle, and skin; and veins to return the blood to the heart (Fig. 3). The aorta branches to form smaller arteries, which in turn branch down to even smaller arteries, until finally the blood reaches the very small capillaries of the vascular bed. These capillaries are so small that the red blood cells must pass single file through them. After passing through the capillaries, where materials being carried by the blood are exchanged with the surrounding tissues, the blood flows to the veins and is returned to the heart.

The flow rate of the blood changes as the blood goes through this system. The cross-sectional area of the vascular bed, which is the product of the cross-sectional area and the number of capillaries, is much greater than the cross-sectional area of the aorta. Because the volume of the blood passing through a cross-sectional area per unit of time is Av, where v is the speed of the blood, we may express the volume flow rate of the blood as

$$\text{Flow rate} = A_{\text{aorta}}v_{\text{aorta}}$$

Furthermore, because the total average flow rate through the aorta and the capillaries must be the same, we have

$$\text{Flow rate} = A_{\text{aorta}}v_{\text{aorta}} = A_{\text{cap}}v_{\text{cap}}$$

EXAMPLE   Flow of Blood in the Aorta and Capillaries

The speed of blood in the aorta is 50 cm/s, and this vessel has a radius of 1.0 cm. (a) What is the rate of flow of blood through this aorta? (b) If the capillaries have a total cross-sectional area of 3000 cm², what is the speed of the blood in them?

Solution

(a) The cross-sectional area of the aorta is

$$A = \pi r^2 = \pi(1.0 \text{ cm})^2 = 3.14 \text{ cm}^2$$

$$\text{Flow rate} = Av = (3.14 \text{ cm}^2)(50 \text{ cm/s}) = 160 \text{ cm}^3/\text{s}$$

(b) The flow rate in the capillaries = 160 cm³/s = $A_c v_c$,

$$v_c = \frac{\text{flow rate}}{A_c} = \frac{160 \text{ cm}^3/\text{s}}{3000 \text{ cm}^2} = 0.053 \text{ cm/s}$$

This low blood speed in the capillaries is necessary to enable the blood to exchange oxygen, carbon dioxide, and other nutrients with the surrounding tissues.

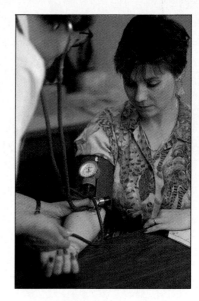

FIGURE 4
(Question 4)   *(Charles Winters)*

Questions and Problems

1. Explain why some individuals tend to black out when they stand up rapidly.
2. If a person is standing at rest, what is the relation between the blood pressure in the left arm and the left leg? What is the relation if the person is in a horizontal position?
3. At what upward acceleration would you expect the blood pressure in the brain to be zero? (Assume that no body mechanisms are operating to compensate for this condition.)
4. When sphygmomanometer is used to measure blood pressure, will the blood pressure readings depend upon the atmospheric pressure? If the atmospheric pressure decreases rapidly, how will this effect the blood pressure readings?
5. Why is it impractical to measure the pulse rate using a vein?
6. Assuming that an artery is clogged such that the effective radius is one-half its normal radius, by what factor must the pressure differential be increased to obtain the normal flow rate through the clogged artery?
7. Determine the average speed of the blood in the aorta if it has a radius of 1.2 cm, and the flow rate is 20 liters/min.
8. If the mean blood pressure in the aorta is 100 mm Hg, (a) determine the blood pressure in the artery located 65 cm above the heart. (b) One cannot apply, without significant error, Bernoulli's principle in the smaller arteries and the capillaries. Why not?
9. When the flow rate is 5.0 liters/min, the blood speed in the capillaries is 0.33 mm/s. Assuming the average diameter of a capillary to be 0.0080 mm, calculate the number of capillaries in the circulatory system.
10. An artery has a length of 20 cm and a radius of 0.50 cm, and blood is flowing at a rate of 6.0 liters/min. What is the difference in the pressure between the ends of the artery?

Thermodynamics PART 2

As we saw in the first part of this book, Newtonian mechanics explains a wide range of phenomena, such as the motions of baseballs, rockets, and planets. We now turn to the study of thermodynamics, which is concerned with the concepts of heat and temperature. As we shall see, thermodynamics is very successful in explaining the bulk properties of matter and the correlation between those properties and the mechanics of atoms and molecules.

Historically, the development of thermodynamics paralleled the development of the atomic theory of matter. By the middle of the 19th century, chemical experiments provided solid evidence for the existence of atoms. At that time, scientists recognized that there must be a connection between the heat-and-temperature theory and the structure of matter. In 1827 the botanist Robert Brown reported that grains of pollen suspended in a liquid moved erratically from one place to another, as if under constant agitation. In 1905 Albert Einstein developed a theory about the cause of this erratic motion, which

today is called Brownian motion. Einstein explained the phenomenon by assuming that the grains of pollen are under constant bombardment by "invisible" molecules in the liquid, which themselves are moving erratically. Einstein's insight gave scientists a means of discovering vital information concerning molecular motion. It also gave reality to the concept of the atomic constituents of matter.

Have you ever wondered how a refrigerator cools, what types of transformations occur in an automobile engine, or what happens to the kinetic energy of a falling object once the object comes to rest? The laws of thermodynamics and the concepts of heat and temperature enable us to answer such practical questions.

Many things can happen to an object when it is heated. Its size changes slightly, but it may also melt, boil, ignite, or even explode. The outcome depends upon the composition of the object and the degree to which it is heated. In general, thermodynamics must concern itself with the physical and chemical transformations of matter in all of its forms: solid, liquid, and gas.

(John Beatty/Tony Stone Images)

10 THERMAL PHYSICS

After the bottle is shaken, the cork is blown off. Contrary to common belief, shaking the champagne bottle does not increase the CO_2 pressure inside. Since the temperature of the bottle and its contents remain constant, the equilibrium pressure does not change, as can be shown by replacing the cork with a pressure gauge. Shaking the bottle displaces some CO_2 from the "head space" to bubbles within the liquid which becomes attached to the walls. If the bubbles remain attached to the walls, when the bottle is opened, the bubbles beneath the liquid level rapidly expand, expelling liquid in the process. So blame the mess on the bubbles. *(Steve Niedorf, The IMAGE Bank)*

Our study thus far has focused exclusively on mechanics. Such concepts as mass, force, and kinetic energy have been carefully defined in order to make the subject quantitative. We now move to a new branch of physics, **thermal physics.** Here we shall find that quantitative descriptions of thermal phenomena require careful definitions of the concepts of temperature, heat, and internal energy. We will seek answers to such questions as: What happens to an object when heat is added to it or removed from it? What physical changes occur in an object when its temperature increases or decreases? One familiar outcome of a temperature increase or decrease for an object is a change in size. We shall examine the details of this process in our discussion of linear expansion.

This chapter concludes with a study of ideal gases. We approach this study on two levels. The first examines ideal gases on the macroscopic scale. Here we are concerned with the relationships among such quantities as pressure, volume, and temperature. On the second level we examine gases on a microscopic scale, using a model that pictures the components of a gas as small particles. This latter approach, called the kinetic theory of gases, helps us understand what happens on the atomic level to affect such macroscopic properties as pressure and temperature.

10.1

TEMPERATURE AND THE ZEROTH LAW OF THERMODYNAMICS

We often associate the concept of temperature with how hot or cold an object feels when we touch it. Thus, our senses provide us with qualitative indications of temperature. However, our senses are unreliable and often misleading. For example, if we remove a metal ice tray and a package of frozen vegetables from the freezer, the ice tray feels colder to the hand than the vegetables even though the two are at the same temperature. This is because metal is a better heat conductor than cardboard, and so the ice tray conducts heat from our hand more efficiently than does the cardboard package. What is needed is a reliable and reproducible method of making quantitative measurements to establish the relative "hotness" or "coldness" of objects. Scientists have developed a variety of thermometers to fulfill this purpose.

We are all familiar with the fact that two objects at different initial temperatures may eventually reach some intermediate temperature when placed in contact with each other. For example, if two soft drinks, one hot and the other cold, are placed in an insulated container, the two eventually reach an equilibrium temperature once the cold one warms up and the hot one cools off. Likewise, if a cup of hot coffee is cooled with an ice cube, the ice eventually melts and the coffee's temperature decreases.

In order to understand the concept of temperature, it is useful to first define two often-used phrases, *thermal contact* and *thermal equilibrium*. To grasp the meaning of thermal contact, imagine two objects placed in an insulated container so that they interact with each other but not with the rest of the world. If the objects are at different temperatures, energy is exchanged between them. The energy exchanged between objects because of a temperature difference is called **heat.** We shall examine the concept of heat in more detail in Chapter 11. For purposes of the current discussion, we shall assume that two objects are in **thermal contact** with each other if heat can be exchanged between them. **Thermal equilibrium** is the situation in which two objects in thermal contact with each other cease to have any exchange of heat.

Now consider two objects, A and B, that are not in thermal contact with each other, and a third object, C, that acts as a thermometer. We wish to determine whether or not A and B would be in thermal equilibrium with each other, once placed in thermal contact. The thermometer (object C) is first placed in thermal contact with A until thermal equilibrium is reached. At that point, the thermometer's reading remains constant, and we record it. The thermometer is then placed in thermal contact with B, and its reading is recorded after thermal equilibrium is reached. If the two readings are the same, then A and B will also be in thermal equilibrium with each other when they are placed in thermal contact.

Molten lava flowing down a mountain in Kilauea, Hawaii. In this case, the hot lava flows smoothly out of a central crater until it cools and solidifies to form the mountains. However, violent eruptions sometimes occur, as in the case of Mount St. Helens in 1980, which can cause both local and global (atmospheric) damage. *(Ken Sakomoto/Black Star)*

We can summarize these results in a statement known as the **zeroth law of thermodynamics** (the law of equilibrium):

> **If bodies A and B are separately in thermal equilibrium with a third body, C, then A and B will be in thermal equilibrium with each other if placed in thermal contact.**

This statement, insignificant and obvious as it may seem, is easily proved experimentally and is very important because it can be used to define temperature. We can think of **temperature** as the property that determines whether or not an object will be in thermal equilibrium with other objects. *Two objects in thermal equilibrium with each other are at the same temperature.*

10.2
THERMOMETERS AND TEMPERATURE SCALES

Thermometers are devices used to measure the temperature of a system. All thermometers make use of a change in some physical property with temperature. Some of the physical properties used are (1) the change in volume of a liquid, (2) the change in length of a solid, (3) the change in pressure of a gas held at constant volume, (4) the change in volume of a gas held at constant pressure, (5) the change in electric resistance of a conductor, and (6) the change in color of a very hot object. For a given substance, a temperature scale can be established based on any one of these physical quantities.

The most common thermometer in everyday use consists of a mass of liquid —usually mercury or alcohol—that expands into a glass capillary tube when heated (Fig. 10.1). In this case the physical property is the change in volume of a liquid, and one can define any temperature change to be proportional to the change in length of the liquid column in the capillary. The thermometer can be calibrated by placing it in thermal contact with some natural systems that remain at constant temperature. One such system is a mixture of water and ice in thermal equilibrium at atmospheric pressure. It is defined to have a temperature of zero degrees Celsius, written 0°C; this temperature is called the ice point of water. Another commonly used system is a mixture of water and steam in thermal equilibrium at atmospheric pressure; its temperature is 100°C, the steam point of water. Once the liquid levels in the thermometer at these two temperatures have been established, the column is divided into 100 equal segments, each corresponding to a change in temperature of one Celsius degree.

Thermometers calibrated in this way present problems when extremely accurate readings are needed. Because mercury and alcohol have different thermal expansion properties, when one indicates a temperature of 50°C, say, the other may indicate a slightly different value. In fact, an alcohol thermometer calibrated at the ice and steam points of water might agree with a mercury thermometer only at the calibration points. The discrepancies between different types of thermometers are especially large when the temperatures to be measured are far from the calibration points.

An additional practical disadvantage of any thermometer is its limited temperature range. A mercury thermometer, for example, cannot be used below the freezing point of mercury, −39°C. To surmount such problems, we need a

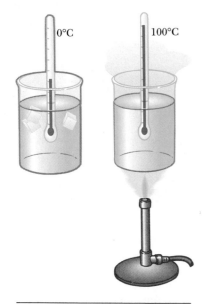

FIGURE 10.1
Schematic diagram of a mercury thermometer. Because of thermal expansion, the level of the mercury rises as the mercury is heated from 0°C (the ice point) to 100°C (the steam point).

universal thermometer whose readings are independent of the substance used. The gas thermometer approaches this requirement.

THE CONSTANT-VOLUME GAS THERMOMETER AND THE KELVIN SCALE

In a **gas thermometer,** the temperature readings are nearly independent of the substance used in the thermometer. One type of gas thermometer is the constant-volume unit shown in Figure 10.2. The physical property used in this device is the pressure variation with temperature of a fixed volume of gas. When the constant-volume gas thermometer was developed, it was calibrated using the ice and steam points of water, as follows. (A different calibration procedure, to be discussed shortly, is now used.) The gas flask was inserted into an ice bath, and mercury reservoir B was raised or lowered until the volume of the confined gas was at some value, indicated by the zero point on the scale. The height h, the difference between the levels in the reservoir and column A, indicated the pressure in the flask at $0°C$. The flask was inserted into water at the steam point, and reservoir B was readjusted until the height in column A was again brought to zero on the scale, ensuring that the gas volume was the same as it had been in the ice bath (hence the designation "constant-volume"). This gave a value for the pressure at $100°C$.

The pressure and temperature values were then plotted on a graph, as in Figure 10.3. The line connecting the two points serves as a calibration curve for measuring unknown temperatures. If we wanted to measure the temperature of a substance, we would place the gas flask in thermal contact with the substance and adjust the column of mercury until the gas took on its specified volume. The height of the mercury column would tell us the pressure of the gas, and we could then find the temperature of the substance from the graph.

Experiments show that the thermometer readings are nearly independent of the type of gas used, so long as the gas pressure is low and the temperature is well above the point at which the gas liquifies (Fig. 10.4). The agreement among thermometers using different gases improves as the pressure is reduced.

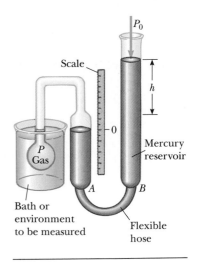

FIGURE 10.2
A constant-volume gas thermometer measures the pressure of the gas contained in the flask on the left. The volume of gas in the flask is kept constant by raising or lowering the column of mercury such that the mercury level remains constant.

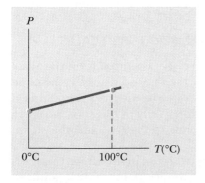

FIGURE 10.3
A typical graph of pressure versus temperature taken with a constant-volume gas thermometer. The dots represent known reference temperatures (the ice point and the steam point).

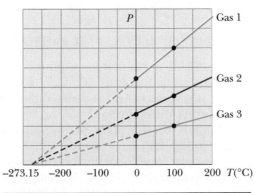

FIGURE 10.4
Pressure versus temperature for dilute gases. Note that, for all gases, the pressure extrapolates to zero at the unique temperature of $-273.15°C$.

If the curves in Figure 10.4 are extended back toward negative temperatures, in every case the pressure is zero when the temperature is $-273.15°C$. This significant temperature is used as the basis for the Kelvin temperature scale, which sets $-273.15°C$ as its zero point (0 K). The size of a Kelvin unit (called a kelvin) is identical to the size of a degree on the Celsius scale. Thus, the relationship of conversion between these temperatures is simply

$$T_C = T - 273.15 \qquad \text{[10.1]}$$

where T_C is the **Celsius temperature** and T is the **Kelvin temperature.**

Early gas thermometers made use of ice and steam points according to the procedure just described. However, these points are experimentally difficult to duplicate because they are pressure-sensitive. Consequently, a new procedure based on a single fixed point was adopted in 1954 by the International Committee on Weights and Measures. The **triple point of water,** which is *the single temperature and pressure at which water, water vapor, and ice can coexist in equilibrium,* was chosen as a convenient and reproducible reference temperature for the Kelvin scale. It occurs at a temperature of $0.01°C$ and a pressure of 4.58 mm of mercury. The temperature at the triple point of water on the Kelvin scale has been assigned a value of 273.16 kelvins (K). Thus, the SI unit of temperature, the **kelvin,** is defined as *1/273.16 of the temperature of the triple point of water.*

Figure 10.5 shows the Kelvin temperatures for various physical processes and structures. The temperature 0 K is often referred to as **absolute zero,** and, as Figure 10.5 shows, this temperature has never been achieved, although laboratory experiments have come close.

What would happen to a substance if its temperature could reach 0 K? As Figure 10.4 indicates, the pressure the substance exerted on the walls of its container would be zero. In Section 10.6 we shall show that the pressure of a gas is proportional to the kinetic energy of the molecules of that gas. Thus, according to classical physics, the kinetic energy of the gas would go to zero, and there would be no motion at all of the individual components of the gas; hence, the molecules would settle out on the bottom of the container. Quantum theory, to be discussed in Chapter 27, modifies this statement to indicate that there would be some residual energy, called the zero-point energy, at this low temperature.

THE CELSIUS, KELVIN, AND FAHRENHEIT TEMPERATURE SCALES

Equation 10.1 shows that the Celsius temperature, T_C, is shifted from the absolute (Kelvin) temperature, T, by 273.15. Because the size of a degree is the same on the two scales, a temperature difference of 5 Celsius degrees is equal to a temperature difference of 5 K. The two scales differ only in the choice of zero point. Thus, the ice point (273.15 K) corresponds to $0.00°C$, and the steam point (373.15 K) is equivalent to $100.00°C$.

The most common temperature scale in everyday use in the United States is the Fahrenheit scale. It sets the temperature of the ice point at $32°F$ and the temperature of the steam point at $212°F$. The relationship between the Celsius and Fahrenheit temperature scales is

$$T_F = \tfrac{9}{5}T_C + 32 \qquad \text{[10.2]}$$

Equation 10.2 can easily be used to find a relationship between changes in temperature on the Celsius and Fahrenheit scales. In an end-of-chapter problem

The kelvin

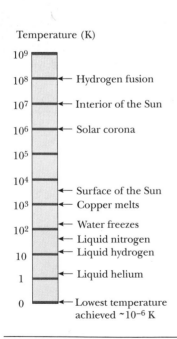

Temperature (K)

10^9	
10^8	← Hydrogen fusion
10^7	← Interior of the Sun
10^6	← Solar corona
10^5	
10^4	
10^3	← Surface of the Sun
	← Copper melts
10^2	← Water freezes
	← Liquid nitrogen
10	← Liquid hydrogen
1	← Liquid helium
0	← Lowest temperature achieved ~10^{-6} K

FIGURE 10.5
Absolute temperature at which various selected physical processes take place.

you will be asked to show that if the Celsius temperature changes by ΔT_C, the Fahrenheit temperature changes by the amount ΔT_F, given by

$$\Delta T_F = \tfrac{9}{5} \Delta T_C \qquad\qquad \text{[10.3]}$$

EXAMPLE 10.1   Converting Temperatures

What is the temperature 50.0°F in degrees Celsius and in kelvins?

Solution   Let us solve Equation 10.2 for T_C and substitute $T_F = 50.0°F$:

$$T_C = \tfrac{5}{9}(T_F - 32.0) = \tfrac{5}{9}(50.0 - 32.0) = \boxed{10°C}$$

From Equation 10.1 we find that

$$T = T_C + 273.15 = \boxed{283\ \text{K}}$$

Exercise   On a hot summer day, the temperature is reported as 30.0°C. What is this temperature in Fahrenheit degrees and in kelvins?

Answer   86°F; 303 K.

EXAMPLE 10.2   Heating a Pan of Water

A pan of water is heated from 25°C to 80°C. What is the change in its temperature on the Kelvin scale and on the Fahrenheit scale?

Solution   From Equation 10.1 we see that the change in temperature on the Celsius scale equals the change on the Kelvin scale. Therefore,

$$\Delta T = \Delta T_C = 80 - 25 = 55°C = \boxed{55\ \text{K}}$$

From Equation 10.3 we find that the change in temperature on the Fahrenheit scale is $\tfrac{9}{5}$ as great as the change on the Celsius scale. That is,

$$\Delta T_F = \tfrac{9}{5}\Delta T_C = \tfrac{9}{5}(80 - 25) = \boxed{99°F}$$

Thermogram of a teakettle showing hot areas in white and cooler areas in purple and black. *(Gary Settles/Science Source/ Photo Researchers)*

10.3
THERMAL EXPANSION OF SOLIDS AND LIQUIDS

Our discussion of the liquid thermometer made use of one of the best-known changes that occurs in a substance: as its temperature increases, its volume increases. (As we shall see shortly, however, in some materials the volume *decreases* when the temperature increases.) This phenomenon, known as **thermal expansion,** plays an important role in numerous applications. For example, thermal expansion joints must be included in buildings, concrete highways, and bridges to compensate for changes in dimensions with temperature variations.

The overall thermal expansion of an object is a consequence of the change in the average separation between its constituent atoms or molecules. To understand this, consider how the atoms in a solid substance behave. At ordinary temperatures, the atoms vibrate about their equilibrium positions with an am-

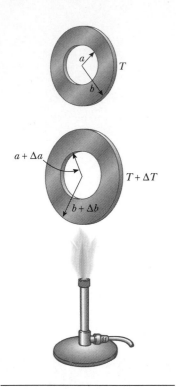

FIGURE 10.6
Thermal expansion of a homogeneous metal washer. As the washer is heated, all dimensions increase. (Note that the expansion is exaggerated in this figure.)

plitude of about 10^{-11} m, and the average spacing between the atoms is about 10^{-10} m. As the temperature of the solid increases, the atoms vibrate with greater amplitudes and the average separation between them increases. Consequently, the solid as a whole expands. If the thermal expansion of an object is sufficiently small compared with the object's initial dimensions, then the change in any dimension is, to a good approximation, dependent on the first power of the temperature change.

Suppose an object has an initial length of L_0 along some direction at some temperature. The length increases by ΔL for the change in temperature ΔT. Experiments show that, when ΔT is small enough, ΔL is proportional to ΔT and to L_0:

$$\Delta L = \alpha L_0 \, \Delta T \qquad\qquad \textbf{[10.4]}$$

or

$$L - L_0 = \alpha L_0 (T - T_0)$$

where L is the final length, T is the final temperature, and the proportionality constant α is called the **average coefficient of linear expansion** for a given material and has units of $(°C)^{-1}$.

It may be helpful to picture a thermal expansion as a magnification or a photographic enlargement. For example, as a metal washer is heated (Fig. 10.6), all dimensions, including the radius of the hole, increase according to Equation 10.4. Table 10.1 lists the average coefficients of linear expansion for various materials. Note that for these materials α is positive, indicating an increase in length with increasing temperature. This is not always the case. For example, some substances, such as calcite ($CaCO_3$), expand along one dimension (positive α) and contract along another (negative α) with increasing temperature.

Because the linear dimensions of an object change with temperature, it follows that surface area and volume also change with temperature. Consider a square having an initial length of side L_0 on a side and therefore an initial area of $A_0 = L_0{}^2$.

TABLE 10.1
Average Coefficients of Linear Expansion for Some Materials Near Room Temperature

Material	Average Coefficient of Linear Expansion [(°C)$^{-1}$]	Material	Average Coefficient of Volume Expansion [(°C)$^{-1}$]
Aluminum	24×10^{-6}	Ethyl alcohol	1.12×10^{-4}
Brass and bronze	19×10^{-6}	Benzene	1.24×10^{-4}
Copper	17×10^{-6}	Acetone	1.5×10^{-4}
Glass (ordinary)	9×10^{-6}	Glycerin	4.85×10^{-4}
Glass (Pyrex)	3.2×10^{-6}	Mercury	1.82×10^{-4}
Lead	29×10^{-6}	Turpentine	9.0×10^{-4}
Steel	11×10^{-6}	Gasoline	9.6×10^{-4}
Invar (Ni-Fe alloy)	0.9×10^{-6}	Air	3.67×10^{-3}
Concrete	12×10^{-6}	Helium	3.665×10^{-3}

Thermal expansion joints are used to separate sections of roadways on bridges. Without these joints, the surfaces would buckle due to thermal expansion on very hot days, or crack due to contraction on very cold days. *(© Frank Siteman, Stock/Boston)*

As the temperature is increased, the length of each side increases to

$$L = L_0 + \alpha L_0 \, \Delta T$$

We are now able to calculate the change in the area of an object as follows. The new area $A = L^2$ is

$$L^2 = (L_0 + \alpha L_0 \, \Delta T)(L_0 + \alpha L_0 \, \Delta T) = L_0^2 + 2\alpha L_0^2 \, \Delta T + \alpha^2 L_0^2 (\Delta T)^2$$

The last term in this expression contains the quantity $\alpha \, \Delta T$ raised to the second power. Because $\alpha \, \Delta T$ is much less than unity, squaring it makes it even smaller. Therefore, we can neglect this term to get a simpler expression:

$$A = L^2 = L_0^2 + 2\alpha L_0^2 \, \Delta T$$

$$A = A_0 + 2\alpha A_0 \, \Delta T$$

or

$$\Delta A = A - A_0 = \gamma A_0 \, \Delta T \qquad \text{[10.5]}$$

where $\gamma = 2\alpha$. The quantity γ (Greek letter gamma) is called the **average coefficient of area expansion.**

By a similar procedure we can show that the *increase in volume* of an object accompanying a change in temperature is

$$\Delta V = \beta V_0 \, \Delta T \qquad \text{[10.6]}$$

where β, the **average coefficient of volume expansion,** is equal to 3α.

As Table 10.1 indicates, each substance has its own characteristic coefficients of expansion. For example, when the temperatures of a brass rod and a steel rod of equal length are raised by the same amount from some common initial value, the brass rod expands more than the steel rod because brass has a larger coefficient of expansion than steel. A simple device that utilizes this principle, called a bimetallic strip, is found in practical devices such as thermostats. The strip is made by securely bonding two different metals together. As the temperature of the strip increases, the two metals expand by different amounts, and the strip bends as in Figure 10.7.

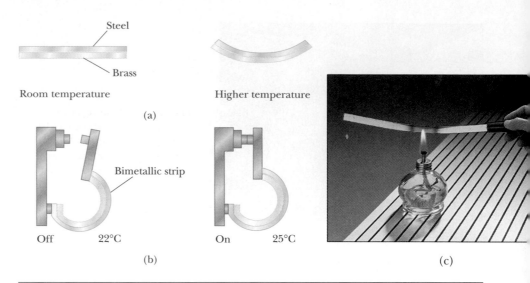

FIGURE 10.7

(a) A bimetallic strip bends as the temperature changes because the two metals have different expansion coefficients. (b) A bimetallic strip used in a thermostat to break or make electrical contact. (c) The blade being heated in the photograph is a bimetallic strip that was straight before being heated. Note that the blade bends when cooled. Which way would it bend if it were heated? *(Courtesy of Central Scientific Company)*

EXAMPLE 10.3   Expansion of a Railroad Track

A steel railroad track has a length of 30.000 m when the temperature is 0°C. What is its length on a hot day when the temperature is 40°C?

Solution   If we use Table 10.1 and Equation 10.4, and note that the change in temperature is 40°C, we find that the *increase* in length is

$$\Delta L = \alpha L_0 \, \Delta T = [11 \times 10^{-6} \, (°C)^{-1}](30.000 \text{ m})(40°C) = 0.013 \text{ m}$$

Therefore, the track's length at 40°C is 30.013 m.

Exercise   What is the length of the same railroad track on a cold winter day when the temperature is 0°F?

Answer   29.994 m.

EXAMPLE 10.4   Does the Hole Get Bigger or Smaller?

A hole of cross-sectional area 100.00 cm² is cut in a piece of steel at 20°C. What is the area of the hole if the steel is heated from 20°C to 100°C?

Solution   A hole in a substance expands in exactly the same way as would a piece of the substance having the same shape as the hole. The change in area of the hole can be found by using Equation 10.5.

$$\Delta A = \gamma A_0 \, \Delta T = [22 \times 10^{-6} \, (°C)^{-1}](100.00 \text{ cm}^2)(80°C) = 0.18 \text{ cm}^2$$

Therefore, the area of the hole at 100°C is

$$A = A_0 + \Delta A = \boxed{100.18 \text{ cm}^2}$$

Thermal expansion: The extreme heat of a July day in Asbury Park, New Jersey, caused these railroad tracks to buckle. *(Wide World Photos)*

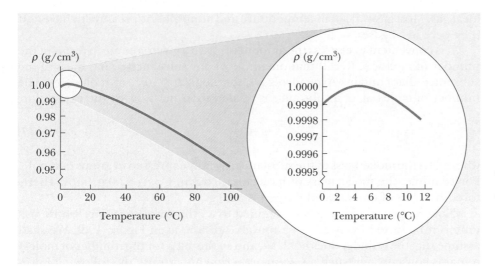

FIGURE 10.8
The density of water as a function of temperature. The inset at the right shows that the maximum density of water occurs at 4°C.

THE UNUSUAL BEHAVIOR OF WATER

Liquids generally increase in volume with increasing temperature and have volume expansion coefficients about ten times greater than those of solids. Water is an exception to this rule, as we can see from its density-versus-temperature curve in Figure 10.8. As the temperature increases from 0°C to 4°C, water contracts and thus its density increases. Above 4°C, water expands with increasing temperature. In other words, the density of water reaches its maximum value (1000 kg/m^3) 4 degrees above the freezing point.

We can use this unusual thermal expansion behavior of water to explain why a pond freezes slowly from the top down. When the atmospheric temperature drops from, say, 7°C to 6°C, the water at the surface of the pond also cools and consequently decreases in volume. This means that the surface water is denser then the water below it, which has not cooled and decreased in volume. As a result, the surface water sinks and warmer water from below is forced to the surface to be cooled. When the atmospheric temperature is between 4°C and 0°C, however, the surface water expands as it cools, becoming less dense than the water below it. The mixing process stops, and eventually the surface water freezes. As the water freezes, the ice remains on the surface because ice is less dense than water. The ice continues to build up on the surface, while water near the bottom of the pool remains at 4°C. If this did not happen, fish and other forms of marine life would not survive.

10.4

MACROSCOPIC DESCRIPTION OF AN IDEAL GAS

In this section we are concerned with the properties of a gas of mass m confined to a container of volume V, pressure P, and temperature T. It is useful to know how these quantities are related. In general, the equation that interrelates them, called the *equation of state,* is very complicated. However, if the gas is maintained at a very low pressure (or low density), the equation of state is found experimentally to be quite simple. Such a low-density gas approximates what is called an

FIGURE 10.9
A gas confined to a cylinder whose volume can be varied with a movable piston.

The universal gas constant

---

ideal gas. Most gases at room temperature and atmospheric pressure behave as if they were ideal gases.

It is convenient to express the amount of gas in a given volume in terms of the number of moles, n. Recall that one mole of any substance is that mass of the substance that contains Avogadro's number, 6.022×10^{23}, of molecules. The number of moles, n, of a substance is related to its mass, m, by the expression

$$n = \frac{m}{M} \tag{10.7}$$

where M is the molar mass of the substance, usually expressed in grams per mole. For example, the molar mass of molecular oxygen, O_2, is 32.0 g/mol. Therefore, the mass of one mole of oxygen is 32.0 g.

Now suppose an ideal gas is confined to a cylindrical container whose volume can be varied by means of a movable piston, as in Figure 10.9. We shall assume that the cylinder does not leak, and so the mass (or the number of moles) remains constant. For such a system, experiments provide the following information. First, when the gas is kept at a constant temperature, its pressure is inversely proportional to the volume (Boyle's law). Second, when the pressure of the gas is kept constant, the volume is directly proportional to the temperature (the law of Charles and Gay-Lussac). These observations can be summarized by the following **equation of state for an ideal gas**:

$$PV = nRT \tag{10.8}$$

In this expression, called the **ideal gas law**, R is a constant for a specific gas that can be determined from experiments, and T is the temperature in kelvins. Experiments on several gases show that, as the pressure approaches zero, the quantity PV/nT approaches the same value of R for all gases. For this reason R is called the **universal gas constant**. In the SI system, where pressure is expressed in pascals and volume in cubic meters, the product PV has units of newton-meters, or joules, and R has the value

$$R = 8.31 \text{ J/mol} \cdot \text{K} \tag{10.9}$$

If the pressure is expressed in atmospheres and the volume in liters (1 L = $10^3 \text{ cm}^3 = 10^{-3} \text{ m}^3$), then R has the value

$$R = 0.0821 \text{ L} \cdot \text{atm/mol} \cdot \text{K}$$

Using this value of R and Equation 10.8, one finds that the volume occupied by 1 mol of any gas at atmospheric pressure and 0°C (273 K) is 22.4 L.

EXAMPLE 10.5   Squeezing a Tank of Gas

Pure helium gas is admitted into a leakproof cylinder containing a movable piston. The initial volume, pressure, and temperature of the gas are 15 L, 2.0 atm, and 300 K. If the volume is decreased to 12 L and the pressure increased to 3.5 atm, find the final temperature of the gas. (Assume that helium behaves as an ideal gas.)

Solution   Because no gas escapes from the cylinder, the number of moles remains constant; therefore, use of $PV = nRT$ at the initial and final points gives

$$\frac{P_i V_i}{T_i} = \frac{P_f V_f}{T_f}$$

where i and f refer to the initial and final values. Solving for T_f, we get

$$T_f = \left(\frac{P_f V_f}{P_i V_i}\right)(T_i) = \frac{(3.5\ \text{atm})(12\ \text{L})}{(2.0\ \text{atm})(15\ \text{L})}(300\ \text{K}) = \boxed{420\ \text{K}}$$

EXAMPLE 10.6   Heating a Bottle of Air

A sealed glass bottle at 27°C contains air at atmospheric pressure and has a volume of 30 cm³. The bottle is then tossed into an open fire. When the temperature of the air in the bottle reaches 200°C, what is the pressure inside the bottle? Assume that any volume changes of the bottle are small enough to be negligible.

Solution   We start with the expression

$$\frac{P_i V_i}{T_i} = \frac{P_f V_f}{T_f}$$

Since the initial and final volumes of the gas are assumed equal, this expression reduces to

$$\frac{P_i}{T_i} = \frac{P_f}{T_f}$$

Before evaluating the final pressure, we must convert the given temperatures to kelvins: $T_i = 27°C = 300\ \text{K}$ and $T_f = 200°C = 473\ \text{K}$. Thus,

$$P_f = \left(\frac{T_f}{T_i}\right)(P_i) = \left(\frac{473\ \text{K}}{300\ \text{K}}\right)(1.0\ \text{atm}) = \boxed{1.6\ \text{atm}}$$

Obviously, the higher the temperature, the higher the pressure exerted by the trapped air. Of course, if the pressure rises high enough, the bottle will shatter.

Exercise   In this example we neglected the change in volume of the bottle. If the coefficient of volume expansion for glass is $27 \times 10^{-6}\ (°C)^{-1}$, find the magnitude of this volume change.

Answer   0.14 cm³.

EXAMPLE 10.7   The Volume of One Mole of Gas

Verify that one mole of oxygen occupies a volume of 22.4 L at 1 atm and 0°C.

Solution   Let us solve the ideal gas equation for V:

$$V = \frac{nRT}{P}$$

In this problem the mass of the gas, m, is assumed to be one mole, M. Thus,

$$n = \frac{m}{M} = 1\ \text{mol}$$

Now let us convert the temperature to kelvins and substitute into the ideal gas equation:

$$V = \frac{nRT}{P} = \frac{(1\ \text{mol})(0.0821\ \text{L}\cdot\text{atm/mol}\cdot\text{K})(273\ \text{K})}{1\ \text{atm}} = \boxed{22.4\ \text{L}}$$

This answer has general validity. *One mole of any gas at standard temperature and pressure (STP) occupies a volume of 22.4 L.*

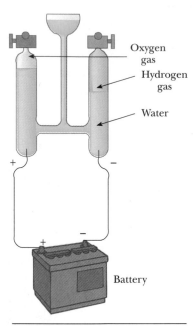

FIGURE 10.10
When water is decomposed by
an electric current, the volume
of hydrogen gas collected is
twice that of oxygen.

Avogadro's number

10.5

AVOGADRO'S NUMBER AND THE IDEAL GAS LAW

In the early 1800s an important field of experimental investigation was created to determine the relative masses of molecules. The equipment used in one such investigation is shown in Figure 10.10. The lower section of the glass vessel contains two electrodes that are connected to a battery so that an electric current can be passed through lightly salted water. (The purpose of the salt is to improve the electrical conductivity of the water.) Bubbles of gas are produced at each electrode and become trapped in the column above it. An analytical examination of the bubbles reveals that the column above the positive electrode contains oxygen gas and the column above the negative electrode contains hydrogen gas. Obviously, the current decomposes the water into its constituent parts.

Experimentally it is found that the volume of hydrogen gas collected is always exactly twice the volume of oxygen gas collected, as we now know in light of the chemical composition of water, H_2O. Additionally, it is found that if 9 g of water are decomposed, 8 g of oxygen and 1 g of hydrogen are collected. From this information it is possible to determine the relative masses of oxygen and hydrogen molecules.

Following such an experimental investigation, Amedeo Avogadro in 1811 stated the following hypothesis:

> **Equal volumes of gas at the same temperature and pressure contain the same numbers of molecules.**

Based on this hypothesis and the fact that the ratio of hydrogen to oxygen molecules collected in the experiment is 2:1, the mass of an oxygen molecule must be 16 times that of a hydrogen molecule.

In Example 10.7, we noted that single moles of all gases occupy the same volume at 1 atm and 0°C. Thus, a corollary to Avogadro's hypothesis is as follows:

> **One mole quantities of all gases at standard temperature and pressure contain the same numbers of molecules.**

Specifically, the number of molecules contained in one mole of any gas is Avogadro's number, 6.02×10^{23} molecules/mol, given the symbol N_A:

$$N_A = 6.02 \times 10^{23} \text{ molecules/mol} \qquad \textbf{[10.10]}$$

It is of historical interest that Avogadro never knew, even approximately, the value for this number.

An alternative method for calculating the number of moles of any gas in a container is to divide the total number of molecules present, N, by Avogadro's number:

$$n = \frac{N}{N_A} \qquad \textbf{[10.11]}$$

With this expression we can rewrite the ideal gas law in the alternative form

$$PV = nRT = \frac{N}{N_A} RT$$

or

$$PV = Nk_B T$$ **[10.12]**   Ideal gas law

where k_B is **Boltzmann's constant** and has the value

$$k_B = \frac{R}{N_A} = 1.38 \times 10^{-23} \, \text{J/K}$$ **[10.13]**   Boltzmann's constant

---

EXAMPLE 10.8   What Is Avogadro's Number?

One mole of hydrogen has a mass of 1.0078 g. In the early 20th century it was found that the mass of a hydrogen atom is approximately 1.673×10^{-24} g. Use these values to find Avogadro's number.

Solution   The number of molecules in 1 mol of hydrogen can be found by dividing the mass of 1 mol by the mass per atom. Hence, we find that

$$N_A = \frac{1.0078 \, \text{g/mol}}{1.673 \times 10^{-24} \, \text{g/atom}} = \boxed{6.020 \times 10^{23} \, \text{atoms/mol}}$$

10.6
THE KINETIC THEORY OF GASES

In Section 10.4 we discussed the properties of an ideal gas, using such quantities as pressure, volume, number of moles, and temperature. In this section we find that pressure and temperature can be understood on the basis of what is happening on the atomic scale. In addition, we re-examine the ideal gas law in terms of the behavior of the individual molecules that make up the gas. Our discussion is restricted to the behavior of gases because the molecular interactions in a gas are much weaker than those in solids and liquids.

The glass vessel contains dry ice (solid carbon dioxide). The white cloud is carbon dioxide vapor, which is denser than air and hence falls from the vessel as shown. *(© R. Folwell/SPL/Photo Researchers)*

MOLECULAR MODEL FOR THE PRESSURE OF AN IDEAL GAS

Let us first use the kinetic theory of gases to show that the pressure a gas exerts on the walls of its container is a consequence of the collisions of the gas molecules with the walls. We make the following assumptions:

Assumptions of kinetic theory for an ideal gas

1. *The number of molecules is large, and the average separation between them is large compared with their dimensions.* This means that the molecules occupy a negligible volume in the container.
2. *The molecules obey Newton's laws of motion, but as a whole they move randomly.* By "randomly" we mean that any molecule can move equally in any direction.
3. *The molecules undergo elastic collisions with each other and with the walls of the container.* Thus, in the collisions *both kinetic energy and momentum are constant.*
4. *The forces between molecules are negligible except during a collision.* The forces between molecules are short-range, so the molecules interact with each other only during collisions.
5. *The gas under consideration is a pure substance;* that is, *all molecules are identical.*

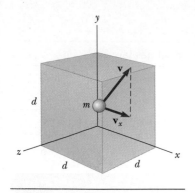

FIGURE 10.11
A cubical box with sides of
length d containing an ideal gas.
The molecule shown moves with
velocity **v**.

Although we often picture an ideal gas as consisting of single atoms, molecular gases at low pressures provide approximations equally as good for ideal gases. Effects of molecular rotations or vibrations have no average effect on the motions considered here.

Now let us derive an expression for the pressure of N molecules of an ideal gas in a container of volume V. The container is a cube with edges of length d (Fig. 10.11). We shall focus our attention on one of these molecules, of mass m and assumed to be moving so that its component of velocity in the x direction is v_x (Fig. 10.12). As the molecule collides elastically with any wall, its velocity is reversed. Since the momentum, p, of the molecule is mv_x before the collision and $-mv_x$ after the collision, the *change in momentum of the molecule* is

$$\Delta p_x = mv_f - mv_i = -mv_x - mv_x = -2\,mv_x$$

Thus, the change in momentum of the wall is $2\,mv_x$. Applying the impulse-momentum theorem (Eq. 6.4) to the wall gives

$$F\,\Delta t = \Delta p = 2\,mv_x$$

In order for the molecule to make another collision with the same wall, it must travel a distance of $2\,d$ in the x direction. Therefore, the time interval between two collisions with the same wall is

$$\Delta t = \frac{2\,d}{v_x}$$

The substitution of this result into the impulse-momentum equation enables us to obtain the force exerted by a molecule on the wall:

$$F_1 = \frac{2\,mv_x}{\Delta t} = \frac{2\,mv_x^2}{2\,d} = \frac{mv_x^2}{d}$$

The total force exerted on the wall by all the molecules is found by adding the forces exerted by all the individual molecules:

$$F = \frac{m}{d}\,(v_{x1}^2 + v_{x2}^2 + \cdots + v_{xN}^2)$$

In this equation, v_{x1} is the x component of velocity of molecule 1, v_{x2} is the x component of velocity of molecule 2, and so on. The summation terminates when we reach molecule N because there are N molecules in the container.

To proceed further, note that the average value of the square of the velocity in the x direction for the N molecules is

$$\overline{v_x^2} = \frac{v_{x1}^2 + v_{x2}^2 + \cdots + v_{xN}^2}{N}$$

Hence, the total force on the wall can be written

$$F = \frac{m}{d}\,N\overline{v_x^2}$$

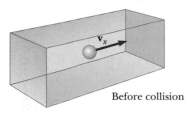

Before collision

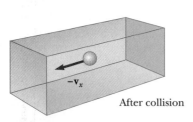

After collision

FIGURE 10.12
A molecule makes an elastic collision with the wall of the container. Its x component of momentum is reversed, and momentum is imparted to the wall.

Now let us focus on one molecule in the container and say that this molecule has velocity components v_x, v_y, and v_z. The Pythagorean theorem relates the square of the velocity to the square of these components:

$$v^2 = v_x^2 + v_y^2 + v_z^2$$

Hence, the average value of v^2 for all the molecules in the container is related to the average values of v_x^2, v_y^2, and v_z^2 according to the expression

$$\overline{v^2} = \overline{v_x^2} + \overline{v_y^2} + \overline{v_z^2}$$

Since all directions of motion are equivalent, the average velocity is the same in any direction. Therefore,

$$\overline{v_x^2} = \overline{v_y^2} = \overline{v_z^2}$$

and we have

$$\overline{v^2} = 3\overline{v_x^2}$$

Thus, the total force on the wall is

$$F = \frac{N}{3}\left(\frac{m\overline{v^2}}{d}\right)$$

This expression allows us to find the pressure exerted on the wall:

$$P = \frac{F}{A} = \frac{F}{d^2} = \frac{1}{3}\left(\frac{N}{d^3}\,m\overline{v^2}\right) = \frac{1}{3}\left(\frac{N}{V}\right)(m\overline{v^2})$$

$$P = \frac{2}{3}\left(\frac{N}{V}\right)(\tfrac{1}{2}\,m\overline{v^2})$$

[10.14]   Pressure of an ideal gas

This result shows that *the pressure is proportional to the number of molecules per unit volume and to the average translational kinetic energy of the molecules, $\frac{1}{2}m\overline{v^2}$*. With this simplified model of an ideal gas, we have arrived at an important result that relates the large-scale quantity of pressure to an atomic quantity, the average value of the square of the molecular speed. Thus, we have a key link between the atomic world and the large-scale world.

Note that Equation 10.14 verifies some features of pressure that are probably familiar to you. One way to increase the pressure inside a container is to increase the number of molecules per unit volume in the container. You do this when you add air to a tire. The pressure in the tire can also be raised by increasing the average translational kinetic energy of the molecules in the tire. As we shall see shortly, this can be accomplished by increasing the temperature of the gas inside the tire. That is why the pressure inside a tire increases as the tire heats up during long trips. The continuous flexing of the tire as it moves along the road surface generates heat that is partially transferred to the air inside, increasing the air's temperature, which in turn produces an increase in pressure.

MOLECULAR INTERPRETATION OF TEMPERATURE

We can gain some insight into the meaning of temperature by first writing Equation 10.14 in the form

$$PV = \tfrac{2}{3}N(\tfrac{1}{2}\,m\overline{v^2})$$

Let us compare this with the equation of state for an ideal gas:

$$PV = Nk_B T$$

Recall that the equation of state is based on experimental facts concerning the macroscopic behavior of gases. Equating the right sides of these expressions, we find that

Temperature is proportional to average kinetic energy

$$T = \frac{2}{3 k_B} \left(\tfrac{1}{2} m \overline{v^2} \right)$$ [10.15]

That is, *temperature is a direct measure of average molecular kinetic energy.*

By rearranging Equation 10.15, we can relate the translational molecular kinetic energy to the temperature:

Average kinetic energy per molecule

$$\tfrac{1}{2} m \overline{v^2} = \tfrac{3}{2} k_B T$$ [10.16]

That is, the average translational kinetic energy per molecule is $\tfrac{3}{2} k_B T$. Since $\overline{v_x^2} = \tfrac{1}{3} \overline{v^2}$, it follows that

$$\tfrac{1}{2} m \overline{v_x^2} = \tfrac{1}{2} k_B T$$ [10.17]

In a similar manner, for the y and z motions it follows that

$$\tfrac{1}{2} m \overline{v_y^2} = \tfrac{1}{2} k_B T \qquad \text{and} \qquad \tfrac{1}{2} m \overline{v_z^2} = \tfrac{1}{2} k_B T$$

Theorem of equipartition of energy

Thus, each translational degree of freedom contributes an equal amount of energy to the gas, namely $\tfrac{1}{2} k_B T$. (In general, "degrees of freedom" refers to the number of independent means by which a molecule can possess energy.) A generalization of this result, known as the **theorem of equipartition of energy**, says that the energy of a system in thermal equilibrium is equally divided among all degrees of freedom.

The total translational kinetic energy of N molecules of gas is simply N times the average energy per molecule, which is given by Equation 10.16:

Total kinetic energy of N molecules

$$E = N \left(\tfrac{1}{2} m \overline{v^2} \right) = \tfrac{3}{2} N k_B T = \tfrac{3}{2} n R T$$ [10.18]

where we have used $k_B = R / N_A$ for Boltzmann's constant and $n = N / N_A$ for the number of moles of gas. This result, together with Equation 10.14, implies that the pressure exerted by an ideal gas depends only on the number of molecules per unit volume and the temperature.

The square root of $\overline{v^2}$ is called the *root-mean-square (rms) speed* of the molecules. From Equation 10.16 we get, for the rms speed,

Root-mean-square speed

$$v_{\mathrm{rms}} = \sqrt{\overline{v^2}} = \sqrt{\frac{3 k_B T}{m}} = \sqrt{\frac{3 R T}{M}}$$ [10.19]

where M is the molar mass in kilograms per mole. This expression shows that, at a given temperature, lighter molecules move faster, on the average, than heav-

TABLE 10.2
Some rms Speeds

Gas	Molar Mass (g/mol)	v_{rms} at 20°C (m/s)
H_2	2.02	1902
He	4.0	1352
H_2O	18	637
Ne	20.1	603
N_2 and CO	28	511
NO	30	494
CO_2	44	408
SO_2	48	390

ier molecules. For example, hydrogen, with a molar mass of 2×10^{-3} kg/mol, moves four times as fast as oxygen, whose molar mass is 32×10^{-3} kg/mol. The rms speed is not the speed at which a gas molecule moves across a room, since such a molecule undergoes several billion collisions per second with other molecules under standard conditions.

Table 10.2 lists the rms speeds for various molecules at 20°C.

EXAMPLE 10.9   A Tank of Helium

A tank contains 2.0 mol of helium gas at 20°C. Assume that the helium behaves as an ideal gas.

(a) Find the total internal energy of the system.

Solution   Using Equation 10.18 with $n = 2.0$ and $T = 293$ K, we get

$$E = \tfrac{3}{2}nRT = \tfrac{3}{2}(2.0 \text{ mol})(8.31 \text{ J/mol·K})(293 \text{ K}) = \boxed{7.3 \times 10^3 \text{ J}}$$

(b) What is the average kinetic energy per molecule?

Solution   From Equation 10.16 we see that the average kinetic energy per molecule is

$$\tfrac{1}{2}m\overline{v^2} = \tfrac{3}{2}k_B T = \tfrac{1}{2}(1.38 \times 10^{-23} \text{ J/K})(293 \text{ K}) = \boxed{6.1 \times 10^{-21} \text{ J}}$$

Exercise   Using the fact that the molar mass of helium is 4.0×10^{-3} kg/mol, determine the rms speed of the atoms at 20°C.

Answer 1.4×10^3 m/s.

SUMMARY

The **zeroth law of thermodynamics** states that if two objects, A and B, are separately in thermal equilibrium with a third object, then A and B are in thermal equilibrium with each other.

The relationship between T_C, the **Celsius** temperature, and T, the **Kelvin (absolute)** temperature, is

$$T_C = T - 273.15 \qquad\qquad \textbf{[10.1]}$$

The relationship between the **Fahrenheit** and **Celsius** temperatures is

$$T_F = \tfrac{9}{5}T_C + 32 \qquad\qquad \textbf{[10.2]}$$

Generally, when a substance is heated, it expands. If an object has an initial length of L_0 at some temperature and undergoes a change in temperature of ΔT, its length changes by the amount ΔL, which is proportional to the object's initial length and the temperature change:

$$\Delta L = \alpha L_0 \Delta T \qquad\qquad \textbf{[10.4]}$$

The parameter α is called the **average coefficient of linear expansion**.

The change in area of a substance is given by

$$\Delta A = \gamma A_0 \, \Delta T \qquad\qquad \textbf{[10.5]}$$

where γ is the **average coefficient of area expansion** and is equal to 2α.

The change in volume of most substances is proportional to the initial volume, V_0, and the temperature change, ΔT:

$$\Delta V = \beta V_0 \, \Delta T \qquad\qquad \textbf{[10.6]}$$

where β is the **average coefficient of volume expansion** and is equal to 3α.

An **ideal gas** is one that obeys the equation

Equation of state for an ideal gas

$$PV = nRT \qquad\qquad \textbf{[10.8]}$$

where P is the pressure of the gas, V is its volume, n is the number of moles of gas, R is the universal gas constant ($8.31 \ \text{J/mol} \cdot \text{K}$), and T is the absolute temperature in kelvins. A real gas at very low pressures behaves approximately as an ideal gas.

The **pressure** of N molecules of an ideal gas contained in a volume V is given by

Pressure and molecular kinetic energy

$$P = \frac{2}{3}\left(\frac{N}{V}\right)\left(\tfrac{1}{2}m\overline{v^2}\right) \qquad\qquad \textbf{[10.14]}$$

where $\tfrac{1}{2}m\overline{v^2}$ is the **average kinetic energy per molecule.**

The average kinetic energy of the molecules of a gas is directly proportional to the absolute temperature of the gas:

Average kinetic energy per molecule

$$\tfrac{1}{2}m\overline{v^2} = \tfrac{3}{2}k_B T \qquad\qquad \textbf{[10.16]}$$

where k_B is **Boltzmann's constant** ($1.38 \times 10^{-23} \ \text{J/K}$).

The **root-mean-square (rms) speed** of the molecules of gas is

$$v_{\text{rms}} = \sqrt{\frac{3 k_B T}{m}} = \sqrt{\frac{3RT}{M}} \qquad\qquad \textbf{[10.19]}$$

ADDITIONAL READING

R. S. Berry, "When the Melting and Freezing Points are not the Same," *Sci. American,* August 1990, p. 68.

T. B. Greenslade, "The Maximum Density of Water," *The Physics Teacher,* November 1985, p. 474.

M. B. Hall, "Robert Boyle," *Sci. American,* August 1967, p. 84.

F. Jones, "Fahrenheit and Celsius, A History," *Physics Today,* 18, 1980, p. 594.

E. Jones and R. Childers, "Observational Evidence for Atoms," *The Physics Teacher,* October 1984, p. 354.

R. H. Romer, "Temperature Scales: Celsius, Fahrenheit, Kelvin, Reaumur, and Romer," *The Physics Teacher,* October 1982, p. 450.

R. E. Wilson, "Standards of Temperature," *Physics Today,* January 1953, p. 10.

CONCEPTUAL QUESTIONS

Example   Markings to indicate length are placed on a steel tape in a room that is at a temperature of 22°C. Measurements are then made with the same tape on a day when the temperature is 27°C. Are the measurements too long, too short, or accurate?

Reasoning   The measurements are too short. At 22°C

the tape would read the width of the object accurately, but an increase in temperature causes the divisions ruled on the tape to be farther apart than they should be. This "too long" ruler will, then, measure objects to be shorter than they really are.

Example   Why do vapor bubbles in a pot of boiling water get larger as they approach the surface?
Reasoning   As a bubble nears the surface, the pressure exerted on it by the liquid decreases. The trapped air inside the bubble then causes it to expand.

Example   Although the average speed of gas molecules in thermal equilibrium at some temperature is greater than zero, the average velocity is zero. Explain.
Reasoning   The gas consists of molecules moving in various directions with a distribution of speeds. There are as many molecules traveling in one direction as in the opposite direction. When all possible directions and speeds are taken into account, the average velocity is found to be zero.

Example   Explain why a column of mercury in a thermometer first descends slightly and then rises when placed in hot water.
Reasoning   The glass surrounding the mercury expands before the mercury does, causing the level of the mercury to drop slightly. The mercury rises after it begins to heat up and approach the temperature of the hot water, because its temperature coefficient of expansion is greater than that for glass.

Example   One container is filled with helium gas and another with argon gas. If both containers are at the same temperature, which molecules have the higher rms speed?
Reasoning   The helium molecules have the higher rms speed since the helium molecule is less massive than the argon molecule. Both molecules have the same average kinetic energy since they are at the same temperature.

Example   If a helium balloon is placed in a freezer, will its volume increase, decrease, or remain the same?
Reasoning   The rubber in a typical balloon stretches easily. The helium inside is nearly at atmospheric pressure. As the temperature drops, the volume will decrease slowly. The rubber wall moves in slowly, just maintaining equality of pressure outside and inside. In the expression $PV = nRT$, V and T decrease by the same factor, while P, n, and R stay constant.

Example   What happens to a helium-filled balloon released into the air? Will it expand or contract? Will it stop rising at some height?
Reasoning   Imagine the balloon rising into air at uniform temperature. The air cannot be uniform in pressure because the lower layers support the weight of all the air above them. The rubber in a typical balloon stretches or contracts until interior and exterior pressures are nearly equal. So as the balloon rises, it expands; this can be considered as an isothermal expansion with P decreasing as V increases by the same factor in $PV = nRT$. If the rubber wall is strong enough, it will eventually contain the helium at a pressure higher than the air outside but at the same density, so that the balloon will stop rising. It is more likely that the rubber will stretch and rupture, releasing the helium, which in turn will "boil out" of the Earth's atmosphere.

Example   When the metal ring and metal sphere in Figure 10.13 are both at room temperature, the sphere does not fit through the ring. After the ring is heated, the sphere can pass through the ring. Why?
Reasoning   The change in dimensions of the ring as it is heated is like a photographic enlargement. Every linear dimension, including the hole diameter, increases by the same factor. Hence, the ring gets larger in diameter as it is heated, enabling the sphere to pass through the ring.

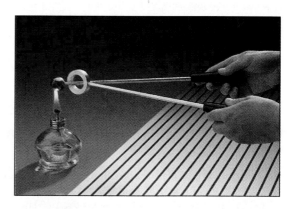

FIGURE 10.13   *(Courtesy of Central Scientific Company)*

1. Is it possible for two objects to be in thermal equilibrium if they are not in thermal contact with each other? Explain.
2. A piece of copper is dropped into a beaker of water. If the water's temperature rises, what happens to the temperature of the copper? When will the water and copper be in thermal equilibrium?
3. What would happen if, upon heating, the glass of a thermometer expanded more than the liquid inside?
4. "If the pressure above a liquid substance is reduced

to the pressure of the triple point, the substance will boil and freeze at the same temperature.'' Is this statement true?

5. A steel wheel bearing is 1 mm smaller in diameter than an axle. How can the bearing be fit onto the axle without removing any material?

6. Why is a power line more likely to break in winter than in summer even if it is loaded with the same weight?

7. On a hot afternoon creaking noises are often heard in the attic of a house. These noises are also frequently heard at night. Why?

8. If a jar lid is screwed on too tightly to remove, it is sometimes possible to loosen the lid by holding it under hot water. Explain.

9. When drinking glasses become stuck together, an old trick is to fill the inner glass with water and then run water of a different temperature over the sides of the outer glass. Which water should be hot and which cold? Why?

10. Chimneys are never used as a weight-bearing part of the structure of a building. Why?

11. Determine the number of grams in 1 mol of (a) hydrogen, (b) helium, and (c) carbon monoxide.

12. Two identical cylinders at the same temperature contain the same kind of gas. If cylinder A contains three times as much gas as cylinder B, what can you say about the relative pressures in the cylinders?

13. A microwave oven is used to heat food in a sealed pouch. Why should the pouch be pricked with a fork before heating begins?

14. The gas from a carbon-dioxide fire extinguisher emerges so rapidly that no immediate heat transfer occurs between the gas and the atmosphere. Use this information to explain why the gas is very cold.

15. During normal breathing the lungs expand upon inhalation and contract upon exhalation. Does the ideal gas law apply to the volume of the lungs under these circumstances? Explain.

16. Milk is a colloidal suspension, which means it contains tiny particles of insoluble matter that float about in the fluid (whey) without settling. Under a high-power microscope the particles are found to be darting around at random (a phenomenon referred to as Brownian motion). How does the molecular model account for this motion?

17. After food is cooked in a pressure cooker, why is it very important to cool the container with cold water before attempting to remove the lid?

18. Moving upward in the atmosphere, should you expect to find the ratio of nitrogen molecules relative to oxygen molecules increasing or decreasing? Explain.

19. An ideal gas is contained in a vessel at a temperature of 300 K. The temperature is increased to 900 K. (a) By what factor is the rms speed of each molecule multiplied? (b) By what factor is the pressure in the vessel multiplied?

PROBLEMS

Section 10.2   Thermometers and Temperature Scales

1. The pressure in a constant-volume gas thermometer is 0.700 atm at 100°C and 0.512 atm at 0°C. (a) What is the temperature when the pressure is 0.0400 atm? (b) What is the pressure at 450°C?

2. A constant-volume gas thermometer is calibrated in dry ice (−80.0°C) and in boiling ethyl alcohol (78.0°C). The two pressures are 0.900 atm and 1.635 atm. (a) What value of absolute zero does the calibration yield? (b) What pressures would be found at the freezing and boiling points of water?

3. Convert the following temperatures to degrees Celsius and Kelvin: (a) the normal human body temperature of 98.6°F; (b) the temperature of a cold day of −5°F.

4. Convert the following temperatures to Fahrenheit and kelvins: (a) the boiling point of liquid hydrogen −252.87°C; (b) the temperature of a room at 20°C.

5. The freezing and boiling points of water on the imaginary "TooHot" temperature scale are selected to be 50 and 200 degrees TH. (a) Derive an equation relating the TooHot scale to the Celsius scale. (b) Calculate the value of absolute zero in degrees TH.

6. At what Fahrenheit temperature are the Kelvin and Fahrenheit temperatures numerically equal?

7. The highest recorded temperature on Earth was 136°F, at Azizia, Libya, in 1922. The lowest recorded temperature was −127°F, at Vostok Station, Antarctica, in 1960. Express these temperature extremes in degrees Celsius.

8. The boiling point of sulfur is 444.60°C. The melting point is 586.1°F below the boiling point. (a) Determine the melting point in degrees Celsius. (b) Find the melting and boiling points in degrees Fahrenheit.

9. Show that the temperature −40° is unique in that it has the same numerical value on the Celsius and Fahrenheit scales.

☐ indicates problems that have full solutions available in the Student Solutions Manual and Study Guide.

10. Show that if the temperature on the Celsius scale changes by ΔT_C, the Fahrenheit temperature changes by the amount $\Delta T_F = (9/5) \Delta T_C$.

Section 10.3 Thermal Expansion of Solids and Liquids

11. The New River Gorge bridge in West Virginia is a 518-m-long steel arch. How much will its length change between temperature extremes of $-20°C$ and $35°C$?

12. A gold ring has an inner diameter of 2.168 cm at a temperature of 15.0°C. Determine its inner diameter at 100°C. ($\alpha_{gold} = 1.42 \times 10^{-5}$ °C^{-1}.)

13. Two concrete spans of a 250-m-long bridge are placed end to end so that there is no room for expansion (Fig. 10.14a). If a temperature increase of 20°C occurs, find the height, y, at which the spans buckle (see Fig. 10.14b). For concrete, $\alpha = 12 \times 10^{-6}$ °C^{-1}.

14. A grandfather clock is controlled by a swinging brass pendulum that is 1.3000 m long at a temperature of 20.0°C. (a) What is the length of the pendulum rod when the temperature drops to 0.0°C? (b) If a pendulum's period is given by $T = 2\pi\sqrt{L/g}$, where L is its length, does the change in length of the rod cause the clock to run fast or slow?

15. Show that the coefficient of volume expansion, β, is related to the coefficient of linear expansion, α, through the expression $\beta = 3\alpha$.

16. A pair of eyeglass frames are made of epoxy plastic (coefficient of linear expansion = 130×10^{-6} °C^{-1}). At room temperature (assume 20.0°C) the frames have circular lens holes 2.20 cm in radius. To what temperature must the frames be heated in order to insert lenses 2.21 cm in radius?

17. A cylindrical brass sleeve is to be shrink-fitted over a brass shaft whose diameter is 3.212 cm at 0°C. The diameter of the sleeve is 3.196 cm at 0°C. (a) To what temperature must the sleeve be heated before it will slip over the shaft? (b) Alternatively, to what temperature must the shaft be cooled before it will slip into the sleeve?

18. At 20.000°C, an aluminum ring has an inner diameter of 5.000 cm, and a brass rod has a diameter of 5.050 cm. (a) To what temperature must the

FIGURE 10.15 (Problem 20)

aluminum ring be heated so that it will just slip over the brass rod? (b) To what temperature must *both* be heated so the aluminum ring will slip off the brass rod? Would this work?

19. An object with coefficient of linear expansion α and Young's modulus Y is raised in temperature by the amount ΔT while its ends are firmly fixed in place. Show that the stress (force per unit area) in the object is $\alpha Y \Delta T$.

20. The band in Figure 10.15 is stainless steel (coefficient of linear expansion = 17.3×10^{-6} °C^{-1}; Young's modulus = 18×10^{10} N/m^2). It is essentially circular with an initial mean radius of 5.0 mm, a height of 4.0 mm, and a thickness of 0.50 mm. If the band just fits snugly over the tooth when heated to a temperature of 80°C, what is the tension in the band when it cools to a temperature of 37°C?

21. A 250-m-long bridge is improperly designed so that it cannot expand with temperature. It is made of concrete with $\alpha = 12 \times 10^{-6}$ °C^{-1}. (a) Assuming that the maximum change in temperature at the site is expected to be 20°C, find the change in length the span would undergo if it were free to expand. (b) If the maximum stress the bridge can withstand without crumbling is 2.0×10^7 Pa, will it crumble because of this temperature increase? Young's modulus for concrete is about 2.0×10^{10} Pa.

22. A construction worker uses a steel tape to measure the length of an aluminum support column. If the measured length is 18.7 m when the temperature is 21.2°C, what is the measured length when the temperature rises to 29.4°C? (*Note:* Do not neglect the expansion of the steel tape.)

23. An automobile fuel tank is filled to the brim with 45 L (12 gal) of gasoline at 10°C. Immediately afterward, the vehicle is parked in the Sun, where the temperature is 35°C. How much gasoline overflows from the tank as a result of the expansion? (Neglect the expansion of the tank.)

24. An underground gasoline tank at 54°F can hold 1000 gallons of gasoline. If the driver of a tanker truck fills the underground tank on a day when the temperature is 90°F, how many gallons, ac-

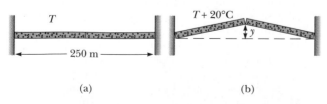

(a)   (b)

FIGURE 10.14 (Problem 13)

cording to his measure on the truck, can he pour in? Assume that the temperature of the gasoline cools to 54°F upon entering the tank.

Section 10.4   Macroscopic Description of an Ideal Gas

25. Gas is contained in an 8.0-L vessel at a temperature of 20°C and a pressure of 9.0 atm. (a) Determine the number of moles of gas in the vessel. (b) How many molecules are in the vessel?

26. (a) An ideal gas occupies a volume of 1.0 cm^3 at 20°C and atmospheric pressure. Determine the number of molecules of gas in the container. (b) If the pressure of the 1.0 cm^3 volume is reduced to 1.0×10^{-11} Pa (an extremely good vacuum) while the temperature remains constant, how many moles of gas remain in the container?

27. The pressure on an ideal gas is cut in half, resulting in a decrease in temperature to three fourths of the original value. Calculate the ratio of the final volume to the original volume of the gas.

28. Gas is confined in a tank at a pressure of 10.0 atm and a temperature of 15.0°C. If half of the gas is withdrawn and the temperature is raised to 65.0°C, what is the new pressure in the tank?

29. A 0.10-cm^3 bubble of gas is formed at the bottom of a 10-cm-deep container of mercury. If the temperature is 27°C at the bottom and 37°C at the surface, what will be the volume of the bubble just beneath the surface of the mercury? Assume that the surface is at atmospheric pressure.

30. A cylinder with a movable piston contains gas at a temperature of 27°C, a volume of 1.5 m^3, and an absolute pressure of 0.20×10^5 Pa. What will be its final temperature if the gas is compressed to 0.70 m^3 and the absolute pressure increases to 0.80×10^5 Pa?

31. One mole of oxygen gas is at a pressure of 6.0 atm and a temperature of 27°C. (a) If the gas is heated at constant volume until the pressure triples, what is the final temperature? (b) If the gas is heated so that both the pressure and volume are doubled, what is the final temperature?

32. A swimmer has 0.820 L of dry air in his lungs when he dives into a lake. Assuming the pressure of the dry air is 95% of the external pressure at all times, what is the volume of the dry air at a depth of 10.0 m? Assume that atmospheric pressure at the surface is 1.013×10^5 Pa.

33. An air bubble has a volume of 1.50 cm^3 when it is released by a submarine 100 m below the surface of a lake. What is the volume of the bubble when it reaches the surface? Assume that the temperature of the air in the bubble remains constant during ascent.

34. The mass of a hot air balloon and its cargo (not including the air inside) is 100 kg. The air outside is at a temperature of 27°C and at a pressure of 1.0 atm. If the volume of the balloon is 400 m^3, what must be the temperature of the air in the balloon for the balloon to begin to rise? Assume that the inside pressure is the same as the outside pressure, and treat air as a gas with a molecular weight of 28.8. (*Hint:* Consider the buoyancy of the outside air.)

35. A weather balloon is designed to expand to a maximum radius of 20 m when in flight at its working altitude, where the air pressure is 0.030 atm and the temperature is 200 K. If the balloon is filled at atmospheric pressure and 300 K, what is its radius at lift-off?

36. (a) Estimate the total mass of air inside a typical-size house on a day when the temperature is 0°F. (Assume a molecular weight of 28.8 g/mol for air.) (b) How much mass must enter or leave the house if the temperature increases to 100°F?

37. A cylindrical diving bell, 3.0 m in diameter and 4.0 m tall with an open bottom, is submerged to a depth of 220 m in the ocean. The surface temperature is 25°C, and the temperature 220 m down is 5.0°C. The density of seawater is 1025 kg/m^3. How high does the seawater rise in the bell when it is submerged?

38. Use the ideal gas equation and the relationship between moles and the mass of a gas to find an expression for the density of a gas.

39. The density of helium gas at $T = 0$°C is $\rho_0 = 0.179$ km/m^3. The temperature is then raised to $T = 100$°C, but the pressure is kept constant. Assuming that helium is ideal, calculate the new density, ρ_f, of the gas.

Section 10.5   Avogadro's Number and the Ideal Gas Law

Section 10.6   The Kinetic Theory of Gases

40. What is the average kinetic energy of a molecule of oxygen at a temperature of 300 K?

41. (a) What is the total random kinetic energy of all the molecules in one mole of hydrogen at a temperature of 300 K? (b) With what speed would a mole of hydrogen have to move so that the kinetic energy of the mass as a whole would be equal to the total random kinetic energy of its molecules?

42. Use Avogadro's number to find the mass of a helium atom.

43. A sealed cubical container 20 cm on a side contains three times Avogadro's number of molecules at a temperature of 20°C. Find the force exerted by the gas on one of the walls of the container.

44. The total random translational kinetic energy of the water molecules in a 20-L container is $4.8 \times$

10^6 J; the total random translational kinetic energy of the water molecules in a 360 000-L swimming pool is 7.3×10^{10} J. Which body of water, the container or the pool, is at the higher temperature?

45. In a period of 1.0 s, 5.0×10^{23} nitrogen molecules strike a wall of area 8.0 cm². If the molecules move at 300 m/s and strike the wall head on in a perfectly elastic collision, find the pressure exerted on the wall. (The mass of one N_2 molecule is 4.68×10^{-26} kg.)

46. Superman leaps in front of Lois Lane to save her from a volley of bullets. In a one-minute interval, an automatic weapon fires 150 bullets, each of mass 8.0 g, at 400 m/s. The bullets strike his mighty chest, which has an area of 0.75 m². Find the average force exerted on Superman's chest if the bullets bounce back after an elastic, head-on collision.

47. Repeat Problem 46 under the following conditions. When Superman steps in front of the bullets he finds they are made of kryptonite, a material that causes them to stick to his chest in an inelastic collision. Find the force exerted on him in this case.

48. (a) Calculate the rms speed of an H_2 molecule when the temperature is 100°C. (b) Repeat the calculation for an N_2 molecule.

49. A cylinder contains a mixture of helium and argon gas in equilibrium at a temperature of 150°C.
(a) What is the average kinetic energy of each type of molecule? (b) What is the rms speed of each type of molecule?

50. The temperature near the top of the atmosphere on Venus is 240 K. (a) Find the rms speed of hydrogen (H_2) at this point in the atmosphere. (b) Repeat for carbon dioxide (CO_2). (c) It has been found that if the rms speed exceeds one sixth of the planet's escape velocity, the gas eventually leaks out of the atmosphere and into outer space. If the escape velocity on Venus is 10.3 km/s, does hydrogen escape? Does carbon dioxide?

51. Three moles of nitrogen gas, N_2, at 27.0°C are contained in a 22.4-L cylinder. Find the pressure the gas exerts on the cylinder walls.

52. At what temperature would the rms speed of helium atoms (mass = 6.66×10^{-27} kg) equal (a) the escape velocity from Earth, 1.12×10^4 m/s? (b) the escape velocity from the Moon, 2.37×10^3 m/s?

ADDITIONAL PROBLEMS

53. The active element of a certain laser is an ordinary glass rod 20 cm long and 1.0 cm in diameter. If the temperature of the rod increases by 75°C, find its increases in (a) length, (b) diameter, and (c) volume.

54. If 2.0 mol of a gas are confined to a 5.0-L vessel at a pressure of 8.0 atm, what is the average kinetic energy of a gas molecule?

55. Absolute zero on the Rankine temperature scale is $T = 0°$ R, and the scale's unit is the same size as the Fahrenheit degree. (a) Determine the freezing and boiling points of water on the Rankine scale. (b) Write a formula that relates the Rankine scale to the Fahrenheit scale. (c) Write a formula that relates the Rankine scale to the Kelvin scale.

56. Before beginning a long trip on a hot day, a driver inflates an automobile tire to a gauge pressure of 1.8 atm at 300 K. At the end of the trip the pressure in the tire has increased to 2.2 atm. (a) Assuming the volume has remained constant, what is the temperature of the air inside the tire? (b) What volume of air (measured at atmospheric pressure) should be released from the tire so that the pressure returns to the initial value? Assume that the air is released during a short time interval during which the temperature remains at the value found in part (a).

57. A liquid with coefficient of volume expansion β just fills a spherical shell of volume V at temperature T (Fig. 10.16). The shell is made of a material that has a coefficient of linear expansion of α. The liquid is free to expand into a capillary of cross-sectional area A at the top. (a) If the temperature increases by ΔT, show that the liquid rises in the capillary by the amount $\Delta h = (V/A)(\beta - 3\alpha) \Delta T$. (b) For a typical system, such as a mercury thermometer, why is it a good approximation to neglect the expansion of the shell?

58. A mercury thermometer is constructed as in Figure 10.16. The capillary tube has a diameter of 0.0050 cm, and the bulb has a diameter of 0.30 cm. Neglecting the expansion of the glass,

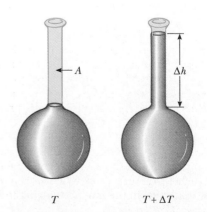

T $T + \Delta T$

FIGURE 10.16 (Problems 57 and 58)

find the change in height of the mercury column for a temperature change of 25°C.

59. The density of gasoline is 730 kg/m³ at 0°C. Its volume expansion coefficient is 9.6×10^{-4} °C^{-1}. If one gallon of gasoline occupies 0.0038 m³, how many extra kilograms of gasoline are obtained when 10 gallons of gasoline are bought at 0°C rather than at 20°C?

60. A 1.5-m-long glass tube, closed at one end, is weighted and lowered to the bottom of a fresh-water lake. When the tube is recovered, an indicator mark shows that water rose to within 0.40 m of the closed end. Determine the depth of the lake. Assume constant temperature.

61. A hollow aluminum cylinder is to be fitted over a steel piston. At 20°C the inside diameter of the cylinder is 99% of the outside diameter of the piston. To what common temperature should the two pieces be heated in order that the cylinder just fit over the piston?

62. Two small containers of equal volume, 100 cm³, contain helium gas at 0°C and 1.0 atm pressure. The two are joined by a small open tube of negligible volume. What pressure is attained by the gas in each container if the temperature of one of the containers is raised to 100°C while the other is kept at 0°C?

63. A steel measuring tape was designed to read correctly at 20°C. A parent uses the tape to measure the height of a 1.1-m-tall child. If the measurement is made on a day when the temperature is 25°C, is the tape reading longer or shorter than the actual height, and by how much?

64. When an ideal gas is held at constant pressure, a small change in volume, ΔV, associated with a small change in temperature, ΔT, is given by $P(\Delta V) = nR(\Delta T)$. (a) Show that the volume coefficient of thermal expansion for an ideal gas at constant pressure is given by $\beta = 1/T$, where T is the Kelvin temperature. (b) What value does this equation predict for β at 0°C?

65. A vertical cylinder of cross-sectional area 0.050 m² is fitted with a tight-fitting, frictionless piston of mass 5.0 kg (Fig. 10.17). If there are 3.0 mol of an ideal gas in the cylinder at 500 K, determine the height, h, at which the piston will be in equilibrium under its own weight.

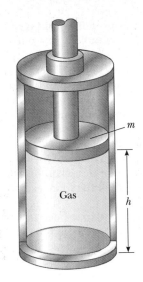

FIGURE 10.17   (Problem 65)

66. An air bubble originating from a deep-sea diver has a radius of 2.0 mm at some depth h. When the bubble reaches the surface of the water, it has a radius of 3.0 mm. Assuming that the temperature of the air in the bubble remains constant, determine (a) the depth, h, of the diver and (b) the absolute pressure at this depth.

HEAT

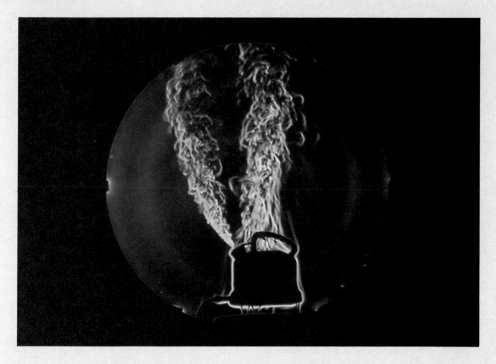

Thermogram of a teakettle showing hot areas in white and cooler areas in purple and black. (Gary Settles/Science Source/Photo Researchers)

It is an experimentally established fact that when two objects at different temperatures are placed in thermal contact with each other, the temperature of the warmer object decreases and the temperature of the cooler object increases. If the two are left in contact for some time, they eventually reach a common equilibrium temperature that is intermediate between the two initial temperatures. When such a process occurs, we say that heat is transferred from the object at the higher temperature to the one at the lower temperature.

Up until about 1850, the subjects of heat and mechanics were considered to be two distinct branches of science, and the principle of conservation of energy seemed to be a rather specialized result that could be used to describe certain kinds of mechanical systems. After the two disciplines were shown to be related, the principle of conservation of energy emerged as a universal principle of nature. From this new perspective, heat is treated as another form of energy that can be transformed into mechanical energy. Experiments performed by the Englishman James Joule (1818–1889) and his contemporaries demonstrated that whenever heat is gained or lost by a system during some process, the gain or loss can be accounted for by an equivalent quantity of mechanical work done on the system. Thus, with the broadening of the concept of energy to include heat as a form of energy, the principle of energy conservation was extended.

The focus of this chapter is to introduce the concept of heat and some of the processes that enable heat to be transferred between a system and its surroundings.

11.1

THE MECHANICAL EQUIVALENT OF HEAT

Heat (or **thermal energy**) is now defined as energy that is transferred between a system and its environment because of a temperature difference between them. Before scientists arrived at a correct understanding of heat, the units in which heat was measured had already been developed. They were chosen because of early misunderstandings about heat. These unusual units are still widely used in many applications, and so we shall discuss them briefly here. One of the most widely used is the **calorie** (cal), defined as *the heat required to raise the temperature of 1 g of water by 1° C.* A related unit is the **kilocalorie** (kcal), *the heat necessary to raise the temperature of 1 kg of water from 14.5° C to 15.5° C* (1 kcal = 10^3 cal). The Calorie with a capital C, used in describing the energy equivalent of foods, is equal to 1 kcal. The unit of heat in the British engineering system is the **British thermal unit** (Btu), defined as *the heat required to raise the temperature of 1 lb of water from 63° F to 64° F.*

Since heat is now recognized as a form of energy, scientists are increasingly using the SI unit of energy, the *joule* (J), for quantities of heat. In this book, heat is most often measured in joules.

Definition of the calorie

JOULE'S EXPERIMENT

When the concept of mechanical energy was introduced in Chapter 5, we found that whenever friction is present in a mechanical system, some mechanical energy is lost. Experiments of various sorts show that this lost mechanical energy

James Prescott Joule (1818–1889)

(Science Photo Library/Photo Researchers)

James Prescott Joule, a British physicist, was born into a wealthy brewing family in Salford, England, on December 24, 1818. He received some formal training in mathematics, philosophy, and chemistry from John Dalton but was in large part self-educated.

Joule's most active research period, from 1837 through 1847, led to the establishment of the principle of conservation of energy and the equivalence of heat and other forms of energy. His study of the quantitative relationships among electrical, mechanical, and chemical effects of heat culminated in his announcement in 1843 of the amount of work required to produce a unit of heat:

> First: that the quantity of heat produced by the friction of bodies, whether solid or liquid, is always proportional to the quantity of energy expended. And second: that the quantity of heat capable of increasing the temperature of water . . . by 1° Fahr requires for its evolution the expenditure of a mechanical energy represented by the fall of 772 lb through the distance of one foot.

This is called the mechanical equivalent of heat (the currently accepted value is 4.186 J/cal).

Much of Joule's later work on the new science of thermodynamics was extended by Lord Kelvin. In 1852 the Joule-Thomson effect showed that when a gas is allowed to expand freely, its temperature drops slightly. This was an important discovery in the field of low-temperature physics.

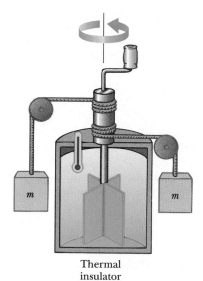

FIGURE 11.1
An illustration of Joule's experiment for measuring the mechanical equivalent of heat. The falling weights rotate the paddles, causing the temperature of the water to rise.

does not simply disappear; instead, it is transformed into thermal energy. Joule (1818–1889) was the first to establish the equivalence of the two forms of energy. Figure 11.1 is a schematic diagram of Joule's most famous experiment. The system of interest is the water in a thermally insulated container. Work is done on the water by a rotating paddle wheel, which is driven by weights falling at a constant speed. The water, stirred by the paddles, warms up because of the friction between it and the paddles. If the energy lost in the bearings and through the walls is neglected, then the loss in potential energy of the weights equals the work done by the paddles on the water. If the two weights fall a distance of h, the loss in potential energy is $2mgh$, and it is this energy that heats the water. By varying the conditions of the experiment, Joule found that the loss in mechanical energy, $2mgh$, is proportional to the increase in temperature of the water, ΔT, and to the mass of water used. The proportionality constant (the specific heat of water) was found to be 4.186 J/g · °C. Hence, 4.186 J of mechanical energy raises the temperature of 1 g of water from 14.5°C to 15.5°C. We use the conversion factor

$$1 \text{ cal} = 4.186 \text{ J} \qquad \textbf{[11.1]}$$

The ratio 4.186 J/cal is known, for purely historical reasons, as the **mechanical equivalent of heat**.

EXAMPLE 11.1   Losing Weight the Hard Way

A student eats a dinner rated at 2000 (food) Calories. He wishes to do an equivalent amount of work in the gymnasium by lifting a 50.0-kg mass. How many times must he raise the weight to expend this much energy? Assume that he raises the weight a distance of 2.00 m each time and that no work is done when the weight is dropped to the floor.

Solution   Since 1 Calorie = 10^3 cal, the work required is 2.00×10^6 cal. Converting this to joules, we have, for the total work required,

$$W = (2.00 \times 10^6 \text{ cal})(4.186 \text{ J/cal}) = 8.37 \times 10^6 \text{ J}$$

The work done in lifting the weight once through the distance h is equal to mgh, and the work done in lifting the weight n times is $nmgh$. If we set $nmgh$ equal to the total work required, we have

$$W = nmgh = 8.37 \times 10^6 \text{ J}$$

Since $m = 50.0$ kg and $h = 2.00$ m, we get

$$n = \frac{8.37 \times 10^6 \text{ J}}{(50.0 \text{ kg})(9.80 \text{ m/s}^2)(2.00 \text{ m})} = \boxed{8.54 \times 10^3 \text{ times}}$$

If the student is in good shape and lifts the weight, say, once every 5.0 s, it will take him about 12 h to perform this feat. Clearly, it is much easier to lose weight by dieting.

This problem is somewhat misleading in that it assumes perfect conversion of chemical energy into mechanical energy. In actual practice, a more realistic assessment of the number of repetitions can be found by dividing the given answer by 6. Thus, about 1400 lifts are required to burn off the calories.

11.2

SPECIFIC HEAT

The quantity of heat energy required to raise the temperature of a given mass of a substance by some amount varies from one substance to another. For example, the heat required to raise the temperature of 1 kg of water by 1°C is 4186 J, but the heat required to raise the temperature of 1 kg of copper by 1°C is only 387 J. Every substance has a unique value for the amount of heat required to change the temperature of 1 kg of it by 1°C, and this number is referred to as the specific heat of the substance. Table 11.1 lists specific heats for a few substances.

Suppose that a quantity, Q, of heat is transferred to m kg of a substance, thereby changing its temperature by ΔT. The **specific heat,** c, of the substance is defined as

Specific heat

$$c \equiv \frac{Q}{m \, \Delta T} \qquad \text{[11.2]}$$

From this definition we can express the heat transferred between a system of mass m and its surroundings for a temperature change of ΔT as

$$Q = mc \, \Delta T \qquad \text{[11.3]}$$

For example, the heat required to raise the temperature of 0.500 kg of water by 3.00°C is equal to $(0.500 \text{ kg})(4186 \text{ J/kg} \cdot °\text{C})(3.00°\text{C}) = 6280$ J. Note that when the temperature increases, ΔT and Q are taken to be *positive,* corresponding to heat flowing *into* the system. Likewise, when the temperature decreases, ΔT and Q are *negative* and heat flows *out* of the system.

Note from Table 11.1 that water has the highest specific heat of the substances we are likely to encounter on a routine basis. This high specific heat is responsible for the moderate temperatures found in regions near large bodies of water. As the temperature of a body of water decreases during winter, the water gives off heat to the air, which carries the heat landward when prevailing winds

TABLE 11.1
Specific Heats of Some Materials at Atmospheric Pressure

Substance	J/kg·°C	cal/g·°C
Aluminum	900	0.215
Beryllium	1820	0.436
Cadmium	230	0.055
Copper	387	0.0924
Germanium	322	0.077
Glass	837	0.200
Gold	129	0.0308
Ice	2090	0.500
Iron	448	0.107
Lead	128	0.0305
Mercury	138	0.033
Silicon	703	0.168
Silver	234	0.056
Steam	2010	0.480
Water	4186	1.00

are favorable. For example, the prevailing winds off the western coast of the United States are toward the land, and the heat liberated by the Pacific Ocean as it cools keeps coastal areas much warmer than they would otherwise be. This explains why the western coastal states generally have more favorable winter weather than the eastern coastal states, where the winds do not carry the heat toward land. The same effect also keeps summers cooler in San Francisco than in a similar region on the East coast.

The fact that the specific heat of water is higher than that of land is responsible for the pattern of air flow at a beach. During the day, the Sun adds roughly equal amounts of energy to beach and water, but the lower specific heat of sand causes the beach to reach a higher temperature than the water. Consequently, the air above the land reaches a higher temperature than that over the water, and cooler air from above the water is drawn in to displace this rising hot air, resulting in a breeze from ocean to land during the day. Because the hot air gradually cools as it rises, it subsequently sinks, setting up the circulating pattern shown in Figure 11.2a. During the night, the land cools more quickly than the water, and the circulating pattern reverses itself because the hotter air is now over the water

Day

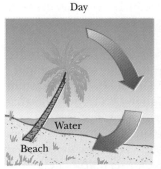

(a)

Night

(b)

FIGURE 11.2
Circulation of air at the beach. (a) On a hot day, the air above the warm sand warms faster than the air above the cooler water. The cooler air over the water moves toward the beach, displacing the rising warmer air. (b) At night, the sand cools more rapidly than the water, and hence the air currents reverse direction.

Hang gliders can soar great distances by taking advantage of air currents and thermal gradients in the atmosphere. *(Superstock)*

(Fig. 11.2b). The offshore and onshore breezes are certainly well known to sailors.

A similar effect produces rising layers of air, called *thermals,* that can help eagles to soar higher and hang gliders to stay in flight longer. A thermal is created when a portion of the Earth reaches a higher temperature than neighboring regions. This often happens to plowed fields, which are heated by the Sun to higher temperatures than nearby fields shaded by vegetation. The cooler, denser air over the vegetation-covered fields pushes the expanding air over the plowed field upward, and a thermal is formed.

11.3

CONSERVATION OF ENERGY: CALORIMETRY

Situations in which mechanical energy is converted to thermal energy occur frequently. We shall look at some of them in examples and in the problems at the end of the chapter, but most of our attention here is directed toward a particular kind of conservation-of-energy situation. In problems using the procedure called *calorimetry,* only the transfer of thermal energy between the system and its surroundings is considered.

One technique for measuring the specific heat of a solid or liquid is simply to heat the substance to some known temperature, place it in a vessel containing water of known mass and temperature, and measure the temperature of the water after equilibrium is reached. Since a negligible amount of mechanical work is done in the process, the law of conservation of energy requires that the heat that leaves the warmer substance (of unknown specific heat) equal the heat that enters the water. Devices in which this heat transfer occurs are called **calorimeters.**

Suppose that m_x is the mass of a substance whose specific heat we wish to measure, c_x its specific heat, and T_x its initial temperature. Let m_w, c_w, and T_w represent the corresponding values for the water. If T is the final equilibrium temperature after everything is mixed, then from Equation 11.2 we find that the heat gained by the water is $m_w c_w (T - T_w)$ and the heat lost by the substance of unknown specific heat, c_x, is $m_x c_x (T_x - T)$. Assuming that the system (water + unknown) neither loses nor gains heat, it follows that the heat gained by the water must equal the heat lost by the unknown (conservation of energy):

$$m_w c_w(T - T_w) = m_x c_x(T_x - T)$$

Solving for c_x gives

$$c_x = \frac{m_w c_w(T - T_w)}{m_x(T_x - T)} \qquad \textbf{[11.4]}$$

Do not attempt to memorize this equation. Instead, always start from first principles in solving calorimetry problems. That is, determine which substances lose heat and which gain heat, and then equate heat loss to heat gain.

A word about sign conventions: In a later chapter we shall find it necessary to use a sign convention in which a positive sign for Q indicates heat gained by a substance and a negative sign indicates heat lost. However, for calorimetry problems it is less confusing if you ignore the positive and negative signs and instead equate heat loss to heat gain. That is, you should always write ΔT as a positive quantity. For example, if T_f is greater than T_i, let $\Delta T = T_f - T_i$. If T_f is less than T_i, let $\Delta T = T_i - T_f$.

EXAMPLE 11.2 Cooling a Hot Ingot

A 0.0500-kg ingot of metal is heated to 200.0°C and then dropped into a beaker containing 0.400 kg of water that is initially at 20.0°C. If the final equilibrium temperature of the mixed system is 22.4°C, find the specific heat of the metal.

Solution Because the heat lost by the ingot equals the heat gained by the water, we can write

$$m_x c_x(T_{ix} - T_{fx}) = m_w c_w(T_{fw} - T_{iw})$$

$$(0.0500 \text{ kg})(c_x)(200.0°C - 22.4°C) =$$
$$(0.400 \text{ kg})(4186 \text{ J/kg} \cdot °C)(22.4°C - 20.0°C)$$

from which we find that

$$c_x = \quad 453 \text{ J/kg} \cdot °C$$

The ingot is most likely iron, as can be seen by comparing this result with the data in Table 11.1.

EXAMPLE 11.3 Fun Time for a Cowboy

A cowboy fires a 2.00-g silver bullet at a muzzle speed of 200 m/s into the pine wall of a saloon. Assume that all the internal energy generated by the impact remains with the bullet. What is the temperature change of the bullet?

Solution The kinetic energy of the bullet is

$$\tfrac{1}{2}mv^2 = \tfrac{1}{2}(2.00 \times 10^{-3} \text{ kg})(200 \text{ m/s})^2 = 40.0 \text{ J}$$

Nothing in the environment is hotter than the bullet, so the bullet gains no thermal energy. Its temperature increases because the 40.0 J of kinetic energy becomes 40.0 J of extra internal energy. The temperature change would be the same as if 40.0 J of thermal energy were transferred from a stove to the bullet, and we imagine this process to compute ΔT from

$$Q = mc\,\Delta T$$

Since the specific heat of silver is 234 J/kg·°C (Table 11.1), we get

$$\Delta T = \frac{Q}{mc} = \frac{40.0\,\text{J}}{(2.00 \times 10^{-3}\,\text{kg})(234\,\text{J/kg·°C})} = \boxed{85.5\text{°C}}$$

Exercise   Suppose the cowboy runs out of silver bullets and fires a lead bullet of the same mass and velocity into the wall. What is the temperature change of the bullet?

Answer   157°C.

11.4

LATENT HEAT AND PHASE CHANGES

A substance usually undergoes a change in temperature when heat is transferred between it and its surroundings. There are situations, however, in which the flow of heat does not result in a change in temperature. This is the case whenever the substance undergoes a physical alteration from one form to another, referred to as a **phase change.** Some common phase changes are solid to liquid (melting), liquid to gas (boiling), and a change in crystalline structure of a solid. Every phase change involves a change in internal energy.

The heat required to change the phase of a given mass, m, of a pure substance is

$$Q = mL \tag{11.5}$$

where L is called the **latent heat** (''hidden'' heat) of the substance and depends on the nature of the phase change as well as on the properties of the substance. **Heat of fusion,** L_f, is the term used when the phase change is from solid to liquid, and **heat of vaporization,** L_v, is the term used when the phase change is from liquid to gas.[1] For example, the heat of fusion for water at atmospheric pressure is 3.33×10^5 J/kg, and the heat of vaporization for water is 2.26×10^6 J/kg. The latent heats of different substances vary considerably, as Table 11.2 shows.

Consider, for example, the heat required to convert a 1.00-g block of ice at −30.0°C to steam (water vapor) at 120.0°C. Figure 11.3 indicates the experimental results obtained when heat is gradually added to the ice. Let us examine each portion of the curve separately.

Part A.   On this portion of the curve the temperature of the ice is changing from −30.0°C to 0.0°C. Since the specific heat of ice is 2090 J/kg·°C, we can calculate the amount of heat added from Equation 11.3:

$$Q = m_i c_i\,\Delta T = (1.00 \times 10^{-3}\,\text{kg})(2090\,\text{J/kg·°C})(30.0\text{°C}) = 62.7\,\text{J}$$

Part B.   When the ice reaches 0°C, the ice/water mixture remains at this temperature—even though heat is being added—until all the ice melts to

---

[1] When a gas cools, it eventually returns to the liquid phase, or *condenses.* The heat per unit mass given up during the process is called the *heat of condensation,* and it equals the heat of vaporization. When a liquid cools, it eventually solidifies, and the *heat of solidification* equals the heat of fusion.

TABLE 11.2
Latent Heats of Fusion and Vaporization

Substance	Melting Point (°C)	Latent Heat of Fusion J/kg	(cal/g)	Boiling Point (°C)	Latent Heat of Vaporization J/kg	(cal/g)
Helium	−269.65	5.23×10^3	(1.25)	−268.93	2.09×10^4	(4.99)
Nitrogen	−209.97	2.55×10^4	(6.09)	−195.81	2.01×10^5	(48.0)
Oxygen	−218.79	1.38×10^4	(3.30)	−182.97	2.13×10^5	(50.9)
Ethyl alcohol	−114	1.04×10^5	(24.9)	78	8.54×10^5	(204)
Water	0.00	3.33×10^5	(79.7)	100.00	2.26×10^6	(540)
Sulfur	119	3.81×10^4	(9.10)	444.60	3.26×10^5	(77.9)
Lead	327.3	2.45×10^4	(5.85)	1750	8.70×10^5	(208)
Aluminum	660	3.97×10^5	(94.8)	2450	1.14×10^7	(2720)
Silver	960.80	8.82×10^4	(21.1)	2193	2.33×10^6	(558)
Gold	1063.00	6.44×10^4	(15.4)	2660	1.58×10^6	(377)
Copper	1083	1.34×10^5	(32.0)	1187	5.06×10^6	(1210)

become water at 0°C. The heat required to melt 1.00 g of ice at 0°C is, from Equation 11.5,

$$Q = mL_f = (1.00 \times 10^{-3}\ \text{kg})(3.33 \times 10^5\ \text{J/kg}) = 333\ \text{J}$$

Part C. Between 0°C and 100°C, nothing surprising happens. No phase change occurs in this region. The heat added to the water is being used to increase its temperature. The amount of heat necessary to increase the temperature from 0°C to 100°C is

$$Q = m_w c_w\, \Delta T = (1.00 \times 10^{-3}\ \text{kg})(4.19 \times 10^3\ \text{J/kg} \cdot \text{°C})(100\text{°C})$$

$$= 4.19 \times 10^2\ \text{J}$$

Part D. At 100°C, another phase change occurs as the water changes from water at 100°C to steam at 100°C. Just as in Part B, the water/steam mixture remains at

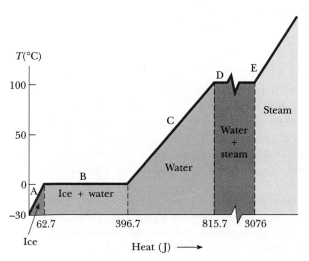

FIGURE 11.3
A plot of temperature versus heat added when 1.00 g of ice, initially at −30.0°C, is converted to steam at 120°C.

100°C—even though heat is being added—until all the liquid has been converted to steam. The heat required to convert 1.00 g of water to steam at 100°C is

$$Q = mL_v = (1.00 \times 10^{-3} \text{ kg})(2.26 \times 10^6 \text{ J/kg}) = 2.26 \times 10^3 \text{ J}$$

Part E.   On this portion of the curve, heat is being added to the steam with no phase change occurring. The heat that must be added to raise the temperature of the steam to 120°C is

$$Q = m_s c_s \Delta T = (1.00 \times 10^{-3} \text{ kg})(2.01 \times 10^3 \text{ J/kg} \cdot ^\circ\text{C})(20^\circ\text{C}) = 40.2 \text{ J}$$

The *total amount of heat* that must be added to change one gram of ice at -30°C to steam at 120°C is therefore about 3.11×10^3 J. Conversely, to cool one gram of steam at 120°C down to the point at which we have ice at -30°C, we must remove 3.11×10^3 J of heat.

Phase changes can be described in terms of rearrangements of molecules when heat is added to or removed from a substance. Consider first the liquid-to-gas phase change. The molecules in a liquid are close together, and the forces between them are stronger than those between the more widely separated molecules of a gas. Therefore, work must be done on the liquid against these attractive molecular forces in order to separate the molecules. The heat of vaporization is the amount of energy that must be added to the liquid to accomplish this.

Similarly, at the melting point of a solid, we imagine that the amplitude of vibration of the atoms about their equilibrium positions becomes great enough to allow the atoms to pass the barriers of adjacent atoms and move to their new positions. The new locations are, on the average, less symmetrical and therefore have higher energy. The heat of fusion is equal to the work required at the molecular level to transform the mass from the ordered solid phase to the disordered liquid phase.

The average distance between atoms is much greater in the gas phase than in either the liquid or the solid phase. Each atom or molecule is removed from its neighbors, without the compensation of attractive forces to new neighbors. Therefore, it is not surprising that more work is required at the molecular level to vaporize a given mass of substance than to melt it; thus the heat of vaporization is much greater than the heat of fusion (Table 11.2).

PROBLEM-SOLVING STRATEGY
Calorimetry Problems

If you are having difficulty with calorimetry problems, make the following considerations.

1. Be sure your units are consistent throughout. That is, if you are using specific heats measured in cal/g · °C, be sure that masses are in grams and temperatures are in Celsius units throughout.
2. Losses and gains in heat are found by using $Q = mc \, \Delta T$ only for those intervals in which no phase changes occur. Likewise, the equations $Q = mL_f$ and $Q = mL_v$ are to be used only when phase changes *are* taking place.
3. Often sign errors occur in heat loss = heat gain equations. One way to determine whether your equation is correct is to examine the signs of all ΔT's that appear in your equation. Every one should be a positive number.

EXAMPLE 11.4 Cooling the Steam

What mass of steam that is initially at 130°C is needed to warm 200 g of water in a 100-g glass container from 20.0° to 50.0°C?

Solution This is a heat transfer problem in which we must equate the heat lost by the steam to the heat gained by the water and glass container. The steam loses heat in three stages. In the first, the steam is cooled to 100°C. The heat liberated in the process is

$$Q_1 = m_x c_s \, \Delta T = m_x (2.01 \times 10^3 \, \text{J/kg} \cdot °\text{C})(30.0°\text{C}) = m_x (6.03 \times 10^4 \, \text{J/kg})$$

In the second stage, the steam is converted to water. In this case, to find the heat removed, we use the heat of vaporization and $Q = mL_v$:

$$Q_2 = m_x (2.26 \times 10^6 \, \text{J/kg})$$

In the last stage, the temperature of the water is reduced to 50.0°C. This liberates heat in the amount of

$$Q_3 = m_x c_w \, \Delta T = m_x (4.19 \times 10^3 \, \text{J/kg} \cdot °\text{C})(50.0°\text{C}) = m_x (2.09 \times 10^5 \, \text{J/kg})$$

If we equate the heat lost by the steam to the heat gained by the water and glass, and use the given information, we find that

$$m_x (6.03 \times 10^4 \, \text{J/kg}) + m_x (2.26 \times 10^6 \, \text{J/kg}) + m_x (2.09 \times 10^5 \, \text{J/kg}) =$$

$$(0.200 \, \text{kg})(4.19 \times 10^3 \, \text{J/kg} \cdot °\text{C})(30.0°\text{C}) + (0.100 \, \text{kg})(837 \, \text{J/kg} \cdot °\text{C})(30.0°\text{C})$$

$$m_x = 1.09 \times 10^{-2} \, \text{kg} = \boxed{10.9 \, \text{g}}$$

EXAMPLE 11.5 Boiling Liquid Helium

Liquid helium has a very low boiling point, 4.2 K, and a very low heat of vaporization, $2.09 \times 10^4 \, \text{J/kg}$ (Table 11.2). A constant power of 10.0 W (1 W = 1 J/s) is transferred to a container of liquid helium from an immersed electric heater. At this rate, how long does it take to boil away 1.00 kg of liquid helium?

Solution Since $L_v = 2.09 \times 10^4 \, \text{J/kg}$, we must supply $2.09 \times 10^4 \, \text{J}$ of energy to boil away 1.00 kg. The power supplied to the helium is 10 W = 10 J/s. That is, in 1.00 s, 10.0 J of energy is transferred to the helium. Therefore, the time it takes to transfer $2.09 \times 10^4 \, \text{J}$ is

$$t = \frac{2.09 \times 10^4 \, \text{J}}{10.0 \, \text{J/s}} = 2.09 \times 10^3 \, \text{s} \approx \boxed{35 \, \text{min}}$$

In contrast, 1.00 kg of liquid nitrogen ($L_v = 2.01 \times 10^5 \, \text{J/kg}$) would boil away in about 5.6 h with the same power input.

Exercise If 10.0 W of power is supplied to 1.00 kg of water at 100°C, how long will it take for the water to completely boil away?

Answer 62.8 h.

11.5

HEAT TRANSFER BY CONDUCTION

There are three ways in which heat energy can be transferred from one location to another: conduction, convection, and radiation. Regardless of the process, however, no net heat transfer takes place between a system and its surroundings

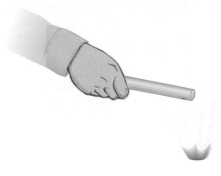

FIGURE 11.4
Heat reaches the hand by conduction through the copper rod.

when the two are at the same temperature. In this section we discuss heat transfer by conduction. Convection and radiation will be discussed in Sections 11.6 and 11.7.

Each of the methods of heat transfer can be examined by considering the ways in which you can warm your hands over an open fire. If you insert a copper rod into the flame, as in Figure 11.4, the temperature of the metal in your hand increases rapidly. **Conduction,** the process by which heat is transferred from the flame through the copper rod to your hand, can be understood by examining what is happening to the atoms of the metal. Initially, before the rod is inserted into the flame, the copper atoms are vibrating about their equilibrium positions. As the flame heats the rod, the copper atoms near the flame begin to vibrate with greater and greater amplitudes. These wildly vibrating atoms collide with their neighbors and transfer some of their energy in the collisions. Gradually, copper atoms progressively farther up the rod increase their amplitudes of vibration until those at the held end are reached. This increased vibration results in an increase in temperature of the metal, and possibly a burned hand.

Although the transfer of heat through a metal can be partially explained by atomic collisions, the rate of heat conduction also depends on the properties of the substance being heated. For example, it is possible to hold a piece of asbestos in a flame indefinitely. This implies that very little heat is being conducted through the asbestos. In general, metals are good conductors of heat, and materials such as asbestos, cork, paper, and fiber glass are poor conductors; gases also are poor heat conductors because of their dilute nature. Metals are good conductors of heat because they contain large numbers of electrons that are relatively free to move through the metal and transport energy from one region to another. Thus, in a good conductor, such as copper, heat conduction takes place both via the vibration of atoms and via the motions of free electrons.

In this section we consider the rate at which heat is transferred from one location to another. If Q is the amount of heat transferred from one location on an object to another in the time Δt, the **heat transfer rate,** H (sometimes called the heat current), is defined as

$$H \equiv \frac{Q}{\Delta t} \qquad\qquad [11.6]$$

Note that H is expressed in watts when Q is in joules and Δt is in seconds (1 W = 1 J/s).

The ceramic material being heated by a flame is a poor conductor of heat but is able to withstand a large temperature gradient; its left side is buried in ice (0°C) while its right side is extremely hot. *(Courtesy of Corning Glass Works)*

The conduction of heat occurs only if a difference in temperature exists between two parts of the conducting medium. Consider a slab of thickness L and cross-sectional area A, as in Figure 11.5. Suppose that one face is maintained at a temperature of T_2 and the other face is held at a lower temperature, T_1. Experimentally, one finds that the rate of heat flow—that is, the heat flow, Q, per unit time, Δt—is proportional to the difference in temperature $T_2 - T_1$ and the area A, and inversely proportional to the thickness of the slab. Specifically, the rate of flow of heat (or heat current) is given by

$$H = \frac{Q}{\Delta t} = kA\left(\frac{T_2 - T_1}{L}\right)$$ [11.7]

where k is a constant called the **thermal conductivity** of the material. This constant is a property of the material. Table 11.3 lists some values of k for metals, gases, and nonmetals. Note that k is large for metals, which are good *heat conductors,* and small for gases and nonmetals, which are poor heat conductors (good *insulators*).

The fact that different materials have different k values should help you understand the following phenomenon, first mentioned in Chapter 10. If you remove a metal ice tray and a package of frozen food from the freezer, which feels colder? Experience tells you that the metal tray feels colder even though it is at the same initial temperature as the cardboard package. This is explained by the fact that metal has a much higher thermal conductivity than cardboard, and so it conducts heat more rapidly and hence removes heat from your hand at a higher rate. Consequently, the metal tray *feels* colder than the carton even though it isn't. By use of a similar argument, you should be able to explain why a tile floor feels colder to bare feet than a carpeted floor does.

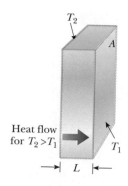

FIGURE 11.5
Heat transfer through a conducting slab of cross-sectional area A and thickness L. The opposite faces are at different temperatures, T_1 and T_2.

TABLE 11.3
Thermal Conductivities

Substance	Thermal Conductivity $(J/s \cdot m \cdot °C)$
Metals (at 25°C)	
Aluminum	238
Copper	397
Gold	314
Iron	79.5
Lead	34.7
Silver	427
Gases (at 20°C)	
Air	0.0234
Helium	0.138
Hydrogen	0.172
Nitrogen	0.0234
Oxygen	0.0238
Nonmetals	
Asbestos	0.25
Concrete	0.80
Glass	0.84
Ice	1.6
Rubber	0.2
Water	0.60
Wood	0.10

EXAMPLE 11.6   Heat Transfer Through a Concrete Wall

Find the amount of heat transferred in 1.00 h by conduction through a concrete wall 2.0 m high, 3.65 m long, and 0.20 m thick if one side of the wall is held at 20°C and the other side is at 5°C.

Solution   Equation 11.7 gives the rate of heat transfer in joules per second. To find the amount of heat transferred in 1.00 h, we rewrite the equation as

$$Q = kA\,\Delta t\left(\frac{T_2 - T_1}{L}\right)$$

If we substitute the values given and consult Table 11.3, we find that

$$Q = (0.80\ \mathrm{J/s \cdot m \cdot °C})(7.3\ \mathrm{m^2})(3600\ \mathrm{s})\left(\frac{15°C}{0.20\ \mathrm{m}}\right) = \boxed{1.6 \times 10^6\ \mathrm{J}}$$

Early houses were insulated by the material of which their walls were constructed—thick masonry blocks. Masonry restricts heat loss by conduction because its k is relatively low. The large thickness, L, also decreases heat loss, as shown by Equation 11.7.

*HOME INSULATION

If you would like to do some calculating to determine whether or not to add insulation to a ceiling or to some other portion of a building, you need to slightly modify what you have just learned about conduction, for two reasons.

1. The insulating properties of materials used in buildings are usually expressed in engineering rather than SI units. For example, measurements stamped on a package of fiber glass insulating board will be in units such as British thermal units, feet, and degrees Fahrenheit.
2. In dealing with the insulation of a building, we must consider heat conduction through a compound slab, with each portion of the slab having a different thickness and a different thermal conductivity. For example, a typical wall in a house consists of an array of materials, such as wood paneling, dry wall, insulation, sheathing, and wood siding.

It is found that the rate of heat transfer through a compound slab is

$$\frac{Q}{\Delta t} = \frac{A(T_2 - T_1)}{\sum_i L_i / k_i}$$

[11.8]

where T_1 and T_2 are the temperatures of the *outer extremities* of the slab and the summation is over all portions of the slab. For example, if the slab consists of three different materials, the denominator is the sum of three terms. In engineering practice, the term L/k for a particular substance is referred to as the *R* **value** of the material. Thus, Equation 11.8 reduces to

$$\frac{Q}{\Delta t} = \frac{A(T_2 - T_1)}{\sum_i R_i}$$

[11.9]

The *R* values for a few common building materials are listed in Table 11.4 (note the units).

This pattern of melted snow on a parking lot indicates the presence of underground steam pipes installed to aid snow removal. Heat from the steam is conducted from the pipes to the pavement, causing the snow to melt. *(Courtesy of Dr. Albert A. Bartlett, University of Colorado, Boulder)*

TABLE 11.4
***R* Values for Some Common Building Materials**

Material	*R* value (ft²·°F·h/Btu)
Hardwood siding (1.0 in. thick)	0.91
Wood shingles (lapped)	0.87
Brick (4.0 in. thick)	4.00
Concrete block (filled cores)	1.93
Styrofoam (1.0 in. thick)	5.0
Fiber glass batting (3.5 in. thick)	10.90
Fiber glass batting (6.0 in. thick)	18.80
Fiber glass board (1.0 in. thick)	4.35
Cellulose fiber (1.0 in. thick)	3.70
Flat glass (0.125 in. thick)	0.89
Insulating glass (0.25-in. space)	1.54
Vertical air space (3.5 in. thick)	1.01
Air film	0.17
Dry wall (0.50 in. thick)	0.45
Sheathing (0.50 in. thick)	1.32

Next to any vertical outside surface is a very thin, stagnant layer of air that must be considered when the total R value for a wall is figured. The thickness of this stagnant layer depends on the speed of the wind. As a result, heat loss from a house on a day when the wind is blowing hard is greater than heat loss on a day when the wind speed is zero. A representative R value for the stagnant layer of air is given in Table 11.4.

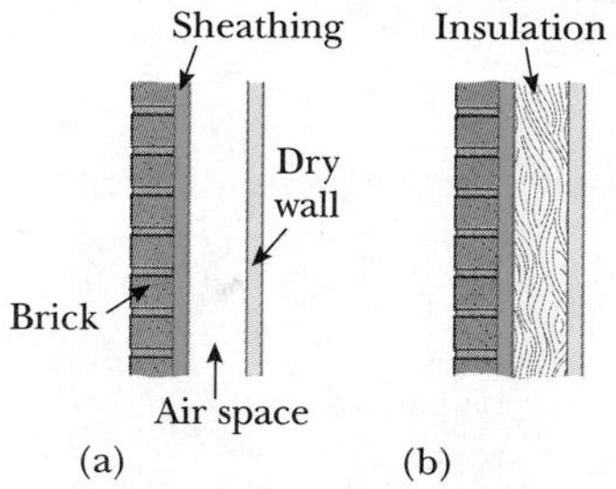

EXAMPLE 11.7   The R Value of a Typical Wall

Calculate the total R value for a wall constructed as shown in Figure 11.6a. Starting outside the house (to the left in the figure) and moving inward, the wall consists of brick, 0.50 in. of sheathing, a vertical air space 3.5 in. thick, and 0.50 in. of dry wall. Do not forget the dead-air layers inside and outside the house.

Solution   Referring to Table 11.4, we find the total R value for the wall as follows:

R_1 (outside air film)	=	$0.17 \ \mathrm{ft^2 \cdot {}^\circ F \cdot h/Btu}$
R_2 (brick)	=	4.00
R_3 (sheathing)	=	1.32
R_4 (air space)	=	1.01
R_5 (dry wall)	=	0.45
R_6 (inside air film)	=	0.17
R_total	=	$7.12 \ \mathrm{ft^2 \cdot {}^\circ F \cdot h/Btu}$

Exercise   If a layer of fiber glass insulation 3.5 in. thick is placed inside the wall to replace the air space, as in Figure 11.6b, what is the total R value of the wall? By what factor is the heat loss reduced?

Answer $R = 17 \ \mathrm{ft^2 \cdot {}^\circ F \cdot h/Btu}$; a factor of 2.5.

FIGURE 11.6
(Example 11.7)   A cross-sectional view of an exterior wall containing (a) an air space and (b) insulation.

11.6
CONVECTION

You have probably warmed your hands by holding them over an open flame as illustrated in Figure 11.7. In this situation, the air directly above the flame is heated and expands. As a result, the density of the air decreases and the air rises. This warmed mass of air heats your hands as it flows by. *Heat transferred by the movement of a heated substance is said to have been transferred by convection.* When the movement results from differences in density, as it does in air around a fire, it is referred to as *natural convection.* When the heated substance is forced to move by a fan or pump, as in some hot-air and hot-water heating systems, the process is called *forced convection.*

The circulating pattern of air flow at a beach (see Fig. 11.2) is an example of convection. Likewise, the mixing that occurs as water is cooled and eventually freezes at its surface (Chapter 10) is an example of convection in nature. Recall that mixing by convection currents ceases when the water temperature reaches 4°C. Because the water in the pool cannot be cooled by convection below 4°C,

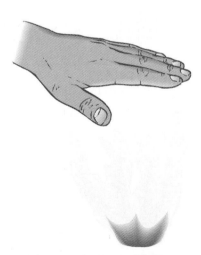

FIGURE 11.7
Heating a hand by convection.

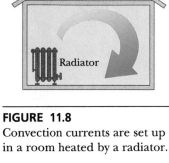

FIGURE 11.8
Convection currents are set up in a room heated by a radiator.

and because water is a relatively poor conductor of heat (see Table 11.3), the water near the bottom remains near 4°C for a long time. As a result, fish live in water of a comfortable temperature even in periods of prolonged cold weather.

If it were not for convection currents, it would be very difficult to boil water. The lower layers of water in a teakettle are warmed first. These heated regions expand and rise to the top because their density is lowered. Meanwhile, denser cool water replaces the warm water at the bottom of the kettle so that it can be heated.

The same process occurs when a room is heated by a radiator. The hot radiator warms the air in the lower regions of the room. The warm air expands and rises to the ceiling because of its lower density. The denser regions of cooler air from above replace the warm air, setting up the continuous air current pattern shown in Figure 11.8.

11.7

RADIATION

The third way of transferring heat is through **radiation.** You have most likely experienced radiant heat when sitting in front of a fireplace. Figure 11.9 shows how you can warm your hands at an open flame by means of radiant heat. The hands are placed to one side of the flame; they are not in physical contact with the flame either directly or indirectly, and therefore conduction cannot account for the heat transfer. Furthermore, convection is not important in this situation since the hands are not above the flame in the path of convection currents. The important process in this case is the radiation of heat energy.

All objects continuously radiate energy in the form of electromagnetic waves, which we shall discuss in Chapter 21. Electromagnetic radiation associated with the transfer of heat energy from one location to another is referred to as *infrared* radiation.

Through electromagnetic radiation, approximately 1340 J of heat energy strikes 1 m² of the top of the Earth's atmosphere every second. Some of this energy is reflected back into space, and some is absorbed by the atmosphere, but enough arrives at the surface of the Earth each day to supply hundreds of times more energy than human technology requires—if it could be captured and used efficiently. The building of an increasing number of solar houses in this country is one attempt to make use of this free energy.

Radiant energy from the Sun affects our day-to-day existence in a number of ways. For example, consider what happens to the atmospheric temperature at night. If there is a cloud cover above the Earth, the water vapor in the clouds reflects back a part of the infrared radiation emitted by the Earth, and consequently the temperature remains moderate. In the absence of a cloud cover, however, there is nothing to prevent this radiation from escaping into space, and thus the temperature drops more on a clear night than when it is cloudy.

The rate at which an object emits radiant energy is proportional to the fourth power of its absolute temperature. This is known as **Stefan's law** and is expressed in equation form as

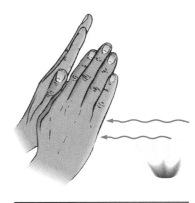

FIGURE 11.9
Warming hands by radiation.

Stefan's law

$$P = \sigma A e T^4 \qquad\qquad [11.10]$$

where P is the power radiated by the object in watts (or joules per second), σ is a constant equal to 5.6696×10^{-8} W/m$^2 \cdot$K^4, A is the surface area of the object in square meters, e is a constant called the **emissivity,** and T is the object's temperature in kelvins. The value of e can vary between zero and unity, depending on the properties of the surface.

An object radiates energy at a rate given by Equation 11.10. At the same time, the object also absorbs electromagnetic radiation. If the latter process did not occur, the object would eventually radiate all of its energy and its temperature would reach absolute zero. The absorbed energy comes from the body's surroundings, which consist of other objects that radiate energy. If an object is at temperature T and its surroundings are at temperature T_0, the net energy gained or lost each second by the object as a result of radiation is

$$P_{\text{net}} = \sigma Ae(T^4 - T_0^4)$$ [11.11]

When an object is in *equilibrium* with its surroundings, *it radiates and absorbs energy at the same rate, and so its temperature remains constant.* When an object is hotter than its surroundings, it radiates more energy than it absorbs, and so it cools. An *ideal absorber* is defined as an object that absorbs all of the energy incident on it; its emissivity is equal to unity. Such an object is often referred to as a **black body.** An ideal absorber is also an ideal radiator of energy. In contrast, an object with an emissivity equal to zero absorbs none of the energy incident on it. Such an object reflects all the incident energy and so is a *perfect reflector.*

White clothing is more comfortable to wear in the heat of the summer than black clothing. Black fabric acts as a good absorber of incoming sunlight and as a good emitter of this absorbed energy. However, about half of the emitted energy travels toward the body, causing the person wearing the garment to feel uncomfortably warm. In contrast, white or light-colored clothing reflects away much of the incoming energy.

The amount of radiant energy emitted by a body can be measured with heat-sensitive recording equipment, using a technique called **thermography.** The radiation emitted by a body is greatest in the body's warmest regions. An image of the pattern formed by varying radiation levels, called a **thermogram,** is brightest in the warmest areas. The photograph below reproduces a thermogram of a house. The center portions of the door and windows are red, signifying

This thermogram of a home, made during cold weather, shows colors ranging from white and yellow (areas of greatest heat loss) to blue and purple (areas of least heat loss). *(Daedalus Enterprises, Inc./Peter Arnold, Inc.)*

temperatures higher than those of surrounding areas. A higher temperature usually means that heat is escaping. Thermograms can be useful for purposes of energy conservation. For example, the owners of this house could conserve energy and reduce their heating costs by adding insulation to the attic area and by installing thermal draperies over the windows.

EXAMPLE 11.8   Who Turned Down the Thermostat?

A student is trying to decide what to wear. The air in his bedroom is at 20.0°C. If the skin temperature of the unclothed student is 37.0°C, how much heat is lost from his body in 10.0 min? Assume that the emissivity of skin is 0.900 and that the surface area of the student is 1.50 m².

Solution   Using Equation 11.11, the rate of heat loss from the skin is

$$P_{net} = \sigma Ae(T^4 - T_0{}^4)$$

$$= (5.67 \times 10^{-8}\ \text{W/m}^2 \cdot \text{K}^4)(1.50\ \text{m}^2)(0.90)[(310\ \text{K})^4 - (293\ \text{K})^4] = 143\ \text{J/s}$$

Note that it was necessary to change the temperature to kelvins. At this rate of heat loss, the total heat lost by the skin in 10.0 min is

$$Q = P_{net} \times \text{time} = (143\ \text{J/s})(600\ \text{s}) = \boxed{8.58 \times 10^4\ \text{J}}$$

---

*11.8

HINDERING HEAT TRANSFER

The Thermos bottle, called a *Dewar flask* (after its inventor) in scientific applications, is designed to minimize heat transfer by conduction, convection, and radiation. It is used to store either cold or hot liquids for long periods of time. The standard vessel (Fig. 11.10) is double-walled Pyrex glass with a silvered inner wall. The space between the walls is evacuated to minimize heat transfer by conduction and convection. By reflecting most of the radiant heat, the silvered surface minimizes heat transfer by radiation. Very little heat is lost through the neck of the flask because Pyrex glass is a poor conductor. A further reduction in heat loss is achieved by reducing the size of the neck. A common scientific application of Dewar flasks is storage of liquid nitrogen (boiling point 77 K) and liquid oxygen (boiling point 90 K). For substances that have very low specific heats, such as liquid helium (boiling point 4.2 K), it is often necessary to use a double Dewar system in which the Dewar flask containing the liquid is surrounded by a second Dewar flask. The space between the two flasks is filled with liquid nitrogen.

Some of the principles of the Thermos bottle are used in the protection of sensitive electronic instruments in orbiting space satellites. In half of its orbit about the Earth, a satellite is exposed to intense radiation from the Sun, and in the other half it is in the Earth's cold shadow. Without protection, its interior would thus be subjected to tremendous extremes of heating and cooling. The interior of the satellite is wrapped with blankets of highly reflective aluminum foil. The foil's shiny surface reflects away much of the Sun's radiation while the

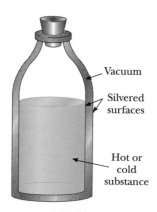

Vacuum

Silvered surfaces

Hot or cold substance

FIGURE 11.10
A cross-sectional view of a Dewar vessel designed to store hot or cold liquids.

satellite is in the unshaded part of the orbit, and helps retain interior heat while the satellite is in the Earth's shadow.

Wool sweaters and down jackets keep us warm by trapping the warmer air in regions close to our bodies and hence reducing heat loss by convection and conduction. In other words, what keeps us warm is not the clothing itself but the air trapped in the clothing.

S P E C I A L   T O P I C
THE GREENHOUSE EFFECT

Many of the principles of heat transfer, and its prevention, can be understood by studying the operation of a glass greenhouse. Glass allows visible light to pass through, but not infrared radiation. During the day, sunlight passes into the greenhouse and is absorbed by the walls, earth, plants, and so on. This absorbed visible light is subsequently re-radiated as infrared radiation, which cannot escape the enclosure. The increasing amount of trapped infrared radiation causes the temperature of the interior to rise.

In addition, convection currents are inhibited in a greenhouse. As a result, heated air cannot rapidly pass over the surfaces of the greenhouse that are exposed to the outside air and thereby cause a heat loss through those surfaces. Many experts consider this to be an even more important effect than the effect of trapped infrared radiation. In fact, experiments have shown that when the glass over a greenhouse is replaced by a special glass that transmits infrared light, the temperature inside is lowered only slightly. Based on this evidence, the primary mechanism that heats a greenhouse is not the absorption of infrared radiation but the inhibition of air flow that occurs under any roof (in an attic, for example).

A phenomenon known as the *greenhouse effect* can also play a major role in determining the Earth's temperature. First note that the Earth's atmosphere is a good transmitter (and hence a poor absorber) of visible radiation and a good absorber of infrared radiation. Carbon dioxide (CO_2) in the Earth's atmosphere acts somewhat like the glass in a greenhouse in that it allows incoming visible radiation from the Sun to pass through more easily than infrared radiation. The visible light that reaches the Earth's surface is absorbed and re-radiated as infrared light, which in turn is absorbed (trapped) by the Earth's atmosphere. An extreme case is the warmest planet, Venus, which has a carbon dioxide–rich atmosphere and temperatures approaching 850°F.

As fossil fuels (coal, oil, and natural gas) are burned, large amounts of carbon dioxide are released into the atmosphere, causing it to retain more heat. This is of great concern to scientists and governments throughout the world. Many scientists are convinced that the 10% increase in the amount of atmospheric carbon dioxide in the last 30 years could lead to drastic changes in world climate. According to one estimate, doubling the carbon dioxide content in the atmosphere will cause temperatures to increase by 2°C! In temperate regions, such as Europe and the United States, the temperature rise would save billions of dollars per year in fuel costs. Unfortunately, the temperature increase would also melt polar ice caps, which could cause flooding and destroy many coastal areas; increase the frequency of droughts; and consequently decrease already low crop yields in tropical and subtropical countries. Even slightly higher average temperatures might make it impossible for certain plants and animals to survive in their customary ranges.

Obviously, the problem is very complex, and the models that have been offered are open to question and further study. Nonetheless, it is an important problem that all nations must address.

SUMMARY

Heat flow is a form of energy transfer that takes place as a consequence of a temperature difference between a system and its surroundings. The **internal energy** of a substance is a function of the state of the substance, and generally increases with increasing temperature.

The **calorie** is the amount of heat necessary to raise the temperature of 1 g of water from 14.5°C to 15.5°C. The **mechanical equivalent of heat** is 4.186 J/cal.

The heat required to change the temperature of a substance by ΔT is

Heat required to raise the temperature of a substance

$$Q = mc\,\Delta T \qquad\qquad [11.3]$$

where m is the mass of the substance and c is its **specific heat.**

The heat required to change the phase of a mass m of a pure substance is

Latent heat

$$Q = mL \qquad\qquad [11.5]$$

The parameter L is called the **latent heat** of the substance and depends on the nature of the phase change and the properties of the substance.

Heat may be transferred by three fundamentally distinct processes: *conduction*, *convection*, and *radiation*. **Conduction** can be viewed as an exchange of kinetic energy between colliding molecules or through the motion of electrons. The rate at which heat flows by conduction through a slab of area A and thickness L is

$$H = \frac{Q}{\Delta t} = kA\left(\frac{T_2 - T_1}{L}\right) \qquad\qquad [11.7]$$

where k is the **thermal conductivity** of the material making up the slab. Heat transferred by the movement of a heated substance is said to have been transferred by **convection.**

All objects **radiate** and absorb energy in the form of electromagnetic waves. An object that is hotter than its surroundings radiates more energy than it absorbs, whereas an object that is cooler than its surroundings absorbs more energy than it radiates. The rate at which an object emits radiant energy is given by **Stefan's law:**

$$P = \sigma A e T^4 \qquad\qquad [11.10]$$

where σ is a constant equal to 5.6696×10^{-8} W/m$^2\cdot$K^4, and e is a constant called the **emissivity.**

ADDITIONAL READING

P. B. Allen, "Conduction of Heat," *The Physics Teacher,* December 1983, p. 582.

E. Barr, "James Prescott Joule and the Quiet Revolution," *The Physics Teacher,* April 1969, p. 199.

F. A. Bazzaz and E. D. Fajer, "Plant Life in a CO_2-Rich World," *Sci. American,* January 1992, p. 68.

B. Chalmers, "How Water Freezes," *Sci. American,* February 1959, p. 144.

J. Dyson, "What Is Heat?" *Sci. American,* September 1954, p. 58.

J. Kelley, "Heat, Cold and Clothing," *Sci. American,* February 1956, p. 194.

M. G. Velarde, "Convection," *Sci. American,* January 1980, p. 92.

M. Wilson, "Count Rumford," *Sci. American,* October 1960, p. 158.

CONCEPTUAL QUESTIONS

Example Concrete has a higher specific heat than does soil. Use this fact to explain (partially) why a city has a higher average temperature than the surrounding countryside. Would you expect breezes to blow from city to country or from country to city? Explain.

Reasoning The large amount of heat stored in the concrete during the day as the sun falls on it is released at night resulting in an overall higher average temperature than the countryside. The heated air in a city rises to be replaced by cooler air drawn in from the countryside. Thus, breezes tend to blow from country to city.

Example A tile floor may feel uncomfortably cold to your bare feet, but a carpeted floor in an adjoining room at the same temperature feels warm. Why?

Reasoning The tile is a better conductor of heat than carpet. Thus, heat is conducted away from your feet more rapidly by the tile than by the carpeted floor.

Example If you hold a paper cup containing water over a flame, you can bring the water to a boil without burning the cup. How is this possible?

Reasoning Because of the small thickness of the bottom of the cup, heat is rapidly conducted through the paper. Convection currents in the water rapidly move this heat away from the surface of the paper.

Example The U.S. penny is now made of copper-coated zinc. Can an experiment be devised to test for the metal content of a collection of one-cent pieces? If so, describe the procedure you would use.

Reasoning A solid copper one-cent piece has a different heat capacity than one which is copper-coated zinc. By measuring the individual heat capacities in a calorimeter, and knowing their masses, you can determine the metal content of the various coins.

Example In a daring lecture demonstration, a professor dips her wetted fingers into molten lead (327°C) and withdraws them quickly without getting burned. How is this possible?

Reasoning The fingers are wetted to create a layer of steam between the fingers and the molten lead. The steam acts as an insulator and prevents serious burns.

1. Ethyl alcohol has about one-half the specific heat of water. If equal masses of alcohol and water in separate beakers are supplied with the same amount of heat, compare the temperature increases of the two liquids.

2. A small crucible is taken from a 200°C oven and immersed in a tub of water at room temperature (in a process often referred to as quenching). What is the approximate final equilibrium temperature?

3. What is wrong with the following statement? "Given any two bodies, the one with the higher temperature contains more thermal energy."

4. Until refrigerators were invented, many people stored fruits and vegetables in underground cellars. Discuss as fully as possible this choice for a storage site.

5. In winter, the people mentioned in Question 4 would set an open barrel of water alongside their produce. Why?

6. During a cold spell, Florida orange growers often spray a mist of water over their trees during the night. Why does this help?

7. Why is it possible to hold a lighted match even when it is burned to within a few millimeters of your fingertips?

8. A piece of paper is wrapped around a rod made half of wood and half of copper. When held over a flame, the paper in contact with the wood burns but the paper in contact with the metal does not. Explain.

9. Why does a piece of metal feel colder than a piece of wood when they are at the same temperature?

10. Why does a potato bake more quickly when a piece of metal has been inserted through it?

11. If your water pipes freeze during the winter, it is almost always the hot water pipes that freeze first. Why? It is often said that if you keep a slow trickle of water running from your pipes they will not freeze. Is this true?

12. A warning sign often seen on highways just before a bridge is "Caution—Bridge Surface Freezes Before Road Surface." Which of the three heat transfer processes is most important in causing a bridge surface to freeze before the road surface on very cold days?

13. Updrafts of air are familiar to all pilots. What causes them?

14. Thermopane windows are usually made with two layers of glass separated by an air space. One manufacturer claims that his panes are farther apart than those of his competitors, which gives them a better insulating factor. Give an argument that would refute his claim.

15. On a very hot day it is possible to cook an egg on the hood of a car. Would you select a black car or a white car on which to cook your egg? Why?

16. Two neighboring houses are identical in all respects save one: one house has a black roof; the other has a white roof. Which house is likely to be a little cooler in the summertime? Explain.

17. Suppose you pour hot coffee for your guests, and one of them chooses to drink the coffee after it has been in the cup for several minutes. Given this delay, in order to have the warmest coffee, should the person

add the (cool) cream just after the coffee is poured or just before drinking? Explain.

18. When insulating a wood-frame house, is it better to place the insulation against the cooler outside wall or against the warmer inside wall? (In either case, there is an air barrier to consider.)

19. When a sealed Thermos bottle full of hot coffee is shaken, what are the changes, if any, in (a) the temperature of the coffee and (b) its internal energy?

20. The air temperature above coastal areas is profoundly influenced by the high specific heat of water. One reason is that the thermal energy released during the cooling of 1 m³ of water by 1°C raises the temperature of an enormously greater volume of air by 1°C. Estimate this volume of air. (The specific heat of air is approximately 1.0 kJ/kg·°C. Take the density of air to be 1.3 kg/m³.)

21. In the winter you might notice that some roofs are uniformly covered with snow while others have areas where the snow has melted, as shown in the photograph. Which buildings are better insulated? Why are there regular spacings between the melted areas in the photograph?

(Question 21)   An alternating pattern of snow-covered and exposed roof. *(Courtesy of Dr. Albert A. Bartlett, University of Colorado, Boulder)*

PROBLEMS

Section 11.2   Specific Heat

1. A 50.0-g sample of copper is at 25°C. If 1200 J of heat energy is added to the copper, what is its final temperature?

2. How many joules of energy are required to raise the temperature of 100 g of gold from 20.0°C to 100°C?

3. A 50-g piece of cadmium is at 20°C. If 400 cal of heat is added to the cadmium, what is its final temperature?

4. A 5.0-g lead bullet traveling at 300 m/s is stopped by a large tree. If half the kinetic energy of the bullet is transformed into heat energy and remains with the bullet while the other half is transmitted to the tree, what is the increase in temperature of the bullet?

5. A 75.0-kg weight-watcher wishes to climb a mountain to work off the equivalent of a large piece of chocolate cake rated at 500 (food) Calories. How high must the person climb?

6. As a part of your exercise routine you climb a 10.0-m rope. How many food calories do you expend in a single climb up the rope?

7. When a driver brakes an automobile, the friction between the brake drums and brake shoes converts the car's kinetic energy to heat. If a 1500-kg automobile traveling at 30 m/s comes to a halt, how much does the temperature rise in each of the four 8.0-kg iron brake drums? (The specific heat of iron is 448 J/kg·°C.)

8. A 3.0-kg rock is initially at rest at the top of a cliff. If the energy equivalent of the rock is sufficient to raise the temperature of 1.0 kg of water by 0.10°C, how high is the cliff?

9. Water at the top of Niagara Falls has a temperature of 10.0°C. If it falls a distance of 50.0 m and all of its potential energy goes into heating the water, calculate the temperature of the water at the bottom of the falls.

10. A hot water heater is operated by solar power. If the solar collector has an area of 6.0 m², and the power delivered by sunlight is 550 W/m², how long will it take to increase the temperature of 1.0 m³ of water from 20°C to 60°C?

11. A 1.5-kg copper block is given an initial speed of 3.0 m/s on a rough horizontal surface. Because of friction, the block finally comes to rest. (a) If the block absorbs 85% of its initial kinetic energy in the form of heat, calculate its increase in temperature. (b) What happens to the remaining energy?

12. A 200-g aluminum cup contains 800 g of water in

□ indicates problems that have full solutions available in the Student Solutions Manual and Study Guide.

thermal equilibrium at 80°C. The combination of cup and water is cooled uniformly so that the temperature decreases by 1.5°C per minute. At what rate is heat energy being removed? Express your answer in watts.

Section 11.3   Conservation of Energy: Calorimetry

13. A 0.40-kg iron horseshoe that is initially at 500°C is dropped into a bucket containing 20 kg of water at 22°C. What is the final equilibrium temperature? Neglect any heat transfer to or from the surroundings.

(Problem 13)   A "red-hot" horseshoe being cooled in water. *(© Chris Sorensen 1993)*

14. A construction worker drops a hot 100-g iron rivet at 500°C into a bucket containing 500 g of mercury at 20°C. Assuming that no heat is lost to the surroundings or the bucket, what is the final temperature of the rivet and mercury?

15. What mass of water at 25°C must be allowed to come to thermal equilibrium with a 3.0-kg gold bar at 100°C in order to lower the temperature of the bar to 50°C?

16. A 200-g block of copper at a temperature of 90°C is dropped into 400 g of water at 27°C. The water is contained in a 300-g glass container. What is the final temperature of the mixture?

17. An aluminum cup contains 225 g of water at 27°C. A 400-g sample of silver at an initial temperature of 87°C is placed in the water. A 40-g copper stirrer is used to stir the mixture until it reaches its final equilibrium temperature of 32°C. Calculate the mass of the aluminum cup.

18. It is desired to cool iron parts from 500°F to 100°F by dropping them into water that is initially at 75°F. Assuming that all the heat from the iron is transferred to the water and that none of the water evaporates, how many kilograms of water are needed per kilogram of iron?

19. Lead pellets, each of mass 1.0 g, are heated to 200°C. How many pellets must be added to 500 g of water that is initially at 20°C to make the equilibrium temperature 25°C? Neglect any heat transfer to or from the container.

20. If 200 g of water is contained in a 300-g aluminum vessel at 10°C and an additional 100 g of water at 100°C is poured into the container, what is the final equilibrium temperature of the mixture?

21. A 100-g aluminum calorimeter contains 250 g of water. The two substances are in thermal equilibrium at 10°C. Two metallic blocks are placed in the water. One is a 50-g piece of copper at 80°C. The other sample has a mass of 70 g and is originally at a temperature of 100°C. The entire system stabilizes at a final temperature of 20°C. Determine the specific heat of the unknown second sample.

22. A student drops two metallic objects into a 120-g steel container holding 150 g of water at 25°C. One object is a 200-g cube of copper that is initially at 85°C, and the other is a chunk of aluminum that is initially at 5°C. To the surprise of the student, the water reaches a final temperature of 25°C, exactly where it started. What is the mass of the aluminum chunk?

Section 11.4   Latent Heat and Phase Changes

23. How much heat is required to change a 40-g ice cube from ice at −10°C to steam at 110°C?

24. A large block of ice at 0°C has a hole chipped in it, and 400 g of aluminum pellets at a temperature of 30°C are poured into the hole. How much of the ice melts?

25. A 50-g ice cube at 0°C is heated until 45 g has become water at 100°C and 5.0 g has become steam at 100°C. How much heat was added to accomplish this?

26. What mass of steam that is initially at 120°C is needed to warm 350 g of water in a 300-g aluminum container from 20°C to 50°C?

27. Assuming no heat loss, determine the mass of water that boils away when 1.94 kg of mercury at a temperature of 200°C is added to 0.050 kg of water at a temperature of 80.0°C.

28. A Styrofoam container used as a picnic cooler contains a block of ice at 0°C. If 225 g of ice melts in 1.0 h, how much heat energy per second passes through the walls of the container?

(Problem 34)   A cross-country skier. *(Nathan Bilow, Leo de Wys, Inc.)*

29. A 100-g ice cube at 0°C is placed in 650 g of water at 25°C. What is the final temperature of the mixture?

30. A 100-g cube of ice at 0°C is dropped into 1.0 kg of water that is originally at 80°C. What is the final temperature of the water after the ice has melted?

31. Steam at 100°C is added to ice at 0°C. (a) Find the amount of ice melted and the final temperature when the mass of steam is 10 g and the mass of ice is 50 g. (b) Repeat with steam of mass 1.0 g and ice of mass 50 g.

32. A 40-g block of ice is cooled to −78°C. It is added to 560 g of water in an 80-g copper calorimeter at a temperature of 25°C. Determine the final temperature. (If not all the ice melts, determine how much ice is left.) Remember that the ice must first warm to 0°C, melt, and then continue warming as water. The specific heat of ice is 0.500 cal/g·°C.

33. A beaker of water sits in the Sun until it reaches an equilibrium temperature of 30°C. The beaker is made of 100 g of aluminum and contains 180 g of water. In an attempt to cool this system down, 100 g of ice at 0°C is added to the water. (a) Determine the final temperature. If $T = 0$°C, determine how much ice remains. (b) Repeat this for 50 g of ice.

34. A 75-kg cross-country skier moves across snow such that the coefficient of friction between skis and snow is 0.20. Assume all the snow beneath his skis is at 0°C and that all the heat generated by friction is added to snow which sticks to his skis until melted. How far would he have to ski to melt 1.0 kg of snow?

Section 11.5   Heat Transfer by Conduction
Section 11.6   Convection
Section 11.7   Radiation

35. (a) Find the rate of heat flow through a copper block of cross-sectional area 15 cm² and length

8.0 cm when a temperature difference of 30°C is established across the block. Repeat the calculation assuming the material is (b) a block of air with these dimensions; (c) a block of wood with these dimensions.

36. A window has a glass surface of 1.6×10^3 cm² and a thickness of 3.0 mm. (a) Find the rate of heat transfer by conduction through this pane when the temperature of the inside surface of the glass is 70°F and the outside temperature is 90°F. (b) Repeat for the same inside temperature and an outside temperature of 0°F.

37. Determine the R value for a wall constructed as follows: The outside of the house consists of lapped wood shingles placed over 0.50-in.-thick sheathing, over 3.0 in. of cellulose fiber, over 0.50 in. of dry wall.

38. A glass window pane has an area of 3.0 m² and a thickness of 0.60 cm. If the temperature difference between its faces is 25°C, how much heat flows through the window per hour?

39. The average thermal conductivity of the walls (including windows) and roof of the house in Figure 11.11 is 4.8×10^{-4} kW/m·°C, and their average thickness is 21.0 cm. The house is heated with natural gas, with a heat of combustion (heat given off per cubic meter of gas burned) of 9300 kcal/m³. How many cubic meters of gas must be burned each day to maintain an inside temperature of 25.0°C if the outside temperature is 0.0°C? Disregard radiation and heat loss through the ground.

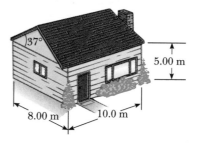

FIGURE 11.11   (Problem 39)

40. An iron rod has a length of 60 cm and a cross-sectional area of 2.0 cm². One end is maintained at 80°C and the other end at 20°C. At steady state, find (a) the temperature gradient (i.e., the change in temperature per unit length along the rod), (b) the rate of heat transfer through the rod, and (c) the temperature in the rod 20 cm from the hot end.

41. A Thermopane window consists of two glass panes, each 0.50 cm thick, with a 1.0-cm-thick sealed layer of air between. If the inside temperature is 23.0°C and the outside temperature is 0.0°C, determine the rate of heat transfer through

1.0 m² of the window. Compare this with the rate of heat transfer through 1.0 m² of a single 1.0-cm-thick pane of glass.

42. A steam pipe is covered with 1.5-cm-thick insulating material of heat conductivity 0.200 cal/cm · °C · s. How much heat is lost every second when the steam is at 200°C and the surrounding air is at 20°C? The pipe has a circumference of 20 cm and a length of 50 m. Neglect losses through the ends of the pipe.

43. Two identical objects are in the same surroundings at 0°C. One is at a temperature of 1200 K, and the other is at 1100 K. Find the ratio of the net power emitted by the hotter object to the net power emitted by the cooler object.

44. A sphere that is to be considered as a perfect black-body radiator has a radius of 0.060 m and is at 200°C in a room where the temperature is 22°C. Calculate the net rate at which the sphere radiates energy.

45. A Styrofoam box has a surface area of 0.80 m² and a wall thickness of 2.0 cm. The temperature inside is 5°C, and that outside is 25°C. If it takes 8.0 h for 5.0 kg of ice to melt in the container, determine the thermal conductivity of the Styrofoam.

46. Calculate the temperature at which a tungsten filament that has an emissivity of 0.25 and a surface area of 2.5×10^{-5} m² will radiate energy at the rate of 25 W in a room where the temperature is 22°C.

47. A copper rod and an aluminum rod of equal diameter are joined end to end in good thermal contact. The temperature of the free end of the copper rod is held constant at 100°C, and that of the far end of the aluminum rod is held at 0°C. If the copper rod is 0.15 m long, what must be the length of the aluminum rod so that the temperature at the junction is 50°C?

ADDITIONAL PROBLEMS

48. In a showdown on the streets of Laredo, the good guy drops a 5.0-g silver bullet, at a temperature of 20°C, into a 100-cm³ cup of water at 90°C. Simultaneously, the bad guy drops a 5.0-g copper bullet, at the same initial temperature, into an identical cup of water. Which one ends the showdown with the coolest cup of water in the west? Neglect any heat transfer into or away from the container.

49. A 1.0-m² solar collector collects radiation from the Sun and focuses it on 250 g of water that is initially at 23°C. The average thermal energy arriving from the Sun at the surface of the Earth at this location is 550 W/m², and we assume that this is collected with 100% efficiency. Find the time re-

quired for the collector to raise the temperature of the water to 100°C.

50. A passive solar home uses a wall of concrete to collect solar energy. The wall has an absorbing area of 20.0 m² and a mass of 1.0×10^4 kg. During a "solar day" the wall is illuminated for 6.0 h, with an average power per unit area of 400 W/m². The wall stores 30% of the energy. (a) If concrete has a specific heat of 920 J/kg · °C, what is the heat capacity of the wall? (Heat capacity is defined as the mass of an object times its specific heat.) (b) How much energy is stored by the wall during a "solar day"? (c) If the initial temperature of the wall is 15°C, what is its temperature at the end of the "solar day"?

51. A solar collector has an effective collecting area of 12 m². The collector is thermally insulated, and so conduction is negligible in comparison with radiation. On a cold but sunny winter's day the temperature outside is −20.0°C, and the Sun irradiates the collector with a power per unit area of 300 W/m². Treating the collector as a black body (i.e., emissivity = 1.0), determine its interior temperature after the collector has achieved a steady-state condition (radiating energy as fast as it is received).

52. The bottom of a copper kettle has a 10.0-cm radius and is 2.0 mm thick. The temperature of the outside surface is 102°C, and the water inside the kettle is boiling at 1 atm of pressure. Find the rate at which heat is being transferred through the bottom of the kettle.

53. A brass statue (60.0% copper, 40.0 zinc) has a mass of 50.0 kg. If its temperature increases by 20.0°C, what is the change of internal energy of the statue? (Specific heat of brass = 380 J/kg · °C = 0.092 cal/g · °C.)

54. A 40-g ice cube floats in 200 g of water in a 100-g copper cup; all are at a temperature of 0°C. A piece of lead at 98°C is dropped into the cup, and the final equilibrium temperature is 12°C. What is the mass of the lead?

55. A 10-g ice cube is in a 200-g copper cup. Initially the ice and cup are at a temperature of −20°C. (a) What happens to the mixture when 500 cal of heat is added? (b) Repeat for 5000 cal.

56. At time $t = 0$, a vessel contains a mixture of 10 kg of water and an unknown mass of ice in equilibrium at 0°C. The temperature of the mixture is measured versus time, with the following results. During the first 50 min, the mixture remains at 0°C. From 50 min to 60 min, the temperature increases to 2°C. Neglecting the heat capacity of the vessel, determine the mass of ice that was initially placed in the vessel. Assume a constant power input to the container.

57. A class of 10 students taking an exam has a power output per student of about 200 W. Assume that the initial temperature of the room is 20°C and that its dimensions are 6.0 m by 15.0 m by 3.0 m. What is the temperature of the room at the end of 1.0 h if all the heat remains in the air in the room and none is added by an outside source? The specific heat of air is 837 J/kg·°C, and its density is about 1.3×10^{-3} g/cm³.

58. A 60-kg runner dissipates 300 W of power while running a marathon. Assuming that 10% of the runner's energy is dissipated in the muscle tissue and that the excess heat is removed from the body primarily by sweating, determine the volume of bodily fluid (assume it is water) lost per hour. (At 37°C the latent heat of vaporization of water is 575 kcal/kg.)

59. An iron plate is held against an iron wheel so that a sliding frictional force of 50 N acts between the two pieces of metal. The relative speed at which the two surfaces slide over each other is 40 m/s. (a) Calculate the rate at which mechanical energy is converted to heat. (b) The plate and the wheel have masses of 5.0 kg each, and each receives 50% of the frictional heat. If the system is run as described for 10 s and each object is then allowed to reach a uniform internal temperature, what is the resultant temperature increase?

60. An automobile has a mass of 1500 kg, and its aluminum brakes have an overall mass of 60 kg. (a) Assuming that all of the frictional heat produced when the car stops is deposited in the brakes, and neglecting heat transfer, how many times could the car be braked to rest starting from 25 m/s (56 mph) before the brakes would begin to melt? (Assume an initial temperature of 20°C.) (b) Identify some effects that are neglected in part (a) but are likely to be important in a more realistic assessment of the heating of brakes.

61. A 1.0-m-long aluminum rod of cross-sectional area 2.0 cm² is inserted vertically into a thermally insulated vessel containing liquid helium at 4.2 K. The rod is initially at 300 K. If half of the rod is inserted into the helium, how many liters of helium boil off by the time the inserted half cools to 4.2 K?

62. A "solar cooker" consists of a curved reflecting mirror that focuses sunlight onto the object to be heated (Fig. 11.12). The solar power per unit area reaching the Earth at the location of a 0.50-m-diameter solar cooker is 600 W/m². Assuming that 50% of the incident energy is converted to heat energy, how long would it take to evaporate 1.0 L of water initially at 20°C? (Neglect the specific heat of the container.)

63. An aluminum rod is 20.000 cm long at 20°C and has a mass of 350 g. If 10 000 J of heat energy is

FIGURE 11.12   (Problem 62)

added to the rod, what is its new length?

64. Three liquids are at temperatures of 10°C, 20°C, and 30°C, respectively. Equal masses of the first two liquids are mixed, and the equilibrium temperature is 17°C. Equal masses of the second and third are then mixed, and the equilibrium temperature is 28°C. Find the equilibrium temperature when equal masses of the first and third are mixed.

65. A *flow calorimeter* is an apparatus used to measure the specific heat of a liquid. The technique is to measure the temperature difference between the input and output points of a flowing stream of the liquid while adding heat at a known rate. (a) Start with the equations $Q = mc\,\Delta T$ and $m = \rho V$, and convince yourself that the rate at which heat is added to the liquid is given by $\rho Q / \Delta t = \rho c\,\Delta T (\Delta V/\Delta t)$. (b) In a particular experiment, a liquid of density 0.72 g/cm³ flows through the calorimeter at the rate of 3.5 cm³/s. At steady state, a temperature difference of 5.8°C is established between the input and output points when heat is supplied at the rate of 40 J/s. What is the specific heat of the liquid?

66. An aluminum rod and an iron rod are joined end to end in a good thermal contact. The two rods have equal lengths and radii. The free end of the aluminum rod is maintained at a temperature of 100°C, and the free end of the iron rod is maintained at 0°C. (a) Determine the temperature of the interface between the two rods. (b) If each rod is 15 cm long and each has a cross-sectional area of 5.0 cm², what quantity of heat energy is conducted across the combination in 30 min?

67. Water is being boiled in an open kettle that has a 0.50-cm-thick circular aluminum bottom with a radius of 12 cm. If the water boils away at rate of 0.50 kg/min, what is the temperature of the lower surface of the bottom of the kettle? Assume that the top surface of the bottom of the kettle is at 100°C.

DAVID GRIFFING   Miami University

ENERGY MANAGEMENT IN THE HUMAN BODY

Why can sprinters run at 10 m/s whereas distance runners can do only a little under 6 m/s? Why does a marathon runner's body temperature often increase by 3°F or more? Why is dieting a better strategy than exercise for losing weight? These and many other questions are readily answered if we examine energy management within the body.

Humans use energy to move, breathe, pump blood, and so forth. Chemical reactions within the body are responsible for storing, releasing, absorbing, and transferring this energy. Some of these reactions *require* energy (endothermic reactions), whereas others *release* energy (exothermic reactions).

The energy source our bodies use to produce endothermic reactions is the food we eat. Exothermic reactions release the energy needed to run, walk, build new cells, and so forth. As an example, locomotion in the human body depends on the cooperation of agonist/antagonist pairs of muscles that alternately contract and relax. Muscle fibers pull but cannot push. With skeletal support, the agonist pulls while the antagonist relaxes, to produce motion. Figure 1 shows the biceps and triceps muscles cooperating to bend the arm at the elbow.

The energy required for muscle contraction is made available in the muscle cells through a network of chemical reactions. Within this network, there are two input and two output ingredients. The input ingredients come from the air we breathe and the food we eat, and the output ingredients are the carbon dioxide we exhale and the water produced as a by-product. In the lungs, the body removes oxygen from the inhaled air and transports it to muscle cells via hemoglobin in the bloodstream. In the mouth, stomach, and intestines, the body digests food, processing some of it into glucose ($C_6H_{12}O_6$). Some of this glucose is transported to the muscle cells. There, oxygen and glucose combine to form water and carbon dioxide in an exothermic reaction:

$$C_6H_{12}O_6 + 6O_2 \Rightarrow 6CO_2 + 6H_2O + E_{out}$$

Because this released energy is linked directly to the oxygen inhaled and carbon dioxide exhaled, it is theoretically simple to measure the energy a person generates. One measures either the net oxygen inhaled or the net carbon dioxide exhaled, and the energy released follows from the glucose oxidation reaction. Such measurements are routinely performed in hospitals and research laboratories.

Energy Production in Muscle Cells

The human body must take time to breathe, transport oxygen, and deliver glucose to the muscle cells. However, muscle fibers need not wait until glucose oxidation supplies the demanded energy because the body has the ability to store energy for future needs. Four processes are dedicated to energy production in a muscle cell; all use a high-energy molecule, adenosine triphosphate (ATP), to store and deliver energy in muscle cells. As long as ATP is available in the cell, immediate muscle contraction can occur without the benefit of breathing.

When the power demand in muscle fibers increases abruptly, the body needs time to adjust to the new level. Energy needed for muscle contraction during this transition time cannot be supplied by glucose oxidation because of the time needed to change to the new equilibrium state. To provide for this temporary energy gap, some "start-

349

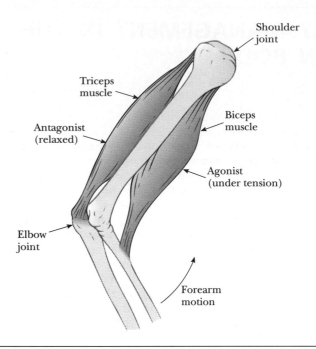

Shoulder
joint

Triceps
muscle

Biceps
muscle

Antagonist
(relaxed)

Agonist
(under tension)

Elbow
joint

Forearm
motion

FIGURE 1

To bend the elbow, the biceps pulls (as agonist) while the triceps relaxes (as antagonist). To straighten the elbow, the roles of biceps and triceps are reversed. The bone is necessary because muscle fibers pull but cannot push.

up'' energy is stored in muscle cells in the form of existing ATP, and in the form of ingredients for anaerobic (without oxygen) production of ATP. After the transition time, respiration will gradually increase ATP production until a steady state is attained. During exercise, therefore, energy is supplied anaerobically for approximately the first two minutes and aerobically thereafter, via respiration.

Energy and Exercise

If either glucose or oxygen is absent, the production of energy in muscle cells stops. If either is in short supply, exercise is limited. Normally, enough glucose is stored in the muscle to supply ordinary energy needs, including vigorous exercise, for a few hours. Enough anaerobic energy is available for the body to function without oxygen for a couple of minutes. As an extreme example, Houdini, the celebrated escape artist of the 1920s, could function under water for 3 or 4 min without breathing.

When the body uses energy faster than respiration can support ATP production, ATP becomes scarce and the available energy declines. This condition is described as **oxygen debt**. At the end of vigorous exercise, when energy demand is back to normal, the debt is ''repaid'' as the exerciser continues to breathe rapidly, as when a sprinter gasps for air after the race.

Experiment has shown that the power, P, required to run a given distance is nearly directly proportional to the running speed v, that is, $P \propto v$. However, the power required to walk a given distance is $P \propto v^x$, with $x > 1$. So to walk a given distance the required energy increases with the walking speed, but to run the same distance the required energy is approximately independent of speed[1]. As a rule of thumb:

$$\text{Running energy/distance} \cong 1.5 \text{ kcal/mi} \cong 0.93 \text{ kcal/km}$$

Using this figure of merit, we can compare the energy expended running with the energy content of food. For example, an 80-kg person burns about 120 kcal running 1 mi at any speed. Burning 1 lb of fat generates about 4200 kcal. Thus, if the goal is to lose 1 lb of fat in 35 days, one could (1) run an extra mile a day, or (2) eat two fewer slices of bread a day.[2] Exercise probably stimulates the appetite enough so that the extra food one eats more than compensates for the extra energy needed for the exercise. Exercise is a poor substitute for dieting as a method to lose weight!

When the body is constructively stressed with exercise, it responds by improving its ability to cope with that stress. In weight lifting, for example, the muscles used to lift weights get stronger, while in training for distance running, the body's ability to ingest and process oxygen improves. In weight lifting, sudden bursts of energy are needed for short times. The exercise is largely anaerobic, and the body's ability to produce anaerobic bursts of energy improves. In distance running, however, a high level of steady-state energy production is needed. The exercise is largely aerobic, and the body responds by improving its capacity for aerobic energy production.

The primary goal of aerobic exercise is to attain good cardiovascular condition. Such exercise works constructively, provided the aerobic stress is regular and not harmfully intensive. Some forms of efficient aerobic activity are potentially harmful to the body in other ways. For example, running puts nonconstructive stress on the skeletal system, and joint problems are not uncommon among long-term distance runners and joggers. When a runner stops a regular exercise routine so that the stress is removed, the body quickly loses its high level of cardiovascular fitness. The time needed to achieve good fitness for distance running is several months or even several years. Other things being equal, runners who stop training during the summer cannot compete with those who train all year long.

To prevent glycogen (a polymer form of glucose) depletion during a marathon, glycogen loading has proven effective. Whereas long training times are needed for the body to develop excellent oxygen-processing in cells, times as short as a week can increase the glycogen stored in the muscle. The body quickly digests carbohydrates and converts them to glycogen. To increase the glycogen stored in muscle cells beyond the normal concentration, the athlete eats food containing no carbohydrates for a week or so and then eats food overconcentrated with carbohydrates for a few days before the race. In effect, the body is tricked into overstocking muscle cells with glycogen, which can provide extra fuel late in the race. Glycogen loading has been effective in helping marathon runners avoid "hitting the wall" after running about 20 mi.

Extreme energy demand, such as that required to swim the English Channel, to compete in the triathalon, or to cycle across the country, pushes the body to its limit both aerobically and anaerobically. Training for such high-stress events normally involves special high-energy diets, and/or sleep deprivation.

Energy production in the body requires both oxygen and glucose. Oxygen cannot be stored, and so the steady-state rate of oxygen ingestion limits the long steady-state power production in muscles. Because glucose can be stored in the muscle cells and liver, the ultimate amount of energy that can be produced in the body without eating depends on how much glucose is stored in the form of glycogen. Beyond this point, the body will support itself by metabolizing body fat.

[1] "Caloric Cost of Walking and Running," Gilbert W. Fellingham et al., *Medicine and Science in Sports* 10:132–136, 1978.
[2] At 60 kcal/slice, 70 slices of bread correspond to 4200 kcal. At 2 slices a day, 35 days are needed.

Work and Heat

A 70-kg person requires, depending on metabolism rate, about 2400 kcal a day for normal bodily functions. The power output of the body rises during exercise and falls during rest. What happens to the energy produced in the body? Some of this energy is used to perform work, and some appears as heat.

When a muscle contracts, a force is exerted through a distance, and so mechanical work is performed. Using only the upper body muscles, a skilled rope climber can climb 5 m in 2 s. For an 80-kg climber, about 2 kW of power is developed. Using leg muscles, up to 5–6 kW may be developed for similar short bursts of activity, such as in stair climbing or jumping. In extended activity, however, the power developed is considerably reduced. For example, in 1932 five men on the Polish Olympic Team climbed 362 m from the 5th to the 102nd floor of the Empire State Building in about 21 min. Again assuming a mass of 80 kg, this works out to a 225-W power output. The longer a person works, the lower the rate at which chemical energy is converted into mechanical work.

In a classic paper[3] the Nobel laureate A. V. Hill described measurements in which the work performed during muscle contraction is accompanied by a temperature increase of the muscle fiber. The energy delivered by ATP in the muscle is converted partly to thermal energy. To prevent the overall body temperature from rising when muscles contract, this thermal energy must be transferred from the body to the environment. A small amount is conducted away, and some is radiated away. Breathing also cools the body because of the water vapor exhaled as a by-product of respiration, and convection is an effective cooling process. Sweating is triggered to provide additional cooling when these cooling processes are saturated.

The body maintains a near-constant temperature by eliminating excess thermal energy as rapidly as possible. Blood vessels dilate so that blood flow increases, pores open so that sweating increases, and respiration increases. In hot weather, however, these mechanisms may be insufficient for extreme aerobic activities such as marathon running, and performance suffers. Athletes can pour water over themselves to increase cooling a little but may have to reduce their activity level to reduce their thermal energy production rate. It is not unusual for marathon runners to experience a rise in body temperature of 3 to 4°F during the race.

Joggers know that fewer clothes are needed to keep warm during exercise than when walking or standing. It is not unusual to observe joggers running in shorts and T-shirts in freezing weather. The heat developed during exercise keeps them warm. Surface blood vessels contract to inhibit heat loss. If any protection is needed against the cold, the most essential garment is a warm hat because up to 40% of the thermal energy lost by the body exits through the head, where a copious supply of blood vessels receives thermal energy from the rest of the body. (In bed in a cold room, the most efficient protection is a nightcap, as our ancestors well knew.)

The efficiency of converting energy to mechanical work is defined to be the ratio of the work output to the energy input. When viewed as an engine the human body's efficiency is probably less than 25%, depending on the basal metabolism rate of the individual. For comparison, this efficiency is about the same as that of an automobile engine.

---

[3] "The Heat of Shortening and the Dynamic Constants of Muscle," A. V. Hill, *Proc. Roy. Soc.* 126:136–195, 1938.

THE LAWS OF THERMODYNAMICS

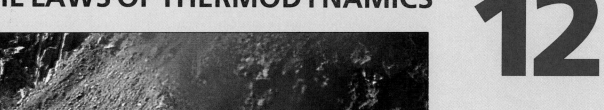

12

This steam-driven locomotive runs from Durango to Silverton, Colorado. Early steam-driven locomotives obtained their energy by burning wood or coal. The generated heat produces the steam, which powers the locomotive. Modern trains use electricity or diesel fuel to power their locomotives. All heat engines extract heat from a burning fuel and convert only a fraction of this energy to mechanical energy. (© Lois Moulton, Tony Stone Worldwide, Ltd.)

The first law of thermodynamics is essentially the principle of conservation of energy generalized to include heat as a form of energy. It tells us that an increase in one form of energy must be accompanied by a decrease in some other form of energy. The law considers both heat and work and places no restrictions on the types of energy conversions that occur. According to the first law, the internal energy of an object (a concept to be discussed shortly) can be increased either by heat added to the object or by work done on it.

The second law of thermodynamics, which can be stated in many equivalent ways, establishes which processes can occur in nature and which cannot. For example, the second law tells us that heat never flows spontaneously from a cold body to a hot body. One important application of this law is in the study of heat engines, such as the internal combustion engine, and the principles that limit their efficiency.

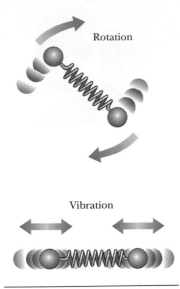

Rotation

Vibration

FIGURE 12.1
The rotational and vibrational energy of a diatomic gas contributes to the internal energy of the gas.

12.1

HEAT AND INTERNAL ENERGY

A major distinction must be made between internal energy and heat. **Internal energy** is all of the energy belonging to a system while it is stationary (neither translating nor rotating), including heat as well as nuclear energy, chemical energy, and strain energy (as in a compressed or stretched spring). **Thermal energy** is the portion of internal energy that changes when the temperature of the system changes. **Heat transfer** is caused by a temperature difference between the system and its surroundings.

In Chapter 10 we showed that the thermal energy of a monatomic ideal gas is associated with the internal motion of its atoms. In this special case, the thermal energy is simply kinetic energy on a microscopic scale; the higher the temperature of the gas, the greater the kinetic energy of the atoms and the greater the thermal energy of the gas. Generally, however, thermal energy includes other forms of molecular energy, such as rotational energy and vibrational kinetic and potential energy (Fig. 12.1).

As an analogy, consider the distinction between work and energy that was discussed in Chapter 5. The work done on (or by) a system is a measure of the energy transferred between the system and its surroundings, whereas the system's mechanical energy (kinetic and/or potential) is a consequence of its motion and coordinates. Thus, when a person does work on a system, energy is transferred from the person to the system. It makes no sense to talk about the work *of* a system—one should refer only to the *work done on or by a system* when some process has occurred in which energy has been transferred to or from the system. Likewise, it makes no sense to use the term *heat* unless energy has been transferred as a result of a temperature difference.

It is also important to recognize that energy can be transferred between two systems even when no *thermal* energy transfer occurs. For example, when a piston compresses a gas, the gas is warmed and its thermal energy increases, but no transfer of heat takes place; if the gas then expands rapidly, it cools and its thermal energy decreases, but there is no transfer of thermal energy to its surroundings. In each case, energy is transferred to or from the system as work but appears within the system as an increase or decrease in thermal energy. The change in internal energy is equal to the change in thermal energy and is measured by a corresponding change in temperature.

12.2

WORK AND HEAT

In the macroscopic approach to thermodynamics, we describe the *state* of a system with the use of such variables as pressure, volume, temperature, and internal energy. The number of macroscopic variables needed to characterize a system depends on the system's nature. For a homogeneous system, such as a gas containing only one type of molecule, usually only two variables are needed. It is important to note that a *macroscopic state* of an isolated system can be specified only if the system is in thermal equilibrium internally. In the case of a gas in a container, internal thermal equilibrium requires that every part of the container be at the same pressure and temperature.

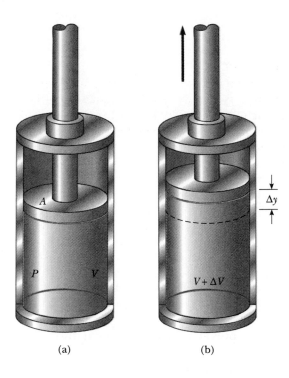

(a) (b)

FIGURE 12.2
(a) A gas in a cylinder occupying a volume V at a pressure P. (b) As the gas expands at constant pressure and the volume increases by the amount ΔV, the work done by the gas is $P\Delta V$.

Consider a gas contained by a cylinder fitted with a movable piston (Fig. 12.2). In equilibrium, the gas occupies a volume of V and exerts a uniform pressure, P, on the cylinder walls and piston. If the piston has cross-sectional area A, the force exerted by the gas on the piston is $F = PA$. Now let us assume that the gas expands slowly enough to allow the system to remain essentially in thermodynamic equilibrium at all times. As the piston moves up a distance of Δy, the work done on the piston by the gas is

$$W = F\,\Delta y = PA\,\Delta y$$

Because $A\,\Delta y$ is the increase in volume of the gas (that is, $\Delta V = A\,\Delta y$), we can express the work done as

$$W = P\,\Delta V \qquad \text{[12.1]}$$

If the gas expands, as in Figure 12.2b, ΔV is positive and the work done by the gas is positive. If the gas is compressed, ΔV is negative and the work done by the gas is negative. (In the latter case, negative work can be interpreted as work being done *on* the system.) Clearly, the work done by (or on) the system is zero when the volume remains constant.

Equation 12.1 can be used to calculate the work done on or by the system only when the pressure of the gas remains constant during the expansion or compression. If the pressure changes, calculus is required to determine the work done. We do not attempt such calculations here.

Consider the process represented by the pressure-volume diagram in Figure 12.3. The gas has expanded from an initial volume, V_i, to a final volume, V_f, at a constant pressure of P. From Equation 12.1 we see that the work done by the gas in this case is $P(V_f - V_i)$. Note that this is just the area under the pressure-volume curve.

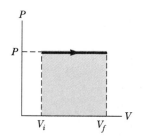

FIGURE 12.3
The PV diagram for a gas expanding at constant pressure. The shaded area represents the work done by the gas.

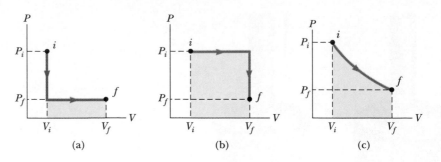

FIGURE 12.4
The work done by a gas as it is taken from an initial state to a final state depends on
the path taken between these states.

> **In general, the work done in an expansion from some initial state to
> some final state is the area under the curve on a *PV* diagram.**

This statement is true whether or not the pressure remains constant during the
process.

As Figure 12.3 shows, the work done in the expansion from the initial state to
the final state depends on the path taken between the two states. To illustrate this
important point, consider several different paths connecting i and f (Fig. 12.4).
In the process depicted in Figure 12.4a, the pressure of the gas is reduced from P_i
to P_f by cooling at constant volume, V_i, and then the gas expands from V_i to V_f
at constant pressure, P_f. The work done along this path is $P_f(V_f - V_i)$. In Figure
12.4b, the gas expands from V_i to V_f at constant pressure, P_i, and then its pressure is reduced to P_f at constant volume, V_f. The work done along this path is
$P_i(V_f - V_i)$, which is greater than the work done in the process of Figure 12.4a.
Finally, in the process depicted in Figure 12.4c, where both P and V change
continuously, the work done has some value intermediate between the values
obtained in the first two processes. To evaluate the work in this case, the shape of
the PV curve must be known. Thus, the work done by a system depends on the
process by which the system goes from the initial to the final state. In other words,
the work done depends on the initial, final, and intermediate states of the
system.

In a similar manner, the heat transferred into or out of a system is also found
to depend on the process. This can be demonstrated by the situations depicted in
Figure 12.5. In all cases, the gases have the same initial volume, temperature,
and pressure and are assumed to be ideal. In Figure 12.5a, the gas is in thermal
contact with a heat reservoir. (A reservoir is a body whose heat capacity is so large
that its temperature remains unchanged when heat is added or extracted from
the body.) If the pressure of the gas is infinitesimally greater than atmospheric
pressure, the gas, because it absorbs heat from the reservoir, expands and causes
the piston to rise. During this expansion to some final volume, V_f, and final
pressure P_f, sufficient heat to maintain a constant temperature of T_i is transferred from the reservoir to the gas.

Now consider the thermally insulated system in Figure 12.5b. When the
membrane is broken, the gas expands rapidly into the vacuum until it occupies a

---

**Work done depends on the
path between the initial and
final states**

Heat reservoir

Free expansion of a gas

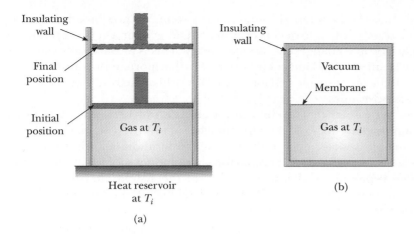

FIGURE 12.5
(a) A gas at temperature T_i expands slowly by absorbing heat from a reservoir at the same temperature. (b) A gas expands rapidly into an evacuated region after a membrane separating it from that region is broken.

volume of V_f and is at a pressure of P_f. In this case the gas does no work, since there is no movable piston. Furthermore, no heat is transferred through the thermally insulated wall, which we call an *adiabatic wall*, and so the temperature remains at T_i. This process is often referred to as **adiabatic free expansion** or simply *free expansion*. In general, an adiabatic process is one in which no heat is transferred between the system and its surroundings.

The initial and final states of the ideal gas in Figure 12.5a are identical to the initial and final states in Figure 12.5b, but the paths are different. In the first case, heat is transferred slowly to the gas, and the gas does work on the piston. In the second case, no heat is transferred, and the work done is zero. Therefore, we conclude that *heat transfer, like work, depends on the initial, final, and intermediate states of the system.*

EXAMPLE 12.1   Work Done by an Expanding Gas

In the system shown in Figure 12.2, the gas in the cylinder is at a pressure of 8000 Pa and the piston has an area of 0.10 m². As heat is slowly added to the gas, the piston is pushed up a distance of 4.0 cm. Calculate the work done on the surroundings by the expanding gas. Assume that the pressure remains constant.

Solution   The change in volume of the gas is

$$\Delta V = A \, \Delta y = (0.10 \text{ m}^2)(4.0 \times 10^{-2} \text{ m}) = 4.0 \times 10^{-3} \text{ m}^3$$

and from Equation 12.1, the work done is

$$W = P \Delta V = (8000 \text{ Pa})(4.0 \times 10^{-3} \text{ m}^3) = \boxed{32 \text{ J}}$$

12.3

THE FIRST LAW OF THERMODYNAMICS

When the principle of conservation of energy was first introduced in Chapter 5, it was stated that the mechanical energy of a system is constant in the absence of nonconservative forces, such as friction. That mechanical model did not encompass changes in the internal energy of the system. We now broaden our scope to use the term ''principle of conservation of energy'' for a generalization encompassing possible changes in internal energy. This is a universally valid law that can be applied to all kinds of processes. Furthermore, it provides a connection between the microscopic and macroscopic worlds. The result will be the **first law of thermodynamics.**

Change in internal energy

We have seen that energy can be transferred between a system and its surroundings in two ways: via work done by (or on) the system, which requires a macroscopic displacement of the point of application of a force (or pressure), and via heat transfer, which occurs through random molecular collisions. Each of these represents a change of energy of the system, and therefore usually results in measurable changes in the macroscopic variables of the system, such as the pressure, temperature, and volume of a gas.

To express these ideas more quantitatively, suppose a system undergoes a change from an initial state to a final state. During this change, positive Q is the heat transferred *to* the system, and positive W is the work done *by* the system. For example, suppose the system is a gas whose pressure and volume change from P_i and V_i to P_f and V_f. If the quantity $Q - W$ is measured for various paths connecting the initial and final equilibrium states (that is, for various *processes*), one finds that it is the same for *all* paths connecting the initial and final states. We conclude that the quantity $Q - W$ is determined completely by the initial and final states of the system, and we call it the *change in the internal energy of the system.* Although Q and W both depend on the path, $Q - W$ *is independent of the path.* If we represent the internal energy function with the letter U, then the *change* in internal energy, $\Delta U = U_f - U_i$, can be expressed as

First-law equation

$$\Delta U = U_f - U_i = Q - W \qquad \textbf{[12.2]}$$

where all quantities must have the same energy units. Equation 12.2 is known as the first law of thermodynamics. This law applies universally to all systems.

Let us examine some special cases in which the only changes in energy are changes in internal energy. First consider an *isolated system,* that is, one that does not interact with its surroundings. In this case, no heat transfer takes place and the work done is zero; hence, the internal energy remains constant. That is, since $Q = W = 0$, $\Delta U = 0$, and so $U_i = U_f$. We conclude that *the internal energy of an isolated system remains constant.*

For isolated systems, *U* remains constant

Next consider the case in which a system (not isolated from its surroundings) is taken through a **cyclic process,** that is, a process that originates and ends at the same state. In this case, the change in the internal energy must again be *zero,* and therefore the heat added to the system must equal the work done during the cycle. That is, in a cyclic process,

Cyclic process

$$\Delta U = 0 \qquad \text{and} \qquad Q = W$$

Note that *the net work done per cycle equals the area enclosed by the path representing the process on a PV diagram.* As we shall see in the next section, cyclic processes are very important for describing the thermodynamics of *heat engines*—devices in which some part of the thermal energy input is converted into mechanical work.

If a process occurs in which the work done is zero, then the change in internal energy equals the heat entering or leaving the system. If heat enters the system, Q is positive and the internal energy increases. For a gas, we can associate this increase in internal energy with an increase in the kinetic energy of the molecules. On the other hand, if a process occurs in which the heat transferred is zero and work is done by the system, then the magnitude of the change in internal energy equals the negative of the work done by the system. That is, the internal energy of the system decreases. For example, if a gas is compressed with no heat transferred (by a moving piston, say), the work done by the gas is negative and the internal energy again increases. This is because kinetic energy is transferred from the moving piston to the gas molecules.

On a microscopic scale, no practical distinction exists between heat transfer and work. Each can produce a change in the internal energy of a system. Although the macroscopic quantities Q and W are *not* properties of a system, they are related to changes of the internal energy of a stationary system through the first law of thermodynamics. Once a process, or path, is defined, Q and W can be either calculated or measured, and the change in internal energy can be found from the first law of thermodynamics.

EXAMPLE 12.2   An Isobaric Process

A gas is enclosed in a container fitted with a piston of cross-sectional area 0.10 m^2. The pressure of the gas is maintained at 8000 Pa while heat is slowly added; as a result, the piston is pushed up a distance of 4.0 cm. (Any process in which the pressure remains constant is called an **isobaric process**.) If 42 J of heat is added to the system during the expansion, what is the change in internal energy of the system?

An isobaric process is one that occurs at constant pressure

Solution   The work done by the gas is

$$W = P \Delta V = (8000 \text{ Pa})(0.10 \text{ m}^2)(4.0 \times 10^{-2} \text{ m}) = 32 \text{ N·m} = 32 \text{ J}$$

The change in internal energy is found from the first law of thermodynamics:

$$\Delta U = Q - W = 42 \text{ J} - 32 \text{ J} = \boxed{10 \text{ J}}$$

We see that *in an isobaric process the work done and the heat transferred are both nonzero.*

Exercise   If 42 J of heat is added to the system with the piston clamped in a *fixed* position, what is the work done by the gas? What is the change in its internal energy?

Answer   No work is done; $\Delta U = 42 \text{ J}$.

EXAMPLE 12.3   An Isovolumetric Process

Water with a mass of 2.0 kg is held at constant volume in a container while 10 000 J of heat is slowly added by a flame. The container is not well insulated, and as a result 2000 J of heat leaks out to the surroundings. What is the temperature increase of the water?

An isovolumetric process is one that takes place at constant volume

Solution   A process that takes place at constant volume is called an **isovolumetric process**. In such a process the work is equal to zero. Thus, the first law of thermodynamics gives

$$\Delta U = Q$$

This indicates that the net heat added to the water goes into increasing the internal energy of the water. The net heat added to the water is

$$Q = 10\,000\,J - 2000\,J = 8000\,J$$

Since $Q = mc\,\Delta T$, the temperature increase of the water is

$$\Delta T = \frac{Q}{mc} = \frac{8000\,J}{(2.0\,\text{kg})(4.186 \times 10^3\,J/\text{kg} \cdot {}^\circ C)} = \boxed{0.96^\circ C}$$

EXAMPLE 12.4   Boiling Water

One gram of water occupies a volume of 1.0 cm³ at atmospheric pressure $(1.013 \times 10^5\,\text{Pa})$. When this water is boiled, it becomes 1671 cm³ of steam. Calculate the change in internal energy for this process.

Solution   Since the heat of vaporization of water is $2.26 \times 10^6\,J/\text{kg}$ at atmospheric pressure, the heat required to boil 1.0 g is

$$Q = mL_v = (1.0 \times 10^{-3}\,\text{kg})(2.26 \times 10^6\,J/\text{kg}) = 2260\,J$$

The work done by the system is positive and equal to

$$W = P(V_{\text{steam}} - V_{\text{water}})$$
$$= (1.013 \times 10^5\,\text{Pa})[(1671 - 1.0) \times 10^{-6}\,\text{m}^3] = 169\,J$$

Hence, the change in internal energy is

$$\Delta U = Q - W = 2260\,J - 169\,J = \boxed{2.1 \times 10^3\,J}$$

The positive ΔU tells us that the internal energy of the system has increased. We see that most of the heat (93%) transferred to the liquid goes into increasing the internal energy. Only a small fraction (7%) goes into external work.

EXAMPLE 12.5   Heat Transferred to a Solid

The internal energy of a solid also increases when heat is transferred to it from its surroundings. To illustrate this point, suppose a 1.0-kg bar of copper is heated at atmospheric pressure. Its temperature increases from 20°C to 50°C.

(a) Find the work done by the copper.

Solution   The change in volume of the copper can be calculated using Equation 10.6 and the linear expansion coefficient for copper taken from Table 10.1 (remembering that $\beta = 3\alpha$):

$$\Delta V = \beta V\,\Delta T = [5.1 \times 10^{-5}\,(^\circ C)^{-1}](50^\circ C - 20^\circ C)\,V = (1.5 \times 10^{-3})\,V$$

The volume is equal to m/ρ, and the density of copper is $8.92 \times 10^3\,\text{kg}/\text{m}^3$. Hence,

$$\Delta V = (1.5 \times 10^{-3})\left(\frac{m}{\rho}\right) = (1.5 \times 10^{-3})\left(\frac{1.0\,\text{kg}}{8.92 \times 10^3\,\text{kg}/\text{m}^3}\right)$$
$$= 1.7 \times 10^{-7}\,\text{m}^3$$

Since the expansion takes place at constant pressure (equal to normal atmospheric pressure), the work done is

$$W = P\,\Delta V = (1.013 \times 10^5 \text{ Pa})(1.7 \times 10^{-7} \text{ m}^3) = \boxed{1.9 \times 10^{-2}\,\text{J}}$$

(b) What quantity of heat is transferred to the copper?

Solution   The specific heat of copper is given in Table 11.1, and from Equation 11.3 we find that the heat transferred is

$$Q = mc\,\Delta T = (1.0 \text{ kg})(387 \text{ J/kg}\cdot{}^\circ\text{C})(30^\circ\text{C}) = \boxed{1.2 \times 10^4\,\text{J}}$$

Exercise   What is the increase in internal energy of the copper?

Answer $\Delta U = 1.2 \times 10^4$ J.

12.4

HEAT ENGINES AND THE SECOND LAW OF THERMODYNAMICS

A heat engine is a device that converts thermal energy to other useful forms, such as electrical and mechanical energy. In a typical process for producing electricity in a power plant, for instance, coal or some other fuel is burned, and the thermal energy produced is used to convert water to steam. This steam is directed at the blades of a turbine, setting it in rotation. Finally, the mechanical energy associated with this rotation is used to drive an electric generator. Another heat engine, the automobile internal combustion engine, extracts heat from a burning fuel and converts a fraction of this energy to mechanical energy.

A heat engine carries some working substance through a cyclic process during which (1) heat is absorbed from a source at a high temperature, (2) work is done by the engine, and (3) heat is expelled by the engine to a reservoir at a lower temperature. As an example, consider the operation of a steam engine in which the working substance is water. The water is carried through a cycle in which it first evaporates into steam in a boiler and then expands against a piston. After the steam is condensed with cooling water, it is returned to the boiler, and the process is repeated.

It is useful to represent a heat engine schematically as in Figure 12.6. The engine absorbs a quantity of heat, Q_h, from the hot reservoir, does work W, and then gives up heat Q_c to the cold reservoir. Because the working substance goes through a cycle, its initial and final internal energies are equal, so $\Delta U = 0$. Hence, from the first law of thermodynamics we see that *the net work, W, done by a heat engine equals the net heat flowing into it.* As we can see from Figure 12.6, $Q_{\text{net}} = Q_h - Q_c$; therefore,

$$W = Q_h - Q_c \qquad [12.3]$$

where Q_h and Q_c are taken to be positive quantities.

If the working substance is a gas, *the net work done for a cyclic process is the area enclosed by the curve representing the process on a PV diagram.* This is shown for an arbitrary cyclic process in Figure 12.7.

The **thermal efficiency, *e*,** of a heat engine is the ratio of the net work done to the heat absorbed at the higher temperature during one cycle:

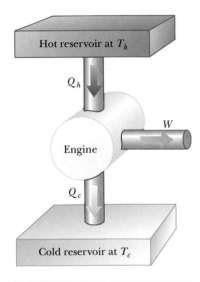

FIGURE 12.6
A schematic representation of a heat engine. The engine receives heat Q_h from the hot reservoir, expels heat Q_c to the cold reservoir, and does work W.

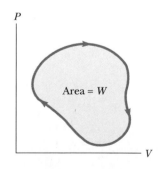

FIGURE 12.7
The *PV* diagram for an arbitrary cyclic process. The area enclosed by the curve equals the net work done.

Thermal efficiency

$$e \equiv \frac{W}{Q_h} = \frac{Q_h - Q_c}{Q_h} = 1 - \frac{Q_c}{Q_h} \qquad \text{[12.4]}$$

We can think of the efficiency as the ratio of what we get (mechanical energy) to what we give (thermal energy at the higher temperature). Equation 12.4 shows that a heat engine has 100% efficiency ($e = 1$) only if $Q_c = 0$—that is, if no heat is expelled to the cold reservoir. In other words, a heat engine with perfect efficiency would have to convert all of the absorbed heat to mechanical work. One of the consequences of the second law of thermodynamics is that this is impossible.

Second law

The **second law of thermodynamics** can be stated as follows: *It is impossible to construct a heat engine that, operating in a cycle, produces no other effect than the absorption of heat from a reservoir and the performance of an equal amount of work.*

This form of the second law is useful for understanding the operation of heat engines. With reference to Equation 12.4, the second law says that, during the operation of a heat engine, W can never be equal to Q_h or, alternatively, that some heat, Q_c, must be rejected to the environment. As a result, it is theoretically impossible to construct an engine that works with 100% efficiency.

Our assessment of the first two laws of thermodynamics can be summed up as follows: the first law says *we cannot get a greater amount of energy out of a cyclic process than we put in*, and the second law says *we cannot break even.*

EXAMPLE 12.6   The Efficiency of an Engine

Find the efficiency of an engine that introduces 2000 J of heat during the combustion phase and loses 1500 J at exhaust.

Solution   The efficiency of the engine is given by Equation 12.4 as

$$e = 1 - \frac{Q_c}{Q_h} = 1 - \frac{1500 \text{ J}}{2000 \text{ J}} = \boxed{0.25, \text{ or } 25\%}$$

Exercise   If an engine has an efficiency of 20% and loses 3000 J at exhaust and to the cooling water, how much work is done by the engine?

Answer   750 J.

12.5

REVERSIBLE AND IRREVERSIBLE PROCESSES

In the next section we shall discuss a theoretical heat engine that is the most efficient engine possible. In order to understand its nature, we must first examine the meanings of reversible and irreversible processes. A **reversible** process is one that can be performed so that, at its conclusion, both the system and its surroundings have been returned exactly to their initial conditions. A process that does not satisfy these requirements is **irreversible.**

All natural processes are known to be irreversible. As an example, let us examine the free expansion of a gas (already discussed in Section 12.2) and show that it cannot be reversible. The gas is contained in an insulated container with a membrane separating it from a vacuum (Fig. 12.8). If the membrane is punc-

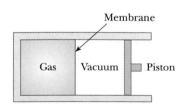

FIGURE 12.8
Free expansion of a gas.

tured, the gas expands freely into the vacuum. Because the gas does not exert a force on its surroundings through a distance, it does no work as it expands. In addition, no heat is transferred to or from the gas since the container is insulated from its surroundings. Thus, in this process the system has changed but the surroundings have not.

Now imagine that we try to reverse the process by first compressing the gas to its original volume. Let's say an engine is being used to force the piston inward. Note, however, that this action is changing both the system and surroundings. The surroundings are changing because work is being done by an outside agent on the system, and the system is changing because the compression is increasing the temperature of the gas. We can lower the temperature of the gas by allowing it to come in contact with an external heat reservoir. Although this second procedure returns the gas to its original state, the surroundings are again affected because thermal energy is added to the surroundings. If this heat could somehow be used to drive the engine and compress the gas, the system and its surroundings could be returned to their initial states. However, our statement of the second law says that this extracted heat cannot be completely converted to mechanical energy isothermally. We must conclude that a reversible process has not occurred.

Although real processes are always irreversible, some are *almost* reversible. If a real process occurs very slowly so that the system is virtually always in equilibrium, the process can be considered reversible. For example, imagine compressing a gas very slowly by dropping some grains of sand onto a frictionless piston, as in Figure 12.9. The compression process can be reversed by the placement of the gas in thermal contact with a heat reservoir. The pressure, volume, and temperature of the gas are well defined during this isothermal compression. Each added grain of sand represents a change to a new equilibrium state. The process can be reversed by the slow removal of grains of sand from the piston.

A general characteristic of a reversible process is that no dissipative effects that convert mechanical energy to thermal energy, such as turbulence or friction, can be present. In reality, such effects are impossible to eliminate completely, and hence it is not surprising that real processes in nature are irreversible.

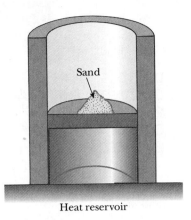

FIGURE 12.9
A gas in thermal contact with a heat reservoir is compressed slowly by grains of sand dropped onto the piston. The compression is isothermal and reversible.

12.6

THE CARNOT ENGINE

In 1824 a French engineer named Sadi Carnot (1796–1832) described a theoretical engine, now called a *Carnot engine,* that is of great importance from both practical and theoretical viewpoints. He showed that a heat engine operating in an ideal, reversible cycle—called a Carnot cycle—between two reservoirs is the most efficient engine possible. **Carnot's theorem** can be stated as follows:

> **No real engine operating between two heat reservoirs can be more efficient than a Carnot engine, operating between the same two reservoirs.**

To describe the Carnot cycle, we assume that the substance working between temperatures T_c and T_h is an ideal gas contained in a cylinder with a movable piston at one end. The cylinder walls and the piston are thermally nonconduct-

PHYSICS IN ACTION

DEVICES THAT CONVERT THERMAL ENERGY INTO OTHER FORMS OF ENERGY

The device in the photograph on the left, called Hero's engine, was invented around 150 B.C. by Hero in Alexandria. When water is boiled in the flask, which is suspended by a cord, steam exits through two tubes at the sides of the flask (in opposite directions), creating a torque that rotates the flask.

The device shown in the center and right-hand photographs, called a thermoelectric converter, uses a series of semiconductor cells to convert thermal energy to electrical energy. In the center photograph, the two "legs" of the device are at the same temperature, and no electrical energy is produced. However, when one leg is at a higher temperature than the other, as in the photograph on the right, electrical energy is produced as the device extracts energy from the hot reservoir and drives a small electric motor. How does this intriguing experiment demonstrate the second law of thermodynamics?

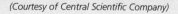

(Courtesy of Central Scientific Company)

(Courtesy of PASCO Scientific Co.)

ing. Figure 12.10 shows four stages of the Carnot cycle, and Figure 12.11 is the *PV* diagram for the cycle. The cycle consists of two adiabatic and two isothermal processes, all reversible.

1. The process $A \rightarrow B$ is an isothermal expansion at temperature T_h, in which the gas is placed in thermal contact with a heat reservoir at temperature T_h (Fig. 12.10a). During the process, the gas absorbs heat Q_h from the reservoir and does work W_{AB} in raising the piston.

2. In the process $B \rightarrow C$, the base of the cylinder is replaced by a thermally nonconducting wall and the gas expands adiabatically; that is, no heat enters or leaves the system (Fig. 12.10b). During the process, the temperature falls from T_h to T_c and the gas does work W_{BC} in raising the piston.

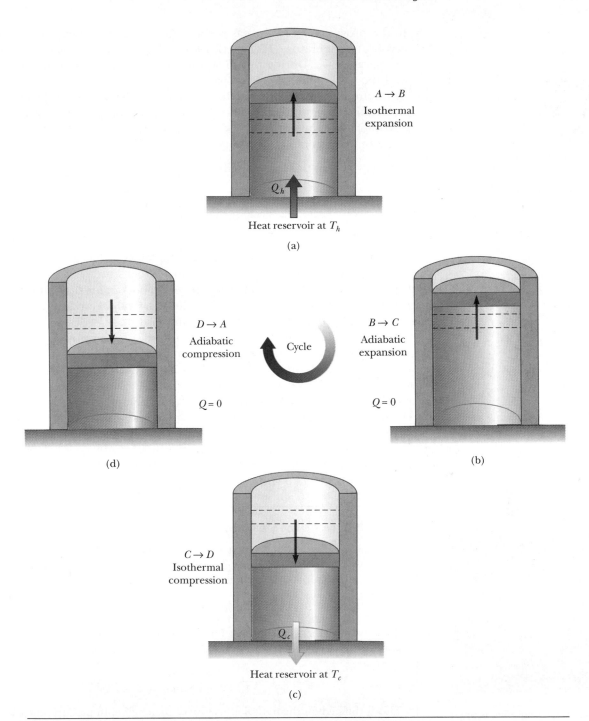

FIGURE 12.10
The Carnot cycle. In process $A \to B$, the gas expands isothermally while in contact with a reservoir at T_h. In process $B \to C$, the gas expands adiabatically ($Q = 0$). In process $C \to D$, the gas is compressed isothermally while in contact with a reservoir at $T_c < T_h$. In process $D \to A$, the gas is compressed adiabatically. The upward arrows on the piston indicate removal of sand during the expansions, and the downward arrows indicate addition of sand during the compressions.

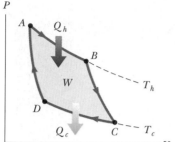

FIGURE 12.11
The *PV* diagram for the Carnot cycle. The net work done, *W*, equals the net heat received in one cycle, $Q_h - Q_c$.

3. In the process $C \rightarrow D$, the gas is placed in thermal contact with a heat reservoir at temperature T_c (Fig. 12.10c) and is compressed isothermally at temperature T_c. During this time, the gas expels heat Q_c to the reservoir, and the work done on the gas is W_{CD}.
4. In the final stage, $D \rightarrow A$, the base of the cylinder is again replaced by a thermally nonconducting wall (Fig. 12.10d) and the gas is compressed adiabatically. The temperature of the gas increases to T_h, and the work done on the gas is W_{DA}.

Carnot showed that the thermal efficiency of a Carnot engine is

$$e_c = \frac{T_h - T_c}{T_h} = 1 - \frac{T_c}{T_h}$$ [12.5]

where T must be in kelvins. From this result, we see that all Carnot engines operating reversibly between the same two temperatures have the same efficiency. Furthermore, the efficiency of any reversible engine operating in a cycle between two temperatures is greater than the efficiency of any irreversible (real) engine operating between the same two temperatures.

Equation 12.5 can be applied to any working substance operating in a Carnot cycle between two heat reservoirs. According to this result, the efficiency is zero if $T_c = T_h$. The efficiency increases as T_c is lowered and as T_h is increased. However, the efficiency can be unity (100%) only if $T_c = 0$ K. Such reservoirs are not available, and so the maximum efficiency is always less than unity. In most practical cases, the cold reservoir is near room temperature, about 300 K. Therefore, one usually strives to increase the efficiency by raising the temperature of the hot reservoir. *All real engines are less efficient than the Carnot engine because they are subject to practical difficulties, including friction, but especially the need to operate irreversibly to complete a cycle in a brief time period.*

A model steam engine equipped with a built-in horizontal boiler. The water is heated electrically, generating steam that is used to power the electric generator at the left. (*Courtesy of Central Scientific Company*)

EXAMPLE 12.7   The Steam Engine

A steam engine has a boiler that operates at 500 K. The heat changes water to steam, which drives the piston. The temperature of the exhaust is that of the outside air, about 300 K. What is the maximum thermal efficiency of this steam engine?

Solution   From the expression for the efficiency of a Carnot engine, we find the maximum thermal efficiency for any engine operating between these temperatures:

$$e_c = 1 - \frac{T_c}{T_h} = 1 - \frac{300 \text{ K}}{500 \text{ K}} = \boxed{0.4, \text{ or } 40\%}$$

This is the highest theoretical efficiency of the engine. In practice, the efficiency is considerably lower.

Exercise   Determine the maximum work the engine can perform in each cycle of operation if it absorbs 200 J of thermal energy from the hot reservoir during each cycle.

Answer   80 J.

Sadi Carnot, a French engineer, was the first to demonstrate the quantitative relationship between work and heat. Carnot was born in Paris on June 1, 1796, and was educated at the École Polytechnique in Paris and at the École Genie in Metz. His many interests included a wide range of study and research in mathematics, tax reform, industrial development, and the fine arts.

In 1824 he published his only work—*Reflections on the Motive Power of Heat*—which reviewed the industrial, political, and economic importance of the steam engine. In it he defined work as "weight lifted through a height." In 1831 Carnot began to study the physical properties of gases, particularly the relationship between temperature and pressure.

On August 24, 1832, Sadi Carnot died suddenly of cholera. In accordance with the health practices of that time, all of his personal effects were burned. Some of his notes that fortunately escaped destruction indicate that Carnot had arrived at the idea that heat is essentially work, or work that has changed its form. For this reason, he is considered the founder of the science of thermodynamics, from which we learn that energy can never disappear; it can only be altered in form. Carnot's notes led Lord Kelvin to confirm and extend the science of thermodynamics in 1850.

**Sadi Carnot
(1796–1832)**

(FPG)

EXAMPLE 12.8 The Carnot Efficiency

The highest theoretical efficiency of a gasoline engine, based on the Carnot cycle, is 30%. If this engine expels its gases into the atmosphere, which has a temperature of 300 K, what is the temperature in the cylinder immediately after combustion?

Solution   The Carnot efficiency is used to find T_h:

$$e_c = 1 - \frac{T_c}{T_h}$$

$$T_h = \frac{T_c}{1 - e_c} = \frac{300 \text{ K}}{1 - 0.30} = \boxed{430 \text{ K}}$$

Actual gas engines operate on a cycle significantly different from the Carnot cycle and therefore have lower maximum possible efficiencies.

Exercise   If the heat engine absorbs 837 J of heat from the hot reservoir during each cycle, how much work can it perform in each cycle?

Answer   251 J.

Lord Kelvin, a British physicist and mathematician (1824–1907). Born William Thomson in Belfast, Kelvin was the first to propose the use of an absolute scale of temperature. His study of Carnot's theory led to the idea that heat cannot pass spontaneously from a colder body to a hotter body; this is known as the second law of thermodynamics. *(J.L. Charmet/SPL/Photo Researchers)*

12.7

HEAT PUMPS AND REFRIGERATORS

A heat pump is a mechanical device that moves thermal energy from a region at lower temperature to a region at higher temperature. Heat pumps have long been popular for cooling and are now used increasingly for heating purposes as

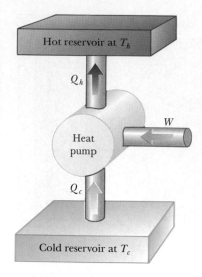

FIGURE 12.12
Schematic diagram of a heat pump, which absorbs heat Q_c from the cold reservoir and expels heat Q_h to the hot reservoir. The work done on the heat pump is W.

well. In the heating mode, a circulating coolant fluid absorbs heat from the outside and releases it to the interior of the structure. The fluid is usually a low-pressure vapor when it is in the coils of the exterior part of the unit, where it absorbs heat from either the air or the ground. This gas is then compressed into a hot, high-pressure vapor and enters the interior part of the unit, where it condenses to a liquid and releases its stored thermal energy. An air conditioner is simply a heat pump installed backward, with "exterior" and "interior" interchanged.

Figure 12.12 is a schematic representation of a heat pump in its heating mode. The outside temperature is T_c, the inside temperature is T_h, and the heat absorbed by the circulating fluid is Q_c. The compressor does work W on the fluid, and the thermal energy transferred from the pump into the structure is Q_h.

The effectiveness of a heat pump, in its heating mode, is described in terms of a number called the **coefficient of performance,** COP. This is defined as the ratio of the heat transferred into the hot reservoir to the work required to transfer that heat:

$$\text{COP(heat pump)} \equiv \frac{\text{heat transferred}}{\text{work done by pump}} = \frac{Q_h}{W} \qquad \textbf{[12.6]}$$

If the outside temperature is 25°F or higher, the COP for a heat pump is about 4. That is, the heat transferred into the house is about four times the work done by the compressor in the heat pump. However, as the outside temperature decreases, it becomes more difficult for the heat pump to extract sufficient heat from the air, and the COP drops. In fact, the COP can fall below unity for temperatures below the midteens.

A Carnot-cycle heat engine run in reverse constitutes the heat pump with the highest possible coefficient of performance for the temperatures between which it operates. The maximum coefficient of performance is

$$\text{COP}_c\text{(heat pump)} = \frac{T_h}{T_h - T_c}$$

Although heat pumps are relatively new products in heating, the refrigerator has been a standard home appliance for decades. The refrigerator works much like a heat pump except that it cools its interior by pumping heat from the food storage compartments into the warmer air outside. During operation a refrigerator removes heat Q_c from its interior, and in the process its motor does work W. The coefficient of performance of a refrigerator or of a heat pump used in its cooling cycle is

$$\text{COP(refrigerator)} = \frac{Q_c}{W} \qquad \textbf{[12.7]}$$

The most efficient refrigerator is one that removes the greatest amount of heat from the cold reservoir with the least amount of work. Thus, a good refrigerator should have a high coefficient of performance, typically 5 or 6.

The highest possible coefficient of performance is again that of a refrigerator whose working substance is carried through the Carnot heat-engine cycle in reverse:

$$\text{COP}_c\text{(refrigerator)} = \frac{T_c}{T_h - T_c}$$

As the difference between temperatures of the two reservoirs approaches zero, the theoretical coefficient of performance of a Carnot heat pump approaches infinity. In practice, the low temperature of the cooling coils and the high temperature at the compressor limit the COP to values below 10.

12.8

ENTROPY

The concept of temperature is involved in the zeroth law of thermodynamics, and the concept of internal energy is involved in the first law. Temperature and internal energy are both state functions; that is, they can be used to describe the thermodynamic state of a system. Another state function related to the second law of thermodynamics is the **entropy function**, *S*.

Consider a reversible process between two equilibrium states. If ΔQ_r is the heat absorbed or expelled by the system during some small interval of the path,

> The **change in entropy, ΔS,** between two equilibrium states is given by the heat transferred, ΔQ_r, divided by the absolute temperature, T, of the system in this interval. That is,
>
> $$\Delta S \equiv \frac{\Delta Q_r}{T} \qquad \text{[12.8]}$$

The subscript *r* on the term ΔQ_r emphasises that the definition applies only to reversible processes. When heat is absorbed by the system, ΔQ_r is positive and hence the entropy increases. When heat is expelled by the system, ΔQ_r is negative and the entropy decreases. Note that Equation 12.8 defines not entropy but the change in entropy. This is consistent with the fact that a change in state always accompanies heat transfer. Hence, the meaningful quantity in a description of a process is the change in entropy.

The concept of entropy was introduced into the study of thermodynamics by Rudolph Clausius in 1865. One reason it became useful and gained wide acceptance is because it provides another variable to describe the state of a system, to go along with pressure, volume, and temperature. The concept of entropy reached a position of even more significance when it was found that

> The entropy of the Universe increases in all natural processes.

This is yet another way of stating the second law of thermodynamics.

The statement of the second law just given must be interpreted with care. Although it says that the entropy of the Universe always increases in all natural processes, there are processes in which the entropy of a system decreases. That is, there are situations in which the entropy of one system (system A) decreases but in correspondence with a net increase in entropy of some other system (system B). In all cases the change in entropy of system B is greater then the change in entropy of system A.

The concept of entropy is satisfying because it enables us to present the second law of thermodynamics in the form of a mathematical statement. In the next section we will find that entropy can also be interpreted in terms of probabilities, a relationship that has profound implications for our world.

Rudolph Clausius (1822–1888). "I propose . . . to call *S* the entropy of a body, after the Greek word 'transformation.' I have designedly coined the word 'entropy' to be similar to energy, for these two quantities are so analogous in their physical significance, that an analogy of denominations seems to be helpful." *(SPL/Photo Researchers)*

EXAMPLE 12.9   **Melting a Piece of Lead**

Calculate the change in entropy when 300 g of lead melts at 327°C (600 K). Lead has a latent heat of fusion of 2.45×10^4 J/kg.

Solution   The amount of heat added to the lead to melt it is

$$Q = mL_f = (0.300 \text{ kg})(2.45 \times 10^4 \text{ J/kg}) = 7.35 \times 10^3 \text{ J}$$

From Equation 12.8, the entropy change of the lead is

$$\Delta S = \frac{Q}{T} = \frac{7.35 \times 10^3 \text{ J}}{600 \text{ K}} = \boxed{12.3 \text{ J/K}}$$

EXAMPLE 12.10   **Which Way Does the Heat Flow?**

A large cold object is at 273 K, and a large hot object is at 373 K. Show that it is impossible for a small amount of heat energy, say 8.00 J, to be transferred from the cold object to the hot object without decreasing the entropy of the Universe and hence violating the second law.

Solution   We assume here that during the heat transfer the two systems do not undergo a temperature change. This is not a necessary assumption; it is used to avoid a need for the techniques of integral calculus. The entropy change of the hot object is

$$\Delta S_h = \frac{Q}{T_h} = \frac{8.00 \text{ J}}{373 \text{ K}} = 0.0214 \text{ J/K}$$

The cold reservoir loses heat, and its entropy change is

$$\Delta S_c = \frac{Q}{T_c} = \frac{-8.00 \text{ J}}{273 \text{ K}} = -0.0293 \text{ J/K}$$

The net entropy change of the Universe is

$$\Delta S_U = \Delta S_c + \Delta S_h = -0.0079 \text{ J/K}$$

This is in violation of the law that the entropy of the Universe always increases in natural processes. That is, *the spontaneous transfer of heat from a cold object to a hot object cannot occur.*

Exercise   In the preceding example, suppose that 8.00 J of heat were transferred from the hot to the cold object. What would be the net change in entropy of the Universe?

Answer   +0.0079 J/K.

12.9

ENTROPY AND DISORDER

As you look around at the beauties of nature, it is easy to recognize that the events of natural processes have in them a large element of chance. For example, the spacing between trees in a natural forest is quite random; if you were to discover a forest where all the trees were equally spaced, you would conclude that it was a planted forest. Likewise, leaves fall to the ground with random arrangements. It would be highly unlikely to find the leaves laid out in perfectly straight rows. We can express the results of such observations by saying that **a disorderly arrangement is much more probable than an orderly one if the laws of nature are allowed to act without interference.**

Entropy originally found its place in thermodynamics, but its importance grew tremendously as the field of statistical mechanics developed. This analytical approach employs an alternative interpretation of entropy. In statistical mechanics, the behavior of a substance is described in terms of the statistical behavior of the atoms and molecules contained in the substance. One of the main products of this approach is the conclusion that

> **Isolated systems tend toward greater disorder, and entropy is a measure of that disorder.**

In light of this new view of entropy, Boltzmann found an alternative method for calculating entropy through use of the relation

$$S = k_B \ln W \qquad \textbf{[12.9]}$$

where k_B is Boltzmann's constant ($k_B = 1.38 \times 10^{-23} \, J/K$) and W is a number proportional to the probability that the system has a particular configuration.

Let us explore the meaning of this equation through a specific example. Imagine that you have a bag of 100 marbles, 50 red and 50 green. You are allowed to draw four marbles from the bag according to the following rules. Draw one marble, record its color, return it to the bag, and draw again. Continue this process until four marbles have been drawn. Note that because each marble is returned to the bag before the next one is drawn, the probability of drawing a red marble is always the same as the probability of drawing a green one.

The results of all possible drawing sequences are shown in Table 12.1. For example, the result RRGR means that a red marble was drawn first, a red one second, a green one third, and a red one fourth. This table indicates that there is only one possible way to draw four red marbles. There are four possible sequences that produce one green and three red marbles, six sequences that produce two green and two red, four sequences that produce three green and one red, and one sequence that produces all green. From Equation 12.9, we see that the state with the greatest disorder (two red and two green marbles) has the highest entropy because it is most probable. In contrast, the most ordered states (all red marbles and all green marbles) are least likely to occur and are states of lowest entropy.

TABLE 12.1
Possible Results of Drawing Four Marbles from a Bag

End Result	Possible Draws	Total Number of Same Results
All R	RRRR	1
1G, 3R	RRRG, RRGR, RGRR, GRRR	4
2G, 2R	RRGG, RGRG, GRRG, RGGR, GRGR, GGRR	6
3G, 1R	GGGR, GGRG, GRGG, RGGG	4
All G	GGGG	1

A full house is a very good hand in the game of poker. Can you calculate the probability of being dealt a full house from a standard deck of 52 cards? *(Tom Mareschel, The IMAGE Bank)*

The outcome of the draw can range between these highly ordered (lowest-entropy) and highly disordered (highest-entropy) states. Thus, entropy can be regarded as an index of how far a system has progressed from an ordered to a disordered state.

The second law of thermodynamics is really a statement of what is most probable rather than of what must be. Imagine placing an ice cube in contact with a hot piece of pizza. There is nothing in nature that absolutely forbids the transfer of heat from the ice to the much warmer pizza. Statistically, it is possible for a slow-moving molecule in the ice to collide with a faster-moving molecule in the pizza so that the slow one transfers some of its energy to the faster one. However, when the great number of molecules present in the ice and pizza are considered, the odds are overwhelmingly in favor of the transfer of energy from the faster-moving molecules to the slower-moving molecules. Furthermore, this example demonstrates that a system naturally tends to move from a state of order to a state of disorder. The initial state, in which all the pizza molecules have high kinetic energy and all the ice molecules have lower kinetic energy, is much more ordered than the final state, after heat transfer has taken place and the ice has melted.

As another example, suppose you were able to measure the velocities of all the air molecules in a room at some instant. It is very unlikely that you would find all molecules moving in the same direction with the same speed—this would, indeed, be a highly ordered state. The most probable situation is a system of molecules moving haphazardly in all directions with a distribution of speeds—a highly disordered state. Let us compare this case to that of drawing marbles from a bag. If a container held 10^{23} molecules of a gas, the probability of finding all of the molecules moving in the same direction with the same speed at some instant would be similar to that of drawing a marble from the bag 10^{23} times and getting a red marble on every draw. This is clearly an unlikely set of events.

The tendency of nature to move toward a state of disorder affects the ability of a system to do work. Consider a ball thrown toward a wall. The ball has kinetic energy, and its state is an ordered one; that is, all of the atoms and molecules of the ball move in unison at the same speed and in the same direction (apart from their random thermal motions). When the ball hits the wall, however, part of the ball's kinetic energy is transformed into the random, disordered, thermal motion of the molecules in the ball and the wall and the temperatures of the ball and the wall both increase slightly. Before the collision, the ball was capable of doing work. It could drive a nail into the wall, for example. With the transformation of part of the ordered energy into disordered thermal energy, this capability of doing work is reduced. That is, the ball rebounds with less kinetic energy than it originally had, because the collision is inelastic.

Various forms of energy can be converted to thermal energy, as in the collision between the ball and the wall, but the reverse transformation is never complete. In general, if two kinds of energy, A and B, can be completely interconverted, we say that they are the *same grade*. However, if form A can be completely converted to form B and the reverse is never complete, then form A is a *higher grade* of energy than form B. In the case of a ball hitting a wall, the kinetic energy of the ball is of a higher grade than the thermal energy contained in the ball and the wall after the collision. Therefore, when high-grade energy is converted to thermal energy, it can never be fully recovered as high-grade energy.

This conversion of high-grade energy to thermal energy is referred to as **degradation of energy.** *The energy is said to be degraded because it takes on a form that is less useful for doing work.* In other words, *in all real processes where heat transfer occurs, the energy available for doing work decreases.*

Finally, note once again that the statement that entropy must increase in all natural processes is true only for isolated systems. There are instances in which the entropy of some system decreases, but with a corresponding net increase in entropy for some other system. When all systems are taken together to form the Universe, *the entropy of the Universe always increases.*

Ultimately, the entropy of the Universe should reach a maximum. At this time the Universe will be in a state of uniform temperature and density. All physical, chemical, and biological processes will have ceased, because a state of perfect disorder implies no available energy for doing work. This gloomy state of affairs is sometimes referred to as an ultimate ''heat death'' of the Universe.

An illustration from Flammarion's novel *La Fin du Monde,* depicting the heat death of the Universe.

SPECIAL TOPIC

ENERGY CONVERSION AND THERMAL POLLUTION

In recent years, many thoughtful people have been concerned with the release of thermal energy into our environment. This **thermal pollution** affects the welfare of our planet, and it is important that we understand its sources and develop methods to control it. There are many sources of thermal pollution, a primary one being the waste heat from electric power plants. In this section we focus on the reasons for the intrinsically low efficiency of power plants and on the acceptable methods currently being used to dispose of thermal energy from these plants.

About 85% of the electric power in the United States is produced by steam engines, which either burn fossil fuel (coal, oil, or natural gas) or use nuclear fuel (uranium-235). The remaining 15% is generated by water in hydroelectric plants. The overall thermal efficiency of a modern fossil-fuel plant is about 40%. The actual efficiency of any power plant is inevitably lower than the theoretical efficiency derived from the second law of thermodynamics. The highest efficiency possible is always sought, for two reasons: first, higher efficiency results in lower fuel costs; second, thermal pollution of the environment is reduced because a highly efficient power plant gives off less waste energy.

The burning of fossil fuels in a electric power plant involves three energy conversion processes: (1) chemical to thermal energy, (2) thermal to mechanical energy, and (3) mechanical to electrical energy. These are presented schematically in Figure 12.13.

During the first step, heat energy is transferred from the burning fuel to water, which is converted to steam. In this process about 12% of the available energy is lost up the chimney. In the second step, thermal energy in the form of steam at high pressure and temperature passes through a turbine and is converted to mechanical energy. A well-designed turbine has an efficiency of about 47%. The steam, which leaves the turbine at a lower pressure, is then condensed into water and gives up heat in the process. Finally, in the third step, the turbine drives an electric generator of very high efficiency, typically 99%. Hence, the overall efficiency is the product of the efficiencies of all steps, which for the figures given is $(0.88)(0.47)(0.99) = 0.41$, or 41%. Therefore, if we account for the 12% of energy lost to the atmosphere, we conclude that the thermal energy transferred to the cooling water amounts to about 47% of the energy theoretically available from the fuel.

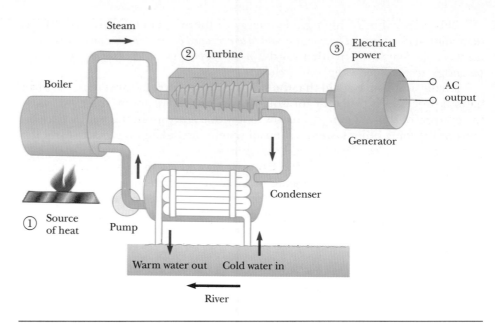

FIGURE 12.13
A schematic diagram of an electric power plant.

The steam generated by a nuclear reactor is at a lower temperature than that generated in a fossil-fuel plant. This is due primarily to material limitations in the reactor. Typical nuclear power plants have an overall efficiency of about 34%.

Cooling towers (Fig. 12.14) are commonly used to dispose of waste heat. They usually use the heat to evaporate water, which is then released to the atmosphere. Cooling towers also present environmental problems since evaporated water can cause increased precipitation, fog, and ice. Another type of tower is the dry (nonevaporative) cooling tower, which transfers heat to the atmosphere by conduction. However, this type is more expensive and cannot cool to as low a temperature as an evaporative tower.

FIGURE 12.14
A cooling tower at a nuclear reactor site. The excess heat produced in the nuclear reactor's core is transferred to steam, which in turn is expelled to the atmosphere through the cooling tower. (G. Nagele/FPG)

SUMMARY

The **work done** as a gas expands or contracts at a constant pressure is

$$W = P \Delta V \qquad \text{[12.1]}$$

The work done is negative if the gas is compressed and positive if the gas expands. In general, the work done in an expansion from some initial state to some final state is the area under the curve on a PV diagram.

From the first law of thermodynamics, when a system undergoes a change from one state to another, the **change in its internal energy, ΔU,** is

$$\Delta U = U_f - U_i = Q - W \qquad \text{[12.2]}$$

where Q is the heat transferred into (or out of) the system and W is the work done by (or on) the system. Q is positive when heat enters the system and negative when the system loses heat.

The **first law of thermodynamics** is the generalization of the law of conservation of energy that includes heat transfer.

In a **cyclic process** (one in which the system returns to its initial state), $\Delta U = 0$ and therefore $Q = W$. That is, the heat transferred into the system equals the work done during the cycle.

An **adiabatic process** is one in which no heat is transferred between the system and its surroundings ($Q = 0$). In this case, the first law gives $\Delta U = -W$. That is, the internal energy changes as a consequence of work being done by (or on) the system.

An **isobaric process** is one that occurs at constant pressure. The work done in such a process is $P \Delta V$.

A **heat engine** is a device that converts thermal energy to other forms of energy, such as mechanical and electrical energy. The work done by a heat engine in carrying a substance through a cyclic process ($\Delta U = 0$) is

$$W = Q_h - Q_c \qquad \text{[12.3]}$$

where Q_h is the heat absorbed from a hot reservoir and Q_c is the heat expelled to a cold reservoir.

The **thermal efficiency** of a heat engine is defined as the ratio of the net work done to the heat absorbed per cycle:

$$e \equiv \frac{W}{Q_h} = 1 - \frac{Q_c}{Q_h} \qquad \text{[12.4]}$$

No real heat engine operating between the kelvin temperatures T_h and T_c can exceed the efficiency of an engine operating between the same two temperatures in a **Carnot cycle**, given by

$$e_c = 1 - \frac{T_c}{T_h} \qquad \text{[12.5]}$$

Real processes proceed in an order governed by the **second law of thermodynamics**, which can be stated in two ways:

1. Heat will not flow spontaneously from a cold object to a hot object.
2. No heat engine operating in a cycle can absorb thermal energy from a reservoir and just perform an equal amount of work.

The second law can also be stated in terms of a quantity called **entropy**. The *change in entropy* of a system is equal to the heat flowing into (or out of) the system as the system changes from one state (A) to another (B), divided by the absolute temperature:

$$\Delta S \equiv \frac{\Delta Q_r}{T} \qquad\qquad [12.8]$$

One of the primary findings of statistical mechanics is that systems tend toward disorder and that entropy is a measure of this disorder. The entropy of the Universe increases in all natural processes; this is an alternative statement of the second law.

ADDITIONAL READING

S. Angrist, "Perpetual Motion Machines," *Sci. American,* January 1968, p. 114.

P. W. Atkins, *The Second Law,* Scientific American Library, W. H. Freeman and Co., New York, 1984.

L. Bryant, "Rudolf Diesel and His Rotational Engine," *Sci. American,* August 1969, p. 108.

D. A. Dicus, J. Letaw, D. Teplitz, and V. Teplitz, "The Future of the Universe," *Sci. American,* March 1983, p. 90.

W. Ehrenburg, "Maxwell's Demon," *Sci. American,* November 1967, p. 107.

K. Ford, "Probability and Entropy in Thermodynamics," *The Physics Teacher,* February 1967, p. 77.

R. Giedel, "Real Otto and Diesel Engine Cycles," *The Physics Teacher,* January 1983, p. 29.

U. Haber-Schaim, "The Role of the Second Law of Thermodynamics in Energy Education," *The Physics Teacher,* January 1983, p. 17.

W. Scaife, "The Parsons Steam Turbine," *Sci. American,* April 1985, p. 132.

G. Walker, "The Stirling Engine," *Sci. American,* August 1973, p. 80.

S. Wilson, "Sadi Carnot," *Sci. American,* August 1981, p. 134.

CONCEPTUAL QUESTIONS

Example   When a sealed Thermos bottle full of hot coffee is shaken, what changes, if any, take place in (a) the temperature of the coffee and (b) its internal energy?
Reasoning   Although no heat is transferred into or out of the system, work is done on the system as the result of the agitation. Consequently, both the temperature and the internal energy of the coffee increase.

Example   Is it possible to cool a room by leaving the door of a refrigerator open? What happens to the temperature of a room in which an air conditioner on a table in the middle of the room is left running?
Reasoning   You cannot cool a room by opening the door of a refrigerator. A refrigerator extracts heat from its interior and rejects it to its surroundings. With the door open, a refrigerator is extracting heat from one portion of the room and rejecting it to another. Thus, the overall temperature of the room does not change. The same argument applies to the air conditioner that is not vented to the outside.

Example   The first law of thermodynamics says we cannot get more out of a process than we put in, but the second law says that we cannot break even. Explain this statement.
Reasoning   The first law is a statement of conservation of energy that says we cannot devise a process that produces more energy than is put into it. The second law, additionally, says that during the operation of a heat engine, some heat must be rejected to the environment. As a result, it is theoretically impossible to construct an engine that will work with 100% efficiency.

Example   Use the first law of thermodynamics to explain why the total energy of an isolated system is always conserved.
Reasoning   An isolated system is one that exchanges neither work nor heat with its surroundings: the rest of the Universe does no work on it and transfers no thermal energy to it as the system goes from some initial to some final state. The first law says that the change in internal energy of the system is zero. Its initial internal energy is equal to its final internal energy. At the same time, however, energy within the system may change from one form

to another. Consider, for example, a bullet being stopped by a board, where the system consists of the bullet plus board. The board does negative work on the bullet, but the outside world does no work on either the bullet or board. Furthermore, the time of the collision is so short that not enough time elapses to allow any significant heat to flow to the outside world. Hence, the total energy of the system remains constant. As mechanical energy disappears, it turns into an equal amount of additional internal energy.

Example If you shake a jar full of jelly beans of two different sizes, the larger jelly beans tend to appear near the top, while the smaller ones tend to settle at the bottom. Why does this occur? Does this process violate the second law of thermodynamics?

Reasoning Shaking opens up spaces between the jelly beans. The smaller ones have a larger chance of falling down into spaces below them. The accumulation of larger beans on top and smaller ones on the bottom implies an increase in order and a decrease in one contribution to the total entropy. However, the second law is not violated. The total entropy increases as the system warms up. The increase in the internal energy of the system comes from the work required to shake the jar of beans and also from the small loss of gravitational potential energy as the beans settle together more compactly.

1. Distinguish clearly between temperature, heat, and internal energy.
2. Is it possible to convert internal energy to mechanical energy?
3. What are some factors that affect the efficiency of automobile engines?
4. A steam-driven turbine is one major component of an electric power plant. Why is it advantageous to increase the temperature of the steam as much as possible?
5. Is it possible to construct a heat engine that creates no thermal pollution?
6. Electrical energy can be converted to heat energy with an efficiency of 100%. Why is this number misleading with regard to heating a home? That is, what other factors must be considered in comparing the cost of electric heating with the cost of hot air or hot water heating?
7. Discuss three common examples of natural processes that involve increases in entropy. Be sure to account for all parts of each system under consideration.
8. Give an example of a process in nature that is nearly reversible.
9. Discuss the change in entropy of a gas that expands (a) at constant temperature; (b) adiabatically.
10. Suppose the waste heat at a power plant is exhausted to a pond of water. Could the efficiency of the plant be increased by refrigerating the water?
11. A heat pump is to be installed in a region where the average outdoor temperature in winter is $-20°C$. In view of this, why would it be advisable to place the outdoor compressor unit deep in the ground? Why are heat pumps not commonly used for heating in cold climates?
12. Can a heat pump have a coefficient of performance less than one? Explain your answer.
13. A designer of an electric heating unit describes his product as being 100% efficient. Is his claim correct?
14. An engineer claims to have developed an engine that takes in 70 000 J of heat at 500 K and expels 20 000 J at 300 K, with 10 000 J of work being done. Would the engine be a good investment? Why or why not?
15. In Israel, solar ponds have been constructed in which the Sun's energy is concentrated near the bottom of a salty pond. With the proper layering of salt in the water, convection is prevented, and temperatures of $100°C$ may be reached. Can you make a guess as to the maximum efficiency with which useful energy can be extracted from the pond?
16. All natural processes are irreversible. Give some examples of irreversible processes that occur in nature.
17. A thermodynamic process occurs in which the entropy of a system changes by $-8.0 J/K$. According to the second law of thermodynamics, what can you conclude about the entropy change of the environment?
18. Coins in the bottom of a box show 10 heads and 10 tails. The box is shaken, and there are 7 heads and 13 tails. Has the second law of thermodynamics been violated? Explain.

PROBLEMS

Section 12.1   Heat and Internal Energy

Section 12.2   Work and Heat

1. Steam moves into the cylinder of a steam engine at a constant pressure of 2.00×10^5 Pa. The diameter of the piston is 16.0 cm, and the piston travels 20.0 cm in one stroke. How much work is done during one stroke?

2. A container of volume 0.40 m³ contains 3.0 mol of argon gas at 30°C. Assuming that argon behaves as an ideal gas, find the total internal energy of the system. (*Hint:* Review the discussion of ideal

☐ indicates problems that have full solutions available in the Student Solutions Manual and Study Guide.

gases in Chapter 10 to find an expression for the energy of an ideal gas.)

3. Sketch a *PV* diagram of the following processes. (a) A gas expands at constant pressure P_1 from volume V_1 to volume V_2. It is then kept at constant volume while the pressure is reduced to P_2. (b) A gas is reduced in pressure from P_1 to P_2 while its volume is held constant at V_1. It is then expanded at constant pressure P_2 to a final volume, V_2. (c) In which of the processes is more work done? Why?

4. A gas expands from *I* to *F* along the three paths indicated in Figure 12.15. Calculate the work done by the gas along paths (a) *IAF*, (b) *IF*, and (c) *IBF*.

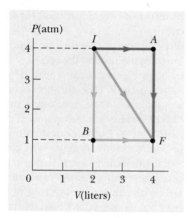

FIGURE 12.15   (Problems 4 and 11)

5. Sketch a *PV* diagram and find the work done by the gas during the following stages. (a) A gas is expanded from a volume of 1.0 L to 3.0 L at a constant pressure of 3.0 atm. (b) The gas is then cooled at constant volume until the pressure falls to 2.0 atm. (c) The gas is then compressed at a constant pressure of 2.0 atm from a volume of 3.0 L to 1.0 L. (*Note:* Be careful of signs.) (d) The gas is heated until its pressure increases from 2.0 atm to 3.0 atm at a constant volume. (e) Find the net work done during the complete cycle.

6. Gas in a container is at a pressure of 1.5 atm and a volume of 4.0 m³. What is the work done by the gas if (a) it expands at constant pressure to twice its initial volume? (b) it is compressed at constant pressure to one-quarter its initial volume?

7. An ideal gas is enclosed under a movable piston in a cylinder. The piston has a mass of 8.0 kg and an area of 5.0 cm²; it is free to slide up and down, keeping the pressure of the gas constant. How much work is done as the temperature of 0.20 mol of the gas is raised from 20°C to 300°C?

8. One mole of an ideal gas initially at a temperature of $T_0 = 0$°C undergoes an expansion, at a constant pressure of one atmosphere, to four times its original volume. (a) Calculate the new temperature of the gas, T_f. (b) Calculate the work done by the gas during the expansion.

9. A force, F, is applied to a metal wire, stretching it by the amount ΔL. Show that the work done on the wire is given by $W = -(\text{stress})(\text{strain})V$, where *V* is the volume of the wire.

Section 12.3   The First Law of Thermodynamics

10. A gas is compressed at a constant pressure of 0.30 atm from a volume of 8.0 L to 3.0 L. In the process, 400 J of heat energy flows out of the gas. (a) What is the work done by the gas? (b) What is the change in its internal energy?

11. A gas expands from *I* to *F* in Figure 12.15. The heat added to the gas is 418 J when the gas goes from *I* to *F* along the diagonal path. (a) What is the change in internal energy of the gas? (b) How much heat must be added to the gas for the indirect path *IAF* to give the same change in internal energy?

12. For a freezer to freeze one liter of water (specific heat, $c = 4186$ J/kg·°C; latent heat of fusion, $L = 3.34 \times 10^5$ J/kg) completely into ice, 500 kJ of work is required. The water is initially at 23.0°C. Calculate the change in internal energy of the water.

13. Two grams of water are sealed in a rigid container; then the water is vaporized by heating. If 1075 cal of heat is needed for vaporization, what is the change in internal energy of the water?

14. A gas is enclosed in a container fitted with a piston of cross-sectional area 0.15 m². The pressure of the gas is maintained at 6000 Pa as the piston moves inward 20 cm. (a) Calculate the work done by the gas. (b) If the internal energy of the gas decreases by 8.0 J, find the amount of heat removed from the system during the compression.

15. Two cm³ of water is boiled at atmospheric pressure to become 3342 cm³ of steam, also at atmospheric pressure. (a) Calculate the work done by the gas during this process. (b) Find the amount of heat added to the water to accomplish this process. (c) From (a) and (b), find the change in internal energy.

16. A gas at a pressure of 100 Pa is in a 2.0-m³ cavity fitted with a movable piston. The gas goes through the following stages: (1) its pressure is increased isovolumetrically to twice its original value; (2) its volume is increased isobarically to four times its original value; (3) its pressure is decreased isovolumetrically back to its original value; (4) its volume

is decreased isobarically back to its original value.
(a) Draw a *PV* diagram of this cyclic process.
(b) Calculate the work done by the gas in each stage and the total work done by the gas.

17. Consider the cyclic process described by Figure 12.16. If *Q* is negative for the process *BC* and Δ*U* is negative for the process *CA*, determine the signs of *Q*, *W*, and Δ*U* associated with each process.

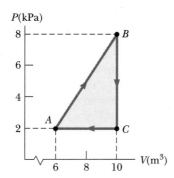

FIGURE 12.16 (Problems 17 and 18)

18. A gas is taken through the cyclic process described by Figure 12.16. (a) Find the net heat transferred to the system during one complete cycle. (b) If the cycle is reversed—that is, the process follows the path *ACBA*—what is the net heat transferred per cycle?

19. A 100-kg steel support rod in a building has a length of 2.0 m at a temperature of 20°C. The rod supports a load of 6000 kg. Find (a) the work done by the rod as the temperature increases to 40°C, (b) the heat added to the rod (assume the specific heat of steel is the same as that for iron), and (c) the change in internal energy of the rod.

20. A 1.0-kg block of aluminum is heated at atmospheric pressure so that its temperature increases from 22°C to 40°C. Find (a) the work done by the aluminum, (b) the heat added to the aluminum, and (c) the change in internal energy of the aluminum.

21. Powdered steel is placed in a container filled with oxygen and fitted with a piston that moves to keep the pressure in the container constant at one atmosphere. A chemical reaction occurs that produces heat. To keep the contents at a constant temperature of 22°C, it is necessary to remove 8.3×10^5 J of heat from the container as it contracts. During the chemical reaction, it is found that 1.5 mol of oxygen is consumed. Find the internal energy change for the system of iron and oxygen.

22. One gram of water changes from liquid to vapor

at a pressure of one atmosphere. In the process, the volume changes from 1.00 cm³ to 1670 cm³. (a) Find the work done and (b) the increase in internal energy of the water.

23. A container is placed in a water bath and held at constant volume as a mixture of fuel and oxygen is burned inside it. The temperature of the water is observed to rise during the burning (the water is also held at constant volume). (a) Consider the burning mixture to be the system. What are the signs of *Q*, Δ*U*, and *W*? (b) What are the signs of these quantities if the water bath is considered to be the system?

24. One mole of gas is initially at a pressure of 2.0 atm and a volume of 0.30 L and has an internal energy equal to 91 J. In its final state the gas is at a pressure of 1.5 atm and a volume of 0.80 L, and its internal energy equals 180 J. For the paths *IAF, IBF,* and *IF* in Figure 12.17, calculate (a) the work done by the gas and (b) the net heat transferred to the gas in the process.

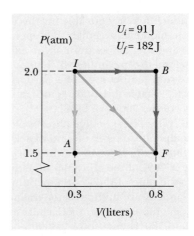

FIGURE 12.17 (Problem 24)

Section 12.4 Heat Engines and the Second Law of Thermodynamics

Section 12.5 Reversible and Irreversible Processes

Section 12.6 The Carnot Engine

25. A student claims that she has constructed a heat engine that operates between the temperatures of 200 K and 100 K with 60% efficiency. The professor does not give her credit for the project. Why not?

26. An engine absorbs 1700 J from a hot reservoir and expels 1200 J to a cold reservoir in each cycle. (a) What is the engine's efficiency? (b) How much work is done in each cycle? (c) What is the power output of the engine if each cycle lasts for 0.30 s?

27. A heat engine performs 200 J of work in each cycle and has an efficiency of 30%. For each cycle of operation, (a) how much heat is absorbed and (b) how much heat is expelled?

28. Suppose that a nuclear power plant operates in a Carnot cycle. Estimate the temperature change, in Celsius, of a river due to the exhausted heat from the plant. Assume that the input power to the boiler in the plant is 25×10^8 W, the efficiency of the use of this power is 30%, and the river flow rate is 9.0×10^6 kg/min.

29. A particular engine has a power output of 5.0 kW and an efficiency of 25%. If the engine expels 8000 J of heat in each cycle, find (a) the heat absorbed in each cycle and (b) the time required for each cycle.

30. The efficiency of a Carnot engine is 30%. The engine absorbs 800 J of heat per cycle from a hot reservoir at 500 K. Determine (a) the heat expelled per cycle and (b) the temperature of the cold reservoir.

31. A heat engine operates between two reservoirs at temperatures of 20°C and 300°C. What is the maximum efficiency possible for this engine?

32. A steam engine has a boiler that operates at 300°F, and the temperature of the exhaust is 150°F. Find the maximum efficiency of this engine.

33. The exhaust temperature of a Carnot heat engine is 300°C. What is the intake temperature if the efficiency of the engine is 30%?

34. In one cycle, a heat engine absorbs 500 J from the high-temperature reservoir and expels 300 J to a low-temperature reservoir. If the efficiency of this engine is 60% of the efficiency of a Carnot engine, what is the ratio of the low temperature to the high temperature in the Carnot engine?

35. A heat engine operates in a Carnot cycle between 80°C and 350°C. It absorbs 21 000 J of heat per cycle from the hot reservoir. The duration of each cycle is 1.0 s. (a) What is the maximum power output of this engine? (b) How much heat does it expel in each cycle?

36. Suppose a heat engine is connected to two heat reservoirs, one a pool of molten aluminum (660°C) and the other a block of solid mercury (−38.9°C). The engine runs by freezing 1.00 g of aluminum and melting 15.0 g of mercury during each cycle. The latent heat of fusion of aluminum is 3.97×10^5 J/kg, and that of mercury is 1.18×10^4 J/kg. (a) What is the efficiency of this engine? (b) How does the efficiency compare with that of a Carnot engine?

37. A power plant that uses the temperature gradient of the ocean has been proposed. The system is to operate between 20°C (surface water temperature) and 5.0°C (water temperature at a depth of about 1 km). (a) What is the maximum efficiency of such a system? (b) If the power output of the plant is 75 MW, how much thermal energy is absorbed per hour? (c) In view of your answer to part (a), do you think such a system is worthwhile?

Section 12.7   Heat Pumps and Refrigerators

38. A heat pump (Fig. 12.18) is essentially an air conditioner run backward. It extracts heat from colder air outside and deposits it in a warmer room. Typically, the ratio of the actual heat entering the room to the work done by the device's motor is 10% of the theoretical maximum. Determine the heat entering the room per joule of work done by the motor if the inside temperature is 20.0°C and the outside temperature is −5.00°C.

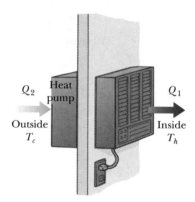

FIGURE 12.18   (Problem 38)

39. What is the coefficient of performance of a refrigerator that operates with Carnot efficiency between temperatures −3.0°C and +27°C?

Section 12.8   Entropy

Section 12.9   Entropy and Disorder

40. What is the change in entropy of 1.0 kg of water at 100°C as it changes to steam at 100°C?

41. A 70-kg log falls from a height of 25 m into a lake. If the log, the lake, and the air are all at 300 K, find the change in entropy of the Universe for this process.

42. Two 2000-kg cars, both traveling at 20 m/s, undergo a head-on collision and stick together. Find the entropy change of the Universe during the collision if the temperature is 23°C.

43. A freezer is used to freeze 1.0 L of water com-

pletely into ice. The water and the freezer remain at a constant temperature of $T = 0°C$. Determine (a) the change in the entropy of the water and (b) the change in the entropy of the freezer.

44. Prepare a table like Table 12.1 for the following occurrence. You toss four coins into the air simultaneously. Record all the possible results of the toss in terms of the numbers of heads and tails that can result. (For example, HHTH and HTHH are two possible ways in which three heads and one tail can be achieved.) (a) On the basis of your table, what is the most probable result of a toss? (b) In terms of entropy, what is the most ordered state and (c) what is the most disordered?

45. Repeat the procedure used to construct Table 12.1 (a) for the case in which you draw three marbles rather than four from your bag and (b) for the case in which you draw five rather than four.

46. Consider a standard deck of 52 playing cards that has been thoroughly shuffled. (a) What is the probability of drawing the ace of spades in one draw? (b) What is the probability of drawing any ace? (c) What is the probability of drawing any spade?

ADDITIONAL PROBLEMS

47. A nuclear power plant has a power output of 1000 MW and operates with an efficiency of 33%. If excess heat is carried away from the plant by a river with a flow rate of 1.0×10^6 kg/s, what is the rise in temperature of the flowing water?

48. A heat engine extracts heat Q_h from a hot reservoir at constant temperature T_h and rejects heat Q_c to a cold reservoir at constant temperature T_c. Find the entropy changes of the (a) hot reservoir, (b) the cold reservoir, (c) the engine, and (d) the complete system.

49. When a gas follows path 123 on the *PV* diagram in Figure 12.19, 418 J of heat flows into the system, and 167 J of work is done. (a) What is the change in the internal energy of the system? (b) How

much heat flows into the system if the process follows path 143? The work done by the gas along this path is 63.0 J. What net work would be done on or by the system if the system followed (c) path 12341? (d) path 14321? (e) What is the change in internal energy of the system in the processes described in parts (c) and (d)?

50. One mole of hydrogen gas is heated from 300 K to 420 K at constant pressure. Calculate (a) the heat energy transferred to the gas, (b) the change in the internal energy of the gas, and (c) the work done by the gas. Note that hydrogen has a specific heat of $c = 28.74$ J/mol·K.

51. A 5.0-kg block of aluminum is heated from 20°C to 90°C at atmospheric pressure. Find (a) the work done by the aluminum, (b) the amount of heat transferred to it, and (c) the increase in its internal energy.

52. One object is at a temperature of T_h and another is at a lower temperature, T_c. Use the second law of thermodynamics to show that heat transfer can only occur from the hotter to the colder object. Assume a constant-temperature process.

53. What is the minimum amount of work that must be done to extract 400 J of heat from a massive object at 0°C while rejecting heat to a hot reservoir at 20°C?

54. The interior of a refrigerator has a surface area of 4.0 m². It is insulated by a 3.0-cm-thick material that has a thermal conductivity of 2.1×10^{-2} J/(m·s·°C). The ratio of the heat extracted from the interior to the work done by the motor is 7.5% of the theoretical maximum. The temperature of the room is 24.0°C, and the temperature inside the refrigerator is 0.0°C. Determine the power required to run the compressor.

55. A refrigerator has a coefficient of performance of 3.0. The ice tray compartment is at $-20°C$, and the room temperature is 22°C. The refrigerator can convert 30 g of water at 22°C to 30 g of ice at $-20°C$ each minute. What input power is required, in watts? Ignore any cooling of the refrigerator.

56. During each cycle, a particular refrigerator absorbs 25 cal from the cold reservoir and expels 32 cal to the high-temperature reservoir. (a) If the refrigerator completes 60 cycles per second, what power is required? (b) What is the COP for this unit?

57. A 1500-kW heat engine operates at 25% efficiency. The heat energy expelled at the low temperature is absorbed by a stream of water that enters the cooling coils at 20°C. If 60 L flows across the coils per second, determine the increase in temperature of the water.

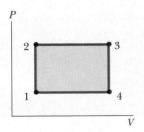

FIGURE 12.19 (Problem 49)

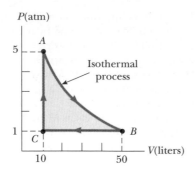

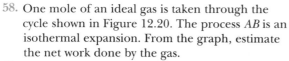

FIGURE 12.20 (Problem 58)

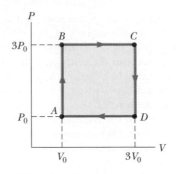

FIGURE 12.22 (Problem 61)

58. One mole of an ideal gas is taken through the cycle shown in Figure 12.20. The process AB is an isothermal expansion. From the graph, estimate the net work done by the gas.

59. One end of a copper rod is in thermal contact with a hot reservoir at $T = 500$ K, and the other end is in thermal contact with a cooler reservoir at $T = 300$ K. If 8000 J of thermal energy is transferred from one end to the other, with no change in the temperature distribution, find the entropy change of each reservoir and the total entropy change of the Universe.

60. One mole of an ideal gas is taken through the reversible cycle shown in Figure 12.21. At point A, the pressure, volume, and temperature are P_0, V_0, and T_0. In terms of R and T_0, find (a) the total heat entering the system per cycle, (b) the total heat leaving the system per cycle, (c) the efficiency of an engine operating in this reversible cycle, and (d) the efficiency of an engine operating in a Carnot cycle between the temperature extremes for this process. (*Hint:* Recall that work equals the area under a PV curve.)

61. An ideal gas initially at pressure P_0, volume V_0, and temperature T_0 is taken through the cycle de-

scribed in Figure 12.22. (a) Find the net work done by the gas per cycle in terms of P_0 and V_0. (b) What is the net heat added to the system per cycle? (c) Obtain a numerical value for the net work done per cycle for 1.00 mol of gas initially at 0°C. (See the hint for Problem 60.)

62. An electrical power plant has an overall efficiency of 15%. The plant is to deliver 150 MW of power to a city, and its turbines use coal as fuel. The burning coal produces steam at 190°C, which drives the turbines. This steam is then condensed into water at 25°C by passing through coils in contact with river water. (a) How many metric tons of coal does the plant consume each day (1 metric ton = 1×10^3 kg)? (b) What is the total cost of the fuel per year if the delivered price is $8 per metric ton? (c) If the river water is delivered at 20°C, at what minimum rate must it flow over the cooling coils in order that its temperature not exceed 25°C? (*Note:* The heat of combustion of coal is 7.8×10^3 cal/g.)

63. Every second at Niagara Falls, some 5000 m³ of

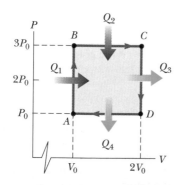

FIGURE 12.21 (Problem 60)

(Problem 63)   A dramatic photograph of Niagara Falls.
(*Jan Kopec, Tony Stone/Worldwide*)

water falls a distance of 50 m. What is the increase in entropy per second due to the falling water? (Assume a 20°C environment.)

64. The work done during an isothermal expansion of an ideal gas from an initial volume V_i to a final volume V_f is

$$W = nRT \ln\left(\frac{V_f}{V_i}\right)$$

Calculate the work done by one mole of an ideal gas at 300°C as it expands isothermally until its volume is tripled.

LAURENT HODGES   Professor of Physics, Iowa State University

ALTERNATIVE SOURCES OF ENERGY

E
S
S
A
Y

The world's major current energy resources are fossil fuels (coal, petroleum, and natural gas) and uranium. However, their disadvantages—rising costs, growing scarcity, vulnerability to political developments, and contributions to air pollution, acid rain, and global warming—have led society to investigate alternatives such as geothermal energy, fusion energy, and solar energy.

Geothermal Energy

Geothermal energy uses the heat in underground rocks or water. It exists everywhere on Earth but is currently being used only in places with natural reservoirs of geothermal steam or hot water: for home heating in Iceland, Boise, Idaho, and Klamath Falls, Oregon, and for generating electricity in Larderello, Italy, and Geysers, California.

In principle, geothermal energy can be extracted almost everywhere in the world from hot dry rocks deep underground by digging two wells, one to bring cold water down from the surface and the other to bring that water back to the surface after it has been heated. The underground rocks are fractured with pressurized water, and the water that passes through the fractures is heated. This method is not currently in commercial use.

Fusion Energy

Nuclear power plants in use around the world today derive their energy from the fissioning of heavy nuclei. An alternative is fusion energy, the production of energy from the fusing of light nuclei into heavier nuclei of lower total mass. The decrease in mass Δm leads to energy release $(\Delta m)c^2$ in accordance with Einstein's relationship, where c is the speed of light. Fusion is the energy source of sunlight and most starlight, as well as of thermonuclear, or "hydrogen," bombs.

There is on Earth an abundance of naturally occurring isotopes suitable for use in fusion energy reactions, notably deuterium (H^2) and lithium. However, controlled fusion has not yet been demonstrated in the laboratory, much less been proved economical. The difficulty is in creating conditions suitable for fusion. In particular, the temperatures must be very high so the kinetic energy of the positively charged nuclei can overcome their electrical repulsion, allowing the strong nuclear interaction to fuse the nuclei. Approaches being studied involve confinement of the highly ionized nuclei by magnetic fields or by bombardment with laser radiation.

Solar Energy

Solar energy, broadly defined, occurs in a variety of forms and currently provides more than 10% of U.S. energy use. In addition to direct solar radiation, which can be used to heat or cool buildings, to heat water, or to produce electricity, there are indirect forms of solar energy, such as water, wind, and photosynthetic power (Figure 1). These indirect forms are properly classified as solar energy because it is solar radiation that drives the hydrological cycle from which water power is derived, that drives the circulation in the atmosphere from which wind power is derived, and that provides the energy for photosynthesis.

384

FIGURE 1
A windmill in south Holland.
This is one example of an indi-
rect source of solar energy.
(David Noble/FPG)

Solar energy is often regarded as the energy source of the future, but in fact it has been the major energy source for humanity throughout almost all of recorded and unrecorded history. The great explorers all used wind energy to travel the world. Agriculture used beasts of burden, whose energy derived from their food. The U.S. industrial revolution began with solar power in the form of mechanical water power, mechanical wind power, and the combustion of wood fuel.

The most cost-effective use of solar energy is for space heating, the main use of energy in American homes. One approach—known as *active solar heating*—uses fans and ducts in air systems or pumps and pipes in liquid systems to distribute the solar energy. Active space heating systems collect solar energy through a flat-plate collector of one or two panes of glass, behind which is a black absorber. In air systems, solar heat is usually stored in a rockbed (a mass of uniformly-sized rocks contained in a volume equal to about the size of a small room), while liquid systems usually store the heat in a large water tank. Active systems were installed on many homes during the 1970s but were never very popular because they tend to be large, complex, and expensive. They require expert design and have many parts requiring constant maintenance. They also tend to be very conspicuous and to make a home look far from conventional.

A much more successful method is *passive solar heating,* in which the energy flows are by natural conduction, convection, and radiation. Passive systems typically collect solar heat through a large expanse of south-facing windows, store it in masonry, and release it at night. The masonry can be located in such places as walls, floors, partitions, or fireplaces. Passive solar homes, which cost little more than conventional homes, work very well in most parts of the United States, particularly in areas that have cold but sunny winters. It is not uncommon for well-insulated passive solar homes in cold northern climates to be heated for a whole winter for only $100 to $200 worth of natural gas heat.

Hydroelectricity

Hydroelectricity accounts for about 12% of the electricity generated in the United States. The energy source for hydroelectricity is the kinetic energy of water, originally evaporated from the oceans, precipitated in the mountains, and then flowing back to the oceans.

(a)   (b)

FIGURE 2

(a) A view of Hoover Dam on the Colorado River. This dam has a height of 221 m and a hydroelectric power capacity of 1.354 MW. *(FPG)* (b) Some of the turbines installed in the Hoover Dam hydroelectric power plant. *(Tom Campbell/FPG)*

Where available—as in the Pacific Northwest and the region served by the Tennessee Valley Authority—hydroelectric power is the cheapest form of electric power, but it has major environmental impacts, since its use requires the damming of a river and often the creation of a large reservoir (Figure 2). The many remaining good hydroelectric sites are unlikely to be developed because vast ecosystems would be adversely affected.

Photosynthetic Energy

Photosynthetic energy is a major source of energy in homes that use wood for space heating. Wood and agricultural wastes are also being used to generate electricity or industrial process heat, usually by small industries. "Gasohol" is 10% ethanol, which can be produced from grain. Wood was the main energy source in the United States during its first century. Not until the 1880s did coal supplant wood as the leading energy source, and wood remained a significant source through the rest of the 19th century.

Photovoltaic Cells

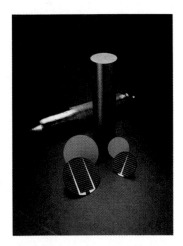

FIGURE 3

Silicon solar cells (foreground). These cells are fabricated from large man-made single crystals of silicon (background). *(© Kai-Dib Films International)*

Photovoltaic cells are made of semiconductor materials that develop a voltage when sunlight strikes them (Figure 3). Originally developed for the space program, to give satellites a source of power while in orbit, they are now widely available on small electronic devices such as "solar" calculators, radios, and battery chargers, but they are also sometimes used to power remote installations or electric fences. A home with photovoltaic cells covering a south-facing roof can collect enough electric energy to meet its needs but requires a battery system to store the energy for use at night. Such a system would currently cost over $10 000, but by the early 21st century it might be cheap enough to be economical. Research is under way, particularly in Europe and Japan, to produce better photovoltaic cells, mainly cells that are either more efficient or cheaper to manufacture.

FIGURE 4
This computer-controlled parabolic mirror tracks the Sun and heats a fluid-filled tube placed at its focus to 745°F. The superheated fluid boils water to drive steam turbine generators. Over 650 000 such mirrors are operating in the Mojave Desert in California, producing 194 MW of electric power. *(Hank Morgan/Rainbow)*

Wind Power

Wind power has long served humanity for transportation, grinding grain, and other purposes, and it is currently an alternative source of electric power (Figure 4). Many wind-driven generators were installed on midwestern farms in the 1920s and 1930s, and a surprising number of them are still in operation. More recently, several large "wind farms" have been built to serve electric utilities in the western United States. Currently they produce electric energy at a cost higher than the current price of coal-generated electricity but less than the cost of some nuclear-generated electric energy.

Questions and Problems

1. What enables glass to act as a solar collector? Would transparent plastic materials work as well?
2. Passive solar homes normally use double- or triple-pane windows. Why not single-pane windows?
3. Does a solar collector have to face exactly south?
4. A passive solar home in Iowa has 40 m² of south-facing double-pane windows and 100 metric tons (100 000 kg) of concrete heat storage. It has a backup furnace whose heat output is 40 MJ/h. Concrete has a specific heat of about 800 J/°C·kg (about one-fifth that of water). Suppose that on a sunny winter day each square meter of windows admits 16 MJ of solar energy into the home. (a) How many hours would the furnace have to run to provide the home with the same amount of heat as was admitted by the windows? (b) How much would the thermal storage warm up if 40% of the solar heat were lost through the windows, walls, and roof of the house and the other 60% went to heat the storage?

PART 3

Vibrations and Wave Motion

(Telegraph Colour Library/FPG)

As we look around us, we find many examples of objects that vibrate or oscillate: a pendulum, the strings of a guitar, an object suspended on a spring, the piston of an engine, the head of a drum, the reed of a saxophone. Most elastic objects vibrate when an impulse is applied to them; that is, once they are distorted, their shape tends to be restored to some equilibrium configuration. Even at the atomic level, the atoms in a solid vibrate about some position as if they were connected to their neighbors by imaginary springs.

Wave motion is closely related to the phenomenon of vibration. Sound waves, earthquake waves, waves on stretched strings, and water waves are all produced by vibrations. As a sound wave travels through a medium (such as air), the molecules of the medium vibrate back and forth; as a water wave travels across a pond, the water molecules vibrate up and down. When any wave travels through a medium, the particles of the medium move in repetitive cycles. Therefore, the motion of the particles bears a strong resemblance to the periodic motion of a vibrating pendulum or a mass attached to a spring.

Many other natural phenomena occur whose explanations require an understanding of vibrations and waves. Although many large structures, such as skyscrapers and bridges, appear to be rigid, they actually vibrate—a fact that must be taken into account by the architects and engineers who design and build them. To understand how radio and television work, we must comprehend the origin and nature of electromagnetic waves and how they propagate through space. Finally, much of what scientists have learned about atomic structure has come from information carried by waves; therefore, to understand the concepts and theories of atomic physics, we must first study waves and vibrations.

13 VIBRATIONS AND WAVES

Large waves such as this travel great distances over the surface of the ocean, yet the water does not flow with the wave. The crests and troughs of the wave often form repetitive patterns.
(Superstock)

This chapter constitutes a brief return to the subject of mechanics as we examine various forms of periodic motion. We concentrate especially on motion that occurs when the force on an object is proportional to the displacement of the object from its equilibrium position. When such a force acts only toward the equilibrium position, the result is a back-and-forth motion called simple harmonic motion—oscillation, or vibration, between two extreme positions for an indefinite period of time with no loss of energy. The terms *harmonic motion* and *periodic motion* are used interchangeably in this chapter. Both refer to back-and-forth motion. You may be familiar with several types of periodic motion, such as the oscillations of a mass on a spring, the motion of a pendulum, and the vibrations of a stringed musical instrument.

Since vibrations can move through a medium, we also study wave motion in this chapter. Many kinds of waves occur in nature, including sound waves, seismic waves, and electromagnetic waves. We end this chapter with a brief discussion of some terms and concepts that are common to all types of waves, and in later chapters we shall focus our attention on specific categories of waves.

13.1
HOOKE'S LAW

One of the simplest types of vibrational motion is that of a mass attached to a spring, as in Figure 13.1. Let us assume that the mass moves on a frictionless horizontal surface. If the spring is stretched or compressed a small distance, x, from its unstretched, or equilibrium, position and then released, it exerts a force on the mass. From experiment this spring force is found to obey the equation

$$F_s = -kx \qquad \text{[13.1]}$$

where x is the displacement of the mass from its unstretched ($x = 0$) position and k is a positive constant called the **spring constant.** This force law for springs was discovered by Robert Hooke in 1678 and is known as **Hooke's law.** The value of k is a measure of the stiffness of the spring. Stiff springs have large k values, and soft springs have small k values.

The negative sign in Equation 13.1 signifies that the force exerted by the spring is always directed *opposite* the displacement of the mass. When the mass is to the right of the equilibrium position, as in Figure 13.1a, x is positive and F_s is negative. This means that the force is in the negative direction, to the left. When the mass is to the left of the equilibrium position, as in Figure 13.1c, x is negative and F_s is positive, indicating that the direction of the force is to the right. Of course, when $x = 0$, as in Figure 13.1b, the spring is unstretched and $F_s = 0$. Because the spring force always acts toward the equilibrium position, it is sometimes called a restoring force. *The direction of the restoring force is such that the mass is being either pulled or pushed toward the equilibrium position.*

Let us examine the motion of the mass if it is initially pulled a distance of A to the right and released from rest. The force exerted on the mass by the spring pulls the mass back toward the equilibrium position. As the object moves toward $x = 0$, the magnitude of the force decreases (because x decreases) and reaches zero at $x = 0$. However, the mass gains speed as it moves toward the equilibrium position. In fact, the speed reaches its maximum value when $x = 0$. The momentum achieved by the mass causes it to overshoot the equilibrium position and to compress the spring. As the object moves to the left of the equilibrium position (negative x values), the force acting on it begins to increase to the right and the speed of the mass begins to decrease. The mass finally comes to a stop at $x = -A$. The process is then repeated, and the mass continues to oscillate back and forth over the same path. This type of motion is called **simple harmonic motion.**

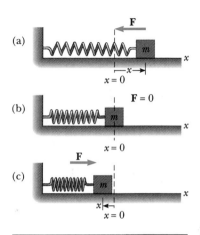

FIGURE 13.1
The force of a spring on a mass varies with the displacement of the mass from the equilibrium position, $x = 0$. (a) When x is positive (stretched spring), the spring force is to the left. (b) When x is zero (unstretched spring), the spring force is zero. (c) When x is negative (compressed spring), the spring force is to the right.

> Simple harmonic motion occurs when the net force along the direction of motion is a Hooke's law type of force — that is, when the net force is proportional to the displacement and in the opposite direction.

Not all repetitive motion over the same path can be classified as simple harmonic motion. For example, a car can start at one end of a city block, accelerate to the other end of the block, stop, back up to its original position, and then repeat the process. The car is moving back and forth over the same path, but the motion is not simple harmonic. Unless the force acting on an object along the direction of motion has the form of Equation 13.1, the object does not exhibit simple harmonic motion.

The motion of a mass suspended from a vertical spring is also simple harmonic. In this case, the force of gravity acting on the attached mass stretches the spring until equilibrium is reached (the mass is suspended and at rest). This position of equilibrium establishes the position of the mass for which $x = 0$. When the mass is stretched a distance of x beyond this equilibrium position and then released, a net force acts (in the form of Hooke's law) toward the equilibrium position. Because the net force is proportional to x, the motion is simple harmonic.

Before we can discuss simple harmonic motion in more detail, it is necessary to define a few terms.

1. The **amplitude,** A, is the *maximum distance traveled by an object away from its equilibrium position.* In the absence of friction, an object continues in simple harmonic motion and reaches a maximum displacement equal to the amplitude on each side of the equilibrium position during each cycle.
2. The **period,** T, is *the time it takes the object to execute one complete cycle of motion.*
3. The **frequency,** f, is *the number of cycles or vibrations per unit of time.*

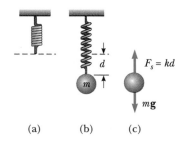

(a)   (b)   (c)

FIGURE 13.2
(Example 13.1)   Determining the spring constant. The elongation, d, of the spring is due to the suspended weight, mg. Since the upward spring force balances the weight when the system is in equilibrium, it follows that $k = mg/d$.

EXAMPLE 13.1   Measuring the Spring Constant

A common technique used to evaluate the spring constant is illustrated in Figure 13.2. The spring is hung vertically (Fig. 13.2a), and a body of mass m is attached to the lower end of the spring (Fig. 13.2b). The spring stretches a distance of d from its initial position under the action of the "load" mg. Since the spring force is upward, it must balance the weight, mg, downward *when the system is at rest.* In this case, we can apply Hooke's law to give

$$F_s = kd = mg$$

$$k = \frac{mg}{d}$$

For example, if a spring is stretched 2.0 cm by a mass of 0.55 kg, the force constant is

$$k = \frac{mg}{d} = \frac{(0.55 \text{ kg})(9.80 \text{ m/s}^2)}{2.0 \times 10^{-2} \text{ m}} = 2.7 \times 10^2 \text{ N/m}$$

EXAMPLE 13.2   Simple Harmonic Motion on a Frictionless Surface

A 0.35-kg mass attached to a spring of spring constant 130 N/m is free to move on a frictionless horizontal surface, as in Figure 13.1. If the mass is released from rest at $x = 0.10$ m, find the force on it and its acceleration at (a) $x = 0.10$ m, (b) $x = 0.050$ m, (c) $x = 0$ m, and (d) $x = -0.050$ m.

Solution   (a) The point at which the mass is released defines the amplitude of the motion. In this case, $A = 0.10$ m. The mass moves continuously between the limits 0.10 m and -0.10 m. When x is a maximum ($x = A$), the force on the mass is a maximum and is calculated as follows:

$$F = -kx$$

$$F_{max} = -kA = -(130 \text{ N/m})(0.10 \text{ m}) = \boxed{-13 \text{ N}}$$

The negative sign indicates that the force acts to the left, in the negative x direction. We can use Newton's second law to calculate the acceleration at this position:

$$F_{max} = ma_{max}$$

$$a_{max} = \frac{F_{max}}{m} = -\frac{13 \text{ N}}{0.35 \text{ kg}} = \boxed{-37 \text{ m/s}^2}$$

Again, the negative sign indicates that the acceleration is to the left.

(b) We can use the same approach to find the force and acceleration at other positions. At $x = 0.05$ m, we have

$$F = -kx = -(130 \text{ N/m})(0.050 \text{ m}) = \boxed{-6.5 \text{ N}}$$

$$a = \frac{F}{m} = -\frac{6.5 \text{ N}}{0.35 \text{ kg}} = \boxed{-19 \text{ m/s}^2}$$

Note that the acceleration of an object moving with simple harmonic motion *is not constant* since F is not constant.

(c) At $x = 0$, the spring force is zero (since $F = -kx = 0$) and the acceleration is zero. In other words, when the spring is unstretched, it exerts no force on the mass attached to it.

(d) At $x = -0.050$ m, we have

$$F = -kx = -(130 \text{ N/m})(-0.050 \text{ m}) = \boxed{6.5 \text{ N}}$$

$$a = \frac{F}{m} = \frac{6.5 \text{ N}}{0.35 \text{ kg}} = \boxed{19 \text{ m/s}^2}$$

This result shows that the force and acceleration are positive when the mass is on the negative side of the equilibrium position. This indicates that the force of the spring on the mass is acting to the right as the spring is being compressed. At the same time, the mass is slowing down as it moves from $x = 0$ to $x = -0.050$ m.

Exercise   Find the force and acceleration when $x = -0.10$ m.

Answer   13 N; 37 m/s^2.

As indicated in Example 13.2, the acceleration of an object moving with simple harmonic motion can be found by using Hooke's law as the force in the equation for Newton's second law, $F = ma$. This gives

$$-kx = ma$$

$$\boxed{a = -\frac{k}{m}x} \qquad \text{[13.2]}$$

Since the maximum value of x is defined to be the amplitude, A, we see that the acceleration ranges over the values $-kA/m$ to $+kA/m$. Equation 13.2 enables us to find the acceleration of the object as a function of its position. In subsequent sections, we shall find equations for velocity as a function of position and position as a function of time.

Earlier we stated that an object moves with simple harmonic motion when the net force acting on it is proportional to its displacement from equilibrium and is directed toward the equilibrium position. Equation 13.2 provides an alternative definition of simple harmonic motion. An object moves with simple harmonic motion if its acceleration is proportional to its displacement and is in the opposite direction to it.

13.2

ELASTIC POTENTIAL ENERGY

So far, for the most part, we have worked with three types of mechanical energy: gravitational potential energy, translational kinetic energy, and rotational kinetic energy. In Chapter 5 we briefly discussed a fourth type of mechanical energy, elastic potential energy. We now consider this form of energy in more detail.

An object has potential energy by virtue of its shape or position. As we learned in Chapter 5, an object of mass m at height h above the ground has gravitational potential energy equal to mgh. This means that the object can do work after it is released. Likewise, a compressed spring has potential energy by virtue of its shape. In this case, the compressed spring, when allowed to expand, can move an object and thus do work on it. As an example, Figure 13.3 shows a ball being projected from a spring-loaded toy gun, where the spring is compressed a distance of x. As the gun is fired, the compressed spring does work on the ball and imparts kinetic energy to it.

> **Elastic potential energy**
>
> The energy stored in a stretched or compressed spring or other elastic material is called **elastic potential energy**, PE_s, defined as
>
> $$PE_s \equiv \tfrac{1}{2}kx^2 \qquad\qquad \textbf{[13.3]}$$

Note that the elastic potential energy stored in a spring is zero when the spring is unstretched or uncompressed ($x = 0$). *Energy is stored in a spring only when it is either stretched or compressed.* Furthermore, *the elastic potential energy is a maximum when a spring has reached its maximum compression or extension.* Finally, the potential energy in a spring is always positive because it is proportional to x^2. We

FIGURE 13.3
A ball projected from a spring-loaded gun. The elastic potential energy stored in the spring is transformed into the kinetic energy of the ball.

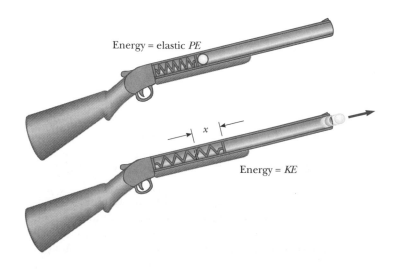

Energy = elastic *PE*

Energy = *KE*

now include this new form of energy in our equation for conservation of mechanical energy:

$$(KE + PE_g + PE_s)_i = (KE + PE_g + PE_s)_f \qquad \textbf{[13.4]}$$

If nonconservative forces such as friction are present, then the final mechanical energy does not equal the initial mechanical energy. In this case, the difference in the two energies must equal the work done by the nonconservative force, W_{nc}. According to the work-energy theorem,

$$W_{nc} = (KE + PE_g + PE_s)_f - (KE + PE_g + PE_s)_i \qquad \textbf{[13.5]}$$

As an example of the energy conversions that take place when a spring is included in the system, consider Figure 13.4. A block of mass m slides on a frictionless horizontal surface with constant velocity v_i and collides with a coiled spring. The description that follows is greatly simplified by assuming that the spring is very light and therefore has negligible kinetic energy. As the spring is compressed, it exerts a leftward force on the block. At maximum compression, the block momentarily stops (Fig. 13.4c). The initial total energy in the system before the collision (block plus spring) is the kinetic energy of the block. After the block collides with the spring and the spring is partially compressed, as in Figure 13.4b, the block has kinetic energy $\frac{1}{2}mv^2$ (where $v < v_i$) and the spring has potential energy $\frac{1}{2}kx^2$. When the block stops momentarily after colliding with the spring, the kinetic energy is zero. Since the spring force is conservative and since

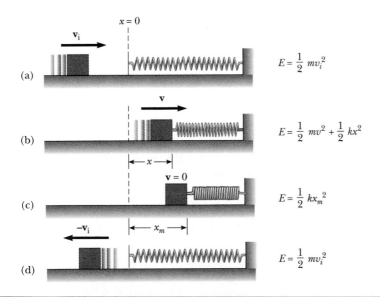

FIGURE 13.4

A block sliding on a frictionless horizontal surface collides with a light spring. (a) Initially, the mechanical energy is entirely the kinetic energy of the block. (b) The mechanical energy at some arbitrary position is the sum of the kinetic energy of the block and the elastic potential energy stored in the spring. (c) When the block comes to rest, the mechanical energy is entirely elastic potential energy stored in the compressed spring. (d) When the block leaves the spring, the mechanical energy is equal to the block's kinetic energy. The total energy remains constant.

FIGURE 13.5
(Example 13.3)   A car starts
from rest on a hill at the position shown. When the car
reaches the bottom of the hill, it
collides with a spring-loaded
guardrail.

there are no external forces that can do work on the system, *the total mechanical energy of the system consisting of the block and spring remains constant.* Thus, energy is transformed from kinetic energy of the block into potential energy stored in the spring. As the spring expands, the block moves in the opposite direction and regains all of its initial kinetic energy, as in Figure 13.4d.

EXAMPLE 13.3   Stop That Car!

A 13 000-N car starts at rest and rolls down a hill from a height of 10 m (Fig. 13.5). It then moves across a level surface and collides with a light spring-loaded guardrail. Neglecting any losses due to friction, find the maximum distance the spring is compressed. Assume a spring constant of 1.0×10^6 N/m.

Solution   The initial potential energy of the car is completely converted to elastic potential energy in the spring at the end of the trip (assuming we neglect any energy losses due to friction during the collision). Thus, conservation of energy gives

$$mgh = \tfrac{1}{2}kx^2$$

Solving for x gives

$$x = \sqrt{\frac{2mgh}{k}} = \sqrt{\frac{2(13\ 000\ \text{N})(10\ \text{m})}{1.0 \times 10^6\ \text{N/m}}} = \boxed{0.50\ \text{m}}$$

Note that it was not necessary at any point to calculate the velocity of the car to obtain a solution. This demonstrates the power of the principle of conservation of energy. One works with the initial and final energy values only, without having to consider all the details in between.

Exercise   What is the speed of the car just before it collides with the guardrail?

Answer   14 m/s.

EXAMPLE 13.4   Motion with and Without Friction

A 1.6-kg block is attached to a spring with a spring constant of 1.0×10^3 N/m (Fig. 13.1). The spring is compressed a distance of 2.0 cm, and the block is released from rest.

(a) Calculate the speed of the block as it passes through the equilibrium position, $x = 0$, if the surface is frictionless.

Solution   By using Equation 13.3, we can find the initial elastic potential energy of the spring when $x = -2.0$ cm $= -2.0 \times 10^{-2}$ m:

$$PE_s = \tfrac{1}{2}kx_i^2 = \tfrac{1}{2}(1.0 \times 10^3\ \text{N/m})(-2.0 \times 10^{-2}\ \text{m})^2 = 0.20\ \text{J}$$

Since the block is always at the same height above the Earth's surface, its gravitational potential energy remains constant. Hence, the initial potential energy stored in the spring is converted to kinetic energy at $x = 0$. That is,

$$\tfrac{1}{2}kx_i^2 = \tfrac{1}{2}mv_f^2$$

$$0.20 \text{ J} = \tfrac{1}{2}(1.6 \text{ kg})(v_f^2)$$

$$v_f = \boxed{0.50 \text{ m/s}}$$

(b) Calculate the speed of the block as it passes through the equilibrium position if a constant frictional force of 4.0 N retards its motion.

Solution   Since sliding friction is present in this situation, we know that the final mechanical energy is less than the initial mechanical energy. The work done by the frictional force for a displacement of 2.0×10^{-2} m is

$$W_{nc} = W_f = -fs = -(4.0 \text{ N})(2.0 \times 10^{-2} \text{ m}) = -0.080 \text{ J}$$

Applying Equation 13.5 to this situation gives

$$-0.080 \text{ J} = \tfrac{1}{2}(1.6 \text{ kg})(v_f^2) - 0.20 \text{ J}$$

$$v_f = \boxed{0.39 \text{ m/s}}$$

Note that this value for v_f is less than that obtained in the frictionless case. Does the result make sense?

Exercise   How far does the block travel before coming to rest? Assume a constant friction force of 4.0 N and the same initial conditions as before.

Answer   5.0 cm.

13.3

VELOCITY AS A FUNCTION OF POSITION

Conservation of energy provides a simple method of deriving an expression for the velocity of a mass undergoing periodic motion as a function of position. The mass in question is initially at its maximum extension, A (Fig. 13.6a), and is then released from rest. In this situation, the initial energy of the system is entirely elastic potential energy stored in the spring, $\tfrac{1}{2}kA^2$. As the mass moves toward the origin to some new position, x (Fig. 13.6b), part of this energy is transformed into kinetic energy, and the potential energy stored in the spring is reduced to $\tfrac{1}{2}kx^2$. Since the total energy of the system is equal to $\tfrac{1}{2}kA^2$ (the initial energy stored in the spring), we can equate this to the sum of the kinetic and potential energies at the final position:

$$\tfrac{1}{2}kA^2 = \tfrac{1}{2}mv^2 + \tfrac{1}{2}kx^2$$

Solving for v, we get

$$\boxed{v = \pm\sqrt{\dfrac{k}{m}(A^2 - x^2)}} \qquad \textbf{[13.6]}$$

This expression shows us that the speed is a maximum at $x = 0$ and zero at the extreme positions $x = \pm A$.

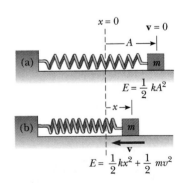

FIGURE 13.6
(a) A mass attached to a spring on a frictionless surface is released from rest with the spring extended a distance of A. Just before release, the total energy is elastic potential energy, $kA^2/2$. (b) When the mass reaches position x, it has kinetic energy $mv^2/2$ and the elastic potential energy has decreased to $kx^2/2$.

The right side of Equation 13.6 is preceded by the $\pm$ sign because the square root of a number can be either positive or negative. The sign of v that is selected depends on the circumstances of the motion. If the mass in Figure 13.6 is moving to the right, v must be positive; if the mass is moving to the left, v must be negative.

EXAMPLE 13.5   The Mass-Spring System Revisited

A 0.50-kg mass connected to a light spring with a spring constant of 20 N/m oscillates on a frictionless horizontal surface.

(a) Calculate the total energy of the system and the maximum speed of the mass if the amplitude of the motion is 3.0 cm.

Solution   From Equation 13.3, we have

$$E = PE_s = \tfrac{1}{2}kA^2 = \tfrac{1}{2}(20 \text{ N/m})(3.0 \times 10^{-2} \text{ m})^2 = \boxed{9.0 \times 10^{-3}\text{ J}}$$

When the mass is at $x = 0$, $PE_s = 0$ and $E = \tfrac{1}{2}mv_{\text{max}}^2$; therefore,

$$\tfrac{1}{2}mv_{\text{max}}^2 = 9.0 \times 10^{-3}\text{ J}$$

$$v_{\text{max}} = \sqrt{\frac{18 \times 10^{-3}\text{ J}}{0.50 \text{ kg}}} = \boxed{0.19 \text{ m/s}}$$

(b) What is the velocity of the mass when the displacement is 2 cm?

Solution   We can apply Equation 13.6 directly:

$$v = \pm\sqrt{\frac{k}{m}(A^2 - x^2)} = \pm\sqrt{\frac{20}{0.50}(3.0^2 - 2.0^2) \times 10^{-4}} = \boxed{\pm 0.14 \text{ m/s}}$$

The $\pm$ sign indicates that the mass could be moving to the right or to the left at this instant.

(c) Compute the kinetic and potential energies of the system when the displacement is 2 cm.

Solution   The results of (b) can be used to give

$$KE = \tfrac{1}{2}mv^2 = \tfrac{1}{2}(0.50 \text{ kg})(0.14 \text{ m/s})^2 = \boxed{5.0 \times 10^{-3}\text{ J}}$$

$$PE_s = \tfrac{1}{2}kx^2 = \tfrac{1}{2}(20 \text{ N/m})(2.0 \times 10^{-2} \text{ m})^2 = \boxed{4.0 \times 10^{-3}\text{ J}}$$

Note that the sum $KE + PE_s$ equals the total energy, E, found in part (a).

Exercise   For what values of x is the speed of the mass 0.10 m/s?

Answer ± 2.6 cm.

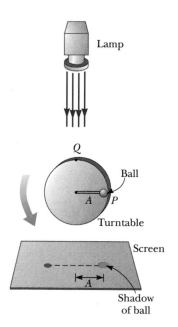

FIGURE 13.7
An experimental setup for demonstrating the connection between simple harmonic motion and uniform circular motion. As the ball rotates on the turntable with constant angular speed, its shadow on the screen moves back and forth with simple harmonic motion.

13.4

COMPARING SIMPLE HARMONIC MOTION WITH UNIFORM CIRCULAR MOTION

We can better understand and visualize many aspects of simple harmonic motion along a straight line by looking at their relationships to uniform circular motion. Figure 13.7 is a top view of an experimental arrangement that is useful for this purpose. A ball is attached to the rim of a phonograph turntable of radius A, illuminated from the side by a lamp. Rather than concentrating on the ball, let us focus our attention on the shadow that the ball casts on the screen. We find that *as the turntable rotates with constant angular speed, the shadow of the ball moves back and forth with simple harmonic motion.*

In order to understand why, let us examine Equation 13.6 more closely. This equation says that the velocity of an object moving with simple harmonic motion is related to the displacement by

$$v = C\sqrt{A^2 - x^2}$$

where C is a constant. To see that the shadow also obeys this relation, consider Figure 13.8, which shows the ball moving with a constant speed, v_0, in a direction tangent to the circular path. At this instant, the velocity of the ball in the x direction is given by $v = v_0 \sin \theta$, or

$$\sin \theta = \frac{v}{v_0}$$

Likewise, the larger triangle containing the angle θ in Figure 13.8 enables us to write

$$\sin \theta = \frac{\sqrt{A^2 - x^2}}{A}$$

Equating the right-hand sides of the last two expressions, we see that v is related to the displacement, x, as

$$\frac{v}{v_0} = \frac{\sqrt{A^2 - x^2}}{A}$$

or

$$v = \frac{v_0}{A}\sqrt{A^2 - x^2} = C\sqrt{A^2 - x^2}$$

The velocity of the ball in the x direction is related to the displacement, x, in exactly the same manner as the velocity of an object undergoing simple harmonic motion. Hence, the shadow moves with simple harmonic motion.

PERIOD AND FREQUENCY

Note that the period, T, of the shadow, which represents the time required for one complete trip back and forth, is also the time it takes the ball to make one complete circular trip on the turntable. Since the ball moves through the distance $2\pi A$ (the circumference of the circle) in the time T, the speed, v_0, of the ball around the circular path is

$$v_0 = \frac{2\pi A}{T}$$

$$T = \frac{2\pi A}{v_0}$$

However, for our purposes, let us consider only a fraction of the complete trip. Imagine that the ball moves from P to Q, a quarter of a revolution, in Figure 13.7. This requires a time interval equal to one fourth of the period, and the distance traveled by the ball is $2\pi A / 4$. Therefore,

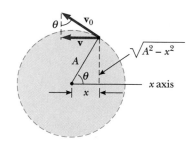

FIGURE 13.8
The ball rotates with constant speed v_0. The x component of the velocity of the ball equals the projection of $\mathbf{v_0}$ on the x axis.

$$\frac{T}{4} = \frac{2\pi A}{4v_0}$$ [13.7]

Now imagine that the motion of the shadow is equivalent to the horizontal motion of a mass on the end of a spring. During this quarter of a cycle, the shadow moves from a point where its energy is solely elastic potential energy to a point where its energy is solely kinetic energy. That is,

$$\tfrac{1}{2} kA^2 = \tfrac{1}{2} mv_0^2$$

$$\frac{A}{v_0} = \sqrt{\frac{m}{k}}$$

Substituting for A/v_0 in Equation 13.7, we find that the **period** is

The period of a mass-spring system moving with simple harmonic motion

$$T = 2\pi \sqrt{\frac{m}{k}}$$ [13.8]

This expression gives the time required for an object to make a complete cycle of its motion. Now recall that the frequency, f, is the number of cycles per unit of time. The symmetry in the units of period and frequency should lead you to see that the two must be related inversely as

$$f = \frac{1}{T}$$ [13.9]

Therefore, the **frequency** of the periodic motion is

Frequency

$$f = \frac{1}{2\pi} \sqrt{\frac{k}{m}}$$ [13.10]

The units of frequency are s^{-1}, or hertz (Hz).

EXAMPLE 13.6 That Car Needs a New Set of Shocks!

A 1300-kg car is constructed on a frame supported by four springs. Each spring has a spring constant of 20 000 N/m. If two people riding in the car have a combined mass of 160 kg, find the frequency of vibration of the car when it is driven over a pothole in the road.

Solution We assume that the weight is evenly distributed; thus, each spring supports one fourth of the load. The total mass supported by the springs is 1460 kg, and therefore each spring supports 365 kg. Hence, the frequency of vibration is

$$f = \frac{1}{2\pi} \sqrt{\frac{k}{m}} = \frac{1}{2\pi} \sqrt{\frac{20\ 000\ \text{N/m}}{365\ \text{kg}}} = \boxed{1.18\ \text{Hz}}$$

Exercise How long does it take the car to execute three complete vibrations?

Answer 2.54 s.

13.5

POSITION AS A FUNCTION OF TIME

We can obtain an expression for the position of an object moving with simple harmonic motion as a function of time by returning to the relationship between simple harmonic motion and uniform circular motion. Again, consider a ball on the rim of a rotating turntable of radius A, as in Figure 13.9. We shall refer to the circle made by the ball as the *reference circle* for the motion. Let us assume that the turntable revolves at a constant angular speed of ω. As the ball rotates on the reference circle, the angle, θ, made by the line OP with the x axis changes with time. Furthermore, as the ball rotates, the projection of P on the x axis, labeled point Q, moves back and forth along the axis with simple harmonic motion.

From the right triangle, OPQ, we see that $\cos \theta = x/A$. Therefore, the x coordinate of the ball is

$$x = A \cos \theta$$

Since the ball rotates with constant angular speed, it follows that $\theta = \omega t$ (see Chapter 7). Therefore,

$$x = A \cos(\omega t) \qquad \text{[13.11]}$$

In one complete revolution, the ball rotates through an angle of 2π rad in the period T. In other words, the motion repeats itself every T seconds. Therefore,

$$\omega = \frac{\Delta \theta}{\Delta t} = \frac{2\pi}{T} = 2\pi f \qquad \text{[13.12]}$$

where f is the frequency of the motion. Consequently, Equation 13.11 can be written

$$x = A \cos(2\pi f t) \qquad \text{[13.13]}$$

This equation represents the position of an object moving with simple harmonic motion as a function of time; it is graphed in Figure 13.10a. The curve should be familiar to you from trigonometry. Note that x varies between A and $-A$ since the cosine function varies between 1 and -1.

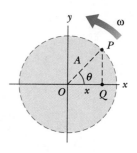

FIGURE 13.9
A reference circle. As the ball at P rotates in a circle with uniform angular speed, its projection, Q, along the x axis moves with simple harmonic motion.

Special Case I

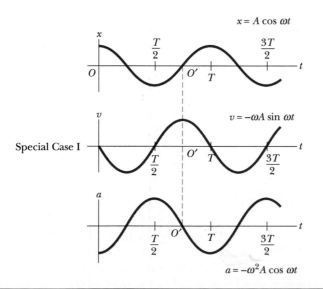

FIGURE 13.10
(a) Displacement, (b) velocity, and (c) acceleration versus time for an object moving with simple harmonic motion under the initial conditions that $x_0 = A$ and $v_0 = 0$ at $t = 0$.

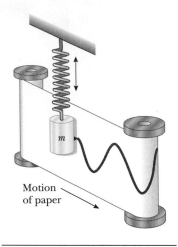

FIGURE 13.11
An experimental apparatus for
demonstrating simple harmonic
motion. A pen attached to the
oscillating mass traces out a sinusoidal wave on the moving
chart paper.

Figures 13.10b and 13.10c represent curves for velocity and acceleration
as a function of time. An end-of-chapter problem will ask you to show that the
velocity and acceleration are sinusoidal functions of time. Note that when x is a
maximum or minimum, the velocity is zero, and when x is zero, the magnitude of
the velocity is a maximum. Furthermore, when x has its maximum positive
value, the acceleration is a maximum but in the negative x direction, and when x
is at its maximum negative position, the acceleration has its maximum value in
the positive direction. These curves are consistent with our earlier discussion of
the points at which v and a reach their maximum, minimum, and zero values.

Figure 13.11 illustrates one experimental arrangement that demonstrates
simple harmonic motion. A mass connected to a spring has a marking pen
attached to it. While the mass vibrates vertically, a sheet of paper is moved
horizontally with constant speed. The pen traces out a sinusoidal pattern.

EXAMPLE 13.7   The Vibrating Mass-Spring System

Find the amplitude, frequency, and period of motion for an object vibrating at
the end of a spring if the equation for its position as a function of time is

$$x = (0.25 \text{ m}) \cos\left(\frac{\pi}{8.0} t\right)$$

Solution   We can find two of our unknowns by comparing this equation with
the general equation for such motion:

$$x = A \cos(2\pi f t)$$

By comparison, we see that

$$A = 0.25 \text{ m}$$

$$2\pi f = \frac{\pi}{8.0} \text{ s}^{-1}$$

$$f = \frac{1}{16} \text{ Hz}$$

Since the period $T = 1/f$, it follows that $T = 1/f = 16 \text{ s}$.

Exercise   What is the position of the object after 2.0 seconds have elapsed?

Answer   0.18 m.

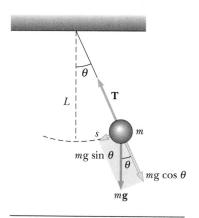

FIGURE 13.12
A simple pendulum consists of a
bob of a mass m suspended by
a light string of length L. (L is
the distance from the pivot to
the center of mass of the bob.)
The restoring force that causes
the pendulum to undergo simple harmonic motion is the component of weight tangent to the
path of motion, $mg \sin \theta$.

13.6
MOTION OF A PENDULUM

A simple pendulum is another mechanical system that exhibits periodic motion.
It consists of a small bob of mass m suspended by a light string of length L fixed at
its upper end, as in Figure 13.12. (By a light string, we mean that the string's mass
is assumed to be very small compared to the mass of the bob and hence can be
ignored.) When released, the mass swings to and fro over the same path; but is its
motion simple harmonic? In order to answer this question, we must examine the

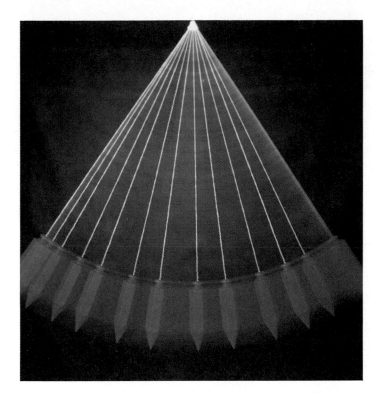

A multiflash photograph of a swinging pendulum. Is the oscillating motion simple harmonic in this case? *(Paul Silverman, Fundamental Photographs)*

force that acts as the restoring force on the pendulum. If this force is proportional to the displacement, s, then the force is of the Hooke's law form, $F = -ks$, and hence the motion is simple harmonic. Furthermore, since $s = L\theta$ in this case, we see that the motion is simple harmonic if F is proportional to the angle θ.

The component of weight tangential to the circular path is the force that acts to restore the pendulum to its equilibrium position. Thus, the restoring force is

$$F_t = -mg \sin \theta$$

From this equation, we see that the restoring force is proportional to $\sin \theta$ rather than to θ. Thus, in general, the motion of a pendulum is *not* simple harmonic. However, for small angles, less than about 15 degrees, the angle θ measured in radians and the sine of the angle are approximately equal. Therefore, if we restrict the motion to small angles, the restoring force can be written as

$$F_t = -mg\theta$$

Because $s = L\theta$, we have

$$F_t = -\left(\frac{mg}{L}\right)s$$

This equation is similar to the general form of the Hooke's law force, given by $F = -ks$, with $k = mg/L$. Thus, we are justified in saying that a pendulum undergoes simple harmonic motion only when it swings back and forth at very small amplitudes (or, in this case, small values of θ, so that $\sin \theta \cong \theta$).

We can find the period of a pendulum by first recalling that the period of a mass-spring system (Eq. 13.8) is

$$T = 2\pi\sqrt{\frac{m}{k}}$$

If we replace k with its equivalent, mg/L, we see that the period of a simple pendulum is

$$T = 2\pi\sqrt{\frac{m}{mg/L}}$$

The period of a simple pendulum depends only on L and g

$$T = 2\pi\sqrt{\frac{L}{g}}$$ [13.14]

This equation reveals the somewhat surprising result that the period of a simple pendulum depends not on mass but only on the pendulum's length and on the free-fall acceleration. Furthermore, the amplitude of the motion is not a factor as long as it is relatively small. The analogy between the motion of a simple pendulum and the mass-spring system is illustrated in Figure 13.13.

FIGURE 13.13
Simple harmonic motion for a mass-spring system and its analogy, the motion of a simple pendulum.

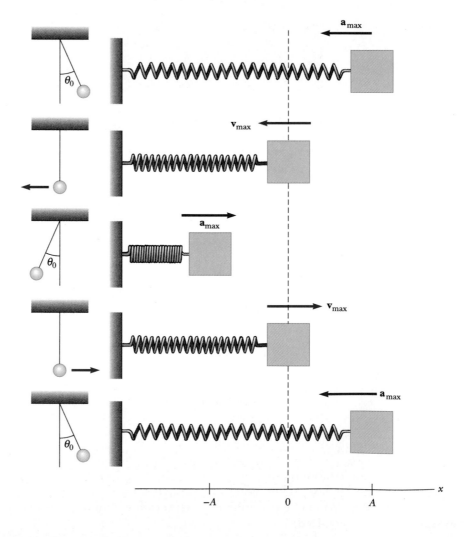

It is of historical interest to point out that it was Galileo who first noted that the motion of a pendulum was independent of its amplitude. He supposedly observed this while attending church services at a cathedral in Pisa. The pendulum he studied was a swinging chandelier that was set in motion when someone bumped it while lighting candles. Galileo was able to measure its frequency, and hence its period, by timing the swings with his pulse.

Geologists often make use of the simple pendulum and Equation 13.14 when prospecting for oil or minerals. Deposits beneath the Earth's surface can produce irregularities in the free-fall acceleration over the region being studied. A specially designed pendulum of known length is used to measure the period, which in turn is used to calculate g. Although such a measurement in itself is inconclusive, it is an important tool for geological surveys.

EXAMPLE 13.8   What Is the Height of That Tower?

A man needs to know the height of a tower, but darkness obscures the ceiling. He does note, however, that a long pendulum extends from the ceiling almost to the floor and that its period is 12.0 s. How tall is the tower?

Solution   If we use $T = 2\pi\sqrt{L/g}$ and solve for L, we get

$$L = \frac{gT^2}{4\pi^2} = \frac{(9.80 \text{ m/s}^2)(12.0 \text{ s})^2}{4\pi^2} = \boxed{35.8 \text{ m}}$$

Exercise   If the length of the pendulum were halved, what would its period of vibration be?

Answer   8.49 s.

---

*13.7

DAMPED OSCILLATIONS

The vibrating motions we have discussed so far have taken place in ideal systems, that is, systems that *oscillate indefinitely* under the action of a linear restoring force. In real systems, forces of friction retard the motion, and consequently the systems do not oscillate indefinitely. The friction reduces the mechanical energy of the system as time passes, and the motion is said to be **damped.**

Shock absorbers in automobiles (Fig. 13.14) make practical application of damped motion. A shock absorber consists of a piston moving through a liquid such as oil. The upper part of the shock absorber is firmly attached to the body of the car, and the piston is attached to a leaf spring that, with the other springs, acts as the main suspension for the car. When the car travels over a bump in the road, holes in the piston allow it to move up and down in the fluid in a damped fashion.

Damped motion varies depending on the fluid used. For example, if the fluid has a relatively low viscosity, the vibrating motion is preserved but the amplitude of vibration decreases in time and the motion ultimately ceases. This is known as *underdamped* oscillation, and its position-time curve appears in Figure 13.15. If the fluid viscosity is increased, the mass returns rapidly to equilibrium after it is released and does not oscillate. In this case, the system is said to be

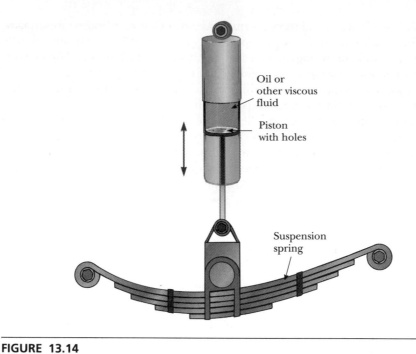

FIGURE 13.14
A cross-sectional view of a shock absorber connected to a spring in the suspension system of an automobile. The upper part of the shock absorber and the ends of the suspension spring are attached to the frame of the automobile (not shown).

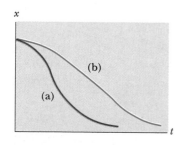

FIGURE 13.15
A graph of displacement versus time for an underdamped oscillator. Note the decrease in amplitude with time.

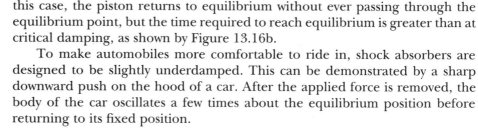

FIGURE 13.16
Plots of displacement versus time for (a) a critically damped oscillator and (b) an overdamped oscillator.

critically damped (Fig. 13.16a), and the piston returns to the equilibrium position in the shortest time possible without once overshooting the equilibrium position. If the viscosity is made greater still, the system is said to be *overdamped*. In this case, the piston returns to equilibrium without ever passing through the equilibrium point, but the time required to reach equilibrium is greater than at critical damping, as shown by Figure 13.16b.

To make automobiles more comfortable to ride in, shock absorbers are designed to be slightly underdamped. This can be demonstrated by a sharp downward push on the hood of a car. After the applied force is removed, the body of the car oscillates a few times about the equilibrium position before returning to its fixed position.

13.8
WAVE MOTION

Most of us first saw waves when, as children, we dropped a pebble into a pool of water. The disturbance created by the pebble generates water waves, which move outward until they finally reach the edge of the pool. There are a wide variety of physical phenomena that have wave-like characteristics. The world is full of waves: sound waves, waves on a string, earthquake waves, and electromagnetic waves, such as visible light, radio waves, television signals, and x-rays. All of these waves have as their source a vibrating object. Thus, we shall use the terminology and concepts of simple harmonic motion as we move into the study of wave motion.

In the case of sound waves, the vibrations that produce waves arise from such sources as a person's vocal cords or a plucked guitar string. The vibrations of electrons in an antenna produce radio or television waves, and the simple up-and-down motion of a hand can produce a wave on a string. Regardless of the type of wave under consideration, there are certain concepts common to all varieties. In the remainder of this chapter, we shall focus our attention on a general study of wave motion. In later chapters we shall study specific types of waves, such as sound and electromagnetic waves.

WHAT IS A WAVE?

As we said before, when you drop a pebble into a pool of water, the disturbance produced by the pebble excites water waves, which move away from the point at which the pebble entered the water. If you carefully examined the motion of a leaf floating near the disturbance, you would see that it moves up and down and back and forth about its original position but does not undergo any net displacement attributable to the disturbance. That is, the water wave (or disturbance) moves from one place to another *but the water is not carried with it.*

Einstein and Infeld made these remarks about wave phenomena:

> A bit of gossip starting in Washington reaches New York very quickly, even though not a single individual who takes part in spreading it travels between these two cities. There are two quite different motions involved, that of the rumor, Washington to New York, and that of the persons who spread the rumor. The wind, passing over a field of grain, sets up a wave which spreads out across the whole field. Here again we must distinguish between the motion of the wave and the motion of the separate plants, which undergo only small oscillations. . . . The particles constituting the medium perform only small vibrations, but the whole motion is that of a progressive wave. The essentially new thing here is that for the first time we consider the motion of something which is not matter, but energy propagated through matter.[1]

When we observe what is called a water wave, what we see is a rearrangement of the water's surface. Without the water there would be no wave. A wave traveling on a string would not exist without the string. Sound waves travel through air as a result of pressure variations from point to point. (We shall discuss sound waves in Chapter 14.) The wave motion we consider in this chapter corresponds to the disturbance of a body or medium. Therefore, we can consider a wave to be *the motion of a disturbance.* In a later chapter we shall discuss electromagnetic waves, which do not require a medium.

The mechanical waves discussed in this chapter require (1) some source of disturbance, (2) a medium that can be disturbed, and (3) some physical connection or mechanism through which adjacent portions of the medium can influence each other. All waves carry energy and momentum. The amount of energy transmitted through a medium and the mechanism responsible for the

[1] Albert Einstein and Leopold Infeld, *The Evolution of Physics.* New York, Simon and Schuster, 1961.

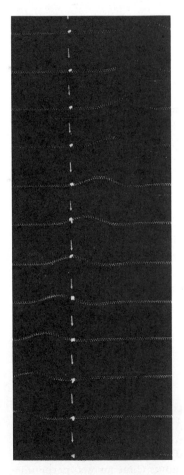

A disturbance traveling from right to left on a stretched spring.

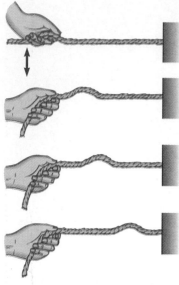

FIGURE 13.17

A wave pulse traveling along a stretched rope. The shape of the pulse is approximately unchanged as it travels.

transport of energy differ from case to case. For instance, the energy carried by ocean waves during a storm is much greater than that carried by a sound wave generated by a single human voice.

13.9

TYPES OF WAVES

One of the simplest ways to demonstrate wave motion is to flip one end of a long rope that is under tension and has its opposite end fixed, as in Figure 13.17. The bump (called a pulse) travels to the right with a definite speed. A disturbance of this type is called a **traveling wave.** Figure 13.17 shows the shape of the rope at three closely spaced times.

As a traveling wave pulse travels along the rope, *each segment of the rope that is disturbed moves perpendicularly to the wave motion.* Figure 13.18 illustrates this point for a particular tiny segment, *P.* Never does the rope move in the direction of the wave. A traveling wave such as this, in which the particles of the disturbed medium move perpendicularly to the wave velocity, is called a **transverse wave.** Figure 13.19a illustrates the formation of transverse waves on a long spring.

In another class of waves, called **longitudinal waves,** the particles of the medium undergo displacements *parallel* to the direction of wave motion. Sound waves in air, for instance, are longitudinal. Their disturbance corresponds to a series of high- and low-pressure regions that may travel through air or through any material medium with a certain speed. A longitudinal pulse can be easily produced in a stretched spring, as in Figure 13.19b. The free end is pumped back and forth along the length of the spring. This action produces compressed and stretched regions of the coil that travel along the spring, parallel to the wave motion.

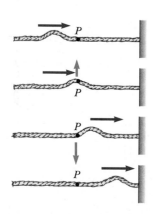

FIGURE 13.18

A pulse traveling on a stretched rope is a transverse wave. That is, any element *P* on the rope moves (blue arrows) perpendicularly to the wave motion (red arrows).

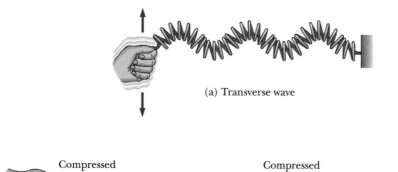

(a) Transverse wave

Compressed   Compressed

Stretched   Stretched

(b) Longitudinal wave

FIGURE 13.19

(a) A transverse wave is set up in a spring by moving one end of the spring perpendicularly to its length. (b) A longitudinal pulse along a stretched spring. The displacement of the coils is in the direction of the wave motion. For the starting motion described in the text, the compressed region is followed by a stretched region.

PICTURE OF A WAVE

Figure 13.20 shows the curved shape of a vibrating string. This pattern, sometimes called a waveform, should be familiar to you from our study of simple harmonic motion; it is a sinusoidal curve. The red curve can be thought of as a snapshot of a traveling wave taken at some instant of time, say $t = 0$; the blue curve, a snapshot of the same traveling wave at a later time. It is not difficult to imagine that this picture can as easily be used to represent a wave on water. In such a case, point A would correspond to the *crest* of the wave and point B to the low point, or *trough,* of the wave.

This same waveform can be used to describe a longitudinal wave. To see how this is done, consider a longitudinal wave traveling on a spring. Figure 13.21a is a snapshot of this wave at some instant, and Figure 13.21b shows the sinusoidal curve that represents the wave. Points where the coils of the spring are compressed correspond to the crests of the waveform, and stretched regions correspond to troughs.

The type of wave represented by the curve in Figure 13.21b is often referred to as a density or pressure wave. This is because the crests, where the spring coils are compressed, are regions of high density, and the troughs, where the coils are stretched, are regions of low density.

An alternative method for representing wave motion along a spring is through the concept of a displacement wave, shown in Figure 13.21c. In this representation, coils that are displaced the greatest distance from equilibrium in one direction are indicated by crests, and coils that are displaced the greatest distance from equilibrium in the opposite direction are represented by troughs. Both of these wave representations will be used in future sections.

13.10
FREQUENCY, AMPLITUDE, AND WAVELENGTH

Figure 13.22 illustrates a method of producing a wave on a very long string. One end of the string is connected to a blade that is set in vibration. As the blade oscillates vertically with simple harmonic motion, a traveling wave moving to the right is set up in the string. Figure 13.22 consists of views of the wave at intervals of one quarter of a period. Note that *each particle of the string, such as P, oscillates vertically in the y direction with simple harmonic motion.* This must be the case, because each particle follows the simple harmonic motion of the blade. Therefore, every segment of the string can be treated as a simple harmonic oscillator vibrating with the same frequency as the blade that drives the string.

The frequencies of the waves we shall study will range from rather low values for waves on strings and waves on water to values between 20 and 20 000 Hz (recall that 1 Hz = 1 s^{-1}) for sound waves, and much higher frequencies for electromagnetic waves.

The horizontal dashed line in Figure 13.22 represents the position of the string if no wave were present. The maximum distance the string is raised above this equilibrium value is called the **amplitude,** A, of the wave. The amplitude can also be designated as the maximum distance the string falls below the equilibrium value. For the waves we work with, the amplitudes at the crest and the trough will be identical.

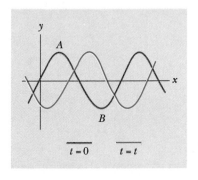

FIGURE 13.20
A one-dimensional harmonic wave traveling to the right with a speed of v. The red curve is a snapshot of the wave at $t = 0$, and the blue curve is another snapshot at some later time t.

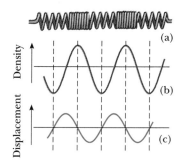

FIGURE 13.21
(a) A longitudinal wave on a spring. (b) The crests of the waveform correspond to compressed regions of the spring, and the troughs correspond to stretched regions of the spring. (c) The displacement wave.

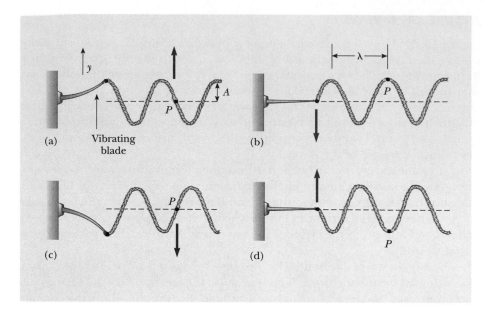

FIGURE 13.22
One method for producing traveling waves on a continuous string. The left end of the string is connected to a blade that is set in vibration. Note that every segment, such as P, oscillates vertically with simple harmonic motion.

Figure 13.22b illustrates another characteristic of a wave. The horizontal arrows show the distance between two successive points that behave identically. This distance is called the **wavelength**, λ (Greek letter lambda).

We can use these definitions to derive an expression for the velocity of a wave. We start with the defining equation for velocity:

$$v = \frac{\Delta x}{\Delta t}$$

A little reflection should convince you that a wave will advance a distance of one wavelength in a time interval equal to one period of the vibration. Thus

$$v = \frac{\lambda}{T}$$

Since the frequency is equal to the reciprocal of the period, we have

Wave velocity

$$v = f\lambda \qquad\qquad \text{[13.15]}$$

We shall apply this equation to many types of waves. For example, we shall use it often in our study of sound and electromagnetic waves.

EXAMPLE 13.9   A Traveling Wave

A wave traveling in the positive x direction is pictured in Figure 13.23. Find the amplitude, wavelength, period, and speed of the wave if it has a frequency of 8.0 Hz.

Solution   The amplitude and wavelength can be read directly off the figure:

$$A = \quad 15 \text{ cm} \qquad \lambda = 40 \text{ cm} = \quad 0.40 \text{ m}$$

The period of the wave is

$$T = \frac{1}{f} = \frac{1}{8.0} \text{ s} = \boxed{0.13 \text{ s}}$$

and the speed is

$$v = f\lambda = (8.0 \text{ s}^{-1})(0.40 \text{ m}) = \boxed{3.2 \text{ m/s}}$$

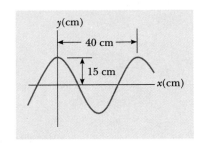

FIGURE 13.23
(Example 13.9)   A harmonic wave of wavelength $\lambda = 40$ cm and amplitude $A = 15$ cm.

EXAMPLE 13.10   Give Me a "C" Note

The note middle C on a piano has a frequency of approximately 264 Hz and a wavelength in air of 1.31 m. Find the speed of sound in air.

Solution   By direct substitution into Equation 13.15, we find that

$$v = f\lambda = (264 \text{ s}^{-1})(1.31 \text{ m}) = \boxed{346 \text{ m/s}}$$

EXAMPLE 13.11   The Speed of Radio Waves

An FM Station broadcasts at a frequency of 100 MHz (M = mega = 10^6), with a radio wave having a wavelength of 3.00 m. Find the speed of the radio wave.

Solution   As in the last example, we use Equation 13.15:

$$v = f\lambda = (100 \times 10^6 \text{ s}^{-1})(3.00 \text{ m}) = \boxed{3.00 \times 10^8 \text{ m/s}}$$

This, in fact, is the speed of *any* electromagnetic wave traveling through empty space.

Exercise   Find the wavelength of an electromagnetic wave whose frequency is 9.0 GHz = 9.0×10^9 Hz (G = giga = 10^9), which is the microwave range.

Answer   0.033 m.

13.11
THE SPEED OF WAVES ON STRINGS

In this section we focus our attention on the speed of a wave on a stretched string. Rather than deriving the equation, we use dimensional analysis to verify that the expression can be valid.

It is easy to understand why the wave speed depends on the tension in the string. If a string under tension is pulled sideways and released, the tension is responsible for accelerating a particular segment back toward its equilibrium position. The acceleration and wave speed increase with increasing tension in the string. Likewise, the wave speed is inversely dependent on the mass per unit length of the string. This is because it is more difficult to accelerate (and impart a large wave speed) to a massive string than to a light string. Thus, wave speed is directly dependent on the tension and inversely dependent on the mass per unit length of the string. The exact relationship of the wave speed, v, the tension, F, and the mass per length, μ, is

$$v = \sqrt{\frac{F}{\mu}}$$ [13.16]

From this we see that the speed of a mechanical wave, such as a wave on a string, depends only on the properties of the medium through which the disturbance travels.

Now let us verify that this expression is dimensionally correct. The dimensions of F are ML/T², and the dimensions of μ are M/L. Therefore, the dimensions of F/μ are L²/T², and those of $\sqrt{F/\mu}$ are L/T, which are indeed the dimensions of velocity. No other combination of F and μ is dimensionally correct, assuming that these are the only variables relevant to the situation.

Equation 13.16 indicates that we can increase the speed of a wave on a stretched string by increasing the tension in the string. It also shows that if we wrap a string with a metallic winding, as is done to the bass strings of pianos and guitars, we decrease the speed of a transmitted wave because the mass per unit length is increased.

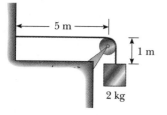

FIGURE 13.24
(Example 13.12)   The tension, F, in the string is maintained by the suspended block. The wave speed is given by the expression $v = \sqrt{F/\mu}$.

EXAMPLE 13.12   A Pulse Traveling on a String

A uniform string has a mass, M, of 0.300 kg and a length, L, of 6.00 m. Tension is maintained in the string by suspending a 2.00-kg block from one end (Fig. 13.24). Find the speed of a pulse on this string.

Solution   The tension, F, in the string is equal to the mass, m, of the block multiplied by the free-fall acceleration:

$$F = mg = (2.00 \text{ kg})(9.80 \text{ m/s}^2) = 19.6 \text{ N}$$

For the string, the mass per unit length, μ, is

$$\mu = \frac{M}{L} = \frac{0.300 \text{ kg}}{6.00 \text{ m}} = 0.0500 \text{ kg/m}$$

Therefore, the wave speed is

$$v = \sqrt{\frac{F}{\mu}} = \sqrt{\frac{19.6 \text{ N}}{0.0500 \text{ kg/m}}} = \boxed{19.8 \text{ m/s}}$$

Exercise   Find the time it takes the pulse to travel from the wall to the pulley.

Answer   0.253 s.

13.12

SUPERPOSITION AND INTERFERENCE OF WAVES

Many interesting wave phenomena in nature are impossible to describe with a single moving wave. Instead, one must analyze what happens when two or more waves attempt to pass through the same region of space. For such analyses one can use the **superposition principle**:

> If two or more traveling waves are moving through a medium, the resultant wave is found by adding together the displacements of the individual waves point by point.

Experiments show that the superposition principle is valid only when the individual waves have small amplitudes of displacement—an assumption we make in all our examples.

One consequence of the superposition principle is that *two traveling waves can pass through each other without being destroyed or even altered.* For instance, when two pebbles are thrown into a pond, the expanding circular waves do not destroy each other. In fact, the ripples pass through each other. Likewise, when sound waves from two sources move through air, they pass through each other. The sound one hears at a given location is the result of both disturbances.

Figures 13.25a and 13.25b show two waves of the same amplitude and frequency. If at some instant of time these two waves attempted to travel through the same region of space, the resultant wave at that instant would have a shape like Figure 13.25c. For example, suppose these are water waves of amplitude 1 m. At the instant they overlap so that crest meets crest and trough meets trough, the resultant wave has an amplitude of 2 m. Waves coming together like this are said to be *in phase* and to undergo **constructive interference**.

Figures 13.26a and 13.26b show two similar waves. In this case, however, the crest of one coincides with the trough of the other; that is, one wave is *inverted* relative to the other. The resultant wave, shown in Figure 13.26c, is seen to be a state of complete cancellation. If these were water waves coming together, one of the waves would be trying to pull an individual drop of water upward at the same instant the other wave was trying to pull it downward. The result would be no motion of the water at all. In such a situation, the two waves are said to be 180°

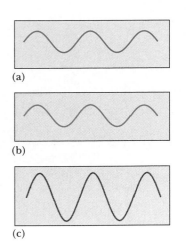

FIGURE 13.25
Constructive interference. If two waves having the same frequency and amplitude are in phase, as in (a) and (b), the resultant wave when they combine (c) has the same frequency as the individual waves but twice their amplitude.

Interference patterns produced by outward-spreading waves from many drops of liquid falling into a body of water. *(Martin Dohrn/Science Photo Library)*

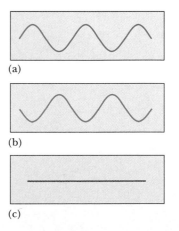

FIGURE 13.26
Destructive interference. When two waves with the same frequency and amplitude are 180° out of phase, as in (a) and (b), the result when they combine (c) is complete cancellation.

FIGURE 13.27
Interference of water waves produced in ripple tank. The sources of the waves are two objects that vibrate perpendicularly to the surface of the tank. *(Courtesy of Central Scientific Company)*

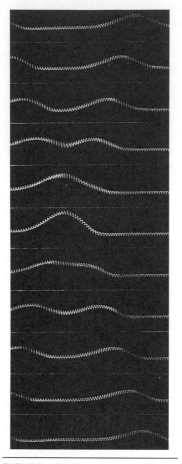

FIGURE 13.28
Constructive interference. Superposition of two equal and symmetric pulses traveling in opposite directions on a stretched spring. *(Courtesy of Education Development Center, Newton, Mass.)*

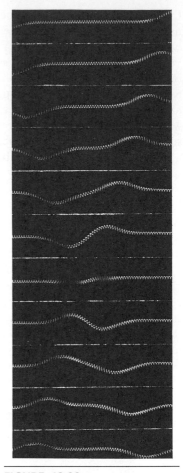

FIGURE 13.29
Destructive interference. Superposition of two symmetric pulses traveling in opposite directions, where one is inverted relative to the other. *(Courtesy of Education Development Center, Newton, Mass.)*

out of phase and to undergo **destructive interference.** Figure 13.27 illustrates the interference of water waves produced in a ripple tank.

Figure 13.28 shows constructive interference in two pulses moving toward each other along a stretched spring; Figure 13.29 shows destructive interference in two pulses. Notice in each case that, when the two pulses separate, their shapes are unchanged, as if they had never met!

13.13

REFLECTION OF WAVES

In our discussion so far, we have assumed that the waves being analyzed could travel indefinitely without striking anything. Obviously, such conditions are not realized in practice. Whenever a traveling wave reaches a boundary, part or all of

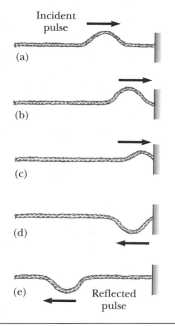

FIGURE 13.30
The reflection of a traveling wave at the fixed end of a stretched string. Note that the reflected pulse is inverted, but its shape remains the same.

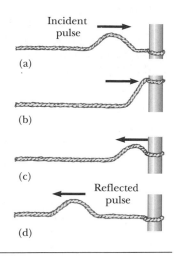

FIGURE 13.31
The reflection of a traveling wave at the free end of a stretched string. In this case, the reflected pulse is not inverted.

the wave is reflected. For example, consider a pulse traveling on a string that is fixed at one end (Fig. 13.30). When the pulse reaches the wall, it is reflected.

Note that *the reflected pulse is inverted*. This can be explained as follows. When the pulse meets the wall, the string exerts an upward force on the wall. By Newton's third law, the wall must exert an equal and opposite (downward) reaction force on the string. This downward force causes the pulse to invert upon reflection.

Now consider a case in which the pulse arrives at the string's end, which is attached to a ring of negligible mass that is free to slide along the post without friction (Fig. 13.31). Again the pulse is reflected, but this time it is not inverted. Upon reaching the post, the pulse exerts a force on the ring, causing it to accelerate upward. The ring is then returned to its original position by the downward component of the tension.

An alternative method of showing that a pulse is reflected without inversion when it strikes a free end of a string is to send a pulse down a string hanging vertically. When the pulse hits the free end, it is reflected without inversion, similarly to the pulse in Figure 13.31.

SUMMARY

Simple harmonic motion occurs when the net force along the direction of motion is a **Hooke's law** type of force—that is, when the net force is proportional to the displacement and in the opposite direction:

$$F_s = -kx \qquad \textbf{[13.1]}$$

The time required for one complete vibration is called the **period** of the motion. The inverse of the period is the **frequency** of the motion, which is the number of oscillations per second.

When an object is moving with simple harmonic motion, its **acceleration** as a function of location is

$$a = -\frac{k}{m}x \qquad \textbf{[13.2]}$$

The energy stored in a stretched or compressed spring or other elastic material is called **elastic potential energy:**

$$PE_s \equiv \tfrac{1}{2}kx^2 \qquad \textbf{[13.3]}$$

The **velocity** of an object as a function of position, when the object is moving with simple harmonic motion, is

$$v = \pm\sqrt{\frac{k}{m}(A^2 - x^2)} \qquad \textbf{[13.6]}$$

The **period** of an object of mass m moving with simple harmonic motion while attached to a spring of spring constant k is

$$T = 2\pi\sqrt{\frac{m}{k}} \qquad \textbf{[13.8]}$$

Because $f = 1/T$, the **frequency** of a mass-spring system is

$$f = \frac{1}{2\pi}\sqrt{\frac{k}{m}} \qquad \textbf{[13.10]}$$

The **position** of an object as a function of time, when the object is moving with simple harmonic motion, is

$$x = A\cos(2\pi f t) \qquad \textbf{[13.13]}$$

A **simple pendulum** of length L moves with simple harmonic motion for small angular displacements from the vertical, with a period of

$$T = 2\pi\sqrt{\frac{L}{g}} \qquad \textbf{[13.14]}$$

The period is independent of the suspended mass.

A **transverse wave** is one in which the particles of the medium move perpendicularly to the direction of the wave velocity. An example is a wave on a stretched string.

A **longitudinal wave** is one in which the particles of the medium move parallel to the direction of the wave velocity. An example is a sound wave.

The relationship of the velocity, wavelength, and frequency of a wave is

$$v = f\lambda \qquad \textbf{[13.15]}$$

The speed of a wave traveling on a stretched string of mass per unit length μ and under tension F is

$$v = \sqrt{\frac{F}{\mu}} \qquad \textbf{[13.16]}$$

The **superposition principle** states that, if two or more traveling waves are moving through a medium, the resultant wave is found by adding the individual waves together point by point. When waves meet crest to crest and trough to trough, they undergo **constructive interference.** When crest meets trough, the waves undergo complete **destructive interference.**

When a wave pulse reflects from a rigid boundary, the pulse is inverted. When the boundary is free, the reflected pulse is not inverted.

ADDITIONAL READING

W. Bascom, ''Ocean Waves,'' *Sci. American,* August 1959, p. 74.

W. Bascom, *Waves and Beaches: The Dynamics of the Ocean Surface,* New York, Doubleday Anchor Books, 1980.

A. Einstein and L. Infeld, *The Evolution of Physics,* New York, Simon and Schuster, 1961.

B. Gilbert and P. Glanz, ''Springs: Distorted and Combined,'' *The Physics Teacher,* October 1983, p. 430.

S. D. Kelby and R. P. Middleton, ''The Vibrations of Hand Bells,'' *Physics Education,* Vol. 15, 1980, p. 320.

J. Oliver, ''Long Earthquake Waves,'' *Sci. American,* March 1959, p. 14.

T. D. Rossing, ''The Physics of Kettledrums,'' *Sci. American,* November 1982, p. 172.

R. A. Waldron, *Waves and Oscillations,* Momentum Series, Princeton, N.J., Van Nostrand, 1964.

CONCEPTUAL QUESTIONS

Example A pendulum bob is made with a ball filled with water. What would happen to the frequency of vibration of this pendulum if there were a hole in the ball that allowed water to leak out slowly?

Reasoning The frequency of a pendulum is given by $f = \dfrac{1}{T} = \dfrac{1}{2\pi}\sqrt{\dfrac{g}{L}}$. Since the frequency depends only on the length of the pendulum and the free-fall acceleration, and not on the mass, the frequency will not change. However, as the water leaks out of the ball, the frequency first decreases (as the distance from the pivot to the center of mass of the ball increases). After the water level in the ball reaches the half-way point, the frequency begins to increase again until the ball is empty. At that point, the frequency is the same as it was when the ball was completely filled with water.

Example Does the acceleration of a simple harmonic oscillator remain constant during its motion? Is the acceleration ever zero? Explain.

Reasoning In simple harmonic motion, the acceleration is not constant. It is zero whenever the object passes through its equilibrium position, to the right whenever the object is to the left of its equilibrium position, and to the left whenever the object is to the right of its equilibrium position. In general, the acceleration is proportional to the displacement but oppositely directed.

Example When all the strings on a guitar are stretched to the same tension, will the speed of a wave along one of the more massive bass strings be greater or less than the speed of a wave on one of the lighter strings?

Reasoning The speed of a wave on a string decreases as the mass of the string is increased. Thus, the wave will move more slowly on the more massive bass strings.

Example Explain why the kinetic and potential energies of a mass-spring system can never be negative.

Reasoning The kinetic energy is proportional to the square of the speed, while the potential energy is proportional to the square of the displacement. The square of either a positive or a negative quantity is positive; thus, both forms of energy must be positive.

1. What is the total distance traveled by an object moving with simple harmonic motion in a time interval equal to its period if its amplitude is A?

2. Given a spring, a clock, and a single known mass, devise an experiment to measure the force constant of the spring.

3. A mass-spring system undergoes simple harmonic motion with an amplitude of A. Does the total energy change if the mass is doubled but the amplitude is not changed? Do the kinetic and potential energies depend on the mass? Explain.

4. Why is a pendulum such a reliable time-keeping aid despite the fact that its oscillations gradually decrease in amplitude with the passing of time?

5. What happens to the period of a simple pendulum if the pendulum's length is doubled? What happens if the suspended mass is doubled?

6. If a pendulum clock keeps perfect time at the base of

a mountain, will it also keep perfect time when moved to the top of the mountain? Explain.

7. If a grandfather clock were running slow, how could we adjust the length of the pendulum to correct the time?

8. Fill your bathroom sink or tub with water. Touch the surface gently with one fingertip and observe the wave pulses. What happens when the wavefronts reach the water's edge? Try touching the surface with the long edge of a ruler. Make a sketch of the resulting wavefronts.

9. If a duck floats on a lake, it bobs up and down but essentially remains in one place as waves pass by. Explain why the duck is not carried along by the wave motion.

10. A flag waves in the breeze. The waves on the flag consist of straight sinusoidal fronts. Describe the motions of various points on the flag. What type of waves are they?

(Question 10)   *(Andre Gallant/The Image Bank)*

11. How do transverse waves differ from longitudinal waves?

12. A solid may transport a longitudinal wave as well as a transverse wave, but a fluid can only transport a longitudinal wave. Why?

13. How would you create a longitudinal wave in a stretched spring? Would it be possible to create a transverse wave in a spring?

14. If you were to shake the end of a stretched rope up and down three times each second, what would be the period of the waves set up in the rope?

15. What happens to the wavelength of a wave on a string when the frequency is doubled? Assume that the tension in the string remains the same.

16. What happens to the speed of a wave on a string when the frequency is doubled? Assume that the tension in the string remains the same.

17. Consider a wave traveling on a stretched rope. What is the difference, if any, between the speed of the wave and the speed of a small section of the rope?

18. By what factor would you have to multiply the tension in a stretched string in order to double the wave speed?

19. If a long rope is hung from a ceiling and waves are sent up the rope from its lower end, the waves do not ascend with constant speed. Explain.

20. Suppose two pulses are moving toward one another on a string. How can you tell whether such pulses reflect off of or pass through one another?

21. When two waves interfere, can the resultant wave be larger than either of the two original waves? Under what conditions?

22. The motion of the Earth going around the Sun is periodic, with a period of 1 year. Is this motion simple harmonic? Explain.

23. A certain pendulum clock runs faster in the winter than in the summer. Explain why. (*Hint:* See Chapter 10.)

PROBLEMS

Section 13.1   Hooke's Law

1. A small ball is set in horizontal motion by rolling it with a speed of 3.00 m/s across a room 12.0 m long, between two walls. Assume that the collisions made with each wall are perfectly elastic and that the motion is perpendicular to the two walls. (a) Show that the motion is periodic, and determine its period. (b) Is this motion simple harmonic? Explain.

2. A 0.40-kg mass is attached to a spring with a spring constant 160 N/m so that the mass is allowed to move on a horizontal frictionless surface.

The mass is released from rest when the spring is compressed 0.15 m. Find (a) the force on the mass and its acceleration at this instant and (b) the maximum values of the force and acceleration.

3. A load of 50 N attached to a spring hanging vertically stretches the spring 5.0 cm. The spring is now placed horizontally on a table and stretched 11 cm. (a) What force is required to stretch the spring by this amount? (b) Plot a graph of force (on the *y* axis) versus spring displacement from the equilibrium position along the *x* axis.

4. A ball dropped from a height of 4.00 m makes a

☐ indicates problems that have full solutions available in the Student Solutions Manual and Study Guide.

perfectly elastic collision with the ground. Assuming no energy lost due to air resistance, (a) show that the motion is periodic and (b) determine the period of the motion. (c) Is the motion simple harmonic? Explain.

Section 13.2   Elastic Potential Energy

5. The mat of a trampoline is held by 32 springs, each having a spring constant of 5000 N/m. A 40.0-kg person jumps from a 1.93-m-high platform onto the trampoline. Determine the stretch of each of the springs. Assume that the springs were initially unstretched and that they stretch equally.

6. An archer pulls her bow string back 0.40 m by exerting a force that increases uniformly from zero to 230 N. (a) What is the equivalent spring constant of the bow? (b) How much work is done in pulling the bow?

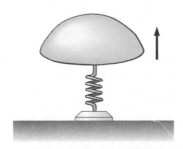

(Problem 6)   *(Eric Lars Baleke/Black Star)*

7. In an arcade game a 0.1-kg disk is shot across a frictionless horizontal surface by compressing it against a spring and releasing it. If the spring has a spring constant of 200 N/m and is compressed from its equilibrium position by 6 cm, find the speed with which the disk slides across the surface.

8. A child's toy consists of a piece of plastic attached to a spring (Fig. 13.32). The spring is compressed

FIGURE 13.32   (Problem 8)

against the floor a distance of 2 cm, and the toy is released. If the toy has a mass of 100 g and rises to a maximum height of 60 cm, estimate the force constant of the spring.

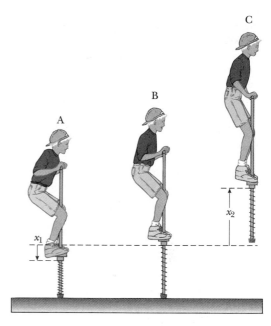

FIGURE 13.33   (Problem 9)

9. A child's pogo stick (Fig. 13.33) stores energy in a spring ($k = 2.5 \times 10^4$ N/m). At position A ($x_1 = -0.10$ m) the spring compression is a maximum, and the child momentarily stops. At position B ($x = 0$) the spring is relaxed, and the child is moving upward. At position C the child again stops at the top of the jump. Assume that the combined mass of the child and pogo stick is 25 kg. (a) Calculate the total energy of the system if both the spring and the gravitational potential energies are zero at $x = 0$. (b) Determine x_2. (c) Calculate the speed of the child at $x = 0$. (d) Determine the acceleration of the child at $x = x_1$.

10. A slingshot consists of a light leather cup, containing a stone, that is pulled back against two rubber bands. It takes a force of 30 N to stretch the bands 1.0 cm. (a) What is the potential energy stored in the bands when a 50-g stone is placed in the cup and pulled back 0.20 m from the equilibrium position? (b) With what speed does the stone leave the slingshot?

11. A simple harmonic oscillator has a total energy of E. (a) Determine the kinetic and potential energies when the displacement is one-half the amplitude. (b) For what value of the displacement does the kinetic energy equal the potential energy?

12. A 1.5-kg block is attached to a spring with a spring constant of 2000 N/m. The spring is then stretched a distance of 0.30 cm and the block is released from rest. (a) Calculate the speed of the block as it passes through the equilibrium position if no friction is present. (b) Calculate the speed of the block as it passes through the equilibrium position if a constant frictional force of 2.0 N retards its motion. (c) What would be the strength of the frictional force if the block reached the equilibrium position the first time with zero velocity?

13. A 10.0-g bullet is fired into and embeds in a 2.00-kg block attached to a spring with a spring constant of 19.6 N/m and whose mass is negligible. How far is the spring compressed if the bullet has a speed of 300 m/s just before it strikes the block, and the block slides on a frictionless surface? (*Hint:* You must use conservation of momentum in this problem. Why?)

14. The spring constant of the spring in Figure 13.34 is 19.6 N/m, and the mass of the object is 1.5 kg. The spring is unstretched and the surface is frictionless. A constant 20-N force is applied horizontally to the object as shown. Find the speed of the object after it has moved a distance of 0.30 m.

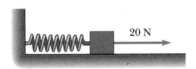

FIGURE 13.34 (Problem 14)

Section 13.3 Velocity as a Function of Position

15. A 50.0-g mass is attached to a horizontal spring with a spring constant of 10.0 N/m and released from rest with an amplitude of 25.0 cm. What is the velocity of the mass when it is halfway to the equilibrium position if the surface is frictionless?

16. A mass of 0.40 kg connected to a light spring with a spring constant of 19.6 N/m oscillates on a frictionless horizontal surface. If the spring is compressed 4.0 cm and released from rest, determine (a) the maximum speed of the mass, (b) the speed of the mass when the spring is compressed 1.5 cm, and (c) the speed of the mass when the spring is stretched 1.5 cm. (d) For what value of x does the speed equal one-half the maximum speed?

17. A particle executes simple harmonic motion with an amplitude of 3.00 cm. (a) At what displacement from the equilibrium position is the particle's speed equal to one-half its maximum speed? (b) What fraction of its maximum speed does the particle have when it is halfway to its amplitude?

18. At an outdoor market, a bunch of bananas is set into oscillatory motion with an amplitude of 20.0 cm on a spring with a spring constant of 16.0 N/m. It is observed that the maximum speed of the bunch of bananas is 40.0 cm/s. What is the weight of the bananas in newtons?

19. A mass-spring system oscillates with an amplitude of 3.5 cm. If the spring constant is 250 N/m and the mass is 0.50 kg, determine (a) the mechanical energy of the system, (b) the maximum speed of the mass, and (c) the maximum acceleration.

Section 13.4 Comparing Simple Harmonic Motion with Uniform Circular Motion

20. A 200-g mass is attached to a spring and executes simple harmonic motion with a period of 0.25 s. If the total energy of the system is 2.0 J, find (a) the force constant of the spring and (b) the amplitude of the motion.

21. A ball moves with constant speed of 5.00 m/s in a circular path of radius 0.400 m (see Fig. 13.8). Find the x component of the velocity of the ball when θ equals (a) 0°, (b) 60°, (c) 90°, (d) 180°, (e) 270°.

22. Consider the simplified single-piston engine in Figure 13.35. If the wheel rotates at a constant angular speed of ω, explain why the piston rod oscillates in simple harmonic motion.

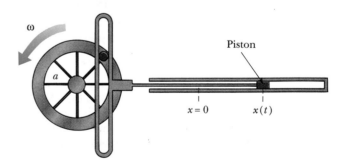

FIGURE 13.35 (Problem 22)

23. While riding behind a car traveling at 3.0 m/s, you notice that one of the car's tires has a small hemispherical boss on its rim, as in Figure 13.36. (a) Explain why the boss, from your viewpoint behind the car, executes simple harmonic motion. (b) If the radii of the car's tires are 0.30 m, what is the boss's period of oscillation?

24. A ball rotates in a circle of radius 20.0 cm, making

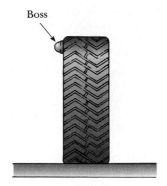

FIGURE 13.36 (Problem 23)

one complete revolution every 2.00 s. What are (a) the speed of the ball, (b) the frequency of motion in hertz, and (c) the angular speed of the ball?

25. The amplitude of a system moving with simple harmonic motion is doubled. Determine the changes in (a) total energy, (b) maximum velocity, (c) maximum acceleration, and (d) period.

26. A spring stretches 3.9 cm when a 10-g mass is hung from it. If a total mass of 25 g attached to this spring oscillates in simple harmonic motion, calculate the period of motion.

27. The frequency of vibration of a mass-spring system is 5.00 Hz when a 4.00-g mass is attached to the spring. What is the force constant of the spring?

28. When four people with a combined mass of 320 kg sit down in a car, they find that the car drops 0.80 cm lower on its springs. Then they get out of the car and bounce it up and down. What is the frequency of the car's vibration if its mass (empty) is 2.0×10^3 kg?

Section 13.5 Position as a Function of Time

29. A 2.00-kg mass on a frictionless horizontal track is attached to the end of a horizontal spring whose force constant is 5.00 N/m. The mass is displaced 3.00 m to the right from its equilibrium position and then released, which initiates simple harmonic motion. (a) What is the force (magnitude and direction) acting on the mass 3.50 s after it is released? (b) How many times does the mass oscillate in 3.50 s?

30. The motion of an object is described by the equation

$$x = (0.3 \text{ m}) \cos(\pi t/3)$$

Find (a) the position of the object at $t = 0$ and $t = 0.20$ s, (b) the amplitude of the motion, (c) the frequency of the motion, and (d) the period of the motion.

31. Given that $x = A \cos(\omega t)$ is a sinusoidal function of time, show that v (velocity) and a (acceleration) are also sinusoidal functions of time. (*Hint:* Use Equations 13.6 and 13.2.)

32. An object on a spring vibrates with a period of 3.00 s. The motion is initiated by releasing the object from its point of maximum displacement, $x = A$. After what time interval is the object first at (a) $x = A/2$? (b) $x = -A/2$? (c) $x = 0$?

33. A spring of negligible mass stretches 3.0 cm from its relaxed length when a force of 7.5 N is applied. A 0.50-kg particle rests on a frictionless horizontal surface and is attached to the free end of the spring. The particle is pulled horizontally so that it stretches the spring 5.0 cm and is then released from rest at $t = 0$. (a) What is the force constant of the spring? (b) What are the period, frequency (f), and angular frequency (ω) of the motion? (c) What is the total energy of the system? (d) What is the amplitude of the motion? (e) What are the maximum velocity and the maximum acceleration of the particle. (f) Determine the displacement, x, of the particle from the equilibrium position at $t = 0.50$ s.

Section 13.6 Motion of a Pendulum

34. A simple 2.00-m-long pendulum oscillates in a location where $g = 9.80$ m/s². How many complete oscillations does it make in 5.00 min?

35. A light balloon filled with helium of density 0.180 kg/m³ is tied to a light string of length $L = 3.00$ m. The string is tied to the ground, forming an "inverted" simple pendulum (Fig. 13.37a). If the balloon is displaced slightly from equilibrium, as in Figure 13.37b, show that the motion is simple harmonic, and determine the period of the motion. Take the density of air to be 1.29 kg/m³. (*Hint:* Use an analogy with the simple pendulum discussed in the text, and see Chapter 9.)

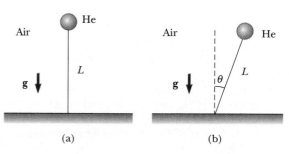

(a)   (b)

FIGURE 13.37 (Problem 35)

36. An aluminum clock pendulum having a period of 1.00 s keeps perfect time at 20.0°C. (a) When placed in a room at a temperature of −5.0°C, will

it gain time or lose time? (b) How much time will it gain or lose every hour? (*Hint:* See Chapter 10.)

37. A pendulum clock that works perfectly on Earth is taken to the Moon. (a) Does it run fast or slow there? (b) If the clock is started at 12:00:00 A.M., what will it read after one Earth day (24.0 h)? Assume that the free-fall acceleration on the Moon is 1.63 m/s^2.

38. A "seconds" pendulum is one that moves through its equilibrium position once each second. (The period of the pendulum is 2.000 s.) The length of a seconds pendulum is 0.9927 m at Tokyo and 0.9942 m at Cambridge, England. What is the ratio of the free-fall acceleration at these two locations?

39. A visitor to a lighthouse wishes to determine the height of the tower. She ties a spool of thread to a small rock to make a simple pendulum, which she hangs down the center of a spiral staircase of the tower. The period of oscillation is 9.40 s. What is the height of the tower?

Section 13.10 Frequency, Amplitude, and Wavelength

40. A bat can detect small objects such as an insect whose size is approximately equal to one wavelength of the sound the bat makes. If bats emit a chirp at a frequency of 60.0 kHz, and if the speed of sound in air is 340 m/s, what is the smallest insect a bat can detect?

41. If the frequency of oscillation of the wave emitted by an FM radio station is 88.0 MHz, determine the wave's (a) period of vibration and (b) wavelength. (Radio waves travel at the speed of light.)

42. Two points, *A* and *B*, on the Earth are at the same longitude and 60.0° apart in latitude. An earthquake at point *A* sends two waves toward *B*. A transverse wave travels along the surface of the Earth at 4.50 km/s, and a longitudinal wave travels through the body of the Earth at 7.80 km/s. (a) Which wave arrives at *B* first? (b) What is the time difference between the arrivals of the two waves at *B*? Take the radius of the Earth to be 6.37 × 10^6 m.

43. A piano emits frequencies that range from a low of about 28 Hz to a high of about 4200 Hz. Find the range of wavelengths spanned by this instrument. The speed of sound in air is approximately 343 m/s.

44. A harmonic wave is traveling along a rope. It is observed that the oscillator that generates the wave completes 40.0 vibrations in 30.0 s. Also, a given maximum travels 425 cm along the rope in 10.0 s. What is the wavelength?

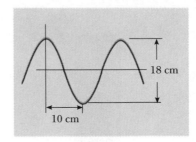

FIGURE 13.38   (Problem 45)

45. A wave traveling in the positive *x* direction is pictured in Figure 13.38. Find the (a) amplitude, (b) wavelength, (c) period, and (d) speed of the wave if it has a frequency of 25.0 Hz.

46. The distance between two successive maxima of a certain transverse wave is 1.20 m. Eight crests, or maxima, pass a given point along the direction of travel every 12.0 s. Calculate the wave speed.

47. (a) How long does it take light to reach us from the Sun, 9.30 × 10^7 mi away? (The speed of light is 3.00 × 10^8 m/s.) (b) An astronaut communicates with Earth from the Moon (3.84 × 10^8 m away). How long does it take his signal to reach us?

48. A 100-car train standing on the siding is started in motion by the engine. If there is 5.00 cm of slack between cars and the engine moves at a constant speed of 40.0 cm/s, how much time is required for the pulse to travel the length of the train?

49. A sound wave, traveling at 343 m/s, is emitted by the foghorn of a tugboat. An echo is heard 2.60 s later. How far away is the reflecting object?

Section 13.11 The Speed of Waves on Strings

50. A circus performer stretches a tightrope between two towers. He strikes one end of the rope and sends a wave along it toward the other tower. He notes that it takes the wave 0.80 s to reach the opposite tower, 20 m away. If one meter of the rope has a mass of 0.35 kg, find the tension in the tightrope.

51. A telephone cord is 4.0 m long. The cord has a mass of 0.20 kg. If a transverse wave pulse travels from the receiver to the telephone box in 0.10 s, what is the tension in the cord?

52. A light string of mass 10.0 g and length *L* = 3.00 m has its ends tied to two walls that are separated by the distance *D* = 2.00 m. Two masses, each of mass *m* = 2.00 kg, are suspended from the string as in Figure 13.39. If a wave pulse is sent from point *A*, how long does it take to travel to point *B*?

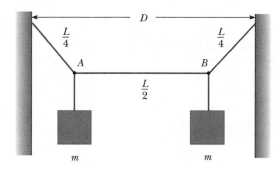

FIGURE 13.39 (Problem 52)

53. Transverse waves with a speed of 50.0 m/s are to be produced on a stretched string. A 5.00-m length of string with a total mass of 0.0600 kg is used. (a) What is the required tension in the string? (b) Calculate the wave speed in the string if the tension is 8.0 N.

54. The elastic limit of a length of steel wire is 2.70×10^9 Pa. What is the maximum speed at which transverse wave pulses can propagate along this wire without exceeding this stress? (The density of the steel is 7.86 g/cm^3.)

55. Transverse waves travel at 20.0 m/s on a string that is under a tension of 6.00 N. What tension is required for a wave speed of 30.0 m/s in the same string?

56. When a steel wire stretched between two clamps is plucked, waves travel along the wire with a speed of 80 m/s. A second wire, of the same material but with twice the length and twice the radius of the first, is stretched between two points under the same tension. At what speed do transverse waves travel along the second wire?

Section 13.12 Superposition and Interference of Waves

Section 13.13 Reflection of Waves

57. A series of pulses of amplitude 0.15 m are sent down a string that is attached to a post at one end. The pulses are reflected at the post and travel back along the string without loss of amplitude. What is the amplitude at a point on the string where two pulses are crossing, (a) if the string is rigidly attached to the post? (b) if the end at which reflection occurs is free to slide up and down?

58. A wave of amplitude 0.30 m interferes with a second wave of amplitude 0.20 m. What are the (a) largest and (b) smallest absolute values of the resultant displacement that can occur, and under what conditions will these maxima and minima occur?

ADDITIONAL PROBLEMS

59. A uniform 100-g, 2.00-m-long rope hangs vertically from a ceiling. A 3.00-kg block is attached to the bottom of the rope. If a wave pulse is set up in the rope, what is its speed when it is 0.5 m from the ceiling?

60. The position of a 0.30-kg object attached to a spring is described by

$$x = (0.25 \text{ m}) \cos(0.4\pi t)$$

Find (a) the amplitude of the motion, (b) the spring constant, (c) the position at $t = 0.30$ s, and (d) the object's velocity at $t = 0.30$ s.

61. A pendulum clock with a period of 1.0000 s works perfectly at sea level. (a) If the clock is moved to the top of Mount Everest (elevation = 8848 m), does it run slower or faster? (b) How much time does it lose or gain in one hour? (*Hint:* Recall from Chapter 7 that g varies with elevation.)

62. A string is 50.0 cm long and has a mass of 3.00 g. A wave travels at 5.00 m/s along this string. A second string has the same length but half the mass of the first. If the two strings are under the same tension, what is the speed of a wave along the second string?

63. Tension is maintained in a string as in Figure 13.40. The observed wave speed is 24 m/s when the suspended mass is 3.0 kg. (a) What is the mass per unit length of the string? (b) What is the wave speed when the suspended mass is 2.0 kg?

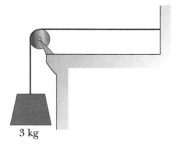

3 kg

FIGURE 13.40 (Problem 63)

64. A spring with a spring constant of 30.0 N/m is stretched 0.200 m from its equilibrium position. How much work must be done to stretch it an additional 0.100 m?

65. A 500-g block is released from rest and slides down a frictionless track that begins 2.00 m above the horizontal, as shown in Figure 13.41. At the bottom of the track, where the surface is horizontal, the block strikes and sticks to a light spring

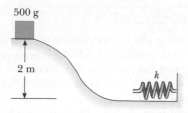

FIGURE 13.41   (Problem 65)

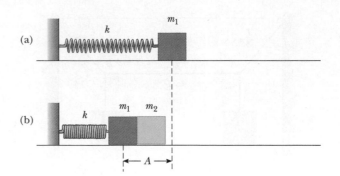

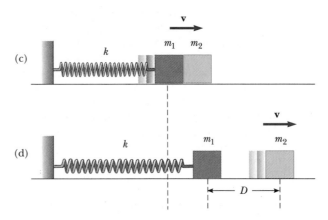

with a spring constant of 20.0 N/m. Find the maximum distance the spring is compressed.

66. A 30-m steel wire and a 20-m copper wire, both with 1.0-mm diameters, are connected end to end and stretched to a tension of 150 N. How long will it take a transverse wave to travel the entire length of the two wires? The density of steel is 7800 kg/m^3, and the density of copper is 8920 kg/m^3.

67. A spring in a toy gun has a spring constant of 9.80 N/m and can be compressed 20.0 cm beyond the equilibrium position. A 1.00-g pellet resting against the spring is propelled forward when the spring is released. (a) Find the muzzle speed of the pellet. (b) If the pellet is fired horizontally from a height of 1.00 m above the floor, what is its range?

68. A mass, m_1 = 9.0 kg, is in equilibrium while connected to a light spring of constant k = 100 N/m that is fastened to a wall as in Figure 13.42a. A second mass, m_2 = 7.0 kg, is slowly pushed up against mass m_1, compressing the spring by the amount A = 0.20 m, as shown in Figure 13.42b. The system is then released, causing both masses to start moving to the right on the frictionless surface. (a) When m_1 reaches the equilibrium point, m_2 loses contact with m_1 (Fig. 13.42c) and moves to the right with speed v. Determine the value of v. (b) How far apart are the masses when the spring is fully stretched for the first time (Fig. 13.42d)? (*Hint:* First determine the period of oscillation and the amplitude of the m_1-spring system after m_2 loses contact with m_1.)

69. A 5.50-g mass is suspended from a cylindrical sample of collagen 3.50 cm long and 2.00 mm in diameter. If the mass vibrates up and down with a frequency of 36.0 Hz, what is the Young's modulus of the collagen?

70. Figure 13.43 shows a crude model of an insect wing. The mass m represents the entire mass of the wing, which pivots about the fulcrum F. The spring represents the surrounding connective tissue. Motion of the wing corresponds to vibration of the spring. Suppose the mass of the wing is

FIGURE 13.42   (Problem 68)

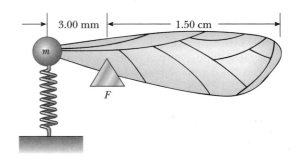

FIGURE 13.43   (Problem 70)

0.30 g and the effective spring constant of the tissue is 4.7×10^{-4} N/m. If the mass m moves up and down a distance of 2.0 mm from its position of equilibrium, what is the maximum speed of the outer tip of the wing?

71. A 5.00-g bullet moving with an initial speed of 400 m/s is fired into and passes through a 1.00-kg block, as in Figure 13.44. The block, initially at rest on a frictionless horizontal surface, is connected to a spring with a spring constant of

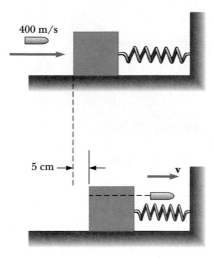

FIGURE 13.44 (Problem 71)

900 N/m. If the block moves 5.00 cm to the right after impact, find (a) the speed at which the bullet emerges from the block and (b) the energy lost in the collision.

72. A particle of mass m slides inside a hemispherical bowl of radius R. Show that for small displacements from equilibrium, the particle exhibits simple harmonic motion like that of a simple pendulum.

73. An 8.00-kg block travels on a rough horizontal surface and collides with a spring. The speed of the block just before the collision is 4.00 m/s. As it rebounds to the left with the spring uncompressed, the block travels at 3.00 m/s. If the coefficient of kinetic friction between the block and the surface is 0.400, determine (a) the loss in mechanical energy due to friction while the block is in contact with the spring and (b) the maximum distance the spring is compressed.

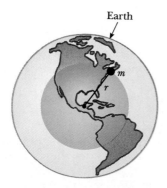

FIGURE 13.45 (Problem 74)

74. Assume that a hole is drilled through the center of the Earth. It can be shown that an object of mass m at a distance of r from the center of the Earth is pulled toward the center of the Earth only by the mass in the shaded portion of Figure 13.45. Write down Newton's law of gravitation for an object at distance r from the center of the Earth, and show that the force on it is of Hooke's law form, $F = kr$, with an effective force constant of $k = (4/3)\pi\rho Gm$, where ρ is the density of the Earth and G is the gravitational constant.

75. A 3.00-kg mass is fastened to a light spring that passes over a pulley (Fig. 13.46). The pulley is frictionless, and its inertia may be neglected. The mass is released from rest when the spring is unstretched. If the mass drops 10.0 cm before stopping, find (a) the spring constant of the spring and (b) the speed of the mass when it is 5.00 cm below its starting point.

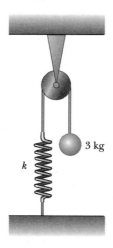

FIGURE 13.46 (Problem 75)

76. A 60.0-kg fire fighter slides down a pole while a constant frictional force of 300 N retards his motion. A horizontal 20.0-kg platform is supported by a spring at the bottom of the pole, to cushion the fall. The fire fighter starts from rest 5.00 m above the platform, and the spring constant is 2500 N/m. Find (a) the fire fighter's speed just before he collides with the platform and (b) the maximum distance the spring is compressed. (Assume that the frictional force acts during the entire motion.)

77. A 2.0-kg block situated on a rough incline is connected to a light spring with a spring constant of 100 N/m (Fig. 13.47). The block is released from rest when the spring is unstretched, and the pulley

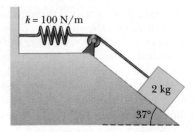

FIGURE 13.47   (Problem 77)

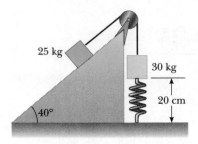

FIGURE 13.48   (Problem 78)

is frictionless. The block moves 20 cm down the incline before stopping. Find the coefficient of kinetic friction between the block and the incline.

78. A 25-kg block is connected to a 30-kg block by a light string that passes over a frictionless pulley. The 30-kg block is connected to a light spring of force constant 200 N/m, as in Figure 13.48. The spring is unstretched when the system is as shown in the figure, and the incline is smooth. The 25-kg block is pulled 20 cm down the incline (so that the 30-kg block is 40 cm above the floor) and is released from rest. Find the speed of each block when the 30-kg block is 20 cm above the floor (that is, when the spring is unstretched).

SOUND

14

Even when silent, this organ at the Mormon Tabernacle in Salt Lake City conveys a sense of the power of sound waves. (Courtesy of Henry Leap)

Sound waves are the most important example of longitudinal waves. In this chapter we discuss the characteristics of sound waves—how they are produced, what they are, and how they travel through matter. We then investigate what happens when sound waves interfere with each other. The insights gained in this chapter will help you understand how we hear what we hear.

14.1

PRODUCING A SOUND WAVE

Whether it conveys the shrill whine of a jet engine or the soft melodies of a pop singer, any sound wave has its source in a vibrating object. Musical instruments produce sounds in a variety of ways. For example, the sound from a clarinet is produced by a vibrating reed, the sound from a drum by the vibration of the taut drum head, the sound from a piano by vibrating strings, and the sound from a singer by vibrating vocal cords.

Sound waves are longitudinal waves traveling through a medium, such as air. In order to investigate how sound waves are produced, we focus our attention on the tuning fork, a common device for producing pure musical notes. A tuning fork consists of two metal prongs, or tines, that vibrate when struck. Their vibration disturbs the air near them, as shown in Figure 14.1. (The amplitude of vibration of the tine in Figure 14.1 has been greatly exaggerated for clarity.) When a tine swings to the right, as in Figure 14.1a, the air molecules in front of its

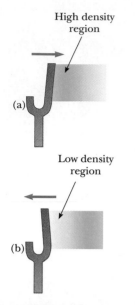

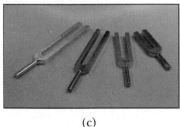

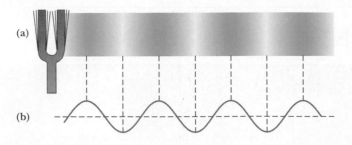

FIGURE 14.2
(a) As the tuning fork vibrates, a series of condensations and rarefactions moves outward, away from the fork. (b) The crests of the wave correspond to condensations, and the troughs correspond to rarefactions.

FIGURE 14.1
A vibrating tuning fork. (a) As the right tine of the fork moves to the right, a high-density region (condensation) of air is formed in front of its movement. (b) As the right tine moves to the left, a low-density region (rarefaction) of air is formed behind it. (c) A set of tuning forks. *(Courtesy of Henry Leap and Jim Lehman)*

movement are forced closer together than normal. Such a region of high molecular density and high air pressure is called a **compression** or **condensation.** This compression moves away from the fork like a ripple on a pond. When the tine swings to the left, as in Figure 14.1b, the molecules to the right of the tine spread apart and the density and air pressure in this region are then lower than normal. Such a region of lower-than-normal density is called a **rarefaction.** Molecules to the right of the rarefaction in the figure move to the left. Hence, the rarefaction itself moves to the right, following the previously produced compression.

As the tuning fork continues to vibrate, a succession of condensations and rarefactions forms and spreads out from it. The resultant pattern in the air is somewhat like that pictured in Figure 14.2a. We can use a sinusoidal curve to represent a sound wave, as in Figure 14.2b. Notice that there are crests in the sinusoidal wave at the points where the sound wave has condensations and troughs where the sound wave has rarefactions. The molecular motion of the sound waves is superposed on the random thermal motion of the atoms and molecules (discussed in Chapter 10).

14.2
CHARACTERISTICS OF SOUND WAVES

As already noted, the general motion of air molecules near a vibrating object is back and forth between regions of compression and rarefaction. Back-and-forth molecular motion in the direction of the disturbance is characteristic of a longitudinal wave. The motion of the medium particles in a **longitudinal sound wave** is *back and forth along the direction in which the wave travels.* In contrast, in a **transverse sound wave** the vibrations of the medium are *at right angles to the direction of travel of the wave.*

CATEGORIES OF SOUND WAVES

Sound waves fall into three categories covering different ranges of frequencies. **Audible waves** are longitudinal waves that lie within the range of sensitivity of the human ear, approximately 20 to 20 000 Hz. **Infrasonic waves** are longitudinal waves with frequencies below the audible range. Earthquake waves are an example. **Ultrasonic waves** are longitudinal waves with frequencies above the audible range for humans. For example, certain types of whistles produce ultrasonic waves. Some animals, such as dogs, can hear the waves emitted by these whistles, even though humans cannot.

14.3

THE SPEED OF SOUND

The speed of a sound wave in a liquid depends on the liquid's compressibility and inertia. If the liquid has a bulk modulus of B and an equilibrium density of ρ, the **speed of sound** is

$$v = \sqrt{\frac{B}{\rho}} \qquad \text{[14.1]}$$

Speed of sound in a liquid

Recall from Chapter 9 that bulk modulus is defined as the ratio of the change in pressure, ΔP, to the resulting fractional change in volume, $\Delta V/V$:

$$B = -\frac{\Delta P}{\Delta V/V} \qquad \text{[14.2]}$$

Note that B is always positive because an increase in pressure (positive ΔP) results in a decrease in volume. Hence the ratio $\Delta P/\Delta V$ is always negative.

It is interesting to compare Equation 14.1 with Equation 13.16 for the speed of transverse waves on a string, $v = \sqrt{F/\mu}$, discussed in Chapter 13. In both cases, the wave speed depends on an elastic property of the medium (B or F) and on an inertial property of the medium (ρ or μ). In fact, the speed of all mechanical waves follows an expression of the general form

$$v = \sqrt{\frac{\text{elastic property}}{\text{inertial property}}}$$

Another example of this general form is the **speed of a longitudinal wave in a solid,** which is

$$v = \sqrt{\frac{Y}{\rho}} \qquad \text{[14.3]}$$

where Y is the Young's modulus of the solid, defined as the longitudinal stress divided by the longitudinal strain (Equation 9.3), and ρ is the density of the solid.

Table 14.1 lists the speeds of sound in various media. As you can see, the speed of sound is much higher in solids than in gases. This makes sense because the molecules in a solid are closer together than those in a gas and hence respond more rapidly to a disturbance. In general, sound travels more slowly in liquids than in solids because liquids are more compressible and hence have smaller bulk moduli.

The speed of sound also depends on the temperature of the medium. For sound traveling through air, the relationship between the speed of sound and temperature is

$$v = (331 \text{ m/s})\sqrt{1 + \frac{T}{273}} \qquad \text{[14.4]}$$

TABLE 14.1
Speeds of Sound in Various Media

Medium	v (m/s)
Gases	
Air (0°C)	331
Air (100°C)	366
Hydrogen (0°C)	1290
Oxygen (0°C)	317
Helium (0°C)	972
Liquids at 25°C	
Water	1490
Methyl alcohol	1140
Seawater	1530
Solids	
Aluminum	5100
Copper	3560
Iron	5130
Lead	1320
Vulcanized rubber	54

where 331 m/s is the speed of sound in air at $0°C$ and T is the temperature in degrees Celsius. Using this equation, one finds that at $20°C$ the speed of sound in air is approximately 343 m/s.

EXAMPLE 14.1   Speed of Sound in a Liquid

Find the speed of sound in water, which has a bulk modulus of about 2.1×10^9 Pa and a density of about 1.0×10^3 kg/m^3.

Solution   From Equation 14.1, we find that

$$v_{\text{water}} = \sqrt{\frac{B}{\rho}} = \sqrt{\frac{2.1 \times 10^9 \text{ Pa}}{1.0 \times 10^3 \text{ kg/m}^3}} \approx \boxed{1500 \text{ m/s}}$$

EXAMPLE 14.2   Sound Waves in a Solid Bar

If a solid bar is struck at one end with a hammer, a longitudinal pulse propagates down the bar. Find the speed of sound in a bar of aluminum, which has a Young's modulus of 7.0×10^{10} Pa and a density of 2.7×10^3 kg/m^3.

Solution   From Equation 14.3 we find that

$$v_{\text{Al}} = \sqrt{\frac{Y}{\rho}} = \sqrt{\frac{7.0 \times 10^{10} \text{ Pa}}{2.7 \times 10^3 \text{ kg/m}^3}} \approx \boxed{5100 \text{ m/s}}$$

This is a typical value for the speed of sound in solids (see Table 14.1).

---

*14.4

ENERGY AND INTENSITY OF SOUND WAVES

As the tines of a tuning fork move back and forth through the air, they exert a force on a layer of air and cause it to move. In other words, the tines do work on the layer of air. The fact that the fork pours energy into the air as sound is one of the reasons that the vibration of the fork slowly dies out. (Other factors, such as the energy lost to friction as the tines bend, also are responsible for the diminution of movement.)

> We define the **intensity, I, of a wave** to be the rate at which energy flows through a unit area, A, perpendicularly to the direction of travel of the wave.

In equation form this is

$$I \equiv \frac{1}{A} \frac{\Delta E}{\Delta t} \qquad \text{[14.5]}$$

Equation 14.5 can be written in an alternative form if you recall that the rate of transfer of energy is defined as power. Thus,

Intensity of a wave

$$I \equiv \frac{\text{power}}{\text{area}} = \frac{P}{A} \qquad \text{[14.6]}$$

This Boeing 747 jet airplane produces sounds at high intensity levels. *(Telegraph Colour Library/ FPG)*

where P is the sound power passing through A, measured in watts, and the intensity has units of watts per square meter.

The faintest sounds the human ear can detect at a frequency of 1000 Hz have an intensity of about 1×10^{-12} W/m². This intensity is called the **threshold of hearing.** The loudest sounds the ear can tolerate have an intensity of about 1 W/m² (the **threshold of pain**). At the threshold of hearing, the increase in pressure in the ear is approximately 3×10^{-5} Pa over normal atmospheric pressure. Since atmospheric pressure is about 1×10^5 Pa, this means the ear can detect pressure fluctuations as small as about 3 parts in 10^{10}! Also, at the threshold of hearing, the maximum displacement of an air molecule is about 1×10^{-11} m. This is a remarkably small number! If we compare this result with the diameter of a molecule (about 10^{-10} m), we see that the ear is an extremely sensitive detector of sound waves.

In a similar manner, one finds that the loudest sounds the human ear can tolerate correspond to a pressure increase of about 29 Pa over normal atmospheric pressure corresponding to a maximum displacement of air molecules of 1×10^{-5} m.

INTENSITY LEVELS IN DECIBELS

As was just mentioned, the human ear can detect a wide range of intensities, with the loudest tolerable sounds having intensities about 1.0×10^{12} times greater than those of the faintest detectable sounds. However, the most intense sound is not perceived as being 1.0×10^{12} times louder than the faintest sound. This is because the sensation of loudness is approximately logarithmic in the human ear. The relative loudness of a sound is called the **intensity level** or **decibel level,** β, and is defined as

$$\beta \equiv 10 \log \left(\frac{I}{I_0} \right)$$ [14.7]   Intensity level

The constant I_0 is the reference intensity level, taken to be the sound intensity at the threshold of hearing ($I_0 = 1.0 \times 10^{-12}$ W/m²), and I is the intensity level β, where β is measured in decibels (dB). (The word *decibel* comes from the name of the inventor of the telephone, Alexander Graham Bell [1847–1922].) On this scale, the threshold of pain ($I = 1.0$ W/m²) corresponds to an intensity level of

TABLE 14.2
**Intensity Levels in Decibels
for Different Sources**

Source of Sound	β (dB)
Nearby jet airplane	150
Jackhammer, ma- chine gun	130
Siren, rock concert	120
Subway, power mower	100
Busy traffic	80
Vacuum cleaner	70
Normal conversation	50
Mosquito buzzing	40
Whisper	30
Rustling leaves	10
Threshold of hearing	0

$\beta = 10 \log(1/1 \times 10^{-12}) = 10 \log(10^{12}) = 120$ dB, and the threshold of hearing corresponds to $\beta = 10 \log(1 \times 10^{-12}/1 \times 10^{-12}) = 0$ dB. Nearby jet airplanes can create intensity levels of 150 dB, and subways and riveting machines have levels of 90 to 100 dB. The electronically amplified sound heard at rock concerts can be at levels of up to 120 dB, the threshold of pain. Prolonged exposure to such high intensity levels can seriously damage the ear. Earplugs are recommended whenever intensity levels exceed 90 dB. Recent evidence suggests that noise pollution, which is common in most large cities and in some industrial environments, may be a contributing factor to high blood pressure, anxiety, and nervousness. Table 14.2 gives some idea of the intensity levels of various sounds.

EXAMPLE 14.3   Intensity Levels of Sound

Calculate the intensity level of a sound wave having an intensity of (a) 1.0×10^{-12} W/m^2; (b) 1.0×10^{-11} W/m^2; (c) 1.0×10^{-10} W/m^2.

Solution   (a) For an intensity of 1.0×10^{-12} W/m^2, the intensity level, in decibels, is

$$\beta = 10 \log \left(\frac{1.0 \times 10^{-12} \text{ W/m}^2}{1.0 \times 10^{-12} \text{ W/m}^2} \right) = 10 \log(1) = \boxed{0 \text{ dB}}$$

This answer should have been obvious without calculation, because an intensity of 1.0×10^{-12} W/m^2 corresponds to the threshold of hearing.

(b) In this case, the intensity is exactly ten times that in part (a). The intensity level is

$$\beta = 10 \log \left(\frac{1.0 \times 10^{-11} \text{ W/m}^2}{1.0 \times 10^{-12} \text{ W/m}^2} \right) = 10 \log(10) = \boxed{10 \text{ dB}}$$

(c) Here the intensity is 100 times greater than at the threshold of hearing, and the intensity level is

$$\beta = 10 \log \left(\frac{1.0 \times 10^{-10} \text{ W/m}^2}{1.0 \times 10^{-12} \text{ W/m}^2} \right) = 10 \log(100) = \boxed{20 \text{ dB}}$$

Note the pattern in these answers. A sound with an intensity level of 10 dB is ten times more intense than the 0-dB sound, and a sound with an intensity level of 20 dB is 100 times more intense than a 0-db sound. This pattern is continued throughout the decibel scale. In short, on the decibel scale *an increase of 10 dB means that the intensity of the sound is multiplied by a factor of 10.* For example, a 50-dB sound is 10 times as intense as a 40-dB sound and a 60-dB sound is 100 times as intense as a 40-dB sound.

Exercise   Determine the intensity level of a sound wave with an intensity of 5.0×10^{-7} W/m^2.

Answer   57 dB.

EXAMPLE 14.4   The Noisy Typewriter

A rather noisy typewriter produces a sound intensity of 1.0×10^{-5} W/m^2. Find the decibel level of this machine, and calculate the new intensity level when a second, identical typewriter is added to the office.

Solution   The intensity level of the single typewriter is

$$\beta = 10 \log \left(\frac{1.0 \times 10^{-5} \text{ W/m}^2}{1.0 \times 10^{-12} \text{ W/m}^2} \right) = 10 \log(10^7) = \boxed{70 \text{ dB}}$$

Adding the second typewriter doubles the energy input into sound and hence doubles the intensity. The new intensity level is

$$\beta = 10 \log \left(\frac{2.0 \times 10^{-5} \text{ W/m}^2}{1.0 \times 10^{-12} \text{ W/m}^2} \right) = \boxed{73 \text{ dB}}$$

Federal regulations now demand that no office or factory worker be exposed to noise levels that average more than 90 dB over an 8-h day. The results in this example read like one of the old jokes that start "There is some good news and some bad news." First the good news. Imagine that you are a manager analyzing the noise conditions in your office. One typewriter in the office produces a noise level of 70 dB. When you add a second typewriter, the noise level increases by only 3 dB. Because of the logarithmic nature of intensity levels, doubling the intensity does not double the intensity level; in fact, it alters it by a surprisingly small amount. This means that additional equipment can be added to an office or factory without appreciably altering the intensity level of an environment.

Now the bad news. The results also work in reverse. As you remove noisy machinery, the intensity level is not lowered appreciably. For example, consider an office with 60 typewriters producing a noise level of 93 dB, which is 3 dB above the maximum allowed. In order to reduce the noise level by 3 dB, half the machines would have to be removed! That is, you would have to remove 30 typewriters to reduce the noise level to 90 dB. To reduce the level another 3 dB, you would have to remove half of the remaining machines, and so on.

*14.5

SPHERICAL AND PLANE WAVES

If a small spherical object oscillates so that its radius changes periodically with time, a spherical sound wave is produced (Fig. 14.3). The wave moves outward from the source at a constant speed.

Because all points on the vibrating sphere behave in the same way, we conclude that the energy in a spherical wave propagates equally in all directions. That is, no one direction is preferred over any other. If P_{av} is the average power emitted by the source, then at any distance r from the source, this power must be distributed over a spherical surface of area $4\pi r^2$, assuming no absorption in the medium. (Recall that $4\pi r^2$ is the surface area of a sphere.) Hence, the **intensity** of the sound at a distance of r from the source is

$$I = \frac{\text{average power}}{\text{area}} = \frac{P_{av}}{A} = \frac{P_{av}}{4\pi r^2} \qquad \textbf{[14.8]}$$

This shows that the intensity of a wave decreases with increasing distance from its source, as you might expect. The fact that I varies as $1/r^2$ is a result of the assumption that the small source (sometimes called a **point source**) emits a spherical wave. Since the average power is the same through any spherical

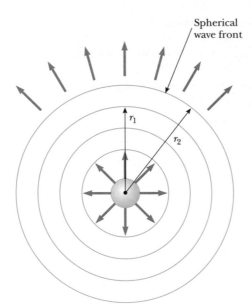

FIGURE 14.3
A spherical wave propagating radially outward from an oscillating sphere. The intensity of the spherical wave varies as $1/r^2$.

surface centered at the source, we see that the intensities at distances r_1 and r_2 (Fig. 14.3) from the center of the source are

$$I_1 = \frac{P_{av}}{4\pi r_1{}^2} \qquad I_2 = \frac{P_{av}}{4\pi r_2{}^2}$$

Therefore, the ratio of intensities at these two spherical surfaces is

$$\frac{I_1}{I_2} = \frac{r_2{}^2}{r_1{}^2}$$

It is useful to represent spherical waves graphically with a series of circular arcs (lines of maximum intensity) concentric with the source representing part of a spherical surface, as in Figure 14.4. We call such an arc a **wavefront**. The distance between adjacent wavefronts equals the wavelength, λ. The radial lines pointing outward from the source and cutting the arcs perpendicularly are called **rays**.

Now consider a small portion of a wavefront that is at a *great* distance (great relative to λ) from the source, as in Figure 14.5. In this case, the rays are nearly parallel to each other and the wavefronts are very close to being planes. Therefore, at distances from the source that are great relative to the wavelength, we can approximate the wavefront with parallel planes. We call such waves plane waves. Any small portion of a spherical wave that is far from the source can be considered a **plane wave**. Figure 14.6 illustrates a plane wave propagating along the x axis. If x is taken to be the direction of the wave motion (or ray) in this figure, then the wavefronts are parallel to the plane containing the y and z axes.

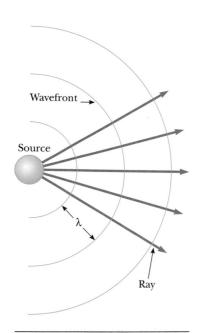

FIGURE 14.4
Spherical waves emitted by a point source. The circular arcs represent the spherical wavefronts concentric with the source. The rays are radial lines pointing outward from the source, perpendicular to the wavefronts.

EXAMPLE 14.5   Intensity Variations of a Point Source

A small source emits sound waves with a power output of 80 W.

(a) Find the intensity 3.0 m from the source.

Solution A small source emits energy in the form of spherical waves (see Fig. 14.3). Let P_{av} be the average power output of the source. At a distance of r from the source, the power is distributed over the surface area of a sphere, $4\pi r^2$. Therefore, the intensity at distance r from the source is given by Equation 14.8. Since $P_{av} = 80$ W and $r = 3.0$ m, we find that

$$I = \frac{P_{av}}{4\pi r^2} = \frac{80 \text{ W}}{4\pi(3.0 \text{ m})^2} = \boxed{0.71 \text{ W/m}^2}$$

which is close to the threshold of pain.

(b) Find the distance at which the sound level is 40 dB.

Solution We can find the intensity at the 40-dB intensity level by using Equation 14.7 with $I_0 = 1.0 \times 10^{-12}$ W/m^2:

$$40 = 10 \log(I/I_0)$$

$$4 = \log(I/I_0)$$

$$10^4 = I/I_0$$

$$I = 10^4 I_0 = 1.0 \times 10^{-8} \text{ W/m}^2$$

When this value for I is used in Equation 14.8, solving for r gives

$$r = \left(\frac{P_{av}}{4\pi I}\right)^{1/2} = \left(\frac{80 \text{ W}}{4\pi \times 10^{-8} \text{ W/m}^2}\right)^{1/2} = \boxed{2.5 \times 10^4 \text{ m}}$$

which is approximately 15 mi!

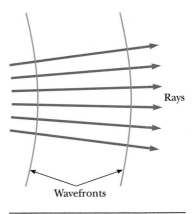

FIGURE 14.5
Far away from a point source, the wavefronts are nearly parallel planes and the rays are nearly parallel lines perpendicular to the planes. Hence, a small segment of a spherical wavefront is approximately a plane wave.

14.6

THE DOPPLER EFFECT

If a car or truck is moving while its horn is blowing, the frequency of the sound you hear is higher as the vehicle approaches you and lower as it moves away from you. This is one example of the Doppler effect, named for the Austrian physicist Christian Doppler (1803–1853), who discovered it.

> In general, a **Doppler effect** is experienced whenever there is relative motion between source and observer. When the source and observer are moving toward each other, the observer hears a frequency higher than the frequency of the source in the absence of relative motion. When the source and observer are moving away from each other, the observer hears a frequency lower than the source frequency.

Although the Doppler effect is most commonly experienced with sound waves, it is a phenomenon common to all waves. For example, the frequencies of light waves are also shifted by the relative motion of source and observer.

First let us consider the case in which the observer is moving and the sound source is stationary. For simplicity, we assume that the air is also stationary and that all velocity measurements are made relative to this stationary medium. Figure 14.7 describes the situation in which the observer is moving with a speed of v_o toward the source (considered a point source), which is at rest ($v_s = 0$).

We shall take the frequency of the source to be f, the wavelength to be λ, and the speed of sound in air to be v. Clearly, if both observer and source were stationary, the observer would detect f wavefronts per second. (That is, when

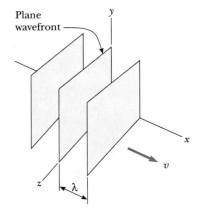

FIGURE 14.6
A representation of a plane wave moving in the positive x direction with a speed of v. The wavefronts are planes parallel to the yz plane.

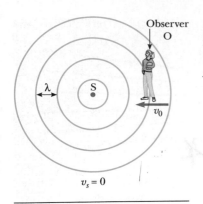

FIGURE 14.7
An observer moving with a speed of v_0 *toward* a stationary point source (S) hears a frequency, f', that is *greater* than the source frequency, f.

$v_o = 0$ and $v_s = 0$, the observed frequency equals the source frequency.) When the observer is moving toward the source, he or she moves a distance of $v_o t$ in t seconds. During this interval, *the observer detects an additional number of wavefronts*. The number of extra wavefronts detected is equal to the distance traveled, $v_o t$, divided by the wavelength, λ. Thus,

$$\text{Additional wavefronts detected} = \frac{v_o t}{\lambda}$$

The number of additional wavefronts detected *per second* is v_o/λ. Hence, the frequency f' heard by the observer is *increased* to

$$f' = f + \frac{v_o}{\lambda}$$

Using the fact that $\lambda = v/f$, we see that $v_o/\lambda = (v_o/v)f$. Hence, f' can be expressed as

$$f' = f\left(\frac{v + v_o}{v}\right) \qquad \textbf{[14.9]}$$

An observer traveling *away* from the source, as in Figure 14.8, *detects fewer wavefronts per second*. Thus, it follows that the frequency heard by the observer in this case is *lowered* to

$$f' = f\left(\frac{v - v_o}{v}\right) \qquad \textbf{[14.10]}$$

We can incorporate these two equations into one:

Observed frequency—observer in motion

$$f' = f\left(\frac{v \pm v_o}{v}\right) \qquad \textbf{[14.11]}$$

This general equation applies when an observer is moving with a speed of v_o relative to a stationary source. *The positive sign is used when the observer is moving toward the source, and the negative sign is used when the observer is moving away from the source.*

Now consider the situation in which the source is in motion and the observer is at rest. If the source is moving directly toward observer A in Figure 14.9a, the wavefronts heard by A are closer together because the source is moving in the direction of the outgoing wave. As a result, the wavelength λ' measured by observer A is shorter than the true wavelength, λ, of the source. During each vibration, which lasts for an interval of T (the period), the source moves a distance of $v_s T = v_s/f$ and *the wavelength is shortened by this amount*. Therefore, the observed wavelength, λ', is given by

$$\lambda' = \lambda - \frac{v_s}{f}$$

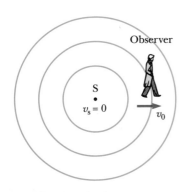

FIGURE 14.8
An observer moving with a speed of v_o *away* from a stationary source hears a frequency, f', that is *lower* than the source frequency.

Since $\lambda = v/f$, the frequency heard by observer A is

$$f' = \frac{v}{\lambda'} = \frac{v}{\lambda - \dfrac{v_s}{f}} = \frac{v}{\dfrac{v}{f} - \dfrac{v_s}{f}} = f\left(\frac{v}{v - v_s}\right) \qquad \textbf{[14.12]}$$

That is, *the observed frequency increases when the source is moving toward the observer.*

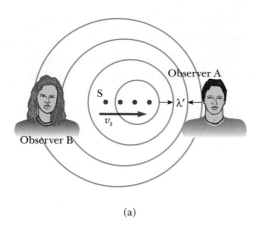

(a)

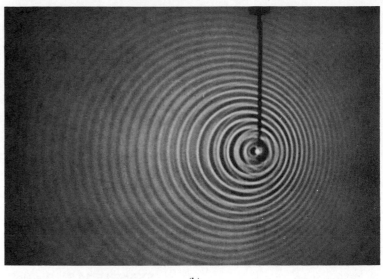

(b)

FIGURE 14.9

(a) A source, S, moving with speed v_s toward stationary observer A and away from stationary observer B. Observer A hears an *increased* frequency, and observer B hears a *decreased* frequency. (b) The Doppler effect in water, observed in a ripple tank. The source producing the water waves is moving to the right. (*Courtesy Educational Development Center, Newton, Mass.*)

In Figure 14.9a the source is moving away from observer B, who is at rest to the left of the source. Thus, observer B measures a wavelength that is *greater* than λ and hears a *decreased* frequency of

$$f' = f\left(\frac{v}{v + v_s}\right)$$ [14.13]

Combining Equations 14.12 and 14.13, we can express the general relationship for the observed frequency when the source is moving and the observer is at rest:

$$f' = f\left(\frac{v}{v \mp v_s}\right)$$ [14.14]

Observed frequency—source in motion

The *negative* sign is used when the source is moving *toward* the observer, and the *positive* sign is used when the source is moving *away from* the observer.

Finally, if both the source and the observer are in motion, one finds the following general relationship for the observed frequency:

$$f' = f\left(\frac{v \pm v_o}{v \mp v_s}\right)$$ [14.15]

Observed frequency—observer and source in motion

In this expression the *upper* signs ($+v_o$ and $-v_s$) refer to motion of one *toward* the other, and the lower signs ($-v_o$ and $+v_s$) refer to motion of one *away* from the other.

EXAMPLE 14.6   Listen, But Don't Stand on the Track

A train moving at a speed of 40.0 m/s sounds its whistle, which has a frequency of 500 Hz. Determine the frequency heard by a stationary observer as the train approaches the observer.

Solution   We can use Equation 14.12 to get the apparent frequency as the train approaches the observer. Taking $v = 345$ m/s as the speed of sound in air, we have

$$f' = f\left(\frac{v}{v - v_s}\right) = (500 \text{ Hz})\left(\frac{345 \text{ m/s}}{345 \text{ m/s} - 40.0 \text{ m/s}}\right) = \boxed{566 \text{ Hz}}$$

Exercise   Determine the frequency heard by the stationary observer as the train recedes from the observer.

Answer   448 Hz.

EXAMPLE 14.7   The Noisy Siren

An ambulance travels down a highway at a speed of 75.0 mi/h, its siren emitting sound at a frequency of 400 Hz. What frequency is heard by a passenger in a car traveling at 55.0 mi/h in the opposite direction as the car (a) approaches? (b) moves away from the ambulance?

Solution   Let us take the velocity of sound in air to be $v = 345$ m/s and use the conversion 1.00 mi/h = 0.447 m/s. Therefore, $v_s = 75.0$ mi/h = 33.5 m/s and $v_o = 55.0$ mi/h = 24.6 m/s. We can use Equation 14.15 in both cases.

(a) As the ambulance and car approach each other, the observed frequency is

$$f' = f\left(\frac{v + v_o}{v - v_s}\right) = (400 \text{ Hz})\left(\frac{345 \text{ m/s} + 24.6 \text{ m/s}}{345 \text{ m/s} - 33.5 \text{ m/s}}\right) = \boxed{475 \text{ Hz}}$$

(b) As the two vehicles recede from each other, the passenger in the car hears a frequency of

$$f' = f\left(\frac{v - v_o}{v + v_s}\right) = (400 \text{ Hz})\left(\frac{345 \text{ m/s} - 24.6 \text{ m/s}}{345 \text{ m/s} + 33.5 \text{ m/s}}\right) = \boxed{339 \text{ Hz}}$$

SHOCK WAVES

Now let us consider what happens when the source speed, v_s, *exceeds* the wave velocity, v. Figure 14.10a describes this situation graphically. The circles represent spherical wavefronts emitted by the source at various times during its motion. At $t = 0$, the source is at point S_0, and at some later time t, the source is at point S_n. In the interval t, the wavefront centered at S_0 reaches a radius of vt. In this same interval, the source travels to S_n, a distance of $v_s t$. At the instant the source is at S_n, the waves just beginning to be generated at this point have wavefronts of zero radius. The line drawn from S_n to the wavefront centered on S_0 is tangent to all other wavefronts generated at intermediate times. All such tangent lines lie on the surface of a cone. The angle, θ, between one of these tangent lines and the direction of travel is given by

$$\sin \theta = \frac{v}{v_s}$$

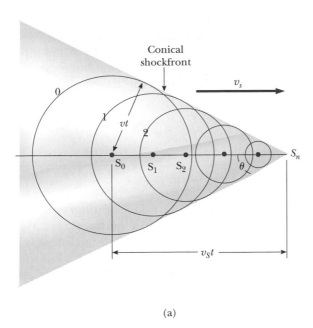

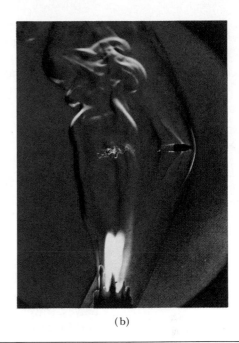

(a)

(b)

FIGURE 14.10

(a) A representation of a shock wave, produced when a source moves from S_0 to S_n with a speed, v_s, that is *greater* than the wave speed, v, in that medium. The envelope of the wavefronts forms a cone whose half-angle is $\sin \theta = v/v_s$. (b) A stroboscopic photograph of a bullet moving at supersonic speed through the hot air above a candle. *(Harold Edgerton, Courtesy of Palm Press, Inc.)*

The ratio v_s/v is referred to as the **Mach number.** The conical wavefront produced when $v_s > v$ (supersonic speeds) is known as a **shock wave.** Figure 14.10b is a photograph of a bullet traveling at supersonic speed through the hot air rising above a candle. Note the shock waves in the vicinity of the bullet. Another interesting example of shock waves is the V-shaped wavefront produced by a boat (the bow wave) when the boat's speed exceeds the speed of the water waves (see the photograph below).

Jet airplanes traveling at supersonic speeds produce shock waves, which are responsible for the loud explosion, or sonic boom, heard on the Earth. A shock

The V-shaped bow wave of a boat is formed because the boat travels at a speed greater than the speed of water waves. A bow wave is analogous to a shock wave formed by an airplane traveling faster than sound. *(Al Satterthwaite/The Image Bank)*

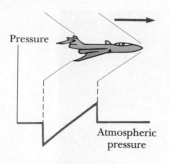

FIGURE 14.11
The two shock waves produced by the nose and tail of a jet airplane traveling at supersonic speed.

wave carries a great deal of energy concentrated on the surface of the cone, with correspondingly great pressure variations. Shock waves are unpleasant to hear and can damage buildings when aircraft fly supersonically at low altitudes. In fact, an airplane flying at supersonic speeds produces a double boom because two shock waves are formed, one from the nose of the plane and one from the tail (Fig. 14.11).

14.7

INTERFERENCE OF SOUND WAVES

Sound waves can be made to interfere with each other. This can be demonstrated with the device shown in Figure 14.12. Sound from a loudspeaker at S is sent into a tube at *P*, where there is a T-shaped junction. Half the sound intensity travels in one direction and half in the opposite direction. Thus, the sound waves that reach the receiver at R travel along two different paths. If the two paths are of the same length, a crest of the wave that enters the junction will separate into two halves, travel the two paths, and then combine again at the receiver. The upper and lower waves reunite at the receiver so that constructive interference takes place, and thus a loud sound is heard at the detector.

Suppose, however, that one of the path lengths is adjusted by sliding the upper U-shaped tube upward so that the upper path is half a wavelength *longer* than the lower path. In this case, an entering sound wave splits and travels the two paths as before, but now the wave along the upper path must travel a distance equivalent to half a wavelength farther than the wave traveling along the lower path. As a result, the crest of one wave meets the trough of the other when they merge at the receiver. Since this is the condition for destructive interference, no sound is detected at the receiver.

You should be able to predict what will be heard if the upper path is adjusted to be one full wavelength longer than the lower path. In this case, constructive interference of the two waves occurs, and a loud sound is detected at the receiver.

Nature provides many other examples of interference phenomena. In a later chapter we shall describe several interesting interference effects involving light waves.

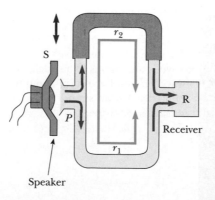

FIGURE 14.12
An acoustical system for demonstrating interference of sound waves. Sound from the speaker enters the tube and splits into two parts at *P*. The two waves combine at the opposite side and are detected at R. The upper path length is varied by the sliding section.

EXAMPLE 14.8   Interference from Two Loudspeakers

Two loudspeakers are placed as in Figure 14.13 and driven by the same source at a frequency of 2000 Hz. The top speaker is then moved left to position A. At this location, an observer at a great distance from the speakers and directly in front of them notices that the intensity of the sound from the two sources has decreased to a minimum. What is the minimum distance the top speaker has been moved? Assume that the speed of sound in air is 345 m/s.

Solution   Initially, both speakers are at the same distance from the observer. Hence, the sound from each speaker must travel the same distance, and the observer hears a loud sound corresponding to constructive interference. When the top speaker is at position A, its sound must travel farther to reach the observer. Since the observer notices a minimum in the sound level when the top speaker is at A, destructive interference is taking place between the two separate sound signals. This means that the top speaker has been moved half a wavelength. With the speed of sound in air equal to 345 m/s, we can calculate the wavelength:

$$\lambda = \frac{v}{f} = \frac{345 \text{ m/s}}{2000 \text{ s}^{-1}} = 0.173 \text{ m}$$

Therefore, the distance moved is half this value, or 0.0865 m.

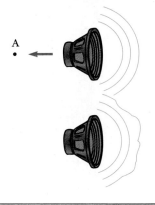

FIGURE 14.13
(Example 14.8)   Two loudspeakers driven by the same source can produce interference.

14.8

STANDING WAVES

If a stretched string is clamped at both ends, traveling waves reflect from the fixed ends, creating waves traveling in both directions. The incident and reflected waves combine according to the superposition principle. For example, if the string is vibrated at exactly the right frequency, a crest moving toward one end and a reflected trough meet at some point along the string. The two waves cancel at this point, which is called a **node.** In the resulting pattern on the string, the wave appears to stand still—hence its name, **standing wave.** There is no motion in the string at the nodes, but midway between two adjacent nodes, at an **antinode,** the string vibrates with the largest amplitude.

Figure 14.14 shows the oscillation of a standing wave during half of a cycle. *Notice that all points on the string oscillate vertically with the same frequency except for the node, which is stationary.* (The points at which the string is attached to the wall are also nodes, labeled N in Figure 14.14a.) Furthermore, different points have different amplitudes of motion.

Consider a string of length L that is fixed at both ends, as in Figure 14.15. The string has a number of natural patterns of vibration, called normal modes. Three of these are pictured in Figures 14.15b, 14.15c, and 14.15d. Each has a characteristic frequency, which we shall now calculate.

First, note that *the ends of the string must be nodes because these points are fixed.* If the string is displaced at its midpoint and released, the vibration shown in Figure 14.15b can be produced, in which case the center of the string is an antinode. For this normal mode, the length of the string equals $\lambda/2$ (the distance between nodes). Thus,

$$L = \frac{\lambda_1}{2} \quad \text{or} \quad \lambda_1 = 2L$$

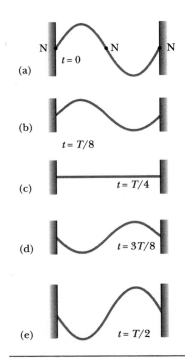

FIGURE 14.14
A standing wave pattern in a stretched string, shown by snapshots of the string during one half of a cycle.

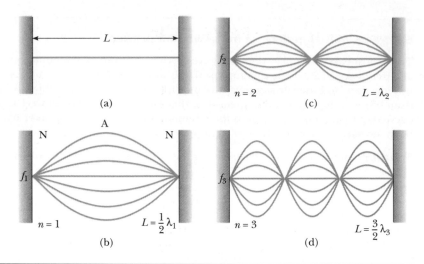

FIGURE 14.15
(a) Standing waves in a stretched string of length L, fixed at both ends. The normal frequencies of vibration form a harmonic series: (b) the fundamental frequency, or first harmonic; (c) the second harmonic; and (d) the third harmonic.

and the frequency of this vibration is

$$f_1 = \frac{v}{\lambda_1} = \frac{v}{2L} \qquad [14.16]$$

In Chapter 13 (Equation 13.16), the speed of a wave on a string was given as $v = \sqrt{F/\mu}$, where F is the tension in the string and μ is its mass per unit length. Thus, we can express Equation 14.16 as

Fundamental frequency

$$f_1 = \frac{1}{2L}\sqrt{\frac{F}{\mu}} \qquad [14.17]$$

This lowest frequency of vibration is called the **fundamental frequency** of the vibrating string.

The next normal mode, of wavelength λ_2 (Fig. 14.15c), occurs when the length of the string equals one wavelength—that is, when $\lambda_2 = L$. Hence,

$$f_2 = \frac{v}{L} = \frac{2v}{2L} = 2f_1 \qquad [14.18]$$

Note that this frequency is equal to *twice* the fundamental frequency. You should convince yourself that the next highest frequency of vibration, shown in Figure 14.15d, is

$$f_3 = \frac{3v}{2L} = 3f_1 \qquad [14.19]$$

In general, the characteristic frequencies are given by

$$f_n = nf_1 = \frac{n}{2L}\sqrt{\frac{F}{\mu}} \qquad [14.20]$$

PHYSICS IN *ACTION*

Photographs of standing waves. As one end of the stretched string is moved from side to side with increasing frequency, patterns with more and more loops are formed. Only certain definite frequencies produce fixed standing-wave patterns. The frequency of the second harmonic, f_2, is twice the fundamental frequency, f_1, and the frequency of the third harmonic, f_3, is three times the fundamental frequency.

STANDING WAVES

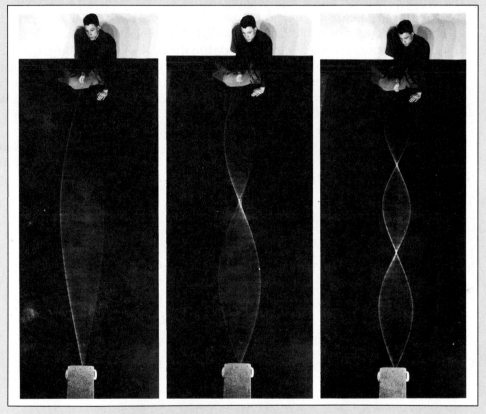

(Courtesy Educational Development Center, Newton, Mass.)

where $n = 1, 2, 3, \ldots$. In other words, the frequencies are integral multiples of the fundamental frequency. The frequencies f_1, $2f_1$, $3f_1$, and so on form a **harmonic series.** The fundamental f_1 corresponds to the **first harmonic;** the frequency $f_2 = 2f_1$, to the **second harmonic;** and so on.

When a stretched string is distorted to a shape that corresponds to any one of its harmonics, after being released it vibrates only at the frequency of that harmonic. If the string is struck or bowed, however, the resulting vibration includes frequencies of various harmonics, including the fundamental. Waves not in the harmonic series are quickly damped out on a string fixed at both ends. In effect, when disturbed, the string "selects" the normal-mode frequencies. As we shall see later, the presence of several harmonics on a string gives stringed

Multiflash photographs of standing wave patterns in a cord with a vibration at the left end. The two-loop pattern at the left represents the second harmonic ($n = 2$), and the three-loop pattern at the right represents the third harmonic ($n = 3$). (© Richard Megna 1991, Fundamental Photographs)

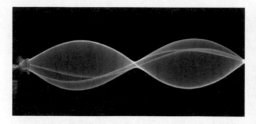

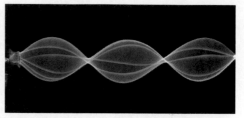

instruments their characteristic sound, which enables us to distinguish one from another even when they are producing identical fundamental frequencies.

The frequency of a string on a musical instrument can be changed either by varying the tension or by changing the length. For example, the tension in guitar and violin strings is varied by turning pegs on the neck of the instrument. As the tension is increased, the frequency of the normal modes increases according to Equation 14.20. Once the instrument is tuned, the musician varies the frequency by pressing the strings against the neck at a variety of positions, thereby changing the effective lengths of the vibrating portions of the strings. As the length is reduced, the frequency increases, as Equation 14.20 indicates.

Finally, Equation 14.20 shows that a string of fixed length can be made to vibrate at a lower fundamental frequency by increasing its mass per unit length. This is achieved in the bass strings of guitars and pianos by wrapping them with metal windings.

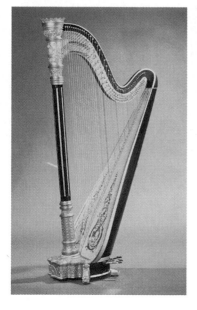

A concert-style harp. (Photo Lyon & Healy Harps, Chicago)

EXAMPLE 14.9 Harmonics of a Stretched String

Find the first four harmonics of a 1.0-m-long string if the string has a mass per unit length of 2.0×10^{-3} kg/m and is under a tension of 80 N.

Solution   The speed of the wave on the string is

$$v = \sqrt{\frac{F}{\mu}} = \sqrt{\frac{80 \text{ N}}{2.0 \times 10^{-3} \text{ kg/m}}} = 200 \text{ m/s}$$

The fundamental frequency can be found using Equation 14.16:

$$f_1 = \frac{v}{2L} = \frac{200 \text{ m/s}}{2(1.0 \text{ m})} = \boxed{100 \text{ Hz}}$$

The frequencies of the next three modes are $f_2 = 2f_1$, $f_3 = 3f_1$, and $f_4 = 4f_1$. Thus, $f_2 = 200$ Hz, $f_3 = 300$ Hz, and $f_4 = 400$ Hz.

Exercise   Find the tension in the string if the fundamental frequency is increased to 120 Hz.

Answer   120 N.

14.9

FORCED VIBRATIONS AND RESONANCE

In Chapter 13 we learned that the energy of a damped oscillator decreases in time because of friction. It is possible to compensate for this energy loss by applying an external force that does positive work on the system.

For example, suppose a mass-spring system having some natural frequency of vibration, f_0, is pushed back and forth with a periodic force whose frequency is f. The system vibrates at the frequency, f, of the driving force. This type of motion is referred to as a **forced vibration.** Its amplitude reaches a maximum when the frequency of the driving force equals the natural frequency of the system, f_0, called the **resonant frequency** of the system. Under this condition, the system is said to be in **resonance.**

In Section 14.8 we learned that a stretched string can vibrate in one or more of its natural modes. Here again, if a periodic force is applied to the string, the amplitude of vibration increases as the frequency of the applied force approaches one of the natural frequencies of vibration.

Resonance vibrations occur in a wide variety of circumstances. Figure 14.16 illustrates one experiment that demonstrates a resonance condition. Several pendulums of different lengths are suspended from a flexible beam. If one of them, such as A, is set in motion, the others begin to oscillate because of vibrations in the flexible beam. Pendulum C, the same length as A, oscillates with the greatest amplitude since its natural frequency matches that of pendulum A (the driving force).

Another apparatus that can easily be set up in the laboratory is a stretched string fixed at one end and connected at the opposite end to a vibrating blade (Fig. 14.17). As the blade oscillates, transverse waves sent down the string are reflected at the fixed end. As we found in Section 14.8, the string has natural frequencies that are determined by its mass per unit length and the tension (Eq. 14.20). When the frequency of the vibrating blade equals one of the natural frequencies of the string, standing waves are produced and the string vibrates with a large amplitude.

Another simple example of resonance is a child being pushed on a swing, which is essentially a pendulum with a natural frequency that depends on the length. The swing is kept in motion by a series of appropriately timed pushes. For its amplitude to be increased, the swing must be pushed each time it returns to the person's hands. This corresponds to a frequency equal to the natural frequency of the swing. If the energy put into the system per cycle of motion exactly equals the energy lost due to friction, the amplitude remains constant.

Opera singers have been known to set crystal goblets in audible vibration with their powerful voices, as shown in the photograph below. This is yet another

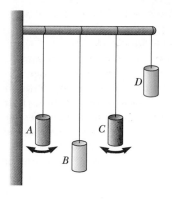

FIGURE 14.16
Resonance. If pendulum A is set in oscillation, only pendulum C, whose length matches that of A, will eventually oscillate with a large amplitude, or resonate. The arrows indicate motion perpendicular to the page.

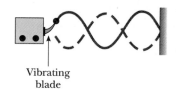

Vibrating blade

FIGURE 14.17
Standing waves can be set up in a stretched string by connecting one end of the string to a vibrating blade. When the blade vibrates at one of the natural frequencies of the string, large-amplitude standing waves are created.

A glass shattered by the amplified sound of a human voice. When the frequency of the singer's sound wave matches one of the natural frequencies of vibration of the glass, resonant vibrations are set up in the glass. *(Ben Rose/The IMAGE Bank)*

FIGURE 14.18
The collapse of the Tacoma Narrows suspension bridge in 1940 was a vivid demonstration of mechanical resonance. High winds set up standing waves in the bridge, causing it to oscillate at one of its natural frequencies. Once established, this resonance condition led to the bridge's collapse. *(United Press International Photo)*

An earthquake in 1989 caused collapse of the I880 freeway near Oakland, California. The damage created by earthquakes is due to waves that travel along the surface of the Earth. *(Peter Menzell/Stock Boston)*

example of resonance: the sound waves emitted by the singer can set up large-amplitude vibrations in the glass. If a highly amplified sound wave has the right frequency, the amplitude of forced vibrations in the glass increases to the point where the glass becomes heavily strained and shatters. In this case, resonance occurs when the wavelength of the emitted sound wave equals the circumference of the glass.

A striking example of structural resonance occurred in 1940, when the Tacoma Narrows bridge in the state of Washington was set in oscillation by the wind (Fig. 14.18). The amplitude of the oscillations increased steadily until the bridge collapsed. A more recent example of destruction by structural resonance occurred during the Loma Prieta earthquake near Oakland, California, in 1989. In one section—almost a mile long—of the double-decker Nimitz Freeway, the upper deck collapsed onto the lower deck, killing several people. The collapse of this particular section of roadway, while other sections escaped serious damage, has been traced to the fact that the earthquake waves had a frequency of approximately 1.5 Hz—very close to the natural resonant frequency of the section of roadway that gave way.

14.10

STANDING WAVES IN AIR COLUMNS

Standing longitudinal waves can be set up in a tube of air, such as an organ pipe, as the result of interference between sound waves traveling in opposite directions. The relationship between the incident wave and the reflected wave depends on whether the reflecting end of the tube is open or closed. A portion of the sound wave is reflected back into the tube at the open end. *If the reflecting end is closed, a node must exist at this end because the movement of air molecules is restricted. If the end is open, the air molecules have complete freedom of motion, and an antinode exists.*

Figure 14.19a shows the first three modes of vibration of a pipe open at both ends. When air is directed against an edge at the left, longitudinal standing waves are formed and the pipe vibrates at its natural frequencies. Note that for the fundamental frequency the wavelength is twice the length of the pipe and hence $f_1 = v/2L$. Similarly, one finds that the frequencies of the second and third harmonics are $2f_1$, $3f_1$, ... Thus,

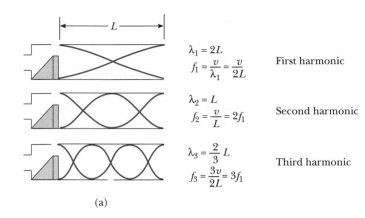

$$\lambda_1 = 2L$$
$$f_1 = \frac{v}{\lambda_1} = \frac{v}{2L}$$
First harmonic

$$\lambda_2 = L$$
$$f_2 = \frac{v}{L} = 2f_1$$
Second harmonic

$$\lambda_3 = \frac{2}{3}L$$
$$f_3 = \frac{3v}{2L} = 3f_1$$
Third harmonic

(a)

$$\lambda_1 = 4L$$
$$f_1 = \frac{v}{\lambda_1} = \frac{v}{4L}$$
First harmonic

$$\lambda_3 = \frac{4}{3}L$$
$$f_3 = \frac{3v}{4L} = 3f_1$$
Third harmonic

$$\lambda_5 = \frac{4}{5}L$$
$$f_5 = \frac{5v}{4L} = 5f_1$$
Fifth harmonic

(b)

FIGURE 14.19
(a) Standing longitudinal waves in an organ pipe open at both ends. The natural frequencies f_1, $2f_1$, $3f_1$ . . . form a harmonic series. (b) Standing longitudinal waves in an organ pipe closed at one end. Only *odd* harmonics are present, and the natural frequencies are f_1, $3f_1$, $5f_1$, and so on.

> In a pipe open at both ends, the natural frequencies of vibration form a series in which *all harmonics are present.* These harmonics are equal to integral multiples of the fundamental frequency.

We can express this harmonic series as

 $$f_n = n\frac{v}{2L} \qquad n = 1, 2, 3, \ldots \qquad \textbf{[14.21]}$$

Pipe open at both ends; all harmonics present

where v is the speed of sound in air.

If a pipe is closed at one end and open at the other, the closed end is a node (Fig. 14.19b). In this case, the wavelength of the fundamental mode is four times the length of the tube. Hence, $f_1 = v/4L$ and the frequencies of the third and fifth harmonics are $3f_1$, $5f_1$, ... That is,

> In a pipe closed at one end and open at the other, *only odd harmonics are present.*

These are given by

$$f_n = n\frac{v}{4L} \qquad n = 1, 3, 5, \ldots \qquad \text{[14.22]}$$

EXAMPLE 14.10   **Harmonics of a Pipe**

A pipe is 2.46 m long.

(a) Determine the frequencies of the first three harmonics if the pipe is open at both ends. Take 345 m/s as the speed of sound in air.

Solution   The fundamental frequency of a pipe open at both ends can be found from Equation 14.21, with $n = 1$.

$$f_1 = \frac{v}{2L} = \frac{345 \text{ m/s}}{2(2.46 \text{ m})} = \boxed{70.0 \text{ Hz}}$$

Since all harmonics are present in a pipe open at both ends, the second and third harmonics have frequencies of $f_2 = 2f_1 = 140$ Hz and $f_3 = 3f_1 = 210$ Hz.

(b) What are the three lowest possible frequencies if the pipe is closed at one end and open at the other?

Solution   The fundamental frequency of a pipe closed at one end can be found from Equation 14.22, with $n = 1$.

$$f_1 = \frac{v}{4L} = \frac{345 \text{ m/s}}{4(2.46 \text{ m})} = \boxed{35.0 \text{ Hz}}$$

In this case, only odd harmonics are present, and so the third and fifth harmonics have frequencies of $f_3 = 3f_1 = 105$ Hz and $f_5 = 5f_1 = 175$ Hz.

Exercise   If the pipe is open at one end, how many harmonics are possible in the normal hearing range, 20 to 20 000 Hz?

Answer   286.

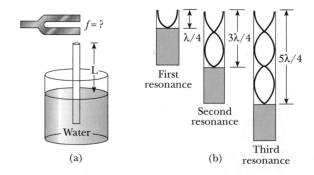

FIGURE 14.20
(Example 14.11)   (a) Apparatus for demonstrating the resonance of sound waves in a tube closed at one end. The length, L, of the air column is varied by moving the tube vertically while it is partially submerged in water. (b) The first three normal frequencies of vibration for the system.

EXAMPLE 14.11 Resonance in a Tube of Variable Length

Figure 14.20 shows a simple apparatus for demonstrating resonance in a tube. A long tube open at both ends is partially submerged in a beaker of water, and a vibrating tuning fork of unknown frequency is placed near the top of the tube. The length of the air column, L, is adjusted by moving the tube vertically. The sound waves generated by the fork are reinforced when the length of the air column corresponds to one of the resonant frequencies of the tube. The smallest value of L for which a peak occurs in the sound intensity is 9.00 cm. For this measurement, determine the frequency of the tuning fork and the value of L for the next two resonant vibrations.

Reasoning and Solution   Once the tube is in the water, this setup represents a pipe closed at one end, and the fundamental has a frequency of $v/4L$ (Fig. 14.20b). If we take $v = 345$ m/s for the speed of sound in air, and $L = 0.0900$ m, we get

$$f_1 = \frac{v}{4L} = \frac{345 \text{ m/s}}{4(0.0900 \text{ m})} = \boxed{958 \text{ Hz}}$$

The fundamental wavelength of the pipe is given by $\lambda = 4L = 0.360$ m. Since the frequency of the source is constant, we see that the next resonance positions (Fig. 14.20b) correspond to lengths of $3\lambda/4 = 0.270$ m and $5\lambda/4 = 0.450$ m.

   This arrangement is often used to measure the speed of sound, in which case the frequency of the tuning fork and the lengths at which resonance occurs must be known.

*14.11

BEATS

The interference phenomena we have been discussing so far have involved the superposition of two or more waves with the same frequency, traveling in opposite directions. Let us now consider another type of interference effect that results from the superposition of two waves with slightly different frequencies. In this situation, the waves at some fixed point are periodically in and out of phase, corresponding to an alternation in time between constructive and destructive interference. In order to understand this phenomenon, consider Figure 14.21. The two waves in Figure 14.21a were emitted by two tuning forks having slightly different frequencies; Figure 14.21b shows the superposition of these two waves. At some time, indicated as t_a in Figure 14.21a, the two waves are out of phase, and

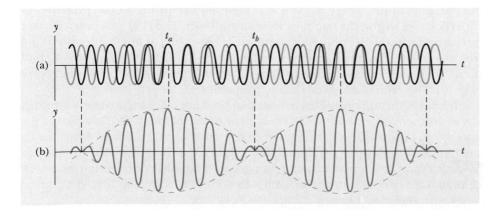

FIGURE 14.21
Beats are formed by the combination of two waves of slightly different frequencies traveling in the same direction. (a) The individual waves. (b) The combined wave has an amplitude (dashed line) that oscillates in time.

destructive interference occurs, as demonstrated by the resultant curve in Figure 14.21b. At some later time, however, the vibrations of the two forks move into step with one another. At time t_b in Figure 14.21a, the two forks simultaneously emit compressions, and constructive interference results, as demonstrated by the curve in Figure 14.21b. As time passes, the vibrations of the two forks move out of phase, then into phase, and so on. Consequently, a listener at some fixed point hears an alternation in loudness, known as **beats.** The number of beats per second, or beat frequency, equals the difference in frequency between the two sources. One can tune a stringed instrument, such as a piano, by beating a note on the instrument against a note of known frequency. The string can then be tuned to the desired frequency by adjusting the tension until no beats are heard.

EXAMPLE 14.12   Sour Notes

A particular piano string is supposed to vibrate at a frequency of 440 Hz. In order to check its frequency, a tuning fork known to vibrate at a frequency of 440 Hz is sounded at the same time the piano key is struck, and a beat frequency of 4 beats per second is heard. Find the possible frequencies at which the string could be vibrating.

Solution   The number of beats per second is equal to the difference in frequency between the two sound sources. In this case, since one of the source frequencies is 440 Hz, 4 beats per second would be heard if the frequency of the string (the second source) were either 444 Hz or 436 Hz.

---

*14.12
QUALITY OF SOUND

The sound-wave patterns produced by most musical instruments are very complex. Figure 14.22 shows characteristic waveforms produced by a tuning fork, a flute, and a clarinet, each playing the same note. Although each instrument has its own characteristic pattern, the figure reveals that each of the waveforms is periodic. Note that the tuning fork produces only one harmonic (the fundamental frequency), but the two instruments emit mixtures of harmonics. Figure 14.23 graphs the harmonics of the waveforms of Figure 14.22. When this note is played on the flute (Fig. 14.23b), part of the sound consists of a vibration at the fundamental frequency, an even higher intensity is contributed by the second harmonic, the fourth harmonic produces about the same intensity as the fundamental, and so on. These sounds add together according to the principle of superposition to give the complex waveform shown. The clarinet emits a certain intensity at a frequency of the first harmonic, about half as much intensity at the frequency of the second harmonic, and so forth. The resultant superposition of these frequencies produces the pattern shown in Figure 14.23c. The tuning fork (Fig. 14.23a) emits sound only at the frequency of the first harmonic.

In music, the mixture of harmonics that produces the characteristic sound of any instrument is referred to as the *quality,* or *timbre,* of the sound. We say that the note C on a flute differs in quality from the same C on a clarinet. Instruments such as the bugle, trumpet, violin, and tuba are rich in harmonics. A musician playing a wind instrument can emphasize one or another of these harmonics by changing the configuration of his or her lips, and can thus play different musical notes with the same valve openings.

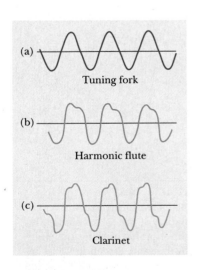

FIGURE 14.22
Waveforms produced by (a) a tuning fork, (b) a harmonic flute, and (c) a clarinet, all at approximately the same frequency. *(Adapted from C. A. Culver, Musical Acoustics)*

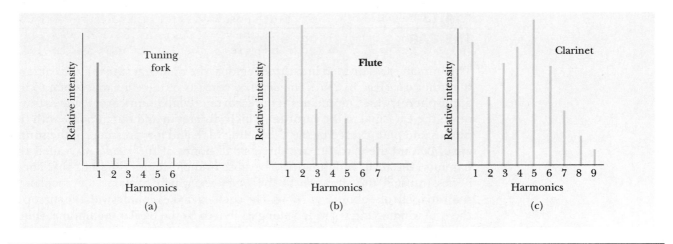

FIGURE 14.23

Harmonics of the waveforms in Figure 14.22. Note their variation in intensity. *(Adapted from C. A. Culver, Musical Acoustics, 4th ed., New York, McGraw-Hill, 1956)*

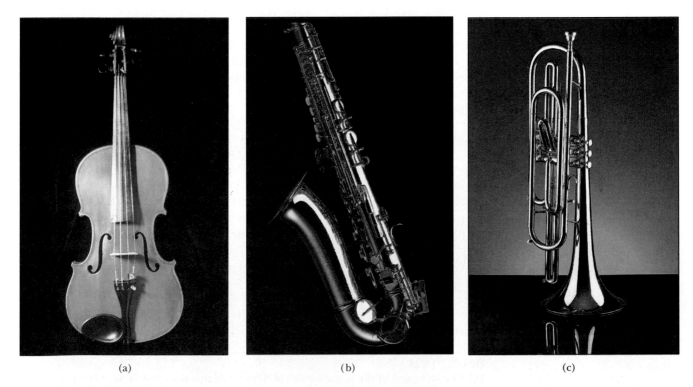

Each musical instrument has its own characteristic sound and mixture of harmonics. Instruments shown are (a) the violin, (b) the saxophone, and (c) the trumpet. *(Photographs courtesy of (a) © 1989 Gary Buss/FPG; (b) and (c) © 1989 Richard Laird/FPG)*

*14.13

THE EAR

The human ear is divided into three regions: the outer ear, the middle ear, and the inner ear (Fig. 14.24a). The *outer ear* consists of the ear canal (open to the atmosphere), which terminates at the eardrum (tympanum). Sound waves travel down the ear canal to the eardrum, which vibrates in and out in phase with the pushes and pulls caused by the alternating high and low pressures of the sound wave. Behind the eardrum are three small bones of the *middle ear,* called the hammer, the anvil, and the stirrup because of their shapes (Fig. 14.24b). These bones transmit the vibration to the *inner ear,* which contains the cochlea, a snail-shaped tube about 2 cm long. The cochlea makes contact with the stirrup at the oval window and is divided along its length by the basilar membrane, which consists of small hairs and nerve fibers. This membrane varies in mass per unit length and in tension along its length, and different portions of it resonate at different frequencies. (Recall that the natural frequency of a string depends on its mass per unit length and on the tension on it.) Along the basilar membrane are numerous nerve endings, which sense the vibration of the membrane and in turn transmit impulses to the brain. The brain interprets the impulses as sounds of varying frequency, depending on the locations along the basilar membrane of the impulse-transmitting nerves and on the rates at which the impulses are transmitted.

Figure 14.25 shows the frequency response curves of an average human ear for sounds of equal loudness, ranging from 0 to 120 dB. To interpret this series of graphs, take the bottom curve as the threshold of hearing. Compare the intensity level on the vertical axis for the two frequencies 100 Hz and 1000 Hz. The vertical axis shows that the 100-Hz sound must be about 38 dB greater than the 1000-Hz sound to be at the threshold of hearing. Thus, we see that the threshold of hearing is very strongly dependent on frequency. The easiest fre-

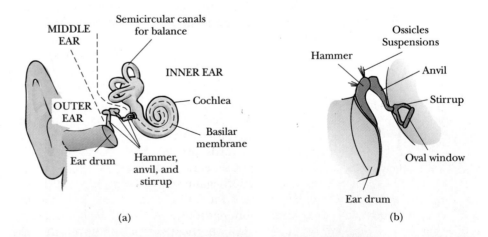

(a)   (b)

FIGURE 14.24
(a) The structure of the human ear. (b) The three tiny bones (ossicles) that connect the eardrum to the window of the cochlea act as a double lever system to decrease the amplitude of vibration and hence increase the pressure on the fluid in the cochlea.

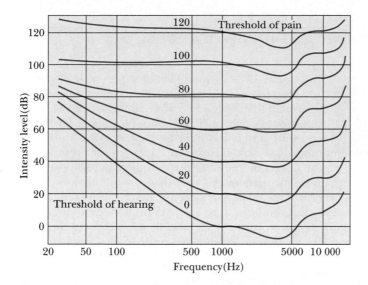

FIGURE 14.25

Curves of intensity level versus frequency for sounds that are perceived to be of equal loudness. Note that the ear is most sensitive at a frequency of about 3300 Hz. The lowest curve corresponds to the threshold of hearing for only about 1% of the population.

quencies to hear are around 3300 Hz, whereas frequencies above 12 000 Hz or below about 50 Hz must be relatively intense to be heard.

Now consider the curve labeled 80. This curve uses as its reference a 1000-Hz tone at an intensity level of 80 dB. The curve shows that a tone of frequency 100 Hz would have to be about 4 dB louder than the 80-dB, 1000-Hz tone in order to sound as loud. Notice that the curves flatten out as the intensity levels of the sounds increase. That is, when sounds are loud, all frequencies can be heard equally well.

The exact mechanism by which sound waves are amplified and detected by the ear is rather complex and not fully understood. We can describe them qualitatively, however. The small bones in the middle ear represent an intricate lever system that increases the force on the oval window. The pressure is greatly magnified because the surface area of the eardrum is about 20 times that of the oval window (in analogy with a hydraulic press). The middle ear, together with the eardrum and oval window, in effect acts as a matching network between the air in the outer ear and the liquid in the inner ear. The overall energy transfer between the outer ear and inner ear is highly efficient, with pressure amplification factors of several thousand. In other words, pressure variations in the inner ear are much greater than those in the outer ear.

The ear has its own built-in protection against loud sounds. The muscles connecting the three middle-ear bones to the walls control the volume of the sound by changing the tension on the bones as sound builds up, thus hindering their ability to transmit vibrations. In addition, the eardrum becomes stiffer as the sound intensity increases. These two occurrences make the ear less sensitive to loud incoming sounds. There is a time delay between the onset of loud sound and the ear's protective reaction, however, so a very sudden loud sound can still damage the ear.

SPECIAL TOPIC

ULTRASOUND AND ITS APPLICATIONS

Ultrasonic waves are sound waves with frequencies from 20 to 100 kHz, frequencies that are beyond the audible range. Because of their high frequency and corresponding short wavelengths, ultrasonic waves can be used to produce images of small objects and are currently in wide use in medical applications, both as a diagnostic tool and in certain treatments. Internal organs can be examined via the images produced by the reflection and absorption of ultrasonic waves. Although ultrasonic waves are far safer than x-rays, their images do not always have as much detail. On the other hand, certain organs, such as the liver and the spleen, are invisible to x-rays but can be imaged with ultrasonic waves.

Medical workers can measure the speed of the blood flow in the body using a device, called an ultrasonic flow meter, that makes use of the Doppler effect. By comparing the frequency of the waves scattered by the blood vessels with the incident frequency, one can obtain the flow speed.

Figure 14.26 illustrates the technique used to produce ultrasonic waves for clinical use. Electrical contacts are made to the opposite faces of a crystal, such as quartz or strontium titanate. If an alternating voltage of very high frequency is applied to these contacts, the crystal vibrates at the same frequency as the applied voltage, emitting a beam of ultrasonic waves. At one time, this was how almost all the headphones used in radio reception produced sound. This method of transforming electrical energy into mechanical energy, called the **piezoelectric effect,** is also reversible. That is, if some external source causes the crystal to vibrate, an alternating voltage is produced across the crystal. Hence, a single crystal can be used to both transmit and receive ultrasonic waves.

The production of electric voltages by a vibrating crystal is a technique that has been used for years in phonograph turntables. A special needle is attached to the crystal, and the vibrations of the needle as it rides in the groove of the record are translated by the crystal into an alternating voltage. This voltage is then amplified and used to drive the system's speakers.

The primary physical principle that makes ultrasound imaging possible is the fact that a sound wave is partially reflected whenever it is incident on a boundary between two materials having different densities. If a sound wave is traveling in a material of density ρ_i and strikes a material of density ρ_t, the percentage of the incident sound wave reflected, PR, is given by

$$PR = \left(\frac{\rho_i - \rho_t}{\rho_i + \rho_t}\right)^2 \times 100 \qquad \textbf{[14.23]}$$

This equation assumes that the incident sound wave travels perpendicularly to the boundary and that the speed of sound is approximately the same in the two materials. The latter assumption holds very well for the human body since the speed of sound does not vary much in the organs of the body.

Physicians commonly use ultrasonic waves to observe fetuses. This technique presents far less risk than x-rays, which can produce birth defects. First the abdomen of the mother is coated with a liquid, such as mineral oil. If this were not done, most of the incident ultrasonic waves from the piezoelectric source would be reflected at the boundary between the air and the mother's skin. Mineral oil has a density similar to that of skin, and as Equation 14.23 indicates, a very small fraction of the incident ultrasonic wave is reflected when $\rho_i \approx \rho_t$. The ultrasound energy is emitted as pulses rather than as a continuous wave, so the same crystal can be used as a detector as well as a transmitter. The source-receiver is then passed over the mother's abdomen. The

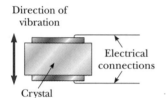

FIGURE 14.26
An alternating voltage applied to the faces of a piezoelectric crystal causes the crystal to vibrate.

reflected sound waves picked up by the receiver are converted to an electric signal, which forms an image along a line on a fluorescent screen. The sound source is then moved a few centimeters on the mother's body, and the process is repeated. The reflected signal produces a second line on the fluorescent screen. In this fashion a complete scan of the fetus can be made. Difficulties such as the likelihood of spontaneous abortion or of breech birth are easily detected with this technique. Also, such fetal abnormalities as spina bifida and water on the brain are readily observable.

Another interesting application of ultrasound is the ultrasonic ranging unit designed by the Polaroid Corporation and used in some of their cameras to provide an almost instantaneous measurement of the distance between the camera and object to be photographed. The principal component of this device is a crystal that acts as both a loudspeaker and a microphone. A pulse of ultrasonic waves is transmitted from the transducer to the object to be photographed. The object reflects part of the signal, producing an echo that is detected by the device. The time interval between the outgoing pulse and the detected echo is then electronically converted to a distance value, since the speed of sound is a known quantity.

A relatively new medical application of ultrasonics is the cavitron ultrasonic surgical aspirator (CUSA). This device has made it possible to surgically remove brain tumors that were previously inoperable. It is a long needle that emits very high-frequency ultrasonic waves (about 23 kHz) at its tip. When the tip touches a tumor, the part of the tumor near the needle is shattered and the residue can be sucked up (aspirated) through the hollow needle.

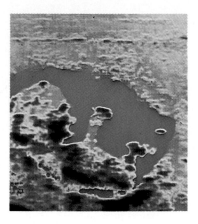

An ultrasound image of a human fetus in the womb after 12 weeks of development, showing the head, body, arms, and legs in profile. *(CNRI/SPL/Photo Researchers)*

SUMMARY

Sound waves are longitudinal waves. **Audible waves** are sound waves with frequencies between 20 and 20 000 Hz. **Infrasonic waves** have frequencies below the audible range, and **ultrasonic waves** have frequencies above the audible range.

The speed of sound in a medium of bulk modulus B and density ρ is

$$v = \sqrt{\frac{B}{\rho}} \qquad \text{[14.1]}$$

The speed of sound also depends on the temperature of the medium. The relationship between temperature and speed for sound in air is

$$v = (331 \text{ m/s})\sqrt{1 + \frac{T}{273}} \qquad \text{[14.4]}$$

where T is the temperature in degrees Celsius and 331 m/s is the speed of sound in air at 0°C.

The **intensity level** of a sound wave, in decibels, is given by

$$\beta \equiv 10 \log\left(\frac{I}{I_0}\right) \qquad \text{[14.7]}$$

The constant I_0 is a reference intensity level, usually taken to be at the threshold of hearing ($I_0 = 1.0 \times 10^{-12}$ W/m²), and I is the intensity level β, where β is measured in **decibels** (dB).

The **intensity** of a *spherical wave* produced by a point source is proportional to the average power emitted and inversely proportional to the square of the distance from the source:

$$I = \frac{P_{av}}{4\pi r^2}$$

[14.8]

The change in frequency heard by an observer whenever there is relative motion between the source and observer is called the **Doppler effect.** If the observer is moving with a speed v_o and the source is at rest, the observed frequency, f', is

$$f' = f\left(\frac{v \pm v_o}{v}\right)$$

[14.11]

The positive sign is used when the observer is moving toward the source, and the negative sign when the observer is moving away from the source.

If the source is moving with a speed v_s and the observer is at rest, the observed frequency is

$$f' = f\left(\frac{v}{v \mp v_s}\right)$$

[14.14]

where $-v_s$ refers to motion toward the observer and $+v_s$ refers to motion away from the observer.

When the observer and source are both moving, the observed frequency is

$$f' = f\left(\frac{v \pm v_o}{v \mp v_s}\right)$$

[14.15]

When waves interfere, the resultant wave is found by adding the individual waves together point by point. When crest meets crest and trough meets trough, the waves undergo **constructive interference.** When crest meets trough, **destructive interference** occurs.

Standing waves are formed when two waves having the same frequency, amplitude, and wavelength travel in opposite directions through a medium. One can set up standing waves of specific frequencies in a stretched string. The natural frequencies of vibration of a stretched string of length L, fixed at both ends, are

$$f_n = \frac{n}{2L}\sqrt{\frac{F}{\mu}} \qquad n = 1, 2, 3, \ldots$$

[14.20]

where F is the tension in the string and μ is its mass per unit length. The natural frequencies of vibration form a **harmonic series;** that is, the frequencies are integral multiples of the fundamental (lowest) frequency.

A system capable of oscillating is said to be in **resonance** with some driving force whenever the frequency of the driving force matches one of the natural frequencies of the system. When the system is resonating, it oscillates with maximum amplitude.

Standing waves can be produced in a tube of air. If the reflecting end of the tube is *open, all* harmonics are present and the natural frequencies of vibration are

$$f_n = n\,\frac{v}{2L} \qquad n = 1, 2, 3, \ldots \qquad \textbf{[14.21]}$$

If the tube is *closed* at the reflecting end, only the *odd* harmonics are present, and the natural frequencies of vibration are

$$f_n = n\,\frac{v}{4L} \qquad n = 1, 3, 5, \ldots \qquad \textbf{[14.22]}$$

The phenomenon of **beats** is an interference effect that occurs when two waves of slightly different frequencies travel in the same direction. For sound waves, the loudness of the resultant sound changes periodically with time.

ADDITIONAL READING

L. N. Baranek, "Noise," *Sci. American,* December 1966, p. 66.

C. A. Culver, *Musical Acoustics,* New York, McGraw-Hill, 1957.

N. H. Fletcher and S. Thwaites, "The Physics of Organ Pipes," *Sci. American,* January 1983, p. 94.

A. J. Hudspeth, "The Hair Cells of the Inner Ear," *Sci. American,* January 1983, p. 54.

C. M. Hutchins, "The Acoustics of Violin Plates," *Sci. American,* October 1981, p. 170.

C. M. Hutchins et al., *The Physics of Music,* New York, Freeman, 1978 (a collection of readings from *Scientific American*).

G. Loeb, "The Functional Replacement of the Ear," *Sci. American,* February 1985, p. 104.

J. Monforte, "The Digital Reproduction of Sound," *Sci. American,* December 1984, p. 78.

B. Patterson, "Musical Dynamics," *Sci. American,* November 1974, p. 78.

C. Shadle, "Experiments on the Acoustics of Whistling," *The Physics Teacher,* March 1983, p. 148.

J. Sundberg, "The Acoustics of the Singing Voice," *Sci. American,* March 1977, p. 82.

G. Von Bekesy, "The Ear," *Sci. American,* August 1957, p. 66.

H. E. White and D. H. White, *Physics and Music,* Philadelphia, Saunders, 1980.

CONCEPTUAL QUESTIONS

Example   As a result of a distant explosion, an observer senses a ground tremor and then hears the explosion. Explain the time lag.

Reasoning   Because it involves compression, the ground tremor can be thought of as a sound wave moving through the ground. Rock is much stiffer against compression than air, hence the ground wave travels at a higher speed than the sound wave traveling through air. Since they start together, the ground tremor reaches the observer before the sound traveling in air.

Example   If a bell is ringing inside a glass container, we cease to hear the sound when the air is pumped out, but we can still see the bell vibrating. What difference does this indicate between the properties of sound waves and light waves?

Reasoning   A sound wave requires a medium such as air or water to propagate; a light wave does not.

Example   Explain why sound waves travel faster in warm air than in cool air.

Reasoning   As temperature increases, the average speed of air molecules increases. Since sound waves are the result of collisions between successive layers of air molecules, and the average speed of the molecules increases with temperature, one expects that the speed of sound should increase with increasing temperature.

Example   Does the wind alter the frequency of sound heard by an observer who is at rest relative to the source of sound? Explain.

Reasoning   Let us assume that the wind is blowing toward the observer with speed v_w, and the source is stationary. As far as the wave motion is concerned, in the reference frame of the air, the effect of the wind is the same as if the observer were moving toward the source at the wind speed. In the Doppler shift frequency equation, both the observer speed v_w and source speed v are positive, and the observed frequency is the same as if no wind were blowing. However, the observer detects the source coming toward him at a speed $v + v_w$. At this larger speed, the observer attributes a larger wavelength, $\lambda' = (v + v_w)/f$, to the wave.

Example   If the wavelength of a sound source is reduced by a factor of 2, what happens to its frequency? its speed?
Reasoning   The speed of sound in a given medium depends only on the physical properties of the medium and temperature. For a harmonic wave, the speed of sound is equal to the product of the frequency and wavelength. Hence, if the wavelength of sound in a medium is reduced by a factor of 2, its frequency must increase by a factor of 2. However, its speed does not change.

Example   A sound wave travels in air at a frequency of 500 Hz. If part of the wave travels from the air into water, does its frequency change? Does its wavelength change? Justify your answers. Note that the speed of sound in air is about 340 m/s, whereas the speed of sound in water is about 1500 m/s.
Reasoning   There is no frequency change when a wave moves across a stationary interface between two media. Every crest in the first medium becomes just one crest in the second, so the number of cycles per second is the same on both sides. As the speed increases from 340 m/s to 1500 m/s, the wavelength ($\lambda = v/f$) increases from 0.68 m to 3.0 m.

1. Why are sound waves characterized as longitudinal?
2. Explain how the distance to a lightning bolt can be determined by counting the seconds between the flash and the sound of the thunder.
3. By listening to a band or orchestra, how can you determine that the speed of sound is the same for all frequencies?
4. Of the following sounds, state which is most likely to have an intensity level of 60 dB: a rock concert, the turning of a page in this text, normal conversation, a cheering crowd at a football game, or background noise at a church?
5. Estimate the decibel level of each of the sounds in Question 4.
6. If the distance from a point source of sound is tripled, by what factor does the sound intensity decrease? Assume there are no reflections from nearby objects to affect your results.
7. Why is the intensity of an echo less than that of the original sound?
8. An airplane mechanic notices that the sound from a twin-engine aircraft rapidly varies in loudness when both engines are running. What could be causing this variation between loud and soft?
9. At certain speeds, an automobile driven on a bumpy road vibrates disastrously and loses traction and braking effectiveness. At other speeds, either lesser or greater, the vibrations are more manageable. Explain. Why are "rumble strips," which work on this same principle, often used just before stop signs?
10. Despite a reasonably steady hand, a person often spills his coffee when carrying it to his seat. Discuss resonance as a possible cause of this difficulty, and devise a means for solving the problem.
11. The radar systems used by police to detect speeders are sensitive to the Doppler shift of a pulse of radio waves. Discuss how this sensitivity can be used to measure the speed of a car.
12. A student records the first ten harmonics for a pipe. Is it possible to determine whether the pipe is open at both ends or closed at one end by studying the difference in frequencies between adjacent harmonics? Explain. What if the ratios of adjacent harmonics are studied?
13. How can an object move with respect to an observer so that the sound from it is not shifted in frequency?
14. Why is it not possible to use sonar (sound waves) to determine the speed of an object traveling faster than the speed of sound in that medium?
15. Why can a duck easily produce a bow wave in water, but an airplane must fly very fast to produce a shock wave in air?
16. How is the natural frequency of vibration of an organ pipe altered as room temperature increases?
17. A person who has just inhaled helium speaks with a high-pitched Donald Duck voice. Why?
18. A soft-drink bottle resonates as air is blown across its top. What happens to the resonant frequency as the level of fluid in the bottle decreases?
19. When two waves interfere constructively or destructively, does any gain or loss in energy occur? Explain.
20. What is the purpose of the slide on a trombone or the valves on a trumpet?
21. Explain why your voice seems to sound better than usual when you sing in the shower.
22. A string of length L, mass per unit length μ, and tension F is vibrating at its fundamental frequency. What effects do the following have on the fundamental frequency? (a) The length of the string is doubled, with all other factors held constant. (b) The mass per unit length is doubled, with all other factors held constant. (c) The tension is doubled, with all other factors held constant.
23. Why does a vibrating guitar string sound louder when placed on the instrument than it would if allowed to vibrate in the air while off the instrument?

PROBLEMS

Section 14.2   Characteristics of Sound Waves

Section 14.3   The Speed of Sound

Unless otherwise stated, use 345 m/s as the speed of sound in air.

1. A sound wave has a frequency of 700 Hz in air and a wavelength of 0.50 m. What is the temperature of the air?

2. The range of human hearing extends from approximately 20 Hz to 20 000 Hz. Find the wavelengths of these extremes at a temperature of $27°C$.

3. A group of hikers hear an echo 3.0 s after they shout. If the temperature is $22°C$, how far away is the mountain that reflected the sound wave?

4. A dolphin located in seawater at a temperature of $25°C$ emits a sound directed toward the bottom of the ocean 150 m below. How much time passes before it hears an echo?

5. A sound wave propagating in air has a frequency of 4000 Hz. Calculate the percent change in wavelength when the wavefront, initially in a region where $T = 27°C$, enters a region where $T = 10°C$.

6. The greatest value ever achieved for the speed of sound in air is about 1.0×10^4 m/s, and the highest frequency ever produced is about 2.0×10^{10} Hz. Find the wavelength of this wave.

7. A stone is dropped from rest into a well. The sound of the splash is heard exactly 2.00 s later. Find the depth of the well if the air temperature is $10.0°C$.

8. If the density of methyl alcohol is 0.80×10^3 kg/m³, find an approximate value for the bulk modulus of this liquid at $25°C$. (See Table 14.1.)

Section 14.4   Energy and Intensity of Sound Waves

9. A microphone in the ocean is sensitive to sounds emitted by porpoises. To produce a usable signal, sound waves striking the microphone must have an intensity of 10 dB. If porpoises emit sound waves with a power of 0.050 W, how far can a porpoise be from the microphone and still be heard? Disregard absorption of sound waves by the water.

10. The intensity level of an orchestra is 85 dB. A single violin achieves a level of 70 dB. How does the intensity of the sound of the full orchestra compare with that of the violin's sound?

11. The area of a typical eardrum is about 5.0×10^{-5} m². Calculate the sound power (the energy per second) incident on an eardrum at (a) the threshold of hearing and (b) the threshold of pain.

12. A noisy machine in a factory produces a decibel rating of 80 dB. How many identical machines could you add to the factory without exceeding the 90-dB limit?

13. Intensity is defined as power per unit area. Show that the decibel level of two sounds can be related to the ratio of two powers as

$$dB = 10 \log \frac{P_1}{P_0}$$

14. On a workday the average decibel level of a busy street is 70 dB, with 100 cars passing a given point every minute. If the number of cars is reduced to 25 every minute on a weekend, what is the decibel level of the street?

Section 14.5   Spherical and Plane Waves

15. A rock group is playing in a bar. Sound emerging from the door spreads uniformly in all directions. If the intensity level of the music is 80.0 dB at a distance of 5.00 m from the door, at what distance is the music just barely audible to a person with a normal threshold of hearing? Disregard absorption.

16. A skyrocket explodes 100 m above the ground (Fig. 14.27). Three observers are spaced 100 m apart, and the first (A) is directly under the point of the explosion. What is the ratio of sound intensities heard by observers A and B? observers A and C?

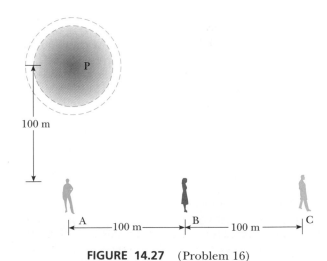

FIGURE 14.27   (Problem 16)

17. A stereo speaker (considered a small source) emits sound waves with a power output of 100 W. (a) Find the intensity 10.0 m from the source. (b) Find the intensity level, in decibels, at this dis-

tance. (c) At what distance would you experience the sound at the threshold of pain, 120 dB?

18. A man shouting loudly produces a 70-dB sound at a distance of 5.0 m. How many watts of power does the man emit? (Treat the man as a point source.)

19. Show that the difference in decibel levels, β_1 and β_2, of a sound source is related to the ratio of its distances, r_1 and r_2, from the receivers by

$$\beta_2 - \beta_1 = 20 \log \frac{r_1}{r_2}$$

20. A stereo speaker is placed between two observers who are 110 m apart, along the line connecting them. If one observer records an intensity level of 60 dB, and the other records an intensity level of 80 dB, how far is the speaker from each observer?

21. A bat flying at 5.0 m/s emits a chirp at 40 kHz. If this sound pulse is reflected by a wall, what is the frequency of the echo received by the bat?

22. An airplane traveling with half the speed of sound ($v = 172$ m/s) emits a sound of frequency 5.00 kHz. At what frequency does a stationary listener hear the sound (a) as the plane approaches? (b) after it passes?

23. At rest, a car's horn sounds the note A (440 Hz). The horn is sounded while the car is moving down the street. A bicyclist moving in the same direction with one-third the car's speed hears a frequency of 440 Hz. How fast is the car moving? Is the cyclist ahead of or behind the car?

24. A fire engine moving to the right at 40 m/s sounds its horn (frequency = 500 Hz) at the two vehicles shown in Figure 14.28. The car is moving to the right at 30 m/s, while the van is at rest. (a) What frequency is heard by the passengers in the car? (*Hint:* See Problem 76.) (b) What is the frequency as heard by the passengers in the van? (c) When the fire engine is 200 m from the car and 250 m from the van, the passengers in the car hear a sound intensity level of 90 dB. At that moment, what intensity level is heard by the passengers in the van?

Fire Engine   Car   Van

FIGURE 14.28   (Problem 24)

25. An ambulance siren emits a note of 500 Hz when the ambulance is at rest. When the ambulance is moving at 10.0 m/s, what frequency is heard by an observer traveling at (a) 15.0 m/s toward the

source? (b) 15.0 m/s, after the observer has passed the ambulance and recedes from it?

26. An alert physics student stands beside the tracks as a train rolls slowly past. He notes that the frequency of the train whistle is 442 Hz when the train is approaching him and 441 Hz when the train is receding from him. From this he can find the speed of the train. What value does he find?

27. Two trains on separate tracks move toward one another. Train 1 has a speed of 130 km/h and train 2 a speed of 90.0 km/h. Train 2 blows its horn, emitting a frequency of 500 Hz. What is the frequency heard by the engineer on train 1?

28. A horn emits a frequency of 1000 Hz. A 14.0-m/s wind is blowing toward a listener. What is the frequency of the sound heard by the listener? (*Hint:* Consider a frame of reference in which the air is at rest. What can you conclude from the result of this problem?)

29. A supersonic jet traveling at Mach 3 at an altitude of 20 000 m is directly overhead at time $t = 0$, as in Figure 14.29. (a) How long will it be before the ground observer encounters the shock wave? (b) Where will the plane be when it is finally heard? (Assume that the speed of sound in air is uniform at 335 m/s.)

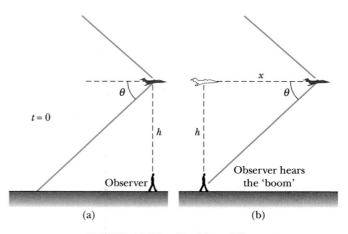

FIGURE 14.29   (Problem 29)

30. The Concorde flies at Mach 1.5, which means the speed of the plane is 1.5 times the speed of sound in air. What is the angle between the direction of propagation of the shock wave and the direction of the plane's velocity?

Section 14.7   Interference of Sound Waves

31. Two loudspeakers are placed above and below one another, as in Figure 14.13, and driven by the same source at a frequency of 500 Hz. (a) What

minimum distance should the top speaker be moved back in order to create destructive interference between the two speakers? (b) If the top speaker is moved back twice the distance calculated in part (a), will constructive or destructive interference occur?

32. A pair of speakers separated by 0.700 m are driven by the same oscillator at a frequency of 700 Hz. An observer, originally positioned at one of the speakers, begins to walk along a line perpendicular to the line joining the speakers. Assuming that the speed of sound in air is 343 m/s, how far must the observer walk before reaching a relative (a) maximum in intensity? (b) minimum in intensity?

33. The sound interferometer shown in Figure 14.12 is driven by a speaker emitting a 400-Hz note. If *destructive* interference occurs at a particular instant, how much must the path length in the U-shaped tube be increased in order to hear (a) constructive interference and (b) destructive interference once again?

34. Two identical speakers separated by 10.0 m are driven by the same oscillator with a frequency of $f = 21.5$ Hz (Fig. 14.30). (a) Explain why a receiver at A records a minimum in sound intensity from the two speakers. (b) If the receiver is moved in the plane of the speakers, what path should it take so that the intensity remains at a minimum? That is, determine the relationship between x and y (the coordinates of the receiver) that causes the receiver to record a minimum in sound intensity. Take the speed of sound to be 344 m/s.

35. Two speakers are driven by a common oscillator at 800 Hz and face each other at a distance of 1.25 m. Locate the points along a line joining the two speakers where relative minima would be expected. (Use $v = 343$ m/s for the speed of sound in air.)

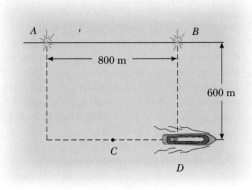

FIGURE 14.31 (Problem 36)

36. The ship in Figure 14.31 travels along a straight line parallel to the shore and 600 m from the shore. The ship's radio receives simultaneous signals of the same frequency from antennas A and B. The signals interfere constructively at point C, which is equidistant from A and B. The signal goes through the first minimum at point D. Determine the wavelength of the radio waves.

Section 14.8 Standing Waves

37. A 12-kg mass hangs in equilibrium from a string of total length $L = 5.0$ m and linear mass density $\mu = 0.0010$ kg/m. The string is wrapped around two light, frictionless pulleys that are separated by the distance $d = 2.0$ m (Fig. 14.32a). (a) Determine the tension in the string. (b) At what frequency must the string between the pulleys vibrate in order to form the standing wave pattern shown in Figure 14.32b?

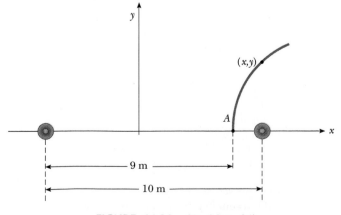

FIGURE 14.30 (Problem 34)

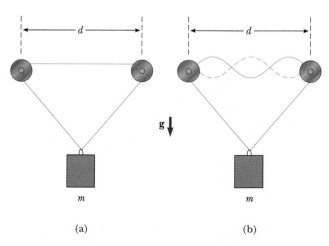

FIGURE 14.32 (Problem 37)

38. A steel wire in a piano has a length of 0.7000 m and a mass of 4.300×10^{-3} kg. To what tension must this wire be stretched in order that the fundamental vibration correspond to middle C ($f_C = 261.6$ Hz on the chromatic musical scale)?

39. How far, and in what direction, should a cellist move his finger to adjust a string's tone from an out-of-tune 449 Hz to an in-tune 440 Hz? The string is 68.0 cm long, and the finger is 20.0 cm from the nut for the 449-Hz tone.

40. A 0.30-g wire is stretched between two points 70 cm apart. If the tension in the wire is 600 N, find the wire's first, second, and third harmonics.

41. In the arrangement shown in Figure 14.33, masses can be hung from a string (with linear mass density $\mu = 0.0020$ kg/m) around a light pulley. The string is connected to a vibrator (of constant frequency, f), and the length of the string between point P and the pulley is $L = 2.0$ m. When the mass m is either 16 kg or 25 kg, standing waves are observed, but no standing waves are observed for any mass between these values. (a) What is the frequency of the vibrator? (*Hint:* The greater the tension in the string, the smaller the number of nodes in the standing wave.) (b) What is the largest mass for which standing waves could be observed?

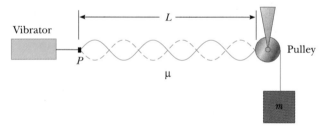

FIGURE 14.33   (Problems 41 and 64)

42. A stretched string fixed at each end has a mass of 40 g and a length of 8.0 m. The tension in the string is 49 N. (a) Determine the positions of the nodes and antinodes for the third harmonic. (b) What is the vibration frequency for this harmonic?

43. Two pieces of steel wire with identical cross-sections have lengths of L and $2L$. Each of the wires is fixed at both ends and stretched so that the tension in the longer wire is four times greater than that in the shorter wire. If the fundamental frequency in the shorter wire is 60 Hz, what is the frequency of the second harmonic in the longer wire?

44. A stretched string of length L is observed to vibrate in five equal segments when driven by a 630-Hz oscillator. What oscillator frequency will set up a standing wave so that the string vibrates in three segments?

Section 14.9   Forced Vibrations and Resonance

45. The chains suspending a child's swing are 2.00 m long. List three frequencies at which this swing could be pushed to set up resonant vibrations in it.

46. A 5.0-kg mass connected to a spring is found to resonate when it is pushed at the frequencies 2.4 Hz, 1.2 Hz, 0.80 Hz, 0.60 Hz, ... Determine the spring constant for the spring.

Section 14.10   Standing Waves in Air Columns

47. The fundamental frequency of an open organ pipe corresponds to middle C (261.6 Hz on the chromatic musical scale). The third resonance of a closed organ pipe has the same frequency. What are the lengths of the two pipes?

48. The human ear canal is about 2.8 cm long. If it is regarded as a tube open at one end and closed at the eardrum, what is the fundamental frequency around which we would expect hearing to be best? Take the speed of sound to be 340 m/s.

49. By adjusting her lips correctly and blowing with the proper pressure, a bugler can cause her instrument to produce a sequence of tones, among which are the following: 440, 660, 880, 1100, . . . Hz—all without changing the length of the air column. (a) Does the bugle behave as an open pipe or a pipe closed at one end? (b) What is the effective length of this bugle? (Use 346 m/s for the speed of sound in air.)

50. A 0.50-m-long brass pipe open at both ends has a fundamental frequency of 350 Hz. (a) Determine the temperature of the air in the pipe. (b) If the temperature is increased by 20°C, what is the new fundamental frequency of the pipe? Be sure to include the effects of temperature on both the speed of sound in air and the length of the pipe.

51. A pipe open at both ends has a fundamental frequency of 300 Hz when the temperature is 0°C. (a) What is the length of the pipe? (b) What is the fundamental frequency at a temperature of 30°C?

52. A tuning fork is sounded above a resonating tube as in Figure 14.20. The first resonant point is 0.080 m from the top of the tube, and the second is at 0.24 m. (a) Where is the third resonant point? (b) What is the frequency of the tuning fork?

53. A 2.00-m-long air column is open at both ends. The frequency of a certain harmonic is 410 Hz, and the frequency of the next higher harmonic is 492 Hz. Determine the speed of sound in the air column.

54. A shower stall measures 86 cm × 86 cm × 210 cm. When a person sings in this shower, which frequencies sound the richest (resonate), given that the temperature in the hot shower stall is 41°C? Assume that, across the width of the stall, the effects are those of a pipe closed at both ends (nodes at both sides), whereas vertically the stall acts as a pipe closed at one end. Assume also that the human voice ranges from 130 to 2000 Hz.

Section 14.11   Beats

55. Two identical mandolin strings under 200 N of tension are sounding tones with frequencies of 523 Hz. The peg of one string slips slightly, and the tension in it drops to 196 N. How many beats are heard?

56. A flute is designed so that it plays a frequency of 261.6 Hz, middle C, when all the holes are covered and the temperature is 20.0°C. (a) Consider the flute to be a pipe open at both ends, and find its length, assuming that the middle-C frequency is the fundamental. (b) A second player, nearby in a colder room, also attempts to play middle C on an identical flute. A beat frequency of 3.00 Hz is heard. What is the temperature of the room?

57. The G string on a violin has a fundamental frequency of 196 Hz. It is 30.0 cm long and has a mass of 0.500 g. As it is being lowered, a nearby violinist fingers a string on her identical violin until a beat frequency of 2.00 Hz is heard between the two. What is the length of the violin string being lowered by the nearby violinist?

Section 14.13   The Ear

58. If a human ear canal can be thought of as resembling an organ pipe, closed at one end, that resonates at a fundamental frequency of 3000 Hz, what is the length of the canal? Use normal body temperature for your determination of the speed of sound in the canal.

59. Some studies indicate that the upper frequency limit of hearing is determined by the diameter of the eardrum. The wavelength of the sound wave and the diameter of the eardrum are approximately equal at this upper limit. If this is so, what is the diameter of the eardrum of a person capable of hearing 20 000 Hz? (Assume a body temperature of 37°C.)

ADDITIONAL PROBLEMS

60. A block with a speaker bolted to it is connected to a spring having spring constant $k = 20.0$ N/m, as in Figure 14.34. The total mass of the block and

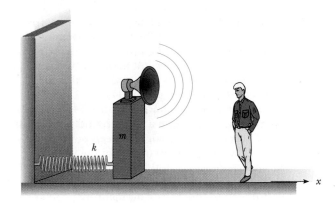

FIGURE 14.34   (Problem 60)

speaker is 5.00 kg, and the amplitude of this unit's motion is 0.500 m. (a) If the speaker emits sound waves of frequency 440 Hz, determine the range in frequencies heard by the person to the right of the speaker. (b) If the maximum intensity level heard by the person is 60 dB when he is *closest* to the speaker, 1.00 m away, what is the minimum intensity heard by the observer? Assume that the speed of sound is 343 m/s.

61. A commuter train passes a passenger platform at a constant speed of 40.0 m/s. The train horn is sounded at a frequency of 320 Hz when the train is at rest. (a) What is the frequency observed by a person on the platform as the train approaches and (b) as the train recedes from him? (c) What wavelength does the observer find in each case?

62. Two train whistles have identical frequencies of 180 Hz. When one train is at rest in the station, sounding its whistle, a beat frequency of 2 Hz is heard from a moving train. What two possible speeds and directions can the moving train have?

"I love hearing that lonesome wail of the train whistle as the magnitude of the frequency of the wave changes due to the Doppler effect."

63. A pipe open at each end has a fundamental frequency of 300 Hz when the speed of sound in air is 333 m/s. (a) How long is the pipe? (b) What is the frequency of the second harmonic when the temperature of the air is increased so that the speed of sound in the pipe is 344 m/s?

64. In the arrangement shown in Figure 14.33, a mass, $m = 5.0$ kg, hangs from a cord around a light pulley. The length of the cord between point P and the pulley is $L = 2.0$ m. (a) When the vibrator is set to a frequency of 150 Hz, a standing wave with six loops is formed. What must be the linear mass density of the cord? (b) How many loops (if any) will result if m is changed to 45 kg? (c) How many loops (if any) will result if m is changed to 10 kg?

65. Two point sound sources have measured intensities of $I_1 = 100$ W/m^2 and $I_2 = 200$ W/m^2. By how many decibels is the level of source 1 lower than the level of source 2? (Assume that the observer is at the same distance from both sources.)

66. A typical sound level for a buzzing mosquito is 40 dB, and normal conversation is approximately 50 dB. How many buzzing mosquitoes will produce a sound intensity equal to that of normal conversation?

67. When high-energy charged particles move through a transparent medium with a speed greater than the speed of light in that medium, a shock wave, or bow wave, of light is produced. This phenomenon is called the *Cerenkov effect* and can be observed in the vicinity of the core of a swimming-pool nuclear reactor due to high-speed electrons moving through the water. In a particular case, the Cerenkov radiation produces a wavefront with a cone angle of 53.0°. Calculate the velocity of the electrons in the water. (Take the speed of light in water to be 2.25×10^8 m/s.)

68. Refer to Table 14.2, which gives decibel values for representative sounds from various sources. Determine the resultant intensity, in decibels, when a vacuum cleaner and a power mower are operated against a background of busy traffic.

69. A variable-length air column is placed just below a vibrating wire that is fixed at both ends. The length of the air column is gradually increased from zero until the first position of resonance is observed at $L = 34$ cm. The wire is 120 cm long and is vibrating in its third harmonic. If the speed of sound in air is 340 m/s, what is the speed of transverse waves in the wire?

70. Two pipes, equal in length, are each open at one end. Each has a fundamental frequency of 480 Hz at 300 K. In one pipe the air temperature is increased to 305 K. If the two pipes are sounded together, what beat frequency results?

71. When at rest, two trains have sirens that emit a frequency of 300 Hz. The two trains travel toward one another and toward an observer stationed between them. One of the trains moves at 30 m/s, and the observer hears a beat frequency of 3 beats per second. What is the velocity of the second train, which travels faster than the first?

72. By proper excitation, it is possible to produce both longitudinal and transverse waves in a long metal rod. In a particular case, the rod is 150 cm long and 0.200 cm in radius and has a mass of 50.9 g. Young's modulus for the material is 6.80×10^{11} dynes/cm^2. Determine the required tension in the rod so that the ratio of the speed of longitudinal waves to the speed of transverse waves is 8.

73. A speaker at the front of a room and an identical speaker at the rear of the room are being driven at 456 Hz by the same sound source. A student walks at a uniform rate of 1.50 m/s along the length of the room. How many beats does the student hear per second?

74. A student stands several meters in front of a smooth reflecting wall, holding a board on which a wire is fixed at each end. The wire, vibrating in its third harmonic, is 75.0 cm long, has a mass of 2.25 g, and is under a tension of 400 N. A second student, moving toward the wall, hears 8.30 beats per second. What is the speed of the student approaching the wall? Use 340 m/s as the speed of sound in air.

75. Two ships are moving along a line due east. The trailing vessel has a speed of 64.0 km/h relative to a land-based observation point, and the leading ship has a speed of 45.0 km/h relative to the same station. The two ships are in a region of the ocean where the current is moving uniformly due west at 10.0 km/h. The trailing ship transmits a sonar signal at a frequency of 1200 Hz. What frequency is monitored by the leading ship? (Use 1520 m/s as the speed of sound in ocean water.)

76. The Doppler equation presented in the text is valid when the motion between the observer and the source occurs on a straight line, so that the source and observer are moving either directly toward or directly away from each other. If this restriction is relaxed, one must use the more general Doppler equation

$$f' = \left[\frac{v + v_o \cos(\theta_o)}{v - v_s \cos(\theta_s)} \right] f$$

where θ_o and θ_s are defined in Figure 14.35a. (a) If both observer and source are moving away from each other, show that the preceding equation re-

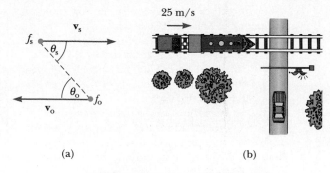

(a)

(b)

FIGURE 14.35 (Problem 76)

duces to Equation 14.15 with lower signs. (b) Use the preceding equation to solve the following problem. A train moves at a constant speed of 25.0 m/s toward the intersection shown in Figure 14.35b. A car is stopped near the intersection, 30.0 m from the tracks. If the train's horn emits a frequency of 500 Hz, what is the frequency heard by the passengers in the car when the train is 40 m from the intersection? Take the speed of sound to be 343 m/s.

77. A duck swims in a pond. Water waves move outward from the duck with a speed of 0.500 m/s. If the wavelength of the waves behind the duck is 50.0% longer than the wavelength of the waves ahead of the duck, how fast is the duck moving?

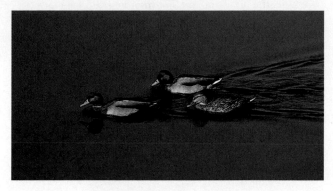

(Problem 77)   *(Dave Gleiter/FPG)*

Electricity and Magnetism

We now begin a study of the branch of physics concerned with electric and magnetic phenomena. The laws of electricity and magnetism play central roles in the operation of many devices such as radios, televisions, electric motors, computers, high-energy accelerators, and a host of electronic devices used in medicine. More fundamentally, we now know that the interatomic and intermolecular forces that are responsible for the formation of solids and liquids are electric in origin. Furthermore, such forces as the pushes and pulls between objects and the elastic force in a spring arise from electric forces at the atomic level.

The ancient Greeks observed electric and magnetic phenomena as early as 700 B.C. They found that a piece of amber, when rubbed, became electrified and attracted pieces of straw or feathers. The existence of magnetic forces was known as a result of observations that a naturally occurring stone called *magnetite* (Fe_2O_3) was attracted to iron. (The word *electric* comes from the Greek word for amber, *elecktron*. The word *magnetic* comes from the name of the country where magnetite was found, *Magnesia*, now Turkey.)

In 1600 William Gilbert discovered that electrification was not limited to amber but is a general phenomenon. Experiments by Charles Coulomb confirmed the inverse-square force law for electricity.

It was not until the early part of the 19th century that scientists established that electricity and magnetism are, in fact, related phenomena. In 1820 Hans Oersted discovered that a compass needle is deflected when placed near a wire carrying an electric current. A few years later, Michael Faraday showed that when a wire is moved near a magnet (or, equivalently, when a magnet is moved near a wire), an electric current is observed in the wire. James Clerk Maxwell used these observations and other experimental facts as bases for formulating the laws of electromagnetism as we now know them. (*Electromagnetism* is a name given to the combined fields of electricity and magnetism.) Shortly thereafter, Heinrich Hertz verified Maxwell's predictions by producing electromagnetic waves in the laboratory. This was followed by such practical developments as radio and television.

Maxwell's contributions to the science of electromagnetism were especially significant because the laws he formulated are basic to *all* forms of electromagnetic phenomena. His work is comparable in importance to Newton's discovery of the laws of motion and the theory of gravitation.

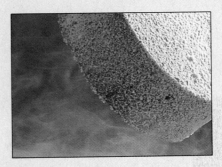

(© Tony Stone/Worldwide)

15 ELECTRIC FORCES AND ELECTRIC FIELDS

This dramatic one-minute exposure captures multiple lightning bolts illuminating Kitt Peak National Observatory in Arizona, an occurrence that illustrates electrical breakdown in the atmosphere.
(© Gary Ladd 1972)

The earliest known study of electricity was conducted by the Greeks about 700 B.C. By modern standards, their contributions to the field were modest. However, from those roots have sprung the enormous electrical distribution systems and sophisticated electronic instruments that are so much a part of our world today. It all began, apparently, when someone noticed that a fossil material called amber would attract small objects after being rubbed with wool. Since then we have learned that this phenomenon is not restricted to amber and wool but occurs (to some degree) when almost any two nonconducting substances are rubbed together.

In this chapter we use this effect, charging by friction, to begin an investigation of electric forces. We then discuss Coulomb's law, which is the fundamental law of force between any two charged particles. The concept of an electric field associated with charges is then introduced, and its effects on other charged particles described. We end with brief discussions of the Van de Graaff generator and the oscilloscope.

15.1

PROPERTIES OF ELECTRIC CHARGES

A number of simple experiments demonstrate the existence of electrostatic forces. For example, after running a plastic comb through your hair, you will find that the comb attracts bits of paper. The attractive force is often strong enough to suspend the paper from the comb. The same effect occurs with other rubbed materials, such as glass and hard rubber.

Another simple experiment is to rub an inflated balloon with wool (or across your hair). On a dry day, the rubbed balloon will then stick to the wall of a room, often for hours. When materials behave in this way, they are said to have become **electrically charged.** You can give your body an electric charge by vigorously rubbing your shoes on a wool rug or by sliding across a car seat. You can then feel, and remove, the charge on your body by lightly touching another person. Under the right conditions, a visible spark can be seen when you touch, and a slight tingle is felt by both parties. (Experiments such as these work best on a dry day because excessive moisture can provide a pathway for charge to leak off a charged object.)

Experiments also demonstrate that there are two kinds of electric charge, which Benjamin Franklin (1706–1790) named **positive** and **negative.** Figure 15.1 illustrates the interaction of the two charges. A hard rubber (or plastic) rod that has been rubbed with fur (or an acrylic material) is suspended by a piece of string. When a glass rod that has been rubbed with silk is brought near the rubber rod, the rubber rod is attracted toward the glass rod (Fig. 15.1a). If two charged rubber rods (or two charged glass rods) are brought near each other, as in Figure 15.1b, the force between them is repulsive. This observation demonstrates that the rubber and glass have different kinds of charge. We use the convention suggested by Franklin, wherein the electric charge on the glass rod is called positive and that on the rubber rod is called negative. On the basis of observations such as these, we conclude that *like charges repel one another and unlike charges attract one another.*

Charge has a natural tendency to be transferred between unlike materials. Rubbing the two materials together serves to increase the area of contact and thus to enhance the charge transfer process.

Like charges repel; unlike charges attract

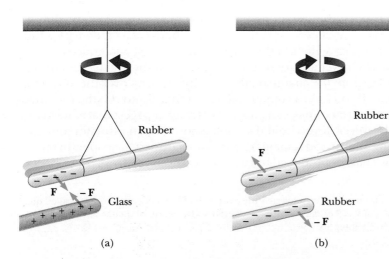

(a)   (b)

FIGURE 15.1
(a) A negatively charged rubber rod, suspended by a thread, is attracted to a positively charged glass rod. (b) A negatively charged rubber rod is repelled by another negatively charged rubber rod.

Another important characteristic of charge is that *electric charge is always conserved.* That is, when two initially neutral objects are charged by being rubbed together, charge is not created in the process. The objects become charged because *negative charge is transferred* from one object to the other. One object gains some amount of negative charge while the other loses an equal amount of negative charge and hence is left with a positive charge. For example, when a glass rod is rubbed with silk, as in Figure 15.2, the silk obtains a negative charge that is equal in magnitude to the positive charge on the glass rod as negatively charged electrons are transferred from the glass to the silk in the rubbing process. Likewise, when rubber is rubbed with fur, electrons are transferred from the fur to the rubber. Any *uncharged object* in our large-scale world contains an enormous number of electrons (on the order of 10^{23}). However, for every negative electron a positively charged proton is present; hence, an uncharged object has no net charge of either sign.

In 1909 Robert Millikan (1886–1953) discovered that if an object is charged, its charge is always a multiple of a fundamental unit of charge, which we designate with the symbol e. In modern terms, the charge is said to be **quantized**. This means that charge occurs as discrete bundles in nature. Thus, an object may have a charge of $\pm e$, or $\pm 2e$, or $\pm 3e$, and so on, but never a fractional charge of $\pm 1.5e$.[1] Other experiments in Millikan's time showed that the electron has a charge of $-e$ and the proton has an equal and opposite charge, $+e$. Some particles, such as a neutron, have no charge. A neutral atom (one with no net charge) contains as many protons as electrons. The value of e is now known to be $1.602\ 19 \times 10^{-19}$ C. (The unit of electric charge, the **coulomb** [C], will be defined more precisely in a later section.)

Charge is conserved; charge is quantized

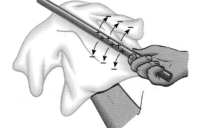

FIGURE 15.2
When a glass rod is rubbed with silk, electrons are transferred from the glass to the silk. Because of conservation of charge, each electron adds negative charge to the silk, and an equal positive charge is left behind on the rod. Also, because the charges are transferred in discrete bundles, the charges on the two objects are $\pm e$, or $\pm 2e$, or $\pm 3e$, and so on.

15.2

INSULATORS AND CONDUCTORS

It is convenient to classify substances in terms of their ability to conduct electric charge.

> **Conductors** are materials in which electric charges move freely, and **insulators** are materials in which electric charges do not move freely.

Glass and rubber are insulators. When such materials are charged by rubbing, only the rubbed area becomes charged, and there is no tendency for the charge to move into other regions of the material. In contrast, materials such as copper, aluminum, and silver are good conductors. When such materials are charged in some small region, the charge readily distributes itself over the entire surface of the material. If you hold a copper rod in your hand and rub the rod with wool or fur, it will not attract a piece of paper. This might suggest that a metal cannot be charged. However, if you hold the copper rod with an insulator and then rub it with wool or fur, the rod remains charged and attracts the paper. In the first case, the electric charges produced by rubbing readily move from the copper through

---

[1] Recent developments have suggested the existence of fundamental particles called **quarks** that have charges of $\pm e/3$ or $\pm 2e/3$. A more complete discussion of quarks and their properties is presented in Chapter 30.

PHYSICS IN *ACTION*

Two balloons, filled with air and suspended from strings, have been rubbed with a cloth and thereby given an electrostatic charge (a). Each carries a net negative charge, and hence they repel each other, as indicated by the fixed separation between them. As their charges are increased by further rubbing, the electrostatic repulsive force between them increases, as does their equilibrium separation.

Two charged balloons are filled with helium, rather than air, and again they are observed to undergo an electrostatic repulsion (b). In this situation, the balloons are in a "floating" position because helium is less dense than air, and the upward buoyant force on each balloon exceeds the weight of the balloon and its contents.

To demonstrate charging by induction (c), two balloons are first electrified by rubbing with a cloth. Each balloon induces a charge of opposite sign on the surface of the insulating wall next to it. The charge of opposite sign causes a net attractive force between each balloon and the wall. This is a common method used to decorate the walls of a room for a party. Why do you suppose this demonstration works better on a day when the humidity is low?

A stream of water is deflected to the right by a charged glass rod (d). Charge of opposite sign has been induced on the stream of water, producing an attractive force between the water and rod and a corresponding deflection.

If a fine water stream is directed upward, and a charged rod is placed next to the stream where it breaks into drops, the drops will mutually repel each other due to their like charge, producing an interesting fountain effect.

ELECTRIFIED BALLOONS AND INDUCED CHARGES

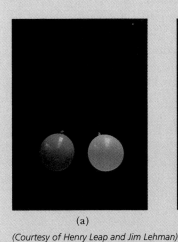

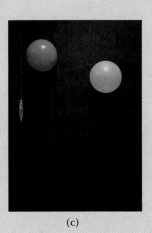

(a) (b) (c) (d)

(Courtesy of Henry Leap and Jim Lehman)

your body and finally to Earth. In the second case, the insulating handle prevents the flow of charge to Earth.

Semiconductors are a third class of materials, and their electrical properties are somewhere between those of insulators and those of conductors. Silicon and germanium are well-known semiconductors that are widely used in the fabrication of a variety of electronic devices. If controlled amounts of certain foreign atoms are added to semiconductors, their electrical properties can be changed by many orders of magnitude.

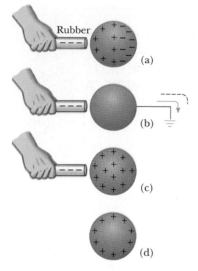

FIGURE 15.3
Charging a metal object by induction. (a) The charge on a neutral metal sphere is redistributed when a charged rubber rod is placed near the sphere.
(b) The sphere is grounded, and some of the electrons leave the sphere through the ground wire. (c) The ground connection is removed, and the sphere is left with excess positive charge. (d) When the rubber rod is moved away, the sphere becomes uniformly charged.

CHARGING BY INDUCTION

When a conductor is connected to Earth by means of a conducting wire or copper pipe, it is said to be **grounded.** The Earth can be considered an infinite reservoir for electrons; this means that it can accept or supply an unlimited number of electrons. With this in mind, we can understand the charging of a conductor by a process known as **induction.**

Consider a negatively charged rubber rod brought near a neutral (uncharged) conducting sphere that is insulated so that there is no conducting path to ground (Fig. 15.3). The repulsive force between the electrons in the rod and those in the sphere causes a redistribution of charge on the sphere so that some electrons move to the side of the sphere farthest away from the rod (Fig. 15.3a). The region of the sphere nearest the negatively charged rod has an excess of positive charge because of the migration of electrons away from this location. If a grounded conducting wire is then connected to the sphere, as in Figure 15.3b, some of the electrons leave the sphere and travel to the Earth. If the wire to ground is then removed (Fig. 15.3c), the conducting sphere is left with an excess of induced positive charge. Finally, when the rubber rod is removed from the vicinity of the sphere (Fig. 15.3d), the induced positive charge remains on the ungrounded sphere. This excess positive charge becomes uniformly distributed over the surface of the ungrounded sphere because of the repulsive forces among the like charges and the high mobility of charge carriers in a metal.

In the process of inducing a charge on the sphere, the charged rubber rod loses none of its negative charge since it never came in contact with the sphere. *Charging an object by induction requires no contact with the object inducing the charge.* This is in contrast to charging an object by rubbing, which does require contact between the two objects.

A process very similar to charging by induction in conductors also takes place in insulators. In most neutral atoms or molecules, the center of positive charge coincides with the center of negative charge. However, in the presence of a charged object, these centers may shift slightly, resulting in more positive charge on one side of the molecule than on the other. This effect is known as **polarization.** The realignment of charge within individual molecules produces an induced charge on the surface of the insulator as shown in Figure 15.4a. With these concepts, you should be able to explain why a comb that has been rubbed through hair will attract bits of neutral paper, or why a balloon that has been rubbed against your clothing can stick to a neutral wall.

FIGURE 15.4
(a) The charged object on the left induces charges on the surface of an insulator. (b) A charged comb attracts bits of paper because charges are displaced in the paper. (© *1968 Fundamental Photographs*)

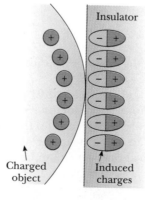

(a)   (b)

Charles Augustin Coulomb, the great French physicist after whom the unit of electric charge the *coulomb* was named, was born in Angoulême in 1736. He was educated at the École du Génie in Mézieres, graduating in 1761 as a military engineer with the rank of first lieutenant. For nine years Coulomb served in the West Indies, where he supervised the building of fortifications in Martinique.

In 1774 Coulomb became a correspondent to the Paris Academy of Science. There he shared the Academy's first prize for his paper on magnetic compasses and also received first prize for his classic work on friction, a study that remained unsurpassed for 150 years. During the next 25 years, he presented 25 papers to the Academy on electricity, magnetism, torsion, and applications of the torsion balance, as well as several hundred committee reports on engineering and civil projects.

Coulomb took full advantage of the variety of positions he held during his lifetime. His experience as an engineer, for example, led him to investigate the strengths of materials and determine the forces that affect objects on beams, thereby contributing to the field of structural mechanics. In the field of ergonomics, his research provided a fundamental understanding of the ways in which people and animals can best do work and greatly influenced the subsequent research of Gaspard Coriolis (1792–1843).

Coulomb's major contribution to science was in the field of electrostatics and magnetism, in which he made use of the torsion balance he had developed (see Fig. 15.9). His paper describing this invention also contained a design for a compass using the principle of torsion suspension. His following paper gave proof of the inverse-square law for the electrostatic force between two charges.

Coulomb died in 1806, five years after becoming president of the Institut de France (formerly the Paris Academy of Science). His research on electricity and magnetism brought that area of physics out of traditional natural philosophy and made it an exact science.

**Charles Coulomb
(1736–1806)**

(Photo courtesy of AIP Niels Bohr Library, E. Scott Barr Collection)

15.3

COULOMB'S LAW

In 1785 Charles Coulomb (1736–1806) established the fundamental law of electric force between two stationary charged particles. Experiments show that

An **electric force** has the following properties:

1. It is inversely proportional to the square of the separation, r, between the two particles and is along the line joining them.
2. It is proportional to the product of the magnitudes of the charges, $|q_1|$ and $|q_2|$, on the two particles.
3. It is attractive if the charges are of opposite sign and repulsive if the charges have the same sign.

From these observations, we can express the magnitude of the electric force between two charges separated by a distance of r as

$$F = k \frac{|q_1||q_2|}{r^2}$$

[15.1]   Coulomb's law

where k is a constant called the *Coulomb constant*.

TABLE 15.1
Charge and Mass of the Electron, Proton, and Neutron

Particle	Charge (C)	Mass (kg)
Electron	-1.60×10^{-19}	9.11×10^{-31}
Proton	$+1.60 \times 10^{-19}$	1.67×10^{-27}
Neutron	0	1.67×10^{-27}

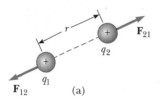

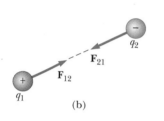

FIGURE 15.5
Two point charges separated by a distance of r exert a force on each other given by Coulomb's law. The force on q_1 is equal in magnitude and opposite in direction to the force on q_2. (a) When the charges are of the same sign, the force is repulsive. (b) When the charges are of opposite sign, the force is attractive.

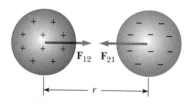

FIGURE 15.6
The attractive force between two oppositely charged spherical charge distributions. Note that r is the distance between the *centers of the spheres.*

The value of the Coulomb constant in Equation 15.1 depends on the choice of units. The SI unit of charge is the **coulomb** (C), which is defined in terms of a current unit called the **ampere** (A), where current is defined as the rate of flow of charge. (The ampere will be defined in Chapter 17.) When the current in a wire is 1 A, the amount of charge that flows past a given point in the wire in 1 s is 1 C. From experiment, we know that the **Coulomb constant** in SI units has the value

$$k = 8.9875 \times 10^9 \ \text{N} \cdot \text{m}^2/\text{C}^2 \qquad \textbf{[15.2]}$$

To simplify our calculations, we shall use the approximate value

$$k \approx 8.99 \times 10^9 \ \text{N} \cdot \text{m}^2/\text{C}^2 \qquad \textbf{[15.3]}$$

The charge on the proton has a magnitude of $e = 1.6 \times 10^{-19}$ C. Therefore, it would take $1/e = 6.3 \times 10^{18}$ protons to create a total charge of $+1$ C. Likewise, 6.3×10^{18} electrons would have a total charge of -1 C. This can be compared with the number of free electrons in 1 cm^3 of copper, which is on the order of 10^{23}. Thus, 1 C is a substantial amount of charge. In typical electrostatic experiments, where a rubber or glass rod is charged by friction, a net charge on the order of 10^{-6} C ($=1 \ \mu$C) is obtained. In other words, only a very small fraction of the total available charge is transferred between the rod and rubbing material. Table 15.1 lists the charges and masses of the electron, proton, and neutron.

When dealing with Coulomb's force law, remember that force is a vector quantity and must be treated accordingly. Furthermore, note that Coulomb's law applies exactly only to point charges or particles and to spherical distributions of charge. Figure 15.5a shows the electric force of repulsion between two positively charged particles. Electric forces obey Newton's third law, and hence the forces $\mathbf{F}_{12}$ and $\mathbf{F}_{21}$ are equal in magnitude but opposite in direction. (The notation $\mathbf{F}_{12}$ denotes the force on particle 1 exerted by particle 2. Likewise, $\mathbf{F}_{21}$ is the force on particle 2 exerted by particle 1.) Figure 15.5b shows the attractive force between two unlike charges. Finally, the attractive force between two spherical distributions of charge is shown in Figure 15.6. Note that in this case, r is the distance between the centers of the spheres.

The Coulomb force is the second example we have seen of a field force—a force exerted by one charge on another even though *there is no physical contact between them.* Recall that another example of a field force is gravitational attraction. The mathematical form of the Coulomb force is the same as that of the gravitational force. That is, they are both inversely proportional to the square of the distance of separation. However, there are some important differences between electric and gravitational forces. Electric forces can be either attractive or repulsive, but gravitational forces are always attractive. Furthermore, gravitational forces are considerably weaker, as shown by the following example.

EXAMPLE 15.1 The Electric Force and the Gravitational Force

The electron and proton of a hydrogen atom are separated (on the average) by a distance of about 5.3×10^{-11} m. Find the magnitudes of the electric force and the gravitational force that each particle exerts on the other.

Solution From Coulomb's law, we find that the attractive electric force has the magnitude

$$F_e = k\frac{|e|^2}{r^2} = \left(8.99 \times 10^9 \ \frac{\text{N} \cdot \text{m}^2}{\text{C}^2}\right)\frac{(1.6 \times 10^{-19} \ \text{C})^2}{(5.3 \times 10^{-11} \ \text{m})^2} = \boxed{8.2 \times 10^{-8} \ \text{N}}$$

From Newton's universal law of gravity and Table 15.1, we find that the gravitational force has the magnitude

$$F_g = G\frac{m_e m_p}{r^2} = \left(6.67 \times 10^{-11} \ \frac{\text{N} \cdot \text{m}^2}{\text{kg}^2}\right)\frac{(9.11 \times 10^{-31} \ \text{kg})(1.67 \times 10^{-27} \ \text{kg})}{(5.3 \times 10^{-11} \ \text{m})^2}$$

$$= \boxed{3.6 \times 10^{-47} \ \text{N}}$$

Since $F_e/F_g \approx 3 \times 10^{39}$, the gravitational force between the charged atomic particles is negligible compared with the electric force.

THE PRINCIPLE OF SUPERPOSITION

Frequently, more than two charges are present and it is necessary to find the net electric force on one of them. This can be accomplished by noting that the electric force between any pair of charges is given by Equation 15.1. Therefore, the resultant force on any one charge equals the vector sum of the forces exerted by the individual charges that are present. This is another example of the **principle of superposition.** For example, if you have three charges and you want to find the force exerted on charge 1 by charges 2 and 3, you first find the force exerted on charge 1 by charge 2 and the force exerted on charge 1 by charge 3. You then add these two forces together vectorially to get the resultant force on charge 1. The following numerical example illustrates this procedure.

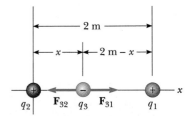

FIGURE 15.7
(Example 15.2) Three point charges are placed along the x axis. The charge q_3 is negative, whereas q_1 and q_2 are positive. If the net force on q_3 is zero, then the force on q_3 due to q_1 must be equal to and opposite the force on q_3 due to q_2.

EXAMPLE 15.2 Where Is the Resultant Force Zero?

Three charges lie along the x axis as in Figure 15.7. The positive charge $q_1 = 15$ μC is at $x = 2.0$ m, and the positive charge $q_2 = 6.0$ μC is at the origin. Where must a *negative* charge, q_3, be placed on the x axis so that the resultant force on it is zero?

Reasoning The only location where the force exerted on q_3 by q_2 is opposite the force exerted on q_3 by q_1 lies on the x axis between q_1 and q_2, as in Figure 15.7. Since we require that the resultant force on q_3 be zero, then F_{32} must equal F_{31}.

Solution If we let x be the coordinate of q_3, then the forces $\mathbf{F}_{31}$ and $\mathbf{F}_{32}$ have the magnitudes

$$F_{31} = k\frac{|q_3|(15 \times 10^{-6} \ \text{C})}{(2.0 - x)^2} \quad \text{and} \quad F_{32} = k\frac{|q_3|(6.0 \times 10^{-6} \ \text{C})}{x^2}$$

If the resultant force on q_3 is zero, then $\mathbf{F}_{32}$ must be equal to and opposite $\mathbf{F}_{31}$, or

$$k\frac{|q_3|\,(15 \times 10^{-6}\ \text{C})}{(2.0 - x)^2} = k\frac{|q_3|\,(6.0 \times 10^{-6}\ \text{C})}{x^2}$$

Because k, 10^{-6}, and q_3 are common to both sides, they can be cancelled from the equation, and we have (after some reduction)

$$(2.0 - x)^2(6.0) = x^2(15)$$

This can be expanded to a quadratic equation, which can be solved for x; but an easier approach is to first take the square root of both sides:

$$(2.0 - x)\sqrt{6.0} = x\sqrt{15}$$

$$(2.0 - x) = x(1.58)$$

$$x = \boxed{0.78\ \text{m}}$$

EXAMPLE 15.3   Using the Superposition Principle

Consider three point charges at the corners of a triangle, as in Figure 15.8, where $q_1 = 6.00 \times 10^{-9}$ C, $q_3 = 5.00 \times 10^{-9}$ C, $q_2 = -2.00 \times 10^{-9}$ C, and the distances of separation are shown in the figure. Find the resultant force on q_3.

Reasoning   It is first necessary to find the direction of the forces exerted on q_3 by q_1 and q_2. The force $\mathbf{F}_{32}$ exerted on q_3 by q_2 is attractive because q_2 and q_3 have opposite signs. The force $\mathbf{F}_{31}$ exerted on q_3 by q_1 is repulsive because both q_1 and q_3 are positive. To find the net force on q_3 it is necessary to find the magnitude of $\mathbf{F}_{32}$ and $\mathbf{F}_{31}$ by use of Coulomb's law and then add the two forces vectorially.

Solution   The magnitude of the force exerted on q_3 by q_2 is

$$F_{32} = k\frac{|q_3|\,|q_2|}{r^2} = (8.99 \times 10^9\ \text{N}\cdot\text{m}^2/\text{C}^2)\frac{(5.00 \times 10^{-9}\ \text{C})(2.00 \times 10^{-9}\ \text{C})}{(4.00\ \text{m})^2}$$

$$= 5.62 \times 10^{-9}\ \text{N}$$

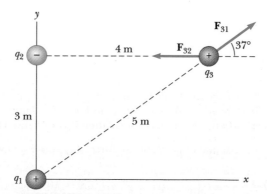

FIGURE 15.8
(Example 15.3)   The force exerted on q_3 by q_1 is $\mathbf{F}_{31}$. The force exerted on q_3 by q_2 is $\mathbf{F}_{32}$. The *total force*, $\mathbf{F}_3$, on q_3 is the *vector* sum $\mathbf{F}_{31} + \mathbf{F}_{32}$.

The magnitude of the force exerted on q_3 by q_1 is

$$F_{31} = k\frac{|q_3||q_1|}{r^3} = (8.99 \times 10^9 \text{ N} \cdot \text{m}^2/\text{C}^2)\frac{(5.00 \times 10^{-9} \text{ C})(6.00 \times 10^{-9} \text{ C})}{(5.00 \text{ m})^2}$$

$$= 1.08 \times 10^{-8} \text{ N}$$

The force $\mathbf{F}_{31}$ makes an angle of 37.0° with the x axis. Therefore, the x component of this force has the magnitude $F_{31} \cos 37.0° = 8.63 \times 10^{-9}$ N, and the y component has the magnitude $F_{31} \sin 37.0° = 6.50 \times 10^{-9}$ N. The force $\mathbf{F}_{32}$ is in the negative x direction. Hence, the x and y components of the resultant force on q_3 are

$$F_x = 8.63 \times 10^{-9} \text{ N} - 5.62 \times 10^{-9} \text{ N} = 3.01 \times 10^{-9} \text{ N}$$

$$F_y = 6.50 \times 10^{-9} \text{ N}$$

The magnitude of the resultant force on the charge q_3 is therefore

$$\sqrt{(3.01 \times 10^{-9} \text{ N})^2 + (6.50 \times 10^{-9} \text{ N})^2} = \boxed{7.16 \times 10^{-9} \text{ N}}$$

and the force vector makes an angle of 65.2° with the x axis.

15.4
EXPERIMENTAL VERIFICATION OF COULOMB'S FORCE LAW

Coulomb measured electric forces between charged objects with a torsion balance (Fig. 15.9). This apparatus consists of two small spheres fixed to the ends of a light horizontal rod made of an insulating material and suspended by a silk thread. Sphere A is given a charge, and charged object B is brought near sphere A. The attractive (or repulsive) force between the two charged objects causes the rod to rotate and to twist the suspension. The angle through which the rod rotates is measured by the deflection of a light beam reflected from a mirror attached to the suspension. The rod rotates through some angle against the restoring force of the twisted thread before reaching equilibrium. The value of the angle of rotation increases as the charge on the objects increases. Thus, the angle of rotation provides a quantitative measure of the electric force of attraction or repulsion. Although Coulomb was unable to establish the inverse square nature of his law with high precision using this apparatus, later scientists using very different methods have shown that the exponent of r is 2 to within a very small uncertainty.

15.5
THE ELECTRIC FIELD

Two different field forces have been introduced into our discussions so far, the gravitational force and the electrostatic force. As pointed out earlier, these forces are capable of acting through space, producing an effect even when there is no physical contact between the objects involved. Field forces can be discussed in a variety of ways, but an approach developed by Michael Faraday (1791–1867) is of such practical value that we shall devote much attention to it in the next several chapters. In this approach, an **electric field** is said to exist in the region of

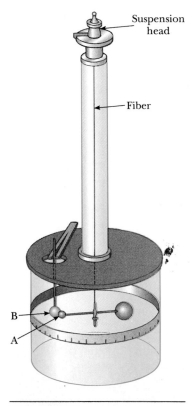

FIGURE 15.9
Coulomb's torsion balance was used to establish the inverse-square law for the electrostatic force between two charges.

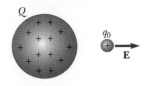

FIGURE 15.10

A small object with a positive charge, q_0, placed near an object with a larger positive charge, Q, experiences an electric field, E, directed as shown. The magnitude of the electric field is defined as the electric force on q_0 divided by the charge q_0.

space around a charged object. When another charged object enters this electric field, forces of an electrical nature arise. As an example, consider Figure 15.10, which shows an object with a small positive charge, q_0, placed near a second object with a larger positive charge, Q.

We define the strength of the electric field at the location of the smaller charge to be the magnitude of the electric force acting on it, divided by the magnitude of its charge:

$$E \equiv \frac{|\mathbf{F}|}{|q_0|} \qquad\qquad [15.4]$$

Note that this is the electric field at the location of q_0 produced by the charge Q, not the field produced by q_0. The electric field is a vector quantity having the SI units newtons per coulomb (N/C). *The direction of* $\mathbf{E}$ *at a point is defined to be the direction of the electric force that would be exerted on a small positive charge placed at that point.* Thus, in Figure 15.10, the direction of the electric field is horizontal and to the right. The electric field at point A in Figure 15.11a is vertical and downward because at this point a positive charge would experience a force of attraction toward the negatively charged sphere.

The definition we have provided for electric field has a serious difficulty. Consider the positively charged conducting sphere in Figure 15.11b. The field in the region surrounding the sphere could be explored by introducing a test charge, q_0, at a point such as P; finding the electric force on this charge; and then dividing this force by the magnitude of the charge on the test charge. Difficulties arise, however, when the magnitude of the test charge is great enough to influence the charge on the conducting sphere. For example, a strong test charge can cause a rearrangement of the charges on the sphere as in Figure 15.11c. As a result, the force exerted on the test charge is different from what it would be if the movement of charge on the sphere had not taken place. Furthermore, the strength of the measured electric field differs from what it would be in the absence of the test charge. To take care of this problem, we simply require that the test charge be small enough to have a negligible effect on the charges on the sphere.

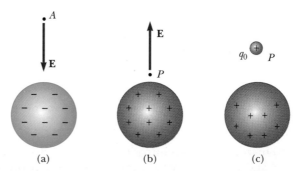

FIGURE 15.11

(a) The electric field at A due to the negatively charged sphere is downward, toward the negative charge. (b) The electric field at P due to the positively charged conducting sphere is upward, away from the positive charge. (c) A test charge, q_0, placed at P will cause a rearrangement of charge on the sphere unless q_0 is very small compared with the charge on the sphere.

Consider a point charge, q, located a distance of r from a test charge, q_0. According to Coulomb's law, the *magnitude* of the force on the test charge is

$$F = k\frac{|q||q_0|}{r^2}$$

Because the magnitude of the electric field at the position of the test charge is defined as $E = F/q_0$, we see that the *magnitude* of the electric field due to the charge q at the position of $|q_0|$ is

$$E = k\frac{|q|}{r^2}$$

[15.5]   Electric field due to a charge q

If q is *positive*, as in Figure 15.12a, the field at P due to this charge is *radially outward* from q. If q is *negative*, as in Figure 15.12b, the field at P is directed *toward* q. Equation 15.5 points out an important property of electric fields that makes them useful quantities for describing electrical phenomena. As the equation indicates, an electric field at a given point depends only on the charge, q, on the object setting up the field and the distance, r, from that object to a specific point in space. As a result, we can say that an electric field exists at point P in Figure 15.12 whether or not there is a charge at P.

The principle of superposition holds when the electric field due to a group of point charges is calculated. We first use Equation 15.5 to calculate the electric field produced by each charge individually at a point, and then add these electric fields together as vectors.

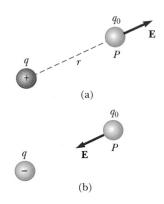

FIGURE 15.12
A test charge, q_0, at P is a distance of r from a point charge, q. (a) If q is positive, the electric field at P points radially *outward* from q. (b) If q is negative, the electric field at P points radially *inward* toward q.

EXAMPLE 15.4   Electric Force on a Proton

Find the electric force on a proton placed in an electric field of 2.0×10^4 N/C that is directed along the positive x axis.

Solution   Because the charge on a proton is $+e = +1.6 \times 10^{-19}$ C, the electric force acting on the proton is

$$F = eE = (1.6 \times 10^{-19}\,\text{C})(2.0 \times 10^4\,\text{N/C}) = \boxed{3.2 \times 10^{-15}\,\text{N}}$$

where the force is in the positive x direction. The weight of the proton has the value $mg = (1.67 \times 10^{-27}\,\text{kg})(9.80\,\text{m/s}^2) = 1.64 \times 10^{-26}$ N. Hence, the magnitude of the gravitational force is negligible compared with that of the electric force.

EXAMPLE 15.5   Electric Field Due to Two Point Charges

Charge $q_1 = 7.00\ \mu$C is at the origin, and charge $q_2 = -5.00\ \mu$C is on the x axis, 0.300 m from the origin (Fig. 15.13). Find the electric field at point P, which has coordinates (0, 0.400) m.

Reasoning   It is first necessary to find the direction of the field at P set up by each charge. The field $\mathbf{E}_1$ at P due to q_1 is vertically upward as in Figure 15.13. Likewise, the field $\mathbf{E}_2$ at P due to q_2 is directed toward q_2 as in Figure 15.13. The magnitudes of the fields can be found from $E = kq/r^2$ and then added together vectorially.

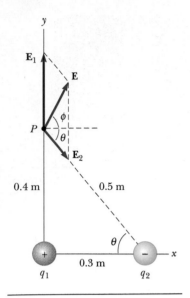

FIGURE 15.13
(Example 15.5)   The total electric field **E** at P equals the vector sum $\mathbf{E}_1 + \mathbf{E}_2$, where $\mathbf{E}_1$ is the field due to the positive charge q_1 and $\mathbf{E}_2$ is the field due to the negative charge q_2.

Solution   The magnitudes of $\mathbf{E}_1$ and $\mathbf{E}_2$ are

$$E_1 = k\frac{|q_1|}{r_1^2} = (8.99 \times 10^9 \text{ N} \cdot \text{m}^2/\text{C}^2)\frac{(7.00 \times 10^{-6} \text{ C})}{(0.400 \text{ m})^2} = 3.93 \times 10^5 \text{ N/C}$$

$$E_2 = k\frac{|q_2|}{r_2^2} = (8.99 \times 10^9 \text{ N} \cdot \text{m}^2/\text{C}^2)\frac{(5.00 \times 10^{-6} \text{ C})}{(0.500 \text{ m})^2} = 1.80 \times 10^5 \text{ N/C}$$

The vector $\mathbf{E}_1$ has an x component of zero. The vector $\mathbf{E}_2$ has an x component given by $E_2 \cos \theta = \frac{3}{5}E_2 = 1.08 \times 10^5$ N/C and a negative y component given by $-E_2 \sin \theta = -\frac{4}{5}E_2 = -1.44 \times 10^5$ N/C. Hence, the resultant component in the x direction is

$$E_x = 1.08 \times 10^5 \text{ N/C}$$

and the resultant component in the y direction is

$$E_y = E_{y1} + E_{y2} = 3.93 \times 10^5 \text{ N/C} - 1.44 \times 10^5 \text{ N/C} = 2.49 \times 10^5 \text{ N/C}$$

From the Pythagorean theorem ($E = \sqrt{E_x^2 + E_y^2}$), we find that **E** has a magnitude of 2.72×10^5 N/C and makes an angle of ϕ of 64.4° with the positive x axis.

Exercise   Find the force on a positive test charge of 2.00×10^{-8} C placed at P.

Answer 5.44×10^{-3} N in the same direction as **E**.

15.6

ELECTRIC FIELD LINES

A convenient aid for visualizing electric field patterns is to draw lines pointing in the direction of the electric field vector at any point. These lines, called **electric field lines,** are related to the electric field in any region of space in the following manner:

1. The electric field vector, **E**, is tangent to the electric field lines at each point.
2. The number of lines per unit area through a surface perpendicular to the lines is proportional to the strength of the electric field in a given region.

Thus, **E** is large when the field lines are close together and small when they are far apart.

Figure 15.14a shows some representative electric field lines for a single positive point charge. Note that this two-dimensional drawing contains only the field lines that lie in the plane containing the point charge. The lines are actually directed radially outward from the charge in *all* directions, somewhat like the quills of a porcupine. Since a positive test charge placed in this field would be repelled by the charge q, the lines are directed radially away from the positive charge. Similarly, the electric field lines for a single negative point charge are directed toward the charge (Fig. 15.14b). In either case, the lines are radial and extend all the way to infinity. Note that the lines are closer together as they get near the charge, indicating that the strength of the field is increasing. Equation 15.5 verifies that this should indeed be the case.

The rules for drawing electric field lines for any charge distribution are as follows:

1. The lines must begin on positive charges (or at infinity) and must terminate on negative charges or, in the case of an excess of charge, at infinity.

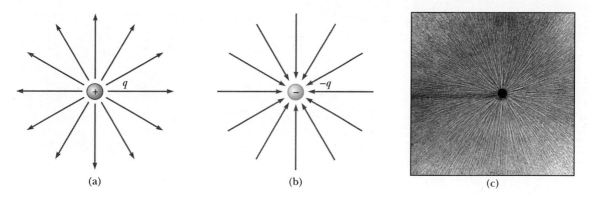

(a)   (b)   (c)

FIGURE 15.14
The electric field lines for a point charge. (a) For a positive point charge, the lines radiate outward. (b) For a negative point charge, the lines converge inward. Note that
the figures show only those field lines that lie in the plane containing the charge.
(c) The dark lines are small pieces of thread suspended in oil, which align with the
electric field produced by a small charged conductor at the center. *(Photo courtesy of
Harold M. Waage, Princeton University)*

2. The number of lines drawn leaving a positive charge or approaching a
   negative charge is proportional to the magnitude of the charge.
3. No two field lines can cross each other.

Figure 15.15 shows the electric field lines for two point charges of equal
magnitude but opposite sign. This charge configuration is called an **electric
dipole.** In this case the number of lines that begin at the positive charge must
equal the number that terminate at the negative charge. At points very near the
charges, the lines are nearly radial. The high density of lines between the charges
indicates a strong electric field in this region.

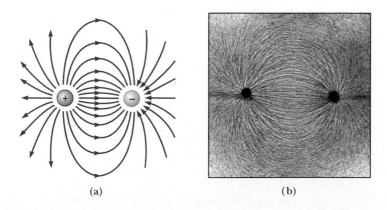

(a)   (b)

FIGURE 15.15
(a) The electric field lines for two equal and opposite point charges (an electric dipole). Note that the number of lines leaving the positive charge equals the number
terminating at the negative charge. (b) The dark lines are small pieces of thread suspended in oil, which align with the electric field produced by two charged conductors.
(Photo courtesy of Harold M. Waage, Princeton University)

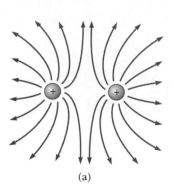 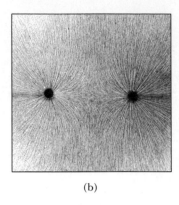

(a)   (b)

FIGURE 15.16
(a) The electric field lines for two positive point charges. (b) The dark lines are small pieces of thread suspended in oil, which align with the electric field produced by two charged conductors. *(Photo courtesy of Harold M. Waage, Princeton University)*

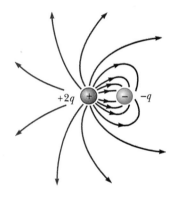

FIGURE 15.17
The electric field lines for a point charge of $+2q$ and a second point charge of $-q$. Note that two lines leave the charge $+2q$ for every line that terminates on $-q$.

Figure 15.16 shows the electric field lines in the vicinity of two equal positive point charges. Again, close to either charge the lines are nearly radial. The same number of lines emerges from each charge because the charges are equal in magnitude. At great distances from the charges, the field is approximately equal to that of a single point charge of magnitude $2q$. The bulging out of the electric field lines between the charges indicates the repulsive nature of the electric force between like charges.

Finally, Figure 15.17 is a sketch of the electric field lines associated with the positive charge $+2q$ and the negative charge $-q$. In this case, the number of lines leaving charge $+2q$ is twice the number terminating on charge $-q$. Hence, only half of the lines that leave the positive charge end at the negative charge. The remaining half terminate on a negative charge that we assume to be located at infinity. At great distances from the charges (great compared with the charge separation), the electric field lines are equivalent to those of a single charge, $+q$.

15.7
CONDUCTORS IN ELECTROSTATIC EQUILIBRIUM

A good electric conductor, such as copper, contains charges (electrons) that are not bound to any atom and are free to move about within the material. When no net motion of charge occurs within a conductor, the conductor is said to be in **electrostatic equilibrium**. As we shall see, an isolated conductor (one that is insulated from ground) has the following properties:

1. **The electric field is zero everywhere inside the conductor.**
2. **Any excess charge on an isolated conductor resides entirely on its surface.**
3. **The electric field just outside a charged conductor is perpendicular to the conductor's surface.**
4. **On an irregularly shaped conductor, the charge tends to accumulate at locations where the radius of curvature of the surface is smallest—that is, at sharp points.**

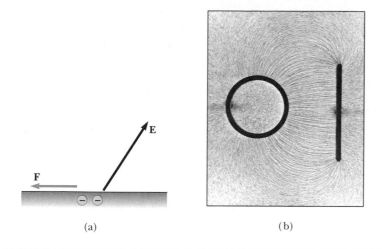

FIGURE 15.18

(a) Negative charges at the surface of a conductor. If the electric field were at an angle to the surface, as shown, an electric force would be exerted on the charges along the surface and they would move to the left. Since the conductor is assumed to be in electrostatic equilibrium, **E** cannot have a component along the surface and hence must be perpendicular to it. (b) The electric field pattern of a charged conducting plate near an oppositely charged conducting cylinder. Small pieces of thread suspended in oil align with the electric field lines. Note that (1) the electric field lines are perpendicular to the conductors and (2) there are no lines inside the cylinder ($\mathbf{E} = 0$). *(Courtesy of Harold M. Waage, Princeton University)*

The first property can be understood by examining what would happen if it were *not* true. If there were an electric field inside a conductor, the free charge there would move and a flow of charge, or current, would be created. However, if there were a net movement of charge, there would no longer be electrostatic equilibrium.

Property 2 is a direct result of the $1/r^2$ repulsion between like charges described by Coulomb's law. If by some means an excess of charge is placed inside a conductor, the repulsive forces arising between the charges force them as far apart as possible, causing them to quickly migrate to the surface. (We shall not prove it here, but it is of interest to note that the excess charge resides on the surface due to the fact that Coulomb's law is an inverse-square law. With any other power law, an excess of charge would exist on the surface, but there would be a distribution of charge, of either the same or opposite sign, inside the conductor.)

Property 3 can be understood by again considering what would happen if it were not true. If the electric field in Figure 15.18a were not perpendicular to the surface, the electric field would have a component along the surface, which would cause the free charges of the conductor to move (to the left in the figure). If the charges moved, however, a current would be created and there would no longer be electrostatic equilibrium. Hence, **E** must be perpendicular to the surface.

To see why property 4 must be true, consider Figure 15.19a, which shows a conductor that is fairly flat at one end and relatively pointed at the other. Any excess charge placed on the object moves to its surface. Figure 15.19b shows the forces between two such charges at the flatter end of the object. These forces are

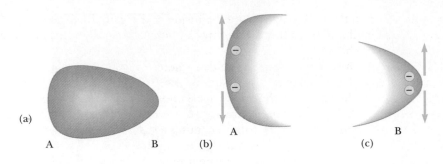

(a)

A   B   (b)   A   (c)   B

FIGURE 15.19
(a) A conductor with a flatter end, A, and a relatively sharp end, B. Excess charge
placed on a conductor resides entirely at its surface and is distributed so that
(b) there is less charge per unit area on the flatter end and (c) there is a large charge
per unit area on the sharper end.

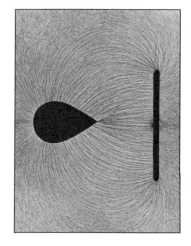

Electric field pattern of a
charged conducting plate near
an oppositely charged pointed
conductor. Small pieces of
thread suspended in oil align
with the electric field lines. Note
that the electric field is most intense near the pointed part of
the conductor where the radius
of curvature is the smallest.
Also, the lines are perpendicular
to the conductors. *(Courtesy of Harold M. Waage, Princeton University)*

predominantly directed parallel to the surface. Thus, the charges move apart
until repulsive forces from other nearby charges create an equilibrium situation.
At the sharp end, however, the forces of repulsion between two charges are
directed predominantly away from the surface, as in Figure 15.19c. As a result,
there is less tendency for the charges to move apart along the surface here, and
the amount of charge per unit area is greater than at the flat end. The cumulative
effect of many such outward forces from nearby charges at the sharp end produces a large force directed away from the surface that can be great enough to
cause charges to leap from the surface into the surrounding air.

Property 4 indicates that if a metal rod having sharp points is attached to a
house, most of the charge sprayed off of or onto the house will pass through these
points, thus eliminating the induced charge on the house produced by storm
clouds. In addition, a lightning discharge striking the house can pass through the
metal rod and be safely carried to the ground through wires leading from the rod
to the Earth. Lightning rods using this principle were first developed by Benjamin Franklin. It is an interesting sidelight to American history to note that some
European countries could not accept the fact that such a worthwhile idea could
have originated in the New World. As a result, they "improved" the design by
eliminating the sharp points. This modification in design drastically reduced the
efficiency of their lightning rods.

Many experiments have shown that the net charge on a conductor resides on
its surface. The experiment described here was first performed by Michael Faraday. A metal ball having a positive charge was lowered at the end of a silk thread
into an uncharged hollow conductor insulated from ground, as in Figure 15.20a.
(This experiment is referred to as **Faraday's ice-pail experiment** because he used
a metal ice pail as the hollow conductor.) As the ball was lowered into the pail,
the needle on an electrometer attached to the outer surface of the pail was
observed to deflect. (An electrometer is a device used to measure charge.) The
needle deflected because the charged ball induced a negative charge on the
inner wall of the pail, which left an equal positive charge on the outer wall (Fig.
15.20b).

Faraday noted that the needle deflection did not change again, either when
the ball touched the inner surface of the pail (Fig. 15.20c) or when it was
removed (Fig. 15.20d). Furthermore, he found that the ball was now uncharged.

Apparently, when the ball had touched the inside of the pail, the excess positive charge on the ball had been neutralized by the induced negative charge on the inner surface of the pail.

Faraday concluded that since the electrometer deflection did not change when the charged ball touched the inside of the pail, the negative charge induced on the inside surface of the pail was just enough to neutralize the positive charge on the ball. As a result of his investigations, he concluded that a charged object suspended inside a metal container causes a rearrangement of charge on the container in such a manner that the sign of the charge on the inside surface of the container is *opposite* the sign of the charge on the suspended object. This produces a charge on the outside surface of the container of the same sign as that on the suspended object.

Faraday also found that if the electrometer was connected to the inside surface of the pail after the experiment had been run, the needle showed no deflection. Thus, the *excess* charge acquired by the pail when contact was made between ball and pail appeared on the outer surface of the pail.

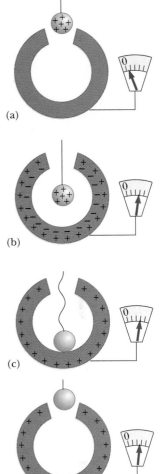

(a)

(b)

(c)

(d)

FIGURE 15.20
An experiment showing that any charge transferred to a conductor resides on its surface in electrostatic equilibrium. The hollow conductor is insulated from ground, and the small metal ball is supported by an insulating thread.

*15.8
THE MILLIKAN OIL-DROP EXPERIMENT

From 1909 to 1913, Robert Andrews Millikan (1868–1953) performed a brilliant set of experiments at the University of Chicago in which he measured the elementary charge, e, on an electron and demonstrated the quantized nature of the electronic charge. The apparatus he used, diagrammed in Figure 15.21, contains

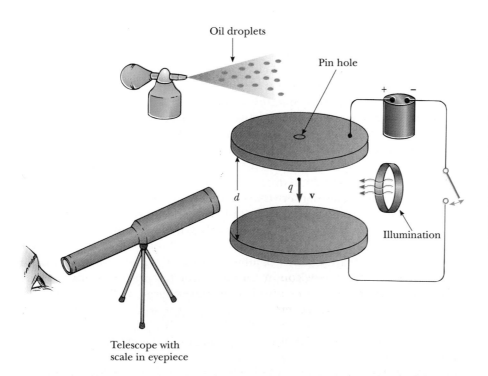

FIGURE 15.21
A schematic view of the Millikan oil-drop apparatus.

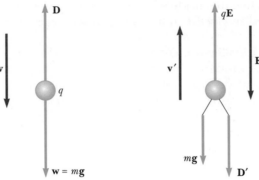

(a) Field off   (b) Field on

FIGURE 15.22
The forces on a charged oil droplet in the Millikan experiment.

two parallel metal plates. Oil droplets that have been charged by friction in an atomizer are allowed to pass through a small hole in the upper plate. A horizontal light beam is used to illuminate the oil droplets, which are viewed by a telescope whose axis is at right angles to the beam. When the droplets are viewed in this manner, they appear as shining stars against a dark background, and the rate of fall of individual drops can be determined.

Let us assume that a single drop having a mass of m and carrying a charge of q is being viewed, and that its charge is negative. If no electric field is present between the plates, the two forces acting on the charge are its weight, mg, acting downward, and an upward viscous drag force, **D** (Fig. 15.22a). The drag force is proportional to the speed of the drop. When the drop reaches its terminal speed, v, the two forces balance each other ($mg = D$).

Now suppose that an electric field is set up between the plates by a battery connected so that the upper plate is positively charged. In this case, a third force, q**E**, acts on the charged drop. Since q is negative and **E** is downward, the electric force is *upward* as in Figure 15.22b. If this force is great enough, the drop moves upward and the drag force **D'** acts downward. When the upward electric force, q**E**, balances the sum of the weight and the drag force, both acting downward, the drop reaches a new terminal speed, v'.

With the field turned on, a drop moves slowly upward, typically at rates of *hundredths* of a centimeter per second. The rate of fall in the absence of a field is comparable. Hence, a single droplet with constant mass and radius can be followed for hours as it alternately rises and falls, by simply turning the electric field on and off.

After making measurements on thousands of droplets, Millikan and his coworkers found that every drop, to within about 1% precision, had a charge equal to some integer multiple of the elementary charge, e. That is,

$$q = ne \qquad n = 0, \pm 1, \pm 2, \pm 3, \ldots \qquad \textbf{[15.6]}$$

where $e = 1.60 \times 10^{-19}$ C. Millikan's experiment is conclusive evidence that charge is quantized. In 1923 he was awarded the Nobel Prize in physics for this work.

*15.9

THE VAN DE GRAAFF GENERATOR

In 1929 Robert J. Van de Graaff designed and built an electrostatic generator that is used extensively in nuclear physics research. The principles of its operation can be understood with the help of the principles of electric fields and charges already presented in this chapter. Figure 15.23 shows the basic construction details of this device. A motor-driven pulley, P, moves a belt past positively charged comb-like metallic needles positioned at A. Negative charges are attracted to these needles from the belt, leaving the left side of the belt with a net positive charge. The positive charges attract electrons onto the belt as it moves past a second comb of needles at B, increasing the excess positive charge on the dome. Since the electric field inside the metal dome is negligible, the positive charge on it can easily be increased regardless of how much charge is already present. The result is that the dome is left with a large amount of positive charge.

This accumulation of charge on the dome cannot continue indefinitely, because eventually an electric discharge through the air takes place. To understand why, consider that, as more and more charge appears on the surface of the dome, the magnitude of the electric field at the surface of the dome is also increasing. Finally, the strength of the field becomes great enough to partially ionize the air near the surface, thus making the air partially conducting. Charges on the dome now have a pathway to leak off into the air, which can produce some spectacular ''lightning bolts'' as the discharge occurs. As noted earlier, charges find it easier to leap off a surface at points where the curvature is great. As a result, one way to inhibit the electric discharge, and to increase the amount of charge that can be stored on the dome, is to increase its radius. Another method for inhibiting discharge is to place the entire system in a container filled with a high-pressure gas, which is significantly more difficult to ionize than air at atmospheric pressure.

If protons (or other charged particles) are introduced into a tube attached to the dome, the large electric field of the dome exerts a repulsive force on the protons, causing them to accelerate to energies high enough to initiate nuclear reactions between the protons and various target nuclei.

*15.10

THE OSCILLOSCOPE

The oscilloscope is an electronic instrument widely used in making electrical measurements. Its main component is the cathode ray tube (CRT), shown in Figure 15.24a. This tube is commonly used to create a visual display of electronic information for other applications, including radar systems, television receivers, and computers. The CRT is essentially a vacuum tube in which electrons are accelerated and deflected under the influence of electric fields.

The electron beam is produced by an assembly called an *electron gun* in the neck of the tube. The electron gun in Figure 15.24a consists of a heater (H), a cathode (C), and a positively charged anode (A). An electric current through the heater causes its temperature to rise, which in turn heats the cathode. The cathode reaches temperatures high enough to cause electrons to be ''boiled

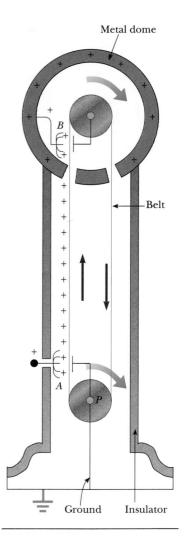

FIGURE 15.23
A diagram of a Van de Graaff generator. Charge is transferred to the dome by means of a rotating belt. The charge is deposited on the belt at point A and transferred to the dome at point B.

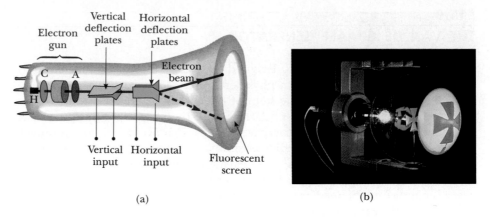

(a)   (b)

FIGURE 15.24

(a) A diagram of a cathode ray tube. Electrons leaving the hot cathode, C, are accelerated to the anode, A, where some pass through a small hole. The electron gun is also
used to focus the beam, and the plates deflect the beam. (b) A ''Maltese cross'' tube
showing the shadow of a beam of cathode rays falling on the tube's luminescent
screen. The hot filament also produces a beam of light and a second shadow of the
cross. *(Courtesy of Central Scientific Company)*

off.'' Although they are not shown in the figure, the electron gun also includes an
element that focuses the electron beam and one that controls the number of
electrons reaching the anode (that is, a brightness control). The anode has a
hole in its center that allows the electrons to pass through without striking the
anode. If left undisturbed, the electrons travel in a straight-line path until they
strike the face of the CRT. The screen at the front of the tube is coated with a
fluorescent material that emits visible light when bombarded with electrons.
This results in a visible spot on the screen of the CRT.

The electrons are moved in a variety of directions by two sets of deflection
plates placed at right angles to each other in the neck of the tube (Fig. 15.24a). In
order to understand how the deflection plates operate, first consider the horizontal ones. External electric circuits can change the amount of charge present
on these plates, with positive charge being placed on one plate and negative on
the other. (In Chapter 16 we shall see that this can be accomplished by applying a
voltage across the plates.) This increasing charge creates an increasing electric
field between the two plates, which deflects the electron beam from its straight-
line path. Slowly increasing the charge on the horizontal plates causes the electron beam to move gradually from the center toward the side of the screen.
Because of the persistence of vision, however, one sees a horizontal line extending across the screen instead of the simple movement of a dot. The horizontal
line can be maintained on the screen by rapid, repetitive tracing.

The vertical deflection plates act in exactly the same way as the horizontal
plates, except that changing the charge on them causes a vertical line on the tube
face. In practice, the horizontal and vertical deflection plates are used simultaneously.

To see how the oscilloscope can display visual information, let us examine
how we can observe the sound wave from a tuning fork on the screen. For this

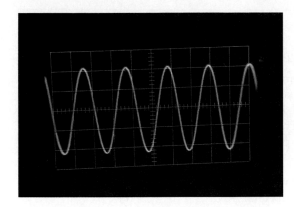

FIGURE 15.25
A sinusoidal wave produced by a wave generator and displayed on the oscilloscope. *(Courtesy of Henry Leap and Jim Lehman)*

purpose, the charge on the horizontal plates changes in such a manner that the beam sweeps across the face of the tube at a constant rate. The tuning fork is then sounded into a microphone, which changes the sound signal to an electric signal that is applied to the vertical plates. The combined effects of the horizontal and vertical plates cause the beam to sweep horizontally and up and down at the same time, with the vertical motion corresponding to the tuning fork signal. A pattern like that in Figure 15.25 is formed on the screen.

PROBLEM-SOLVING STRATEGY
Electric Forces and Fields

1. **Units. When performing calculations that use the Coulomb constant, k, charges must be in coulombs and distances in meters. If they are given in other units, you must convert them.**
2. **Applying Coulomb's law to point charges. It is important to use the superposition principle properly when dealing with a collection of interacting point charges. If several charges are present, the resultant force on any one of them is found by determining the individual force exerted on it by every other charge and then determining the vector sum of all these forces. The magnitude of the force that any charged object exerts on another is given by Coulomb's law, and the direction of the force is found by noting that the forces are repulsive between like charges and attractive between unlike charges.**
3. **Calculating the electric field of point charges. Remember that the superposition principle can be applied to electric fields, which are also vector quantities. To find the total electric field at a given point, first calculate the electric field at the point due to each individual charge. The vector sum of the fields due to all the individual charges is the resultant field at the point.**

SUMMARY

Electric charges have the following important properties:

1. Unlike charges attract one another, and like charges repel one another.
2. Electric charge is always conserved.
3. Charge is quantized; that is, it exists in discrete packets that are integral multiples of the electronic charge.
4. The force between charged particles varies as the inverse square of their separation.

Conductors are materials in which charges move freely. **Insulators** are materials that do not readily transport charge.

Coulomb's law states that the electric force between two stationary charged particles separated by a distance of r has the magnitude

$$F = k \frac{|q_1||q_2|}{r^2} \qquad \text{[15.1]}$$

where $|q_1|$ and $|q_2|$ are the magnitudes of the charges on the particles in coulombs and k is the **Coulomb constant,** which has the value

$$k \approx 8.99 \times 10^9 \ \text{N} \cdot \text{m}^2/\text{C}^2 \qquad \text{[15.3]}$$

The magnitude of the **electric field, E,** at some point in space is defined as the magnitude of the electric force that acts on a small positive charge placed at that point, divided by the magnitude of its charge, $|q_0|$:

$$E \equiv \frac{|\mathbf{F}|}{|q_0|} \qquad \text{[15.4]}$$

The direction of the electric field at a point in space is defined to be the direction of the electric force that would be exerted on a small positive charge placed at that point.

The magnitude of the electric field due to a *point charge, q,* at distance r from the point charge is

$$E = k \frac{|q|}{r^2} \qquad \text{[15.5]}$$

Electric field lines are useful for describing the electric field in any region of space. The electric field vector, **E**, is tangent to the electric field lines at every point. Furthermore, the number of electric field lines per unit area through a surface perpendicular to the lines is proportional to the strength of the electric field in that region.

A **conductor in electrostatic equilibrium** has the following properties:

1. The electric field is zero everywhere inside the conductor.
2. Any excess charge on an isolated conductor must reside entirely on its surface.
3. The electric field just outside a charged conductor is perpendicular to the conductor's surface.
4. On an irregularly shaped conductor, charge tends to accumulate where the radius of curvature of the surface is smallest, that is, at sharp points.

ADDITIONAL READING

R. B. Kaner and A. G. MacDiarmid, "Plastics That Conduct Electricity," *Sci. American,* February 1988, p. 106.

W. F. Magie, *Source Book in Physics,* Cambridge, Mass., Harvard University Press, 1963. This includes extracts from the works of Coulomb and others.

H. W. Meyer, *History of Electricity and Magnetism,* Cambridge, Mass., MIT Press, 1971.

A. D. Moore, "Electrostatics," *Sci. American,* March 1972, p. 46.

D. Roller and D. H. D. Roller, *The Development of the Concept of Electric Charge,* Cambridge, Mass., Harvard University Press, 1954.

E. R. Williams, "The Electrification of Thunderstorms," *Sci. American,* November 1988, p. 88.

CONCEPTUAL QUESTIONS

Example If a metal object receives a positive charge, does its mass increase, decrease, or stay the same? What happens to its mass if the object receives a negative charge?
Reasoning An object's mass decreases very slightly (immeasurably) when it is given a positive charge, because it loses electrons. When the object is given a negative charge, its mass increases slightly because it gains electrons.

Example Would life be different if the electron were positively charged and the proton were negatively charged?
Reasoning No. The assignment of positive and negative charge is completely arbitrary.

Example If a suspended object, A, is attracted to object B, which is charged, can we conclude that object A is charged? Explain.
Reasoning Object A might have a charge opposite in sign to that of B, but it also might be a neutral conductor. Let us assume that B has a positive charge. If object A is a neutral conductor, the nonuniform electric field created by B would polarize A, displacing negative charge to the near face and pushing an equal amount of positive charge to the other face. Then the force of attraction on the near side would be slightly larger than the force of repulsion on the far side, so the net force on A is toward B.

Example When defining the electric field, why is it necessary to specify that the magnitude of the test charge be very small?
Reasoning If the charges creating the field are located on movable objects, or if they are on conductors, then a large test charge would exert a force on the objects. This would cause the objects to either move or (in the case of conductors) their charge distribution would be altered. This, in turn, would change the electric field at the location where we are trying to measure it.

Example An uncharged, metallic-coated Ping-Pong ball is placed in the region between two vertical parallel metal plates. If the two plates are charged, one positive and one negative, describe the motion the ball undergoes.

Reasoning The two oppositely charged plates create a region of uniform electric field between them, pointing from the positive to the negative plate. An uncharged weightless ball could hang at rest in this region, feeling no total force. However, if it is disturbed so as to touch one plate, say the negative one, it will acquire some negative charge and experience a force that will accelerate it to the positive plate. When it touches the positive plate, it will acquire a positive charge and start to accelerate back to the negative plate. The ball will rattle back and forth between the charged plates (acting as a transporter of charge) until it has transferred all their charges to make both plates neutral.

1. Explain from an atomic viewpoint why charge is usually transferred by electrons.

2. In very dry air, sparks are often observed (or heard) when clothes are removed in the dark. Explain.

3. Because of a higher moisture content, air is a better conductor of charge in the summer than in the winter. Would you expect the shocks from static electricity to be more severe in summer or winter? Discuss.

4. A balloon is negatively charged by rubbing and then clings to a wall. Does this mean that the wall is positively charged? Why does the balloon eventually fall?

5. A charged comb will attract small bits of dry paper that fly away when they touch the comb. Explain.

6. A light piece of aluminum foil is draped over a wooden rod. When a rod with a positive charge is brought close to the foil, the foil leaves stand apart. Why? What kind of charge is on the foil?

7. What is the difference between charging an object by induction and charging it by conduction?

8. Operating room personnel must wear special conducting shoes while working around oxygen. Why? Contrast the possible effect of wearing rubber shoes.

9. A student stands on a piece of insulating material, places her hand on top of a Van de Graaff generator, and then turns on the generator. Is she shocked? Why or why not? She finds that her hair stands on end as in the photograph. Why does this occur? Why does a second student, who is also on an insulated stand, find that her hair stands up when she touches the first student?

10. In some instances, people who happened to be near the point where a lightning bolt struck the Earth have reported that their clothes were thrown off as a result of the strike. Why might this happen?

11. Compare and contrast Newton's law of universal gravitation with Coulomb's force law.

12. Imagine that someone proposes a theory that says people are bound to the Earth by electric forces rather than by gravity. How could you prove this theory wrong?

13. Is it possible for an electric field to exist in empty space? Explain.

14. Explain why electric field lines never cross. (*Hint:* **E** must have a unique direction at every point.)

15. A "free" electron and "free" proton are placed in an identical electric field. Compare the electric force on each particle. Compare their accelerations.

16. In Figure 15.17, where do the extra lines leaving the charge $+2q$ end?

17. Are people in a steel-frame building safer than those in a wood-frame house during an electrical storm, or vice versa? Explain.

18. Suppose someone told you that it is the rubber tires on a car that make a car safe during a lightning storm. What would be your response? Do the relatively new steel-belted tires on a car affect the safety of those in the car?

19. Why should a ground wire be connected to the metal support rod for a television antenna?

20. Why should you get out of a swimming pool during a lightning storm?

21. Why is it not a good idea to seek shelter under a tree during a lightning storm?

22. Suggest a method by which Coulomb might have placed exactly any fraction of a total charge on one of two metal balls, with the remaining fraction on the other ball. (*Hint:* Use the size of each ball as a way of distributing the total charge.)

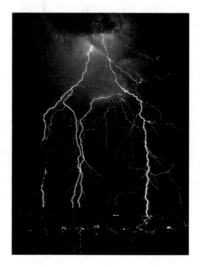

(Question 21)　　*(Gordon Garradd/Photo Researchers)*

(Question 9)　　*(Mark C. Burnett/Photo Researchers)*

PROBLEMS

Section 15.3　Coulomb's Law

1. Calculate the net charge on a substance consisting of (a) 5×10^{14} electrons; (b) a combination of 7×10^{13} protons and 4×10^{13} electrons.

2. A typical lead-acid storage battery contains sulfuric acid, H_2SO_4, which breaks down into $2H + SO_4$ and each molecule delivers two electrons to the external circuit. If the battery delivers a total charge of 2.0×10^5 C, how many grams of sulfuric acid are used up?

3. A 4.5×10^{-9} C charge is located 3.2 m from a -2.8×10^{-9} C charge. Find the electrostatic force exerted by one charge on the other.

□ indicates problems that have full solutions available in the Student Solutions Manual and Study Guide.

4. Two identical conducting spheres are placed with their centers 0.30 m apart. One is given a charge of 12×10^{-9} C and the other a charge of -18×10^{-9} C. (a) Find the electrostatic force exerted on one sphere by the other. (b) The spheres are connected by a conducting wire. After equilibrium has occurred, find the electrostatic force between the two.

5. Suppose that 1.00 g of hydrogen is separated into electrons and protons. Suppose also that the protons are placed at the Earth's north pole and the electrons are placed at the south pole. What is the resulting compressional force on the Earth?

6. At the point of fission, a nucleus of ^{238}U, with 92 protons, is divided into two smaller spheres, each with 46 protons and a radius of 5.9×10^{-15} m. What is the repulsive force pushing the two spheres apart?

7. The Moon and Earth are bound together by gravity. If, instead, the force of attraction were the result of each having a charge of the same magnitude but opposite in sign, find the quantity of charge that would have to be placed on each to produce the required force.

8. An electron is released above the surface of the Earth. A second electron directly below it exerts an electrostatic force on the first electron just great enough to cancel the gravitational force on it. How far below the first electron is the second?

9. A 2.2×10^{-9} C charge is on the x axis at $x = -1.5$ m, a 5.4×10^{-9} C charge is on the x axis at $x = 2.0$ m, and a 3.5×10^{-9} C charge is at the origin. Find the net force on the 3.5×10^{-9} C charge.

10. Calculate the magnitude and direction of the Coulomb force on each of the three charges in Figure 15.26.

11. A molecule of DNA (deoxyribonucleic acid) is 2.17 μm long. The ends of the molecule become singly ionized—negative on one end, positive on the other. The helical molecule acts as a spring and compresses 1.00% upon becoming charged. Determine the effective spring constant of the molecule.

12. Three charges are arranged as shown in Figure 15.27. Find the magnitude and direction of the electrostatic force on the charge at the origin.

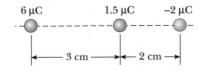

6 μC   1.5 μC   −2 μC

|← 3 cm →|← 2 cm →|

FIGURE 15.26   (Problems 10 and 27)

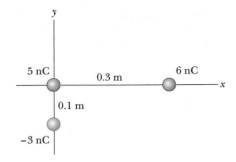

FIGURE 15.27   (Problems 12 and 36)

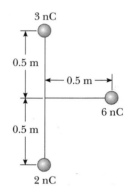

FIGURE 15.28   (Problem 13)

13. Three charges are arranged as shown in Figure 15.28. Find the magnitude and direction of the electrostatic force on the 6.00-nC charge.

14. In the Bohr theory of the hydrogen atom, an electron moves in a circular orbit about a proton, where the radius of the orbit is 0.51×10^{-10} m. (a) Find the electrostatic force between the two. (b) If this force serves as the centripetal force on the electron, what is the speed of the electron?

15. An alpha particle (charge $= +2.0e$) is sent at high speed toward a gold nucleus (charge $= +79e$). What is the electrical force acting on the alpha particle when it is 2.0×10^{-14} m from the gold nucleus?

16. Find (a) the total number of electrons in one gram of copper and (b) the total charge of all these electrons.

17. A charge of 2.00×10^{-9} C is placed at the origin, and a charge of 4.00×10^{-9} C is placed at $x = 1.5$ m. Locate the point between the two charges where a charge of 3.00×10^{-9} C should be placed so that the net electric force on it is zero.

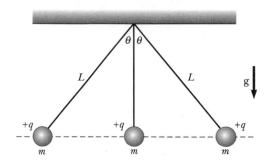

FIGURE 15.29   (Problem 20)

18. Two small metallic spheres, each of mass 0.20 g, are suspended as pendulums by light strings from a common point. They are given the same electric charge, and it is found that the two come to equilibrium when each string is at an angle of 5.0° with the vertical. If each string is 30.0 cm long, what is the magnitude of the charge on each sphere?

19. A charge of 6.00×10^{-9} C and a charge of -3.00×10^{-9} C are separated by a distance of 60.0 cm. Find the position at which a third charge, of 12.0×10^{-9} C, can be placed so that the net electrostatic force on it is zero.

20. Three identical point charges, each of mass $m = 0.10$ kg, hang from three strings, as shown in Figure 15.29. If the lengths of the left and right strings are each $L = 30.0$ cm, and if the angle θ is 45°, determine the value of q.

21. A $+2.7$-μC point charge is on the x axis at $x = -3.0$ m, and a $+2.0$-μC point charge is on the x axis at $x = +1.0$ m. Determine the net electric field (magnitude and direction) on the y axis at $y = +2.0$ m.

22. An electron and a proton are each placed at rest in an external electric field of 520 N/C. Calculate the speed of each particle after 48 ns.

23. The nucleus of ^{208}Pb has 82 protons within a sphere of radius 6.34×10^{-15} m. Calculate the electric field at the surface of the nucleus.

24. An electron and a proton both start from rest and from the same point in a uniform electric field of 370 N/C. How far apart are they 1.0 μs after they are released? Ignore the attraction between the electron and the proton. (*Hint:* Imagine the experiment performed with the proton only, and then repeat with the electron only.)

25. What are the magnitude and direction of the electric field set up by the proton at the position of the electron in the hydrogen atom? (See Problem 14.)

26. Find the electric field at a point midway between two charges of $+30.0 \times 10^{-9}$ C and (a) $+60.0 \times 10^{-9}$ C, separated by 30.0 cm; (b) -60.0×10^{-9} C, separated by 30.0 m.

27. (a) Determine the electric field strength at a point 1.00 cm to the left of the middle charge shown in Figure 15.26. (b) If a charge of -2.00 μC is placed at this point, what are the magnitude and direction of the force on it?

28. What are the magnitude and direction of the electric field that will balance the weight of (a) an electron? (b) a proton? (Use the data in Table 15.1.)

29. A piece of aluminum foil of mass 5.00×10^{-2} kg is suspended by a string in an electric field directed vertically upward. If the charge on the foil is 3.00 μC, find the strength of the field that will reduce the tension in the string to zero.

30. An object with a net charge of 24 μC is placed in a uniform electric field of 610 N/C, directed vertically. What is the mass of this object if it "floats" in this electric field?

31. An electron is accelerated by a constant electric field of magnitude 300 N/C. (a) Find the acceleration of the electron. (b) Use the equations of motion with constant acceleration to find the electron's speed after 1.00×10^{-8} s, assuming it starts from rest.

32. A constant electric field directed along the positive x axis has a strength of 2000 N/C. Find (a) the force exerted on the proton by the field, (b) the acceleration of the proton, and (c) the time required for the proton to reach a speed of 1.00×10^{6} m/s, assuming it starts from rest.

33. A proton accelerates from rest in a uniform electric field of 640 N/C. At some later time, its speed is 1.20×10^{6} m/s. (a) Find the magnitude of the acceleration of the proton. (b) How long does it take the proton to reach this speed? (c) How far has it moved in this interval? (d) What is its kinetic energy at the later time?

34. Consider an electron that is released from rest in a uniform electric field. (a) If the electron is accelerated to 1.0% of the speed of light after traveling 2.0 mm, what is the strength of the electric field? (b) What speed does the electron have after traveling 4.0 mm from rest?

35. Each of the protons in a particle beam has a kinetic energy of 3.25×10^{-15} J. What are the magnitude and direction of the electric field that will stop these protons in a distance of 1.25 m?

36. Find the net electric field exerted on the charge at the origin in Figure 15.27.

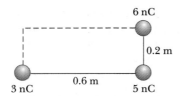

FIGURE 15.30   (Problem 37)

37. Positive charges are situated at three corners of a rectangle, as shown in Figure 15.30. Find the electric field at the fourth corner.

38. A point particle having charge q is at point (x_0, y_0) in the xy plane. Show that the x and y components of the electric field at point (x, y) due to the charge q are

$$E_x = \frac{kq(x - x_0)}{[(x - x_0)^2 + (y - y_0)^2)]^{3/2}}$$

$$E_y = \frac{kq(y - y_0)}{[(x - x_0)^2 + (y - y_0)^2]^{3/2}}$$

39. Three identical charges ($q = -5.0\ \mu$C) are along a circle of 2.0-m radius at angles of 30°, 150°, and 270°, as shown in Figure 15.31. What is the resultant electric field at the center of the circle?

40. Three identical point charges ($q = +2.7\ \mu$C) are placed on the corners of an equilateral triangle with sides of length $a = 35$ cm, as shown in Figure 15.32. (a) At what point in the plane of the charges (other than infinity) is the electric field zero? (b) What are the magnitude and direction

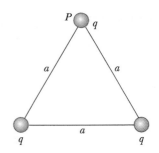

FIGURE 15.32   (Problem 40)

of the electric field at point P *due to the two charges at the base of the triangle?*

41. Three charges are at the corners of an equilateral triangle, as shown in Figure 15.33. Calculate the electric field at a point midway between the two charges on the x axis.

42. Each of the electrons in a particle beam has a kinetic energy of 1.60×10^{-17} J. (a) What is the magnitude of the uniform electric field (pointing in the direction of the electrons' movement) that will stop these electrons in a distance of 10.0 cm? (b) How long will it take to stop the electrons? (c) After the electrons stop, what will they do? Explain.

43. In Figure 15.34, determine the point (other than infinity) at which the total electric field is zero.

44. Repeat Problem 43 for the case in which the 6.00-μC charge is replaced by a charge of $-6.00\ \mu$C.

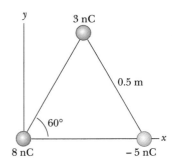

FIGURE 15.33   (Problem 41)

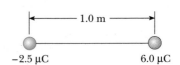

FIGURE 15.34   (Problem 43)

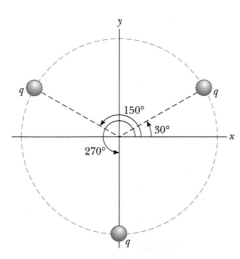

FIGURE 15.31   (Problem 39)

Section 15.6 Electric Field Lines

Section 15.7 Conductors in Electrostatic Equilibrium

45. (a) Sketch the electric field lines around an isolated 1-μC positive point charge. (b) Sketch the electric field pattern around an isolated negative point charge of magnitude -2 μC.

46. (a) Sketch the electric field pattern around two positive point charges of magnitude 1 μC, placed close together. (b) Sketch the electric field pattern around two negative point charges of magnitude -2 μC, placed close together. (c) Sketch the pattern around a 1-μC positive point charge and a -2-μC charge placed close together.

47. Figure 15.35 shows the electric field lines for two point charges separated by a small distance. (a) Determine the ratio q_1/q_2. (b) What are the signs of q_1 and q_2?

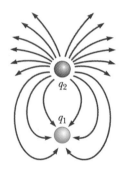

FIGURE 15.35 (Problem 47)

48. (a) Sketch the electric field pattern set up by a positively charged hollow sphere. Include the lines both inside and outside the sphere. (b) A conducting cube is given a positive charge. Sketch the electric field pattern both inside and outside the cube.

49. Consider a rod of finite length having a uniform negative charge. Sketch the pattern of the electric field lines in a plane containing the rod.

50. Two point charges are a small distance apart. (a) Sketch the electric field lines for the two if one has a charge four times that of the other and both charges are positive. (b) Repeat for the case that both charges are negative.

51. Refer to Figure 15.20. The charge lowered into the center of the hollow conductor has a magnitude of 5 μC. Find the magnitude and sign of the charge on the inside and outside of the hollow conductor when the charge is as shown in (a) Figure 15.20a; (b) Figure 15.20b; (c) Figure 15.20c; (d) Figure 15.20d.

52. A positive point charge is placed a distance of $R/2$ from the center of an uncharged thin conducting spherical shell of radius R. Sketch the electric field patterns set up by this arrangement, both inside and outside the shell.

Section 15.9 The Van de Graaff Generator

53. The dome of a Van de Graaff generator receives a charge of 2×10^{-4} C. Find the strength of the electric field (a) inside the dome; (b) at the surface of the dome, assuming it has a radius of 1 m; (c) 4 m from the center of the dome. (*Hint:* See Section 15.7 to review properties of conductors in electrostatic equilibrium. Also use the fact that the points on the surface are outside a spherically symmetric charge distribution; the total charge may be considered as located at the center of the sphere.)

54. If the electric field strength in air exceeds 3.0×10^6 N/C, the air becomes a conductor. Using this fact, determine the maximum amount of charge that can be carried by a metal sphere 2.0 m in radius. (See the hint in Problem 53.)

55. Air breaks down (loses its insulating quality) and sparking results if the field strength is increased to about 3.0×10^6 N/C. (a) What acceleration does an electron experience in such a field? (b) If the electron starts from rest, in what distance does it acquire a speed equal to 10% of the speed of light?

56. A Van de Graaff generator is charged so that the electric field at its surface is 3.0×10^4 N/C. Find (a) the electric force exerted on a proton released at its surface and (b) the acceleration of the proton at this instant of time.

ADDITIONAL PROBLEMS

57. You are told that an electron remains suspended between the ground of the Earth (assumed neutral) and a fixed positive point charge, q, at a distance of 7.62 m from the point charge. Is this observation possible? Explain.

58. Five equal negative point charges, $-q$, are placed symmetrically around a circle of radius R as in Figure 15.36. Calculate the electric field, **E**, at the center of the circle.

59. Three point charges are aligned along the x axis as shown in Figure 15.37. Find the electric field at (a) the position (2.0, 0) m and (b) the position (0, 2.0) m.

60. Three point charges lie along the y axis. A charge of $q_1 = -9.0$ μC is at $y = 6.0$ m, and a charge of

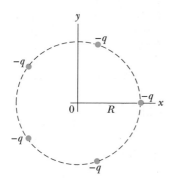

FIGURE 15.36   (Problem 58)

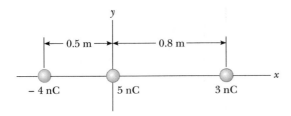

FIGURE 15.37   (Problem 59)

$q_2 = -8.0$ μC is at $y = -4.0$ m. Where must a third positive charge, q_3, be placed so that the resultant force on it is zero?

61. A small 2.00-g plastic ball is suspended by a 20.0-cm-long string in a uniform electric field, as shown in Figure 15.38. If the ball is in equilibrium when the string makes a 15.0° angle with the vertical as indicated, what is the net charge on the ball?

62. Protons are projected with an initial speed of $v_0 = 9550$ m/s into a region where a uniform electric field, $E = 720$ N/C, is present (Fig. 15.39). The protons are to hit a target that lies a horizontal distance of 1.27 mm from the point where the

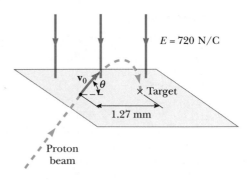

FIGURE 15.39   (Problem 62)

protons are launched. Find (a) the two projection angles, θ, that will result in a hit and (b) the total duration of flight for each of these two trajectories.

63. Three point charges lie along the y axis. A charge of $q_1 = -3.0$ μC is at $y = 5.0$ m, and a charge of $q_2 = 8.0$ μC is at $y = 2.0$ m. Where must a positive charge, q_3, be placed so that the resultant force on it is zero?

64. A uniformly charged rod of length L carries a total positive charge of Q. The electric field at a point a distance of d away from one end of the rod and on the line containing the rod is directed away from the rod, and its magnitude is

$$E = \frac{kQ}{d(L + d)}$$

If a 2.00-μC point charge is placed 2.00 m from the left end, and on the line containing a uniform 3-m rod having a charge of 5.00 μC, determine the force *on the rod* due to the point charge.

65. Two 2.0-g spheres are suspended by 10.0-cm-long light strings (Fig. 15.40). A uniform electric field is applied in the x direction. If the spheres have charges of -5.0×10^{-8} C and $+5.0 \times 10^{-8}$ C, determine the electric field intensity that enables the spheres to be in equilibrium at $\theta = 10°$.

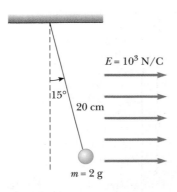

FIGURE 15.38   (Problem 61)

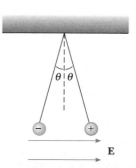

FIGURE 15.40   (Problem 65)

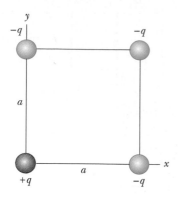

FIGURE 15.41 (Problem 66)

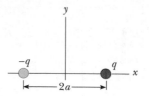

FIGURE 15.42 (Problem 68)

66. Four point charges are situated at the corners of a square with sides of length a, as in Figure 15.41. Find the expression for the resultant force on the positive charge q.

67. An electron traveling with an initial speed of 4.0×10^6 m/s enters a region with a uniform electric field of magnitude 2.5×10^4 N/C. The direction of travel of the electron is the direction of the field. (a) Find the acceleration of the electron. (b) Determine the time it takes for the electron to stop after it enters the field. (c) How far does the electron move in the electric field before stopping?

68. Two point charges like those in Figure 15.42 are called an electric dipole. Show that the electric field at a distant point along the x axis is given by the expression $E_x = 4\,kqa/x^3$.

69. A 2.00-μC charged 1.00-g cork ball is suspended vertically on a 0.500-m-long light string in the presence of a uniform downward-directed electric field of magnitude $E = 1.00 \times 10^5$ N/C. If the ball is displaced slightly from the vertical, it oscillates like a simple pendulum. (a) Determine the period of this oscillation. (b) Should gravity be included in the calculation for part (a)? Explain.

70. Two equal positive charges, q, are on the x axis at $x = a$ and $x = -a$. Show that the field along the positive y axis is in the y direction and is given by the relation $E_y = 2\,kqy(y^2 + a^2)^{-3/2}$.

ELECTRICAL ENERGY AND CAPACITANCE

Jennifer holds onto a charged sphere that reaches a potential of about 100 000 volts. The device that generates this high potential is called a Van de Graaff generator. Why do you suppose Jennifer's hair stands on end like the needles of a porcupine? Why is it important that she stand on a pedestal, insulated from the ground? (Courtesy of Henry Leap and Jim Lehman)

T he concept of potential energy was first introduced in Chapter 5. A potential energy function can be defined for any conservative force, such as the force of gravity. By using the principle of conservation of energy, we were often able to avoid working directly with forces when solving problems. In this chapter we discover that the potential energy concept is also useful in the study of electricity. Because the Coulomb force is conservative, we can define an electrical potential energy corresponding to the Coulomb force. This concept of potential energy is of value, but perhaps even more valuable is a quantity called electric potential, defined as potential energy per unit charge.

Electric potential is of great practical value for dealing with electric circuits. For example, when we speak of a voltage applied between two points, we are actually referring to an electric potential difference between those points. We take our first steps toward circuits with a discussion of electric potential, carried forward by an investigation of a common circuit element called a capacitor.

16.1

POTENTIAL DIFFERENCE AND ELECTRIC POTENTIAL

In Chapter 5 we showed that the gravitational force is a conservative force. As you may recall, this means that the work done on an object by this force depends only on the initial and final positions of the object and not on the path connecting the two positions. Furthermore, since the gravitational force is conservative, it is possible to define a potential energy function, which we call gravitational potential energy. Because the Coulomb force law is of the same form as the universal law of gravity, it follows that *the electrostatic force is also conservative.* Therefore, it is possible to define an electrical potential energy function associated with this force.

Let us consider potential energy from the point of view of the particular situation shown in Figure 16.1. Imagine a small positive charge placed at point A in a uniform electric field of magnitude E. As the charge moves from point A to point B under the influence of the electric force exerted on it, qE, the work done on the charge by the electric force is

$$W = Fd = qEd$$

where d is the distance between A and B.

By definition, *the work done by a conservative force equals the negative of the change in potential energy,* ΔPE. The change in electrical potential energy is therefore

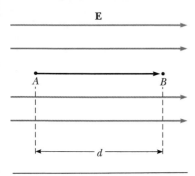

FIGURE 16.1
When a charge, q, moves in a uniform electric field, **E**, from point A to point B, the work done on the charge by the electric force is qEd.

<table>
<tr><td>Change in potential energy</td></tr>
</table>

$$\Delta PE = -W = -qEd \qquad \text{[16.1]}$$

Note that although potential energy can be defined for any electric field, *Equation 16.1 is valid only for the case of a uniform electric field.* In subsequent sections we shall examine situations in which the electric field is not uniform.

In the coming pages we shall often have occasion to use electrical potential energy, but of even more practical importance in the study of electricity is the concept of electric potential.

<table>
<tr><td>Potential difference between two points</td></tr>
</table>

> The potential difference between points A and B, $V_B - V_A$, is defined as the change in potential energy (final value minus initial value) of a charge, q, moved from A to B, divided by the charge.

$$\Delta V \equiv V_B - V_A = \frac{\Delta PE}{q} \qquad \text{[16.2]}$$

Potential difference should not be confused with potential energy. The change in potential between two points is proportional to the change in potential energy of a charge as it moves between the points, and we see from Equation 16.2 that the two are related as $\Delta PE = q\,\Delta V$. Because potential energy is a scalar quantity, *electric potential is also a scalar quantity.* From Equation 16.2 we see that potential difference is a measure of energy per unit charge. Alternatively, electrical potential difference is the work done to move a charge from a point A to a point B divided by the magnitude of the charge. Thus, the SI units of potential are joules per coulomb, called volts (V):[1]

$$1\text{ V} \equiv 1\text{ J}/\text{C} \qquad\qquad \textbf{[16.3]}$$

This says that 1 J of work must be done to move a 1-C charge between two points that are at a potential difference of 1 V. In the process of moving through a potential difference of 1 V, the 1-C charge gains (or loses) 1 J of energy. Dividing Equation 16.1 by q gives

$$\frac{\Delta PE}{q} = V_B - V_A = -Ed \qquad\qquad \textbf{[16.4]}$$

This equation shows that potential difference also has units of electric field times distance. From this, it follows that the SI units of electric field, newtons per coulomb, can also be expressed as volts per meter:

$$1\text{ N}/\text{C} = 1\text{ V}/\text{m}$$

Since Equation 16.4 is directly related to Equation 16.1, it, too, is valid only for the case of a uniform field.

Let us examine the changes in energy associated with movements of charge in the electric field pictured in Figure 16.2a. Because the positive charge q tends to move in the direction of the electric field, we must apply an upward external

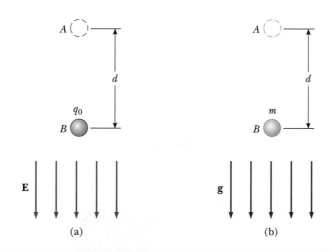

(a)   (b)

FIGURE 16.2
(a) When the electric field, **E**, is directed downward, point B is at a lower electric potential than point A. A positive test charge that moves from A to B loses electric potential energy. (b) A mass, m, moving downward in the direction of the gravitational field, **g**, loses gravitational potential energy.

---

[1] Note that the symbol V (italic) represents potential, whereas V (nonitalic) is the symbol for the unit of this quantity—volts. Try not to confuse these two symbols.

force on the charge to move it from *A* to *B*. Work is done on the charge, and this means that *a positive charge gains electrical potential energy when it is moved in a direction opposite the electric field.* This is analogous to a mass gaining gravitational potential energy when it rises to higher elevations in the presence of gravity, as in Figure 16.2b. If a positive charge is released from rest at point *A*, it experiences a force, *qE*, in the direction of the field (downward in Figure 16.2a). Therefore, it accelerates downward, gaining kinetic energy. *As it gains kinetic energy, it loses an equal amount of potential energy.*

On the other hand, if the test charge *q* is negative, the situation is reversed. *A negative charge loses electrical potential energy when it moves in the direction opposite the electric field.* That is, a negative charge released from rest in the field **E** accelerates in a direction *opposite* the field.

Let us pause briefly to discuss a situation that illustrates the concept of potential difference. Consider the common 12-V automobile battery. Such a battery maintains a potential difference across its terminals, where the positive terminal is 12 V higher in potential than the negative terminal. In practice, the negative terminal is usually connected to the metal body of the car, which can be considered at a potential of zero volts. The battery becomes a useful device when it is connected by conducting wires to such things as lightbulbs, a radio, power windows, motors, and so forth. Now consider a charge of +1 C, to be moved around a circuit that contains the battery connected to some of these external devices. As the charge is moved inside the battery from the negative terminal (at 0 V) to the positive terminal (at 12 V), the work done on the charge by the battery is 12 J. Thus, every coulomb of positive charge that leaves the positive terminal of the battery carries an energy of 12 J. As this charge moves through the external circuit toward the negative terminal, it gives up its 12 J of electrical energy to the external devices. When the charge reaches the negative terminal, its electrical energy is zero. At this point, the battery takes over and restores 12 J of energy to the charge as it is moved from the negative to the positive terminal, enabling it to make another transit of the circuit. The actual amount of charge that leaves the battery and traverses the circuit depends on the properties of the external devices, as we shall see in the next chapter.

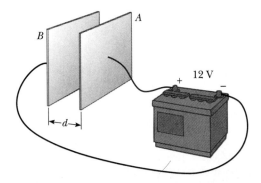

FIGURE 16.3

(Example 16.1)   A 12-V battery connected to two parallel plates. The electric field between the plates has a magnitude given by the potential difference divided by the plate separation, *d*.

EXAMPLE 16.1   The Field Between Two Parallel Plates of Opposite Charge

Figure 16.3 illustrates a situation in which a constant electric field can be set up. A 12-V battery is connected between two parallel metal plates separated by 0.30 cm. Find the strength of the electric field.

Reasoning   The electric field is uniform (except near the edges of the metal plates), and thus the relationship between potential difference and the magnitude of the field is given by Equation 16.4:

$$V_B - V_A = -Ed$$

As already noted above, chemical forces inside a battery maintain one electrode, called the positive terminal, at a higher potential than a second electrode, the negative terminal. Thus, in Figure 16.3, plate B, which is connected to the negative terminal, must be at a lower potential than plate A, which is connected to the positive terminal.

Solution   We have

$$V_B - V_A = -12 \text{ V}$$

This gives a value for E of

$$E = -\frac{(V_B - V_A)}{d} = -\frac{(-12 \text{ V})}{0.30 \times 10^{-2} \text{ m}} = \boxed{4.0 \times 10^3 \text{ V/m}}$$

The direction of this field is from the positive plate to the negative plate. A device consisting of two plates separated by a small distance is called a *parallel-plate capacitor* (to be discussed later in this chapter).

EXAMPLE 16.2   Motion of a Proton in a Uniform Electric Field

A proton is released from *rest* in a uniform electric field of magnitude 8.0×10^4 V/m, directed along the positive x axis (Fig. 16.4). The proton undergoes a displacement of 0.50 m in the direction of the field.

(a) Find the *change* in electric potential of the proton as a result of this displacement.

Solution   From Equation 16.4, we have

$$\Delta V = V_B - V_A = -Ed = -(8.0 \times 10^4 \text{ V/m})(0.50 \text{ m}) = \boxed{-4.0 \times 10^4 \text{ V}}$$

Thus, the electric potential of the proton *decreases* as it moves from A to B.

(b) Find the change in potential energy of the proton for this displacement.

Solution

$$\Delta PE = q \, \Delta V = e \, \Delta V = (1.6 \times 10^{-19} \text{ C})(-4.0 \times 10^4 \text{ V}) = \boxed{-6.4 \times 10^{-15} \text{ J}}$$

The negative sign here means that the potential energy of the proton decreases as it moves in the direction of the electric field. This makes sense since, as the proton *accelerates* in the direction of the field, it gains kinetic energy and at the same time loses potential energy (mechanical energy is conserved).

(c) Find the speed of the proton after it has moved 0.50 m, starting from rest.

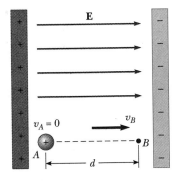

FIGURE 16.4
(Example 16.2)   A proton accelerates from A to B in the direction of the uniform electric field.

Solution   If no forces other than the conservative electrical force are acting on the proton, we can apply the principle of conservation of mechanical energy in the form

$$KE_i + PE_i = KE_f + PE_f$$

In our case, $KE_i = 0$; hence, the preceding expression gives

$$KE_f = PE_i - PE_f = -\Delta PE$$

With this equation and the results of part (b), we find that

$$\tfrac{1}{2} m v_f^2 = 6.4 \times 10^{-15}\,\mathrm{J}$$

and

$$v_f^2 = \frac{2(6.4 \times 10^{-15}\,\mathrm{J})}{1.67 \times 10^{-27}\,\mathrm{kg}} = 7.66 \times 10^{12}\,\mathrm{m^2/s^2}$$

$$v_f = \boxed{2.8 \times 10^6\,\mathrm{m/s}}$$

16.2

ELECTRIC POTENTIAL AND POTENTIAL ENERGY DUE TO POINT CHARGES

In electric circuits, a point of zero potential is often defined by grounding (connecting to Earth) some point in the circuit. For example, if the negative plate in Example 16.1 were grounded, it would be considered to have a potential of zero, and the positive plate to have a potential of 12 V. It is also possible to define the potential due to a point charge at a point in space. In this case, the point of zero potential is taken to be at an infinite distance from the charge. With this choice, the methods of calculus can be used to show that *the electric potential due to a point charge, q, at any distance r from the charge is given by*

Electric potential due to a point charge

$$V = k \frac{q}{r} \qquad \text{[16.5]}$$

Equation 16.5 points out a significant property of electric potential that makes it an important quantity in the study of electricity: the potential at a given point depends only on the charge, q, on the object setting up the potential and the distance r from that object to a specific point in space. As a result, we can say that a potential exists at some point in space whether or not there is a charge at that point.

Superposition principle

The electric potential of two or more charges is obtained by applying the **superposition principle.** That is, *the total potential at some point P due to several point charges is the algebraic sum of the potentials due to the individual charges.* This is similar to the method used in Chapter 15 to find the resultant electric field at a point in space. However, note that in the case of potentials, one must evaluate an *algebraic sum* of individual potentials to obtain the total, because *potentials are scalar quantities.* Thus, it is much easier to evaluate the electric potential at some point due to several charges than to evaluate the electric field, which is a vector quantity.

We now consider the potential energy of interaction of a system of two charged particles. If V_1 is the electric potential due to charge q_1 at a point, P, then the work required to bring charge q_2 from infinity to P without acceleration is $q_2 V_1$. By definition, this work equals the potential energy, PE, of the two-particle system when the particles are separated by a distance of r (Fig. 16.5).

Therefore, we can express the potential energy of the *pair* of charges as

$$PE = q_2 V_1 = k \frac{q_1 q_2}{r} \qquad \text{[16.6]}$$

Note that if the charges are of the *same* sign, PE is positive. This is consistent with the fact that like charges repel, and so positive work must be done on the system to bring two charges near one another. Conversely, if the charges are of *opposite* sign, the force is attractive and PE is negative. This means that negative work must be done to bring unlike charges close together.

Potential energy of a pair of charges

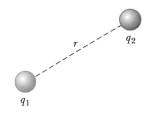

FIGURE 16.5
If two point charges are separated by the distance r, the potential energy of the pair is $k_e q_1 q_2 / r$.

PROBLEM-SOLVING STRATEGY
Electric Potential

1. **When you work problems involving electric potential, remember that potential is a *scalar quantity* (rather than a vector quantity, like the electric field), so there are no components to worry about. Therefore, when using the superposition principle to evaluate the electric potential due to a system of point charges at a point, simply take the algebraic sum of the potentials due to all charges. You must keep track of signs, however. The potential due to each positive charge is positive, and the potential due to each negative charge is negative. Use the basic equation $V = kq/r$.**
2. **As in mechanics, only changes in potential are significant; hence, the point you choose for zero potential is arbitrary.**

EXAMPLE 16.3   Finding the Electric Potential

A 5.0-μC point charge is at the origin, and a point charge of -2.0 μC is on the x axis at (3.0, 0) m, as in Figure 16.6.

(a) If the potential is taken to be zero at infinity, find the total electric potential due to these charges at point P, with coordinates (0, 4.0) m.

Reasoning   The potential at P due to each charge can be calculated from $V = kq/r$. The total electric potential is the scalar sum of these two potentials.

Solution   The potential at P due to the 5.0-μC charge is

$$V_1 = k \frac{q_1}{r_1} = \left(8.99 \times 10^9 \, \frac{\text{N} \cdot \text{m}^2}{\text{C}^2} \right) \left(\frac{5.0 \times 10^{-6} \, \text{C}}{4.0 \, \text{m}} \right) = 1.12 \times 10^4 \, \text{V}$$

and the potential due to the -2.0-μC charge is

$$V_2 = k \frac{q_2}{r_2} = \left(8.99 \times 10^9 \, \frac{\text{N} \cdot \text{m}^2}{\text{C}^2} \right) \left(\frac{-2.0 \times 10^{-6} \, \text{C}}{5.0 \, \text{m}} \right) = -0.360 \times 10^4 \, \text{V}$$

and

$$V_P = V_1 + V_2 = 7.6 \times 10^3 \, \text{V}$$

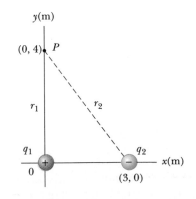

FIGURE 16.6
(Example 16.3)   The electric potential at point P due to the point charges q_1 and q_2 is the algebraic sum of the potentials due to the individual charges.

(b) How much work is required to bring a third point charge of 4.0 μC from infinity to P?

Solution

$$W = q_3 V_P = (4.0 \times 10^{-6} \text{ C})(7.6 \times 10^3 \text{ V})$$

Since $1 \text{ V} = 1 \text{ J/C}$, W reduces to

$$W = \boxed{3.1 \times 10^{-2} \text{ J}}$$

Exercise   Find the magnitude and direction of the electric field at point P.

Answer $2.3 \times 10^3 \text{ N/C}$ at an angle of 79° with the x axis.

16.3

POTENTIALS AND CHARGED CONDUCTORS

In order to determine the potential at all points on a charged conductor, let us combine Equations 16.1 and 16.2. From Equation 16.1 we see that the work done on a charge by electric forces is related to the change in electrical potential energy of the charge by

$$W = -\Delta PE$$

Furthermore, from Equation 16.2 we see that the change in potential energy between two points, A and B, is related to the potential difference between these points by

$$\Delta PE = q(V_B - V_A)$$

Combining these two equations, we find that

$$\boxed{W = -q(V_B - V_A)} \qquad \text{[16.7]}$$

As we see from this result,

> No work is required to move a charge between two points that are at the same potential. That is, $W = 0$ when $V_B = V_A$.

In Chapter 15 we found that, when a conductor is in electrostatic equilibrium, a net charge placed on it resides entirely on its surface. Furthermore, we showed that the electric field just outside the surface of a charged conductor in electrostatic equilibrium is perpendicular to the surface and that the field inside the conductor is zero. We shall now show that *all points on the surface of a charged conductor in electrostatic equilibrium are at the same potential.*

Consider a surface path connecting any points A and B on a charged conductor, as in Figure 16.7. The electric field, **E**, is always perpendicular to the displacement along this path; therefore, no work is done by the electric field if a charge is moved between these points. From Equation 16.7 we see that if the work done is zero, the difference in potential, $V_B - V_A$, is also zero. Therefore,

> The electric potential is a constant everywhere on the surface of a charged conductor in equilibrium.

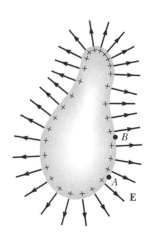

FIGURE 16.7
An arbitrarily shaped conductor with an excess positive charge. When the conductor is in electrostatic equilibrium, all of the charge resides at the surface, **E** = 0 inside the conductor, and the electric field just outside the conductor is perpendicular to the surface. The potential is constant inside the conductor and is equal to the potential at the surface.

Furthermore, because the electric field inside a conductor is zero, no work is required to move a charge between two points inside the conductor. Again, Equation 16.7 shows that if the work done is zero, the difference in potential between any two points inside a conductor must also be zero. Thus, we conclude that the electric potential is constant everywhere inside a conductor.

Finally, since one of the points could be arbitrarily close to the surface of the conductor, we conclude that

Properties of a charged conductor in equilibrium

> **The electric potential is constant everywhere inside a conductor and equal to its value at the surface.**

Consequently, no work is required to move a charge from the interior of a charged conductor to its surface. (Note that the potential inside a conductor is not necessarily zero even though the interior electric field is zero.)

THE ELECTRON VOLT

A unit of energy commonly used in atomic and nuclear physics is the electron volt (eV).

Definition of the electron volt

> **The electron volt is defined as the energy that an electron (or proton) gains when accelerated through a potential difference of 1 V.**

Since $1 \text{ V} = 1 \text{ J/C}$ and since the magnitude of charge on the electron or proton is 1.60×10^{-19} C, we see that the electron volt is related to the joule by

$$1 \text{ eV} = 1.60 \times 10^{-19} \text{ C} \cdot \text{V} = 1.60 \times 10^{-19} \text{ J} \qquad \textbf{[16.8]}$$

EXAMPLE 16.4   The Bohr Atom

In the Bohr model of the hydrogen atom, an electron in its lowest energy state moves in a circular orbit about the nucleus (a single proton) at a distance of 5.29×10^{-11} m.

(a) Find the Coulomb force of attraction exerted on the electron by the proton.

Solution   The magnitude of the Coulomb force is

$$F = k \frac{q_1 q_2}{r^2} = (8.99 \times 10^9 \text{ N} \cdot \text{m}^2/\text{C}^2) \frac{(1.60 \times 10^{-19} \text{ C})^2}{(5.29 \times 10^{-11} \text{ m})^2} = \boxed{8.22 \times 10^{-8} \text{ N}}$$

(b) Find the speed of the electron and its kinetic energy in electron volts.

Solution   The force found in part (a) is the centripetal force acting on the electron directed toward the proton, so we know $F_c = (mv^2)/r$, from which we can find v:

$$v = \sqrt{\frac{F_c r}{m}} = \sqrt{\frac{(8.22 \times 10^{-8} \text{ N})(5.29 \times 10^{-11} \text{ m})}{9.11 \times 10^{-31} \text{ kg}}} = \boxed{2.18 \times 10^6 \text{ m/s}}$$

and the kinetic energy is

$$KE = \tfrac{1}{2} mv^2 = \tfrac{1}{2}(9.11 \times 10^{-31} \text{ kg})(2.18 \times 10^6 \text{ m/s})^2$$

$$= 2.17 \times 10^{-18} \text{ J} = \boxed{13.6 \text{ eV}}$$

(c) Find the electrical potential energy of the electron.

Solution   The potential set up by the proton at the location of the electron is

$$V = k\frac{q}{r} = (8.99 \times 10^9 \text{ N} \cdot \text{m}^2/\text{C}^2) \frac{1.60 \times 10^{-19} \text{ C}}{5.29 \times 10^{-11} \text{ m}} = 27.2 \text{ V}$$

and the electrical potential energy of the electron is

$$PE = q_e V = (-1.60 \times 10^{-19} \text{ C})(27.2 \text{ V}) = 4.36 \times 10^{-18} \text{ J} = \boxed{-27.2 \text{ eV}}$$

(d) Find the total energy of the electron.

Solution

$$E = KE + PE = 13.6 \text{ eV} - 27.2 \text{ eV} = \boxed{-13.6 \text{ eV}}$$

The negative sign for the total energy is characteristic of the total energy for bound systems. The value is interpreted to mean that 13.6 eV of energy must be added to the electron to free it from the proton. Thus, the binding energy of the hydrogen atom is 13.6 eV. As we shall see later, the energy of the system is quantized and the energy of the electron in any of the possible Bohr orbits is

$$E = -\frac{13.6 \text{ eV}}{n^2}$$

where n is a quantum number that can take on integer values from one to infinity.

16.4

EQUIPOTENTIAL SURFACES

A surface on which all points are at the same potential is called an **equipotential surface.** The potential difference between any two points on an equipotential surface is zero. Hence, *no work is required to move a charge at constant speed on an equipotential surface.* Equipotential surfaces have a simple relationship to the electric field. *The electric field at every point of an equipotential surface is perpendicular to the surface.* If the electric field, **E**, had a component parallel to the surface, this component would produce an electric force on a charge placed on the surface. This force would do work on the charge as it moved from one point to another, in contradiction to the definition of an equipotential surface.

It is convenient to represent equipotential surfaces on a diagram by drawing **equipotential lines,** which are two-dimensional views of the intersections of the equipotential surfaces with the plane of the drawing. Figure 16.8a shows the equipotential lines (in blue) associated with a positive point charge. Note that the equipotential lines are perpendicular to the electric field lines (in red) at all points. Recall that the potential due to a point charge q is given by $V = kq/r$. This relation shows that, for a single point charge, the potential is constant on any surface in which r is constant. Therefore, the equipotential surfaces of a point charge are a family of spheres centered on the point charge. Figure 16.8b shows the equipotential lines associated with two charges of equal magnitude but opposite sign.

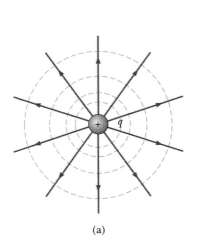

(a)

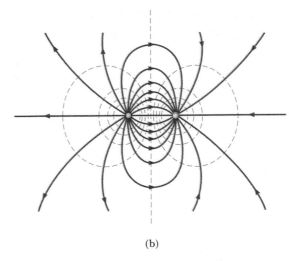

(b)

FIGURE 16.8
Equipotential surfaces (dashed blue lines) and electric field lines (red lines) for (a) a positive point charge and (b) two equal but opposite point charges. In all cases, the equipotential surfaces are *perpendicular* to the electric field lines at every point.

SPECIAL TOPIC

THE ELECTROSTATIC PRECIPITATOR

One important application of electric discharge in gases is a device called an *electrostatic precipitator*. It is used to remove particulate matter from combustion gases, thereby reducing air pollution. It is especially useful in coal-burning power plants and in industrial operations that generate large quantities of smoke. Systems currently in use can eliminate approximately 90% (by mass) of the ash and dust from the smoke. Unfortunately, a very high percentage of the lighter particles still escape, and these contribute significantly to smog and haze.

Figure 16.9 illustrates the basic idea of the electrostatic precipitator. A high voltage (typically 40 to 100 kV) is maintained between a wire running down the center of a duct and the outer wall, which is grounded. The wire is maintained at a negative potential with respect to the wall, and so the electric field is directed toward the wire. The electric field near the wire reaches a high enough value to cause a discharge around the wire and the formation of positive ions, electrons, and negative ions, such as O_2^-. As the electrons and negative ions are accelerated toward the outer wall by the nonuniform electric field, the dirt particles in the streaming gas become charged by collisions and ion capture. Since most of the charged dirt particles are negative, they are also drawn to the outer wall by the electric field. When the duct is shaken, the particles fall loose and are collected at the bottom.

In addition to reducing the amounts of harmful gases and particulate matter in the atmosphere, the electrostatic precipitator recovers valuable metal oxides from the stack.

An *electrostatic air cleaner,* used in homes to relieve the discomfort of allergy sufferers, utilizes many of the same principles as the precipitator. Air laden with dust and pollen is drawn into the device across a positively charged mesh screen. The airborne particles become positively charged when they make intimate contact with the screen. The particles then pass through a second, negatively charged mesh screen. The electrostatic force of attraction between the positively charged particles in the air and the negatively charged screen causes the particles to precipitate out on the surface of the screen. In this fashion, a very high percentage of contaminants are removed from the air stream.

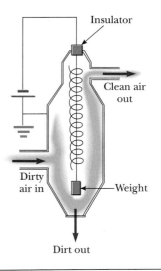

FIGURE 16.9
A schematic diagram of an electrostatic precipitator. The high voltage maintained on the central wires creates an electric discharge in the vicinity of the wire.

SPECIAL TOPIC

XEROGRAPHY AND LASER PRINTERS

The process of xerography is widely used for making photocopies of printed materials. The basic idea behind the process was developed by Chester Carlson, who was granted a patent for his invention in 1940. In 1947 the Xerox Corporation launched a full-scale program to develop automated duplicating machines using Carlson's process. The huge success of that development is quite evident; today, practically all offices and libraries have one or more duplicating machines, and the capabilities of modern machines are still expanding.

Some features of the xerographic process involve simple concepts from electrostatics and optics. However, the one idea that makes the process unique is the use of photoconductive material to form an image. (A photoconductor is a material that is a poor conductor of electricity in the dark but becomes a reasonably good electric conductor when exposed to light.)

Figure 16.10 illustrates the steps in the xerographic process. First, the surface of a plate or drum is coated with a thin film of the photoconductive material (usually selenium or some compound of selenium), and the photoconductive surface is given a positive electrostatic charge in the dark (Fig. 16.10a). The page to be copied is then projected onto the charged surface (Fig. 16.10b). The photoconducting surface becomes conducting only in areas where light strikes; there the light produces charge carriers in the photoconductor, which neutralize the positively charged surface. The

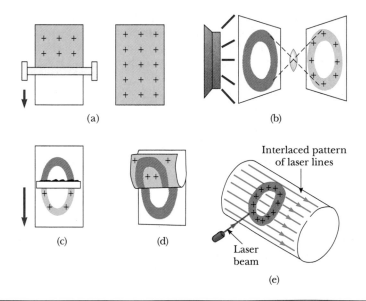

FIGURE 16.10

The xerographic process. (a) The photoconductive surface is positively charged. (b) Through the use of a light source and lens, a hidden image is formed on the charged surface in the form of positive charges. (c) The surface containing the image is covered with a negatively charged powder, which adheres only to the image area. (d) A piece of paper is placed over the surface and given a charge. This transfers the image to the paper, which is then heated to "fix" the powder to the paper. (e) The image on the drum of a laser printer is produced by turning a laser beam on and off as it sweeps across the selenium-coated drum.

charges remain on those areas of the photoconductor not exposed to light, however, leaving a hidden image of the object in the form of a positive surface charge distribution.

Next, a negatively charged powder called a *toner* is dusted onto the photoconducting surface (Fig. 16.10c). The charged powder adheres only to the areas that contain the positively charged image. At this point, the image becomes visible. It is then transferred to the surface of a sheet of positively charged paper. Finally, the toner is "fixed" to the surface of the paper by heat (Fig. 16.10d). This results in a permanent copy of the original.

The steps for producing a document on a laser printer are similar to those used in a photocopy machine, in that parts (a), (c), and (d) of Figure 16.10 remain essentially the same. The difference between the two techniques lies in the way the image is formed on the selenium-coated drum. In a laser printer, the command to print the letter O, for instance, is sent to a laser from the memory of a computer. A rotating mirror inside the printer causes the beam of the laser to sweep across the selenium-coated drum in an interlaced pattern (Fig. 16.10e). Electrical signals generated by the printer turn the laser beam on and off in a pattern that traces out the letter O in the form of positive charges on the selenium. Toner is then applied to the drum, and the transfer to paper is accomplished as in a photocopy machine.

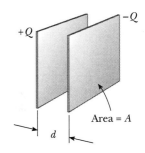

16.5

THE DEFINITION OF CAPACITANCE

A **capacitor** is a device used in a variety of electric circuits—for example, to tune the frequency of radio receivers, eliminate sparking in automobile ignition systems, or store short-term energy in electronic flash units. Figure 16.11 shows a typical design for a capacitor. It consists of two parallel metal plates separated by a distance of d. When used in an electric circuit, the plates are connected to the positive and negative terminals of a battery or some other voltage source. When this connection is made, electrons are pulled off one of the plates, leaving it with a charge of $+Q$, and transferred through the battery to the other plate, leaving it with a charge of $-Q$, as shown in the figure. This charge transfer stops when the potential difference across the plates equals the potential difference of the battery. Thus, a charged capacitor acts as a storehouse of charge and energy that can be reclaimed when needed for a specific application.

FIGURE 16.11
A parallel-plate capacitor consists of two parallel plates, each of area A, separated by a distance d. The plates carry equal and opposite charges.

Capacitance of a pair of conductors

> The capacitance, C, of a capacitor is defined as the ratio of the magnitude of the charge on either conductor to the magnitude of the potential difference between the conductors:

$$C \equiv \frac{Q}{V} \qquad \text{[16.9]}$$

From this equation we see that a large capacitance is needed to store a large amount of charge for a given applied voltage. Also, we see that capacitance has the SI units coulombs per volt, called **farads** (F) in honor of Michael Faraday. That is,

$$1 \text{ F} \equiv 1 \text{ C/V}$$

The farad is a very large unit of capacitance. In practice, most typical capacitors have capacitances ranging from microfarads ($1\ \mu\text{F} = 1 \times 10^{-6}\ \text{F}$) to picofarads ($1\ \text{pF} = 1 \times 10^{-12}\ \text{F}$).

EXAMPLE 16.5   The Charge on the Plates of a Capacitor

A 3.0-μF capacitor is connected to a 12-V battery. What is the magnitude of the charge on each plate of the capacitor?

Solution   The definition of capacitance (Eq. 16.9) gives

$$Q = CV = (3.0 \times 10^{-6}\ \text{F})(12\ \text{V}) = \boxed{36\ \mu\text{C}}$$

16.6

THE PARALLEL-PLATE CAPACITOR

The capacitance of a device depends on the geometric arrangement of the conductors. For example, the capacitance of a parallel-plate capacitor whose plates are separated by air (see Fig. 16.11) is

Capacitance of a parallel-plate capacitor

$$C = \epsilon_0 \frac{A}{d} \qquad \textbf{[16.10]}$$

where A is the area of one of the plates, d is the distance of separation of the plates, and ϵ_0 (ϵ is the Greek letter epsilon) is a constant called the **permittivity of free space,** with the value

$$\epsilon_0 = 8.85 \times 10^{-12}\ \text{C}^2/\text{N}\cdot\text{m}^2$$

The permittivity of free space is related to the Coulomb constant, k, by

$$k = \frac{1}{4\pi\epsilon_0}$$

Although we shall not derive Equation 16.10, we shall attempt to make it seem plausible. As you can see from the definition of capacitance, $C \equiv Q/V$, the amount of charge a given capacitor can store for a given potential difference across its plates increases as the capacitance increases. Therefore, it seems reasonable that a capacitor constructed from plates with large areas should be able to store a large charge. Furthermore, if the oppositely charged plates are close together, the attractive force between them will be large. In fact, for a given potential difference, the charge on the plates increases with decreasing plate separation.

One practical device that uses a capacitor is the flash attachment on a camera. A battery is used to charge the capacitor, and this stored charge is then released when the shutter-release button is pressed to take a picture. The stored charge is delivered to a flash tube very quickly, illuminating the subject at the instant more light is needed.

Computers make use of capacitors in many ways. For example, one type of computer keyboard has capacitors at the bases of its keys, as in Figure 16.12. Each key is connected to a movable plate, which represents one side of the capacitor;

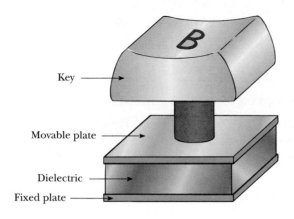

FIGURE 16.12
When the key of one type of keyboard is pressed, the capacitance of a parallel-plate capacitor increases as the plate spacing decreases.

Key

Movable plate

Dielectric

Fixed plate

the fixed plate on the bottom of the keyboard represents the other side of the capacitor. When a key is pressed, the capacitor spacing decreases, causing an increase in capacitance. External electronic circuits recognize each key by the *change* in its capacitance when it is pressed.

EXAMPLE 16.6   Calculating C for a Parallel-Plate Capacitor

A parallel-plate capacitor has an area of $A = 2.00$ cm$^2 = 2.00 \times 10^{-4}$ m^2 and a plate separation of $d = 1.00$ mm $= 1.00 \times 10^{-3}$ m. Find its capacitance.

Solution   From $C = \epsilon_0 A/d$ we find that

$$C = \epsilon_0 \frac{A}{d} = (8.85 \times 10^{-12} \text{ C}^2/\text{N} \cdot \text{m}^2) \left(\frac{2.00 \times 10^{-4} \text{ m}^2}{1.00 \times 10^{-3} \text{ m}} \right)$$

$$= 1.77 \times 10^{-10} \text{ F} = \boxed{177 \text{ pF}}$$

Exercise   Show that 1 C^2/N·m equals 1 F.

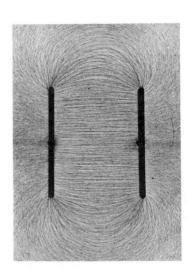

The electric field pattern of two oppositely charged conducting parallel plates. Note the nonuniform nature of the electric field at the ends of the plates. *(Courtesy of Harold M. Waage, Princeton University)*

Symbols for Circuit Elements

The symbol that is commonly used to represent a capacitor in a circuit is ——||——, or sometimes ——)|——. Do not confuse this with the circuit symbol ——=||+—— used to designate a battery (or any other direct current source). The positive terminal of the battery is at the higher potential and is represented by the longer vertical line in the battery symbol. In the next chapter we shall discuss another circuit element, called a resistor, represented by the symbol ——ᴧᴧᴧ—. The wires in a circuit that do not have appreciable resistance compared to other elements in the circuit will be represented by straight lines.

16.7

COMBINATIONS OF CAPACITORS

Two or more capacitors can be combined in circuits in several ways. The equivalent capacitances of certain combinations can be calculated with methods described in this section.

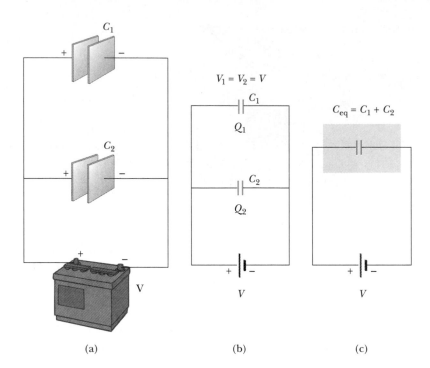

FIGURE 16.13
(a) A parallel connection of two capacitors. (b) The circuit diagram for the parallel combination. (c) The potential differences across the capacitors are the same, and the equivalent capacitance is $C_{eq} = C_1 + C_2$.

PARALLEL COMBINATION

Two capacitors connected as shown in Figure 16.13a are known as a *parallel combination* of capacitors. The left plate of each capacitor is connected by a conducting wire to the positive terminal of the battery, and the left plates are therefore at the same potential. Likewise, the right plates are connected to the negative terminal of the battery. When the capacitors are first connected in the circuit, electrons are transferred from the left plates through the battery to the right plates, leaving the left plates positively charged and the right plates negatively charged. The energy source for this charge transfer is the internal chemical energy stored in the battery, which is converted to electrical energy. The flow of charge ceases when the voltage across the capacitors equals that of the battery. The capacitors reach their maximum charge when the flow of charge ceases. Let us call the maximum charges on the two capacitors Q_1 and Q_2. Then the *total charge, Q,* stored by the two capacitors is

$$Q = Q_1 + Q_2 \qquad \text{[16.11]}$$

We can replace these two capacitors with one equivalent capacitor having a capacitance of C_{eq}. This equivalent capacitor must have exactly the same external effect on the circuit as the original two. That is, it must store Q units of charge. We also see, from Figure 16.13b, that

V is the same across capacitors connected in parallel

> The potential differences across the capacitors in a parallel circuit are the same; each is equal to the voltage of the battery, *V*.

From Figure 16.13c, we see that the voltage across the equivalent capacitor is also *V*. Thus, we have

$$Q_1 = C_1 V \qquad Q_2 = C_2 V$$

and, for the equivalent capacitor,

$$Q = C_{eq} V$$

Substituting these relations into Equation 16.11 gives

$$C_{eq} V = C_1 V + C_2 V$$

or

$$C_{eq} = C_1 + C_2 \quad \left(\begin{array}{l} \text{parallel} \\ \text{combination} \end{array}\right) \qquad \text{[16.12]}$$

If we extend this treatment to three or more capacitors connected in parallel, the equivalent capacitance is found to be

$$C_{eq} = C_1 + C_2 + C_3 + \cdots \quad \left(\begin{array}{l} \text{parallel} \\ \text{combination} \end{array}\right) \qquad \text{[16.13]}$$

Thus, we see that *the equivalent capacitance of a parallel combination of capacitors is greater than any of the individual capacitances.*

EXAMPLE 16.7   Four Capacitors Connected in Parallel

Determine the capacitance of the single capacitor that is equivalent to the parallel combination of capacitors shown in Figure 16.14, and find the charge on the 12.0-μF capacitor.

Solution   The equivalent capacitance is found by use of Equation 16.13:

$$C_{eq} = C_1 + C_2 + C_3 + C_4$$

$$= 3.00 \ \mu\text{F} + 6.00 \ \mu\text{F} + 12.0 \ \mu\text{F} + 24.0 \ \mu\text{F} = \boxed{45.0 \ \mu\text{F}}$$

The potential difference across the 12.0-μF capacitor (and all other capacitors in this case) is equal to the voltage of the battery, and so

$$Q = CV = (12.0 \times 10^{-6} \ \text{F})(18.0 \ \text{V}) = 216 \times 10^{-6} \ \text{C} = \boxed{216 \ \mu\text{C}}$$

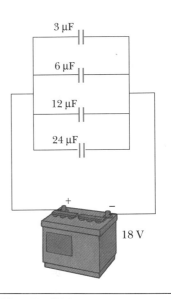

FIGURE 16.14
(Example 16.7)   Four capacitors connected in parallel.

SERIES COMBINATION

Now consider two capacitors connected in *series,* as illustrated in Figure 16.15a.

> For a series combination of capacitors, the magnitude of the charge must be the same on all the plates.

Q is the same for all capacitors connected in series

To see why this must be true, let us consider the charge transfer process in some detail. We start with uncharged capacitors. When a battery is connected to the circuit, electrons are transferred from the left plate of C_1 to the right plate of C_2 through the battery. As this negative charge accumulates on the right plate of C_2, an equivalent amount of negative charge is removed from the left plate of C_2, leaving it with an excess positive charge. The negative charge leaving the left plate of C_2 accumulates on the right plate of C_1, where again an equivalent amount of negative charge is removed from the left plate. The result of this is that *all of the right plates gain charges of $-Q$ while all the left plates have charges of $+Q$.* (This is a consequence of the conservation of charge.)

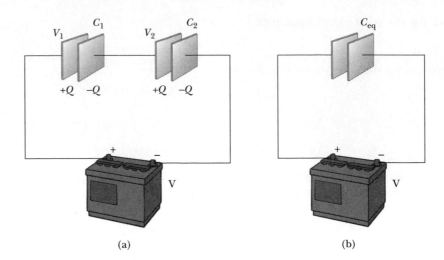

FIGURE 16.15

A series combination of two capacitors. The charges on the capacitors are the same, and the equivalent capacitance can be calculated from the reciprocal relationship

$1/C_{eq} = (1/C_1) + (1/C_2)$.

(a)   (b)

One can find an equivalent capacitor that performs the same function as the series combination. After it is fully charged, *the equivalent capacitor must end up with a charge of* $-Q$ *on its right plate and a charge of* $+Q$ *on its left plate.* By applying the definition of capacitance to the circuit in Figure 16.15b, we have

$$V = \frac{Q}{C_{eq}}$$

where V is the potential difference between the terminals of the battery and C_{eq} is the equivalent capacitance. From Figure 16.15a we see that

$$V = V_1 + V_2 \qquad\qquad [16.14]$$

where V_1 and V_2 are the potential differences across capacitors C_1 and C_2. (This is a consequence of the conservation of energy.)

> **The potential difference across any number of capacitors (or other circuit elements) in series equals the sum of the potential differences across the individual capacitors.**

Since $Q = CV$ can be applied to each capacitor, the potential differences across them are given by

$$V_1 = \frac{Q}{C_1} \qquad V_2 = \frac{Q}{C_2}$$

Substituting these expressions into Equation 16.14, and noting that $V = Q/C_{eq}$, we have

$$\frac{Q}{C_{eq}} = \frac{Q}{C_1} + \frac{Q}{C_2}$$

Cancelling Q, we arrive at the relationship

$$\frac{1}{C_{eq}} = \frac{1}{C_1} + \frac{1}{C_2} \qquad \begin{pmatrix} \text{series} \\ \text{combination} \end{pmatrix} \qquad [16.15]$$

If this analysis is applied to three or more capacitors connected in series, the equivalent capacitance is found to be

$$\frac{1}{C_{eq}} = \frac{1}{C_1} + \frac{1}{C_2} + \frac{1}{C_3} + \cdots \qquad \left(\begin{array}{c}\text{series} \\ \text{combination}\end{array}\right) \qquad \textbf{[16.16]}$$

As we shall demonstrate in Example 16.8, this implies that *the equivalent capacitance of a series combination is always less than any individual capacitance in the combination.*

PROBLEM-SOLVING STRATEGY
Capacitors

1. **Be careful with your choice of units. To calculate the capacitance of a device in farads, make sure that distances are in meters and use the SI value of ϵ_0.**

2. **When two or more unequal capacitors are connected in *series*, they carry the same charge, but the potential differences across them are not the same. Their capacitances add as reciprocals, and the equivalent capacitance of the combination is always *less* than the smallest individual capacitor.**

3. **When two or more capacitors are connected in *parallel*, the potential differences across them are the same. The charge on each capacitor is proportional to its capacitance; hence, the capacitances add directly to give the equivalent capacitance of the parallel combination.**

4. **A complicated circuit consisting of capacitors can often be reduced to a simple circuit containing only one capacitor. To do this, examine your initial circuit and replace any capacitors in series or any in parallel with equivalent capacitors, using the rules in Steps 2 and 3. After making these changes, sketch your new circuit. Examine it and replace any series or parallel combinations again. Continue this process until a single, equivalent capacitor is found.**

5. **To find the charge on or the potential difference across one of the capacitors in the complicated circuit, start with the final circuit found in Step 4 and gradually work your way back through the circuits using $C = Q/V$ and the rules given in Steps 2 and 3.**

EXAMPLE 16.8   Four Capacitors Connected in Series

Four capacitors are connected in series with a battery, as in Figure 16.16.

(a) Find the capacitance of the equivalent capacitor.

Solution   (a) The equivalent capacitance is found from Equation 16.16:

$$\frac{1}{C_{eq}} = \frac{1}{3.0\ \mu\text{F}} + \frac{1}{6.0\ \mu\text{F}} + \frac{1}{12\ \mu\text{F}} + \frac{1}{24\ \mu\text{F}}$$

$$C_{eq} = \boxed{1.6\ \mu\text{F}}$$

Note that the equivalent capacitance is less than the capacitance of any of the individual capacitors in the combination.

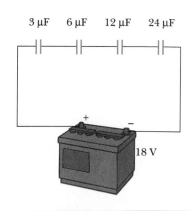

FIGURE 16.16
(Example 16.8)   Four capacitors connected in series.

(b) Find the charge on the 12-μF capacitor.

Solution   We find the charge on the equivalent cap⌐

$$Q = C_{eq} V = (1.6 \times 10^{-6} \text{ F})(18 \text{ V}) = \boxed{29 \ \mu\text{C}}$$

This is also the charge on each of the capacitors it replaced. Thus, the charge on the 12-μF capacitor in the original circuit is 29 μC.

EXAMPLE 16.9   Equivalent Capacitance

Find the equivalent capacitance between a and b for the combination of capacitors shown in Figure 16.17a. All capacitances are in microfarads.

Solution   Using Equations 16.13 and 16.16, we reduce the combination step by step as indicated in the figure. The 1.0-μF and 3.0-μF capacitors are in *parallel* and combine according to $C_{eq} = C_1 + C_2$. Their equivalent capacitance is 4.0 μF. Likewise, the 2.0-μF and 6.0-μF capacitors are also in *parallel* and have an equivalent capacitance of 8.0 μF. The upper branch in Figure 16.17b now consists of two 4.0-μF capacitors in *series*, which combine according to

$$\frac{1}{C_{eq}} = \frac{1}{C_1} + \frac{1}{C_2} = \frac{1}{4.0 \ \mu\text{F}} + \frac{1}{4.0 \ \mu\text{F}} = \frac{1}{2.0 \ \mu\text{F}}$$

$$C_{eq} = 2.0 \ \mu\text{F}$$

Likewise, the lower branch in Figure 16.17b consists of two 8.0-μF capacitors in *series* with an equivalent capacitance of 4.0 μF. Finally, the 2.0-μF and 4.0-μF capacitors in Figure 16.17c are in *parallel* and have an equivalent capacitance of 6.0 μF. Hence, the equivalent capacitance of the circuit is 6.0 μF.

FIGURE 16.17
(Example 16.9)   To find the equivalent capacitance of the circuit in (a), the circuit is reduced in steps—as indicated in (b), (c), and (d)—using the series and parallel rules described in the text.

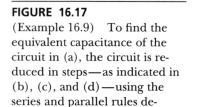 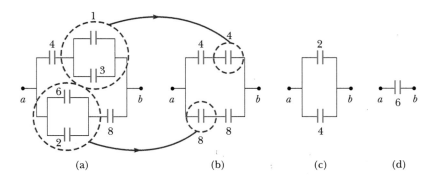

(a)   (b)   (c)   (d)

16.8

ENERGY STORED IN A CHARGED CAPACITOR

Almost everyone who works with electronic equipment has at some time verified that a capacitor can store energy. If the plates of a charged capacitor are connected by a conductor, such as a wire, charge transfers from one plate to the other until the two are uncharged. The discharge can often be observed as a visible spark. If you accidentally touched the opposite plates of a charged capacitor, your fingers would act as a pathway by which the capacitor could discharge, inflicting an electric shock. The degree of shock would depend on the capacitance and voltage applied to the capacitor. Where high voltages and large quantities of charge are present, as in the power supply of a television set, such a shock can be fatal.

If a capacitor is initially uncharged (both plates neutral), so that the plates are at the same potential, almost no work is required to transfer a small amount of charge, ΔQ, from one plate to the other. However, once this charge has been transferred, a small potential difference, $\Delta V = \Delta Q / C$, appears between the plates. Therefore, work must be done to transfer additional charge through this potential difference. As more charge is transferred from one plate to the other, the potential difference increases in proportion. If the potential difference at any instant during the charging process is V, the work required to move more charge, ΔQ, through this potential difference is $V \Delta Q$; that is,

$$\Delta W = V \Delta Q$$

We know that $V = Q/C$ for a capacitor that has a total charge of Q. Therefore, a plot of voltage versus charge gives a straight line with a slope of $1/C$, as shown in Figure 16.18. Since the work ΔW is the sum of the areas of the shaded rectangles, the total work done in charging the capacitor to a final voltage, V, is the area under the voltage-charge curve, which in this case equals the area under the straight line. Since the area under this line is the area of a triangle (which is one-half the product of the base and height), the total work done is

$$W = \tfrac{1}{2} QV \qquad \text{[16.17]}$$

Note that this is also the energy stored in the capacitor, since the work required to charge the capacitor equals the energy stored in the capacitor after it is charged. From the definition of capacitance, we find $Q = CV$; hence, we can express the energy stored as

$$\text{Energy stored} = \tfrac{1}{2} QV = \tfrac{1}{2} CV^2 = \frac{Q^2}{2C} \qquad \text{[16.18]}$$

This result applies to any capacitor. In practice, there is a limit to the maximum energy (or charge) that can be stored, because electrical breakdown ultimately occurs between the plates of the capacitor at a sufficiently large value of V. For this reason, capacitors are usually labeled with a maximum operating voltage.

Large capacitors can store enough electrical energy to cause severe burns or even death if they are discharged so that the flow of charge can pass through the heart. Under the proper conditions, however, they can be used to sustain life by stopping cardiac fibrillation in heart attack victims. When fibrillation occurs, the heart produces a rapid, unregulated pattern of beats. A fast discharge of electrical energy through the heart can return the organ to its normal beat pattern. Emergency medical teams use defibrillators—batteries capable of charging a capacitor to a high voltage. (The circuit actually permits the capacitor to be charged to a much higher voltage than the battery.) The stored electrical energy is released through the heart by conducting electrodes, called paddles, that are placed on both sides of the victim's chest.

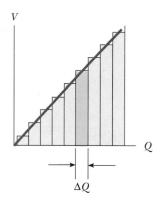

FIGURE 16.18
A plot of voltage versus charge for a capacitor is a straight line with the slope $1/C$. The work required to move a charge of ΔQ through a potential difference of V across the capacitor plates is $\Delta W = V \Delta Q$, which equals the area of the blue rectangle. The *total work* required to charge the capacitor to a final charge of Q is the area under the straight line, which equals $QV/2$.

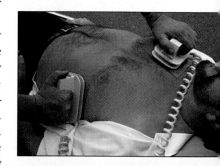

A paramedic uses a portable defibrillator machine in an attempt to revive a man who has suffered a heart attack. The defibrillator delivers a controlled electric shock to restore normal heart rhythm. (*Adam Hart-Davis/ SPL/Photo Researchers*)

EXAMPLE 16.10 Energy Stored in a Charged Capacitor

Find the amount of energy stored in a 5.0-μF capacitor when it is connected across a 120-V battery.

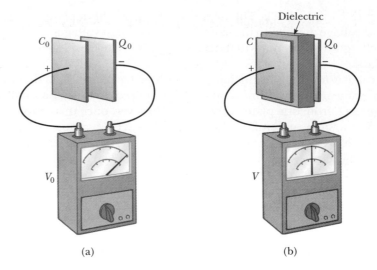

FIGURE 16.19
(a) With air between the plates, the voltage across the capacitor is V_0, the capacitance is C_0, and the charge is Q_0. (b) With a dielectric between the plates, the charge remains at Q_0, but the voltage and capacitance both change.

(a)　　　　　　(b)

Solution　Using Equation 16.18, we have

$$\text{Energy stored} = \tfrac{1}{2}CV^2 = \tfrac{1}{2}(5.0 \times 10^{-6}\,\text{F})(120\,\text{V})^2 = \boxed{3.6 \times 10^{-2}\,\text{J}}$$

Very large capacitors are used as components in high-voltage power generators. Here a technician installs a safety switch in the "oil section" that contains the generators seen at the right side of the picture. This is the main high-voltage mechanical switch linking the generators to a particle accelerator. *(Courtesy of Sandia National Laboratories)*

*16.9
CAPACITORS WITH DIELECTRICS

A **dielectric** is an insulating material, such as rubber, glass, or waxed paper. When a dielectric is inserted between the plates of a capacitor, the capacitance increases. If the dielectric completely fills the space between the plates, the capacitance is multiplied by the factor κ, called the **dielectric constant.**

The following experiment can be performed to illustrate the effect of a dielectric in a capacitor. Consider a parallel-plate capacitor of charge Q_0 and capacitance C_0 in the absence of a dielectric. The potential difference across the capacitor plates can be measured, and it is given by $V_0 = Q_0/C_0$ (Fig. 16.19a). Because the capacitor is not connected to an external circuit, there is no pathway for charge to leave or be added to the plates. If a dielectric is now inserted between the plates as in Figure 16.19b, it is found that the voltage across the plates *is reduced* by the factor κ (>1) to the value V, where

$$V = \frac{V_0}{\kappa}$$

Since $\kappa > 1$, V is less than V_0. Because the charge Q_0 on the capacitor does not change, we conclude that the capacitance in the presence of the dielectric, C, must change to the value

$$C = \frac{Q_0}{V} = \frac{Q_0}{V_0/\kappa} = \frac{\kappa Q_0}{V_0}$$

or

$$C = \kappa C_0 \qquad\qquad \textbf{[16.19]}$$

TABLE 16.1
Dielectric Constants and Dielectric Strengths of Various Materials at Room Temperature

Material	Dielectric Constant, κ	Dielectric Strength (V/m)
Vacuum	1.000 00	—
Air	1.000 59	3×10^6
Bakelite	4.9	24×10^6
Fused quartz	3.78	8×10^6
Pyrex glass	5.6	14×10^6
Polystyrene	2.56	24×10^6
Teflon	2.1	60×10^6
Neoprene rubber	6.7	12×10^6
Nylon	3.4	14×10^6
Paper	3.7	16×10^6
Strontium titanate	233	8×10^6
Water	80	—
Silicone oil	2.5	15×10^6

According to this result, the capacitance is *multiplied* by the factor κ when the dielectric completely fills the region between the plates. For a parallel-plate capacitor, where the capacitance in the absence of a dielectric is $C_0 = \epsilon_0 A/d$, we can express the capacitance in the presence of a dielectric as

$$C = \kappa \epsilon_0 \frac{A}{d} \qquad \text{[16.20]}$$

From this result it appears that the capacitance could be made very large by decreasing d, the distance between the plates. In practice, the lowest value of d is limited by the electric discharge that can occur through the dielectric material separating the plates. For any given plate separation, there is a maximum electric field that can be produced in the dielectric before it breaks down and begins to conduct. This maximum electric field is called the **dielectric strength**, and for air its value is about 3×10^6 V/m. Most insulating materials have dielectric strengths greater than that of air, as indicated by the values in Table 16.1.

Dielectric breakdown in air. Sparks are produced when a large alternating voltage is applied across the electrodes using a high-voltage induction coil power supply. *(Courtesy of Central Scientific Company)*

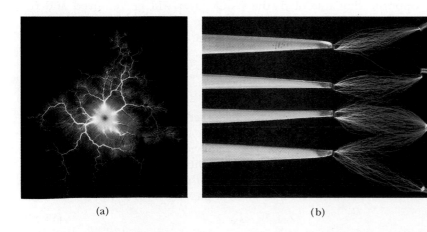

(a)　　　　　(b)

(a) A Kirlian photograph created by dropping a steel ball into a high-energy electric field. This technique is also known as electrophotography. *(Henry Dakin/Science Photo Library)* (b) Sparks from static electricity discharge between a fork and four electrodes. Many discharges were used to make this image, because only one spark forms for a given discharge. Each spark follows the line of least resistance through the air at the time. Note that the bottom prong of the fork forms discharges to both electrodes at bottom right. The light of each spark is created by the ionization of gas atoms along its path. *(Adam Hart-Davis/Science Photo Library)*

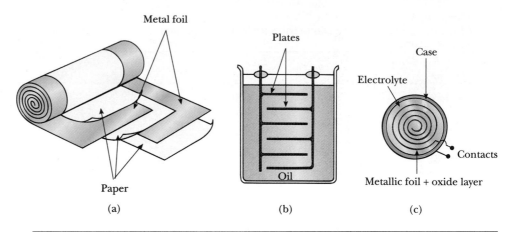

FIGURE 16.20

Three commercial capacitor designs: (a) a tubular capacitor whose plates are separated by paper and then rolled into a cylinder, (b) a high-voltage capacitor consisting of many parallel plates separated by oil, and (c) an electrolytic capacitor.

A collection of capacitors used in a variety of applications. *(Courtesy of Henry Leap and Jim Lehman)*

Commercial capacitors are often made using metal foil interlaced with thin sheets of paraffin-impregnated paper or mylar, which serves as the dielectric material. These alternate layers of metal foil and dielectric are then rolled into a small cylinder (Fig. 16.20a). A high-voltage capacitor commonly consists of a number of interwoven metal plates immersed in silicone oil (Fig. 16.20b). Small capacitors are often constructed from ceramic materials. Variable capacitors (typically 10 to 500 pF) usually consist of two interwoven sets of metal plates, one fixed and the other movable, with air as the dielectric.

An electrolytic capacitor (Fig. 16.20c) is often used to store large amounts of charge at relatively low voltages. It consists of a metal foil in contact with an electrolyte—a solution that conducts charge by virtue of the motion of the ions contained in it. When a voltage is applied between the foil and the electrolyte, a thin layer of metal oxide (an insulator) is formed on the foil, and this layer serves as the dielectric. Enormous capacitances can be attained because the dielectric layer is very thin.

When electrolytic capacitors are used in circuits, the polarity (the plus and minus signs on the device) must be observed. If the polarity of the applied voltage is opposite that intended, the oxide layer will be removed and the capacitor will conduct rather than store charge. Furthermore, reversing the polarity can result in such a large current that the capacitor may either burn or produce steam and explode.

THE STUD FINDER

If you have ever tried to hang a picture on a wall securely, you know that it can be difficult to locate a wooden stud in which to anchor your nail. The principles discussed in this section are fundamental to a device called a stud finder, shown in Figure 16.21. The primary element in this device is a capacitor with its plates arranged side by side instead of facing one another. As the detector is moved along a wall, its capacitance changes when it passes across a stud because the

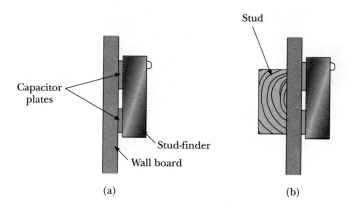

Stud

Capacitor plates

Stud-finder

Wall board

(a)

(b)

FIGURE 16.21
A stud finder. (a) The materials between the plates of the capacitor are the drywall and the air behind it. (b) The materials become drywall and wood when the detector moves across a stud in the wall. The change in the dielectric constant causes a signal light to illuminate.

dielectric constant of the material "between" the plates changes. The change in capacitance can be used to cause a light to come on, signalling the presence of the stud.

EXAMPLE 16.11   A Paper-Filled Capacitor

A parallel-plate capacitor has plates 2.0 cm by 3.0 cm. The plates are separated by a 1.0-mm thickness of paper.

(a) Find the capacitance of this device.

Solution   (a) Since $\kappa = 3.7$ for paper (Table 16.1), we get

$$C = \kappa \epsilon_0 \frac{A}{d} = 3.7 \left(8.85 \times 10^{-12} \frac{C^2}{N \cdot m^2} \right) \left(\frac{6.0 \times 10^{-4} \, m^2}{1.0 \times 10^{-3} \, m} \right)$$

$$= 20 \times 10^{-12} \, F = \boxed{20 \, pF}$$

(b) Find the maximum charge that can be placed on the capacitor.

Solution   From Table 16.1 we see that the dielectric strength of paper is 16×10^6 V/m. Because the paper thickness is 1.0 mm, the maximum voltage that can be applied before electrical breakdown occurs can be calculated using Equation 16.4:

$$V_{max} = E_{max} d = (16 \times 10^6 \, V/m)(1.0 \times 10^{-3} \, m) = 16 \times 10^3 \, V$$

Hence, the maximum charge that can be placed on the capacitor is

$$Q_{max} = CV_{max} = (20 \times 10^{-12} \, F)(16 \times 10^3 \, V) = \boxed{0.32 \, \mu C}$$

AN ATOMIC DESCRIPTION OF DIELECTRICS

The explanation of why a dielectric increases the capacitance of a capacitor is based on an atomic description of the material, which in turn involves a property of some molecules called **polarization**. A molecule is said to be polarized when there is a separation between the "centers of gravity" of its negative charge and its positive charge. In some molecules, such as water, this condition is always present. To see why, consider the geometry of a water molecule (Fig. 16.22). The

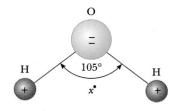

FIGURE 16.22
The water molecule, H_2O, has a permanent polarization resulting from its bent geometry.

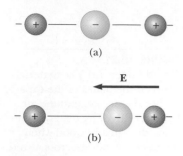

FIGURE 16.23
(a) A symmetric molecule has no permanent polarization. (b) An external electric field induces a polarization in the molecule.

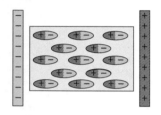

FIGURE 16.24
When a dielectric is placed between the plates of a charged parallel-plate capacitor, the dielectric becomes polarized. This creates a net positive induced charge on the left side of the dielectric and a net negative induced charge on the right side. As a result, the capacitance of the device is multiplied by the factor κ.

molecule is arranged so that the negative oxygen atom is bonded to the positively charged hydrogen atoms with a 105° angle between the two bonds. The center of negative charge is at the oxygen atom, and the center of positive charge lies at a point midway along the line joining the hydrogen atoms (point x in the diagram). Materials composed of molecules that are permanently polarized in this fashion have large dielectric constants, and, indeed, Table 16.1 shows that the dielectric constant of water is quite large ($\kappa = 80$).

A symmetric molecule (Fig. 16.23a) can have no permanent polarization, but a polarization can be induced by an external electric field. A field directed to the left, as in Figure 16.23b, would cause the center of positive charge to shift to the left from its initial position, and the center of negative charge to shift to the right. This *induced polarization* is the effect that predominates in most materials used as dielectrics in capacitors.

To understand why the polarization of a dielectric can affect capacitance, consider Figure 16.24, which shows a slab of dielectric placed between the plates of a parallel-plate capacitor. The dielectric becomes polarized as shown because it is in the electric field that exists between the metal plates. Notice that a net positive charge appears on the dielectric surface adjacent to the negatively charged metal plate. The presence of this positive charge on the dielectric effectively reduces some of the negative charge on the metal, allowing more negative charge to be stored on the capacitor plates for a given applied voltage. From the definition of capacitance, $C = Q/V$, we see that, since the plates can store more charge for a given voltage, the capacitance must increase.

SPECIAL TOPIC
LIVING CELLS AS CAPACITORS

We shall examine electrical phenomena in the human body in detail in Chapter 18. For now, let us consider a feature of living cells that gives them capacitor-like characteristics.

As shown in Figure 16.25, the presence of charged ions in a cell and in the fluid surrounding the cell sets up a charge distribution across the membrane wall. With this charge distribution, the cell is equivalent to a small capacitor separated by a dielectric, with the membrane wall acting as the dielectric. The potential difference across this "capacitor" can be measured via the technique depicted in Figure 16.26. A tiny probe is forced through the cell wall, and a second probe is placed in the extracellular fluid, revealing that potential differences across cell walls are typically on the order of 100 mV.

The charge distribution of a cell can be understood on the basis of the theory of transport through selectively permeable membranes. The primary ionic constituents of a cell are potassium ions (K^+) and chloride ions (Cl^-). The cell wall is highly permeable to K^+ ions but only moderately permeable to Cl^- ions. Since the K^+ ions can cross the cell wall with ease, the charge distribution on a cell wall is determined primarily by their movement.

To explain the process, let us assume that the cellular fluid is electrically neutral at some time. Under normal circumstances, the concentration of K^+ ions outside the cell is much smaller than that inside the cell. Therefore, K^+ ions diffuse out of the cell, leaving behind a net negative charge. As diffusion commences, electrostatic forces of attraction across the membrane wall produce a layer of positive charge on the exterior surface of the wall and a layer of negative charge on the interior surface.

At the same time, the electric field set up by these charges *impedes* the flow of additional K^+ ions diffusing out of the cell. Because of this impedence, an equilibrium is finally established so that there is no net movement of K^+ ions through the cell wall. When equilibrium is reached, the potential difference across the cell wall is given by the so-called **Nernst potential,**

$$V_N = \frac{kT}{q} \ln\left(\frac{c_i}{c_o}\right) \qquad \textbf{[16.21]}$$

where k is Boltzmann's constant (1.38×10^{-23} J/K), T is the Kelvin temperature, q is 1.6×10^{-19} C, and c_i and c_o are the concentrations of K^+ ions inside and outside the cell. At normal body temperatures ($T = 310$ K), this equation becomes

$$V_N = (26.7 \text{ mV}) \ln\left(\frac{c_i}{c_o}\right) \qquad \textbf{[16.22]}$$

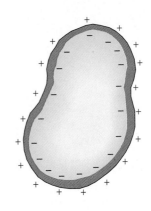

FIGURE 16.25
A living cell is equivalent to a small capacitor separated by a dielectric (the membrane wall of the cell).

The electrical balance achieved in the cell is necessary for it to function properly. A threat to this balance occurs when the extracellular fluid contains a higher concentration of sodium ions (Na^+) than is normally found inside a cell. Although the cell wall is not highly permeable to these ions, a fraction succeed in penetrating the wall. This encroachment by Na^+ ions inside the cell gradually depletes the negative charge on the interior of the cell wall. If this occurs, more K^+ ions escape, thus endangering the cell. Fortunately, the cell can prevent this by effectively "pumping" Na^+ ions out of itself while "pumping" K^+ ions into itself. The mechanism for these processes is not fully understood.

EXAMPLE   Concentration Ratio

Find the ratio of the concentration of K^+ ions inside a cell to the concentration outside the cell if the Nernst potential is measured to be 90.0 mV.

Solution   Solving Equation 16.22 for $\ln(c_i/c_o)$, we obtain

$$\ln\left(\frac{c_i}{c_o}\right) = \frac{V_N}{26.7 \times 10^{-3}} = \frac{90.0 \times 10^{-3} \text{ V}}{26.7 \times 10^{-3} \text{ V}} = 3.37$$

The number that has a natural logarithm of 3.37 is 29.1. (Use your calculator to verify this.) Thus, the required concentration ratio is

$$\frac{c_i}{c_o} = \boxed{29.1}$$

FIGURE 16.26
An experimental technique for measuring the potential difference across the walls of a living cell.

SUMMARY

The **difference in potential** between two points, A and B, is

$$V_B - V_A \equiv \frac{\Delta PE}{q} \qquad \textbf{[16.2]}$$

where ΔPE is the *change* in electrical potential energy experienced by a charge, q, as it moves between A and B. The units of potential difference are joules per coulomb, or **volts**; 1 J/C = 1 V.

The **potential difference** between two points, A and B, in a *uniform electric field*, **E**, is

$$V_B - V_A = -Ed \qquad \textbf{[16.4]}$$

where d is the distance between A and B, and E is the strength of the electric field in that region.

The **electric potential** due to a point charge, q, at distance r from the point charge is

$$V = k\frac{q}{r} \qquad \textbf{[16.5]}$$

The **electrical potential energy** of a pair of point charges separated by distance r is

$$PE = k\frac{q_1 q_2}{r} \qquad \textbf{[16.6]}$$

Every point on the surface of a charged conductor in electrostatic equilibrium is at the same potential. Furthermore, the potential is constant everywhere inside the conductor and equals its value on the surface.

The **electron volt** is defined as the energy that an electron (or proton) gains when accelerated through a potential difference of 1 V. The conversion between electron volts and joules is

$$1\ \text{eV} = 1.60 \times 10^{-19}\ \text{J} \qquad \textbf{[16.8]}$$

A **capacitor** consists of two metal plates with charges that are equal in magnitude but opposite in sign. The **capacitance** (C) of any capacitor is the ratio of the magnitude of the charge, Q, on either plate to the potential difference, V, between them:

$$C \equiv \frac{Q}{V} \qquad \textbf{[16.9]}$$

Capacitance has the units coulombs per volt, or **farads**; $1\ \text{C/V} \equiv 1\ \text{F}$.

The capacitance of two *parallel metal plates* of area A separated by distance d is

$$C = \epsilon_0 \frac{A}{d} \qquad \textbf{[16.10]}$$

where ϵ_0 is a constant called the **permittivity of free space**, with the value

$$\epsilon_0 = 8.85 \times 10^{-12}\ \text{C}^2/\text{N}\cdot\text{m}^2$$

The **equivalent capacitance of a parallel combination** of capacitors is

$$C_{eq} = C_1 + C_2 + C_3 + \cdots \qquad \textbf{[16.13]}$$

If two or more capacitors are connected in series, the **equivalent capacitance of the series combination** is

$$\frac{1}{C_{eq}} = \frac{1}{C_1} + \frac{1}{C_2} + \frac{1}{C_3} + \cdots \qquad \textbf{[16.16]}$$

Three equivalent expressions for calculating the **energy stored** in a charged capacitor are

$$\text{Energy stored} = \tfrac{1}{2}QV = \tfrac{1}{2}CV^2 = \frac{Q^2}{2C} \qquad \textbf{[16.18]}$$

When a nonconducting material, called a **dielectric,** is placed between the plates of a capacitor, the capacitance is multiplied by the factor κ, which is called the **dielectric constant** and is a property of the dielectric material. The capacitance of a parallel-plate capacitor filled with a dielectric is

$$C = \kappa\epsilon_0\,\frac{A}{d} \qquad \textbf{[16.20]}$$

ADDITIONAL READING

Alan H. Cromer, *Physics for the Life Sciences,* New York, McGraw-Hill, 1974, Chap. 17.

A. Einstein and L. Infeld, *The Evolution of Physics,* New York, Simon and Schuster, 1938.

A. L. Stanford Jr., *Foundations of Biophysics,* New York, Academic Press, 1975. Electric dipoles.

CONCEPTUAL QUESTIONS

Example   If the electric potential at some point is zero, can you conclude that there are no charges in the vicinity of that point? Explain.

Reasoning   No. The electric potential at some point may be due to a collection of positive and negative charges whose individual potentials cancel at the point in question. For example, the potential at the midpoint of an isolated electric dipole is zero.

Example   Explain why, under static conditions, all points in a conductor must be at the same electric potential.

Reasoning   If two points on a conducting object were at different potentials, then free charges in the object would move, and we would have a nonstatic situation, in contradiction to the initial assumption. (Free positive charges would migrate from higher to lower potential locations; free electrons would rapidly move from lower to higher potential locations.) The charges would continue to move until the potential is equal everywhere in the conductor.

Example   What happens to the charge on a capacitor if the potential difference between its plates is doubled?

Reasoning   Since the charge on the capacitor is given by $Q = CV$, and the capacitance is constant, doubling the potential difference would double the charge.

Example   Why is it dangerous to touch the terminals of a high-voltage capacitor even after the voltage source has been removed? What could be done to make the capacitor safe to handle?

Reasoning   The capacitor may remain charged long after the voltage source is removed. The residual charge can be lethal if it passes through the body to ground. The capacitor may be safely handled after discharging the plates by short-circuiting the device with a conductor, such as a screwdriver with an insulating handle.

Example   If you want to increase the maximum operating voltage of a parallel-plate capacitor, describe how you can do this for a fixed plate separation.

Reasoning   Changing the area will change the capacitance and maximum charge, but not the maximum voltage. The question does not allow you to increase the plate separation. You can increase the maximum operating voltage by inserting a material with higher dielectric strength between the plates, or by evacuating the space between the plates.

Example   A parallel-plate capacitor is charged by a battery, and the battery is then disconnected from the capacitor. Since the charges on the capacitor plates are equal and opposite, they attract each other. Hence, it takes positive work to increase the plate separation. What happens to the external work when the plate separation is increased?

Reasoning   The work done in pulling the capacitor plates further apart is transferred into additional electric energy stored in the capacitor. The charge is constant and the capacitance decreases but the potential difference between the plates increases, which results in an increase in the stored electric energy.

1. A constant electric field is parallel to the *x* axis. In what direction can a charge be displaced in this field without any external work being done on the charge?

2. If the potential is constant in a certain region, what is the electric field in that region?

3. Explain why equipotential surfaces are always perpendicular to the electric field lines. Do equipotential surfaces ever intersect?

4. If a proton is released from rest in a uniform electric field, does the corresponding electric potential increase or decrease? What about its potential energy?

5. If an electron is released from rest in a uniform electric field, does the corresponding electric potential increase or decrease? What about its potential energy?

6. In your own words, distinguish between electric potential and electric potential energy.

7. Give a physical explanation of the fact that the potential energy of a pair of like charges is positive, whereas the potential energy of a pair of unlike charges is negative.

8. Two capacitors have the same potential difference across them, but one has a large capacitance and one a small capacitance. You would get a greater shock from touching the leads to the one with high capacitance than from touching the leads to the one with low capacitance. From this information, explain what causes the sensation of shock.

9. The plates of a capacitor are connected to a battery. What happens to the charge on the plates if the connecting wires are removed from the battery? What happens to the charge if the wires are removed from the battery and connected to each other?

10. A capacitor is designed so that one plate is large and the other is small. Do the plates have the same charge when connected to a battery?

11. Since the net charge in a capacitor is always zero, what does a capacitor store?

12. If the potential difference across a capacitor is doubled, by what factor is the energy stored by the capacitor multiplied?

13. Two parallel plates are uncharged. Does the set of plates have a capacitance? Explain.

14. If you were asked to design a small capacitor with high capacitance, what factors would be important in your design?

15. Explain why a dielectric increases the maximum operating voltage of a capacitor even though the physical size of the capacitor does not change.

16. A pair of capacitors is connected in parallel, and an identical pair is connected in series. Which pair would be more dangerous to handle after being connected to the same voltage source?

17. What advantage might there be in using two identical capacitors in parallel, connected in series with another identical parallel pair, rather than using a single capacitor by itself?

18. Is it always possible to reduce a combination of capacitors to one equivalent capacitor with the rules developed in this chapter? Explain.

19. If the area of the plates of a parallel-plate capacitor is doubled while the spacing between the plates is halved, how is the capacitance affected?

20. A purchasing director gets a quantity discount for buying identical capacitors in large numbers. How many different values of effective capacitance can a technician produce using all combinations of the capacitors? Evaluate the capacitance for each combination in terms of the capacitance, *C*, of a single capacitor.

21. A parallel-plate capacitor is charged and then disconnected from a battery. How much does the stored energy change (increase or decrease) when the plate separation is doubled?

PROBLEMS

Section 16.1   Potential Difference and Electric Potential

1. (a) How much work is done on a proton by a uniform electric field of 200 N/C as the charge moves a distance of 2.0 cm in the field? (b) What is the difference in potential energy between these two points?

2. The gap between electrodes in a spark plug is 0.060 cm. To produce an electric spark in a gasoline-air mixture, an electric field of 3.0×10^6 V/m must be achieved. On starting a car, what minimum voltage must be supplied by the ignition circuit?

3. A uniform electric field of magnitude 250 V/m is directed in the positive *x* direction. A +12-μC

charge moves from the origin to the point $(x, y) = (20$ cm, 50 cm$)$. (a) What was the change in the potential energy of this charge? (b) Through what potential difference did the charge move?

4. The difference in potential between the accelerating plates of a T.V. set is about 25 000 V. If the distance between these plates is 1.5 cm, find the magnitude of the uniform electric field in this region.

5. To recharge a 12-V battery, a battery charger must move 3.6×10^5 C of charge from the negative terminal to the positive terminal. How much work is

□ indicates problems that have full solutions available in the Student Solutions Manual and Study Guide.

(Problem 5) *(Superstock)*

done by the battery charger? Express your answer in joules.

6. A pair of oppositely charged, parallel plates are separated by 5.33 mm. A potential difference of 600 V exists between the plates. (a) What is the magnitude of the electric field strength between the plates? (b) What is the magnitude of the force on an electron between the plates? Make a sketch showing the direction of the force on the electron. (c) How much work must be done on the electron to move it to the negative plate if it is initially positioned 2.90 mm from the positive plate?

7. How much work is done (by a battery, generator, or some other source of electrical energy) in moving Avogadro's number of electrons from a point where the electric potential is 6.0 V to a point where the electric potential is −10 V?

8. A spherical cell is 3.6 μm in diameter, and its outer membrane is 0.11 μm thick. The potential difference between the inner and outer surfaces of the cell is 90 mV, and the inner surface is negative. How much work is required to eject a positive sodium ion (Na^+) from the interior of the cell?

9. An ion, after being accelerated through a potential difference of 60.0 V, experiences an increase of potential energy of 1.92×10^{-17} J. Calculate the charge on the ion.

10. Two positively charged particles, accelerated from rest through the same potential difference, are found to have the same speed throughout the region of the electric field. What can you say about the charge-to-mass ratios for both particles? Explain.

11. (a) Calculate the speed of a proton that is accelerated from rest through a potential difference of 120 V. (b) Calculate the speed of an electron that is accelerated through the same potential difference.

12. A positron (a particle with charge $+e$ and a mass equal to that of an electron), accelerated from rest between two points at a fixed potential difference, acquires a speed of 30% that of light. What speed is achieved by a *proton* accelerated from rest between the same two points? (Neglect relativistic effects.)

13. (a) Through what potential difference would an electron need to accelerate to achieve a speed of 60% of the speed of light, starting from rest? (The speed of light is 3.00×10^8 m/s.) (b) Repeat this calculation for a proton. (Do not consider relativistic effects.)

14. An electron moves from one plate to another across which there is a potential difference of 2000 V. (a) Find the speed with which the electron strikes the positive plate. (b) Repeat part (a) for a proton moving from the positive to the negative plate.

Section 16.2   Electric Potential and Potential Energy Due to Point Charges

Section 16.3   Potentials and Charged Conductors

15. An electron in the beam of a typical television picture tube is accelerated through a potential difference of 20 000 V before it strikes the face of the tube. What is the energy of this electron, in electron volts, and what is its speed when it strikes the screen?

16. (a) Find the potential 1.0 cm from a proton. (b) What is the potential difference between two points that are 1.0 cm and 2.0 cm from a proton? (c) Repeat parts (a) and (b) for an electron.

17. An electron starts from rest 3.00 cm from the center of a uniformly charged sphere of radius 2.00 cm. If the sphere carries a total charge of 1.00×10^{-9} C, how fast will the electron be moving when it reaches the surface of the sphere?

18. Two point charges are on the y axis—one, of magnitude 3.0×10^{-9} C, at the origin and a second, of magnitude 6.0×10^{-9} C, at the point $y = 30$ cm. Calculate the potential (a) at $y = 60$ cm and (b) at $y = -60$ cm. (c) Repeat this problem, assuming that the 6.0×10^{-9} C charge is replaced by a -6.0×10^{-9} C charge.

19. Four particles with charges +5.00 μC, +3.00 μC, +3.00 μC, and −5.00 μC are placed at the corners of a 2.00 m × 2.00 m square. (a) Determine the electric potential at the center of the

square, assuming zero potential at infinity. (b) If one of the $+3.00$-μC charges is replaced by a -4.00-μC charge, what is the electric potential at the square's center?

20. A charge of -3.00×10^{-9} C is at the origin of a coordinate system, and a charge of 8.00×10^{-9} C is on the x axis at $x = 2.00$ m. At what two locations on the x axis is the electric potential equal to zero?

21. A point charge of 9.00×10^{-9} C is located at the origin. How much work is required to bring a positive charge of 3.00×10^{-9} C from infinity to the location $x = 30.0$ cm?

22. Two point charges, $Q_1 = +5.0$ nC and $Q_2 = -3.0$ nC, are separated by 35 cm. (a) What is the electric potential at a point midway between the charges? (b) What is the potential energy of the pair of charges? What is the significance of the algebraic sign of your answer?

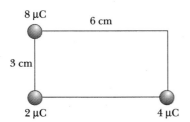

FIGURE 16.27   (Problem 23)

23. Three charges are situated at corners of a rectangle as in Figure 16.27. How much energy would be expended in moving the 8.0-μC charge to infinity?

24. How much work is required to move an electron from a point 50 cm away from a proton to a point 100 cm away from the proton?

25. Calculate the speed of (a) an electron that has a kinetic energy of 1.00 eV and (b) a proton that has a kinetic energy of 1.00 eV.

26. A proton is accelerated from rest through a potential difference of 25 000 V. (a) What is its kinetic energy in electron volts? (b) What is the speed of the proton after this acceleration?

27. In Rutherford's famous scattering experiments that led to the "planetary model" of the atom, alpha particles (having charges of $+2e$ and masses of 6.6×10^{-27} kg) were fired toward a "fixed" gold nucleus with charge $+79e$. An alpha particle, initially very far from the gold nucleus, is fired at

2.0×10^7 m/s directly toward the gold nucleus as in Figure 16.28. How close does the alpha particle get to the gold nucleus before turning around?

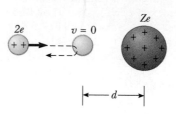

FIGURE 16.28   (Problem 27)

Section 16.5   The Definition of Capacitance

Section 16.6   The Parallel-Plate Capacitor

28. An air-filled parallel-plate capacitor is to have a capacitance of 1.00 F. If the distance between the plates is 1.00 mm, calculate the required surface area of each plate. Convert your answer to square miles.

29. (a) How much charge is on each plate of a 4.00-μF capacitor when it is connected to a 12.0-V battery? (b) If this same capacitor is connected to a 1.50-V battery, what charge is stored?

30. Consider various combinations of three capacitors, each with a capacitance of 2.0 μF. (a) Sketch the arrangement that would give a circuit the largest equivalent capacitance. (b) Sketch the arrangement that would give a circuit the smallest equivalent capacitance. (c) Sketch the arrangement that would give an equivalent capacitance of 3.0 μF.

31. The plates of a parallel-plate capacitor are separated by 0.100 mm. If the material between the plates is air, what plate area is required to provide a capacitance of 2.00 pF?

32. The potential difference between a pair of oppositely charged parallel plates is 400 V. (a) If the spacing between the plates is doubled without altering the charge on the plates, what is the new potential difference between the plates? (b) If the plate spacing is doubled while the potential difference between the plates is kept constant, what is the ratio of the final charge on one of the plates to the original charge?

33. A parallel-plate capacitor has an area of 2.0 cm², and the plates are separated by 2.0 mm with air

between them. How much charge does this capacitor store when connected to a 6.0-V battery?

34. A circular parallel-plate capacitor with a spacing of 3.0 mm is charged to produce an electric field strength of 3.0×10^6 V/m. What plate radius is required if the stored charge is 1.0 μC?

35. A parallel-plate capacitor has an area of 5.00 cm^2, and the plates are separated by 1.00 mm with air between them. It stores a charge of 400 pC. (a) What is the potential difference across the plates of the capacitor? (b) What is the magnitude of the uniform electric field in the region between the plates?

36. Consider the Earth and a cloud layer 800 m above the Earth to be the plates of a parallel-plate capacitor. (a) If the cloud layer has an area of 1.0 km$^2 = 1.00 \times 10^6$ m^2, what is the capacitance? (b) If an electric field strength greater than 2.0×10^6 N/C causes the air to break down and conduct charge (lightning), what is the maximum charge the cloud can hold?

Section 16.7 Combinations of Capacitors

37. Two capacitors, $C_1 = 5.00$ μF and $C_2 = 12.0$ μF, are connected in parallel, and the resulting combination is connected to a 9.00-V battery. What is the value of the equivalent capacitance of the combination?

38. (a) What is the potential difference across each capacitor in Problem 37? (b) What is the charge stored on each capacitor?

39. A 25-μF capacitor and a 40-μF capacitor are charged by being connected across separate 50-V batteries. (a) Determine the resulting charge on each capacitor. (b) The capacitors are then disconnected from their batteries and connected to each other, with each negative plate connected to the other positive plate. What is the final charge of each capacitor, and what is the final potential difference across the 40-μF capacitor?

40. A series circuit consists of a 0.050-μF capacitor, a 0.10-μF capacitor, and a 400-V battery. Find the charge (a) on each of the capacitors; (b) on each of the capacitors if they are reconnected in parallel across the battery.

41. Three capacitors, $C_1 = 5.00$ μF, $C_2 = 4.00$ μF, and $C_3 = 9.00$ μF, are connected together. (a) Find the effective capacitance of the group if they are all in parallel. (b) Find the effective capacitance of the group if they are all in series.

42. Four capacitors ($C_1 = 1.0$ μF, $C_2 = 2.0$ μF, $C_3 = 3.0$ μF, and $C_4 = 4.0$ μF) are connected to a 6.0-V battery as in Figure 16.29. Determine the energy stored in the 2.0-μF capacitor.

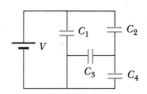

FIGURE 16.29 (Problem 42)

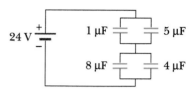

FIGURE 16.30 (Problem 43)

43. Find the charge on each of the capacitors in Figure 16.30.

44. (a) Find the equivalent capacitance of the group of capacitors in Figure 16.31. (b) Find the charge on and the potential difference across each.

45. Consider the circuit shown in Figure 16.32, which consists of four capacitors ($C = 5.00$ μF) that are initially uncharged and a 13.0-V battery. (a) If the switch is thrown to position a, what are the

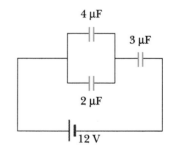

FIGURE 16.31 (Problem 44)

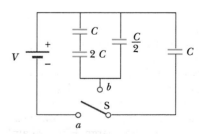

FIGURE 16.32 (Problem 45)

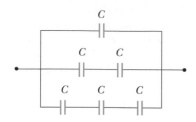

FIGURE 16.33   (Problem 46)

charges on all four capacitors? (b) If the switch is then thrown to position *b*, what are the charges on all four capacitors?

46. Evaluate the effective capacitance of the configuration in Figure 16.33. The capacitors are identical, and each has capacitance *C*.

47. A 10.0-μF capacitor is fully charged across a 12.0-V battery. The capacitor is then disconnected from the battery and connected across an initially uncharged capacitor, *C*. The voltage across each capacitor is 3.00 V. What is the capacitance *C*?

48. How should four 2.0-μF capacitors be connected to have a total capacitance of (a) 8.0 μF? (b) 2.0 μF? (c) 1.5 μF? (d) 0.50 μF?

49. Figure 16.34 shows a network of capacitors between terminals *a* and *b*, with $C_1 = 5.00$ μF, $C_2 = 10.0$ μF, $C_3 = 2.00$ μF. If the potential difference between points *a* and *b* is 60.0 V, what charge is stored by C_3?

50. Consider the combination of capacitors in Figure 16.35. (a) What is the equivalent capacitance of the group? (b) Determine the charge on each capacitor.

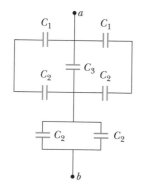

FIGURE 16.34   (Problem 49)

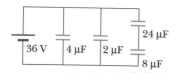

FIGURE 16.35   (Problem 50)

51. A 1.00-μF capacitor is first charged by being connected across a 10.0-V battery. It is then disconnected from the battery and connected across an uncharged 2.00-μF capacitor. Determine the charge on each capacitor.

Section 16.8   Energy Stored in a Charged Capacitor

52. Two capacitors, $C_1 = 25$ μF and $C_2 = 5.0$ μF, are connected in parallel and charged with a 100-V power supply. (a) Calculate the total energy stored in the two capacitors. (b) What potential difference would be required across the same two capacitors connected in *series* in order that the combination store the same energy as in (a)?

53. The energy stored in a 52.0-μF capacitor is used to melt a 6.00-mg sample of lead. To what voltage must the capacitor be initially charged, assuming that the initial temperature of the lead is 20.0°C? Lead has a specific heat of 128 J/kg·°C, a melting point of 327.3°C, and a heat of fusion of 2.45 × 10^4 J/kg.

54. A storm cloud has a potential difference relative to a tree of 1.0 × 10^8 V. If, during a lightning stroke, 50 C of charge is transferred through this potential difference and 1% of the energy is absorbed by the tree, how much water (sap in the tree) can be boiled away, starting at 30°C? Water has a specific heat of 4186 J/kg·°C, a boiling point of 100°C, and a heat of vaporization of 2.26 × 10^6 J/kg.

55. A parallel-plate capacitor has 2.00-cm² plates that are separated by 5.00 mm with air between them. If a 12.0-V battery is connected to this capacitor, how much energy does it store?

56. Consider a parallel-plate capacitor with charge *Q* and area *A*, filled with dielectric material having dielectric constant κ. It can be shown that the magnitude of the attractive force exerted on each plate by the other is given by $F = Q^2/(2\kappa\epsilon_0 A)$. When a potential difference of 100 V exists between the plates of an air-filled 20-μF parallel-plate capacitor, what force does each plate exert on the other if they are separated by 2.01 mm?

Section 16.9   Capacitors with Dielectrics

57. Two parallel plates, each of area 2.00 cm², are separated by 2.00 mm with water between them. A voltage of 6.00 V is applied between the plates. Calculate (a) the magnitude of the electric field between the plates, (b) the charge stored on each plate, and (c) the charge stored on each plate if the water is removed and replaced with air.

58. A capacitor that has air between its plates is connected across a potential difference of 12 V and stores 48 μC of charge. It is then disconnected from the source while still charged. (a) Find its capacitance. (b) A piece of Teflon is then inserted between the plates. Find the voltage and charge on the capacitor. (c) Find its new capacitance.

59. A 3.0-μF capacitor is connected to a 12-V battery. If the material between the plates is initially air, by what factor is the stored energy multiplied when a sheet of nylon is placed between the plates?

60. A capacitor with air between its plates is charged to 100 V and then disconnected from the battery. When a piece of glass is placed between the plates, the voltage across the capacitor drops to 25 V. What is the dielectric constant of the glass? (Assume the glass completely fills the space between the plates.)

61. Determine (a) the capacitance and (b) the maximum voltage that can be applied to a Teflon-filled parallel-plate capacitor having a plate area of 175 cm^2 and insulation thickness of 0.0400 mm.

62. A model of a red blood cell portrays the cell as a spherical capacitor—a positively charged liquid sphere of surface area A, separated by a membrane of thickness t from the surrounding, negatively charged fluid. Tiny electrodes introduced into the interior of the cell show a potential difference of 100 mV across the membrane. The membrane's thickness is estimated to be 100 nm and its dielectric constant to be 5.00. (a) If an average red blood cell has a mass of 1.00×10^{-12} kg, estimate the volume of the cell and thus find its surface area. The density of blood is 1100 kg/m^3. (b) Estimate the capacitance of the cell. (c) Calculate the charge on the surface of the membrane. How many electronic charges does this represent?

ADDITIONAL PROBLEMS

63. An isolated capacitor of unknown capacitance has been charged to a potential difference of 100 V. When the charged capacitor is then connected in parallel to an uncharged 10-μF capacitor, the voltage across the combination is measured to be 30 V. Calculate the unknown capacitance.

64. At distance r from a point charge, q, the electric potential is $V = 600$ V and the magnitude of the electric field is $E = 200$ N/C. Determine the values of q and r.

65. Find the potential at point P for the rectangular grouping of charges shown in Figure 16.36.

66. The three charges shown in Figure 16.37 are at the vertices of an isosceles triangle. Calculate the electric potential at the midpoint of the base, taking the charge $q = 5.0 \times 10^{-9}$ C.

67. Find the equivalent capacitance of the group of capacitors in Figure 16.38.

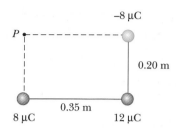

FIGURE 16.36 (Problem 65)

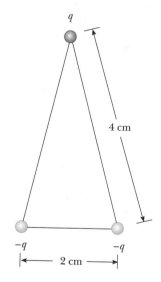

FIGURE 16.37 (Problem 66)

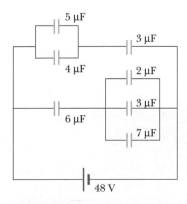

FIGURE 16.38 (Problem 67)

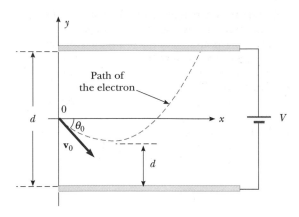

FIGURE 16.39 (Problem 68)

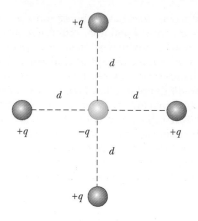

FIGURE 16.41 (Problem 71)

68. An electron is fired at a speed of $v_0 = 5.6 \times 10^6$ m/s, and at an angle of $\theta_0 = -45°$, between two parallel conducting plates that are $D = 2.0$ mm apart, as in Figure 16.39. If the voltage difference between the plates is $V = 100$ V, determine (a) how close, d, the electron will get to the bottom plate and (b) where the electron will strike the top plate.

69. A parallel-plate capacitor is constructed using three dielectric materials, as in Figure 16.40. Determine the capacitance of the device, given that the plate area is 1.0 cm^2, $d = 2.0$ mm, $\kappa_1 = 4.9$, $\kappa_2 = 5.6$, and $\kappa_3 = 2.1$. (*Hint:* Think of the device as consisting of three capacitors, two of them being in a series combination that is in parallel with the third capacitor.)

70. Suppose that the extracellular fluid of a cell has a potassium ion concentration of 0.00450 mol/L and the intracellular fluid has a potassium ion concentration of 0.138 mol/L. Calculate the potential difference across the cell membrane. Neglect the diffusion of any other ions across the membrane and assume a temperature of 310 K.

71. Determine (in terms of q, d, and k) the energy stored in the point-charge system in Figure 16.41.

(*Hint:* First determine the amount of work required to construct the system, one charge at a time.)

72. When a certain air-filled parallel-plate capacitor is connected across a battery, it acquires a charge (on each plate) of 150 μC. While the battery connection is maintained, a dielectric slab is inserted into and fills the region between the plates. This results in the accumulation of an additional charge of 200 μC on each plate. What is the dielectric constant of the dielectric slab?

73. It is possible to create large potential differences by first charging a group of capacitors connected in parallel and then activating a switching arrangement that in effect disconnects the capacitors from the charging source and reconnects them in series. The group of charged capacitors is then discharged in series. What is the maximum potential difference that can be achieved in this manner, using ten 500-μF capacitors and a charging source of 800 V?

74. A parallel-plate capacitor is to be constructed using Pyrex glass as a dielectric. If the capacitance of the device is to be 0.20 μF and it is to be operated at 6000 V, (a) calculate the minimum plate area required. (b) What is the energy stored in the capacitor at the operating voltage?

75. A 2.00-μF capacitor charged to 200 V and a 4.00-μF capacitor charged to 400 V are connected to each other, with the positive plate of each connected to the negative plate of the other. (a) What is the final value of the charge that resides on each capacitor? (b) What is the potential difference across each capacitor after they have been connected?

76. Capacitors $C_1 = 4.0$ μF and $C_2 = 2.0$ μF are charged as a series combination across a 100-V

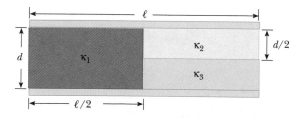

FIGURE 16.40 (Problem 69)

battery. The two capacitors are disconnected from the battery and from each other. They are then connected positive plate to positive plate and negative plate to negative plate. Calculate the resulting charge on each capacitor.

77. A pair of oppositely charged, parallel plates are separated by a distance of 5.0 cm with a potential difference of 500 V between the plates. A proton is released from rest at the positive plate, and at the same time an electron is released from rest at the negative plate. Neglect any interaction between the proton and the electron. (a) After what interval (at what time) will their paths cross? (b) How fast will each particle be going when their paths cross? (c) At what time will the electron reach the opposite plate? (d) At what time will the proton reach the opposite plate?

78. Capacitors $C_1 = 6.0 \ \mu F$ and $C_2 = 2.0 \ \mu F$ are charged as a parallel combination across a 250-V battery. The capacitors are disconnected from the battery and from each other. They are then connected positive plate to negative plate and negative plate to positive plate. Calculate the resulting charge on each capacitor.

CURRENT AND RESISTANCE

A small permanent magnet floats freely above a ceramic disk of the superconductor YBa$_2$Cu$_3$O$_7$ cooled by liquid nitrogen at 77 K. The superconductor has zero electric resistance at temperatures below 92 K and expels any applied magnetic field. (D.O.E./ Science Source/Photo Researchers)

Many practical applications and devices are based on the principles of static electricity, but electricity truly became an inseparable part of our daily lives when scientists learned how to control the flow of electric charges. Electric currents power our lights, radios, television sets, air conditioners, and refrigerators; they ignite the gasoline in automobile engines, travel through miniature components making up the chips of microcomputers, and perform countless other invaluable tasks.

In this chapter we define current and discuss some of the factors that contribute to the resistance to flow of charge in conductors. We also discuss energy transformations in electric circuits. (These topics will be the foundation for additional work with circuits in later chapters.) Finally, an interesting essay on the exciting topic of superconductivity follows this chapter.

17.1

ELECTRIC CURRENT

Whenever electric charges of like sign move, a *current* is said to exist. To define current more precisely, suppose the charges are moving perpendicularly to a surface of area A, as in Figure 17.1. (This area could be the cross-sectional area of a wire, for example.) The **current** is *the rate at which charge flows through this surface*. If ΔQ is the amount of charge that passes through this area in a time interval of Δt, the current, I, is equal to the ratio of the charge to the time interval:

$$I \equiv \frac{\Delta Q}{\Delta t}$$

[17.1]   Direction of current

The SI unit of current is the **ampere** (A):

$$1\text{ A} = 1\text{ C/s}$$

[17.2]

Thus, 1 A of current is equivalent to 1 C of charge passing through the cross-sectional area in a time interval of 1 s.

When charges flow through a surface as in Figure 17.1, they can be positive, negative, or both. *It is conventional to give the current the same direction as the flow of positive charge.* In a common conductor, such as copper, the current is due to the motion of the negatively charged electrons. Therefore, when we speak of current in such a conductor, the direction of the current is opposite the direction of flow of electrons. On the other hand, if one considers a beam of positively charged protons in an accelerator, the current is in the direction of motion of the protons. In some cases—gases and electrolytes, for example—the current is the result of the flows of both positive and negative charges. It is common to refer to a moving charge (whether it is positive or negative) as a mobile *charge carrier*. In a metal, for example, the charge carriers are electrons.

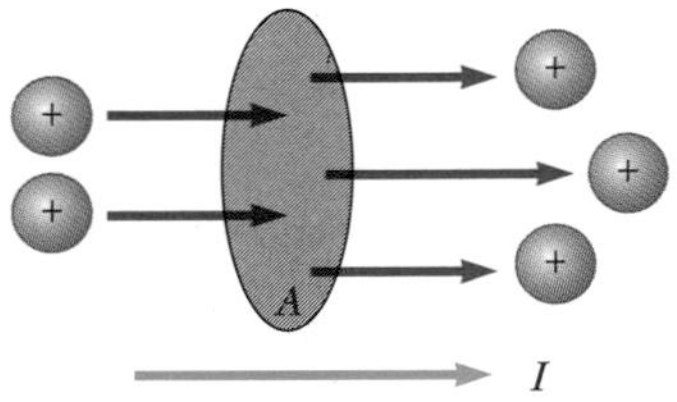

FIGURE 17.1
Charges in motion through an area of A. The time rate of flow of charge through the area is defined as the current I. The direction of the current is the direction of flow of positive charges.

EXAMPLE 17.1   The Current in a Lightbulb

The amount of charge that passes through the filament of a certain lightbulb in 2.00 s is 1.67 C. Find (a) the current in the lightbulb and (b) the number of electrons that pass through the filament in one second.

Solution   (a) From Equation 17.1 we have

$$I = \frac{\Delta Q}{\Delta t} = \frac{1.67\text{ C}}{2.00\text{ s}} = \boxed{0.835\text{ A}}$$

(b) In one second, 0.835 C of charge must pass the cross-sectional area of the filament. This total charge per second is equal to the number of electrons, N, times the charge on a single electron.

$$Nq = N(1.60 \times 10^{-19}\text{ C/electron}) = 0.835\text{ C}$$

$$N = \boxed{5.22 \times 10^{18}\text{ electrons}}$$

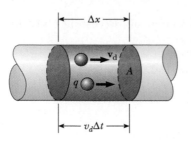

FIGURE 17.2

A section of a uniform conductor of cross-sectional area A. The charge carriers move with a speed of v_d, and the distance they travel in the time Δt is given by $\Delta x = v_d \Delta t$. The number of mobile charge carriers in the section of length Δx is given by $nAv_d \Delta t$, where n is the number of mobile carriers per unit volume.

17.2
CURRENT AND DRIFT SPEED

It is instructive to relate current to the motion of the charged particles. Consider the current in a conductor of cross-sectional area A (Fig. 17.2). The volume of an element of length Δx of the conductor is $A \Delta x$. If n represents the number of mobile charge carriers per unit volume, then the number of carriers in the volume element is $nA \Delta x$. Therefore, the charge, ΔQ, in this element is

$$\Delta Q = \text{number of carriers} \times \text{charge per carrier} = (nA \Delta x) q$$

where q is the charge on each carrier. If the carriers move with a speed of v_d, the distance they move in time Δt is $\Delta x = v_d \Delta t$. Therefore, we can write ΔQ as

$$\Delta Q = (nAv_d \Delta t) q$$

If we divide both sides of this equation by Δt, we see that the current in the conductor is

Current is proportional to the drift velocity	$$I = \frac{\Delta Q}{\Delta t} = nqv_d A \qquad \text{[17.3]}$$

The speed of the charge carriers, v_d, is an average speed called the **drift speed**. To understand its meaning, consider a conductor in which the charge carriers are free electrons. If the conductor is isolated, these electrons undergo random motion similar to that of gas molecules. When a potential difference is applied across the conductor (say, by means of a battery), an electric field is set up in the conductor, creating an electric force on the electrons and hence a current. In reality, the electrons do not simply move in straight lines along the conductor. Instead, they undergo repeated collisions with the metal atoms, and the result is a complicated zigzag motion (Fig. 17.3). The energy transferred from the electrons to the metal atoms during collision increases the vibrational energy of the atoms and causes a corresponding increase in the temperature of the conductor. However, despite the collisions, the electrons move slowly along the conductor (in a direction opposite **E**) with the drift velocity, $\mathbf{v}_d$. The work done by the field on the electrons exceeds the average loss in energy due to collisions, and this work provides a steady current. One can think of the collisions within a conductor as being an effective internal friction (or drag) force similar to that experienced by the molecules of a liquid flowing through a pipe stuffed with steel wool.

Before we close our discussion of current, it is wise to clarify what might appear to be a discrepancy regarding a property of electric fields in conductors. We just stated that an electric field is set up in a conductor when a voltage is applied across it, but in Chapter 15 we stated that the electric field inside a

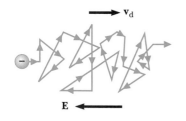

FIGURE 17.3

A schematic representation of the zigzag motion of a charge carrier in a conductor. The changes in direction are due to collisions with atoms in the conductor. Note that the net motion of electrons is opposite the direction of the electric field.

conductor is zero. There is no conflict between these two statements because an electric field *can* exist in a conductor when charges are in motion, in contrast with the discussion of Chapter 15, where electrostatic equilibrium conditions (charges at rest) were assumed. In the latter case, the electric field inside the conductor must indeed be zero.

EXAMPLE 17.2   The Drift Speed in a Copper Wire

A copper wire of cross-sectional area 3.00×10^{-6} m^2 carries a current of 10.0 A. Assuming that each copper atom contributes one free electron to the metal, find the drift speed of the electrons in this wire. The density of copper is 8.95 g/cm^3.

Reasoning   All the variables in Equation 17.3 are known except n, the number of free charge carriers per unit volume. We can find n by recalling that one atomic mass of any substance contains Avogadro's number (6.02×10^{23}) of atoms, and each atom contributes one charge carrier to the metal. The volume of one atomic mass can be found from copper's known density and its atomic mass.

Solution   From the periodic table of the elements, we find that the atomic mass of copper is 63.5 g/mol. Knowing the density of copper enables us to calculate the volume occupied by 63.5 g of copper:

$$V = \frac{m}{\rho} = \frac{63.5 \text{ g}}{8.95 \text{ g/cm}^3} = 7.09 \text{ cm}^3$$

If we now assume that each copper atom contributes one free electron to the body of the material, we have

$$n = \frac{6.02 \times 10^{23} \text{ electrons}}{7.09 \text{ cm}^3} = 8.48 \times 10^{22} \text{ electrons/cm}^3$$

$$= \left(8.48 \times 10^{22} \frac{\text{electrons}}{\text{cm}^3} \right) \left(10^6 \frac{\text{cm}^3}{\text{m}^3} \right) = 8.48 \times 10^{28} \text{ electrons/m}^3$$

From Equation 17.3, we find that the drift speed is

$$v_d = \frac{I}{nqA} = \frac{10.0 \text{ C/s}}{(8.48 \times 10^{28} \text{ electrons/m}^3)(1.60 \times 10^{-19} \text{ C})(3.00 \times 10^{-6} \text{ m}^2)}$$

$$= \boxed{2.46 \times 10^{-4} \text{ m/s}}$$

Example 17.2 shows that drift speeds are typically very small. In fact, the drift speed is much smaller than the average speed between collisions; for instance, electrons traveling at 2.46×10^{-4} m/s would take about 68 min to travel 1 m! In view of this low speed, you might wonder why a light turns on almost instantaneously when a switch is thrown. Think of the flow of water through a pipe. If a drop of water is forced into one end of a pipe that is already filled with water, a drop must be pushed out the other end of the pipe. Although it may take an individual drop a long time to make it through the pipe, a flow initiated at one end produces a similar flow at the other end very quickly. In a conductor, the electric field that drives the free electrons travels with a speed close to that of light. Thus, when you flip a light switch, the message for the electrons to start moving through the wire (the electric field) reaches them at a speed on the order of 10^8 m/s.

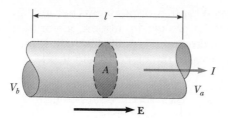

FIGURE 17.4
A uniform conductor of length l and cross-sectional area A. The current, I, in the conductor is proportional to the applied voltage, $V = V_b - V_a$. The electric field, **E**, set up in the conductor is also proportional to the current.

17.3

RESISTANCE AND OHM'S LAW

When a voltage (potential difference), V, is applied across the ends of a metallic conductor as in Figure 17.4, the current in the conductor is found to be proportional to the applied voltage; that is, $I \propto V$. If the proportionality is exact, we can write $V = IR$, where the proportionality constant R is called the resistance of the conductor. In fact, we define this **resistance** as the ratio of the voltage across the conductor to the current it carries:

Resistance

$$R \equiv \frac{V}{I} \qquad \text{[17.4]}$$

Resistance has the SI units volts per ampere, called **ohms** (Ω). Thus, if a potential difference of 1 V across a conductor produces a current of 1 A, the resistance of the conductor is 1 Ω. For example, if an electrical appliance connected to a 120-V source carries a current of 6 A, its resistance is 20 Ω.

It is useful to compare the concepts of electric current, voltage, and resistance with the flow of water in a river. As water flows downhill in a river of constant width and depth, the flow rate (water current) depends on the angle of flow and the effects of rocks, the river bank, and other obstructions. Likewise, electric current in a uniform conductor depends on the applied voltage and the resistance of the conductor caused by collisions of the electrons with atoms in the conductor.

For many materials, including most metals, experiments show that *the resistance is constant over a wide range of applied voltages.* This statement is known as **Ohm's law** after Georg Simon Ohm (1789–1854), who was the first to conduct a systematic study of electrical resistance.

Ohm's law is not a fundamental law of nature, but an empirical relationship that is valid only for certain materials. Materials that obey Ohm's law, and hence have a constant resistance over a wide range of voltages, are said to be *ohmic*. Materials that do not obey Ohm's law are *nonohmic*. Ohmic materials have a linear current-voltage relationship over a large range of applied voltages (Fig. 17.5a). Nonohmic materials have a nonlinear current-voltage relationship (Fig. 17.5b). One common semiconducting device that is nonohmic is the diode. Its resistance is small for currents in one direction (positive V) and large for currents in the reverse direction (negative V). Most modern electronic devices, such as transistors, have nonlinear current-voltage relationships; their operation depends on the particular ways in which they violate Ohm's law.

It is common practice to express Ohm's law as

Ohm's law

$$V = IR \qquad \text{[17.5]}$$

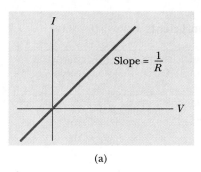

(a)

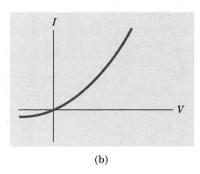

(b)

FIGURE 17.5
(a) The current-voltage curve for an ohmic material. The curve is linear, and the slope gives the resistance of the conductor. (b) A nonlinear current-voltage curve for a semiconducting diode. This device does not obey Ohm's law.

where R is understood to be independent of V. We shall continue to use this traditional form of Ohm's law when discussing electrical circuits. A **resistor** is a simple circuit element that provides a specified resistance in an electric circuit. The symbol for a resistor in circuit diagrams is a zigzag line, ⎓⋀⋀⎓ .

EXAMPLE 17.3   The Resistance of a Steam Iron

All electric devices are required to have identifying plates that specify their electrical characteristics. The plate on a certain steam iron states that the iron carries a current of 6.4 A when connected to a 120-V source. What is the resistance of the steam iron?

Solution   From Ohm's law, we find the resistance to be

$$R = \frac{V}{I} = \frac{120 \text{ V}}{6.4 \text{ A}} = \boxed{19 \ \Omega}$$

Exercise   The resistance of a hot plate is 48 Ω. How much current does the plate carry when connected to a 120-V source?

Answer   2.5 A.

17.4
RESISTIVITY

In an earlier section we pointed out that electrons do not move in straight-line paths through a conductor. Instead, they undergo repeated collisions with the metal atoms. Consider a conductor with a voltage applied between its ends. An electron gains speed as the electric force associated with the internal electric field accelerates it, giving it a velocity in the direction opposite that of the electric field. A collision with an atom randomizes the electron's velocity, thus reducing its velocity in the direction opposite the field. The process then repeats itself. Together these collisions affect the electron somewhat as a force of internal friction would. This is the origin of a material's resistance. The resistance of an ohmic conductor is proportional to its length, l, and inversely proportional to its cross-sectional area, A. That is,

$$R = \rho \frac{l}{A} \qquad\qquad \textbf{[17.6]}$$

TABLE 17.1
Resistivities and Temperature Coefficients of Resistivity for Various Materials[a]

Material	Resistivity $(\Omega \cdot m)$	Temperature Coefficient of Resistivity $[(°C)^{-1}]$
Silver	1.59×10^{-8}	3.8×10^{-3}
Copper	1.7×10^{-8}	3.9×10^{-3}
Gold	2.44×10^{-8}	3.4×10^{-3}
Aluminum	2.82×10^{-8}	3.9×10^{-3}
Tungsten	5.6×10^{-8}	4.5×10^{-3}
Iron	10.0×10^{-8}	5.0×10^{-3}
Platinum	11×10^{-8}	3.92×10^{-3}
Lead	22×10^{-8}	3.9×10^{-3}
Nichrome[b]	150×10^{-8}	0.4×10^{-3}
Carbon	3.5×10^{5}	-0.5×10^{-3}
Germanium	0.46	-48×10^{-3}
Silicon	640	-75×10^{-3}
Glass	10^{10}–10^{14}	
Hard rubber	$\approx 10^{13}$	
Sulfur	10^{15}	
Quartz (fused)	75×10^{16}	

[a] All values are at 20°C.
[b] A nickel-chromium alloy commonly used in heating elements.

An assortment of resistors used for a variety of applications in electronic circuits. *(Courtesy of Henry Leap and Jim Lehman)*

where the constant of proportionality, ρ, is called the **resistivity** of the material.[1] Every material has a characteristic resistivity that depends on its electronic structure and on temperature. Good electric conductors have very low resistivities, and good insulators have very high resistivities. Table 17.1 lists the resistivities of a variety of materials at 20°C. Because resistance values are in ohms, resistivity values must be in ohm-meters.

Equation 17.6 shows that the resistance of a cylindrical conductor is proportional to its length and inversely proportional to its cross-sectional area. This is analogous to the flow of liquid through a pipe. As the length of the pipe is increased, the resistance to liquid flow increases because of a gain in friction between the fluid and the walls of the pipe. As its cross-sectional area is increased, the pipe can transport more fluid in a given time interval, so its resistance drops.

EXAMPLE 17.4   The Resistance of Nichrome Wire

(a) Calculate the resistance per unit length of a 22-gauge nichrome wire of radius 0.321 mm.

Solution   The cross-sectional area of this wire is

$$A = \pi r^2 = \pi(0.321 \times 10^{-3} \text{ m})^2 = 3.24 \times 10^{-7} \text{ m}^2$$

---

[1] The symbol ρ used for resistivity should not be confused with the same symbol used earlier in the book for density. Very often, a single symbol is used to represent different quantities.

The resistivity of nichrome is $1.5 \times 10^{-6} \; \Omega \cdot m$ (Table 17.1). Thus, we can use Equation 17.6 to find the resistance per unit length:

$$\frac{R}{l} = \frac{\rho}{A} = \frac{1.5 \times 10^{-6} \; \Omega \cdot m}{3.24 \times 10^{-7} \; m^2} = \boxed{4.6 \; \Omega/m}$$

(b) If a potential difference of 10.0 V is maintained across a 1.0-m length of the nichrome wire, what is the current in the wire?

Solution   Since a 1.0-m length of this wire has a resistance of 4.6 Ω, Ohm's law gives

$$I = \frac{V}{R} = \frac{10.0 \; V}{4.6 \; \Omega} = \boxed{2.2 \; A}$$

Note from Table 17.1 that the resistivity of nichrome is about 100 times that of copper, a typical "good" conductor. Therefore, a copper wire of the same radius would have a resistance per unit length of only 0.052 Ω/m, and a 1.0-m length of copper wire of the same radius would carry the same current (2.2 A) with an applied voltage of only 0.11 V.

Because of its high resistivity and its resistance to oxidation, nichrome is often used for heating elements in toasters, irons, and electric heaters.

Exercise   What is the resistance of a 6.0-m length of 22-gauge nichrome wire? How much current does it carry when connected to a 120-V source?

Answer   28 Ω; 4.3 A.

Exercise   Calculate the current density and electric field in the wire, assuming that it carries a current of 2.2 A.

Answer $6.7 \times 10^6 \; A/m^2$; 10 N/C.

An old-fashioned carbon filament incandescent lamp. The resistance of such a lamp is typically 10 Ω, but it changes with temperature. Does the resistance increase or decrease as the filament is heated? *(Courtesy of Central Scientific Company)*

17.5

TEMPERATURE VARIATION OF RESISTANCE

The resistivity, and hence the resistance, of a conductor depends on a number of factors. One of the most important is the temperature of the metal. For most metals, resistivity increases with increasing temperature. This correlation can be understood as follows. As the temperature of the material increases, its constituent atoms vibrate with increasingly greater amplitudes. Just as it is more difficult to weave one's way through a crowded room when the people are in motion than when they are standing still, so do the electrons find it more difficult to pass atoms moving with large amplitudes.

For most metals, resistivity increases approximately linearly with temperature over a limited temperature range, according to the expression

$$\rho = \rho_0[1 + \alpha(T - T_0)] \qquad \text{[17.7]}$$

where ρ is the resistivity at some temperature, T (in Celsius degrees); ρ_0 is the resistivity at some reference temperature, T_0 (usually taken to be 20°C); and α is a parameter called the **temperature coefficient of resistivity.** The temperature coefficients for various materials are provided in Table 17.1.

Since the resistance of a conductor with uniform cross-section is proportional to the resistivity according to Equation 17.6 ($R = \rho l / A$), the temperature variation of resistance can be written

$$R = R_0[1 + \alpha(T - T_0)]$$ [17.8]

Precise temperature measurements are often made using this property, as shown by the following example.

EXAMPLE 17.5   A Platinum Resistance Thermometer

A resistance thermometer, which measures temperature by measuring the change in resistance of a conductor, is made of platinum and has a resistance of 50.0 Ω at 20.0°C. When the device is immersed in a vessel containing melting indium, its resistance increases to 76.8 Ω. From this information, find the melting point of indium.

Solution   If we solve Equation 17.8 for $T - T_0$, and get α for platinum from Table 17.1, we obtain

$$T - T_0 = \frac{R - R_0}{\alpha R_0} = \frac{76.8 \ \Omega - 50.0 \ \Omega}{[3.92 \times 10^{-3} \ (°C)^{-1}][50.0 \ \Omega]} = 137°C$$

Since $T_0 = 20.0°C$, we find that the melting point of indium is

$$T = \quad 157°C$$

---

*THE CARBON MICROPHONE

Figure 17.6 illustrates the construction of a carbon microphone, commonly used in the mouthpiece of a telephone. A flexible steel diaphragm is placed in contact with carbon granules inside a container. The carbon granules serve as the primary resistance medium in a circuit containing a source of current—here, a battery—and a transformer (described in a later chapter).

The magnitude of the current in the circuit changes when a sound wave strikes the diaphragm. When a compression arrives at the microphone, the diaphragm flexes inward, causing the carbon granules to press together into a smaller-than-normal volume, corresponding to a decrease in the length of the resistance medium. This results in a lower circuit resistance and hence a greater

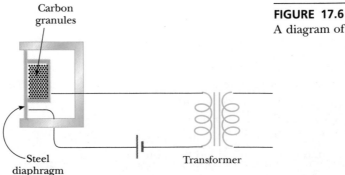

Carbon
granules

Steel
diaphragm

Transformer

FIGURE 17.6

A diagram of a carbon microphone.

current in the circuit. When a rarefaction arrives at the microphone, the diaphragm relaxes, and the carbon granules become more loosely packed, causing an increase in the circuit resistance and a corresponding decrease in current. These variations in current, following the changes of the sound wave, are sent through the transformer into the telephone company's transmission line. A speaker in the listener's earpiece then converts electric signals back to a sound wave.

The carbon microphone has a very poor frequency response. It adequately reproduces frequencies below 4000 Hz and is therefore suitable for speech transmission, since the critical frequencies in normal conversation are usually below this value. However, its capabilities fall off dramatically at higher frequencies, rendering it useless for high-fidelity purposes, which require reliable sound reproduction at all frequencies between 20 Hz and 20 000 Hz.

17.6

SUPERCONDUCTORS

There is a class of metals and compounds whose resistances go virtually to *zero* below a certain temperature, T_c, called the *critical temperature*. These materials are known as **superconductors**. The resistance-temperature graph for a superconductor follows that of a normal metal at temperatures above T_c (Fig. 17.7). When the temperature is at or below T_c, the resistance suddenly drops to zero. This phenomenon was discovered in 1911 by the Dutch physicist H. Kamerlingh Onnes as he and a graduate student worked with mercury, which is a superconductor below 4.2 K. Recent measurements have shown that the resistivities of superconductors below T_c are less than 4×10^{-25} $\Omega \cdot$m—around 10^{17} times smaller than the resistivity of copper, and in practice considered to be zero.

Today thousands of superconductors are known, including such common metals as aluminum, tin, lead, zinc, and indium. Table 17.2 lists the critical temperatures of several superconductors. The value of T_c is sensitive to chemical composition, pressure, and crystalline structure. Interestingly, copper, silver, and gold, which are excellent conductors, do not exhibit superconductivity.

One of the truly remarkable features of superconductors is the fact that, once a current is set up in them, it persists *without any applied voltage* (since $R = 0$). In fact, steady currents in superconducting loops have been observed to persist for many years with no apparent decay!

An important recent development in physics that has created much excitement in the scientific community is the discovery of high-temperature copper-oxide-based superconductors. The excitement began with a 1986 publication by J. Georg Bednorz and K. Alex Müller, scientists at the IBM Zurich Research Laboratory in Switzerland, in which they reported evidence for superconductivity at a temperature near 30 K in an oxide of barium, lanthanum, and copper. Bednorz and Müller were awarded the Nobel Prize in 1987 for their remarkable discovery. Shortly thereafter, a new family of compounds was open for investigation, and research activity in the field of superconductivity proceeded vigorously. In early 1987, groups at the University of Alabama at Huntsville and the University of Houston announced the discovery of superconductivity at about 92 K in an oxide of yttrium, barium, and copper ($YBa_2Cu_3O_7$). Late in 1987,

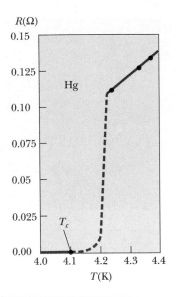

FIGURE 17.7
Resistance versus temperature for a sample of mercury. The graph follows that of a normal metal above the critical temperature, T_c. The resistance drops to zero at the critical temperature, which is 4.2 K for mercury, and remains at zero for lower temperatures.

TABLE 17.2
Critical Temperatures for Various Superconductors

Material	T_c (K)
Zn	0.88
Al	1.19
Sn	3.72
Hg	4.15
Pb	7.18
Nb	9.46
Nb_3Sn	18.05
Nb_3Ge	23.2
$YBa_2Cu_3O_7$	90
Bi–Sr–Ca–Cu–O	105
Tl–Ba–Ca–Cu–O	125

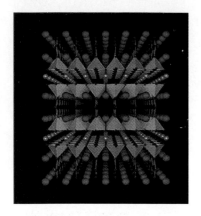

A computer-generated model of the high-temperature superconductor $YBa_2Cu_3O_7$, which is a member of a crystal structure family known as perovskites. *(Courtesy of IBM Research)*

teams of scientists from Japan and the United States reported superconductivity at 105 K in an oxide of bismuth, strontium, calcium, and copper. More recently, scientists have reported superconductivity at temperatures as high as 150 K in an oxide containing mercury. At this point one cannot rule out the possibility of room-temperature superconductivity, and the search for novel superconducting materials continues. It is an important search both for scientific reasons and because practical applications become more probable and widespread as the critical temperature is raised. The essay at the end of this chapter discusses superconductivity and the new generation of superconductors in more detail.

An important and useful application is superconducting magnets in which the magnetic field intensities are about ten times greater than those of the best normal electromagnets. Such magnets are being considered as a means of storing energy. The idea of using superconducting power lines for transmitting power efficiently is also receiving some consideration. Modern superconducting electronic devices consisting of two thin-film superconductors separated by a thin insulator have been constructed. They include magnetometers (magnetic-field measuring devices) and various microwave devices.

17.7

ELECTRICAL ENERGY AND POWER

If a battery is used to establish an electric current in a conductor, chemical energy stored in the battery is continuously transformed into kinetic energy of the charge carriers. This kinetic energy is quickly lost as a result of collisions between the charge carriers and the lattice ions, causing an increase in the temperature of the conductor. Thus, the chemical energy stored in the battery is continuously transformed into thermal energy.

In order to understand the process of energy transfer in a simple circuit, consider a battery whose terminals are connected to a resistor, indicated by the symbol —⋀⋀— (Fig. 17.8). (Remember that the positive terminal of the battery is always at the higher potential.) Now imagine following a quantity of positive charge ΔQ around the circuit from point A through the battery and resistor and back to A. Point A is a reference point that is grounded (the ground symbol is ⏚), and its potential is taken to be zero. As the charge moves from A to B through the battery, its electrical potential energy increases by the amount $V \Delta Q$ (where V is the potential at B) while the chemical potential energy in the battery decreases by the same amount. (Recall from Chapter 16 that $\Delta PE = q \Delta V$.) However, as the charge moves from C to D through the resistor, it loses this electrical potential energy during collisions with atoms in the resistor, thereby producing thermal energy. Note that, if we neglect the resistance of the interconnecting wires, no loss in energy occurs for paths BC and DA. When the charge returns to point A, it must have the same potential energy (zero) as it had at the start.

The rate at which the charge ΔQ loses potential energy as it passes through the resistor is

$$\frac{\Delta Q}{\Delta t} V = IV$$

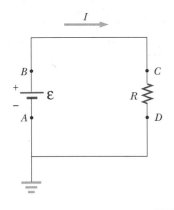

FIGURE 17.8

A circuit consisting of a battery and resistance R. Positive charge flows clockwise from the positive to the negative terminal of the battery. Point A is grounded.

where I is the current in the circuit. Of course, the charge regains this energy when it passes through the battery. Since the rate at which the charge loses energy equals the power, P, dissipated in the resistor, we have

$$P = IV \qquad\qquad \text{[17.9]}$$

Power

In this case, the power is supplied to a resistor by a battery. However, Equation 17.9 can be used to determine the power transferred from a battery to *any* device carrying a current, I, and having a potential difference, V, between its terminals.

Using Equation 17.9 and the fact that $V = IR$ for a resistor, we can express the power dissipated by the resistor in the alternative form

$$P = I^2R = \frac{V^2}{R} \qquad\qquad \text{[17.10]}$$

Power dissipated by a resistor

When I is in amperes, V in volts, and R in ohms, the SI unit of power is the watt (introduced in Chapter 5). The dissipation of power as heat in a conductor of resistance R is called *joule heating*. It is also often referred to as an I^2R *loss*.

EXAMPLE 17.6   The Power Consumed by an Electric Heater

An electric heater is operated by applying a potential difference of 50.0 V to a nichrome wire of total resistance 8.00 Ω. Find the current carried by the wire and the power rating of the heater.

Solution   Since $V = IR$, we have

$$I = \frac{V}{R} = \frac{50.0\ \text{V}}{8.00\ \Omega} = \boxed{6.25\ \text{A}}$$

We can find the power rating using $P = I^2R$:

$$P = I^2R = (6.25\ \text{A})^2(8.00\ \Omega) = \boxed{313\ \text{W}}$$

Exercise   If we doubled the applied voltage to the heater, what would happen to the current and power?

Answer   The current would double, and the power would quadruple.

High-voltage power lines in Austria. (*Index/Stock International*)

EXAMPLE 17.7   Electrical Rating of a Lightbulb

A lightbulb is rated at 120 V and 75 W. That is, its operating voltage is 120 V and it has a power rating of 75 W. The bulb is powered by a 120-V direct current power supply. Find the current in the bulb and its resistance.

Solution   Since we know that the power rating of the bulb is 75 W and the operating voltage is 120 V, we can use $P = IV$ to find the current:

$$I = \frac{P}{V} = \frac{75\ \text{W}}{120\ \text{V}} = \boxed{0.63\ \text{A}}$$

(Top) A small permanent magnet levitated above a disk of the superconductor $YBa_2Cu_3O_{7-\delta}$, which is at 77 K. *(Tony Stone/Worldwide) (Bottom)* This high-speed express train in Tokyo, Japan, levitates a few inches above the track and is capable of speeds exceeding 225 km/h. *(Joseph Brignolo/The Image Bank)*

Using Ohm's law, $V = IR$, the resistance is calculated to be

$$R = \frac{V}{I} = \frac{120 \text{ V}}{0.63 \text{ A}} = \boxed{190 \ \Omega}$$

Exercise   What would the resistance be in a lamp rated at 120 V and 100 W?

Answer   140 Ω.

*17.8

ENERGY CONVERSION IN HOUSEHOLD CIRCUITS

The heat generated when current passes through a resistive material is used in many common devices. Figure 17.9a is a cross-sectional view of the spiral heating element of an electric range. The material through which the current passes is surrounded by an insulating substance; this prevents the charge from flowing through the cook to the Earth when he or she touches the range. A material that is a good conductor of heat surrounds the insulator.

Figure 17.9b shows a common hair dryer, in which a fan blows air past heating coils. In this case the warm air is used to dry hair, but the same principle is used to dry clothes and to heat buildings.

In a steam iron (Fig. 17.9c), a heating coil warms the bottom of the iron and simultaneously turns water to steam, which is sprayed from jets in the bottom of the iron.

Regardless of the ways in which you use electrical energy in your home, you ultimately must pay for it or risk having your power turned off. The unit of energy used by electric companies to calculate consumption, the **kilowatt-hour**, is defined in terms of the unit of power. One kilowatt-hour (kWh) is the energy

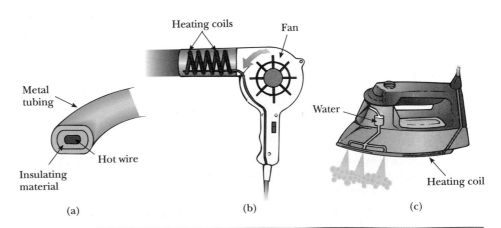

FIGURE 17.9
(a) A cross-section of the heating element used in an electric range. (b) In a hair dryer, air is warmed by being blown from a fan past the heating coils. (c) In a steam iron, water is turned into steam by heat from a heating coil.

converted or consumed in 1 h at the constant rate of 1 kW. It has the numerical value

$$1 \text{ kWh} = (10^3 \text{ W})(3600 \text{ s}) = 3.60 \times 10^6 \text{ J} \qquad \textbf{[17.11]}$$

The kilowatt-hour is a unit of energy

On an electric bill, the amount of electricity used in a given period is usually stated in multiples of kilowatt-hours.

EXAMPLE 17.8 The Cost of Operating a Lightbulb

How much does it cost to burn a 100-W lightbulb for 24 h if electric energy costs $0.080 per kilowatt-hour?

Solution   A 100-W lightbulb is equivalent to a 0.10-kW bulb. Since energy consumed equals power × time, the amount of energy you must pay for, expressed in kilowatt-hours, is

$$\text{Energy} = (0.10 \text{ kW})(24 \text{ h}) = 2.4 \text{ kWh}$$

If energy is purchased at $0.080 per kilowatt-hour, the 24-h cost is

$$\text{Cost} = (2.4 \text{ kWh})(\$0.080/\text{kWh}) = \quad \$0.19$$

This is a small amount of money, but when larger and more complex electric devices are used, the cost goes up rapidly.

Exercise   If electric energy costs $0.080/kWh, what does it cost to operate an electric oven, which operates at 20.0 A and 220 V, for 5.0 h?

Answer $1.80.

SUMMARY

The **electric current,** I, in a conductor is defined as

$$I \equiv \frac{\Delta Q}{\Delta t} \qquad \textbf{[17.1]}$$

where ΔQ is the charge that passes through a cross-section of the conductor in time Δt. The SI unit of current is the **ampere** (A); 1 A = 1 C/s. By convention, the direction of current is in the direction of flow of positive charge.

The current in a conductor is related to the motion of the charge carriers by

$$I = nqv_d A \qquad \textbf{[17.3]}$$

where n is the number of mobile charge carriers per unit volume, q is the charge on each carrier, v_d is the drift speed of the charges, and A is the cross-sectional area of the conductor.

The **resistance,** R, of a conductor is defined as the ratio of the potential difference across the conductor to the current:

$$R \equiv \frac{V}{I} \qquad \textbf{[17.4]}$$

The SI units of resistance are volts per ampere, or **ohms** (Ω); 1 Ω = 1 V/A.

Ohm's law provides a relationship between the voltage drop across a conductor (V), the current through it (I), and the resistance of the conductor (R):

$$V = IR \qquad\qquad \textbf{[17.5]}$$

If a conductor has length l and cross-sectional area A, its **resistance** is

$$R = \rho \frac{l}{A} \qquad\qquad \textbf{[17.6]}$$

where ρ is an intrinsic property of the conductor called the **electrical resistivity.** The SI unit of resistivity is the **ohm-meter** ($\Omega \cdot$m).

The resistivity of a conductor varies with temperature over a limited temperature range, according to the expression

$$\rho = \rho_0[1 + \alpha(T - T_0)] \qquad\qquad \textbf{[17.7]}$$

where α is the **temperature coefficient of resistivity** and ρ_0 is the resistivity at some reference temperature, T_0 (usually taken to be 20°C).

The resistance of a conductor varies with temperature according to the expression

$$R = R_0[1 + \alpha(T - T_0)] \qquad\qquad \textbf{[17.8]}$$

If a potential difference, V, is maintained across a resistor, the **power,** or rate at which energy is supplied to the resistor, is

$$P = IV \qquad\qquad \textbf{[17.9]}$$

Since the potential difference across a resistor is $V = IR$, the power dissipated by a resistor can be expressed as

$$P = I^2R = \frac{V^2}{R} \qquad\qquad \textbf{[17.10]}$$

A **kilowatt-hour** is the amount of energy converted or consumed in one hour by a device supplied with power at the rate of 1 kW. This is equivalent to

$$1 \text{ kWh} = 3.60 \times 10^6 \text{ J} \qquad\qquad \textbf{[17.11]}$$

ADDITIONAL READING

M. Azbel et al., "Conduction Electrons in Metals," *Sci. American,* January 1973, p. 88.

R. J. Cava, "Superconductors Beyond 1-2-3," *Sci. American,* August 1990, p. 60.

H. Ehrenreich, "The Electrical Properties of Materials," *Sci. American,* September 1967, p. 194.

R. M. Hazen, *The Breakthrough: The Race for the Superconductor,* New York, Summit Books, 1988. An account of recent discoveries in high-temperature superconductivity.

L. P. Williams, "André-Marie Ampère," *Sci. American,* January 1989, p. 90.

A. M. Wolsky, R. F. Giese, and E. J. Daniels, "The New Superconductors: Prospects for Applications," *Sci. American,* February 1989, p. 60.

CONCEPTUAL QUESTIONS

Example Two wires, A and B, are made of the same metal and have equal lengths, but the resistance of wire A is three times that of wire B. What is the ratio of their cross-sectional areas? How do their radii compare?
Reasoning Since $R = \rho l/A$, the ratio of their resistances is given by $R_1/R_2 = A_2/A_1$. Hence the ratio of their areas is three. That is, the area of wire B is three times that of wire A. The radius of wire B is $\sqrt{3}$ times the radius of wire A.
Example When the voltage across a certain conductor is doubled, the current is observed to triple. What can you conclude about the conductor?
Reasoning Since the current is not proportional to the applied voltage, the conductor does not obey Ohm's law. This occurs, for example, in an electronic device known as a diode.
Example Two lightbulbs operate from 120 V, but one has a power rating of 25 W and the other has a power rating of 100 W. Which bulb has the higher resistance? Which carries the greater current?
Reasoning To analyze this situation, it is important to note that the voltage V across each lightbulb is the same. Since the power is $P = V^2/R$, the 25-W bulb would have the higher resistance. Furthermore, since $P = IV$, we see that the 100-W bulb carries the greater current.
Example Two conductors of the same length and radius are connected across the same voltage source. One conductor has twice the resistance of the other. Which conductor dissipates more power?
Reasoning The voltage (potential difference) across each conductor is the same. Since the power dissipated in a conductor is $P = V^2/R$, the conductor with the lower resistance will dissipate more power.
Example If you were to design an electric heater using nichrome wire as the heating element, what parameters of the wire would you vary to meet a specific power output?
Reasoning Having decided on the power desired (say 1000 W), choose the operating voltage of the heater (say 120 V). This determines the current ($I = P/V = 8.33$ A) and required resistance ($R = V/I = 14.4\ \Omega$). The material and operating temperature determine the resistivity of the heater, so what remains is to choose a length and cross-sectional area for the heater wire to satisfy $R = \rho l/A$. If you choose the area A to be very small, the wire will be fragile; if A is large, you will have to buy more material.
Example Car batteries are often rated in ampere-hours. Does this designate the amount of current, power, energy, or charge that can be drawn from the battery?
Reasoning Because an ampere is a coulomb per second, an ampere-hour is equal to 3600 C. This is the amount of charge that the battery can "push" through itself, increasing the energy of each bit of this charge by the advertised voltage.

1. In an analogy between traffic flow and electrical current, what would correspond to the charge, Q? What would correspond to the current, I?
2. Suppose you charge a pocket comb by friction and then walk around the room carrying the comb. Are you producing an electric current?
3. We have seen that an electric field must exist inside a conductor that carries a current. How is this possible in view of the fact that, in our study of electrostatics, we concluded that the electric field inside a conductor must be zero?
4. What is the difference between the drift speed of an electron and its velocity?
5. If charges flow very slowly through a metal, why doesn't it take several hours for a light to come on after you throw a switch?
6. What factors affect the resistance of a conductor?
7. What is the difference between resistance and resistivity?
8. Use the atomic theory of matter to explain why the resistance of a material should increase as its temperature increases.
9. How does resistance change with temperature in copper and in silicon? Why are they different?
10. To reduce static electricity, a certain material is to be sprayed on a nylon rug. Describe the necessary electrical properties of this material.
11. Edison's original lightbulb contained a carbon filament. How should carbon be described electrically?
12. When an incandescent lamp burns out, it typically does so just after being switched on. Why?
13. Some homes have light dimmers that are operated by rotation of a knob. What is being changed in the electric circuit when the knob is rotated?
14. In the water analogy for an electric circuit, what corresponds to the power supply, resistor, charge, and potential difference?
15. What single experimental requirement makes superconducting devices expensive to operate? In principle, can this limitation be overcome?
16. If materials could be produced that would be superconducting at room temperature, list some ways in which such materials could benefit mankind.
17. A heating pad for sore, aching muscles carries a label indicating the proper voltage and current. How are these specifications related to the amount of heat produced by the heating pad? What would happen if the heating pad were accidentally plugged into the wrong voltage? Consider both cases in which the voltage is greater and cases in which it is less than the proper voltage.

18. If electrical power is transmitted over long distances, the resistance of the wires becomes significant. Why? Which mode of transmission would result in less energy loss—high current and low voltage or low current and high voltage? Discuss.

19. A typical monthly utility rate structure might be: $1.60 for the first 16 kWh, 7.05 cents/kWh for the next 34 kWh used, 5.02 cents/kWh for the next 50 kWh, 3.25 cents/kWh for the next 100 kWh, 2.95 cents/kWh for the next 200 kWh, and 2.35 cents/kWh for all in excess of 400 kWh. Based on these rates, what would be the charge for 527 kWh? From the standpoint of encouraging resource conservation, what is wrong with this pricing method?

20. Suppose a uniform wire of resistance R is stretched uniformly to three times its original length. What is its new resistance, assuming that its density and resistivity remain constant?

21. Two conductors made of the same material are connected across a common potential difference. Conductor A has twice the diameter and twice the length of conductor B. What is the ratio of the power delivered to the two conductors?

22. If electricity costs $0.080/kWh, estimate how much it costs a person to dry her hair with a 1500-W blow dryer during a year's time.

23. Two conductors of the same length and same material are connected in series (one after the other) across a difference in potential. The wires have different cross-sectional areas. (a) When the wires are in series, their currents are the same. Does the wire of greater cross-sectional area dissipate more or less power than the smaller wire? (b) Repeat the exercise with the wires connected in parallel (so that the voltages across them are the same).

PROBLEMS

Section 17.1   Electric Current

Section 17.2   Current and Drift Speed

1. The compressor on an air conditioner draws 90 A when it starts up. If the start-up time is about 0.50 s, how much charge passes a cross-sectional area of the circuit in this time?

2. A total charge of 6.0 mC passes through a cross-sectional area of a wire in 2.0 s. What is the current in the wire?

3. In the Bohr model of the hydrogen atom, an electron in the lowest energy state follows a circular path 5.29×10^{-11} m from the proton. (a) Show that the speed of the electron is 2.19×10^6 m/s. (b) What is the effective current associated with this orbiting electron?

4. A single positive point charge of 1.67 μC travels with a speed of 80.0 km/s in a circular orbit of radius 2.15 m. Determine the time-averaged current corresponding to the circulating charge.

5. In a particular television picture tube, the measured beam current is 60.0 μA. How many electrons strike the screen every second?

6. If a current of 80.0 mA exists in a metal wire, how many electrons flow past a given cross-section of the wire in 10.0 min? Sketch the directions of the current and the electrons' motion.

7. If 3.25×10^{-3} kg of gold is deposited on the negative electrode of an electrolytic cell in a period of 2.78 h, what is the current through the cell in this period? Assume that the gold ions carry one elementary unit of positive charge.

8. A teapot with a surface area of 700 cm² is to be silver plated. It is attached to the negative elec-trode of an electrolytic cell containing silver nitrate ($Ag^+NO_3^-$). If the cell is powered by a 12.0-V battery and has a resistance of 1.80 Ω, how long does it take to build up a 0.133-mm layer of silver on the teapot? (Density of silver = 10.5×10^3 kg/m³.)

9. Calculate the number of free electrons per cubic meter for gold, assuming one free electron per atom. (Density of gold = 19.3×10^3 kg/m³.)

10. An aluminum wire with a cross-sectional area of 4.0×10^{-6} m² carries a current of 5.0 A. Find the drift speed of the electrons in the wire. The density of aluminum is 2.7 g/cm³. (Assume that one electron is supplied by each atom.)

11. A 200-km-long high-voltage transmission line 2.0 cm in diameter carries a steady current of 1000 A. If the conductor is copper with a free charge density of 8.5×10^{28} electrons per cubic meter, how long (in years) does it take one electron to travel the full length of the cable?

12. A typical aluminum wire has about 1.5×10^{20} free electrons in every centimeter of its length, and the drift speed of these electrons is about 0.03 cm/s. (a) How many electrons pass through a cross-sectional area of the wire every second? (b) What is the current in the wire?

Section 17.3   Resistance and Ohm's Law

Section 17.4   Resistivity

13. (a) A 34.5-m length of copper wire at 20.0°C has a radius of 0.25 mm. If a potential difference of 9.0 V is applied across the length of the wire, determine the current in the wire. (b) If the wire is

heated to 30.0°C, while the 9.0-V potential difference is maintained, what is the resulting current in the wire?

14. An electrician finds that a 1.0-m length of a certain type of wire has a resistance of 0.20 Ω. What is the total resistance of the 120 m of this wire he plans to use to wire a house?

15. Calculate the diameter of a 2.0-cm length of tungsten filament in a small lightbulb if its resistance is 0.050 Ω.

16. Eighteen-gauge wire has a diameter of 1.024 mm. Calculate the resistance of 15 m of 18-gauge copper wire at 20°C.

17. A person notices a mild shock if the current along a path through the thumb and index finger exceeds 80 μA. Compare the maximum allowable voltage without shock across the thumb and index finger with a dry-skin resistance of 4.0×10^5 Ω and a wet-skin resistance of 2000 Ω.

18. A wire of uniform cross-section is stretched along a meter stick, and a potential difference of 0.50 V is maintained between the 0-cm and 100-cm marks. How far apart on the wire are two points that differ in potential by 80.0 mV?

19. A typical color television draws about 2.5 A when connected to a 120-V source. What is the effective resistance of the T.V. set?

20. When operating at 120 V, a resistor carries a current of 0.50 A. What current is carried if (a) the operating voltage is lowered to 90 V? (b) the voltage is raised to 130 V?

21. A metal wire, 50.0 m long and 2.00 mm in diameter, is connected to a source with a potential difference of 9.11 V, and the current is found to be 36.0 A. Identify the metal, using Table 17.1.

22. (a) Calculate the resistance of a 200-cm-long piece of nichrome wire with a cross-sectional area of 0.020 cm². (b) What current does the wire carry when connected to a 3.0-V flashlight battery?

23. Suppose you wish to fabricate a uniform wire out of 1 g of copper. If the wire is to have a resistance of 0.50 Ω and all of the copper is to be used, what must be the (a) length and (b) diameter of this wire?

24. The heating element of a toaster is made of nichrome wire 1.0 mm in diameter. If the toaster has a resistance of 8.5 Ω, how long is the wire?

25. A potential difference of 12 V is found to produce a current of 0.40 A in a 3.2-m length of wire with a uniform radius of 0.40 cm. What is (a) the resistivity of the wire? (b) the resistance of the wire?

26. A 2.0-m piece of iron wire carries a current of 0.25 A when connected to a 6.0-V battery. What length of gold wire, with the same radius, would carry the same current when connected to the battery?

27. A rectangular block of copper has sides of length 10 cm, 20 cm, and 40 cm. If the block is connected to a 6-V source across opposite faces of the rectangular block, what are (a) the maximum current and (b) minimum current that can be carried?

28. A battery establishes a uniform electric field of 100 V/m in a wire of radius 0.5 cm. If the current in the wire is 5 A, what is the resistivity of the material?

Section 17.5 Temperature Variation of Resistance

29. If a silver wire has a resistance of 10.0 Ω at 20.0°C, what resistance does it have at 40.0°C? Neglect any change in length or cross-sectional area resulting from the change in temperature.

30. The copper wire used in a house has a cross-sectional area of 3.00 mm². If 10.0 m of this wire is used to wire a circuit in the house at 20°C, find the resistance of the wire at temperatures of (a) 30.0°C and (b) 10.0°C.

31. At 40.0°C, the resistance of a segment of gold wire is 100.0 Ω. When the wire is placed in a liquid bath, the resistance decreases to 97.0 Ω. What is the temperature of the bath?

32. At 20°C the carbon resistor in an electric circuit, connected to a 5-V battery, has a resistance of 200 Ω. What is the current in the circuit when the temperature of the carbon rises to 80°C?

33. A wire 3.00 m long and 0.450 mm² in cross-sectional area has a resistance of 41 Ω at 20°C. If its resistance increases to 41.4 Ω at 29.0°C, what is the temperature coefficient of resistivity?

34. A platinum resistance thermometer has resistances of 200.0 Ω when placed in a 0°C ice bath and 253.8 Ω when immersed in a crucible containing melting potassium. What is the melting point of potassium? (*Hint:* First determine the resistance of the platinum resistance thermometer at room temperature, 20°C.)

35. In one form of plethysmograph (a device for measuring volume), a rubber capillary tube with an inside diameter of 1.00 mm is filled with mercury at 20°C. The resistance of the mercury is measured with the aid of electrodes sealed into the ends of the tube. If 100.00 cm of the tube is wound in a spiral around a patient's upper arm, the blood flow during a heartbeat causes the arm to expand, stretching the tube to a length of 100.04 cm. From this observation (assuming cylindrical symmetry) you can find the change in volume, which gives an indication of blood flow. (a) Calculate the resistance of the mercury. (b) Calculate the fractional change in resistance during the heartbeat. (*Hint:* The fraction by which the cross-sectional area of the mercury thread decreases is the fraction by

which the length increases, since the volume of mercury is constant.) $\rho_{Hg} = 9.4 \times 10^{-7}\ \Omega \cdot m$.

36. A certain lightbulb has a tungsten filament with a resistance of $19.0\ \Omega$ when cold and $140\ \Omega$ when hot. Assume that Equation 17.8 can be used over the large temperature range involved here, and find the temperature of the filament when it is hot. Assume an initial temperature of $20°C$.

37. A 100-cm-long copper wire 0.50 cm in radius has a potential difference across it sufficient to produce a current of 3.0 A at $20°C$. (a) What is the potential difference? (b) If the temperature of the wire is increased to $200°C$, what potential difference is now required to produce a current of 3.0 A?

38. The temperature coefficients of resistivity in Table 17.1 are at $20°C$. What would they be at $0°C$? (*Hint:* The temperature coefficient of resistivity at $20°C$ satisfies $R = R_0[1 + \alpha(T - T_0)]$ where R_0 is the resistance of the material at $T_0 = 20°C$. The temperature coefficient of resistivity, α', at $0°C$ must satisfy $R = R'(1 + \alpha' T)$, where R' is the resistance of the material at $0°C$.)

Section 17.7 Electrical Energy and Power

39. (a) Determine the resistance of a 100-W lightbulb in your home. (b) Assuming that the filament is tungsten and has a cross-sectional area of $0.010\ mm^2$, determine the length of the wire inside the bulb when the bulb is turned on. (c) Why do you think the wire inside the bulb is tightly coiled? (d) If the temperature of the tungsten wire is $2600°C$ when the bulb is turned on, what is the length of the wire when the bulb is turned off and has cooled to $20°C$? (See Chapter 10, and use $4.5 \times 10^{-6}/°C$ as the coefficient of linear expansion for tungsten.)

40. The power supplied to a typical black-and-white television set is 90 W when the set is connected to 120 V. (a) How much electric energy does this set consume in one hour? (b) A color television set draws about 2.5 A when connected to 120 V. How much time is required for it to consume the same energy as the black-and-white model consumes in one hour?

41. A high-voltage transmission line carries 1000 A at 700 kV for a distance of 160 km. If the resistance in the wire is $0.31\ \Omega/km$, what is the power dissipated due to resistive losses?

42. The potential difference across a resting neuron is about 75 mV, and the current through it is about $200\ \mu A$. How much power does the neuron release?

43. The tungsten heating element in a 1500-W heater is 3.00 m long, and the resistor is to be connected to a 120-V source. What is the cross-sectional area of the wire?

44. Determine the loss of electrical power per meter of a copper wire 2.00 mm in diameter if a current of 40.0 A exists in the wire.

45. Suppose that a voltage surge produces 140 V for a moment. By what percentage will the output of a 120-V, 100-W lightbulb increase, assuming its resistance does not change?

46. How many 100-W lightbulbs can you use in a 120-V circuit without tripping a 15-A circuit breaker? (The bulbs are connected in parallel.)

47. A copper cable is designed to carry a current of 300 A with a power loss of 2.00 W/m. What is the required radius of this cable?

48. What is the required resistance of an immersion heater that will increase the temperature of 1.5 kg of water from $10.0°C$ to $50.0°C$ in 10.0 min while operating at 120 V?

49. In a hydroelectric installation, a turbine delivers 2000 hp to a generator, which in turn converts 90% of the mechanical energy to electrical energy. Under these conditions, what current does the generator deliver at a potential difference of 3000 V?

50. In a certain stereo system, each speaker has a resistance of $4.00\ \Omega$. The system is rated at 60.0 W in each channel. Each speaker circuit includes a fuse rated at a maximum current of 4.00 A. Is this system adequately protected against overload?

Section 17.8 Energy Conversion in Household Circuits

51. How much does it cost to watch a complete 21-hour-long World Series on a 90.0-W black-and-white television set? Assume that electricity costs $0.0700/kWh.

52. A certain household uses an average of 2500 kWh of energy per month. If coal supplies 7.0×10^6 cal/kg, how much coal must be burned in a 40% efficient generator to supply this much energy?

53. An 11-W energy-efficient fluorescent lamp is designed to produce the same illumination as a conventional 40-W lamp. How much does the energy-efficient lamp save during 100 hours of use? Assume a cost of $0.080/kWh for electrical energy.

54. A small motor draws a current of 1.75 A from a 120-V line. The output power of the motor is 0.20 hp. (a) At a rate of $0.060/kWh, what is the cost of operating the motor for 4.0 h? (b) What is the efficiency of the motor?

55. An electric resistance heater is to deliver 1500 kcal/h to a room using 110-V electricity. If fuses come in 10-A, 20-A, and 30-A sizes, what is the smallest fuse that can safely be used in the heater circuit?

56. An advertisement for a vaporizer that operates on 110 V states that it converts 400 cm^3 of water to

steam in an hour. Ignoring heat losses, determine the resistance of the vaporizer.

57. A 110-V motor produces 2.5 hp of mechanical power. If this motor is 90% efficient in converting electrical power to mechanical power, find (a) the current drawn by the motor and (b) the total electrical energy used by this motor in running for one hour. (c) If electrical energy costs $0.080/kWh, what does it cost to run the motor for this hour?

58. The heating coil of a hot water heater has a resistance of 20 Ω and operates at 210 V. If electrical energy costs $0.080/kWh, what does it cost to raise the 200 kg of water in the tank from 15°C to 80°C? (See Chapter 11.)

ADDITIONAL PROBLEMS

59. Storage batteries are often rated in terms of the amounts of charge they can deliver. How much charge can a 90-ampere-hour battery deliver?

60. A steam iron draws 6.0 A from a 120-V line. (a) How many joules of thermal energy are produced in 20 min? (b) How much does it cost, at $0.080/kWh, to run the steam iron for 20 min?

61. An x-ray tube used for cancer therapy operates at 4.0 MV, with a beam current of 25 mA striking the metal target. The power in this beam is transferred to a stream of water flowing through holes drilled in the target. What rate of flow, in kilograms per second, is needed if the temperature rise of the water is not to exceed 50°C?

62. The headlights on a car are rated at 80 W. If they are connected to a fully charged 90-A·h, 12-V battery, how long does it take the battery to completely discharge?

63. A particular wire has a resistivity of 3.0×10^{-8} $\Omega \cdot m$ and a cross-sectional area of 4.0×10^{-6} m². A length of this wire is to be used as a resistor that will develop 48 W of power when connected across a 20-V battery. What length of wire is required?

64. Birds resting on high-voltage power lines are a common sight. The copper wire on which a bird stands is 2.2 cm in radius and carries a current of 50 A. If the bird's feet are 4.0 cm apart, calculate the potential difference across its body.

65. A length of metal wire has a radius of 5.00×10^{-3} m and a resistance of 0.100 Ω. When the potential difference across the wire is 15.0 V, the electron drift speed is found to be 3.17×10^{-4} m/s. Based on these data, calculate the density of free electrons in the wire.

66. In some television sets built in the 1970s, the picture-tube filament was kept warm at all times, using 10.0 W of power, to provide "instant-on"

viewing. If such a set is idle 16 h/day, (a) calculate the dollar cost per year, at $0.080/kWh, for this convenience; (b) calculate the ecological cost, in kg/year of bituminous coal (fuel value 30.0 MJ/kg) burned in a 35% efficient power plant.

67. A small sphere that carries a charge of 8.00 nC is whirled in a circle at the end of an insulating string. The rotation frequency is 100π rad/s. What average current does this rotating charge represent?

68. It is estimated that in the United States (population 250 million) there is one electric clock per person, with each clock using energy at a rate of 2.5 W. To supply this energy, about how many metric tons of coal are burned per hour in coal-fired electric generating plants that are, on the average, 25% efficient? The heat of combustion for coal is 33 MJ/kg.

69. (a) A 115-g mass of aluminum is formed into a right circular cylinder, shaped so that its diameter equals its height. Calculate the resistance between the top and bottom faces of the cylinder at 20°C. (b) Calculate the resistance between opposite faces if the same mass of aluminum is formed into a cube.

70. A resistor is constructed by forming a material of resistivity 3.5×10^5 $\Omega \cdot m$ into the shape of a hollow cylinder of length 4.0 cm and inner and outer radii of 0.50 cm and 1.2 cm, respectively. In use, a potential difference is applied between the ends of the cylinder, producing a current parallel to the length of the cylinder. Find the resistance of the cylinder.

71. The current in a conductor varies in time as shown in Figure 17.10. (a) How many coulombs of charge pass through a cross-section of the conductor in the interval $t = 0$ to $t = 5.0$ s? (b) What constant current would transport the same total charge during the 5.0-s interval as does the actual current?

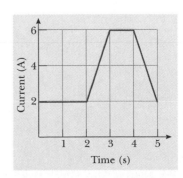

FIGURE 17.10 (Problem 71)

72. A wire of initial length L_0 and radius r_0 has a measured resistance of 1.0 Ω. The wire is drawn under tensile stress to a new uniform radius of $r = 0.25r_0$. What is the new resistance of the wire?

73. A 50.0-g sample of a conducting material is all that is available. The resistivity of the material is measured to be 11×10^{-8} $\Omega \cdot$m, and the density is 7.86 g/cm^3. The material is to be shaped into a wire that has a total resistance of 1.5 Ω. (a) What length is required? (b) What must be the diameter of the wire?

74. (a) A sheet of copper ($\rho = 1.7 \times 10^{-8}$ $\Omega \cdot$m) is 2.0 mm thick and has surface dimensions of 8.0 cm $\times$ 24 cm. If the long edges are joined to form a tube 24 cm in length, what is the resistance between the ends? (b) What mass of copper is required to manufacture a 1500-m-long spool of copper cable with a total resistance of 4.5 Ω?

75. When a straight wire is heated, its resistance changes according to Equation 17.8:

$$R = R_0[1 + \alpha(T - T_0)]$$

where α is the temperature coefficient of resistivity. (a) Show that a more precise result, which includes the fact that the length and area of a wire change when it is heated, is

$$R = \frac{R_0[1 + \alpha(T - T_0)][1 + \alpha'(T - T_0)]}{[1 + 2\alpha'(T - T_0)]}$$

where α' is the coefficient of linear expansion (see Chapter 10). (b) Compare these two results for a 2.00-m-long copper wire of radius 0.100 mm, both at 20.0°C and heated to 100.0°C.

"I'M AFRAID MR. EDISON IS TOO BUSY TO DISCUSS YOUR IDEA WITH YOU AT THIS TIME. PLEASE CALL AGAIN, MR. FLUORESCENT."

VOLTAGE MEASUREMENTS IN MEDICINE

Electrocardiograms

Every action involving the body's muscles is initiated by electrical activity. The voltages produced by muscular action in the heart are particularly important to physicians. Voltage pulses cause the heart to beat, and the waves of electrical excitation associated with the heart beat are conducted through the body via the body fluids. These voltage pulses are large enough to be detected by suitable monitoring equipment attached to the skin. Standard electric devices can be used to record these voltage pulses because the amplitude of a typical pulse associated with heart activity is of the order of 1 mV. These voltage pulses are recorded on an instrument called an **electrocardiograph,** and the pattern recorded by this instrument is called an **electrocardiogram (EKG).** In order to understand the information contained in an EKG pattern, it is useful to first describe the underlying principles concerning electrical activity in the heart.

The right atrium of the heart contains a specialized set of muscle fibers called the SA (sinoatrial) node, which initiate the heartbeat (Fig. 1). Electric impulses that originate in these fibers gradually spread from cell to cell throughout the right and left atrial muscles, causing them to contract. The pulse that passes through the muscle cells is often called a *depolarization wave* because of its effect on individual cells. If an individual muscle cell were examined, an electric charge distribution would be found on its surface, as shown in Figure 2a. (See Living Cells as Capacitors in Ch. 16 for an explanation of how this charge distribution arises.) The impulse generated by the SA node momentarily changes the cell's charge distribution to that shown in Figure 2b. The positively charged ions on the surface of the cell are temporarily able to diffuse through the membrane wall so that the cell attains an excess positive charge on its in-

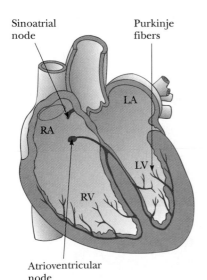

Sinoatrial node

Purkinje fibers

LA

RA

LV

RV

Atrioventricular node

FIGURE 1
The electrical conduction system of the human heart. (RA: right atrium; LA: left atrium; RV: right ventricle; LV: left ventricle.)

557

FIGURE 2
(a) Charge distribution of a muscle cell in the atrium before a depolarization wave has passed through the cell. (b) Charge distribution as the wave passes.

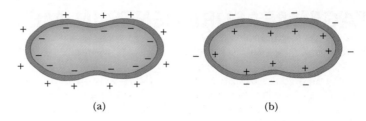

(a)   (b)

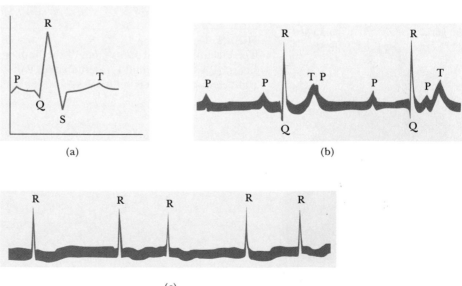

FIGURE 3

An EKG response for a normal heart.

side surface. As the depolarization wave travels from cell to cell throughout the atria, the cells recover to the charge distribution shown in Figure 2a. When the impulse reaches the AV (atrioventricular) node (Fig. 1), the muscles of the atria begin to relax, and the pulse is directed by the AV node to the ventricular muscles. The muscles of the ventricles then contract as the depolarization wave spreads through the ventricles along a group of fibers called the *Purkinje fibers*. The ventricles then relax after the pulse has passed through. At this point, the SA node is again triggered and the cycle is repeated.

A sketch of the electrical activity registered on an EKG for one beat of a normal heart is shown in Figure 3. The pulse indicated by P occurs just before the atria begin to contract. The QRS pulse occurs in the ventricles just before they contract, and the T pulse occurs when the cells in the ventricles begin to recover. EKGs for an abnormal heart are shown in Figure 4. The QRS portion of the pattern shown in Figure 4a is wider than normal. This indicates that the patient may have an enlarged heart. Figure 4b indicates that there is no relationship between the P pulse and the QRS pulse. This suggests a blockage in the electrical conduction path between the SA and AV nodes. This can occur when the atria and ventricles beat independently. Finally, Figure 4c shows a situation in which there is no P pulse and an irregular spacing between the QRS pulses. This is symptomatic of irregular atrial contraction, which is called fibrillation. In this situation, the atrial and ventricular contractions are irregular.

FIGURE 4
Abnormal EKGs.

Electroencephalography

The electrical activity of the brain can be measured with an instrument called an **electroencephalograph** in much the same way that an electrocardiograph measures the electrical activity of the heart. The voltage pattern measured by an electroencephalograph is referred to as an **electroencephalogram (EEG)**. An EEG pattern is recorded by placing electrodes on the patient's scalp. While an EKG voltage pulse is typically 1 mV, a typical voltage amplitude associated with brain activity is only a few *micro*volts and is therefore more difficult to measure.

The EEGs in Figure 5 represent the brain wave pattern of a patient awake and then in various stages of sleep. In Figure 5b the patient begins to fall into a light sleep, and in Figure 5d the patient is in a deep sleep. Figure 5c is the EEG during a type of sleep called REM (rapid eye movement) sleep, which occurs approximately every 2 h. In this stage, the brain activity is quite similar to that of the patient while awake. It is interesting to note that a person in this stage of sleep is extremely difficult to awaken. Apparently, REM sleep is necessary for psychological well-being. Anyone who is deprived of this stage of sleep for an extended period of time becomes extremely fatigued and irritable.

The EEG is an extremely important diagnostic tool for detecting epilepsy, brain tumors, brain hemorrhage, meningitis, and so forth. For example, the EEG pattern of a patient suffering an epileptic seizure would show very little structure, indicating that the brain activity is greatly reduced.

Brain waves are often discussed in terms of their frequencies. Frequencies of about 10 Hz are referred to as *alpha waves*, frequencies between 10 Hz and 60 Hz are called *beta waves*, and those below 10 Hz are called *delta waves*. As a person falls asleep, the frequency of brain wave activity generally decreases. For example, the brain wave frequency for a person in a very deep sleep may drop as low as 1 or 2 Hz.

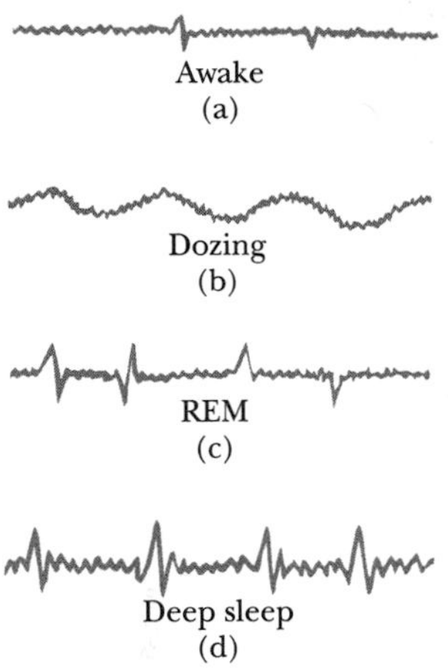

Awake
(a)

Dozing
(b)

REM
(c)

Deep sleep
(d)

FIGURE 5
Brain waves from an individual in various stages of sleep.

DAVID MARKOWITZ   University of Connecticut

SUPERCONDUCTIVITY

At a temperature typically between 1 and 10 K, many metals and alloys undergo a change of phase. At temperatures above this range they exhibit ordinary resistance R to electric current (Ohm's law), but below ≈ 10 K the resistance vanishes. The high-T phase is called the normal state of the electrons, and the low-T phase is called the superconducting state. When you plot R vs. T, you find a sharp transition at a critical temperature T_c (Fig. 1). The purer the material, the sharper the transition.

The discovery of this phenomenon came in 1911 at the laboratory of Heike Kamerlingh Onnes in Leiden, Holland. Several aspects of this amazing behavior were realized early on. First, there is a close analogy to the phase transition from water to ice. Pure water freezes at $0°C$, but impure water is still liquid below $0°C$. The clearest property showing the sharp liquid-solid transition for water is density, since water abruptly expands upon freezing. The higher-T phase has the higher entropy (liquid water), and the lower-T phase has lower entropy (ice). The same idea applies to normal and superconducting electrons.

Second, any unusual electrical property is accompanied by an equally unusual magnetic property because the same mobile charges are responsible for both. In the 1930s it was established that at T below T_c, a moderate magnetic flux is expelled from the interior of a superconductor. A pure superconductor expels 100% of the flux except for a narrow surface layer. However, a large magnetic flux destroys superconductivity, and the metal reverts to the normal state even at T below T_c. The large flux penetrates the metal in the normal way.

By the 1950s numerous electromagnetic and thermal properties of the superconductivity state were known. They were explained by the theory of superconductivity due to John Bardeen, Leon Cooper, and Robert Schrieffer (''the BCS theory'') at the University of Illinois and by mathematical and conceptual extensions due to many others.

Explanation

The BCS theory treats the conduction electrons as an ''electron fluid.'' This fluid is capable of freezing at a low enough T, but the freezing process is not the same as for water. The orderly arrangement of molecules in ice takes place in what we call ''position space.'' This is the space we live in and look at. We place axes labeled x, y, and z in this space and use them to designate the position of any particle. A regular arrangement of positions is a low-entropy state.

An alternative transition for a fluid takes place in an abstract space called ''velocity space,'' the space conventionally used to describe motions of particles (or motions of the centers of mass of bodies). We place axes labeled v_x, v_y, and v_z in this space and use them to designate the velocity of any particle. A traffic engineer could study the distribution of automobile velocities on a highway. This would show the fraction of drivers obeying the posted speed limits. Automobiles travel at a variety of speeds. In my state 55 mph is the limit on an interstate highway. Some cars go much faster than that. It is inconceivable that all cars would go at 55, even tolerating slight variations and accepting 54 or 56 mph. If they did (and stayed in their lanes), the number of accidents would be greatly reduced.

What is inconceivable for automobiles is achievable by electrons that conspire to form a superconducting phase. This is the state of the electron fluid in which all electrons do exactly the same thing. What they do is form bound pairs, and all pairs behave in exactly the same way. They tolerate no variation at all, and they never have an accident. The process of forming this state from previously unpaired electrons is "freezing in velocity space." The regular arrangement of velocities is a low-entropy (hence, low-T) state.

Some questions arise. Why do electrons want to form identical pairs? Answer: Pairing lowers their potential energy of interaction. They pair for the same reason a released ball falls to the Earth. Why are electrons able to form identical pairs? Answer: Only the fact that modern physics governs electrons allows them to do that. In classical physics properties of motion are continuous, but in modern physics they are quantized. An analogy may help. Throw a number of darts at a target. They will not all hit the same point because there are an infinite number of nearby points, but they may all hit the same ring of the target. The rings are quantized; neighboring rings are clearly separated.

Classifications of Superconductors

There are different classifications of superconductors, all are electron systems but not all are regarded as fluids. Still, it appears that they all rely on the pairing mechanism, and it is pairs that carry current without resistance.

The simplest class of superconductors is elemental metals, and the next simplest is alloys formed from them. Apparently, any structure is permissible to a metallic superconductor as long as the material satisfies two criteria: large enough number of valence electrons per atom and strong enough interaction between electrons and the polarizable lattice. Since these properties are widespread among elements in the periodic table, there are many metallic superconductors and many alloy superconductors.

Some materials are superconducting when they are amorphous but not when they are crystalline. Others are semiconducting in one crystalline form, metallic superconductors in another. This suggests that structure influences electronic behavior. Then too, some materials have strikingly different superconducting properties when they are fashioned into thin films rather than bars or wires. A film that is thin enough constrains electrons to move in two dimensions instead of the usual three, implying that dimensionality affects electronic behavior.

The Present Scene

Physicists greet a new record for critical temperature as eagerly as sports fans greet new athletic records. Often the record is pushed gradually. Occasionally it jumps as a result of a technological breakthrough, analogous to the effect the fiberglass pole had on the polevault.

Gradually, critical temperatures have risen past 130 K. Experimentalists have tried many metal and oxide combinations, numerous structure types, and various dimensionalities, including films and filaments. In 1993 a layered compound containing mercury, alkaline earths, and copper oxide of different configurations achieved 133.5 K at atmospheric pressure and 164 K at very high pressure. These are the highest confirmed critical temperatures, and they came with heroic chemical and physical effort.

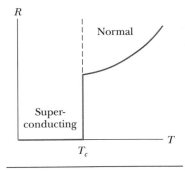

FIGURE 1
Resistance versus temperature for a pure superconductor. The metal behaves normally down to the critical temperature, T_c, below which the electrical resistance drops to zero.

Now for the technological leap. In December 1993, within days of each other, two French laboratories independently reported evidence suggesting that superconductivity may occur above 250 K, perhaps as high as 280 K, the temperature inside a home refrigerator or outside on a winter day. The research articles are technical; they are remarkable for what they omit as well as what they include. The articles fail to provide details of preparation and measurement as if the authors wish to keep some secrets. The best place to read about the work is Paul Grant's page in the British journal *Nature*, Vol. 367, 6 Jan. 94, p. 16. He describes the novel materials and methods of preparation and emphasizes the points of evidence and their inconclusive nature. Of greatest interest to us are a report by one group of a sharp drop in electrical resistance in a sample near 250 K and a report by the other of a sharp drop to zero resistance at 235 K in a markedly different sample. Extensions of the work suggest that T_c, if that is what is being observed, can be pushed to higher values. In Grant's metaphor, we may call the claims the new French Revolution.

Disappointingly, the superconductors with very high reported values of T_c are not superconducting at very low values of current, as well as at the anticipated higher currents. Electric current destroys superconductivity. The phenomenon abolishes itself. This means that the promise of this widespread material property to be technologically useful remains largely unfulfilled.

As an example, in a computer, photons and electrons can play the same roles. For example, the storage of one bit of information can be a tiny electric current or a tiny ray of light. As another example, the switching of information from one value to another can be a change in current or in illumination. It is not known which means will eventually be more reliable and economical. In this rivalry between electronics and photonics, high T_c superconductors may play the decisive role.

Questions and Problems

1. An electron inside a sample possesses values of energy, linear momentum, and "spin" angular momentum. No two electrons coincide in all these values. What scheme might electrons follow in their pairing to arrange that all pairs have the same values of these decisive properties?
2. What value of binding energy for an electron pair corresponds to temperatures of 1 K and 100 K? Express your answers in electron-volts and compare them with other energies of electrons or atoms you know.
3. What pairing condition or conditions would be ruined by a large magnetic flux penetrating a sample?
4. If the magnetic energy inside a sample exceeds the binding energy of all pairs in the sample, then superconductivity is destroyed. Magnetic energy "wins." From your answer to Question 2, find the critical value of magnetic field strength. Is it orbital or spin motion that is damaged first by an increasing magnetic field?
5. What pairing condition gives rise to a "supercurrent"? Why does a large electric current destroy superconductivity? In essence, the large current "self-destructs."
6. From your answer to Question 4, find the value of the "critical current density" J_c.

DIRECT CURRENT CIRCUITS

18

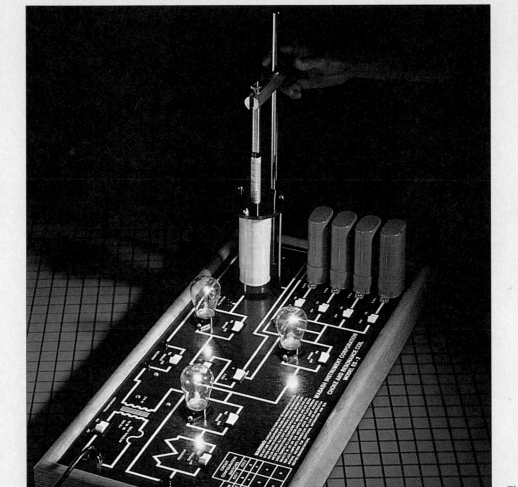

This versatile circuit enables the experimenter to examine the properties of circuit elements such as capacitors and resistors and their effects on circuit behavior.
(Courtesy of Central Scientific Company)

This chapter analyzes some simple circuits whose elements include batteries, resistors, and capacitors in varied combinations. Such analysis is simplified by the use of two rules known as Kirchhoff's rules, which follow from the principle of conservation of energy and the law of conservation of charge. Most of the circuits are assumed to be in *steady state,* which means that the currents are constant in magnitude and direction. We close the chapter with a discussion of circuits containing resistors and capacitors, in which current varies with time.

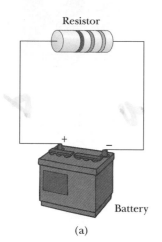

Resistor

Battery

(a)

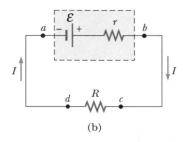

(b)

FIGURE 18.1

(a) A circuit consisting of a resistor connected to the terminals of a battery. (b) A circuit diagram of a source of emf, $\mathcal{E}$, of internal resistance r connected to an external resistor, R.

An assortment of batteries. (*Courtesy of Henry Leap and Jim Lehman*)

18.1
SOURCES OF emf

The source that maintains the constant current in a closed circuit is called a source of "emf."[1] Any devices (such as batteries and generators) that increase the potential energy of charges circulating in circuits are sources of emf. One can think of such a source as a "charge pump" that forces electrons to move in a direction opposite the electrostatic field inside the source. The emf, $\mathcal{E}$, of a source is the work done per unit charge, and hence the SI unit of emf is the volt.

Consider the circuit in Figure 18.1a, consisting of a battery connected to a resistor. We assume that the connecting wires have no resistance. If we neglect the internal resistance of the battery, the potential drop across the battery (the terminal voltage) equals the emf of the battery. However, because a real battery always has some internal resistance, r, the terminal voltage is not equal to the emf. The circuit of Figure 18.1a can be described schematically by the diagram in Figure 18.1b. The battery, represented by the dashed rectangle, consists of a source of emf, $\mathcal{E}$, in series with an internal resistance, r. Now imagine a positive charge moving from a to b in Figure 18.1b. As the charge passes from the negative to the positive terminal of the battery, the potential of the charge increases by $\mathcal{E}$. However, as the charge moves through the resistance, r, its potential decreases by the amount Ir, where I is the current in the circuit. Thus, the terminal voltage of the battery, $V = V_b - V_a$, is

$$V = \mathcal{E} - Ir \qquad [18.1]$$

Note from this expression that $\mathcal{E}$ is equal to the *terminal voltage when the current is zero,* called the open-circuit voltage. By inspecting Figure 18.1b, we see that the terminal voltage, V, must also equal the potential difference across the external resistance, R, often called the **load resistance;** that is, $V = IR$. Combining this with Equation 18.1, we see that

$$\mathcal{E} = IR + Ir \qquad [18.2]$$

Solving for the current gives

$$I = \frac{\mathcal{E}}{R + r}$$

This shows that the current in this simple circuit depends on both the resistance external to the battery and the internal resistance. If R is much greater than r, we can neglect r in our analysis and we do, for many circuits.

If we multiply Equation 18.2 by the current, I, we get

$$I\mathcal{E} = I^2R + I^2r$$

This equation tells us that the total power output of the source of emf, $I\mathcal{E}$, is converted to power that is dissipated as joule heat in the load resistance, I^2R, *plus* power that is dissipated in the internal resistance, I^2r. Again, if $r \ll R$, most of the power delivered by the battery is transferred to the load resistance.

Unless otherwise stated, we will assume in our examples and end-of-chapter problems that the internal resistance of a battery in a circuit is negligible.

---

[1] The term was originally an abbreviation for *electromotive force,* but emf is not really a force, so the long form is discouraged.

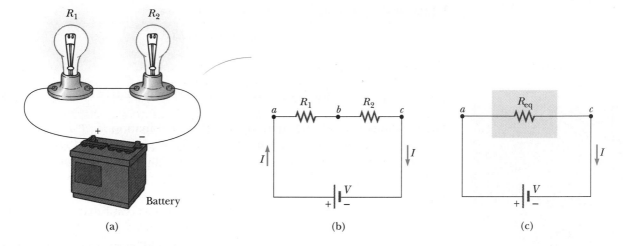

FIGURE 18.2
A series connection of two resistors, R_1 and R_2. The currents in the resistors are the same, and the equivalent resistance of the combination is given by $R_{eq} = R_1 + R_2$.

18.2
RESISTORS IN SERIES

The conservation laws are grand unifying principles that serve as bases for many outcomes in the study of physics. In particular, the next three sections of this text use two of these laws, the conservation of energy and the conservation of charge, as a framework. Rather than point out the places where these laws appear in our derivations, we will ask you, as an exercise, to identify their appearances yourself.

Figure 18.2 shows two resistors, R_1 and R_2, connected to a battery in a circuit called a series circuit, in which there is only one pathway for the current. Charges must pass through both resistors and the battery as they traverse the circuit. Hence, all charges in a series circuit must follow the same conducting path.

> Note that the currents through all resistors in a series circuit are the same, because any charge that flows through R_1 must also flow through R_2.

For a series connection of resistors, the current is the same in all the resistors

This is analogous to water flowing through a pipe with two constrictions, corresponding to R_1 and R_2. Whatever volume of water flows in one end in a given time interval must exit the opposite end.

Since the potential drop from a to b in Figure 18.2b equals IR_1 and the potential drop from b to c equals IR_2, the potential drop from a to c is

$$V = IR_1 + IR_2$$

Figure 18.2c shows an equivalent resistor, R_{eq}, that can replace the two resistors of the original circuit. Applying Ohm's law to this resistor, we have

$$V = IR_{eq}$$

Equating the preceding two expressions, we have

$$IR_{eq} = IR_1 + IR_2$$

or

$$R_{eq} = R_1 + R_2 \qquad \text{(series combination)} \qquad \text{[18.3]}$$

The resistance R_{eq} is equivalent to the series combination $R_1 + R_2$ in the sense that the circuit current is unchanged when R_{eq} replaces R_1 and R_2.

An extension of this analysis shows that the equivalent resistance of three or more resistors connected in series is

$$R_{eq} = R_1 + R_2 + R_3 + \cdots \qquad \text{(series combination)} \qquad \text{[18.4]}$$

From this, we see that *the equivalent resistance of a series combination of resistors is always greater than any individual resistance.*

Note that if the filament of one lightbulb in Figure 18.2 were to break, or "burn out," the circuit would no longer be complete (an open-circuit condition would exist) and the second bulb would also go out. Some Christmas-tree light sets (especially older ones) are connected in this way, and the task of determining which bulb is burned out is a tedious one.

In many circuits, fuses are used in series with other circuit elements for safety purposes. The conductor in the fuse is designed to melt and open the circuit at some maximum current, the value of which depends on the nature of the circuit. If a fuse were not used, excessive currents could damage circuit elements, overheat wires, and perhaps cause a fire. In modern home construction, circuit breakers are used in place of fuses. When the current in a circuit exceeds some value (typically 15 A), the circuit breaker acts as a switch and opens the circuit.

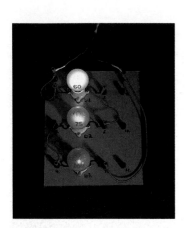

A series connection of three lamps, all rated at 120 V, with power ratings of 60 W, 75 W, and 150 W. Why do the intensities of the lamps differ? Which lamp has the greatest resistance? How would their relative intensities differ if they were connected in parallel? *(Courtesy of Henry Leap and Jim Lehman)*

EXAMPLE 18.1   Four Resistors in Series

Four resistors are arranged as shown in Figure 18.3a. Find (a) the equivalent resistance and (b) the current in the circuit if the emf of the battery is 6.0 V.

Solution   (a) The equivalent resistance is found from Equation 18.4:

$$R_{eq} = R_1 + R_2 + R_3 + R_4 = 2.0\ \Omega + 4.0\ \Omega + 5.0\ \Omega + 7.0\ \Omega = \boxed{18\ \Omega}$$

FIGURE 18.3
(Example 18.1)   (a) Four resistors connected in series. (b) The equivalent resistance of the circuit in (a).

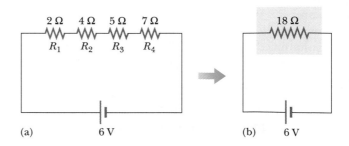

(a)   6 V   (b)   6 V

(b) If we apply Ohm's law to the equivalent resistor in Figure 18.3b, we find the current in the circuit to be

$$I = \frac{V}{R_{eq}} = \frac{6.0 \text{ V}}{18 \; \Omega} = \boxed{\tfrac{1}{3} \text{ A}}$$

Exercise   Since the current in the equivalent resistor is $\tfrac{1}{3}$ A, this must also be the current in each resistor of the original circuit. Find the voltage drop across each resistor.

Answer $V_{2\Omega} = \tfrac{2}{3}$ V, $V_{4\Omega} = \tfrac{4}{3}$ V, $V_{5\Omega} = \tfrac{5}{3}$ V, $V_{7\Omega} = \tfrac{7}{3}$ V.

18.3

RESISTORS IN PARALLEL

Now consider two resistors connected in parallel, as in Figure 18.4.

> **When resistors are connected in parallel, the potential differences across them are the same.**

This must be true because the left sides of the resistors are connected to a common point, the positive side of the battery, point a in Figure 18.4b, and the right sides are connected to a common point, the negative terminal of the battery in Figure 18.4b.

The currents, however, are generally not the same. They are the same only if the resistors have the same resistance. When the current, I, reaches point a (called a junction) in Figure 18.4b, it splits into two parts, I_1 going through R_1

Georg Simon Ohm (1787–1854), German physicist. *(Courtesy of AIP Niels Bohr Library, E. Scott Barr Collection)*

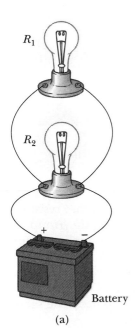

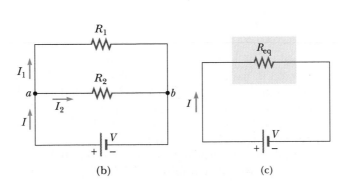

(a)   (b)   (c)

FIGURE 18.4
(a) A parallel connection of two resistors. (b) A circuit diagram for the parallel combination. (c) The voltages across the resistors are the same, and the equivalent resistance of the combination is given by the reciprocal relationship

$$1/R_{eq} = 1/R_1 + 1/R_2.$$

Three incandescent lamps with power ratings of 25 W, 75 W, and 150 W, connected in parallel to a voltage source of about 100 V. All lamps are rated at the same voltage. Why do the intensities of the lamps differ? Which lamp draws the most current? Which has the least resistance? *(Courtesy of Henry Leap and Jim Lehman)*

and I_2 going through R_2. If R_1 is greater than R_2, then I_1 is less than I_2. That is, the charge tends to follow the path of least resistance. *Since charge is conserved, the current, I, that enters point a must equal the total current leaving that point, $I_1 + I_2$.* That is,

$$I = I_1 + I_2$$

The potential drop must be the same for the two resistors and must also equal the potential drop across the battery. Ohm's law applied to each resistor gives

$$I_1 = \frac{V}{R_1} \qquad I_2 = \frac{V}{R_2}$$

Ohm's law applied to the equivalent resistor in Figure 18.4c gives

$$I = \frac{V}{R_{eq}}$$

When these expressions for the current are substituted into the equation $I = I_1 + I_2$, and V is cancelled, we obtain

$$\frac{1}{R_{eq}} = \frac{1}{R_1} + \frac{1}{R_2} \qquad \text{(parallel combination)} \qquad \textbf{[18.5]}$$

An extension of this analysis to three or more resistors in parallel produces the following general expression for the equivalent resistance:

$$\frac{1}{R_{eq}} = \frac{1}{R_1} + \frac{1}{R_2} + \frac{1}{R_3} + \cdots \qquad \text{(parallel combination)} \qquad \textbf{[18.6]}$$

From this it can be shown that the equivalent resistance of two or more resistors connected in parallel is always *less* than the smallest resistance in the group.

Household circuits are always wired so that the lightbulbs (or appliances, or whatever) are connected in parallel, as in Figure 18.4a. In this manner, each device operates independently of the others, so that if one is switched off, the others remain on. Equally important, each device gets the same voltage.

Finally, it is interesting to note that parallel resistors combine in the same way series capacitors combine.

PROBLEM-SOLVING STRATEGY
Resistors

1. **When two or more unequal resistors are connected in** *series,* **they carry the same current, but the potential differences across them are not the same. The resistors add directly to give the equivalent resistance of the series combination.**

2. **When two or more unequal resistors are connected in** *parallel,* **the potential differences across them are the same. Since the current is inversely proportional to the resistance, the currents through them are not the same. The equivalent resistance of a parallel combination of resistors is found through reciprocal addition, and the equivalent resistance is always** *less* **than the smallest individual resistor in the combination.**

3. **A complicated circuit consisting of resistors can often be reduced to a simple circuit with only one resistor. To do so, examine the initial circuit and replace any resistors in series or any in parallel using the procedures outlined in Steps 1 and 2. Sketch the new circuit after these changes have been made. Examine the new circuit and replace any series or parallel combinations. Continue this process until a single equivalent resistance is found.**

4. **If the current through or the potential difference across a resistor in the complicated circuit is to be identified, start with the final circuit found in Step 3 and gradually work back through the circuits, using** $V = IR$ **and the rules of Steps 1 and 2.**

EXAMPLE 18.2   Three Resistors in Parallel

Three resistors are connected in parallel as in Figure 18.5. A potential difference of 18 V is maintained between points a and b.

(a) Find the current in each resistor.

Solution   The resistors are in parallel, and so the potential difference across each is 18 V. Let us apply $V = IR$ to find the current in each resistor:

$$I_1 = \frac{V}{R_1} = \frac{18 \text{ V}}{3.0 \text{ }\Omega} = \boxed{6.0 \text{ A}}$$

$$I_2 = \frac{V}{R_2} = \frac{18 \text{ V}}{6.0 \text{ }\Omega} = \boxed{3.0 \text{ A}}$$

$$I_3 = \frac{V}{R_3} = \frac{18 \text{ V}}{9.0 \text{ }\Omega} = \boxed{2.0 \text{ A}}$$

(b) Calculate the power dissipated by each resistor and the total power dissipated by the three resistors.

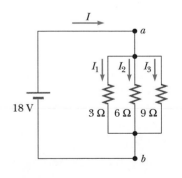

FIGURE 18.5
(Example 18.2)   Three resistors connected in parallel. The voltage across each resistor is 18 V.

Solution Applying $P = I^2R$ to each resistor gives

$$3\ \Omega: \qquad P_1 = I_1^2R_1 = (6.0\ \text{A})^2(3.0\ \Omega) = \boxed{110\ \text{W}}$$

$$6\ \Omega: \qquad P_2 = I_2^2R_2 = (3.0\ \text{A})^2(6.0\ \Omega) = \boxed{54\ \text{W}}$$

$$9\ \Omega: \qquad P_3 = I_3^2R_3 = (2.0\ \text{A})^2(9.0\ \Omega) = \boxed{36\ \text{W}}$$

(Note that you can also use $P = V^2/R$ to find the power dissipated by each resistor.) Summing the three quantities gives a total power of 200 W.

Exercise Calculate the equivalent resistance of the three resistors, and from this result find the total power dissipated.

Answer $\frac{18}{11}\ \Omega$; 200 W.

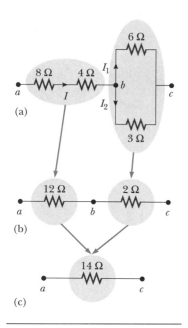

FIGURE 18.6

(Example 18.3) The four resistors shown in (a) can be reduced in steps to an equivalent 14-Ω resistor.

EXAMPLE 18.3 Equivalent Resistance

Four resistors are connected as shown in Figure 18.6a.

(a) Find the equivalent resistance between points a and c.

Solution The circuit can be reduced in steps, as shown in Figures 18.6b and 18.6c. The 8.0-Ω and 4.0-Ω resistors are in series, and so the equivalent resistance between a and b is 12 Ω (Eq. 18.4). The 6.0-Ω and 3.0-Ω resistors are in parallel, and so from Equation 18.6 we find that the equivalent resistance from b to c is 2 Ω. Hence, the equivalent resistance from a to c is 14 Ω.

(b) What is the current in each resistor if a 42-V battery is placed between a and c?

Solution The current, I, is the same in the 8.0-Ω and 4.0-Ω resistors since they are in series. Using Ohm's law and the results of (a), we get

$$I = \frac{V_{ac}}{R_{\text{eq}}} = \frac{42\ \text{V}}{14\ \Omega} = 3.0\ \text{A}$$

When this current enters the junction at b, it splits; part of it passes through the 6.0-Ω resistor (I_1), and part passes through the 3.0-Ω resistor (I_2). Since the potential difference across these resistors, V_{bc}, is the *same* (they are in parallel), $(6\ \Omega)\,I_1 = (3\ \Omega)\,I_2$, or $I_2 = 2I_1$. Using this result and the fact that $I_1 + I_2 = 3.0$ A, we find that $I_1 = 1.0$ A and $I_2 = 2.0$ A. We could have guessed this from the start by noting that the current through the 3.0-Ω resistor has to be twice the current through the 6.0-Ω resistor in view of their relative resistances and the fact that the same voltage is applied to both.

As a final check, note that $V_{bc} = (6\ \Omega)\,I_1 = (3\ \Omega)\,I_2 = 6.0$ V and $V_{ab} = (12\ \Omega)\,I_1 = 36$ V; therefore, $V_{ac} = V_{ab} + V_{bc} = 42$ V, as expected.

18.4

KIRCHHOFF'S RULES AND COMPLEX DC CIRCUITS

As demonstrated in the preceding section, we can analyze simple circuits using Ohm's law and the rules for series and parallel combinations of resistors. However, there are many ways in which resistors can be connected so that the circuits formed cannot be reduced to a single equivalent resistor. The procedure for analyzing more complex circuits is greatly simplified by the use of two simple rules called **Kirchhoff's rules:**

> 1. The sum of the currents entering any junction must equal the sum of the currents leaving that junction. (This rule is often referred to as the **junction rule.**)
> 2. The sum of the potential differences across all the elements around any closed circuit loop must be zero. (This rule is usually called the **loop rule.**)

The junction rule is a statement of *conservation of charge.* Whatever current enters a given point in a circuit must leave that point because charge cannot build up or disappear at a point. If we apply this rule to the junction in Figure 18.7a, we get

$$I_1 = I_2 + I_3$$

Figure 18.7b represents a mechanical analog to this situation, in which water flows through a branched pipe with no leaks. The flow rate into the pipe equals the total flow rate out of the two branches.

The loop rule is equivalent to the principle of *conservation of energy.* Any charge that moves around any closed loop in a circuit (it starts and ends at the same point) must gain as much energy as it loses. Its energy may decrease in the form of a potential drop, $-IR$, across a resistor or as a result of flowing backward through a source of emf, that is, from the positive to the negative terminal inside the battery. In the latter case, electrical energy is converted to chemical energy as the battery is charged.

When applying Kirchhoff's rules, you must make two decisions at the beginning of the problem.

1. You must assign symbols and directions to the currents in all branches of the circuit. If you should happen to guess the wrong direction for a current, the end result for that current will be negative but its magnitude will be correct.
2. When applying the loop rule, you must choose a direction (clockwise or counterclockwise) for going around the loop. As you traverse the loop, record voltage drops and rises according to the following rules:

When you apply the loop rule, it is helpful to keep the following points in mind. They are summarized in Figure 18.8, where it is assumed that movement is from point a toward point b:

1. If a resistor is traversed in the direction of the current, the change in potential across the resistor is $-IR$ (Fig. 18.8a).
2. If a resistor is traversed in the direction opposite the current, the change in potential across the resistor is $+IR$ (Fig. 18.8b).
3. If a source of emf is traversed in the direction of the emf (from $-$ to $+$ on the terminals), the change in potential is $+\mathcal{E}$ (Fig. 18.8c).
4. If a source of emf is traversed in the direction opposite the emf (from $+$ to $-$ on the terminals), the change in potential is $-\mathcal{E}$ (Fig. 18.8d).

There are limits to the numbers of times the junction rule and the loop rule can be used. You can use the junction rule as often as needed so long as, each time you write an equation, you include in it a current that has not been used in a

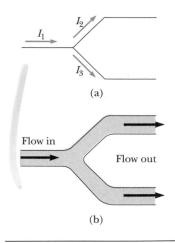

FIGURE 18.7
(a) A schematic diagram illustrating Kirchhoff's junction rule. Conservation of charge requires that whatever current enters a junction must leave that junction. Therefore, in this case, $I_1 = I_2 + I_3$. (b) A hydraulic analog of the junction rule: the net flow out must equal the net flow in.

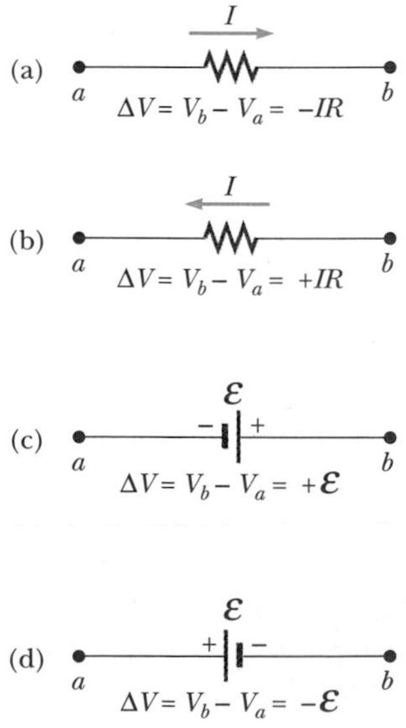

(a) $\Delta V = V_b - V_a = -IR$

(b) $\Delta V = V_b - V_a = +IR$

(c) $\Delta V = V_b - V_a = +\mathcal{E}$

(d) $\Delta V = V_b - V_a = -\mathcal{E}$

FIGURE 18.8
Rules for determining the potential changes across a resistor and a battery, assuming the battery has no internal resistance.

previous junction-rule equation. In general, the number of times the junction rule can be used is one fewer than the number of junction points in the circuit. The loop rule can be used as often as needed so long as a new circuit element (resistor or battery) or a new current appears in each new equation. In general, *to solve a particular circuit problem, you need as many independent equations as you have unknowns.*

Voltages, currents, and resistances are frequently measured by digital multimeters like this one. *(Courtesy of Henry Leap and Jim Lehman)*

PROBLEM-SOLVING STRATEGY
Kirchhoff's Rules

1. **First, draw the circuit diagram and assign labels and symbols to all the known and unknown quantities. You must assign *directions* to the currents in each part of the circuit. Do not be alarmed if you guess the direction of a current incorrectly; the resulting value will be negative, but *its magnitude will be correct*. Although the assignment of current directions is arbitrary, you must stick with it throughout as you apply Kirchhoff's rules.**
2. **Apply the junction rule to any junction in the circuit. The junction rule may be applied as many times as a new current (one not used in a previous application) appears in the resulting equation.**
3. **Now apply Kirchhoff's loop rule to as many loops in the circuit as are needed to solve for the unknowns. In order to apply this rule, you must correctly identify the change in potential as you cross each element in traversing the closed loop. Watch out for signs!**
4. **Solve the equations simultaneously for the unknown quantities. Be careful in your algebraic steps, and check your numerical answers for consistency.**

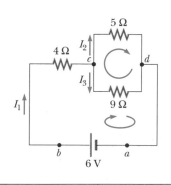

FIGURE 18.9
(Example 18.4)   A multiloop circuit.

EXAMPLE 18.4   Applying Kirchhoff's Rules

Find the currents in the circuit shown in Figure 18.9.

Reasoning   There are three unknown currents in this circuit, and so we must obtain three independent equations. We can find the equations with one application of the junction rule and two applications of the loop rule.

Solution   The first step is to assign a current to each branch of the circuit; these are our unknowns and are labeled I_1, I_2, and I_3 in Figure 18.9. It is also necessary to guess directions for the currents. Your experience with circuits such as this should tell you that the directions of all three have been chosen correctly. (However, recall that if a current direction is chosen incorrectly, the numerical answer will turn out negative, but the magnitude will be correct. This point will be demonstrated in the next example.)

We now apply Kirchhoff's rules. First we can apply the junction rule using either *c* or *d*, the only two junctions in the circuit. Let us choose junction *c*. The net current into this junction is I_1, and the net current leaving it is $I_2 + I_3$. Thus, the junction rule applied to *c* gives

$$I_1 = I_2 + I_3$$

Recall that you may apply the junction rule over and over until you reach a situation in which no new currents appear in an equation. In this example we have reached that point with one application. If we apply the junction rule at d, we find that $I_1 = I_2 + I_3$, exactly the same equation.

We have three unknowns in our problem, I_1, I_2, and I_3; thus, we need two more independent equations before we can find a solution. We obtain these equations by applying the loop rule to the two loops indicated in the figure. Note that there are actually three loops in the circuit, but these two are sufficient to complete the problem. (Where is the loop that we do not use?)

When applying the loop rule, we must first choose the loops to be traversed, and then the directions in which to traverse them. We have selected the two loops indicated in the figure, and have decided to traverse both of them clockwise. Other choices could be made, but the final result would be the same.

Starting at point a and moving clockwise around the large loop, we encounter the following voltage changes (see Fig. 18.8 for the basic rules):

1. From a to b, we encounter a voltage change of 6.0 V.
2. From b to c through the 4-Ω resistor, we encounter a voltage change of $-(4\,\Omega)\,I_1$.
3. From c to d through the 9-Ω resistor, we encounter a voltage change of $-(9\,\Omega)\,I_3$.

No voltage change occurs from d back to a. Now that we have made a complete traversal of the loop, we can equate the sum of the voltage changes to zero:

$$6\,V - (4\,\Omega)\,I_1 - (9\,\Omega)\,I_3 = 0$$

Moving clockwise around the small loop from point c, we encounter the following:

1. From c to d through the 5.0-Ω resistor, a voltage change of $-(5\,\Omega)\,I_2$
2. From d to c through the 9.0-Ω resistor, a voltage change of $+(9\,\Omega)\,I_3$

We find that

$$-(5\,\Omega)\,I_2 + (9\,\Omega)\,I_3 = 0$$

Thus, we have the following three equations to be solved for the three unknowns:

$$I_1 = I_2 + I_3$$
$$6\,V - (4\,\Omega)\,I_1 - (9\,\Omega)\,I_3 = 0$$
$$-(5\,\Omega)\,I_2 + (9\,\Omega)\,I_3 = 0$$

If you need help in solving three equations with three unknowns, see Example 18.5. You should be able to obtain the following answers:

$$I_1 = \boxed{0.83\text{ A}} \qquad I_2 = \boxed{0.53\text{ A}} \qquad I_3 = \boxed{0.30\text{ A}}$$

Exercise Solve this same problem by using the methods learned earlier for series and parallel combinations of resistors. First find the equivalent resistance of the circuit, which you can then use to obtain I_1.

EXAMPLE 18.5   Another Application of Kirchhoff's Rules

Find I_1, I_2, and I_3 in Figure 18.10.

Reasoning To find the three unknown currents, we apply the junction rule once and the loop rule twice.

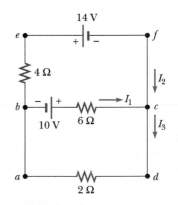

FIGURE 18.10
(Example 18.5)   A circuit containing three loops.

Solution   We choose the directions of the currents as shown in the figure. Applying Kirchhoff's first rule to junction c gives

$$(1) \qquad\qquad I_1 + I_2 = I_3$$

The circuit has three loops: *abcda*, *befcb*, and *aefda*. We need only two loop equations to determine the unknown currents. The third loop equation would give no new information. Applying Kirchhoff's second rule to loops *abcda* and *befcb* and traversing these loops clockwise, we obtain the following expressions:

$$(2) \quad \text{Loop } abcda: \qquad 10\text{ V} - (6\ \Omega)I_1 - (2\ \Omega)I_3 = 0$$

$$(3) \quad \text{Loop } befcb: \quad -14\text{ V} + (6\ \Omega)I_1 - 10\text{ V} - (4\ \Omega)I_2 = 0$$

Note that in loop *befcb*, a positive sign is obtained when the 6-Ω resistor is traversed, since the direction of the path is opposite the direction of the current I_1. A third loop equation for *aefda* gives $-14\text{ V} - (2\ \Omega) I_3 - (4\ \Omega) I_2 = 0$, which is just the sum of (2) and (3). Expressions (1), (2), and (3) represent three linear, independent equations with three unknowns.

We can solve the problem as follows: Substitution of (1) into (2) gives, with units ignored for the moment,

$$10 - 6I_1 - 2(I_1 + I_2) = 0$$

$$(4) \qquad\qquad 10 = 8I_1 + 2I_2$$

Dividing each term in (3) by 2 and rearranging the equation gives

$$(5) \qquad\qquad -12 = -3I_1 + 2I_2$$

Subtracting (5) from (4) eliminates I_2, giving

$$22 = 11 I_1$$

$$I_1 = 2.0\text{ A}$$

Using this value of I_1 in (5) yields a value for I_2:

$$2I_2 = 3I_1 - 12 = 3(2) - 12 = -6$$

$$I_2 = -3.0\text{ A}$$

Finally, $I_3 = I_1 + I_2 = -1$ A. Hence, the currents have the values

$$I_1 = \boxed{2.0\text{ A}} \qquad I_2 = \boxed{-3.0\text{ A}} \qquad I_3 = \boxed{-1.0\text{ A}}$$

The fact that I_2 and I_3 are both negative indicates only that we chose the wrong directions for these currents. The numerical values are correct.

Exercise   Find the potential difference between junctions b and c.

Answer $V_b - V_c = 2.0$ V.

18.5

RC CIRCUITS

So far, we have been concerned with circuits with constant currents. We now consider direct-current circuits containing capacitors, in which the currents vary with time. Consider the series circuit in Figure 18.11a. Let us assume that the capacitor is initially uncharged with the switch opened. After the switch is closed, the battery begins to charge the plates of the capacitor and a current passes through the resistor. The charging process continues until the capacitor is

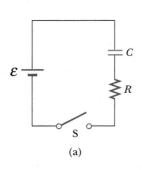

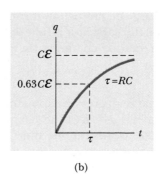

(a)   (b)

FIGURE 18.11
(a) A capacitor in series with a resistor, a battery, and a switch. (b) A plot of the charge on the capacitor versus time after the switch for the circuit is closed. After one time constant, τ, the charge is 63% of the maximum value, $C\mathcal{E}$. The charge approaches its maximum value as t approaches infinity.

charged to its maximum equilibrium value, $Q = C\mathcal{E}$, where $\mathcal{E}$ is the maximum voltage across the capacitor. Once the capacitor is fully charged, the current in the circuit is zero. If we assume that the capacitor is uncharged before the switch is closed, and the switch is closed at $t = 0$, we can show that the charge on the capacitor varies with time according to the expression

$$q = Q(1 - e^{-t/RC}) \qquad \text{[18.7]}$$

where $e = 2.718 \ldots$ is the base of the natural logarithms. Figure 18.11b is a graph of this expression. Note that the charge is zero at $t = 0$ and approaches its maximum value, Q, as t approaches infinity. The voltage, V, across the capacitor at any time is obtained by dividing the charge by the capacitance. That is, $V = q/C$.

As you can see from Equation 18.7, it takes an infinite amount of time for the capacitor to become fully charged. The term RC that appears in Equation 18.7, called the **time constant,** τ (Greek letter tau), is:

$$\tau = RC \qquad \text{[18.8]}$$

The time constant represents the time required for the charge to increase from zero to 63.2% of its maximum equilibrium value. That is, in one time constant, the charge on the capacitor increases from zero to $0.632Q$. (This can be seen by substituting $t = \tau = RC$ in Equation 18.7 and solving for q.) It is important to note that a capacitor charges very slowly in a circuit with a long time constant, whereas it charges very rapidly in a circuit with a short time constant.

Now consider the circuit in Figure 18.12a, consisting of a capacitor with an initial charge of Q, a resistor, and a switch. Before the switch is closed, the potential difference across the charged capacitor is Q/C. Once the switch is

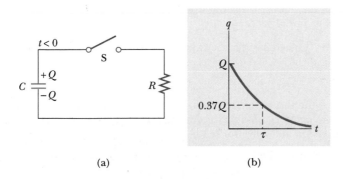

(a)   (b)

FIGURE 18.12
(a) A charged capacitor connected to a resistor and a switch. (b) A graph of the charge on the capacitor versus time after the switch is closed.

closed, the charge begins to flow through the resistor from one capacitor plate to the other until the capacitor is fully discharged. If the switch is closed at $t = 0$, it can be shown that the charge, q, on the capacitor varies with time according to the expression

$$q = Qe^{-t/RC} \qquad\qquad \textbf{[18.9]}$$

That is, the charge decreases exponentially with time as shown in Figure 18.12b. In the interval $t = \tau = RC$, the charge decreases from its initial value, Q, to $0.368Q$. In other words, in one time constant, the capacitor loses 63.2% of its initial charge. Since $V = q/C$, we see that the voltage across the capacitor also decreases exponentially with time according to the expression $V = \mathcal{E}e^{-t/RC}$, where $\mathcal{E}$ (which equals Q/C) is the initial voltage across the fully charged capacitor.

EXAMPLE 18.6   Charging a Capacitor in an *RC* Circuit

An uncharged capacitor and a resistor are connected in series to a battery, as in Figure 18.11a. If $\mathcal{E} = 12$ V, $C = 5.0\ \mu$F, and $R = 8.0 \times 10^5\ \Omega$, find the time constant of the circuit, the maximum charge on the capacitor, and the charge on the capacitor after one time constant.

Solution   The time constant of the circuit is

$$\tau = RC = (8.0 \times 10^5\ \Omega)(5.0 \times 10^{-6}\ \text{F}) = \boxed{4.0\ \text{s}}$$

The maximum charge on the capacitor is

$$Q = C\mathcal{E} = (5.0 \times 10^{-6}\ \text{F})(12\ \text{V}) = \boxed{60\ \mu\text{C}}$$

After one time constant, the charge on the capacitor is 63.2% of its maximum value:

$$q = 0.632Q = 0.632(60 \times 10^{-6}\ \text{C}) = \boxed{38\ \mu\text{C}}$$

Exercise   Find the charge on the capacitor and the voltage across the capacitor after time t has elapsed.

Answer $q = 60\ \mu\text{C}(1 - e^{-t/4})$; $V = 12\ \text{V}(1 - e^{-t/4})$.

EXAMPLE 18.7   Discharging a Capacitor in an *RC* Circuit

Consider a capacitor, C, being discharged through a resistor, R, as in Figure 18.12a. After how many time constants does the charge on the capacitor drop to one fourth of its initial value?

Solution   The charge on the capacitor varies with time according to Equation 18.9:

$$q(t) = Qe^{-t/RC}$$

where Q is the initial charge on the capacitor. To find the time it takes the charge q to drop to one fourth of its initial value, we substitute $q(t) = Q/4$ into this expression and solve for t:

$$\tfrac{1}{4}Q = Qe^{-t/RC}$$

$$\tfrac{1}{4} = e^{-t/RC}$$

Taking logarithms of both sides, we find that

$$-\ln 4 = -\frac{t}{RC}$$

$$t = RC \ln 4 = \boxed{1.39\ RC}$$

Exercise   If $R = 8.0 \times 10^5\ \Omega$, $C = 5.0\ \mu F$, and the initial voltage across the capacitor is 6.0 V, what is the voltage across the capacitor after time t has elapsed?

Answer $V = (6.0\ V)e^{-t/4}$.

*18.6

HOUSEHOLD CIRCUITS

Household circuits are a very practical application of some of the ideas presented in this chapter. In a typical installation, the utility company distributes electric power to individual houses with a pair of wires, or power lines. Electrical devices in a house are then connected in parallel to these lines, as shown in Figure 18.13. The potential drop between the two wires is about 120 V. (These are actually alternating currents and voltages, but for the present discussion we shall assume that they are direct currents and voltages.) One of the wires is connected to ground, and the other wire, sometimes called the "hot" wire, has a potential of 120 V. A meter and a circuit breaker or fuse are connected in series with the wire entering the house, as indicated in Figure 18.13. Figure 18.14 is a cutaway view of a fuse. The fuse is a small metallic strip, which melts if the current exceeds a certain value. If a circuit did not include a fuse, excessive currents could damage circuit elements, overheat wires, and perhaps cause a fire.

In modern homes, circuit breakers are used in place of fuses. When the current in a circuit exceeds some value (typically 15 A), the circuit breaker acts as a switch and opens the circuit. Figure 18.15 is one design for a circuit breaker. Current passes through a bimetallic strip, the top of which bends to the left when excessive current heats it. If the strip bends far enough to the left, it settles into a groove in the spring-loaded metal bar. When this occurs, the bar drops enough to open the circuit at the contact point. The bar also flips a switch that indicates that the circuit breaker is not operational. (After the overload is removed, the switch can be flipped back on.) Circuit breakers based on this design have the disadvantage that some time is required for the heating of the strip, and thus the circuit is not opened rapidly enough when it is overloaded. Consequently, many circuit breakers are now designed to use electromagnets, which we shall discuss later.

The wire and circuit breaker are carefully selected to meet the current demands of a circuit. If the circuit is to carry currents as large as 30 A, a heavy-duty wire and appropriate circuit breaker must be used. Household circuits that are normally used to power lamps and small appliances often require only 15 A. Each circuit has its own circuit breaker to accommodate its expected load.

As an example, consider a circuit that powers a toaster, a microwave oven, and a heater (represented by R_1, R_2, . . . in Fig. 18.13). We can calculate the current through each appliance using the equation $P = IV$. The toaster, rated at 1000 W, draws a current of $1000/120 = 8.33$ A. The microwave oven, rated at 800 W, draws a current of 6.67 A, and the heater, rated at 1300 W, draws a

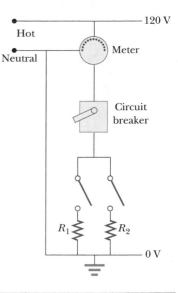

FIGURE 18.13
A wiring diagram for a household circuit. The resistances R_1 and R_2 represent appliances or other electrical devices that operate at an applied voltage of 120 V.

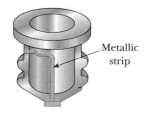

FIGURE 18.14
A cutaway view of a fuse.

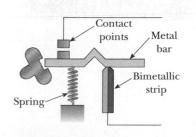

FIGURE 18.15
A circuit breaker uses a bimetallic strip for its operation.

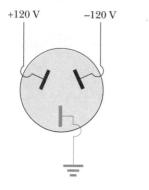

+120 V   −120 V

FIGURE 18.16
Power connections for a 240-V
appliance.

current of 10.8 A. If the three appliances are operated simultaneously, they draw a total current of 25.8 A. Therefore, the breaker should be able to handle at least this much current, or else it will be tripped. Alternatively, one could operate the toaster and microwave oven on one 15-A circuit and the heater on a separate 15-A circuit.

Many heavy-duty appliances, such as electric ranges and clothes dryers, require 240 V to operate. The power company supplies this voltage by providing, in addition to a live wire that is 120 V above ground potential, a wire, also considered live, that is 120 V below ground potential (Fig. 18.16). Therefore, the potential drop across the two live wires is 240 V. An appliance operating from a 240-V line requires half the current of one operating from a 120-V line; therefore, smaller wires can be used in the higher-voltage circuit without becoming overheated.

*18.7

ELECTRICAL SAFETY

A person can be electrocuted by touching a live wire (which commonly is live because of a frayed cord and exposed conductors) while in contact with ground. The ground contact might be made by touching a water pipe (which is normally at ground potential) or by standing on the ground with wet feet, since impure water is a good conductor. Obviously such situations should be avoided at all costs.

Electric shock can result in fatal burns, or it can cause the muscles of vital organs, such as the heart, to malfunction. The degree of damage to the body depends on the magnitude of the current, the length of time it acts, and the part of the body through which it passes. Currents of 5 mA or less can cause a sensation of shock but ordinarily do little or no damage. If the current is larger than about 10 mA, the hand muscles contract and the person may be unable to let go of the live wire. If a current of about 100 mA passes through the body for just a few seconds, it can be fatal. Such large currents paralyze the respiratory muscles. In some cases, currents of about 1 A through the body produce serious (and sometimes fatal) burns.

As an additional safety feature for consumers, electrical equipment manufacturers now use electrical cords that have a third wire, called a case ground. To understand how this works, consider the drill being used in Figure 18.17. Figure 18.17a shows a two-wire device that has one wire, called the ''hot'' wire, connected to the high-potential (120-V) side of the input power line, while the second wire is connected to ground (0 V). Under normal operating conditions, the path of the current through the drill is like that shown in Figure 18.17a. However, if the high-voltage wire comes in contact with the case of the drill (Fig. 18.17b), a ''short circuit'' can occur. In this undesirable circumstance, the pathway for the current is from the high-voltage wire through the person holding the drill and to Earth—a pathway that can kill. Protection is provided by a third wire, connected to the case of the drill (Fig. 18.17c). In this case, if a short occurs, the path of least resistance for the current is from the high-voltage wire through the case and back to ground through the third wire. The resulting high current produced will blow a fuse or trip a circuit breaker before the consumer is injured.

Special power outlets called ground-fault interrupters (GFIs) are now being used in kitchens, bathrooms, basements, and other hazardous areas of new

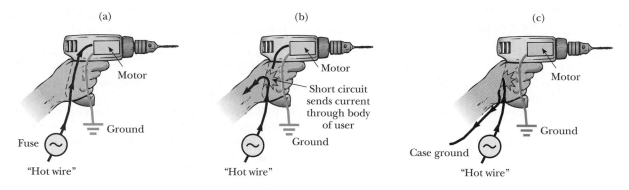

FIGURE 18.17
(a) When the drill is operated with two wires, the "hot wire," at 120 V, is always fused on this side of the circuit for safety. (b) A two-wire connection is potentially dangerous. If the high-voltage side comes in contact with the drill case, the person holding the drill receives an electrical shock. (c) Shock can be prevented by a third wire running from the drill case to the ground.

homes. They are designed to protect people from electrical shock by sensing small currents—approximately 5 mA and greater—leaking to ground. When current above this level is detected, the device shuts off (interrupts) the current in less than a millisecond. Ground-fault interrupters will be discussed in Chapter 19.

SUMMARY

A **source of emf** is any device that transforms nonelectrical energy into electrical energy.

The **equivalent resistance** of a set of resistors connected in **series** is

$$R_{eq} = R_1 + R_2 + R_3 + \cdots \qquad \text{[18.4]}$$

The **equivalent resistance** of a set of resistors connected in **parallel** is

$$\frac{1}{R_{eq}} = \frac{1}{R_1} + \frac{1}{R_2} + \frac{1}{R_3} + \cdots \qquad \text{[18.6]}$$

Complex circuits are conveniently analyzed by using **Kirchhoff's rules:**

1. The sum of the currents entering any junction must equal the sum of the currents leaving that junction.
2. The sum of the potential differences across all the elements around any closed circuit loop must be zero.

The first rule is a statement of **conservation of charge.** The second is a statement of **conservation of energy.**

As a capacitor is charged by a battery through a resistor, the current drops from a maximum value to zero. The **time constant,** $\tau = RC$, represents the time it takes the charge on the capacitor to increase from zero to 63% of its maximum value.

ADDITIONAL READING

P. F. Baker, "The Nerve Axon," *Sci. American,* March 1966, p. 74.

P. Davidovits, *Physics in Biology and Medicine,* Englewood Cliffs, N.J., Prentice-Hall, 1977.

B. Katz, "The Nerve Impulse," *Sci. American,* November 1952, p. 55.

K. Kordesch and K. Tomantschger, "Primary Batteries," *The Physics Teacher,* January 1981, p. 12.

T. F. Robinson, S. M. Factor, and E. H. Sonnenblick,

"The Heart as a Suction Pump," *Sci. American,* June 1986, p. 84.

A. M. Scher, "The Electrocardiogram," *Sci. American,* November 1961, p. 132.

G. M. Shepherd, "Microcircuits in the Nervous System," *Sci. American,* February 1978, p. 92.

A. K. Solomon, "Pumps in the Living Cell," *Sci. American,* August 1962, p. 100.

CONCEPTUAL QUESTIONS

Example   How can you tell that the headlights on a car are wired in parallel, and not in series? What would happen if you rewired your headlights in series?
Reasoning   If they were wired in series and one burned out, the other would go out also. If you rewired your headlights to be in series across a 12-V battery, the voltage across each would only be 6 V, and they would not glow as brightly.

Example   Embodied in Kirchhoff's rules are two conservation laws. What are they?
Reasoning   The junction rule is a statement of conservation of charge. It says that the amount of charge that enters a junction in some time interval must equal the charge that leaves the junction in that time interval. The loop rule is a statement of conservation of energy. It says that the potential increases and decreases around a closed loop in a circuit must add to zero.

Example   Suppose you are flying a kite when it strikes a high-voltage wire (a very dangerous situation). What factors determine how great a shock you will receive?
Reasoning   A few of the factors involved are as follows: the conductivity of the string (is it wet or dry?); how well you are insulated from ground (are you wearing thick rubber or leather shoes?); the magnitude of the potential difference between you and the kite; the type and condition of the soil under your feet.

Example   Why is it possible for a bird to be perched on a high-voltage wire without being electrocuted?
Reasoning   The bird is resting on a wire whose electric potential has some constant value. For the bird to be electrocuted, a potential difference must exist across two points of the bird's body. There is no potential difference between the bird's feet, so it is safe.

Example   Suppose a parachutist lands on a high-voltage wire, and grabs the wire as she prepares to be rescued. Will she be electrocuted? If the wire then breaks, should she continue to hold onto the wire as she falls to the ground?

Reasoning   She will not be electrocuted if she holds onto only the one high-voltage wire, because she is not completing a circuit. There is no potential difference across her body as long as she clings to only one wire. However, she should immediately release the wire once it breaks, since she will become part of a closed circuit when she reaches the ground or comes into contact with another object.

Birds on a high-voltage wire. *(Superstock)*

1. Is the direction of current through a battery always from negative to positive on the terminals?
2. Car batteries are often rated in ampere-hours. Does this designate the amount of current, power, energy, or charge that can be drawn from the battery?
3. Two sets of Christmas tree lights are available. In set A, when one bulb is removed or burns out, all the others remain illuminated. In set B, when one bulb is removed, the remaining bulbs will not operate. Explain the difference in the wiring of the two sets.

4. How would you connect resistors in order for the equivalent resistance to be greater than any of the individual resistances? Give an example.

5. How would you connect resistors in order for the equivalent resistance to be smaller than any of the individual resistances? Give an example.

6. When resistors are connected in series, which of the following is (are) the same for all the resistors: potential difference, current, or power?

7. When resistors are connected in parallel, which of the following is (are) the same for all the resistors: potential difference, current, or power?

8. Sketch as many different electric circuits as you can, using three lightbulbs and a battery.

9. The circuits in a car are "single-wire" circuits. For them to work properly, a ground wire must be attached from the chassis of the car to one of the battery posts. Why? What happens if the ground wire corrodes?

10. An incandescent lamp connected to a 120-V source by a short extension cord provides more illumination than the same lamp connected to the same source with a long extension cord. Explain.

11. In the circuits considered in this chapter, the wires connecting the elements are idealized to be perfect conductors (represented schematically by lines between the elements). Discuss the effect of the real wires in a circuit. Consider the resistance of the wires in relation to resistive elements of the circuit.

12. A "short circuit" is a circuit containing a path of very low resistance in parallel with some other part of the circuit. Discuss the effect of a short circuit on the portion of the circuit with which it is in parallel. Use a lamp with a frayed line cord as an example.

13. A series circuit consists of three identical lamps connected to a battery, as shown in Figure 18.18. Imagine that switch S is closed. (a) What happens to the intensities of lamps A and B? (b) What happens to the intensity of lamp C? (c) What happens to the current in the circuit? (d) Does the power dissipated in the circuit increase, decrease, or remain the same?

14. What advantage might there be in using two identical resistors in parallel, connected in series with another identical parallel pair, rather than just using a single resistor?

15. When can the potential difference across a resistor be positive?

16. Since charged particles have mass, can Kirchhoff's junction rule be based on conservation of mass rather than conservation of charge? Explain.

17. Would a fuse work successfully if it were placed in parallel with the device it was supposed to protect?

18. Does it matter whether a fuse is placed in the high-voltage line of a household circuit or in the ground line?

19. What advantage does 120-V operation offer over 240 V? What disadvantages?

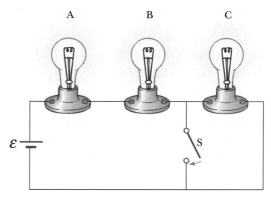

FIGURE 18.18 (Question 13)

20. If it is the current flowing through the body that determines the seriousness of a shock, why do we see warnings of high *voltage* rather than high *current* near electric equipment?

21. Why is it dangerous to turn on a light when you are in the bathtub?

22. When electricians work with potentially live wires, they often use the backs of their hands or fingers to move the wires. Why do you suppose they do this?

23. What procedure would you use to try to save a person "frozen" to a live high-voltage wire without endangering your own life?

24. At what level of current do you experience (a) the sensation of shock? (b) involuntary muscle contractions? (c) paralysis of respiratory muscles? (d) serious and possibly fatal burns? In practice, what current level is regarded as safe?

25. Three identical lightbulbs are connected to a battery as in Figure 18.19. Compare the levels of brightness of the bulbs when all are illuminated. What happens to the brightness of each bulb when (a) A is removed from its socket? (b) C is removed from its socket? (c) a wire is connected between points a and b? (d) a wire is connected between points a and c?

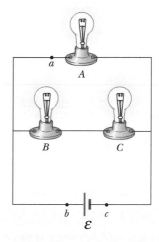

FIGURE 18.19 (Question 25)

PROBLEMS

Section 18.2   Resistors in Series

Section 18.3   Resistors in Parallel

1. A 4.0-Ω resistor, an 8.0-Ω resistor, and a 12-Ω resistor are connected in series with a 24-V battery. What are (a) the equivalent resistance and (b) the current in each resistor?

2. A 9.00-V battery delivers 117 mA when connected to a 72.0-Ω load. Determine the internal resistance of the battery.

3. The resistors of Problem 1 are connected in parallel across a 24-V battery. Find (a) the equivalent resistance and (b) the current in each resistor.

4. A length of wire is cut into five equal pieces. The five pieces are then connected in parallel, with the resulting resistance being 2.00 Ω. What was the resistance of the original length of wire?

5. A technician has a box full of resistors, all with the same resistance, R. How many different values of effective resistance can the technician achieve, using all the possible combinations of one to three separate resistors? Express the effective resistance of each combination in terms of R.

6. The resistance between points a and b in Figure 18.20 drops to one-half its original value when switch S is closed. Determine the value of R.

7. (a) You need a 45-Ω resistor, but the stockroom has only 20-Ω and 50-Ω resistors. How can the desired resistance be achieved under these circumstances? (b) What can you do if you need a 35-Ω resistor?

8. A battery with an internal resistance of 10.0 Ω produces an open-circuit voltage of 12.0 V. A variable load resistance with a range of 0 to 30.0 Ω is connected across the battery. (*Note:* A battery has a resistance which depends on the condition of its chemicals and increases as the battery ages. This so-called internal resistance can be represented in

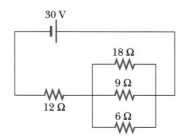

FIGURE 18.21   (Problem 9)

a simple circuit diagram as a resistor in series with the battery.) (a) Graph the power dissipated in the load resistor as a function of the load resistance. (b) With your graph, demonstrate the following important theorem: *The power delivered to a load is maximum if the load resistance equals the internal resistance of the source.*

9. Find the equivalent resistance of the circuit in Figure 18.21.

10. An 18-Ω resistor and a 6.0-Ω resistor are connected in series across an 18-V battery. (a) Find the current through each resistor and the voltage drop across each resistor. (b) Repeat part (a) for the situation in which the resistors are connected in parallel.

11. Find the equivalent resistance of the circuit in Figure 18.22.

12. A 9.0-Ω resistor and a 6.0-Ω resistor are connected in series with a power supply. (a) The voltage drop across the 6.0-Ω resistor is measured to be 12 V. Find the voltage output of the power supply. (b) The two resistors are connected in parallel across a power supply, and the current through the 9.0-Ω resistor is found to be 0.25 A. Find the voltage setting of the power supply.

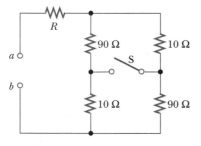

FIGURE 18.20   (Problem 6)

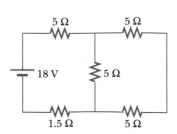

FIGURE 18.22   (Problem 11)

□ indicates problems that have full solutions available in the Student Solutions Manual and Study Guide.

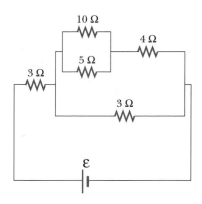

FIGURE 18.23 (Problem 13)

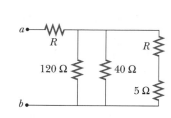

FIGURE 18.24 (Problem 14)

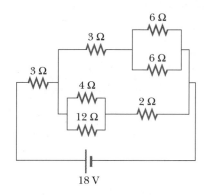

FIGURE 18.25 (Problem 15)

13. (a) Find the equivalent resistance of the circuit in Figure 18.23. (b) If the total power supplied to the circuit is 4.00 W, find the emf of the battery.

14. The resistance between terminals a and b in Figure 18.24 is 75 Ω. If the resistors labeled R have the same value, determine R.

15. Find the current in the 12-Ω resistor in Figure 18.25.

16. Two resistors, A and B, are connected in series to a 6.0-V battery. A voltmeter connected across resistor A measures a voltage of 4.0 V. When the two resistors are connected in parallel across the 6.0-V battery, the current through B is found to be 2.0 A. Find the resistances of A and B.

Section 18.4 Kirchhoff's Rules and Simple DC Circuits

17. Find the current through each of the three resistors of Figure 18.26 (a) by the rules for resistors in series and parallel and (b) by the use of Kirchhoff's rules.

18. Determine the potential on terminals A, B, and C in Figure 18.27. (a) if terminal A is grounded; (b) if terminal B is grounded; (c) if terminal C is grounded.

19. Four resistors are connected to a battery with a terminal voltage of 12 V, as shown in Figure 18.28. Determine the power lost in the 50-Ω resistor.

20. Two 1.50-V batteries—with their positive terminals in the same direction—are inserted in series into the barrel of a flashlight. One battery has an internal resistance of 0.255 Ω, the other an internal resistance of 0.153 Ω. When the switch is closed, a current of 0.600 A passes through the lamp. (See the note in Problem 8.) (a) What is the lamp's resistance? (b) What fraction of the power dissipated is dissipated in the batteries?

21. An unmarked battery has an unknown internal resistance. If the battery is connected to a fresh 5.60-V battery (negligible internal resistance) positive to positive and negative to negative, the current through the circuit is 10.0 mA. If the polarity of the unknown battery is reversed, the current increases to 25.0 mA. Determine the source voltage and internal resistance of the unknown battery. Assume that in each case the direction of the current is negative to positive in the 5.60-V battery. (See the note in Problem 8.)

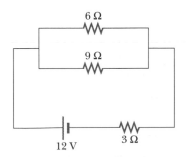

FIGURE 18.26 (Problem 17)

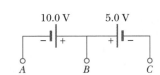

FIGURE 18.27 (Problem 18)

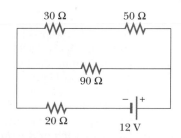

FIGURE 18.28 (Problem 19)

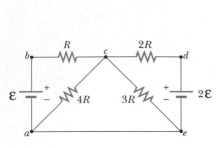

FIGURE 18.29 (Problem 22)

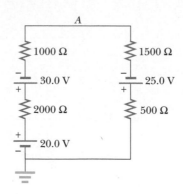

FIGURE 18.30 (Problem 23)

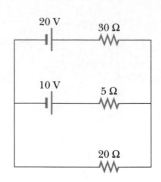

FIGURE 18.31 (Problem 24)

22. If $R = 1.0$ kΩ and $\mathcal{E} = 250$ V in Figure 18.29, determine the direction and magnitude of the current in the horizontal wire between a and e.

23. Figure 18.30 shows a circuit diagram. Determine (a) the current, (b) the potential of wire A relative to ground, and (c) the voltage drop across the 1500-Ω resistor.

24. Find the current in each resistor in Figure 18.31.

25. For the network in Figure 18.32, show that the resistance between points a and b is $R_{ab} = \frac{27}{17}$ Ω. (*Hint:* Connect a battery with emf $\mathcal{E}$ across points a and b and determine $\mathcal{E}/I$, where I is the current through the battery.)

29. What is the emf, $\mathcal{E}_1$, of the battery in the circuit of Figure 18.36?

30. Find the potential difference across each resistor in Figure 18.37.

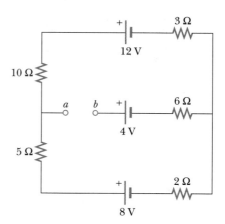

FIGURE 18.33 (Problem 26)

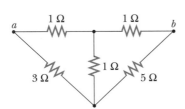

FIGURE 18.32 (Problem 25)

26. Determine the potential difference, V_{ab}, for the circuit in Figure 18.33.

27. (a) Determine the potential difference, V_{ab}, for the circuit in Figure 18.34. Note that each battery has an internal resistance, as indicated in the figure. (b) If points a and b are connected by a 7.0-Ω resistor, what is the current through this resistor?

28. Calculate each of the unknown currents I_1, I_2, and I_3 for the circuit of Figure 18.35.

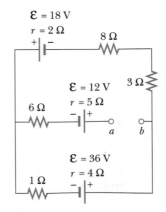

FIGURE 18.34 (Problem 27)

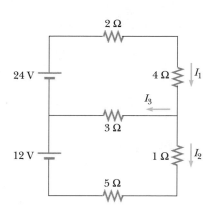

FIGURE 18.35   (Problem 28)

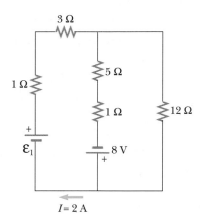

FIGURE 18.36   (Problem 29)

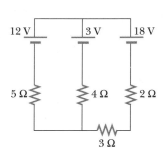

FIGURE 18.37   (Problem 30)

Section 18.5   *RC* Circuits

31. Consider a series *RC* circuit for which $C = 6.0\ \mu F$, $R = 2.0 \times 10^6\ \Omega$, and $\mathcal{E} = 20$ V. Find (a) the time constant of the circuit and (b) the maximum charge on the capacitor after a switch in the circuit is closed.

32. An uncharged capacitor and a resistor are connected in series to a source of emf. If $\mathcal{E} = 9.0$ V, $C = 20\ \mu F$, and $R = 100\ \Omega$, find (a) the time constant of the circuit, (b) the maximum charge on the capacitor, and (c) the charge on the capacitor after one time constant.

33. Consider an *RC* circuit in which the capacitor is being charged by a battery connected in the circuit. In a time equal to two time constants, what percentage of the *final* charge is left on the capacitor?

34. Consider a series *RC* circuit (see Fig. 18.11) for which $R = 1.0$ MΩ, $C = 5.0\ \mu F$, and $\mathcal{E} = 30$ V. Find the charge on the capacitor 10 s after the switch is closed.

35. An uncharged capacitor, $C = 10\ \mu F$, and a resistor are connected in series to an emf of 9.0 V. (a) What should the resistance be to give a time constant of 5.0 s? (b) What is the charge on the capacitor after two time constants have elapsed, if the resistance has the value found in part (a)?

36. Show that $\tau = RC$ has units of time.

Section 18.6   Household Circuits

37. An electric heater is rated at 1300 W, a toaster is rated at 1000 W, and an electric grill is rated at 1500 W. The three appliances are connected in parallel to a common 120-V circuit. (a) How much current does each appliance draw? (b) Is a 30.0-A circuit breaker sufficient in this situation? Explain.

38. Your toaster oven and coffee maker each dissipate 1200 W of power. Can you operate them together if the 120-V line that feeds them has a circuit breaker rated at 15 A? Explain.

39. A heating element in a stove is designed to dissipate 3000 W when connected to 240 V. (a) Assuming that the resistance is constant, calculate the current in this element if it is connected to 120 V. (b) Calculate the power it dissipates at this voltage.

40. Three 2.0-Ω resistors are connected as shown in Figure 18.38. Each can dissipate a maximum power of 32 W without being excessively heated. Determine the maximum power the network can dissipate.

41. Aluminum wiring has been used in the past instead of copper for economic reasons. According to the National Electrical Code, the maximum allowable current for 12-gauge copper wire with rubber insulation is 20 A. What should be the maximum allowable current in a 12-gauge aluminum wire if it is to dissipate the same power per unit length as the copper wire?

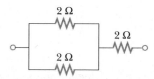

FIGURE 18.38   (Problem 40)

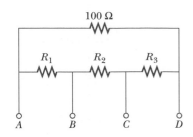

FIGURE 18.39 (Problem 45)

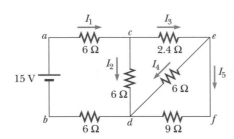

FIGURE 18.40 (Problems 46 and 47)

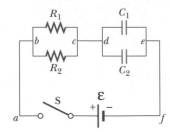

FIGURE 18.41 (Problem 48)

42. A lamp ($R = 150\ \Omega$), an electric heater ($R = 25\ \Omega$), and a fan ($R = 50\ \Omega$) are connected in parallel across a 120-V line. (a) What total current is supplied to the circuit? (b) What is the voltage across the fan? (c) What is the current in the lamp? (d) What power is expended in the heater?

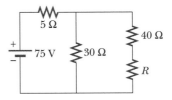

FIGURE 18.42 (Problem 50)

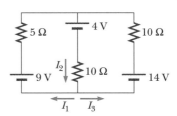

FIGURE 18.43 (Problem 51)

ADDITIONAL PROBLEMS

43. Arrange nine 100-Ω resistors in a series-parallel network so that the total resistance of the network is also 100 Ω. All nine resistors must be used.

44. A series combination of a 12-kΩ resistor and an unknown capacitor is connected to a 12-V battery. One second after the circuit is completed, the voltage across the capacitor is 10 V. Determine the capacitance of the capacitor.

45. In Figure 18.39, $R_1 = 0.100\ \Omega$, $R_2 = 1.00\ \Omega$, and $R_3 = 10.0\ \Omega$. Find the equivalent resistance of the circuit and the current in each resistor when a 5.00-V power supply is connected between (a) points A and B; (b) points A and C; (c) points A and D.

46. Find the equivalent resistance of the circuit in Figure 18.40.

47. For the circuit of Problem 46, find (a) each current in the circuit, (b) the potential difference across each resistor, and (c) the power dissipated by each resistor.

48. The circuit in Figure 18.41 contains two resistors, $R_1 = 2.0$ kΩ and $R_2 = 3.0$ kΩ, and two capacitors, $C_1 = 2.0\ \mu$F and $C_2 = 3.0\ \mu$F, connected to a battery with emf $\mathcal{E} = 120$ V. If there are no charges on the capacitors before switch S is closed, determine the charges q_1 and q_2 on capacitors C_1 and C_2, respectively, after the switch is closed. (*Hint:* First reconstruct the circuit so that it becomes a simple *RC* circuit containing a single resistor and single capacitor in series, connected to the battery, and then determine the total charged, q, stored in the circuit.)

49. A series *RC* circuit has a time constant of 0.960 s. The battery has an emf of 48.0 V, and the maxi-

mum current in the circuit is 500 mA. What are (a) the value of the capacitance and (b) the charge stored in the capacitor 1.92 s after the switch is closed?

50. The resistor R in Figure 18.42 dissipates 20.0 W of power. Determine the value of R.

51. Find the values of I_1, I_2, and I_3 for the circuit in Figure 18.43.

52. Two resistors, R_1 and R_2, have an equivalent resistance of 690 Ω when they are connected in series and an equivalent resistance of 150 Ω when they are connected in parallel. What are R_1 and R_2?

53. The student engineer of a campus radio station wishes to verify the effectiveness of the lightning rod on the antenna mast (Fig. 18.44). The unknown resistance R_x is between points C and E. Point E is a "true ground" but is inaccessible for direct measurement since this stratum is several meters below the Earth's surface. Two identical rods are driven into the ground at A and B, introducing an unknown resistance, R_y. The procedure is as follows: Measure resistance R_1 between points A and B, then connect A and B with a heavy con-

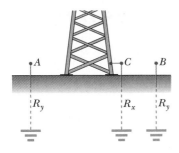

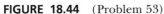

FIGURE 18.44 (Problem 53)

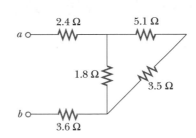

FIGURE 18.45 (Problem 54)

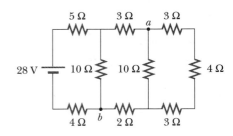

FIGURE 18.46 (Problem 55)

ducting wire and measure resistance R_2 between points A and C. (a) Derive a formula for R_x in terms of the observable resistances, R_1 and R_2. (b) A satisfactory ground resistance would be $R_x < 2.0\ \Omega$. Is the grounding of the station adequate if measurements give $R_1 = 13\ \Omega$ and $R_2 = 6.0\ \Omega$?

54. Find the equivalent resistance between points a and b in Figure 18.45.

55. For the circuit in Figure 18.46, calculate (a) the equivalent resistance of the circuit and (b) the power dissipated by the entire circuit. (c) Find the current in the 5.0-Ω resistor.

56. An emf of 10 V is connected to a series RC circuit consisting of a resistor of $2.0 \times 10^6\ \Omega$ and a capacitor of 3.0 μF. Find the time required for the charge on the capacitor to reach 90% of it final value.

57. In the circuit in Figure 18.47, the 4.0-V battery and the 2.0-V battery have internal resistances of 0.5 Ω and 0.25 Ω, respectively. (See the note in Problem 8.) (a) Find the current in the circuit. (b) Find the power dissipated as heat in the circuit. (c) Find the power absorbed by the 2.0-V battery as it charges, including the power dissipated as heat within the battery.

58. When a battery of unknown emf is connected to a 5.0-Ω resistor, the current in the circuit is 0.30 A. If the battery is now connected to an 8.0-Ω resistor, the current is 0.20 A. What are the emf of the battery and its internal resistance? (See the note in Problem 8.)

59. An automobile battery has an emf of 12.60 V and an internal resistance of 0.080 Ω. The headlights have total resistance 5.00 Ω (assumed constant). What is the potential difference across the headlight bulbs (a) when they are the only load on the battery? (b) when the starter motor is operated, taking an additional 35.0 A from the battery?

60. (a) Apply Kirchhoff's loop rule to the RC circuit in Figure 18.11a and show that $\mathcal{E} - q/C - IR = 0$. Use this equation and Equation 18.7 to determine the current in the circuit as a function of time. (b) Apply Kirchhoff's loop rule to the RC circuit in Figure 18.12a and show that $q/C - IR = 0$. Use

this equation and Equation 18.9 to determine the current in the circuit as a function of time.

61. What are the expected readings of the ammeter and voltmeter for the circuit in Figure 18.48?

62. A voltage, V, is applied to a series configuration of n resistors, each of value R. The circuit components are reconnected in a parallel configuration, and voltage V is again applied. Show that the power consumed by the series configuration is $1/n^2$ times the power consumed by the parallel configuration.

63. A generator has a terminal voltage of 110 V when it delivers 10.0 A, and 106 V when it delivers 30.0 A. Calculate the emf and the internal resistance of the generator.

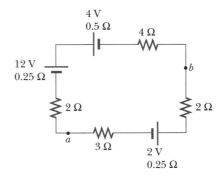

FIGURE 18.47 (Problem 57)

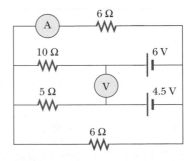

FIGURE 18.48 (Problem 61)

PAUL DAVIDOVITS   Boston College

CURRENT IN THE NERVOUS SYSTEM

The most remarkable use of electrical phenomena in living organisms is found in the nervous system of animals. Specialized cells in the body called **neurons** form a complex network that receives, processes, and transmits information from one part of the body to another. The center of this network is located in the brain, which has the ability to store and analyze information. Based on this information, the nervous system controls parts of the body.

The nervous system is very complex: the human nervous system, for example, consists of about 10^{10} interconnected neurons. Some aspects of the nervous system are well known. During the past 38 years, the method of signal propagation through the nervous system has been firmly established. The messages are electric pulses transmitted by neurons. When a neuron receives an appropriate stimulus, it produces electric pulses that are propagated along its cable-like structure. The strength of the stimulus is conveyed by the number of pulses produced. When the pulses reach the end of the ''cable,'' they activate either neurons or muscle cells.

The neurons, which are the basic units of the nervous system, can be divided into three classes: sensory neurons, motoneurons, and interneurons. The sensory neurons receive stimuli from sensory organs that monitor the external and internal environment of the body. Depending on their specialized functions, the sensory neurons convey messages about factors such as heat, light, pressure, muscle tension, and odor to higher centers in the nervous system. The motoneurons carry messages that control the muscle cells. The messages are based on the information provided by the sensory neurons and by the brain. The interneurons transmit information from one neuron to another.

Each neuron consists of a cell body to which are attached input ends called **dendrites** and a long tail called the **axon,** which propagates the signal away from the cell (Fig. 1). The far end of the axon branches into nerve endings that transmit the signal across small gaps to other neurons or to muscle cells. A simple sensory-motoneuron circuit is shown in Figure 2. A stimulus from a muscle produces nerve impulses that travel to the spine. Here the signal is transmitted to a motoneuron, which in turn sends impulses to control the muscle.

The axon, which is an extension of the neuron cell, conducts the electric impulses away from the cell body. Some axons are extremely long. In humans, for example, the axons connecting the spine with the fingers and toes are more than 1 m long. The neuron can transmit messages because of the special electrical characteristics of the axon. Most of the information about the electrical and chemical properties of the axon is obtained by inserting small needle-like probes into the axon. With such probes it is possible to measure currents in the axon and to sample its chemical composition. Such experiments are usually difficult to run because the diameter of most axons is very small. Even the largest axons in the human nervous system have a diameter of only about 20×10^{-4} cm. The giant squid, however, has an axon with a diameter of about 0.5 mm, which is large enough for the convenient insertion of probes. Much of the information about signal transmission in the nervous system has come from experiments with the squid axon.

In the aqueous environment of the body, salts and other molecules dissociate into positive and negative ions. As a result, body fluids are relatively good conductors of electricity. The inside of the axon is filled with an ionic fluid that is separated from the surrounding body fluid by a thin membrane that is only about 5 to 10 nm thick.

The resistivities of the internal and external fluids are about the same, but their chemical compositions are substantially different. The external fluid is similar to sea-

588

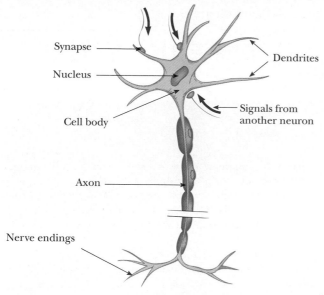

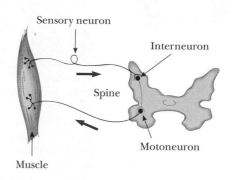

FIGURE 1
Diagram of a neuron.

FIGURE 2
A simple neural circuit.

water. Its ionic solutes are mostly positive sodium ions and negative chloride ions. Inside the axon, the positive ions are mostly potassium ions and the negative ions are mostly large organic ions.

Since there is a large concentration of sodium ions outside the axon and a large concentration of potassium ions inside, we may ask why the concentrations are not equalized by diffusion. In other words, why don't the sodium ions leak into the axon and the potassium ions leak out of it? The answer lies in the properties of the axon membrane.

In the resting condition, when the axon is not conducting an electric pulse, the axon membrane is highly permeable to potassium ions, slightly permeable to sodium ions, and impermeable to large organic ions. Thus, while sodium ions cannot easily leak into the axon, potassium ions can certainly leak out of it. As the potassium ions leak out of the axon, however, they leave behind the large negative organic ions, which cannot follow them through the membrane. As a result, a negative potential is produced inside the axon with respect to the outside. The negative potential, which has been measured to about 70 mV, holds back the outflow of potassium ions so that, at equilibrium, the concentration of ions is as we have stated.

The mechanism for the production of an electric signal by the neuron is conceptually remarkably simple. When a neuron receives an appropriate stimulus, which may be heat, pressure, or a signal from another neuron, the properties of its membrane change. As a result, sodium ions rush into the cell while potassium ions flow out of it. This flow of charged particle constitutes an electric current signal which propagates along the axon to its destination.

Although the axon is a highly complex structure, its main electrical properties can be represented by the standard electric circuit concepts of resistance and capacitance. The propagation of the signal along the axon is then well described by the techniques of electric circuit analysis discussed in the text.

FIGURE 3
Stellate neuron from human cortex. (© *Dr. Dennis Kunkel/Phototake*)

Questions

1. Why has the nervous system developed to utilize sodium and potassium ions to conduct electrical signals?
2. In the nervous system, the strength of a stimulus is conveyed by the number of pulses produced rather than by the amplitude of the signal. What is the advantage of this arrangement?

19

MAGNETISM

Oxygen, a paramagnetic substance, is attracted to a magnetic field. The liquid oxygen in this photograph is suspended between the poles of a permanent magnet. Paramagnetic substances contain atoms (or ions) that have permanent magnetic dipole moments. These dipoles interact weakly with each other and are randomly oriented in the absence of an external magnetic field. When the substance is placed in an external magnetic field, its atomic dipoles tend to line up with the field. (Courtesy of Leon Lewandowski)

T he list of important technological applications of magnetism is very long. For instance, large electromagnets are used to pick up heavy loads. Magnets are also used in such devices as meters, motors, and loudspeakers. Magnetic tapes are routinely used in sound and video recording equipment and for computer memory, and magnetic recording material is used on computer disks. Intense magnetic fields generated by superconducting magnets are currently being used to contain the plasmas (heated to temperatures on the order of 10^8 K) used in controlled nuclear fusion research.

As we investigate magnetism in this chapter, you will find that the subject cannot be divorced from electricity. For example, magnetic fields affect moving charges and moving charges produce magnetic fields. The ultimate source of all magnetic fields is electric current, whether it be the current in a wire or the current produced by the motion of charges within atoms or molecules.

19.1

MAGNETS

Most people have had experience with some form of magnet. You are most likely familiar with the common iron horseshoe magnet that can pick up iron-containing objects such as paper clips and nails. In the discussion that follows, we shall assume that the magnet has the shape of a bar. Iron objects are most strongly attracted to the ends of such a bar magnet, called its **poles.** One end is called the **north pole,** and the other the **south pole.** The names come from the behavior of a magnet in the presence of the Earth's magnetic field. If a bar magnet is suspended from its midpoint by a piece of string so that it can swing freely in a horizontal plane, it will rotate until its north pole points to the north of the Earth and its south pole points to the south of the Earth. The same idea is used to construct a simple compass. Magnetic poles also exert attractive or repulsive forces on each other similar to the electrical forces between charged objects. In fact, simple experiments with two bar magnets show that

> **Like poles repel each other and unlike poles attract each other.**

Although the force between two magnetic poles is similar to the force between two electric charges, there is an important difference. Electric charges can be isolated (witness the proton and the electron), but magnetic poles cannot be isolated. In fact, no matter how many times a permanent magnet is cut, each piece always has a north pole and a south pole. Thus, magnetic poles always occur in pairs. There is some theoretical basis for the speculation that magnetic monopoles (isolated north or south poles) may exist in nature, and attempts to detect them are currently an active experimental field of investigation. However, none of these attempts has proven successful.

There is yet another similarity between electric and magnetic effects, which concerns methods for making permanent magnets. In Chapter 15 we learned that when two materials such as rubber and wool are rubbed together, each becomes charged, one positively and the other negatively. In a somewhat analogous fashion, an unmagnetized piece of iron can be magnetized by stroking with a magnet. Magnetism can be induced in iron (and other materials) by other means. For example, if a piece of unmagnetized iron is placed near a strong permanent magnet, the piece of iron eventually becomes magnetized. The process can be accelerated by either heating and cooling the iron or by hammering. Naturally occurring magnetic materials, such as magnetite, achieve their magnetism in this manner, since they have been subjected to the Earth's magnetic field over very long periods of time. The extent to which a piece of material retains its magnetism depends on whether it is classified as being magnetically hard or soft. **Soft** magnetic materials, such as iron, are easily magnetized but also tend to easily lose their magnetism. In contrast, **hard** magnetic materials such as cobalt and nickel are difficult to magnetize but tend to retain their magnetism.

In earlier chapters we found it convenient to describe the interaction between charged objects in terms of electric fields. Recall that an electric field surrounds any electric charge. The region of space surrounding a *moving* charge also includes a magnetic field. In addition, a magnetic field surrounds any magnetized material.

An assortment of commercially available magnets. The four red magnets and the large black magnet on the left are made of an alloy of iron, aluminum, and cobalt. The six horseshoe magnets on the right are made of different nickel alloy steels. The rectangular magnets on the lower right are ceramics made of iron, nickel, and beryllium oxides. *(Courtesy of Central Scientific Company)*

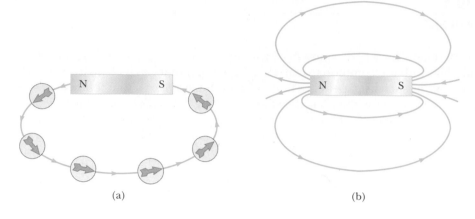

(a)

(b)

FIGURE 19.1
(a) Tracing the magnetic field of a bar magnet. (b) Several magnetic field lines of a bar magnet.

To describe any type of field, we must define its magnitude, or strength, and its direction. The direction of the magnetic field, **B**, at any location is the direction in which the north pole of a compass needle points at that location. Figure 19.1a shows how the magnetic field of a bar magnet can be traced with the aid of a compass. Several magnetic field lines of a bar magnet traced out in this manner appear in Figure 19.1b. Magnetic field patterns can be displayed by small iron filings, as shown in Figure 19.2.

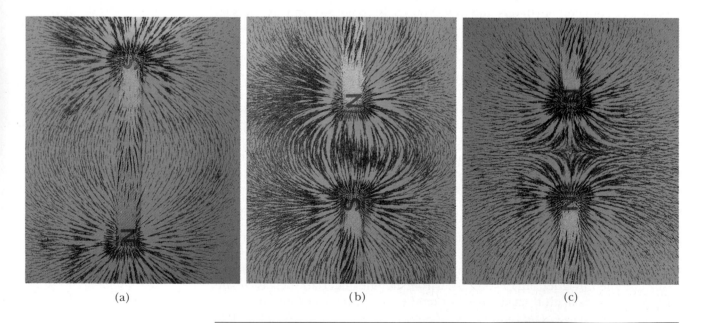

(a) (b) (c)

FIGURE 19.2
(a) The magnetic field pattern of a bar magnet, displayed by iron filings on a sheet of paper. (b) The magnetic field pattern between *unlike* poles of two bar magnets, displayed by iron filings. (c) The magnetic field pattern between two *like* poles. *(Courtesy of Henry Leap and Jim Lehman)*

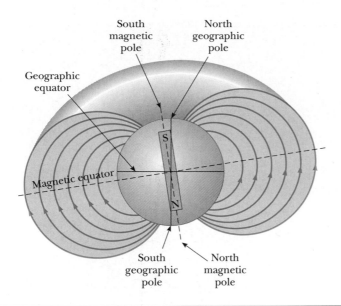

South
magnetic
pole

North
geographic
pole

Geographic
equator

Magnetic equator

South
geographic
pole

North
magnetic
pole

FIGURE 19.3

The Earth's magnetic field lines. Note that magnetic south is at the north geographic pole, and magnetic north is at the south geographic pole.

19.2

MAGNETIC FIELD OF THE EARTH

When we speak of a small bar magnet as having north and south poles, we should more properly say that it has a "north-seeking" pole and a "south-seeking" pole. By this we mean that if such a magnet is used as a compass, one end will seek, or point to, the north geographic pole of the Earth. Thus, we conclude that

> **The magnetic north pole corresponds to the south geographic pole, and the magnetic south pole corresponds to the north geographic pole.**

In fact, the configuration of the Earth's magnetic field, pictured in Figure 19.3, very much resembles what would be achieved by burying a bar magnet deep in the interior of the Earth.

If a compass needle is suspended in bearings that allow it to rotate in the vertical plane as well as in the horizontal plane, the needle is horizontal with respect to the Earth's surface only near the equator. As the device is moved northward, the needle rotates so that it points more and more toward the surface of the Earth. Finally, at a point just north of Hudson Bay in Canada, the north pole of the needle points directly downward. This site, first found in 1832, is considered to be the location of the south magnetic pole of the Earth. It is approximately 1300 mi from the Earth's geographic north pole and varies with time. Similarly, the magnetic north pole of the Earth is about 1200 miles from the geographic south pole. Thus, it is only approximately correct to say that a compass needle points north. The difference between true north, defined as the geographic north pole, and north indicated by a compass varies from point to point on the Earth, and the difference is referred to as *magnetic declination*. For

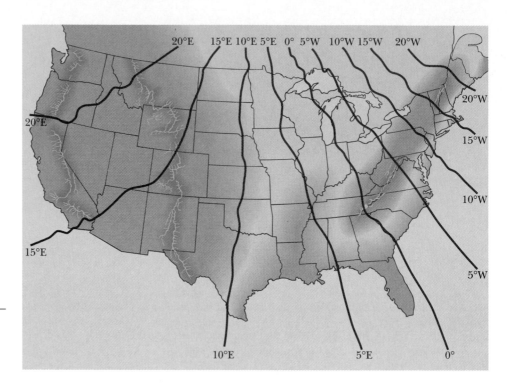

FIGURE 19.4
A map of the United States showing the declination of a compass from true north.

example, along a line through South Carolina and the Great Lakes, a compass indicates true north, whereas in Washington state it aligns 25° east of true north (Fig. 19.4).

Although the magnetic field pattern of the Earth is similar to the pattern that would be set up by a bar magnet deep in the Earth, it is easy to understand why the source of the Earth's field cannot be large masses of permanently magnetized material. The Earth does have large deposits of iron ore deep beneath its surface, but the high temperatures in the Earth's core prevent the iron from retaining any permanent magnetization. It is considered more likely that the true source of the Earth's field is charge-carrying convection currents in its core. Charged ions or electrons circling in the liquid interior could produce a magnetic field. There is also some evidence that the strength of a planet's field is related to the planet's rate of rotation. For example, Jupiter rotates faster than the Earth, and recent space probes indicate that Jupiter's magnetic field is stronger than ours. Venus, on the other hand, rotates more slowly than the Earth, and its magnetic field is found to be weaker. Investigation into the cause of the Earth's magnetism continues.

An interesting sidelight concerns the Earth's magnetic field. It has been found that the direction of the field has reversed several times during the last million years. Evidence for this is provided by basalt (an iron-containing rock) that is sometimes spewed forth by volcanic activity on the ocean floor. As the lava cools, it solidifies and retains a picture of the Earth's magnetic field direction. When the basalt deposits are dated, they provide evidence for periodic reversals of the magnetic field.

It has long been speculated that some animals, such as birds, use the magnetic field of the Earth to guide their migrations. Studies have shown that a type of anaerobic bacterium that lives in swamps has a magnetized chain of magnetite as part of its internal structure. (The term *anaerobic* means that these bacteria live and grow without oxygen; in fact, oxygen is toxic to them.) The magnetized chain acts as a compass needle that enables the bacteria to align with the Earth's magnetic field. When they find themselves out of the mud on the bottom of the swamp, they return to their oxygen-free environment by following the magnetic field lines of the Earth. Further evidence for their magnetic sensing ability is the fact that bacteria found in the Northern Hemisphere have internal magnetite chains that are opposite in polarity to those of similar bacteria in the Southern Hemisphere. This is consistent with the fact that in the Northern Hemisphere the Earth's field has a downward component, whereas in the Southern Hemisphere it has an upward component.

19.3

MAGNETIC FIELDS

Experiments show that a stationary charged particle does not interact with a static magnetic field. However, *when moving through a magnetic field a charged particle experiences a force.* This force has its maximum value when the charge moves perpendicularly to the magnetic field lines, decreases in value at other angles, and becomes zero when the particle moves along the field lines. We shall make use of these observations in describing the magnetic field.

In our discussion of electricity, the electric field at some point in space was defined as the electric force per unit charge acting on some test charge placed at that point. In a similar manner, we can describe the properties of the magnetic field, **B**, at some point in terms of the magnetic force exerted on a test charge at that point. Our test object is assumed to be a charge, q, moving with velocity v. It is found experimentally that the strength of the magnetic force on the particle is proportional to the magnitude of the charge, q, the magnitude of the velocity, v, the strength of the external magnetic field, **B**, and the sine of the angle θ between the direction of v and the direction of **B**. These observations can be summarized by writing the magnitude of the magnetic force as

$$F = qvB \sin \theta$$

[19.1]   Magnetic force

This expression is used to define the magnitude of the magnetic field as

$$B \equiv \frac{F}{qv \sin \theta}$$

[19.2]   Magnetic field defined

If F is in newtons, q in coulombs, and v in meters per second, the SI unit of magnetic field is the **tesla** (T), also called the **weber** (Wb) **per square meter** (that is, $1 \text{ T} = 1 \text{ Wb/m}^2$). Thus, if a 1-C charge moves through a magnetic field of magnitude 1 T with a velocity of 1 m/s, perpendicularly to the field ($\sin \theta = 1$), the magnetic force exerted on the charge is 1 N. We can express the units of B as

$$[B] = T = \frac{Wb}{m^2} = \frac{N}{C \cdot m/s} = \frac{N}{A \cdot m}$$

[19.3]

PHYSICS IN *ACTION*

THE MOTION OF CHARGED PARTICLES IN MAGNETIC FIELDS

(*Left*) The white arc in this photograph indicates the circular path followed by an electron beam in a magnetic field. The vessel contains gas at very low pressure, and the beam is made visible as the electrons collide with the gas atoms, which emit visible light. The magnetic field is produced by two coils (not shown). The apparatus can be used to measure the ratio of e/m for the electron.

(*Right*) Aurora borealis (the northern lights), photographed near Fairbanks, Alaska. Auroras occur when cosmic rays—electrically charged particles, mainly from the Sun—become trapped in the Earth's atmosphere over the magnetic poles and collide with other atoms, resulting in the emission of visible light.

(*Courtesy of Central Scientific Company*)

(*Jack Finch/SPL/Photo Researchers*)

In practice, the cgs unit for magnetic field, the **gauss** (G), is often used. The gauss is related to the tesla through the conversion

$$1 \text{ T} = 10^4 \text{ G}$$

Conventional laboratory magnets can produce magnetic fields as large as about 25 000 G, or 2.5 T. Superconducting magnets that can generate magnetic fields as great as 3×10^5 G, or 30 T, have been constructed. These values can be compared with the value of the Earth's magnetic field near its surface, which is about 0.5 G, or 0.5×10^{-4} T.

From Equation 19.1 we see that the force on a charged particle moving in a magnetic field has its maximum value when the particle moves *perpendicularly* to the magnetic field, corresponding to $\theta = 90°$, so that $\sin \theta = 1$. The magnitude of this maximum force has the value

$$F_{\text{max}} = qvB \qquad \textbf{[19.4]}$$

Also, note from Equation 19.1 that F is zero when **v** is parallel to **B** (corresponding to $\theta = 0°$ or $180°$). Thus, no magnetic force is exerted on a charged particle when it moves in the direction of the magnetic field or opposite the field.

Experiments show that the direction of the magnetic force is always perpendicular to both **v** and **B**, as shown in Figure 19.5. To determine the direction of the force, we employ the following right-hand rule:

> Hold your right hand open as illustrated in Figure 19.6, and then place your fingers in the direction of **B** with your thumb pointing in the direction of **v**. The force, **F**, on a positive charge is directed *out* of the palm of your hand.

If the charge is negative rather than positive, the force is directed *opposite* that shown in Figure 19.6. That is, if q is negative, simply use the right-hand rule to find the direction of **F** for positive q, and then reverse this direction for the negative charge.

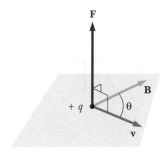

FIGURE 19.5
The direction of the magnetic force on a charged particle moving with a velocity of **v** in the presence of a magnetic field. When **v** is at an angle of θ with **B**, the magnetic force is perpendicular to both **v** and **B**.

EXAMPLE 19.1   A Proton Traveling in the Earth's Magnetic Field

A proton moves with a speed of 1.0×10^5 m/s through the Earth's magnetic field, which has a value of 55 μT at a particular location. When the proton moves eastward, the magnetic force acting on it is a maximum, and when it moves northward, no magnetic force acts on it. What is the strength of the magnetic force, and what is the direction of the magnetic field?

Solution   The magnitude of the force can be found from Equation 19.4:

$$F_{max} = qvB = (1.6 \times 10^{-19}\ \text{C})(1.0 \times 10^5\ \text{m/s})(55 \times 10^{-6}\ \text{T})$$

$$= \boxed{8.8 \times 10^{-19}\ \text{N}}$$

The direction of the magnetic field cannot be determined precisely from the information given in the problem. Since no magnetic force acts on a charged particle when it is moving parallel to the field, all that we can say for sure is that the magnetic field is directed either northward or southward.

Exercise   Calculate the gravitational force on the proton, and compare it with the magnetic force. Note that the mass of the proton is 1.67×10^{-27} kg.

Answer 1.6×10^{-26} N. $F_{grav}/F_{max} \approx 1.8 \times 10^{-8}$

EXAMPLE 19.2   A Proton Moving in a Strong Magnetic Field

A proton moves at 8.0×10^6 m/s along the x axis. It enters a region where there is a magnetic field of magnitude 2.5 T, directed at an angle of 60° with the x axis and lying in the xy plane (Fig. 19.7). Calculate the initial force on and acceleration of the proton.

Solution   From Equation 19.1 we get

$$F = qvB \sin \theta = (1.6 \times 10^{-19}\ \text{C})(8.0 \times 10^6\ \text{m/s})(2.5\ \text{T})(\sin 60°)$$

$$= \boxed{2.8 \times 10^{-12}\ \text{N}}$$

Use the right-hand rule, noting that the charge is positive, to see that the force is in the positive z direction. Verify that the units of F in the calculation reduce to newtons.

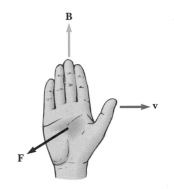

FIGURE 19.6
The right-hand rule for determining the direction of the magnetic force on a positive charge moving with a velocity of **v** in a magnetic field, **B**. With your thumb in the direction of **v** and your four fingers in the direction of **B**, the force is directed out of the palm of your hand.

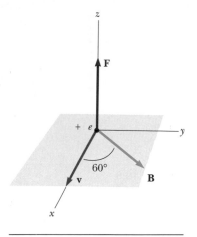

FIGURE 19.7
(Example 19.2)   The magnetic force, **F**, on a proton is in the positive z direction when **v** and **B** lie in the xy plane.

Since the mass of the proton is 1.67×10^{-27} kg, its initial acceleration is

$$a = \frac{F}{m} = \frac{2.8 \times 10^{-12} \text{ N}}{1.67 \times 10^{-27} \text{ kg}} = \boxed{1.7 \times 10^{15} \text{ m/s}^2}$$

in the positive z direction.

Exercise   Calculate the acceleration of an electron that moves through the same magnetic field at the same speed as the proton. The mass of an electron is 9.11×10^{-31} kg.

Answer 3.0×10^{18} m/s^2 in the negative z direction.

19.4
MAGNETIC FORCE ON A CURRENT-CARRYING CONDUCTOR

If a force is exerted on a single charged particle when it moves through a magnetic field, it should be no surprise that a current-carrying wire also experiences a force when placed in a magnetic field. This follows from the fact that the current is a collection of many charged particles in motion; hence, the resultant force on the wire is due to the sum of the individual forces on the charged particles. The force on the particles is transmitted to the "bulk" of the wire through collisions with the atoms making up the wire.

Before we continue, some explanation is in order concerning notation in many of the figures. To indicate the direction of **B**, we use the following convention. If **B** is directed into the page, as in Figure 19.8, we use a series of blue crosses, representing the tails of arrows. If **B** is directed out of the page, we use a series of blue dots, representing the tips of arrows. If **B** lies in the plane of the page, we use a series of blue field lines with arrowheads.

The force on a current-carrying conductor can be demonstrated by hanging a wire between the faces of a magnet as in Figure 19.8. In this figure, the magnetic field is directed into the page and covers the region within the shaded circle. The wire deflects to the right or left when a current is passed through it.

Let us quantify this discussion by considering a straight segment of wire of length ℓ and cross-sectional area A, carrying current I in a uniform external

FIGURE 19.8
A flexible vertical wire partially stretched between the faces of a magnet, with the field (blue crosses) directed into the page. (a) When there is no current in the wire, it remains vertical. (b) When the current is upward, the wire deflects to the left. (c) When the current is downward, the wire deflects to the right.

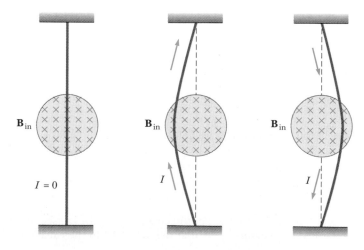

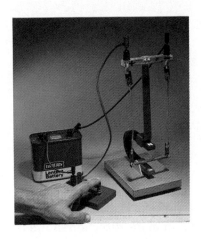

This apparatus demonstrates the force on a current-carrying conductor in an external magnetic field. Why does the bar swing *away* from the magnet after the switch is closed? *(Courtesy of Henry Leap and Jim Lehman)*

magnetic field, **B**, as in Figure 19.9. We assume that the magnetic field is perpendicular to the wire and is directed into the page. Each charge carrier in the wire experiences a force of magnitude $F_{max} = qv_d B$, where v_d is the drift velocity of the charge. To find the total force on the wire, we multiply the force on one charge carrier by the number of carriers in the segment. Since the volume of the segment is $A\ell$, the number of carriers is $nA\ell$, where n is the number of carriers per unit volume. Hence, the magnitude of the total magnetic force on the wire of length ℓ is

   Total force = (force on each charge carrier)(total number of carriers)

$$F_{max} = (qv_d B)(nA\ell)$$

From Chapter 17, however, we know that the current in the wire is given by $I = nqv_d A$. Therefore, F_{max} can be expressed as

$$F_{max} = BI\ell \qquad \text{[19.5]}$$

This equation can be used only when *the current and the magnetic field are at right angles to each other.*

   If the wire is not perpendicular to the field but is at some arbitrary angle, as in Figure 19.10, the magnitude of the magnetic force on the wire is

$$F = BI\ell \sin \theta \qquad \text{[19.6]}$$

where θ is the angle between **B** and the direction of the current. The direction of this force can be obtained by use of the right-hand rule. However, in this case, you must place your thumb in the direction of the current rather than in the direction of v. In Figure 19.10, the direction of the magnetic force on the wire is out of the page.

   Finally, when the current is either in the direction of the field or opposite the direction of the field, the magnetic force on the wire is zero.

   The fact that a magnetic force acts on a current-carrying wire in a magnetic field is the operating principle of most loudspeakers in sound systems. One speaker design, shown in Figure 19.11, consists of a coil of wire, called the voice coil; a flexible paper cone that acts as the speaker; and a permanent magnet. The coil of wire surrounding the north pole of the magnet is shaped so that the

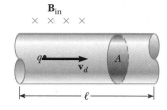

FIGURE 19.9
A section of a wire containing moving charges in an external magnetic field, **B**.

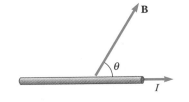

FIGURE 19.10
A wire carrying a current, I, in the presence of an external magnetic field, **B**, that makes an angle of θ with the wire.

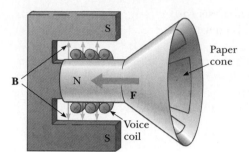

FIGURE 19.11
A cross-sectional view of a loudspeaker.

magnetic field lines are directed radially outward from the coil's axis. When an electrical signal is sent to the coil, producing a current in the direction shown in Figure 19.11, a magnetic force to the left acts on the coil. (This can be seen by applying the right-hand rule to each turn of wire.) When the current reverses direction, as it would for a sinusoidally varying current, the magnetic force on the coil also reverses direction, and the cone accelerates to the right. An alternating current through the coils causes an alternating force on the speaker, which results in vibrations of the cone. The cone creates sound waves as it pushes and pulls on the air in front of it. In this way an electrical signal is converted to a sound wave.

EXAMPLE 19.3   A Current-Carrying Wire in the Earth's Magnetic Field

A wire carries a current of 22 A from east to west. Assume that at this location the magnetic field of the Earth is horizontal and directed from south to north, and that it has a magnitude of 0.50×10^{-4} T. Find the force on a 36-m length of wire. How does the force change if the current runs west to east?

Solution   Because the directions of the current and magnetic field are at right angles, we can use Equation 19.5. The magnitude of the force is

$$F_{\text{max}} = BI\ell = (0.50 \times 10^{-4}\,\text{T})(22\,\text{A})(36\,\text{m}) = \boxed{4.0 \times 10^{-2}\,\text{N}}$$

The right-hand rule shows that the force on the wire is directed toward the Earth.

   If the current is directed from west to east, the force has the same magnitude but its direction is upward, away from the Earth.

Exercise   If the current is directed north to south, what is the magnetic force on the wire?

Answer   Zero.

19.5

TORQUE ON A CURRENT LOOP

In the preceding section we showed how a force is exerted on a current-carrying conductor when the conductor is placed in an external magnetic field. With this as a starting point, we now show that a torque is exerted on a current loop placed in a magnetic field. The results of this analysis will be of great practical value when we discuss the galvanometer (in this chapter) and generators (in Chapter 20).

Consider a rectangular loop carrying current I in the presence of an external uniform magnetic field in the plane of the loop, as shown in Figure 19.12a. The forces on the sides of length a are zero because these wires are parallel to the field. The magnitude of the forces on the sides of length b, however, is

$$F_1 = F_2 = BIb$$

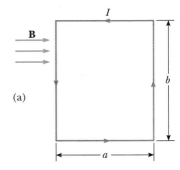

(a)

The direction of $\mathbf{F}_1$, the force on the left side of the loop, is out of the page, and that of $\mathbf{F}_2$, the force on the right side of the loop, is into the page. If we view the loop from an end, as in Figure 19.12b, the forces are directed as shown. If we assume that the loop is pivoted so that it can rotate about point O, we see that these two forces produce a torque about O that rotates the loop clockwise. The magnitude of this torque, τ_{max}, is

$$\tau_{max} = F_1 \frac{a}{2} + F_2 \frac{a}{2} = (BIb)\frac{a}{2} + (BIb)\frac{a}{2} = BIab$$

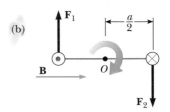

(b)

where the moment arm about O is $a/2$ for both forces. Since the area of the loop is $A = ab$, the torque can be expressed as

$$\tau_{max} = BIA \qquad \text{[19.7]}$$

Note that this result is valid only when the magnetic field is parallel to the plane of the loop, as in the view in Figure 19.12b. If the field makes an angle of θ with a line perpendicular to the plane of the loop, as in Figure 19.12c, the moment arm for each force is given by $(a/2)\sin\theta$. An analysis like that just used produces, for the magnitude of the torque,

$$\tau = BIA \sin\theta \qquad \text{[19.8]}$$

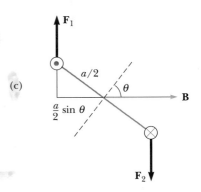

(c)

This result shows that the torque has the *maximum* value, BIA, when the field is parallel to the plane of the loop ($\theta = 90°$) and is *zero* when the field is perpendicular to the plane of the loop ($\theta = 0$). As seen in Figure 19.12c, the loop tends to rotate to smaller values of θ (so that the normal to the plane of the loop rotates toward the direction of the magnetic field).

Although this analysis is for a rectangular loop, a more general derivation would indicate that Equation 19.8 applies regardless of the shape of the loop. Furthermore, the torque on a loop with N turns is

$$\tau = NBIA \sin\theta \qquad \text{[19.9]}$$

FIGURE 19.12
(a) A rectangular loop in a uniform magnetic field, **B**. There are no magnetic forces on the sides of length a parallel to **B**, but there are forces acting on the sides of length b. (b) An end view of the rectangular loop shows that the forces $\mathbf{F}_1$ and $\mathbf{F}_2$ on the sides of length b create a torque that tends to twist the loop clockwise. (c) If **B** is at an angle of θ with a line perpendicular to the plane of the loop, the torque is given by $BIA \sin\theta$.

EXAMPLE 19.4   The Torque on a Circular Loop in a Magnetic Field

A circular wire loop of radius 50.0 cm is oriented at an angle of 30.0° to a magnetic field of 0.50 T, as shown in an edge view in Figure 19.13. The current in the loop is 2.0 A in the direction shown. Find the magnitude of the torque at this instant.

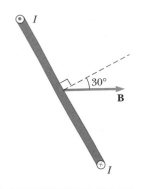

FIGURE 19.13
(Example 19.4) An edge view of a circular current loop in an external magnetic field, **B**.

Solution Regardless of the shape of the loop, Equation 19.8 is valid:

$$\tau = BIA \sin \theta = (0.50 \text{ T})(2.0 \text{ A})[\pi(0.50 \text{ m})^2](\sin 30.0°) = \boxed{0.39 \text{ N·m}}$$

Exercise Find the torque on the loop if it has three turns rather than one.

Answer The torque is three times that on the one-turn loop, or 1.2 N·m.

19.6

THE GALVANOMETER AND ITS APPLICATIONS

THE GALVANOMETER

A *galvanometer* is a device used in the construction of both ammeters and volt-meters. Its basic operation makes use of the fact that a torque acts on a current loop in the presence of a magnetic field. Figure 19.14 shows the main compo-nents of a galvanometer. It consists of a coil of wire mounted so that it is free to rotate on a pivot in a magnetic field provided by a permanent magnet. The torque experienced by the coil is proportional to the current through it. This means that the larger the current, the greater the torque and the more the coil will rotate before the spring tightens enough to stop the movement. Hence, the amount of deflection is proportional to the current. Once the instrument is properly calibrated, it can be used in conjunction with other circuit elements to measure either currents or potential differences. Figure 19.14b is a photograph of a large-scale galvanometer movement.

A GALVANOMETER IS THE BASIS OF AN AMMETER

A typical off-the-shelf galvanometer is usually not suitable for use as an ammeter (a current-measuring device). One of the main reasons is that a typical galva-nometer has a resistance of about 60 Ω, and an ammeter resistance this large can considerably alter the current in the circuit in which it is placed. To understand this, consider the following case. Suppose you construct a simple series circuit containing a 3-V battery and a 3-Ω resistor. The current in such a circuit is 1 A.

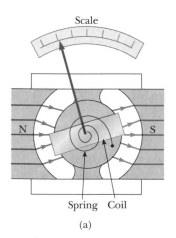

Scale

N S

Spring Coil

(a)

(b)

FIGURE 19.14
(a) The principal components of a galvanometer. When current passes through the coil, situated in a magnetic field, the magnetic torque causes the coil to twist. The angle through which the coil ro-tates is proportional to the current through it. (b) A large-scale demonstration model of a galva-nometer movement. Why does the coil rotate about the vertical axis after the switch is closed? *(Courtesy of Jim Lehman)*

However, if you include a 60-Ω galvanometer in the circuit in an attempt to measure the current, the total resistance of the circuit is now 63 Ω and the current is reduced to 0.048 A.

A second factor that limits the use of a galvanometer as an ammeter is the fact that a typical galvanometer gives a full-scale deflection for very low currents, on the order of 1 mA and less. Consequently, such a galvanometer cannot be used directly to measure currents greater than 1 mA.

Now suppose we wish to convert a 60-Ω, 1-mA galvanometer to an ammeter that deflects full-scale when 2 A passes through it. In spite of the factors just described, this can be accomplished by simply placing a resistor, R_p, in *parallel* with the galvanometer, as in Figure 19.15. (The combination of the galvanometer and parallel resistor constitutes an ammeter.) The size of the resistor must be selected so that when 2 A passes through the ammeter, only 0.001 A passes through the galvanometer and the remaining 1.999 A passes through the resistor, R_p, sometimes called the *shunt resistor*. Because the galvanometer and shunt resistor are in parallel, the potential differences across them are the same. Thus, using Ohm's law, we get

$$(0.001 \text{ A})(60 \text{ }\Omega) = (1.999 \text{ A}) R_p$$

$$R_p = 0.03 \text{ }\Omega$$

Notice that the shunt resistance, R_p, is extremely small. Thus, the configuration in Figure 19.15 solves both problems associated with converting a galvanometer to an ammeter. The ammeter just described can measure a large current (2 A) and has a low resistance, on the order of 0.03 Ω. (Recall that the equivalent resistance of two resistors in parallel is always *less* than the value of either of the individual resistors.)

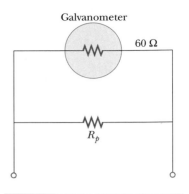

FIGURE 19.15
When a galvanometer is to be used as an ammeter, a resistor (R_p) is connected in parallel with the galvanometer.

A GALVANOMETER IS THE BASIS OF A VOLTMETER

With the proper modification, the basic galvanometer can also be used to measure potential differences in a circuit. To understand how to accomplish this, let us first calculate the largest voltage that can be measured with a galvanometer. If the galvanometer has a resistance of 60 Ω and gives a maximum deflection for a current of 1 mA, the largest voltage it can measure is

$$V_{\text{max}} = (0.001 \text{ A})(60 \text{ }\Omega) = 0.06 \text{ V}$$

From this you can see that some modification is required to enable this device to measure larger voltages. Furthermore, a voltmeter must have a very high resistance to ensure that it will not disturb the circuit in which it is placed. The basic galvanometer, with a resistance of only 60 Ω, is not acceptable for direct voltage measurements.

The circuit in Figure 19.16 has the basic modification that is necessary to convert a galvanometer to a voltmeter. Suppose we want to construct a voltmeter capable of measuring a maximum voltage of 100 V. In this situation, a resistor, R_s, is placed in *series* with the galvanometer. The value of R_s is found by noting that a current of 1 mA must pass through the galvanometer when the voltmeter

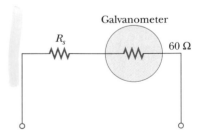

FIGURE 19.16
When a galvanometer is to be used as a voltmeter, a resistor (R_s) is connected in series with the galvanometer.

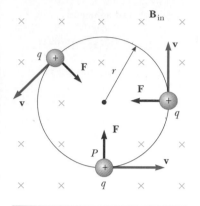

FIGURE 19.17

When the velocity of a charged particle is perpendicular to a uniform magnetic field, the particle moves in a circle whose plane is perpendicular to **B**, which is directed into the page. (The crosses represent the tails of the magnetic field vectors.) The magnetic force, **F**, on the charge is always directed toward the center of the circle.

is connected across a potential difference of 100 V. Application of Ohm's law to this circuit gives

$$100 \text{ V} = (0.001 \text{ A})(R_s + 60 \text{ } \Omega)$$

$$R_s = 99\,940 \text{ } \Omega$$

demonstrating that this voltmeter has a very high resistance.

When a voltmeter is constructed with several available ranges, values of R_s may be selected by use of a switch that can be connected to a preselected set of resistors. The required value of R_s increases as the maximum voltage to be measured increases.

19.7
MOTION OF A CHARGED PARTICLE IN A MAGNETIC FIELD

Consider the case of a positively charged particle moving in a uniform magnetic field so that the direction of the particle's velocity is *perpendicular to the field,* as in Figure 19.17. The label $\mathbf{B}_{in}$ indicates that **B** is directed into the page. Application of the right-hand rule at point *P* shows that the direction of the magnetic force, **F**, at this location is upward. This causes the particle to alter its direction of travel and to follow a curved path. Application of the right-hand rule at any point shows that *the magnetic force is always toward the center of the circular path;* therefore, the magnetic force is effectively a centripetal force that changes only the direction of **v** and not its magnitude. Since **F** is a centripetal force, we can equate its magnitude, qvB in this case, to the mass of the particle multiplied by the centripetal acceleration, v^2/r. From Newton's second law, we find that

$$F = qvB = \frac{mv^2}{r}$$

which gives

$$r = \frac{mv}{qB} \qquad \qquad \textbf{[19.10]}$$

This says that the radius of the path is proportional to the momentum, mv, of the particle and is inversely proportional to the magnetic field.

A magnet placed near the screen of a television tube distorts the picture as shown. This is due to the deflection of the electron beam in the tube in the presence of the magnetic field.
(Courtesy of Henry Leap and Jim Lehman)

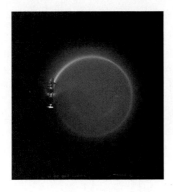

The bending of an electron beam in an external magnetic field. The tube contains gas at very low pressure, and the beam is made visible as the electrons collide with the gas atoms, which in turn emit visible light. The apparatus used to take this photograph is part of a system used to measure the ratio *e/m*. *(Courtesy of Henry Leap and Jim Lehman)*

If the initial direction of the velocity of the charged particle is not perpendicular to the magnetic field but instead is directed at an angle to the field, as shown in Figure 19.18, the path followed by the particle is a spiral (called a helix) along the magnetic field lines.

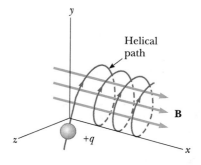

FIGURE 19.18
A charged particle that has a velocity vector with a component parallel to a uniform magnetic field moves in a helical path.

EXAMPLE 19.5   A Proton Moving Perpendicularly to a Uniform Magnetic Field

A proton is moving in a circular orbit of radius 14 cm in a uniform magnetic field of magnitude 0.35 T, directed perpendicularly to the velocity of the proton. Find the orbital speed of the proton.

Solution   From Equation 19.10, we get

$$v = \frac{qBr}{m} = \frac{(1.6 \times 10^{-19} \text{ C})(0.35 \text{ T})(14 \times 10^{-2} \text{ m})}{1.67 \times 10^{-27} \text{ kg}} = \boxed{4.7 \times 10^6 \text{ m/s}}$$

Exercise   If an electron moves perpendicularly to the same magnetic field with this speed, what is the radius of its circular orbit?

Answer 7.6×10^{-5} m.

EXAMPLE 19.6   The Mass Spectrometer

Two singly ionized atoms move out of a slit at point *S* in Figure 19.19 and into a magnetic field of 0.10 T. Each has a speed of 1.0×10^6 m/s. The nucleus of the first atom contains one proton and has a mass of 1.68×10^{-27} kg, and the nucleus of the second atom contains a proton and a neutron and has a mass of 3.36×10^{-27} kg. Atoms with the same chemical properties but different masses are called isotopes. The two isotopes here are hydrogen and deuterium. Find their distance of separation when they strike a photographic plate at *P*.

Solution   The radius of the circular path followed by the lighter isotope, hydrogen, is

$$r_1 = \frac{m_1 v}{qB} = \frac{(1.68 \times 10^{-27} \text{ kg})(1.0 \times 10^6 \text{ m/s})}{(1.6 \times 10^{-19} \text{ C})(0.10 \text{ T})} = 0.11 \text{ m}$$

The radius of the path of the heavier isotope, deuterium, is

$$r_2 = \frac{m_2 v}{qB} = \frac{(3.36 \times 10^{-27} \text{ kg})(1.0 \times 10^6 \text{ m/s})}{(1.6 \times 10^{-19} \text{ C})(0.10 \text{ T})} = 0.21 \text{ m}$$

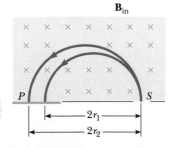

FIGURE 19.19
(Example 19.6)   Two isotopes leave the slit at point *S* and travel in different circular paths before striking a photographic plate at *P*.

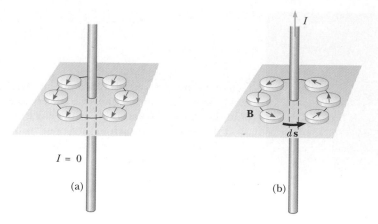

FIGURE 19.20
(a) When there is no current in the vertical wire, all compass needles point in the same direction. (b) When the wire carries a strong current, the compass needles deflect in directions tangent to the circle, pointing in the direction of **B** due to the current.

The distance of separation is

$$x = 2r_2 - 2r_1 = \quad 0.20 \text{ m}$$

The concepts used in this example underlie the operation of a device called a **mass spectrometer,** which is sometimes used to separate isotopes according to their mass-to-charge ratios, but more often is used to measure masses.

19.8

MAGNETIC FIELD OF A LONG, STRAIGHT WIRE AND AMPÈRE'S LAW

Hans Christian Oersted (1777–1851), Danish physicist. *(North Wind Picture Archives)*

During a lecture demonstration in 1819, the Danish scientist Hans Oersted (1777–1851) found that an electric current in a wire deflected a nearby compass needle. This discovery, linking a magnetic field with an electric current, was the beginning of our understanding of the origin of magnetism.

A simple experiment first carried out by Oersted in 1820 clearly demonstrates that a current-carrying conductor produces a magnetic field. In this experiment, several compass needles are placed in a horizontal plane near a long vertical wire, as in Figure 19.20a. When there is no current in the wire, all needles point in the same direction (that of the Earth's field), as one would expect. However, when the wire carries a strong, steady current, the needles all deflect in directions tangent to the circle, as in Figure 19.20b. These observations show that the direction of **B** is consistent with the following convenient rule:

If the wire is grasped in the right hand with the thumb in the direction of the current, as in Figure 19.21a, the fingers will curl in the direction of **B**.

When the current is reversed, the needles in Figure 19.20b also reverse.

Since the needles point in the direction of **B**, we conclude that the lines of **B** form circles about the wire. By symmetry, the magnitude of **B** is the same everywhere on a circular path centered on the wire and lying in a plane perpendicular to the wire. By varying the current and distance from the wire, one finds that **B** is proportional to the current and inversely proportional to the distance from the wire.

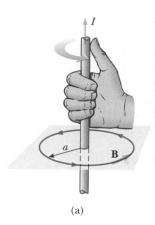

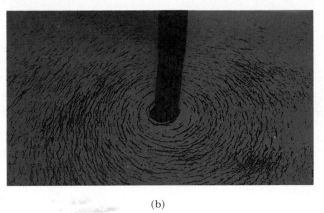

(a)   (b)

FIGURE 19.21
(a) The right-hand rule for determining the direction of the
magnetic field due to a long,
straight wire carrying a current.
Note that the magnetic field
lines form circles around the
wire. (b) Circular magnetic field
lines surrounding a current-
carrying wire, displayed by iron
filings. *(Courtesy of Henry Leap and
Jim Lehman)*

Shortly after Oersted's discovery, scientists arrived at an expression for the strength of the magnetic field due to the current in a long, straight wire. The magnetic field strength at distance r from a wire carrying current I is

$$B = \frac{\mu_0 I}{2\pi r} \qquad \text{[19.11]}$$

Magnetic field due to a long straight wire

This result shows that the magnitude of the magnetic field is proportional to the current and decreases as the distance from the wire increases, as one might intuitively expect. The proportionality constant μ_0, called the **permeability of free space,** is defined to have the value

$$\mu_0 \equiv 4\pi \times 10^{-7} \ \text{T}\cdot\text{m/A} \qquad \text{[19.12]}$$

*AMPÈRE'S LAW

Equation 19.11 enables us to calculate the magnetic field due to a long, straight wire carrying a current. A general procedure for deriving such equations was proposed by the French scientist André-Marie Ampère (1775–1836); it provides a relation between the current in an arbitrarily shaped wire and the magnetic field produced by the wire.

Consider a circular path surrounding a current, as in Figure 19.21a. The path can be divided into many short segments, each of length $\Delta \ell$. Let us now multiply one of these lengths by the component of the magnetic field parallel to that segment, where the product is labeled $B_{\parallel} \Delta \ell$. According to Ampère, the sum of all such products over the closed path is equal to μ_0 times the net current, I, that passes through the surface bounded by the closed path. This statement, known as **Ampère's circuital law,** can be written

$$\sum B_{\parallel} \Delta \ell = \mu_0 I \qquad \text{[19.13]}$$

Ampère's circuital law

where $\Sigma B_{\parallel} \Delta \ell$ means that we take the sum over all the products $B_{\parallel} \Delta \ell$ around the closed path.

We can use Ampère's circuital law to derive the magnetic field due to a long, straight wire carrying a current, I. As discussed earlier, the magnetic field lines of this configuration form circles with the wire at their centers, as shown in Figure 19.21a. The magnetic field is tangent to this circle at every point and has the same value, $B_{||}$, over the entire circumference of a circle of radius r. We now calculate the sum $\Sigma\, B_{||}\, \Delta\ell$ over a circular path and note that $B_{||}$ can be removed from the sum (since it has the same value for each element on the circle). Equation 19.13 then gives

$$\sum B_{||}\, \Delta\ell = B_{||} \sum \Delta\ell = B_{||}\, (2\,\pi r) = \mu_0\, I$$

Dividing both sides by $2\,\pi r$, we obtain

$$B = \frac{\mu_0 I}{2\,\pi r}$$

This is identical to Equation 19.11, which is the magnetic field of a long, straight current.

Ampère's circuital law is extremely important because it provides an elegant and simple method for calculating the magnetic fields of highly symmetric current configurations. However, it cannot be used to calculate magnetic fields for complex current configurations that lack symmetry.

EXAMPLE 19.7   The Magnetic Field of a Long Wire

A long, straight wire carries a current of 5.00 A. At one instant, a proton, 4.00 mm from the wire, travels at 1.50×10^3 m/s parallel to the wire and in the same direction as the current (Fig. 19.22). Find the magnitude and direction of the magnetic force that is acting on the proton because of the field produced by the wire.

Solution   From Equation 19.11, the magnitude of the magnetic field at a point 4.00 mm from the wire is

$$B = \frac{\mu_0 I}{2\,\pi r} = \frac{(4\pi \times 10^{-7}\ \text{T}\cdot\text{m/A})(5.00\ \text{A})}{2\,\pi(4.00 \times 10^{-3}\ \text{m})} = 2.50 \times 10^{-4}\ \text{T}$$

This field is directed into the page at the location of the proton, as shown by the right-hand rule for a long, straight wire (see Fig. 19.21).

FIGURE 19.22
(Example 19.7)   The magnetic field due to the current is into the page at the location of the proton, and the magnetic force on the proton is to the left.

André-Marie Ampère was a French mathematician, chemist, and philosopher who founded the science of electrodynamics. The unit of measure for electric current was named in his honor.

Ampère's genius, particularly in mathematics, became evident early in his life: he had mastered advanced mathematics by the age of 12. In his first publication, *Considerations on the Mathematical Theory of Games,* an early contribution to the theory of probability, he proposed the virtual inevitability of a player's losing a continuing game of chance (played with money) to a player with much greater financial resources.

Ampère is credited with the discovery of electromagnetism—the relationship between electric currents and magnetic fields. His work in this field was influenced by the findings of Danish physicist Hans Christian Oersted. Ampère presented a series of papers expounding the theory and basic laws of electromagnetism, which he called electrodynamics to differentiate it from the study of stationary electric forces, which he called electrostatics. The culmination of his studies came in 1827 with the publication of *Mathematical Theory of Electrodynamic Phenomena Deduced Solely from Experiment*, in which he derived precise mathematical formulations of electromagnetism, notably Ampère's law.

Many stories are told of Ampère's absent-mindedness, a trait he shared with Newton. On one occasion, he forgot to honor an invitation to dine with the Emperor Napoleon.

Ampère's personal life was filled with tragedy. His father, a wealthy city official, was guillotined during the French Revolution, and his wife died young, in 1803. Ampère died at the age of 61 of pneumonia. His judgment of his life is clear from the epitaph he chose for his gravestone: *Tandem felix* (Happy at last).

André-Marie Ampère (1775–1836)

(North Wind Picture Archives)

The magnetic force on the proton is

$$F = qvB = (1.60 \times 10^{-19}\ \text{C})(1.50 \times 10^3\ \text{m/s})(2.50 \times 10^{-4}\ \text{T})$$

$$= 6.00 \times 10^{-20}\ \text{N}$$

The force is directed toward the wire, as shown by the right-hand rule for the force on a moving charge (see Fig. 19.6).

19.9

MAGNETIC FORCE BETWEEN TWO PARALLEL CONDUCTORS

As we have seen, a magnetic force acts on a current-carrying conductor when the conductor is placed in an external magnetic field. Since a current in a conductor creates its own magnetic field, it is easy to understand that two current-carrying wires placed close together exert magnetic forces on each other. Consider two long, straight, parallel wires separated by the distance d and carrying currents I_1 and I_2 in the same direction, as shown in Figure 19.23. Let us determine the force on one wire due to a magnetic field set up by the other wire.

Wire 2, which carries current I_2, sets up magnetic field $\mathbf{B}_2$ at wire 1. The direction of $\mathbf{B}_2$ is perpendicular to the wire, as shown in the figure. Using Equation 19.11, we see that the magnitude of this magnetic field is

$$B_2 = \frac{\mu_0 I_2}{2\pi d}$$

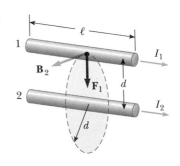

FIGURE 19.23
Two parallel wires, each carrying a steady current, exert forces on each other. The field at wire 1 due to wire 2, $\mathbf{B}_2$, produces a force on wire 1 given by $F_1 = B_2 I_1 \ell$. The force is attractive if the currents have the same direction, as shown, and repulsive if the two currents have opposite directions.

According to Equation 19.5, the magnetic force on wire 1 in the presence of field $\mathbf{B}_2$ due to I_2 is

$$F_1 = B_2 I_1 \ell = \left(\frac{\mu_0 I_2}{2\pi d}\right) I_1 \ell = \frac{\mu_0 I_1 I_2 \ell}{2\pi d}$$

We can rewrite this in terms of the force per unit length:

$$\frac{F_1}{\ell} = \frac{\mu_0 I_1 I_2}{2\pi d} \qquad \text{[19.14]}$$

The direction of $\mathbf{F}_1$ is downward, toward wire 2, as indicated by the right-hand rule. If one considers the field set up at wire 2 due to wire 1, the force $\mathbf{F}_2$ on wire 2 is found to be equal to and opposite $\mathbf{F}_1$. This is what one would expect from Newton's third law of action-reaction.

We have already shown that parallel conductors carrying currents in the same direction *attract* each other. Now we use the approach indicated by Figure 19.23 and the steps leading to Equation 19.14 to show that parallel conductors carrying currents in opposite directions *repel* each other.

The force between two parallel wires carrying a current is used to define the SI unit of current, the **ampere** (A), as follows:

> If two long, parallel wires 1 m apart carry the same current, and the force per unit length on each wire is 2×10^{-7} N/m, then the current is defined to be 1 A.

The SI unit of charge, the **coulomb** (C), can now be defined in terms of the ampere as follows:

> If a conductor carries a steady current of 1 A, then the quantity of charge that flows through any cross-section in 1 s is 1 C.

EXAMPLE 19.8   Levitating a Wire

Two wires, each having a weight per unit length of 1.0×10^{-4} N/m, are strung parallel to one another above the surface of the Earth, one directly above the other. The wires are aligned in a north-south direction so that the Earth's magnetic field will not affect them. When their distance of separation is 0.10 m, what must be the current in each in order for the lower wire to levitate the upper wire? Assume that the wires carry the same currents, traveling in opposite directions.

Solution   If the upper wire is to float, it must be in equilibrium under the action of two forces: its weight and magnetic repulsion. The weight per unit length—here 1.0×10^{-4} N/m—must be equal and opposite the magnetic force per unit length given in Equation 19.14. Since the currents are the same, we have

$$\frac{F_1}{\ell} = \frac{mg}{\ell} = \frac{\mu_0 I^2}{2\pi d}$$

$$1.0 \times 10^{-4} \text{ N/m} = \frac{(4\pi \times 10^{-7} \text{ T·m/A})(I^2)}{(2\pi)(0.10 \text{ m})}$$

We solve for the current to find

$$I = \boxed{7.1 \text{ A}}$$

Exercise   If the current in each wire is doubled, what is the equilibrium separation of the two wires?

Answer   0.40 m.

19.10

MAGNETIC FIELD OF A CURRENT LOOP

The strength of the magnetic field set up by a piece of wire carrying a current can be enhanced at a specific location if the wire is formed into a loop. You can understand this by considering the effect of several small segments of the current loop, as in Figure 19.24. The small segment at the top of the loop, labeled Δx_1, produces at the loop's center a magnetic field of magnitude B_1, directed out of the page. The direction of **B** can be verified using the right-hand rule for a long, straight wire. Grab the wire with your right hand, with your thumb pointing in the direction of the current. Your fingers curl around in the direction of **B**.

A segment at the bottom of the loop, Δx_2, also contributes to the field at the center, thus increasing its strength. The field produced at the center of the current loop by the segment Δx_2 has the same magnitude as B_1 and is also directed out of the page. Similarly, all other such segments of the current loop contribute to the field. The net effect is a magnetic field for the current loop as pictured in Figure 19.25a.

Notice in Figure 19.25a that the magnetic field lines enter at the left side of the current loop and exit at the right. Thus, one side of the loop acts as though it were the north pole of a magnet, and the other acts as a south pole. The fact that the field set up by such a current loop bears a striking resemblance to the field of a bar magnet (Fig. 19.25c) will be of more than casual interest to us in a future section.

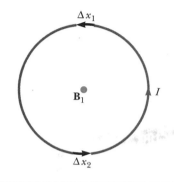

FIGURE 19.24

All segments of the current loop produce a magnetic field at the center of the loop, directed *out of the page.*

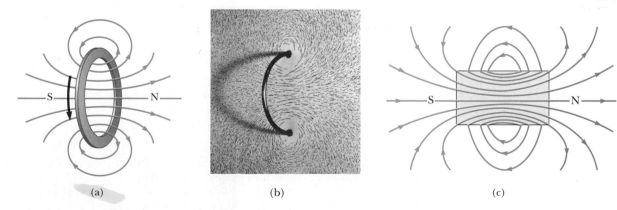

(a)   (b)   (c)

FIGURE 19.25

(a) Magnetic field lines for a current loop. Note that the magnetic field lines of the current loop resemble those of a bar magnet. (b) Field lines of a current loop, displayed by iron filings. *(Education Development Center, Newton, Mass.)* (c) The magnetic field of a bar magnet is similar to that of a current loop.

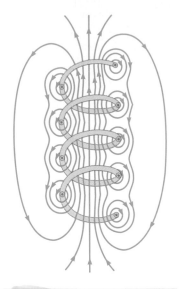

FIGURE 19.26

The magnetic field lines for a loosely wound solenoid. *(Adapted from D. Halliday and R. Resnick, Physics, New York, Wiley, 1978)*

The field inside a solenoid

19.11

MAGNETIC FIELD OF A SOLENOID

If a long, straight wire is bent into a coil of several closely spaced loops, the resulting device is a **solenoid,** often called an **electromagnet.** This device is important in many applications since it acts as a magnet only when it carries a current. As we shall see, the magnetic field inside a solenoid increases with the current and is proportional to the number of coils per unit length.

Figure 19.26 shows the magnetic field lines of a loosely wound solenoid of length ℓ and total number of turns N. Note that the field lines inside the solenoid are nearly parallel, uniformly spaced, and close together. This indicates that the field inside the solenoid is nearly uniform and strong. The exterior field at the sides of the solenoid is nonuniform and is much weaker than the interior field.

If the turns are closely spaced, the field lines are as shown in Figure 19.27a, entering at one end of the solenoid and emerging at the other. This means that one end of the solenoid acts as a north pole and the other end acts as a south pole. If the length of the solenoid is much greater than its radius, the lines that leave the north end of the solenoid spread out over a wide region before returning to enter the south end. Hence, as you can see in Figure 19.27a, the magnetic field lines outside are widely separated, indicative of a weak field. This is in contrast to a much stronger field inside the solenoid, where the lines are close together. Also, the field inside the solenoid has a constant magnitude at all points far from its ends. The expression for the field inside the solenoid is

$$B = \mu_0 nI \qquad \textbf{[19.15]}$$

where $n = N/\ell$ is the number of turns per unit length.

FIGURE 19.27

(a) Magnetic field lines for a tightly wound solenoid of finite length carrying a steady current. The field inside the solenoid is nearly uniform and strong. Note that the field lines resemble those of a bar magnet, so the solenoid effectively has north and south poles. (b) The magnetic field pattern of a bar magnet, displayed by small iron filings on a sheet of paper. *(Courtesy of Henry Leap and Jim Lehman)*

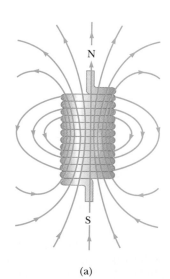

(a)

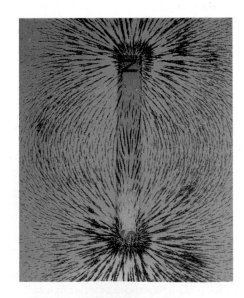

(b)

EXAMPLE 19.9   **The Magnetic Field Inside a Solenoid**

A certain solenoid consists of 100 turns of wire and has a length of 10.0 cm.

(a) Find the magnetic field inside the solenoid when it carries a current of 0.500 A.

Solution   The number of turns per unit length is

$$n = \frac{N}{\ell} = \frac{100 \text{ turns}}{0.10 \text{ m}} = 1000 \text{ turns/m}$$

so

$$B = \mu_0 nI = (4\pi \times 10^{-7} \text{ T·m/A})(1000 \text{ turns/m})(0.500 \text{ A}) = \quad 6.28 \times 10^{-4} \text{ T}$$

(b) Assume that the field still has almost this value just outside the solenoid, like the point labeled N in Figure 19.27a. Find the magnitude and direction of the magnetic force acting on an electron that is moving from right to left in the figure, through point N, at 375 m/s.

Solution   The magnitude of the magnetic force on the electron is

$$F = qvB = (1.60 \times 10^{-19} \text{ C})(375 \text{ m/s})(6.28 \times 10^{-4} \text{ T}) = \quad 3.77 \times 10^{-20} \text{ N}$$

By use of the right-hand rule in Figure 19.6, the direction of this force is found to be out of the page. Since the electron has a negative charge, we must reverse the direction of the magnetic force after applying the right-hand rule. This force will deflect the electron from its original direction of motion. So-called steering magnets placed along the neck of the picture tube in a television set, as in Figure 19.28, are used to make the electron beam move to the desired locations on the screen, thus tracing out the images of your favorite program.

Exercise   How many turns should the solenoid have (assuming it carries the same current) if the field inside is to be five times as great?

Answer   500 turns.

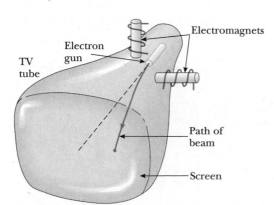

Electromagnets

Electron
gun

TV
tube

Path of
beam

Screen

FIGURE 19.28
Electromagnets are used to deflect electrons to desired positions on the screen of a television tube.

*19.12

MAGNETIC DOMAINS

The magnetic field produced by a current in a coil of wire gives us a hint as to what might cause certain materials to exhibit strong magnetic properties. A single coil like that in Figure 19.25 has a north pole and a south pole, but if this is true for a coil of wire, it should also be true for any current confined to a circular path. In particular, an individual atom should act as a magnet because of the motion of the electrons about the nucleus. Each electron, with its charge of 1.6×10^{-19} C, circles the atom once in about 10^{-16} s. If we divide the electronic charge by this time interval, we see that the orbiting electron is equivalent to a current of 1.6×10^{-3} A. Such a current produces a magnetic field on the order of 20 T at the center of the circular path. From this we see that a very strong magnetic field would be produced if several of these atomic magnets could be aligned inside a material. This does not occur, however, because the simple model we have described is not the complete story. A thorough analysis of atomic structure shows that the magnetic field produced by one electron in an atom is often canceled by an oppositely revolving electron in the same atom. The net result is that *the magnetic effect produced by the electrons orbiting the nucleus is either zero or very small for most materials.*

The magnetic properties of many materials are explained by the fact that an electron not only circles in an orbit but also spins on its axis like a top (Fig. 19.29). (This classical description should not be taken literally. The property of *spin* can be understood only with the methods of quantum mechanics, which we shall not discuss here.) The spinning electron represents a charge in motion that produces a magnetic field. The field due to the spinning is generally stronger than the field due to the orbital motion. In atoms containing many electrons, the electrons usually pair up with their spins opposite each other, so that their fields cancel each other. That is why most substances are not magnets. However, in certain strongly magnetic materials such as iron, cobalt, and nickel, the magnetic fields produced by the electron spins do not cancel completely. Such materials are said to be **ferromagnetic.** In ferromagnetic materials, strong coupling occurs between neighboring atoms, to form large groups of atoms whose spins are aligned, called **domains.** Typically, the sizes of these groups range from about 10^{-4} cm to 0.1 cm. In an unmagnetized substance the domains are randomly oriented, as shown in Figure 19.30a. When an external field is applied, as in Figures 19.30b, 19.31b, and 19.31c, the magnetic field of each domain tends to come nearer alignment with the external field, resulting in magnetization.

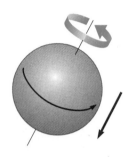

FIGURE 19.29
A model of a spinning electron.

FIGURE 19.30
(a) Random orientation of domains in an unmagnetized substance. (b) When an external magnetic field, $\mathbf{B}_0$, is applied, the domains tend to align with the magnetic field.

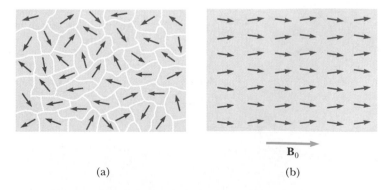

(a)

(b)

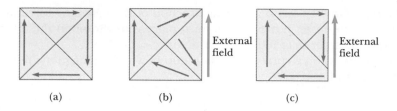

FIGURE 19.31
(a) An unmagnetized sample of four domains. (b) Magnetization due to rotation of the domains. (c) Magnetization due to a shift in domain boundaries.

In some substances a second effect occurs: domains that are already aligned with the field tend to grow at the expense of the others (Fig. 19.31c). Thus, two effects, both dependent on the domains, are responsible for a material's becoming magnetized.

In what are called hard magnetic materials, domain alignment persists after the external field is removed; the result is a **permanent magnet.** In soft magnetic materials, such as iron, once the external field is removed, thermal agitation produces domain motion and the material quickly returns to an unmagnetized state.

The alignment of domains explains why the strength of an electromagnet is increased dramatically by the insertion of an iron core into the magnet's center. The magnetic field produced by the current in the loops causes alignment of the domains, thus producing a large net external field. The use of iron as a core is also advantageous because it is a soft magnetic material and loses its magnetism almost instantaneously after the current in the coils is turned off.

SUMMARY

The **magnetic force** that acts on a charge, q, moving with velocity **v** in a magnetic field, **B**, has the magnitude

$$F = qvB \sin \theta \qquad \text{[19.1]}$$

where θ is the angle between **v** and **B**.

To find the direction of this force, you can use the **right-hand rule:** Place the fingers of your open right hand in the direction of **B** and point your thumb in the direction of the velocity, **v**. The force, **F**, on a positive charge is directed out of the palm of your hand.

If the charge is *negative* rather than positive, the force is directed opposite the force given by the right-hand rule.

The SI unit of magnetic field is the **tesla** (T), or weber per square meter (Wb/m²). An additional commonly used unit for magnetic field is the **gauss** (G); $1 \text{ T} = 10^4 \text{ G}$.

If a straight conductor of length ℓ carries current I, the magnetic force on that conductor when it is placed in a uniform external magnetic field, **B**, is

$$F = BI\ell \sin \theta \qquad \text{[19.6]}$$

The right-hand rule also gives the direction of the magnetic force on the conductor. In this case, however, you must place your thumb in the direction of the current rather than in the direction of **v**.

The torque, τ, on a current-carrying loop of wire in a magnetic field, **B**, has the magnitude

$$\tau = BIA \sin \theta \qquad\qquad \textbf{[19.8]}$$

where I is the current in the loop and A is its cross-sectional area. The angle between **B** and a line drawn perpendicularly to the plane of the loop is θ.

The galvanometer can be used in the construction of both ammeters and voltmeters.

If a charged particle moves in a uniform magnetic field so that its initial velocity is perpendicular to the field, it will move in a circular path whose plane is perpendicular to the magnetic field. The radius, r, of the circular path is

$$r = \frac{mv}{qB} \qquad\qquad \textbf{[19.10]}$$

where m is the mass of the particle and q is its charge.

The magnetic field at distance r from a **long, straight wire** carrying current I has the magnitude

$$B = \frac{\mu_0 I}{2\pi r} \qquad\qquad \textbf{[19.11]}$$

where $\mu_0 = 4\pi \times 10^{-7}\ \text{T} \cdot \text{m/A}$ is the **permeability of free space.** The magnetic field lines around a long, straight wire are circles concentric with the wire.

Ampère's law can be used to find the magnetic field around certain simple current-carrying conductors. It can be written

$$\sum B_{\parallel}\, \Delta \ell = \mu_0 I \qquad\qquad \textbf{[19.13]}$$

where $B_{\parallel}$ is the component of **B** tangent to a small current element of length $\Delta \ell$ that is part of a closed path, and I is the total current that penetrates the closed path.

The force per unit length on each of two parallel wires separated by the distance d and carrying currents I_1 and I_2 has the magnitude

$$\frac{F}{\ell} = \frac{\mu_0 I_1 I_2}{2\pi d} \qquad\qquad \textbf{[19.14]}$$

The forces are attractive if the currents are in the same direction and repulsive if they are in opposite directions.

The magnetic field inside a solenoid has the magnitude

$$B = \mu_0 n I \qquad\qquad \textbf{[19.15]}$$

where n is the number of turns of wire per unit length, $n = N/\ell$.

ADDITIONAL READING

S. Akasofu, "The Dynamic Aurora," *Sci. American,* May 1989, p. 90.

S. Banerjee, "Polar Flip-Flops," *The Sciences,* November/ December 1984, p. 24.

F. Bitter, *Magnets: The Education of a Physicist,* Science Study Series, Garden City, N.Y., Doubleday, 1959.

J. Bloxham and D. Gubbins, "The Evolution of the Earth's Magnetic Field," *Sci. American,* December 1989, p. 68.

R. A. Carrigan, Jr., and W. P. Trower, "Superheavy Magnetic Monopoles," *Sci. American,* April 1982, p. 106.

B. Dibner, *Oersted and the Discovery of Electromagnetism,* Blaisdell, 1962.

S. Felch, "Searches for Magnetic Monopoles and Fractional Electric Charge," *The Physics Teacher,* March 1984, p. 142.

K. A. Hoffman, "Ancient Magnetic Reversals: Clues to the Geodynamo," *Sci. American,* May 1988, p. 76.

H. H. Kolm and A. J. Freeman, "Intense Magnetic Fields," *Sci. American,* April 1965, p. 66.

A. Nier, "The Mass Spectrometer," *Sci. American,* March 1953, p. 68.

CONCEPTUAL QUESTIONS

Example   Why does the picture on a television screen become distorted when a magnet is brought near the screen?
Reasoning   The magnetic field produces an additional force on the electrons moving toward the screen. This magnetic force causes the electrons to be deflected to a spot other than the one to which they are supposed to go. The result is a distorted image.

Example   It is found that charged particles from outer space, called cosmic rays, strike the Earth more frequently at the poles than at the equator. Why?
Reasoning   At the poles, the magnetic field of the Earth points almost straight downward, in the direction the charges are moving. As a result, there is little or no magnetic force exerted on the charged particles at the poles to deflect them away from the Earth.

Example   If a charged particle moves in a straight line through some region of space, can you say that the magnetic field in that region is zero?
Reasoning   No. There may be another field such as an electric field or a gravitational field that produces a force on the charged particle that is strong enough to cancel out the magnetic force. Also, a charged particle moving parallel to a magnetic field would experience no magnetic force.

Example   A charged particle moves in a circular path in the presence of a magnetic field applied perpendicular to the particle's velocity vector. Does the particle gain energy from the magnetic field? Justify your answer.
Reasoning   No. The magnetic field produces a magnetic force which is directed toward the center of the circular path of the particle, and this causes a change in direction of the particle's velocity. However, since the magnetic force is always perpendicular to the displacement of the particle, no work is done on the particle, and its speed remains constant. Since the particle's speed does not change, its kinetic energy remains constant.

Example   How can a current loop be used to determine the presence of a magnetic field in a given region of space?
Reasoning   The loop can be mounted on an axle that can rotate. The current loop will rotate when placed in an external magnetic field for some arbitrary orientation of the field relative to the loop. As the current in the loop is increased, the torque on it will also increase.

Example   Is it possible to orient a current loop in a uniform magnetic field such that the loop will not tend to rotate?
Reasoning   Yes. If the magnetic field is directed perpendicular to the plane of the loop, the forces on opposite sides of the loop will be equal and opposite, but will produce no net torque on the loop.

Example   How can the motion of a charged particle be used to distinguish between a magnetic field and an electric field? Give a specific example to justify your argument.
Reasoning   The magnetic force on a moving charged particle is always perpendicular to the direction of motion. There is no magnetic force on the charge when it moves in the direction of the magnetic field. On the other hand, the force on a charged particle moving in an electric field is never zero. Therefore, by projecting the charged particle in different directions, it is possible to determine the nature of the field.

1. The north-seeking pole of a magnet is attracted toward the geographic north pole of the Earth; yet like poles repel. What is the way out of this dilemma?
2. Which way would a compass point if you were at the north magnetic pole?
3. Two charged particles are projected into a region where there is a magnetic field perpendicular to their velocities. If the particles are deflected in opposite directions, what can you say about them?

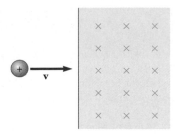

FIGURE 19.32   (Question 6)

4. Suppose an electron is chasing a proton up this page when suddenly a magnetic field is applied perpendicularly to the page. What happens to the particles?

5. At a given instant, a proton moves in the positive x direction in a region where there is a magnetic field in the negative z direction. What is the direction of the magnetic force? Does the proton continue to move in the positive x direction? Explain.

6. A proton moving horizontally enters a region where there is a uniform magnetic field perpendicular to the proton's velocity, as shown in Figure 19.32. Describe the proton's subsequent motion. How would an electron behave under the same circumstances?

7. List several similarities and differences between electric and magnetic forces.

8. Can a magnetic field set a resting electron in motion? If so, how?

9. Justify the following statement: "It is impossible for a constant (i.e., time-independent) magnetic field to alter the speed of a charged particle."

10. Explain why it is not possible to determine the charge and mass of a charged particle separately, by electric and magnetic forces.

11. If a solenoid were suspended by a string so that it could rotate freely, could it be used as a magnet when it carried a direct current? Could it also be used if the current were alternating in direction?

12. What would be the internal resistances of an ideal ammeter and an ideal voltmeter? Why should these be called *ideal* meters?

13. A *bubble chamber* is a device used for observing tracks of particles that pass through the chamber, which is immersed in a magnetic field. If some of the tracks are spirals and others are straight lines, what can you say about the particles?

14. Explain why two parallel wires carrying currents in opposite directions repel each other.

15. Two wires carrying equal but opposite currents are twisted together in the construction of a circuit. Why does this technique reduce stray magnetic fields?

16. Describe the change in the magnetic field inside a solenoid carrying a steady current if (a) the length of the solenoid is doubled but the number of turns re-

FIGURE 19.33   (Question 17) *(Courtesy of Central Scientific Company)*

mains the same; (b) the number of turns is doubled but the length remains the same.

17. The electron beam in Figure 19.33 is projected to the right. The beam deflects downward in the presence of a magnetic field produced by a pair of current-carrying coils. (a) What is the direction of the magnetic field? (b) What would happen to the beam if the current in the coils were reversed?

18. A magnet attracts a piece of iron. The iron can then attract another piece of iron. Explain, on the basis of alignment of the domains, what happens in each piece of iron.

19. Will a nail be attracted to either pole of a magnet? Explain what is happening inside the nail.

20. You are an astronaut stranded on a planet with no test equipment or minerals around. The planet does not even have a magnetic field. You have two iron bars in your possession; one is magnetized, one is not. How can you determine which is magnetized?

21. An electron travels in a circular orbit of radius 1.77 m in a region of uniform magnetic field. If the magnetic field strength doubles, what is the new radius of the orbit?

22. (a) Determine the units for the combination

$$\frac{1}{\sqrt{\mu_0 \epsilon_0}}$$

where ϵ_0 is the permittivity of vacuum and μ_0 is the permeability of vacuum. (b) How does this value compare to the speed of light in vacuum?

23. Explain why, if the current loop in Figure 19.12 were set in motion, the motion would be periodic. Under what conditions would it be simple harmonic?

24. The photograph shows two permanent magnets, with holes bored through their centers, placed one over the other. Because the poles of the upper magnet are reversed from those of the lower, the upper magnet levitates above the lower magnet. If the upper magnet were displaced slightly, either up or down, would the resulting motion be periodic? Explain. What would happen if the upper magnet were inverted?

25. As we have seen in this chapter, moving charges (electric currents) create magnetic fields. Because of the mathematical similarity between the magnitude of the electric force, $F = k|q_1||q_2|/r^2$, and the magnitude of the gravitational force, $F = Gm_1 m_2/r^2$, one might postulate that moving masses (gravitational currents) create a gravitational analog to the magnetic field. Comment on this possibility.

(Question 24)   *(Courtesy of Central Scientific Company)*

PROBLEMS

Section 19.3   Magnetic Fields

1. A duck flying due north at 15 m/s passes over Atlanta, where the magnetic field of the Earth is 5.0×10^{-5} T in a direction 60° below a horizontal line running north and south. The duck has a positive charge of 4.0×10^{-8} C. What is the magnetic force acting on the duck?

2. A proton moves at right angles to a magnetic field of 0.10 T with a speed of 2.0×10^7 m/s. Find the magnitude of the acceleration of the proton.

3. A proton travels with a speed of 3.0×10^6 m/s at an angle of 37° with the direction of a magnetic field of 0.30 T in the $+y$ direction. What are (a) the magnitude of the magnetic force on the proton and (b) the proton's acceleration?

4. A proton moves at 2.50×10^6 m/s horizontally, at right angles to a magnetic field. (a) What is the strength of the field that is required to just balance the weight of the proton and keep it moving horizontally? (b) Should the direction of the magnetic field be in a horizontal or a vertical plane?

5. (a) Find the direction of the force on a proton moving through the magnetic fields in Figure 19.34, as shown. (b) Repeat part (a), assuming the moving particle is an electron.

6. A proton moves eastward in the plane of the Earth's magnetic equator so that its distance from the ground remains constant. What is the speed of the proton?

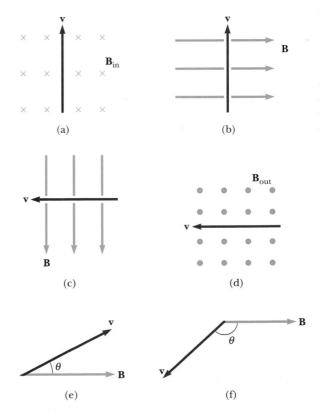

FIGURE 19.34 (Problems 5 and 17). For Problem 17, replace the velocity vector with a current in that direction.

□ indicates problems that have full solutions available in the Student Solutions Manual and Study Guide.

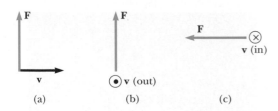

FIGURE 19.35 (Problems 7 and 18). For Problem 18, replace the velocity vector with a current in that direction.

7. Find the direction of the magnetic field on the positively charged particle moving in the various situations shown in Figure 19.35, if the direction of the magnetic force acting on it is as indicated.

8. Sodium ions (Na^+) move at 0.851 m/s through a bloodstream in the arm of a person standing near a large magnet. The magnetic field has a strength of 0.254 T and makes an angle of 51.0° with the motion of the sodium ions. The arm contains 100 cm³ of blood with 3.00×10^{20} Na^+ ions per cubic centimeter. If no other ions were present in the arm, what would be the magnetic force on the arm?

9. What speed would a proton need to circle the Earth 1000 km above the magnetic equator, where the Earth's magnetic field is directed on a line between magnetic north and south and has an intensity of 4.00×10^{-8} T?

10. An electron is accelerated through 2400 V from rest and then enters a region where there is a uniform 1.70-T magnetic field. What are the (a) maximum and (b) minimum values of the magnetic force this charge can experience?

11. A proton moves perpendicularly to a uniform magnetic field, **B**, at 1.0×10^7 m/s and experiences an acceleration of 2.0×10^{13} m/s² in the $+x$ direction when its velocity is in the $+z$ direction. Determine the magnitude and direction of the field.

12. Show that the work done by the magnetic force on a charged particle moving in a uniform magnetic field is zero for any displacement of the particle.

Section 19.4 Magnetic Force on a Current-Carrying Conductor

13. Imagine a very long, uniform wire with a linear mass density of 0.0010 kg/m that encircles the Earth at its magnetic equator. What are the magnitude and direction of the current in the wire that will keep it levitated above the ground?

14. A wire carries a current of 10.0 A in a direction that makes an angle of 30.0° with the direction of

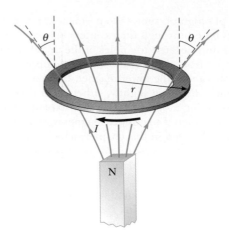

FIGURE 19.36 (Problem 15)

a magnetic field of strength 0.300 T. Find the magnetic force on a 5.00-m length of the wire.

15. A strong magnet is placed under a horizontal conducting ring of radius $r = 2.0$ cm that carries a current of $I = 2.0$ A, as shown in Figure 19.36. (a) If the magnetic field lines have magnitude 0.010 T and make an angle of $\theta = 30°$ with the vertical at the ring's location, what are the magnitude and direction of the resultant force on the ring? (b) If this ring has a uniform linear mass density of 0.010 kg/m, what current in the ring will keep it levitated at its present location?

16. A current, $I = 15$ A, is directed along the positive x axis and perpendicularly to a magnetic field. The conductor experiences a magnetic force per unit length of 0.12 N/m in the negative y direction. Calculate the magnitude and direction of the magnetic field in the region through which the current passes.

17. In Figure 19.34, assume that in each case the velocity vector shown is replaced with a wire carrying a current in the direction of the velocity vector. For each case, find the direction of the magnetic force acting on the wire.

18. In Figure 19.35, assume that in each case the velocity vector shown is replaced with a wire carrying a current in the direction of the velocity vector. For each case, find the direction of the magnetic field that will produce the magnetic force shown.

19. A wire with a mass of 1.00 g/cm is placed on a horizontal surface with a coefficient of friction of 0.200. The wire carries a current of 1.50 A eastward and moves horizontally to the north. What are the magnitude and the direction of the *smallest*

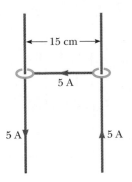

FIGURE 19.37 (Problem 22)

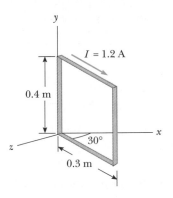

FIGURE 19.38 (Problem 24)

magnetic field that enables the wire to move in this fashion?

20. A uniform horizontal wire with a linear mass density of 0.50 g/m carries a 2.0-A current. It is placed in a constant magnetic field, with a strength of 4.0×10^{-3} T, that is perpendicular to the wire. As the wire moves upward starting from rest, (a) what is its acceleration and (b) how long does it take to rise 50 cm? Neglect the magnetic field of the Earth.

21. A thin, 1.00-m-long copper rod has a mass of 50.0 g. What is the minimum current in the rod that will cause it to float in a magnetic field of 2.00 T?

22. An unusual message delivery system is pictured in Figure 19.37. A 15-cm length of conductor that is free to move is held in place between two thin conductors. When a 5.0-A current is directed as shown in the figure, the wire segment moves upward at a constant velocity. If the mass of the wire is 0.15 kg, find the magnitude and direction of the minimum magnetic field that is required to move the wire. (The wire slides without friction on the two vertical conductors.)

23. The Earth has a magnetic field of 0.60×10^{-4} T, pointing 75° below the horizontal in a north-south plane. A 10.0-m-long straight wire carries a 15-A current. (a) If the current is directed horizontally toward the east, what are the magnitude and direction of the magnetic force on the wire? (b) What are the magnitude and direction of the force if the current is directed vertically upward?

Section 19.5   Torque on a Current Loop

24. A rectangular loop consists of 100 closely wrapped turns and has dimensions 0.40 m by 0.30 m. The loop is hinged along the y axis, and the plane of the coil makes an angle of 30.0° with the x axis

(Fig. 19.38). What is the magnitude of the torque exerted on the loop by a uniform magnetic field of 0.80 T directed along the x axis, when the current in the windings has a value of 1.2 A in the direction shown? What is the expected direction of rotation of the loop?

25. A 2.00-m-long wire loop carrying a current of 2.00 A is in the shape of an equilateral triangle. If the loop is placed in a constant magnetic field of magnitude 0.500 T, determine the *maximum* torque that acts on it.

26. A single circular wire loop of radius 50.0 cm, carrying a current of 2.00 A, is in a magnetic field of 0.4 T. (a) Find the maximum torque that acts on this loop. (b) Find the angle that the plane of the loop makes with the field when the torque is one-half the value found in part (a).

Section 19.6   The Galvanometer and Its Applications

27. A 50.0-Ω, 10.0-mA galvanometer is to be converted to an ammeter that reads 3.00 A at full-scale deflection. What value of R_p should be placed in parallel with the coil?

28. An ammeter reads 10.0 A at full-scale deflection. The meter was constructed by inserting a resistor in parallel with a 50.0-Ω galvanometer coil that deflects full-scale when the voltage across it is 50.0 mV. What is the value of the parallel resistance that should be used?

29. Consider a galvanometer with an internal resistance of 60.0 Ω. If it deflects full-scale when it carries a current of 0.500 mA, what is the value of the series resistance that must be connected to it if this combination is to be used as a voltmeter having a full-scale deflection for a potential difference of 1.00 V?

30. A 40.0-Ω, 2.0-mA galvanometer is to be converted

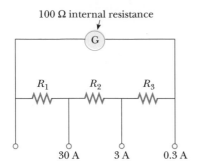

FIGURE 19.39   (Problem 31)

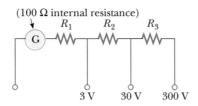

FIGURE 19.40   (Problem 32)

to a voltmeter that reads 150 V at full-scale deflection. What value of series resistance should be used with the galvanometer coil?

31. The galvanometer of Problem 32 is to be converted to a multirange ammeter using the circuit shown in Figure 19.39. Find the values of R_1, R_2, and R_3 that will give the full-scale readings in the figure.

32. A galvanometer has an internal resistance of 100 Ω and deflects full-scale for a current of 100 μA. This galvanometer is to be used to construct the multirange voltmeter in Figure 19.40. Find the values of R_1, R_2, and R_3 that will enable the meter to give the full-scale readings in the figure.

Section 19.7   Motion of a Charged Particle in a Magnetic Field

33. A singly charged positive ion with a mass of 6.68×10^{-27} kg moves clockwise in a circular path of radius 3.00 cm with a speed of 1.00×10^4 m/s. Find the direction and strength of the magnetic field.

34. A +2.0-μC charged particle with a kinetic energy of 0.09 J is placed in a uniform magnetic field of magnitude 0.10 T. If the particle moves in a circular path of radius 3.0 m, determine its mass.

35. A proton moves in a circular orbit perpendicularly to a uniform magnetic field of 0.758 T. Find the time it takes the proton to make one pass around the orbit.

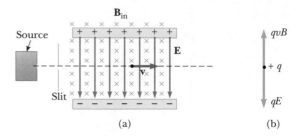

FIGURE 19.41   (Problem 38)

36. A proton moves in a circular path perpendicularly to a constant magnetic field so that it takes 1.00 μs to complete one revolution. Determine the strength of the magnetic field.

37. An electron moves in a circular path perpendicularly to a constant magnetic field of magnitude 1.00 mT. If the angular momentum of the electron about the center of the circle is 4.00×10^{-25} J·s, determine (a) the radius of the circular path and (b) the speed of the electron.

38. Figure 19.41a is a diagram of a device called a velocity selector, in which particles of a specific velocity pass through undeflected while those with greater or lesser velocities are deflected either upward or downward. An electric field is directed perpendicularly to a magnetic field. This produces on the charged particle an electric force and a magnetic force that are equal in magnitude and opposite in direction (Fig. 19.41b), and hence cancel. Show that particles with a speed of $v = E/B$ will pass through undeflected.

39. A proton (charge $+e$, mass m_p), a deuteron (charge $+e$, mass $2m_p$), and an alpha particle (charge $+2e$, mass $4m_p$) are accelerated through a common potential difference, V. The particles enter a uniform magnetic field, **B**, in a direction perpendicular to **B**. The proton moves in a circular path of radius r_p. Determine the values of the radii of the circular orbits for the deuteron, r_d, and the alpha particle, r_α, in terms of r_p.

 40. A singly charged positive ion has a mass of 2.5×10^{-26} kg. After being accelerated through a potential difference of 250 V, the ion enters a magnetic field of 0.50 T, in a direction perpendicular to the field. Calculate the radius of the path of the ion in the field.

41. Consider the mass spectrometer shown schematically in Figure 19.42. The electric field between the plates of the velocity selector is 950 V/m, and the magnetic fields in both the velocity selector and the deflection chamber have magnitudes of 0.93 T. Calculate the radius of the path in the sys-

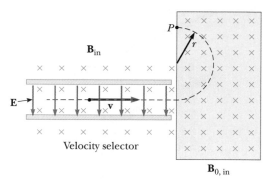

FIGURE 19.42 (Problem 41). A mass spectrometer. Charged particles are first sent through a velocity selector. They then enter a region where a magnetic field, $\mathbf{B}_0$ (inward), causes positive ions to move in a semicircular path and strike a photographic film at P.

tem for a singly charged ion with mass $m = 2.18 \times 10^{-26}$ kg. (*Hint:* See Problem 38.)

42. Two ions with masses of 6.64×10^{-27} kg move out of the slit of a mass spectrometer and into a region where the magnetic field is 0.20 T. Each has a speed of 1.0×10^6 m/s, but one ion is singly charged and the other is doubly charged. Find (a) the radius of the circular path followed by each in the field and (b) the distance of separation when they have moved through one-half their circular path and strike a piece of photographic paper.

43. A mass spectrometer is used to examine the isotopes of uranium. Ions in the beam emerge from the velocity selector at a speed of 3.00×10^5 m/s and enter a uniform magnetic field of 0.600 T directed perpendicularly to the velocity of the ions. What is the distance between the impact points formed on the photographic plate by singly charged ions of ^{235}U and ^{238}U?

44. A cosmic-ray proton in interstellar space has an energy of 10 MeV and executes a circular orbit with a radius equal to that of Mercury's orbit around the Sun (5.8×10^{10} m). What is the galactic magnetic field in that region of space?

Section 19.8 Magnetic Field of a Long, Straight Wire and Ampère's Law

45. Figure 19.43 is a cross-sectional view of a coaxial cable (such as a VCR cable). The center conductor is surrounded by a rubber layer, which is surrounded by an outer conductor, which is surrounded by another rubber layer. The current in the inner conductor is 1.00 A out of the page, and the current in the outer conductor is 3.00 A into the page. Using Ampère's law, determine the

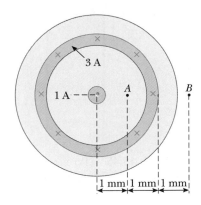

FIGURE 19.43 (Problem 45)

magnitude and direction of the magnetic field at points A and B.

46. At what distance from a long, straight wire carrying a current of 5.0 A is the magnetic field due to the wire equal to the strength of the Earth's field, approximately 5.0×10^{-5} T?

47. Niobium metal becomes a superconductor (with zero electrical resistance) when cooled below 9 K. If superconductivity is destroyed when the surface magnetic field exceeds 0.100 T, determine the maximum current a 2.00-mm-diameter niobium wire can carry and remain superconducting.

48. Find the direction of the current in the wire in Figure 19.44 that would produce a magnetic field directed as shown, in each case.

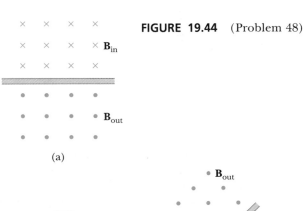

FIGURE 19.44 (Problem 48)

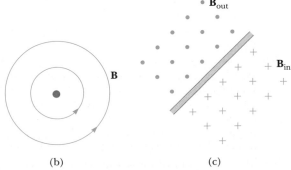

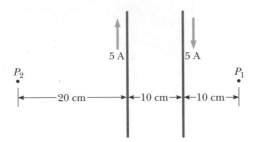

FIGURE 19.45   (Problem 49)

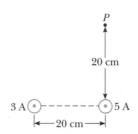

FIGURE 19.46   (Problem 50)

49. The two wires shown in Figure 19.45 carry currents of 5.00 A in opposite directions and are separated by 10.0 cm. Find the direction and magnitude of the net magnetic field (a) at a point midway between the wires; (b) at point P_1, that is, 10.0 cm to the right of the wire on the right; and (c) at point P_2, that is, 20.0 cm to the left of the wire on the left.

50. The two wires in Figure 19.46 carry currents of 3.00 A and 5.00 A in the direction indicated. (a) Find the direction and magnitude of the magnetic field at a point midway between the wires. (b) Find the magnitude and direction of the magnetic field at a point 20.0 cm above the wire carrying the 5.00-A current.

Section 19.9   Magnetic Force Between Two Parallel Conductors

51. Two long, parallel conductors are carrying currents in the same direction, as in Figure 19.47. Conductor A carries a current of 150 A and is held firmly in position; conductor B carries current I_B and is allowed to slide freely up and down (parallel to A) between a set of nonconducting guides. If the linear mass density of conductor B is 0.10 g/cm, what value of current I_B will result in equilibrium when the distance between the two conductors is 2.5 cm?

52. Two long, parallel wires, each with a mass per unit length of 40 g/m, are supported in a horizontal

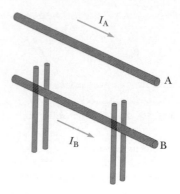

FIGURE 19.47   (Problem 51)

plane by 6.0-cm-long strings, as shown in Figure 19.48. Each wire carries the same current, I, causing the wires to repel each other so that the angle, θ, between the supporting strings is 16°. (a) Are the currents in the same or opposite directions? (b) Determine the magnitude of each current.

53. Two parallel wires are 10.0 cm apart, and each carries a current of 10.0 A. (a) If the currents are in the same direction, find the force per unit length exerted on one of the wires by the other. Are the wires attracted or repelled? (b) Repeat the problem with the currents in opposite directions.

54. A 2.0-m-long wire weighing 0.080 N/m is suspended directly above a second wire. The top wire carries a current of 30.0 A and the bottom wire carries a current of 60.0 A. Find the distance of separation between the wires so that the top wire will be held in place by magnetic repulsion.

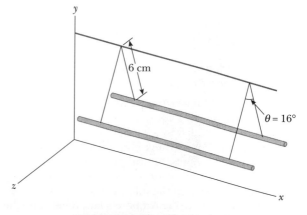

FIGURE 19.48   (Problem 52)

FIGURE 19.49   (Problem 56)

Section 19.10   Magnetic Field of a Current Loop

Section 19.11   Magnetic Field of a Solenoid

55. An 8.0-m-long copper wire with a cross-sectional area of 1.0×10^{-4} m², in the shape of a square loop, is connected to a 0.10-V battery. If the loop is placed in a uniform magnetic field of magnitude 0.4 T, what is the maximum torque that can act on it? The resistivity of copper is 1.7×10^{-8} Ω·m.

56. The magnetic field at the center of a circular wire loop carrying current I is perpendicular to the plane of the loop, conforms to the right-hand rule illustrated in Figure 19.21, and has magnitude $B = \mu_0 I/(2a)$, where a is the radius of the loop. A conductor consists of a circular loop of radius $a = 0.100$ m and two straight, very long sections, as shown in Figure 19.49. The current in the wire is 7.00 A, and the wire is in the plane of the paper. Calculate the direction and magnitude of the magnetic field at the center of the loop.

57. A single-turn square loop of wire, 2.00 cm on a side, carries a current of 0.200 A. The loop is inside a solenoid, with the plane of the loop perpendicular to the magnetic field of the solenoid. The solenoid has 30 turns per centimeter and carries a current of 15.0 A. Find the force on each side of the loop and the torque acting on it.

58. Use the path shown in Figure 19.50 and Ampère's law to derive the expression for the magnetic field inside a solenoid of n turns per unit length and current I. (*Hint:* Assume that the strength of the magnetic field is zero outside the solenoid—a good assumption.)

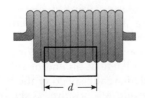

FIGURE 19.50   (Problem 58)

ADDITIONAL PROBLEMS

59. A cosmic-ray proton traveling at half the speed of light heads directly toward the center of the Earth in the plane of the Earth's magnetic equator. Will it hit the Earth? As an estimate, assume that the Earth's magnetic field is 5.0×10^{-5} T and extends out one Earth diameter, or 1.3×10^{7} m. (*Hint:* Calculate the radius of curvature of the proton in this magnetic field.)

60. The planetary model of the hydrogen atom consists of an electron in a circular orbit of radius 5.3×10^{-11} m about a proton. The motion of the electron creates an electric current. (a) Show that the electron's speed is 2.2×10^{6} m/s. (b) Determine the magnetic field strength at the location of the proton. (*Hint:* See Problem 56.)

61. What magnetic field is required to constrain an electron with a kinetic energy of 400 eV to a circular path of radius 0.80 m?

62. A circular coil consisting of a single loop of wire has a radius of 30.0 cm and carries a current of 25 A. It is placed in an external magnetic field of 0.30 T. Find the torque on the wire when the plane of the coil makes an angle of 35° with the direction of the field.

63. Two species of singly charged positive ions of masses 20.0×10^{-27} kg and 23.4×10^{-27} kg enter a magnetic field at the same location with a speed of 1.00×10^{5} m/s. If the strength of the field is 0.200 T, and they move perpendicularly to the field, find their distance of separation after they complete one half of their circular path.

64. Three long, parallel conductors carry currents of $I = 2.0$ A. Figure 19.51 is an end view of the conductors, with each current coming out of the page. Given that $a = 1.0$ cm, determine the magnitude and direction of the magnetic field at points (a) A, (b) B, and (c) C.

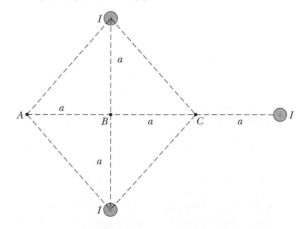

FIGURE 19.51   (Problem 64)

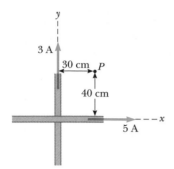

FIGURE 19.52 (Problem 65)

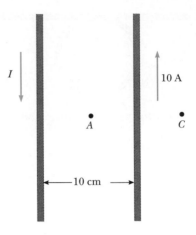

FIGURE 19.53 (Problem 69)

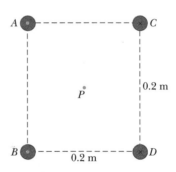

FIGURE 19.54 (Problem 70)

65. Two long, straight wires cross each other at right angles, as shown in Figure 19.52. (a) Find the direction and magnitude of the magnetic field at point *P*, which is in the same plane as the two wires. (b) Find the magnetic field at a point 30.0 cm above the point of intersection (30.0 cm out of the page, toward you).

66. A rectangular coil of 150 turns and area 0.12 m² is in a uniform magnetic field of 0.15 T. Measurements indicate that the maximum torque exerted on the loop by the field is 6.0×10^{-4} N·m. (a) Calculate the current in the coil. (b) Would the value for the required current be different if the 150 turns of wire were used to form a single-turn coil of greater area? Explain.

67. A rectangular loop of wire carrying a current of 2.00 A is suspended vertically and attached to the right arm of a balance. After the system is balanced, an external magnetic field, **B**, is introduced. The field threads the lower end of the loop in a direction perpendicular to the wire. If the width of the loop is 20.0 cm and it takes 13.5 g of added mass on the left arm to rebalance the system, determine **B**.

68. An electron circles at a speed of 10^4 m/s in a radius of 2 cm in a solenoid. The magnetic field of the solenoid is perpendicular to the plane of the electron's path. Find (a) the strength of the magnetic field inside the solenoid and (b) the current in the solenoid if it has 25 turns per centimeter.

69. Two parallel conductors carry currents in opposite directions, as shown in Figure 19.53. One conductor carries a current of 10.0 A. Point *A* is the midpoint between the wires, and point *C* is 5.00 cm to the right of the 10.0-A current. *I* is adjusted so that the magnetic field at *C* is zero. Find (a) the value of the current *I* and (b) the value of the magnetic field at *A*.

70. Four long, parallel conductors all carry currents of 4.00 A. Figure 19.54 is an end view of the conductors. The current direction is out of the page at points *A* and *B* (indicated by dots) and into the page at *C* and *D* (indicated by crosses). Calculate the magnitude and direction of the magnetic field at point *P*, at the center of the square of edge length 0.200 m.

71. A singly charged heavy ion is observed to complete 5 revolutions in a uniform magnetic field of magnitude 0.050 T in 1.50 ms. Calculate (approximately) the mass of the ion.

72. A heart surgeon monitors the flow rate of blood through an artery using an electromagnetic flow-meter (Fig. 19.55) in which electrodes *A* and *B* are attached to the outer surface of a blood vessel of inside diameter 3.0 mm. (a) For a magnetic field strength of 0.040 T, an emf of 160 μV is developed. Calculate the speed of the blood. (b) Verify that electrode *A* is positive, as shown. Does the

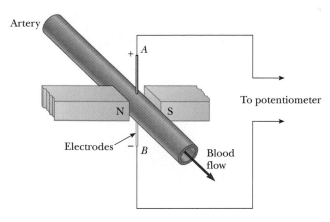

Artery

+ *A*

N S

To potentiometer

Electrodes

− *B*

Blood
flow

FIGURE 19.55 (Problem 72)

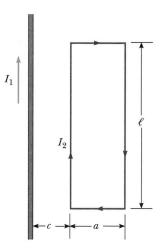

I_1

I_2

ℓ

$\leftarrow c \rightarrow \leftarrow a \rightarrow$

FIGURE 19.56 (Problem 73)

sign of the emf depend on whether the ions in the
blood are positively or negatively charged? Explain.

73. For the arrangement shown in Figure 19.56, the
current in the straight conductor has the value
$I_1 = 5.0$ A and lies in the plane of the rectangular
loop, which carries current $I_2 = 10.0$ A. The dimensions are $c = 0.10$ m, $a = 0.15$ m, and $\ell =$
0.45 m. Find the magnitude and direction of the
net force exerted on the rectangle by the magnetic
field of the straight current-carrying conductor.

74. A straight wire of mass 10.0 g and length 5.0 cm is
suspended from two identical springs that, in turn,
form a closed circuit (Fig. 19.57). The springs
stretch a distance of 0.50 cm under the weight of
the wire. The circuit has a total resistance of 12 Ω.
When a magnetic field is turned on, directed out
of the page (indicated by the dots in Fig. 19.57),
the springs are observed to stretch an additional
0.30 cm. What is the strength of the magnetic
field? (The upper portion of the circuit is fixed.)

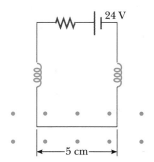

24 V

5 cm

FIGURE 19.57 (Problem 74)

20

INDUCED VOLTAGES AND INDUCTANCE

This photograph of water-driven generators was taken at the Bonneville Dam in Oregon. Hydroelectric power is generated when water from a dam passes through the generators under the influence of gravity, which causes turbines in the generator to rotate. The mechanical energy of the rotating turbines is transformed into electrical energy using the principle of electromagnetic induction, which you will study in this chapter. (David Weintraub, Photo Researchers, Inc.)

In 1819 Hans Christian Oersted discovered that a magnetic compass experiences a force in the vicinity of an electric current. Although there had long been speculation that such a relationship existed, this was the first evidence of a link between electricity and magnetism. Because nature is often symmetric, the discovery that electric currents produce magnetic fields led scientists to suspect that magnetic fields could produce electric currents. Indeed, experiments conducted by Michael Faraday in England and, independently, by Joseph Henry in the United States in 1831 showed that a changing magnetic field could induce an electric current in a circuit. The results of these experiments led to a very basic and important law known as Faraday's law. In this chapter we discuss several practical applications of Faraday's law, one of which is the production of electrical energy in power generation plants throughout the world.

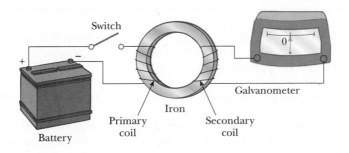

FIGURE 20.1
Faraday's experiment. When the switch in the primary circuit at the left is closed, the galvanometer in the secondary circuit at the right deflects momentarily. The emf in the secondary circuit is induced by the changing magnetic flux through the coil in this circuit.

20.1

INDUCED emf AND MAGNETIC FLUX

INDUCED emf

We begin this chapter by describing an experiment, first conducted by Faraday, that demonstrates that a current can be produced by a changing magnetic field. The apparatus shown in Figure 20.1 consists of a coil connected to a switch and a battery. We shall refer to this coil as the *primary coil* and to the corresponding circuit as the primary circuit. The coil is wrapped around an iron ring to intensify the magnetic field produced by the current through it. A second coil, at the right, is wrapped around the iron ring and is connected to a galvanometer. We shall refer to this as the *secondary coil* and to the corresponding circuit as the secondary circuit. Note that there is no battery in the secondary circuit. The only purpose of this circuit is to detect any current that might be produced by a magnetic field.

At first glance, you might guess that no current would ever be detected in the secondary circuit. However, when the switch in the primary circuit is suddenly closed or opened, something quite amazing happens. Just after the switch is closed, the galvanometer in the secondary circuit deflects in one direction and then returns to zero. When the switch is opened, the galvanometer deflects in the opposite direction and again returns to zero. Finally, when there is a steady current in the primary circuit, the galvanometer reads zero.

From observations such as these, Faraday concluded that an electric current can be produced by a changing magnetic field. (A steady magnetic field cannot produce a current.) The current produced in the secondary circuit occurs only for an instant while the magnetic field through the secondary coil is changing. In effect, the secondary circuit behaves as though a source of emf were connected to it for a short instant. It is customary to say that *an induced emf is produced in the secondary circuit by the changing magnetic field.*

MAGNETIC FLUX

In order to quantitatively evaluate induced emfs, it is first necessary to fully understand what factors affect the phenomenon. As you will see later, the emf is

Michael Faraday (1791–1867). "It appeared very extraordinary, that as every electric current was accompanied by a corresponding intensity of magnetic action at right angles to the current, good conductors of electricity, when placed within the sphere of this action, should have any current induced through them, or some sensible effect produced equivalent in force to such a current." *(By kind permission of the President and Council of the Royal Society)*

Michael Faraday (1791–1867)

Michael Faraday was a British physicist and chemist who is often regarded as the greatest experimental scientist of the 1800s. His many contributions to the study of electricity include the invention of the electric motor, the electric generator, and the transformer as well as the discovery of electromagnetic induction, the laws of electrolysis, and the observation that the plane of polarization of light is rotated by an electric field.

Faraday was born in 1791 in rural England, but his family moved to London shortly thereafter. One of ten children of a blacksmith, Faraday received minimal formal education and was apprenticed to a bookbinder at age 14. He was fascinated by articles on electricity and chemistry and was fortunate to have an employer who allowed him to read books and attend scientific lectures. He received some education in science from the City Philosophical Society.

When Faraday finished his apprenticeship in 1812, he expected to devote himself to bookbinding rather than to science. That same year, he attended a lecture by Humphry Davy, who had made many contributions in the field of heat and thermodynamics. Faraday sent 386 pages of notes, bound in leather, to the scientist; Davy was impressed and appointed Faraday as his permanent assistant at the Royal Institution. From 1813 to 1815, Faraday toured France and Italy with Davy, visiting leading scientists of the time such as Volta and Vauquelin.

Despite his limited mathematical skills, Faraday succeeded in making the basic discoveries on which virtually all our uses of electricity depend. He conceived the fundamental natures of magnetism and, to a degree, electricity and light.

A modest man who was content to serve science as best he could, Faraday declined a knighthood and a position as president of the Royal Society. He was also a moral man; he refused to participate in the preparation of poison gas for use in the Crimean War.

Faraday died in 1867. His many achievements are acknowledged through the use of his name: the Faraday constant is the quantity of charge required to deliver a standard amount of substance in electrolysis, and the farad is the SI unit of capacitance.

(North Wind Picture Archives)

induced by a change in a quantity called the magnetic flux rather than simply by a change in the magnetic field.

Consider a loop of wire in the presence of a uniform magnetic field, **B**. If the loop has an area of A, the **magnetic flux,** Φ, through the loop is defined as

Magnetic flux

$$\Phi \equiv B_{\perp} A = BA \cos \theta \qquad \text{[20.1]}$$

where $B_{\perp}$ is the component of **B** perpendicular to the plane of the loop, as in Figure 20.2a, and θ is the angle between **B** and the normal (perpendicular) to the plane of the loop. Figure 20.2b is an end view of the loop and the penetrating magnetic field lines. When the field is perpendicular to the plane of the loop, as in Figure 20.3a, $\theta = 0$ and Φ has a maximum value, $\Phi_{\max} = BA$. When the plane of the loop is parallel to **B**, as in Figure 20.3b, $\theta = 90°$ and $\Phi = 0$. Since B is in teslas, or webers per square meter, the units of flux are $T \cdot m^2$, or webers.

We can emphasize the significance of Equation 20.1 by first drawing magnetic field lines as in Figure 20.3. The number of lines per unit area increases as the field strength increases.

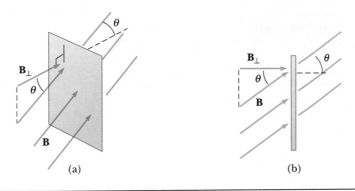

FIGURE 20.2

(a) A uniform magnetic field, **B**, making an angle of θ with the normal to the plane of a wire loop of area A. (b) An edge view of the loop.

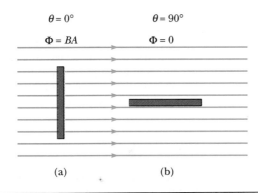

FIGURE 20.3

An edge view of a loop in a uniform magnetic field. (a) When the field lines are perpendicular to the plane of the loop, the magnetic flux through the loop is a maximum and equal to $\Phi = BA$. (a) When the field lines are parallel to the plane of the loop, the magnetic flux through the loop is zero.

> The value of the magnetic flux is proportional to the total number of lines passing through the loop.

Thus, we see that the most lines pass through the loop when its plane is perpendicular to the field, as in Figure 20.3a, and so the flux has its maximum value. As Figure 20.3b shows, no lines pass through the loop when its plane is parallel to the field, and so in this case $\Phi = 0$.

20.2

FARADAY'S LAW OF INDUCTION

The usefulness of the concept of magnetic flux can be made obvious by another simple experiment that demonstrates the basic idea of electromagnetic induc-

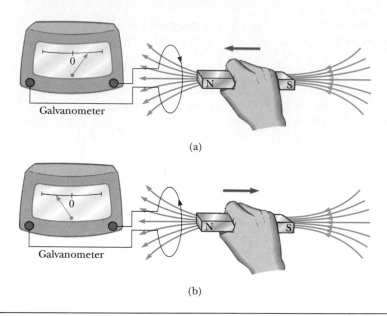

(a)

(b)

FIGURE 20.4
(a) When a magnet is moved toward a wire loop connected to a galvanometer, the galvanometer deflects as shown. This shows that a current is induced in the loop.
(b) When the magnet is moved away from the loop of wire, the galvanometer deflects in the opposite direction, indicating that the induced current is opposite that shown in (a).

tion. Consider a wire loop connected to a galvanometer, as in Figure 20.4. If a magnet is moved toward the loop, the galvanometer needle deflects in one direction, as in Figure 20.4a. If the magnet is moved away from the loop, the galvanometer needle deflects in the opposite direction, as in Figure 20.4b. If the magnet is held stationary and the loop is moved either toward or away from the magnet, the needle also deflects. From these observations, it can be concluded that *a current is set up in the circuit as long as there is relative motion between the magnet and the loop.* These results are quite remarkable in view of the fact that the circuit contains no batteries! We call such a current an *induced current* because it is produced by an induced emf.

This experiment has something in common with the Faraday experiment discussed in Section 20.1. In each case, an emf is induced in a current when the magnetic flux through the circuit changes with time. In fact, we can make the following general summary of such experiments involving induced currents and emfs:

> **The instantaneous emf induced in a circuit equals the rate of change of magnetic flux through the circuit.**

If a circuit contains N tightly wound loops and the flux through each loop changes by the amount $\Delta\Phi$ during the interval Δt, the average emf induced in the circuit during time Δt is

Faraday's law

$$\varepsilon = -N\frac{\Delta\Phi}{\Delta t} \qquad [20.2]$$

This is a statement of **Faraday's law of magnetic induction.** The minus sign is included to indicate the polarity of the induced emf, which can be found by use of **Lenz's law:**

> The polarity of the induced emf is such that it produces a current whose magnetic field opposes the change in magnetic flux through the loop. That is, the induced current tends to maintain the original flux through the circuit.

Lenz's law

We shall consider several applications of Lenz's law in Section 20.4.

The ground fault interrupter (GFI) is an interesting safety device that protects users of electrical power against electric shock when they touch appliances. Its operation makes use of Faraday's law. Figure 20.5 shows the essential parts of a ground fault interrupter. Wire 1 leads from the wall outlet to the appliance to be protected, and wire 2 leads from the appliance back to the wall outlet. An iron ring surrounds the two wires so as to confine the magnetic field set up by each wire. A sensing coil, which can activate a circuit breaker when changes in magnetic flux occur, is wrapped around part of the iron ring. Because the currents in the wires are in opposite directions, the net magnetic field through the sensing coil due to the currents is zero. However, if a short circuit occurs in the appliance so that there is no returning current, the net magnetic field through the sensing coil is no longer zero. (This can happen, for example, if one of the wires loses its insulation and accidentally touches the metal case of the appliance, providing a direct path to ground.) Because the current is alternating, the magnetic flux through the sensing coil changes with time, producing an induced voltage in the coil. This induced voltage is used to trigger a circuit breaker, stopping the current before it reaches a level that might be harmful to the person using the appliance.

Another interesting application of Faraday's law is the production of sound in an electric guitar. A vibrating string induces an emf in a coil (Fig. 20.6). The pickup coil is placed near the vibrating guitar string, which is made of a metal that can be magnetized. The permanent magnet inside the coil magnetizes the

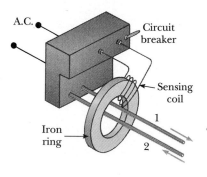

FIGURE 20.5
Essential components of a ground fault interrupter.

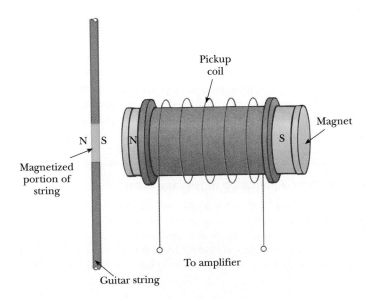

FIGURE 20.6
In an electric guitar, a vibrating string induces a voltage in the pickup coil.

PHYSICS IN *ACTION*

DEMONSTRATIONS OF ELECTROMAGNETIC INDUCTION

(Left) To demonstrate electromagnetic induction, an ac voltage is applied to the lower coil in the apparatus. A voltage is induced in the upper coil, as indicated by the illuminated lamp connected to this coil. What do you think happens to the lamp's intensity as the upper coil is moved over the vertical tube? To answer this question, note that the magnetic field associated with the lower coil varies along the axis of the tube. *(Right)* This modern electric range cooks food using the principle of induction. An oscillating current is passed through a coil placed below the cooking surface made of a special glass. The current produces an oscillating magnetic field, which induces a current in the cooking utensil. Since the cooking utensil has some electrical resistance, the electrical energy associated with the induced current transforms into thermal energy, causing the utensil and its contents to heat up.

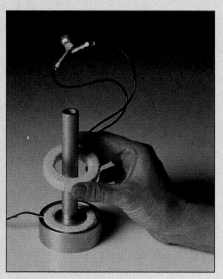

(Courtesy of Central Scientific Company)

(Corning, Inc.)

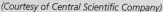

portion of the string nearest the coil. When the guitar string vibrates at some frequency, its magnetized segment produces a changing magnetic flux through the pickup coil. The changing flux induces a voltage in the coil; the voltage is fed to an amplifier. The output of the amplifier is sent to the loudspeakers, producing the sound waves that we hear.

EXAMPLE 20.1   Application of Faraday's Law

A coil with 200 turns of wire is wrapped on an 18.0-cm-square frame. Each turn has the same area, equal to that of the frame, and the total resistance of the coil is 2.00 Ω. A uniform magnetic field is applied perpendicularly to the plane of the coil. If the field changes uniformly from 0 to 0.500 T in 0.800 s, find the magnitude of the induced emf in the coil while the field is changed.

Reasoning The magnitude of the induced emf can be found from Faraday's law of induction, $\varepsilon = -N\dfrac{\Delta\Phi}{\Delta t}$. In this equation, $\Delta\Phi$ is the difference between the final and initial fluxes, where $\Phi = BA$.

Solution The area of the coil is $(0.180 \text{ m})^2 = 0.0324 \text{ m}^2$. The magnetic flux through the coil at $t = 0$ is zero because $B = 0$. At $t = 0.800$ s, the magnetic flux through the coil is

$$\Phi_f = BA = (0.500 \text{ T})(0.0324 \text{ m}^2) = 0.0162 \text{ T}\cdot\text{m}^2$$

Therefore, the *change* in flux through the coil during the 0.800-s interval is

$$\Delta\Phi = \Phi_f - \Phi_i = 0.0162 \text{ T}\cdot\text{m}^2$$

Faraday's law of induction enables us to find the magnitude of the induced emf:

$$|\varepsilon| = N\frac{\Delta\Phi}{\Delta t} = (200 \text{ turns})\left(\frac{0.0162 \text{ T}\cdot\text{m}^2}{0.800 \text{ s}}\right) = \boxed{4.05 \text{ V}}$$

(Note that $1 \text{ T}\cdot\text{m}^2 = 1 \text{ V}\cdot\text{s}$.)

Exercise Find the magnitude of the induced current in the coil while the field is changing.

Answer 2.03 A.

20.3

MOTIONAL emf

In Section 20.2, we considered a situation in which an emf is induced in a circuit when the magnetic field changes with time. In this section we describe a particular application of Faraday's law in which a so-called **motional emf** is produced. This is the emf induced in a conductor moving through a magnetic field.

First consider a straight conductor of length ℓ moving with constant velocity through a uniform magnetic field directed into the paper, as in Figure 20.7. For simplicity, we assume that the conductor is moving perpendicularly to the field. The electrons in the conductor experience a force of magnitude $F = qvB$ directed downward along the conductor. Because of this magnetic force, the electrons move to the lower end and accumulate there, leaving a net positive charge at the upper end. As a result of this charge separation, an electric field is produced in the conductor. The charge at the ends builds up until the downward magnetic force, qvB, is balanced by the upward electric force, qE. At this point, charge stops flowing and the condition for equilibrium requires that

$$qE = qvB \quad \text{or} \quad E = vB$$

Since the electric field is constant, the field produced in the conductor is related to the potential difference across the ends by $V = E\ell$. Thus,

$$V = E\ell = B\ell v \qquad \qquad \textbf{[20.3]}$$

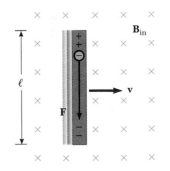

FIGURE 20.7
A straight conductor of length ℓ moving with velocity **v** through a uniform magnetic field, **B**, directed perpendicular to **v**. The vector **F** is the force on an electron in the conductor. An emf of $B\ell v$ is induced between the ends of the bar.

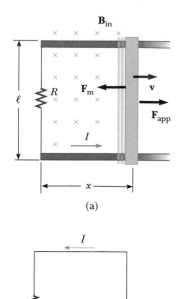

(a)

(b)

FIGURE 20.8

(a) A conducting bar sliding with velocity **v** along two conducting rails under the action of an applied force, $\mathbf{F}_{app}$. A counterclockwise current is induced in the loop. (b) The equivalent circuit of (a).

Since there is an excess of positive charge at the upper end of the conductor and an excess of negative charge at the lower end, the upper end is at a higher potential than the lower end. Thus,

> A potential difference is maintained across the conductor as long as there is motion through the field. If the motion is reversed, the polarity of the potential difference is also reversed.

A more interesting situation occurs if the moving conductor is part of a closed conducting path. This situation is particularly useful for illustrating how a changing magnetic flux induces a current in a closed circuit. Consider a circuit consisting of a conducting bar of length ℓ, sliding along two fixed parallel conducting rails, as in Figure 20.8a. For simplicity, we assume that the moving bar has zero resistance and that the stationary part of the circuit has resistance R. A uniform and constant magnetic field, **B**, is applied perpendicularly to the plane of the circuit. As the bar is pulled to the right with velocity **v** under the influence of an applied force, $\mathbf{F}_{app}$, the free charges in the bar experience a magnetic force along the length of the bar. This force in turn sets up an induced current because the charges are free to move in a closed conducting path. In this case, the changing magnetic flux through the loop and the corresponding induced emf across the moving bar arise from the change in area of the loop as the bar moves through the magnetic field.

Let us assume that the bar moves a distance of Δx in time Δt, as shown in Figure 20.9. The increase in flux, $\Delta\Phi$, through the loop in that time is the amount of flux that now passes through the portion of the circuit that has area $\ell\,\Delta x$:

$$\Delta\Phi = BA = B\ell\,\Delta x$$

Using Faraday's law and noting that there is one loop ($N = 1$), we find that the induced emf has the magnitude

$$\mathcal{E} = \frac{\Delta\Phi}{\Delta t} = B\ell\,\frac{\Delta x}{\Delta t} = B\ell v \qquad \textbf{[20.4]}$$

This induced emf is often called a *motional emf* because it arises from the motion of a conductor through a magnetic field.

Furthermore, if the resistance of the circuit is R, the magnitude of the induced current is

$$I = \frac{\mathcal{E}}{R} = \frac{B\ell v}{R} \qquad \textbf{[20.5]}$$

Figure 20.8b is the equivalent circuit diagram for this example.

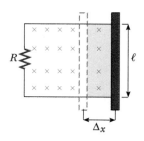

FIGURE 20.9

As the bar moves to the right, the area of the loop increases by the amount $\ell\Delta x$, and the magnetic flux through the loop increases by $B\ell\Delta x$.

EXAMPLE 20.2   The Electrified Airplane Wing

An airplane with a wing span of 30.0 m flies parallel to the Earth's surface at a location where the downward component of the Earth's magnetic field is 0.60×10^{-4} T. Find the difference in potential between the wing tips when the speed of the plane is 250 m/s.

Solution   Because the plane is flying horizontally, we do not have to concern ourselves with the horizontal component of the Earth's field. Thus, we find that

$$\varepsilon = B\ell v = (0.60 \times 10^{-4}\ \text{T})(30.0\ \text{m})(250\ \text{m/s}) = \boxed{0.45\ \text{V}}$$

In an application similar to this, NASA plans to deploy a long conducting tether from a shuttle and use the motion through the Earth's magnetic field to generate power.

EXAMPLE 20.3   Where Is the Energy Source?

(a) The sliding bar in Figure 20.8a has a length of 0.50 m and moves at 2.0 m/s in a magnetic field of magnitude 0.25 T. Find the induced voltage in the moving rod.

Solution   We use Equation 20.3 and find that

$$\varepsilon = B\ell v = (0.25\ \text{T})(0.50\ \text{m})(2.0\ \text{m/s}) = \boxed{0.25\ \text{V}}$$

(b) If the resistance in the circuit is 0.50 Ω, find the current in the circuit.

Solution   The current is found from Ohm's law to be

$$I = \frac{\varepsilon}{R} = \frac{0.25\ \text{V}}{0.50\ \Omega} = \boxed{0.50\ \text{A}}$$

(c) Find the amount of energy delivered to the 0.50 Ω resistor in one second.

Solution   The power dissipated by the resistor is

$$P = IV = (0.50\ \text{A})(0.25\ \text{V}) = 0.13\ \text{W}$$

Because power is defined as the rate at which energy is converted in a device, the energy, W, dissipated in the resistor in one second is

$$W = Pt = (0.125\ \text{W})(1.0\ \text{s}) = \boxed{0.13\ \text{J}}$$

(d) The source of the energy calculated in part (c) is some external agent that keeps the bar moving at a constant speed of 2.0 m/s by exerting an applied force, F_{app}. Find the value of F_{app}.

Solution   From part (c), we know that the work done by the applied force in one second is 0.13 J. In one second, the bar moves a distance of

$$d = vt = (2.0\ \text{m/s})(1.0\ \text{s}) = 2.0\ \text{m}$$

Thus, from the definition of work, we find that $W = F_{\text{app}}\, d$, or

$$F_{\text{app}} = \frac{W}{d} = \frac{0.13\ \text{J}}{2.0\ \text{m}} = \boxed{0.063\ \text{N}}$$

Exercise   If the rod is to move at constant speed, the applied force must be equal in magnitude to the retarding magnetic force, $I\ell B$. Show that this approach also gives $F_{\text{app}} = 0.063$ N, as found in part (d).

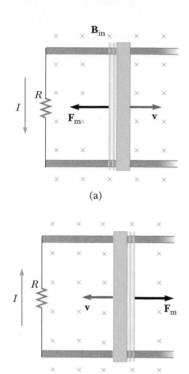

FIGURE 20.10
(a) As the conducting bar slides on the two fixed conducting rails, the magnetic flux through the loop increases in time. By Lenz's law, the induced current must be *counterclockwise* so as to produce a counteracting flux *out of the paper*. (b) When the bar moves to the left, the induced current must be *clockwise*. Why?

20.4

LENZ'S LAW REVISITED

To attain a better understanding of Lenz's law, let us return to the example of a bar moving to the right on two parallel rails in the presence of a uniform magnetic field directed into the paper (Fig. 20.10a). As the bar moves to the right, the magnetic flux through the circuit increases with time because the area of the loop increases. Lenz's law says that the induced current must be in a direction such that the flux *it* produces opposes the change in the external magnetic flux. Since the flux due to the external field is increasing *into* the paper, the induced current, to oppose the change, must produce a flux *out* of the paper. Hence, the induced current must be counterclockwise when the bar moves to the right. (Use the right-hand rule to verify this direction.) On the other hand, if the bar is moving to the left, as in Figure 20.10b, the magnetic flux through the loop decreases with time. Since the flux is into the paper, the induced current has to be clockwise to produce a flux into the paper. In either case, the induced current tends to maintain the original flux through the circuit.

Let us examine this situation from the viewpoint of energy considerations. Suppose that the bar is given a slight push to the right. In the preceding analysis, we found that this motion led to a counterclockwise current in the loop. Let's see what would happen if we assumed that the current was clockwise. For a clockwise current I, the direction of the magnetic force, $BI\ell$, on the sliding bar would be to the right. This force would accelerate the rod and increase its velocity. This, in turn, would cause the area of the loop to increase more rapidly, thereby increasing the induced current, which would increase the force, which would increase the current, would. . . . In effect, the system would acquire energy with zero input energy. This is clearly inconsistent with all experience and with the law of conservation of energy. We are forced to conclude that the current must be counterclockwise.

Consider another situation. A bar magnet is moved to the right toward a stationary loop of wire, as in Figure 20.11a. As the magnet moves, the magnetic flux through the loop increases with time. To counteract this, the induced current produces a flux to the left, as in Figure 20.11b; hence, the induced

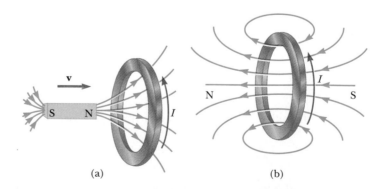

(a)   (b)

FIGURE 20.11
(a) When the magnet is moved toward the stationary conducting loop, a current is induced in the direction shown. (b) This induced current produces its own flux to the left to counteract the increasing external flux to the right.

current is in the direction shown. Note that the magnetic field lines associated with the induced current oppose the motion of the magnet. Therefore, the left face of the current loop is a north pole and the right face is a south pole.

On the other hand, if the magnet were moving to the left, its flux through the loop, which is toward the right, would decrease in time. Under these circumstances, the induced current in the loop would be in a direction so as to set up a field directed from left to right through the loop, in an effort to maintain a constant number of flux lines. Hence, the induced current in the loop would be opposite that shown in Figure 20.11b. In this case, the left face of the loop would be a south pole and the right face would be a north pole. In both cases, we have to do work to move the magnet, and it is this mechanical work that is transformed into electrical energy.

EXAMPLE 20.4   Application of Lenz's Law

A coil of wire is placed near an electromagnet as in Figure 20.12a. Find the direction of the induced current in the coil (a) at the instant the switch is closed, (b) after the switch has been closed for several seconds, and (c) when the switch is opened.

Reasoning and Solution   (a) When the switch is closed, the situation changes from a condition in which no lines of flux pass through the coil to one in which lines of flux pass through in the direction shown in Figure 20.12b. To counteract this change in the number of lines, the coil must set up a field from left to right in the figure. This requires a current directed as shown in Figure 20.12b.

(b) After the switch has been closed for several seconds, there is no change in the number of lines through the loop; hence, the induced current is zero.

(c) Opening the switch causes the magnetic field to change from a condition in which flux lines thread through the coil from right to left to a condition of zero flux. The induced current must then be as shown in Figure 20.12c, so as to set up its own field from right to left.

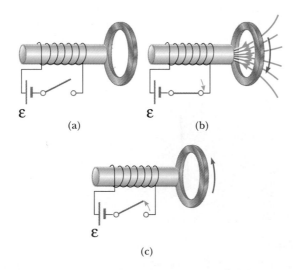

(a)

(b)

(c)

FIGURE 20.12
(Example 20.4)

SPECIAL TOPIC

TAPE RECORDERS

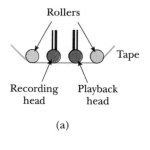

(a)

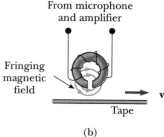

(b)

FIGURE 20.13
(a) The head of a magnetic tape recorder. (b) The fringing magnetic field magnetizes the tape during recording.

One common practical use of induced currents and emfs is in the tape recorder. Many different types of tape recorders are made, but the basic principles are the same for all. A magnetic tape moves past a recording head and a playback head, as in Figure 20.13a. The tape is a plastic ribbon coated with iron oxide or chromium oxide.

The recording process uses the fact that a current passing through an electromagnet produces a magnetic field. Figure 20.13b illustrates the steps in the process. A sound wave sent into a microphone is transformed into an electric current, amplified, and allowed to pass through a wire coiled around a doughnut-shaped piece of iron, which functions as the recording head. The iron ring and the wire constitute an electromagnet, in which the lines of the magnetic field are contained completely inside the iron except at the point where a slot is cut in the ring. Here the magnetic field fringes out of the iron and magnetizes the small pieces of iron oxide embedded in the tape. Thus, as the tape moves past the slot, it becomes magnetized in a pattern that reproduces both the frequency and the intensity of the sound signal entering the microphone.

To reconstruct the sound signal, the tape is allowed to pass through a recorder with the playback head in operation. This head is very similar to the recording head in that it consists of wire-wound doughnut-shaped piece of iron with a slot in it. When the tape moves past this head, the varying magnetic fields on the tape produce changing field lines through the wire coil. The changing lines induce a current in the coil that corresponds to the current in the recording head that originally produced the tape. This changing electric current can be amplified and used to drive a speaker. Playback is thus an example of induction of a current by a moving magnet.

20.5

GENERATORS

Generators and motors are important practical devices that operate on the principle of electromagnetic induction. First, let us consider the **alternating current** (ac) **generator,** a device that converts mechanical energy to electrical energy. In its simplest form, the ac generator consists of a wire loop rotated in a magnetic field by some external means (Fig. 20.14a). In commercial power

FIGURE 20.14
(a) A schematic diagram of an ac generator. An emf is induced in a coil, which rotates by some external means in a magnetic field. (b) A plot of the alternating emf induced in the loop versus time.

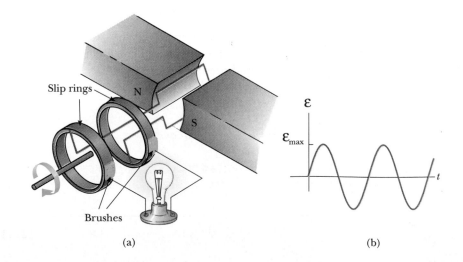

(a)

(b)

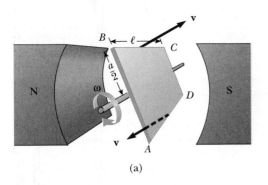

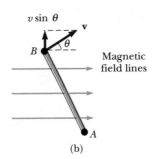

FIGURE 20.15
(a) A loop rotating at constant angular velocity in an external magnetic field. The emf induced in the loop varies sinusoidally with time. (b) An edge view of the rotating loop.

plants, the energy required to rotate the loop can be derived from a variety of sources. For example, in a hydroelectric plant, falling water directed against the blades of a turbine produces the rotary motion; in a coal-fired plant, heat produced by burning coal is used to convert water to steam, and this steam is directed against the turbine blades. As the loop rotates, the magnetic flux through it changes with time, inducing an emf and a current in an external circuit. The ends of the loop are connected to slip rings that rotate with the loop. Connections to the external circuit are made by stationary brushes in contact with the slip rings.

We can derive an expression for the emf generated in the rotating loop by making use of the expression for motional emf, $\varepsilon = B\ell v$. Figure 20.15a shows a loop of wire rotating clockwise in a uniform magnetic field directed to the right. The magnetic force (qvB) on the charges in wires AB and CD is not along the lengths of the wires. (The force on the electrons in these wires is perpendicular to the wires.) Hence, an emf is generated only in wires BC and AD. At any instant, wire BC has velocity **v** at an angle of θ with the magnetic field, as shown in Figure 20.15b. (Note that the component of velocity parallel to the field has no effect on the charges in the wire, whereas the component of velocity perpendicular to the field produces a magnetic force on the charges that moves electrons from C to B.) The emf generated in wire BC equals $B\ell v_\perp$, where ℓ is the length of the wire and $v_\perp$ is the component of velocity perpendicular to the field. An emf of $B\ell v_\perp$ is also generated in wire DA, and the sense of this emf is the same as in wire BC. Since $v_\perp = v \sin \theta$, the total emf is

$$\varepsilon = 2B\ell v_\perp = 2B\ell v \sin \theta \qquad\qquad \textbf{[20.6]}$$

If the loop rotates with a constant angular velocity of ω, we can use the relation $\theta = \omega t$ in Equation 20.6. Furthermore, since every point on wires BC and DA rotates in a circle about the axis of rotation with the same angular velocity, ω, we have $v = r\omega = (a/2)\omega$, where a is the length of sides AB and CD. Therefore, Equation 20.6 reduces to

$$\varepsilon = 2B\ell\left(\frac{a}{2}\right)\omega \sin \omega t = B\ell a\omega \sin \omega t$$

If the loop has N turns, the emf is N times as large because each loop has the same emf induced in it. Furthermore, since the area of the loop is $A = \ell a$, the total emf is

$$\varepsilon = NBA\omega \sin \omega t \qquad\qquad \textbf{[20.7]}$$

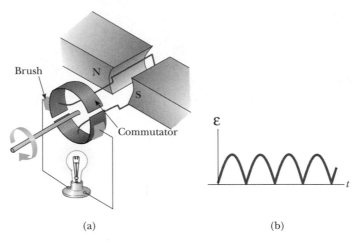

FIGURE 20.16
(a) A schematic diagram of a dc generator. (b) The emf fluctuates in magnitude but always has the same polarity.

(a)

(b)

This result shows that the emf varies sinusoidally with time, as plotted in Figure 20.14b. Note that the maximum emf has the value

$$\mathcal{E}_{max} = NBA\omega \qquad \text{[20.8]}$$

which occurs when $\omega t = 90°$ or $270°$. In other words, $\mathcal{E} = \mathcal{E}_{max}$ when the plane of the loop is parallel to the magnetic field. Furthermore, the emf is zero when $\omega t = 0$ or $180°$, that is, when the magnetic field is perpendicular to the plane of the loop. In the United States and Canada, the frequency of rotation for commercial generators is 60 Hz, whereas in some European countries 50 Hz is used. (Recall that $\omega = 2\pi f$, where f is the frequency in hertz.)

The **direct current** (dc) **generator** is illustrated in Figure 20.16a. The components are essentially the same as those of the ac generator, except that the contacts to the rotating loop are made by a split ring, or commutator. In this design, the output voltage always has the same polarity and the current is a pulsating direct current, as in Figure 20.16b. This can be understood by noting that the contacts to the split ring reverse their roles every half cycle. At the same time, the polarity of the induced emf reverses. Hence, the polarity of the split ring remains the same.

A pulsating dc current is not suitable for most applications. To produce a steady dc current, commercial dc generators use many loops and commutators distributed around the axis of rotation so that the sinusoidal pulses from the loops overlap phase. When these pulses are superimposed, the dc output is almost free of fluctuations.

Turbines that generate electrical energy at a hydroelectric power plant. *(Luis Castaneda/ The IMAGE Bank)*

EXAMPLE 20.5   Emf Induced in an ac Generator

An ac generator consists of eight turns of wire of area $A = 0.0900$ m² with a total resistance of 12.0 Ω. The loop rotates in a magnetic field of 0.500 T at a constant frequency of 60.0 Hz.

(a) Find the maximum induced emf.

Solution   First note that $\omega = 2\pi f = 2\pi(60.0 \text{ Hz}) = 377$ rad/s. When we substitute the appropriate numerical values into Equation 20.8, we obtain

$$\mathcal{E}_{max} = NAB\omega = 8(0.0900 \text{ m}^2)(0.500 \text{ T})(377 \text{ rad/s}) = \boxed{136 \text{ V}}$$

(b) What is the maximum induced current?

Solution   From Ohm's law and the result of (a), we find that

$$I_{max} = \frac{\mathcal{E}_{max}}{R} = \frac{136\ V}{12.0\ \Omega} = \boxed{11.3\ A}$$

(c) Determine the time variation of the induced emf.

Solution   We can use Equation 20.7 to obtain the time variation of $\mathcal{E}$:

$$\mathcal{E} = \mathcal{E}_{max} \sin \omega t = \boxed{(136\ V)\ \sin 377 t}$$

where t is in seconds.

Exercise   Determine the time variation of the induced current.

Answer $I = (11.3\ A)\ \sin 377t.$

*MOTORS AND BACK emf

Motors are devices that convert electrical energy to mechanical energy. Essentially, *a motor is a generator run in reverse.* Instead of a current being generated by a rotating loop, a current is supplied to the loop by a source of emf, and the magnetic force on the current-carrying loop causes it to rotate.

A motor can perform useful mechanical work when a shaft connected to its rotating coil is attached to some external device. As the coil in the motor rotates, however, the changing magnetic flux through it induces an emf which acts to reduce the current in the coil. If this were not the case, Lenz's law would be violated. The phrase **back emf** is used for an emf that tends to reduce the applied current. The back emf increases in magnitude as the rotational speed of the coil increases. We can picture this state of affairs as the equivalent circuit in Figure 20.17. For illustrative purposes, assume that the external power source attempting to drive current through the coil of the motor has a voltage of 120 V, that the coil has a resistance of 10 Ω, and that the back emf induced in the coil at this instant is 70 V. Thus, the voltage available to supply current equals the difference between the applied voltage and the back emf, 50 V in this case. It is clear that the current is limited by the back emf.

When a motor is turned on, there is no back emf initially and the current is very large because it is limited only by the resistance of the coil. As the coil begins to rotate, the induced back emf opposes the applied voltage and the current in the coil is reduced. If the mechanical load increases, the motor slows down, which decreases the back emf. This reduction in the back emf increases the current in the coil and therefore also increases the power needed from the external voltage source. Consequently, the power requirements for starting a motor and for running it under heavy loads are greater than those for running the motor under average loads. If the motor is allowed to run under no mechanical load, the back emf reduces the current to a value just large enough to overcome energy losses by heat and friction.

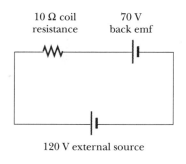

FIGURE 20.17
A motor can be represented as a resistance plus a back emf.

EXAMPLE 20.6   The Induced Current in a Motor

A motor has coils with a resistance of 10 Ω and is supplied by a voltage of 120 V. When the motor is running at its maximum speed, the back emf is 70 V. Find the current in the coils (a) when the motor is first turned on; (b) when the motor has reached maximum speed.

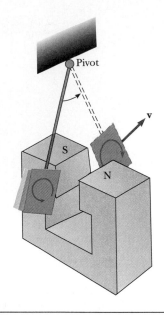

FIGURE 20.18

An apparatus that demonstrates the formation of eddy currents in a conductor moving through a magnetic field. As the plate enters or leaves the field, the changing magnetic flux sets up an induced emf, which causes the eddy currents in the plate.

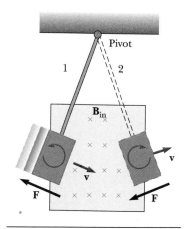

FIGURE 20.19

As the conducting plate enters the magnetic field in position 1, the eddy currents are counter-clockwise. At position 2, however, the currents are clockwise. In either case, the magnetic force retards the motion of the plate.

Solution   (a) When the motor is first turned on, the back emf is zero. (The coils are motionless.) Thus, the current in the coils is a maximum and is

$$I = \frac{\mathcal{E}}{R} = \frac{120 \text{ V}}{10 \text{ } \Omega} = \boxed{12 \text{ A}}$$

(b) At the maximum speed, the back emf has its maximum value. Thus, the effective supply voltage is now that of the external source minus the back emf, and the current is reduced to

$$I = \frac{\mathcal{E} - \mathcal{E}_{\text{back}}}{R} = \frac{120 \text{ V} - 70 \text{ V}}{10 \text{ } \Omega} = \frac{50 \text{ V}}{10 \text{ } \Omega} = \boxed{5.0 \text{ A}}$$

Exercise   If the current in the motor is 8.0 A at some instant, what is the back emf at this time?

Answer   40 V.

*20.6

EDDY CURRENTS

As we have seen, an emf and a current are induced in a circuit by a changing magnetic flux. In the same manner, circulating currents called **eddy currents** are set up in pieces of metal moving through a magnetic field. This can easily be demonstrated by allowing a flat metal plate at the end of a bar to swing through a magnetic field (Fig. 20.18). The metal should be a nonmagnetic material, such as aluminum or copper. As the plate enters the field, the changing flux creates an induced emf in the plate, which in turn causes the free electrons in the metal to move, producing swirling eddy currents. According to Lenz's law, the direction of the eddy currents must oppose the change that causes them. Thus, the eddy currents must produce effective magnetic poles on the plate, which are repelled by the poles of the magnet, giving rise to a repulsive force that opposes the swinging motion of the plate.

As indicated in Figure 20.19, when the magnetic field is into the paper, the eddy current is counterclockwise as the swinging plate enters the field at position 1. This is because the external flux into the paper is increasing, and hence, by Lenz's law, the induced current must provide a flux out of the paper. The opposite is true as the plate leaves the field at position 2, where the current is clockwise. Since the induced eddy current always produces a retarding force when the plate enters or leaves the field, the swinging plate quickly comes to rest.

If slots are cut in the metal plate, as in Figure 20.20, the eddy currents and the corresponding retarding force are greatly reduced. The cuts in the plate are open circuits for any large current loops that might otherwise be formed.

The braking systems on many rapid-transit cars make use of electromagnetic induction and eddy currents. An electromagnet, which can be energized with a current, is positioned near the steel rails. The braking action occurs when a large current is passed through the electromagnet. The relative motion of the magnet and rails induces eddy currents in the rails, and the direction of these currents produces a drag force on the moving vehicle. The vehicle's loss in mechanical energy is transformed into heat. Because the eddy currents decrease steadily in magnitude as the vehicle slows, the braking effect is quite smooth.

Eddy currents are undesirable in motors and transformers because they dissipate energy in the form of heat. To reduce this energy loss, moving conducting parts are often laminated—that is, built up in thin layers separated by a nonconducting material, such as lacquer or metal oxide. This layered structure increases the resistance of the possible paths of the eddy currents and effectively confines the currents to individual layers. Lamination is used in the cores of transformers and motors to minimize eddy currents and thereby increase efficiency.

20.7

SELF-INDUCTANCE

Consider a circuit consisting of a switch, a resistor, and a source of emf, as in Figure 20.21. When the switch is closed, the current does not immediately change from zero to its maximum value, $\mathcal{E}/R$. The law of electromagnetic induction, Faraday's law, prevents this. What happens instead is the following. As the current increases with time, the magnetic flux through the loop due to this current also increases. The increasing flux induces an emf in the circuit that opposes the change in magnetic flux. By Lenz's law, the induced electric field in the loop must therefore be opposite the direction of the current. That is, the induced emf is in the direction indicated by the dashed battery in Figure 20.21. The net potential difference across the resistor is the emf of the battery minus the opposing induced emf. As the magnitude of the current increases, the *rate* of increase lessens and hence the induced emf decreases. This opposing emf results in a gradual increase in the current. For the same reason, when the switch is opened, the current gradually decreases to zero. This effect is called **self-induction** because the changing flux through the circuit arises from the circuit itself. The emf that is set up in this case is called a **self-induced emf.**

As a second example of self-inductance, consider Figure 20.22, which shows a coil wound on a cylindrical iron core. (A practical device would have several hundred turns.) Assume that the current changes with time. When the current is in the direction shown, a magnetic field is set up inside the coil, directed from right to left. As a result, some lines of magnetic flux pass through the cross-sectional area of the coil. As the current changes with time, the flux through the coil changes and induces an emf in the coil. Application of Lenz's law shows that this

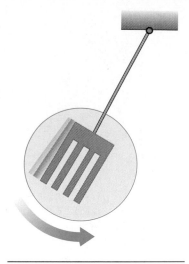

FIGURE 20.20
When slots are cut in the conducting plate, the eddy currents are reduced and the plate swings more freely through the magnetic field.

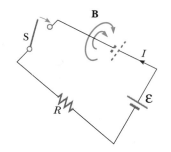

FIGURE 20.21
After the switch in the circuit is closed, the current produces its own magnetic flux through the loop. As the current increases toward its equilibrium value, the flux changes in time and induces an emf in the loop.

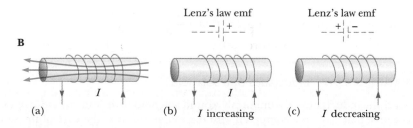

(a) (b) *I* increasing (c) *I* decreasing

FIGURE 20.22
(a) A current in the coil produces a magnetic field directed to the left. (b) If the current increases, the coil acts as a source of emf directed as shown by the dashed battery. (c) The emf of the coil changes its polarity if the current decreases.

induced emf has a direction so as to oppose the change in the current. That is, if the current is increasing, the induced emf is as pictured in Figure 20.22b, and if the current is decreasing, the induced emf is as shown in Figure 20.22c.

To evaluate self-inductance quantitatively, first note that, according to Faraday's law, the induced emf is as given by Equation 20.2:

$$\varepsilon = -N\frac{\Delta\Phi}{\Delta t}$$

The magnetic flux is proportional to the magnetic field, which is proportional to the current in the circuit. Thus, *the self-induced emf must be proportional to the time rate of change of the current:*

Self-induced emf

$$\varepsilon \equiv -L\frac{\Delta I}{\Delta t}$$ [20.9]

where L is a proportionality constant called the **inductance** of the device. The negative sign indicates that a changing current induces an emf in opposition to that change. This means that if the current is increasing (ΔI positive), the induced emf is negative, indicative of opposition to the increase in current. Likewise, if the current is decreasing (ΔI negative), the sign of the induced emf is positive to indicate that the emf is acting to oppose the decrease.

The inductance of a coil depends on the cross-sectional area of the coil and other quantities, all of which can be grouped under the general heading of geometric factors. The SI unit of inductance is the **henry** (H), which, from Equation 20.9, is equal to 1 volt-second per ampere:

$$1\text{ H} = 1\text{ V}\cdot\text{s/A}$$

Examples 20.7 and 20.8 discuss some simple situations for which self-inductances are easily evaluated. In the process, it is often convenient to equate Equations 20.2 and 20.9 to find an expression for L:

$$N\frac{\Delta\Phi}{\Delta t} = L\frac{\Delta I}{\Delta t}$$

Inductance

$$L = N\frac{\Delta\Phi}{\Delta I} = \frac{N\Phi}{I}$$ [20.10]

where we assume that $\Phi = 0$ and $I = 0$ at $t = 0$.

EXAMPLE 20.7   Inductance of a Solenoid

Find the inductance of a uniformly wound solenoid with N turns and length ℓ. Assume that ℓ is large compared with the radius and that the core of the solenoid is air.

Reasoning   The inductance can be found from $L = N\Phi/I$. The flux through each turn is $\Phi = BA$, and $B = \mu_0 nI$.

Joseph Henry, an American physicist who carried out early experiments in electromagnetic induction, was born in Albany, New York, in 1797. The son of a laborer, Henry had little schooling and was forced to find employment at a very young age. After working his way through Albany Academy to study medicine, then engineering, Henry became professor of mathematics and physics in 1826. He later served as professor of natural philosophy at New Jersey College (now Princeton University).

In 1848, Henry became the first director of the Smithsonian Institute, where he introduced a weather-forecasting system based on meteorological information received by the electric telegraph. He was also the first president of the Academy of Natural Science, a position he held until his death in 1878.

Many of Henry's early experiments dealt with electromagnetism. He improved the electromagnet of William Sturgeon and made one of the first electromagnetic motors. By 1830 Henry had made powerful electromagnets by using many turns of fine insulated wire wound around iron cores. He discovered the phenomenon of self-induction but failed to publish his findings; as a result, credit was given to Michael Faraday.

Henry's contribution to science was formally acknowledged in 1893, when the unit of inductance was named the henry.

Joseph Henry (1797–1878)

(North Wind Picture Archives)

Solution We take the interior field to be uniform and given by Equation 19.15:

$$B = \mu_0 n I = \mu_0 \frac{N}{\ell} I$$

where $n = N/\ell$ is the number of turns per unit length. The flux through each turn is

$$\Phi = BA = \mu_0 \frac{N}{\ell} AI$$

where A is the cross-sectional area of the solenoid. From this expression and Equation 20.10, we find that

$$L = \frac{N\Phi}{I} = \frac{\mu_0 N^2 A}{\ell} \qquad \textbf{[20.11]}$$

This shows that L depends on the geometric factors ℓ and A and on μ_0 and is proportional to the square of the number of turns. Since $N = n\ell$, we can also express the result in the form

$$L = \mu_0 \frac{(n\ell)^2}{\ell} A = \mu_0 n^2 A\ell = \mu_0 n^2 V \qquad \textbf{[20.12]}$$

where $V = A\ell$ is the volume of the solenoid.

EXAMPLE 20.8   Calculating Inductance and Self-Induced emf

(a) Calculate the inductance of a solenoid containing 300 turns if the length of the solenoid is 25.0 cm and its cross-sectional area is $4.00 \text{ cm}^2 = 4.00 \times 10^{-4} \text{ m}^2$.

Solution Using Equation 20.11, we get

$$L = \frac{\mu_0 N^2 A}{\ell} = (4\pi \times 10^{-7}\,\text{T}\cdot\text{m/A})\,\frac{(300)^2(4.00 \times 10^{-4}\,\text{m}^2)}{25.0 \times 10^{-2}\,\text{m}}$$

$$= 1.81 \times 10^{-4}\,\text{T}\cdot\text{m}^2/\text{A} = \boxed{0.181\,\text{mH}}$$

(b) Calculate the self-induced emf in the solenoid described in (a) if the current through it is decreasing at the rate of 50.0 A/s.

Solution Equation 20.9 can be combined with $\Delta I/\Delta t = -50.0$ A/s to give

$$\varepsilon = -L\frac{\Delta I}{\Delta t} = -(1.81 \times 10^{-4}\,\text{H})(-50.0\,\text{A/s}) = \boxed{9.05\,\text{mV}}$$

*20.8

RL CIRCUITS

A circuit element that has a large inductance, such as a closely wrapped coil of many turns, is called an **inductor**. The circuit symbol for an inductor is ... . We shall always assume that the self-inductance of the remainder of the circuit is negligible compared with that of the inductor in the circuit.

To gain some insight into the effect of an inductor in a circuit, consider the two circuits in Figure 20.23. Figure 20.23a shows a resistor connected to the terminals of a battery. Ohm's law applied to this circuit gives

$$\varepsilon = RI \tag{20.13}$$

In the past, we have interpreted resistance as a measure of opposition to the current. Now consider the circuit in Figure 20.23b, consisting of an inductor connected to the terminals of a battery. At the instant the switch in this circuit is closed, the emf of the battery equals the back emf generated in the coil. Thus, we have

$$\varepsilon = -L\frac{\Delta I}{\Delta t} \tag{20.14}$$

From this expression, we can interpret L as a measure of *opposition to the rate of change in current.*

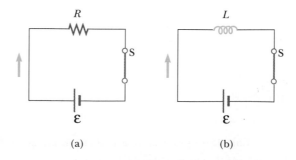

(a)   (b)

FIGURE 20.23
A comparison of the effect of a resistor to that of an inductor in a simple circuit.

Figure 20.24 shows a circuit consisting of a resistor, inductor, and battery. Suppose the switch is closed at $t = 0$. The current begins to increase, but the inductor produces an emf that opposes the increasing current. Thus, the current is unable to change from zero to its maximum value of $\mathcal{E}/R$ instantaneously. Equation 20.14 shows that the induced emf is a maximum when the current is changing most rapidly, which occurs when the switch is first closed. As the current approaches its steady-state value, the back emf of the coil falls off because the current is changing more slowly. Finally, when the current reaches its steady-state value, the rate of change is zero and the back emf is also zero. Figure 20.25 plots current in the circuit as a function of time.[1] This plot is very similar to that of the charge on a capacitor as a function of time, discussed in Chapter 18, Section 18.5. In that case, we found it convenient to introduce a quantity called the time constant of the circuit, which told us something about the time required for the capacitor to approach its steady-state charge. In the same fashion, time constants are defined for circuits containing resistors and inductors. The **time constant** τ for an *RL* circuit is the time required for the current in the circuit to reach 63.2% of its final value, $\mathcal{E}/R$; the time constant is given by

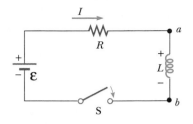

FIGURE 20.24
A series *RL* circuit. As the current increases toward its maximum value, the inductor produces an emf that opposes the increasing current.

$$\tau = \frac{L}{R}$$ [20.15]

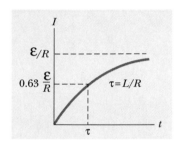

FIGURE 20.25
A plot of current versus time for the *RL* circuit shown in Figure 20.24. The switch is closed at $t = 0$, and the current increases toward its maximum value, $\mathcal{E}/R$. The time constant, τ, is the time it takes the current to reach 63.2% of its maximum value.

EXAMPLE 20.9   The Time Constant for an *RL* Circuit

The circuit shown in Figure 20.24 consists of a 30-mH inductor, a 6.0-Ω resistor, and a 12-V battery. The switch is closed at $t = 0$.

(a) Find the time constant of the circuit.

Solution   The time constant is given by Equation 20.15:

$$\tau = \frac{L}{R} = \frac{30 \times 10^{-3}\,\text{H}}{6.0\,\Omega} = \boxed{5.0\ \text{ms}}$$

(b) Find the current after one time constant has elapsed.

Solution   After one time constant, the current in the circuit has risen to 63.2% of its final value. Thus, the current is

$$I = 0.632\,\frac{\mathcal{E}}{R} = (0.632)\left(\frac{12\,\text{V}}{6.0\,\Omega}\right) = \boxed{1.3\ \text{A}}$$

Exercise   What is the voltage drop across the resistor (a) at $t = 0$? (b) after one time constant?

Answer   (a) 0; (b) 7.6 V.

---

[1] The equation for the current in the circuit as a function of time is

$$I = \frac{\mathcal{E}}{R}\left(1 - e^{-Rt/L}\right)$$

20.9

ENERGY STORED IN A MAGNETIC FIELD

The emf induced by an inductor prevents a battery from establishing a current in a circuit instantaneously. The battery has to do work to produce a current. We can think of this needed work as energy stored by the inductor in its magnetic field. In a manner quite similar to that used in Section 16.8 to find the energy stored by a capacitor, we find that the energy stored by an inductor is

Energy stored in an inductor

$$PE_L = \tfrac{1}{2}LI^2 \qquad \text{[20.16]}$$

Note that the result is similar in form to the expression for the energy stored in a capacitor:

Energy stored in a capacitor

$$PE_C = \tfrac{1}{2}CV^2$$

SUMMARY

The **magnetic flux, Φ,** through a closed loop is defined as

$$\Phi \equiv BA \cos \theta \qquad \text{[20.1]}$$

where B is the strength of the uniform magnetic field, A is the cross-sectional area of the loop, and θ is the angle between **B** and the direction perpendicular to the plane of the loop.

Faraday's law of induction states that the instantaneous emf induced in a circuit equals the rate of change of magnetic flux through the circuit:

$$\varepsilon = -N\frac{\Delta\Phi}{\Delta t} \qquad \text{[20.2]}$$

where N is the number of loops in the circuit.

Lenz's law states that the polarity of the induced emf is such that it produces a current whose magnetic field opposes the *change* in magnetic flux through a circuit.

If a conducting bar of length ℓ moves through a magnetic field with a speed, v, so that **B** is perpendicular to the bar, the emf induced in the bar, often called a **motional emf,** is

$$\varepsilon = B\ell v \qquad \text{[20.4]}$$

When a coil of wire with N turns, each of area A, rotates with constant angular velocity ω in a uniform magnetic field **B** as in Figure 20.15, the emf induced in the coil is

$$\varepsilon = NAB\omega \sin \omega t \qquad \text{[20.7]}$$

When the current in a coil changes with time, an emf is induced in the coil according to Faraday's law. This **self-induced emf** is defined by the expression

$$\varepsilon \equiv -L\frac{\Delta I}{\Delta t} \qquad \text{[20.9]}$$

where L is the inductance of the coil. The SI unit for inductance is the henry (H); 1 H = 1 V·s/A.

The **inductance** of a coil can be found from the expression

$$L = \frac{N\Phi}{I} \qquad \textbf{[20.10]}$$

where N is the number of turns on the coil, I is the current in the coil, and Φ is the magnetic flux through the coil.

If a resistor and inductor are connected in series to a battery and a switch is closed at $t = 0$, the current in the circuit does not rise instantly to its maximum value. After one **time constant**, $\tau = L/R$, the current in the circuit is 63.2% of its final value, $\mathcal{E}/R$.

The **energy stored** in the magnetic field of an inductor carrying current I is

$$PE_L = \tfrac{1}{2}LI^2 \qquad \textbf{[20.16]}$$

ADDITIONAL READING

H. Kondo, "Michael Faraday," *Sci. American,* October 1953, p. 90.

D. K. C. McDonald, *Faraday, Maxwell and Kelvin,* New York, Doubleday Anchor, 1964.

H. L. Sharlin, "From Faraday to the Dynamo," *Sci. American,* May 1961, p. 107.

G. Shiers, "The Induction Coil," *Sci. American,* May 1971, p. 80.

CONCEPTUAL QUESTIONS

Example A circular current loop is located in a uniform and constant magnetic field. Describe how an emf can be induced in the loop.

Reasoning According to Faraday's law, an emf is induced in a current loop if the magnetic flux through the loop changes with time. An emf can be induced in the loop in this situation by either rotating the loop about an arbitrary axis or by changing the shape of the loop.

Example A spacecraft orbiting the Earth has a coil of wire in it. An astronaut measures a small current in the coil although there is no battery connected to it and there are no magnets on the spacecraft. What is causing the current?

Reasoning As the spacecraft passes through space, it is apparently moving from a region of one magnetic field strength to a region of a different magnetic field strength. The changing field through the coil induces a current in the coil.

Example As the conducting bar in Figure 20.26 moves to the right, an electric field directed downward is set up in the conductor. If the bar were moving to the left, explain why the electric field would be upward.

Reasoning If the bar were moving to the left, the magnetic force on the negative charge carriers in the bar would be upward, causing an accumulation of negative charge on the top and positive charge on the bottom. Hence, the electric field in the bar would be upward.

Example As the bar in Figure 20.26 moves perpendicularly to the magnetic field, is an external force required to keep it moving with constant velocity? Explain.

Reasoning No. Once the bar is in motion and the charges are separated, no external force is required to maintain the motion. An applied external force in the x direction will cause the bar to accelerate in that direction.

Example A circuit containing a coil, resistor, and battery connected in series is in steady state; that is, the current has reached a constant value. Does the coil have an inductance? Does the coil affect the value of the final current in the circuit? Explain.

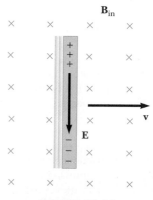

FIGURE 20.26

Reasoning   The coil has an inductance regardless of the nature of the current in the circuit. Inductance is only a function of the coil geometry and the nature of the material inside the coil. Since the current is constant, the self-induced emf of the coil is zero, so the coil (neglecting its resistance) has no effect on the steady-state current. The coil influences only the rate at which the current increases after the circuit is closed.

Example   Suppose the switch in a circuit containing an inductor, a battery, and a resistor all connected in series has been closed for a long time and is suddenly opened. Does the current instantaneously drop to zero? Why does a spark sometimes appear at the switch contacts when the switch is opened?

Reasoning   No. The current decays to zero exponentially with time due to the inductance in the circuit. The characteristic time of the decay depends on the values of both the inductance and resistance. A spark tends to appear at the switch when it is opened since the back emf associated with the inductor has its maximum value at this instant, which in turn can cause electrical breakdown in the air between the switch contacts.

Example   If the current in an inductor is doubled, by what factor is the stored energy multiplied?

Reasoning   The energy stored in an inductor carrying a current I is given by $PE_L = \frac{1}{2}LI^2$. Doubling the current will multiply the stored energy by a factor of four.

Example   How is electrical energy produced in dams (that is, how is the energy of motion of the water converted to electric energy)?

Reasoning   The falling water pushes on the vanes of a turbine functioning as a waterwheel causing it to turn, and with it, the rotating coil of a generator. The emf induced in the coil of the generator is the source of voltage driving current throughout the electric power grid. When current passes through the generator coil, a retarding magnetic force acts on each wire. The water must do work against this magnetic force.

1. What is the difference between magnetic flux and magnetic field?
2. A wire loop is placed in a uniform magnetic field. For what orientation of the loop is the magnetic flux a maximum? For what orientation is the flux zero?
3. We saw in the last chapter that a steady magnetic field does not change the energy of a charge moving in it. Can a changing magnetic field affect the energy of a charge?
4. Does dropping a magnet down a long copper tube produce a current in the tube? Explain.
5. Explain why an external force is necessary to keep the bar in Figure 20.27 moving with a constant velocity.
6. The bar in Figure 20.27 moves to the right with velocity **v**, and a uniform, constant magnetic field is di-

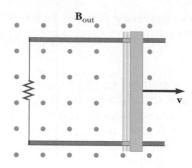

FIGURE 20.27   (Questions 5 and 6)

rected *outward* (represented by dots). Why is the induced current clockwise? If the bar were moving to the left, what would be the direction of the induced current?

7. Magnetic storms on the Sun (sunspots) can cause difficulties for communications here on the Earth. Why?
8. Would you expect the tape from a tape recorder to be attracted to a magnet? (Try it, but not with a recording you wish to save.)
9. Wearing a metal bracelet in a region of strong magnetic field could be hazardous. Discuss.
10. What happens when the rate of rotation of a generator coil is increased?
11. Could a current be induced in a coil by rotating a magnet inside it? If so, how?
12. "If the shaft of a dc motor is cranked manually, the motor becomes a dc generator." Is this statement true? Discuss.
13. A piece of aluminum is dropped vertically between the poles of an electromagnet. Does the magnetic field affect the velocity of the aluminum?
14. A bar magnet is dropped toward a conducting ring lying on the floor. As the magnet falls toward the ring, does it move as a freely falling body?
15. In a beam balance scale, an aluminum plate is sometimes used to slow the oscillations of the beam near equilibrium. The plate is mounted at the end of the beam and moves between the poles of a small horseshoe magnet attached to the frame. Why are the oscillations of the beam strongly damped near equilibrium?
16. Why is the induced emf that appears in an inductor called a back (counter) emf?
17. Does the inductance of a coil depend on the current in the coil? What parameters affect the inductance of a coil?
18. If a small bar magnet is placed at the center of a sphere, what is the total magnetic flux through the sphere? (*Hint:* Think about the symmetry in this problem.) Would your answer be any different if the magnet were not at the center of the sphere? What if the magnet were outside the sphere?

19. When the switch in the circuit in Figure 20.28a is closed, a current is set up in the coil and the metal ring springs upward (see Fig. 20.28b). Explain this behavior.

20. Assume that the battery in Figure 20.28a is replaced by an alternating current source and the switch S is held closed. If the metal ring on top of the solenoid is held down, it becomes *hot*. Why?

FIGURE 20.28 (Questions 19 and 20) *(Photo courtesy of Central Scientific Company)*

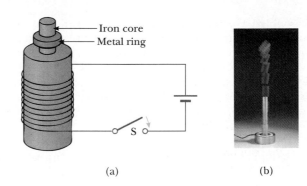

(a)   (b)

PROBLEMS

Section 20.1   Induced emf and Magnetic Flux

1. A wire in the shape of an equilateral triangle 10.0 cm on a side is placed in a uniform magnetic field of magnitude 0.10 T. If the field makes an angle of 45° with the plane of the triangle, determine the magnetic flux through the triangle.

2. Find the flux of the Earth's magnetic field, of magnitude 5.00×10^{-5} T, through a square loop of area 20.0 cm², (a) when the field is perpendicular to the plane of the loop; (b) when the field makes a 30.0° angle with the normal to the plane of the loop; (c) when the field makes a 90.0° angle with the normal to the plane.

3. A square loop 2.00 m on a side is placed in a magnetic field of strength 0.300 T. If the field makes an angle of 50.0° with the normal to the plane of the loop, as in Figure 20.2, determine the magnetic flux through the loop.

4. A long, straight wire lies in the plane of a circular coil with a radius of 0.010 m. The wire carries a current of 2.0 A and is placed along a diameter of the coil. (a) What is the net flux through the coil? (b) If the wire passes through the center of the coil and is perpendicular to the plane of the coil, find the net flux through the coil.

5. A solenoid 4.00 cm in diameter and 20.0 cm long has 250 turns and carries a current of 15.0 A. Calculate the magnetic flux through the circular cross-sectional area of the solenoid.

6. The solenoid of Problem 5 is surrounded by a single loop of wire 10.0 cm in diameter. The loop is positioned perpendicularly to and centered on the axis of the solenoid, as shown in Figure 20.29. Find the magnetic flux through the loop of wire. (The field outside the solenoid is small enough to be negligible.)

7. A long, straight wire carrying a current of 2.00 A is placed along the axis of a cylinder of radius 0.500 m and a length of 3.00 m. Determine the total magnetic flux through the cylinder.

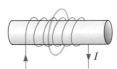

FIGURE 20.29 (Problem 6)

8. A circular wire loop of radius R is placed in a uniform magnetic field, **B**, and is then spun at a constant angular velocity ω about an axis through its diameter. Determine the magnetic flux through the loop as a function of time if the axis of rotation is (a) perpendicular to **B**; (b) parallel to **B**.

Section 20.2   Faraday's Law of Induction

9. A circular wire loop of radius 0.50 m lies in a plane perpendicular to a uniform magnetic field of magnitude 0.40 T. If in 0.10 s the wire is re-shaped from a circle into a square, but remains in the same plane, what is the magnitude of the average induced emf in the wire during this time?

10. A coil of radius 20 cm is placed in an external magnetic field of strength 0.20 T so that the plane of the coil is perpendicular to the field. The coil is pulled out of the field in 0.30 s. Find the average induced emf during this interval.

11. The bolt of lightning depicted in Figure 20.30 strikes the ground 200 m from a 100-turn coil

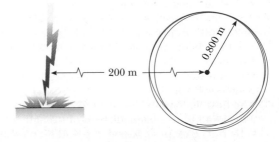

FIGURE 20.30 (Problem 11)

□ indicates problems that have full solutions available in the Student Solutions Manual and Study Guide.

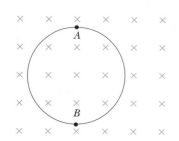

FIGURE 20.31 (Problem 17)

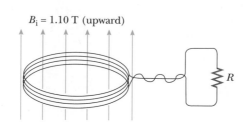

FIGURE 20.32 (Problem 19)

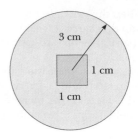

FIGURE 20.33 (Problem 21)

oriented as shown. If the current in the lightning bolt falls from 6.02×10^6 A to zero in 10.5 μs, what is the average voltage induced in the coil? Assume that the distance to the center of the coil determines the average magnetic induction at the coil's position. Treat the lightning bolt as a vertical wire.

12. A powerful electromagnet has a field of 1.6 T and a cross-sectional area of 0.20 m². If a coil of 200 turns with a total resistance of 20 Ω is placed around it, and then the power to the electromagnet is turned off in 0.020 s, what current is induced in the coil?

13. A two-turn circular wire loop of radius 0.500 m lies in a plane perpendicular to a uniform magnetic field of magnitude 0.40 T. If the wire is reshaped from a two-turn circle to a one-turn circle in 0.10 s (while remaining in the same plane), what is the magnitude of the average induced emf in the wire during this time? (*Hint:* Use Faraday's law in the form $\mathcal{E} = -\Delta(N\Phi)/\Delta t$.)

14. A wire loop of radius 0.30 m lies so that an external magnetic field of strength +0.30 T is perpendicular to the loop. The field changes to −0.20 T in 1.5 s. (The plus and minus signs here refer to opposite directions through the coil.) Find the magnitude of the average induced emf in the coil during this time.

15. A 500-turn circular-loop coil 15.0 cm in diameter is initially aligned so that its axis is parallel to the Earth's magnetic field. In 2.77 ms the coil is flipped so that its axis is perpendicular to the Earth's magnetic field. If a voltage of 0.166 V is thereby induced in the coil, what is the value of the Earth's magnetic field?

16. A square, single-turn coil 0.20 m on a side is placed with its plane perpendicular to a constant magnetic field. An emf of 18 mV is induced in the winding when the area of the coil decreases at a rate of 0.10 m²/s. What is the magnitude of the magnetic field?

17. The flexible loop in Figure 20.31 has a radius of 12 cm and is in a magnetic field of strength 0.15 T. The loop is grasped at points A and B and

stretched until it closes. If it takes 0.20 s to close the loop, find the magnitude of the average induced emf in it during this time.

18. The plane of a rectangular coil, 5.0 cm by 8.0 cm, is perpendicular to the direction of a magnetic field, **B**. If the coil has 75 turns and a total resistance of 8.0 Ω, at what rate must the magnitude of **B** change to induce a current of 0.10 A in the windings of the coil?

19. A circular coil, enclosing an area of 100 cm², is made of 200 turns of copper wire, as shown in Figure 20.32. Initially, a 1.1-T uniform magnetic field points perpendicularly upward through the plane of the coil. The direction of the field then reverses so that the final magnetic field has a magnitude of 1.1 T and points downward through the coil. During the period in which the field is changing direction, how much charge flows through the coil if the coil is connected to a 5.0-Ω resistor as shown?

20. A 300-turn solenoid with a length of 20 cm and a radius of 1.5 cm carries a current of 2.0 A. A second coil of four turns is wrapped tightly about this solenoid so that it can be considered to have the same radius as the solenoid. Find (a) the change in the magnetic flux through the coil and (b) the magnitude of the average induced emf in the coil when the current in the solenoid increases to 5.0 A in a period of 0.90 s.

21. A square, single-turn wire coil 1.00 cm on a side is placed inside a solenoid that has a circular cross-section of radius 3.00 cm, as shown in Figure 20.33. The solenoid is 20.0 cm long and wound with 100 turns of wire. (a) If the current in the solenoid is 3.00 A, find the flux through the coil. (b) If the current in the solenoid is reduced to zero in 3.00 s, find the magnitude of the average induced emf in the coil.

22. The wire shown in Figure 20.34 is bent in the shape of a "tent," with $\theta = 60°$ and $L = 1.5$ m, and is placed in a uniform magnetic field of 0.30 T perpendicular to the tabletop. The wire is "hinged" at points a and b. If the "tent" is flattened out on the table in 0.10 s, what is the average induced emf in the wire during this time?

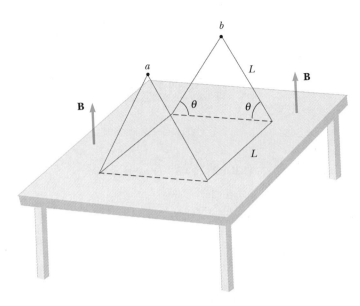

FIGURE 20.34 (Problem 22)

Section 20.3 Motional emf

23. A car with a 1.00-m-long radio antenna travels at 80.0 km/h in a place where the Earth's magnetic field is 5.00×10^{-5} T. What is the *maximum* possible induced emf in the antenna as it moves through the Earth's magnetic field?

24. An airplane with a wingspan of 40 m flies parallel to the Earth's surface, at a location where the downward component of the Earth's magnetic field is 0.60×10^{-4} T. At what speed would the plane have to fly to produce a difference in potential of 1.5 V between the tips of its wings? Is this likely to occur during a typical airplane flight?

25. Over a region where the *vertical* component of the Earth's magnetic field is 40.0 μT directed downward, a 5.00-m length of wire is held in an east-west direction and moved horizontally to the north with a speed of 10.0 m/s. Calculate the potential difference between the ends of the wire, and determine which end is positive.

26. Consider the arrangement shown in Figure 20.35. Assume that $R = 6.0$ Ω and $\ell = 1.2$ m, and that a uniform 2.5-T magnetic field is directed *into* the page. At what speed should the bar be moved to produce a current of 0.50 A in the resistor?

27. Over a region where the vertical component of the Earth's magnetic field is 40.0 μT, a 0.500-m length of wire held in an east-west direction moves at 5.00 m/s along parallel rails, as shown in Figure 20.9. Find (a) the induced emf in the circuit and (b) the induced current in the circuit if there is a

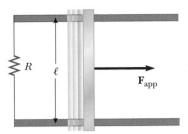

FIGURE 20.35 (Problem 26)

resistor of 5.00 Ω connected between the parallel rails. The resistance of the wire and track can be considered negligible.

28. For the circuit described in Problem 27, find (a) the average power supplied to the circuit by the induced emf, (b) the power dissipated in the 5.00-Ω resistor, (c) the force required to keep the length of wire moving at a constant speed, and (d) the power supplied to the circuit by the agent exerting the force calculated in part (c).

29. A 12.0-m-long steel beam is accidentally dropped by a construction crane from a height of 9.00 m. The horizontal component of the Earth's magnetic field over the region is 18.0 μT. What is the induced emf in the beam just before impact with the Earth, assuming its long dimension remains in a horizontal plane, oriented perpendicularly to the horizontal component of the Earth's magnetic field?

Section 20.4 Lenz's Law Revisited

30. A bar magnet is held above the center of a wire loop in a horizontal plane, as shown in Figure 20.36. The south end of the magnet is toward the loop. The magnet is dropped. Find the direction of the current through the resistor (a) while the magnet is falling toward the loop and (b) after the magnet has passed through the loop and moves away from it.

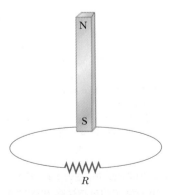

FIGURE 20.36 (Problem 30)

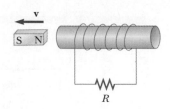

FIGURE 20.37   (Problem 31)

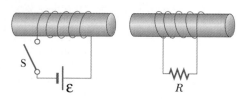

FIGURE 20.38   (Problem 32)

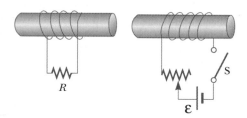

FIGURE 20.39   (Problem 33)

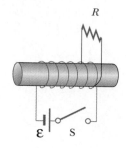

FIGURE 20.40   (Problem 34)

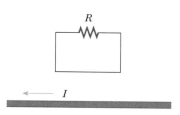

FIGURE 20.41   (Problem 35)

31. A bar magnet is positioned near a coil of wire as shown in Figure 20.37. What is the direction of the current through the resistor when the magnet is moved (a) to the left? (b) to the right?

32. Find the direction of the current through the resistor in Figure 20.38, (a) at the instant the switch is closed, (b) after the switch has been closed for several minutes, and (c) at the instant the switch is opened.

33. Find the direction of the current in resistor R in Figure 20.39 after each of the following steps, (taken in the order given). (a) The switch is closed. (b) The variable resistance in series with the battery is decreased. (c) The circuit containing resistor R is moved to the left. (d) The switch is opened.

34. In Figure 20.40, what is the direction of the current induced in the resistor at the instant the switch is closed?

35. What is the direction of the current induced in the resistor when the current in the long, straight wire in Figure 20.41 decreases rapidly to zero?

36. Two flat coils of wire are lying on a table, with the small one (the secondary) completely inside the larger coil (the primary), as shown in Figure 20.42. (a) If the switch is closed, what is the direction of the current in the primary? (b) What is the

direction of the induced current in the secondary just after the switch is closed?

37. The magnetic field shown in Figure 20.43 has a uniform magnitude of 25.0 mT directed out of the paper. The initial diameter of the kink is 2.00 cm. (a) The wire is quickly pulled taut, and the kink shrinks to a diameter of zero in 50.0 ms. Determine the average voltage induced between endpoints A and B. Include the polarity. (b) Suppose the kink is undisturbed, but the magnetic field increases to 100 mT in 4.00×10^{-3} s. Determine the average voltage across terminals A and B, including polarity, during this period.

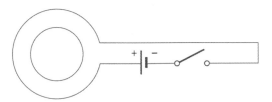

FIGURE 20.42   (Problem 36)

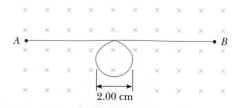

FIGURE 20.43   (Problem 37)

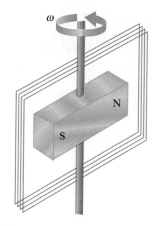

FIGURE 20.44   (Problem 38)

38. A bar magnet is spun at constant angular speed ω about an axis, as shown in Figure 20.44. A flat rectangular conducting loop surrounds the magnet, and at $t = 0$ the magnet is oriented as shown. Sketch the induced current in the loop as a function of time, plotting counterclockwise currents as positive and clockwise currents as negative.

Section 20.5   Generators

39. The alternating voltage of a generator is represented by the equation $\mathcal{E} = (240 \text{ V}) \sin 500t$, where $\mathcal{E}$ is in volts and t is in seconds. Find the frequency of the voltage and the maximum voltage output of the source.

40. A semicircular conductor of radius $R = 0.25$ m is rotated about axis AC at a constant rate of 120 revolutions per minute (Fig. 20.45). A uniform magnetic field in all of the lower half of the figure is directed out of the page and has a magnitude of 1.3 T. Determine the maximum value of the emf induced in the conductor.

41. A 100-turn square wire coil of area 0.040 m² rotates about a vertical axis at 1500 rpm, as indicated in Figure 20.46. The horizontal component of the Earth's magnetic field at the location of the loop is 2.0×10^{-4} T. Calculate the maximum emf induced in the coil by the Earth's field.

42. A loop of area 0.10 m² is rotating at 60 rev/s with its axis of rotation perpendicular to a 0.20-T magnetic field. (a) If there are 1000 turns on the loop,

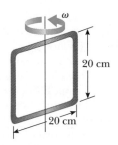

FIGURE 20.46   (Problem 41)

what is the maximum voltage induced in the loop? (b) When the maximum induced voltage occurs, what is the orientation of the loop with respect to the magnetic field?

43. An ac generator with its terminals shorted together consists of 40 turns of wire with an area of 0.12 m² and a total resistance of 30 Ω. The loop rotates in a magnetic field of 0.10 T at a constant frequency of 60.0 Hz. (a) Find the maximum induced emf, (b) the maximum induced current, (c) an expression for the time variation of $\mathcal{E}$, and (d) an expression for the time variation of the induced current.

44. A motor has coils with a resistance of 30 Ω and operates from a voltage of 240 V. When the motor is operating at its maximum speed, the back emf is 145 V. Find the current in the coils (a) when the motor is first turned on and (b) when the motor has reached maximum speed. (c) If the current in the motor is 6.0 A at some instant, what is the back emf at that time?

45. When the coil of a motor is rotating at maximum speed, the current in the windings is 4.0 A. When the motor is first turned on, the current in the windings is 11 A. If the motor is operated at 120 V, find the back emf in the coil and the resistance of the windings.

46. Figure 20.47 represents an electromagnetic brake that utilizes eddy currents. An electromagnet

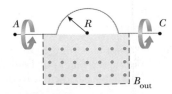

FIGURE 20.45   (Problem 40)

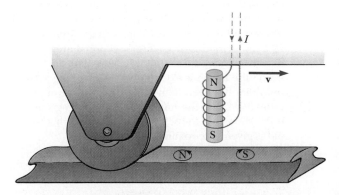

FIGURE 20.47   (Problem 46)

hangs from the car near the rails; to stop the car, a large current is sent through the coils of the electromagnet. The moving electromagnet induces eddy currents in the rails whose fields oppose the field of the electromagnet, and thereby slows the car to a stop. From the direction of the car's motion and the direction of the current shown, determine which of the two eddy currents shown on the rails is correct. Explain.

47. In a model ac generator, a 500-turn rectangular coil, 8.0 cm by 20 cm, rotates at 120 rev/min in a uniform magnetic field of 0.60 T. (a) What is the maximum emf induced in the coil? (b) What is the instantaneous value of the emf in the coil at $t = (\pi/32)$ s? Assume that the emf is zero at $t = 0$. (c) What is the smallest value of t for which the emf will have its maximum value?

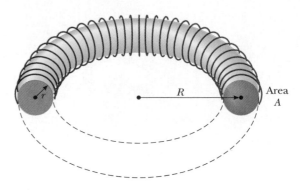

FIGURE 20.48   (Problem 53)

Section 20.7   Self-Inductance

48. A coil has an inductance of 3.0 mH, and the current through it changes from 0.20 A to 1.5 A in 0.20 s. Find the magnitude of the average induced emf in the coil during this period.

49. A Slinky toy spring has a radius of 4.00 cm and an inductance of 275 μH when extended to a length of 1.50 m. What is the total number of turns in the spring?

50. A solenoid of radius 2.5 cm has 400 turns and a length of 20 cm. Find (a) its inductance and (b) the rate at which current must change through it to produce an emf of 75 mV.

51. Show that the two expressions for inductance given by

$$L = \frac{N\Phi}{I} \quad \text{and} \quad L = \frac{-\mathcal{E}}{\Delta I/\Delta t}$$

have the same units.

52. An emf of 24.0 mV is induced in a 500-turn coil at an instant when the current is 4.00 A and is changing at a rate of 10.0 A/s. What is the magnetic flux through each turn of the coil?

53. The toroid in Figure 20.48 has a major radius of R, a minor radius of r, a cross-sectional area of $A = \pi r^2$, and N turns. If $R \gg r$, the magnetic field inside the toroid is essentially that of a long solenoid that has been bent into a large circle of radius R. Using the uniform field of a long solenoid, show that the self-inductance of such a toroid is given (approximately) by

$$L \approx \frac{\mu_0 N^2 A}{2\pi R}$$

Section 20.8   RL Circuits

54. Show that the SI units for the inductive time constant τ are seconds.

55. A 6.0-V battery is connected in series with a resistor and an inductor. The series circuit has a time constant of 600 μs, and the maximum current is 300 mA. What is the value of the inductance?

56. A 25-mH inductor, an 8.0-Ω resistor, and a 6.0-V battery are connected in series. The switch is closed at $t = 0$. Find the voltage drop across the resistor (a) at $t = 0$ and (b) after one time constant has passed. Also, find the voltage drop across the inductor (c) at $t = 0$ and (d) after one time constant has elapsed.

57. The switch in a series RL circuit in which $R = 6.00 \, \Omega$, $L = 3.00$ H, and $\mathcal{E} = 24.00$ V is closed at $t = 0$. (a) What is the maximum current in the circuit? (b) What is the current when $t = 0.500$ s?

58. An RL circuit with $L = 3.00$ H and an RC circuit with $C = 3.00 \, \mu$F have the same time constant. If the two circuits have the same resistance, R, (a) what is the value of R, and (b) what is this common time constant?

Section 20.9   Energy Stored in a Magnetic Field

59. How much energy is stored in a 70.0-mH inductor at the instant when the current is 2.00 A?

60. A 24-V battery is connected in series with a resistor and an inductor, where $R = 8.0 \, \Omega$ and $L = 4.0$ H. Find the energy stored in the inductor (a) when the current reaches its maximum value and (b) one time constant after the switch is closed.

ADDITIONAL PROBLEMS

61. A 50-turn rectangular coil, 0.20 m by 0.30 m, is rotated at 90 rad/s in a magnetic field so that the axis of rotation is perpendicular to the direction

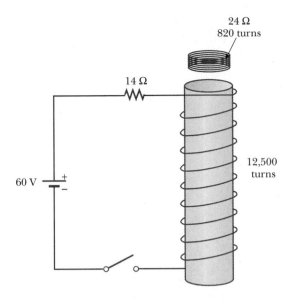

24 Ω
820 turns

14 Ω

60 V

12,500
turns

FIGURE 20.49   (Problem 62)

of the field. The maximum emf induced in the coil is 0.50 V. What is the magnitude of the field?

62. An 820-turn wire coil of resistance 24.0 Ω is placed on top of a 12 500-turn, 7.00-cm-long solenoid, as in Figure 20.49. Both coil and solenoid have cross-sectional areas of 1.00×10^{-4} m². (a) How long does it take the solenoid current to reach 0.632 times its maximum value? (b) Determine the average back emf caused by the self-inductance of the solenoid during this interval. (c) Determine the average rate of change in magnetic flux through the coil during this interval. (d) Find the magnitude of the average induced current in the coil.

63. A tightly wound circular coil has 50 turns, each of radius 0.20 m. A uniform magnetic field is introduced perpendicularly to the plane of the coil. If the field increases in strength from 0 to 0.30 T in 0.40 s, what average emf is induced in the windings of the coil?

64. A coiled telephone cord has 70 turns, a cross-sectional diameter of 1.3 cm, and an unstretched length of 60 cm. Determine the self-inductance of the unstretched cord.

65. An automobile starter motor draws a current of 3.5 A from a 12-V battery when operating at normal speed. A broken pulley locks the armature in position, and the current increases to 18 A. What was the back emf of the motor when operating normally?

66. A five-turn circular coil of radius 15 cm is oriented with its plane perpendicular to a uniform magnetic field of 0.15 T. This field increases at a uniform rate of 0.20 T in 3.0 s. If the resistance of the coil is 8.0 Ω, find the amount of charge that passes through the coil during this 3.0-s interval.

67. In Figure 20.50, the bar magnet is moved toward the loop. Is $V_a - V_b$ positive, negative, or zero? Explain.

68. A 300-turn inductor has a radius of 5.00 cm and a length of 20.0 cm. Find the energy stored in it when the current is 0.500 A.

69. An open hemispherical surface of radius 0.10 m is in a magnetic field of 0.15 T. The circular cross-section of the surface is perpendicular to the direction of the field. Calculate the magnetic flux through the surface.

70. An aluminum ring of radius 5.0 cm and resistance 3.0×10^{-4} Ω is placed on top of a long air-core

FIGURE 20.50   (Problem 67)

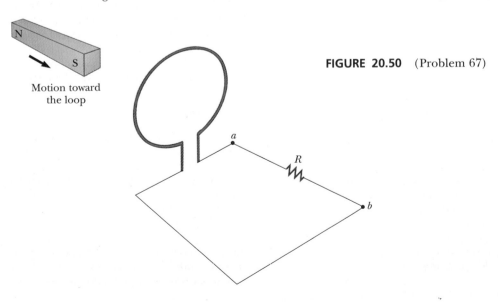

N

S

Motion toward
the loop

a

R

b

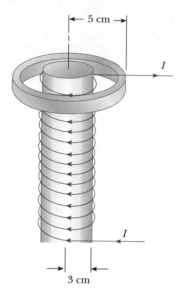

FIGURE 20.51 (Problem 70)

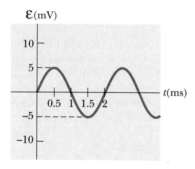

FIGURE 20.52 (Problem 71)

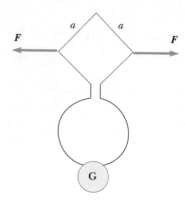

FIGURE 20.53 (Problem 72)

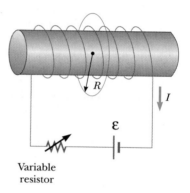

FIGURE 20.54 (Problem 73)

solenoid with 1000 turns per meter and radius 3 cm, as in Figure 20.51. The magnetic field due to the current in the solenoid at the location of the ring is one-half that at the center of the solenoid. If the current in the solenoid is increasing at a constant rate of 270 A/s, what is the induced current in the ring?

71. Figure 20.52 is a graph of induced emf versus time for a coil of N turns rotating with angular speed ω in a uniform magnetic field directed perpendicularly to the axis of rotation of the coil. Copy this sketch (increasing the scale), and on the same set of axes show the graph of emf versus t when (a) the number of turns in the coil is doubled; (b) the angular speed is doubled; (c) the angular speed is doubled while the number of turns in the coil is halved.

72. A square loop of wire with edge length $a = 0.20$ m is perpendicular to the Earth's magnetic field at a point where $B = 15\ \mu$T, as in Figure 20.53. The total resistance of the loop and the wires connecting the loop to the galvanometer is 0.50 Ω. If the loop is suddenly collapsed by horizontal forces, as shown, what total charge passes through the galvanometer?

73. A single-turn circular loop of radius 0.20 m is coaxial with a long 1600-turn solenoid of radius 0.050 m and length 0.80 m, as in Figure 20.54. The variable resistor is changed so that the solenoid current decreases linearly from 6.0 A to 1.5 A in 0.20 s. Calculate the induced emf in the circular loop. (The field just outside the solenoid is small enough to be negligible.)

74. A small circular washer of radius 0.500 cm is held directly below a long, straight wire carrying a current of 10.0 A, and is 0.500 meter above the top of a table (Fig. 20.55). (a) If this washer is dropped from rest, what is the magnitude of the

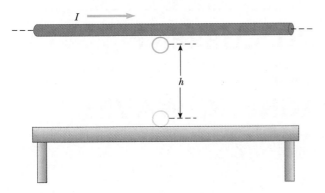

FIGURE 20.55 (Problem 74)

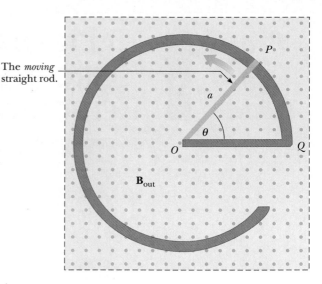

FIGURE 20.56 (Problem 75)

average induced emf in the washer from the time it is released to the moment it hits the tabletop? You may assume that the magnetic field through the entire washer is the same as the magnetic field at the center of the washer. (b) In what direction does current flow through the wire?

75. Figure 20.56 shows a stationary conductor, whose shape is similar to the letter "e" ($a = 50.0$ cm), that is placed in a constant magnetic field of magnitude $B = 0.500$ T directed out of the page. A 50.0-cm-long, straight conducting rod pivoted about point O rotates with a constant angular speed of 2 rad/s. (a) Determine the induced emf in loop POQ. (*Hint:* The area of loop POQ is $A = \theta a^2/2$.) (b) If the conducting material has a resistance per unit length of 5.00 Ω/m, what is the induced current in loop POQ at 0.250 s? (*Hint:* The length of arc PQ is $a\theta$.)

76. A horizontal wire is free to slide on the vertical rails of a conducting frame, as in Figure 20.57. The wire has mass m and length ℓ, and the resistance of the circuit is R. If a uniform magnetic field is directed perpendicularly to the frame, what is the terminal speed of the wire as it falls under the force of gravity? (Neglect friction.)

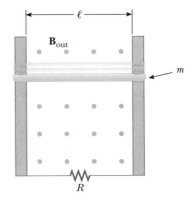

FIGURE 20.57 (Problem 76)

21

ALTERNATING CURRENT CIRCUITS AND ELECTROMAGNETIC WAVES

A nighttime view of the satellite receiver-transmitter dish at the Land Earth Station at Goonhilly in Cornwall, United Kingdom. This station serves the INMARSAT (International Maritime Satellite) organization, providing telephone, telex, data, and facsimile (fax) operations to the shipping, aviation, offshore, and land mobile industries. (Photo Researchers, Inc.)

It is important to understand the basic principles of alternating current (ac) circuits because they are so much a part of our everyday life. Every time we turn on a television set or a stereo, or any of a multitude of other electric appliances, we are calling on alternating currents to provide the power to operate them. We begin our study of ac circuits by examining the characteristics of a circuit containing a source of emf and a single circuit element: a resistor, a capacitor, or an inductor. Then we examine what happens when these elements are connected in combination with each other. Our discussion is limited to situations in which the elements are arranged in simple series configurations.

We conclude this chapter with a discussion of **electromagnetic waves,** which are composed of fluctuating electric and magnetic fields. Electromagnetic waves in the form of visible light enable us to view the world around us; infrared waves warm our environment; radio-frequency waves carry our favorite television and radio programs; the list goes on and on.

21.1

RESISTORS IN AN ac CIRCUIT

An ac circuit consists of combinations of circuit elements and an ac generator, which provides the alternating current. We have seen that the output of an ac generator is sinusoidal and varies with time according to

$$v = V_m \sin 2\pi ft \qquad \text{[21.1]}$$

where v is the instantaneous voltage, V_m is the maximum voltage of the ac generator, and f is the frequency at which the voltage changes, measured in hertz. We first consider a simple circuit consisting of a resistor and an ac generator (designated by the symbol —Ⓐ—), as in Figure 21.1. The current and the voltage across the resistor are shown in Figure 21.2.

Let us briefly discuss the current-versus-time curve in Figure 21.2. At point a on the curve, the current has a maximum value in one direction, arbitrarily called the positive direction. Between points a and b, the current is decreasing in magnitude but is still in the positive direction. At point b, the current is momentarily zero; it then begins to increase in the opposite (negative) direction between points b and c. At point c, the current has reached its maximum value in the negative direction.

Note that the current and voltage are in step with each other since they vary identically with time.

> Since the current and the voltage reach their maximum values at the same time, they are said to be *in phase*.

Note that *the average value of the current over one cycle is zero*. That is, the current is maintained in one direction (the positive direction) for the same amount of time and at the same magnitude as it is in the opposite direction (the negative direction). However, the direction of the current has no effect on the behavior of the resistor in the circuit. This can be understood by realizing that collisions between electrons and the fixed atoms of the resistor result in an increase in the temperature of the resistor. Although this temperature increase depends on the magnitude of the current, it is independent of its direction.

We can quantify this discussion by recalling that the rate at which electrical energy is converted to heat in a resistor, which is the power P, is

$$P = i^2 R$$

where i is the instantaneous current in the resistor. Since the heating effect of a current is proportional to the *square* of the current, it makes no difference whether the current is direct or alternating — that is, whether the sign associated with the current is positive or negative. However, the heating effect produced by an alternating current with a maximum value of I_m *is not the same* as that produced by a direct current of the same value. This is because the alternating current is at this maximum value for only a very brief instant of time during a cycle. What is important in an ac circuit is an average value of current, referred to as the rms current. The **rms current** is the direct current that would dissipate the same amount of energy in a resistor as is dissipated by the actual alternating current. The term *rms* stands for *root mean square*, which simply means that the square root of the average value of the square of the current is taken. Since i^2

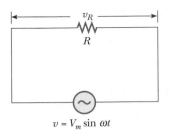

FIGURE 21.1
A series circuit consisting of a resistor, R, connected to an ac generator, designated by the symbol —Ⓐ—.

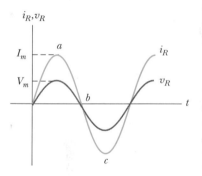

FIGURE 21.2
A plot of current and voltage across a resistor versus time.

varies as $\sin^2 2\pi ft$, one can show that the average value of i^2 is $\frac{1}{2}I_m{}^2$ (Fig. 21.3).[1] Therefore, the rms current, I, is related to the maximum value of the alternating current, I_m, by

rms current

$$I = \frac{I_m}{\sqrt{2}} = 0.707 I_m \qquad \text{[21.2]}$$

This equation says that an alternating current with a maximum value of 2 A produces the same heating effect in a resistor as a direct current of $(2/\sqrt{2})$ A. Thus, we can say that the average power dissipated in a resistor that carries alternating current I is $P_{av} = I^2 R$, where I is the rms current.

Alternating voltages are also best discussed in terms of rms voltages, with the relationship being identical to the preceding one; that is, the rms voltage, V, is related to the maximum value of the alternating voltage, V_m, by

rms voltage

$$V = \frac{V_m}{\sqrt{2}} = 0.707 V_m \qquad \text{[21.3]}$$

When we speak of measuring an ac voltage of 120 V from an electric outlet, we really mean an *rms* voltage of 120 V. A quick calculation using Equation 21.3 shows that such an ac voltage actually has a peak value of about 170 V. In this chapter we use rms values when discussing alternating currents and voltages. One reason is that ac ammeters and voltmeters are designed to read rms values. Furthermore, if we use rms values, many of the equations we rely on will have the same form as those used in the study of direct current (dc) circuits. Table 21.1 summarizes the notations used in this chapter.

Consider the series circuit in Figure 21.1, consisting of a resistor connected to an ac generator. A resistor limits the current in an ac circuit just as it does in a dc circuit. Therefore, Ohm's law is valid for an ac circuit, and we have

$$V_R = IR \qquad \text{[21.4]}$$

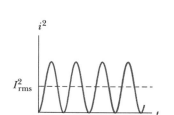

FIGURE 21.3
A plot of the square of the current in a resistor versus time. The rms current is the square root of the average of the square of the current.

That is, *the rms voltage across a resistor is equal to the rms current in the circuit times the resistance*. This equation also applies if maximum values of current and voltage are used. That is, the maximum voltage drop across a resistor equals the maximum current in the resistor times the resistance.

---

[1] The fact that the square root of the average value of the square of the current equals $I_m/\sqrt{2}$ can be shown as follows. The current in the circuit varies with time according to the expression $i = I_m \sin 2\pi ft$, and so $i^2 = I_m{}^2 \sin^2 2\pi ft$. Therefore, we can find the average value of i^2 by calculating the average value of $\sin^2 2\pi ft$. Note that a graph of $\cos^2 2\pi ft$ versus time is identical to a graph of $\sin^2 2\pi ft$ versus time, except that the points are shifted on the time axis. Thus, the time average of $\sin^2 2\pi ft$ is equal to the time average of $\cos^2 2\pi ft$ when taken over one or more cycles. That is,

$$(\sin^2 2\pi ft)_{av} = (\cos^2 2\pi ft)_{av}$$

With this fact and the trigonometric identity $\sin^2 \theta + \cos^2 \theta = 1$, we get

$$(\sin^2 2\pi ft)_{av} + (\cos^2 2\pi ft)_{av} = 2(\sin^2 2\pi ft)_{av} = 1$$

$$(\sin^2 2\pi ft)_{av} = \tfrac{1}{2}$$

When this result is substituted into the expression $i^2 = I_m{}^2 \sin^2 2\pi ft$, we get $(i^2)_{av} = I^2 = I_m{}^2/2$, or $I = I_m/\sqrt{2}$, where I is the rms current.

TABLE 21.1
Notation Used in This Chapter

	Voltage	Current
Instantaneous value	v	i
Maximum value	V_m	I_m
rms value	V	I

EXAMPLE 21.1 What Is the rms Current?

An ac voltage source has an output of $v = (200 \text{ V}) \sin 2\pi ft$. This source is connected to a 100-Ω resistor as in Figure 21.1. Find the rms current through the resistor.

Reasoning Compare the expression for the voltage output just given with the general form, $v = V_m \sin 2\pi ft$.

Solution By comparison, we see that the maximum output voltage of the device is 200 V. Thus, the rms voltage output of the source is

$$V = \frac{V_m}{\sqrt{2}} = \frac{200 \text{ V}}{\sqrt{2}} = 141 \text{ V}$$

Ohm's law can be used in resistive ac circuits as well as in dc circuits. The calculated rms voltage can be used with Ohm's law to find the rms current in the circuit:

$$I = \frac{V}{R} = \frac{141 \text{ V}}{100 \text{ } \Omega} = \boxed{1.41 \text{ A}}$$

Exercise Find the maximum current in the circuit.

Answer 2.00 A.

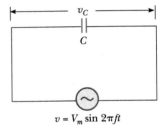

FIGURE 21.4
A series circuit consisting of a capacitor, C, connected to an ac generator.

21.2

CAPACITORS IN AN ac CIRCUIT

To understand the effect of a capacitor on the behavior of a circuit containing an ac voltage source, let us first recall what happens when a capacitor is placed in a circuit containing a dc source, such as a battery. At the instant a switch is closed in a series circuit containing a battery, a resistor, and a capacitor, there is zero charge on the plates of the capacitor. Therefore, the motion of charge through the circuit is relatively free, and initially there is a large current in the circuit. As more charge accumulates on the capacitor, the voltage across it increases, opposing the current. After some time interval—which depends on the time constant, RC—has elapsed, the current approaches zero. From this, we see that a capacitor in a dc circuit limits, or impedes, the current so that it approaches zero after a brief time.

Now consider the simple series circuit in Figure 21.4, consisting of a capacitor connected to an ac generator. Let us sketch a curve of current versus time and one of voltage versus time, and then attempt to make the graphs seem reasonable. The curves are shown in Figure 21.5. First, note that the segment of the current curve from a to b indicates that the current starts out at a rather large

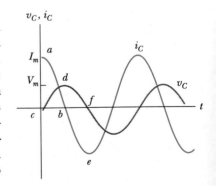

FIGURE 21.5
Plots of current and voltage across a capacitor versus time, in an ac circuit. The voltage lags the current by 90°.

value. This can be understood by recognizing that there is no charge on the capacitor at $t = 0$; consequently, there is nothing in the circuit except the resistance of the wires to hinder the flow of charge at this instant. However, the current decreases as the voltage across the capacitor increases from c to d on the voltage curve. When the voltage is at point d, the current reverses and begins to increase in the opposite direction (from b to e on the current curve). During this time, the voltage across the capacitor decreases from d to f because the plates are now losing the charge they accumulated earlier. The remainder of the cycle for both voltage and current is a repeat of what happened during the first half of the cycle. The current reaches a maximum value in the opposite direction at point e on the current curve and then decreases as the voltage across the capacitor builds up.

Note that the current and voltage are not in step with each other, as they are in a purely resistive circuit. The curves of Figure 21.5 indicate that, when an alternating voltage is applied across a capacitor, the voltage reaches its maximum value one quarter of a cycle after the current reaches its maximum value. In this situation, it is common to say that

The voltage across a capacitor lags behind the current by 90°

the voltage always lags behind the current by 90°.

The impeding effect of a capacitor on the current in an ac circuit is expressed in terms of a factor called the **capacitive reactance**, X_C, defined as

Capacitive reactance

$$X_C \equiv \frac{1}{2\pi f C} \qquad \text{[21.5]}$$

You will be asked in an end-of-chapter problem to show that when C is in farads and f is in hertz, the unit of X_C is the ohm.

Let us examine whether Equation 21.5 is reasonable. With a dc source (a dc source can be considered an ac source with zero frequency), X_C is infinitely large. This means that a capacitor impedes the direct current the same way a resistor of infinitely large resistance would. The current in such a circuit is zero. Indeed, we found that to be the case in Chapter 16. On the other hand, Equation 21.5 predicts that, as the frequency increases, the capacitive reactance decreases. This means that, before the charge on a capacitor has time to build up to the point where the current is zero, the direction of the current has reversed.

The analogy between capacitive reactance and resistance allows us to write an equation of the same form as Ohm's law to describe ac circuits containing capacitors. This equation relates the rms voltage and rms current in the circuit to the reactance as

$$V_C = IX_C \qquad \text{[21.6]}$$

EXAMPLE 21.2   A Purely Capacitive ac Circuit

An 8.00-μF capacitor is connected to the terminals of an ac generator with an rms voltage of 150 V and a frequency of 60.0 Hz. Find the capacitive reactance and the rms current in the circuit.

Solution   From Equation 21.5 and the fact that $2\pi f = 377\text{ s}^{-1}$, we have

$$X_C = \frac{1}{2\pi fC} = \frac{1}{(377\text{ s}^{-1})(8.00\times 10^{-6}\text{ F})} = \boxed{332\ \Omega}$$

If we substitute this result into Equation 21.6, we find that

$$I = \frac{V_C}{X_C} = \frac{150\text{ V}}{332\ \Omega} = \boxed{0.452\text{ A}}$$

Exercise   If the frequency is doubled, what happens to the capacitive reactance and the current?

Answer X_C is halved, and I is doubled.

21.3
INDUCTORS IN AN ac CIRCUIT

Now consider an ac circuit consisting only of an inductor connected to the terminals of an ac generator, as in Figure 21.6. (In any real circuit, there is some resistance in the wire forming the inductive coil, but we ignore this for now.) The changing current output of the generator produces a back emf in the coil of magnitude

$$v_L = L\frac{\Delta I}{\Delta t} \tag{21.7}$$

Thus, the current in the circuit is impeded by the back emf of the inductor. The effective resistance of the coil in an ac circuit is measured by a quantity called the **inductive reactance**, X_L:

$$X_L \equiv 2\pi fL \tag{21.8}$$

You will be asked in an end-of-chapter problem to show that when f is in hertz and L is in henries, the unit of X_L is the ohm. Note that the inductive reactance increases with increasing frequency and increasing inductance.

To understand the meaning of inductive reactance, let us compare this equation for X_L with Equation 21.7. First, note from Equation 21.8 that the inductive reactance depends on the inductance, L. This seems reasonable because the back emf (Eq. 21.7) is large for large values of L. Second, note that the inductive reactance depends on the frequency, f. This, too, seems reasonable because the back emf depends on $\Delta I/\Delta t$, a quantity that is large when the current changes rapidly, as it would for large frequencies.

With inductive reactance defined in this manner, we can write an equation of the same form as Ohm's law for the voltage across the coil or inductor:

$$V_L = IX_L \tag{21.9}$$

where V_L is the rms voltage drop across the coil and I is the rms current in the coil.

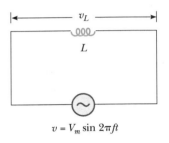

$v = V_m \sin 2\pi ft$

FIGURE 21.6
A series circuit consisting of an inductor, L, connected to an ac generator.

Inductive reactance

The voltage across an inductor leads the current by 90°

Figure 21.7 shows the instantaneous voltage and instantaneous current across the coil as functions of time. When a sinusoidal voltage is applied across an inductor, the voltage reaches its maximum value one quarter of an oscillation period before the current reaches its maximum value. In this situation, we say that

The voltage always leads the current by 90°.

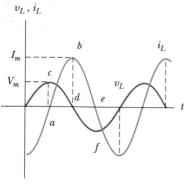

To see why this phase relationship between voltage and current should exist, let us examine a few points on the curves of Figure 21.7. Note that at point a on the current curve, the current is beginning to increase in the positive direction. At this instant, the rate of change of current is at a maximum, and we see from Equation 21.7 that the voltage across the inductor is consequently also at a maximum at this time. As the current rises between points a and b on the curve, $\Delta I/\Delta t$ (the slope of the current curve) gradually decreases until it reaches zero at point b. As a result, the voltage across the inductor is decreasing during this same time interval, as the segment between c and d on the voltage curve indicates. Immediately after point b, the current begins to decrease, although it still has the same direction it had during the previous quarter cycle. As the current decreases to zero (from b to e on the curve), a voltage is again induced in the coil (d to f), but the sense of this voltage is opposite the sense of the voltage induced between c and d. This occurs because back emfs are always directed to oppose the change in the current.

We could continue to examine other segments of the curves, but no new information would be gained since the current and voltage variations are repetitive.

FIGURE 21.7

Plots of current and voltage across an inductor versus time in an ac circuit. The voltage leads the current by 90°.

EXAMPLE 21.3   A Purely Inductive ac Circuit

In a purely inductive ac circuit (see Fig. 21.6), $L = 25.0$ mH and the rms voltage is 150 V. Find the inductive reactance and rms current in the circuit if the frequency is 60.0 Hz.

Solution   First, note that $2\pi f = 2\pi(60.0) = 377$ s^{-1}. Equation 21.8 then gives

$$X_L = 2\pi fL = (377 \text{ s}^{-1})(25.0 \times 10^{-3} \text{ H}) = \boxed{9.43 \ \Omega}$$

Substituting this result into Equation 21.9 gives

$$I = \frac{V_L}{X_L} = \frac{150 \text{ V}}{9.43 \ \Omega} = \boxed{15.9 \text{ A}}$$

Exercise   Calculate the inductive reactance and rms current in the circuit if the frequency is 6 kHz.

Answer $X_L = 943 \ \Omega$; $I_m = 0.159$ A.

21.4

THE *RLC* SERIES CIRCUIT

In the foregoing sections, we examined the effects of an inductor, a capacitor, and a resistor when they are connected separately across an ac voltage source. We now consider what happens when these devices are combined.

Figure 21.8 shows a circuit containing a resistor, an inductor, and a capacitor connected in series across an ac generator. The current in the circuit varies sinusoidally with time, as indicated in Figure 21.9a. Thus,

$$i = I_m \sin 2\pi ft$$

Earlier we learned that the voltage across each element may or may not be in phase with the current. The instantaneous voltages across the three elements, shown in Figure 21.9, have the following phase relations to the instantaneous current.

1. The instantaneous voltage across the resistor, v_R, is *in phase* with the instantaneous current. (See Fig. 21.9b.)
2. The instantaneous voltage across the inductor, v_L, *leads* the current by 90°. (See Fig. 21.9c.)
3. The instantaneous voltage across the capacitor, v_C, *lags behind* the current by 90°. (See Fig. 21.9d.)

The net instantaneous voltage, v, across all three elements is the sum of the instantaneous voltages across the separate elements:

$$v = v_R + v_C + v_L \qquad \text{[21.10]}$$

This net voltage can be obtained graphically as shown in Figure 21.9e. Note that the instantaneous voltage across the resistor (Fig. 21.9b) is in phase with the current, the instantaneous voltage across the inductor (Fig. 21.9c) leads the current by a quarter cycle, and the instantaneous voltage across the capacitor (Fig. 21.9d) lags behind the current by a quarter cycle. Because both v_L and v_C are out of phase with v_R, the net voltage is a complicated function of the voltage amplitudes across the individual elements.

Rather than following the analytical procedure of adding the individual instantaneous voltages to find the net voltage, it is helpful to use another technique involving vectors. We represent the voltage across each element with a rotating vector, as in Figure 21.10. The rotating vectors are referred to as **phasors,** and the diagram is called a **phasor diagram**. This particular diagram represents the circuit voltage given by the expression $v = V_m \sin(2\pi ft + \phi)$, where V_m is the maximum voltage (the amplitude of the phasor) and ϕ is the angle between the phasor and the $+x$ axis. The phasor can be viewed as a vector of magnitude V_m rotating at a constant frequency, f, so that its projection along the y axis is the instantaneous voltage in the circuit. Since ϕ is the phase angle between the voltage and current in the circuit, the phasor for the current (not shown in Figure 21.10) lies along the $+x$ axis and is expressed by the relation $i = I_m \sin(2\pi ft)$.

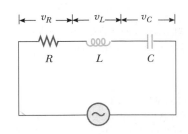

FIGURE 21.8
A series circuit consisting of a resistor, an inductor, and a capacitor connected to an ac generator.

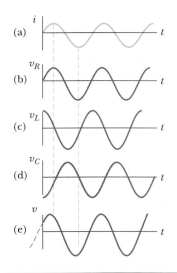

FIGURE 21.9
Phase relations in the series *RLC* circuit shown in Figure 21.8.

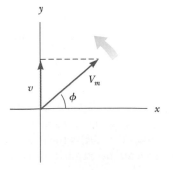

FIGURE 21.10
A phasor diagram for the voltage in an ac circuit, where ϕ is the phase angle between the voltage and current, and v is the instantaneous voltage.

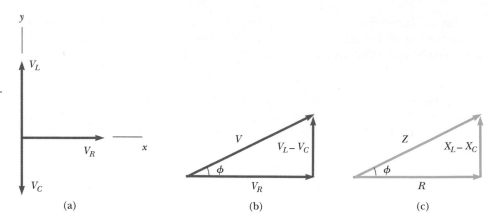

FIGURE 21.11
(a) A phasor diagram for the *RLC* circuit. (b) Addition of the phasors as vectors gives $V = \sqrt{V_R^2 + (V_L - V_C)^2}$. (c) The reactance triangle that gives the impedance relation,
$Z = \sqrt{R^2 + (X_L - X_C)^2}$.

The phasor diagrams in Figure 21.11 are useful for analyzing the *series RLC* circuit. Voltages in phase with the current are represented by vectors along the $+x$ axis, and voltages out of phase with the current lie along other axes. Thus, V_R is horizontal and to the right because it is in phase with the current. Likewise, V_L is represented by a phasor along the $+y$ axis because it leads the current by 90°. Finally, V_C is along the $-y$ axis since it lags behind the current by 90°. If the phasors are added as vector quantities, Figure 21.11a shows that the only x component for the voltages is V_R, and the net y component is $V_L - V_C$. It is convenient to now add the phasors vectorially (Fig. 21.11b), where V is the total rms voltage in the circuit. The right triangle in Figure 21.11b gives the following equations for the total voltage and phase angle:

$$V = \sqrt{V_R^2 + (V_L - V_C)^2} \qquad \text{[21.11]}$$

$$\tan \phi = \frac{V_L - V_C}{V_R} \qquad \text{[21.12]}$$

where all voltages are rms values. Note that although we choose to use rms voltages in our analysis, the equations above apply equally well to peak voltages, because the two quantities are related to each other by the same factor for all circuit elements. The result for the total voltage, V, as given by Equation 21.11, reinforces the fact that *the rms voltages across the resistor, capacitor, and inductor are not in phase, so one cannot simply add them to get the voltage across the combination of elements.*

We can write Equation 21.11 in the form of Ohm's law, using the relations $V_R = IR$, $V_L = IX_L$, and $V_C = IX_C$, where I is the rms current in the circuit:

$$V = I\sqrt{R^2 + (X_L - X_C)^2} \qquad \text{[21.13]}$$

It is convenient to define a parameter called the **impedance,** Z, of the circuit as

Impedance

$$Z \equiv \sqrt{R^2 + (X_L - X_C)^2} \qquad \text{[21.14]}$$

so that Equation 21.13 becomes

$$V = IZ \qquad \text{[21.15]}$$

Note that Equation 21.15 is in the form of Ohm's law, $V = IR$, where R is replaced by the impedance, in ohms. Equation 21.15 can be regarded as a

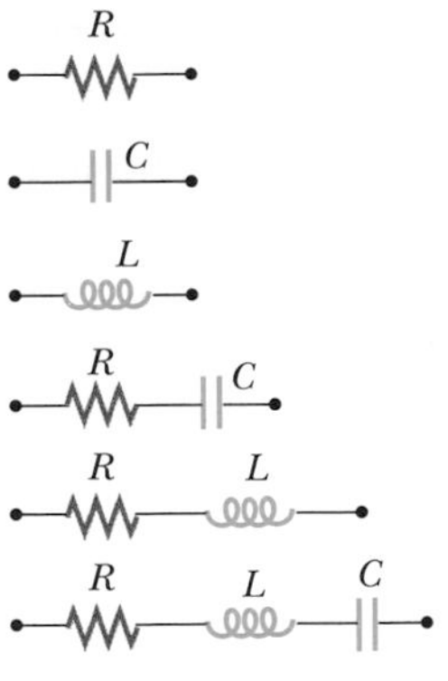

Circuit Elements	Impedance, Z	Phase angle, ϕ
R —WW—	R	$0°$
C —||—	X_C	$-90°$
L —000—	X_L	$+90°$
R —WW— C —||—	$\sqrt{R^2 + X_C^2}$	Negative, between $-90°$ and $0°$
R —WW— L —000—	$\sqrt{R^2 + X_L^2}$	Positive, between $0°$ and $90°$
R —WW— L —000— C —||—	$\sqrt{R^2 + (X_L - X_C)^2}$	Negative if $X_C > X_L$ Positive if $X_C < X_L$

FIGURE 21.12
The impedance values and phase angles for various combinations of circuit elements. In each case, an ac voltage (not shown) is applied across the combination of elements (that is, across the dots).

generalized form of Ohm's law applied to a series ac circuit. Note that the current in the circuit depends upon the resistance, the inductance, the capacitance, *and* the frequency since the reactances are frequency-dependent.

It is useful to represent the impedance, Z, with a vector diagram like Figure 21.11c. A right triangle is constructed; the right side is the quantity $X_L - X_C$, the base is R, and the hypotenuse is Z. Applying the Pythagorean theorem to this triangle, we see that

$$Z = \sqrt{R^2 + (X_L - X_C)^2}$$

which is consistent with Equation 21.14. Furthermore, we see from the vector diagram that the phase angle, ϕ, between the current and the voltage is given by

$$\tan \phi = \frac{X_L - X_C}{R}$$ **[21.16]**   Phase angle, ϕ

The physical significance of the phase angle will become apparent in Section 21.5.

Figure 21.12 provides impedance values and phase angles for some series circuits containing different combinations of circuit elements.

Many parallel alternating current circuits are also useful in everyday applications. We shall not discuss them here, however, because their analysis is beyond the scope of this book.

PROBLEM-SOLVING STRATEGY
Alternating Current

The following procedures are recommended for solving alternating-current problems:

1. **The first step in analyzing alternating-current circuits is to calculate as many of the unknown quantities, such as X_L and X_C, as possible. (When you calculate X_C, express the capacitance in farads rather than, say, microfarads.)**
2. **Apply the equation $V = IZ$ to the portion of the circuit that is of interest. For example, if you want to know the voltage drop across the combination of an inductor and a resistor, the equation for the voltage drop reduces to $V = I\sqrt{R^2 + X_L^2}$.**

EXAMPLE 21.4   Analyzing a Series *RLC* ac Circuit

Analyze a series *RLC* ac circuit for which $R = 250\ \Omega$, $L = 0.600\ H$, $C = 3.50\ \mu F$, $f = 60\ Hz$, and $V = 150\ V$.

Solution   The reactances are given by $X_L = 2\pi f L = 226\ \Omega$ and $X_C = 1/2\pi f C = 758\ \Omega$. Therefore, the impedance is

$$Z = \sqrt{R^2 + (X_L - X_C)^2} = \sqrt{(250\ \Omega)^2 + (226\ \Omega - 758\ \Omega)^2} = 588\ \Omega$$

The rms current is

$$I = \frac{V}{Z} = \frac{150\ V}{588\ \Omega} = 0.255\ A$$

The phase angle between the current and voltage is

$$\phi = \tan^{-1}\left(\frac{X_L - X_C}{R}\right) = \tan^{-1}\left(\frac{226\ \Omega - 758\ \Omega}{250\ \Omega}\right) = -64.8°$$

Since the circuit is more capacitive than inductive (that is, $X_C > X_L$), ϕ is negative. A negative phase angle means that the current leads the applied voltage.

The rms voltages across the elements are

$$V_R = IR = (0.255\ A)(250\ \Omega) = 63.8\ V$$

$$V_L = IX_L = (0.255\ A)(226\ \Omega) = 57.6\ V$$

$$V_C = IX_C = (0.255\ A)(758\ \Omega) = 193\ V$$

Note that the sum of the three rms voltages, $V_R + V_L + V_C$, is 314 V, which is much greater than the rms voltage of the generator, 150 V. The sum 314 V is a meaningless quantity because, when alternating voltages are added, *both their amplitudes and their phases* must be taken into account. That is, the voltages must be added in a way that takes account of the different phases. The relationship among V, V_R, and V_C is given by Equation 21.11. You should use the values found above to verify this equation.

21.5

POWER IN AN ac CIRCUIT

Nikola Tesla (1856–1943) was born in Croatia but spent most of his professional life as an inventor in the United States. He was a key figure in the development of alternating-current electricity, high-voltage transformers, and the transport of electrical power using ac transmission lines. Tesla's viewpoint was at odds with the ideas of Edison, who committed himself to the use of direct current in power transmission. Tesla's ac approach won out. *(UPI/ Bettmann)*

No power losses are associated with capacitors and pure inductors in an ac circuit. (A pure inductor is defined as one with no resistance or capacitance.) Let us begin by analyzing the power dissipated in an ac circuit that contains only a generator and a capacitor.

When the current begins to increase in one direction in an ac circuit, charge begins to accumulate on the capacitor and a voltage drop appears across it. When this voltage drop reaches its maximum value, the energy stored in the capacitor is

$$PE_C = \tfrac{1}{2}CV_m^{\ 2}$$

However, this energy storage is only momentary. When the current reverses direction, the charge leaves the capacitor plates and returns to the voltage source. Thus, during one half of each cycle the capacitor is being charged, and during the other half the charge is being returned to the voltage source. Therefore, the average power supplied by the source is zero. In other words, *a capacitor in an ac circuit does not dissipate energy.*

Similarly, the source must do work against the back emf of an inductor, which carries a current. When the current reaches its maximum value, the energy stored in the inductor is a maximum and is given by

$$PE_L = \tfrac{1}{2}LI_m^2$$

When the current begins to decrease in the circuit, this stored energy is returned to the source as the inductor attempts to maintain the current in the circuit. The only element in an *RLC* circuit that dissipates energy is the resistor. The average power lost in a resistor is

The resistor is the only element in an *RLC* circuit that dissipates energy

$$P_{\text{av}} = I^2 R \qquad\qquad \textbf{[21.17]}$$

where *I* is the rms current in the circuit. An alternative equation for the average power dissipated in an ac circuit can be found by substituting (from Ohm's law) $R = V_R/I$ into Equation 21.17:

$$P_{\text{av}} = IV_R$$

It is convenient to refer to a voltage triangle that shows the relationship among *V*, V_R, and $V_L - V_C$, such as Figure 21.11b. From this figure, we see that the voltage drop across a resistor can be written in terms of the voltage of the source:

$$V_R = V \cos \phi$$

Hence, the average power dissipated in an ac circuit is

$$P_{\text{av}} = IV \cos \phi \qquad\qquad \textbf{[21.18]} \qquad \text{Average power}$$

where the quantity $\cos \phi$ is called the **power factor**.

EXAMPLE 21.5   Calculate the Average Power

Calculate the average power delivered to the series *RLC* circuit described in Example 21.4.

Reasoning and Solution   We are given 150 V for the rms voltage supplied to the circuit, and we have calculated that the rms current in the circuit is 0.255 A and the phase angle, ϕ, is $-64.8°$. Thus, the power factor, $\cos \phi$, is 0.426.

Solution   From these values, we calculate the average power using Equation 21.18:

$$P_{\text{av}} = IV \cos \phi = (0.255 \text{ A})(150 \text{ V})(0.426) = \boxed{16.3 \text{ W}}$$

The same result can be obtained using Equation 21.17.

21.6

RESONANCE IN A SERIES *RLC* CIRCUIT

In general, the current in a series *RLC* circuit can be written

$$I = \frac{V}{Z} = \frac{V}{\sqrt{R^2 + (X_L - X_C)^2}} \qquad\qquad \textbf{[21.19]}$$

From this we see that the current has its *maximum* value when the impedance has its *minimum* value. This occurs when $X_L = X_C$. In such a circumstance, the impedance of the circuit reduces to $Z = R$. The frequency, f_0, at which this happens is called the **resonance frequency** of the circuit. To find f_0, we set $X_L = X_C$, which gives, from Equations 21.5 and 21.8,

$$2\pi f_0 L = \frac{1}{2\pi f_0 C}$$

Resonance frequency

$$f_0 = \frac{1}{2\pi\sqrt{LC}} \qquad\qquad [21.20]$$

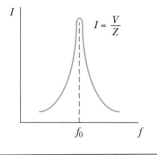

FIGURE 21.13
A plot of current amplitude in a series *RLC* circuit versus frequency of the generator voltage. Note that the current reaches its maximum value at the resonance frequency, f_0.

Figure 21.13 is a plot of current as a function of frequency for a circuit containing a fixed value for both the capacitance and inductance. From Equation 21.19 it must be concluded that the current would become infinite at resonance when $R = 0$. Although Equation 21.19 predicts this result, real circuits always have some resistance, which limits the value of the current.

The receiving circuit of a radio is an important application of a series resonance circuit. The radio is tuned to a particular station (which transmits a specific radio-frequency signal) by varying a capacitor, which changes the resonance frequency of the receiving circuit. When this resonance frequency matches that of the incoming radio wave, the current in the receiving circuit increases.

EXAMPLE 21.6   The Capacitance of a Circuit in Resonance

Consider a series *RLC* circuit for which $R = 150\ \Omega$, $L = 20$ mH, $V = 20$ V, and $2\pi f = 5.0 \times 10^3\ \text{s}^{-1}$. Determine the value of the capacitance for which the rms current is a maximum.

Reasoning   The current is a maximum at the resonance frequency, f_0, which should be made to match the driving frequency, $5.0 \times 10^3\ \text{s}^{-1}$.

Solution   In this problem,

$$2\pi f_0 = 5.0 \times 10^3\ \text{s}^{-1} = \frac{1}{\sqrt{LC}}$$

$$C = \frac{1}{(25 \times 10^6\ \text{s}^{-2})\,L} = \frac{1}{(25 \times 10^6\ \text{s}^{-2})(20.0 \times 10^{-3}\ \text{H})} = \boxed{2.0\ \mu\text{F}}$$

Exercise   Calculate the maximum rms current in the circuit.

Answer   0.13 A.

---

*21.7

THE TRANSFORMER

In many situations it is desirable or necessary to change a small ac voltage to a larger one or vice versa. Before we examine a few such cases, let us consider the device that makes these conversions possible, the ac transformer.

In its simplest form, the **ac transformer** consists of two coils of wire wound around a core of soft iron, as in Figure 21.14. The coil on the left, which is connected to the input ac voltage source and has N_1 turns, is called the primary winding, or the *primary*. The coil on the right, which is connected to a resistor R and consists of N_2 turns, is the *secondary*. The purpose of the common iron core is to increase the magnetic flux and to provide a medium in which nearly all the flux through one coil passes through the other.

When an input ac voltage, V_1, is applied to the primary, the induced voltage across it is given by

$$V_1 = -N_1 \frac{\Delta \Phi}{\Delta t} \qquad \text{[21.21]}$$

where Φ is the magnetic flux through each turn. If we assume that no flux leaks from the iron core, then the flux through each turn of the primary equals the flux through each turn of the secondary. Hence, the voltage across the secondary coil is

$$V_2 = -N_2 \frac{\Delta \Phi}{\Delta t} \qquad \text{[21.22]}$$

The term $\Delta \Phi / \Delta t$ is common to Equations 21.21 and 21.22. Therefore, we see that

$$V_2 = \frac{N_2}{N_1} V_1 \qquad \text{[21.23]}$$

When N_2 is greater than N_1, and thus V_2 exceeds V_1, the transformer is referred to as a *step-up transformer*. When N_2 is less than N_1, making V_2 less than V_1, we speak of a *step-down transformer*.

It should be clear that a voltage is generated across the secondary only when there is a *change* in the number of flux lines passing through the secondary. Thus, the input current in the primary must change with time, which is what happens when an alternating current is used. However, when the input at the primary is a direct current, a voltage output occurs at the secondary only at the instant a switch in the primary circuit is opened or closed. Once the current in the primary reaches a steady value, the output voltage at the secondary is zero.

It may seem that a transformer is a device in which it is possible to get something for nothing. For example, a step-up transformer can change an input voltage from, say, 10 V to 100 V. This means that each 1 coulomb of charge leaving the secondary has 100 J of energy, whereas each coulomb of charge entering the primary has only 10 J of energy. However, this is not an example of a breakdown in the principle of conservation of energy, because *the power input to the primary equals the power output at the secondary;* that is,

$$I_1 V_1 = I_2 V_2 \qquad \text{[21.24]}$$

Thus, if the voltage at the secondary is ten times that at the primary, the current at the secondary is reduced by a factor of 10. Equation 21.24 assumes an **ideal transformer,** in which there are no power losses between the primary and the secondary. Real transformers typically have power efficiencies ranging from 90% to 99%. Power losses occur because of such factors as eddy currents induced in the iron core of the transformer, which dissipate energy in the form of I^2R losses.

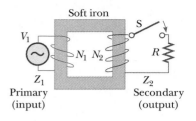

FIGURE 21.14
An ideal transformer consists of two coils wound on the same soft iron core. An ac voltage, V_1, is applied to the primary coil, and the output voltage, V_2, is observed across the load resistance, R.

In an ideal transformer, the input power equals the output power

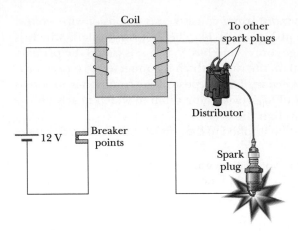

FIGURE 21.15
A circuit diagram of an automobile ignition system.

When electric power is transmitted over large distances, it is economical to use a high voltage and a low current because the power lost via resistive heating in the transmission lines varies as $I^2 R$. This means that if a utility company can reduce the current by a factor of 10, for example, the power loss is reduced by a factor of 100. In practice, the voltage is stepped up to around 230 000 V at the generating station, then stepped down to around 20 000 V at a distribution station, and finally stepped down to 120 V at the customer's utility pole.

In the electrical systems of automobiles, transformers are used with dc sources. In the engine, spark plugs produce a spark to ignite a gasoline-air mixture in the cylinders. For a plug to work, there must be a voltage high enough to cause the spark to jump across the plug's gap. The technique for creating this high voltage is indicated in Figure 21.15. The primary coil of the transformer is connected to the 12-V battery of the car. The secondary is connected to the spark plugs through the distributor. At the instant the plug is supposed to fire, the distributor mechanically connects, in turn, each spark plug to the secondary coil. A switch (called the "breaker points" or simply the "points") breaks the circuit between the battery and the primary. This interrupts the current in the primary and induces a voltage in the secondary. The transformer used is a step-up transformer, and the resulting voltage in the secondary is high enough (about 20 000 V) to cause a spark to jump when the voltage is applied across the gap of the spark plug.

EXAMPLE 21.7   Distributing Power to a City

A generator at a utility company produces 100 A of current at 4000 V. The voltage is stepped up to 240 000 V by a transformer before it is sent on high-voltage transmission lines across a rural area to a city. Assume that the effective resistance of the power line is 30.0 Ω.

(a) Determine the percentage of power lost.

Solution   From Equation 21.24, the current in the transmission line is

$$I_2 = \frac{I_1 V_1}{V_2} = \frac{(100 \text{ A})(4000 \text{ V})}{2.40 \times 10^5 \text{ V}} = 1.67 \text{ A}$$

and the power lost in the transmission line is

$$P_{\text{lost}} = I_2^2 R = (1.67 \text{ A})^2 (30.0 \text{ Ω}) = 83.7 \text{ W}$$

The power output of the generator is

$$P = IV = (100 \text{ A})(4000 \text{ V}) = 4.00 \times 10^5 \text{ W}$$

From this we can find the percentage of power lost as

$$\% \text{ power lost} = \left(\frac{83.7 \text{ W}}{4.00 \times 10^5 \text{ W}} \right) \times 100 = 0.0209\%$$

(b) What percentage of the original power would be lost in the transmission line if the voltage were not stepped up?

Solution If the voltage were not stepped up, the current in the transmission line would be 100 A and the power lost in the line would be

$$P_{\text{lost}} = I^2R = (100 \text{ A})^2(30.0 \text{ }\Omega) = 3.00 \times 10^5 \text{ W}$$

In this case, the percentage of power lost would be

$$\% \text{ power lost} = \left(\frac{3.00 \times 10^5 \text{ W}}{4.00 \times 10^5 \text{ W}} \right) \times 100 = 75\%$$

This example illustrates the advantage of high-voltage transmission lines. At the city, a transformer at a substation steps the voltage back down to about 4000 V, and this voltage is sent through utility lines throughout the city. When the power is to be used at a home or business, a transformer on a utility pole near the establishment reduces the voltage to 240 V or 120 V.

Exercise If the transmission line is cooled so that the resistance is reduced to 5.0 Ω, how much power is lost in the line if it carries a current of 0.89 A?

Answer 4.0 W.

21.8

MAXWELL'S PREDICTIONS

During the early stages of their study and development, electric and magnetic phenomena were thought to be unrelated. In 1865, however, James Clerk Maxwell (1831–1879) provided a mathematical theory that showed a close relationship between all electric and magnetic phenomena. Additionally, his theory predicted that electric and magnetic fields can move through space as waves. The theory he developed is based upon the following four pieces of information:

1. Electric fields originate on positive charges and terminate on negative charges. The electric field due to a point charge can be determined at a location by applying Coulomb's force law to a test charge placed at that location.
2. Magnetic field lines always form closed loops; that is, they do not begin or end anywhere.
3. A varying magnetic field induces an emf and hence an electric field. This is a statement of Faraday's law (Chapter 20).
4. Magnetic fields are generated by moving charges (or currents), as summarized in Ampère's law (Chapter 19).

Let us examine these statements further in order to understand their significance and Maxwell's contributions to the theory of electromagnetism. The first statement is a consequence of the nature of the electrostatic force between charged particles, given by Coulomb's law. It embodies a recognition of the fact that *free charges (electric monopoles) exist in nature.*

**James Clerk Maxwell
(1831–1879)**

James Clerk Maxwell is generally regarded as the greatest theoretical physicist of the 19th century. Born in Edinburgh to a well-known Scottish family, he entered the University of Edinburgh at age 15, around the time he developed an original method of drawing a perfect oval. Maxwell was appointed to his first professorship in 1856 at Aberdeen. This was the beginning of a career during which he developed the electromagnetic theory of light, the kinetic theory of gases, and explanations of the natures of Saturn's rings and color vision.

Maxwell's development of the electromagnetic theory of light took many years and began with the paper "On Faraday's Lines of Force," in which he expanded upon Faraday's theory that electric and magnetic effects result from fields of lines of force surrounding conductors and magnets. His next publication, *On Physical Lines of Force,* included a series of papers on the nature of electromagnetism. By considering how the motions of the vortices and cells could produce magnetic and electric effects, Maxwell successfully explained all the known effects of electromagnetism. He effectively showed that the lines of force behaved in a similar way.

Maxwell's other important contributions to theoretical physics were made in the area of the kinetic theory of gases. Here, he furthered the work of Rudolf Clausius, who in 1858 had shown that a gas must consist of molecules in constant motion, colliding with one another and the walls of the container. This resulted in Maxwell's distribution of molecular velocities and made important applications of the theory to viscosity, conduction of heat, and diffusion of gases.

Maxwell's successful interpretation of Faraday's concept of the electromagnetic field resulted in the field equations that bear Maxwell's name. Formidable mathematical ability combined with great insight enabled him to lead the way in the study of the two most important areas of physics at that time. He died of cancer before he was 50.

(North Wind Picture Archives)

The second statement—that magnetic fields form continuous loops—is exemplified by the magnetic field lines around a long, straight wire, which are closed circles, and the magnetic field lines of a bar magnet, which form closed loops.

The third statement is equivalent to Faraday's law of induction, and the fourth statement is equivalent to Ampère's law.

In one of the greatest theoretical developments of the 19th century, Maxwell used these four statements within a corresponding mathematical framework to prove that electric and magnetic fields play symmetric roles in nature. It was already known from experiments that a changing magnetic field produced an electric field according to Faraday's law. Maxwell suspected that nature should be symmetric, and he therefore hypothesized that a changing electric field should produce a changing magnetic field. This hypothesis could not be proven experimentally at the time it was developed, because the magnetic fields generated by changing electric fields are generally very weak and therefore difficult to detect.

| A changing electric field produces a changing magnetic field |

To justify his hypothesis, Maxwell searched for other phenomena that might be explained by it. He turned his attention to the motion of rapidly oscillating charges, such as those in a conducting rod connected to an alternating voltage. Such charges experience accelerations and, according to Maxwell's predictions, generate changing electric and magnetic fields. The changing fields cause electromagnetic disturbances that travel through space as waves, similar to the spreading water waves created by a pebble thrown into a pool. The waves sent out by the oscillating charges are fluctuating electric and magnetic fields, and so they are called *electromagnetic waves*. Maxwell calculated their speed to be equal to the speed of light, $c = 3 \times 10^8$ m/s. He concluded that light waves are electromagnetic in nature—that light waves and other electromagnetic waves consist of fluctuating electric and magnetic fields traveling through space with a speed of 3×10^8 m/s! This was truly one of the greatest discoveries of science. It had a profound influence on later developments.

21.9

HERTZ'S DISCOVERIES

In 1887, Heinrich Hertz (1857–1894) was the first to generate and detect electromagnetic waves in a laboratory setting. To appreciate the details of his experiment, let us re-examine the properties of an *LC* circuit. In such a circuit, a charged capacitor is connected to an inductor, as in Figure 21.16. When the switch is closed, oscillations occur in the current in the circuit and in the charge on the capacitor. If the resistance of the circuit is neglected, no energy is lost to heat, and the oscillations continue.

In the following analysis, we shall neglect the resistance in the circuit. Let us assume that the capacitor has an initial charge of Q_m and that the switch is closed at $t = 0$. It is convenient to describe what ensues from an energy viewpoint. When the capacitor is fully charged, the total energy in the circuit is stored in the electric field of the capacitor and is equal to $Q_m^2/2C$. At this time, the current is zero and so no energy is stored in the inductor. As the capacitor begins to discharge, the energy stored in its electric field decreases. At the same time, the current increases and energy equal to $LI^2/2$ is now stored in the magnetic field of the inductor. Thus, energy is transferred from the electric field of the capacitor to the magnetic field of the inductor. When the capacitor is fully discharged, it stores no energy. At this time, the current reaches its maximum value, and all of the energy is stored in the inductor. The process then repeats in the reverse direction. The energy continues to transfer between the inductor and the capacitor, corresponding to oscillations in the current and charge.

Figure 21.17 is a representation of this energy transfer. The circuit behavior is analogous to that of the oscillating mass-spring system studied in Chapter 13. The potential energy stored in a stretched spring, $kx^2/2$, corresponds to the potential energy stored in the capacitor, $Q_m^2/2C$; the kinetic energy of the moving mass, $mv^2/2$, corresponds to the energy stored in the inductor, $LI^2/2$, which requires the presence of moving charges. In Figure 21.17a, all of the energy is stored as potential energy in the capacitor at $t = 0$ (because $I = 0$). In Figure 21.17b, all of the energy is stored as "kinetic" energy in the inductor, $LI_m^2/2$, where I_m is the maximum current. At intermediate points, part of the energy is potential energy and part is kinetic energy.

As we saw in Section 21.6, the frequency of oscillation of an *LC* circuit is called the *resonance frequency* of the circuit and is given by

$$f_0 = \frac{1}{2\pi\sqrt{LC}}$$

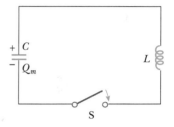

FIGURE 21.16
A simple *LC* circuit. The capacitor has an initial charge of Q_m, and the switch is closed at $t = 0$.

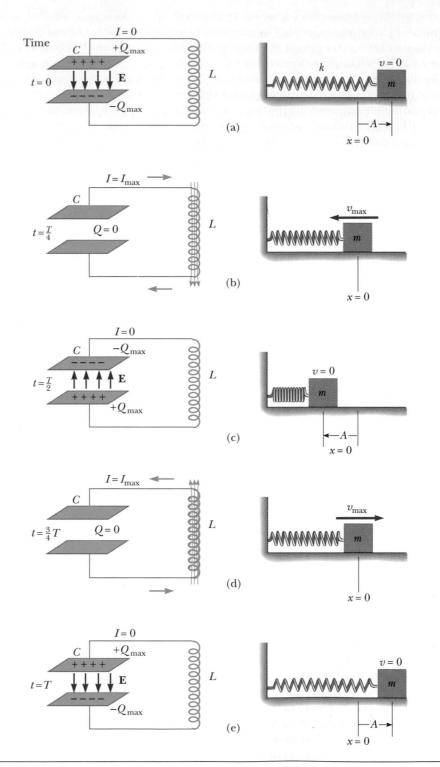

FIGURE 21.17
Stages of energy transfer in an LC circuit with zero resistance. The capacitor has a charge of Q_m at $t = 0$, when the switch is closed. The mechanical analog of this circuit, the mass-spring system, is shown at the right.

The circuit Hertz used in his investigations of electromagnetic waves is similar to that just discussed and is shown schematically in Figure 21.18. An induction coil (a large coil of wire) is connected to two metal spheres with a narrow gap between them to form a capacitor. Oscillations are initiated in the circuit by short voltage pulses sent via the coil to the spheres, charging one positive, the other negative. Because L and C are quite small in this circuit, the frequency of oscillation is quite high, $f \approx 100$ MHz. This circuit is called a transmitter because it produces electromagnetic waves.

Several meters from the transmitter circuit, Hertz placed a second circuit, the receiver, which consisted of a single loop of wire connected to two spheres. It had its own effective inductance, capacitance, and natural frequency of oscillation. Hertz found that energy was being sent from the transmitter to the receiver when the resonance frequency of the receiver was adjusted to match that of the transmitter. The energy transfer was detected when the voltage across the spheres in the receiver circuit became high enough to produce ionization in the air, which caused sparks to appear in the air gap separating the spheres. Hertz's experiment is analogous to the mechanical phenomenon in which a tuning fork picks up the vibrations from another, identical tuning fork.

Hertz assumed that the energy transferred from the transmitter to the receiver is carried in the form of waves, which are now known to be electromagnetic waves. In a series of experiments, he also showed that the radiation generated by the transmitter exhibits wave properties: interference, diffraction, reflection, refraction, and polarization. As you will see shortly, all of these properties are exhibited by light. Thus, it became evident that these waves had properties similar to those of light waves and differed only in frequency and wavelength.

Perhaps the most convincing experiment Hertz performed was the measurement of the speed of waves from the transmitter, accomplished as follows. Waves of known frequency from the transmitter were reflected from a metal sheet so that an interference pattern was set up, much like the standing wave pattern on a stretched string. As we saw in our discussion of standing waves, the distance between nodes is $\lambda/2$, so Hertz was able to determine the wavelength, λ. Using the relationship $v = \lambda f$, he found that v was close to 3×10^8 m/s, the known speed of visible light. Hertz's experiments thus provided the first evidence in support of Maxwell's theory.

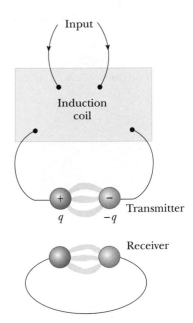

FIGURE 21.18
A schematic diagram of Hertz's apparatus for generating and detecting electromagnetic waves. The transmitter consists of two spherical electrodes connected to an induction coil, which provides short voltage surges to the spheres, setting up oscillations in the discharge. The receiver is a nearby single loop of wire.

21.10

PRODUCTION OF ELECTROMAGNETIC WAVES BY AN ANTENNA

In the last section, we found that the energy stored in an LC circuit is continually transferred between the electric field of the capacitor and the magnetic field of the inductor. However, this energy transfer continues for prolonged periods of time only when the changes occur slowly. If the current alternates rapidly, the circuit loses some of its energy in the form of electromagnetic waves. In fact, electromagnetic waves are radiated by *any* circuit carrying an alternating current. The fundamental mechanism responsible for this radiation is the acceleration of a charged particle. Whenever a charged particle undergoes an acceleration, it must radiate energy.

An alternating voltage applied to the wires of an antenna forces an electric charge in the antenna to oscillate. This is a common technique for accelerating charged particles and is the source of the radio waves emitted by the antenna of a radio station.

An accelerating charge radiates energy

**Heinrich Rudolf Hertz
(1857–1894)**

(The Bettmann Archive)

Heinrich Hertz was born in 1857 in Hamburg, Germany. He studied physics under Helmholtz and Kirchhoff at the University of Berlin. In 1885 Hertz accepted the position of professor of physics at Karlsruhe; it was here that he made his discovery of radio waves, his most important accomplishment, in 1888.

Discovering radio waves, demonstrating their generation, and determining their speed were among Hertz's many achievements. After finding that the speed of a radio wave was the same as that of light, Hertz showed that radio waves, like light waves, could be reflected, refracted, and diffracted.

Hertz died of blood poisoning at the age of 36. During his short life, he made many contributions to science. The hertz, equal to one complete vibration or cycle per second, is named after him.

Figure 21.19 illustrates the production of an electromagnetic wave by oscillating electric charges in an antenna. Two metal rods are connected to an ac generator, which causes charges to oscillate between the two rods. The output voltage of the generator is sinusoidal. At $t = 0$, the upper rod is given a maximum positive charge and the bottom rod an equal negative charge, as in Figure 21.19a. The electric field near the antenna at this instant is also shown in Figure 21.19a. As the charges oscillate, the rods become less charged, the field near the rods decreases in strength, and the downward-directed maximum electric field produced at $t = 0$ moves away from the rod. When the charges are neutralized, as in Figure 21.19b, the electric field has dropped to zero. This occurs after an

FIGURE 21.19
An electric field set up by oscillating charges in an antenna. The field moves away from the antenna at the speed of light.

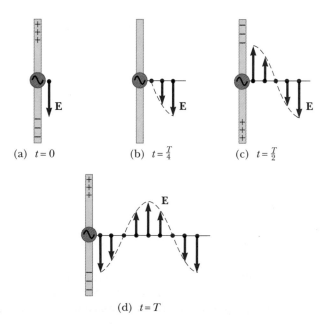

(a) $t = 0$ (b) $t = \frac{T}{4}$ (c) $t = \frac{T}{2}$

(d) $t = T$

interval equal to one quarter of the period of oscillation. Continuing in this fashion, the upper rod soon obtains a maximum negative charge and the lower rod becomes positive, as in Figure 21.19c, resulting in an electric field directed upward. This occurs after an interval equal to one-half the period of oscillation. The oscillations continue as indicated in Figure 21.19d. Note that the electric field near the antenna oscillates in phase with the charge distribution. That is, the field points down when the upper rod is positive and up when the upper rod is negative. Furthermore, the magnitude of the field at any instant depends on the amount of charge on the rods at that instant.

As the charges continue to oscillate (and accelerate) between the rods, the electric field set up by the charges moves away from the antenna at the speed of light. Figure 21.19 shows the electric field pattern at certain times during the oscillation cycle. As you can see, one cycle of charge oscillation produces one full wavelength in the electric field pattern.

Since the oscillating charges create a current in the rods, a magnetic field is also generated when the current in the rods is upward, as shown in Figure 21.20. The magnetic field lines circle the antenna and are *perpendicular to the electric field at all points.* As the current changes with time, the magnetic field lines spread out from the antenna. At great distances from the antenna, the strengths of the electric and magnetic fields become very weak. However, at these distances it is necessary to take into account the facts that (1) a changing magnetic field produces a changing electric field and (2) a changing electric field produces a changing magnetic field, as predicted by Maxwell. These induced electric and magnetic fields are in phase: at any point, the two fields reach their maximum values at the same instant. This is illustrated at one instant of time in Figure 21.21. Note that (1) these fields are perpendicular to each other and (2) both fields are perpendicular to the direction of motion of the wave. This second property is characteristic of transverse waves. Hence, we see that *an electromagnetic wave is a transverse wave.*

FIGURE 21.20
Magnetic field lines around an antenna carrying a changing current. Why do the circles have different radii?

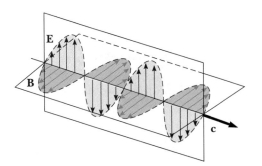

FIGURE 21.21
An electromagnetic wave sent out by oscillating charges in an antenna, represented at one instant of time. Note that the electric field is perpendicular to the magnetic field, and both are perpendicular to the direction of wave propagation.

21.11

PROPERTIES OF ELECTROMAGNETIC WAVES

We have seen that Maxwell's detailed analysis predicted the existence and properties of electromagnetic waves. We have already examined some of those properties. In this section we summarize what we have found out about electromagnetic waves thus far and consider some additional properties. In our discussion here and in future sections, we shall often make reference to a type of wave called

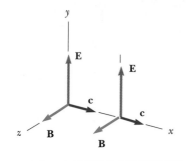

FIGURE 21.22
A plane electromagnetic wave traveling in the positive x direction. The electric field is along the y direction, and the magnetic field is along the z direction.

a **plane wave.** A plane electromagnetic wave is a wave in which the oscillating fields associated with the wave are uniform over a plane at any given time. Figure 21.22 pictures such a wave at a given instant of time. In this case, the oscillations of the electric and magnetic fields take place in planes perpendicular to the x axis and thus to the direction of travel for the wave.

Since the electric and magnetic fields are perpendicular to the direction of travel of the wave, electromagnetic waves are transverse waves. In Figure 21.22, the vibration of the electric field portion of the wave is taken to be in the y direction and the vibration of the magnetic field is in the z direction. Both vibrations are perpendicular to the direction of travel of the wave, which is along the x axis. Figure 21.23 is a "snapshot"—a representation at some instant of time—of an electromagnetic wave moving in the x direction. In this representation, the strength of the electric field is a maximum at the origin, and the direction of the electric field is the positive y direction. As movement proceeds along the x axis, the magnitude of **E** decreases sinusoidally to zero. With further movement along the x axis, **E** reverses and points in the negative y direction, and the magnitude begins to increase. The magnetic field follows a similar pattern, except its vibrations take place along the z axis in Figure 21.23.

Electromagnetic waves travel with the speed of light. In fact, it can be shown that the speed of an electromagnetic wave is related to the permeability and permittivity of the medium through which it travels. Maxwell found this relationship for free space to be

Speed of light

$$c = \frac{1}{\sqrt{\mu_0 \epsilon_0}} \qquad [21.25]$$

where c is the speed of light, $\mu_0 = 4\pi \times 10^{-7} \, \text{N} \cdot \text{s}^2/\text{C}^2$ is the permeability constant of free space, and $\epsilon_0 = 8.85 \times 10^{-12} \, \text{C}^2/\text{N} \cdot \text{m}^2$ is the permittivity of free space. Substituting these values into Equation 21.25, we find that

$$c = 2.99792 \times 10^8 \, \text{m/s} \qquad [21.26]$$

Since electromagnetic waves travel at a speed that is precisely the same as the speed of light in vacuum, one is led to believe (correctly) that *light is an electromagnetic wave.*

The ratio of the electric to the magnetic field in an electromagnetic field equals the speed of light. That is,

Light is an electromagnetic wave and transports energy and momentum

$$\frac{E}{B} = c \qquad [21.27]$$

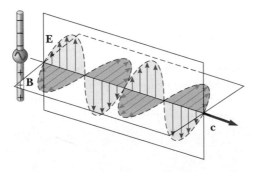

FIGURE 21.23
An electromagnetic wave sent out by oscillating charges in an antenna, represented at one instant of time. Note that the electric field is perpendicular to the magnetic field, and both are perpendicular to the direction of wave propagation.

Electromagnetic waves carry energy as they travel through space, and this energy can be transferred to objects placed in their paths. The average rate at which energy passes through an area perpendicular to the direction of travel of a wave, or the average power per unit area, is given by

$$\text{Average power per unit area} = \frac{E_m B_m}{2\mu_0} \qquad \textbf{[21.28]}$$

Since $E = cB = B/\sqrt{\mu_0 \epsilon_0}$, this can also be expressed as

$$\text{Average power per unit area} = \frac{E_m^2}{2\mu_0 c} = \frac{c}{2\mu_0} B_m^2 \qquad \textbf{[21.29]}$$

Note that in these expressions we use the *average* power per unit area. Also note that the values to be used for E and B are the *maximum* values. Interestingly, a detailed analysis would show that the energy carried by an electromagnetic wave is shared equally by the electric and magnetic fields.

Electromagnetic waves transport momentum as well as energy. This is demonstrated by the fact that radiation pressure is exerted on a surface when an electromagnetic wave impinges upon it. Although radiation pressures are very small (about 5×10^{-6} N/m^2 for direct sunlight), they have been measured with a device like that in Figure 21.24. Light is allowed to strike a mirror and a black disk that are connected to each other by a horizontal bar suspended from a fine fiber. Light striking the black disk is completely absorbed, and so *all* of the momentum of the light is transferred to the disk. Light striking the mirror head on is totally reflected; hence, the momentum transfer to the mirror is twice that transmitted to the disk. As a result, the horizontal bar supporting the disks twists counterclockwise as seen from above. The bar comes to equilibrium at some angle under the action of the torques caused by radiation pressure and the twisting of the fiber. The radiation pressure can be determined by measuring the angle at which equilibrium occurs. The apparatus must be placed in a high vacuum to eliminate the effects of air currents.

In summary, electromagnetic waves traveling through free space have the following properties:

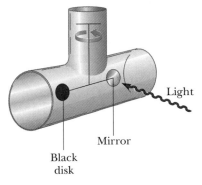

FIGURE 21.24
An apparatus for measuring the radiation pressure of light. In practice, the system is contained in a high vacuum.

1. Electromagnetic waves travel at the speed of light.
2. Electromagnetic waves are transverse waves, since the electric and magnetic fields are perpendicular to the direction of propagation of the wave and to each other.
3. The ratio of the electric field to the magnetic field in an electromagnetic wave equals the speed of light.
4. Electromagnetic waves carry both energy and momentum, which can be delivered to a surface.

EXAMPLE 21.8 Solar Energy

The Sun delivers an average power per unit area of about 700 W/m^2 to the Earth's surface. Calculate the total power incident on a roof 8.00 m by 20.0 m. Assume that the radiation is incident *normal* to the roof (the Sun is directly overhead).

A solar home in Oregon. (*John Neal/ Photo Researchers, Inc.*)

Solution   The power per unit area, or light intensity, is 700 W/m². For normal incidence we get

$$\text{Power} = (700 \text{ W/m}^2)(8.00 \times 20.0 \text{ m}^2) = \boxed{1.12 \times 10^5 \text{ W}}$$

Note that if this power could *all* be converted to electric power, it would be more than enough for the average home. Unfortunately, solar energy is not easily harnessed, and the prospects for large-scale conversion are not as bright as they may appear from this simple calculation. For example, the conversion efficiency from solar to electrical energy is far less than 100%; 10% is typical for photovoltaic cells. Roof systems for converting solar energy to thermal energy with efficiencies of around 50% have been built. However, other practical problems must be considered, such as overcast days, geographic location, and energy storage.

Exercise   How much solar energy (in joules) is incident on the roof in 1.00 h?

Answer 4.03×10^8 J.

21.12

THE SPECTRUM OF ELECTROMAGNETIC WAVES

We have seen that all electromagnetic waves travel in a vacuum with the speed of light, c. These waves transport energy and momentum from some source to a receiver. In 1887 Hertz successfully generated and detected the radio-frequency electromagnetic waves predicted by Maxwell. Maxwell himself had recognized as electromagnetic waves both visible light and the infrared radiation discovered in 1800 by William Herschel. It is now known that other forms of electromagnetic waves exist that are distinguished by their frequencies and wavelengths.

Because all electromagnetic waves travel through vacuum with a speed of c, their frequency, f, and wavelength, λ, are related by the important expression

$$c = f\lambda \tag{21.30}$$

The types of electromagnetic waves are presented in Figure 21.25. Note the wide range of frequencies and wavelengths. For instance, a radio wave with a frequency of 5.00 MHz (a typical value) has a wavelength of

$$\lambda = \frac{c}{f} = \frac{3.00 \times 10^8 \text{ m/s}}{5.00 \times 10^6 \text{ s}^{-1}} = 60.0 \text{ m}$$

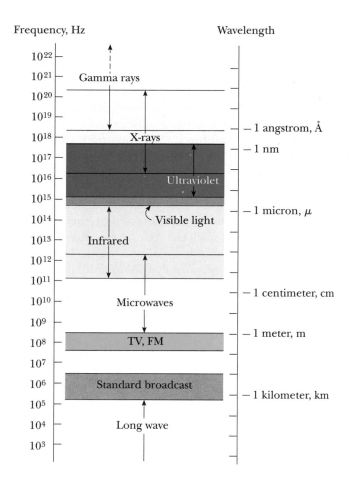

FIGURE 21.25
The electromagnetic spectrum. Note the overlap between one type of wave and the next. There is no sharp division between the types.

The following abbreviations are often used to designate short wavelengths and distances:

$$1 \text{ micrometer } (\mu\text{m}) = 10^{-6} \text{ m}$$

$$1 \text{ nanometer } (\text{nm}) = 10^{-9} \text{ m}$$

$$1 \text{ angstrom } (\text{Å}) = 10^{-10} \text{ m}$$

The wavelengths of visible light, for example, range from 0.4 to 0.7 μm, or 400 to 700 nm, or 4000 to 7000 Å.

Brief descriptions of these wave types follow, in order of decreasing wavelength. There is no sharp division between one kind of wave and the next. Note that all forms of radiation are produced by accelerating charges.

Radio waves, which were discussed in the preceding section, are the result of charges accelerating through conducting wires. They are used in radio and television communication systems.

Microwaves (short-wavelength radio waves) have wavelengths ranging between about 1 mm and 30 cm and are generated by electronic devices. Their short wavelengths make them well suited for the radar systems used in aircraft navigation and for the study of atomic and molecular properties of matter.

Microwave ovens are an interesting domestic application of these waves. It has been suggested that solar energy might be harnessed by beaming microwaves to Earth from a solar collector in space.

Infrared waves (sometimes called heat waves), produced by hot bodies and molecules, have wavelengths ranging from about 1 mm to the longest wavelength of visible light, 7×10^{-7} m. They are readily absorbed by most materials. The infrared energy absorbed by a substance appears as heat. This is because the energy agitates the atoms of the object, increasing their vibrational or translational motion, and the result is a temperature rise. Infrared radiation has many practical and scientific applications including physical therapy, infrared photography, and the study of the vibrations of atoms.

Visible light, the most familiar form of electromagnetic waves, may be defined as the part of the spectrum that is detected by the human eye. Light is produced by the rearrangement of electrons in atoms and molecules. The wavelengths of visible light are classified as colors ranging from violet ($\lambda \approx 4 \times 10^{-7}$ m) to red ($\lambda \approx 7 \times 10^{-7}$ m). The eye's sensitivity is a function of wavelength and is greatest at a wavelength of about 5.6×10^{-7} m (yellow-green).

Ultraviolet (uv) light covers wavelengths ranging from about 3.8×10^{-7} m (380 nm) down to 6×10^{-8} m (60 nm). The Sun is an important source of ultraviolet light (which is the main cause of suntans). Most of the ultraviolet light from the Sun is absorbed by atoms in the upper atmosphere, or stratosphere. This is fortunate, since uv light in large quantities has harmful effects on humans. One important constituent of the stratosphere is ozone (O_3) from reactions of oxygen with ultraviolet radiation. This ozone shield converts lethal high-energy ultraviolet radiation to heat, which warms the stratosphere.

X-rays are electromagnetic waves with wavelengths from about 10^{-8} m (10 nm) down to 10^{-13} m (10^{-4} nm). The most common source of x-rays is the acceleration of high-energy electrons bombarding a metal target. X-rays are used as a diagnostic tool in medicine and as a treatment for certain forms of cancer. Since x-rays damage or destroy living tissues and organisms, care must be taken to avoid unnecessary exposure and overexposure.

Gamma rays, electromagnetic waves emitted by radioactive nuclei, have wavelengths ranging from about 10^{-10} m to less than 10^{-14} m. They are highly penetrating and cause serious damage when absorbed by living tissues. Consequently, those working near such radiation must be protected by garments containing heavily absorbing materials, such as layers of lead.

SUMMARY

If an ac circuit consists of a generator and a resistor, the current in the circuit is in phase with the voltage. That is, the current and voltage reach their maximum values at the same time.

In discussions of voltages and currents in ac circuits, **rms values** of voltages are usually used. One reason is that ac ammeters and voltmeters are designed to read rms values. The rms values of currents and voltage (I and V) are related to the maximum values of these quantities (I_m and V_m) as follows:

$$I = \frac{I_m}{\sqrt{2}} \qquad V = \frac{V_m}{\sqrt{2}} \qquad \text{[21.2, 21.3]}$$

The rms voltage across a resistor is related to the rms current through the resistor by **Ohm's law:**

$$V_R = IR \qquad \text{[21.4]}$$

If an ac circuit consists of a generator and a capacitor, the voltage lags behind the current by 90°. That is, the voltage reaches its maximum value one quarter of a period after the current reaches its maximum value.

The impeding effect of a capacitor on current in an ac circuit is given by the **capacitive reactance,** X_C, defined as

$$X_C \equiv \frac{1}{2\pi f C} \qquad \text{[21.5]}$$

where f is the frequency of the ac generator.

The rms voltage across and the rms current through a capacitor are related by

$$V_C = IX_C \qquad \text{[21.6]}$$

If an ac circuit consists of a generator and an inductor, the voltage leads the current by 90°. That is, the voltage reaches its maximum value one quarter of a period before the current reaches its maximum value.

The effective impedance of a coil in an ac circuit is measured by a quantity called the **inductive reactance,** X_L, defined as

$$X_L \equiv 2\pi f L \qquad \text{[21.8]}$$

The rms voltage drop across a coil is related to the rms current through the coil by

$$V_L = IX_L \qquad \text{[21.9]}$$

In an RLC series ac circuit, the applied rms voltage, V, is related to the rms voltages across the resistor (V_R), capacitor (V_C), and inductor (V_L) by

$$V = \sqrt{V_R{}^2 + (V_L - V_C)^2} \qquad \text{[21.11]}$$

If an ac circuit contains a resistor, an inductor, and a capacitor, the effective resistance of the circuit is given by the **impedance,** Z, of the circuit, defined as

$$Z \equiv \sqrt{R^2 + (X_L - X_C)^2} \qquad \text{[21.14]}$$

The relationship between the rms voltage supplied to an RLC circuit and the rms current in the circuit is

$$V = IZ \qquad \text{[21.15]}$$

In an RLC series ac circuit, the applied rms voltage and current are out of phase. The **phase angle,** ϕ, between the current and voltage is given by

$$\tan \phi = \frac{X_L - X_C}{R} \qquad \text{[21.16]}$$

The **average power** delivered by the generator in an RLC ac circuit is

$$P_{\text{av}} = IV \cos \phi \qquad \text{[21.18]}$$

where the constant $\cos \phi$ is called the **power factor.**

Electromagnetic waves were predicted by James Clerk Maxwell and later generated and detected by Heinrich Hertz. These waves have the following properties:

1. Electromagnetic waves are transverse waves, since the electric and magnetic fields are perpendicular to the direction of travel.
2. Electromagnetic waves travel with the speed of light.
3. The ratio of the electric field to the magnetic field in an electromagnetic wave equals the speed of light; that is,

$$\frac{E}{B} = c \qquad\qquad\qquad \text{[21.27]}$$

4. Electromagnetic waves carry energy as they travel through space. The average power per unit area is

$$\frac{E_m B_m}{2\mu_0} = \frac{E_m^2}{2\mu_0 c} = \frac{c}{2\mu_0} B_m^2 \qquad\qquad \text{[21.28, 21.29]}$$

where E_m and B_m are the maximum values of the electric and magnetic fields.
5. Electromagnetic waves transport momentum as well as energy. The speed, c, frequency, f, and wavelength, λ, of an electromagnetic wave are related by

$$c = f\lambda \qquad\qquad\qquad \text{[21.30]}$$

The **electromagnetic spectrum** includes waves covering a broad range of frequencies and wavelengths. These waves have a variety of applications and characteristics, depending on their frequencies or wavelengths.

ADDITIONAL READING

L. Barthold and H. G. Pfeiffer, "High-Voltage Transmission," *Sci. American*, May 1964, p. 38.

A. H. W. Beck, *Words and Waves*, World University Library, New York, McGraw-Hill, 1967.

J. W. Coltman, "The Transformer," *Sci. American*, January 1988, p. 86.

O. Kedem and U. Ganiel, "Solar Energy, How Much Do We Receive?" *The Physics Teacher*, December 1983, p. 573.

D. K. C. McDonald, *Faraday, Maxwell and Kelvin*, Science Study Series, Garden City, N.Y., Doubleday, 1964.

J. R. Pierce, *Electrons and Waves*, Science Study Series, Garden City, N.Y., Doubleday, 1964.

R. S. Stolarski, "The Antarctic Ozone Hole," *Sci. American*, January 1988, p. 30.

CONCEPTUAL QUESTIONS

Example   How can the average value of an alternating current be zero yet the square root of the average value squared not be zero?

Reasoning   The average value of an alternating current is zero because its direction is positive as often as it is negative, and its time average is zero. The average value of the square of the current is not zero, however, since the square of positive and negative values are always positive.

Example   What is meant by the statement that "the voltage across an inductor leads the current by 90°"?

Reasoning   This means that the voltage across an inductor reaches its peak value one quarter of a cycle before the current reaches its peak value.

Example   Would an inductor and a capacitor used together in an ac circuit dissipate any energy?

Reasoning   No. The only element that dissipates energy in an ac circuit is a resistor. Inductors and capacitors store energy during one half of a cycle and release that energy during the other half of a cycle, so they dissipate no net energy.

Example   Is the voltage applied to a series *RLC* circuit always in phase with the current in the resistor?
Reasoning   No. The voltage and current are in phase when $L = 0$ and $C = 0$, or when the frequency corresponds to the resonance frequency of the circuit.

Example   Will a transformer operate if a battery is used for the input voltage across the primary?
Reasoning   No. When a battery is used in the primary, the current in the primary coil is constant, so the flux through the core remains constant. A voltage can only be induced in the secondary coil if the flux through the core changes with time.

Example   If you charge a comb by running it through your hair and then hold the comb next to a bar magnet, do the electric and magnetic fields produced constitute an electromagnetic wave?
Reasoning   An electric field is set up by the charges on the comb, and the bar magnet sets up a magnetic field. However, these fields are constant in magnitude, while the fields of an electromagnetic wave are continually changing in magnitude and direction. Hence, no electromagnetic wave is produced. (Electromagnetic waves are produced by accelerating charges.)

Example   Does a wire connected to a battery emit an electromagnetic wave?
Reasoning   No. The wire will emit electromagnetic waves only if the current varies in time. The radiation is the result of accelerating charges, which can only occur when the current is not constant.

1. Do ac ammeters and voltmeters read peak, rms, or average values?
2. Explain why the reactance of a capacitor decreases with increasing frequency, whereas the reactance of an inductor increases with increasing frequency.
3. A memory device often used with alternating current circuits is ELI the ICE man. ICE means that in a capacitive circuit, the current leads the voltage. What does ELI mean?
4. "At high frequencies a capacitor looks like a short circuit." Is this statement true? Discuss. Explain what is meant by high frequencies.
5. If it were not for the eye's persistence of vision, one could see a lightbulb operated with an ac voltage flicker on and off. Would you see 60 or 120 flickers per second?
6. If the frequency in a series *RLC* circuit is doubled, what happens to the resistance, the inductive reactance, and the capacitive reactance?
7. In a series *RLC* circuit, what is the possible range of values for the phase angle?
8. Does the phase angle depend on frequency? What is the phase angle when the inductive reactance equals the capacitive reactance?
9. What is the impedance of an *RLC* circuit at the resonance frequency?
10. Show that the maximum current in an *RLC* circuit occurs when the circuit is in resonance.
11. How would you make a lamp dimmer with a coil of a variable inductance? Would the dimmer work with dc?
12. A series *RLC* circuit models a driven harmonic oscillator. If the inductance represents the mass of the oscillator, what do the capacitance and resistance represent? Discuss.
13. Why are the primary and secondary coils of a transformer wrapped on an iron core that passes through both coils?
14. If the fundamental source of a sound wave is a vibrating object, what is the fundamental source of an electromagnetic wave?
15. When light (or other electromagnetic radiation) travels across a given region, what is it that moves?
16. List as many similarities and differences as you can between sound waves and light waves.
17. What does a radio wave do to the charges in the receiving antenna to provide a signal for your car radio?
18. For a given incident energy of an electromagnetic wave, why is the radiation pressure on a perfectly reflecting surface twice as great as the pressure on a perfectly absorbing surface?
19. Is there any difference between 1.0-nm ultraviolet light and 1.0-nm x-rays? Discuss.
20. Explain what is meant by monochromatic x-rays.
21. Suppose a creature from another planet had eyes that were sensitive to infrared radiation. Describe what he would see if he looked around the room you are in now. That is, what would be bright and what would be dim?

PROBLEMS

Section 21.1   Resistors in an ac Circuit

1. If a toaster is plugged into a source of alternating voltage with an rms value of 110 V, an rms current of 7.5 A occurs in the heating element. What are (a) the resistance of the toaster's heating element and (b) the average power input to the toaster?

2. (a) What is the resistance of a lightbulb that uses an average power of 75 W when connected to a 60-Hz power source with an rms voltage of 120 V? (b) What is the resistance of a 100-W bulb?

□ indicates problems that have full solutions available in the Student Solutions Manual and Study Guide.

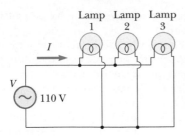

FIGURE 21.26   (Problem 3)

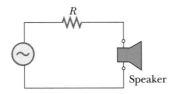

FIGURE 21.27   (Problem 4)

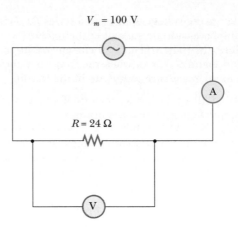

FIGURE 21.28   (Problem 6)

3. Figure 21.26 shows three lamps connected to the 110-V ac (rms) household supply voltage. Lamps 1 and 2 have 150-W bulbs; lamp 3 has a 100-W bulb (average power). Find the rms value of the current I and the resistance of each bulb.

4. An audio amplifier, represented by the ac source and the resistor R in Figure 21.27, delivers alternating voltages at audio frequencies to the speaker. If the source puts out an alternating voltage with an amplitude of 15.0 V, resistance R is 8.20 Ω, and the speaker is equivalent to a resistance of 10.4 Ω, what is the time-averaged power input to the speaker?

5. An ac voltage source has an output of $v = 150 \sin 377t$. Find (a) the rms voltage output, (b) the frequency of the source, and (c) the voltage at $t = 1/120$ s. (d) Find the maximum current in the circuit when the generator is connected to a 50.0-Ω resistor.

Section 21.2   Capacitors in an ac Circuit

6. An ac power supply produces a peak voltage of $V_m = 100$ V. This power supply is connected to a 24-Ω resistor, and the current and resistor voltage are measured with an ideal ac ammeter and an ideal ac voltmeter, as shown in Figure 21.28. What does each meter read? Recall that an ideal ammeter has zero resistance and an ideal voltmeter has infinite resistance.

7. A certain capacitor in a circuit has a capacitive reactance of 30.0 Ω when the frequency is 120 Hz.

What capacitive reactance does the capacitor have at a frequency of 10 000 Hz?

8. A variable-frequency ac generator with $V_m = 18$ V is connected across a 9.4×10^{-8} F capacitor. At what frequency should the generator be operated to produce a peak current of 5.0 A?

9. The generator in a purely capacitive ac circuit has an angular frequency of 120 π rad/s. If $V_m = 140$ V and $C = 6$ μF, what is the rms current in the circuit?

10. What value of capacitor must be inserted in a 60-Hz circuit in series with a generator of 170 V maximum output voltage to produce an rms current output of 0.75 A?

Section 21.3   Inductors in an ac Circuit

11. The generator in a purely inductive ac circuit has an angular frequency of 120π rad/s. If $V_m = 140$ V and $L = 0.100$ H, what is the rms current in the circuit?

12. A 2.4-μF capacitor is connected across an alternating voltage with an rms value of 9.0 V. The rms current through the capacitance is 25.0 mA. (a) What is the source frequency? (b) If the capacitor is replaced by an ideal coil with an inductance of 0.160 H, what is the rms current through the coil?

13. An inductor is connected to a 20.0-Hz power supply that produces a 50.0-V rms voltage. What inductance is needed to keep the maximum current in the circuit below 80.0 mA?

14. In a purely inductive circuit, the rms voltage is 120 V. (a) If the rms current is 10.0 A at a frequency of 60.0 Hz, calculate the inductance. (b) At what frequency is the rms current 5.00 A?

15. In a purely inductive circuit, $L = 30.0$ mH, the frequency is 60.0 Hz, and the maximum voltage is

200 V. (a) Find the rms current in the circuit. (b) Repeat for a frequency of 6000 Hz.

16. An inductor has a 54-Ω reactance at 60.0 Hz. What will be the peak current if this inductor is connected to a 50.0-Hz source that produces a 100-V rms voltage?

Section 21.4 The *RLC* Series Circuit

17. A 40.0-μF capacitor is connected to a 50.0-Ω resistor and a generator whose rms output is 30.0 V at 60.0 Hz. Find (a) the rms current in the circuit, (b) the voltage drop across the resistor, (c) the voltage drop across the capacitor, and (d) the phase angle for the circuit. (e) Sketch the phasor diagram for this circuit.

18. A 10.0-μF capacitor and a 2.00-H inductor are connected in series with a 60.0-Hz source whose rms output is 100 V. Find (a) the rms current in the circuit, (b) the voltage drop across the inductor, (c) the voltage drop across the capacitor, and (d) the phase angle for the circuit. (e) Sketch the phasor diagram for this circuit.

19. A pure 20.0-mH inductor is connected in series with a 20.0-Ω resistor and a 60.0-Hz, 100-V rms source. Find (a) the rms current in the circuit, (b) the voltage drop across the inductor, (c) the voltage drop across the resistor, and (d) the phase angle for this circuit. (e) Sketch the phasor diagram for this circuit.

20. A 100-μF capacitor is connected in series with an 18-Ω resistor and a 60.0-Hz ac voltage source. The voltage supplied to the circuit is 120 V. Determine the rms current in the circuit.

21. A 50.0-Ω resistor, a 0.100-H inductor, and a 10.0-μF capacitor are connected in series to a 60.0-Hz source. The rms current in the circuit is 2.75 A. Find the rms voltages across (a) the resistor, (b) the inductor, (c) the capacitor, and (d) the *RLC* combination. (e) Sketch the phasor diagram for this circuit.

22. A 60.0-Ω resistor, a 3.0-μF capacitor, and a 0.40-H inductor are connected in series to a 90.0-V, 60.0-Hz source. Find (a) the voltage drop across the *LC* combination and (b) the voltage drop across the *RC* combination.

23. A person is working near the secondary of a transformer, as shown in Figure 21.29. The primary voltage is 110 V at 60.0 Hz. The stray capacitance between the person's finger and the secondary winding, C_s, is 20.0 pF. Assuming that the person has a body resistance to ground, R_b, of 50.0 kΩ (as indicated), determine the rms voltage across the person's body. (*Hint:* Redraw the circuit with the secondary of the transformer as a simple ac source.)

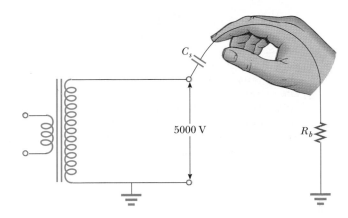

FIGURE 21.29 (Problem 23)

24. An ac source with a peak voltage of 150 V and $f = 50.0$ Hz is connected between points *a* and *d* in Figure 21.30. Calculate the *peak* voltages between points (a) *a* and *b*, (b) *b* and *c*, (c) *c* and *d*, (d) *b* and *d*.

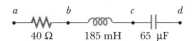

FIGURE 21.30 (Problem 24)

25. A 50.0-Ω resistor is connected in series with a 15.0-μF capacitor and a 60.0-Hz, 120-V source. (a) Find the current in the circuit. (b) What is the value of the inductor that must be inserted in the circuit to reduce the current to one-half that found in (a)?

26. A series *RLC* circuit has a phase angle of 30.0° and an impedance of 100 Ω. (a) Determine the resistance in the circuit. (b) If $C = 5.0$ μF and $L = 7.0$ mH, determine the reactances in the capacitor and the inductor. (c) What is the frequency of the circuit?

27. A resistor ($R = 900$ Ω), a capacitor ($C = 0.25$ μF), and an inductor ($L = 2.5$ H) are connected in series across a 240-Hz ac source for which $V_m = 140$ V. Calculate the (a) impedance of the circuit, (b) peak current delivered by the source, and (c) phase angle between the current and voltage. (d) Is the current leading or lagging behind the voltage?

Section 21.5 Power in an ac Circuit

28. A multimeter in an *RL* circuit records an rms current of 0.500 A and a 60.0-Hz rms generator voltage of 104 V. A wattmeter shows that the average thermal power developed in the resistor is

10.0 W. Determine (a) the impedance in the circuit, (b) the resistance, R, and (c) the inductance, L.

29. A 50.0-Ω resistor is connected to a 30.0-μF capacitor and to a 60.0-Hz, 100-V rms source. (a) Find the power factor and the average power delivered to the circuit. (b) Repeat part (a) when the capacitor is replaced with a 0.300-H inductor.

30. In a certain RLC circuit, the rms current is 6.0 A, the rms voltage is 240 V, and the current leads the voltage by 53°. (a) What is the total resistance of the circuit? (b) Calculate the total reactance, $X_L - X_C$. (c) Find the average power dissipated in the circuit.

31. An ac voltage with an amplitude of 100 V is applied to a series combination of a 200-μF capacitor, a 100-mH inductor, and a 20.0-Ω resistor. Calculate the power dissipation and the power factor for frequencies of (a) 60.0 Hz and (b) 50.0 Hz.

32. An inductor and a resistor are connected in series. When connected to a 60-Hz, 90-V source, the voltage drop across the resistor is found to be 50 V and the power dissipated in the circuit is 14 W. Find (a) the value of the resistance and (b) the value of the inductance.

Section 21.6   Resonance in a Series RLC Circuit

33. A circuit consists of a source of alternating voltage with a constant amplitude of 5.0 V and a variable frequency in series with a resistance of 100 Ω, a capacitance of 2.25 μF, and an inductance of 0.25 H. (a) Determine the rms current at frequencies of 10.0 Hz, 100 Hz, 1000 Hz, and 10 000 Hz. (b) Determine the phase of the source voltage relative to the current at these same frequencies. (c) Determine the frequency at which the rms current is at a maximum.

34. Consider an LC circuit at resonance. Show that the maximum current (I_m) in the circuit is related to the maximum charge (Q_m) on the capacitor by the expression $I_m = 2\pi f_0 Q_m$. (*Hint:* Use conservation of energy.)

35. A resonant circuit in a radio receiver is tuned to a certain station when the inductor has a value of 0.200 mH and the capacitor has a value of 30.0 pF. Find the frequency of the radio station and the wavelength sent out by the station.

36. The AM band extends from approximately 500 kHz to 1600 kHz. If a 2.0-μH inductor is used in a tuning circuit for a radio, what are the extremes that a capacitor must reach in order to cover the complete band of frequencies?

37. Figure 21.31 shows a tuning circuit for the AM broadcast frequencies, 500–1600 kHz. If the inductance of the coil is 100 μH and the fixed ca-

FIGURE 21.31   (Problem 37)

pacitance, C_1, is 80.0 pF, what should be the range of the variable capacitance, C_2?

38. A series circuit contains a 3.00-H inductor, a 3.00-μF capacitor, and a 30.0-Ω resistor connected to a 120-V rms source of variable frequency. Find the power delivered to the circuit when the frequency of the source is (a) the resonance frequency; (b) one-half the resonance frequency; (c) one-fourth the resonance frequency; (d) two times the resonance frequency; (e) four times the resonance frequency. From your calculations, can you draw a conclusion about the frequency at which the maximum power is delivered to the circuit?

39. The Q value of an RLC circuit is defined as the voltage drop across the inductor (or capacitor) at resonance, divided by the voltage drop across the resistor. The greater the Q value, the sharper, or narrower, is the curve of power versus frequency. Figure 21.32 shows such curves for small R and large R. (a) Show that the Q value is given by the expression $Q = 2\pi f_0 L / R$, where f_0 is the resonance frequency. (b) Calculate the Q value for the circuit of Problem 38.

40. The curves of Figure 21.32 are characterized by the width at half-maximum, Δf. This width is the difference in frequency between the two points on the curve where the power is one-half the maximum value. It can be shown that this width is related to the Q value (when the Q value is large) by $\Delta f = f_0 / Q$. (a) Find Δf for the circuit in Problem 39. (b) Find the Q value and Δf for the circuit in Problem 39 if R is replaced by a 300-Ω resistor.

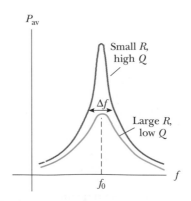

FIGURE 21.32   (Problems 39 and 40)

Section 21.7   The Transformer

41. An ac power generator produces 50 A (rms) at 3600 V. The voltage is stepped up to 100 000 V by an ideal transformer, and the energy is transmitted through a long-distance power line that has a resistance of 100 Ω. What percentage of the power delivered by the generator is dissipated as heat in the power line?

42. A transformer on a pole near a factory steps the voltage down from 3600 V to 120 V. The transformer is to deliver 1000 kW to the factory at 90% efficiency. Find (a) the power delivered to the primary, (b) the current in the primary, and (c) the current in the secondary.

43. At a given moment, every inhabitant of a city of 20 000 people turns on a 100-W lightbulb. Assume no other power in the city is being used. (a) If the utility company furnishes this total power at 120 V, calculate the current in the power lines from the utility to the city. (b) Calculate this current if the power company first steps up the voltage to 200 000 V. (c) How much heat is lost in each 1.0-m length of the power lines if the resistance of the lines is 5.0×10^{-4} Ω/m? Repeat this calculation for situation (a) and for situation (b). (d) If an individual line from the utility can handle only 100 A, how many lines are required to handle the current in each situation described?

Section 21.9   Hertz's Discoveries

44. If the coil in the resonant circuit of a radio has an inductance of 2.0 μH, what range of values must the tuning capacitor have in order to cover the complete range of FM frequencies? (The FM range of frequencies goes from 88 to 108 MHz.)

45. A standing wave interference pattern is set up by radio waves between two metal sheets positioned 2.00 m apart. This is the smallest distance between the plates that will produce a standing wave pattern. What is the fundamental frequency?

Section 21.10   Production of Electromagnetic Waves by an Antenna

Section 21.11   Properties of Electromagnetic Waves

46. Experimenters at the National Bureau of Standards have made precise measurements of the speed of light using the fact that, in vacuum, the speed of electromagnetic waves is $c = 1/\sqrt{\mu_0 \epsilon_0}$, where $\mu_0 = 4\pi \times 10^{-7}$ N·s²/C² and $\epsilon_0 = 8.854 \times 10^{-12}$ C²/N·m². What value (to four significant figures) does this give for the speed of light in vacuum?

47. Near the surface of the Earth, the magnitude of the Earth's magnetic field is about 0.50×10^{-4} T.

How much energy is stored in 1.0 m³ of Earth's atmosphere?

48. An incandescent lamp is radiating uniformly in all directions at 15 W. Calculate the maximum values of the electric and magnetic fields (a) 1.0 m and (b) 5.0 m from the source. (*Hint:* Use the formula for the surface area of a sphere, $A = 4\pi r^2$, to find the area on which the total energy is incident in each case, and then use Equation 21.29.)

49. Assume that the solar radiation incident on the Earth is 1340 W/m² (at the top of the Earth's atmosphere). Calculate the total power radiated by the Sun, taking the average separation between the Earth and the Sun to be 1.49×10^{11} m.

50. The Sun delivers an average power of 1340 W/m² to the top of the Earth's atmosphere. Find the magnitudes of $\mathbf{E}_m$ and $\mathbf{B}_m$ for the electromagnetic waves at the top of the atmosphere.

51. A coil with a resistance of $R = 16.0$ Ω and an inductance of $L = 0.0500$ H is connected in series with a variable capacitor, C, as in Figure 21.33. If the frequency of the circuit is $f = 76.4$ Hz, and an rms voltage of 100 V exists across ab, find (a) the peak voltage across ab, (b) the inductive reactance of the coil, and (c) the impedance of the coil. (d) By what angle does the current in the coil lag behind the voltage across it? (e) Beginning with very low values, C is increased until the rms current is 5.00 A. Determine the total impedance and capacitive reactance for this new adjustment.

52. The U.S. Navy has long proposed the construction of extremely low-frequency (ELF waves) communications systems; such waves could penetrate the oceans to reach distant submarines. Calculate the length of a quarter-wavelength antenna for a transmitter generating ELF waves of frequency 75 Hz. How practical is this?

53. A community plans to build a facility to convert solar radiation to electrical power. They require 1.0 MW of power, and the system to be installed has an efficiency of 30% (that is, 30% of the solar energy incident on the surface is converted to electrical energy). What must be the effective area of a perfectly absorbing surface used in such an installation, assuming a constant power per unit area of 1000 W/m² in the incident solar radiation?

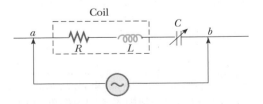

FIGURE 21.33   (Problem 51)

54. A plane electromagnetic wave has an average power per unit area of 300 W/m². A flat, rectangular surface, 20.0 cm by 40.0 cm, is placed perpendicularly to the direction of the plane wave. If the surface absorbs half the energy and reflects half, calculate the net energy delivered to the surface in 1.0 min.

Section 21.12 The Spectrum of Electromagnetic Waves

55. A diathermy machine, used in physiotherapy, generates electromagnetic radiation that gives the effect of "deep heat" when absorbed in tissue. One assigned frequency for diathermy is 27.33 MHz. What is the wavelength of this radiation?

56. Locate the positions in the electromagnetic spectrum of the waves that have the following frequencies, and find the wavelength associated with each frequency: (a) 10^6 Hz, (b) 10^8 Hz, (c) 10^{10} Hz, (d) 10^{13} Hz, (e) 10^{15} Hz, (f) 10^{17} Hz, (g) 10^{21} Hz. (*Hint:* See Fig. 21.25.)

57. What are the wavelength ranges in the (a) AM radio band (540–1600 kHz) and (b) the FM radio band (88–108 MHz)?

58. A wave has a wavelength of 4.0×10^{-7} m and a speed of 2.5×10^8 m/s in a particular material. (a) What is its frequency in this material? (b) The frequency of a wave does not change as the wave moves from one material to another. Find the wavelength of this wave in vacuum.

59. If the North Star, Polaris, were to burn out today, in what year would it disappear from our vision? The distance from the Earth to Polaris is approximately 6.44×10^{18} m.

60. A singer's voice is transmitted by a radio wave to a person 100 km away. (a) How much time passes before the distant listener hears the sound? (b) By the time the radio message reaches a listener, how far from the singer has the sound wave moved in the auditorium? Assume that the speed of sound is 345 m/s.

61. An important news announcement is transmitted by radio waves to people who are 100 km away, sitting next to their radios, and by sound waves to people sitting across the newsroom, 3.0 m from the newscaster. Who receives the news first? Explain. Take the speed of sound in air to be 343 m/s.

ADDITIONAL PROBLEMS

62. (a) What capacitance will resonate with a one-turn loop of inductance 400 pH to give a radar wave of wavelength 3.0 cm? (b) If the capacitor has square parallel plates separated by 1.0 mm of air, what should the edge length of the plates be? (c) What

is the common reactance of the loop and capacitor at resonance?

63. A 0.250-H inductor is connected to a capacitor and a 30.0-Ω resistor along with a 60.0-Hz, 30.0-V generator. (a) To what value would the capacitor have to be adjusted to produce resonance? (b) Find the voltage drop across the capacitor-and-inductor combination at resonance.

64. The primary coil of a certain transformer has an inductance of 2.5 H and a resistance of 80.0 Ω (a) If the primary coil is connected to an ac source with a frequency of 60.0 Hz and a voltage of 110 V rms, what is the rms current through the primary? (b) If the primary is connected to 110 V dc, what is the current through the primary? Disregard initial effects. (c) In each case, compare the power dissipated in the resistance.

65. The electromagnetic power, P, radiated by a moving point charge, q, with acceleration a is

$$P = \frac{2 k q^2 a^2}{3 c^3}$$

where k is the Coulomb constant and c is the speed of light in vacuum. If an electron is placed in a constant electric field of 100 N/C, determine (a) the acceleration of the electron and (b) the power radiated by this electron. (c) Show that the right side of this equation is in watts.

66. A particular inductor has appreciable resistance. When the inductor is connected to a 12-V battery, the current through the inductor is 3.0 A. When it is connected to an ac source with an rms output of 12 V and a frequency of 60.0 Hz, the current drops to 2.0 A. What are (a) the impedance at 60.0 Hz and (b) the inductance of the inductor?

67. A 0.700-H inductor is connected in series with a fluorescent lamp to limit the current drawn by the lamp. If the combination is connected to a 60.0-Hz, 120-V line, and if the voltage across the lamp is to be 40.0 V, what is the current in the circuit? (The lamp is a pure resistive load.)

68. A small transformer is used to supply an ac voltage of 6.0 V to a model-railroad lighting circuit. The primary has 220 turns and is connected to a standard 110-V, 60-Hz line. Although the resistance of the primary may be neglected, it has an inductance of 150 mH. (a) How many turns are required on the secondary winding? (b) If the transformer is left plugged in, what current is drawn by the primary when the secondary is open? (c) What power is drawn by the primary when the secondary is open?

69. A 200-Ω resistor is connected in series with a 5.0-μF capacitor and a 60-Hz, 120-V rms line. If electrical energy costs $0.080 kWh, how much does it cost to leave this circuit connected for 24 h?

70. A series *RLC* circuit has a resonance frequency of $2000/\pi$ Hz. When it is operating at a frequency of $\omega > \omega_0$, $X_L = 12$ Ω and $X_C = 8.0$ Ω. Calculate the values of L and C for the circuit.

71. An inductor is to be made of a 5.0-m length of 1.0-mm-diameter copper wire wound on a coil of radius 3.0 cm. (The coil has an air core.) (a) Estimate the length of the completed solenoid and number of turns on it. Find (b) the inductance of the completed coil and (c) the resistance of the completed coil. (d) Find the current that is drawn when this device is connected to a 60-Hz, 20-V rms source.

72. Two connections allow contact with two circuit elements in series inside a box, but it is not known whether the circuit elements are *R*, *L*, or *C*. In an attempt to find what is inside the box, you make some measurements, with the following results. When a 3.0-V dc power supply is connected across the terminals, there is a direct current of 300 mA in the circuit. When a 3.0-V, 60-Hz source is connected, the current becomes 200 mA. (a) What are the two elements in the box? (b) What are their values of *R*, *L*, or *C*?

73. A transmission line with a resistance per unit length of 4.5×10^{-4} Ω/m is to be used to transmit 5000 kW of power over a distance of 400 miles $(6.44 \times 10^5$ m). The output voltage of the generator is 4500 V. (a) What is the line loss if a transformer is used to step up the voltage to 500 kV? (b) What fraction of the input power is lost to the line under these circumstances? (c) What difficulties would be encountered in an attempt to transmit the 5000 kW of power at the generator voltage of 4500 V?

74. Suppose you wish to use a transformer as an impedance-matching device between an audio amplifier that has an output impedance of 8000 Ω and a speaker that has an input impedance of 8.0 Ω. What should be the ratio of primary to secondary turns on the transformer?

75. What value of inductance should be used in series with a capacitor of 1.50 pF to form an oscillating circuit that will radiate a wavelength of 5.25 m?

76. The nearest star to us is approximately 4.0 lightyears away (1 lightyear is the distance light travels in one year). Find the distance to this star in meters.

77. A possible means of space flight is to place a perfectly reflecting aluminized sheet into Earth's orbit and use the light from the Sun to push this

FIGURE 21.34   (Problem 80)

solar sail. Suppose such a sail, of area 6.00×10^4 m^2 and mass 6000 kg, is placed in orbit facing the Sun. (a) What force is exerted on the sail? (b) What is the sail's acceleration? (c) How long does it take for this sail to reach the Moon, 3.84×10^8 m away? Ignore all gravitational effects, and assume a solar intensity of 1380 W/m^2. (*Hint:* The radiation pressure by a reflected wave is given by 2 (average power per area)/c.)

78. What power must be radiated uniformly in all directions by a source if the amplitude of the magnetic field is 7.0×10^{-8} T at a distance of 2.0 m?

79. A microwave transmitter emits electromagnetic waves of a single wavelength. The maximum electric field 1.0 km from the transmitter is 6.0 V/m. Assuming that the transmitter is a point source and neglecting waves reflected from the Earth, calculate (a) the maximum magnetic field at this distance and (b) the total power emitted by the transmitter.

80. A dish antenna with a diameter of 20.0 m receives (at normal incidence) a radio signal from a distant source, as shown in Figure 21.34. The radio signal is a continuous sinusoidal wave with amplitude $E_m = 0.20$ μV/m. Assume the antenna absorbs all the radiation that falls on the dish. (a) What is the amplitude of the magnetic field in this wave? (b) What is the intensity of the radiation received by this antenna? (c) What is the power received by the antenna?

81. Compute the average energy content of a liter of sunlight as it reaches the top of the Earth's atmosphere, where its intensity is 1340 W/m^2.

Light and Optics PART 5

Scientists have long been intrigued by the nature of light, and philosophers have argued endlessly concerning the proper definition and perception of light. It is important to understand the nature of this basic ingredient of life on Earth. Plants convert light energy from the Sun to chemical energy through photosynthesis. Light is the means by which we transmit and receive information from objects around us and throughout the Universe.

The Greeks believed that light consisted of tiny particles (corpuscles) that were emitted by a light source, then stimulated the perception of vision upon striking the observer's eye. Newton used this corpuscular theory to explain the reflection and refraction of light. In 1670 one of Newton's contemporaries, the Dutch scientist Christian Huygens, succeeded in explaining many properties of light by proposing that light was wave-like. In 1801 Thomas Young gave strong support to the wave theory by showing that light beams can interfere with one another. In 1865 Maxwell developed a brilliant theory that electromagnetic waves travel with the speed of light (Chap. 21). By that time, the wave theory of light seemed to be on firm ground.

However, at the beginning of the 20th century, Max Planck introduced the notion of quantization of electromagnetic radiation, and Albert Einstein returned to the corpuscular theory of light in order to explain the radiation emitted by hot objects and the electrons emitted by a metal exposed to light (the photoelectric effect). We shall discuss those and other modern topics in the last part of this book.

Today scientists view light as having a dual nature. Experiments can be devised that will display either its particle-like or its wave-like nature. In this part of the book, we concentrate on the aspects of light that are best understood through the wave model. First we discuss the reflection of light at the boundary between two media and the refraction (bending) of light as it travels from one medium into another. We use these ideas to study the refraction of light as it passes through lenses and the reflection of light from mirrored surfaces. Finally, we describe how lenses and mirrors can be used to view objects with telescopes and microscopes and how lenses are used in photography.

(John McDermott/ Tony Stone Images)

22 REFLECTION AND REFRACTION OF LIGHT

This photograph taken in Salamanca, Spain, shows the reflection of the New Cathedral in the Tormes River. Are you able to distinguish the cathedral from its image? (David Parker/SPL/Photo Researchers)

22.1

THE NATURE OF LIGHT

Until the beginning of the 19th century, light was considered to be a stream of particles, emitted by a light source, that stimulated the sense of sight upon entering the eye. The chief architect of the particle theory of light was Newton. With this theory he provided simple explanations of some known experimental facts concerning the nature of light, namely the laws of reflection and refraction.

Christian Huygens was a Dutch physicist and astronomer who is best known for his contributions to the fields of dynamics and optics. His accomplishments in physics include the invention of the pendulum clock and the first exposition of a wave theory of light. As an astronomer, Huygens was the first to recognize the rings around Saturn and to discover Titan, a satellite of Saturn.

Huygens was born into a prominent family in The Hague in 1629. He was the son of Constantin Huygens, one of the most important figures of the Renaissance in Holland. Educated at the University of Leyden, Christian became a close friend of the French philosopher René Descartes, who was a frequent guest at the Huygens home. Huygens published his first paper in 1651, on the quadrature of curves.

Huygens' reputation in optics and dynamics spread throughout Europe, and in 1663 he was elected a charter member of the Royal Society. In 1666 he was lured to France by Louis XIV, who surrounded himself with scholars for the glory of his regime. While in France, Huygens became one of the founders of the French Academy of Science.

In 1673, in Paris, Huygens published *Horologium Oscillatorium*. In this work he described a solution to the problem of the compound pendulum, for which he calculated the equivalent simple pendulum length. In the same publication he also derived a formula for computing the period of oscillation of a simple pendulum and explained the force governing uniform motion in a circle.

Huygens returned to Holland in 1681, constructed some lenses of large focal lengths, and invented the achromatic eyepiece for telescopes. Shortly after a visit to England, where he had met Isaac Newton, Huygens published his treatise on the wave theory of light. To Huygens, light was a vibratory motion in the ether, spreading out and producing the sensation of light when impinging on the eye. On the basis of this theory, he deduced the laws of reflection and refraction and explained the phenomenon of double refraction.

Huygens, second only to Newton among scientists in the second half of the 17th century, was the first to proceed beyond the point reached by Galileo and Descartes in the field of dynamics. A solitary man, Huygens did not attract students or disciples and was very slow in publishing his findings. He died in 1695 after a long illness.

**Christian Huygens
(1629–1695)**

Christian Huygens (1629–1695). *(Courtesy of Rijksmuseum voor de Geschiedenis der Natuurwetenschappen. Courtesy AIP Niels Bohr Library)*

Most scientists accepted Newton's particle theory of light. However, during Newton's lifetime another theory was proposed. In 1678 a Dutch physicist and astronomer, Christian Huygens (1629–1695), showed that a wave theory of light could also explain the laws of reflection and refraction. The wave theory did not receive immediate acceptance for several reasons. All the waves known at the time (sound, water, and so on) traveled through some sort of medium, but light from the Sun could travel to Earth through empty space. Furthermore, it was argued that if light were some form of wave, it would bend around obstacles; hence, we should be able to see around corners. It is now known that light does indeed bend around the edges of objects. This phenomenon, known as *diffraction*, is not easy to observe because light waves have such short wavelengths. Even though experimental evidence for the diffraction of light was discovered by Francesco Grimaldi (1618–1663) around 1660, for more than a century most scientists rejected the wave theory and adhered to Newton's particle theory. This was, for the most part, due to Newton's great reputation as a scientist.

The first clear demonstration of the wave nature of light was provided in 1801 by Thomas Young (1773–1829), who showed that, under appropriate conditions, light exhibits interference behavior. That is, at certain points in the vicinity of two sources, light waves can combine and cancel each other by destructive interference. Such behavior could not be explained at that time by a particle theory. Several years later, a French physicist, Augustin Fresnel (1788–1829), performed a number of detailed experiments dealing with interference and diffraction phenomena. In 1850 Jean Foucault (1791–1868) provided further evidence of the inadequacy of the particle theory by showing that the speed of light is lower in liquids than in air; according to the particle model, the speed of light would be higher in glasses and liquids than in air. Additional developments during the 19th century led to the general acceptance of the wave theory of light.

The most important development concerning the theory of light was the work of Maxwell, who in 1865 predicted that light was a form of high-frequency electron agnetic wave (Chapter 21). His theory predicted that these waves should have a speed of about 3×10^8 m/s, in agreement with measured values. As discussed in Chapter 21, Hertz provided experimental confirmation of Maxwell's theory in 1887 by producing and detecting electromagnetic waves. Furthermore, Hertz and other investigators showed that these waves exhibited reflection, refraction, and all the other characteristic properties of waves.

Although the classical theory of electricity and magnetism explained most known properties of light, some subsequent experiments could not be explained by the assumption that light was a wave. The most striking of these was the *photoelectric effect* (which we shall examine more closely in Chapter 27), also discovered by Hertz. Hertz found that clean metal surfaces emit charges when exposed to ultraviolet light. In 1888 Hallwachs discovered that the emitted charges are negative, and in 1899 J. J. Thomson showed that the emitted charges are electrons.

In 1902 Philip Lenard found that the maximum kinetic energy of photoelectrons (electrons emitted by light) does not depend on the intensity of the incoming radiation. Although he was unable to establish the precise relationship, Lenard found that the maximum kinetic energy increases with light frequency. The lack of dependence on intensity was completely unexpected and could not be explained by classical physics. In 1905 Einstein published a paper that formulated the theory of light quanta and explained the photoelectric effect. He reached the conclusion that light is composed of "corpuscles," or discontinuous quanta of energy. Furthermore, he asserted that light interacting with matter also consists of quanta, and he brilliantly worked out the implications of the photoelectric process. More specifically, Einstein showed that the energy of a photon is proportional to the frequency of the electromagnetic wave:

Energy of a photon

$$E = hf \qquad \text{[22.1]}$$

where $h = 6.63 \times 10^{-34}$ J·s is *Planck's constant*. This theory retains some features of both the wave and particle theories of light. As we shall discuss later, the photoelectric effect is the result of energy transfer from a single photon to an electron in the metal. That is, the electron interacts with one photon of light as if it, the electron, had been struck by a particle. Yet the photon has wave-like characteristics as implied by the fact that light exhibits interference phenomena.

In view of these developments, light must be regarded as having a *dual nature*. That is, *in some cases light acts as a wave and in others it acts as a particle.* Classical electromagnetic wave theory provides adequate explanations of light propagation and of the effects of interference, whereas the photoelectric effect and other experiments involving the interaction of light with matter are best explained by assuming that light is a particle. Light is light, to be sure. However, the question "Is light a wave or a particle?" is inappropriate; sometimes it acts as one, sometimes as the other. Fortunately, it never acts as both in the same experiment.

22.2
MEASUREMENTS OF THE SPEED OF LIGHT

Light travels so fast ($c \approx 3 \times 10^8$ m/s) that early attempts to measure its speed were unsuccessful. Galileo attempted to measure the speed of light by positioning two observers in towers separated by about 5 miles. Each observer carried a shuttered lantern. One observer opened his lantern first, and then the other opened his lantern at the moment he saw the light from the first lantern. In principle, the speed could then be obtained from the transit time of the light beams between lanterns. However, at a speed of 186 000 mi/s, light would travel the 10-mile round trip in approximately 54 μs. Because the transit time is so small compared with the reaction time of the observers, it is impossible to measure the speed of light in this manner.

ROEMER'S METHOD

The first successful estimate of the speed of light was made in 1675 by the Danish astronomer Ole Roemer (1644–1710). His technique involved astronomical observations of one of the moons of Jupiter, Io. It is of interest that Roemer did not set out to measure the speed of light. He was, instead, trying to explain a perplexing problem related to the prediction of eclipses of Io and a few other moons. After realizing that the underlying difficulty with predicting the eclipses had to do with the finite speed of light, he estimated that speed.

At the time of Roemer, only four of Jupiter's moons had been discovered, and the periods of their orbits were known. Io, the innermost moon, has a period of 42.5 h, and its orbit, the orbit of Jupiter, and the orbit of the Earth all lie in approximately the same plane. As a result, Io goes into eclipse behind Jupiter every 42.5 h, as seen from Earth. Using the orbital motion of Io as a clock, Roemer expected to find a constant period over long time intervals. Instead, he observed a systematic variation in Io's period. He found that the periods were longer than average when Earth receded from Jupiter and shorter than average when Earth approached Jupiter. If Io had a constant period, Roemer should have been able to observe a particular eclipse and be able to predict when future eclipses would occur. But he found that his predictions often did not agree with the actual occurrences of the eclipses. For example, consider the situation diagrammed in Figure 22.1 Suppose an eclipse occurred when the Earth was at position E_1. Knowing the period of Io, Roemer could predict when an eclipse should occur three months later, with the Earth at position E_2. However, the actual eclipse at E_2 occurred approximately 600 s later than the predicted time.

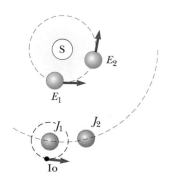

FIGURE 22.1
As the Earth moves from E_1 to E_2, Jupiter moves only from J_1 to J_2.

Roemer attributed this discrepancy to the fact that the distance between the Earth and Jupiter was changing between the two observations. In three months (one quarter of the Earth's period), the Earth moved through one fourth of its orbit, from E_1 to E_2, as shown in Figure 22.1. In the same time interval, Jupiter, whose period is about 12 years, moved a much shorter distance, from J_1 to J_2. Therefore, as the Earth moved from E_1 to E_2, light from Jupiter had to travel an additional distance equal to the radius of the Earth's orbit.

Using the data available at that time, Roemer estimated the speed of light to be about 2.1×10^8 m/s. The large discrepancy between this value and the currently accepted value, 3×10^8 m/s, is due to a large error in the assumed radius of the Earth's orbit. Roemer's experiment is important historically because it demonstrated that light does have a finite speed and established a rough estimate of the magnitude of that speed.

FIZEAU'S TECHNIQUE

The first successful method of measuring the speed of light using purely Earth-bound techniques was developed in 1849 by Armand H. L. Fizeau (1819–1896). Figure 22.2 is a simplified diagram of his apparatus. The basic idea is to measure the total time it takes light to travel from some point to a distant mirror and back. If d is the distance between the light source and the mirror and if the transmit time for one round trip is t, then the speed of light is $c = 2d/t$. To measure the transit time, Fizeau used a rotating toothed wheel, which converts an otherwise continuous beam of light to a series of light pulses. Additionally, the rotation of the wheel controls what an observer at the light source sees. For example, if the light passing the opening at point A in Figure 22.2 returned at the instant tooth B had rotated into position to cover the return path, the light would not reach the observer. At a faster rate of rotation, the opening at point C could move into position to allow the reflected beam to pass and reach the observer. Knowing the distance, d, the number of teeth in the wheel, and the angular speed of the wheel, Fizeau arrived at a value of $c = 3.1 \times 10^8$ m/s. Similar measurements made by subsequent investigators yielded more accurate values for c, approximately 2.9977×10^8 m/s.

A variety of more accurate measurements have since been reported for c. A recent value, obtained using a laser technique, is

$$c = 2.997\ 924\ 574\ (12) \times 10^8 \text{ m/s}$$

where the (12) indicates the uncertainty in the last two digits. The number of significant figures here is certainly impressive. In fact, the speed of light has been

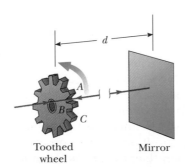

FIGURE 22.2
Fizeau's method for measuring the speed of light using a rotating toothed wheel.

determined with such high accuracy that it is now used to define the SI unit of length, the meter. As noted in Chapter 1, the meter is defined as the distance traveled by light in a vacuum during an interval of 1/299 792 458 s.

EXAMPLE 22.1   Measuring the Speed of Light with Fizeau's Toothed Wheel

Assume that the toothed wheel of the Fizeau experiment has 360 teeth and is rotating at a speed of 27.5 rev/s when the light from the source is extinguished—that is, when a burst of light passing through opening A in Figure 22.2 is blocked by tooth B on return. If the distance to the mirror is 7500 m, find the speed of light.

Reasoning and Solution   If the wheel has 360 teeth, it turns through an angle of 1/720 rev in the time that it takes the light to make its round trip. From the definition of angular velocity, we see that the time is

$$t = \frac{\theta}{\omega} = \frac{(1/720)\ \text{rev}}{27.5\ \text{rev/s}} = 5.05 \times 10^{-5}\ \text{s}$$

Hence, the speed of light is

$$c = \frac{2d}{t} = \frac{2(7500\ \text{m})}{5.05 \times 10^{-5}\ \text{s}} = \boxed{2.97 \times 10^{8}\ \text{m/s}}$$

22.3

HUYGENS' PRINCIPLE

The laws of reflection and refraction can be developed using a geometric method proposed by Huygens in 1678. Huygens assumed that light is some form of wave motion rather than a stream of particles. He had no knowledge of the nature of light or of its electromagnetic character. Nevertheless, his simplified wave model is adequate for understanding many practical aspects of the propagation of light.

Huygens' principle is a geometric construction for determining at some instant the position of a new wavefront from the knowledge of the wavefront that preceded it. (A wavefront is a surface passing through those points of a wave that have the same phase and amplitude. For instance, a wavefront could be a surface passing through the crests of waves.) In Huygens' construction,

> All points on a given wavefront are taken as point sources for the production of spherical secondary waves, called wavelets, which propagate outward with speeds characteristic of waves in that medium. After some time has elapsed, the new position of the wavefront is the surface tangent to the wavelets.

Figure 22.3 illustrates two simple examples of Huygens' construction. First, consider a plane wave moving through free space, as in Figure 22.3a. At $t = 0$, the wavefront is indicated by the plane labeled AA'. In Huygens' construction, each point on this wavefront is considered a point source. For clarity, only a few points on AA' are shown. With these points as sources for the wavelets, we draw circles each of radius ct, where c is the speed of light in vacuum and t is the

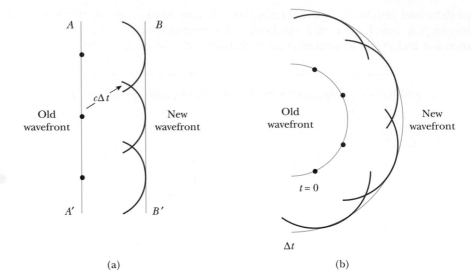

FIGURE 22.3
Huygens' constructions for (a) a plane wave propagating to the right and (b) a spherical wave.

(a)   (b)

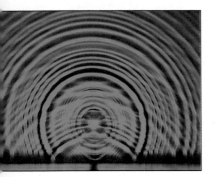

FIGURE 22.4
Water waves in a ripple tank demonstrate Huygens' wavelets. A plane wave at the bottom is incident on a barrier with a small opening. The opening acts as a source of circular wavelets.
(Erich Schrempp/ Photo Researchers)

period of propagation from one wavefront to the next. The surface drawn tangent to these wavelets is the plane BB', which is parallel to AA'. In a similar manner, Figure 22.3b shows Huygens' construction for an outgoing spherical wave.

A convincing demonstration of Huygens' principle is performed with water waves in a shallow tank (called a ripple tank), as in Figure 22.4. Plane waves at the bottom are incident on a barrier that contains a small opening. The opening acts as a source of two-dimensional circular waves propagating outward.

THE RAY APPROXIMATION IN GEOMETRIC OPTICS

In studying geometric optics here and in Chapter 23, we shall use what is called the *ray approximation*. To understand this approximation, first recall that the direction of energy flow of a wave, corresponding to the direction of wave propagation, is called a ray. The rays of a given wave are straight lines perpendicular to the wavefronts, as illustrated in Figure 22.5 for a plane wave. In the ray approximation, we assume that a wave moving through a medium travels in a

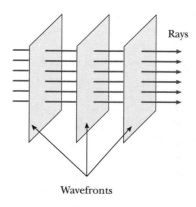

FIGURE 22.5
A plane wave traveling to the right. Note that the rays, corresponding to the direction of wave motion, are straight lines perpendicular to the wavefronts.

straight line in the direction of its rays. That is, a ray is a line drawn in the direction in which the light is traveling. For example, a beam of sunlight passing through a darkened room traces out the path of a light ray.

22.4

REFLECTION AND REFRACTION

REFLECTION OF LIGHT

When a light ray traveling in a transparent medium encounters a boundary leading into a second medium, part of the incident ray is reflected back into the first medium. Figure 22.6a shows several rays of a beam of light incident on a smooth, mirror-like, reflecting surface. The reflected rays are parallel to each other, as indicated in the figure. Reflection of light from such a smooth surface is called *specular reflection*. On the other hand, if the reflecting surface is rough, as in Figure 22.6b, the surface reflects the rays in a variety of directions. Reflection from any rough surface is known as *diffuse reflection*. A surface behaves as a smooth surface as long as the surface variations are small compared with the wavelength of the incident light. Figures 22.6c and 22.6d are photographs of specular and diffuse reflection of laser light.

For instance, consider the two types of reflection from a road surface that one sees while driving at night. When the road is dry, light from oncoming vehicles is scattered off the road in different directions (diffuse reflection) and the road is quite visible. On a rainy night, when the road is wet, the road irregularities are filled with water. Because the wet surface is quite smooth, the light undergoes specular reflection. This means that the light is reflected straight ahead, and the driver of a car sees only what is directly in front of him. Light from

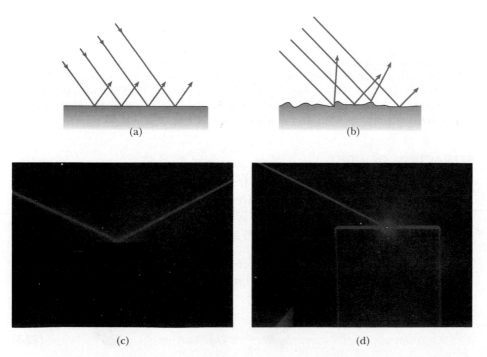

(a) (b)

(c) (d)

FIGURE 22.6
A schematic representation of (a) specular reflection, where the reflected rays are all parallel to each other, and (b) diffuse reflection, where the reflected rays travel in random directions. (c, d) Photographs of specular and diffuse reflection, using laser light. *(Photographs courtesy of Henry Leap and Jim Lehman)*

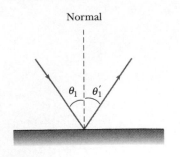

FIGURE 22.7
According to the law of reflection, $\theta_1 = \theta_1'$.

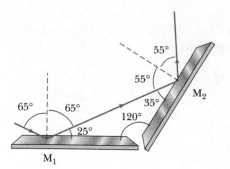

FIGURE 22.8
(Example 22.2)   Mirrors M_1 and M_2 make an angle of $120°$ with each other.

the side never reaches his eye. In this book we concern ourselves only with specular reflection, and we use the term *reflection* to mean specular reflection.

Consider a light ray traveling in air and incident at some angle on a flat, smooth surface, as in Figure 22.7. The incident and reflected rays make angles θ_1 and θ_1', respectively, with a line perpendicular to the surface at the point where the incident ray strikes the surface. We call this line the *normal* to the surface. Experiments show that *the angle of reflection equals the angle of incidence;* that is,

$$\theta_1' = \theta_1 \qquad \text{[22.2]}$$

EXAMPLE 22.2   The Double-Reflecting Light Ray

Two mirrors make an angle of $120°$ with each other, as in Figure 22.8. A ray is incident on mirror M_1 at an angle of $65°$ to the normal. Find the direction of the ray after it is reflected from mirror M_2.

Reasoning and Solution   From the law of reflection, we see that the first reflected ray also makes an angle of $65°$ with the normal. It follows that this same ray makes an angle of $90° - 65°$, or $25°$, with the horizontal. From the triangle made by the first reflected ray and the two mirrors, we see that the first reflected ray makes an angle of $35°$ with M_2 (since the sum of the interior angles of any triangle is $180°$). This means that this ray makes an angle of $55°$ with the normal to M_2. From the law of reflection, it follows that the second reflected ray makes an angle of $55°$ with the normal to M_2.

REFRACTION OF LIGHT

When a ray of light traveling through a transparent medium encounters a boundary leading into another transparent medium, as in Figure 22.9a, part of the ray is reflected and part enters the second medium. The ray that enters the second medium is bent at the boundary and is said to be *refracted*. The incident ray, the reflected ray, the refracted ray and the normal at the point of incidence all lie in the same plane. The **angle of refraction,** θ_2 in Figure 22.9a, depends on the properties of the two media and on the angle of incidence, through the relationship

$$\frac{\sin \theta_2}{\sin \theta_1} = \frac{v_2}{v_1} = \text{constant} \qquad \text{[22.3]}$$

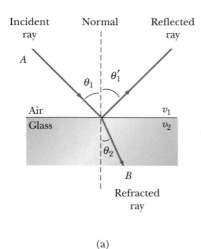

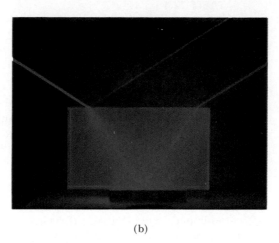

(a)

(b)

FIGURE 22.9
(a) A ray obliquely incident on an air-glass interface. The refracted ray is bent toward the normal since $v_2 < v_1$. (b) Light incident on the Lucite block bends both when it enters the block and when it leaves the block. *(Courtesy of Henry Leap and Jim Lehman)*

where v_1 is the speed of light in medium 1 and v_2 is the speed of light in medium 2. Willebrord Snell (1591–1626) is usually credited with the experimental discovery of this relationship, which is therefore known as Snell's law. In Section 22.8 we shall derive the laws of reflection and refraction using Huygens' principle.

Experiment shows that *the path of a light ray through a refracting surface is reversible.* For example, the ray in Figure 22.9a travels from point *A* to point *B*. If the ray originated at *B*, it would follow the same path to reach point *A*, but the reflected ray would be in the glass.

When light moves from a material in which its speed is high to a material in which its speed is lower, the angle of refraction, θ_2, is less than the angle of incidence, as shown in Figure 22.10a. If the ray moves from a material in which its speed is low to a material in which its speed is higher, it is bent away from the normal, as in Figure 22.10b.

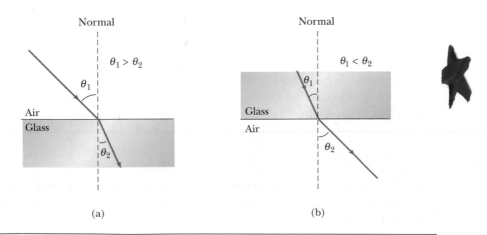

(a)

(b)

FIGURE 22.10
(a) When the light beam moves from air into glass, its path is bent toward the normal.
(b) When the beam moves from glass into air, its path is bent away from the normal.

TABLE 22.1
Indices of Refraction for Various Substances, Measured with Light of Vacuum Wavelength $\lambda_0 = 589$ nm

Substance	Index of Refraction	Substance	Index of Refraction
Solids at 20°C		Liquids at 20°C	
Cubic Zirconia	2.20	Benzene	1.501
Diamond (C)	2.419	Carbon disulfide	1.628
Fluorite (CaF_2)	1.434	Carbon tetrachloride	1.461
Fused quartz (SiO_2)	1.458	Ethyl alcohol	1.361
Glass, crown	1.52	Glycerine	1.473
Glass, flint	1.66	Water	1.333
Ice (H_2O) (at 0°C)	1.309		
Polystyrene	1.49	Gases at 0°C, 1 atm	
Sodium chloride (NaCl)	1.544	Air	1.000 293
Zircon	1.923	Carbon dioxide	1.000 45

22.5

THE LAW OF REFRACTION

When light passes from one transparent medium to another, it is refracted because the speed of light is different in the two media. It is convenient to define the **index of fraction,** n, of a medium as the ratio

Index of refraction

$$n \equiv \frac{\text{speed of light in vacuum}}{\text{speed of light in a medium}} = \frac{c}{v}$$ **[22.4]**

From this definition, we see that the index of refraction is a dimensionless number that is greater than unity since v is always less than c. Furthermore, n equals unity for a vacuum. Table 22.1 lists the indices of refraction for some representative substances.

As light travels from one medium to another, *its frequency does not change.* To see why, consider Figure 22.11. Wavefronts pass an observer at point A in medium 1 with a certain frequency and are incident on the boundary between medium 1 and medium 2. The frequency with which the wavefronts pass an observer at point B in medium 2 must equal the frequency at which they arrive at point A in medium 1. If this were not the case, either wavefronts would pile up at the boundary or they would be destroyed or created at the boundary. Since there is no mechanism for this to occur, the frequency must be a constant as a light ray passes from one medium into another.

Therefore, because the relation $v = f\lambda$ must be valid in both media and because $f_1 = f_2 = f$, we see that

$$v_1 = f\lambda_1 \qquad \text{and} \qquad v_2 = f\lambda_2$$

where the subscripts refer to the two media. A relationship between index of refraction and wavelength can be obtained by dividing these two equations and making use of the definition of index of refraction provided by Equation 22.4:

$$\frac{\lambda_1}{\lambda_2} = \frac{v_1}{v_2} = \frac{c/n_1}{c/n_2} = \frac{n_2}{n_1}$$ **[22.5]**

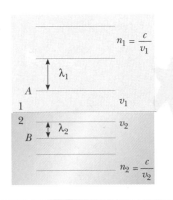

FIGURE 22.11
As the wave moves from medium 1 to medium 2, its wavelength changes but its frequency remains constant.

which gives

$$\lambda_1 n_1 = \lambda_2 n_2 \qquad [22.6]$$

Let medium 1 be the vacuum so that $n_1 = 1$. From Equation 22.6, it follows that the index of refraction of any medium can be expressed as the ratio

$$n = \frac{\lambda_0}{\lambda_n} \qquad [22.7]$$

where λ_0 is the wavelength of light in vacuum and λ_n is the wavelength in a medium with index of refraction n. Figure 22.12 is a schematic representation of this reduction in wavelength when light passes from vacuum into a transparent medium.

We are now in a position to express Snell's law (Eq. 22.3) in an alternative form. If we substitute Equation 22.5 into Equation 22.3, we get

$$n_1 \sin \theta_1 = n_2 \sin \theta_2 \qquad [22.8]$$

This is the most widely used and practical form of Snell's law.

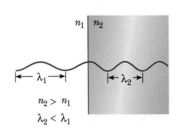

FIGURE 22.12
A schematic diagram of the *reduction* in wavelength when light travels from a medium with a low index of refraction to one with a higher index of refraction.

Snell's law

EXAMPLE 22.3   An Index of Refraction Measurement

A beam of light of wavelength 550 nm, traveling in air, is incident on a slab of transparent material. The incident beam makes an angle of 40.0° with the normal, and the refracted beam makes an angle of 26.0° with the normal. Find the index of refraction of the material.

Solution   Snell's law of refraction (Eq. 22.8), together with the given data—$\theta_1 = 40.0°$, $n_1 = 1.00$ for air, and $\theta_2 = 26.0°$—gives

$$n_1 \sin \theta_1 = n_2 \sin \theta_2$$

$$n_2 = \frac{n_1 \sin \theta_1}{\sin \theta_2} = (1.00) \frac{\sin 40.0°}{\sin 26.0°} = \frac{0.643}{0.438} = \boxed{1.47}$$

If we compare this value with the data in Table 22.1, we see that the material is probably fused quartz.

Exercise   What is the wavelength of this light in the slab?

Answer   374 nm.

EXAMPLE 22.4   Angle of Refraction for Glass

A light ray of wavelength 589 nm (produced by a sodium lamp) traveling through air is incident on a smooth, flat slab of crown glass at an angle of 30.0° to the normal, as sketched in Figure 22.13. Find the angle of refraction, θ_2.

Solution   Snell's law (Eq. 22.8) can be rearranged as

$$\sin \theta_2 = \frac{n_1}{n_2} \sin \theta_1$$

From Table 22.1, we find that $n_1 = 1.00$ for air and $n_2 = 1.52$ for crown glass. Therefore, the unknown refraction angle is determined by

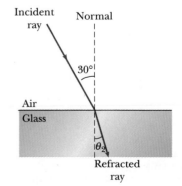

FIGURE 22.13
(Example 22.4)   Refraction of light by glass.

$$\sin \theta_2 = \left(\frac{1.00}{1.52} \right) (\sin 30.0°) = 0.329$$

$$\theta_2 = \sin^{-1}(0.329) = \boxed{19.2°}$$

We see that the ray is bent *toward* the normal, as expected.

Exercise If the light ray moves from inside the glass toward the glass-air interface at an angle of 30.0° to the normal, determine the angle of refraction.

Answer 49.5° *away* from the normal.

EXAMPLE 22.5 The Speed of Light in Fused Quartz

Light of wavelength 589 nm in vacuum passes through a piece of fused quartz of index of refraction $n = 1.458$.

(a) Find the speed of light in fused quartz.

Solution The speed of light in fused quartz can be obtained from Equation 22.4:

$$v = \frac{c}{n} = \frac{3.00 \times 10^8 \text{ m/s}}{1.458} = \boxed{2.06 \times 10^8 \text{ m/s}}$$

It is interesting to note that the speed of light in vacuum, 3.00×10^8 m/s, is an upper limit for the speed of material objects. In our treatment of relativity in Chapter 26, we shall find that this upper limit is consistent with experimental observations. However, it is possible for a particle moving in a medium to have a speed that exceeds the speed of light in that medium. For example, it is theoretically possible for a particle to travel through fused quartz at a speed greater than 2.06×10^8 m/s, but it must have a speed less than 3.00×10^8 m/s in a vacuum.

(b) What is the wavelength of this light in fused quartz?

Solution We can use $\lambda_n = \lambda_0/n$ (Eq. 22.7) to calculate the wavelength in fused quartz, noting that we are given $\lambda_0 = 589$ nm $= 589 \times 10^{-9}$ m:

$$\lambda_n = \frac{\lambda}{n} = \frac{589 \text{ nm}}{1.458} = \boxed{404 \text{ nm}}$$

Exercise Find the frequency of the light passing through the fused quartz.

Answer 5.09×10^{14} Hz.

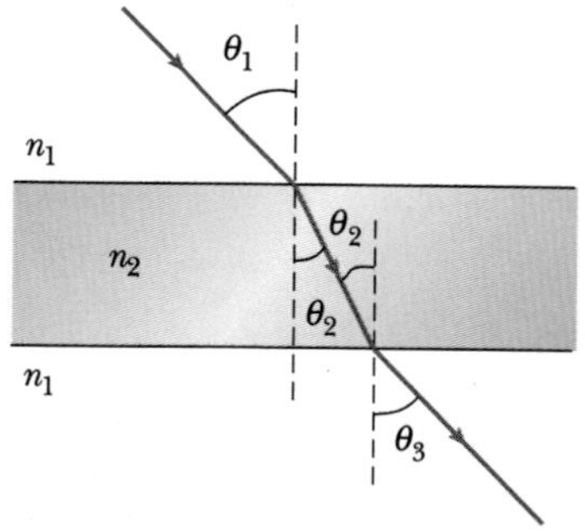

EXAMPLE 22.6 Light Passing Through a Slab

A light beam traveling through a transparent medium of index of refraction n_1 passes through a thick transparent slab with parallel faces and index of refraction n_2 (Fig. 22.14). Show that the emerging beam is parallel to the incident beam.

Reasoning To solve this problem, it is necessary to apply Snell's law twice, once at the upper surface and once at the lower surface. The two equations will be related because the angle of refraction at the upper surface equals the angle of incidence at the lower surface. This procedure will enable us to compare angles θ_1 and θ_3.

FIGURE 22.14
(Example 22.6) When light passes through a flat slab of material, the emerging beam is parallel to the incident beam, and therefore $\theta_1 = \theta_3$.

Solution   First, let us apply Snell's law to the upper surface:

$$(1) \qquad \sin \theta_2 = \frac{n_1}{n_2} \sin \theta_1$$

Applying Snell's law to the lower surface gives

$$(2) \qquad \sin \theta_3 = \frac{n_2}{n_1} \sin \theta_2$$

Substituting (1) into (2) gives

$$\sin \theta_3 = \frac{n_2}{n_1} \left(\frac{n_1}{n_2} \sin \theta_1 \right) = \sin \theta_1$$

That is, $\theta_3 = \theta_1$, and so the slab does not alter the direction of the beam. It does, however, displace the beam. The same result is obtained when light passes through multiple layers of materials.

22.6

DISPERSION AND PRISMS

An important property of the index of refraction is that its value in anything but vacuum depends on the wavelength of light. This phenomenon is called **dispersion** (Fig. 22.15). Since n is a function of wavelength, Snell's law indicates that light of *different wavelengths* is bent at *different angles* when incident on a refracting material. As seen in Figure 22.15, the index of refraction decreases with increasing wavelength. This means that blue light ($\lambda \cong 470$ nm) bends more than red light ($\lambda \cong 650$ nm) when passing into a refracting material.

To understand how dispersion can affect light, let us consider what happens when light strikes a prism, as in Figure 22.16a. A single ray of light that is incident on the prism from the left emerges bent away from its original direction of travel by an angle of δ, called the **angle of deviation.** Now suppose a beam of white light (a combination of all visible wavelengths) is incident on a prism, as in Figure 22.16b. Because of dispersion, the blue component of the incident beam is bent

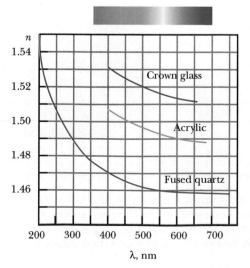

FIGURE 22.15
Variations of index of refraction with wavelength for three materials.

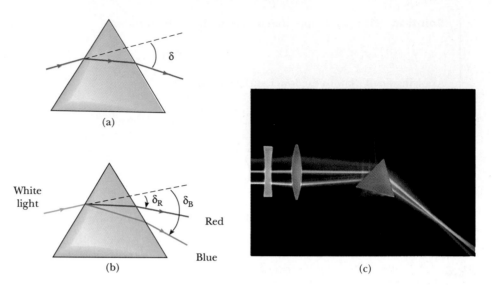

FIGURE 22.16
(a) A prism refracts a light ray and deviates the light through the angle δ. (b) When light is incident on a prism, the blue light is bent more than the red. (c) Light of different colors passes through a prism and two lenses. Note that as the light passes through the prism, different wavelengths are refracted at different angles. *(David Parker/ SPL/ Photo Researchers)*

more than the red component, and the rays that emerge from the second face of the prism fan out in a series of colors known as a visible **spectrum,** as shown in Figure 22.17. These colors, in order of decreasing wavelength, are red, orange, yellow, green, blue, and violet. Clearly, the angle of deviation, δ, depends on wavelength. Violet light ($\lambda \cong 400$ nm) deviates the most, red light ($\lambda \cong 650$ nm) deviates the least, and the remaining colors in the visible spectrum fall between these extremes.

Prisms are often used in an instrument known as a **prism spectrometer,** the essential elements of which are shown in Figure 22.18a. This instrument is commonly used to study the wavelengths emitted by a light source, such as a sodium vapor lamp. Light from the source is sent through a narrow, adjustable slit and lens to produce a parallel, or collimated, beam. The light then passes through the prism and is dispersed into a spectrum. The refracted light is observed through a telescope. The experimenter sees an image of the slit through

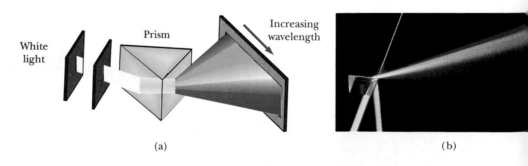

FIGURE 22.17
(a) Dispersion of white light by a prism. Since *n* varies with wavelength, the prism disperses the white light into its various spectral components. (b) Different colors of light that pass through a prism are refracted at different angles because the index of refraction of the glass depends on wavelength. The blue light bends the most; red light bends the least. *(Age/ Peter Arnold, Inc.)*

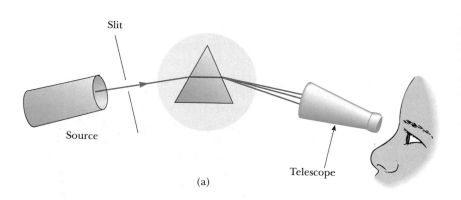

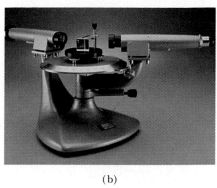

(a)

(b)

FIGURE 22.18
(a) A diagram of a prism spectrometer. The colors in the spectrum are viewed through a telescope. (b) A photograph of a prism spectrometer. *(Courtesy of Central Scientific Company)*

the eyepiece of the telescope. The telescope can be moved or the prism can be rotated in order to view the various wavelengths, which have different angles of deviation. Figure 22.18b shows the type of prism spectrometer used in undergraduate laboratories.

All hot, low-pressure gases emit their own characteristic spectra. Thus, one use of a prism spectrometer is to identify gases. For example, sodium emits only two wavelengths in the visible spectrum: two closely spaced yellow lines. A gas emitting these and only these colors can thus be identified as sodium. Likewise, mercury vapor has its own characteristic spectrum, consisting of four prominent wavelengths—orange, green, blue, and violet lines—along with some wavelengths of lower intensity. The particular wavelengths emitted by a gas serve as "fingerprints" of that gas.

*22.7

THE RAINBOW

The dispersion of light into a spectrum is demonstrated most vividly in nature through the formation of a rainbow, often seen by an observer positioned between the Sun and a rain shower. To understand how a rainbow is formed,

A complete circular rainbow as seen from an airplane with clouds in the background. Can you explain why a complete circle is observed, rather than the half-circle that we normally observe on Earth? *(Ron Chapple/FPG)*

Sunlight

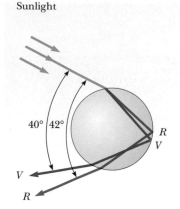

FIGURE 22.19
Refraction of sunlight by a spherical raindrop.

consider Figure 22.19. A ray of light passing overhead strikes a drop of water in the atmosphere and is refracted and reflected as follows. It is first refracted at the front surface of the drop, with the violet light deviating the most and the red light the least. At the back surface of the drop, the light is reflected and returns to the front surface, where it again undergoes refraction as it moves from water into air. The rays leave the drop so that the angle between the incident white light and the returning violet ray is 40°, and the angle between the white light and the returning red ray is 42° (Fig. 22.19). This small angular difference between the returning rays causes us to see the bow.

Now consider an observer viewing a rainbow, as in Figure 22.20. If a raindrop high in the sky is being observed, the red light returning from the drop can reach

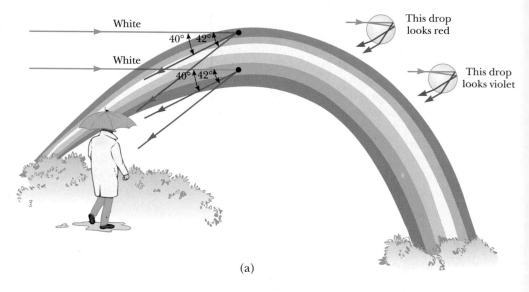

(a)

(b)

FIGURE 22.20
(a) The formation of a rainbow. (b) Dramatic photograph of a rainbow over Niagara Falls in Ontario, Canada. (*John Edwards/ Tony Stone Images*)

the observer because it is deviated the most, but the violet light passes over the observer because it is deviated the least. Hence, the observer sees this drop as being red. Similarly, a drop lower in the sky would direct violet light toward the observer and appear to be violet. (The red light from this drop would strike the ground and not be seen.) The remaining colors of the spectrum would reach the observer from raindrops lying between these two extreme positions. Figure 22.20b shows a rainbow.

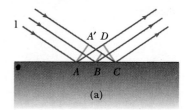

(a)

22.8

HUYGENS' PRINCIPLE APPLIED TO REFLECTION AND REFRACTION

The laws of reflection and refraction were stated earlier in this chapter without proof. We shall now derive these laws using Huygens' principle. For the law of reflection, refer to Figure 22.21a. The line AA' represents a wavefront of the incident light. As ray 3 travels from A' to C, ray 1 reflects from A and produces a spherical wavelet of radius AD. (Recall that the radius of a Huygens wavelet is vt.) Since the two wavelets having radii $A'C$ and AD are in the same medium, they have the same speed, v, and thus $AD = A'C$. Meanwhile, the spherical wavelet centered at B has spread only half as far as the one centered at A, since ray 2 strikes the surface later than ray 1.

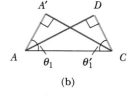

(b)

FIGURE 22.21
(a) Huygens' construction for proving the law of reflection. (b) Triangle ADC is identical to triangle $AA'C$.

From Huygens' principle, we find that the reflected wavefront is CD, a line tangent to all the outgoing spherical wavelets. The remainder of our analysis depends upon geometry, as summarized in Figure 22.21b. Note that the right triangles ADC and $AA'C$ are congruent because they have the same hypotenuse, AC, and because $AD = A'C$. From Figure 22.21b we have

$$\sin \theta_1 = \frac{A'C}{AC} \quad \text{and} \quad \sin \theta_1' = \frac{AD}{AC}$$

Thus,

$$\sin \theta = \sin \theta_1'$$
$$\theta_1 = \theta_1'$$

which is the law of reflection.

Now let us use Huygens' principle and Figure 22.22a to derive Snell's law of refraction. Note that in the time interval Δt, ray 1 moves from A to B and ray 2 moves from A' to C. The radius of the outgoing spherical wavelet centered at A is equal to $v_2 \Delta t$. The distance $A'C$ is equal to $v_1 \Delta t$. Geometric considerations show that angle $A'AC$ equals θ_1 and angle ACB equals θ_2. From triangles $AA'C$ and ACB, we find that

$$\sin \theta_1 = \frac{v_1 \Delta t}{AC} \quad \text{and} \quad \sin \theta_2 = \frac{v_2 \Delta t}{AC}$$

If we divide these two equations, we get

$$\frac{\sin \theta_1}{\sin \theta_2} = \frac{v_1}{v_2}$$

But from Equation 22.4 we know that $v_1 = c/n_1$ and $v_2 = c/n_2$. Therefore,

$$\frac{\sin \theta_1}{\sin \theta_2} = \frac{c/n_1}{c/n_2} = \frac{n_2}{n_1}$$
$$n_1 \sin \theta_1 = n_2 \sin \theta_2$$

which is the law of refraction.

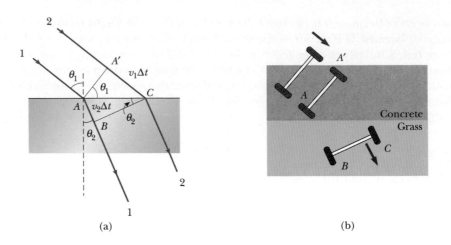

FIGURE 22.22
(a) Huygens' construction for proving the law of refraction. (b) A mechanical analog
of refraction.

A mechanical analog of refraction is shown in Figure 22.22b. The wheels
change direction as they move from a concrete surface to a grass surface.

22.9

TOTAL INTERNAL REFLECTION

An interesting effect called *total internal reflection* can occur when light attempts to
move from a medium with a *high* index of refraction to one with a *lower* index of
refraction. Consider a light beam traveling in medium 1 and meeting the boundary between medium 1 and medium 2, where n_1 is greater than n_2 (Fig. 22.23).

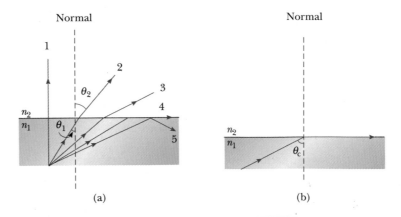

FIGURE 22.23
(a) A ray from a medium with index of refraction n_1 to a medium with index of refraction n_2, where $n_1 > n_2$. As the angle of incidence increases, the angle of refraction increases until θ_2 is 90° (ray 4). For even larger angles of incidence, total internal reflection occurs (ray 5). (b) The angle of incidence producing a 90° angle of refraction is
often called the *critical angle*, θ_c.

Possible directions of the beam are indicated by rays 1 through 5. Note that the refracted rays are bent away from the normal because n_1 is greater than n_2. At some particular angle of incidence, θ_c, called the **critical angle,** the refracted light ray moves parallel to the boundary so that $\theta_2 = 90°$ (Fig. 22.23b). *For angles of incidence greater than* θ_c, the beam is entirely reflected at the boundary, as is ray 5 in Figure 22.23a. This ray is reflected at the boundary as though it had struck a perfectly reflecting surface. It and all rays like it obey the law of reflection; that is, the angle of incidence equals the angle of reflection.

We can use Snell's law to find the critical angle. When $\theta_1 = \theta_c$, $\theta_2 = 90°$ and Snell's law (Eq. 22.8) gives

$$n_1 \sin \theta_c = n_2 \sin 90° = n_2$$

$$\sin \theta_c = \frac{n_2}{n_1} \qquad \text{for } n_1 > n_2 \qquad \text{[22.9]}$$

Note that this equation can be used only when n_1 is greater than n_2. That is,

> **Total internal reflection occurs only when light attempts to move from a medium of high index of refraction to a medium of lower index of refraction.**

If n_1 were less than n_2, Equation 22.9 would give $\sin \theta_c > 1$, which is an absurd result because the sine of an angle can never be greater than unity.

When medium 2 is air, the critical angle is small for substances with large indices of refraction, such as diamond, where $n = 2.42$ and $\theta_c = 24.0°$. By comparison for crown glass, $n = 1.52$ and $\theta_c = 41.0°$. This property, combined with proper faceting, causes diamonds and crystal glass to sparkle brilliantly.

One can use a prism and the phenomenon of total internal reflection to alter the direction of travel of a light beam. Figure 22.24 illustrates two such possibilities. In one case the light beam is deflected by 90° (Fig. 22.24a), and in the second case the path of the beam is reversed (Fig. 22.24b). A common application of total internal reflection is a submarine periscope. In this device, two prisms are arranged as in Figure 22.24c so that an incident beam of light follows the path shown and the user can "see around corners."

This photograph shows nonparallel light rays entering a glass prism. The bottom two rays undergo total internal reflection at the longest side of the prism. The top three rays are refracted at the longest side as they leave the prism. *(Courtesy of Henry Leap and Jim Lehman)*

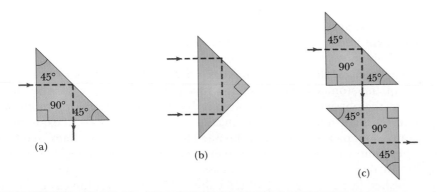

FIGURE 22.24

Internal reflection in a prism. (a) The ray is deviated by 90°. (b) The direction of the ray is reversed. (c) Two prisms used as a periscope.

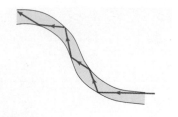

FIGURE 22.25
Light travels in a curved transparent rod by multiple internal reflections.

EXAMPLE 22.7   A View from the Fish's Eye

(a) Find the critical angle for a water-air boundary if the index of refraction of water is 1.33.

Solution   Applying Equation 22.9, we find the critical angle to be

$$\sin \theta_c = \frac{n_2}{n_1} = \frac{1.00}{1.33} = 0.752$$

$$\theta_c = \boxed{48.8°}$$

(b) Use the results of (a) to predict what a fish will see if it looks up toward the water surface at angles of 40.0°, 48.8°, and 60.0°.

Reasoning and Solution   Because the path of a light ray is reversible, the fish can see out of the water if it looks toward the surface at an angle less than the critical angle. Thus, at 40.0°, the fish can see into the air above the water. At an angle of 48.8°, the critical angle for water, the light that reaches the fish has to skim along the water surface before being refracted to the fish's eye. At angles greater than the critical angle, the light reaching the fish comes via internal reflection at the surface. Thus, at 60.0°, the fish sees a reflection of some object on the bottom of the pool.

FIBER OPTICS

Another interesting application of total internal reflection is the use of glass or transparent plastic rods to "pipe" light from one place to another. As indicated in Figure 22.25, light is confined to traveling within the rods, even around gentle curves, as a result of successive internal reflections. Such a "light pipe" can be flexible if thin fibers are used rather than thick rods. If a bundle of parallel fibers is used to construct an optical transmission line, images can be transferred from one point to another.

This technique is used in an industry known as *fiber optics*. Very little light intensity is lost in these fibers as a result of reflections on the sides. Any loss of intensity is due essentially to reflections from the two ends and absorption by the fiber material. Fiber-optic devices are particularly useful for viewing images produced at inaccessible locations. Physicians often use fiber-optic cables to aid in the diagnosis and correction of certain medical problems without the intrusion of major surgery. For example, a fiber-optic cable can be threaded through the esophagus and into the stomach to look for ulcers. In this application, the cable actually consists of two fiber-optic lines, one to transmit a beam of light into the stomach for illumination and the other to allow this light to be transmitted out of the stomach. The resulting image can, in some cases, be viewed directly by the physician but most often is displayed on a television monitor or captured on film. In a similar fashion, the cables can be used to examine the colon or to do repair work without the need for large incisions. Damaged knees and other joints can sometimes be repaired using a process called arthroscopic surgery. In this technique, a small incision is made into the joint. Repair is accomplished by inserting a small fiber-optic cable through the cut to provide illumination and then trimming cartilage or damaged tissue with a small knife at the end of a second cable. The field of fiber optics is also finding increasing use in telecommunications, since the fibers can carry much higher volumes of tele-

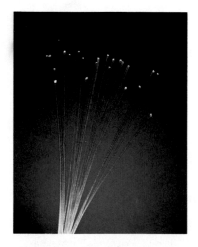

Strands of glass optical fibers are used to carry voice, video, and data signals in telecommunication networks. Typical fibers have diameters of 60 μm.
(© *Richard Megna 1983, Fundamental Photographs*)

phone calls and other forms of communication than electrical wires because of the higher frequency of light.

SUMMARY

Huygens' principle states that all points on a wavefront are point sources for the production of spherical secondary waves called wavelets. These wavelets propagate outward at a speed characteristic of waves in a particular medium. After some time has elapsed, the new position of the wavefront is the surface tangent to the wavelets.

The **index of refraction** of a material, n, is defined as

$$n \equiv \frac{c}{v} \qquad \text{[22.4]}$$

where c is the speed of light in a vacuum and v is the speed of light in the material. The index of refraction of a material is

$$n = \frac{\lambda_0}{\lambda_n} \qquad \text{[22.7]}$$

where λ_0 is the wavelength of the light in vacuum and λ_n is its wavelength in the material.

The **law of reflection** states that a wave reflects from a surface so that the *angle of reflection, θ_1',* equals the *angle of incidence, θ_1.*

The **law of refraction,** or **Snell's law,** states that

$$n_1 \sin \theta_1 = n_2 \sin \theta_2 \qquad \text{[22.8]}$$

Total internal reflection can occur when light attempts to move from a material with a high index of refraction to one with a lower index of refraction. The *maximum angle of incidence, θ_c,* for which light can move from a medium with index n_1 into a medium with index n_2, where n_1 is greater than n_2, is called the **critical angle** and is given by

$$\sin \theta_c = \frac{n_2}{n_1} \qquad \text{for } n_1 > n_2 \qquad \text{[22.9]}$$

ADDITIONAL READING

W. S. Boyle, "Light-Wave Communications," *Sci. American,* August 1977, p. 40.

W. Bragg, *The Universe of Light,* Dover, 1959.

A. B. Fraser and W. H. Mach, "Mirages," *Sci. American,* January 1976, p. 102.

N. S. Kapany, "Fiber Optics," *Sci. American,* November 1960, p. 72.

E. A. Lacy, *Fiber Optics,* Englewood Cliffs, N.J., Prentice-Hall, 1982.

E. H. Land, "The Retinex Theory of Color Vision," *Sci. American,* December 1977, p. 108.

"Light," *Sci. American,* September 1968 (entire issue).

D. F. Mandoli and W. R. Briggs, "Fiber Optics in Plants," *Sci. American,* August 1984, p. 90.

E. W. Stark, "Diffuse Reflection: Uses That Affect Our Lives," *The Physics Teacher,* March 1986, p. 144.

A. C. S. van Heel and C. H. F. Velzel, *What Is Light,* World University Library, New York, McGraw-Hill, 1968, Chap. 1.

S. Williamson and H. Cummins, *Light and Color in Nature and Art,* New York, John Wiley and Son, 1983.

CONCEPTUAL QUESTIONS

Example   Why do astronomers observing distant galaxies talk about looking backward in time?

Reasoning   To understand this, consider light traveling to us from the Sun. Since the speed of light is 3×10^8 m/s, and the distance from the Sun to Earth is about 1.5×10^{11} m, a light signal takes 500 s (about 8 min) to travel from the Sun to the Earth. Thus, when we view the Sun, the light we see is light that left the Sun 8 min earlier. The effect is even greater for more distant objects. In fact, when you look at objects deep in space, the light we see may have left them millions, or perhaps even billions of years ago.

Example   Suppose you are told that only two colors of light (X and Y) are sent through a glass prism and that X is bent more than Y. Which color travels more slowly in the prism?

Reasoning   The color traveling slowest is bent the most. Thus, X travels more slowly in the glass prism.

Example   As light travels from vacuum ($n = 1$) to a medium such as glass ($n > 1$), does its wavelength change? Does its frequency change? Does its velocity change?

Reasoning   The wavelength decreases since it depends on the index of refraction of the medium ($\lambda_n = \lambda_0/n$). The frequency does not change; it depends only on the nature of the light source. The velocity decreases since it also depends on the index of refraction of the medium according to the relation $v = c/n$.

Example   Under certain circumstances, sound can be heard from extremely far away. This frequently happens over a body of water, where the air near the water surface is cooler than the air at higher altitudes. Explain how the refraction of sound waves could increase the distance over which sound can be heard.

Reasoning   A portion of the sound is bent back toward the Earth by refraction. This means that the sound can reach the listener by this path as well as by a direct path. Thus, the sound is louder.

Example   The level of water in a clear, colorless glass container is easily observed with the naked eye. The level of liquid helium in a clear glass vessel is extremely difficult to see with the naked eye. Explain.

Reasoning   The index of refraction of water is 1.333, which is quite different from that of air (about 1). On the other hand, the index of refraction of liquid helium is closer to that of air. Consequently, viewing a vessel of liquid helium is similar to viewing an empty vessel.

1. Explain why we see lightning before we hear thunder.
2. Would Huygens' wavelets be spherical in a nonisotropic medium (a medium in which the wave speed depends on the direction of the wave propagation)? Discuss.
3. A solar eclipse occurs when the Moon gets between the Earth and the Sun. Use a diagram to show why some areas of the Earth see a total eclipse, other areas see a partial eclipse, and most areas see no eclipse.
4. Some department stores have their windows slanted slightly inward at the bottom. This is to decrease the glare from streetlights or the Sun, which would make it difficult for shoppers to see the display inside. Sketch a light ray reflecting off such a window to show how this technique works.
5. You can make a corner reflector by placing three plane mirrors in the corner of a room where the ceiling meets the walls. Show that no matter where you are in the room, you can see yourself reflected in the mirrors—upside down.
6. Several corner reflectors were left on the Moon's Sea of Tranquility by the astronauts of Apollo 11. How can scientists utilize a laser beam sent from Earth even today to determine the precise distance from the Earth to the Moon?
7. The rectangular aquarium sketched in Figure 22.26 contains only one goldfish. When the fish is near a corner of the aquarium and is viewed along a direction that makes an equal angle with two adjacent faces, one observes two fish that are mirror images of each other, as shown. Explain this observation.

FIGURE 22.26   (Question 7)

8. Does a light ray traveling from one medium into another always bend toward the normal as in Figure 22.13?
9. Explain why an oar partially in water appears to be bent.
10. A laser beam passing through a nonhomogeneous sugar solution is observed to follow a curved path. Explain.
11. In Figure 22.27, light from a helium-neon laser beam

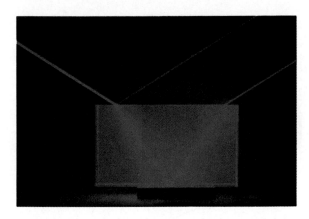

FIGURE 22.27 (Questions 11 and 12) *(Courtesy of Henry Leap and Jim Lehman)*

($\lambda = 632.8$ nm) is incident on a block of Lucite. Can you identify the incident, reflected, and refracted rays? From this photograph, estimate the index of refraction of Lucite at this wavelength.

12. Suppose blue light were used instead of red in the experiment shown in Figure 22.27. Would the refracted beam be bent at a larger or smaller angle?

13. What are the conditions for the production of a mirage? On a hot day, what is it that we are seeing when we observe "water on the road"?

14. Sound waves have much in common with light waves, including the properties of reflection and refraction. Give examples of such phenomena for sound waves.

15. Explain why a diamond shows more "sparkle" than a glass crystal of the same shape and size.

16. A color-television image is formed by superimposing three separate images of red, green, and blue phosphors. Discuss the effect of viewing the composite image through a prism.

17. Why does a diamond show flashes of color when observed under ordinary white light?

18. Why does the arc of a rainbow appear with red colors on top and violet hues on the bottom?

19. How is it possible that a complete *circular* rainbow can sometimes be seen from an airplane?

20. Is it possible to have total internal reflection for light incident from air on water? Explain.

21. What would happen to a light pipe if the coating separating individual fibers were to wear off?

22. Redesign the periscope of Figure 22.24c so that it can show you where you have been rather than where you are going.

23. As visible light travels from one medium to another, does its "color" change? Explain.

24. The following photograph shows multiple reflections that were created with two plane mirrors. What can you conclude about the relative orientation of the mirrors?

(Question 24) *(Richard Megna 1986/ Fundamental Photographs)*

PROBLEMS

Section 22.1 The Nature of Light

Section 22.2 Measurements of the Speed of Light

1. As a result of his observations, Roemer concluded that the time interval between successive eclipses of the moon Io by the planet Jupiter increased by 22 minutes during a six-month period as the Earth moved from a point in its orbit on the side of the Sun nearer Jupiter to a point on the side opposite Jupiter. Using 1.5×10^8 km as the average radius of the Earth's orbit about the Sun, calculate the speed of light from these data.

2. During the Apollo XI Moon landing, a highly reflecting screen was erected on the Moon's surface. The speed of light is found by measuring the time it takes a laser beam to travel from Earth, reflect from the screen, and return to Earth. If this interval is measured to be 2.51 s, what is the measured speed of light? Take the center-to-center distance from Earth to Moon to be 3.84×10^8 m, and do not neglect the sizes of the Earth and Moon.

3. The Fizeau experiment is performed so that the round-trip distance for the light is 40 m. (a) Find the two lowest speeds of rotation that allow the

□ indicates problems that have full solutions available in the Student Solutions Manual and Study Guide.

light to pass through the notches. Assume that the wheel has 360 teeth and that the speed of light is 3.00×10^8 m/s. (b) Repeat for a round-trip distance of 4000 m.

4. Albert A. Michelson very carefully measured the speed of light using an alternative version of the technique developed by Fizeau. Figure 22.28 shows the approach he used. Light was reflected from one face of a rotating eight-sided mirror toward a stationary mirror 35 km away. At certain rates of rotation, the returning beam of light was directed toward the eye of an observer as shown. (a) What minimum angular speed must the rotating mirror have in order that side A will have rotated to position B, causing the light to be reflected to the eye? (b) What is the next highest angular velocity that will enable the source of light to be seen?

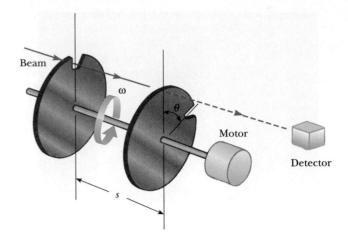

FIGURE 22.29　(Problem 5)

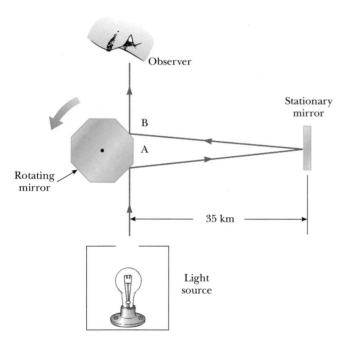

FIGURE 22.28　(Problem 4)

5. Figure 22.29 shows an apparatus used to measure the speed distribution of gas molecules. It consists of two slotted rotating disks separated by a distance s, with the slots displaced by the angle θ. Suppose the speed of light is measured by sending a light beam toward the right disk of this apparatus. (a) Show that a light beam will be seen in the detector (that is, will make it through both slots) only if its speed is given by $c = s\omega/\theta$, where

ω is the angular speed of the disks and θ is measured in radians. (b) What is the measured speed of light if the distance between the two slotted rotating disks is 2.5 m, the slot in the second disk is displaced $\frac{1}{60}$ of one degree from the slot in the first disk, and the disks are rotating at 5555 rev/s?

Section 22.4　Reflection and Refraction

Section 22.5　The Law of Refraction

6. You are standing with a mirror at the center of a giant clock. Someone at 12 o'clock shines a beam of light toward you, and you want to use the mirror to reflect the beam toward an observer at 5 o'clock. What should the angle of incidence be to achieve this?

7. The angle between the two mirrors in Figure 22.30 is a right angle. The beam of light in the vertical plane P strikes mirror 1 as shown. (a) Determine the distance the reflected light beam travels be-

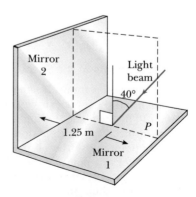

FIGURE 22.30　(Problem 7)

fore striking mirror 2. (b) In what direction does the light beam travel after being reflected from mirror 2?

8. How many times will the incident beam shown in Figure 22.31 be reflected by each of the parallel mirrors?

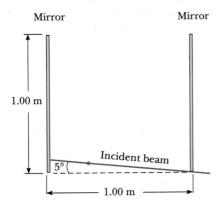

FIGURE 22.31   (Problem 8)

9. A beam of light enters a glass of water at an angle of 36° with the vertical. What is the angle between the refracted ray and the vertical?

10. Find the speeds of light in (a) flint glass, (b) water, and (c) zircon.

11. The reflecting surfaces of two intersecting plane mirrors are at an angle of θ ($0° < \theta < 90°$), as in Figure 22.32. If a light ray strikes the horizontal mirror, show that the emerging ray will intersect the incident ray at an angle of $\beta = 180° - 2\theta$.

12. Light of wavelength 436 nm in air enters a fishbowl filled with water, then exits through the crown-glass wall of the container. Find the wavelengths of the light (a) in the water and (b) in the glass.

13. A light ray follows the path shown in Figure 22.33.

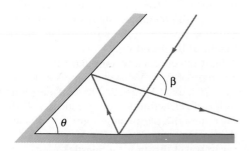

FIGURE 22.32   (Problem 11)

Given that $n_1 = 1.7$, $n_2 = 1.5$, $n_3 = 1.3$, and $\theta = 60°$, determine the angles θ_1, θ_2, θ_3, and θ_4. Assume that the index of refraction for air is 1.0 and path AB is parallel to the base of the figure.

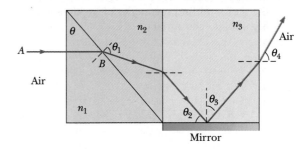

FIGURE 22.33   (Problem 13)

14. A beam of light strikes the surface of mineral oil at an angle of 23.1° with the normal to the surface. If the light travels at 2.17×10^8 m/s through the oil, what is the angle of refraction?

15. A narrow beam of sodium yellow light ($\lambda_0 = 589$ nm) is incident from air on a smooth surface of water at an angle of $\theta_1 = 35.0°$. Determine the angle of refraction, θ_2, and the wavelength of the light in water.

16. The laws of refraction and reflection are the same for sound as for light. The speed of sound is 340 m/s in air and 1510 m/s in water. If a sound wave traveling in air approaches a plane water surface at an angle of incidence of 12.0°, what is the angle of refraction?

17. A narrow beam of ultrasonic waves reflects off the liver tumor in Figure 22.34. If the speed of the wave is 10.0% less in the liver than in the surrounding medium, determine the depth of the tumor.

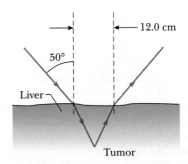

FIGURE 22.34   (Problem 17)

18. A ray of light is incident on the surface of a block of clear ice at an angle of 40.0° with the normal. Part of the light is reflected and part is refracted.

Find the angle between the reflected and refracted light.

19. A ray of light strikes a flat 2.00-cm-thick block of glass ($n = 1.50$) at an angle of 30.0° with the normal (Fig. 22.35). Trace the light beam through the glass and find the angles of incidence and refraction at each surface.

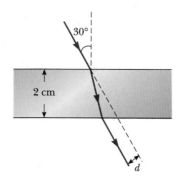

FIGURE 22.35   (Problems 19 and 20)

20. When the light ray in Problem 19 passes through the glass block, it is shifted laterally by a distance d (Fig. 22.35). Find the value of d.

21. A material with index of refraction $n = 2.0$ is in the shape of a quarter circle of radius $R = 10$ cm and is surrounded by a vacuum (Fig. 22.36). A light ray, parallel to the base of the material, is incident from the left at a distance of $L = 5.0$ cm above the base and emerges out of the material at the angle θ. Determine the value of θ.

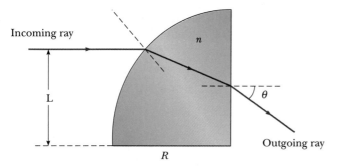

FIGURE 22.36   (Problem 21)

22. A flashlight on the bottom of a 4.00-m-deep swimming pool sends a ray upward and at an angle so that the ray strikes the surface of the water 2.00 m from the point directly above the flashlight. What angle (in air) does the emerging ray make with the water's surface?

23. A ray of light strikes the midpoint of one face of an equiangular glass prism ($n = 1.5$) at an angle

of incidence of 30°. (a) Trace the path of the light ray through the glass, and find the angles of incidence and refraction at each surface. (b) If a small fraction of light is also reflected at each surface, find the angles of incidence and reflection at these surfaces.

24. The light beam shown in Figure 22.37 makes an angle of 20.0° with the normal line NN' in the linseed oil. Determine the angles θ and θ'. (The refractive index for linseed oil is 1.48.)

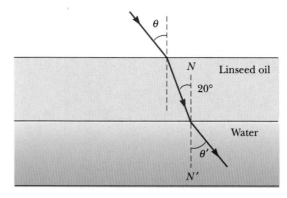

FIGURE 22.37   (Problem 24)

25. Three sheets of plastic have unknown indices of refraction. Sheet 1 is placed on top of sheet 2, and a laser beam is directed onto the sheets from above so that it strikes the interface at an angle of 26.5° with the normal. The refracted beam in sheet 2 makes an angle of 31.7° with the normal. The experiment is repeated with sheet 3 on top of sheet 2 and, with the same angle of incidence, the refracted beam makes an angle of 36.7° with the normal. If the experiment is repeated again with sheet 1 on top of sheet 3, what is the expected angle of refraction in sheet 3? Assume the same angle of incidence.

26. A submarine is 300 m horizontally out from the shore and 100 m beneath the surface of the water. A laser beam is sent from the sub so that it strikes the surface of the water at a point 210 m from the shore. If the beam just strikes the top of a building standing directly at the water's edge, find the height of the building.

27. A light ray enters a rectangular block of plastic at an angle of $\theta_1 = 45°$ and emerges at an angle of $\theta_2 = 76°$, as in Figure 22.38. (a) Determine the index of refraction for the plastic. (b) If the light ray enters the plastic at a point $L = 50$ cm from the bottom edge, how long does it take the light ray to travel through the plastic?

28. A cylindrical cistern, constructed below ground level, is 3.0 m in diameter and 2.0 m deep and is

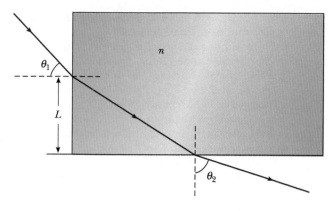

FIGURE 22.38 (Problem 27)

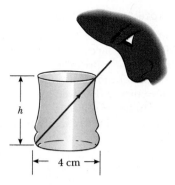

FIGURE 22.40 (Problem 30)

filled to the brim with a liquid whose index of refraction is 1.5. A small object rests on the bottom of the cistern at its center. How far from the edge of the cistern can a girl whose eyes are 1.2 m from the ground stand and still see the object?

29. A light ray traveling in air is incident normally on the base of a right-angle prism of index of refraction $n = 1.5$. The ray follows the path shown in Figure 22.39. If $\theta = 60°$ and the base of the prism is mirrored, determine the angle, ϕ, made by the outgoing ray with the normal to the right face of the prism.

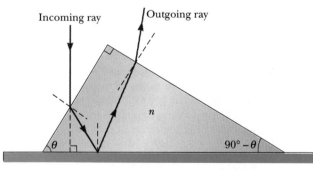

FIGURE 22.39 (Problem 29)

30. A drinking glass is 4.0 cm wide at the bottom, as shown in Figure 22.40. When an observer's eye is positioned as shown, the observer sees the edge of the bottom of the glass. When this glass is filled with water, the observer sees the center of the bottom of the glass. Find the height of the glass.

31. A cylindrical tank with an open top has a diameter of 3.0 m and is completely filled with water. When the setting Sun reaches an angle of 28° above the horizon, sunlight ceases to illuminate the bottom of the tank. How deep is the tank?

Section 22.6 Dispersion and Prisms

32. A certain kind of glass has an index of refraction of 1.650 for blue light of wavelength 430 nm and an index of refraction of 1.615 for red light of wavelength 680 nm. If a beam containing these two colors is incident at an angle of 30.00° on a piece of this glass, what is the angle between the two beams inside the glass?

33. Light of wavelength 400 nm is incident at an angle of 45° on acrylic and is refracted as it passes into the material. What wavelength of light incident on fused quartz at an angle of 45° would be refracted at exactly this same angle? (See Fig. 22.15.)

34. The index of refraction for red light in water is 1.331, and that for blue light is 1.340. If a ray of white light enters the water at an angle of incidence of 83.00°, what are the underwater angles of refraction for the blue and red components of the light?

35. The index of refraction for violet light in silica flint glass is 1.66, and that for red light is 1.62. What is the angular dispersion of visible light passing through an equilateral prism of apex angle 60.0° if the angle of incidence is 50.0°? (See Fig. 22.41.)

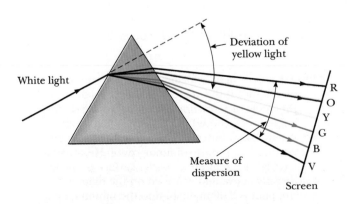

FIGURE 22.41 (Problem 35)

Section 22.9   Total Internal Reflection

36. As shown in Figure 22.42, a light ray is incident normally on one face of a 30°-60°-90° block of dense flint glass (a prism) that is immersed in water. (a) Determine the exit angle, θ_4, of the ray. (b) A substance is dissolved in the water to increase the index of refraction. At what value of n_2 does total internal reflection cease at point P?

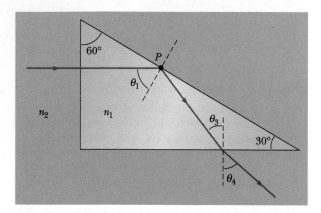

FIGURE 22.42   (Problem 36)

37. Calculate the critical angles for the following materials when surrounded by air: (a) zircon, (b) fluorite, (c) ice. Assume that $\lambda = 589$ nm.
38. A laser beam strikes one end of a slab of material, as in Figure 22.43. The index of refraction of the slab is 1.48. Determine the number of internal reflections of the beam before it emerges from the opposite end of the slab.

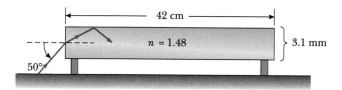

FIGURE 22.43   (Problem 38)

39. A beam of light is incident from air on the surface of a liquid. If the angle of incidence is 30.0° and the angle of refraction is 22.0°, find the critical angle for the liquid when surrounded by air.
40. A jewel thief hides a diamond by placing it on the bottom of a public swimming pool. He places a circular piece of wood on the surface of the water directly above and centered on the diamond. If the pool is 2.00 m deep, find the minimum diame-

ter of the piece of wood that would prevent the diamond from being seen.
41. From within a diamond, a light ray is incident on the interface between the diamond and air. What is the critical angle for total internal reflection? Use Table 22.1. (The smallness of θ_c for diamond means that light is easily "trapped" within a diamond and eventually emerges from the many cut faces after many internal reflections; this makes a diamond more "brilliant" than stones with smaller n and larger θ_c.)
42. The light beam in Figure 22.44 strikes surface 2 at the critical angle. Determine the angle of incidence, θ_i.

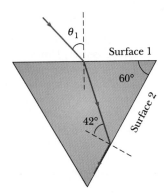

FIGURE 22.44   (Problem 42)

43. A light ray is incident normally to the long face (the hypotenuse) of a 45°-45°-90° prism surrounded by air, as shown in Figure 22.24b. Calculate the minimum index of refraction of the prism for which the ray will follow the path shown.
44. Three adjacent faces (that all share a corner) of a plastic cube of index of refraction n are painted black, with a clear spot at the painted corner serving as a source of diverging rays when light comes through the clear spot. Show that a ray from this corner to the center of a clear face is totally reflected if $n \geq \sqrt{3}$.
45. A large Lucite cube ($n = 1.59$) has a small air bubble (a defect in the casting process) below the surface of one of its faces. When a penny (diameter 1.90 cm) is placed directly over the bubble, it cannot be seen at any angle from the opposite face of the cube. However, when a dime (diameter 1.75 cm) is placed directly over the bubble, it can be seen from the opposite face of the cube. What is the range of possible depths of the air bubble beneath the surface?
46. A plastic light pipe has an index of refraction of

1.53. For total internal reflection, what is the minimum angle of incidence to the wall of the pipe if the pipe is in (a) air? (b) water?

47. A light pipe consists of a central strand of material surrounded by an outer coating. The interior portion of the pipe has an index of refraction of 1.60. If all rays striking the interior walls of the pipe with incident angles greater than 59.5° are subject to total internal reflection, what is the index of refraction of the coating?

48. Determine the maximum angle, θ, for which the light rays incident on the end of the pipe in Figure 22.45 are subject to total internal reflection along the walls of the pipe. Assume that the pipe has an index of refraction of 1.36 and that the outside medium is air.

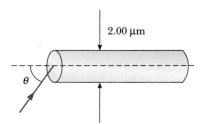

2.00 μm

θ

FIGURE 22.45   (Problem 48)

ADDITIONAL PROBLEMS

49. Light is incident normally on a 1.00-cm layer of water that lies on top of a flat Lucite plate with a thickness of 0.500 cm. How much more time is required for light to pass through this double layer than is required to traverse the same distance in air? ($n_{\text{Lucite}} = 1.59$)

50. A layer of ice floats on water. If light is incident on the upper surface of the ice at an angle of incidence of 30.0°, what is the angle of refraction in the water?

51. Figure 22.46 shows the path of a beam of light through several layers of different indices of refraction. (a) If $\theta_1 = 30.0°$, what is the angle, θ_2, of the emerging beam? (b) What must the incident angle, θ_1, be in order to have total internal reflection at the surface between the $n = 1.20$ medium and the $n = 1.00$ medium?

52. Light is incident on the surface of a prism, $n = 1.8$, as shown in Figure 22.24a. If the prism is surrounded by a fluid, what is the maximum index of refraction of the fluid that will still cause total internal reflection?

53. A cylindrical material of radius $R = 2.00$ m has a mirrored surface on its right half, as in Figure 22.47. A light ray traveling in air is incident on the left side of the cylinder. If the incident light ray and exiting light ray are parallel and $d = 2.00$ m, determine the index of refraction of the material.

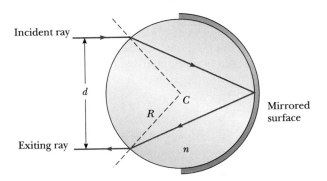

Incident ray

d

C

R

Mirrored surface

Exiting ray

n

FIGURE 22.47   (Problem 53)

54. A narrow beam of light is incident from air onto a glass surface with index of refraction 1.56. Find the angle of incidence for which the corresponding angle of refraction is one-half the angle of incidence. (*Hint:* You might want to use the trigonometric identity sin $2\theta = 2$ sin θ cos θ.)

55. A thick plate of flint glass ($n = 1.66$) rests on top of a thick plate of transparent acrylic ($n = 1.50$). A beam of light is incident on the top surface of the flint glass at an angle θ_i. The beam passes through the glass and the acrylic and emerges from the acrylic at an angle of 40.0° with respect to the normal. Calculate the value of θ_i. A sketch of the light path through the two plates of refracting material would be helpful.

56. One technique to measure the angle of a prism is shown in Figure 22.48. A parallel beam of light is directed on the apex of the prism so that the

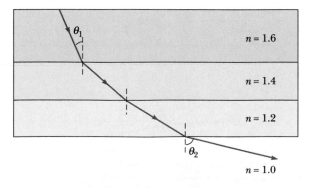

θ_1

$n = 1.6$

$n = 1.4$

$n = 1.2$

θ_2

$n = 1.0$

FIGURE 22.46   (Problem 51)

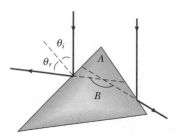

FIGURE 22.48   (Problem 56)

beam reflects from opposite faces of the prism. Show that the angular separation of the two beams is given by $B = 2A$.

57. A. H. Pfund's method for measuring the index of refraction of glass is illustrated in Figure 22.49. One face of a slab of thickness t is painted white, and a small hole scraped clear at point P serves as a source of diverging rays when the slab is illuminated from below. Ray PBB' strikes the clear surface at the critical angle and is totally reflected, as are rays such as PCC'. Rays such as PAA' emerge from the clear surface. On the painted surface there appears a dark circle of diameter d, surrounded by an illuminated region, or halo. (a) Derive a formula for n in terms of the measured quantities d and t. (b) What is the diameter of the dark circle if $n = 1.52$ for a slab 0.600 cm thick? (c) If white light is used, the critical angle depends on color due to dispersion. Is the inner edge of the white halo tinged with red light or violet light? Explain.

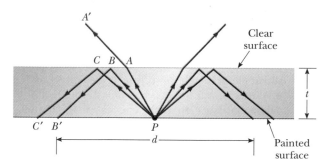

FIGURE 22.49   (Problem 57)

58. Repeat Example 22.2 for the case in which the angle is 90.0°. Show that the ray of light is always reflected from the second mirror so that it travels opposite the original direction.

59. Light of wavelength λ_0 in vacuum has a wavelength of 438 nm in water and a wavelength of 390 nm in benzene. What is the index of refraction of water relative to benzene at wavelength λ_0?

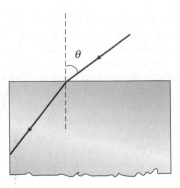

FIGURE 22.50   (Problem 60)

60. A light ray of wavelength 589 nm is incident at an angle θ on the top surface of a block of polystyrene, as shown in Figure 22.50. (a) Find the maximum value of θ for which the refracted ray will undergo total internal reflection at the left vertical face of the block. (b) Repeat the calculation for the case in which the polystyrene block is immersed in water. (c) What happens if the block is immersed in carbon disulfide?

61. For this problem, refer to Figure 22.16. For various angles of incidence, it can be shown that the angle δ is a minimum when the ray passes through the glass so that the ray is parallel to the base of the prism. A measurement of this minimum angle of deviation enables one to find the index of refraction of the prism material. Show that n is given by the expression

$$n = \frac{\sin\left[\frac{1}{2}\left(A + \delta_{\min}\right)\right]}{\sin\left(\dfrac{A}{2}\right)}$$

where A is the apex angle of the prism.

62. A light ray is incident on a prism and refracted at the first surface, as shown in Figure 22.51. Let ϕ represent the apex angle of the prism and n its index of refraction. Find, in terms of n and ϕ, the smallest allowed value of the angle of incidence at the first surface for which the refracted ray will not undergo internal reflection at the second surface.

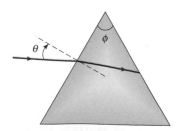

FIGURE 22.51   (Problem 62)

MIRRORS AND LENSES

Three beams of light are incident on a biconvex lens. Part of each beam is reflected, and part refracted, by the glass lens at the left air-glass interface. The refracted beams entering the lens are then partially reflected and partially refracted at the opposite glass-air interface. How do you explain the fact that most of the red beam is reflected, while the yellow and blue beams are mostly refracted, at the second glass-air interface? (Richard Megna, Fundamental Photographs)

This chapter is concerned with the formation of images when plane and spherical waves fall on plane and spherical surfaces. Images can be formed by either reflection or refraction. Mirrors and lenses form images in both ways. In our study of mirrors and lenses, we continue to use the ray approximation and to assume that light travels in straight lines (in other words, we ignore diffraction). This chapter completes our study of geometric optics. In Chapter 25 we shall examine the construction and properties of some optical instruments that use mirrors and lenses.

23.1
PLANE MIRRORS

We begin our investigation by examining the simplest possible mirror, the plane mirror. Consider a point source of light placed at O in Figure 23.1, a distance of p in front of a plane mirror. The distance p is called the **object distance.** Light rays

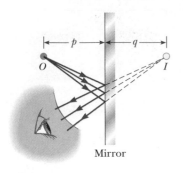

FIGURE 23.1
An image formed by reflection from a plane mirror. The image point, I, is behind the mirror at distance q, which is equal to the object distance, p.

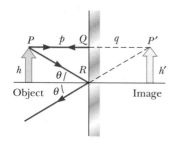

FIGURE 23.2
A geometric construction to locate the image of an object placed in front of a plane mirror. Because the triangles PQR and $P'QR$ are identical, $p = q$, and $h = h'$.

leave the source and are reflected from the mirror. After reflection, the rays diverge (spread apart), but they appear to the viewer to come from a point, I, behind the mirror. Point I is called the **image** of the object at O. Regardless of the system under study, *images are formed at the point where rays of light actually intersect or at which they appear to originate*. Since the rays in Figure 23.1 appear to originate at I, which is a distance of q behind the mirror, this is the location of the image. The distance q is called the **image distance**.

Images are classified as real or virtual. A *real image* is one in which light actually intersects, or passes through, the image point; a *virtual image* is one in which the light does not really pass through the image point but appears to come (diverge) from that point. The image formed by the plane mirror in Figure 23.1 is a virtual image. In fact, the images seen in plane mirrors are always virtual for real objects. Real images can be displayed on a screen (as at a movie), but virtual images cannot.

We shall examine some of the properties of the images formed by plane mirrors by using the simple geometric techniques shown in Figure 23.2. To find out where an image is formed, it is always necessary to follow at least two rays of light as they reflect from the mirror. One of those rays starts at P, follows a horizontal path, PQ, to the mirror, and reflects back on itself. The second ray follows the oblique path PR and reflects as shown. An observer to the left of the mirror would trace the two reflected rays back to the point from which they appear to have originated, that is, point P'. A continuation of this process for points other than P on the object would result in a virtual image (drawn as a yellow arrow) to the right of the mirror. Since triangles PQR and $P'QR$ are identical, $PQ = P'Q$. Hence, we conclude that *the image formed by an object placed in front of a plane mirror is as far behind the mirror as the object is in front of the mirror*. Geometry also shows that the object height, h, equals the image height, h'. **Lateral magnification,** M, is defined as follows:

$$M \equiv \frac{\text{image height}}{\text{object height}} = \frac{h'}{h} \qquad \text{[23.1]}$$

This is a general definition of the magnification of any type of mirror. For a plane mirror, $M = 1$ because $h' = h$.

The observer assumes that the image formed by a plane mirror has right-left reversal. This can be seen by standing in front of a mirror and raising your right hand. The image you see raises its left hand. Likewise, your hair appears to be parted on the opposite side and a mole on your right cheek appears to be on your left cheek.

In summary, the image formed by a plane mirror has the following properties:

1. The image is as far behind the mirror as the object is in front.
2. The image is unmagnified, virtual, and erect. (By *erect* we mean that, if the object arrow points upward as in Figure 23.2, so does the image arrow. The opposite of an erect image is an inverted image.)

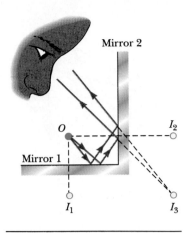

EXAMPLE 23.1   Multiple Images Formed by Two Mirrors

Two plane mirrors are at right angles to each other, as in Figure 23.3, and an object is placed at point O. In this situation, multiple images are formed. Locate the positions of these images.

Reasoning and Solution   The image of the object is at I_1 in mirror 1 and at I_2 in mirror 2. In addition, a third image is formed at I_3 that is considered to be the image of I_1 in mirror 2 or, equivalently, the image of I_2 in mirror 1. That is, the image at I_1 (or I_2) serves as the object for I_3. When viewing I_3, note that the rays reflect twice after leaving the object at O.

Exercise   Sketch the rays that correspond to viewing the images at I_1 and I_2, and show that the light is reflected only once in each of these cases.

FIGURE 23.3
(Example 23.1)   When an object is placed in front of two perpendicular mirrors as shown, three images are formed.

Most rearview mirrors in cars have a day setting and a night setting. The night setting greatly diminishes the intensity of the image so that lights from trailing cars will not blind the driver. To understand how such a mirror works, consider Figure 23.4. The mirror is a wedge of glass with a reflecting mirror on the back side. When the mirror is in the day setting, as in Figure 23.4a, light from an object behind the car strikes the mirror at point 1. Most of the light enters the wedge, is refracted, and reflects from the back of the mirror to return to the front surface, where it is refracted again as it re-enters the air as ray B (for *bright*). In addition, a small portion of the light is reflected at the front surface, as indicated by ray D (for *dim*). This dim reflected light is responsible for the image observed when the mirror is in the night setting, as in Figure 23.4b. In this case, the wedge is rotated so that the path followed by the bright light (ray B) does not lead to the eye. Instead, the dim light reflected from the front surface travels to the eye, and the brightness of trailing headlights does not become a hazard.

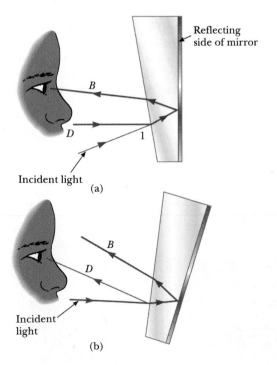

FIGURE 23.4
A cross-sectional view of a rearview mirror. (a) The day setting forms a bright image, *B*. (b) The night setting forms a dim image, *D*.

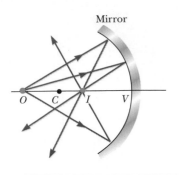

Mirror

FIGURE 23.5
A point object placed at *O*, outside the center of curvature of a concave spherical mirror, forms a real image at *I* as shown. If the rays diverge from *O* at small angles, they all reflect through the same image point.

23.2

IMAGES FORMED BY SPHERICAL MIRRORS

CONCAVE MIRRORS

A **spherical mirror,** as its name implies, has the shape of a segment of a sphere. Figure 23.5 shows a spherical mirror with light reflecting from its silvered inner, concave surface; this is called a **concave mirror.** The mirror has radius of curvature *R*, and its center of curvature is at point *C*. Point *V* is the center of the spherical segment, and a line drawn from *C* to *V* is called the **principal axis** of the mirror.

Now consider a point source of light placed at point *O* in Figure 23.5, on the principal axis and outside point *C*. Several diverging rays originating at *O* are shown. After reflecting from the mirror, these rays converge to meet at *I*, called the **image point.** The rays then continue and diverge from *I* as if there were an object there. As a result, a real image is formed. *Whenever reflected light actually passes through a point, any image formed there is real.*

We assume that all rays that diverge from the object make small angles with the principal axis. All such rays reflect through the image point, as in Figure 23.5. Rays that are far from the principal axis, as in Figure 23.6, converge to other points on the principal axis, producing a blurred image. This effect, called **spherical aberration,** is present to some extent with any spherical mirror and will be discussed in Section 23.7.

We can use the geometry shown in Figure 23.7 to calculate the image distance, *q*, from the object distance, *p*, and radius of curvature, *R*. By convention, these distances are measured from point *V*. Figure 23.7 shows two rays of light leaving the tip of the object. One ray passes through the center of curvature, *C*, of the mirror, hitting the mirror head on (perpendicularly to the mirror surface) and reflecting back on itself. The second ray strikes the mirror at point *V* and reflects as shown, obeying the law of reflection. The image of the tip of the arrow is at the point where the two rays intersect. From the largest triangle in Figure 23.7 we see that $\tan \theta = h/p$; the light blue triangle gives $\tan \theta = -h'/q$. The negative sign signifies that the image is inverted, and so *h'* is negative. Thus, from Equation 23.1 and these results, we find that the magnification of the mirror is

$$M = \frac{h'}{h} = -\frac{q}{p} \qquad [23.2]$$

We also note, from two other triangles in the figure, that

$$\tan \alpha = \frac{h}{p - R} \qquad \text{and} \qquad \tan \alpha = -\frac{h'}{R - q}$$

from which we find that

$$\frac{h'}{h} = -\frac{R - q}{p - R} \qquad [23.3]$$

If we compare Equation 23.2 to Equation 23.3, we see that

$$\frac{R - q}{p - R} = \frac{q}{p}$$

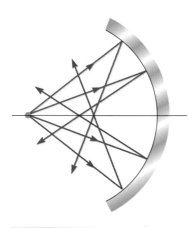

FIGURE 23.6
Rays at large angles from the horizontal axis reflect from a spherical, concave mirror to intersect the optic axis at different points, resulting in a blurred image. This is called *spherical aberration.*

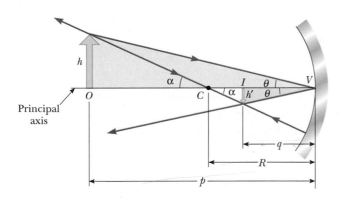

FIGURE 23.7
The image formed by a spherical concave mirror, where the object, at O, lies outside the center of curvature, C.

Simple algebra reduces this to

$$\frac{1}{p} + \frac{1}{q} = \frac{2}{R}$$

[23.4] Mirror equation

This expression is called the **mirror equation.**

If the object is very far from the mirror—that is, if the object distance, p, is great enough compared with R that p can be said to approach infinity—then $1/p \approx 0$, and we see from Equation 23.4 that $q \approx R/2$. In other words, when the object is very far from the mirror, *the image point is halfway between the center of curvature and the center of the mirror,* as in Figure 23.8a. The incoming rays are essentially parallel in this figure because the source is assumed to be very far from

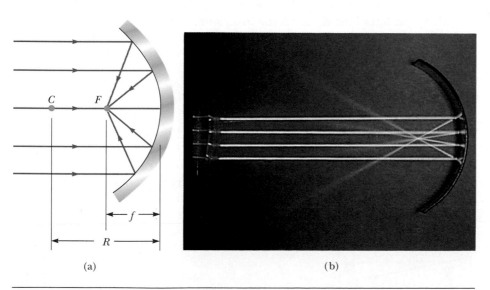

(a) (b)

FIGURE 23.8
(a) Light rays from a distant object ($p = \infty$) reflect from a concave mirror through the focal point, F. In this case, the image distance $q = R/2 = f$, where f is the focal length of the mirror. (b) A photograph of the reflection of parallel rays from a concave mirror. *(Courtesy of Jim Lehman, James Madison University)*

the mirror. In this special case we call the image point the **focal point,** F, and the image distance the **focal length,** f, where

$$f = \frac{R}{2} \qquad\qquad \text{[23.5]}$$

The mirror equation can therefore be expressed in terms of the focal length:

$$\frac{1}{p} + \frac{1}{q} = \frac{1}{f} \qquad\qquad \text{[23.6]}$$

Note that rays from objects at infinity are always focused at the focal point.

23.3

CONVEX MIRRORS AND SIGN CONVENTIONS

Figure 23.9 shows the formation of an image by a **convex mirror,** which is silvered so that light is reflected from the outer, convex surface. This is sometimes called a **diverging mirror** because the rays from any point on the object diverge after reflection as though they were coming from some point behind the mirror. The image in Figure 23.9 is virtual rather than real because it lies behind the mirror at the point where the reflected rays appear to originate. In general, as shown in the figure, the image formed by a convex mirror is erect, virtual, and smaller than the object.

We shall not derive any equations for convex spherical mirrors. If we did, we would find that the equations developed for concave mirrors can be used with convex mirrors if a particular sign convention is used. Let us call the region in which light rays move the *front side* of the mirror, and the other side, where virtual images are formed, the *back side*. For example, in Figures 23.7 and 23.9, the side to the left of the mirror is the front side, and the side to the right is the back side. Figure 23.10 is helpful for understanding the rules for object and image distances, and Table 23.1 summarizes the sign conventions for all the necessary quantities.

RAY DIAGRAMS FOR MIRRORS

We can conveniently determine the positions and sizes of images formed by mirrors by constructing *ray diagrams* similar to the ones we have been using. This kind of graphical construction tells us the overall nature of the image and can be used to check parameters calculated from the mirror and magnification equations. To make a ray diagram, one needs to know the position of the object and

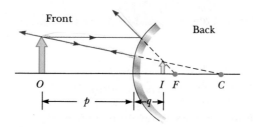

FIGURE 23.9
Formation of an image by a spherical convex mirror. Note that the image is virtual and erect.

TABLE 23.1
Sign Conventions for Mirrors

p is + if the object is in front of the mirror (real object).
p is − if the object is in back of the mirror (virtual object).

q is + if the image is in front of the mirror (real image).
q is − if the image is in back of the mirror (virtual image).

Both f and R are + if the center of curvature is in front of the mirror (concave mirror).
Both f and R are − if the center of curvature is in back of the mirror (convex mirror).

If M is positive, the image is erect.
If M is negative, the image is inverted.

p = object distance; p' = image distance; f = focal length; R = radius of curvature;
M = lateral magnification.

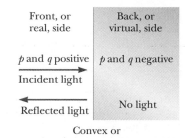

FIGURE 23.10
A diagram describing the signs of p and q for convex and concave mirrors.

the location of the center of curvature. To locate the image, three rays are constructed (rather than just the two we have been constructing so far), as shown by the examples in Figure 23.11. All three rays start from the same object point; for these examples the tip of the arrow was chosen. For the concave mirrors in Figure 23.11a and b, the rays are drawn as follows:

1. Ray 1 is drawn parallel to the principal axis and is reflected back through the focal point, *F*.
2. Ray 2 is drawn through the focal point. Thus, it is reflected parallel to the principal axis.
3. Ray 3 is drawn through the center of curvature, *C*, and is reflected back on itself.

The intersection of any *two* of these rays at a point locates the image. The third ray serves as a check of construction. The image point obtained in this fashion must always agree with the value of *q* calculated from the mirror formula.

In the case of a concave mirror, note what happens as the object is moved closer to the mirror. The real, inverted image in Figure 23.11a moves to the left as the object approaches the focal point. When the object is at the focal point, the image is infinitely far to the left. However, when the object lies between the focal point and the mirror surface, as in Figure 23.11b, the image is virtual and erect.

With the convex mirror shown in Figure 23.11c, the image of a real object is always virtual and erect. As the object distance increases, the virtual image shrinks and approaches the focal point as *p* approaches infinity. You should construct a ray diagram to verify this.

The image-forming characteristics of curved mirrors obviously determine their uses. For example, suppose you want to design a mirror that will help people shave or apply cosmetics, such as the one in Figure 23.11b. That is, you need a concave mirror that puts the user inside the focal point. In such a situation, the image is erect and greatly enlarged. In contrast, suppose that the primary purpose of a mirror is to observe a large field of view, in which case you need a convex mirror such as the one in Figure 23.11c. The diminished size of the

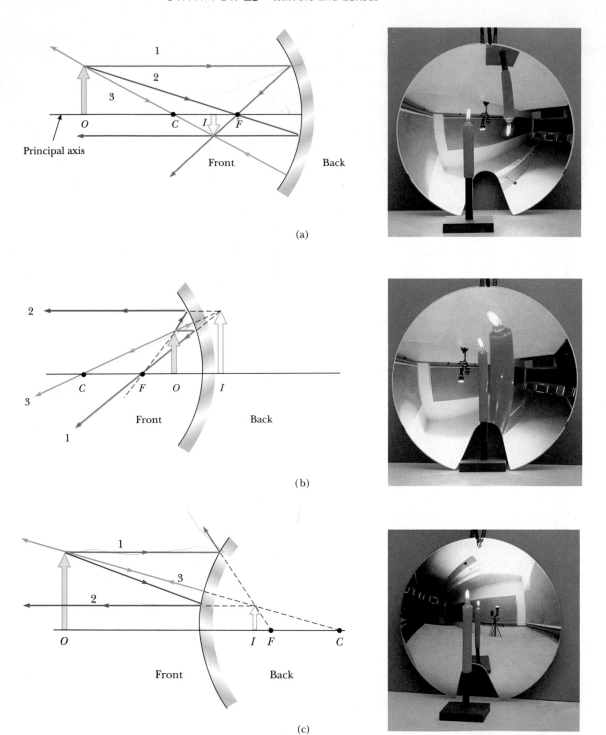

Principal axis

(a)

(b)

(c)

FIGURE 23.11
Ray diagrams for spherical mirrors, and corresponding photographs of the images of candles. (a) When an object is outside the center of curvature of a concave mirror, the image is real, inverted, and reduced in size. (b) When an object is between a concave mirror and the focal point, the image is virtual, erect, and magnified. (c) When an object is in front of a convex mirror, the image is virtual, erect, and reduced in size.
(Courtesy of David Rogers)

image means that a fairly large field of view is seen in the mirror. Mirrors such as this are often placed in stores to help employees watch for shoplifting. A second use is as a sideview mirror on a car (Fig. 23.12). This kind of mirror is usually placed on the passenger side of the car and carries the warning "Objects are closer than they appear." Without this warning, a driver might think he is looking into a plane mirror, which does not alter the size of the image. Thus, he could be fooled into believing that a truck is far away because it looks small, when it is actually a large semi very close behind him but diminished in size because of the image formation characteristics of the convex mirror.

FIGURE 23.12
A convex sideview mirror on a vehicle produces an erect image that is smaller than the object. (*© Junebug Clark 1988/Photo Researchers, Inc.*)

EXAMPLE 23.2 Images Formed by a Concave Mirror

Assume that a certain concave spherical mirror has a focal length of 10.0 cm. Locate the images for object distances of (a) 25.0 cm, (b) 10.0 cm, and (c) 5.00 cm. Describe the image in each case.

Solution (a) For an object distance of 25.0 cm, we find the image distance using the mirror equation:

$$\frac{1}{p} + \frac{1}{q} = \frac{1}{f}$$

$$\frac{1}{25.0 \text{ cm}} + \frac{1}{q} = \frac{1}{10.0 \text{ cm}}$$

$$q = \boxed{16.7 \text{ cm}}$$

The magnification is given by Equation 23.2:

$$M = -\frac{q}{p} = -\frac{16.7 \text{ cm}}{25.0 \text{ cm}} = \boxed{-0.667}$$

Thus, the image is smaller than the object. Furthermore, the image is inverted because M is negative. Finally, because q is positive, the image is on the front side of the mirror and is real. This situation is pictured in Figure 23.11a.

(b) When the object distance is 10.0 cm, the object is at the focal point. Substituting the values $p = 10.0$ cm and $f = 10.0$ cm into the mirror equation, we find that

$$\frac{1}{10.0 \text{ cm}} + \frac{1}{q} = \frac{1}{10.0 \text{ cm}}$$

$$q = \boxed{\infty}$$

Thus, we see that rays of light originating from an object at the focal point of a concave mirror are reflected so that the image is formed an infinite distance from the mirror; that is, the rays travel parallel to one another after reflection.

(c) When the object is at 5.00 cm, inside the focal point of the mirror, the mirror equation gives

$$\frac{1}{5.00 \text{ cm}} + \frac{1}{q} = \frac{1}{10.0 \text{ cm}}$$

$$q = \boxed{-10.0 \text{ cm}}$$

That is, the image is virtual since it is behind the mirror. The magnification is

$$M = -\frac{q}{p} = -\left(\frac{-10.0 \text{ cm}}{5.00 \text{ cm}}\right) = \boxed{2.00}$$

We see that the image height is magnified by a factor of 2, and the positive sign indicates that the image is erect (Fig. 23.11b).

Note the characteristics of an image formed by a concave spherical mirror. When the object is outside the focal point, the image is inverted and real; at the focal point, the image is formed at infinity; inside the focal point, the image is erect and virtual.

Exercise   If the object distance is 20.0 cm, find the image distance and the magnification of the mirror.

Answer $q = 20.0$ cm, $M = -1.00$.

EXAMPLE 23.3   Images Formed by a Convex Mirror

An object 3.00 cm high is placed 20.0 cm from a convex mirror with a focal length of 8.00 cm. Find (a) the position of the final image and (b) the magnification of the mirror.

Solution   (a) Since the mirror is convex, its focal length is negative. To find the image position, we use the mirror equation:

$$\frac{1}{p} + \frac{1}{q} = \frac{1}{f}$$

$$\frac{1}{20.0 \text{ cm}} + \frac{1}{q} = \frac{1}{-8.00 \text{ cm}}$$

$$q = \boxed{-5.71 \text{ cm}}$$

The negative value of q indicates that the image is virtual, or behind the mirror, as in Figure 23.11c.

(b) The magnification of the mirror is

$$M = -\frac{q}{p} = -\left(\frac{-5.71 \text{ cm}}{20.0 \text{ cm}}\right) = \boxed{0.286}$$

The image is erect because M is positive.

Exercise   Find the height of the image.

Answer   0.857 cm.

EXAMPLE 23.4   An Enlarged Image

When a woman stands with her face 40.0 cm from a cosmetic mirror, the erect image is twice as tall as her face. What is the focal length of the mirror?

Reasoning   Most of the problems we have encountered so far have been simple applications of the mirror equation. However, to find f in this example, we must first find q, the image distance. Since the problem states that the image is erect, the magnification must be positive (in this case, $M = +2$), and because $M = -q/p$, we can determine q.

Solution   The magnification equation gives us a relationship between the object and image distances:

$$M = -\frac{q}{p} = 2$$

$$q = -2p = -2(40.0 \text{ cm}) = -80.0 \text{ cm}$$

First, note that a virtual image is formed because the woman is able to see her erect image in the mirror. This explains why the image distance is negative. Substitute $q = -80.0$ cm into the mirror equation to obtain

$$\frac{1}{40.0 \text{ cm}} - \frac{1}{80.0 \text{ cm}} = \frac{1}{f}$$

$$f = \boxed{80.0 \text{ cm}}$$

The positive sign for the focal length indicates that the mirror is concave, a fact that we already knew because the mirror magnified the object (a convex mirror would have produced a diminished image).

23.4

IMAGES FORMED BY REFRACTION

In this section we describe how images are formed by refraction at a spherical surface. Consider two transparent media with indices of refraction n_1 and n_2, where the boundary between the two media is a spherical surface of radius R (Fig. 23.13). Let us assume that the medium to the right has a higher index of refraction than the one to the left; that is, $n_2 > n_1$. This would be the case for light entering a curved piece of glass from air or for light entering the water in a fishbowl from air. The rays originating at the object location, O, are refracted at the spherical surface and then converge to the image point, I. We can begin with Snell's law of refraction and use simple geometric techniques to show that the object distance, image distance, and radius of curvature are related by the equation

$$\frac{n_1}{p} + \frac{n_2}{q} = \frac{n_2 - n_1}{R} \qquad \text{[23.7]}$$

Furthermore, the magnification of a refracting surface is

$$M = \frac{h'}{h} = -\frac{n_1 q}{n_2 p} \qquad \text{[23.8]}$$

As with mirrors, we must use a sign convention if we are to apply these equations to a variety of circumstances. First note that real images are formed on the side of the surface *opposite* the side from which the light comes. This is in contrast with mirrors, where real images are formed on the *same* side of the reflecting surface. Therefore, *the sign convention for spherical refracting surfaces is the same as for mirrors, recognizing the change in sides of the surface for real and virtual images.* For example, in Figure 23.13, p, q, and R are all positive.

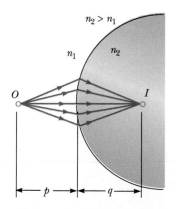

FIGURE 23.13
An image formed by refraction at a spherical surface. Rays making small angles with the optic axis diverge from a point object at O and pass through the image point, I.

TABLE 23.2
Sign Conventions for Refracting Surfaces

p is + if the object is in front of the surface (real object).
p is − if the object is in back of the surface (virtual object).

q is + if the image is in back of the surface (real image).
q is − if the image is in front of the surface (virtual image).

R is + if the center of curvature is in back of the surface.
R is − if the center of curvature is in front of the surface.

p = object distance; p' = image distance; R = radius of curvature.

The sign conventions for spherical refracting surfaces are summarized in Table 23.2. (The same conventions are used for thin lenses, which will be discussed in the next section.) As with mirrors, we assume that the front of the refracting surface is the side from which the light approaches the surface.

PLANE REFRACTING SURFACES

If the refracting surface is a plane, then R approaches infinity and Equation 23.7 reduces to

$$\frac{n_1}{p} = -\frac{n_2}{q}$$

$$q = -\frac{n_2}{n_1} p \qquad \text{[23.9]}$$

From Equation 23.9 we see that the sign of q is opposite that of p. Thus, *the image formed by a plane refracting surface is on the same side of the surface as the object.* This is illustrated in Figure 23.14 for the situation in which n_1 is greater than n_2, where a virtual image is formed between the object and the surface. Note that the refracted ray bends *away* from the normal in this case, because $n_1 > n_2$.

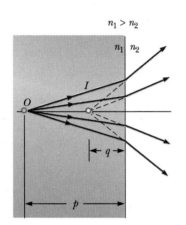

$n_1 > n_2$

FIGURE 23.14
The image formed by a plane refracting surface is virtual; that is, it forms to the left of the refracting surface.

EXAMPLE 23.5 Gaze into the Crystal Ball

A coin 2.00 cm in diameter is embedded in a solid glass ball of radius 30.0 cm (Fig. 23.15). The index of refraction of the ball is 1.50, and the coin is 20.0 cm from the surface. Find the position and height of the image.

Solution Because they are moving from a medium of high index of refraction to a medium of lower index of refraction, the rays originating at the object are refracted away from the normal at the surface and diverge outward. The image is formed in the glass and is virtual. Applying Equation 23.7 and taking $n_1 = 1.500$, $n_2 = 1.000$, $p = 20.0$ cm, and $R = -30.0$ cm, we get

$$\frac{n_1}{p} + \frac{n_2}{q} = \frac{n_2 - n_1}{R}$$

$$\frac{1.500}{20.0 \text{ cm}} + \frac{1.000}{q} = \frac{1.000 - 1.500}{-30.0 \text{ cm}}$$

$$-17.1 \text{ cm}$$

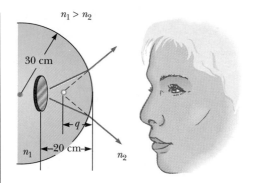

FIGURE 23.15
(Example 23.5) A coin embedded in a glass ball forms a virtual image between the coin and the glass surface.

The negative sign indicates that the image is in the same medium as the object (the side of incident light), in agreement with our ray diagram, and therefore must be virtual.

To find the image height, we first use Equation 23.8 for the magnification:

$$M = -\frac{n_1 q}{n_2 p} = -\frac{1.500(-17.1 \text{ cm})}{1.000(20.0 \text{ cm})} = \frac{h'}{h} = 1.28$$

Therefore,

$$h' = 1.28h = (1.28)(2.00 \text{ cm}) = \quad 2.56 \text{ cm}$$

The positive value for M indicates an erect image.

EXAMPLE 23.6   The One That Got Away

A small fish is swimming at a depth of d below the surface of a pond (Fig. 23.16). What is the *apparent depth* of the fish as viewed from directly overhead?

Reasoning   In this example, the refracting surface is a plane, and so R is infinite. Hence, we can use Equation 23.9 to determine the location of the image.

Solution   The facts that $n_1 = 1.33$ for water and $p = d$ give us

$$q = -\frac{n_2}{n_1} p = -\frac{1}{1.33} d = \quad -0.752 \, d$$

Again, since q is negative, the image is virtual, as indicated in Figure 23.16. The apparent depth is three-fourths the actual depth. For instance, if $d = 4.0$ m, $q = -3.0$ m.

Exercise   If the fish is 12 cm long, how long is its image?

Answer   12 cm.

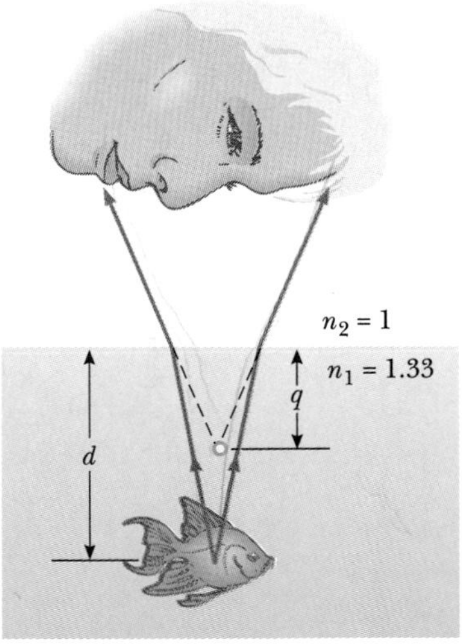

FIGURE 23.16
(Example 23.6)   The apparent depth, q, of the fish is less than the true depth, d.

*23.5

ATMOSPHERIC REFRACTION

Images formed by refraction in our atmosphere lead to some interesting results. In this section we look at two examples. A situation that occurs daily is the visibility of the Sun at dusk even though it has passed below the horizon. Figure

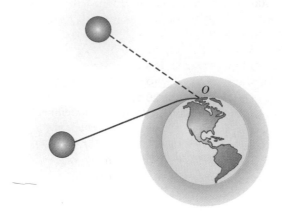

FIGURE 23.17
Because of refraction, an observer at O sees the Sun even though it has fallen below the horizon.

23.17 shows why this occurs. Rays of light from the Sun strike the Earth's atmosphere (represented by the shaded area around the Earth) and are bent as they pass into a medium that has an index of refraction different from that of the almost empty space in which they have been traveling. The bending in this situation differs somewhat from the bending we have considered previously in being gradual and continuous as the light moves through the atmosphere toward an observer at point O. This is because the light moves through layers of air that have a continuously changing index of refraction. When the rays reach the observer, the eye follows them back along the direction from which they appear to have come (indicated by the dashed path in the figure). The end result is that the Sun is seen to be above the horizon even after it has fallen below it.

The **mirage** is another phenomenon of nature produced by refraction in the atmosphere. A mirage can be observed when the ground is so hot that the air directly above it is warmer than the air at higher elevations. The desert is, of

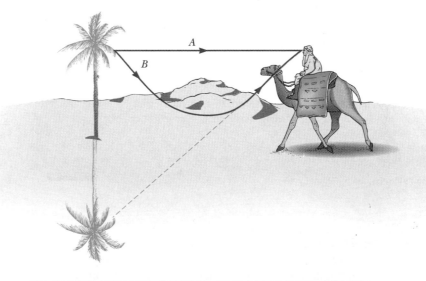

FIGURE 23.18
A mirage is produced by the bending of light rays in the atmosphere when there are large temperature differences between the ground and the air.

course, a region in which such circumstances prevail, but mirages are also seen on heated roadways during the summer. The layers of air at different heights above the Earth have different densities and different refractive indices. The effect this can have is pictured in Figure 23.18. In this situation the observer sees a tree in two different ways. One group of light rays reaches the observer by the straight-line path *A*, and the eye traces these rays back to see the tree in the normal fashion. In addition, a second group of rays travels along the curved path *B*. These rays are directed toward the ground and are then bent as a result of refraction. Consequently, the observer also sees an inverted image of the tree as he traces these rays back to the point at which they appear to have originated. Because an erect image and an inverted image are seen when the image of a tree is observed in a reflecting pool of water, the observer unconsciously calls upon this past experience and concludes that a pool of water must be in front of the tree.

23.6

THIN LENSES

A typical **thin lens** consists of a piece of glass or plastic, ground so that each of its two refracting surfaces is a segment of either a sphere or a plane. Lenses are commonly used to form images by refraction in optical instruments, such as cameras, telescopes, and microscopes. The equation that relates object and image distances for a lens is virtually identical to the mirror equation derived earlier, and the method used to derive it is also similar.

Figure 23.19 shows some representative shapes of lenses. Notice that we have placed these lenses in two groups. Those in Figure 23.19a are thicker at the center than at the rim, and those in Figure 23.19b are thinner at the center than at the rim. The lenses in the first group are examples of **converging lenses,** and those in the second group are **diverging lenses.** The reason for these names will become apparent shortly.

As we did for mirrors, it is convenient to define a point called the **focal point** for a lens. For example, in Figure 23.20a, a group of rays parallel to the axis passes through the focal point, *F*, after being converged by the lens. The distance from the focal point to the lens is called the **focal length,** *f. The focal length is the image distance that corresponds to an infinite object distance.* Recall that we are considering the lens to be very thin. As a result, it makes no difference whether we take the focal length to be the distance from the focal point to the surface of the lens or the distance from the focal point to the center of the lens, because the difference between these two lengths is negligible. A thin lens has *two* focal points, as illustrated in Figure 23.20, corresponding to parallel rays traveling from the left and from the right.

Rays parallel to the axis diverge after passing through a lens of biconcave shape in Figure 23.20b. In this case, the focal point is defined to be the point at which the diverged rays appear to originate, labeled *F* in the figure. Figures 23.20a and 23.20b indicate why the names *converging* and *diverging* are applied to these lenses.

Consider a ray of light passing through the center of a lens, labeled ray 1 in Figure 23.21. If we apply Snell's law at both surfaces, we find that this ray is deflected from its original direction of travel by a distance of δ, shown in Figure

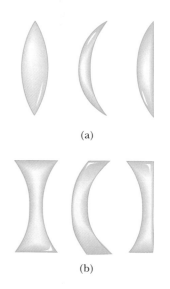

(a)

(b)

FIGURE 23.19
Lens shapes. (a) Converging lenses have positive focal lengths and are thickest at the middle. From left to right, these are biconvex, convex-concave, and plano-convex. (b) Diverging lenses have negative focal lengths and are thickest at the edges. From left to right are biconcave, convex-concave, and plano-convex lenses.

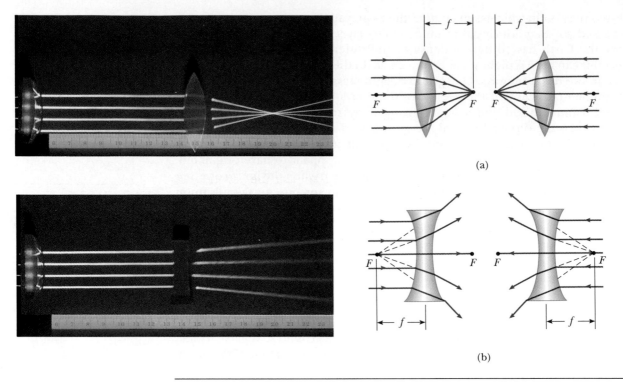

FIGURE 23.20
(Left) Photographs of the effects of converging and diverging lenses on parallel rays. *(Courtesy of Jim Lehman, James Madison University)* *(Right)* The focal points of (a) the bioconvex lens and (b) the biconcave lens.

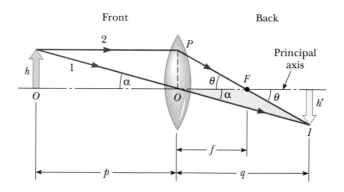

FIGURE 23.21
A geometric construction for developing the thin-lens equation.

23.22. To avoid the complications arising from this deflection, we make what is called the *thin-lens approximation: the thickness of the lens is assumed to be negligible.* As a result, δ becomes vanishingly small and we see that the ray passes through the lens undeflected. Ray 2 in Figure 23.21 is parallel to the principal axis of the lens (the horizontal axis passing through *O*), and as a result it passes through the focal point, *F*, after refraction. The point at which rays 1 and 2 intersect is the image point.

We first note that the tangent of the angle α can be found by using the shaded triangles in Figure 23.21:

$$\tan \alpha = \frac{h}{p} \quad \text{or} \quad \tan \alpha = -\frac{h'}{q}$$

From this we find that

$$M = \frac{h'}{h} = -\frac{q}{p} \qquad \text{[23.10]}$$

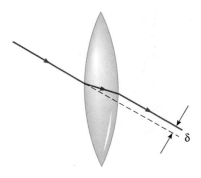

Thus, the equation for magnification by a lens is the same as the equation for magnification by a mirror. We also note from Figure 23.21 that the tangent of θ is

$$\tan \theta = \frac{PO}{f} \quad \text{or} \quad \tan \theta = -\frac{h'}{q - f}$$

However, the height PO used in the first of these equations is the same as h, the height of the object. Therefore,

$$\frac{h}{f} = -\frac{h'}{q - f}$$

$$\frac{h'}{h} = -\frac{q - f}{f}$$

Using this in combination with Equation 23.10 gives

$$\frac{q}{p} = \frac{q - f}{f}$$

which reduces to

$$\frac{1}{p} + \frac{1}{q} = \frac{1}{f} \qquad \text{[23.11]}$$

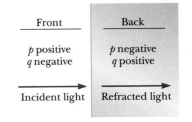

FIGURE 23.23
A diagram for obtaining the signs of p and q for a thin lens or a refracting surface.

This equation, called the **thin-lens equation,** can be used with both converging and diverging lenses if we adhere to a set of sign conventions. Figure 23.23 is useful for obtaining the signs of p and q, and Table 23.3 gives the complete sign

TABLE 23.3
Sign Conventions for Thin Lenses

p is $+$ if the object is in front of the lens.
p is $-$ if the object is in back of the lens.

q is $+$ if the image is in back of the lens.
q is $-$ if the image is in front of the lens.

R_1 and R_2 are $+$ if the center of curvature for each surface is in back of the lens.
R_1 and R_2 are $-$ if the center of curvature for each surface is in front of the lens.

f is $+$ for a converging lens.
f is $-$ for a diverging lens.

p = object distance; q = image distance; R_1 = radius of curvature of front surface;
R_2 = radius of curvature of back surface; f = focal length.

conventions for lenses. Note that *a converging lens has a positive focal length* under this convention, and *a diverging lens has a negative focal length*. Hence the names *positive* and *negative* are often given to these lenses.

The focal length for a lens in air is related to the curvatures of its front and back surfaces and to the index of refraction, *n*, of the lens material by

Lens maker's equation

$$\frac{1}{f} = (n - 1)\left(\frac{1}{R_1} - \frac{1}{R_2}\right) \qquad \textbf{[23.12]}$$

where R_1 is the radius of curvature of the front surface of the lens and R_2 is the radius of curvature of the back surface. (As with mirrors, we arbitrarily call the side from which the light approaches the *front* of the lens.) Equation 23.12 enables us to calculate the focal length from the known properties of the lens. It is called the **lens maker's equation.**

RAY DIAGRAMS FOR THIN LENSES

Ray diagrams are very convenient for determining the image formed by a thin lens or a system of lenses. They should also help clarify the sign conventions we have already discussed. Figure 23.24 illustrates this method for three single-lens situations. To locate the image formed by a converging lens (Fig. 23.24a and b), the following three rays are drawn from the top of the object:

FIGURE 23.24
Ray diagrams for locating the image of an object. (a) The object is outside the focal point of a converging lens. (b) The object is inside the focal point of a converging lens. (c) The object is outside the focal point of a diverging lens.

1. The first ray is drawn parallel to the principal axis. After being refracted by the lens, this ray passes through (or appears to come from) one of the focal points.
2. The second ray is drawn through the center of the lens. This ray continues in a straight line.
3. The third ray is drawn through the focal point, *F*, and emerges from the lens parallel to the principal axis.

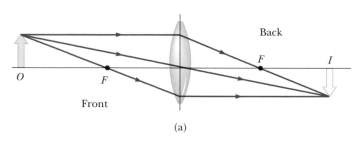

(a)

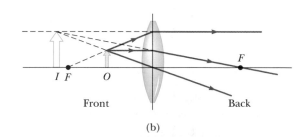

(b)

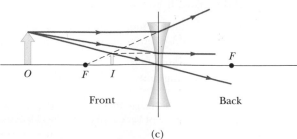

(c)

A similar construction is used to locate the image formed by a diverging lens, as shown in Figure 23.24c. The point of intersection of *any two* of the rays in these diagrams can be used to locate the image. The third ray serves as a check on construction.

For the converging lens in Figure 23.24a, where the object is *outside* the front focal point ($p > f$), the image is real and inverted. When the real object is *inside* the front focal point ($p < f$), as in Figure 23.24b, the image is virtual and erect. For the diverging lens of Figure 23.24c, the image is virtual and erect.

PROBLEM-SOLVING STRATEGY
Lenses and Mirrors

Your success or failure in working lens and mirror problems will be determined largely by whether or not you make sign errors when substituting into the lens and mirror equations. The only way to ensure that you don't make sign errors is to become adept at using the sign conventions. The best way to do this is to work a multitude of problems on your own. Watching an instructor or reading the example problems is no substitute for practice.

EXAMPLE 23.7   The Lens Maker's Equation

The double convex lens of Figure 23.25 has an index of refraction of 1.50. The radius of curvature of the front surface is $R_1 = 10.0$ cm, and that of the back surface is $R_2 = -15.0$ cm. Find the focal length of the lens.

Solution   From the sign conventions in Table 23.3 we find that $R_1 = +10.0$ cm and $R_2 = -15.0$ cm. Thus, using the lens maker's equation, we have

$$\frac{1}{f} = (n - 1)\left(\frac{1}{R_1} - \frac{1}{R_2}\right) = (1.50 - 1)\left(\frac{1}{10.0\ \text{cm}} - \frac{1}{-15.0\ \text{cm}}\right)$$

$$f = \boxed{12\ \text{cm}}$$

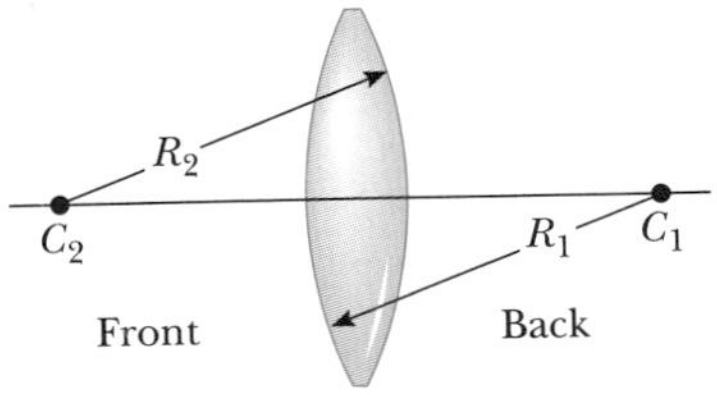

FIGURE 23.25
This converging lens has two curved surfaces with radii of curvature R_1 and R_2. The center of curvature R_1 lies to the right of the lens, and the center of R_2 lies to the left.

EXAMPLE 23.8   Images Formed by a Converging Lens

A converging lens of focal length 10.0 cm forms images of objects placed (a) 30.0 cm, (b) 10.0 cm, and (c) 5.0 cm from the lens. In each case, find the image distance and describe the image.

Solution   (a) The thin-lens equation, Equation 23.11, can be used to find the image distance:

$$\frac{1}{p} + \frac{1}{q} = \frac{1}{f}$$

$$\frac{1}{30.0\ \text{cm}} + \frac{1}{q} = \frac{1}{10.0\ \text{cm}}$$

$$q = \boxed{15\ \text{cm}}$$

The positive sign for the image distance tells us that the image is on the real side of the lens (Fig. 23.23). The magnification of the lens is

$$M = -\frac{q}{p} = -\frac{15 \text{ cm}}{30.0 \text{ cm}} = \boxed{-0.50}$$

Thus, the image is reduced in height by one half, and the negative sign for M tells us that the image is inverted. The situation is like that in Figure 23.24a.

(b) No calculation is necessary for this case because we know that, when the object is placed at the focal point, the image is formed at infinity. This is readily verified by substituting $p = 10.0$ cm into the lens equation.

(c) We now move inside the focal point, to an object distance of 5.0 cm. In this case, the lens equation gives

$$\frac{1}{5.0 \text{ cm}} + \frac{1}{q} = \frac{1}{10.0 \text{ cm}}$$

$$q = \boxed{-10.0 \text{ cm}}$$

$$M = -\frac{q}{p} = -\left(\frac{-10.0 \text{ cm}}{5.0 \text{ cm}}\right) = \boxed{2.0}$$

The negative image distance tells us that the image is formed on the side of the lens from which the light is incident, the virtual side (Fig. 23.23). The image is enlarged, and the positive sign for M tells us that the image is erect, as shown in Figure 23.24b.

EXAMPLE 23.9   The Case of a Diverging Lens

Repeat the problem of Example 23.8 for a *diverging* lens of focal length 10.0 cm.

Solution   (a) Let us apply the lens equation with an object distance of 30.0 cm:

$$\frac{1}{p} + \frac{1}{q} = \frac{1}{f}$$

$$\frac{1}{30.0 \text{ cm}} + \frac{1}{q} = -\frac{1}{10.0 \text{ cm}}$$

$$q = \boxed{-7.5 \text{ cm}}$$

The magnification is

$$M = -\frac{q}{p} = -\left(\frac{-7.5 \text{ cm}}{30.0 \text{ cm}}\right) = \boxed{0.25}$$

Thus, the image is virtual, smaller than the object, and erect.

(b) When the object is at the focal point, $p = 10.0$ cm, we have

$$\frac{1}{10.0 \text{ cm}} + \frac{1}{q} = -\frac{1}{10.0 \text{ cm}}$$

$$q = \boxed{-5.0 \text{ cm}}$$

$$M = -\frac{q}{p} = -\left(\frac{-5.0 \text{ cm}}{10.0 \text{ cm}}\right) = \boxed{0.50}$$

(c) When the object is inside the focal point, at 5.0 cm, we have

$$\frac{1}{5.0 \text{ cm}} + \frac{1}{q} = -\frac{1}{10.0 \text{ cm}}$$

$$q = \boxed{-3.3 \text{ cm}}$$

$$M = -\left(\frac{-3.3 \text{ cm}}{5.0 \text{ cm}}\right) = \boxed{0.66}$$

Again, we have a virtual image that is smaller than the object and erect, as in Figure 23.24c.

COMBINATION OF THIN LENSES

If two thin lenses are used to form an image, the system can be treated in the following manner. First, the image of the first lens is calculated as though the second lens were not present. The light then approaches the second lens *as if* it had come from the image formed by the first lens. Hence, **the image formed by the first lens is treated as the object for the second lens.** The image formed by the second lens is the final image of the system. If the image formed by the first lens lies on the back side of the second lens, then the image is treated as a virtual object for the second lens (that is, p is negative). The same procedure can be extended to a system of three or more lenses. The overall magnification of a system of thin lenses is the *product* of the magnifications of the separate lenses.

EXAMPLE 23.10   Two Lenses in a Row

Two converging lenses are placed 20.0 cm apart as shown in Figure 23.26. If the first lens has a focal length of 10.0 cm and the second has a focal length of 20.0 cm, locate the final image formed of an object 30.0 cm in front of the first lens. Find the magnification of the system.

Reasoning   We apply the thin-lens equation to both lenses. The image formed by the first lens is treated as the object for the second lens. Also, we use the fact that the total magnification of the system is the product of the magnifications produced by the separate lenses.

Solution   The location of the image formed by the first lens is found via the thin-lens equation:

$$\frac{1}{30.0 \text{ cm}} + \frac{1}{q} = \frac{1}{10.0 \text{ cm}}$$

$$q = 15 \text{ cm}$$

The magnification of this lens is

$$M_1 = -\frac{q}{p} = -\frac{15 \text{ cm}}{30.0 \text{ cm}} = -\frac{1}{2}$$

The image formed by this lens becomes the object for the second lens. Thus, the object distance for the second lens is 5.0 cm. We again apply the thin-lens equation to find the location of the final image.

$$\frac{1}{5.0 \text{ cm}} + \frac{1}{q} = \frac{1}{20.0 \text{ cm}}$$

$$q = \boxed{-6.7 \text{ cm}}$$

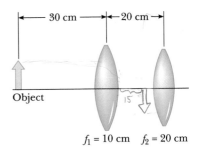

FIGURE 23.26
(Example 23.10)

The magnification of the second lens is

$$M_2 = -\frac{q}{p} = -\frac{(-6.7 \text{ cm})}{5.0 \text{ cm}} = 1.3$$

Thus, the final image is 6.7 cm to the left of the second lens, and the overall magnification of the system is

$$M = M_1 M_2 = (-\tfrac{1}{2})(1.3) = \boxed{-0.67}$$

The negative sign indicates that the final image is inverted with respect to the initial object.

Exercise   If the two lenses in Figure 23.26 are separated by 10.0 cm, locate the final image and find the magnification of the system.

Answer   4 cm behind the second lens; $M = -0.40$.

PHYSICS IN *ACTION*

MIRRORS AND LENSES

(Left) The "professor in the box" appears to be balancing himself on a few fingers with both of his feet elevated from the floor. Since the professor can maintain this position for a long time, he appears to defy gravity. This is one example of an optical illusion, used by magicians, that makes use of a mirror. Where in the box was the mirror placed to create this illusion? (The same strange professor and some of his friends from Virginia Military Institute have been known to "fly" above a lecture bench.)

(Right) This photograph shows two converging lenses separated by about 5 cm. Parallel light rays are incident from the left on the larger lens, whose focal length was measured by an independent experiment to be 11 cm. The final image is formed to the right of the smaller lens, as shown. Using the image formed by the larger lens as the object for the smaller lens, and the information given in the photograph, you should be able to show that the focal length of the smaller lens is about 5.2 cm.

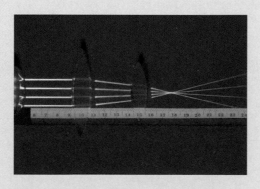

(Courtesy of Henry Leap and Jim Lehman)

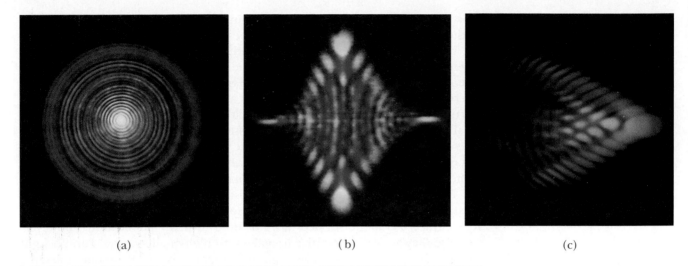

(a)   (b)   (c)

FIGURE 23.27

Lenses can produce varied forms of aberrations, as shown by these blurred photographic images of a point source. (a) Spherical aberration occurs when light passing through the lens at different distances from the optical axis is focused at different points. (b) Astigmatism is an aberration that occurs when the object is not on the optical axis of the lens. (c) Coma. This aberration occurs when light passing through the lens far from the optical axis focuses at a different part of the focal plane from light passing near the center of the lens. *(Photos by Norman Goldberg)*

*23.7
LENS ABERRATIONS

One of the basic problems of lenses and lens systems is the imperfect quality of the images, which is largely the result of defects in shape and form. The simple theory of mirrors and lenses assumes that rays make small angles with the principal axis and that all rays reaching the lens or mirror from a point source are focused at a single point, producing a sharp image. Clearly, this is not always true in the real world. Where the approximations used in this theory do not hold, imperfect images are formed.

If one wishes to precisely analyze image formation, it is necessary to trace each ray, using Snell's law, at each refracting surface. This procedure shows that there is no single point image; instead, the image is blurred. The departures of real (imperfect) images from the ideal predicted by the simple theory are called **aberrations.** Two common types of aberrations are spherical aberration and chromatic aberration. Photographs of three forms of lens aberrations are shown in Figure 23.27.

SPHERICAL ABERRATION

Spherical aberration results from the fact that the focal points of light rays far from the principal axis of a spherical lens (or mirror) are different from the focal points of rays with the same wavelength passing near the axis. Figure 23.28 illustrates spherical aberration for parallel rays passing through a converging

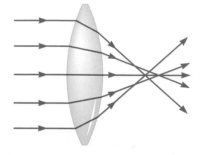

FIGURE 23.28

A spherical aberration produced by a converging lens. Does a diverging lens produce spherical aberration? (Angles are greatly exaggerated for clarity.)

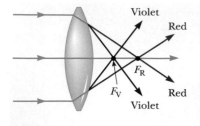

FIGURE 23.29

A chromatic aberration produced by a converging lens. Rays of different wavelengths focus at different points. (Angles are greatly exaggerated for clarity.)

lens. Rays near the middle of the lens are imaged farther from the lens than rays at the edges. Hence, there is no single focal length for a lens.

Most cameras are equipped with an adjustable aperture to control the light intensity and, when possible, reduce spherical aberration. (An aperture is an opening that controls the amount of light transmitted through the lens.) As the aperture size is reduced, sharper images are produced, since only the central portion of the lens is exposed to the incident light when the aperture is very small. At the same time, however, progressively less light is imaged. To compensate for this loss, a longer exposure time is used. An example of the results obtained with small apertures is the sharp image produced by a "pinhole" camera, with an aperture size of approximately 1 mm.

In the case of mirrors used for very distant objects, one can eliminate, or at least minimize, spherical aberration by employing a parabolic rather than spherical surface. Parabolic surfaces are not used in many applications, however, because they are very expensive to make with high-quality optics. Parallel light rays incident on such a surface focus at a common point. Parabolic reflecting surfaces are used in many astronomical telescopes to enhance the image quality. They are also used in searchlights, in which a nearly parallel light beam is produced from a small lamp placed at the focus of the reflecting surface.

CHROMATIC ABERRATION

The fact that different wavelengths of light refracted by a lens focus at different points gives rise to chromatic aberration. In Chapter 22 we described how the index of refraction of a material varies with wavelength. When white light passes through a lens, one finds, for example, that violet light rays are refracted more than red light rays (Fig. 23.29); thus, the focal length for red light is greater than that for violet light. Other wavelengths (not shown in Fig. 23.29) would have intermediate focal points. The chromatic aberration for a diverging lens is opposite that for a converging lens. Chromatic aberration can be greatly reduced by the use of a combination of converging and diverging lenses made from two different types of glass.

SUMMARY

Images are formed where rays of light intersect or where they appear to originate. A **real image** is one in which light intersects, or passes through, an image point. A **virtual image** is one in which the light does not pass through the image point but appears to diverge from that point.

The image formed by a plane mirror has the following properties:

1. The image is as far behind the mirror as the object is in front.
2. The image is unmagnified, virtual, and erect.

The **magnification,** M, of a mirror is defined as the ratio of **image height,** h', to **object height,** h, which is the negative of the ratio of image distance, q, to object distance, p:

$$M = \frac{h'}{h} = -\frac{q}{p} \qquad \textbf{[23.2]}$$

The **object distance** and **image distance** for a spherical mirror of radius R are related by the **mirror equation**:

$$\frac{1}{p} + \frac{1}{q} = \frac{1}{f} \qquad \textbf{[23.6]}$$

where $f = R/2$ is the **focal length** of the mirror.

An image can be formed by refraction at a spherical surface of radius R. The object and image distances for refraction from such a surface are related by

$$\frac{n_1}{p} + \frac{n_2}{q} = \frac{n_2 - n_1}{R} \qquad \textbf{[23.7]}$$

The **magnification of a refracting surface** is

$$M = \frac{h'}{h} = -\frac{n_1 q}{n_2 p} \qquad \textbf{[23.8]}$$

where the object is located in the medium with index of refraction n_1 and the image is formed in the medium with index of refraction n_2.

The **magnification for a thin lens** is

$$M = \frac{h'}{h} = -\frac{q}{p} \qquad \textbf{[23.10]}$$

and the object and image distances are related by the **thin-lens equation**:

$$\frac{1}{p} + \frac{1}{q} = \frac{1}{f} \qquad \textbf{[23.11]}$$

Aberrations are responsible for the formation of imperfect images by lenses and mirrors. **Spherical aberration** results from the fact that the focal points of light rays far from the principal axis of a spherical lens or mirror are different from those of rays passing through the center. **Chromatic aberration** arises from the fact that light rays of different wavelengths focus at different points when refracted by a lens.

ADDITIONAL READING

A. B. Fraser and M. W. Hirsch, *Mirages,* Springer-Verlag, 1975.

T. B. Greenslade, "Multiple Images in Plane Mirrors," *The Physics Teacher,* January 1982, p. 29.

R. C. Jones. "How Images Are Detected," *Sci. American,* September 1968, p. 111.

F. D. Smith, "How Images Are Formed," *Sci. American,* September 1968, p. 97.

W. Tape, "The Topology of Mirages," *Sci. American,* June 1985, p. 120.

D. E. Thomas, "Mirror Images," *Sci. American,* December 1980, p. 206.

A. C. S. van Heel and C. H. F. Velzel, *What Is Light?* World University Library, New York, McGraw-Hill, 1968, Chaps. 1 and 2.

W. B. Veldkamp and T. J. McHugh, "Binary Optics," *Sci. American,* May 1992, p. 92.

R. Winston, "Nonimaging Optics," *Sci. American,* March 1991, p. 76.

CONCEPTUAL QUESTIONS

Example A person spear fishing from a boat sees a fish located 3 m from the boat at an apparent depth of 1 m. To spear the fish, should the person aim at, above, or below the image of the fish?

Reasoning Because of refraction, the fish appears to be at a depth that is less than its actual depth. Therefore, the person should aim below the image of the fish to hit it.

Example What are the conditions for the production of a mirage? On a hot day, what is it that we are seeing when we observe "water on the road"?

Reasoning A mirage occurs when light changes direction as it moves between regions of air having different indices of refraction because of their different densities at different temperatures. When a blacktop road is warmed by the Sun, an apparent wet spot is observed due to refraction of light from the bright sky. This light, headed originally a little below the horizontal, bends up as it first enters and leaves the hotter, less-dense, lower-index of refraction of layers of air closer to the road surface.

Example Describe the type of lens that can be used to start a fire.

Reasoning The burning glass must make a real image of the distant Sun, causing nearly parallel rays to converge. Any converging lens will do. A large-diameter lens will gather more power to start the fire more quickly.

Example Explain why a mirror cannot give rise to chromatic aberration.

Reasoning Chromatic aberration is produced when light passes *through* a material, as it does when passing through the glass of a lens. A mirror, silvered on its front surface, never has light passing through it, so this aberration cannot occur. This is only one of many reasons why large telescopes use mirrors rather than lenses for their optical elements.

Example A solar furnace can be constructed by using a concave mirror to reflect and focus sunlight into a furnace enclosure. What factors in the design of the reflecting mirror will guarantee that very high temperatures can be achieved?

Reasoning Make the mirror an efficient reflector (shiny); use a parabolic-shaped mirror so that it reflects all rays to the image point, even those far from the axis; most important, use a large-diameter mirror in order to collect more solar power.

1. Why do some emergency vehicles have the symbol ƎƆИAⅬUBMA written on the front?

2. When you look in a mirror, your left and right sides are reversed, yet your head and legs have not traded places. Explain.

3. The rearview mirror on a late-model car warns the user that objects may be closer than they appear. What kind of mirror is being used, and why was that type selected?

4. Consider a concave spherical mirror with a real object. Is the image always inverted? Is the image always real? Give conditions for your answers.

5. Repeat the preceding question for a convex spherical mirror.

6. Consider a spherical concave mirror with the object to the left of the mirror, beyond the focal point. Using ray diagrams, show that the image of the object moves to the left as the object approaches the focal point.

7. What is the magnification of a plane mirror? What is its focal length?

8. Why does a clear stream always appear to be shallower than it actually is?

9. Explain why a fish in a spherical fishbowl appears larger than it really is.

10. A cylinder of solid glass or clear plastic is placed above the words LEAD OXIDE, and the words are viewed from above, as in Figure 23.30. LEAD appears inverted but OXIDE does not. Explain.

FIGURE 23.30 (Question 10)
(Courtesy of Henry Leap and Jim Lehman)

11. Explain this statement: "The focal point of a lens is the image of a point object at infinity." Discuss the notion of infinity in real terms as it applies to object distances. Based on this statement, can you think of a "quick and dirty" method for determining the focal length of a positive lens?

12. One method for determining the position of an image, either real or virtual, is by means of *parallax.* If a finger or other object is placed at the position of the image, as in Figure 23.31, and the finger and

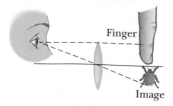

FIGURE 23.31 (Question 12)

image are viewed simultaneously (the image through the lens if it is virtual), the finger and image have the same parallax; that is, if it is viewed from different positions, the image will appear to move along with the finger. Use this method to locate the image formed by a lens. Explain why the method works.

13. It is well known that distant objects viewed underwater with the naked eye appear blurred and out of focus. On the other hand, goggles provide the swimmer with clear vision. Explain, using the fact that the indices of refraction of the cornea, water, and air are 1.376, 1.333, and 1.000 29, respectively.

14. Lenses used in eyeglasses, whether converging or diverging, are always designed so that the middle of the lens curves away from the eye, like the center lenses in Figure 23.19a and b. Why?

15. In a Jules Verne novel, a piece of ice is shaped into the form of a magnifying lens to focus sunlight and thereby start a fire. Is this possible?

16. Consider the image formed by a thin converging lens. Under what conditions will the image be (a) inverted? (b) erect? (c) real? (d) virtual? (e) larger than the object? (f) smaller than the object?

17. Repeat Question 16 for a thin diverging lens.

18. Discuss the proper position of a slide relative to the lens in a slide projector. What type of lens must the slide projector have?

19. Discuss why the following statements are true or false: (a) A virtual image can be a virtual object. (b) A virtual image can be a real object. (c) A real image can be a virtual object. (d) A real image can be a real object.

20. Discuss the type of aberration involved in each of the following situations. (a) The edges of the image appear reddish. (b) The image's central portion cannot be clearly focused. (c) The image's outer portion cannot be clearly focused. (d) The central portion of the image is enlarged relative to the outer portions.

21. You are standing in a room that has two adjacent walls made up of plane mirrors. In addition, the ceiling is a plane mirror. What is the maximum number of images you can see of yourself? Explain.

22. Figure 23.32 shows a lithograph by M. C. Escher titled *Hand with Reflection Sphere (Self-Portrait in Spherical*

FIGURE 23.32   (Question 22) *(© 1994 M.C. Escher/Cordon Art–Baarn–Holland. All rights reserved.)*

Mirror). Escher had this to say about the work: ''The picture shows a spherical mirror, resting on a left hand. But as a print is the reverse of the original drawing on stone, it was my right hand that you see depicted. (Being left-handed, I needed my left hand to make the drawing.) Such a globe reflection collects almost one's whole surroundings in one disk-shaped image. The whole room, four walls, the floor, and the ceiling, everything, albeit distorted, is compressed into that one small circle. Your own head, or more exactly the point between your eyes, is the absolute center. No matter how you turn or twist yourself, you can't get out of that central point. You are immovably the focus, the unshakable core, of your world.'' Comment on the accuracy of Escher's description.

PROBLEMS

Section 23.1   Plane Mirrors

1. Use Figure 23.2 to give a geometric proof that the virtual image formed by a plane mirror is the same distance behind the mirror as the object is in front of it.

2. (a) What is the minimum length for a plane mirror that would allow a person 6.00 ft tall to see his full height? (b) Does the distance from the mirror affect your answer? (*Hint:* A diagram showing the path followed by the rays that leave the

☐ indicates problems that have full solutions available in the Student Solutions Manual and Study Guide.

person would help. Follow those that leave the top of the head and the feet, strike the mirror, and then enter the eye.)

3. A person walks into a room that has, on opposite walls, two plane mirrors producing multiple images. When the person is 5.00 ft from the mirror on the left wall and 10.0 ft from the mirror on the right wall, find the distances from the person to the first three images seen in the left-hand mirror.

4. A light beam enters a small hole in a rectangular box that has three mirrored sides, as in Figure 23.33. (a) For what angle θ will the light beam emerge from the hole, assuming that it hits each mirror only once? (b) Show that your answer does not depend on the location of the hole.

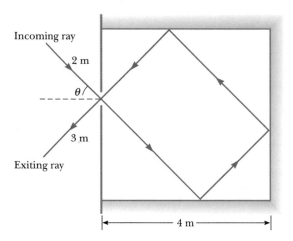

FIGURE 23.33   (Problem 4)

Section 23.2   Images Formed by Spherical Mirrors

Section 23.3   Convex Mirrors and Sign Conventions

In the following problems, algebraic signs are not given. We leave it to you to determine the correct sign to use with each quantity, based on an analysis of the problem and the sign conventions in Table 23.1.

5. (a) Prove, by using Equations 23.2 and 23.6, that the image of a real object formed by a convex mirror is *always* erect, virtual, and smaller than the object. (b) Prove, by using Equation 23.6, that the image of a real object placed in front of a spherical concave mirror is always virtual and erect when $p < |f|$.

6. A concave spherical mirror has a radius of curvature of 20.0 cm. Locate the images for object distances of (a) 40.0 cm, (b) 20.0 cm, and (c) 10.0 cm. In each case, state whether the image

is real or virtual and erect or inverted, and find the magnification.

7. A rectangle, 10.0 cm × 20.0 cm, is placed so that its right edge is 40.0 cm to the left of a concave spherical mirror, as in Figure 23.34. The radius of curvature of the mirror is 20.0 cm. (a) Draw the image seen through this mirror. (b) What is the area of the image?

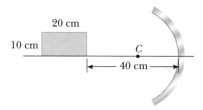

FIGURE 23.34   (Problem 7)

8. A ball is dropped from rest 3.00 m directly above the vertex of a concave mirror of radius 1.00 m. The mirror lies in a horizontal plane. (a) Describe the motion of the ball's image in the mirror. (b) At what time will the ball and its image coincide?

9. A dentist uses a mirror to examine a tooth. The tooth is 1.00 cm in front of the mirror, and the image is formed 10.0 cm behind the mirror. Determine (a) the mirror's radius of curvature and (b) the magnification of the image.

10. A convex mirror with a radius of curvature of 0.55 m monitors the aisles in a store. Locate and describe the image of a customer 10.0 m from the mirror. Determine the magnification.

11. A dedicated sports car enthusiast polishes the inside and outside surfaces of a hubcap that is a section of a sphere. When he looks into one side of the hubcap, he sees an image of his face 30.0 cm in back of the hubcap. He then turns the hubcap over and sees another image of his face 10.0 cm in back of the hubcap. (a) How far is his face from the hubcap? (b) What is the radius of curvature of the hubcap?

12. A spherical mirror is to be used to form an image, five times as tall as an object, on a screen positioned 5.0 m from the object. (a) Describe the type of mirror required. (b) Where should the mirror be positioned relative to the object?

13. A spherical Christmas tree ornament is 6.00 cm in diameter. What is the magnification of an object placed 10.0 cm away from the ornament?

14. A convex mirror has a focal length of 20.0 cm. Determine the object location for which the image will be one-half as tall as the object.

15. The real image height of a concave mirror is observed to be four times greater than the object height when the object is 30.0 cm in front of a mirror. What is the radius of curvature of the mirror?

16. A concave makeup mirror is designed so that a person 25 cm in front of it sees an erect image magnified by a factor of two. What is the radius of curvature of the mirror?

17. A 2.00-cm-high object is placed 10.0 cm in front of a mirror. What type of mirror and what radius of curvature are needed to create an upright image that is 4.00 cm high?

18. A man standing 1.52 m in front of a shaving mirror produces an inverted image 18.0 cm in front of it. How close to the mirror should he stand if he wants to form an upright image of his chin that is twice the chin's actual size?

19. A convex spherical mirror with a radius of curvature of 10.0 cm produces a virtual image one-third the size of the object. Where is the object?

20. A child holds a candy bar 10.0 cm in front of a convex mirror and notices that the image height is reduced by one half. What is the radius of curvature of the mirror?

21. A concave mirror has a focal length of 40.0 cm. Determine the object position that results in an erect image four times the height of the object.

22. A 2.00-cm-high object is placed 3.0 cm in front of a concave mirror. If the image is 5.0 cm high and virtual, what is the focal length of the mirror?

23. What type of mirror is required to form, on a wall 2.00 m from the mirror, an image of an object placed 10.0 cm in front of the mirror? What is the magnification of the image?

24. Under certain limiting conditions Equation 23.4 can be used for a plane mirror. (a) What value must be assumed for R if the mirror is plane? (b) What is the relationship between the object and image distances in this limiting case? (c) Does the result of (b) agree with the foregoing discussion of the plane mirror?

Section 23.4   Images Formed by Refraction

25. A swimming pool is 2.00 m deep. How deep does it appear to be (a) when completely filled with water? (b) when filled halfway with water?

26. A cubical block of ice 50.0 cm on an edge is placed on a level floor over a speck of dust. Locate the image of the speck if the index of refraction of ice is 1.309.

27. A colored marble is dropped into a large tank filled with benzene ($n = 1.50$). What is the depth of the tank if the apparent depth of the marble, viewed from directly above the tank, is 35.0 cm?

28. A transparent sphere of unknown composition is observed to form an image of the Sun on its surface opposite the Sun. What is the refractive index of the sphere material?

29. A goldfish is swimming in water inside a spherical plastic bowl of index of refraction 1.33. If the goldfish is 10.0 cm from the wall of the 15.0-cm-radius bowl, where does the goldfish appear to an observer outside the bowl?

30. A paperweight is made of a solid glass hemisphere of index of refraction 1.50. The radius of the circular cross-section is 4.0 cm. The center of the hemisphere is placed directly over a 2.5-mm-long line drawn on a sheet of paper. What length of line is seen by someone looking vertically down on the hemisphere?

31. One end of a long glass rod ($n = 1.50$) is formed into the shape of a convex surface of radius 8.00 cm. An object is positioned in air along the axis of the rod. Find the image position that corresponds to each of the following object positions: (a) 20.0 cm, (b) 8.00 cm, (c) 4.00 cm, (d) 2.00 cm.

32. Repeat Problem 31, assuming that the rod is placed in water instead of in air.

33. A flint glass plate ($n = 1.66$) rests on the bottom of an aquarium tank. The plate is 8.00 cm thick (vertical dimension) and is covered with water ($n = 1.33$) to a depth of 12.0 cm. Calculate the apparent thickness of the plate as viewed from above the water. (Assume nearly normal incidence.)

34. A goldfish is swimming at 2.00 cm/s toward the right side of a rectangularly shaped aquarium tank. What is the apparent speed of the goldfish as measured by an observer looking in from outside the right side of the tank? The index of refraction for water is 1.33.

Section 23.6   Thin Lenses

35. A contact lens is made of plastic with an index of refraction of 1.58. The lens has a focal length of +25.0 cm, and its inner surface has a radius of curvature of +18.0 mm. What is the radius of curvature of the outer surface?

36. A lens with radii of curvature of 52.5 cm and −61.9 cm has a focal length of +60.0 cm. Find its index of refraction.

37. A converging lens has a focal length of 20.0 cm. Locate the images for object distances of (a) 40.0 cm, (b) 20.0 cm, and (c) 10.0 cm. For each case, state whether the image is real or virtual and erect or inverted, and find the magnification.

38. A glass converging lens ($n = 1.50$) is designed to look like the lens in Figure 23.24a. The radius of the first surface is 15.0 cm, and the radius of the second surface is 10.0 cm. (a) Find the focal length of the lens. Determine the positions of the images for object distances of (b) infinity, (c) $3f$, (d) f, and (e) $f/2$.

39. A magnifying glass has a convex lens of focal length 15.0 cm. At what distance from a postage stamp should you hold this lens to get a magnification of $+2.00$?

40. Where must an object be placed to have no magnification ($|m| = 1.00$) (a) for a converging lens of focal length 12.0 cm? (b) for a diverging lens of focal length 12.0 cm?

41. A diverging lens has a focal length of 20.0 cm. Locate the images for object distances of (a) 40.0 cm, (b) 20.0 cm, and (c) 10.0 cm. For each case, state whether the image is real or virtual and erect or inverted, and find the magnification.

42. A diverging lens ($n = 1.50$) is shaped like that in Figure 23.24c. The radius of the first surface is 15.0 cm, and that of the second surface is 10.0 cm. (a) Find the focal length of the lens. Determine the positions of the images for object distances of (b) infinity, (c) $3|f|$, (d) $|f|$, and (e) $|f|/2$.

43. When an object is moved along the principal axis of a thin lens, the height of the image is five times the height of the object if the object is at point A (to the left of the lens) and also if the object is at point B, 20.0 cm farther from the lens. (a) Is the lens converging or diverging? (b) What is the focal length of the lens?

44. A microscope slide is placed in front of a converging lens with a focal length of 2.44 cm. The lens forms an image of the slide 12.9 cm from the slide. How far is the lens from the slide if the image is (a) real? (b) virtual?

45. The nickel's image in Figure 23.35 has twice the

FIGURE 23.35 (Problem 45)

diameter of the nickel and is 2.84 cm from the lens. Determine the focal length of the lens.

46. Figure 23.36 shows a thin glass ($n = 1.5$) converging lens with radii of curvature $R_1 = 15.0$ cm and $R_2 = -12.0$ cm. To the left of the lens is a square object of area 100.0 cm², with its base on the axis of the lens and with the right end of the square 20.0 cm to the left of the lens. (a) Determine the focal length of the lens. (b) Draw the image of the square as seen through the lens. What type of geometric figure is this? (c) Determine the area of the image.

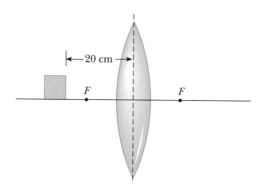

FIGURE 23.36 (Problem 46)

47. A diverging lens is used to form a virtual image of an object. The object is 80.0 cm to the left of the lens, and the image is 40.0 cm to the left of the lens. Determine the focal length of the lens.

48. We want to form an image 30.0 cm from a diverging lens with a focal length of -40.0 cm. Where must we place the object? Determine the magnification.

49. An object's distance from a converging lens is ten times the focal length. How far is the image from the focal point? Express the answer as a fraction of the focal length.

50. A person looks at a gem with a converging lens with a focal length of 12.5 cm. The lens forms a virtual image 30.0 cm from the lens. Determine the magnification. Is the image upright or inverted?

51. An object is placed 50.0 cm from a screen. Where should a converging lens with a 10.0-cm focal length be placed to form an image on the screen? Find the magnification(s).

52. A diverging lens is to be used to produce a virtual image one-third as tall as the object. Where should the object be placed?

53. Two converging lenses, each of focal length 15.0 cm, are placed 40.0 cm apart, and an object

is placed 30.0 cm in front of the first. Where is the final image formed, and what is the magnification of the system?

54. A converging lens is placed 30.0 cm to the right of a diverging lens of focal length 10.0 cm. A beam of parallel light enters the diverging lens from the left, and the beam is again parallel when it emerges from the converging lens. Calculate the focal length of the converging lens.

55. An object is placed 20 cm to the left of a converging lens of focal length 25 cm. A diverging lens of focal length 10 cm is 25 cm to the right of the converging lens. Find the position and magnification of the final image.

56. An object is 5.00 m to the left of a flat screen. A converging lens with focal length $f = 0.800$ m is placed between the object and the screen.
(a) Show that there are two positions for the lens that will cause an image to form on the screen, and determine how far these positions are from the object. (b) In what way would the two resultant images differ?

57. Lens L_1 in Figure 23.37 has a focal length of 15.0 cm, and lens L_2 has a focal length of 13.3 cm. The distance, d, of lens L_2 from the film plane can be varied from 5.00 cm to 10.0 cm. Determine the range of distances for which objects can be focused on the film.

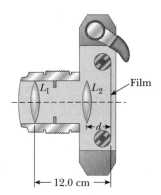

FIGURE 23.37 (Problem 57)

58. A 1.00-cm-high object is placed 4.00 cm to the left of a converging lens of focal length 8.00 cm. A diverging lens of focal length -16.00 cm is 6.00 cm to the right of the converging lens. Find the position and height of the final image. Is the image inverted or erect? Real or virtual?

59. Two converging lenses having focal lengths of 10.0 cm and 20.0 cm are placed 50.0 cm apart, as shown in Figure 23.38. The final image is to be located between the lenses, at the position indi-

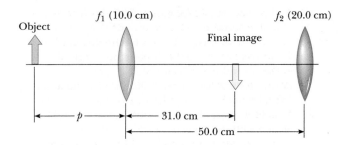

FIGURE 23.38 (Problem 59)

cated. (a) How far to the left of the first lens should the object be positioned? (b) What is the overall magnification? (c) Is the final image erect or inverted?

ADDITIONAL PROBLEMS

60. The lens and mirror in Figure 23.39 have focal lengths of $+80.0$ cm and -50.0 cm, respectively. An object is placed 1.00 m to the left of the lens, as shown. Locate the final image. State whether the image is erect or inverted, and determine the overall magnification.

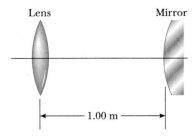

FIGURE 23.39 (Problem 60)

61. The object in Figure 23.40 is midway between the lens and the mirror. The mirror's radius of curvature is 20.0 cm, and the lens has a focal length of

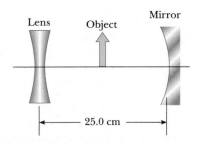

FIGURE 23.40 (Problem 61)

-16.7 cm. Considering only the light that leaves the object and travels first toward the mirror, locate the final image formed by this system. Is this image real or virtual? Is it erect or inverted? What is the overall magnification?

62. A thin lens of focal length 20.0 cm lies on a horizontal front-surfaced mirror. How far above the lens should an object be held if its image is to coincide with the object?

63. Figure 23.41 shows a converging lens with radii $R_1 = 9.00$ cm and $R_2 = -11.0$ cm, in front of a concave spherical mirror of radius $R = 8.00$ cm. The focal points (F_1 and F_2) for the thin lens and the center of curvature (C) of the mirror are also shown. (a) If the focal points F_1 and F_2 are 5.00 cm from the vertex of the thin lens, determine the index of refraction for the lens. (b) If the lens and mirror are 20.0 cm apart, and an object is placed 8.00 cm to the left of the lens, determine the position of the final image and its magnification as seen by the eye in the figure. (c) Is the final image inverted or upright? Explain.

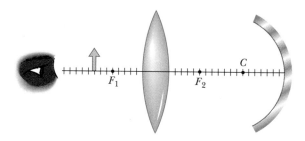

FIGURE 23.41   (Problem 63)

64. The disk of the Sun subtends an angle of 0.50° at the Earth. What are the position and diameter of a solar image formed by a concave spherical mirror of radius 3.0 m? (*Hint:* The angle subtended, θ, is given by $\tan \theta = -h'/q$.)

65. A converging lens has a focal length of 20.0 cm. Find the positions of the image for object distances of (a) 50.0 cm, (b) 30.0 cm, and (c) 10.0 cm. (d) For each object distance, determine the magnification of the lens and whether the image is erect or inverted.

66. An object placed 10.0 cm from a concave spherical mirror produces a real image 8.00 cm from the mirror. If the object is moved to a new position 20.0 cm from the mirror, what is the position of the image? Is the final image real or virtual?

67. A converging lens of focal length 20.0 cm is separated by 50.0 cm from a converging lens of focal length 5.00 cm. (a) Find the final position of the image of an object placed 40.0 cm in front of the

first lens. (b) If the height of the object is 2.00 cm, what is the height of the final image? Is it real or virtual? (c) Determine the image position of an object placed 5.00 cm in front of the two lenses in contact.

68. A real object is placed at the zero end of a meter stick. A large concave mirror at the 100-cm end of the meter stick forms an image of the object at the 70.0-cm position. A small convex mirror placed at the 20.0-cm position forms a final image at the 10.0-cm point. What is the radius of curvature of the convex mirror?

69. An object is positioned 36.0 cm to the left of a biconvex lens of index of refraction 1.50. The left surface of the lens has a radius of curvature of 20.0 cm. The right surface of the lens is to be shaped so that a real image will be formed 72.0 cm to the right of the lens. What is the required radius of curvature of the second surface?

70. An object is placed 12 cm to the left of a diverging lens of focal length -6.0 cm. A converging lens of focal length 12 cm is placed a distance of d to the right of the diverging lens. Find the distance d that places the final image at infinity.

71. A "floating coin" illusion consists of two parabolic mirrors, each with a focal length of 7.5 cm, facing each other so that their centers are 7.5 cm apart (Fig. 23.42). If a few coins are placed on the lower mirror, an image of the coins forms at the small opening at the center of the top mirror. Show that the final image forms at that location, and describe its characteristics. (*Note:* A flashlight beam shone on these *images* has a very startling effect. Even at a glancing angle, the incoming light beam is seemingly reflected off the *images* of the coins! Do you understand why?)

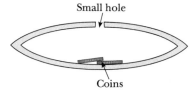

FIGURE 23.42   (Problem 71)

72. Object O_1 is 15.0 cm to the left of a converging lens of 10.0-cm focal length. A second lens is positioned 10.0 cm to the right of the first lens and is observed to form a final image at the position of the original object, O_1. (a) What is the focal length of the second lens? (b) What is the overall magnification of this system? (c) What is the nature (i.e., real or virtual, erect or inverted) of the final image?

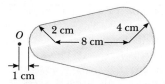

FIGURE 23.43   (Problem 73)

73. To work this problem, use the fact that the image formed by the first surface becomes the object for the second surface. Figure 23.43 shows a piece of glass with index of refraction 1.50. The ends are hemispheres with radii 2.00 cm and 4.00 cm, and the centers of the hemispherical ends are separated by a distance of 8.00 cm. A point object is in air, 1.00 cm from the left end of the glass. Locate the image of the object due to refraction at the two spherical surfaces.

74. The lens maker's equation for a lens with index n_1 immersed in a medium with index n_2 takes the form

$$\frac{1}{f} = \left(\frac{n_1}{n_2} - 1\right)\left(\frac{1}{R_1} - \frac{1}{R_2}\right)$$

A thin diverging glass (index = 1.5) lens with $R_1 = -3.0$ m and $R_2 = -6.0$ m is surrounded by air. An arrow is placed 10.0 m to the left of the lens. (a) Determine the position of the image. Repeat part (a) with the arrow and lens immersed in (b) water (index = 1.33); (c) a medium with an index of refraction of 2.0. (d) How can a lens that is diverging in air be changed into a converging lens?

75. Find the object distances (in terms of f) of a thin converging lens of focal length f if (a) the image is real and the image distance is four times the focal length; (b) the image is virtual and the image distance is three times the focal length. (c) Calculate the magnification of the lens for cases (a) and (b).

24 WAVE OPTICS

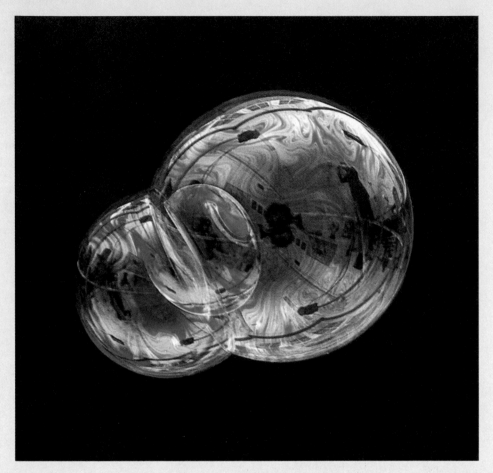

A layer of soap-film bubbles on water. The colors, produced just before the bubbles burst, are due to interference between light rays reflected from the front and back of the thin film of soap making up the bubble. The color depends on the thickness of the film, ranging from black where the film is at its thinnest to magenta where it is thickest. (Dr. Jeremy Burgess/Science Photo Library)

O ur discussion of light has thus far been concerned with what happens when light passes through a lens or reflects from a mirror. Because explanations of such phenomena rely on a geometric analysis of light rays, that part of optics is often called geometric optics. We now expand our study of light into an area called *wave optics*. The three primary topics we examine in this chapter are interference, diffraction, and polarization. These phenomena cannot be adequately explained with ray optics, but the wave theory leads us to satisfying descriptions.

24.1
CONDITIONS FOR INTERFERENCE

In our discussion of interference of mechanical waves in Chapter 13, we found that two waves could add together either constructively or destructively. In constructive interference, the amplitude of the resultant wave is greater than that of either of the individual waves, whereas in destructive interference, the resultant amplitude is less than that of either individual wave. Electromagnetic waves also undergo interference. Fundamentally, all interference associated with electromagnetic waves arises from the combining of the electromagnetic fields that constitute the individual waves.

Interference effects in light waves are not easy to observe because of the short wavelengths involved (about 4×10^{-7} m to about 7×10^{-7} m). For sustained interference between two sources of light to be observed, the following conditions must be met:

1. The sources must be **coherent;** that is, they must maintain a constant phase with respect to each other.
2. The sources must have identical wavelengths.
3. The superposition principle must apply.

Conditions for interference

Let us examine the characteristics of coherent sources. Two sources (producing two traveling waves) are needed to create interference. To produce a stable interference pattern, the individual waves must maintain a constant phase with one another. When this situation prevails, the sources are said to be coherent. The sound waves emitted by two side-by-side loudspeakers driven by a single amplifier can produce interference because the two speakers respond to the amplifier in the same way at the same time—i.e., are in phase.

If two light sources are placed side by side, however, no interference effects are observed, because the light waves from one source are emitted independently of the waves from the other source; hence, the emissions from the two sources do not maintain a constant phase relationship with each other during the time of observation. An ordinary light source undergoes random changes about once every 10^{-8} s. Therefore, the conditions for constructive interference, destructive interference, and intermediate states have durations on the order of 10^{-8} s. The result is that no interference effects are observed, since the eye cannot follow such short-term changes. Such light sources are said to be **noncoherent.**

A common method for producing two coherent light sources is to use a single-wavelength source to illuminate a screen containing two small slits. The light emerging from the two slits is coherent because a single source produces the original light beam and the slits serve only to separate the original beam into two parts (which is exactly what was done to the sound signal just mentioned). Any random change in the light emitted by the source will occur in the two separate beams at the same time, and interference effects can be observed.

24.2

YOUNG'S DOUBLE-SLIT INTERFERENCE

Interference in light waves from two sources was first demonstrated by Thomas Young in 1801. Figure 24.1a is a schematic diagram of the apparatus used in this experiment. (Young used pinholes rather than slits in his original experiments.) Light is incident on a screen in which there is a narrow slit, S_0. The light waves emerging from this slit arrive at a second screen that contains two narrow, parallel slits, S_1 and S_2. These slits serve as a pair of coherent light sources because waves emerging from them originate at the same wavefront and there-

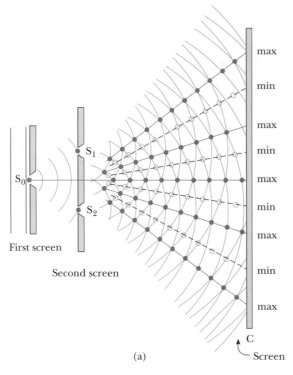

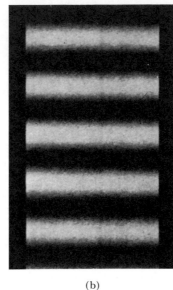

max
min
max
min
max
min
max
min
max

S₀

S₁

S₂

First screen

Second screen

C

Screen

(a)

(b)

FIGURE 24.1

(a) A schematic diagram of Young's double-slit experiment. The narrow slits act as sources of waves. Slits S_1 and S_2 behave as coherent sources that produce an interference pattern on screen C. (This drawing is not to scale.) (b) The fringe pattern formed on screen C could look like this.

FIGURE 24.2

An interference pattern involving water waves is produced by two vibrating sources at the water's surface. The pattern is analogous to that observed in Young's double-slit experiment. Note the regions of constructive and destructive interference.

(Richard Megna, Fundamental Photographs)

fore are always in phase. The light from the two slits produces on screen C a visible pattern consisting of a series of bright and dark parallel bands called **fringes** (Fig. 24.1b). When the light from slits S_1 and S_2 arrives at a point on the screen so that constructive interference occurs at that location, a bright line appears. When the light from the two slits combines destructively at any location on the screen, a dark line results. Figure 24.2 is a photograph of an interference pattern produced by two coherent vibrating sources in a water tank.

Figure 24.3 is a schematic diagram of some of the ways in which the two waves can combine at screen C. In Figure 24.3a, the two waves, which leave the two slits in phase, strike the screen at the central point, P. Since these waves travel equal distances, they arrive in phase at P, and as a result constructive interference occurs there and a bright fringe is observed. In Figure 24.3b, the two light waves again start in phase, but the upper wave has to travel one wavelength farther to reach point Q on the screen. Since the upper wave falls behind the lower one by exactly one wavelength, the two waves still arrive in phase at Q, and so a second bright fringe appears at that location. Now consider point R, midway between P and Q in Figure 24.3c. There the upper wave has fallen half a wavelength behind the lower wave. This means that the trough of the bottom wave overlaps the crest of the upper wave, giving rise to destructive interference at R. Consequently, a dark region can be observed at R. Figure 24.3d shows the intensity distribution on the screen. Notice that the central fringe is most intense and that the intensity decreases for higher-order fringes.

We can describe Young's experiment quantitatively with the help of Figure 24.4. Consider point P on the viewing screen; the screen is positioned a perpendicular distance of L from the screen containing slits S_1 and S_2, which are

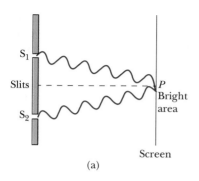

(a)

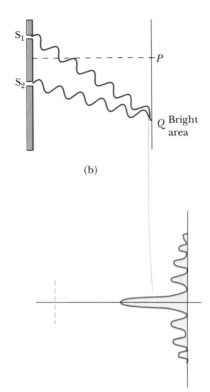

(b)

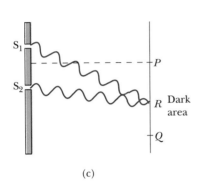

(c)

(d)

FIGURE 24.3
(a) Constructive interference occurs at *P* when the waves combine. (b) Constructive interference also occurs at *Q*. (c) Destructive interference occurs at *R* when the wave from the upper slit falls one-half wavelength behind the wave from the lower slit. (d) The intensity of the fringes decreases with movement to higher orders. (These figures are not drawn to scale.)

separated by distance d, and r_1 and r_2 are the distances the secondary waves travel from slit to screen. Let us assume that the source is monochromatic. Under these conditions, the waves emerging from S_1 and S_2 have the same frequency and amplitude and are in phase. The light intensity on the screen at P is the resultant of the light from both slits. Note that a wave from the lower slit travels farther than a wave from the upper slit by the amount $d \sin \theta$. This distance is called the **path difference, δ** (lowercase Greek delta), where

$$\delta = r_2 - r_1 = d \sin \theta \qquad \textbf{[24.1]} \qquad \text{Path difference}$$

This equation assumes that the two waves travel in parallel lines, which is approximately true, because L is much greater than d. As noted earlier, the value of this

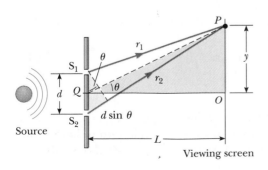

FIGURE 24.4
A geometric construction to describe Young's double-slit experiment. The path difference between the two rays is $r_2 - r_1 = d \sin \theta$. (This figure is not drawn to scale.)

path difference determines whether the two waves are in phase when they arrive at P. If the path difference is either zero or some integral multiple of the wavelength, the two waves are in phase at P and constructive interference results. Therefore, the condition for bright fringes, or **constructive interference,** at P is

Condition for constructive interference

$$\delta = d \sin \theta = m\lambda \qquad m = 0, \pm 1, \pm 2, \ldots \qquad \text{[24.2]}$$

The number m is called the **order number** of the fringe. The central bright fringe at $\theta = 0$ ($m = 0$) is called the *zeroth-order maximum;* the first maximum on either side, when $m = \pm 1$, is called the *first-order maximum;* and so forth.

Similarly, when the path difference is an odd multiple of $\lambda/2$, the two waves arriving at P are 180° out of phase and give rise to destructive interference. Therefore, the condition for dark fringes, or **destructive interference,** at P is

Condition for destructive interference

$$\delta = d \sin \theta = (m + \tfrac{1}{2})\lambda \qquad m = 0, \pm 1, \pm 2, \ldots \qquad \text{[24.3]}$$

If $m = 0$ in this equation, the path difference is $\delta = \lambda/2$, which is the condition for the location of the first dark line on either side of the central (bright) maximum. Likewise, if $m = 1$, $\delta = 3\lambda/2$, which is the condition for the second dark line on each side, and so forth.

It is useful to obtain expressions for the positions of the bright and dark fringes measured vertically from O to P. We assume that $L \gg d$ (Fig. 24.4) and that $d \gg \lambda$. The first assumption says that the distance from the slits to the screen is much greater than the distance between the two slits. The second says that the distance between the two slits is much greater than the wavelength. This situation can prevail in practice because L is often on the order of 1 m, whereas d is a fraction of a millimeter and λ is less than a micrometer for visible light. Under these conditions θ is small for the first several orders, and so we can use the approximation $\sin \theta \cong \tan \theta$. From the triangle OPQ in Figure 24.4, we see that

$$\sin \theta \approx \tan \theta = \frac{y}{L} \qquad \text{[24.4]}$$

Using this result together with the substitution $\sin \theta = m\lambda/d$ from Equation 24.2, we see that the positions of the *bright fringes,* measured from O, are

$$y_{\text{bright}} = \frac{\lambda L}{d} m \qquad \text{[24.5]}$$

Similarly, using Equations 24.3 and 24.4, we find that the *dark fringes* are located at

$$y_{\text{dark}} = \frac{\lambda L}{d}\left(m + \frac{1}{2}\right) \qquad \text{[24.6]}$$

As we shall demonstrate in Example 24.1, Young's double-slit experiment provides a method for measuring the wavelength of light. In fact, Young used this technique to make the first measurement of the wavelength of light. Additionally, his experiment gave the wave model of light a great deal of credibility.

Reflection, interference, and diffraction can be seen in this aerial photograph of waves in the sea. As the waves pass through a narrow gap, they spread out (diffract), and the interference of two waveforms is manifested in cross-patterned areas. (*John S. Shelton*)

Today we still use the phenomenon of interference to explain many observations of wave-like behavior.

EXAMPLE 24.1   Measuring the Wavelength of a Light Source

A screen is separated from a double-slit source by 1.2 m. The distance between the two slits is 0.030 mm. The second-order bright fringe ($m = 2$) is measured to be 4.5 cm from the centerline. Determine (a) the wavelength of the light and (b) the distance between adjacent bright fringes.

Solution   (a) We can use Equation 24.5 with $m = 2$, $y_2 = 4.5 \times 10^{-2}$ m, $L = 1.2$ m, and $d = 3.0 \times 10^{-5}$ m:

$$\lambda = \frac{y_2 d}{mL} = \frac{(4.5 \times 10^{-2}\ \text{m})(3.0 \times 10^{-5}\ \text{m})}{2(1.2\ \text{m})} = 5.6 \times 10^{-7}\ \text{m} = \boxed{560\ \text{nm}}$$

Reasoning and Solution   (b) Since the positions of the bright fringes are given by Equation 24.5, we see that the distance between *any* adjacent bright fringes (say, those characterized by m and $m + 1$) is

$$\Delta y = y_{m+1} - y_m = \frac{\lambda L}{d}(m + 1) - \frac{\lambda L}{d}m = \frac{\lambda L}{d}$$

$$= \frac{(5.6 \times 10^{-7}\ m)\left(1.2\ \dfrac{d}{m}\right)}{3.0 \times 10^{-5}\ m} = \boxed{2.2\ \text{cm}}$$

24.3

CHANGE OF PHASE DUE TO REFLECTION

Young's method of producing two coherent light sources involves illuminating a pair of slits with a single source. Another simple, although ingenious, arrangement for producing an interference pattern with a single light source is known as *Lloyd's mirror*. A light source is placed at point S, close to a mirror, as illustrated in Figure 24.5. Waves can reach the viewing point, P, either by the direct path SP or by the path involving reflection from the mirror. The reflected ray can be treated as a ray originating at the source S', behind the mirror. Source S', which is the image of S, can be considered a virtual source.

At points far from the source, one would expect an interference pattern due to waves from S and S', just as is observed for two real coherent sources. An interference pattern is indeed observed. However, the positions of the dark and bright fringes are *reversed* relative to the pattern of two real coherent sources (Young's experiment). This is because the coherent sources S and S' differ in phase by 180°. This 180° phase change is produced by reflection.

To illustrate this further, consider point P', where the mirror meets the screen. This point is equidistant from S and S'. If path difference alone were responsible for the phase difference, one would expect to see a bright fringe at P' (since the path difference is zero for this point), corresponding to the central fringe of the two-slit interference pattern. Instead, one observes a *dark* fringe at P' because of the 180° phase change produced by reflection. In general, an electromagnetic wave undergoes a phase change of 180° upon reflection from a medium of higher index of refraction than the one in which it was traveling.

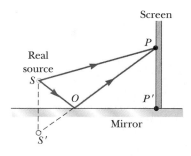

FIGURE 24.5
Lloyd's mirror. An interference pattern is produced on a screen at P as a result of the combination of the direct ray and the reflected ray. The reflected ray undergoes a phase change of 180°.

INTERFERENCE

The photograph on the left shows a layer of soap-film bubbles on water. The colors, produced just before the bubbles burst, are due to interference between light rays reflected from the front and back of the thin film of soap making up the bubble. The color depends on the thickness of the film, ranging from black where the film is at its thinnest to magenta where it is at its thickest.

In the photograph on the right, a thin film of oil on water displays interference, evidenced by the pattern of colors when white light is incident on the film. The film thickness varies in the vicinity of the blade, thereby producing the interesting color pattern.

(Dr. Jeremy Burgess/Science Photo Library)

(Peter Aprahamian/Science Photo Library)

It is useful to draw an analogy between reflected light waves and the reflections of a transverse wave on a stretched string when the wave meets a boundary, as in Figure 24.6. The reflected pulse on a string undergoes a phase change of 180° when it is reflected from the boundary of a denser medium, such as a heavier string, and no phase change when it is reflected from the boundary of a less dense medium. Similarly, an electromagnetic wave undergoes a 180° phase change when reflected from the boundary of a medium of higher index of refraction than the one in which it has been traveling. There is no phase change when the wave is reflected from a boundary leading to a medium of lower index of refraction. The part of the wave that crosses the boundary also undergoes no phase change.

24.4

INTERFERENCE IN THIN FILMS

Interference effects are commonly observed in thin films, such as soap bubbles and thin layers of oil on water. The varied colors observed with ordinary white light result from the interference of waves reflected from the opposite surfaces of the film.

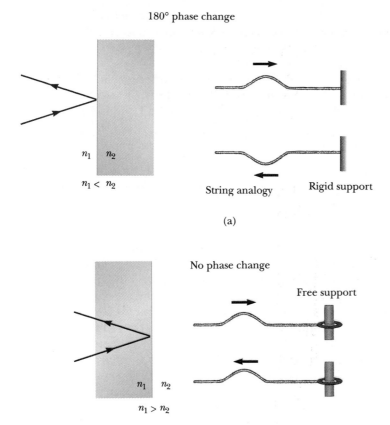

180° phase change

n_1 n_2

$n_1 < n_2$

String analogy Rigid support

(a)

No phase change

Free support

n_1 n_2

$n_1 > n_2$

(b)

FIGURE 24.6

(a) A ray reflecting from a medium of higher refractive index undergoes a 180° phase change. The right side shows the analogy with a reflected pulse on a string. (b) A ray reflecting from a medium of lower refractive index undergoes *no* phase change.

Consider a film of uniform thickness t and index of refraction n, as in Figure 24.7. Let us assume that the light rays traveling in air are nearly normal to the two surfaces of the film. To determine whether the reflected rays interfere constructively or destructively, we must first note the following facts:

1. An electromagnetic wave traveling from a medium of index of refraction n_1 toward a medium of index of refraction n_2 undergoes a 180° phase change upon reflection when $n_2 > n_1$. There is no phase change in the reflected wave if $n_2 < n_1$.
2. The wavelength of light, λ_n, in a medium with index of refraction n is

$$\lambda_n = \frac{\lambda}{n} \qquad \text{[24.7]}$$

where λ is the wavelength of light in vacuum.

We apply these rules to the film of Figure 24.7. According to the first rule, ray 1, which is reflected from the upper surface (A), undergoes a phase change of 180° with respect to the incident wave. Ray 2, which is reflected from the lower surface (B), undergoes no phase change with respect to the incident wave. Therefore, ray 1 is 180° out of phase with respect to ray 2, a situation that is equivalent to a path difference of $\lambda_n/2$. However, we must also consider the fact that ray 2 travels an extra distance of $2t$ before the waves recombine. For exam-

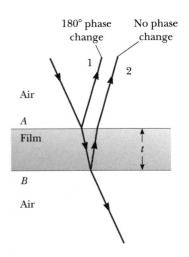

180° phase change No phase change

1 2

Air

A

Film t

B

Air

FIGURE 24.7

Interference observed in light reflected from a thin film is due to a combination of rays reflected from the upper and lower surfaces.

ple, if $2t = \lambda_n/2$, rays 1 and 2 recombine in phase and constructive interference results. In general, the condition for constructive interference is

$$2t = (m + \tfrac{1}{2})\lambda_n \qquad (m = 0, 1, 2, \ldots) \qquad \textbf{[24.8]}$$

This condition takes into account two factors: (a) the difference in optical path length for the two rays (the term $m\lambda_n$) and (b) the 180° phase change upon reflection (the term $\lambda_n/2$). Since $\lambda_n = \lambda/n$, we can write Equation 24.8 in the form

$$2nt = (m + \tfrac{1}{2})\lambda \qquad (m = 0, 1, 2, \ldots) \qquad \textbf{[24.9]}$$

If the extra distance $2t$ traveled by ray 2 is a multiple of λ_n, the two waves combine out of phase and destructive interference results. The general equation for destructive interference is

$$2nt = m\lambda \qquad (m = 0, 1, 2, \ldots) \qquad \textbf{[24.10]}$$

It is important to realize that two factors influence interference: (1) phase reversals on reflection and (2) differences in travel distance. The foregoing conditions for constructive and destructive interference are valid only when the medium above the top surface of the film is the same as the medium below the bottom surface. The surrounding medium may have a refractive index less than or greater than that of the film. In either case, the rays reflected from the two surfaces will be out of phase by 180°. If the film is placed between two *different* media, one of lower refractive index and one of higher refractive index, the conditions for constructive and destructive interference are reversed. In this case, either there is a phase change of 180° for both ray 1 reflecting from surface *A* and ray 2 reflecting from surface *B*, or there is no phase change for either ray; hence, the net change in relative phase due to the reflections is *zero*.

NEWTON'S RINGS

Another method for observing interference of light waves is to place a plano-convex lens on top of a flat glass surface, as in Figure 24.8a. With this arrangement, the air film between the glass surfaces varies in thickness from zero at the point of contact to some value t at P. If the radius of curvature of the lens, R, is very large compared with the distance r, and if the system is viewed from above using light of wavelength λ, a pattern of light and dark rings is observed (Fig. 24.8b). These circular fringes are called **Newton's rings** after their discoverer. Newton's particle model of light could not explain the origin of the rings.

The interference is due to the combination of ray 1, reflected from the plate, with ray 2, reflected from the lower surface of the lens. Ray 1 undergoes a phase change of 180° upon reflection, because it is reflected from a boundary leading into a medium of higher refractive index, whereas ray 2 undergoes no phase change. Hence, the conditions for constructive and destructive interference are given by Equations 24.9 and 24.10, respectively, with $n = 1$ since the "film" is air. Here again, one might guess that the contact point, O, would be bright, corresponding to constructive interference. Instead, it is dark, as seen in Figure 24.8b, because ray 1, reflected from the plate, undergoes a 180° phase change with respect to ray 2. Using the geometry shown in Figure 24.8a, one can obtain expressions for the radii of the bright and dark bands in terms of the radius of curvature, R, and vacuum wavelength, λ. For example, the dark rings have radii of $r \approx \sqrt{m\lambda R/n}$. In Problem 37 at the end of the chapter, you will be asked to supply the details.

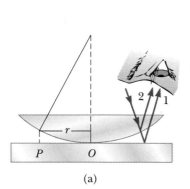
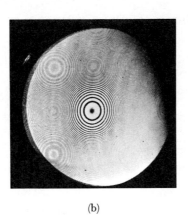
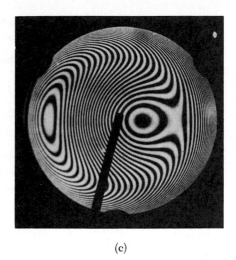

(a)   (b)   (c)

FIGURE 24.8
(a) The combination of rays reflected from the glass plate and the curved surface of the lens gives rise to an interference pattern known as Newton's rings. (b) A photograph of Newton's rings. *(Courtesy of Bausch & Lomb Optical Co.)* (c) This asymmetric interference pattern indicates imperfections in the lens. *(From Physical Science Study Committee, College Physics, Lexington, Mass., D. C. Heath and Co., 1968)*

One of the important uses of Newton's rings is in the testing of optical lenses. A circular pattern like that in Figure 24.8b is achieved only when the lens is ground to a perfectly spherical curvature. Variations from such symmetry might produce a pattern like that in Figure 24.8c. These variations give an indication of how the lens must be ground and polished to remove the imperfections.

PROBLEM-SOLVING STRATEGY
Thin-Film Interference

The following features should be kept in mind when you work thin-film interference problems:

1. **Identify the thin film causing the interference.**
2. **The type of interference that occurs is determined by the phase relationship between the portion of the wave reflected at the upper surface of the film and the portion reflected at the lower surface.**
3. **Phase differences between the two portions of the wave have two causes: (a) differences in the distances traveled by the two portions and (b) phase changes occurring upon reflection. *Both* causes must be considered when you are determining which type of interference occurs.**
4. **When distance and phase changes upon reflection are both taken into account, the interference is constructive if the path difference between the two waves is an integral multiple of λ, and destructive if the equivalent path difference is $\lambda/2$, $3\lambda/2$, $5\lambda/2$, and so forth.**

EXAMPLE 24.2   Interference in a Soap Film

Calculate the minimum thickness of a soap-bubble film ($n = 1.33$) that will result in constructive interference in the reflected light, if the film is illuminated by light with a wavelength in free space of 602 nm.

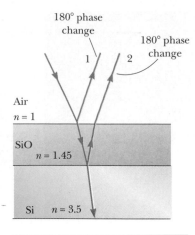

FIGURE 24.9
(Example 24.3)   Reflective losses from a silicon solar cell are minimized by coating it with a thin film of silicon monoxide, SiO.

Reasoning   The minimum film thickness for constructive interference corresponds to $m = 0$ in Equation 24.9. This gives $2nt = \lambda/2$.

Solution   Since $2nt = \lambda/2$, we have

$$t = \frac{\lambda}{4n} = \frac{602 \text{ nm}}{4(1.33)} = \boxed{113 \text{ nm}}$$

Exercise   What other film thicknesses will produce constructive interference?

Answer   338 nm, 564 nm, 789 nm, and so on.

EXAMPLE 24.3   Nonreflecting Coatings for Solar Cells

Semiconductors such as silicon are used to fabricate solar cells—devices that generate electric energy when exposed to sunlight. Solar cells are often coated with a transparent thin film, such as silicon monoxide (SiO; $n = 1.45$) to minimize reflective losses (Fig. 24.9). A silicon solar cell ($n = 3.50$) is coated with a thin film of silicon monoxide for this purpose. Assuming normal incidence, determine the minimum thickness of the film that will produce the least reflection at a wavelength of 552 nm.

Reasoning   Reflection is least when rays 1 and 2 in Figure 24.9 meet the condition of destructive interference. Note that both rays undergo 180° phase changes upon reflection. Hence, the net change in phase due to reflection is zero, and the condition for a reflection *minimum* is a path difference of $\lambda_n/2$; therefore, $2t = \lambda/2n$.

Solution   Since $2t = \lambda/2n$, the required thickness is

$$t = \frac{\lambda}{4n} = \frac{552 \text{ nm}}{4(1.45)} = \boxed{95.2 \text{ nm}}$$

Typically, such coatings reduce the reflective loss from 30% (with no coating) to 10% (with coating), thereby increasing the cell's efficiency, since more light is available to create charge carriers in the cell. In reality, the coating is never perfectly nonreflecting, because the required thickness is wavelength-dependent and the incident light covers a wide range of wavelengths.

Glass lenses used in cameras and other optical instruments are usually coated with a transparent thin film, such as magnesium fluoride (MgF_2), to reduce or eliminate unwanted reflection. More important, such coatings enhance the transmission of light through the lenses.

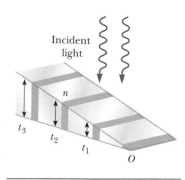

FIGURE 24.10
(Example 24.4)   Interference bands in reflected light can be observed by illuminating a wedge-shaped film with monochromatic light. The dark areas correspond to positions of destructive interference.

EXAMPLE 24.4   Interference in a Wedge-Shaped Film

A thin, wedge-shaped film of refractive index n is illuminated with monochromatic light of wavelength λ, as illustrated in Figure 24.10. Describe the interference pattern observed in this case.

Reasoning and Solution   The interference pattern is that of a thin film of variable thickness surrounded by air. Hence, the pattern is a series of alternating bright and dark parallel bands. A dark band corresponding to destructive interference appears at point O, the apex, since the upper reflected ray undergoes a 180° phase change while the lower one does not. According to Equation 24.10, other dark bands appear when $2nt = m\lambda$, so that $t_1 = \lambda/2n$, $t_2 = \lambda/n$,

$t_3 = 3\lambda/2n$, and so on. Similarly, bright bands are observed when the thickness satisfies the condition $2nt = (m + \frac{1}{2})\lambda$, corresponding to thicknesses of $\lambda/4n$, $3\lambda/4n$, $5\lambda/4n$, and so on. If white light is used, bands of different colors are observed at different points, corresponding to the different wavelengths of light present.

24.5

DIFFRACTION

Suppose a light beam is incident on two slits, as in Young's double-slit experiment. If the light truly traveled in straight-line paths after passing through the slits, as in Figure 24.11a, the waves would not overlap and no interference pattern would be seen. Instead, Huygens' principle requires that the waves spread out from the slits, as shown in Figure 24.11b. In other words, the light deviates from a straight-line path and enters the region that would otherwise be shadowed. This divergence of light from its initial line of travel is called **diffraction.**

In general, diffraction occurs when waves pass through small openings, around obstacles, or by sharp edges. For example, when a narrow slit is placed between a distant light source (or a laser beam) and a screen, the light produces a diffraction pattern like that in Figure 24.12. The pattern consists of a broad, intense central band, the **central maximum,** flanked by a series of narrower, less intense secondary bands (called **secondary maxima**) and a series of dark bands, or **minima.** This cannot be explained within the framework of geometric optics, which says that light rays traveling in straight lines should cast a sharp image of the slit on the screen.

Figure 24.13 shows the diffraction pattern and shadow of a penny. The pattern consists of the shadow, a bright spot at its center, and a series of bright and dark bands of light near the edge of the shadow. The bright spot at the center (called the *fresnel bright spot* after its discoverer, Augustin Fresnel) can be ex-

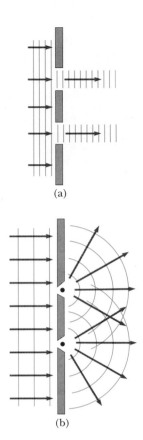

(a)

(b)

FIGURE 24.11
(a) If light did not spread out after passing through the slits, no interference would occur. (b) The light from the two slits overlaps as it spreads out, producing interference fringes.

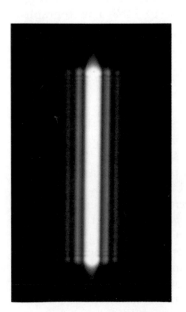

FIGURE 24.12
The diffraction pattern that appears on a screen when light passes through a narrow vertical slit. The pattern consists of a broad central band and a series of less intense and narrower side bands.

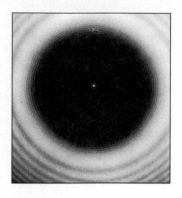

FIGURE 24.13
The diffraction pattern of a penny placed midway between the screen and the source. (*Courtesy of P. M. Rinard, from* Am. J. Phys., *44:70, 1976*)

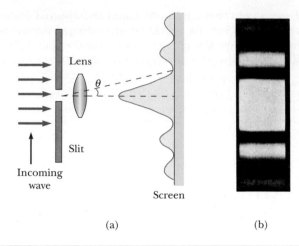

(a)   (b)

FIGURE 24.14
(a) The Fraunhofer diffraction pattern of a single slit. The parallel rays are brought into focus on the screen with a converging lens. The pattern consists of a central bright region flanked by much weaker maxima. (This drawing is not to scale.) (b) A photograph of a single-slit Fraunhofer diffraction pattern. (*From M. Cagnet, M. Francon, and J. C. Thierr,* Atlas of Optical Phenomena, *Berlin, Springer-Verlag, 1962, plate 18*)

plained by the wave theory of light, which predicts constructive interference at this point. In contrast, from the viewpoint of geometric optics, the center of the pattern would be completely screened by the penny, and so one would never observe a central bright spot.

One type of diffraction, called **Fraunhofer diffraction,** occurs when the rays reaching the observing screen are approximately parallel. This can be achieved experimentally either by placing the observing screen far from the slit or by using a converging lens to focus the parallel rays on the screen, as in Figure 24.14a. A bright fringe is observed along the axis at $\theta = 0$, with alternating dark and bright fringes on each side of the central bright fringe. Figure 24.14b is a photograph of a single-slit Fraunhofer diffraction pattern.

24.6

SINGLE-SLIT DIFFRACTION

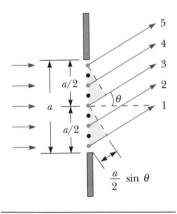

FIGURE 24.15
Diffraction of light by a narrow slit of width a. Each portion of the slit acts as a point source of waves. The path difference between rays 1 and 3 or between rays 2 and 4 is equal to $(a/2)$ $\sin \theta$ (This drawing is not to scale, and the rays are assumed to converge at a distant point.)

Until now we have assumed that slits are point sources of light. In this section we determine how their finite widths are the basis for understanding the nature of the Fraunhofer diffraction pattern produced by a single slit.

We can deduce some important features of this problem by examining waves coming from various portions of the slit, as shown in Figure 24.15. According to Huygens' principle, *each portion of the slit acts as a source of waves.* Hence, *light from one portion of the slit can interfere with light from another portion,* and the resultant intensity on the screen depends on the direction θ.

To analyze the diffraction pattern, it is convenient to divide the slit into halves, as in Figure 24.15. All the waves that originate at the slit are in phase. Consider waves 1 and 3, which originate at the bottom and center of the slit, respectively. Wave 1 travels farther than wave 3 by an amount equal to the path difference $(a/2) \sin \theta$, where a is the width of the slit. Similarly, the path difference between waves 3 and 5 is also $(a/2) \sin \theta$. If this path difference is exactly half of a wavelength (corresponding to a phase difference of 180°), the two waves cancel each other and destructive interference results. This is true, in fact, for any two waves that originate at points separated by half the slit width, because the phase difference between two such points is 180°. Therefore, waves from the upper half of the slit interfere *destructively* with waves from the lower half of the slit when

$$\frac{a}{2} \sin \theta = \frac{\lambda}{2}$$

or when

$$\sin \theta = \frac{\lambda}{a}$$

If we divide the slit into four parts rather than two, and use similar reasoning, we find that the screen is also dark when

$$\sin \theta = \frac{2\lambda}{a}$$

Likewise, we can divide the slit into six parts and show that darkness occurs on the screen when

$$\sin \theta = \frac{3\lambda}{a}$$

Therefore, the general condition for **destructive interference** is

$$\sin \theta = m \frac{\lambda}{a} \qquad m = \pm 1, \pm 2, \pm 3, \ldots \qquad \textbf{[24.11]}$$

Equation 24.11 gives the values of θ for which the diffraction pattern has zero intensity—that is, a dark fringe is formed. However, this equation tells us nothing about the variation in intensity along the screen. The general features of the intensity distribution along the screen are shown in Figure 24.16. A broad central bright fringe is observed, flanked by much weaker, alternating bright fringes. The various dark fringes (points of zero intensity) occur at the values of θ

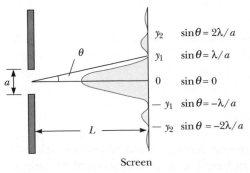

y_2 $\sin \theta = 2\lambda/a$

y_1 $\sin \theta = \lambda/a$

0   $\sin \theta = 0$

$-y_1$ $\sin \theta = -\lambda/a$

$-y_2$ $\sin \theta = -2\lambda/a$

Screen

FIGURE 24.16
Positions of the minima for the Fraunhofer diffraction pattern of a single slit of width a. (This is not to scale.)

A compact disc acts as a diffraction grating when observed under white light. The colors that are observed in the reflected light and their intensities depend on the orientation of the disc relative to the eye and to the light source. Can you explain how this works? *(Kristen Brochmann/Fundamental Photographs)*

that satisfy Equation 24.11. The points of constructive interference lie approximately halfway between the dark fringes. Note that the central bright fringe is twice as wide as the weaker maxima.

EXAMPLE 24.5   Where Are the Dark Fringes?

Light of wavelength 580 nm is incident on a slit of width 0.30 mm. The observing screen is placed 2.0 m from the slit. Find the positions of the first dark fringes and the width of the central bright fringe.

Solution   The first dark fringes that flank the central bright fringe correspond to $m = \pm 1$ in Equation 24.11:

$$\sin \theta = \pm \frac{\lambda}{a} = \pm \frac{5.8 \times 10^{-7}\ \text{m}}{0.30 \times 10^{-3}\ \text{m}} = \pm 1.9 \times 10^{-3}$$

From the triangle in Figure 24.16, we see that $\tan \theta = y_1/L$. Since θ is very small, we can use the approximation $\sin \theta \approx \tan \theta$, so that $\sin \theta \approx y_1/L$. Therefore, the positions of the first minima, measured from central axis, are

$$y_1 \approx L \sin \theta = \pm L \frac{\lambda}{a} = \quad \pm 3.9 \times 10^{-3}\ \text{m}$$

The positive and negative signs correspond to the first dark fringes on either side of the central bright fringe. Hence, the width of the central bright fringe is given by $2|y_1| = 7.8 \times 10^{-3}\ \text{m} = 7.8\ \text{mm}$. Note that this value is much greater than the width of the slit. However, as the width of the slit is *increased,* the diffraction pattern *narrows,* corresponding to smaller values of θ. In fact, for large values of a, the maxima and minima are so closely spaced that the only observable pattern is a large central bright area resembling the geometric image of the slit.

Exercise   Determine the width of the first-order bright fringe.

Answer   3.9 mm.

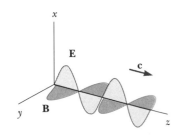

FIGURE 24.17
A schematic diagram of a polarized electromagentic wave propagating in the z direction. The electric field vector, **E**, vibrates in the xz plane, and the magnetic field vector, **B**, vibrates in the yz plane.

24.7

POLARIZATION OF LIGHT WAVES

In Chapter 21 we described the transverse nature of electromagnetic waves. Figure 24.17 shows that the electric and magnetic vectors associated with an electromagnetic wave are at right angles to each other and also to the direction of wave propagation. The phenomenon of polarization, described in this section, is firm evidence of the transverse nature of electromagnetic waves.

An ordinary beam of light consists of a large number of waves emitted by the atoms or molecules of the light source. Each atom produces a wave with its own orientation, **E**, as in Figure 24.17, corresponding to the direction of atomic vibration. However, since all directions of vibration are possible, the resultant electromagnetic wave is a superposition of waves produced by the individual atomic sources. The result is an **unpolarized** light wave, represented schematically in Figure 24.18a. The direction of wave propagation in this figure is perpendicular to the page. Note that *all* directions of the electric field vector are equally probable and lie in a plane (such as the plane of this page) perpendicular to the direction of propagation. At any given point and at some instant of time, there is only one resultant electric field; do not be misled by Figure 24.18a.

A wave is said to be **linearly polarized** if **E** vibrates in the same direction *at all times* at a particular point, as in Figure 24.18b. (Sometimes such a wave is described as *plane-polarized* or simply *polarized*.) The wave in Figure 24.17 is an example of a wave linearly polarized in the *y* direction. As the field propagates in the *x* direction, **E** is always in the *y* direction. The plane formed by **E** and the direction of propagation is called the *plane of polarization* of the wave. In Figure 24.17, the plane of polarization is the *xy* plane. It is possible to obtain a linearly polarized wave from an unpolarized wave by removing from the unpolarized wave all components except those whose electric field vectors oscillate in a single plane. We shall now discuss three processes for doing this: (1) selective absorption, (2) reflection, and (3) scattering.

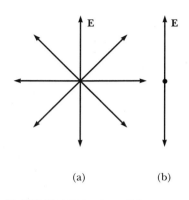

(a)   (b)

FIGURE 24.18
(a) An unpolarized light beam viewed along the direction of propagation (perpendicular to the page). The transverse electric field vector can vibrate in any direction with equal probability. (b) A linearly polarized light beam with the electric field vector vibrating in the vertical direction.

POLARIZATION BY SELECTIVE ABSORPTION

The most common technique for polarizing light is to use a material that (1) transmits waves whose electric field vectors vibrate in a plane parallel to a certain direction and (2) absorbs those waves whose electric field vectors vibrate in directions perpendicular to that direction.

In 1932 E. H. Land discovered a material, which he called **polaroid,** that polarizes light through selective absorption by oriented molecules. This material is fabricated in thin sheets of long-chain hydrocarbons, which are stretched during manufacture so that the molecules align. After a sheet is dipped into a solution containing iodine, the molecules become good electrical conductors. However, the conduction takes place primarily along the hydrocarbon chains, since the valence electrons of the molecules can move easily only along the chains (recall that valence electrons are "free" electrons that can readily move through the conductor). As a result, the molecules readily *absorb* light whose electric field vector is parallel to their lengths, and *transmit* light whose electric field vector is perpendicular to their lengths. It is common to refer to the direction perpendicular to the molecular chains as the **transmission axis.** In an ideal polarizer, all light with **E** parallel to the transmission axis is transmitted, and all light with **E** perpendicular to the transmission axis is absorbed.

Let us now describe the intensity of light that passes through a polarizing material. In Figure 24.19, an unpolarized light beam is incident on the first polarizing sheet, called the **polarizer,** where the transmission axis is as indicated. The light that is passing through this sheet is polarized vertically, and the transmitted electric field vector is $\mathbf{E}_0$. A second polarizing sheet, called the **analyzer,** intercepts this beam with its transmission axis at an angle of θ to the axis of the

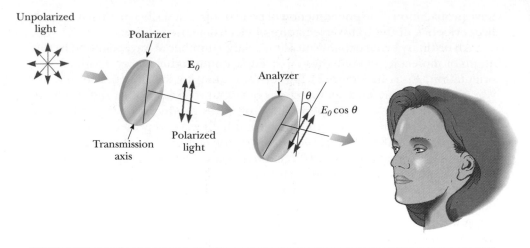

FIGURE 24.19
Two polarizing sheets whose transmission axes make an angle of θ with each other. Only a fraction of the polarized light incident on the analyzer is transmitted.

polarizer. The component of $\mathbf{E}_0$ that is perpendicular to the axis of the analyzer is completely absorbed, and the component parallel to the axis, $E_0 \cos \theta$, passes through. The transmitted intensity varies as the *square* of the transmitted amplitude, and the intensity of the transmitted (polarized) light varies as

$$I = I_0 \cos^2 \theta \qquad \text{[24.12]}$$

where I_0 is the intensity of the polarized wave incident on the analyzer. This expression, known as **Malus's law,** applies to any two polarizing materials whose transmission axes are at an angle of θ to each other. From this expression, note that the transmitted intensity is a maximum when the transmission axes are parallel ($\theta = 0$ or $180°$) and zero (complete absorption by the analyzer) when the transmission axes are perpendicular to each other. This variation in transmitted intensity through a pair of polarizing sheets is illustrated in Figure 24.20.

POLARIZATION BY REFLECTION

When an unpolarized light beam is reflected from a surface, the reflected light is completely polarized, partially polarized, or unpolarized, depending on the angle of incidence. If the angle of incidence is either $0°$ or $90°$ (a normal or grazing angle), the reflected beam is unpolarized. However, for angles of incidence between $0°$ and $90°$, the reflected light is polarized to some extent. For one particular angle of incidence, the reflected beam is completely polarized. Let us now investigate that special angle.

Suppose an unpolarized light beam is incident on a surface, as in Figure 24.21a. The beam can be described by two electric field components, one parallel to the surface (represented by dots) and the other perpendicular to the first component and to the direction of propagation (represented by red arrows). It is found that the parallel component reflects more strongly than the other components, and this results in a partially polarized beam. Furthermore, the refracted beam is also partially polarized.

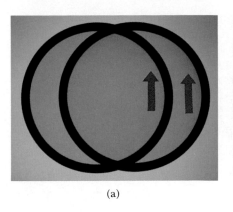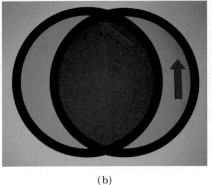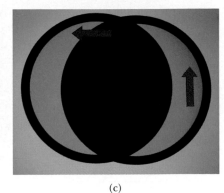

(a) (b) (c)

FIGURE 24.20

The intensity of light transmitted through two polarizers depends on the relative orientations of their transmission axes. (a) The transmitted light has *maximum* intensity when the transmission axes are *aligned* with each other. (b) The transmitted light intensity diminishes when the transmission axes are at an angle of 45° with each other. (c) The transmitted light intensity is a *minimum* when the transmission axes are at *right angles* to each other. *(Photos courtesy of Henry Leap)*

Now suppose that the angle of incidence, θ_1, is varied until the angle between the reflected and refracted beams is 90° (Fig. 24.21b). At this particular angle of incidence, the reflected beam is completely polarized, with its electric field vector parallel to the surface, while the refracted beam is partially polarized. The angle of incidence at which this occurs is called the **polarizing angle**, θ_p.

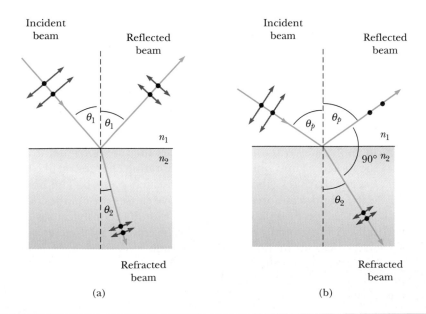

(a) (b)

FIGURE 24.21

(a) When unpolarized light is incident on a reflecting surface, the reflected and refracted beams are partially polarized. (b) The reflected beam is completely polarized when the angle of incidence equals the polarizing angle, θ_p, satisfying the equation $n = \tan \theta_p$.

An expression relating the polarizing angle to the index of refraction of the reflecting surface can be obtained by use of Figure 24.21b. From this figure we see that, at the polarizing angle, $\theta_p + 90° + \theta_2 = 180°$, so that $\theta_2 = 90° - \theta_p$. Using Snell's law, and taking $n_1 = n_{air} = 1.00$ and $n_2 = n$,

$$n = \frac{\sin \theta_1}{\sin \theta_2} = \frac{\sin \theta_p}{\sin \theta_2}$$

Because $\sin \theta_2 = \sin(90° - \theta_p) = \cos \theta_p$, the expression for n can be written

Brewster's law

$$n = \frac{\sin \theta_p}{\cos \theta_p} = \tan \theta_p \qquad \text{[24.13]}$$

This expression is called **Brewster's law,** and the polarizing angle θ_p is sometimes called **Brewster's angle** after its discoverer, Sir David Brewster (1781–1868). For example, Brewster's angle for crown glass ($n = 1.52$) has the value $\theta_p = \tan^{-1}(1.52) = 56.7°$. Because n varies with wavelength for a given substance, Brewster's angle is also a function of wavelength.

Polarization by reflection is a common phenomenon. Sunlight reflected from water, glass, or snow is partially polarized. If the surface is horizontal, the electric field vector of the reflected light has a strong horizontal component. Sunglasses made of polarizing material reduce the glare of reflected light. The transmission axes of the lenses are oriented vertically to absorb the strong horizontal component of the reflected light.

POLARIZATION BY SCATTERING

When light is incident on a system of particles, such as a gas, the electrons in the medium can absorb and reradiate part of the light. The absorption and reradiation of light by the medium, called **scattering,** is what causes sunlight reaching an observer on the Earth from straight overhead to be polarized. You can observe this effect by looking directly up through a pair of sunglasses made of polarizing glass. Less light passes through at certain orientations of the lenses than at others.

Figure 24.22 illustrates how the sunlight becomes polarized. The left side of the figure shows an incident unpolarized beam of sunlight on the verge of striking an air molecule. When the beam strikes the air molecule, it sets the electrons of the molecule into vibration. These vibrating charges act like those in an antenna except that they vibrate in a complicated pattern. The horizontal part of the electric field vector in the incident wave causes the charges to vibrate horizontally, and the vertical part of the vector simultaneously causes them to vibrate vertically. A horizontally polarized wave is emitted by the electrons as a result of their horizontal motion, and a vertically polarized wave is emitted parallel to the Earth as a result of their vertical motion.

Scientists have found that bees and homing pigeons use the polarization of sunlight as a navigational aid.

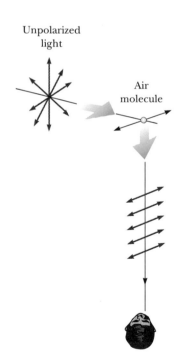

Unpolarized light

Air molecule

FIGURE 24.22
The scattering of unpolarized sunlight by air molecules. The light observed at right angles is plane polarized because the vibrating molecule has a horizontal component of vibration.

OPTICAL ACTIVITY

Many important practical applications of polarized light involve the use of certain materials that display the property of **optical activity**. A substance is said to be optically active if it rotates the plane of polarization of transmitted light. To

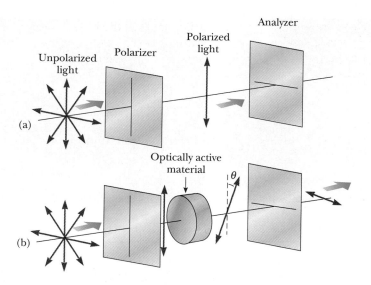

FIGURE 24.23

(a) When crossed polarizers are used, none of the polarized light can pass through the analyzer. (b) An optically active material rotates the direction of polarization through the angle θ, enabling some of this light to pass through the analyzer.

understand how this process occurs, suppose unpolarized light is incident on a polarizer from the left, as in Figure 24.23a. The transmitted light is polarized vertically, as shown. If this light is then incident on an analyzer with its axis perpendicular to that of the polarizer, no light emerges from it. If an optically active material is placed between the polarizer and analyzer as in Figure 24.23b, the material causes the direction of the polarized beam to rotate through the angle θ. As a result, some light is able to pass through the analyzer. The angle through which the light is rotated by the material can be found by rotating the polarizer until the light is again extinguished. It is found that the angle of rotation depends on the length of the sample and, if the substance is in solution, on the concentration. One optically active material is a solution of common sugar, dextrose. A standard method for determining the concentration of sugar solutions is to measure the rotation produced by a fixed length of the solution.

Optical activity occurs in a material because of an asymmetry in the shape of its constituent molecules. For example, some proteins are optically active because of their spiral shapes. Other materials, such as glass and plastic, become optically active when placed under stress. If polarized light is passed through an unstressed piece of plastic and then through an analyzer with an axis perpendicular to that of the polarizer, none of the polarized light is transmitted. However, if the plastic is placed under stress, the regions of greatest stress produce the largest angles of rotation of polarized light. Hence, one observes a series of light and dark bands in the transmitted light. Engineers often use this procedure in the design of structures ranging from bridges to small tools. A plastic model is built and analyzed under different load conditions to determine positions of potential weakness and failure under stress. If the design is poor, patterns of light and dark bands will indicate the points of greatest weakness, and the design can be corrected at an early stage. Figure 24.24 shows examples of stress patterns in two plastic models.

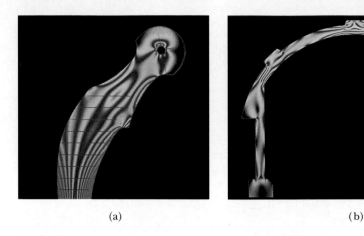

(a)   (b)

FIGURE 24.24

(a) Strain distribution in a plastic model of a hip replacement used in a medical research laboratory. The pattern is produced when the plastic model is placed between two crossed polarizers. (b) A plastic model of an arch structure under load conditions observed between two crossed polarizers. The stress pattern is produced in the regions where the stresses are greatest. Such patterns and models are useful for the optimal design of architectural components. (a *and* b, *Peter Aprahamian/Sharples Stress Engineers Ltd./Science Photo Library*)

SPECIAL TOPIC

LIQUID CRYSTALS

An effect similar to rotation of the plane of polarization is used to create the familiar displays on pocket calculators, wristwatches, laptop computers, and so forth. The properties of a unique type of substance called a liquid crystal make these displays (correspondingly called LCDs for *liquid crystal displays*) possible. As its name implies, a **liquid crystal** is a substance with properties intermediate between those of a crystalline solid and those of a liquid; that is, the molecules of the substance are more orderly than those in a liquid but less than those in a pure crystalline solid. The forces that hold the molecules together in such a state are just barely strong enough to enable the substance to maintain a definite shape, so it is reasonable to call it a solid. However, small imputs of mechanical or electrical energy can disrupt these weak bonds and make the substance flow, rotate, or twist.

To see how liquid crystals can be used to create a display, consider Figure 24.25a. The liquid crystal is placed between two glass plates in the pattern shown, and electrical contacts, indicated by the thin lines, are made to the liquid crystal. When a voltage is applied across any segment in the display, that segment turns dark. In this fashion, any number between 0 and 9 can be formed by the pattern, depending on the voltages applied to the eight segments.

To see why a segment can be changed from dark to light by the application of a voltage, consider Figure 24.25b, which shows the basic construction of a portion of the display. The liquid crystal is placed between two glass substrates that are packaged between two pieces of Polaroid material with their transmission axes perpendicular. A reflecting surface is placed behind one of the pieces of Polaroid. First consider what happens when light falls on this package and no voltages are applied to the liquid crystal, as shown in Figure 24.25b. Incoming light is polarized by the first polarizer

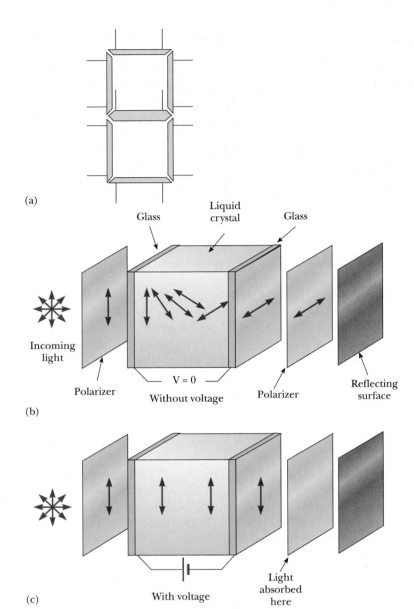

(a)

(b)

(c)

FIGURE 24.25
(a) The light-segment pattern of a liquid crystal display. (b) Rotation of a polarized light beam by a liquid crystal when the applied voltage is zero. (c) Molecules of the liquid crystal align with the electric field when a voltage is applied.

and then falls on the liquid crystal. A voltage is applied to the liquid crystal so that its molecules gradually change orientation by 90° from the first piece of glass to the second. This causes a rotation of the plane of polarization of the light as it moves through the crystal. Thus, when light strikes the second polarizer, it has rotated 90° and is allowed to pass through to the reflecting surface. The light reflects and retraces its path, and the segment appears light to a viewer at the left of the figure. When a voltage is applied as in Figure 24.25c, the molecules of the liquid crystal rotate to align with the electric field—that is, parallel to the direction of travel of the light. As a result, no rotation of the plane of polarization occurs, and the light is absorbed by the second polarizer. Thus, there is no light to be reflected, and a viewer at the left sees this segment of the liquid crystal as black. Changing the voltage applied to the crystal in a precise pattern and at precise times can make the pattern tick off the seconds on a watch, display a letter on a computer display, and so forth.

SUMMARY

Interference occurs when two or more light waves overlap at a given point. A sustained interference pattern is observed if (1) the sources are coherent (that is, they maintain a constant phase relationship with one another), (2) the sources have identical wavelengths, and (3) the superposition principle is applicable.

In **Young's double-slit experiment**, two slits separated by distance d are illuminated by a single-wavelength light source. An interference pattern consisting of bright and dark fringes is observed on a screen a distance of L from the slits. The condition for **bright fringes** (constructive interference) is

$$d \sin \theta = m\lambda \qquad m = 0, \pm 1, \pm 2, \ldots \qquad \textbf{[24.2]}$$

The condition for **dark fringes** (destructive interference) is

$$d \sin \theta = (m + \tfrac{1}{2})\lambda \qquad m = 0, \pm 1, \pm 2, \ldots \qquad \textbf{[24.3]}$$

The number m is called the **order number** of the fringe.

An electromagnetic wave undergoes a phase change of 180° upon reflection from a medium with an index of refraction higher than that of the medium in which the wave is traveling.

The wavelength of light, λ_n, in a medium with index of refraction n is

$$\lambda_n = \frac{\lambda}{n} \qquad \textbf{[24.7]}$$

where λ is the wavelength of the light in free space.

The **diffraction pattern** produced by a single slit on a distant screen consists of a central bright maximum and alternating bright and dark regions of lower intensity. The angles θ at which the diffraction pattern has zero intensity (regions of destructive interference) are

$$\sin \theta = \frac{m\lambda}{a} \qquad m = \pm 1, \pm 2, \pm 3, \ldots \qquad \textbf{[24.11]}$$

where a is the width of the slit and λ is the wavelength of the light incident on the slit.

Unpolarized light can be polarized by selective absorption, reflection, and scattering.

In general, light reflected from an amorphous material, such as glass, is partially polarized. Reflected light is completely polarized with its electric field parallel to the surface when the angle of incidence produces a 90° angle between the reflected and refracted beams. This angle of incidence, called the **polarizing angle**, θ_p, satisfies **Brewster's law**, given by

$$n = \tan \theta_p \qquad \textbf{[24.13]}$$

where n is the index of refraction of the reflecting medium.

ADDITIONAL READING

P. Baumeister and G. Pincus, "Optical Interference Coatings," *Sci. American,* December 1970, p. 59.

C. Bohrens and A. Fraser, "Colors of the Sky," *The Physics Teacher,* May 1985, p. 267.

H. M. Nussensvieg, "The Theory of the Rainbow," *Sci. American,* April 1977, p. 116.

Scientific American, September 1968. The entire issue is devoted to light.

W. A. Shurcliffe and S. S. Ballard, *Polarized Light,* Van Nostrand, 1964.

R. E. Slusher and B. Yurke, "Squeezed Light," *Sci. American,* May 1988, p. 50.

A. C. S. van Heel and C. H. F. Velzel, *What Is Light?* World University Library, New York, McGraw-Hill, 1968, Chaps. 3, 4, and 5.

H. C. Von Baeyer, "Rainbows, Whirlpools, and Clouds," *The Scientist,* July/August 1984, p. 24.

R. Wehner, "Polarized Light Navigation by Insects," *Sci. American,* July 1976, p. 106.

A. Wood and F. Oldham, *Thomas Young, Natural Philosopher, 1773–1829,* Cambridge, England, Cambridge University Press, 1954.

CONCEPTUAL QUESTIONS

Example   If Young's double-slit experiment were performed under water, how would the observed interference pattern be affected?

Reasoning   The wavelength of light traveling in water would decrease, since the wavelength of light in a medium is given by $\lambda_0 = \lambda / n$, where λ is the wavelength in vacuum and n is the index of refraction of the medium. Since the positions of the bright and dark fringes are proportional to the wavelength, the fringe separations would decrease.

Example   If white light is used in Young's double-slit experiment rather than monochromatic light, how does the interference pattern change?

Reasoning   Every color produces its own interference pattern and we see them superposed. The central maximum is white. The first side maximum is a full spectrum with violet on the inside and red on the outside. The second maximum is also a full spectrum, with red in it overlapping with violet in the third maximum. At larger angles the light soon starts mixing to white again.

Example   What is the necessary condition on path length difference between two waves that interfere (a) constructively and (b) destructively?

Reasoning   (a) Two waves interfere constructively if their path difference is either zero or some integral multiple of the wavelength; that is, if the path difference is $m\lambda$, where m is an integer. (b) Two waves interfere destructively if their path difference is an odd multiple of one-half of a wavelength; that is, if the path difference equals $(m + \frac{1}{2})\lambda$.

Example   Describe the change in width of the central maximum of the single-slit diffraction pattern as the width of the slit is made smaller.

Reasoning   The width of the central maximum increases as the width of the slit is made smaller. This can be seen by inspecting the condition for destructive interference, $\sin \theta = \lambda / a$. Since the width of the central maximum is proportional to $\sin \theta$, as the slit-width a is reduced, the width of the central maximum increases.

Example   Suppose reflected white light is used to observe a thin, transparent coating on glass as the coating material is gradually deposited by evaporation in a vacuum. Describe possible color changes that might occur during the process of building up the thickness of the coating.

Reasoning   Suppose the index of refraction of the coating is intermediate between vacuum and the glass. When the coating is very thin, light reflected from its top and bottom surfaces will interfere constructively, so you see the surface white and brighter. Once the thickness reaches one-quarter of the wavelength of violet light in the coating, destructive interference for violet light will make the surface look red. Then other colors in spectral order (blue, green, yellow, orange, and red) will interfere destructively, making the surface look red, violet, and then blue. As the coating gets thicker, constructive interference is observed for violet and then for other colors in spectral order. Even thicker coatings give constructive and destructive interference for several visible wavelengths, so the reflected light starts looking white again.

Example   Certain sunglasses use a polarizing material to reduce the intensity of light reflected from shiny surfaces, such as water or the hood of a car. What orientation of the transmission axis should the material have to be most effective?

Reasoning   The reflected light is partially polarized, with the component parallel to the reflecting surface being the most intense. Therefore, the polarizing material should have its transmission axis oriented in the vertical direction in order to minimize the intensity of the reflected light.

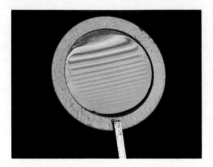

FIGURE 24.26   (Question 5)

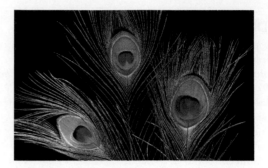

FIGURE 24.27   (Question 19) *(Diane Schiumo/Fundamental Photographs)*

1. A simple way of observing an interference pattern is to look at a distant light source through a stretched handkerchief or an open umbrella. Explain how this works.
2. Why is it so much easier to perform interference experiments with a laser than with an ordinary light source?
3. A lens with outer radius of curvature R and index of refraction n rests on a flat glass plate. It is illuminated with white light from above. Is there a dark spot or a light spot at the center of the lens? What does it mean if the observed rings are noncircular?
4. In the process of evaporation, a soap bubble appears black just before it breaks. Explain this phenomenon in terms of the phase changes that occur upon reflection from the two surfaces.
5. If a soap film on a wire loop is held in air, as in Figure 24.26, and observed by reflected light, it appears black in the thinnest regions and shows a variety of colors in thicker regions. Explain.
6. If an oil film is observed on water, the film appears brightest at the outer regions, where it is thinnest. From this information, what can you say about the index of refraction of the oil relative to that of water?
7. Lenses with a "nonreflective" coating usually appear to have a purplish cast. Why?
8. Would it be possible to place a nonreflective coating on an airplane to cancel radar waves of wavelength 3 cm?
9. To observe interference in a thin film, why must the film be thin compared with the wavelengths of visible light? (*Hint:* How far apart are the two reflected

waves when they attempt to interfere, if the film is thick?)
10. Washed dishes that are not rinsed well often have colored bands or rings across them. Discuss the interference effect that causes this.
11. When the battery of a calculator that uses an LCD is low, the display is easier to see if it is at an angle to one's eyes (other than 90°). Explain.
12. Explain why you can hear someone around an open doorway, but you cannot see the person.
13. A point source of light occurs inside a container that is opaque except for a single hole. Discuss what happens to the image on the screen as the diameter of the hole is reduced (a) in the ray approximation and (b) in an actual situation.
14. Light falls on a slit that is twice as wide as it is high. What kind of pattern is formed on a screen? In which direction is the beam the most spread out?
15. Show, by a sketch, that the waves emitted by a radio antenna are polarized.
16. Can a sound wave be polarized? Explain.
17. Why is the sky black when viewed from the Moon?
18. Is light from the sky polarized? Why is it that clouds seen through Polaroid glasses stand out in bold contrast to the sky?
19. The brilliant colors of peacock feathers (Fig. 24.27) are due to a phenomenon known as *iridescence*. The melanin fibers in the feathers act as a series of narrow slits. How do you explain the differing colors? Why do the colors often change as the bird moves?

PROBLEMS

Section 24.2   Young's Double-Slit Interference

1. A radio transmitter, A, operating at 60.0 MHz is 10.0 m from another similar transmitter, B, that is 180° out of phase with A. How far must an observer move from A toward B along the line connecting A and B, to reach the nearest point where the two beams are in phase?
2. If the distance between two slits is 0.050 mm and the distance to a screen is 2.5 m, find the spacing

□ indicates problems that have full solutions available in the Student Solutions Manual and Study Guide.

between the first- and second-order bright fringes for yellow light of 600-nm wavelength.

3. A pair of narrow, parallel slits separated by 0.250 mm are illuminated by the green component from a mercury vapor lamp ($\lambda = 546.1$ nm). The interference pattern is observed on a screen 1.20 m from the plane of the parallel slits. Calculate the distance (a) from the central maximum to the first bright region on either side of the central maximum and (b) between the first and second dark bands in the interference pattern.

4. A double slit with a spacing of 0.083 mm between the slits is 2.5 m from a screen. (a) If yellow light of wavelength 570 nm strikes the double slit, what is the separation between the zeroth- and first-order maxima on the screen? (b) If blue light of wavelength 410 nm strikes the double slit, what is the separation between the second- and fourth-order maxima? (c) Repeat parts (a) and (b) for the minima.

5. A riverside warehouse has two open doors, as in Figure 24.28. A boat on the river sounds its horn. To person A the sound is loud and clear. To person B the sound is barely audible. The principal wavelength of the sound waves is 3.00 m. Assuming person B is at the position of the first minimum, determine the distance between the doors, center to center.

6. Light of wavelength 575 nm falls on a double slit, and the first bright fringe is seen at an angle of 16.5°. Find the distance between the slits.

7. Light of wavelength 460 nm falls on two slits spaced 0.30 mm apart. What is the required distance from the slits to a screen if the spacing between the first and second dark fringes is to be 4.0 mm?

8. White light spans the wavelength range between about 400 nm and 700 nm. If white light passes through two slits 0.30 mm apart and falls on a screen 1.5 m from the slits, find the distance between the first-order violet and red fringes.

9. The yellow component of light from a helium discharge tube ($\lambda = 587.5$ nm) is allowed to fall on a plate containing parallel slits that are 0.200 mm apart. A screen is positioned so that the second bright fringe in the interference pattern is a distance equal to 10 slit spacings from the central maximum. What is the distance between the plate and the screen?

10. A Young's interference experiment is performed with blue-green argon laser light. The separation between the slits is 0.50 mm, and the interference pattern on a screen 3.3 m away shows the first maximum 3.4 mm from the center of the pattern. What is the wavelength of argon laser light?

11. Two radio antennas separated by 300 m, as shown in Figure 24.29, simultaneously transmit identical signals (assume waves) of the same wavelength. A radio in a car traveling due north receives the signals. (a) If the car is at the position of the second maximum, what is the wavelength of the signals? (b) How much farther must the car travel to encounter the next minimum in reception? (*Caution:* Avoid small angle approximations in this problem.)

12. Waves from a radio station have a wavelength of 300 m. They travel by two paths to a home receiver 20.0 km from the transmitter. One path is a direct path, and the second is by reflection from a mountain directly behind the home receiver. What is the minimum distance from the mountain

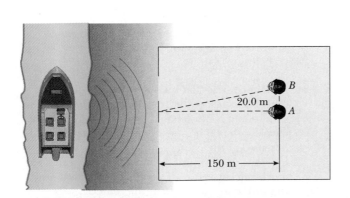

FIGURE 24.28 (Problem 5)

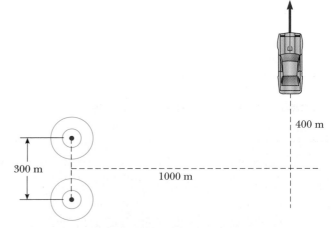

FIGURE 24.29 (Problem 11)

to the receiver that produces destructive interference at the receiver? (Assume that no phase change occurs on reflection from the mountain.)

13. Waves broadcast by a 1500-kHz radio station arrive at a home receiver by two paths. One is a direct path, and the second is by reflection off an airplane directly above the home receiver. The airplane is approximately 100 m above the home receiver, and the direct distance from the station to the home is 20 km. What is the exact height of the airplane if destructive interference is occurring? (Assume that no phase change occurs on reflection from the plane.)

14. The waves from a radio station can reach a home receiver by two different paths. One is a straight-line path from the transmitter to the home, a distance of 30.0 km. The second path is by reflection from the ionosphere (a layer of ionized air molecules near the top of the atmosphere). Assume that this reflection takes place at a point midway between receiver and transmitter. If the wavelength broadcast by the radio station is 350 m, find the minimum height of the ionospheric layer that will produce destructive interference between the direct and reflected beams. (Assume no phase changes on reflection.)

15. Radio waves from a star, of wavelength 250.0 m, reach a radio telescope by two separate paths. One is a direct path to the receiver, which is situated on the edge of a cliff by the ocean. The second is by reflection off the water. The first minimum of destructive interference occurs when the star is 25.0° above the horizon. Find the height of the cliff. (Assume no phase change on reflection.)

16. One of the bright bands in Young's interference pattern is 12.0 mm from the central maximum. The screen is 119 cm from the pair of slits that serve as sources. The slits are 0.241 mm apart and are illuminated by the blue light from a hydrogen discharge tube (λ = 486 nm). How many bright lines are observed between the central maximum and the 12.0-mm position?

Section 24.3   Change of Phase Due to Reflection

Section 24.4   Interference in Thin Films

17. Suppose the film shown in Figure 24.7 has an index of refraction of 1.36 and is surrounded by air on both sides. Find the minimum thickness, other than zero, that will produce constructive interference in the reflected light when the film is illuminated by light of wavelength 500 nm.

18. Nonreflective coatings on camera lenses reduce the loss of light at the surfaces of multi-lens systems and prevent internal reflections that might mar the image. Find the minimum thickness of a layer of magnesium fluoride (n = 1.38) on flint glass (n = 1.66) that will cause destructive interference of reflected light of wavelength 550 nm near the middle of the visible spectrum.

19. A thin layer of liquid methylene iodide (n = 1.756) is sandwiched between two flat parallel plates of glass. What must be the thickness of the liquid layer if normally incident light with λ = 600 nm in air is to be strongly reflected?

20. A thin film of oil (n = 1.38) floats on water. What color is the film at points where the oil thickness is 300 nm? Assume that the film is viewed from above with white light.

21. Two parallel glass plates are placed in contact and illuminated from above with light of wavelength 580 nm. As the plates are slowly moved apart and the reflected light is observed, darkness occurs at certain separations. What are the distances (other than, perhaps, zero) of the first three of these separations?

22. When a liquid is introduced into the air space between the lens and the plate in a Newton's-rings apparatus, the diameter of the tenth ring changes from 1.50 to 1.31 cm. Find the index of refraction of the liquid. (See Problem 37.)

23. A beam of light of wavelength 580 nm passes through two closely spaced glass plates, as shown in Figure 24.30. For what minimum nonzero value of the plate separation, d, will the transmitted light be bright?

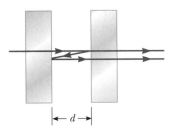

FIGURE 24.30   (Problem 23)

24. A planoconvex (flat on one side, convex on the other) lens rests with its curved side on a flat glass surface and is illuminated from above by light of wavelength 500 nm (see Fig. 24.8). A dark spot is observed at the center, surrounded by 19 concentric dark rings (with bright rings in between). How much thicker is the air wedge at the position of the 19th dark ring than at the center?

25. A planoconvex lens (see Problem 24) with radius of curvature R = 3.0 m is in contact with a flat plate of glass. A light source and the observer's

eye are both close to the normal, as shown in Figure 24.8. The radius of the 50th bright Newton's ring is found to be 9.8 mm. What is the wavelength of the light produced by the source?

26. A transparent oil of index of refraction 1.29 spills on the surface of water (index of refraction 1.33), producing a maximum of reflection with normally incident orange light (wavelength 600 nm in air). Assuming the maximum occurs in the first order, determine the thickness of the oil slick.

27. In the Newton's-rings arrangement, how much greater (as a fraction) is the radius of the ninth dark ring to that of the fourth dark ring? (See Problem 37.)

28. A thin film of glass ($n = 1.50$) floats on a liquid of $n = 1.35$. When the glass is illuminated by light of $\lambda = 580$ nm incident from air above it, find the minimum thickness, other than zero, that will produce destructive interference in the reflected light.

29. A thin 1.00×10^{-5}-cm-thick film of MgF_2 ($n = 1.38$) is used to coat a camera lens. Are any wavelengths in the visible spectrum intensified in the reflected light?

30. A coating is applied to a lens to minimize reflections. The index of refraction of the coating is 1.550, and that of the lens is 1.480. If the coating is 177.4 nm thick, which reflected wavelengths are minimally reflected for normal incidence in the lowest order?

31. Determine the minimum thickness of a soap film ($n = 1.330$) that will result in constructive interference of (a) the red H_α line ($\lambda = 656.3$ nm); (b) the blue H_γ line ($\lambda = 434.0$ nm).

32. A soap bubble of index of refraction 1.40 strongly reflects both red and green colors in white light. What thickness of soap bubble allows this to happen? (In air, $\lambda_{red} = 700$ nm, $\lambda_{green} = 500$ nm.)

33. Two rectangular optically flat plates ($n = 1.52$) are in contact along one end and are separated along the other end by a sheet of paper that is 4.00×10^{-3} cm thick (Fig. 24.31). The top plate is illuminated by monochromatic light of wavelength 546.1 nm. Calculate the number of dark parallel bands crossing the top plate (including the dark band at zero thickness along the edge of contact between the two plates).

34. The condition for constructive interference by reflection from a thin film in air, as developed in Section 24.4, assumes nearly normal incidence. (a) Show that if the light is incident on the film at an angle of $\theta_1 \gg 0$ (relative to the normal), then the condition for constructive interference is given by $2nt/\cos\theta_2 = (m + 1/2)\lambda$, where θ_2 is the angle of refraction. (b) Calculate the minimum thickness for constructive interference if sodium light ($\lambda = 590$ nm) is incident at an angle of 30.0° on a film with index of refraction 1.38.

35. An air wedge is formed between two thick glass plates in a manner like that described in Problem 33. Light of wavelength 434 nm is incident vertically on the top plate. In this case, there are 20 bright parallel interference fringes across the top plate. Calculate the thickness of the paper separating the plates.

36. A flat piece of glass is supported horizontally above the flat end of a 10.0-cm-long metal rod that has its lower end rigidly fixed. The thin film of air between the rod and glass is observed to be bright when illuminated by light of wavelength 500 nm. As the temperature is slowly increased by 25.0° C, the film changes from bright to dark and back to bright 200 times. What is the coefficient of linear expansion of the metal?

37. A planoconvex lens (flat on one side, convex on the other) with index of refraction n rests with its curved side (radius of curvature R) on a flat glass surface of the same index of refraction with a film of index n_{film} between them. The lens is illuminated from above by light of wavelength λ. Using the geometry in Figure 24.8a, show that the dark Newton rings have radii of

$$r \approx \sqrt{m\lambda R / n_{film}}$$

where m is an integer.

Section 24.6   Single-Slit Diffraction

38. Microwaves of wavelength 5.00 cm enter a long, narrow window in a building that is otherwise essentially opaque to the microwaves. If the window is 36.0 cm wide, what is the distance from the central maximum to the first-order minimum along a wall 6.50 m from the window?

39. Prove that if the wavelength of light is equal to or greater than the width of a slit, light striking the slit perpendicularly passes through without forming any dark interference bands.

40. Light of wavelength 600 nm falls on a 0.40-mm-wide slit and forms a diffraction pattern on a screen 1.5 m away. (a) Find the position of the first dark band on each side of the central maximum. (b) Find the width of the central maximum.

41. A screen is placed 50.0 cm from a single slit,

FIGURE 24.31   (Problems 33 and 35)

which is illuminated with light of wavelength 680 nm. If the distance between the first and third minima in the diffraction pattern is 3.00 mm, what is the width of the slit?

42. If the light in Figure 24.15 strikes the single slit at an angle of β from the perpendicular direction, show that Equation 24.11, the condition for destructive interference, must be modified to read

$$\sin \theta = m \left(\frac{\lambda}{a} \right) - \sin \beta$$

43. Light of wavelength 587.5 nm illuminates a single 0.75-mm-wide slit. (a) At what distance from the slit should a screen be placed if the first minimum in the diffraction pattern is to be 0.85 mm from the central maximum? (b) Calculate the width of the central maximum.

44. A slit of width 0.50-mm is illuminated with light of wavelength 500 nm, and a screen is placed 120 cm in front of the slit. Find the widths of the first and second maxima on each side of the central maximum.

Section 24.7   Polarization of Light Waves

45. (a) If light is incident at an angle of θ from a medium of index n_1 on a medium of index n_2 so that the angle between the reflected ray and refracted ray is β, show that

$$\tan \theta = \frac{n_2 \sin \beta}{n_1 - n_2 \cos \beta}$$

(*Hint:* Use the following identity.)

$$\sin(A + B) = \sin A \cos B + \cos A \sin B$$

(b) Show that the foregoing equation for $\tan \theta$ reduces to Brewster's law when $\beta = 90°$, $n_1 = 1$, and $n_2 = n$.

46. The index of refraction of a glass plate is 1.52. What is the Brewster's angle when the plate is (a) in air? (b) in water? (See Problem 48.)

47. A light beam is incident on heavy flint glass ($n = 1.65$) at the polarizing angle. Calculate the angle of refraction for the transmitted ray.

48. Equation 24.13 assumes that the incident light is in air. If the light is incident from a medium of index n_1 on a medium of index n_2, follow the procedure used to derive Equation 24.13 to show that $\tan \theta_p = n_2 / n_1$.

49. Use the result of Problem 48 to find Brewster's angle when light is reflected off a piece of glass ($n = 1.50$) submerged in water.

50. At what angle above the horizon is the Sun if light from it is completely polarized upon reflection from water?

51. The angle of incidence of a light beam in air onto a reflecting surface is continuously variable. The reflected ray is found to be completely polarized when the angle of incidence is 48.0°. (a) What is the index of refraction of the reflecting material? (b) If some of the light passes into the material, what is the angle of refraction?

52. Light with a wavelength in vacuum of 546.1 nm falls perpendicularly on a biological specimen that is 1.000 μm thick. The light splits into two beams polarized at right angles, for which the indices of refraction are 1.320 and 1.333. (a) Calculate the wavelength of each component of the light while it is traversing the specimen. (b) Calculate the phase difference between the two beams when they emerge from the specimen.

53. The critical angle for total internal reflection for sapphire surrounded by air is 34.4°. Calculate the Brewster angle for sapphire if the light is incident from the air.

54. For a particular transparent medium surrounded by air, show that the critical angle for total internal reflection and Brewster's angle are related by the expression $\cot \theta_p = \sin \theta_c$.

ADDITIONAL PROBLEMS

55. When a monochromatic beam of light is incident from air at an angle of 37.0° with the normal on the surface of a glass block, it is observed that the refracted ray is directed at 22.0° with the normal. What angle of incidence from air would result in total polarization of the reflected beam?

56. A thin layer of oil ($n = 1.25$) is floating on water. How thick is the oil in the region that reflects green light ($\lambda = 525$ nm)?

57. Figure 24.32 shows a radio wave transmitter and a receiver, both $h = 30.0$ m above the ground and $d = 600$ m apart. The receiver can receive signals both directly, from the transmitter, and indirectly,

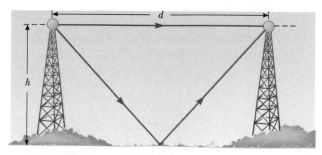

Transmitter   Receiver

FIGURE 24.32   (Problem 57)

from signals that bounce off the ground. Assuming that the ground is level between the transmitter and receiver and that a $\lambda/2$ phase shift occurs upon reflection, determine the longest wavelengths that interfere (a) constructively and (b) destructively.

58. The transmitting antenna on a submarine is 5.00 m above the water when the ship surfaces. The captain wishes to transmit a message to a receiver on a 90.0-m-tall cliff at the ocean shore. If the signal is to be completely polarized by reflection off the ocean surface, how far must the ship be from the shore?

59. Coherent light rays from a great distance strike a pair of slits separated by distance d at an angle of θ_1, as in Figure 24.33. If an interference maximum is formed at an angle of θ_2 a great distance from the slits, show that

$$d(\sin \theta_1 - \sin \theta_2) = m\lambda$$

where m is an integer.

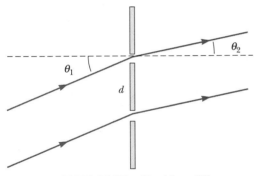

FIGURE 24.33 (Problem 59)

60. Light of wavelength 546 nm (the intense green line from a mercury source) produces a Young's interference pattern in which the second minimum from the central maximum is along a direction that makes an angle of 18.0 min of arc with the axis through the central maximum. What is the distance between the parallel slits?

61. A beam containing light of wavelengths λ_1 and λ_2 is incident on a set of parallel slits. In the interference pattern, the fourth bright line of the λ_1 light occurs at the same position as the fifth bright line of the λ_2 light. If λ_1 is known to be 540 nm, what is the value of λ_2?

62. When a beam of light passes through a pair of polarizing plates (polarizer and analyzer), the intensity of the transmitted beam is a maximum when the transmission axes of the polarizer and the analyzer are parallel to each other ($\theta = 0°$ or $180°$)

and zero when the axes are perpendicular ($\theta = 90°$). For any value of θ between $0°$ and $90°$, the transmitted intensity is given by **Malus's law:**

$$I = I_i \cos^2 \theta$$

where I_i is the intensity of the light incident on the analyzer after the light has passed through the polarizer. (a) Polarized light is incident on the analyzer so that the direction of polarization makes an angle of $45°$ with respect to the transmission axes. What is the ratio I/I_i? (b) What should the angle between the transmission axes be to make $I/I_i = 1/3$?

63. Three polarizing plates whose planes are parallel are centered on a common axis. The directions of the transmission axes relative to the common vertical direction are shown in Figure 24.34. A plane-polarized beam of light with the plane of polarization parallel to the vertical reference direction is incident from the left on the first disk with intensity $I_i = 10.0$ units (arbitrary). Calculate the transmitted intensity, I_f, when $\theta_1 = 20.0°$, $\theta_2 = 40.0°$, and $\theta_3 = 60.0°$. (*Hint:* Refer to Problem 62.)

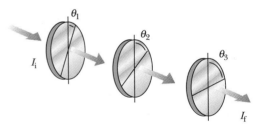

FIGURE 24.34 (Problem 63)

64. Suppose that a slit 6.0 cm wide is placed in front of a microwave source operating at a frequency of 7.5 GHz. Calculate the angle (measured from the central maximum) where the first minimum in the diffraction pattern occurs.

65. A diffraction pattern is produced on a screen 140 cm from a single slit, using monochromatic light of wavelength 500 nm. The distance from the center of the central maximum to the first-order maximum is 3.00 mm. Calculate the slit width. (*Hint:* Assume that the first-order maximum is halfway between the first- and second-order minima.)

66. In a double-slit arrangement, $d = 0.150$ mm, $L = 140$ cm, $\lambda = 643$ nm, and $y = 1.80$ cm. (a) What is the path difference, δ, for the two slits at this y location? (b) Express this path difference in terms of the wavelength. (c) Will the interference corre-

spond to a maximum, a minimum, or an intermediate condition?

67. A glass plate ($n = 1.61$) is covered with a thin,
uniform layer of oil ($n = 1.20$). A light beam of
variable wavelength from air is incident normally
on the oil surface. Observation of the reflected
beam shows destructive interference at 500 nm
and constructive interference at 750 nm. From
this information, calculate the thickness of the oil
film.

FIGURE 24.35 (Problem 68)

68. A piece of transparent material with index of refraction n is cut into the shape of a wedge, as
shown in Figure 24.35. The angle of the wedge is
small, and monochromatic light of wavelength λ
is normally incident from above. If the height of
the wedge is h and its length is ℓ, show that bright
fringes occur at the positions

$$x = \frac{\lambda \ell \left(m + \frac{1}{2}\right)}{2hn}$$

and dark fringes occur at the positions $x = \lambda \ell m / 2hn$, where $m = 0, 1, 2, \ldots$ where x is measured as shown.

69. Figure 24.36 illustrates the formation of an interference pattern by the Lloyd's mirror method.
Light from source S reaches the screen via two different pathways. One is a direct path, and the second is by reflection from a horizontal mirror. The
interference is as though light from two different
sources, S and S', had interfered as in the Young's
double-slit arrangement. Assume that the actual
source, S, and the virtual source, S', are in a plane
25 cm to the left of the mirror, and the screen is
a distance of $L = 120$ cm to the right of this plane.
Source S is a distance of $d = 2.5$ mm above the top
surface of the mirror, and the light is monochromatic with $\lambda = 620$ nm. Determine the distance of
the first bright fringe above the surface of the
mirror.

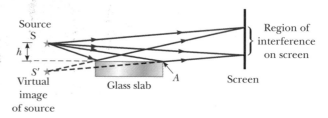

FIGURE 24.36 (Problem 69)

OPTICAL INSTRUMENTS

25

Some common navigational instruments. (Left) A magnifying glass is used to view fine print and small objects. (Bottom right) Binoculars are used to view distant objects. *(Murray Alcosser/The Image Bank)*

We use devices made from lenses, mirrors, or other optical components every time we put on a pair of eyeglasses, take a photograph, look at the sky through a telescope, and so on. In this chapter we examine how these and other optical instruments work. For the most part, our analyses will involve the laws of reflection and refraction and the procedures of geometric optics. However, to explain certain phenomena we must use the wave nature of light.

25.1

THE CAMERA

The single-lens photographic **camera** is a simple optical instrument whose essential features are shown in Figure 25.1a. It consists of a lighttight box, a converging lens that produces a real image, and a film behind the lens to receive the image. Focusing is accomplished by varying the distance between lens and film—with an adjustable bellows in old-style cameras, and with some other mechanical arrangement in newer models. For proper focusing, or sharp images, the lens-to-film distance will depend on the object distance as well as on the focal length of the lens. The shutter, located behind the lens, is a mechanical device that is opened for selected time intervals. With this arrangement, moving objects can be photographed with the use of short exposure times, and dark scenes (low light levels) with the use of long exposure times. Without this con-

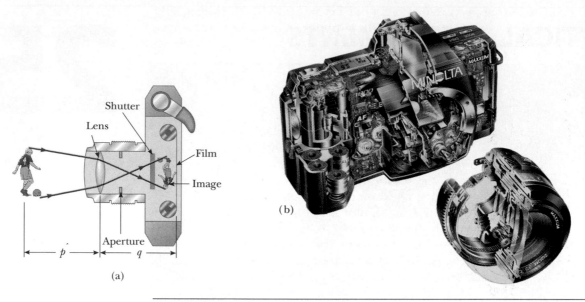

Shutter

Lens

Film

Image

Aperture

p q

(a)

(b)

FIGURE 25.1

(a) A cross-sectional view of a simple camera. (b) A cutaway view of a modern 35-mm camera. *(Courtesy of Minolta Corporation)*

trol, it would be impossible to take stop-action photographs. For example, a speeding race car would move far enough while the shutter was open to produce a blurred image. Typical shutter "speeds" are 1/30, 1/60, 1/125, and 1/250 s. Stationary objects are often shot with a shutter speed of 1/60 s. More expensive cameras, like that shown in Figure 25.1b, have a second adjustable aperture either behind or in front of the lens, to provide further control of the intensity of light reaching the film.

The brightness of the image focused on the film depends on the diameter and focal length of the lens. Clearly, the amount of light reaching the film, and hence the brightness of the image formed on the film, increases with the size of the lens. The focal length of the lens also affects the brightness of the image. We can see this by considering the magnification equation for a thin lens:

$$M = \frac{h'}{h} = \frac{q}{p}$$

$$h' = h\frac{q}{p}$$

where h and h' are the object and image heights, respectively, and p and q are the object and image distances. When p is large, q is approximately equal to the focal length, f. Thus, we have

$$h' = \frac{h}{p}f$$

From this result, we see that a lens with a short focal length produces a small image, corresponding to a small value of h'.

A small image is brighter than a larger one because all of the incoming light is concentrated in a much smaller area. Because the brightness of the image depends on f and on D, the diameter of the lens, a quantity called the *f*-**number** is defined as

$$f\text{-number} \equiv \frac{f}{D} \qquad\qquad \textbf{[25.1]}$$

The f-number is a measure of the "light-concentrating" power of a lens and determines what is called the speed of the lens. A fast lens has a small f-number and usually a small focal length and large diameter. Camera lenses are often marked with a range of f-numbers such as $f/2.8$, $f/4$, $f/5.6$, $f/8$, $f/11$, $f/16$. They are selected by adjusting the aperture, which effectively changes D. When the f-number is increased by one position, or one "stop," the light admitted decreases by a factor of 2. Likewise, the shutter speed is changed in steps whose factor is 2. The smallest f-number corresponds to the case where the aperture is wide open and as much as possible of the lens area is in use. Fast lenses, with f-numbers as low as 1.2, are relatively expensive because it is more difficult to keep aberrations acceptably small. A simple camera for routine snapshots usually has a fixed focal length and fixed aperture size, with an f-number of about $f/11$.

EXAMPLE 25.1   Choosing the *f*-Number

Suppose you are using a single-lens 35-mm camera (35 mm is the width of the film strip) with only two f-stops, $f/2.8$ and $f/22$. Which f-number would you use on a cloudy day, and why?

Solution   Substituting the given f-numbers into Equation 25.1, we have

$$2.8 = \frac{f_1}{D_1} \qquad \text{and} \qquad 22 = \frac{f_2}{D_2}$$

The focal length of the camera is fixed ($f_1 = f_2$ in the above), but the diameter of the aperture is not. On a cloudy day, you should make the shutter opening as large as possible. As these equations indicate, the largest value of D produces the smallest f-number. Thus, you should use the 2.8 setting.

25.2
THE EYE

The eye is a remarkable and extremely complex organ. Because of this complexity, defects sometimes arise that impair vision. To compensate for the defects, external aids, such as eyeglasses, are often used. In this section we describe the parts of the eye, their purposes, and some of the corrections that can be made when the eye does not function properly. You will find that the eye has much in common with the camera. Like the camera, the eye gathers light and produces a sharp image. However, the mechanisms by which the eye controls the amount of light admitted and adjusts itself to produce correctly focused images are far more complex, intricate, and effective than those in even the most sophisticated camera. In all respects, the eye is a wonder of design.

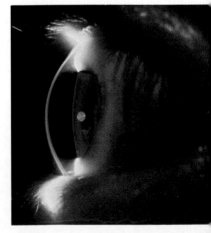

Close-up photograph of the cornea of a human eye. (©*Lennart Nilsson*, Behold Man, *Little Brown and Company*)

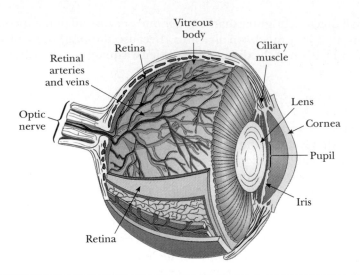

FIGURE 25.2

Essential parts of the eye. Can you correlate the essential parts of the eye with those of the simple camera in Figure 25.1a?

Figure 25.2 shows the essential parts of the eye. The front is covered by a transparent membrane called the *cornea*. Inward from the cornea are a clear liquid region (the *aqueous humor*), a variable aperture (the *iris* and *pupil*), and the *crystalline lens*. Most of the refraction occurs in the cornea, because the liquid medium surrounding the lens has an average index of refraction close to that of the lens. The iris, the colored portion of the eye, is a muscular diaphragm that regulates the amount of light entering the eye by dilating the pupil (increasing its diameter) in light of low intensity and contracting the pupil in high-intensity light. The *f*-number range of the eye is about $f/2.8$ to $f/16$.

Light entering the eye is focused by the cornea-lens system onto the back surface of the eye, called the *retina*. The surface of the retina consists of millions of sensitive receptors called *rods* and *cones*. When stimulated by light, these structures send impulses via the optic nerve to the brain, where a distinct image of an object is perceived.

The eye focuses on a given object by varying the shape of the pliable crystalline lens through an amazing process called **accommodation.** An important component in accommodation is the *ciliary muscle,* which is attached to the lens. When the eye is focused on distant objects, the ciliary muscle is relaxed. For an object distance of infinity, the focal length of the eye (the distance between the lens and the retina) is about 1.7 cm. The eye focuses on nearby objects by tensing the ciliary muscle. This action effectively reduces the focal length by slightly decreasing the radius of curvature of the lens, which allows the image to be focused on the retina. This lens adjustment takes place so swiftly that we are not aware of the change. In this respect, as in others, even the finest electronic camera is a toy compared with the eye. It is evident that there is a limit to accommodation, because objects that are very close to the eye produce blurred images.

The **near point** is the smallest distance for which the lens will produce a sharp image on the retina. This distance usually increases with age.

Typically, the near point of the eye is about 18 cm at age 10, about 25 cm at age 20, 50 cm at age 40, and 500 cm or greater at age 60.

DEFECTS OF THE EYE

An eye can have several abnormalities that keep it from functioning properly. These can often be corrected with eyeglasses, contact lenses, or surgery.

When the relaxed eye produces an image of a distant object *behind* the retina, as in Figure 25.3a, the abnormality is known as **hyperopia,** and the person is said to be *farsighted.* With this defect, distant objects are seen clearly but near objects are blurred. Either the hyperopic eye is too short or the ciliary muscle cannot change the shape of the lens enough to properly focus the image. The condition can be corrected with a converging lens, as shown in Figure 25.3b.

Another condition, known as **myopia,** or *nearsightedness,* occurs either when the eye is longer than normal or when the maximum focal length of the lens is insufficient to produce a clearly formed image on the retina. In this case, light from a distant object is focused in front of the retina (Fig. 25.4a). The distinguishing feature of this imperfection is that distant objects are not seen clearly. Nearsightedness can be corrected with a diverging lens, as in Figure 25.4b.

Beginning with middle age, most people lose some of their accommodation power, usually as a result of hardening of the crystalline lens. This causes farsightedness, which can be corrected with converging lenses.

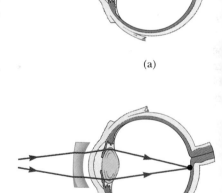

(a)

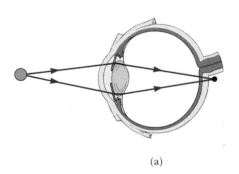

(a)

FIGURE 25.3
(a) A farsighted eye is slightly shorter than normal; hence, the image of a nearby object focuses *behind* the retina. (b) The condition can be corrected with a converging lens. (The object is assumed to be very small in these figures.)

(b)

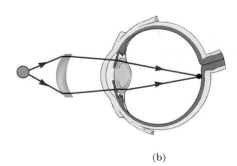

(b)

FIGURE 25.4
(a) A nearsighted eye is slightly longer than normal; hence, the image of a distant object focuses *in front of* the retina. (b) The condition can be corrected with a diverging lens. (The object is assumed to be very small in these figures.)

A common eye defect is **astigmatism,** in which light from a point source produces a line image on the retina. This occurs when the cornea or the crystalline lens or both are not perfectly spherical. Astigmatism can be corrected by lenses with different curvatures in two mutually perpendicular directions. A cylindrical lens (a segment of a cylinder) is typically used for this purpose.

The eye is also subject to several diseases. One, which usually occurs in old age, is the formation of **cataracts,** which make the lens partially or totally opaque. The common remedy for cataracts is surgical removal of the lens. Another disease, called **glaucoma,** arises from an abnormal increase in fluid pressure inside the eyeball. This pressure increase can cause a reduction in blood supply to the retina, which can eventually lead to blindness when the nerve fibers of the retina die. If the disease is discovered early enough, it can be treated with medicine or surgery.

Optometrists and ophthalmologists usually prescribe corrective lenses measured in diopters.

> The **power, P,** of a lens in **diopters** equals the inverse of the focal length in meters; that is, $P = 1/f$.

For example, a converging lens whose focal length is $+20$ cm has a power of $+5$ diopters, and a diverging lens whose focal length is -40 cm has a power of -2.5 diopters.

EXAMPLE 25.2 Prescribing a Lens

The near point of an eye is 50.0 cm. (a) What focal length must a corrective lens have to enable the eye to clearly see an object 25.0 cm away?

Reasoning The thin-lens equation (Eq. 23.11, Chapter 23) enables us to solve this problem. We have placed an object at 25 cm, and we want the lens to form an image at the closest point that the eye can see clearly. This corresponds to the near point, 50.0 cm.

Solution Applying the thin-lens equation, we have

$$\frac{1}{25.0 \text{ cm}} + \frac{1}{(-50.0 \text{ cm})} = \frac{1}{f}$$

$$f = \boxed{50.0 \text{ cm}}$$

Why did we use a negative sign for the image distance? Notice that the focal length is positive, indicating the need for a converging lens to correct farsightedness such as this.

(b) What is the power of this lens?

Solution The power is the reciprocal of the focal length in meters:

$$P = \frac{1}{f} = \frac{1}{0.500 \text{ m}} = \boxed{2.00 \text{ diopters}}$$

EXAMPLE 25.3 A Case of Nearsightedness

A particular nearsighted person cannot see objects clearly when they are beyond 50.0 cm (the far point of the eye). What focal length should the prescribed lens have to correct this problem?

Reasoning The purpose of the lens in this instance is to "move" an object from infinity to a distance where it can be seen clearly.

Solution From the thin-lens equation, we have

$$\frac{1}{p} + \frac{1}{q} = \frac{1}{\infty} + \frac{1}{(-50.0 \text{ cm})} = \frac{1}{f}$$

$$f = \boxed{-50.0 \text{ cm}}$$

Why did we use a negative sign for the image distance? As you should have suspected, the lens must be diverging (have a negative focal length) to correct nearsightedness.

Exercise What is the power of this lens?

Answer -2.00 diopters.

25.3
THE SIMPLE MAGNIFIER

The **simple magnifier** is one of the simplest and most basic of all optical instruments because it consists of only a single converging lens. As the name implies, this device is used to increase the apparent size of an object. Suppose an object is viewed at some distance, p, from the eye, as in Figure 25.5. Clearly, the size of the image formed at the retina depends on the angle, θ, subtended by the object at the eye. As the object moves closer to the eye, θ increases and a larger image is observed. However, a normal eye cannot focus on an object closer than about 25 cm, the near point (Fig. 25.6a). (Try it!) Therefore, θ is maximum at the near point.

To further increase the apparent angular size of an object, a converging lens can be placed in front of the eye with the object positioned at point O, just inside the focal point of the lens, as in Figure 25.6b. At this location, the lens forms a virtual, erect, and enlarged image, as shown. Clearly, the lens increases the angular size of the object. We define the **angular magnification, m,** as the ratio of the angle subtended by the object when the lens is in use (angle θ in Fig. 25.6b) to that subtended by the object when it is placed at the near point with no lens (angle θ_0 in Fig. 25.6a):

$$m \equiv \frac{\theta}{\theta_0} \qquad \text{[25.2]}$$

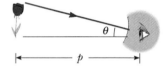

FIGURE 25.5
The size of the image formed on the retina depends on the angle, θ, subtended at the eye.

Angular magnification

The angular magnification is a maximum when the image is at the near point of the eye, that is, when $p = -25$ cm (see Fig. 25.6b). The object distance that corresponds to this image distance can be calculated from the thin-lens formula:

$$\frac{1}{p} + \frac{1}{-25 \text{ cm}} = \frac{1}{f}$$

[25.3]

$$p = \frac{25f}{25 + f}$$

where f is the focal length in centimeters. The small-angle approximation is made as follows:

$$\tan \theta_0 \approx \theta_0 \approx \frac{h}{25} \qquad \text{and} \qquad \tan \theta \approx \theta \approx \frac{h}{p}$$

[25.4]

Thus, Equation 25.2 becomes

$$m = \frac{\theta}{\theta_0} = \frac{h/p}{h/25} = \frac{25}{p} = \frac{25}{25f/(25+f)}$$

$$m = 1 + \frac{25 \text{ cm}}{f}$$

[25.5]

The magnification given by Equation 25.5 is the ratio of the angular size seen with the lens to the angular size seen without the lens, with the object at the near point of the eye. The eye can actually focus on an image formed anywhere between the near point and infinity, and is most relaxed when the image is at infinity (Sec. 25.2). In order for the image formed by the magnifying lens to appear at infinity, the object must be placed at the focal point of the lens, that is, $p = f$. In this case, Equation 25.4 becomes

$$\theta_0 \approx \frac{h}{25} \qquad \text{and} \qquad \theta \approx \frac{h}{f}$$

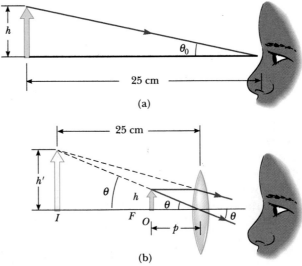

(a)

(b)

FIGURE 25.6
(a) An object placed at the near point ($p = 25$ cm) subtends an angle of $\theta_0 \approx h/25$ at the eye. (b) An object placed near the focal point of a converging lens produces a magnified image, which subtends an angle of $\theta \approx h'/25$ at the eye.

and the magnification is

$$m = \frac{\theta}{\theta_0} = \frac{25 \text{ cm}}{f}$$ [25.6]

With a single lens, it is possible to achieve angular magnifications up to about 4 without serious aberrations. Magnifications up to about 20 can be achieved by using a second lens to correct for aberrations.

EXAMPLE 25.4 Maximum Magnification of a Lens

What is the maximum magnification of a lens with a focal length of 10.0 cm, and what is the magnification of this lens when the eye is relaxed?

Reasoning The maximum magnification occurs when the image formed by the lens is at the near point of the eye. Under these circumstances, Equation 25.5 gives us the maximum magnification.

Solution Using Equation 25.5, we have

$$m = 1 + \frac{25 \text{ cm}}{f} = 1 + \frac{25 \text{ cm}}{10.0 \text{ cm}} = \boxed{3.5}$$

When the eye is relaxed, the image is at infinity. In this case, we use Equation 25.6:

$$m = \frac{25}{f} = \frac{25 \text{ cm}}{10.0 \text{ cm}} = \boxed{2.5}$$

25.4
THE COMPOUND MICROSCOPE

A simple magnifier provides only limited assistance with inspection of the minute details of an object. Greater magnification can be achieved by combining two lenses in a device called a compound microscope, a schematic diagram of which is shown in Figure 25.7a. It consists of two lenses: an objective with a very short focal length, f_o (where $f_o < 1$ cm), and an ocular lens, or eyepiece, with a focal length, f_e, of a few centimeters. The two lenses are separated by distance L, which is much greater than either f_o or f_e. The object, O, placed just outside the focal

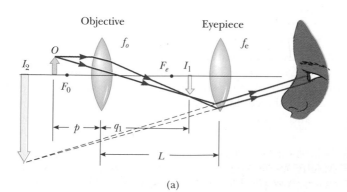

(a)

(b)

FIGURE 25.7
(a) A diagram of a compound microscope, which consists of an objective and an eyepiece, or ocular lens. (b) This simple microscope is accompanied by a laser beam. *(Garry Gay/The Image Bank)*

length of the objective, forms a real, inverted image at I_1, which is at or just inside the focal point of the eyepiece. This image is real and much enlarged. (For clarity, the enlargement of I_1 is not shown in Fig. 25.7a.) The eyepiece, which serves as a simple magnifier, uses the image at I_1 as its object and produces an image at I_2. The image seen by the eye at I_2 is virtual, inverted, and very much enlarged.

The lateral magnification, M_1, of the first image is q_1/p_1. Note in Figure 25.7a that q_1 is approximately equal to L. This occurs because the object is placed close to the focal point of the objective lens, which ensures that the image formed will be far from the objective lens. Furthermore, since the object is very close to the focal point of the objective lens, $q_1 \approx f_o$. This gives a magnification of

$$M_1 = -\frac{q_1}{p_1} \approx -\frac{L}{f_o}$$

for the objective. The angular magnification of the eyepiece for an object (corresponding to the image at I_1) placed at the focal point is found from Equation 25.6 to be

$$m_e = \frac{25 \text{ cm}}{f_e}$$

The overall magnification of the compound microscope is defined as the product of the lateral and angular magnifications:

Magnification of a microscope

$$M = M_1 m_e = -\frac{L}{f_o}\left(\frac{25 \text{ cm}}{f_e}\right) \qquad [25.7]$$

The negative sign indicates that the image is inverted with respect to the object.

The microscope has extended our vision into the previously unknown realm of incredibly small objects, and the capabilities of this instrument have steadily increased with improved techniques in precision grinding of lenses. A question that is often asked about microscopes is, "With extreme patience and care, would it be possible to construct a microscope that would enable us to see an atom?" The answer to this question is no, as long as visible light is used to illuminate the object. The reason is that, in order to be seen, the object under a microscope must be at least as large as a wavelength of light. An atom is many times smaller than the wavelength of visible light, and so its mysteries must be probed via other techniques.

The wavelength dependence of the "seeing" ability of a wave can be illustrated by water waves set up in a bathtub in the following manner. Imagine that you vibrate your hand in the water until waves with a wavelength of about 6 in. are moving along the surface. If you fix a small object, such as a toothpick, in the path of the waves, you will find that the waves are not appreciably disturbed by the toothpick but continue along their path. Now suppose you fix a larger object, such as a toy sailboat, in the path of the waves. In this case, the waves are considerably "disturbed" by the object. The toothpick was much smaller than the wavelength of the waves, and as a result the waves did not "see" it. The toy sailboat, in contrast, is about the same size as the wavelength of the waves and hence creates a disturbance. Light waves behave in this same general way. The ability of an optical microscope to view an object depends on the size of the object relative to the wavelength of the light used to observe it. Hence, it will never be possible to observe atoms or molecules with such a microscope, since their dimensions are so small (≈ 0.1 nm) relative to the wavelength of the light (≈ 500 nm).

EXAMPLE 25.5   Magnifications of a Microscope

A certain microscope has two possible objectives. One has a focal length of 20.0 mm, and the other has a focal length of 2.0 mm. Also available are two eyepieces of focal lengths 2.5 cm and 5.0 cm. If the length of the microscope is 18 cm, what magnifications are possible?

Reasoning and Solution   The solution consists of applying Equation 25.7 to four different combinations of lenses. For the combination of the two long focal lengths, we have

$$M = -\frac{L}{f_o}\left(\frac{25 \text{ cm}}{f_e}\right) = -\frac{18}{2.0}\left(\frac{25}{5.0}\right) = \boxed{-45}$$

The combination of the 20.0-mm objective and the 2.5-cm eyepiece gives

$$M = -\frac{18}{2.0}\left(\frac{25}{2.5}\right) = \boxed{-90}$$

The 2.0-mm and 5.0-cm combination produces

$$M = -\frac{18}{0.20}\left(\frac{25}{5.0}\right) = \boxed{-450}$$

Finally, the two short focal lengths give

$$M = -\frac{18}{0.20}\left(\frac{25}{2.5}\right) = \boxed{-900}$$

25.5

THE TELESCOPE

There are two fundamentally different types of telescopes, both designed to help us view distant objects such as the planets in our Solar System. These two types are (1) the **refracting telescope,** which uses a combination of lenses to form an image, and (2) the **reflecting telescope,** which uses a curved mirror and a lens to form an image.

Let us first consider the refracting telescope. In this device, two lenses are arranged so that the objective forms a real, inverted image of the distant object very near the focal point of the eyepiece (Fig. 25.8a). Furthermore, the image at I_1 is formed at the focal point of the objective because the object is essentially at infinity. Hence, the two lenses are separated by the distance $f_o + f_e$, which corresponds to the length of the telescope's tube. The eyepiece finally forms, at I_2, an enlarged, inverted image of the image at I_1.

The angular magnification of the telescope is given by θ/θ_o, where θ_o is the angle subtended by the object at the objective and θ is the angle subtended by the final image. From the triangles in Figure 25.8a, and for small angles, we have

$$\theta \approx \frac{h'}{f_e} \quad \text{and} \quad \theta_o \approx \frac{h'}{f_o}$$

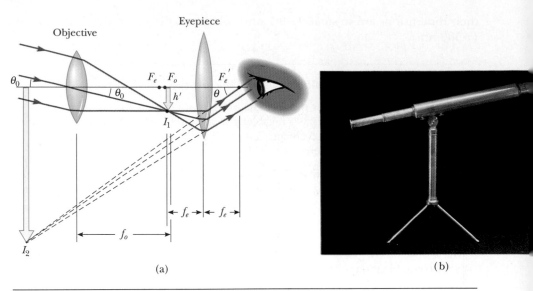

FIGURE 25.8
(a) A diagram of a refracting telescope, with the object at infinity. (b) A photograph of a refracting telescope. *(Photo courtesy of Henry Leap and Jim Lehman)*

Therefore, the angular magnification of the telescope can be expressed as

Angular magnification of a telescope

$$m = \frac{\theta}{\theta_o} = \frac{h'/f_e}{h'/f_o} = \frac{f_o}{f_e} \qquad \text{[25.8]}$$

This says that the angular magnification of a telescope equals the ratio of the objective focal length to the eyepiece focal length. Here again, the magnification is the ratio of the angular size seen with the telescope to the angular size seen with the unaided eye.

In some applications—for instance, the observation of nearby objects such as the Sun, Moon, or planets—magnification is important. Stars, in contrast, are so far away that they always appear as small points of light regardless of how much magnification is used. The large research telescopes used to study very distant objects must have great diameters to gather as much light as possible. It is difficult and expensive to manufacture such large lenses for refracting telescopes. Additionally, the heaviness of large lenses leads to sagging, which is another source of aberration.

These problems can be partially overcome by replacing the objective lens with a reflecting, concave mirror. Figure 25.9 shows the design for a typical reflecting telescope. Incoming light rays pass down the barrel of the telescope and are reflected by a parabolic mirror at the base. These rays converge toward point *A* in the figure, where an image would be formed. However, before this image is formed, a small flat mirror at point *M* reflects the light toward an opening in the side of the tube that passes into an eyepiece. This design is said to have a *Newtonian focus* because Newton developed it. Note that in the reflecting telescope the light never passes through glass (except the small eyepiece). As a result, problems associated with chromatic aberration are virtually eliminated. Also, difficulties arising from spherical aberration are reduced by the parabolic shape of the mirror.

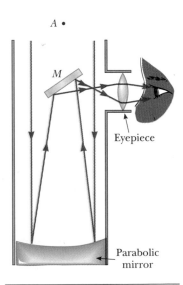

FIGURE 25.9
A reflecting telescope with a Newtonian focus.

The largest single-mirrored telescope in the world is the 6-m-diameter reflecting telescope on Mount Pastukhov in the Caucasus. The largest single-mirrored reflecting telescope in the United States is the 5-m-diameter instrument on Mount Palomar in California. In contrast, the largest refracting telescope in the world, at the Yerkes Observatory in Williams Bay, Wisconsin, has a diameter of only 1 m.

EXAMPLE 25.6   Magnification of a Reflecting Telescope

A reflecting telescope has an 8-in.-diameter objective mirror with a focal length of 1500 mm. What is the magnification of this telescope when an eyepiece with an 18-mm focal length is used?

Solution   The equation for finding the magnification of a reflector is the same as that for a refractor. Thus, Equation 25.8 gives

$$m = \frac{f_o}{f_e} = \frac{1500 \text{ mm}}{18 \text{ mm}} = \boxed{83}$$

25.6

RESOLUTION OF SINGLE-SLIT AND CIRCULAR APERTURES

The ability of an optical system such as a microscope or telescope to distinguish between closely spaced objects is limited because of the wave nature of light. To understand this difficulty, consider Figure 25.10, which shows two light sources far from a narrow slit of width a. The sources can be taken as two point sources, S_1 and S_2, that are *not* coherent. For example, they could be two distant stars. If no

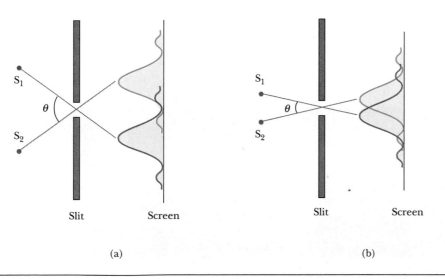

(a)   (b)

FIGURE 25.10
Each of two point sources at some distance from a small aperture produces a diffraction pattern. (a) The angle subtended by the sources at the aperture is large enough so that the diffraction patterns are distinguishable. (b) The angle subtended by the sources is so small that the diffraction patterns are not distinguishable. (Note that the angles are greatly exaggerated.)

diffraction occurred, one would observe two distinct bright spots (or images) on the screen at the right in the figure. However, because of diffraction, each source is imaged as a bright central region flanked by weaker bright and dark rings. What is observed on the screen is the sum of two diffraction patterns, one from S_1 and the other from S_2.

If the two sources are separated so that their central maxima do not overlap, as in Figure 25.10a, their images can be distinguished and are said to be *resolved*. If the sources are close together, however, as in Figure 25.10b, the two central maxima may overlap and the images are *not resolved*. To decide whether two images are resolved, the following condition is often applied to their diffraction patterns:

> When the central maximum of one image falls on the first minimum of another image, the images are said to be just resolved. This limiting condition of resolution is known as **Rayleigh's criterion.**

Figure 25.11 shows diffraction patterns in three situations. When the objects are far apart, their images are well resolved (Fig. 25.11a). The images are just

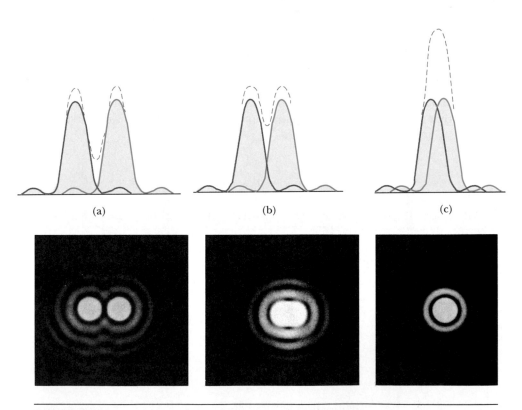

(a)   (b)   (c)

FIGURE 25.11
The diffraction patterns of two point sources (solid curves) and the resultant pattern (dashed curve) for three angular separations of the sources. (a) The sources are far apart, and their patterns are well resolved. (b) The sources are closer together, and their patterns are just resolved. (c) The sources are so close together that their patterns are not resolved. (*From M. Cagnet, M. Francon, and J. C. Thierr,* Atlas of Optical Phenomena, *Berlin, Springer-Verlag, 1962, plate 16*)

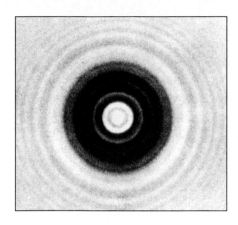

FIGURE 25.12
The Fresnel diffraction pattern of a circular aperture consists of a central bright disk surrounded by concentric bright and dark rings. (*From M. Cagnet, M. Francon, and J. C. Thierr,* Atlas of Optical Phenomena, *Berlin, Springer-Verlag, 1962, plate 34*)

resolved when their angular separation satisfies Rayleigh's criterion (Fig. 25.11b). Finally, the images in Figure 25.11c are not resolved.

From Rayleigh's criterion, we can determine the minimum angular separation, θ_m, subtended by the source at the slit so that the images will be just resolved. In Chapter 24 we found that the first minimum in a single-slit diffraction pattern occurs at the angle that satisfies the relationship

$$\sin \theta = \frac{\lambda}{a}$$

where a is the width of the slit. According to Rayleigh's criterion, this expression gives the smallest angular separation for which two images can be resolved. Because $\lambda \ll a$ in most situations, $\sin \theta$ is small and we can use the approximation $\sin \theta \approx \theta$. Therefore, the limiting angle of resolution for a slit of width a is

$$\theta_m \approx \frac{\lambda}{a} \qquad \text{[25.9]}$$

Limiting angle for a slit

where θ_m is in radians. Hence, the angle subtended by the two sources at the slit must be *greater* than λ / a if the images are to be resolved.

Many optical systems use circular apertures rather than slits. The diffraction pattern of a circular aperture (Fig. 25.12) consists of a central circular bright disk surrounded by progressively fainter rings. Analysis shows that the limiting angle of resolution of the circular aperture is

$$\theta_m = 1.22 \frac{\lambda}{D} \qquad \text{[25.10]}$$

Limiting angle for a circular aperture

where D is the diameter of the aperture. Note that Equation 25.10 is similar to Equation 25.9 except for the factor 1.22, which arises from a complex mathematical analysis of diffraction from a circular aperture.

EXAMPLE 25.7 **Limiting Resolution of a Microscope**

Sodium light of wavelength 589 nm is used to view an object under a microscope. The aperture of the objective is 0.90 cm. (a) Find the limiting angle of resolution. (b) Using visible light of any wavelength you desire, what is the maximum limit of resolution for this microscope? (c) Suppose water of index of refraction 1.33 filled the space between the object and the objective. What effect would this have on the resolving power of the microscope?

A large reflecting telescope used at Mount Palomar Observatory.
(*J.L.T. Rhodes/The Image Bank*)

Solution   (a) From Equation 25.10, we find the limiting angle of resolution to be

$$\theta_m = 1.22 \left(\frac{589 \times 10^{-9} \text{ m}}{0.90 \times 10^{-2} \text{ m}} \right) = \boxed{8.0 \times 10^{-5} \text{ rad}}$$

This means that any two points on the object subtending an angle of less than 8.0×10^{-5} rad at the objective cannot be distinguished in the image.

(b) To obtain the maximum resolution, we have to use the shortest wavelength available in the visible spectrum. Violet light of wavelength 400 nm gives us a limiting angle of resolution of

$$\theta_m = 1.22 \left(\frac{400 \times 10^{-9} \text{ m}}{0.90 \times 10^{-2} \text{ m}} \right) = \boxed{5.4 \times 10^{-5} \text{ rad}}$$

(c) In this case, the wavelength of the sodium light in the water is found by $\lambda_w = \lambda_a / n$ (Chapter 22). Thus, we have

$$\lambda_w = \frac{\lambda_a}{n} = \frac{589 \text{ nm}}{1.33} = 443 \text{ nm}$$

The limiting angle of resolution at this wavelength is

$$\theta_m = 1.22 \left(\frac{443 \times 10^{-9} \text{ m}}{0.90 \times 10^{-2} \text{ m}} \right) = \boxed{6.0 \times 10^{-5} \text{ rad}}$$

EXAMPLE 25.8   Resolution of a Telescope

The Hale telescope at Mount Palomar has a diameter of 200 in. What is its limiting angle of resolution at a wavelength of 600 nm?

Solution   Because $D = 200$ in. $= 5.08$ m and $\lambda = 6.00 \times 10^{-7}$ m, Equation 25.10 gives

$$\theta_m = 1.22 \frac{\lambda}{D} = 1.22 \left(\frac{6.00 \times 10^{-7} \text{ m}}{5.08 \text{ m}} \right) = \boxed{1.44 \times 10^{-7} \text{ rad}}$$

Therefore, any two stars that subtend an angle greater than or equal to this value will be resolved (assuming ideal atmospheric conditions).

It is interesting to compare this value with the resolution of a large radio-telescope, such as the system at Arecibo, Puerto Rico, which has a diameter of 1000 ft (305 m). This telescope detects radio waves at a wavelength of 0.75 m. The corresponding minimum angle of resolution is calculated to be 3.0×10^{-3} rad (10 min 19 s of arc), which is more than 10 000 times larger than the calculated minimum angle for the Hale telescope.

EXAMPLE 25.9   Comparing Two Telescopes

Two telescopes have the following properties:

Telescope	Diameter of Objective (in.)	Focal Length of Objective (mm)	Focal Length of Eyepiece (mm)
A	6.00	1000	6.0
B	8.00	1250	25.0

(a) Which has the better resolving power? (b) Which has the greater light-gathering ability? (c) Which produces a greater magnification?

Solution (a) The telescope with the larger objective has the greater ability to discriminate between nearby objects. This is telescope B.

(b) The telescope with the larger objective can collect more light. Hence, B is again the choice.

(c) The magnification of telescope A is

$$m = \frac{f_o}{f_e} = \frac{1000 \text{ mm}}{6.00 \text{ mm}} = 167$$

That of telescope B is

$$m = \frac{1250 \text{ mm}}{25.0 \text{ mm}} = 50.0$$

Thus, telescope A has the greater magnification.

25.7

THE MICHELSON INTERFEROMETER

The camera and the telescope are examples of commonly used optical instruments. In contrast, the Michelson interferometer is an optical instrument that is unfamiliar to most people. It has great scientific importance, however. Invented by the American physicist A. A. Michelson (1852–1931), it is an ingenious device that splits a light beam into two parts and then recombines them to form an interference pattern. The interferometer is used to make accurate length measurements.

Figure 25.13 is a schematic diagram of an interferometer. A beam of light provided by a monochromatic source is split into two rays by a partially silvered mirror, M, inclined at an angle of 45° relative to the incident light beam. One ray is reflected vertically upward to mirror M_1, and the other ray is transmitted horizontally through mirror M to mirror M_2. Hence, the two rays travel separate paths, L_1 and L_2. After reflecting from mirrors M_1 and M_2, the two rays eventually recombine to produce an interference pattern, which can be viewed through a telescope. The glass plate, P, equal in thickness to mirror M, is placed in the path of the horizontal ray to ensure that the two rays travel the same distance through glass.

The interference pattern for the two rays is determined by the difference in their path lengths. When the two rays are viewed as shown, the image of M_2 is at M_2' parallel to M_1. Hence, the space between M_2' and M_1 forms the equivalent of a parallel air film. The effective thickness of the air film is varied by using a finely threaded screw to move mirror M_1 in the direction indicated by the arrows in Figure 25.13. If one of the mirrors is tipped slightly with respect to the other, the thin film between the two is wedge-shaped, and an interference pattern consisting of parallel fringes is set up, as described in Chapter 24, Example 24.4. Now suppose we focus on one of the dark lines with the crosshairs of a telescope. As the mirror M_1 is moved to lengthen the path L_1, the thickness of the wedge increases. When the thickness increases by $\lambda/4$, the destructive interference that initially produced the dark fringe has changed to constructive interference, and we now observe a bright fringe at the location of the crosshairs. The term *fringe*

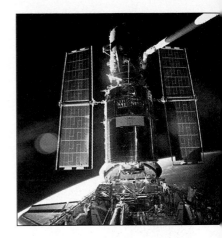

Sunlight reflects off Endeavour's aft windows and the shiny Hubble Space Telescope (HST) prior to its post-servicing deployment near the end of the 11-day STS-61 mission. A hand-held Hasselblad camera was used inside Endeavour's cabin to record the image. (*NASA*)

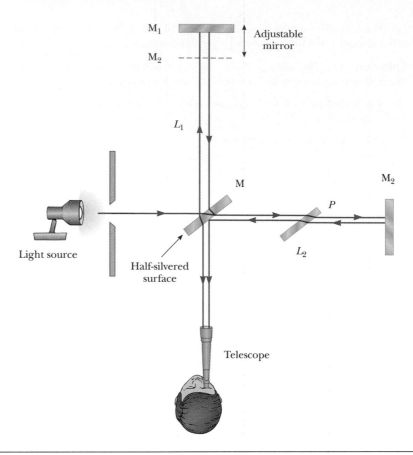

FIGURE 25.13
A diagram of the Michelson interferometer. A single beam is split into two rays by the
half-silvered mirror, M. The path difference between the two rays is varied with the adjustable mirror, M_1.

shift is used to describe the change in a fringe from dark to light or light to dark.
Thus, successive light and dark fringes are formed each time M_1 is moved a
distance of $\lambda/4$. The wavelength of light can be measured by counting the
number of fringe shifts for a measured displacement of M_1. Conversely, if the
wavelength is accurately known (as with a laser beam), the mirror displacement
can be determined to within a fraction of the wavelength. Because the interferometer can measure displacements precisely, it is often used to make highly
accurate measurements of the dimensions of mechanical components.

If the mirrors are perfectly aligned, rather than tipped with respect to one
another, the path difference differs slightly for different angles of view. This
results in an interference pattern that resembles Newton's rings. The pattern can
be used in a fashion similar to that for tipped mirrors. One concentrates on the
center spot in the interference pattern. For example, suppose the spot is initially
dark, indicating that destructive interference is occurring. If M_1 is now moved by
a distance of $\lambda/4$, this central spot changes to a light region, corresponding to a
fringe shift.

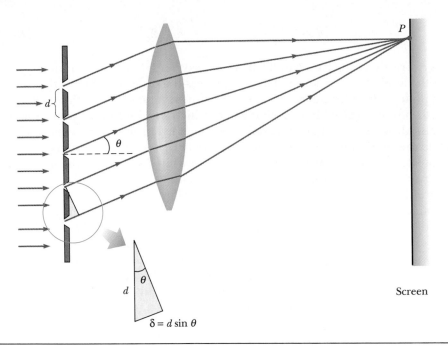

FIGURE 25.14
A side view of a diffraction grating. The slit separation is d, and the path difference between adjacent slits is $d \sin \theta$.

25.8

THE DIFFRACTION GRATING

The diffraction grating, a very useful device for analyzing light sources, consists of many equally spaced parallel slits. A grating can be made by scratching parallel lines on a glass plate with a precision machining technique. The spaces between the scratches are transparent to the light and hence act as separate slits. A typical grating contains several thousand lines per centimeter. For example, a grating ruled with 5000 lines/cm has a slit spacing, d, equal to the inverse of this number; hence, $d = (1/5000)$ cm $= 2 \times 10^{-4}$ cm.

Figure 25.14 is a schematic diagram of a section of a plane diffraction grating. A plane wave is incident from the left, normal to the plane of the grating. A converging lens can be used to bring the rays together at point P. The intensity of the pattern on the screen is the result of the combined effects of interference and diffraction. Each slit produces diffraction, and the diffracted beams in turn interfere with one another to produce the pattern. Moreover, each slit acts as a source of waves, and all waves start at the slits in phase. However, for some arbitrary direction θ measured from the horizontal, the waves must travel *different* path lengths before reaching a particular point P on the screen. From Figure 25.14, note that the path difference between waves from any two adjacent slits is $d \sin \theta$. If this path difference equals one wavelength or some integral multiple of a wavelength, waves from all slits will be in phase at P and a bright line will be observed. Therefore, the condition for **maxima** in the interference pattern at the angle θ is

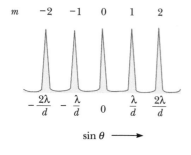

FIGURE 25.15
Intensity versus $\sin \theta$ for the diffraction grating. The zeroth-, first-, and second-order principal maxima are shown.

$$d \sin \theta = m\lambda \qquad m = 0, 1, 2, \ldots \qquad [25.11]$$

This expression can be used to calculate the wavelength from the grating spacing and the angle of deviation, θ. The integer m is the **order number** of the diffraction pattern. If the incident radiation contains several wavelengths, each wavelength deviates through a specific angle, which can be found from Equation 25.11. All wavelengths are focused at $\theta = 0$, corresponding to $m = 0$. This is called the *zeroth-order maximum*. The *first-order maximum*, corresponding to $m = 1$, is observed at an angle that satisfies the relationship $\sin \theta = \lambda/d$; the *second-order maximum*, corresponding to $m = 2$, is observed at a larger angle θ, and so on. Figure 25.15 is a sketch of the intensity distribution for some of the orders produced by a diffraction grating. Note the sharpness of the principal maxima and the broad range of the dark areas. This is in contrast to the broad bright fringes characteristic of the two-slit interference pattern.

A simple arrangement that can be used to measure the orders of a diffraction pattern is shown in Figure 25.16. This is a form of diffraction grating spectrometer. The light to be analyzed passes through a slit and is formed into a parallel beam by a lens. The light then strikes the grating at a 90° angle. The diffracted light leaves the grating at angles that satisfy Equation 25.11. A telescope is used to view the image of the slit. The wavelength can be determined by measuring the angles at which the images of the slit appear for the various orders.

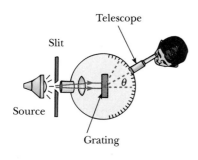

FIGURE 25.16
A diagram of a diffraction grating spectrometer. The collimated beam incident on the grating is diffracted into the various orders at the angles θ that satisfy the equation $d \sin \theta = m\lambda$, where $m = 0, 1, 2, \ldots$.

EXAMPLE 25.10   The Orders of a Diffraction Grating

Monochromatic light from a helium-neon laser ($\lambda = 632.8$ nm) is incident normally on a diffraction grating containing 6000 lines/cm. Find the angles at which one would observe the first-order maximum, the second-order maximum, and so forth.

Solution   First we must calculate the slit separation, which is the inverse of the number of lines per centimeter:

$$d = \frac{1}{6000} \text{ cm} = 1.667 \times 10^{-4} \text{ cm} = 1667 \text{ nm}$$

For the first-order maximum ($m = 1$), we get

$$\sin \theta_1 = \frac{\lambda}{d} = \frac{632.8 \text{ nm}}{1667 \text{ nm}} = 0.3796$$

$$\theta_1 = \;\; 22.31°$$

For $m = 2$ we find that

$$\sin \theta_2 = \frac{2\lambda}{d} = \frac{2(632.8 \text{ nm})}{1667 \text{ nm}} = 0.7592$$

$$\theta_2 = \;\; 49.39°$$

However, for $m = 3$ we find that $\sin \theta_3 = 1.139$. Since $\sin \theta$ cannot exceed unity, this is not a realistic solution. Hence, only zeroth-, first-, and second-order maxima would be observed in this situation.

RESOLVING POWER OF THE DIFFRACTION GRATING

The diffraction grating is most useful for making accurate wavelength measurements. Like the prism, it can be used to disperse a spectrum into its components. Of the two devices, the grating is more precise if one wants to distinguish between two closely spaced wavelengths. We say that the grating spectrometer has a higher *resolution* than the prism spectrometer. If λ_1 and λ_2 are two nearly equal wavelengths between which the spectrometer can just barely distinguish, the **resolving power,** R, of the grating is defined as

$$R \equiv \frac{\lambda}{\lambda_2 - \lambda_1} = \frac{\lambda}{\Delta\lambda} \qquad \text{[25.12]}$$

where $\lambda \approx \lambda_1 \approx \lambda_2$ and $\Delta\lambda = \lambda_2 - \lambda_1$. Thus, we see that a grating with a high resolving power can distinguish small differences in wavelength. Furthermore, if N lines of the grating are illuminated, it can be shown that the resolving power in the mth-order diffraction equals the product Nm:

$$R = Nm \qquad \text{[25.13]}$$

Resolving power of a grating

Thus, the resolving power increases with order number. Furthermore, R is large for a grating with a great number of illuminated slits. Note that for $m = 0$, $R = 0$, which signifies that *all wavelengths are indistinguishable* for the zeroth-order maximum (all wavelengths fall at the same point on the screen). However, consider the second-order diffraction pattern of a grating that has 5000 rulings illuminated by the light source. The resolving power of such a grating in second order is $R = 5000 \times 2 = 10\,000$. Therefore, the *minimum* wavelength separation between two spectral lines that can be just resolved, assuming a mean wavelength of 600 nm, is calculated from Equation 25.12 to be $\Delta\lambda = \lambda/R = 6 \times 10^{-2}$ nm. For the third-order principal maximum $R = 15\,000$ and $\Delta\lambda = 4 \times 10^{-2}$ nm, and so on.

EXAMPLE 25.11   Resolving the Sodium Spectral Lines

Two strong lines in the spectrum of sodium have wavelengths of 589.00 nm and 589.59 nm.

(a) What must the resolving power of a grating be in order to distinguish these wavelengths?

Solution   From Equation 25.12, we find that

$$R = \frac{\lambda}{\Delta\lambda} = \frac{589 \text{ nm}}{589.59 \text{ nm} - 589.00 \text{ nm}} = \frac{589}{0.59} = \boxed{998}$$

(b) To resolve these lines in the second-order spectrum, how many lines of the grating must be illuminated?

Solution   From Equation 25.13 and the result of (a), we find that

$$N = \frac{R}{m} = \frac{998}{2} = \boxed{499 \text{ lines}}$$

SUMMARY

The light-concentrating power of a lens of focal length f and diameter D is determined by the f-**number,** defined as

$$f\text{-number} \equiv \frac{f}{D} \qquad \textbf{[25.1]}$$

The smaller the f-number of a lens, the brighter the image formed.

Hyperopia (farsightedness) is a defect of the eye that occurs either when the eyeball is too short or when the ciliary muscle cannot change the shape of the lens enough to form a properly focused image. **Myopia** (nearsightedness) occurs either when the eye is longer than normal or when the maximum focal length of the lens is insufficient to produce a clearly focused image on the retina.

The **power** of a lens in **diopters** is the inverse of the focal length in meters. The **angular magnification of a lens** is defined as

$$m \equiv \frac{\theta}{\theta_0} \qquad \textbf{[25.2]}$$

where θ is the angle subtended by an object at the eye with a lens in use and θ_0 is the angle subtended by the object when it is placed at the near point of the eye and no lens is used. The **maximum magnification of a lens** is

$$m = 1 + \frac{25 \text{ cm}}{f} \qquad \textbf{[25.5]}$$

When the eye is relaxed, the magnification is

$$m = \frac{25 \text{ cm}}{f} \qquad \textbf{[25.6]}$$

The overall **magnification of a compound microscope** of length L is the product of the magnification produced by the objective, of focal length f_o, and the magnification produced by the eyepiece, of focal length f_e:

$$M = -\frac{L}{f_o}\left(\frac{25 \text{ cm}}{f_e}\right) \qquad \textbf{[25.7]}$$

The **magnification of a telescope** is

$$m = \frac{f_o}{f_e} \qquad \textbf{[25.8]}$$

where f_o is the focal length of the objective and f_e is the focal length of the eyepiece.

Two images are said to be **resolved** when the central maximum of the diffraction pattern for one image falls on the first minimum of the other image. This limiting condition of resolution is known as **Rayleigh's criterion.** The limiting angle of resolution for a **slit** of width a is

$$\theta_m \approx \frac{\lambda}{a} \qquad \textbf{[25.9]}$$

The limiting angle of resolution of a **circular aperture** is

$$\theta_m = 1.22 \frac{\lambda}{D} \qquad\qquad [27.10]$$

where D is the diameter of the aperture.

A **diffraction grating** consists of many equally spaced, identical slits. The condition for **maximum intensity** in the interference pattern of a diffraction grating is

$$d \sin \theta = m\lambda \qquad m = 0, 1, 2, \ldots \qquad [25.11]$$

where d is the spacing between adjacent slits and m is the order number of the diffraction pattern. The **resolving power** of a diffraction grating in the mth order is

$$R = Nm \qquad\qquad [25.13]$$

where N is the number of illuminated rulings on the grating.

ADDITIONAL READING

J. Bahcall and L. Spitzer, Jr., "The Space Telescope," *Sci. American,* July 1982, p. 40.

E. Bandana, "The Mystery of Myopia," *The Sciences,* Nov./Dec. 1985, p. 46.

D. Marr, *Vision,* New York, W. H. Freeman, 1982.

C. R. Michael, "Retinal Processing of Visual Images," *Sci. American,* May 1969, p. 105.

U. Neisser, "The Processes of Vision," *Sci. American,* September 1968, p. 204.

W. H. Price, "Photographic Lens," *Sci. American,* August 1976, p. 72.

M. Ruiz, "Camera Optics," *The Physics Teacher,* September 1982, p. 372.

R. S. Shankland, "Michelson and His Interferometer," *Physics Today,* April 1976, p. 72.

G. Wald, "Eye and Camera," *Sci. American,* August 1950, p. 32.

T. D. Walker, *Light and Its Uses,* New York, W. H. Freeman, 1980.

CONCEPTUAL QUESTIONS

Example A pinhole camera can be constructed by punching a small hole in one side of a cardboard box. If the opposite side is cut out and replaced with tissue paper, the image of a distant object is formed by light passing through the pinhole onto the tissue paper (the screen). No lens is involved here. In effect, the pinhole replaces the lens of the camera. Explain why an image is formed, and describe the nature of the image.

Reasoning Only light from the top of the distant object can pass through the hole and reach the bottom of the screen, while only light from the bottom of the object can pass through the hole and reach the top of the screen. Light from intermediate points on the object falls on the screen between these two limits. Thus, an inverted image of the object is formed on the screen. You should construct a simple ray diagram to confirm these statements.

Example If you want to examine the fine detail of an object with a magnifying lens of focal length 15 cm, where should the object be placed in order to observe a magnified image of the object?

Reasoning A magnified image of an object is produced by a converging lens when the object is placed somewhere between the focal point and the lens. Hence, the distance of the object from the lens should be less than 15 cm.

Example Large telescopes are usually reflecting rather than refracting. List some reasons for this choice.

Reasoning Large lenses are difficult to manufacture and machine with accuracy. Their large weight leads to sagging, which produces a distorted image. In reflecting telescopes, light does not pass through glass; hence, problems associated with chromatic aberrations are eliminated. Large-diameter reflecting telescopes are also technically easier to construct. Some designs use a rotating pool of mercury as the reflecting surface.

Example   A classic science-fiction story, The Invisible Man, tells of a person who becomes invisible by changing the index of refraction of his body to that of air. This story has been criticized by students who know how the eye works; they claim the invisible man would be unable to see. On the basis of your knowledge of the eye, could he see or not?

Reasoning   He would not be able to see. In order for the eye to ''see'' an object, incoming light must be refracted at the cornea and lens to form an image on the retina. If the cornea and lens had the same index of refraction as air, refraction would not occur, and an image would not be formed.

Example   Explain why it is theoretically impossible to see an object as small as an atom regardless of the quality of the light microscope being used.

Reasoning   In order to ''see'' an object, the wavelength of the light in the microscope must be comparable to the size of the object. An atom is much smaller than the wavelength of light in the visible spectrum, so an atom can never be seen using visible light.

Example   The diffraction grating effect is easily observed with everyday equipment. For example, a compact disc can be held so that light is reflected from it a glancing angle (Fig. 25.17), and various colors in the reflected light can be seen. Furthermore, the observation depends on the orientation of the disc relative to the eye and light source. Explain how this works.

Reasoning   A row of pits in the compact disc forms a narrow groove which should by itself scatter light over many directions. The disc has many closely and equally spaced grooves. The light scattered by these grooves interferes constructively only in certain directions that depend on the wavelength and on the direction of the incident light. Thus, any one section of the disc functions as a diffraction grating for white light, sending different colors of light off in different directions. The different spectral colors you see when looking at one section of the disc changes as the light source, the disc, or you move to change the angles of incidence or diffraction.

FIGURE 25.17   Compact discs act as diffraction gratings when observed under white light. (© *Bobbie Kingsley/Photo Researchers, Inc.*)

1. A photoflash unit makes a very bright light for a very short time. Discuss differences in the exposure of film with and without a flash unit. Consider the *f*-stop and shutter speed in your deliberation. Which is important: the intensity of the light, or the product of the intensity and the time of the flash?

2. Estimate the shutter speed and *f*-number necessary to photograph a center driving for the goal in a well-lighted basketball gymnasium. Repeat for a race car during the final lap on a cloudy day. Why will the latter probably be blurred anyway?

3. Compare and contrast the eye and a camera. What parts of the camera correspond to the iris, the retina, and the cornea of the eye?

4. In a darkened room we can see reflected images of a luminous object in a person's eye (Fig. 25.18). Upon accommodation, image *b* shifts and changes size, and the other two images do not. Discuss the origins of the three images.

FIGURE 25.18   (Question 4)

5. Suppose we look through a person's eyeglasses, with the lenses a few centimeters from our eyes. If we move the glasses from side to side, the image appears to move. Which way does the image move if the person has (a) myopia? (b) hyperopia? Discuss.

6. The optic nerve and the brain invert the image formed on the retina. Why do we not see everything upside down?

7. If you want to use a converging lens to set fire to a piece of paper, why should the light source be farther away from the focal point of the lens?

8. Explain why you could not use a diverging lens as a simple magnifier.

9. In what respect are the optical systems of a microscope and a telescope (a) similar? (b) different?

10. How should a telescope or microscope be altered to photograph the image? Discuss.

11. What difficulty would astronauts encounter in finding their landing site on the Moon if astronomers were not aware of the properties of the telescope?

12. Assuming that the headlights of a car are point sources, estimate the maximum distance from an ob-

server to the car at which the headlights are distinguishable from each other.

13. A laser beam is incident at a shallow angle on a machinist's ruler that has a finely calibrated scale. The rulings on the scale give rise to a diffraction pattern on a screen. Discuss how you can use this technique to measure the wavelength of the laser light.

14. Describe the zeroth-order spectrum of a grating spectroscope. Include the angular location relative to the source and the color(s) of the image, in the event that the source emits white light.

15. Short of looking inside a spectroscope, how do we know whether the instrument is a grating or prism type?

16. Can a diffraction grating be used for infrared rays, ultraviolet rays, or x-rays? Can a prism be used to examine the spectra of these parts of the electromagnetic spectrum?

17. Discuss the physical principles involved in each of the following instruments: (a) an endoscope (a device for examining the inside of a person's stomach), (b) a telephoto lens, (c) a spectrum analyzer, (d) a pair of binoculars.

18. Many modern cameras with autofocus capability use an infrared beam that bounces off the subject and is reflected back to the camera. By electronic measurement of the angle of the beam and the beam's duration of travel, the distance to the subject can be determined, and hence the camera can be focused. A person uses such a camera to take a picture of herself through a mirror and is confused when the picture comes out blurred. Explain this observation.

PROBLEMS

Section 25.1   The Camera

1. The focal length of the lens in a simple camera is 10.0 cm, and the image formed on the film is to be 35 mm high. How far from the camera does a 2.0-m-tall person have to stand so that the person's image fits on the film?

2. A camera used by a professional photographer to shoot portraits has a focal length of 25.0 cm. The photographer takes a portrait of a person 1.50 m in front of the camera. Where is the image formed, and what is the magnification?

3. A photographic image of a building is 0.0920 m high. The image was made with a lens with a focal length of 52.0 mm. If the lens was 100 m from the building when the photograph was made, determine the height of the building.

4. When the f-number of a camera is increased by one position, the admitted light decreases by a factor of 2. Using this result, explain why the f-number is given as $f/2^{\text{Integer}/2}$. For example, $f/1.4 = f/2^{1/2}$, $f/2.0 = f/2^{2/2}$, $f/2.8 = f/2^{3/2}$, $f/4 = f/2^{4/2}$, $f/5.6 = f/2^{5/2}$, and so on.

5. A photograph of the full Moon is shot with a camera lens of focal length 120 mm. Determine the diameter of the Moon's image on the film. (*Note:* The radius of the Moon is 1.74×10^6 m, and the Earth-Moon distance is 3.84×10^8 m.)

6. If a certain type of film requires an exposure time of 0.010 s with an f-stop of 11.0, and another type of film requires twice the light energy for the same exposure, what f-stop does the second type of film need with the 0.010-s exposure time?

7. Assume that the camera in Figure 25.1a has a fixed focal length of 6.50 cm and is adjusted to properly focus the image of a distant object. How far and in what direction must the lens be moved to focus the image of an object that is 2.00 m away?

8. A camera is being used with the correct exposure at $f/4$ and a shutter speed of $1/32$ s. In order to "stop" a fast-moving subject, the shutter speed is changed to $1/256$ s. Find the new f-stop that should be used to maintain satisfactory exposure, assuming no change in lighting conditions.

Section 25.2   The Eye

9. The near point of an eye is 100 cm. (a) What focal length should the lens have so that the eye can clearly see an object 25.0 cm in front of it? (b) What is the power of the lens?

10. A certain young girl can adjust the power of her eye's lens-cornea combination between limits of +57 diopters and +65 diopters. With the lens relaxed, she can see a distant star clearly. (a) How far is this girl's near point from her eye? (b) How far is her retina from her eye lens?

11. A certain child's near point is 10.0 cm; her far point (with eye lenses relaxed) is 125 cm. Each eye lens is 2.00 cm from the retina. (a) Between what limits, measured in diopters, does the power of this lens-cornea combination vary? (b) Calculate the power of the eyeglass lens this child should use for relaxed distance vision. Is the lens converging or diverging?

□ indicates problems that have full solutions available in the Student Solutions Manual and Study Guide.

12. A person has far points 8.44 cm from the right eye and 12.2 cm from the left eye. Write a prescription for the powers of the corrective lenses.

13. An artificial lens is implanted in a person's eye to replace a diseased lens. The distance between the artificial lens and the retina is 2.80 cm. In the absence of the lens, the image of a distant object falls 2.53 cm behind the retina. The lens is designed to put the image of the distant object on the retina. What is the power of the implanted lens? (*Hint:* Consider the image formed by the eye as a virtual object.)

14. An individual is nearsighted; his near point is 13.0 cm and his far point is 20.0 cm. (a) What lens power is needed to correct his nearsightedness? (b) When the lenses are in use, what is this person's near point?

15. A retired bank president can easily read the fine print of the financial page when the newspaper is held at arm's length, 60.0 cm from the eye. What should be the focal length of an eyeglass lens that will allow her to read at the more comfortable distance of 24.0 cm?

16. A person is to be fitted with bifocals. She can see clearly when the object is between 30 cm and 1.5 m from the eye. (a) The upper portions of the bifocals should be designed to enable her to see distant objects clearly. What power should they have? (b) The lower portions of the bifocals should enable her to see objects comfortably at 25 cm. What power should they have (Fig. 25.19)?

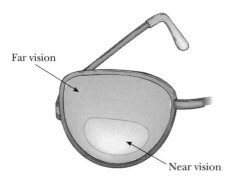

FIGURE 25.19   (Problems 16 and 17)

17. An older person has far points of 1.00 m and near points of 0.667 m in both eyes. Bifocal lenses (Fig. 25.19) are prescribed. Determine the powers of the upper and lower portions of the bifocal lens.

18. A person wearing contact lenses finds that the correction is imperfect. With the lenses in place, the near points are 50.5 cm from the eyes, whereas without the lenses they are 84.0 cm away.

Determine the effective focal length of the lenses when they are in place on the eyes.

Section 25.3   The Simple Magnifier

19. A leaf of length h is positioned 71.0 cm in front of a converging lens with a focal length of 39.0 cm. An observer views the image of the leaf from a position 1.26 m behind the lens, as shown in Figure 25.20. (a) What magnification (ratio of image size to object size) is produced by the lens? (b) What angular magnification is achieved by viewing the image of the leaf rather than viewing the leaf directly?

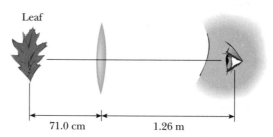

Leaf

71.0 cm   1.26 m

FIGURE 25.20   (Problem 19)

20. In 1675 the Dutch biologist Anton van Leeuwenhoek, using a single lens probably of focal length 1.25 mm, discovered bacteria. What was the magnifying power of this early simple microscope?

21. A biology student uses a simple magnifier to examine the structural features of the wing of an insect. The lens is mounted in a frame 3.50 cm above the work surface, and the image is formed 25.0 cm from the eye. What is the focal length of the lens?

22. A stamp collector uses a lens with 7.5-cm focal length as a simple magnifier. The virtual image is produced at the normal near point (25 cm). (a) How far from the lens should the stamp be placed? (b) What is the expected magnification?

23. (a) What is the angular magnification of a lens having a focal length of 25 cm? (b) What is the magnification of this lens when the eye is relaxed?

24. A boy scout starts a fire by using a lens from his eyeglasses to focus sunlight on kindling 5.0 cm from the lens. The boy scout has a near point of 15 cm. When the lens is used as a simple magnifier, (a) what is the maximum magnification that can be achieved, and (b) what is the magnification when the eye is relaxed? (*Caution:* The equations derived in the text for a simple magnifier assume a "normal" eye.)

Section 25.4 The Compound Microscope

Section 25.5 The Telescope

25. An elderly sailor is shipwrecked on a desert island but manages to save his eyeglasses. The lens for one eye has a power of +1.20 diopters, and the other lens has a power of +9.00 diopters. (a) What is the magnifying power of the telescope he can construct with these lenses? (b) How far apart are the lenses when the telescope is adjusted for minimum eyestrain?

26. The length of a microscope tube is 15 cm. The focal length of the objective is 1.0 cm, and the focal length of the eyepiece is 2.5 cm. What is the magnification of the microscope, assuming it is adjusted so that the eye is relaxed?

27. The text discusses the astronomical telescope. Another type is the Galilean telescope, in which an objective lens gathers light (Fig. 25.21), but the converging rays are not allowed to reach their focus. Instead, a strong diverging lens changes the curvature to -4.00 m^{-1}, so that a virtual, erect image is formed at B (Fig. 25.21a), 25.0 cm from the eye. When adjusted for minimum eyestrain, the rays of the telescope emerge parallel to each other, as in Figure 25.21b, and the virtual image is at infinity. An opera glass, which is a Galilean telescope, is used to view a 30.0-cm-tall singer's head that is 40.0 m from the objective lens. The focal length of the objective is +8.00 cm, and that of the eyepiece is −2.00 cm. The telescope is adjusted so that parallel rays enter the eye. Compute (a) the size of the real image that would have been formed by the objective, (b) the virtual object distance for the negative lens, (c) the distance between the lenses, and (d) the approximate overall magnifying power.

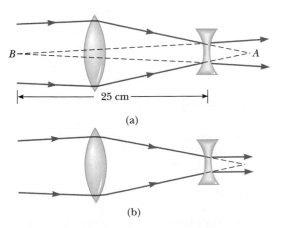

(a)

(b)

FIGURE 25.21 (Problem 27)

28. The objective lens in a microscope with a 20.0-cm-long tube has a magnification of 50.0 and the eyepiece has a magnification of 20.0. What are the focal lengths of (a) the objective and (b) the eyepiece? (c) What is the overall magnification of the microscope?

29. A microscope has an objective lens with a focal length of 16.22 mm and an eyepiece with a focal length of 9.50 mm. With the length of the barrel set at 29.0 cm, the diameter of a red blood cell's image subtends an angle of 1.43 mrad with the eye. If the final image distance is 29.0 cm from the eyepiece, what is the actual diameter of the red blood cell?

30. A certain telescope has a 5.0-in. diameter and an objective lens of focal length 1250 mm. (a) What is the f-number of this lens? (b) What is the magnification of this telescope when it is used with a 25-mm eyepiece?

31. The lenses of an astronomical telescope are 92 cm apart when adjusted for viewing a distant object with minimum eyestrain. The magnifying power is 45. Compute the focal length of each lens.

32. An astronomical telescope has an objective with a focal length of 75 cm and an eyepiece with a focal length of 4.0 cm. What is the magnifying power of this instrument?

33. A telescope adjusted for very distant objects has an objective with a focal length of 100 cm and an eyepiece with a focal length of 10.0 cm. A bee buzzes in front of the telescope, 150 cm from the objective. (a) Use the thin-lens equation twice (once for each lens) to find the location of the final image of the bee. (b) On the basis of your answer to (a), could you see the bee through the telescope? Why or why not?

34. The objective lens of a refracting telescope has a focal length of 1.75 m. The eyepiece has a focal length of −9.73 cm. How much larger does a distant tree appear to be when viewed through the telescope?

35. A certain telescope has an objective of focal length 1500 cm. If the Moon is used as an object, a 1.0-cm length of the image formed by the objective corresponds to what distance, in miles, on the Moon? Assume 3.8×10^8 m for the Earth-Moon distance.

36. A person decides to use an old pair of eyeglasses to make some optical instruments. He knows that the near point in his left eye is 50.0 cm and the near point in his right eye is 100 cm. (a) What is the maximum magnification he can produce in a telescope? (b) If he places the lenses 10.0 cm apart, what is the maximum magnification he can

produce in a microscope? [Go back to basics and use the thin-lens equation to solve part (b).]

Section 25.6   Resolution of Single-Slit and Circular Apertures

37. If the distance from the Earth to the Moon is 3.8×10^8 m, what diameter would be required for a telescope objective to resolve a Moon crater 300 m in diameter? Assume a wavelength of 500 nm.

38. Two telescopes have the following characteristics. Telescope A has a diameter of 5.0 in. and a focal length for the objective of 1200 mm, and is used with an 18-mm eyepiece. Telescope B has a diameter of 3.0 in. and a focal length for the objective of 1000 mm, and is used with a 25-mm eyepiece. (a) What is the limiting angle of resolution for each telescope when it is used with 600-nm light? (b) What is the magnification of each telescope?

39. Suppose a 5.00-m-diameter telescope were constructed on the dark side of the Moon. The viewing there (except for brief periods of sunlight) would be excellent because of no atmospheric distortion. As an example, what would be the separation between two objects that could just be resolved on the planet Mars in 500-nm light? The distance from the dark side of the Moon to Mars at closest approach is 8.0×10^7 km.

40. A positive lens with a diameter of 30.0 cm forms an image of a satellite passing overhead. The satellite has two green lights (wavelength 500 nm) spaced 1.0 m apart. If the lights can just be resolved according to the Rayleigh criterion, what is the altitude of the satellite?

41. A spy satellite circles the Earth at an altitude of 200 km and carries out surveillance with a special high-resolution telescopic camera having a lens diameter of 35 cm. If the angular resolution of this camera is limited by diffraction, estimate the separation of two small objects on the Earth's surface that are just resolved in yellow-green light ($\lambda = 550$ nm).

42. Two stars are 8.0 lightyears away and can just be resolved by a 20-in. telescope equipped with a filter that only allows light of wavelength 500 nm to pass. What is the distance between the two stars?

43. To increase the resolving power of a microscope, the object and the objective are immersed in oil ($n = 1.5$). If the limiting angle of resolution without the oil is 0.60 μrad, what is the limiting angle of resolution with the oil?

44. Two radio telescopes are used in a special technique called long-baseline interferometry, which causes the two to act as a single telescope with an effective diameter of 1000 mi. At what distance from this telescope would two objects 250 000 mi apart be at the limit of resolution? Assume that the wavelength of the radio waves from these objects is about 1.0 m.

45. (a) Calculate the limiting angle of resolution for the eye, assuming a pupil diameter of 2.00 mm, a wavelength of 500 nm in air, and an index of refraction for the eye of 1.33. (b) What is the maximum distance from the eye at which two points separated by 1.00 cm could be resolved?

46. Two motorcycles, separated laterally by 2.0 m, are approaching an observer holding an infrared detector that is sensitive to radiation of wavelength 885 nm. What aperture diameter is required in the detector if the two headlights are to be resolved at a distance of 10.0 km?

47. The photosensitive screen of a television camera is 4.0 cm wide and 3.0 cm high. It contains 2.0×10^5 light-sensitive elements (pixels) arranged in a square lattice of 387 rows and 516 columns. The camera lens has a focal length of 15 cm. (a) What is the spacing between centers of adjacent pixels on the screen? (b) What is the minimum diameter of the camera lens if the resolution of the camera is to be limited by the pixel spacing on the screen (assume a 500-nm wavelength)? (c) If the leaves of a tree have a typical linear dimension of the order of 9.0 cm, what is the maximum distance from the camera at which individual leaves of the tree can be resolved?

48. The separate stars of a double star are just resolved by a lens with a diameter of 10.0 cm and a focal length of 1.40 m. If the image is viewed from a distance of 30.0 cm, what is the angle subtended by the lines of sight of the two stars in the image? (Assume a 500-nm wavelength.)

Section 25.7   The Michelson Interferometer

49. One arm of a Michelson interferometer contains a cell that is 5.00 cm long and parallel to the arm. How many fringe shifts would be observed if all the air were evacuated from the cell? The wavelength of the light source is 590 nm and the refractive index of air is 1.000 29.

50. An interferometer is used to measure the length of an insect. The wavelength of the light used is 650 nm. As one arm of the interferometer is moved from one end of the insect to the other, 310 fringe shifts are counted. How long is the insect?

51. Light of wavelength 550 nm is used to calibrate a Michelson interferometer. By use of a micrometer screw, the platform on which one mirror is mounted is moved 0.18 mm. How many fringe shifts are counted?

52. The Michelson interferometer can be used to measure the index of refraction of a gas by placing an evacuated transparent tube along one arm of the device. Fringe shifts occur as the gas is slowly added to the tube. Assume that the tube is 5.0 cm long and that 60 fringe shifts occur as the pressure of the gas in the tube increases to atmospheric pressure. What is the index of refraction of the gas? (*Hint:* The fringe shifts occur because the wavelength of the light changes inside the gas-filled tube.)

53. A thin sheet of transparent material has an index of refraction of 1.40 and is 15.0 μm thick. When it is inserted along one arm of an interferometer in the path of 600-nm light, how many fringe shifts occur in the pattern?

Section 25.8   The Diffraction Grating

54. A diffraction grating with 2500 lines/cm is used to examine the sodium spectrum. Calculate the angular separation of the two closely spaced sodium-yellow lines (588.995 nm and 589.592 nm) in each of the first three orders.

55. A diffraction grating is calibrated by using the 546.1-nm line of mercury vapor. It is found that the first-order line is at an angle of 21.0°. Calculate the number of lines per centimeter on this grating.

56. A grating with 1500 slits per centimeter is illuminated with light of wavelength 500 nm. (a) What is the highest order number that can be observed with this grating? (b) Repeat for a grating of 15 000 slits per centimeter.

57. Three discrete spectral lines occur at angles of 10.09°, 13.71°, and 14.77° in the first-order spectrum of a grating spectroscope. (a) If the grating has 3660 slits/cm, what are the wavelengths of the light? (b) At what angles are these lines found in the second-order spectra?

58. A diffraction grating with 4000 lines/cm is illuminated by light from the Sun. The first-order solar spectrum is spread out on a white wall across the room. (a) At what angle from the centerline is blue light (400 nm)? (b) At what angle from the centerline does red light (650 nm) appear?

59. Light from a hydrogen source is incident on a diffraction grating. The incident light contains four wavelengths: $\lambda_1 = 410.1$ nm, $\lambda_2 = 434.0$ nm, $\lambda_3 = 486.1$ nm, and $\lambda_4 = 656.3$ nm. The diffraction grating has 410 lines/mm. Calculate the angles between (a) λ_1 and λ_4 in the first-order spectrum and (b) λ_1 and λ_3 in the third-order spectrum.

60. Visible light ranging from 400 nm to 700 nm is focused on a diffraction grating. The entire first order is observed, but no second-order spectrum is seen. What is the maximum spacing between lines on this grating?

61. Light containing two different wavelengths passes through a diffraction grating with 1200 slits/cm. On a screen 15.0 cm from the grating, the third-order maximum of the shorter wavelength falls on top of the first-order minimum of the longer wavelength. If the neighboring maxima of the longer wavelength are 8.44 mm apart on the screen, what are the wavelengths in the light? (*Hint:* Use the small angle approximation.)

62. A light source contains two major spectral lines, an orange line of wavelength 610 nm and a blue-green line of wavelength 480 nm. If the spectrum is resolved by a diffraction grating having 5000 lines/cm and viewed on a screen 2.00 m from the grating, what is the distance between the two spectral lines in the second-order spectrum?

63. A 15.0-cm-long grating has 6000 slits per centimeter. When it is used in the second order, can two lines of wavelengths 600.000 nm and 600.003 nm be separated? Explain.

64. The H$_\alpha$ line in hydrogen has a wavelength of 656.2 nm. This line differs in wavelength from the corresponding spectral line in deuterium (the heavy stable isotope of hydrogen) by 0.18 nm. (a) Determine the minimum number of lines a grating must have to resolve these two wavelengths in the first order. (b) Repeat part (a) for the second order.

ADDITIONAL PROBLEMS

65. A telescope has an objective of focal length 100 cm and an eyepiece of focal length 1.50 cm. (a) What is the distance between the two lenses when the telescope is used to view a distant object? (b) What is the angular magnification of the telescope?

66. The near point of an eye is 75 cm. (a) What lens power should be prescribed to enable the eye to see an object clearly at 25 cm? (b) If the user can see an object clearly at 26 cm but not 25 cm, by how many diopters did the lens grinder miss the prescription?

67. The 546.1-nm line in mercury is measured at an angle of 81.0° in the third-order spectrum of a diffraction grating. Calculate the number of lines per millimeter for the grating.

68. A laboratory (astronomical) telescope is used to view a scale that is 300 cm from the objective, which has a focal length of 20.0 cm; the eyepiece

has a focal length of 2.00 cm. Calculate the angular magnification when the telescope is adjusted for minimum eyestrain. (*Note:* The object is not at infinity, and so the simple expression $m = f_o/f_e$ is not sufficiently accurate for this problem. Also, assume small angles so that tan $\theta \approx \theta$.)

69. A source emits three lines: a violet line of wavelength 400 nm, a green line of wavelength 550 nm, and a red line of wavelength 700 nm. Images of these lines are formed by a diffraction grating with a line spacing of 2500 nm. Calculate sin θ for all observable orders of each line, and arrange the images (13 in all) in sequence according to their sin θ values—thus, V_1, G_1, R_1, V_2, ... The resulting sequence illustrates a phenomenon known as the "overlapping of orders" in a grating spectrum of visible light.

70. The wavelengths of the sodium spectrum are $\lambda_1 = 589.0$ nm and $\lambda_2 = 589.6$ nm. Determine the minimum number of lines in a grating that will allow resolution of the sodium spectrum in (a) the first order and (b) the third order.

71. Light consisting of two wavelength components is incident on a grating. The shorter wavelength component has a wavelength of 440.0 nm. The third-order image of this component is coincident with the second-order image of the longer wavelength component. Determine the wavelength of the longer wavelength component.

72. Sunlight is incident on a diffraction grating that has 2750 lines/cm. The second-order spectrum over the visible range (400–700 nm) is to be limited to 1.75 cm along a screen that is distance L from the grating. What is the required value of L?

73. If a typical eyeball is 2.0 cm long and has a pupil opening that can range from about 2.0 mm to 6.0 mm, what is (a) the focal length of the eye when it is focused on objects 1.00 m away, (b) the smallest f-number of the eye when it is focused on objects 1.0 m away, and (c) the largest f-number of the eye when it is focused on objects 1.0 m away?

74. The 501.5-nm line in helium is observed at an angle of 30.0° in the second-order spectrum of a diffraction grating. Calculate the angular deviation of the 667.8-nm line in helium in the first-order spectrum for the same grating.

75. A fringe pattern is established in the field of view of a Michelson interferometer, using light of wavelength 580 nm. A parallel-faced sheet of transparent material 2.5 μm thick is placed in front of one of the mirrors, perpendicularly to the incident and reflected light beams. An observer counts 12 fringe shifts. What is the index of refraction of the sheet?

76. A microwave version of the Michelson interferometer uses waves of wavelength 4.00 cm. The detector receiving the two interfering microwave beams is sensitive to phase differences corresponding to path differences of one tenth of a wavelength. What is the smallest displacement that can be detected with the interferometer?

77. If the aqueous humor of the eye has an index of refraction of 1.34 and the distance from the vertex of the cornea to the retina is 2.00 cm, what is the radius of curvature of the cornea for which distant objects will be focused on the retina? (For simplicity, assume that all refraction occurs in the aqueous humor.)

"JUST CHECKING."

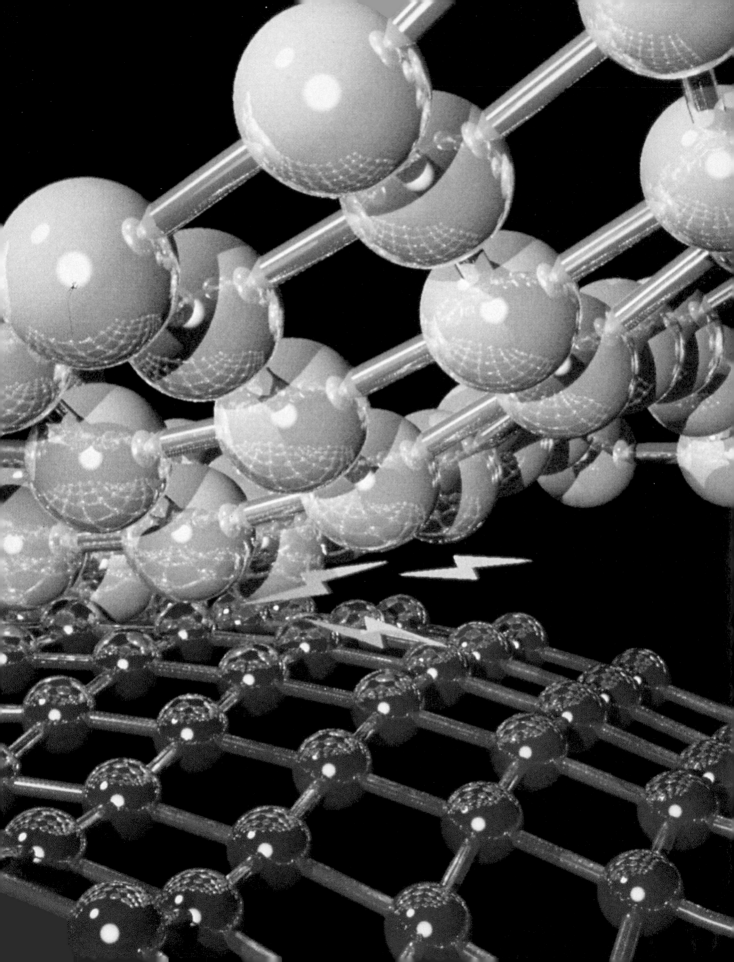

Modern Physics

PART 6

At the end of the 19th century, scientists believed that they had learned most of what there was to know about physics. Newton's laws of motion and his universal theory of gravitation, Maxwell's theoretical work in unifying electricity and magnetism, and the laws of thermodynamics and kinetic theory were highly successful in explaining a wide variety of phenomena.

However, at the turn of the 20th century a major revolution shook the world of physics. In 1900 Planck provided the basic ideas that led to the formulation of the quantum theory, and in 1905 Einstein formulated his brilliant special theory of relativity. The excitement of the times is captured in Einstein's own words: "It was a marvelous time to be alive." Both theories were to have a profound effect on our understanding of nature, and within a few decades they inspired new developments and theories in the fields of atomic physics, nuclear physics, and condensed matter physics.

Our discussion of modern physics will begin with a treatment of the special theory of relativity in Chapter 26. Although its underlying concepts often violate our common sense, the theory provides us with a new and deeper view of physical laws. In Chapter 27 we shall discuss various developments in quantum theory, which provides us with a successful model for understanding electrons, atoms, and molecules. The last three chapters of the text are concerned with applications of quantum theory. Chapter 28 discusses the structure and properties of atoms using concepts from quantum mechanics. Chapter 29 is concerned with the structure and properties of the atomic nucleus. Chapter 30 discusses many practical applications of nuclear physics and concludes with a discussion of elementary particles.

Keep in mind that, although modern physics has been developed during this century and has led to a multitude of important technological achievements, the story is still incomplete. New dis-

Mel Pruitt

coveries will be made during our lifetime, many of which will deepen or refine our understanding of nature and the world around us. It is still "a marvelous time to be alive."

RELATIVITY

26.1

INTRODUCTION

Most of our everyday experiences and observations deal with objects that move at speeds much lower than the speed of light. Newtonian mechanics and the early ideas on space and time were formulated to describe the motion of such objects. As we saw in the chapters on mechanics, this formalism is very successful in describing a wide range of phenomena. Although Newtonian mechanics works very well at low speeds, it fails when applied to particles whose speeds approach that of light. Experimentally, the predictions of Newtonian theory at high speeds can be tested by accelerating an electron through a large electric potential difference. For example, it is possible to accelerate an electron to a speed of

$0.99c$ by using a potential difference of several million volts. According to Newtonian mechanics, if the potential difference (as well as the corresponding energy) is increased by a factor of 4, then the speed of the electron should be doubled to $1.98c$. However, experiments show that the speed of the electron always remains *lower* than the speed of light, regardless of the size of the accelerating voltage. Since Newtonian mechanics places no upper limit on the speed that a particle can attain, it is contrary to modern experimental results and is clearly a limited theory.

In 1905, at the age of only 26, Einstein published his special theory of relativity. Regarding the theory, Einstein wrote

> The relativity theory arose from necessity, from serious and deep contradictions in the old theory from which there seemed no escape. The strength of the new theory lies in the consistency and simplicity with which it solves all these difficulties, using only a few very convincing assumptions.[1]

Although Einstein made many other important contributions to science, his theory of relativity alone represents one of the greatest intellectual achievements of the 20th century. With this theory, experimental observations over the range from $v = 0$ to velocities approaching the speed of light can be predicted. Newtonian mechanics, which was accepted for more than 200 years, is in fact a specialized case of Einstein's theory. This chapter introduces the special theory of relativity, with emphasis on some of the consequences of the theory. A discussion of general relativity and some of its consequences is presented in the essay that follows this chapter.

As we shall see, the special theory of relativity is based on two basic postulates:

1. The laws of physics are the same in all inertial reference systems.
2. The speed of light in a vacuum is always measured to be 3×10^8 m/s, and the measured value is independent of the motion of the observer or of the motion of the source of light.

Special relativity covers such phenomena as the slowing down of moving clocks and the contraction of moving rods as measured by a stationary observer. In addition to these topics, we also discuss the relativistic forms of momentum and energy, terminating the chapter with the famous mass-energy equivalence formula, $E = mc^2$.

26.2

THE PRINCIPLE OF RELATIVITY

In order to describe a physical event, it is necessary to choose a *frame of reference*. For example, when you perform an experiment in a laboratory, you must select a coordinate system, or frame of reference, that is at rest with respect to the laboratory. However, suppose an observer in a passing car moving at a constant velocity with respect to the lab were to observe your experiment. Would the

---

[1] A. Einstein and L. Infeld, *The Evolution of Physics*, New York, Simon and Schuster, 1961.

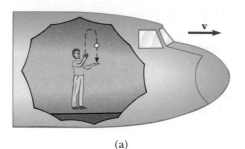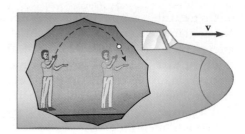

(a)

(b)

FIGURE 26.1
(a) The observer on the airplane sees the ball move in a vertical path when thrown upward. (b) The Earth observer views the path of the ball to be a parabola.

observations made by the moving observer differ dramatically from yours? That is, if you found Newton's first law to be valid in your frame of reference, would the moving observer agree with you?

> According to the principle of Newtonian relativity, the laws of mechanics are the same in all inertial frames of reference. Inertial frames of reference are those reference frames in which Newton's first law, the law of inertia, is valid.

For the situation described above, the laboratory coordinate system and the coordinate system of the moving car are both inertial frames of reference. Consequently, if the laws of mechanics are found to be true in the lab, the person in the car must also observe the same laws.

Let us describe a common observation to illustrate the equivalence of the laws of mechanics in different inertial frames. Consider an airplane in flight, moving with a constant velocity as in Figure 26.1a. If a passenger in the airplane throws a ball straight up in the air, the passenger observes that the ball moves in a vertical path. The motion of the ball is precisely the same as it would be if the ball were thrown while at rest on Earth. The law of gravity and the equations of motion under constant acceleration are obeyed whether the airplane is at rest or in uniform motion. Now consider the same experiment when viewed from another observer at rest on the Earth. This stationary observer views the path of the ball to be a parabola as in Figure 26.1b. Furthermore, according to this observer, the ball has a velocity to the right equal to the velocity of the plane. Although the two observers disagree on certain aspects of the experiment, both agree that the motion of the ball obeys the law of gravity and Newton's laws of motion. Thus, we draw the following important conclusion:

> There is no preferred frame of reference for describing the laws of mechanics.

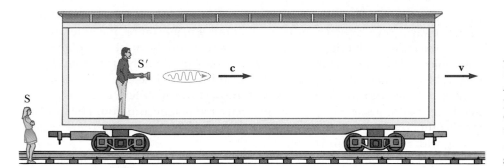

FIGURE 26.2
A pulse of light is sent out by a person in a moving boxcar. According to Newtonian relativity, the speed of the pulse should be $c + v$ relative to a stationary observer.

26.3

THE SPEED OF LIGHT

It is quite natural to ask whether the concept of Newtonian relativity in mechanics also applies to experiments in electricity, magnetism, optics, and other areas. For example, if we assume that the laws of electricity and magnetism are the same in all inertial frames, a paradox concerning the speed of light immediately arises. This can be understood by recalling that according to electromagnetic theory, the speed of light always has the fixed value of $2.997\ 924\ 58 \times 10^8$ m/s in free space. But this is in direct contradiction to common sense. For example, suppose a light pulse is sent out by an observer in a boxcar moving with a velocity **v** (Fig. 26.2). The light pulse has a velocity **c** relative to observer S′ in the boxcar. According to Newtonian relativity, the velocity of the pulse relative to the stationary observer S outside the boxcar should be $\mathbf{c} + \mathbf{v}$. This obviously contradicts Einstein's theory, which postulates that the velocity of the light pulse is the same for all observers.

In order to resolve this paradox, we must conclude either that (1) the addition law for velocities is incorrect or that (2) the laws of electricity and magnetism are not the same in all inertial frames. If the Newtonian addition law for velocities were incorrect, we would be forced to abandon the seemingly "obvious" notions of absolute time and absolute length that form the basis for this law.

If instead we assume that the second conclusion is true, then a preferred reference frame must exist in which the speed of light has the value c, while in any other reference frame the speed of light must have a value that is greater or less than c. It is useful to draw an analogy with sound waves, which propagate through a medium such as air. The speed of sound in air is about 330 m/s when measured in a reference frame in which the air is stationary. However, the speed of sound is greater or less than this value when measured from a reference frame that is moving with respect to the air.

In the case of light signals (electromagnetic waves), recall that electromagnetic theory predicted that such waves must propagate through free space with a speed equal to the speed of light. However, the theory does not require the presence of a medium for wave propagation. This is in contrast to other types of waves that we have studied, such as water and sound waves, which do require a medium to support the disturbances. In the 19th century, physicists thought that electromagnetic waves also required a medium in order to propagate. They proposed that such a medium existed, and they gave it the name **luminiferous**

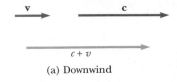

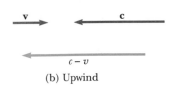

(a) Downwind

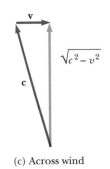

(b) Upwind

(c) Across wind

FIGURE 26.3
If the velocity of the ether wind relative to the Earth is v, and c is the velocity of light relative to the ether, the speed of light relative to the Earth is (a) $c + v$ in the downwind direction,
(b) $c - v$ in the upwind direction, and (c) $(c^2 - v^2)^{1/2}$ in the direction perpendicular to the wind.

ether. The ether was assumed to be present everywhere, even in empty space, and light waves were viewed as ether oscillations. Furthermore, the ether would have to be a massless but rigid medium with no effect on the motion of planets or other objects. These are indeed strange concepts. Additionally, it was found that the troublesome laws of electricity and magnetism would take on their simplest forms in a frame of reference at *rest* with respect to the ether. This frame was called the *absolute frame*. The laws of electricity and magnetism would be valid in this absolute frame, but they would have to be modified in any reference frame moving with respect to this frame.

As a result of the importance attached to this absolute frame, it became of considerable interest in physics to prove by experiment that it existed. A direct method for detecting the ether wind was to measure its influence on the speed of light relative to a frame of reference on Earth. If v is the velocity of the ether relative to the Earth, then the speed of light should have its maximum value, $c + v$, when propagating downwind as shown in Figure 26.3a. Likewise, the speed of light should have its minimum value, $c - v$, when propagating upwind as in Figure 26.3b, and some intermediate value, $(c^2 - v^2)^{1/2}$, in the direction perpendicular to the ether wind as in Figure 26.3c. If the Sun is assumed to be at rest in the ether, then the velocity of the ether wind would be equal to the orbital velocity of the Earth around the Sun, which has a magnitude of about 3×10^4 m/s. Since $c = 3 \times 10^8$ m/s, a change in speed of about 1 part in 10^4 m/s for measurements in the upwind or downwind directions should be detectable. However, as we shall see in the next section, all attempts to detect such changes and establish the existence of the ether (and hence the absolute frame) proved futile!

26.4

THE MICHELSON–MORLEY EXPERIMENT

The most famous experiment designed to detect small changes in the speed of light was performed in 1887 by A. A. Michelson (1852–1931) and E. W. Morley (1838–1923). We should state at the outset that the outcome of the experiment was *negative*, thus contradicting the ether hypothesis. The experiment was designed to determine the velocity of the Earth with respect to the hypothetical ether. The tool used was the Michelson interferometer shown in Figure 26.4. When one of the arms of the interferometer was aligned along the direction of the Earth's motion through space, the motion of the Earth through the ether would have been equivalent to the ether flowing past the Earth in the opposite direction. This ether wind blowing in the opposite direction should have caused the speed of light as measured in the Earth's frame of reference to be $c - v$ as it approached the mirror M_2 in Figure 26.4, and $c + v$ after reflection. The speed v is the speed of the Earth through space, and hence the speed of the ether wind, while c is the speed of light in the absolute ether frame. In the experiment the two beams of light reflected from M_1 and M_2 recombined, and an interference pattern consisting of alternating dark and bright bands or fringes was formed. During the experiment, the interference pattern was observed while the interferometer was rotated through an angle of 90 degrees. The effect of this rotation should have been to cause a slight but measurable shift in the fringe pattern. Measurements failed to show any change in the interference pattern!

The Michelson–Morley experiment was repeated by other researchers under various conditions and at different locations, but the results were always the same: *No fringe shift of the magnitude required by the ether hypothesis was ever observed.*

The negative results of the Michelson–Morley experiment meant that it was impossible to measure the absolute orbital velocity of the Earth with respect to the ether frame. However, as we shall see in the next section, Einstein developed a postulate for his theory of relativity that places quite a different interpretation on these results. In later years, when more was known about the nature of light, the idea of an ether that permeates all of space was relegated to the ash heap of worn-out concepts. Light is now understood to be an electromagnetic wave that requires no medium for its propagation. As a result, the idea of having an ether in which electromagnetic waves travel became unnecessary.

*DETAILS OF THE MICHELSON–MORLEY EXPERIMENT

As we mentioned earlier, the Michelson–Morley experiment was designed to detect the motion of the Earth with respect to the ether. Before we examine the details of this important, historical experiment, it is instructive to first consider a race between two airplanes, as shown in Figure 26.5a. One airplane flies from point O to point A perpendicular to the direction of the wind, while the second airplane flies from point O to point B parallel to the wind. We shall assume that they start at O at the same time, travel the same distance L with the same cruising speed c with respect to the wind, and return to O. Which airplane will win the race? In order to answer this question, we shall first calculate the time of flight for both airplanes.

First, consider the airplane that moves along path I parallel to the wind. As it moves to the right, its speed is enhanced by the wind, and its velocity with respect to the Earth is $c + v$. As it moves to the left on its return journey, it must fly

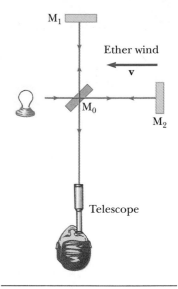

FIGURE 26.4
According to the ether wind theory, the speed of light should be $c - v$ as the beam approaches mirror M_2 and $c + v$ after reflection.

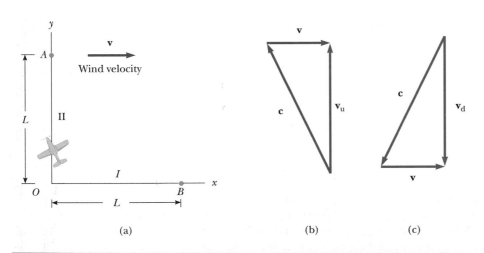

(a)   (b)   (c)

FIGURE 26.5
(a) If an airplane wishes to travel from O to A with a wind blowing to the right, it must head into the wind at some angle. (b) Vector diagram for determining the airplane's direction for the trip from O to A. (c) Vector diagram for determining its direction for the trip from A to O.

Albert A. Michelson (1852–1931). *(SPL/Photo Researchers)*

opposite the wind; hence its speed with respect to the Earth is $c - v$. Thus, the times of flight to the right and to the left are, respectively,

$$t_R = \frac{L}{c + v} \quad \text{and} \quad t_L = \frac{L}{c - v}$$

and the total time of flight for the airplane moving along path I is

$$t_1 = t_R + t_L = \frac{L}{c + v} + \frac{L}{c - v} = \frac{2Lc}{c^2 - v^2}$$

$$= \frac{2L}{c\left(1 - \dfrac{v^2}{c^2}\right)} \qquad \text{[26.1]}$$

Now consider the airplane flying along path II. If the pilot aims the airplane directly toward point A, it will be blown off course by the wind and will not reach its destination. To compensate for the wind, the pilot must point the airplane into the wind at some angle as shown in Figure 26.5a. This angle must be selected so that the vector sum of **c** and **v** leads to a velocity vector pointed directly toward A. The resultant vector diagram is shown in Figure 26.5b, where $\mathbf{v_u}$ is the velocity of the airplane with respect to the ground as it moves from O to A. From the Pythagorean theorem, the magnitude of the vector $\mathbf{v_u}$ is

$$v_u = \sqrt{c^2 - v^2} = c\sqrt{1 - \frac{v^2}{c^2}}$$

Likewise, on the return trip from A to O, the pilot must again head into the wind so that the airplane's velocity with respect to the Earth, $\mathbf{v_d}$, will be directed toward O, as shown in Figure 26.5c. From this figure, we see that

$$v_d = \sqrt{c^2 - v^2} = c\sqrt{1 - \frac{v^2}{c^2}}$$

Thus, the total time of flight for the trip along path II is

$$t_2 = \frac{L}{v_u} + \frac{L}{v_d} = \frac{L}{c\sqrt{1 - \dfrac{v^2}{c^2}}} + \frac{L}{c\sqrt{1 - \dfrac{v^2}{c^2}}}$$

$$= \frac{2L}{c\sqrt{1 - \dfrac{v^2}{c^2}}} \qquad \text{[26.2]}$$

Comparing Equations 26.1 and 26.2, we see that the airplane flying along path II wins the race. The difference in flight times is given by

$$\Delta t = t_1 - t_2 = \frac{2L}{c}\left[\frac{1}{\left(1 - \dfrac{v^2}{c^2}\right)} - \frac{1}{\sqrt{1 - \dfrac{v^2}{c^2}}}\right]$$

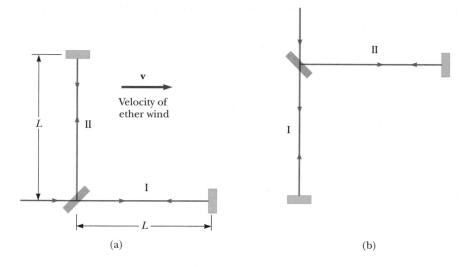

FIGURE 26.6
(a) Top view of the Michelson–Morley interferometer, where v is the velocity of the ether and L is the length of each arm.
(b) When the interferometer is rotated by 90°, the role of each arm is reversed.

This expression can be simplified using the following binomial expansions in v/c (assumed to be much smaller than 1) after dropping all terms higher than second order:

$$\left(1 - \frac{v^2}{c^2}\right)^{-1} \approx 1 + \frac{v^2}{c^2}$$

and

$$\left(1 - \frac{v^2}{c^2}\right)^{-1/2} \approx 1 + \frac{1}{2}\frac{v^2}{c^2}$$

This gives

$$\Delta t = \frac{Lv^2}{c^3} \qquad \text{[26.3]}$$

The analogy between this airplane race and the Michelson–Morley experiment is shown in Figure 26.6a. Two beams of light travel along two arms of an interferometer. In this case, the "wind" is the ether blowing across the Earth from left to right as the Earth moves through the ether from right to left. Because the speed of the Earth in its orbital path is approximately equal to 3×10^4 m/s, the speed of the wind should be at least this great. The two light beams start out in phase and return to form an interference pattern. Let us assume that the interferometer is adjusted for parallel fringes and that a telescope is focused on one of these fringes. The time difference between the two light beams gives rise to a phase difference between the beams, producing an interference pattern when they combine at the position of the telescope. The difference in the pattern is detected by rotating the interferometer through 90° in a horizontal plane, so that the two beams exchange roles (Fig. 26.6b). This results in a net time shift of twice the time difference given by Equation 26.3. Thus, the net time difference is

$$\Delta t_{\text{net}} = 2\,\Delta t = \frac{2Lv^2}{c^3} \qquad \text{[26.4]}$$

The corresponding path difference is

$$\Delta d = c\,\Delta t_{\text{net}} = \frac{2Lv^2}{c^2} \qquad \text{[26.5]}$$

In the first experiments by Michelson and Morley, each light beam was reflected by the mirrors many times to give an increased effective path length L

of about 11 meters. Using this value and taking v to be equal to 3×10^4 m/s gives a path difference of

$$\Delta d = \frac{2(11 \text{ m})(3.0 \times 10^4 \text{ m/s})^2}{(3.0 \times 10^8 \text{ m/s})^2} = 2.2 \times 10^{-7} \text{ m}$$

This extra travel distance should produce a noticeable shift in the fringe pattern. Specifically, calculations show that if the pattern is viewed while the interferometer is rotated through 90°, a shift of about 0.4 fringes should be observed. The instrument used by Michelson and Morley was capable of detecting a shift in the fringe pattern as small as 0.01 fringes. However, *they detected no shift in the fringe pattern*. Since then, the experiment has been repeated many times by various scientists under various conditions and no fringe shift has ever been detected. Thus, it was concluded that the motion of the Earth with respect to the ether cannot be detected.

Many efforts were made to explain the null results of the Michelson–Morley experiment. For example, perhaps the Earth drags the ether with it in its motion through space. To test this assumption, interferometer measurements were made at various altitudes, but again no fringe shift was detected. In the 1890s G. F. Fitzgerald and H. A. Lorentz tried to explain the null results by making the following ad hoc assumption. They proposed that the length of an object moving along the direction of the ether wind would contract by a factor of $\sqrt{1 - v^2/c^2}$. The net result of this contraction would be a change in length of one of the arms of the interferometer such that no path difference would occur as the interferometer was rotated.

No other experiment in the history of physics has received such valiant efforts to explain the absence of an expected result as has the Michelson–Morley experiment. The stage was set for the brilliant Albert Einstein, who solved the problem in 1905 with his special theory of relativity.

26.5

EINSTEIN'S PRINCIPLE OF RELATIVITY

In the previous section we noted the serious contradiction between the Newtonian addition law for velocities and the fact that the speed of light is the same for all observers. In 1905 Albert Einstein proposed a theory that would resolve this contradiction but at the same time would completely alter our notions of space and time. Einstein based his special theory of relativity on the following general hypothesis, which is called the **principle of relativity**:

Postulates of relativity

> **All the laws of physics are the same in all inertial frames.**

An immediate consequence of the principle of relativity is that:

> **The speed of light in a vacuum has the same value, $c = 2.997\ 924\ 58 \times 10^8$ m/s, in all inertial reference frames.**

In other words, anyone who measures the speed of light will get the same value, c. This implies that the ether does not exist. Together, the principle of relativity and its immediate consequence are often referred to as the two postulates of special relativity.

Albert Einstein, one of the greatest physicists of all times, was born in Ulm, Germany. He showed little intellectual promise as a youngster and left the highly disciplined German school system after one teacher stated "You will never amount to anything, Einstein." Following a vacation in Italy, he completed his education at the Swiss Federal Polytechnic School in 1901. Although Einstein attended very few lectures, he was able to pass the courses with the help of excellent lecture notes taken by a friend. Unable to find an academic position, Einstein accepted a position as a junior official in the Swiss Patent Office in Berne. In this setting, and during his "spare time," he continued his independent studies in theoretical physics. In 1905, at the age of 26, he earned his Ph.D. and published four scientific papers that revolutionized physics. One of these papers, for which he was awarded the 1921 Nobel prize in physics, dealt with the photoelectric effect. Another was concerned with Brownian motion, the irregular motion of small particles suspended in a liquid. The remaining two papers were concerned with what is now considered his most important contribution of all, the special theory of relativity. In 1915 Einstein published his work on the general theory of relativity, which relates gravity to the structure of space and time. The most dramatic prediction of this theory was that light would be deflected by a gravitational field. Measurements made by astronomers on bright stars in the vicinity of the eclipsed Sun in 1919 confirmed Einstein's prediction, and he suddenly became a world celebrity. This and other predictions of the general theory of relativity are discussed in the delightful essay by Clifford Will that follows this chapter.

In 1913, following academic appointments in Switzerland and Czechoslovakia, Einstein accepted a special position created for him at the Kaiser Wilhelm Institute in Berlin. This made it possible for him to devote all of his time to research, free of financial troubles and routine duties. Einstein left Germany in 1933, which was then under Hitler's power, thereby escaping the fate of millions of other European Jews. In the same year he accepted a special position at the Institute for Advanced Study in Princeton, where he remained for the rest of his life. He became an American citizen in 1940. Although a pacifist, Einstein was persuaded by Leo Szilard to write a letter to President Franklin D. Roosevelt urging him to initiate a program to develop a nuclear bomb. The result was the successful six-year Manhattan project and the two nuclear explosions in Japan that ended World War II in 1945.

Einstein made many important contributions to the development of modern physics, including the concept of the light quantum and the idea of stimulated emission of radiation, which led to the invention of the laser 40 years later. Einstein was deeply disturbed by the development of quantum mechanics in the 1920s despite his own role in this theory. In particular, he could never accept the probabilistic view of events in nature that is a central feature of the highly successful quantum theory. He once said, "God does not play dice with nature." The last few decades of his life were devoted to an unsuccessful search for a unified theory that would combine gravitation and electromagnetism into one picture.

Albert Einstein (1879–1955)
"Imagination is more important than knowledge."

(AIP Niels Bohr Library)

The null result of the Michelson–Morley experiment can be readily understood within the framework of Einstein's theory. According to his principle of relativity, the premises of the Michelson–Morley experiment were incorrect. In the process of trying to explain the expected results, we stated that when light traveled against the ether wind its speed was $c - v$. However, if the state of motion of the observer or of the source has no influence on the value found for the speed of light, the measured value will always be c. Likewise, the light makes the return trip after reflection from the mirror at a speed of c, not the speed of

$c + v$. Thus, the motion of the Earth should not influence the fringe pattern observed in the Michelson–Morley experiment and a null result should be expected.

If we accept Einstein's theory of relativity, we must conclude that relative motion is unimportant when measuring the speed of light. At the same time, we must alter our common-sense notions of space and time and be prepared for some rather bizarre consequences.

26.6
CONSEQUENCES OF SPECIAL RELATIVITY

Almost everyone who has dabbled even superficially in science is aware of some of the startling predictions that arise because of Einstein's approach to relative motion. As we examine some of the consequences of relativity in this section, we shall find that they conflict with some of our basic notions of space and time. We shall restrict our discussion to the concepts of length, time, and simultaneity, which are quite different in relativistic mechanics from what they are in Newtonian mechanics. For example, we shall see that *the distance between two points and the time interval between two events depend on the frame of reference in which they are measured.* That is, *in relativity, there is no such thing as absolute length or absolute time. Furthermore, events at different locations that occur simultaneously in one frame are not simultaneous in another frame.*

Absolute length and absolute time intervals are meaningless in relativity

SIMULTANEITY AND THE RELATIVITY OF TIME

A basic premise of Newtonian mechanics is that there is a universal time scale that is the same for all observers. In fact, Newton wrote, "Absolute, true, and mathematical time, of itself, and from its own nature, flows equably without relation to anything external." In his special theory of relativity, Einstein abandoned this assumption. According to Einstein, *time interval measurements depend on the reference frame in which they are made.*

Einstein devised the following thought experiment to illustrate this point. A boxcar moves with uniform velocity, and two lightning bolts strike its ends, as in Figure 26.7a, leaving marks on the boxcar and the ground. The marks left on the boxcar are labeled A' and B', and those on the ground are labeled A and B. An observer at O' moving with the boxcar is midway between A' and B', and an observer on the ground at O is midway between A and B. The events recorded by the observers are the light signals from the lightning bolts.

Let us assume that the two light signals reach the observer at O at the same time, as indicated in Figure 26.7b. This observer realizes that the light signals have traveled at the same speed over distances of equal length. Thus, the observer at O concludes that the events at A and B occurred simultaneously. Now consider the same events as viewed by the observer on the boxcar at O'. By the time the light has reached the observer at O, the observer at O' has moved, as indicated in Figure 26.7b. Thus, the light signal from B' has already swept past O', while the light from A' has not yet reached O'. According to Einstein's second postulate, the observer at O' must find that light travels at the same speed as that measured by the observer at O. Therefore, the observer at O' concludes that the lightning struck the front of the boxcar before it struck the back. This thought experiment clearly demonstrates that the two events which appear to be simulta-

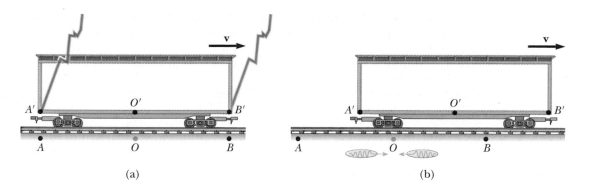

(a)   (b)

FIGURE 26.7
Two lightning bolts strike the ends of a moving boxcar.
(a) The events appear to be simultaneous to the stationary observer at O, who is midway between A and B. (b) The events do not appear to be simultaneous to the observer at O', who claims that the front of the train is struck *before* the rear.

neous to the observer at O do not appear to be simultaneous to the observer at O'. In other words,

> Two events that are simultaneous in one reference frame are in general not simultaneous in a second frame moving with respect to the first. That is, simultaneity is not an absolute concept.

At this point, you might wonder which observer is right concerning the two events. The answer is that *both are correct* because the principle of relativity states that *there is no preferred inertial frame of reference*. Although the two observers reach different conclusions, both are correct in their own reference frames because the concept of simultaneity is not absolute.

TIME DILATION

Consider a vehicle moving to the right with a speed v, as in Figure 26.8a. A perfectly reflecting mirror is fixed to the ceiling of the vehicle, and an observer

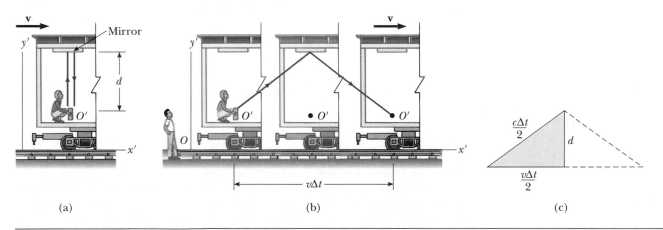

(a)   (b)   (c)

FIGURE 26.8
(a) A mirror is fixed to a moving vehicle, and a light pulse leaves O' at rest in the vehicle. (b) Relative to a stationary observer on Earth, the mirror and O' move with a speed v. Note that the distance the pulse travels is greater than $2d$ as measured by the stationary observer. (c) The right triangle for calculating the relationship between Δt and $\Delta t'$.

at O' at rest in this system holds a flash gun a distance d below the mirror. At some instant, the flash gun goes off and a pulse of light is released. Because the light pulse has a speed c, the time it takes it to travel from the observer to the mirror and back again can be found from the definition of velocity,

$$\Delta t' = \frac{\text{distance traveled}}{\text{velocity}} = \frac{2d}{c} \qquad \text{[26.6]}$$

where the prime notation indicates that this is the time measured by the observer in the reference frame of the moving vehicle.

Now consider the same set of events as viewed by an observer at O in a stationary frame (Fig. 26.8b). According to this observer, the mirror and flash gun are moving to the right with a speed of v. The sequence of events just described would appear entirely different to this stationary observer. By the time the light from the flash gun reaches the mirror, the mirror will have moved a distance of $v\,\Delta t/2$, where Δt is the time it takes the light pulse to travel from O' to the mirror and back, as measured by the stationary observer. In other words, the stationary observer concludes that, because of the motion of the system, the light, if it is to hit the mirror, will leave the flash gun at an angle with respect to the vertical. Comparing Figures 26.8a and 26.8b, we see that the light must travel farther in the stationary frame than in the moving frame.

Now, according to Einstein's second postulate, the speed of light must be c as measured by both observers. Therefore, it follows that the time interval Δt, measured by the observer in the stationary frame, is *longer* than the time interval $\Delta t'$, measured by the observer in the moving frame. To obtain a relationship between Δt and $\Delta t'$, it is convenient to use the right triangle shown in Figure 26.7c. The Pythagorean theorem applied to this triangle gives

$$\left(\frac{c\Delta t}{2}\right)^2 = \left(\frac{v\Delta t}{2}\right)^2 + d^2$$

Solving for Δt gives

$$\Delta t = \frac{2d}{\sqrt{c^2 - v^2}} = \frac{2d}{c\sqrt{1 - v^2/c^2}} \qquad \text{[26.7]}$$

Since $\Delta t' = 2d/c$, we can express Equation 26.7 as

Time dilation

$$\Delta t = \frac{\Delta t'}{\sqrt{1 - v^2/c^2}} = \gamma \Delta t' \qquad \text{[26.8]}$$

where $\gamma = 1/\sqrt{1 - v^2/c^2}$. This result says that the time interval measured by the observer in the stationary frame is *longer* than that measured by the observer in the moving frame (γ is always greater than unity).

For example, suppose an observer in a moving vehicle has a clock that he uses to measure the time required for the light flash to leave the gun and return. Let us assume that the measured time interval in this frame of reference, $\Delta t'$, is one second. (This would require a very tall vehicle.) Now let us find the time interval as measured by a stationary observer using an identical clock. If the vehicle is traveling at half the speed of light ($v = 0.500c$), then $\gamma = 1.15$, and according to Equation 26.6 $\Delta t = \gamma \Delta t' = 1.15(1.00 \text{ s}) = 1.15$ s. Thus, when the observer on the moving vehicle claims that 1.00 s has passed, a stationary observer claims that 1.15 s has passed. From this we may conclude that

> **According to a stationary observer, a moving clock runs more slowly than an identical stationary clock by a factor of γ^{-1}. This effect is known as time dilation.**

A moving clock runs more slowly than an identical stationary clock

The time interval $\Delta t'$ in Equation 26.8 is called the proper time. In general, **proper time** is defined as *the time interval between two events as measured by an observer who sees the events occur at the same place.* In our case, the observer at O' measures the proper time. That is, *proper time is always the time measured by an observer moving along with the clock.*

We have seen that moving clocks run slow by a factor of γ^{-1}. This is true for ordinary mechanical clocks as well as for the light clock just described. In fact, we can generalize these results by stating that *all physical processes, including chemical and biological reactions, slow down relative to a stationary clock when they occur in a moving frame.* For example, the heartbeat of an astronaut moving through space has to keep time with a clock inside the spaceship. Both the spaceship clock and the heartbeat are slowed down relative to a stationary clock. The astronaut would not, however, have any sensation of life slowing down in the spaceship.

Time dilation is a very real phenomenon that has been verified by various experiments. Muons are unstable elementary particles with a charge equal to that of the electron and a mass 207 times that of the electron. They can be produced by the absorption of cosmic radiation high in the atmosphere. These unstable particles have a lifetime of only 2.2 μs when measured in a reference frame at rest with respect to them. If we take 2.2 μs as the average lifetime of a muon and assume that their speed is close to the speed of light, we find that these particles can travel only about 600 m before they decay (Fig. 26.9a). Hence, they could never reach the Earth from the upper atmosphere where they are produced. However, experiments show that a large number of muons *do* reach the Earth, and the phenomenon of time dilation explains how. Relative to an observer on Earth, the muons have a lifetime equal to $\gamma \tau$, where $\tau = 2.2$ μs is the lifetime in a frame of reference traveling with the muons. For example, for $v = 0.99c$, $\gamma \approx 7.1$ and $\gamma \tau \approx 16$ μs. Hence, the average distance traveled as measured by an observer on Earth is $\gamma v \tau \approx 4800$ m, as indicated in Figure 26.9b.

In 1976 experiments with muons were conducted at the laboratory of the European Council for Nuclear Research (CERN) in Geneva. Muons were injected into a large storage ring, reaching speeds of about $0.9994c$. Electrons produced by the decaying muons were detected by counters around the ring, enabling scientists to measure the decay rate, and hence the lifetime of the muons. The lifetime of the moving muons was measured to be about 30 times as long as that of stationary muons to within two parts in a thousand, in agreement with the prediction of relativity.

The results of an experiment reported by Hafele and Keating provided direct evidence for the phenomenon of time dilation.[2] The experiment involved

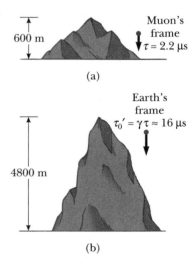

FIGURE 26.9

(a) The muons travel only about 600 m as measured in the muons' reference frame, where their lifetime is about 2.2 μs. Because of time dilation, the muons' lifetime is longer as measured by the observer on Earth. (b) Muons traveling with a speed of $0.99c$ travel a distance of about 4800 m as measured by an observer on Earth.

---

[2] J. C. Hafele and R. E. Keating, "Around the World Atomic Clocks: Relativistic Time Gains Observed," *Science,* July 14, 1972, p. 168.

the use of very stable cesium-beam atomic clocks. Time intervals measured with four such clocks in jet flight were compared with time intervals measured by reference atomic clocks at the U.S. Naval Observatory. (Because of the Earth's rotation about its axis, a ground-based clock is not in a true inertial frame.) Time intervals measured with the flying clocks were compared to time intervals measured with the Earth-based reference clocks. In order to compare the results with the theory, many factors had to be considered, including periods of acceleration and deceleration relative to the Earth, variations in direction of travel, and the weaker gravitational field experienced by the flying clocks. Their results were in good agreement with the predictions of the special theory of relativity. In their paper, Hafele and Keating report the following: "Relative to the atomic time scale of the U.S. Naval Observatory, the flying clocks lost 59 ± 10 ns during the eastward trip and gained 273 ± 7 ns during the westward trip. . . . These results provide an unambiguous empirical resolution of the famous clock paradox with macroscopic clocks."

EXAMPLE 26.1 What Is the Period of the Pendulum?

The period of a pendulum is measured to be 3.0 s in the inertial frame of the pendulum. What is the period when measured by an observer moving at a speed of $0.95c$ with respect to the pendulum?

Reasoning and Solution   In this case, the proper time is 3.0 s. We can use Equation 26.8 to calculate the period measured by the moving observer:

$$T = \gamma T' = \frac{1}{\sqrt{1 - \frac{(0.95c)^2}{c^2}}} \, T' = (3.2)(3.0 \text{ s}) = \boxed{9.6 \text{ s}}$$

That is, the observer moving with a speed of $0.95c$ observes that the pendulum slows down.

THE TWIN PARADOX

An interesting consequence of time dilation is the so-called twin paradox. Consider a controlled experiment involving 20-year-old twin brothers Speedo and Goslo. Speedo, the more venturesome twin, sets out on a journey toward a star located 30 lightyears from Earth. His spaceship is able to accelerate to a speed close to the speed of light. After reaching the star, Speedo becomes very homesick and immediately returns to Earth at the same high speed. Upon his return, he is shocked to find that many things have changed. Old cities have expanded and new cities have appeared. Lifestyles, fashions, and transportation systems have changed dramatically. Speedo's twin brother, Goslo, has aged to about 80 years old and is now wiser, feeble, and somewhat hard-of-hearing. Speedo, on the other hand, has aged only about ten years. This is because his bodily processes slowed down during his travels in space.

It is quite natural to raise the question, "Which twin actually travels at a speed close to the speed of light, and therefore does not age as much?" Herein lies the paradox: From Goslo's frame of reference, he is at rest while his brother Speedo travels at a high velocity. On the other hand, according to the space traveler Speedo, it is he who is at rest while his brother zooms away from him on

The space traveler ages more slowly than his twin who remains on Earth

Earth and then returns. This leads to confusion about which twin actually ages more.

In order to resolve this paradox, it should be pointed out that the trip is not as symmetrical as we may have led you to believe. Speedo, the space traveler, experiences a series of accelerations and decelerations during his journey to the star and back home, and therefore is not always in uniform motion. This means that Speedo is in a noninertial frame during part of his trip, so that predictions based on special relativity are not valid in his frame. On the other hand, the brother on Earth is in an inertial frame and can make reliable predictions based on the special theory. The situation is not symmetrical since Speedo experiences accelerations when his spaceship turns around, whereas Goslo is not subject to such accelerations. Therefore, the space traveler is indeed younger upon returning to Earth.

LENGTH CONTRACTION

We have seen that measured time intervals are not absolute; that is, the time interval between two events depends on the frame of reference in which it is measured. Likewise, the measured distance between two points depends on the frame of reference. The **proper length** of an object is defined as *the length of the object measured in the reference frame in which the object is at rest*. The length of an object measured in a reference frame in which the object is moving is always less than the proper length. This effect is known as **relativistic length contraction**.

To understand relativistic length contraction quantitatively, let us consider a spaceship traveling with a speed v from one star to another, as seen by two observers. An observer at rest on Earth (and also assumed to be at rest with respect to the two stars) measures the distance between the stars to be L' (where L' is the proper length). According to this observer, it takes a time $\Delta t = L'/v$ for the spaceship to complete the voyage. What does an observer in the spaceship measure? Because of time dilation, the space traveler measures a smaller time of travel: $\Delta t' = \Delta t/\gamma$. The observer in the spaceship claims to be at rest and sees the destination star as moving toward the ship with speed v. Since the space traveler reaches the star in the time $\Delta t'$, she concludes that the distance, L, between the stars is shorter than L'. This distance is given by

$$L = v\,\Delta t' = v\frac{\Delta t}{\gamma}$$

Since $L' = v\,\Delta t$, we see that

$$L = \frac{L'}{\gamma}$$

or,

$$L = L'\sqrt{1 - v^2/c^2} \qquad \text{[26.9]} \qquad \text{Length contraction}$$

According to this result, illustrated in Figure 26.10, if an observer at rest with respect to an object measures its length to be L', an observer moving at a relative speed v with respect to the object will find it to be shorter than its rest length by the factor $\sqrt{1 - v^2/c^2}$. You should note that *the length contraction takes place only along the direction of motion*.

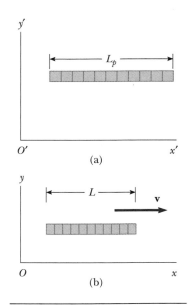

FIGURE 26.10
A meter stick moves to the right with a speed v. (a) The meter stick as viewed by an observer at rest with respect to the meter stick. (b) The meter stick as seen by an observer moving with a speed v with respect to the meter stick. The moving meter stick is always measured to be *shorter* than in its own rest frame by a factor of $\sqrt{1 - v^2/c^2}$.

EXAMPLE 26.2   The Contraction of a Spaceship

A spaceship is measured to be 120 m long while it is at rest with respect to an observer. If this spaceship now flies past the observer with a speed of $0.99c$, what length will the observer measure for the spaceship?

Solution   From Equation 26.9, the length measured by the observer is

$$L = L' \sqrt{1 - v^2/c^2} = (120 \text{ m}) \sqrt{1 - \frac{(0.99c)^2}{c^2}} = \boxed{17 \text{ m}}$$

Exercise   If the ship moves past the observer with a speed of $0.01000c$, what length will the observer measure?

Answer   99.99 m.

EXAMPLE 26.3   How High is the Spaceship?

An observer on Earth sees a spaceship at an altitude of 435 m moving downward toward the Earth with a speed of $0.970c$. What is the altitude of the spaceship as measured by an observer in the spaceship?

Solution   The moving observer in the spaceship finds the altitude to be

$$L = L' \sqrt{1 - v^2/c^2} = (435 \text{ m}) \sqrt{1 - \frac{(0.970c)^2}{c^2}} = \boxed{106 \text{ m}}$$

EXAMPLE 26.4   The Triangular Spaceship

A spaceship in the form of a triangle flies by an observer with a speed of $0.95c$. When the spaceship is at rest (Fig. 26.11a), the distances x and y are found to be 52 m and 25 m, respectively. What is the shape of the spaceship as seen by an observer at rest when the spaceship is in motion along the direction shown in Figure 26.11b?

Solution   The observer sees the horizontal length of the spaceship to be contracted to a length of

$$L = L' \sqrt{1 - v^2/c^2} = (52 \text{ m}) \sqrt{1 - \frac{(0.95c)^2}{c^2}} = \boxed{16 \text{ m}}$$

The 25-m vertical height is unchanged because it is perpendicular to the direction of relative motion between the observer and the spaceship. Figure 26.11b represents the shape of the spaceship as seen by the observer at rest.

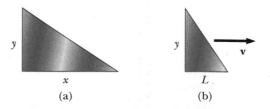

(a)   (b)

FIGURE 26.11
(Example 26.4)   (a) When the spaceship is at rest, its shape is as shown.
(b) The spaceship appears to look like this when it moves to the right with a speed v. Note that only its x dimension is contracted in this case.

26.7

RELATIVISTIC MOMENTUM

In order to properly describe the motion of particles within the framework of special relativity, we must generalize Newton's laws of motion and the definitions of momentum and energy. As we shall see, these generalized definitions reduce to the classical (nonrelativistic) definitions when v is much less than c.

First, recall that conservation of momentum states that when two objects collide, the total momentum of the system remains constant, assuming that the objects are isolated (that is, they interact only with each other). If such a collision is analyzed within the framework of Einstein's postulates of relativity, it is found that momentum is not conserved if the classical definition of momentum, $p = m_0 v$, is used. (In discussions of relativity, the subscript on m_0 indicates the mass of an object at rest or moving at low speed.) However, according to the principle of relativity, momentum must be conserved in all reference systems. In view of this condition, it is necessary to modify the definition of momentum to satisfy the following conditions:

1. The relativistic momentum must be conserved in all collisions.
2. The relativistic momentum must approach the classical value $m_0 v$ as the quantity v/c approaches zero.

The correct relativistic equation for momentum that satisfies these conditions is

$$p \equiv \frac{m_0 v}{\sqrt{1 - v^2/c^2}} = \gamma m_0 v \qquad [26.10]$$

Momentum

where m_0 is the rest mass of the particle and v is its velocity. The theoretical derivation of this generalized expression for momentum is beyond the scope of this text. Note that when v is much less than c, the denominator of Equation 26.10 approaches unity and so p approaches $m_0 v$. Therefore, the relativistic equation for momentum reduces to the classical expression when v is small compared with c.

EXAMPLE 26.5 The Relativistic Momentum of an Electron

An electron, which has a rest mass of 9.11×10^{-31} kg, moves with a speed of $0.75c$. Find its relativistic momentum and compare this value to the momentum calculated from the classical expression.

Solution   From Equation 26.10, with $v = 0.75c$, we have

$$p = \frac{m_0 v}{\sqrt{1 - v^2/c^2}}$$

$$= \frac{(9.11 \times 10^{-31} \text{ kg})(0.75 \times 3 \times 10^8 \text{ m/s})}{\sqrt{1 - (0.75c)^2/c^2}} = 3.1 \times 10^{-22} \text{ kg} \cdot \text{m/s}$$

The classical expression gives

$$\text{Momentum} = m_0 v = 2.1 \times 10^{-22} \text{ kg} \cdot \text{m/s}$$

The (correct) relativistic result is 50% greater than the classical result!

26.8

MASS AND THE ULTIMATE SPEED

We have seen that the values of length and time intervals depend on the reference frame in which they are measured. Einstein showed that *the mass also depends on the frame of reference and that observed mass of an object increases with speed according to the expression*

$$m \equiv \frac{m_0}{\sqrt{1 - v^2/c^2}}$$

[26.11]

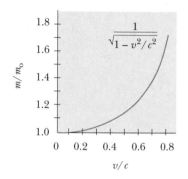

FIGURE 26.12

Variation of mass with speed. This theoretical curve is in excellent agreement with experimental data on particles such as electrons.

where m is the mass of the object as measured by an observer moving with a speed v, and m_0 is the mass of the object as measured by an observer at rest with respect to the object. We shall refer to m_0 as the **rest mass** of the object. From this definition, the relativistic momentum given by Equation 26.10 can be interpreted as being the product of the relativistic mass, m, and the velocity of the object, v; that is, $p = mv$. Equation 26.11 indicates that *the mass of an object increases as its speed increases,* as shown in Figure 26.12. This prediction has been borne out through observations on elementary particles, such as energetic electrons.

It is also possible to see from Equation 26.11 that the speed of light is the ultimate speed of an object. The equation indicates that as an object is accelerated, its mass increases, and hence the unbalanced force on the object has to increase in proportion in order to maintain a constant acceleration. Finally, when $v = c$, the mass becomes infinite. This means that an infinite amount of energy would be required to accelerate an object to the speed of light. Thus, *the speed of light is the ultimate speed for any material object.*

EXAMPLE 26.6   **The Mass of a Speedy Ball**

Superman, who has an exceptionally strong arm, throws a fastball with a speed of $0.90c$. If the rest mass of the ball is 0.50 kg, what is its mass in flight?

Solution   This is a straightforward application of Equation 26.11. The relativistic mass is

$$m = \frac{m_0}{\sqrt{1 - v^2/c^2}} = \frac{0.50 \text{ kg}}{\sqrt{1 - \frac{(0.90c)^2}{c^2}}} = 1.2 \text{ kg}$$

26.9

RELATIVISTIC ADDITION OF VELOCITIES

Imagine a motorcycle rider moving with a speed of $0.80c$ past a stationary observer, as shown in Figure 26.13. If the rider tosses a ball in the forward direction with a speed of $0.70c$ relative to himself, what is the speed of the ball as seen by the stationary observer at the side of the road? Common sense and the ideas of Newtonian relativity say that the speed should be the sum of the two speeds, or $1.5c$. This answer must be incorrect since it contradicts the assertion that no material object can travel faster than the speed of light.

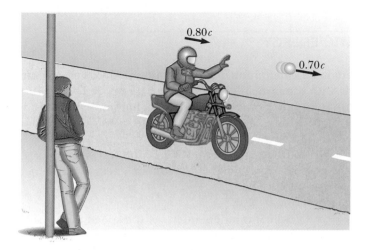

FIGURE 26.13
A motorcycle moves past a stationary observer with a speed of $0.80c$; the motorcyclist throws a ball in the direction of motion with a speed of $0.70c$ relative to himself.

Einstein resolved this dilemma by deriving an equation for the relativistic addition of velocities. For one-dimensional motion, this equation is

$$u = \frac{v + u'}{1 + \dfrac{vu'}{c^2}}$$

[26.12]   Velocity addition

where u is the velocity of an object as seen in one frame of reference (frame A), u' is the velocity of the same object as measured in a different frame of reference (frame B), and v is the velocity of frame B relative to frame A. Let us apply this equation to the case of the speedy motorcycle rider and the stationary observer.

The velocity of the motorcycle with respect to the stationary observer is $0.80c$. Thus, $v = 0.80c$. The velocity of the ball in the frame of the motorcyclist is $0.70c$. That is, $u' = 0.70c$. Therefore, the velocity, u, of the ball relative to the stationary observer is

$$u = \frac{0.80c + 0.70c}{1 + \dfrac{(0.80c)(0.70c)}{c^2}} = 0.96c$$

EXAMPLE 26.7   Measuring the Speed of a Light Beam

Suppose that the motorcyclist moving with a speed of $0.80c$ turns on a beam of light that moves away from the motorcycle with a speed of c in the same direction as the moving motorcycle. What speed would the stationary observer measure for the beam of light?

Solution   In this case, $v = 0.80c$ and $u' = c$. Substituting these values into Equation 26.12, we find

$$u = \frac{0.80c + c}{1 + \dfrac{(0.80c)(c)}{c^2}} = c$$

This is consistent with the statement made earlier that *all observers measure the speed of light to be c regardless of the motion of the source of light.*

26.10

RELATIVISTIC ENERGY

We have seen that the definition of momentum required generalization to make it compatible with the principle of relativity. Likewise, the definition of kinetic energy requires modification in relativistic mechanics. In view of the definition of relativistic momentum, we might guess that the classical expression for kinetic energy, $mv^2/2$, could be translated to a relativistic expression by substituting the value for m as given in Equation 26.11. This, however, is not correct. Einstein found that the correct expression for the **kinetic energy** of an object is

Kinetic energy

$$KE = mc^2 - m_0c^2 \qquad \text{[26.13]}$$

The term $m_0c^2 = E_0$, which is independent of the speed of the object, is called the **rest energy** of the object. The term mc^2, which depends on the object speed, is therefore the sum of the kinetic and rest energies. We define mc^2 to be the **total energy**, E, that is,

$$E = mc^2 = KE + m_0c^2 \qquad \text{[26.14]}$$

This is Einstein's famous *mass-energy equivalence equation*. If we make use of Equation 26.11, we can express the total energy as[3]

Total energy

$$E = \frac{m_0c^2}{\sqrt{1 - v^2/c^2}} \qquad \text{[26.15]}$$

The relation $E_0 = m_0c^2$ (obtained from Eq. 26.15 by setting $v = 0$) shows that *mass is a form of energy*. Furthermore, this result shows that *even a small mass corresponds to an enormous amount of energy*. The concept stated in Equation 26.15 has revolutionized the field of nuclear physics. The validity of the relationship between mass and energy has been proved beyond question.

In many situations, the momentum or energy of a particle is known rather than its speed. It is therefore useful to have an expression relating the total

---

[3] At first glance, Equation 26.13 does not look even remotely similar to the classical expression for kinetic energy. However, we shall show that the two equations are equivalent in the limit of low velocities. First, let us substitute the expression for total energy from Equation 26.15 into Equation 26.13 to find

$$KE = mc^2 - m_0c^2 = \frac{m_0c^2}{\sqrt{1 - \dfrac{v^2}{c^2}}} - m_0c^2$$

$$= m_0c^2 \left[\left(1 - \frac{v^2}{c^2} \right)^{-1/2} - 1 \right]$$

For small velocities, v becomes very small compared to c; hence the first term in the brackets can be approximated by the first two terms in the binomial series expansion for $\left(1 - \dfrac{v^2}{c^2} \right)^{-1/2}$ to give

$$KE \approx m_0c^2 \left[1 + \frac{1}{2}\frac{v^2}{c^2} - 1 \right] = \frac{1}{2}m_0v^2$$

energy, E, to the relativistic momentum, p. This is accomplished by using the expressions $E = mc^2$ and $p = mv$.[4] The result is

$$E^2 = p^2 c^2 + (m_0 c^2)^2 \qquad \text{[26.16]}$$

When the particle is at rest, $p = 0$, and so $E_0 = m_0 c^2$. That is, by definition, *the total energy of a particle at rest equals its rest energy*. As we shall discuss in Chapter 27, it is well established that there are particles with zero rest mass, such as photons. If we set m_0 equal to zero in Equation 26.16, we see that the *total energy of a photon* can be expressed as

$$E = pc \qquad \text{[26.17]}$$

This equation is an exact expression relating energy and momentum for photons, which always travel at the speed of light.

When dealing with electrons or other subatomic particles, it is convenient to express the energy in electron volts (eV) because the particles are usually given energy by acceleration through an electrostatic potential difference. Recall that

$$1 \text{ eV} = 1.60 \times 10^{-19} \text{ J}$$

For example, the mass of an electron is 9.11×10^{-31} kg. Hence, its rest energy is

$$m_0 c^2 = (9.11 \times 10^{-31} \text{ kg})(3.00 \times 10^8 \text{ m/s})^2 = 8.20 \times 10^{-14} \text{ J}$$

Converting this to electron volts, we have

$$m_0 c^2 = (8.20 \times 10^{-14} \text{ J}) \left(\frac{1 \text{ eV}}{1.60 \times 10^{-19} \text{ J}} \right) = 0.511 \text{ MeV}$$

where $1 \text{ MeV} = 10^6$ eV.

EXAMPLE 26.8   The Energy Contained in a Baseball

If a 0.50-kg baseball could be converted completely to energy of forms other than mass, how much energy of other forms would be released?

---

[4] The expression relating the total energy, E, to the momentum, p, can be found by evaluating the quantity $m^2 c^4 - m_0^2 c^4$ as

$$m^2 c^4 - m_0^2 c^4 = \frac{m_0^2 c^4}{1 - v^2/c^2} - m_0^2 c^4 = m_0^2 c^4 \left(\frac{v^2/c^2}{1 - v^2/c^2} \right) = \frac{m_0^2 v^2 c^2}{1 - v^2/c^2}$$

But,

$$p = \frac{m_0 v}{\sqrt{1 - v^2/c^2}}$$

so,

$$m^2 c^4 - m_0^2 c^4 = p^2 c^2$$

Thus,

$$m^2 c^4 = p^2 c^2 + (m_0 c^2)^2$$

Since $E^2 = m^2 c^4$, we have

$$E^2 = p^2 c^2 + (m_0 c^2)^2$$

Solution   The energy equivalent of the baseball is found from Equation 26.14 (with $KE = 0$):

$$E_0 = m_0 c^2 = (0.50 \text{ kg})(3.0 \times 10^8 \text{ m/s})^2 = \boxed{4.5 \times 10^{16} \text{ J}}$$

This is enough energy to keep a 100-W lightbulb burning for approximately ten million years. However, it is impossible to achieve complete conversion from mass to energy of other forms in practical situations. For example, mass is converted to energy in nuclear power plants, but only a small fraction of the mass actually undergoes conversion.

EXAMPLE 26.9   **The Energy of a Speedy Electron**

An electron moves with a speed of $v = 0.850\,c$. Find its total energy and kinetic energy in electron volts.

Solution   The fact that the rest energy of an electron is 0.511 MeV, along with Equation 26.15, gives

$$E = \frac{m_0 c^2}{\sqrt{1 - v^2/c^2}} = \frac{0.511 \text{ MeV}}{\sqrt{1 - \dfrac{(0.850\,c)^2}{c^2}}}$$

$$= 1.90(0.511 \text{ MeV}) = \boxed{0.970 \text{ MeV}}$$

The kinetic energy is obtained by subtracting the rest energy from the total energy:

$$KE = E - m_0 c^2 = 0.970 \text{ MeV} - 0.511 \text{ MeV} = \boxed{0.459 \text{ MeV}}$$

EXAMPLE 26.10   **The Energy of a Speedy Proton**

The total energy of a proton is three times its rest energy.

(a) Find the proton's rest energy in electron volts.

Solution

$$\text{Rest energy} = m_0 c^2 = (1.67 \times 10^{-27} \text{ kg})(3.00 \times 10^8 \text{ m/s})^2$$

$$= (1.50 \times 10^{-10} \text{ J}) \left(\frac{1 \text{ eV}}{1.60 \times 10^{-19} \text{ J}} \right) = \boxed{938 \text{ MeV}}$$

(b) With what speed is the proton moving?

Solution   Since the total energy, E, is three times the rest energy, Equation 26.15 gives

$$E = mc^2 = 3m_0 c^2 = \frac{m_0 c^2}{\sqrt{1 - v^2/c^2}}$$

$$3 = \frac{1}{\sqrt{1 - v^2/c^2}}$$

Solving for v gives

$$1 - \frac{v^2}{c^2} = \frac{1}{9}$$

$$\frac{v^2}{c^2} = \frac{8}{9}$$

$$v = \frac{\sqrt{8}}{3} c = \boxed{2.83 \times 10^8 \text{ m/s}}$$

(c) Determine the kinetic energy of the proton in electron volts.

Solution

$$KE = E - m_0 c^2 = 3 m_0 c^2 - m_0 c^2 = 2 m_0 c^2$$

Since $m_0 c^2 = 938$ MeV, $KE = \boxed{1880 \text{ MeV}}$

SUMMARY

The two basic postulates of the **special theory of relativity** are

1. The laws of physics are the same in all inertial frames of reference.
2. The speed of light is the same for all inertial observers, independent of their motion or of the motion of the source of light.

Some of the consequences of the special theory of relativity are as follows:

1. Clocks in motion relative to an observer slow down. This is known as **time dilation.** The relationship between time intervals in the moving and at-rest systems is

$$\Delta t = \gamma \Delta t' \qquad \text{[26.8]}$$

where Δt is the time interval in the system in relative motion with respect to the clock, $\gamma = 1/\sqrt{1 - v^2/c^2}$, and $\Delta t'$ is the time interval in the system moving with the clock.

2. The length of an object in motion is *contracted* in the direction of motion. The equation for **length contraction** is

$$L = L' \sqrt{1 - v^2/c^2} \qquad \text{[26.9]}$$

where L is the length in the system in motion relative to the object, and L' is the length in the system in which the object is at rest.

3. Events that are simultaneous for one observer are not simultaneous for another observer in motion relative to the first.

The relativistic expression for the **momentum** of a particle moving with a velocity v is

$$p \equiv \frac{m_0 v}{\sqrt{1 - v^2/c^2}} = \gamma m_0 v \qquad \text{[26.10]}$$

The **mass** of an object increases as its speed increases according to

$$m \equiv \frac{m_0}{\sqrt{1 - v^2/c^2}}$$ **[26.11]**

where m_0 is the rest mass of the object, that is, its mass as measured by an observer at rest with respect to it.

The relativistic expression for the addition of velocities is

$$u = \frac{v + u'}{1 + \dfrac{vu'}{c^2}}$$ **[26.12]**

where u is the velocity of an object as seen in one frame of reference, u' is the velocity of the same object in a different frame of reference, and v is the relative velocity of the two frames.

The relativistic expression for the **kinetic energy** of an object is

$$KE = mc^2 - m_0 c^2$$ **[26.13]**

where $m_0 c^2$ is the **rest energy** of the object.

The **total energy** of a particle is

$$E = \frac{m_0 c^2}{\sqrt{1 - v^2/c^2}}$$ **[26.15]**

This is Einstein's famous mass-energy equivalence equation.

The relativistic momentum is related to the total energy through the equation

$$E^2 = p^2 c^2 + (m_0 c^2)^2$$ **[26.16]**

ADDITIONAL READING

H. Bondi, *Relativity and Common Sense,* Science Study Series, Garden City, N.Y., Doubleday, 1964.

J. Bronowski, "The Clock Paradox," *Sci. American,* February 1963, p. 134.

R. W. Clark, *Einstein: The Life and Times,* New York, World Publishing, 1971.

J. Crelinsten, "Relativity, Einstein, Physicists, and the Public," *The Physics Teacher,* February 1980, p. 115.

J. Crelinsten, "Physicists Receive Relativity: Revolution and Reaction," *The Physics Teacher,* February 1980, p. 187.

A. Einstein, *Ideas and Opinions,* New York, Crown, 1954.

A. Einstein, *Out of My Later Years,* Secaucus, N.J., Citadel Press, 1973.

G. Gamow, *Mr. Tomkins in Wonderland,* New York, Cambridge University Press, 1939.

R. D. Henry, "Special Relativity Made Transparent," *The Physics Teacher,* December 1985, p. 536.

L. Infeld, *Albert Einstein,* New York, Scribner's, 1950.

P. K. MacKeown, "Gravity is Geometry," *The Physics Teacher,* December 1984, p. 557.

J. Schwinger, *Einstein's Legacy,* Scientific American Library, New York, W. H. Freeman and Co., 1985.

R. S. Shankland, "The Michelson-Morley Experiment," *Sci. American,* November 1964, p. 107.

G. J. Whitrow, ed., *Einstein, The Man and His Achievement,* New York, Dover, 1967.

CONCEPTUAL QUESTIONS

Example It is said that Einstein, in his teenage years, asked the question, "What would I see in a mirror if I carried it in my hands and ran at the speed of light?" How would you answer this question?

Reasoning You would see the same thing that you see when looking at a mirror when at rest. The theory of relativity tells us that all experiments will give the same results in all inertial frames of reference.

Example Suppose astronauts were paid according to the time spent traveling in space. After a long voyage traveling at a speed near that of light, a crew of astronauts returns and open their pay envelopes. What will their reaction be?

Reasoning Assuming that their on-duty time was kept on Earth, they will be pleasantly surprised with a large paycheck. Less time will have passed for the astronauts in their frame of reference than for their employer back on Earth.

Example Two identically constructed clocks are synchronized. One is put in orbit around the Earth while the other remains on Earth. Which clock runs more slowly? When the moving clock returns to Earth, will the two clocks still be synchronized?

Reasoning The clock in orbit will run more slowly. The experiment for clocks carried in airplanes is described in Section 26.6. As with the twin "paradox," the extra centripetal acceleration of the orbiting clock makes its history fundamentally different from that of the clock on Earth.

Example The equation $E = mc^2$ is often given in popular descriptions of Einstein's theory of relativity. Is this expression strictly correct? For example, does it accurately account for the kinetic energy of a moving mass?

Reasoning The equation $E = mc^2$, better written $\Delta E = \Delta mc^2$, relates any energy transferred into or removed from a system to the change in the mass of the system. When work is done on an object to increase its speed, we can say that the increase in its kinetic energy corresponds to an increase in its effective mass from m to γm. $KE = (\gamma m - m)c^2$ is one version of $\Delta E = \Delta mc^2$. Another is $Q = \Delta mc^2$ for the energy output of an exothermic nuclear reaction.

Example Imagine an astronaut on a trip to Sirius, which lies 8 lightyears from the Earth. Upon arrival at Sirius, the astronaut finds that the trip lasted 6 years. If the trip was made at a constant speed of $0.8c$, how can the 8-lightyear distance be reconciled with the 6-year duration?

Reasoning The 8 lightyears represents the proper length of a rod from here to Sirius, measured by an observer seeing both nearly at rest. The astronauts sees Sirius coming toward her at $0.8c$, but also sees the distance contracted to

$$\frac{8 \text{ lightyears}}{\gamma} = (8 \text{ lightyears}) \sqrt{1 - \frac{v^2}{c^2}}$$

$$= (8 \text{ lightyears}) \sqrt{1 - 0.8^2}$$

$$= 5 \text{ lightyears}$$

So the travel time as measured on her clocks is

$$t = \frac{d}{v} = \frac{5 \text{ lightyears}}{0.8c} = 6 \text{ years}$$

1. What one measurement will two observers in relative motion always agree upon?

2. Some of the distant stars, called quasars, are receding from us at half the speed of light or faster. What is the speed of light we receive from these quasars?

3. The speed of light in a transparent material medium is lower than the speed of light in a vacuum. Is this a violation of relativity? Discuss.

4. High-energy particles sometimes travel through a material medium faster than the speed of light in that medium. Does this violate the principles of relativity?

5. A spaceship in the shape of a sphere moves past an observer on Earth with a speed of $0.50c$. What shape will the observer see as the spaceship moves past?

6. When defining length, why is it necessary to specify that the positions of the ends of a rod are to be determined simultaneously?

7. Suppose two observers who are moving relativistically each have meter sticks that they wish to use as standards. How can the observers be sure that their sticks have the same proper length? Discuss.

8. An astronaut moves away from the Earth at a speed close to the speed of light. If an observer on Earth could measure the astronaut's size and pulse rate, what changes (if any) would he measure? Would the astronaut measure any changes?

9. Assuming that the Universe started with a Big Bang 10 billion years ago and that evolution proceeded at the same pace everywhere, would it make any sense to listen for intelligent life elsewhere in the Universe? Discuss.

10. Causality means that cause precedes effect. What limitations does special relativity place on causality in the event that a cause occurring in one location produces an effect an another location? Discuss.

11. Imagine a rocket with a flasher on top that flashes a sequence of numbers, 1, 2, 3, . . . at regular proper time intervals. Discuss the sequence in which the numbers are emitted, the position of the flasher, and the time interval between the emission of each number in a frame of reference in which the rocket is moving with speed $0.99c$.

12. Give a physical argument showing that it is impossible to accelerate an object of mass m to the speed of light, even with a continuous force acting on it.

13. What happens to the density of an object as its speed increases?

14. Since mass is a form of energy, can we conclude that a compressed spring has more mass than the same spring when not compressed? On the basis of your answer, which has more mass, a spinning planet or an otherwise identical but nonspinning planet?

15. Consider the incorrect statement, "Matter can neither be created nor destroyed." How would you correct this statement in view of the special theory of relativity?

16. List some ways in which our day-to-day lives would change if the speed of light were 100 mi/h.

17. Relativistic quantities must make a smooth transition to their Newtonian counterparts as the velocity of a system becomes small with respect to the speed of light. Explain.

18. Why did Einstein postulate that the speed of light in a **vacuum**, rather than in some medium, such as water, is independent of the motion of the observer and of the motion of the source of light?

19. Suppose the distance between you and a 500-Hz sound source is decreasing in time in such a way that either you are standing still and the source is moving toward you, or the source is stationary and you are moving toward it. If the medium between you and the sound source is air, and if you measure the frequency of the sound to be 544.4 Hz, can you determine who is moving? Explain.

20. If you are doing relativistic calculations dealing with the motion of objects in water, what should you use for c in the expression for γ?

PROBLEMS

Section 26.4 The Michelson–Morley Experiment

1. Two airplanes fly paths I and II specified in Figure 26.5a. Both planes have air speeds of 100 m/s and fly a distance of 200 km. The wind blows at 20.0 m/s in the direction shown in the figure. Find (a) the time of flight to each city, (b) the time to return, and (c) the difference in total flight times.

2. In one version of the Michelson–Morley experiment, the lengths L in Figure 26.4 were 28 m. Take v to be 3.0×10^4 m/s and find (a) the time difference caused by rotation of the interferometer and (b) the expected fringe shift, assuming that the light used has a wavelength of 550 nm.

Section 26.6 Consequences of Special Relativity

3. An astronaut at rest on Earth has a heartbeat rate of 70 beats/min. When the astronaut is traveling in a spaceship at $0.90c$, what will this rate be as measured by (a) an observer also in the ship and (b) an observer at rest on the Earth?

4. The proper length of one spaceship is three times that of another. The two spaceships are traveling in the same direction and, while both are passing overhead, an Earth observer measures the two spaceships to have the same length. If the slower spaceship is moving with a speed of $0.35c$, determine the speed of the faster spaceship.

5. With what speed must a clock move in order to run at a rate that is one-half the rate of a clock at rest?

6. A deep-space probe moves away from Earth with a speed of $0.80c$. An antenna on the probe requires 3.0 s, probe time, to rotate through 1.0 rev. How much time is required for 1.0 rev according to an observer on Earth?

7. If astronauts could travel at $v = 0.950c$, we on Earth would say it takes $(4.20/0.950) = 4.42$ years to reach Alpha Centauri, 4.2 lightyears away. The astronauts disagree. (a) How much time passes on the astronaut's clocks? (b) What is the distance to Alpha Centauri as measured by the astronauts?

8. The average lifetime of a pi meson in its own frame of reference is 2.6×10^{-8} s. (This is the proper lifetime.) If the meson moves with a speed of $0.98c$, what is (a) its mean lifetime as measured by an observer on Earth and (b) the average distance it travels before decaying as measured by an observer on Earth? (c) What distance would it travel if time dilation did not occur?

9. A supertrain of proper length 100 m travels at a speed of $0.95c$ as it passes through a tunnel having proper length 50 m. As seen by a trackside observer, is the train ever completely within the tunnel? If so, by how much?

10. A cube is 2.0 cm on a side when at rest. (a) What shape does it take on when moving past an observer at 2.5×10^8 m/s, and (b) what is the length of each side?

11. An unidentified flying object (UFO) flashes across the night sky. A UFO enthusiast on the top of Pike's Peak determines that the length of the UFO is 100 m along the direction of motion. If the

□ indicates problems that have full solutions available in the Student Solutions Manual and Study Guide.

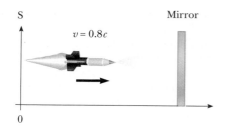

FIGURE 26.14   (Problem 14)

UFO is moving with a speed of $0.90c$, how long is the UFO according to its pilot?

12. A muon formed high in the Earth's atmosphere travels at speed $v = 0.99c$ for a distance of 4.6 km before it decays into an electron, a neutrino, and an anti-neutrino ($\mu^- \rightarrow e^- + \nu + \bar{\nu}$). (a) How long does the muon live, as measured in its reference frame? (b) How far does the muon travel, as measured in its frame?

13. A spaceship of proper length 300 m takes 0.75 μs to pass an Earth observer. Determine the speed of this spaceship as measured by the Earth observer.

14. An observer in a rocket moves toward a mirror at a speed of $0.80c$ with respect to the reference frame labeled by S in Figure 26.14. The mirror is stationary with respect to S. A light pulse emitted by the rocket travels toward the mirror and is reflected back to the rocket. The front of the rocket is 1.8×10^{12} m from the mirror (as measured by observers in S) at the moment the light pulse leaves the rocket. What is the total travel time of the pulse as measured by observers in (a) the frame S and (b) the front of the rocket?

15. A friend in a spaceship travels past you at a high speed. He tells you that his ship is 20 m long and that the identical ship you are sitting in is 19 m long. According to your observations, (a) how long is your ship, (b) how long is his ship, and (c) what is the speed of your friend's ship?

16. A moving rod is 2.00 m long, and its length is oriented at an angle of $30.0°$ with respect to the direction of motion. The rod has a speed of $0.995c$. (a) What is the proper length of the rod? (b) What is the orientation angle in the frame moving with the rod?

17. Observer A measures the length of two rods, one stationary, the other moving with a speed of $0.955c$. She finds that the rods have the same length. A second observer B travels along with the moving rod. What is the ratio of the length of A's rod to the length of B's rod according to observer B?

18. The nearest star to Earth is approximately 4.0 lightyears away. If you travel at 2.5×10^8 m/s in a spaceship, how long does it take to get there (a) according to an Earthbound observer and (b) according to an observer on the spaceship?

19. An atomic clock is placed on a jet airplane. The clock measures a time interval of 3600 s when the jet moves at 400 m/s. What corresponding time interval does an identical clock held by an observer on the ground measure? (*Hint:* For $v/c \ll 1$, use the approximation $\gamma \approx 1 + v^2/2c^2$.)

20. A 2.0-m long bobsled is traveling at 65 mi/h. What is the decrease in its length as seen by a stationary observer? (See the hint in Problem 19.)

Section 26.7   Relativistic Momentum

21. Calculate the momentum of an electron moving with a speed of (a) $0.010c$, (b) $0.50c$, (c) $0.90c$.

22. An electron has a speed $v = 0.90c$. At what speed will a proton have a momentum equal to that of the electron?

23. An electron has a momentum that is 90% larger than its classical momentum. (a) Find the speed of the electron. (b) How would your result change if the particle were a proton?

24. Show that the speed of an object having momentum p and rest mass m_0 is given by

$$v = \frac{c}{\sqrt{1 + (m_0 c/p)^2}}$$

25. An unstable particle at rest breaks up into two fragments of *unequal mass*. The rest mass of the lighter fragment is 2.50×10^{-28} kg, and that of the heavier fragment is 1.67×10^{-27} kg. If the lighter fragment has a speed of $0.893c$ after the breakup, what is the speed of the heavier fragment?

Section 26.8   Mass and the Ultimate Speed

26. What is the speed of an electron that has been accelerated to the point at which its mass is ten times its rest mass?

27. A small particle with a charge of $+3.0$ μC and a mass of 1.0×10^{-6} kg accelerates upward in an electric field with a strength of 10 N/C. At what speed will the downward gravitational force equal the upward electrical force?

28. At what speed must an electron move for its mass to equal a proton's rest mass?

29. At low speeds, a charged object in a magnetic field moves in a circular path of radius $r = mv/qB$. If an electron moves in an orbit of radius 10 cm with a speed of 1.0×10^5 m/s, what will the radius be when its speed is $0.96c$?

30. A meter stick with a rest mass of 0.50 kg is found to be 40 cm long. Determine its mass while moving.

31. A cube of mass 8.0 kg is 0.50 m on a side.
(a) What is its density (mass per unit volume) as seen by an observer when the cube is moving away from the observer with a speed of $0.90c$? (b) What is the density as measured by an observer moving with the cube?

32. An electron is accelerated along a straight path in a linear accelerator. The mass of the electron becomes 10 000 times its rest mass when it reaches its final speed. (a) What is its final speed? (b) If the length of the accelerator is 3500 m, what is its length in the electron's frame of reference?

Section 26.9   Relativistic Addition of Velocities

33. A space vehicle is moving at a speed of $0.75c$ with respect to an external observer. An atomic particle is projected at $0.90c$ in the same direction as the spaceship's velocity with respect to an observer inside the vehicle. What is the speed of the projectile as seen by the external observer?

34. Ted and Mary are playing a game of catch in frame S', which is moving with a speed of $0.60c$; Jim in frame S is watching (Fig. 26.15). Ted throws the ball to Mary with a speed of $0.80c$ (according to Ted) and their separation (measured in S') is 1.8×10^{12} m. (a) According to Mary, how fast is the ball moving? (b) According to Mary, how long will it take the ball to reach her? (c) According to Jim, how far apart are Ted and Mary, and how fast is the ball moving? (d) According to Jim, how long will it take the ball to reach Mary?

35. Spaceship I, which contains students taking a physics exam, approaches Earth with a speed of $0.60c$, while spaceship II, which contains instructors proctoring the exam, moves at $0.28c$ (relative to Earth) directly toward the students. If the instructors in spaceship II stop the exam after 50 min have passed on *their clock*, how long does the exam last as measured by (a) the students? (b) an observer on Earth?

36. Consider two inertial reference frames, S and S' where S' is moving to the right with a constant speed of $0.60c$ as measured by observers in S. A stick of proper length 1.0 m is moving to the left toward the origins of both S and S', and the length of the stick is 50 cm as measured by observers in S'. (a) Determine the speed of the stick as measured by observers in S and S'. (b) What is the length of the stick as measured by observers in S?

37. An electron moves to the right with a speed of $0.90c$ relative to the laboratory frame. A proton moves to the right with a speed of $0.70c$ relative to the electron. Find the speed of the proton relative to the laboratory frame.

38. Spaceship R is moving to the right at a speed of $0.70c$ with respect to the Earth. A second spaceship, L, moves to the left at the same speed with respect to the Earth. What is the speed of L with respect to R?

39. A pulsar is a stellar object that emits light in short bursts. Suppose a pulsar with a speed of $0.950c$ approaches the Earth, and a rocket with a speed of $0.995c$ heads toward the pulsar (both speeds measured in the Earth's frame of reference). If the pulsar emits 10.0 pulses per second in its own frame of reference, at what rate are the pulses emitted in the rocket's frame of reference?

40. A rocket moves with a velocity of $0.92c$ to the right with respect to a stationary observer A. An observer B moving relative to observer A finds that the rocket is moving with a velocity of $0.95c$ to the left. What is the velocity of observer B relative to observer A? (*Hint:* Consider observer B's velocity in the frame of reference of the rocket.)

41. According to an observer A, a rocket moves at $0.91c$ to the left. In A's frame of reference the rocket has a mass of 1.2×10^4 kg. According to a second observer B, the rocket has a mass of 2.7×10^4 kg. (a) What is the *speed* of the rocket according to observer B? (b) What is the *velocity* of B relative to observer A? (*Caution:* There are two possible answers to part (b) of this problem.)

Section 26.10   Relativistic Energy

42. What is the speed of a particle whose kinetic energy is equal to its own rest energy?

43. A proton moves with a speed of $0.950c$. Calculate its (a) rest energy, (b) total energy, and (c) kinetic energy.

44. In a color television tube, electrons are accelerated through a potential difference of 20 000 volts. With what speed do the electrons strike the screen?

45. If it takes 3750 MeV of work to accelerate a proton from rest to a speed of v, determine v.

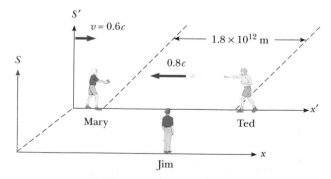

FIGURE 26.15   (Problem 34)

46. A mass of 0.50 kg is converted completely into energy of other forms. (a) How much energy of other forms is produced and (b) how long will this much energy keep a 100-W lightbulb burning?

47. The Sun radiates approximately 4.0×10^{26} J of energy into space each second. (a) How much mass is converted into energy of other forms each second? (b) If the mass of the Sun is 2.0×10^{30} kg, how long can the Sun survive if the energy transformation continues at the present rate?

48. Show that the total energy, E, momentum, p, and speed, v, of an object are all related by $v = pc^2/E$.

49. (a) Show that a potential difference of 1.02×10^6 V would be sufficient to give an electron a speed equal to twice the speed of light if Newtonian mechanics remained valid at high speeds. (b) What speed would an electron acquire in falling through a potential difference of 1.02×10^6 V?

50. Electrons are accelerated to an energy of 2.0×10^{10} eV in the 3.0-km-long Stanford Linear Accelerator. (a) What is the γ factor for the electrons? (b) What is the speed of the 20-GeV electrons? (c) How long does the accelerator appear to a 20-GeV electron?

51. A proton in a high-energy accelerator is given a kinetic energy of 50.0 GeV. Determine (a) the momentum and (b) the speed of the proton.

52. An unstable particle with a rest mass equal to 3.34×10^{-27} kg is initially at rest. The particle decays into two fragments that fly off with velocities of $0.987c$ and $-0.868c$. Find the rest masses of the fragments. (*Hint:* Conserve both mass-energy and momentum.)

ADDITIONAL PROBLEMS

53. Two futuristic rockets are on a collision course. The rockets are moving with speeds of $0.800c$ and $0.600c$ and are initially 2.52×10^{12} m apart as measured by Liz, an Earth observer, as in Figure 26.16. Both rockets are 50.0 m in length as measured by Liz. (a) What are their respective proper lengths? (b) What is the length of each rocket as measured by an observer in the other rocket? (c) According to Liz, how long before the rockets collide? (d) According to rocket 1, how long before they collide? (e) According to rocket 2, how long before they collide? (f) If both rocket crews are capable of total evacuation within 90 minutes (their own time), will there by any casualties?

54. What are the mass and speed of a proton that has been accelerated from rest through a difference of potential of (a) 500 V and (b) 5.0×10^8 V?

55. A radioactive nucleus moves with a speed of v relative to a laboratory observer. The nucleus emits

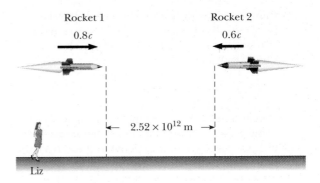

FIGURE 26.16 (Problem 53)

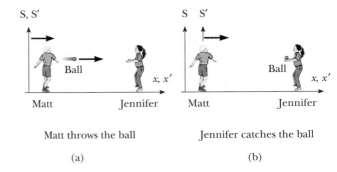

FIGURE 26.17 (Problem 56)

an electron in the positive x direction with a speed of $0.70c$ relative to the decaying nucleus and a speed of $0.85c$ in the $+x$ direction relative to the laboratory observer. What is the value of v?

56. Consider two inertial reference frames, S and S', where S' is moving to the right with constant speed $0.6c$ as measured by observers in S. Jennifer is located 1.8×10^{11} m to the right of the origin of S and is fixed in S (as measured by observers in S), while Matt is fixed in S' at the origin in S' (as measured by observers in S'). At the instant their origins coincide, Matt throws a ball toward Jennifer at constant speed $0.80c$ as measured by Matt (Fig. 26.17). (a) What is the speed of the ball as measured by Jennifer? How long before Jennifer catches the ball, as measured by (b) Jennifer, (c) the ball, and (d) Matt?

57. An electron has a total energy equal to five times its rest energy. (a) What is its momentum? (b) Repeat for a proton.

58. A physics professor on Earth gives an exam to her students who are on a rocket ship traveling at speed of v with respect to Earth. The moment the

ship passes the professor, she signals the start of the exam. If she wishes her students to have T_0 (rocket time) to complete the exam, show that she should wait a time of

$$T = T_0 \sqrt{\frac{1 - v/c}{1 + v/c}}$$

(Earth time) before sending a light signal telling them to stop. (*Hint:* Remember that it takes some time for the second light signal to travel from the professor to the students.)

59. An alarm clock is set to sound in 10 h. At $t = 0$ the clock is placed in a spaceship moving with a speed of $0.75c$ (relative to the Earth). What distance, as determined by an Earth observer, does the spaceship travel before the alarm clock sounds?

60. Determine the energy required to accelerate an electron from (a) $0.50c$ to $0.75c$ and (b) $0.90c$ to $0.99c$.

61. The muon is an unstable particle that spontaneously decays into an electron and two neutrinos. If the number of muons at $t = 0$ is N_0, the number at time t is given by $N = N_0 e^{-t/\tau}$ where τ is the mean lifetime, equal to 2.2 μs. Suppose that the muons move at a speed of $0.95c$ and that there are 5.0×10^4 muons at $t = 0$. (a) What is the observed lifetime of the muons? (b) How many muons remain after traveling a distance of 3.0 km?

62. A rod of length L_0 moves with a speed of v along the horizontal direction. The rod makes an angle of θ_0 with respect to the axis of a coordinate system moving with the rod. (a) Show that the length of the rod as measured by a stationary observer is given by

$$L = L_0 \left[1 - \left(\frac{v^2}{c^2} \right) \cos^2 \theta_0 \right]^{1/2}$$

(b) Show that the angle the rod makes with the axis as seen by the stationary observer is given by the expression $\tan \theta = \gamma \tan \theta_0$. These results show that the rod is both contracted and rotated. (Take the lower end of the rod to be at the origin of the moving coordinate system.)

63. If a light source moves with a speed of v relative to an observer, there is a shift in the observed frequency analogous to the Doppler effect for sound waves. Show that the observed frequency, f_0, is related to the true frequency through the expression

$$f_0 = \sqrt{\frac{c \pm v_s}{c \mp v_s}} f$$

where the upper signs correspond to the source approaching the observer and the lower signs correspond to the source receding from the observer. (*Hint:* In the moving frame the period is the proper time interval and is given by $T = 1/f$. Furthermore, the wavelength measured by the observer is given by $\lambda_0 = (c - v_s) T_0$, where T_0 is the period measured in the stationary frame.)

CLIFFORD M. WILL
McDonnell Center for the Space Sciences, Washington University

THE RENAISSANCE OF GENERAL RELATIVITY

Despite its enormous influence on scientific thought in its early years, general relativity had become a sterile, formalistic subject by the late 1950s, cut off from the mainstream of physics. Yet by 1970, it had become one of the most active and exciting branches of physics. It took on new roles both as a theoretical tool of the astrophysicist and as a playground for the elementary-particle physicist. New experiments verified its predictions in unheard-of ways and to remarkable levels of precision. One of the most remarkable and important aspects of this renaissance was the degree to which experiment and observation motivated and complemented theoretical advances.

This was not always the case. In deriving general relativity during the final months of 1915, Einstein was not particularly motivated by a desire to account for observational results. Instead, he was driven by purely theoretical ideas of elegance and simplicity. His goal was to produce a theory of gravitation that incorporated both the special theory of relativity, which deals with physics in inertial frames, and the principle of equivalence, the proposal that physics in a frame falling freely in a gravitational field is in some sense equivalent to physics in an inertial frame.

Once he formulated the general theory, however, Einstein proposed three tests. One was an immediate success: the explanation of the anomalous advance in the perihelion of Mercury of 43 arcseconds per century, a problem that had bedeviled celestial mechanicians of the latter part of the 19th century. The next test, the deflection of light by the Sun, was such a success that it produced what today would be called a "media event." The third test, the gravitational redshift of light, remained unfulfilled until 1960, by which time it was no longer viewed as a true test of general relativity.

The turning point for general relativity came in the early 1960s, when discoveries of unusual astronomical objects such as quasars demonstrated that the theory has important applications in astrophysics. After 1960, the pace of research in general relativity and in an emerging field called "relativistic astrophysics" accelerated.

Einstein's Three Tests

Perihelion Shift.   The explanation of the anomalous perihelion shift of Mercury's orbit was an early triumph of general relativity. This had been an unsolved problem in celestial mechanics for over half a century, since the announcement by Le Verrier in 1859 that, after the perturbing effects of the planets on Mercury's orbit had been accounted for, there remained in the data an unexplained advance in the perihelion of Mercury. The modern value for this discrepancy is 42.98 arcseconds per century, and general relativity accounted for it in a natural way. Radar measurements of the orbit of Mercury since 1966 have led to improved accuracy, so that the predicted relativistic advance agrees with the observations to about 0.1%.

Deflection of Light.   Einstein's second test concerns the deflection of light. According to general relativity, a light ray which passes the Sun at a distance d is deflected by an angle $\Delta\theta = 1.75''/d$, where d is measured in units of the solar radius, and the notation $''$ denotes seconds of arc. Confirmation by the British astronomers Eddington and Crommelin of the bending of optical starlight observed during a total solar eclipse in the first months following World War I helped make Einstein a celebrity. However, those measurements had only 30% accuracy, and succeeding eclipse experiments weren't much better.

859

The development of long-baseline radio interferometry produced a method for greatly improved determinations of the deflection of light. Coupled with this technological advance is a series of heavenly coincidences: each year groups of strong quasars pass near the Sun (as seen from the Earth). The idea is to measure the differential deflection of radio waves from one quasar relative to those from another as they pass near the Sun. A number of measurements of this kind occurred almost annually over the period from 1969 to 1975, yielding a confirmation of the predicted deflection to 1.5%. Recent measurements have improved this to 0.1%.

Gravitational Redshift. Another consequence of Einstein's insight is the gravitational redshift effect, which is a frequency shift between two identical clocks placed at different heights in a gravitational field. For small differences in height h between clocks, the shift in the frequency Δf is given by

$$\frac{\Delta f}{f} = \frac{gh}{c^2}$$

where g is the local gravitational acceleration and c is the speed of light. If the receiver clock is at a lower height than the emitter clock, the received signal is shifted to higher frequencies ("blueshift"); if the receiver is higher, the signal is shifted to lower frequencies ("redshift").

The most precise gravitational redshift experiment performed to date was a rocket experiment carried out in June 1976. A "hydrogen maser" atomic clock was flown on a Scout D rocket to an altitude of 10 000 km, and its frequency compared to a similar clock on the ground using radio signals. After the effects of the rocket's motion were taken into account, the observations confirmed the gravitational redshift to 0.02%.

Applications

General relativity has passed every experimental test to which it has been put, and many alternative theories have fallen by the wayside. Most physicists now take the theory for granted and look to see how it can be used as a practical tool in physics and astronomy.

Gravitational Radiation. One of these new tools is gravitational radiation, a subject almost as old as general relativity. By 1916, Einstein had succeeded in showing that the field equations of general relativity admitted wave-like solutions analogous to those of electromagnetic theory. For example, a dumbbell rotating about an axis passing at right angles through its handle will emit gravitational waves that travel at the speed of light. They also carry energy away from the dumbbell, just as electromagnetic waves carry energy away from a light source.

In 1968, Joseph Weber made the stunning announcement that he had detected gravitational radiation of extraterrestrial origin using massive aluminum bars as detectors. A passing gravitational wave acts as an oscillating gravitational force field that alternately compresses and extends the bar lengthwise. However, subsequent observations by other researchers using bars with sensitivity that was claimed to be better than Weber's failed to confirm Weber's results. His results are now generally regarded as a false alarm, although there is still no good explanation for the "events" that he recorded in his bars.

Nevertheless, Weber's experiments did initiate the program of gravitational-wave detection and inspired other groups to build better detectors. Currently a dozen laboratories around the world are engaged in building and improving upon the basic "Weber bar" detector, striving to reduce noise from thermal, electrical, and environmental sources in order to detect the very weak oscillations produced by a gravitational wave. For a bar of one meter length, the challenge is to detect a variation in length smaller than 10^{-20} meters, or 10^{-5} of the radius of a proton.

Although gravitational radiation has not been detected directly, we know that it exists, through a remarkable system known as the binary pulsar. Discovered in 1974 by radio astronomers Russell Hulse and Joseph Taylor, it consists of a pulsar (which is a rapidly spinning neutron star) and a companion star in orbit around each other. Although the companion has not been seen directly, it is also believed to be a neutron star. The pulsar acts as an extremely stable clock, its pulse period of approximately 59 milliseconds drifting by only 0.25 ns/year. By measuring the arrival times of radio pulses at Earth, observers were able to determine the motion of the pulsar about its companion with amazing accuracy. For example, the accurate value for the orbital period is 27 906.980 895 s, and the orbital eccentricity is 0.617 131. Like a rotating dumbbell, an orbiting binary system should emit gravitational radiation, and in the process lose some of its orbital energy. This energy loss will cause the pulsar and its companion to spiral in toward each other, and the orbital period to shorten. According to general relativity, the predicted decrease in the orbital period is 75.8 μs/year. The observed decrease rate is in agreement with the prediction to better than 1%. This confirms the existence of gravitational radiation and the general relativistic equations that describe it.

Black Holes.   The first glimmerings of the black hole idea date to the 18th century, in the writings of a British amateur astronomer, the Reverend John Michell. Reasoning on the basis of the corpuscular theory that light would be attracted by gravity, he noted that the speed of light emitted from the surface of a massive body would be reduced by the time the light was very far from the source. (Michell of course did not know special relativity.) It would therefore be possible for a body to be sufficiently massive and compact to prevent light from escaping from its surface.

Although the general relativistic solution for a nonrotating black hole was discovered by Karl Schwarzschild in 1916 and a calculation of gravitational collapse to a black hole state was performed by J. Robert Oppenheimer and Hartland Snyder in 1939, black hole physics didn't really begin until the middle 1960s, when astronomers confronted the problem of the energy output of the quasars and started to take black holes seriously.

A black hole is formed when a star has exhausted the thermonuclear fuel necessary to produce the heat and pressure that support it against gravity. The star begins to collapse, and if it is massive enough, it continues to collapse until its radius approaches a value called the gravitational radius or Schwarzschild radius. In the nonrotating spherical case, this radius is $2GM/c^2$, where M is the mass of the star. For a body of one solar mass, the gravitational radius is about 3 km; for a body of the mass of the Earth, it is about 9 mm. An observer sitting on the surface of the star sees the collapse continue to smaller and smaller radii, until both star and observer reach the origin $r = 0$, with consequences too horrible to describe. On the other hand, an observer at great distances observes the collapse to slow down as the radius approaches the gravitational radius, a result of the gravitational redshift of the light signals sent outward. The distant observer never sees any signals emitted by the falling observer once the latter is inside the gravitational radius. This is because any signal emitted inside can never escape the sphere bounded by the gravitational radius, called the "event horizon."

In 1974, Stephen Hawking discovered that the laws of quantum mechanics require a black hole to evaporate by the creation of particles with a thermal energy spectrum and to have an associated temperature and entropy. The temperature of a Schwarzschild black hole is $T = hc^3/8\pi k_B GM$, where h is Planck's constant and k is Boltzmann's constant. This discovery demonstrated a remarkable connection between gravity, thermodynamics, and quantum mechanics that helped renew the theoretical quest for a grand synthesis of all the fundamental interactions. On the other hand, for black holes of astronomical masses, the evaporation is completely negligible, since for a solar-mass black hole, $T \approx 10^{-6}$ K.

Although a great deal is known about black holes in theory, rather less is known about them observationally. There are several instances in which the evidence for the existence of black holes is impressive, but in all cases it is indirect. For instance, in the x-ray source Cygnus XI, the source of the x-rays is believed to be a black hole with a mass larger than about six solar masses in orbit around a giant star. The x-rays are emitted by matter pulled from the surface of the companion star and sent into a spiralling orbit around the black hole.

Cosmology.   Although Einstein in 1917 first used general relativity to calculate a model for the Universe, the subject was not considered a serious branch of physics until the 1960s, when astronomical observations lent credence to the idea that the Universe is expanding from a ''Big Bang.'' In 1965 came the discovery of the cosmic background radiation by Arno Penzias and Robert Wilson. This radiation is the remains of the hot electromagnetic black-body radiation that once dominated the Universe in its earlier phase, now cooled to 3 K by the subsequent expansion of the Universe. Next came calculations of the amount of helium synthesized from hydrogen by thermonuclear fusion in the very early Universe, around 1000 s after the Big Bang. The amount, approximately 25% by mass, was in agreement with the abundances of helium observed in stars and in interstellar space. This was an important confirmation of the hot Big-Bang picture because the amount of helium believed to be produced by fusion in the interiors of stars is woefully inadequate to explain the observed abundances.

Today, the general relativistic hot Big-Bang model of the Universe has broad acceptance, and cosmologists now focus their attention on more detailed issues, such as how galaxies and other large-scale structures formed out of the hot primordial soup, and on what the Universe might have been like earlier than 1000 s.

Questions and Problems

1. Two stars are a certain distance from each other as seen in the night sky. Half a year later, the stars are now overhead during the day, and the Sun is now located midway between the two stars. Because of the deflection of light by the Sun, do the stars now appear closer together or farther apart?
2. When gravitational waves are emitted by the binary pulsar, the double-star system loses energy. As a consequence, its orbital velocity increases, and thus its kinetic energy increases. How is this possible?
3. What is the gravitational frequency shift between two atomic clocks separated by 1000 km, both at sea level?
4. Calculate the deflection of light for light rays that pass the Sun at distances of 2 solar radii, 10 solar radii, and 100 solar radii.

QUANTUM PHYSICS

27

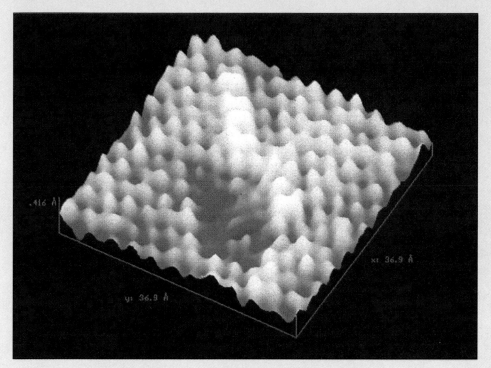

In the previous chapter, we discussed why Newtonian mechanics must be replaced by Einstein's special theory of relativity when dealing with particles whose speeds are comparable to the speed of light. Although many problems were indeed resolved by the theory of relativity in the early part of the 20th century, many experimental and theoretical problems remained unsolved. Attempts to explain the behavior of matter on the atomic level with the laws of classical physics were totally unsuccessful. Various phenomena, such as blackbody radiation, the photoelectric effect, and the emission of sharp spectral lines by atoms in a gas discharge tube, could not be understood within the framework of classical physics. We shall describe these phenomena because of their importance in subsequent developments.

Another revolution took place in physics between 1900 and 1930. This was the era of a new and more general formulation called *quantum mechanics*. This new approach was highly successful in explaining the behavior of atoms, molecules, and nuclei. As with relativity, the quantum theory requires a modification of our ideas concerning the physical world.

The earliest and most basic ideas of quantum theory were introduced by Planck, and most of the subsequent mathematical developments, interpretations, and improvements were made by a number of distinguished physicists, including Einstein, Bohr, Schrödinger, de Broglie, Heisenberg, Born, and Dirac. An extensive study of quantum theory is certainly beyond the scope of this book.

This chapter is simply an introduction to the underlying ideas of quantum theory and the wave-particle nature of matter. We also discuss some simple applications of quantum theory, including the photoelectric effect, the Compton effect, and x-rays.

27.1
BLACKBODY RADIATION AND PLANCK'S HYPOTHESIS

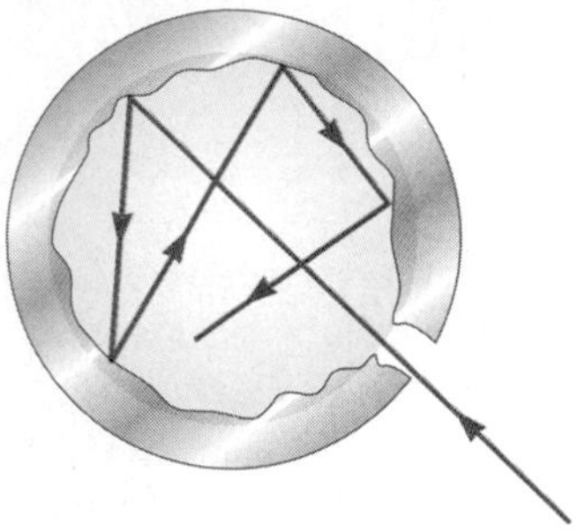

An object at any temperature is known to emit radiation, sometimes referred to as **thermal radiation.** The characteristics of this radiation depend on the temperature and properties of the object. At low temperatures, thermal radiation wavelengths are mainly in the infrared region and hence not observable by the eye. As the temperature of an object increases, it eventually begins to glow red. At sufficiently high temperatures, it appears to be white, as in the glow of the hot tungsten filament of a lightbulb. A careful study of thermal radiation shows that it consists of a continuous distribution of wavelengths from the infrared, visible, and ultraviolet portions of the spectrum.

From a classical viewpoint, thermal radiation (electromagnetic waves) originates from accelerated charged particles near the surface of an object, which emit radiation much like small antennas. The thermally agitated charges can have a distribution of accelerations, which accounts for the continuous spectrum of radiation emitted by the object. By the end of the 19th century, it had become apparent that the classical theory of thermal radiation was inadequate. The basic problem was in understanding the observed distribution of wavelengths in the radiation emitted by a black body. By definition, a black body is an ideal system that absorbs *all* radiation incident on it. A good approximation to a black body is the inside of a hollow object, as shown in Figure 27.1. The nature of the radiation emitted through a small hole leading to the cavity depends only on the temperature of the cavity walls.

Experimental data for the distribution of energy for blackbody radiation at three different temperatures are shown in Figure 27.2. Note that the radiated energy varies with wavelength and temperature. As the temperature of the black body increases, the total amount of energy it emits increases. Also, as temperatures increase, the peak of the distribution shifts to shorter wavelengths. This shift was found to obey the following relationship, called **Wien's displacement law:**

$$\lambda_{\max} T = 0.2898 \times 10^{-2} \text{ m} \cdot \text{K} \qquad \textbf{[27.1]}$$

where $\lambda_{\max}$ is the wavelength at which the curve peaks and T is the absolute temperature of the object emitting the radiation.

Early attempts to use classical ideas to explain the shape of the curves shown in Figure 27.2 failed. Figure 27.3 shows an experimental plot of the blackbody radiation spectrum (red curve) together with the theoretical picture of what this curve should look like based on classical theories (blue curve). At long wavelengths, classical theory is in good agreement with the experimental data. However, at short wavelengths there is major disagreement. This can be seen by noting that as λ approaches zero, classical theory predicts that the amount of energy being radiated should increase. In fact, theory predicts that the intensity

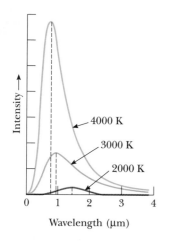

FIGURE 27.1
The opening in the cavity of a body is a good approximation of a black body. As light enters the cavity through the small opening, part is reflected and part is absorbed on each reflection from the interior walls. After many reflections, essentially all of the incident energy is absorbed.

FIGURE 27.2
Intensity of blackbody radiation versus wavelength at three different temperatures. Note that the total radiation emitted (the area under a curve) increases with increasing temperature.

Max Planck was born in Kiel and received his college education in Munich and Berlin. He joined the faculty at Munich in 1880 and five years later received a professorship at Kiel University. He replaced Kirchhoff at the University of Berlin in 1889 and remained there until 1926. He introduced the concept of a "quantum of action" (Planck's constant h) in an attempt to explain the spectral distribution of blackbody radiation, which laid the foundations for quantum theory. He was awarded the Nobel prize in 1918 for this discovery of the quantized nature of energy. The work leading to the "lucky" blackbody radiation formula was described by Planck in his 1920 Nobel prize acceptance speech: "But even if the radiation formula proved to be perfectly correct, it would after all have been only an interpolation formula found by lucky guesswork and, thus, would have left us rather unsatisfied. I therefore strived from the day of its discovery, to give it a real physical interpretation and this led me to consider the relations between entropy and probability according to Boltzmann's ideas."

Planck's life was filled with personal tragedies. One of his sons was killed in action in World War I, and two daughters were lost in childbirth in the same period. His house was destroyed by bombs in World War II, and in 1944 his son Erwin was executed by the Nazis after being accused of plotting to assassinate Hitler.

Planck became president of the Kaiser Wilhelm Institute of Berlin in 1930. The institute was renamed the Max Planck Institute in his honor after World War II. Although Planck remained in Germany during the Hitler regime, he openly protested the Nazi treatment of his Jewish colleagues and consequently was forced to resign his presidency in 1937. Following World War II, he was renamed the president of the Max Planck Institute. He spent the last two years of his life in Göttingen as an honored and respected scientist and humanitarian.

**Max Planck
(1858–1947)**

(Science Photo Library)

should be infinite. This is contrary to the experimental data, which show that as λ approaches zero, the amount of energy carried by short-wavelength radiation also approaches zero. This contradiction is often called the **ultraviolet catastrophe.**

In 1900 Planck developed a formula for blackbody radiation that was in complete agreement with experiments at all wavelengths. His analysis led to a curve that is shown by the red line in Figure 27.3. Planck's original theoretical approach is rather abstract in that it involves arguments based on entropy and thermodynamics. We shall present arguments that are easier to visualize physically, while attempting to convey the spirit and revolutionary impact of Planck's original work.

Planck was convinced that blackbody radiation was produced by submicroscopic electric oscillators, which he called resonators. He assumed that the walls of a glowing cavity were composed of literally billions of these resonators (whose exact nature was unknown). The resonators could have only certain discrete amounts of energy, E_n, given by

$$E_n = nhf \qquad \text{[27.2]}$$

where n is a positive integer called a **quantum number** and f is the frequency of vibration of the resonators. The energies of the molecule are said to be

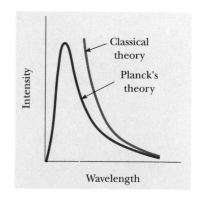

FIGURE 27.3
Comparison of the Planck theory with the classical theory for the distribution of blackbody radiation.

quantized, and the allowed energy states are called *quantum states.* The factor *h* is a constant, known as **Planck's constant,** given by

$$h = 6.626 \times 10^{-34} \, \text{J} \cdot \text{s} \qquad\qquad \textbf{[27.3]}$$

The resonators emit energy in discrete units of light energy called **quanta** (or **photons,** as they are now called). They do so by "jumping" from one quantum state to another. If the quantum number changes by one unit, Equation 27.2 shows that the amount of energy radiated by the molecule equals *hf.* Hence, the energy of a light quantum, corresponding to the energy difference between two adjacent levels, is given by

$$E = hf \qquad\qquad \textbf{[27.4]}$$

The molecule will radiate or absorb energy only when it changes quantum states. If it remains in one quantum state, no energy is absorbed or emitted.

The key point in Planck's theory is the radical assumption of quantized energy states. This development marked the birth of the quantum theory. At that time, most scientists, including Planck, did not consider the quantum concept to be realistic. Hence, Planck and others continued to search for a more rational explanation of blackbody radiation. However, subsequent developments showed that a theory based on the quantum concept (rather than on classical concepts) had to be used to explain a number of atomic phenomena.

EXAMPLE 27.1   Thermal Radiation from the Human Body

The temperature of the skin is approximately 35°C. At what wavelength does the radiation emitted from the skin reach its peak?

Solution   From Wien's displacement law, we have

$$\lambda_{\text{max}} \, T = 0.2898 \times 10^{-2} \, \text{m} \cdot \text{K}$$

Solving for λ_{max}, noting that 35°C corresponds to an absolute temperature of 308 K, we have

$$\lambda_{\text{max}} = \frac{0.2898 \times 10^{-2} \, \text{m} \cdot \text{K}}{308 \, \text{K}} = \boxed{940 \; \mu\text{m}}$$

This radiation is in the infrared region of the spectrum.

Exercise   (a) Find the wavelength corresponding to the peak of the radiation curve for the heating element of an electric oven at a temperature of 1.20×10^3 K. Note that although this radiation peak lies in the infrared, there is enough visible radiation at this temperature to give the element a red glow. (b) Calculate the wavelength corresponding to the peak of the radiation curve for an object whose temperature is 5.00×10^3 K, an approximate temperature for the surface of the Sun.

Answer   (a) 2.415 μm; (b) 580 nm, in the visible region.

EXAMPLE 27.2 The Quantized Oscillator

A 2.0-kg mass is attached to a massless spring of force constant $K = 25$ N/m. The spring is stretched 0.40 m from its equilibrium position and released.

(a) Find the total energy and frequency of oscillation according to classical calculations.

(b) Assume that the energy is quantized and find the quantum number, n, for the system.

(c) How much energy would be carried away in a one-quantum change?

Solution   (a) The total energy of a simple harmonic oscillator having an amplitude A is $\frac{1}{2} KA^2$. Therefore,

$$E = \frac{1}{2} KA^2 = \frac{1}{2} (25 \text{ N/m})(0.40 \text{ m})^2 = \boxed{2.0 \text{ J}}$$

The frequency of oscillation is

$$f = \frac{1}{2\pi} \sqrt{\frac{K}{m}} = \frac{1}{2\pi} \sqrt{\frac{25 \text{ N/m}}{2.0 \text{ kg}}} = \boxed{0.56 \text{ Hz}}$$

(b) If the energy is quantized, we have $E_n = nhf$, and from the result of (a), we have

$$E_n = nhf = n(6.63 \times 10^{-34} \text{ J·s})(0.56 \text{ Hz}) = 2.0 \text{ J}$$

Therefore,

$$n = \boxed{5.4 \times 10^{33}}$$

(c) The energy carried away in a one-quantum change of energy is

$$E = hf = (6.63 \times 10^{-34} \text{ J·s})(0.56 \text{ Hz}) = \boxed{3.7 \times 10^{-34} \text{ J}}$$

The energy carried away by a one-quantum change is such a small fraction of the total energy of the oscillator that we could not expect to see it. Thus, even though the decrease in energy of a spring-mass system is quantized and does decrease by small quantum jumps, our senses perceive the decrease as continuous. Quantum effects become important and measurable only on the submicroscopic level of atoms and molecules.

EXAMPLE 27.3   The Energy of a "Yellow" Photon

Yellow light with a frequency of approximately 6.0×10^{14} Hz is the predominant frequency in sunlight. What is the energy carried by a quantum of this light?

Solution   The energy carried by one quantum of light is given by Equation 27.4:

$$E = hf = (6.63 \times 10^{-34} \text{ J·s})(6.0 \times 10^{14} \text{ Hz}) = 4.0 \times 10^{-19} \text{ J} = \boxed{2.5 \text{ eV}}$$

Light

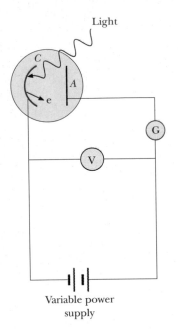

FIGURE 27.4
Circuit diagram for observing the photoelectric effect. When light strikes plate C, photoelectrons are ejected from the plate. Electrons collected at A and passing through the galvanometer constitute a current in the circuit.

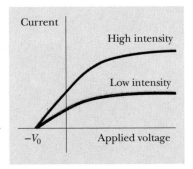

FIGURE 27.5
Photoelectric current versus voltage for two light intensities. The current increases with intensity but reaches a saturation level for large values of V. At voltages equal to or less than $-V_0$, the current is zero.

27.2

THE PHOTOELECTRIC EFFECT

In the latter part of the 19th century, experiments showed that, when light is incident on certain metallic surfaces, electrons are emitted from the surfaces. This phenomenon is known as the **photoelectric effect,** and the emitted electrons are called **photoelectrons.** The first discovery of this phenomenon was made by Hertz, who was also the first to produce the electromagnetic waves predicted by Maxwell.

Figure 27.4 is a schematic diagram of an apparatus in which the photoelectric effect can occur. An evacuated glass or quartz tube contains a metal plate, C, connected to the negative terminal of a battery, and a second metal plate, A, maintained at a positive potential by the battery. When the tube is kept in the dark, the galvanometer, G, reads zero, indicating that there is no current in the circuit. However, when monochromatic light of an appropriate wavelength shines on plate C, a current is detected by the galvanometer, indicating a flow of charges across the gap between C and A. The current associated with this process arises from electrons emitted from the negative plate and collected at the positive plate.

A plot of the photoelectric current versus the potential difference, V, between A and C is shown in Figure 27.5 for two light intensities. Note that for large values of V, the current reaches a maximum value, corresponding to the case where all photoelectrons are collected at A. In addition, the current increases as the incident light intensity increases, as you might expect. Finally, when V is negative, that is, when the battery in the circuit is reversed to make C positive and A negative, the photoelectrons are repelled by the negative plate A. Only those electrons having a kinetic energy greater than eV will reach A, where e is the charge on the electron. Furthermore, if V is less than or equal to V_0, called the **stopping potential,** no electrons will reach A and the current will be zero. The stopping potential is *independent* of the radiation intensity. The maximum kinetic energy of the photoelectrons is related to the stopping potential through the relation

$$KE_{max} = eV_0 \qquad \text{[27.5]}$$

In 1905 Einstein developed a theory that explained the photoelectric effect and made several predictions that were in disagreement with the wave theory of light. The major features of the theory that could not be explained with a wave model are as follows:

1. No electrons are emitted if the incident light frequency falls below some **cutoff frequency,** f_c, which depends on the material being illuminated. For example, in the case of sodium, $f_c = 5.50 \times 10^{14}$ Hz. This is inconsistent with the wave theory, which predicts that the photoelectric effect should occur at any frequency, provided the light intensity is high enough.
2. If the light frequency exceeds the cutoff frequency, a photoelectric effect is observed and the number of photoelectrons emitted is proportional to the light intensity. However, the maximum kinetic energy of the photoelectrons is independent of light intensity, a fact that cannot be explained by the concepts of classical physics.

3. The maximum kinetic energy of the photoelectrons increases with increasing light frequency.

4. Electrons are emitted from the surface almost instantaneously, even at low light intensities. Classically, it would be expected that the electrons would require some time to absorb the incident radiation before acquiring enough kinetic energy to escape from the metal.

In his photoelectric paper, for which he received the Nobel prize in 1921, Einstein extended Planck's concept of quantization to electromagnetic waves. He assumed that light (or any electromagnetic wave) of frequency f can be considered a stream of photons. Each photon has an energy, E, given by

$$E = hf$$

[27.6]

Energy of a photon

where h is Planck's constant. Thus, Einstein considered light to be much like a stream of particles (rather than a wave) traveling through space, where each "particle" could be absorbed as a unit by an electron. Furthermore, Einstein argued that when the photon's energy is transferred to an electron in a metal, the energy acquired by the electron must be hf. However, the electron must also pass through the metal surface in order to be emitted, and some energy is required to overcome this barrier. The amount of energy, ϕ, required to escape the metal is known as the **work function** of the substance and is on the order of a few electron volts for metals. For example, the work function for zinc is about 3.0 eV. Hence, in order to conserve energy, the maximum kinetic energy of the ejected photoelectrons is the difference between the photon energy and the work function of the metal, or

$$KE_{max} = hf - \phi$$

[27.7]

Photoelectric effect equation

That is, the excess energy, $hf - \phi$, equals the maximum kinetic energy that the liberated electron can have outside the surface.

With the photon theory of light, the features of the photoelectric effect that cannot be understood using classical concepts can be explained. These are briefly described in the order they were introduced earlier:

1. The fact that the photoelectric effect is not observed below a certain cutoff frequency follows from the fact that the energy of the photon must be greater than or equal to ϕ. If the energy of the incoming photon is not equal to or greater than ϕ, the electrons will never be ejected from the surface, regardless of the intensity of the light.

2. The fact that KE_{max} is independent of the light intensity can be understood with the following argument. If the light intensity is doubled, the number of photons is doubled, which doubles the number of photoelectrons emitted. However, their kinetic energy, which equals $hf - \phi$, depends only on the light frequency and the work function, not on the light intensity.

3. The fact that KE_{max} increases with increasing frequency is easily understood with Equation 27.7.

4. Finally, the fact that the electrons are emitted almost instantaneously is consistent with the particle theory of light, in which the incident energy appears in small packets and there is a one-to-one interaction between

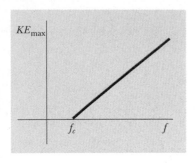

FIGURE 27.6
A sketch of KE_{max} versus frequency of incident light for photoelectrons in a typical photoelectric effect experiment. Photons with frequency less than f_c do not have sufficient energy to eject an electron from the metal.

photons and electrons. This is in contrast to having the energy of the radiation distributed uniformly over a large area.

A final confirmation of Einstein's theory is a test of the prediction of a linear relationship between f and KE_{max}. Indeed, such a linear relationship is observed, as sketched in Figure 27.6. The slope of such a curve gives a value for h. The intercept on the horizontal axis gives the cutoff frequency, which is related to the work function through the relation $f_c = \phi/h$. This corresponds to a **cutoff wavelength** of

$$\lambda_c = \frac{c}{f_c} = \frac{c}{\phi/h} = \frac{hc}{\phi} \qquad \text{[27.8]}$$

where c is the speed of light (3.00×10^8 m/s). Wavelengths *greater* than λ_c incident on a material with a work function ϕ do not result in the emission of photoelectrons.

EXAMPLE 27.4   The Photoelectric Effect for Sodium

A sodium surface is illuminated with light of wavelength 3.00×10^{-7} m. The work function for sodium is 2.46 eV. Find (a) the kinetic energy of the ejected photoelectrons and (b) the cutoff wavelength for sodium.

Solution   (a) The energy of each photon of the illuminating light beam is

$$E = hf = \frac{hc}{\lambda} = \frac{(6.63 \times 10^{-34}\,\text{J·s})(3.00 \times 10^8\,\text{m/s})}{3.00 \times 10^{-7}\,\text{m}}$$

$$= 6.63 \times 10^{-19}\,\text{J} = \frac{6.63 \times 10^{-19}\,\text{J}}{1.60 \times 10^{-19}\,\text{J/eV}} = 4.14\,\text{eV}$$

where we have used the conversion 1 eV = 1.6×10^{-19} J. Using Equation 27.7 gives

$$KE_{max} = hf - \phi = 4.14\,\text{eV} - 2.46\,\text{eV} = \boxed{1.68\,\text{eV}}$$

(b) The cutoff wavelength can be calculated from Equation 27.8 after we convert ϕ from electron volts to joules:

$$\phi = 2.46\,\text{eV} = (2.46\,\text{eV})(1.6 \times 10^{-19}\,\text{J/eV}) = 3.94 \times 10^{-19}\,\text{J}$$

Hence

$$\lambda_c = \frac{hc}{\phi} = \frac{(6.63 \times 10^{-34}\,\text{J·s})(3.00 \times 10^8\,\text{m/s})}{3.94 \times 10^{-19}\,\text{J}}$$

$$= 5.05 \times 10^{-7}\,\text{m} = \boxed{505\,\text{nm}}$$

This wavelength is in the green region of the visible spectrum.

27.3

APPLICATIONS OF THE PHOTOELECTRIC EFFECT

The photoelectric cell shown in Figure 27.4 acts much like a switch in an electric circuit in that it produces a current in an external circuit when light of sufficiently high frequency falls on the cell, but it does not allow a current in the dark. Many

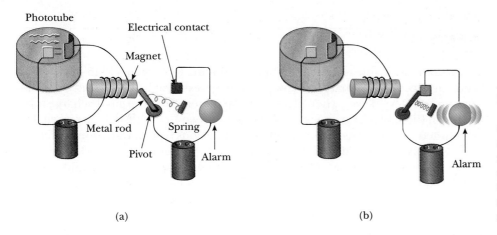

(a)

(b)

FIGURE 27.7
(a) When light strikes the photo-tube, the current in the circuit on the left energizes the mag-net, breaking the burglar alarm circuit on the right. (b) When the light source is removed, the alarm circuit is closed, and the alarm sounds.

practical devices in our everyday lives depend on the photoelectric effect. One of the first practical uses was as the detector in the light meter of a camera. Light reflected from the object to be photographed struck a photoelectric surface, causing it to emit electrons, which then passed through a very sensitive ammeter. The magnitude of the current depended on the intensity of the light. Modern solid-state devices have now replaced light meters that used the photoelectric effect.

A second example of the photoelectric effect in action is the burglar alarm. This device often uses ultraviolet rather than visible light in order to make the beam less obvious. A beam of light passes from the source to a photosensitive surface; the current produced is then amplified and used to energize an electro-magnet that attracts a metal rod, as in Figure 27.7a. If an intruder breaks the light beam, the electromagnet switches off and the spring pulls the iron rod to the right (Fig. 27.7b). In this position, a completed circuit allows current to pass and activate the alarm system.

Figure 27.8 shows how the photoelectric effect is used to produce the sound on a movie film. The soundtrack is located along the side of the film in the form of an optical pattern of light and dark lines. A beam of light in the projector is directed through the soundtrack toward a phototube. The variation in shading on the soundtrack varies the light intensity falling on the plate of the phototube, thus changing the current in the circuit. This changing current electrically simu-lates the original sound wave and reproduces it in the speaker.

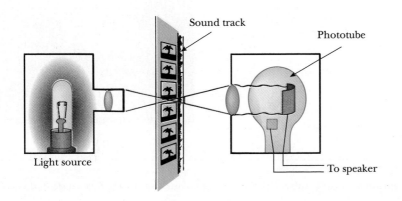

FIGURE 27.8
The shading on the soundtrack varies the light intensity reach-ing the phototube and hence the current to the speaker.

27.4

X-RAYS

In 1895 at the University of Wurzburg, Wilhelm Roentgen (1845–1923) was studying electrical discharges in low-pressure gases when he noted that a fluorescent screen glowed even when placed several meters from the gas discharge tube and even when black cardboard was placed between the tube and the screen. He concluded that the effect was caused by a mysterious type of radiation, which he called **x-rays** because of their unknown nature. Subsequent study showed that these rays traveled at or near the speed of light and that they could not be deflected by either electric or magnetic fields. This last fact indicated that x-rays did not consist of beams of charged particles, although the possibility that they were beams of uncharged particles remained.

In 1912 Max von Laue (1879–1960) suggested that one should be able to diffract x-rays by using the regular atomic spacings of a crystal lattice as a diffraction grating, just as visible light is diffracted by a ruled grating. Shortly thereafter, researchers demonstrated that such a diffraction pattern could be observed, similar to that shown in Figure 27.9 for NaCl. The wavelengths of the x-rays were then determined from the diffraction data and the known values of the spacing between atoms in the crystal. X-ray diffraction has proved to be an invaluable technique for understanding the structure of matter. We shall discuss this subject in more detail in the next section.

Typical x-ray wavelengths are about 0.1 nm, which is on the order of the atomic spacing in a solid. We now know that x-rays are a part of the electromagnetic spectrum, characterized by frequencies higher than those of ultraviolet radiation and having the ability to penetrate most materials with relative ease.

X-rays are produced when high-speed electrons are suddenly decelerated, for example, when a metal target is struck by electrons that have been accelerated through a potential difference of several thousand volts. Figure 27.10a shows a schematic diagram of an x-ray tube. A current in the filament causes electrons to be boiled off, and these freed electrons are accelerated toward a dense metal target, such as tungsten, which is held at a higher potential than the filament.

FIGURE 27.9
X-ray diffraction pattern of NaCl.

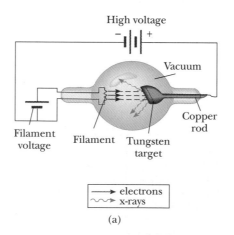

(a)

(b)

FIGURE 27.10
(a) Diagram of an x-ray tube. (b) Photograph of an x-ray tube. *(Courtesy of GE Medical Systems)*

Figure 27.11 represents a plot of x-ray intensity versus wavelength for the spectrum of radiation emitted by an x-ray tube. Note that there are two distinct patterns. One pattern is a continuous broad spectrum that depends on the voltage applied to the tube. Superimposed on this pattern is a series of sharp, intense lines that depend on the nature of the target material. The accelerating voltage must exceed a certain value, called the **threshold voltage,** in order to observe these sharp lines, which represent radiation emitted by the target atoms as their electrons undergo rearrangements. We shall discuss this further in Chapter 28. The continuous radiation is sometimes called **bremsstrahlung,** a German word meaning "braking radiation." The term arises from the nature of the mechanism responsible for the radiation. That is, electrons emit radiation when they undergo a deceleration inside the target.

The deceleration of the electrons by the target produces an effect similar to an inverse photoelectric effect. In the photoelectric process, a quantum of radiant energy is absorbed by an electron in a metal and the electron gains enough energy to escape the metal. In the case of x-ray production, the inverse of this process occurs, as shown in Figure 27.12. As an electron passes close to a positively charged nucleus contained in a target material, it is deflected from its path because of its electrical attraction to the nucleus, and hence experiences an acceleration. An analysis from classical physics shows that any charged particle will radiate energy in the form of electromagnetic radiation when it is accelerated. (An example of this is the production of electromagnetic waves by accelerated charges in a radio antenna, as described in Chapter 21.) According to quantum theory, this radiation must appear in the form of photons. Since the radiated photon shown in Figure 27.12 carries energy, the electron must lose kinetic energy because of its encounter with the target nucleus. Let us consider an extreme example in which the electron loses all of its energy in a single collision. In this case, the initial energy of the electron (eV) is transformed completely into the energy of the photon (hf_{max}). In equation form we have

$$eV = hf_{max} = \frac{hc}{\lambda_{min}} \qquad \textbf{[27.9]}$$

where eV is the energy of the electron after it has been accelerated through a potential difference of V volts and e is the charge on the electron. This says that the shortest-wavelength radiation that can be produced is

$$\lambda_{min} = \frac{hc}{eV} \qquad \textbf{[27.10]}$$

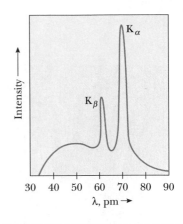

FIGURE 27.11
The x-ray spectrum of a metal target consists of a broad continuous spectrum plus a number of sharp lines, which are due to *characteristic x-rays*. The data shown were obtained when 35-keV electrons bombarded a molybdenum target. Note that $1 \text{ pm} = 10^{-12} \text{ m} = 10^{-3} \text{ nm}$.

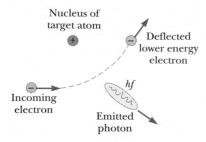

FIGURE 27.12
An electron passing near a charged target atom experiences an acceleration, and a photon is emitted in the process.

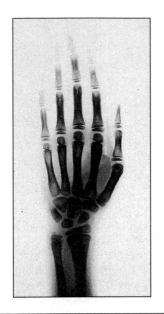

FIGURE 27.13
X-ray photograph of a human hand.

The reason that all the radiation produced does not have this particular wavelength is because many of the electrons are not stopped in a single collision. This results in the production of the continuous spectrum of wavelengths.

As noted earlier, x-rays are extremely penetrating and can produce burns or other complications if proper precautions are not taken. Between 1930 and 1950 an x-ray device called a fluoroscope was widely used in shoe stores to examine the bones of the foot. Such devices are no longer in use since they are now known to be health hazards although physicians use similar devices to study the skeletal structures of their patients. An x-ray photograph of a human hand is shown in Figure 27.13.

EXAMPLE 27.5 **The Minimum X-Ray Wavelength**

Calculate the minimum wavelength produced when electrons are accelerated through a potential difference of 100 000 V, a not-uncommon voltage for an x-ray tube.

Solution From Equation 27.10, we have

$$\lambda_{min} = \frac{(6.63 \times 10^{-34}\,\text{J}\cdot\text{s})(3.00 \times 10^{8}\,\text{m/s})}{(1.60 \times 10^{-19}\,\text{C})(10^{5}\,\text{V})} = \boxed{1.24 \times 10^{-11}\,\text{m}}$$

---

*27.5

DIFFRACTION OF X-RAYS BY CRYSTALS

In Chapter 25 we described how a diffraction grating can be used to measure the wavelength of light. In principle, the wavelength of *any* electromagnetic wave can be measured if a grating having the proper line spacing can be found. The spacing between lines must be approximately equal to the wavelength of the radiation to be measured. X-rays are electromagnetic waves with wavelengths on the order of 0.1 nm. It would be impossible to construct a grating with such a small spacing. As noted in the previous section, Max von Laue suggested that the regular array of atoms in a crystal could act as a three-dimensional grating for observing the diffraction of x-rays.

One experimental arrangement for observing x-ray diffraction is shown in Figure 27.14. A narrow beam of x-rays with a continuous wavelength range is incident on a crystal such as sodium chloride. The diffracted radiation is very

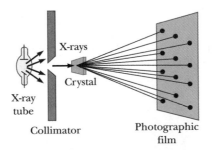

FIGURE 27.14
Schematic diagram of the technique used to observe the diffraction of x-rays by a single crystal. The array of spots formed on the film by the diffracted beams is called a Laue pattern.

intense in certain directions, corresponding to constructive interference from waves reflected from layers of atoms in the crystal. The diffracted radiation is detected by a photographic film and forms an array of spots known as a "Laue pattern." The crystal structure is determined by analyzing the positions and intensities of the various spots in the pattern.

The arrangement of atoms in a crystal of NaCl is shown in Figure 27.15. The smaller red spheres represent Na^+ ions and the larger blue spheres represent Cl^- ions. The spacing between successive Na^+ (or Cl^-) ions in this cubic structure, denoted by the symbol a in Figure 27.15, is approximately 0.563 nm.

A careful examination of the NaCl structure shows that the ions lie in various planes. The shaded areas in Figure 27.15 represent one example in which the atoms lie in equally spaced planes. Now suppose an x-ray beam is incident at grazing angle θ on one of the planes, as in Figure 27.16. The beam can be reflected from both the upper and lower plane of atoms. However, the geometric construction in Figure 27.16 shows that the beam reflected from the lower surface travels farther than the beam reflected from the upper surface by a distance of $2d \sin \theta$. The two portions of the reflected beam will combine to produce constructive interference when this path difference equals some integral multiple of the wavelength λ. The same is true for reflection from the entire family of parallel planes. (Note the similarity between this analysis and that used to describe thin film interference.) The condition for constructive interference is given by

$$2d \sin \theta = m\lambda \qquad (m = 1, 2, 3, \ldots)$$ **[27.11]**   Bragg's law

This condition is known as **Bragg's law** after W. L. Bragg (1890–1971), who first derived the relationship. If the wavelength and diffraction angle are measured, Equation 27.11 can be used to calculate the spacing between atomic planes.

The method of x-ray diffraction to determine crystalline structures was thoroughly developed in England by W. H. Bragg and his son W. L. Bragg, who shared a Nobel prize in 1915 for their work. Since then, thousands of crystalline structures have been investigated. Recently, the technique of x-ray structural analysis has been used to unravel the mysteries of such complex organic systems as the important DNA molecule.

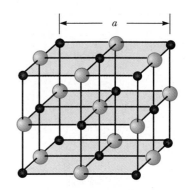

FIGURE 27.15
A model of the cubic crystalline structure of sodium chloride. The blue spheres represent the Cl^- ions, and the red spheres represent the Na^+ ions. The length of the cube edge is $a = 0.563$ nm.

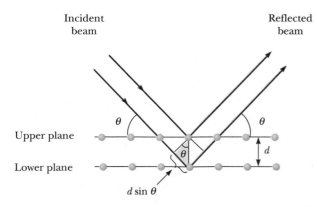

FIGURE 27.16
A two-dimensional description of the reflection of an x-ray beam from two parallel crystalline planes separated by a distance d. The beam reflected from the lower plane travels farther than the one reflected from the upper plane by an amount equal to $2d \sin \theta$.

EXAMPLE 27.6   **Reflection from Calcite**

If the spacing between certain planes in a crystal of calcite is 0.314 nm, find the angles of incidence at which first- and third-order interference will occur for x-rays of wavelength 0.070 nm.

Solution   For first-order interference, the value of m in Equation 27.11 is 1. Thus, the angle of incidence corresponding to this order of interference is found as follows:

$$\sin \theta = \frac{m\lambda}{2d} = \frac{(0.0700 \text{ nm})}{2(0.314 \text{ nm})} = 0.111$$

$$\theta = \boxed{6.37°}$$

In third-order interference, $m = 3$, and we find

$$\sin \theta = \frac{m\lambda}{2d} = \frac{3(0.0700 \text{ nm})}{2(0.314 \text{ nm})} = 0.334$$

$$\theta = \boxed{19.5°}$$

27.6

COMPTON SCATTERING

Further justification for the photon theory of light came from an experiment conducted by Arthur H. Compton in 1923. In his experiment, Compton directed an x-ray beam of wavelength λ_0 toward a block of graphite. He found that the scattered x-rays had a slightly longer wavelength, λ, than the incident x-rays, and hence the energies of the scattered rays were lower. The amount of energy reduction depended on the angle at which the x-rays were scattered. The change in wavelength, $\Delta\lambda$, between a scattered x-ray and an incident x-ray is called the **Compton shift**.

In order to explain this effect, Compton assumed that if a photon behaves like a particle, its collision with other particles is similar to that between two billiard balls. Hence, both energy and momentum must be conserved. If the incident photon collides with an electron initially at rest, as in Figure 27.17, the

FIGURE 27.17

Diagram representing Compton scattering of a photon by an electron. The scattered photon has less energy (or longer wavelength) than the incident photon.

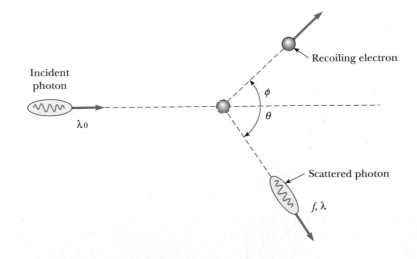

photon transfers some of its energy and momentum to the electron. Consequently, the energy and frequency of the scattered photon are lowered and its wavelength increases. Applying relativistic energy and momentum conservation to the collision described in Figure 27.17, the shift in wavelength of the scattered photon is given by

$$\Delta\lambda = \lambda - \lambda_0 = \frac{h}{m_0 c}(1 - \cos\theta)$$

[27.12]   The Compton shift formula

where m_0 is the rest mass of the electron and θ is the angle between the directions of the scattered and incident photons. The quantity $h/m_0 c$ is called the **Compton wavelength** and has a value of $h/m_0 c = 0.002\,43$ nm. Note that the Compton wavelength is very small relative to the wavelengths of visible light and hence would be very difficult to detect if visible light were used. Furthermore, note that the Compton shift depends on the scattering angle, θ, and not on the wavelength. Experimental results for x-rays scattered from various targets strongly support the photon concept.

EXAMPLE 27.7   The Compton Shift for Carbon

X-rays of wavelength $\lambda_0 = 0.200$ nm are scattered from a block of carbon. The scattered x-rays are observed at an angle of 45.0° to the incident beam. Calculate the wavelength of the scattered x-rays at this angle.

Solution   The shift in wavelength of the scattered x-rays is given by Equation 27.12. Taking $\theta = 45.0°$, we find that

$$\Delta\lambda = \frac{h}{m_0 c}(1 - \cos\theta)$$

$$= \frac{6.63 \times 10^{-34}\,\text{J}\cdot\text{s}}{(9.11 \times 10^{-31}\,\text{kg})(3.00 \times 10^8\,\text{m/s})}(1 - \cos 45.0°)$$

$$= 7.11 \times 10^{-13}\,\text{m} = 0.000\,711\,\text{nm}$$

Hence, the wavelength of the scattered x-ray at this angle is

$$\lambda = \Delta\lambda + \lambda_0 = \boxed{0.200\,711\,\text{nm}}$$

27.7

PAIR PRODUCTION AND ANNIHILATION

In the photoelectric and Compton effects, the energy of a photon is transformed into the kinetic and potential energy of an electron. When a photon interacts with matter through the photoelectric effect, an electron is removed from an atom and the photon disappears. In the Compton effect, a photon is scattered off an electron (or a nucleus) and loses some energy in the process. We shall now describe a process in which the energy of a photon is converted completely into rest mass. This is a striking verification of the equivalence of mass and other forms of energy as predicted by Einstein's theory of relativity.

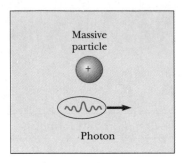

Before

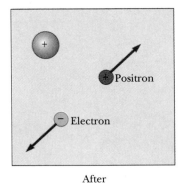

After

FIGURE 27.18
Representation of the process of pair production.

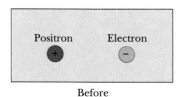

Before

Photon   Photon

After

FIGURE 27.19
Representation of the process of pair annihilation.

Light has a dual nature

A common process in which a photon creates matter is called **pair production,** illustrated in Figure 27.18. In this process, an electron and a positron are simultaneously produced, while the photon disappears. (Note that the positron is a positively charged particle having the same mass as an electron. The positron is often called the *antiparticle* of the electron.) In order for pair production to occur, energy, momentum, and charge must all be conserved during the process. Note that it is impossible for a photon to produce a single electron because the photon has zero charge, and charge would not be conserved in the process. The *minimum* energy that a photon must have to produce an electron-positron pair can be found using conservation of energy by equating the photon energy to the total rest energy of the pair. That is,

$$hf_{min} = 2 m_0 c^2 \qquad \text{[27.13]}$$

Since the rest energy of an electron is $m_0 c^2 = 0.51$ MeV (see Chapter 28), the minimum energy required for pair production is 1.02 MeV. The wavelength of a photon carrying this much energy is 0.0012 nm. Photons with such short wavelengths are in the gamma-ray (or very short x-ray) region of the spectrum.

Pair production cannot occur in a vacuum but can only take place in the presence of a massive particle such as an atomic nucleus. The massive particle must participate in the interaction in order that energy and momentum be conserved simultaneously.

Pair annihilation is a process in which an electron-positron pair produces two photons, the inverse of pair production. Figure 27.19 is one example of pair annihilation in which an electron and positron initially at rest combine with each other, disappear, and create two photons. Since the initial momentum of the pair is zero, it is impossible to produce a single photon. Momentum can be conserved only if two photons moving in opposite directions, both with the same energy and magnitude of momentum, are produced. We shall discuss particles and their antiparticles further in Chapter 30.

27.8

PHOTONS AND ELECTROMAGNETIC WAVES

An explanation of a phenomenon such as the photoelectric effect presents very convincing evidence in support of the photon (or particle) concept of light. An obvious question that arises at this point is, "How can light be considered a photon when it exhibits wave-like properties?" On the one hand, we describe light in terms of photons having energy and momentum. On the other hand, we must also recognize that light and other electromagnetic waves exhibit interference and diffraction effects that are consistent only with a wave interpretation. Which model is correct? Is light a wave or a particle? The answer depends on the specific phenomenon being observed. Some experiments can be better (or solely) explained on the basis of the photon concept, whereas others are best (or solely) described with a wave model. The end result is that *we must accept both models and admit that the true nature of light is not describable in terms of a single classical picture— light has a dual nature.*

We can perhaps understand why photons are compatible with electromagnetic waves in the following manner. We may suspect that long-wavelength radio waves do not exhibit particle characteristics. Consider, for instance, radio waves at a frequency of 2.5 MHz. The energy of a photon having this frequency is only about 10^{-8} eV. From a practical viewpoint, this energy is too small to be detected as a single photon. A sensitive radio receiver might require as many as 10^{10} of these photons to produce a detectable signal. With such a large number of photons reaching the detector every second, it would be unlikely for any graininess to appear in the detected signal; hence it would appear as a continuous wave. That is, we would not be able to detect the individual photons striking the antenna.

Now consider what happens as we go to higher frequencies, or shorter wavelengths. In the visible region, it is possible to observe both the photon and the wave characteristics of light. As we mentioned earlier, a light beam shows interference phenomena and at the same time can produce photoelectrons, which can be understood best by using Einstein's photon concept. At even higher frequencies and correspondingly shorter wavelengths, the momentum and energy of the photons increase. Consequently, the photon nature of light becomes more evident than its wave nature. For example, an x-ray photon is easily detected as a single event. However, as the wavelength decreases, wave effects, such as interference and diffraction, become more difficult to observe. Very indirect methods are required to detect the wave nature of very-high-frequency radiation, such as gamma rays.

All forms of electromagnetic radiation can be described from two points of view. At one extreme, the electromagnetic waves description suits the overall interference pattern formed by a large number of photons. At the other extreme, the photon description is natural when dealing with highly energetic photons of very short wavelength. Hence,

Light has a dual nature; it exhibits both wave and photon characteristics.

27.9
THE WAVE PROPERTIES OF PARTICLES

Students first introduced to the dual nature of light often find it very difficult to accept. In the world around us, we are accustomed to regarding such things as a thrown baseball solely as particles and such things as sound waves solely as forms of wave motion. Every large-scale observation can be interpreted by considering either a wave explanation or a particle explanation, but in the world of photons and electrons, such distinctions are not as sharply drawn. Even more disconcerting is the fact that, under certain conditions, *particles such as electrons also exhibit wave characteristics.*

In 1924 the French physicist Louis de Broglie wrote a doctoral dissertation proposing that *because photons have wave and particle characteristics, perhaps all forms of matter have wave as well as particle properties.* This was a highly revolutionary idea with no experimental confirmation. It was later to play an important role in the development of quantum mechanics.

In Chapter 26 we found that the relationship between energy and momentum for a photon, which has a rest mass of zero, is $p = E/c$. We also know that the energy of a photon is

$$E = hf = \frac{hc}{\lambda}$$

[27.14]   Energy of a photon

Thus, the momentum of a photon can be expressed as

Momentum of a photon

$$p = \frac{E}{c} = \frac{hc}{c\lambda} = \frac{h}{\lambda} \qquad [27.15]$$

From this equation we see that the photon wavelength can be specified by its momentum, $\lambda = h/p$. De Broglie suggested that

de Broglie's hypothesis

material particles of momentum p should also have wave properties and a corresponding wavelength.

Because the momentum of a particle of mass m and velocity v is $p = mv$, the **de Broglie wavelength** of the particle is

de Broglie wavelength

$$\lambda = \frac{h}{p} = \frac{h}{mv} \qquad [27.16]$$

Furthermore, in analogy with photons, de Broglie postulated that the frequencies of matter waves (that is, waves associated with real particles) obey the Einstein relation $E = hf$, so that

$$f = \frac{E}{h} \qquad [27.17]$$

The dual nature of matter is quite apparent in Equations 27.16 and 27.17. That is, each equation contains both particle concepts (mv and E) and wave concepts (λ and f).

De Broglie's proposal that all particles exhibit both wave and particle properties was first regarded as pure speculation. If particles such as electrons had wave properties, then under the correct conditions they should exhibit interference phenomena. Three years later, in 1927, C. J. Davisson and L. Germer of the United States discovered that electrons can be diffracted. This important discovery provided the first experimental confirmation of the matter waves proposed by de Broglie. In their original experiment, low-energy electrons (about 54 eV) were scattered from a single crystal of nickel. The electrons were scattered in preferred directions, as evidenced by intensity peaks at certain scattering angles. The results were explained by recognizing that the regularly spaced planes of atoms in the crystal acted as a diffraction grating for the electron waves, as discussed in Section 27.5.

Louis de Broglie (1892–1987), a French physicist, was awarded the Nobel prize in 1929 for his discovery of the wave nature of electrons. He was the scion of an aristocratic French family that produced marshals, ambassadors, foreign ministers, and at least one duke, his older brother Maurice de Broglie. Louis de Broglie came rather late to theoretical physics, as he first studied history. Only after serving as a radio operator in World War I did he follow the lead of his older brother and begin his studies of physics. Maurice de Broglie was an outstanding experimental physicist in his own right and conducted experiments in the family mansion in Paris. *(AIP Niels Bohr Library)*

EXAMPLE 27.8   **The Wavelength of an Electron**

Calculate the de Broglie wavelength for an electron ($m = 9.11 \times 10^{-31}$ kg) moving at 1.00×10^7 m/s.

Solution   Equation 27.16 gives

$$\lambda = \frac{h}{mv} = \frac{6.63 \times 10^{-34}\,\text{J}\cdot\text{s}}{(9.11 \times 10^{-31}\,\text{kg})(1.00 \times 10^7\,\text{m/s})} = \boxed{7.28 \times 10^{-11}\,\text{m}}$$

This wavelength corresponds to that of x-rays in the electromagnetic spectrum.

Exercise   Find the de Broglie wavelength of a proton ($m = 1.67 \times 10^{-27}$ kg) moving with a speed of 1.00×10^7 m/s.

Answer 3.97×10^{-14} m.

EXAMPLE 27.9   The Wavelength of a Rock

A rock of mass 50.0 g is thrown with a speed of 40.0 m/s. What is the de Broglie wavelength of the rock?

Solution   From Equation 27.16, we have

$$\lambda = \frac{h}{mv} = \frac{6.63 \times 10^{-34}\,\text{J·s}}{(50.0 \times 10^{-3}\,\text{kg})(40.0\,\text{m/s})} = 3.32 \times 10^{-34}\,\text{m}$$

Notice that this wavelength is much smaller than any possible aperture through which the rock could pass. This means that we could not observe diffraction effects, and as a result the wave properties of large-scale objects cannot be observed.

SPECIAL TOPIC

THE ELECTRON MICROSCOPE

A practical device that relies on the wave characteristics of electrons is the **electron microscope** (Fig. 27.20a), which is in many respects similar to an ordinary compound microscope. One important difference is that the electron microscope has a much greater resolving power because electrons can be accelerated to very high kinetic energies, giving them a very short wavelength. Any microscope is capable of detecting details that are comparable in size to the wavelength of the radiation used to illuminate the object. Typically, the wavelengths of electrons are about 100 times shorter than those of the visible light used in optical microscopes. As a result, electron microscopes are able to distinguish details about 100 times smaller.

In operation, a beam of electrons falls on a thin slice of the material to be examined. The sample must be very thin, typically a few hundred angstroms, in order to minimize undesirable effects such as absorption or scattering of the electrons. The electron beam is controlled by electrostatic or magnetic deflection, which acts on the charges to focus the beam to an image. Rather than examining the image through an eyepiece as in an ordinary microscope, a magnetic lens forms an image on a fluorescent screen. The fluorescent screen is necessary because the image produced would not otherwise be visible. An example of a photograph taken by an electron microscope is shown in Figure 27.20b.

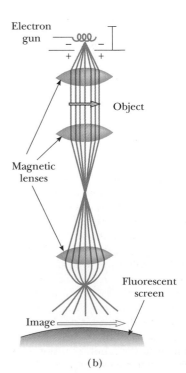

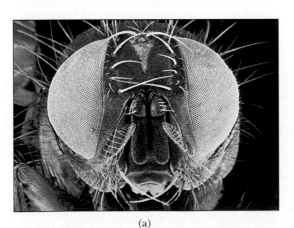

FIGURE 27.20

(a) Scanning electron micrograph of the head of a female housefly. The head is dominated by large eyes that contain around 4000 image-forming elements known as ommatidia. Each of these elements contains a lens and an array of light-sensitive cells. *(Dr. Jeremy Burgess/SPL/ Photo Researchers).* (b) Diagram of an electron microscope. The "lenses" that control the electron beam are magnetic deflection coils.

(a)

(b)

27.10

THE WAVE FUNCTION

De Broglie's revolutionary idea that particles should have a wave nature soon moved out of the realm of skepticism to the point where it was viewed as a necessary concept in understanding the subatomic world. In 1926 the Austrian-German physicist Erwin Schrödinger proposed a wave equation that described the manner in which matter waves change in space and time. The Schrödinger wave equation represents a key element in the theory of quantum mechanics. Its role is as important in quantum mechanics as that played by Newton's laws of motion in classical mechanics. Schrödinger's equation has been successfully applied to the hydrogen atom and to many other microscopic systems. Its importance in most aspects of modern physics cannot be overemphasized.

We shall not go through a mathematical derivation of Schrödinger's wave equation, nor shall we even state the equation here since it involves mathematical operations beyond the scope of this text. When we attempt to solve the Schrödinger equation, the basic entity we seek to determine is a quantity, Ψ, called the **wave function**. Each particle is represented by a wave function Ψ that depends both on the position of the object and on time. Once Ψ is found, what information about the particle does it give? To answer this question, let us consider an analogy with light.

In Chapter 24 we discussed Young's double-slit experiment and explained experimental observations of the interference pattern solely in terms of the wave nature of light. Let us now discuss this same experiment in terms of both the wave and particle nature of light.

First, recall from Chapter 21 that the intensity of a light beam is proportional to the square of the electric field strength, E, associated with the beam. That is, $I \propto E^2$. According to the wave model of light, there are certain points on the viewing screen where the net electric field is zero due to destructive interference of waves from the two slits. Because E is zero at these points, the intensity is also zero, and the screen is dark at these locations. Likewise, at points on the screen where constructive interference occurs, E is large, as is the intensity; hence these locations are bright.

Now consider the same experiment when light is viewed as having a particle nature. It seems reasonable to assume that the number of photons reaching a point on the screen per second increases as the intensity (brightness) increases. Thus, the number of photons that strikes a unit area on the screen each second is proportional to the square of the electric field, or $N \propto E^2$. Now let us consider the behavior of a single photon. What will be the fate of the photon as it moves through the slits in Young's experiment? From a probabilistic point of view, a photon has a high probability of striking the screen at a point where the intensity (and E^2) is high, and a low probability of striking the screen where the intensity is low.

When describing particles rather than photons, Ψ rather than E plays the role of the amplitude. Using an analogy with the description of light, we make the following interpretation of Ψ for particles: If Ψ is a wave function used to describe a single particle, the value of Ψ^2 at some location at a given time is proportional to the probability of finding the particle at that location at that time.

This photograph of Werner Heisenberg (1901–1976) was taken around 1924 (University of Hamburg). Heisenberg obtained his Ph.D. in 1923 at the University of Munich, where he studied under Arnold Sommerfeld and became an enthusiastic mountain climber and skier. Later he worked as an assistant to Max Born at Göttingen and Niels Bohr in Copenhagen. While physicists such as de Broglie and Schrödinger tried to develop pictorialized models of the atom, Heisenberg, with the help of Born and Pascual Jordan, developed an abstract mathematical model called matrix mechanics to explain the wavelengths of spectral lines. The more successful wave mechanics model published by Schrödinger a few months later was shown to be equivalent to Heisenberg's approach. Heisenberg made many other significant contributions to physics, including his famous uncertainty principle, which earned him the Nobel Prize in 1932, his prediction of two forms of molecular hydrogen, and theoretical models of the nucleus. During World War II he was director of the Max Planck Institute at Berlin, where he was in charge of German research on atomic weapons. Following the war, he moved to West Germany and became director of the Max Planck Institute for Physics at Göttingen.

**Werner Heisenberg
(1901–1976)**

27.11

THE UNCERTAINTY PRINCIPLE

If you were to measure the position and velocity of a particle at any instant, you would always be faced with reducing the experimental uncertainties in your measurements as much as possible. According to classical mechanics, there is no fundamental barrier to an ultimate refinement of the apparatus and/or experimental procedures. That is, in principle it would be possible to make such measurements with arbitrarily small uncertainty or with infinite accuracy. Quantum theory predicts, however, that

> **It is fundamentally impossible to make simultaneous measurements of a particle's position and velocity with infinite accuracy.**

This statement, known as the **uncertainty principle,** was first derived by Werner Heisenberg in 1927.

Consider a particle moving along the x axis and suppose that Δx and Δp represent the uncertainty in the measured values of the particle's position and momentum, respectively, at some instant. The uncertainty principle says that the product $\Delta x \, \Delta p$ is never less than a number of the order of Planck's constant. More specifically,

$$\Delta x \, \Delta p \geq \frac{h}{4\pi}$$

[27.18]   Uncertainty principle

**Erwin Schrödinger
(1887–1961)**

Erwin Schrödinger (1887–1961) was an Austrian theoretical physicist best known as the creator of wave mechanics. He also produced important papers in the fields of statistical mechanics, color vision, and general relativity. Schrödinger did much to hasten the universal acceptance of quantum theory by demonstrating the mathematical equivalence between his wave mechanics and the more abstract matrix mechanics developed by Heisenberg.

Although Schrödinger's wave theory was generally based on clear physical ideas, one of its major shortcomings in 1926 was the physical interpretation of the wave function Ψ. Schrödinger felt that the electron was ultimately a wave, Ψ was the vibration amplitude of this wave, and Ψ^2 was the electric charge density. Born, Bohr, Heisenberg, and others pointed out the problems with this interpretation and presented the currently accepted view that Ψ^2 is a probability, and that the electron is ultimately no more a wave than a particle. Schrödinger never accepted this view, but registered his "concern and disappointment" that this "transcendental, almost psychical interpretation" had become "universally accepted dogma."

In 1927 Schrödinger accepted the chair of theoretical physics at the University of Berlin, where he formed a close friendship with Max Planck. He left Germany in 1933 and eventually settled at the Dublin Institute of Advanced Study. After spending 17 happy, creative years there working on problems in general relativity, cosmology, and the application of quantum physics to biology, he returned home in 1956 to his beloved Tirolean mountains in Austria, where he died in 1961.

That is, *it is physically impossible to measure simultaneously the exact position and exact momentum of a particle.* If Δx is made very small, Δp will be large, and vice versa.

In order to understand the uncertainty principle, consider the following thought experiment. Suppose you wish to measure the position and momentum of an electron as accurately as possible. You might be able to do this by viewing the electron with a powerful microscope. In order for you to see the electron and thus determine its location, at least one photon of light must bounce off the electron and pass through the microscope into your eye. This incident photon is shown moving toward the electron in Figure 27.21a. When the photon strikes the electron, as in Figure 27.21b, it transfers some of its energy and momentum to the electron. Thus, in the process of attempting to locate the electron very accurately (that is, by making Δx very small), we have caused a rather large uncertainty in its momentum. In other words, *the measurement procedure itself limits the accuracy to which we can determine position and momentum simultaneously.*

Let us analyze the collision between the photon and the electron by first noting that the incoming photon has a momentum of h/λ. As a result of the collision, the photon transfers part or all of its momentum to the electron. Thus, the uncertainty in the electron's momentum after the collision is at least as great as the momentum of the incoming photon. That is, $\Delta p = h/\lambda$. Furthermore, since light also has wave properties, we would expect the uncertainty in the position of the electron to be on the order of one wavelength of the light being

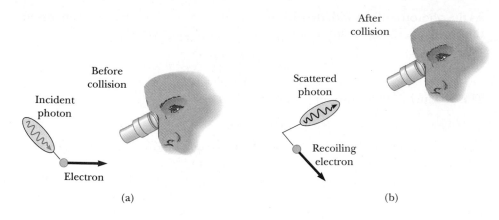

After
collision

Before
collision

Scattered
photon

Incident
photon

Recoiling
electron

Electron

(a)   (b)

FIGURE 27.21
A thought experiment for viewing an electron with a powerful
microscope. (a) The electron is
viewed before colliding with the
photon. (b) The electron recoils
(is disturbed) as the result of
the collision with the photon.

used to view it, because of diffraction effects. Thus, $\Delta x = \lambda$. Multiplying these
two uncertainties gives

$$\Delta x \, \Delta p = \lambda \left(\frac{h}{\lambda} \right) = h$$

This represents the minimum in the products of the uncertainties. Since the
uncertainty can always be greater than this minimum, we have

$$\Delta x \, \Delta p \geqslant h$$

This agrees with Equation 27.18 (apart from a numerical factor introduced by
Heisenberg's more precise analysis).

Another form of the uncertainty principle that applies to the simultaneous
measurement of energy and time is

$$\Delta E \, \Delta t \geqslant \frac{h}{4\pi} \qquad\qquad \textbf{[27.19]}$$

where ΔE is the uncertainty in a measurement of the energy and Δt is the
time it takes to make the measurement. It can be inferred from this relationship
that the energy of a particle cannot be measured with complete precision in an
infinitely short interval of time. Thus, when an electron is viewed as a particle,
the uncertainty principle tells us that (a) its position and velocity cannot both be
known precisely at the same time and (b) its energy can be uncertain (or may not
be conserved) for a period of time given by $\Delta t = h/(4\pi \, \Delta E)$.

We can arrive at the uncertainty principle in the form of Equation 27.19 by
returning to Figure 27.21. Recall that the uncertainty in the position of the
electron being viewed by the microscope is on the order of the wavelength of the
light being used to detect the electron. The photon used to detect the electron
travels with a speed c; therefore the time it takes the photon to travel a distance
Δx is given by

$$\Delta t \approx \frac{\Delta x}{c} \approx \frac{\lambda}{c}$$

As the photon collides with the electron, it can transfer all or part of its energy to the electron. Thus, the uncertainty in the energy transferred is

$$\Delta E \approx hf \approx \frac{hc}{\lambda}$$

The product of these two uncertainties is

$$\Delta E\, \Delta t \approx h$$

Again, the result of our approximate analysis agrees with Heisenberg's more accurate and detailed analysis within a small numerical factor.

Heisenberg's uncertainty principle enables us to better understand the dual wave-particle nature of light and matter. We have seen that the wave description is quite different from the particle description. Therefore, if an experiment is designed to reveal the particle character of an electron (such as the photoelectric effect), its wave character will become fuzzy. Likewise, if the experiment is designed to accurately measure the electron's wave properties (such as diffraction from a crystal), its particle character will become fuzzy.

EXAMPLE 27.10   Locating an Electron

The speed of an electron is measured to be 5.00×10^3 m/s to an accuracy of $0.003\ 00\%$. Find the uncertainty in determining the position of this electron.

Solution   The momentum of the electron is

$$p = mv = (9.11 \times 10^{-31}\ \text{kg})(5.00 \times 10^3\ \text{m/s}) = 4.56 \times 10^{-27}\ \text{kg·m/s}$$

Because the uncertainty in p is $0.00\ 300\%$ of this value, we get

$$\Delta p = 0.000\ 0300p = (0.000\ 0300)(4.56 \times 10^{-27}\ \text{kg·m/s})$$
$$= 1.37 \times 10^{-31}\ \text{kg·m/s}$$

The uncertainty in position can now be calculated by using this value of Δp and Equation 27.18:

$$\Delta x\, \Delta p \geqslant \frac{h}{4\pi}$$

$$\Delta x \geqslant \frac{h}{4\pi\, \Delta p} = \frac{6.63 \times 10^{-34}\ \text{J·s}}{4\pi(1.37 \times 10^{-31}\ \text{kg·m/s})}$$

$$= 0.385 \times 10^{-3}\ \text{m} = \boxed{0.385\ \text{mm}}$$

EXAMPLE 27.11   Excited States of Atoms

As we shall see in the next chapter, electrons in atoms can be found in higher states of energy called excited states for short periods of time. If the average time that an electron exists in one of these states is 1.00×10^{-8} s, what is the minimum uncertainty in energy of the excited state?

Solution   From the uncertainty principle in the form of Equation 27.19, we find that the minimum uncertainty in energy is

$$\Delta E = \frac{h}{4\pi\, \Delta t} = \frac{(6.63 \times 10^{-34}\ \text{J·s})}{4\pi(1.00 \times 10^{-8}\ \text{s})} = 5.28 \times 10^{-27}\ \text{J} = \boxed{3.30 \times 10^{-8}\ \text{eV}}$$

SUMMARY

The characteristics of **blackbody radiation** cannot be explained using classical concepts. The peak of a blackbody radiation curve is given by **Wien's displacement law:**

$$\lambda_{\max} T = 0.2898 \times 10^{-2} \ \text{m} \cdot \text{K} \qquad \textbf{[27.1]}$$

where $\lambda_{\max}$ is the wavelength at which the curve peaks and T is the absolute temperature of the object emitting the radiation.

Planck first introduced the quantum concept when he assumed that the vibrating molecules responsible for blackbody radiation could have only discrete amounts of energy given by

$$E_n = nhf \qquad \textbf{[27.2]}$$

where n is a positive integer called a **quantum number** and f is the frequency of vibration of the molecule.

The **photoelectric effect** is a process whereby electrons are ejected from a metal surface when light is incident on that surface. Einstein provided a successful explanation of this effect by extending Planck's quantum hypothesis to electromagnetic waves. In this model, light is viewed as a stream of particles called photons, each with energy $E = hf$, where f is the frequency and h is **Planck's constant.** The maximum kinetic energy of the ejected photoelectrons is

$$KE_{\max} = hf - \phi \qquad \textbf{[27.7]}$$

where ϕ is the **work function** of the metal.

X-rays are produced when high-speed electrons are suddenly decelerated. When electrons are accelerated through a voltage V, the shortest-wavelength radiation that can be produced is

$$\lambda_{\min} = \frac{hc}{eV} \qquad \textbf{[27.10]}$$

The regular array of atoms in a crystal can act as a diffraction grating for x-rays. The condition for constructive interference of the diffracted rays is given by **Bragg's law:**

$$2d \sin \theta = m\lambda \qquad (m = 1, 2, 3, \ . \ . \ .) \qquad \textbf{[27.11]}$$

X-rays from an incident beam are scattered at various angles by electrons in a target such as carbon. In such a scattering event, a shift in wavelength is observed for the scattered x-rays. This phenomenon is known as the **Compton shift.** Conservation of momentum applied to a photon-electron collision yields the following expression for the shift in wavelength of the scattered x-rays:

$$\Delta \lambda = \lambda - \lambda_0 = \frac{h}{m_0 c} (1 - \cos \theta) \qquad \textbf{[27.12]}$$

where m_0 is the rest mass of the electron, c is the speed of light, and θ is the scattering angle.

Pair production is a process in which the energy of a photon is converted into rest mass. In this process, the photon disappears as an electron-positron pair is created. Likewise, the energy of an electron-positron pair can be converted into electromagnetic radiation by the process of **pair annihilation.**

De Broglie proposed that all matter has both a particle and a wave nature. The **de Broglie wavelength** of any particle of mass m and speed v is

$$\lambda = \frac{h}{p} = \frac{h}{mv}$$ [27.16]

De Broglie also proposed that the frequencies of the waves associated with particles obey the Einstein relationship, $E = hf$.

In the theory of **quantum mechanics,** each particle is described by a quantity Ψ called the **wave function.** The probability of finding the particle at a particular point at some instant is proportional to Ψ^2. Quantum mechanics is very successful in describing the behavior of atomic and molecular processes.

According to Heisenberg's **uncertainty principle,** it is impossible to measure simultaneously the exact position and exact momentum of a particle. If Δx is the uncertainty in the measured position and Δp the uncertainty in the momentum, the product $\Delta x \, \Delta p$ is given by

$$\Delta x \, \Delta p \geqslant \frac{h}{4\pi}$$ [27.18]

or by

$$\Delta E \, \Delta t \geqslant \frac{h}{4\pi}$$ [27.19]

where ΔE is the uncertainty in the energy of the particle and Δt is the uncertainty in the time it takes to measure the energy.

ADDITIONAL READING

A. Baker, *Modern Physics and Antiphysics,* Reading, Mass., Addison-Wesley, 1970.

D. C. Cassidy, "Heisenberg, Uncertainty and the Quantum Revolution," *Sci. American,* May 1992, p. 44.

L. de Broglie, "The Revolution in Physics: A Non-Mathematical Survey of Quanta," Noonday Press, 1953.

G. Gamow, *Thirty Years That Shook Physics,* New York, 1966. Doubleday, Anchor Books, 1966.

T. Goldman, R. J. Hughes, and M. M. Nieto, "Gravity and Antimatter," *Sci. American,* March 1988, p. 48.

W. Heisenberg, *Physics and Beyond,* New York, Harper and Row, 1971 (a "biography" of quantum mechanics).

B. Hoffman, *Strange Story of the Quantum,* New York, Dover, 1959.

M. R. Howells, J. Kirz, and D. Sayre, "X-ray Microscopes," *Sci. American,* February 1991, p. 88.

A. Shimony, "The Reality of the Quantum World," *Sci. American,* January 1988, p. 46.

B. Wheaton, "Louis de Broglie and the Origin of Wave Mechanics," *The Physics Teacher,* May 1984, p. 297.

H. K. Wickramasinghe, "Scanned-Probe Microscopes," *Sci. American,* October 1989, p. 98.

CONCEPTUAL QUESTIONS

Example   All objects radiate energy. Why, then, are we not able to see all objects in a dark room?

Reasoning   All objects do radiate energy, but at room temperature, this energy is primarily in the infrared region of the spectrum, which our eyes cannot detect. Some snakes have a membrane located above their eyes that is sensitive to infrared radiation; thus they can seek out their prey in what we would consider to be complete darkness.

Example   The brightest star in the constellation Lyra is the bluish star Vega, while the brightest star in Boötes is the reddish star Arcturus. How do you account for the difference in color of the two stars?

Reasoning In general, the stars with the highest surface temperature produce photons with the highest energy and frequency. (See Wien's law). Thus, the color of a star is an indication of its surface temperature. You might want to spend a few moments scanning the night sky with binoculars or a small telescope to see how many different colors of stars you can find.

Example An x-ray photon is scattered by an electron. What happens to the frequency of the scattered photon relative to that of the incident photon?

Reasoning Part of the energy of the x-ray photon is transferred to the electron. Thus, the energy of the scattered photon decreases, and its frequency decreases relative to that of the incident photon.

Example In the photoelectric effect, explain why the stopping potential depends on the frequency of light but not on the intensity.

Reasoning We can picture higher frequency light as a stream of photons of higher energy. In a collision, one photon gives all of its energy to a single electron. The kinetic energy of such an electron is measured by the stopping potential. The reverse voltage (stopping potential) required to stop the current is proportional to the frequency of the incoming light. More intense light consists of more photons striking a unit area each second, but atoms are so small that one emitted electron never gets a "kick" from more than one photon. Increasing the light intensity will generally increase the size of the current but will not change the energy of the individual ejected electrons. Thus, the stopping potential stays constant.

Example Why is an electron microscope more suitable than an optical microscope for "seeing" objects of an atomic size?

Reasoning A microscope can see details no smaller than the wavelength of the waves it uses to produce images. Electrons with kinetic energies of several electron volts have wavelengths of less than a nanometer, which is much smaller than the wavelength of visible light (having wavelengths ranging from about 400 to 700 nm.) Therefore, an electron microscope can resolve details of much smaller sizes as compared to an optical microscope.

Example If the photoelectric effect is observed for one metal using light of a certain wavelength, can you conclude that the effect will also be observed for another metal under the same conditions?

Reasoning No. The second metal may have a higher work function than the first metal, in which case the incident photons may not have enough energy to eject photoelectrons.

1. When wood is stacked on a special elevated grate in a fireplace, a pocket of burning wood that is hotter than the burning wood at the top of the stack forms beneath the grate. Explain how this device provides more heat to the room than does a conventional fire, thus increasing the efficiency of the fireplace.

2. What assumptions were made by Planck in dealing with the problem of blackbody radiation? Discuss the consequences of these assumptions.

3. Which has more energy, a photon of ultraviolet radiation or a photon of yellow light?

4. What effect, if any, would you expect the temperature of a material to have on the ease with which electrons can be ejected from it in the photoelectric effect?

5. In the photoelectric effect, explain why the photocurrent depends on the intensity of the light source but not on the frequency.

6. If a photon is deflected by the Compton effect, can its wavelength ever become shorter?

7. What assumptions were made by Compton in dealing with the scattering of a photon from an electron?

8. Is light a wave or a particle? Support your answer by citing specific experimental evidence.

9. Suppose a photograph were made of a person's face using only a few photons. Would the result be simply a very faint image of the face? Discuss.

10. An electron and a proton are accelerated from rest through the same potential difference. Which particle has the longer wavelength?

11. If matter has a wave nature, why is this wave-like character not observable in our daily experiences?

12. Is an electron a particle or a wave? Support your answer by citing some experimental results.

13. Why is it impossible to simultaneously measure the position and velocity of a particle with infinite accuracy?

14. Suppose there are two electrons inside a hollow box. Can we determine which electron is which? Discuss this in terms of the attributes that uniquely identify each electron and the uncertainty principle.

15. Discuss whether the behavior of an electron is mainly wave-like or particle-like in each of the following situations: (a) traversing a circular orbit in a magnetic field; (b) absorbing a photon and being photoelectrically ejected from the surface of a metal; (c) forming an interference pattern.

16. The Wien displacement law says that the peak of a blackbody radiation curve, when viewed as a function of the *wavelength* of radiation emitted, occurs at λ_{max}, where $\lambda_{max} T = 2.898$ mm K. If this blackbody radiation curve is viewed as a function of the *frequency* of the radiation emitted, it peaks at f_{max}, where $f_{max} / T = 58.79$ GHz/K. Computing $\lambda_{max} f_{max}$ leads to $\lambda_{max} f_{max} = 0.568c \neq c$. Comment on why this product does not give c.

17. If a photon travels from a vacuum to a medium having index of refraction n, which of the following are changed: (a) its energy, (b) its momentum. Explain.

18. Why is the photoelectric effect studied using conservation of energy and not conservation of momentum?

19. Does the process of pair production violate conservation of mass?

20. Figure 27.16 shows one x-ray beam reflecting off the surface of the crystal (upper plane) and one penetrating the crystal and reflecting off the lower plane. The derivation of Bragg's law looks similar to the derivation of the thin film equations from Chapter 24. Why isn't the index of refraction for the crystal considered in the calculation of Bragg's law?

PROBLEMS

Section 27.1   Blackbody Radiation and Planck's Hypothesis

1. (a) Assuming that the tungsten filament of a lightbulb is a black body, determine its peak wavelength if its temperature is 2900 K. (b) Why does your answer to part (a) suggest that more energy from a lightbulb goes into heat than into light?

2. Calculate the energy in electron volts of a photon having a wavelength in (a) the microwave range, 5.00 cm, (b) the visible light range, 5.00×10^{-7} m, and (c) the x-ray range, 1.00×10^{-8} m.

3. The threshold of dark-adapted (scotopic) vision is 4.0×10^{-11} W/m² at a central wavelength of 500 nm. If light with this intensity and wavelength enters the eye when the pupil is open to its maximum diameter of 8.5 mm, how many photons per second enter the eye?

4. Figure 27.22 shows the spectrum of light emitted by a firefly. Determine the temperature of a black body that would emit radiation peaked at the same frequency. Based on your result, would you say firefly radiation is blackbody radiation?

5. If the surface temperature of the Sun is 5800 K, find the wavelength that corresponds to the maximum rate of energy emission from the Sun.

6. A certain light source is found to emit radiation whose peak value has a frequency of 1.00×10^{15} Hz. Find the temperature of the source.

7. A quantum of electromagnetic radiation has an energy of 2.0 keV. What is its wavelength?

8. A 0.50-kg mass falls from a height of 3.0 m. If all of the energy of this mass could be converted to visible light of wavelength 5.0×10^{-7} m, how many photons would be produced?

9. A star that is moving away from the Earth at a speed of $0.280\,c$ is observed to emit radiation with a peak wavelength value of 500 nm. Determine the surface temperature of this star. (*Hint:* See Problem 63 in Chapter 26.)

10. A 1.5-kg mass vibrates at an amplitude of 3.0 cm on the end of a spring of spring constant 20.0 N/m. (a) If the energy of the spring is quantized, find its quantum number. (b) If n changes by 1, find the fractional change in energy of the spring.

Section 27.2   The Photoelectric Effect

11. From the scattering of sunlight, Thomson calculated that the classical radius of the electron has a value of 2.82×10^{-15} m. If sunlight having an intensity of 500 W/m² falls on a disk with this radius, estimate the time required to accumulate 1.0 eV of energy. Assume that light is a classical wave and that the light striking the disk is completely absorbed. How does your estimate compare with the observation that photoelectrons are promptly (within 10^{-9} s) emitted?

12. Ultraviolet light is incident normally on the surface of a certain substance. The binding energy of the electrons in this substance is 3.44 eV. The incident light has an intensity of 0.055 W/m². The electrons are photoelectrically emitted with a maximum speed of 4.2×10^5 m/s. How many electrons are emitted from a square centimeter of the surface? Assume that 100% of the photons are absorbed.

13. When light of wavelength 445 nm strikes a certain metal surface, the stopping potential is 70.0% of that which results when light of wavelength 410 nm strikes the same metal surface. Based on this information and the following table of work functions, identify the metal involved in the experiment.

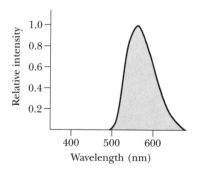

FIGURE 27.22   (Problem 4)

Metal	Work Function (eV)
Cesium	1.90
Potassium	2.23
Silver	4.73
Tungsten	4.58

14. When light of wavelength 350 nm falls on a potassium surface, electrons are emitted that have a maximum kinetic energy of 1.31 eV. Find (a) the work function of potassium, (b) the cutoff wavelength, and (c) the frequency corresponding to the cutoff wavelength.

15. Electrons are ejected from a metallic surface with speeds ranging up to 4.6×10^5 m/s when light with a wavelength of $\lambda = 625$ nm is used. (a) What is the work function of the surface? (b) What is the cutoff frequency for this surface?

16. When a certain metal is illuminated with light of frequency 3.0×10^{15} Hz, a stopping potential of 7.0 V is required to stop the ejected electrons. What is the work function of this metal?

17. Consider the metals lithium, iron, and mercury, which have work functions of 2.3 eV, 3.9 eV, and 4.5 eV, respectively. If light of wavelength 3.0×10^{-7} m is incident on each of these metals, determine (a) which metals exhibit the photoelectric effect and (b) the maximum kinetic energy for the photoelectrons for those that exhibit the effect.

18. When light of wavelength 253.7 nm falls on cesium, the required stopping potential is 3.00 V. If light of wavelength 435.8 nm is used, the stopping potential is 0.900 V. Use this information to plot a graph like that shown in Figure 27.6, and from the graph determine the cutoff frequency for cesium and its work function.

19. A light source emitting radiation at a frequency of 7.0×10^{14} Hz is incapable of ejecting photoelectrons from a certain metal. In an attempt to use this source to eject photoelectrons from the metal, the source is given a velocity toward the metal. (a) Explain how this procedure produces photoelectrons. (*Hint:* See Problem 63 in Chapter 26.) (b) When the speed of the light source is equal to $0.28c$, photoelectrons just begin to be ejected from the metal. What is the work function of the metal? (c) When the speed of the light source is increased to $0.90c$, determine the maximum kinetic energy of the photoelectrons.

20. Red light of wavelength 670.0 nm produces photoelectrons from a certain photoemissive material. Green light of wavelength 520.0 nm produces photoelectrons from the same material with 1.50 times the maximum kinetic energy. What is the material's work function?

Section 27.4   X-Rays

21. How fast must an electron be moving if all its kinetic energy is lost to a single x-ray photon (a) at the high end of the x-ray electromagnetic spectrum with a wavelength of 1.00×10^{-8} m; (b) at the low end of the x-ray electromagnetic spectrum with a wavelength of 1.00×10^{-13} m?

22. What minimum accelerating voltage would be required to produce an x-ray with a wavelength of 0.0300 nm?

23. Calculate the minimum wavelength x-ray that can be produced when a target is struck by an electron that has been accelerated through a potential difference of (a) 15.0 kV; (b) 1.00×10^5 V.

24. The extremes of the x-ray portion of the electromagnetic spectrum range from approximately 1.0×10^{-8} m to 1.0×10^{-13} m. Find the minimum accelerating voltages required to produce wavelengths at these two extremes.

Section 27.5   Diffraction of X-Rays by Crystals

25. The highest-quality diffraction gratings have about 25 000 grooves/cm. Suppose that such a grating were used in the standard transmission mode (Fig. 25.11) to analyze x-rays with a wavelength of 0.50 Å. At what angle θ would the first-order maximum appear? Comment on your result.

26. The spacing between planes of nickel atoms in a nickel crystal is 0.352 nm. At what angle does a second-order Bragg reflection occur in nickel for 11.3-keV x-rays?

27. As a single crystal is rotated in an x-ray spectrometer (Fig. 27.14), many parallel planes of atoms besides *AA* and *BB* produce strong diffraction beams. Two such planes are shown in Figure 27.23. Determine geometrically the interplanar spacings d_1 and d_2 in terms of d_0.

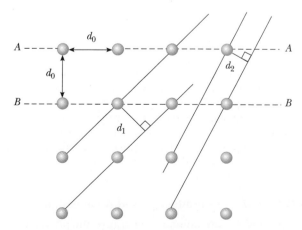

FIGURE 27.23 (Problem 27)

28. Potassium iodide has an interplanar spacing of $d = 0.296$ nm. A monochromatic x-ray beam shows a first-order diffraction maximum when the angle of incidence is 7.6°. Calculate the x-ray wavelength.

29. A monochromatic x-ray beam is incident on a NaCl crystal surface where $d = 0.353$ nm. The second-order maximum in the reflected beam is found when the angle between the incident beam and the surface is 20.5°. Determine the wavelength of the x-rays.

30. X-rays of wavelength 0.140 nm are reflected from a certain crystal, and the first-order maximum occurs at an angle of 14.4°. What value does this give for the interplanar spacing of this crystal?

Section 27.6   Compton Scattering

31. An electron initially at rest recoils after a head-on collision with a 6.2-keV photon. Determine the kinetic energy acquired by the electron.

32. Show that the Compton wavelength has the numerical value 0.002 43 nm.

33. An incident photon with energy 0.88 MeV is scattered by a free electron initially at rest in such a way that the scattering angle of the scattered electron equals that of the scattered photon ($\phi = \theta$ in Fig. 27.17). (a) Determine the angles ϕ and θ. (b) Determine the energy and momentum of the scattered photon. (c) Determine the kinetic energy and momentum of the scattered electron. (*Hint:* See Problem 63.)

34. X-rays are scattered from electrons in a carbon target. The measured wavelength shift is 0.0012 nm. Calculate the scattering angle.

35. A beam of 0.68-nm photons undergoes Compton scattering from free electrons. What are the energy and momentum of the photons that emerge at a 45° angle with respect to the incident beam?

36. A 0.45-nm x-ray photon is deflected through a 23° angle after scattering from a free electron. (a) What is the kinetic energy of the recoiling electron? (b) What is its speed?

37. After a 0.80 nm x-ray photon scatters from a free electron, the electron recoils with a speed equal to 1.4×10^6 m/s. (a) What was the Compton shift in the photon's wavelength? (b) Through what angle was the photon scattered?

38. A 0.0016-nm photon scatters from a free electron. For what (photon) scattering angle will the recoiling electron and scattered photon have the same kinetic energy?

Section 27.7   Pair Production and Annihilation

39. Find (a) the minimum energy of the photon required to produce a proton-antiproton pair and (b) the wavelength of this radiation.

40. An electron moving at a speed of $0.60c$ collides head on with a positron also moving at $0.60c$. Determine the energy and momentum of each photon produced in the process.

41. How much total kinetic energy will an electron-positron pair have if produced by a photon of energy 3.00 MeV?

42. If an electron-positron pair with a total kinetic energy of 2.50 MeV is produced, find (a) the energy of the photon that produced the pair and (b) its frequency.

43. Two photons are produced when a proton and an antiproton annihilate each other. What is the minimum frequency and corresponding wavelength of each photon?

Section 27.9   The Wave Properties of Particles

44. (a) If the wavelength of an electron is equal to 5.00×10^{-7} m, how fast is it moving? (b) If the electron has a speed of 1.00×10^7 m/s, what is its wavelength?

45. After learning about de Broglie's hypothesis that particles of momentum p have wave characteristics with wavelength $\lambda = h/p$, an 80-kg student has grown concerned about being diffracted when passing through a 75-cm-wide doorway. Assuming that significant diffraction occurs when the width of the diffraction aperature is less that 10 times the wavelength of the wave being diffracted, (a) determine the maximum speed at which the student can pass through the doorway in order to be significantly diffracted. (b) With that speed, how long will it take the student to pass through the doorway if it is 15 cm thick? Compare your result to the currently accepted age of the Universe, which is 4×10^{17} s. (c) Should this student worry about being diffracted?

46. Calculate the de Broglie wavelength of a proton moving at (a) 1.00×10^4 m/s; (b) 1.00×10^7 m/s.

47. At what speed must an electron move so that its de Broglie wavelength equals its Compton wavelength? (*Hint:* If you get an answer of c, see Problem 63.)

48. The resolving power of a microscope is proportional to the wavelength used. A resolution of approximately 1.0×10^{-11} m (0.10 Å) would be required in order to "see" an atom. (a) If electrons were used (electron microscope), what minimum kinetic energy would be required for the electrons? (b) If photons were used, what minimum photon energy would be needed to obtain 1.0×10^{-11} m resolution?

49. Find the de Broglie wavelength of a ball whose mass is 0.200 kg just before it strikes the Earth after being dropped from a building 50.0 m tall.

50. Calculate the kinetic energy in MeV of electrons having a de Broglie wavelength of 1.0×10^{-3} nm.

51. Through what potential difference would an electron have to be accelerated from rest to give it a de Broglie wavelength of 1.0×10^{-10} m?

52. Show that the de Broglie wavelength of a nonrelativistic electron accelerated from rest through a potential difference V is given by $\lambda = 1.228/\sqrt{V}$ nm, where V is in volts.

53. A monoenergetic beam of electrons is incident on a single slit of width 0.500 nm. A diffraction pattern is formed on a screen 20.0 cm from the slit. If the distance between successive minima of the diffraction pattern is 2.10 cm, what is the energy of the incident electrons?

54. De Broglie postulated that the relationship $\lambda = h/p$ is valid for relativistic particles. What is the de Broglie wavelength for a (relativistic) electron whose kinetic energy is 3.00 MeV?

Section 27.11 The Uncertainty Principle

55. A 50.0-g ball moves at 30.0 m/s. If its speed is measured to an accuracy of 0.10%, what is the minimum uncertainty in its position?

56. A 0.50-kg block rests on the icy surface of a frozen pond, which we can assume to be frictionless. If the location of the block is measured to a precision of 0.50 cm, what speed must the block acquire because of the measurement process?

57. An electron is located on a pinpoint having a diameter of 2.5 μm. What is the minimum uncertainty in the speed of the electron?

58. In the ground state of hydrogen, the uncertainty in the position of the electron is roughly 0.10 nm. If the speed of the electron is on the order of the uncertainty in the speed, how fast is the electron moving?

59. Show that an electron confined to a nucleus of diameter 2.0×10^{-15} m must be studied using Einsteinian relativity, but a proton confined to the same nucleus can be studied using Newtonian relativity.

60. Suppose optical radiation ($\lambda = 5.00 \times 10^{-7}$ m) is used to determine the position of an electron to within the wavelength of the light. What will be the resulting uncertainty in the electron's velocity?

61. (a) Show that the kinetic energy of a nonrelativistic particle can be written in terms of its momentum as $KE = p^2/2m$. (b) Use the results of (a) to find the smallest kinetic energy of a proton confined to a 1.0×10^{-15}-m nucleus.

ADDITIONAL PROBLEMS

62. A photon strikes a metal with a work function of ϕ and produces a photoelectron with a de Broglie wavelength equal to the wavelength of the original photon. (a) Show that the energy of this photon must have been given by $E = \dfrac{\phi(mc^2 - \phi/2)}{mc^2 - \phi}$, where m is the rest mass of the electron. (*Hint:* Begin with conservation of energy, $E + mc^2 = \phi + \sqrt{(pc)^2 + (mc^2)^2}$.) (b) If one of these photons strikes platinum ($\phi = 6.2$ eV), determine the resulting maximum speed of the photoelectron.

63. Show that the speed of a particle having de Broglie wavelength λ and Compton wavelength $\lambda_C = h/(mc)$ is given by $v = \dfrac{c}{\sqrt{1 + (\lambda/\lambda_C)^2}}$.

Hint: Use $p = \dfrac{mv}{\sqrt{1 - v^2/c^2}}$.

64. A 70.0-kg jungle hero swings at the end of a vine at a frequency of 0.50 Hz at 2.0 m/s as he moves through the lowest point on his arc. (a) Assume the energy is quantized and find the quantum number n for the system. (b) Find the energy carried away in a one-quantum change in his energy.

65. How many photons are emitted per second by a 100.0-W sodium lamp if the wavelength of sodium light is 589.3 nm?

66. An x-ray tube is operated at 50 000 V. (a) Find the minimum wavelength of the radiation emitted by this tube. (b) If this radiation is directed at a crystal, the first-order maximum in the reflected radiation occurs when the angle of incidence is 2.5°. What is the spacing between reflecting planes in the crystal?

67. A light source of wavelength λ illuminates a metal and ejects photoelectrons with a maximum kinetic energy of 1.00 eV. A second light source of wavelength $\lambda/2$ ejects photoelectrons with a maximum kinetic energy of 4.00 eV. What is the work function of the metal?

68. An electron and a proton each have a thermal kinetic energy of $3k_BT/2$. Calculate the de Broglie wavelength of each particle at a temperature of 2000 K. (Recall that k_B is Boltzmann's constant, which has the value $k_B = 1.38 \times 10^{-23}$ J/K.)

69. Photons of wavelength 450 nm are incident on a metal. The most energetic electrons ejected from the metal are bent into a circular arc of radius 20.0 cm by a magnetic field whose strength is 2.00×10^{-5} T. What is the work function of the metal?

70. In the Compton scattering event illustrated in Figure 27.17, the scattered photon has an energy of 120.0 keV and the recoiling electron has a kinetic energy of 40.0 keV. Find (a) the wavelength of the incident photon, (b) the angle θ at which the photon is scattered, and (c) the recoil angle of the electron. (*Hint:* Conserve both mass-energy and relativistic momentum.)

Roger A. Freedman and Paul K. Hansma
University of California, Santa Barbara

The Scanning Tunneling Microscope

The basic idea of quantum mechanics, that particles have wave-like properties and vice versa, is among the strangest found anywhere in science. Because of this strangeness, and because quantum mechanics mostly deals with the very small, it might seem to have little practical application. In fact, however, one of the basic phenomena of quantum mechanics—tunneling—is at the heart of a very practical device, the *scanning tunneling microscope,* or *STM,* which enables us to get highly detailed images of surfaces with resolution comparable to the size of a *single atom.*

Figure 1, an image of the surface of a piece of graphite, shows what the STM can do. Note the high quality of the image and the recognizable rings of carbon atoms. What makes this image so remarkable is that its *resolution*—the size of the smallest detail that can be discerned—is about 0.2 nm. For an ordinary microscope, the resolution is limited by the wavelength of the waves used to make the image. Thus, an optical microscope has a resolution no better than 200 nm, about half the wavelength of visible light, and so could never show the detail displayed in Figure 1. Electron microscopes can have a resolution of 0.2 nm by using electron waves of this wavelength, given by the de Broglie formula $\lambda = h/p$. The electron momentum p required to give this wavelength is 10 000 eV$/c$, corresponding to an electron speed of 2% of the speed of light. Electrons traveling at this speed would penetrate into the interior of the piece of graphite in Figure 1 and so couldn't give us information about individual surface atoms.

The STM achieves its very fine resolution by using the basic idea shown in Figure 2. A conducting probe with a very sharp tip is brought near the surface to be studied. Because it is attracted to the positive ions in the surface, an electron in the surface has a lower total energy than an electron in the empty space between surface and tip. The same thing is true for an electron in the probe tip, which is attracted to the positive ions in the tip. In Newtonian mechanics, this means that electrons cannot move between surface and tip because they lack the energy to escape either material. Because the electrons obey quantum mechanics, however, they can "tunnel" across the barrier of empty space. By applying a voltage between surface and tip, the electrons can be made to tunnel preferentially from surface to tip. In this way the tip samples the distribution of electrons just above the surface.

Because of the nature of tunneling, the STM is very sensitive to the distance z from tip to surface. The reason is that in the empty space between tip and surface, the electron wave function falls off exponentially with a decay length of order 0.1 nm, that is, the wave function decreases by $1/e$ over that distance. For distances z greater than 1 nm (that is, beyond a few atomic diameters), essentially no tunneling takes place. This exponential behavior causes the current of electrons tunneling from surface to tip to depend very strongly on z. This sensitivity is the basis of the operation of the STM: By monitoring the tunneling current as the tip is scanned over the surface, scientists obtain a sensitive measure of the topography of the electron distribution on the surface. The result of this scan is used to make images like that in Figure 1. In this way the STM can measure the height of surface features to within 0.001 nm, approximately $1/100$ of an atomic diameter!

You can see just how sensitive the STM is by looking carefully at Figure 1. You may be able to see that of the six carbon atoms in each ring, three *appear* lower than the other three. In fact all six atoms are at the same level, but they all have slightly different electron distributions. The three atoms that appear lower are bonded to other carbon atoms directly beneath them in the underlying atomic layer, and so their electron distributions—which are responsible for the bonding—extend downward beneath the surface. The atoms in the surface layer that appear higher do not lie directly

894

over subsurface atoms and hence are not bonded to carbon atoms beneath them. For these higher-appearing atoms, the electron distribution extends upward into the space above the surface. This extra electron density is what makes these electrons appear higher in Figure 1, since what the STM maps is the topography of the electron distribution.

The STM has, however, one serious limitation: it depends on electrical conductivity of the sample and the tip. Unfortunately, most materials are not electrically conductive at their surface. Even metals such as aluminum are covered with nonconductive oxides. A newer microscope, the atomic force microscope, or AFM, overcomes this limitation. It measures the force between a tip and the sample rather than an electrical current. This force, which is typically a result of the exclusion principle, depends very strongly on the tip–sample separation just as the electron tunneling current does for the STM. Thus the AFM has comparable sensitivity for measuring topography and has become widely used for technological applications.

Perhaps the most remarkable thing about the STM is that its operation is based on a quantum-mechanical phenomenon—tunneling—that was well understood in the 1920s, even though the first STM was not built until the 1980s. What other applications of quantum mechanics may yet be waiting to be discovered?

Questions

1. Our discussion of STM operation describes only a single tip. In fact, a probe may have a number of protrusions on its end, each of which acts as a tip. Hence it might be expected that such a probe would give multiple STM images of the surface, greatly complicating the analysis. Explain why there is no problem with multiple images provided the multiple tips differ in proximity to the surface by 2 nm or more.

2. The STM, while using physical concepts that date from the 1920s, was not developed until the 1980s. Suggest some reasons why the STM was not invented half a century earlier.

3. It is stated in the essay that the resolution for conventional microscopes is limited by the wavelength used to make the image. To see whether this guideline applies to the STM, estimate the wavelength of an electron in the surface of a conductor and compare your estimate to the vertical resolution (about 0.001 nm) and lateral resolution (about 0.2 nm). Does the guideline apply to these two resolutions? Why or why not?

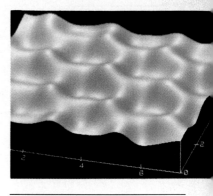

FIGURE 1

The surface of graphite as "viewed" with a scanning tunneling microscope. This technique enables scientists to see small details on surfaces with a lateral resolution of about 0.2 nm and a vertical resolution of 0.001 nm. The contours seen here represent the arrangement of individual carbon atoms on the crystal surface.

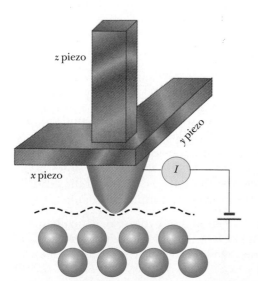

FIGURE 2

A schematic view of an STM. The tip, shown as a rounded cone, is mounted on a piezoelectric x, y, z scanner. A scan of the tip over the sample can reveal contours of the surface down to the atomic level. An STM image is composed of a series of scans displaced laterally from each other. *(Based on a drawing from P. K. Hansma, V. B. Elings, O. Marti, and C. Bracker,* Science *242:209, 1988. Copyright 1988 by the AAAS.)*

28 ATOMIC PHYSICS

An artist's rendition of "bucky-balls," short for buckminsterful-lerene. These nearly spherical molecular structures that look like soccer balls were named for R. Buckminster Fuller, inventor of the geodesic dome. This new form of carbon, C_{60}, was discovered by astrophysicists while investigating the carbon gas that exists between stars. Scientists are actively studying the properties and potential uses of buckminsterfullerene.

A large portion of this chapter is concerned with the study of the hydrogen atom. Although the hydrogen atom is the simplest atomic system, it is especially important for several reasons:

1. Much of what is learned about the hydrogen atom with its single electron can be extended to such single-electron ions as He^+ and Li^{2+}.
2. The hydrogen atom is an ideal system for performing precise comparisons of theory and experiment and for improving our overall understanding of atomic structure.
3. The quantum numbers used to characterize the allowed states of hydrogen can also be used to describe (approximately) the allowed states of more complex atoms. This enables us to understand the periodic table of the elements, one of the greatest triumphs of quantum mechanics.

In this chapter we first discuss the Bohr model of hydrogen, which helps us understand many features of hydrogen but fails to explain many finer details of atomic structure. Next we examine the hydrogen atom from the viewpoint of quantum mechanics and the quantum numbers used to characterize various atomic states. Additionally, we examine the physical significance of the quantum numbers and the effect of a magnetic field on certain quantum states. The Pauli exclusion principle is also presented. This physical principle is extremely important in understanding the properties of complex atoms and the arrangement of elements in the periodic table. Finally, we apply our knowledge of atomic structure to describe the mechanisms involved in the production of x-rays and the operation of a laser.

28.1

EARLY MODELS OF THE ATOM

The model of the atom in the days of Newton was that of a tiny, hard, indestructible sphere, and was a good basis for the kinetic theory of gases. However, new models had to be devised when later experiments revealed the electrical nature of atoms. J. J. Thomson (1856–1940) suggested a model of the atom as a volume of positive charge with electrons embedded throughout the volume, much like the seeds in a watermelon (Fig. 28.1).

In 1911 Geiger and Marsden, under the supervision of Ernest Rutherford (1871–1937), performed a critical experiment showing that Thompson's model could not be correct. In this experiment, a beam of positively charged **alpha particles,** the nuclei of helium atoms, was projected against a thin metal foil, as in Figure 28.2a. The results of the experiment were astounding. Most of the alpha particles passed through the foil as if it were empty space! Furthermore, some of the alpha particles that were deflected from their original direction of travel were scattered through very large angles. Some particles were even deflected backwards reversing their directions. As Rutherford wrote, "It was quite the most incredible event that has ever happened to me in my life. It was almost as incredible as if you fired a 15-inch shell at a piece of tissue paper and it came back and hit you."

Such large deflections were not expected on the basis of Thomson's model, in which a positively charged alpha particle would never come close enough to a large enough positive charge to cause any large-angle deflections. Rutherford

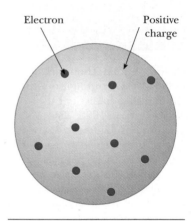

FIGURE 28.1
Thomson's model of the atom, with the electrons embedded inside the positive charge like seeds in a watermelon.

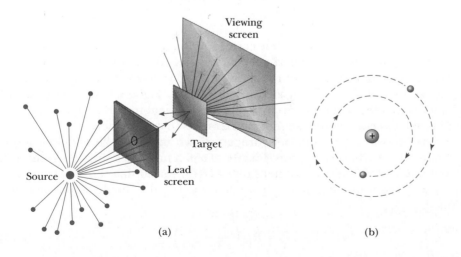

(a)   (b)

FIGURE 28.2
(a) Geiger and Marsden's technique for observing the scattering of alpha particles from a thin foil target. The source is a naturally occurring radioactive substance, such as radium. (b) Rutherford's planetary model of the atom.

Sir Joseph John Thomson (1856–1940), an English physicist and the recipient of the Nobel Prize in 1906. Thomson, usually considered the discoverer of the electron, opened up the field of subatomic particle physics with his extensive work on the deflection of cathode rays (electrons) in an electric field. *(Stock Montage, Inc.)*

explained these observations by assuming that the positive charge in an atom was concentrated in a region that was small relative to the size of the atom. He called this concentration of positive charge the **nucleus** of the atom. Any electrons belonging to the atom were assumed to be in the relatively large volume outside the nucleus. In order to explain why electrons in this outer region of the atom were not pulled into the nucleus, Rutherford viewed them as moving in orbits about the positively charged nucleus in the same manner that the planets orbit the Sun (see Fig. 28.2b).

There are two basic difficulties with Rutherford's planetary model. As we shall see in the next section, an atom emits certain discrete characteristic frequencies of electromagnetic radiation and no others; the Rutherford model is unable to explain this phenomenon. A second difficulty is that Rutherford's electrons are undergoing a centripetal acceleration. According to Maxwell's theory of electromagnetism, centripetally accelerated charges revolving with frequency f should radiate electromagnetic waves of frequency f. Unfortunately, this classical model leads to disaster when applied to the atom. As the electron radiates energy, the radius of its orbit steadily decreases and its frequency of revolution increases. This leads to an ever-increasing frequency of emitted radiation and a rapid collapse of the atom as the electron plunges into the nucleus.

28.2

ATOMIC SPECTRA

As you may have already learned in chemistry, the hydrogen atom is the simplest known atomic system, and an especially important one to understand. Much of what we know about the hydrogen atom (which consists of one proton and one electron) can be extended to other single-electron ions such as He^+ and Li^{2+}. Furthermore, a thorough understanding of the physics underlying the hydrogen atom can then be used to describe more complex atoms and the periodic table of the elements.

Suppose an evacuated glass tube is filled with hydrogen (or some other gas) at very low pressure. If a voltage applied between metal electrodes in the tube is great enough to produce an electric current in the gas, the tube emits light whose color is characteristic of the gas in the tube (this is how a neon sign works). When the emitted light is analyzed with a device called a spectroscope, a series of discrete lines is observed, each line corresponding to a different wavelength, or color, of light. Such a series of spectral lines is commonly referred to as an **emission spectrum.** The wavelengths contained in a given line spectrum are characteristic of the element emitting the light. (Fig. 28.3). Because no two elements emit the same line spectrum, this phenomenon represents a marvelous and reliable technique for identifying elements in a substance.

The emission spectrum of hydrogen shown in Figure 28.4 includes four prominent lines that occur at wavelengths of 656.3 nm, 486.1 nm, 434.1 nm, and 410.2 nm. In 1885 Johann Balmer (1825–1898) found that the wavelengths of these and less prominent lines can be described by the simple empirical equation:

Balmer series

$$\frac{1}{\lambda} = R_H \left(\frac{1}{2^2} - \frac{1}{n^2} \right)$$

[28.1]

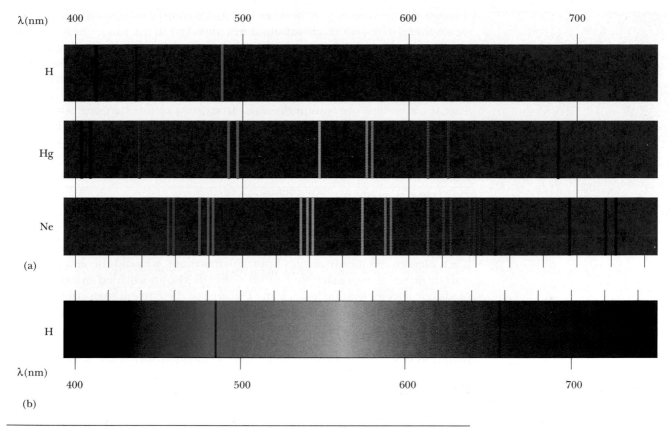

FIGURE 28.3

Visible spectra. (a) Line spectra produced by emission in the visible range for the elements hydrogen, mercury, and neon. (b) The absorption spectrum for hydrogen. The dark absorption lines occur at the same wavelengths as the emission lines for hydrogen shown in (a). *(K. W. Whitten, K. D. Gailey, and R. E. Davis,* General Chemistry, *3rd ed., Philadelphia, Saunders College Publishing, 1987)*

where n may have integral values of 3, 4, 5, . . . , and R_H is a constant, called the **Rydberg constant.** If the wavelength is in meters, R_H has the value

$$R_H = 1.097\ 373\ 2 \times 10^7 \text{ m}^{-1} \qquad \textbf{[28.2]}$$

Rydberg constant

The first line in the Balmer series, at 656.3 nm, corresponds to $n = 3$ in Equation 28.1; the line of 486.1 nm corresponds to $n = 4$; and so on.

In addition to emitting light at specific wavelengths, an element can also absorb light at specific wavelengths. The spectral lines corresponding to this process form what is known as an **absorption spectrum.** An absorption spectrum can be obtained by passing a continuous radiation spectrum (one containing all wavelengths) through a vapor of the element being analyzed. The absorption spectrum consists of a series of dark lines superimposed on the otherwise continuous spectrum. Each line in the absorption spectrum of a given element coincides with a line in the emission spectrum of the element. That is, if hydrogen is the absorbing vapor, dark lines will appear at the visible wavelengths 656.3 mn, 486.1 nm, 434.1 nm, and 410.2 nm, as shown in Figures 28.3b and 28.4.

The absorption spectrum of an element has many practical applications. For example, the continuous spectrum of radiation emitted by the Sun must pass

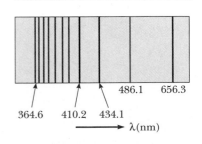

FIGURE 28.4

A series of spectral lines for atomic hydrogen. The prominent labeled lines are part of the Balmer series.

through the cooler gases of the solar atmosphere and then through the Earth's atmosphere. The various absorption lines observed in the solar spectrum have been used to identify elements in the solar atmosphere. It is interesting to note that, when the solar spectrum was first being studied, some lines were found that did not correspond to any known element. A new element had been discovered! Since the Greek word for Sun is *helios,* the new element was named helium. Scientist are able to examine the light from stars other than our Sun in this fashion, but elements other than those present on Earth have never been detected.

28.3

THE BOHR THEORY OF HYDROGEN

At the beginning of the 20th century, scientists were perplexed by the failure of classical physics to explain the characteristics of spectra. Why did atoms of a given element emit only certain lines? Furthermore, why did the atoms absorb only those wavelengths that they emitted? In 1913 Bohr provided an explanation of atomic spectra that includes some features of the currently accepted theory. Using the simplest atom, hydrogen, Bohr developed a model of what he thought must be the atom's structure in an attempt to explain why the atom was stable. His model of the hydrogen atom contains some classical features as well as some revolutionary postulates that could not be justified within the framework of classical physics. The basic assumptions of the Bohr theory as it applies to the hydrogen atom are as follows:

Assumptions of the Bohr theory

1. The electron moves in circular orbits about the proton under the influence of the Coulomb force of attraction, as in Figure 28.5. In this case, the Coulomb force is the centripetal force.
2. Only certain electron orbits are stable. These are orbits in which the hydrogen atom does not emit energy in the form of radiation. Hence, the total energy of the atom remains constant, and classical mechanics can be used to describe the electron's motion.
3. Radiation is emitted by the hydrogen atom when the electron "jumps" from a more energetic initial state to a lower state. The "jump" cannot be visualized or treated classically. In particular, the frequency, f, of the radiation emitted in the jump is related to the change in the atom's energy and is *independent of the frequency of the electron's orbital motion.* The frequency of the emitted radiation is

$$E_i - E_f = hf \qquad \text{[28.3]}$$

where E_i is the energy of the initial state, E_f is the energy of the final state, h is Planck's constant, and $E_i > E_f$.

4. The size of the allowed electron orbits is determined by a condition imposed on the electron's orbital angular momentum: the allowed orbits are those for which the electron's orbital angular momentum about the nucleus is an integral multiple of $\hbar = h/2\pi$.

$$mvr = n\hbar \qquad n = 1, 2, 3, \ldots \qquad \text{[28.4]}$$

With these assumptions, we can calculate the allowed energies and emission wavelengths of the hydrogen atom. We shall use the model pictured in Figure

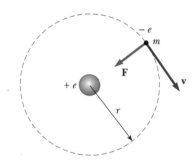

FIGURE 28.5
Diagram representing Bohr's model of the hydrogen atom. In this model, the orbiting electron is allowed only in specific orbits of discrete radius.

Niels Bohr, a Danish physicist, proposed the first quantum model of the atom. He was an active participant in the early development of quantum mechanics and provided much of its philosophical framework. He made many other important contributions to theoretical nuclear physics, including the development of the liquid-drop model of the nucleus and work in nuclear fission. He was awarded the 1922 Nobel prize in physics for his investigation of the structure of atoms and of the radiation emanating from them.*

Bohr spent most of his life in Copenhagen, Denmark, and received his doctorate at the University of Copenhagen in 1911. The following year he traveled to England, where he worked first under J. J. Thomson at Cambridge, and later under Ernest Rutherford in Manchester. He married in 1912 and returned to the University of Copenhagen in 1916 as professor of physics. During the 1920s and 1930s, Bohr headed the Institute for Advanced Studies in Copenhagen, under the support of the Carlsberg brewery. (This was certainly beer's greatest gift to the field of theoretical physics.) The institute was a magnet for many of the world's best physicists and provided a forum for the exchange of ideas. Bohr, a firm believer in doing physics on a ''person-to-person'' basis, was always the initiator of probing questions, reflections, and discussions with his guests.

When Bohr visited the United States in 1939 to attend a scientific conference, he brought news that the fission of uranium had been observed by Hahn and Strassman in Berlin. The results, confirmed by other scientists shortly thereafter, were the foundations of the atomic bomb developed in the United States during World War II. He returned to Denmark and was there during the German occupation in 1940. He escaped to Sweden in 1943 to avoid imprisonment, and helped to arrange the escape of many imperiled Danish citizens.

Although Bohr himself worked on the Manhattan project at Los Alamos until 1945, he felt strongly that openness between nations concerning nuclear weapons should be the first step in establishing their control. After the war, Bohr committed himself to many human issues, including the development of peaceful uses of atomic energy. He was awarded the Atoms for Peace award in 1957. As John Archibald Wheeler summarizes, ''Bohr was a great scientist. He was a great citizen of Denmark, of the World. He was a great human being.''

*For several interesting articles concerning Bohr, read the special Niels Bohr centennial issue of *Physics Today,* October 1985.

**Niels Bohr
(1885–1962)**

28.5, in which the electron travels in a circular orbit of radius r with an orbit speed v.

The electrical potential energy of the atom is

$$PE = k\frac{q_1 q_2}{r} = k\frac{(-e)(e)}{r} = -k\frac{e^2}{r}$$

where k is the Coulomb constant. Assuming the nucleus is at rest the total energy, E, of the atom is the sum of the kinetic energy and the potential energy of the electron:

$$E = KE + PE = \tfrac{1}{2}mv^2 - k\frac{e^2}{r} \qquad \text{[28.5]}$$

Let us apply Newton's second law to the electron. We know that the electric force of attraction on the electron, ke^2/r^2, must equal ma_r, where $a_r = v^2/r$ is the centripetal acceleration of the electron. Thus,

$$k\frac{e^2}{r^2} = m\frac{v^2}{r} \qquad \text{[28.6]}$$

From this equation, we see that the kinetic energy of the electron is

$$\tfrac{1}{2}mv^2 = \frac{ke^2}{2r} \qquad \text{[28.7]}$$

We can combine this result with Equation 28.5 and express the **total energy** of the atom as

Total energy of the hydrogen atom

$$E = -\frac{ke^2}{2r} \qquad \text{[28.8]}$$

An expression for r is obtained by solving Equations 28.4 and 28.6 for v and equating the results:

$$v^2 = \frac{n^2\hbar^2}{m^2 r^2} = \frac{ke^2}{mr}$$

The radii of the Bohr orbits are quantized

$$r_n = \frac{n^2\hbar^2}{mke^2} \qquad n = 1, 2, 3, \ldots \qquad \text{[28.9]}$$

This equation is based on the assumption that *the electron can exist only in certain allowed orbits determined by the integer n.*

The orbit with the smallest radius, called the **Bohr radius,** a_0, corresponds to $n = 1$ and has the value

Bohr radius

$$a_0 = \frac{\hbar^2}{mke^2} = 0.0529 \text{ nm} \qquad \text{[28.10]}$$

A general expression for the radius of any orbit in the hydrogen atom is obtained by substituting Equation 28.10 into Equation 28.9:

$$r_n = n^2 a_0 = n^2(0.0529 \text{ nm}) \qquad \text{[28.11]}$$

The first three Bohr orbits for hydrogen are shown in Figure 28.6.

Equation 28.9 may be substituted into Equation 28.8 to give the following expression for the energies of the quantum states:

Allowed energies of the hydrogen atom

$$E_n = -\frac{mk^2 e^4}{2\hbar^2}\left(\frac{1}{n^2}\right) \qquad n = 1, 2, 3, \ldots \qquad \text{[28.12]}$$

If we insert numerical values into Equation 28.12, we have

$$E_n = -\frac{13.6}{n^2} \text{ eV} \qquad \text{[28.13]}$$

The lowest stationary energy state, or **ground state,** corresponds to $n = 1$ and has an energy $E_1 = -mk^2 e^4/2\hbar = -13.6$ eV. The next state, corresponding to $n =$

2, has an energy $E_2 = E_1/4 = -3.4$ eV, and so on. An energy level diagram showing the energies of these stationary states and the corresponding quantum numbers is shown in Figure 28.7. The uppermost level shown, corresponding to $n \to \infty$, represents the state for which the electron is completely removed from the atom. In this case, $E = 0$ for $r = \infty$. The minimum energy required to ionize the atom, that is, to completely remove the electron, is called the **ionization energy**. The ionization energy for hydrogen is 13.6 eV.

Equations 28.3 and 28.12 and the third Bohr postulate show that if the electron jumps from one orbit, whose quantum number is n_i, to a second orbit, whose quantum number is n_f, it emits a photon of frequency f, given by

$$f = \frac{E_i - E_f}{h} = \frac{mk^2 e^4}{4\pi\hbar^3}\left(\frac{1}{n_f^2} - \frac{1}{n_i^2}\right) \qquad \textbf{[28.14]}$$

Finally, to compare this result with the empirical formulas for the various spectral series, we use the fact that $\lambda f = c$ and Equation 28.14 to get

$$\frac{1}{\lambda} = \frac{f}{c} = \frac{mk^2 e^4}{4\pi c\hbar^3}\left(\frac{1}{n_f^2} - \frac{1}{n_i^2}\right) \qquad \textbf{[28.15]}$$

A comparison of this result with Equation 28.1 gives the following expression for the Rydberg constant:

$$R = \frac{mk^2 e^4}{4\pi c\hbar^3} \qquad \textbf{[28.16]}$$

If we insert the known values of m, k, e, c, and $\hbar$ into this expression, the resulting theoretical value for R is found to be in excellent agreement with the value determined experimentally for the Rydberg constant. When Bohr demonstrated this agreement, it was recognized as a major accomplishment of his theory.

In order to compare Equation 28.15 with spectroscopic data, it is convenient to express it in the form

$$\frac{1}{\lambda} = R\left(\frac{1}{n_f^2} - \frac{1}{n_i^2}\right) \qquad \textbf{[28.17]}$$

We can use this expression to evaluate the wavelengths for the various series in the hydrogen spectrum. For example, in the Balmer series, $n_f = 2$ and $n_i = 3, 4, 5, \ldots$ (Eq. 28.1). For the Lyman series, we take $n_f = 1$ and $n_i = 2, 3, 4, \ldots$. The energy level diagram for hydrogen, shown in Figure 28.7, indicates the origin of the spectral lines described above. The transitions between levels are represented by vertical arrows. Note that whenever a transition occurs between a state designated by n_i to one designated by n_f (where $n_i > n_f$), a photon with a frequency of $(E_i - E_f)/h$ is emitted. This can be interpreted as follows. The lines in the visible part of the hydrogen spectrum arise when the electron jumps from the third, fourth, or even higher orbit to the second orbit. Likewise, the lines of the Lyman series (in the ultraviolet) arise when the electron jumps from the second, third, or even higher orbit to the innermost ($n_f = 1$) orbit. Hence, the Bohr theory successfully predicts the wavelengths of all observed spectral lines of hydrogen.

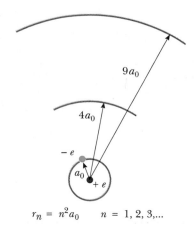

$$r_n = n^2 a_0 \qquad n = 1, 2, 3, \ldots$$

FIGURE 28.6

The first three circular orbits predicted by the Bohr model of the hydrogen atom.

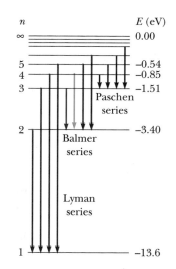

FIGURE 28.7

An energy level diagram for hydrogen. In such diagrams the discrete allowed energies are plotted on the vertical axis. Nothing is plotted on the horizontal axis, but the horizontal extent of the diagram is made large enough to show allowed transitions. Note that the quantum numbers are given on the left and the energies (in eV) are on the right.

EXAMPLE 28.1   An Electronic Transition in Hydrogen

The electron in the hydrogen atom makes a transition from the $n = 2$ energy state to the ground state (corresponding to $n = 1$). Find the wavelength and frequency of the emitted photon.

Solution   We can use Equation 28.17 directly to obtain λ, with $n_i = 2$ and $n_f = 1$:

$$\frac{1}{\lambda} = R \left(\frac{1}{n_f^2} - \frac{1}{n_i^2} \right)$$

$$\frac{1}{\lambda} = R \left(\frac{1}{1^2} - \frac{1}{2^2} \right) = \frac{3R}{4}$$

$$\lambda = \frac{4}{3R} = \frac{4}{3(1.097 \times 10^7 \text{ m}^{-1})} = 1.215 \times 10^{-7} \text{ m} = \boxed{121.5 \text{ nm}}$$

This wavelength lies in the ultraviolet region.

Since $c = f\lambda$, the frequency of the photon is

$$f = \frac{c}{\lambda} = \frac{3.00 \times 10^8 \text{ m/s}}{1.215 \times 10^{-7} \text{ m}} = \boxed{2.47 \times 10^{15} \text{ Hz}}$$

Exercise   What is the wavelength of the photon emitted by hydrogen when the electron makes a transition from the $n = 3$ state to the $n = 1$ state?

Answer $9/8R = 102.6$ nm.

EXAMPLE 28.2   The Balmer Series for Hydrogen

The Balmer series for the hydrogen atom corresponds to electronic transitions that terminate in the state of quantum number $n = 2$, as shown in Figure 28.8.

(a) Find the longest-wavelength photon emitted and determine its energy.

Solution   The longest-wavelength photon in the Balmer series results from the transition from $n = 3$ to $n = 2$. Using Equation 28.17 gives

$$\frac{1}{\lambda} = R \left(\frac{1}{n_f^2} - \frac{1}{n_i^2} \right)$$

$$\frac{1}{\lambda_{\text{max}}} = R \left(\frac{1}{2^2} - \frac{1}{3^2} \right) = \frac{5}{36} R$$

$$\lambda_{\text{max}} = \frac{36}{5R} = \frac{36}{5(1.097 \times 10^7 \text{ m}^{-1})} = \boxed{656.3 \text{ nm}}$$

This wavelength is in the red region of the visible spectrum.

The energy of this photon is

$$E_{\text{photon}} = hf = \frac{hc}{\lambda_{\text{max}}}$$

$$= \frac{(6.626 \times 10^{-34} \text{ J} \cdot \text{s})(3.00 \times 10^8 \text{ m/s})}{656.3 \times 10^{-9} \text{ m}}$$

$$= 3.03 \times 10^{-19} \text{ J} = \boxed{1.89 \text{ eV}}$$

We could also obtain the energy of the photon by using Equation 28.3 in the form $hf = E_3 - E_2$, where E_2 and E_3 are the energy levels of the hydrogen

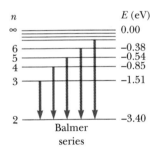

n	E (eV)
∞	0.00
6	-0.38
5	-0.54
4	-0.85
3	-1.51
2	-3.40

Balmer series

FIGURE 28.8
(Example 28.2)   Transitions responsible for the Balmer series for the hydrogen atom. All transitions terminate at the $n = 2$ level.

atom, calculated from Equation 28.13. Note that this is the lowest-energy photon in this series since it involves the smallest energy change.

(b) Find the shortest-wavelength photon emitted in the Balmer series.

Solution   The shortest-wavelength photon in the Balmer series is emitted when the electron makes a transition from $n = \infty$ to $n = 2$. Therefore

$$\frac{1}{\lambda_{min}} = R \left(\frac{1}{2^2} - \frac{1}{\infty} \right) = \frac{R}{4}$$

$$\lambda_{min} = \frac{4}{R} = \frac{4}{1.097 \times 10^7 \text{ m}^{-1}} = \boxed{364.6 \text{ nm}}$$

This wavelength is in the ultraviolet region and corresponds to the series limit.

Exercise   Find the energy of the shortest-wavelength photon emitted in the Balmer series for hydrogen.

Answer   3.40 eV.

BOHR'S CORRESPONDENCE PRINCIPLE

In our study of relativity in Chapter 26, we found that newtonian mechanics cannot be used to describe phenomena that occur at speeds approaching the speed of light. Newtonian mechanics is a special case of relativistic mechanics and is usable only when v is much smaller than c. Similarly, *quantum mechanics is in agreement with classical physics when the energy differences between quantized levels are very small.* This principle, first set forth by Bohr, is called the **correspondence principle.**

For example, consider the hydrogen atom with $n > 10\ 000$. For such large values of n, the energy differences between adjacent levels approach zero and the levels are nearly continuous. Consequently, the classical model is reasonably accurate in describing the system for large values of n. According to the classical model, the frequency of the light emitted by the atom is equal to the frequency of revolution of the electron in its orbit about the nucleus. Calculations show that for $n > 10\ 000$, this frequency is different from that predicted by quantum mechanics by less than 0.015%.

28.4

MODIFICATION OF THE BOHR THEORY

The Bohr theory of the hydrogen atom was a tremendous success in certain areas because it explained several features of the spectra of hydrogen that had previously defied explanation. It accounted for the Balmer series and other series, it predicted a value for the Rydberg constant that is in excellent agreement with the experimental value, it gave an expression for the radius of the atom, and it predicted the energy levels of hydrogen. Although these successes were important to scientists, it is perhaps even more significant that the Bohr theory gave us a model of what the atom looks like and how it behaves. Once a basic model is constructed, refinements and modifications can be made to enlarge upon the concept and to explain finer details.

The analysis used in the Bohr theory is also successful when applied to *hydrogen-like* atoms. An atom is said to be hydrogen-like when it contains only one

TABLE 28.1
Shell and Subshell Notations

n	Shell Symbol	ℓ	Subshell Symbol
1	K	0	*s*
2	L	1	*p*
3	M	2	*d*
4	N	3	*f*
5	O	4	*g*
6	P	5	*h*
. . .		. . .	

electron. Examples are singly ionized helium, doubly ionized lithium, triply ionized beryllium, etc. The results of the Bohr theory for hydrogen can be extended to hydrogen-like atoms by substituting Ze^2 for e^2 in the hydrogen equations, where Z is the atomic number of the element. For example, Equations 28.12 and 28.15 become

$$E_n = -\frac{mk^2Z^2e^4}{2\hbar^2}\left(\frac{1}{n^2}\right) \qquad n = 1, 2, 3, \ldots \qquad \text{[28.18]}$$

and

$$\frac{1}{\lambda} = \frac{mk^2Z^2e^4}{4\pi c\hbar^3}\left(\frac{1}{n_f^2} - \frac{1}{n_i^2}\right) \qquad \text{[28.19]}$$

Although many attempts were made to extend the Bohr theory to more complex (multi-electron) atoms, the results were unsuccessful. Even today, only approximate methods are available for treating multi-electron atoms.

Within a few months following the publication of Bohr's theory, Arnold Sommerfeld (1868–1951) extended the results to include elliptical orbits. We shall examine his model briefly because much of the nomenclature used in this treatment is still in use today. Bohr's concept of quantization of angular momentum led to the **principal quantum number,** n, which determines the energy of the allowed states of hydrogen. Sommerfeld's theory retained n, but also introduced a new quantum number, ℓ, called the **orbital quantum number,** where the value of ℓ ranges from 0 to $n - 1$ in integer steps. According to this model, an electron in any one of the allowed energy states of a hydrogen atom may move in any one of a number of orbits. For each value of n there are n possible orbits. Since $n = 1$ and $\ell = 0$ for the first energy level (ground state), there is only one possible orbit for this state. The second energy level, with $n = 2$, has two possible orbits corresponding to $\ell = 0$ and $\ell = 1$. The third energy level, with $n = 3$, has three possible orbits corresponding to $\ell = 0$, $\ell = 1$, and $\ell = 2$.

It became customary to use a specific nomenclature when referring to these states and orbits. **All states with the same principal quantum number are said to form a shell.** These shells are identified by the letters K, L, M, . . . , which designate the states for which $n = 1, 2, 3, ...$ Likewise, **the states having the same values of n and ℓ are said to form a subshell.** The letters $s, p, d, f, g,$ . . . are used to designate the states for which $\ell = 0, 1, 2, 3, 4, ...$ These notations are summarized in Table 28.1.

Note that the maximum number of electrons allowed in any given subshell is given by $2(2\ell + 1)$. For example, the p subshell ($\ell = 1$) is filled when it contains

6 electrons. This fact will be important to us later when we discuss the *Pauli exclusion principle.*

Another modification of the Bohr theory arose when it was discovered that the spectral lines of a gas are split into several closely spaced lines when the gas is placed in a strong magnetic field. (This is called the Zeeman effect after its discoverer.) Figure 28.9 shows a single spectral line being split into three closely spaced lines. This observation indicates that the energy of an electron is slightly modified when the atom is immersed in a magnetic field. In order to explain this observation, a new quantum number, m_ℓ, called the **orbital magnetic quantum number,** was introduced. The theory is in accord with experimental results when m_ℓ is restricted to values ranging from $-\ell$ to $+\ell$, in integer steps.

Finally, very high resolution spectrometers revealed that spectral lines of gases are in fact two very closely spaced lines even in the absence of an external magnetic field. This splitting was referred to as **fine structure.** In 1925 Samuel Goudsmit and George Uhlenbeck introduced the idea of an electron spinning about its own axis to explain the origin of fine structure. The results of their work introduced yet another quantum number, m_s, called the **spin magnetic quantum number.** We shall save further discussion of this quantum number for a later section. It is interesting to note that each of the new concepts introduced into the original Bohr theory added new quantum numbers and improved the original model. However, the most profound step forward in our current understanding of atomic structure came with the development of quantum mechanics.

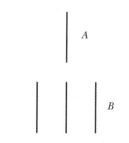

FIGURE 28.9
A single line (*A*) can split into three separate lines (*B*) in a magnetic field.

EXAMPLE 28.3   Singly Ionized Helium

Singly ionized helium, He$^+$, a hydrogen-like system, has one electron in the $1s$ orbit when the atom is in its ground state. Find (a) the energy of the electron in the ground state and (b) the radius of the ground-state orbit.

Solution   (a) From Equation 28.18, the energy of a level whose principal quantum number is n is given by

$$E_n = -\frac{mk^2 Z^2 e^4}{2\hbar^2}\left(\frac{1}{n^2}\right)$$

This can be expressed in eV units as

$$E_n = -\frac{Z^2(13.6)}{n^2}\ \text{eV}$$

Since $Z = 2$ for helium, and $n = 1$ in the ground state, we have

$$E_1 = -4(13.6)\ \text{eV} = \boxed{-54.4\ \text{eV}}$$

(b) The radius of the ground state orbit can be found with the help of Equation 28.9. This equation must be modified in the case of a hydrogen-like atom by substituting Ze^2 for e^2 to obtain

$$r_n = \frac{n^2\hbar^2}{mkZe^2} = \frac{n^2}{Z}(0.0529\ \text{nm})$$

For our case, $n = 1$ and $Z = 2$, and the result is

$$r_1 = \boxed{0.0265\ \text{nm}}$$

FIGURE 28.10
(a) Standing wave pattern for an electron wave in a stable orbit of hydrogen. There are three full wavelengths in this orbit.
(b) Standing wave pattern for a vibrating stretched string fixed at its ends. This pattern has three full wavelengths.

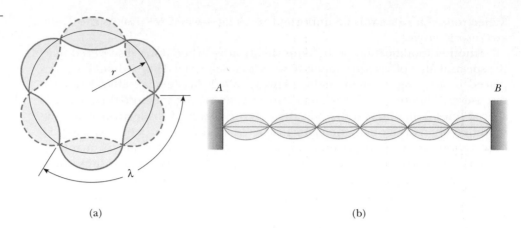

(a)   (b)

28.5

DE BROGLIE WAVES AND THE HYDROGEN ATOM

One of the postulates made by Bohr in this theory of the hydrogen atom was that angular momentum of the electron is quantized in units of $\hbar = h/2\pi$, or

$$mvr = n\hbar$$

For more than a decade following Bohr's publication, no one was able to explain why the angular momentum of the electron was restricted to these discrete values. Finally, de Broglie recognized a connection between his theory of the wave character of material properties and the quantization condition given above. De Broglie assumed that an electron orbit would be stable (allowed) only if it contained an integral number of electron wavelengths. Figure 28.10a demonstrates this point when three complete wavelengths are contained in one circumference of the orbit. Similar patterns can be drawn for orbits containing two wavelengths, four wavelengths, five wavelengths, and so forth. This situation is analogous to that of standing waves on a string discussed in Chapter 14. There we found that strings have preferred (resonant) frequencies of vibration. Figure 28.10b shows a standing wave pattern containing three wavelengths for a string fixed at each end. Now imagine that the vibrating string is removed from its supports at A and B and bent into a circular shape that brings points A and B together. The end result is a pattern like that shown in Figure 28.10a.

In general, the condition for a de Broglie standing wave in an electron orbit is that the circumference must contain an integral multiple of electron wavelengths. We can express this condition as

$$2\pi r = n\lambda \qquad n = 1, 2, 3, \ldots \qquad \text{[28.20]}$$

De Broglie's equation for the wavelength of an electron in terms of its momentum is

De Broglie wavelength

$$\lambda = \frac{h}{mv} \qquad \text{[28.21]}$$

When this expression for λ is substituted into Equation 28.20, we find

$$2\pi r = \frac{nh}{mv}$$

or

$$mvr = n\hbar$$

This is precisely the quantization of angular momentum condition imposed by Bohr in his original theory of hydrogen.

The electron orbit shown in Figure 28.10a contains three complete wavelengths and corresponds to the case in which the principal quantum number n equals three. The orbit with one complete wavelength in its circumference corresponds to the first Bohr orbit, $n = 1$; the orbit with two complete wavelengths corresponds to the second Bohr orbit, $n = 2$, and so forth.

By applying the wave theory of matter to electrons in atoms, de Broglie was able to explain the appearance of integers in the Bohr theory as a natural consequence of interference. This was the first convincing argument that the wave nature of matter was at the heart of the behavior of atomic systems. Although the analysis provided by de Broglie was a promising first step, gigantic strides were subsequently made with the development of Schrödinger's wave equation and its application to atomic systems.

28.6

QUANTUM MECHANICS AND THE HYDROGEN ATOM

One of the first great achievements of quantum mechanics was the solution of the wave equation for the hydrogen atom. We shall not attempt to carry out this solution. Rather, we shall simply describe its properties and some of its implications with regard to atomic structure.

According to quantum mechanics, the energies of the allowed states are in exact agreement with the values obtained by the Bohr theory (Eq. 28.12), when the allowed energies depend only on the principal quantum number, n.

In addition to the principal quantum number, two other quantum numbers emerged from the solution of the wave equation, ℓ and m_ℓ. The quantum number ℓ is called the **orbital quantum number,** and m_ℓ is called the **orbital magnetic quantum number.** As pointed out in Section 28.4, these quantum numbers had already appeared in modifications made to the Bohr theory. The significance of quantum mechanics is that these quantum numbers and the restrictions placed on their values arose directly from mathematics and not from any ad hoc assumptions to make the theory consistent with experimental observation. Because we shall need to make use of the various quantum numbers in the next several sections, the ranges of their values are repeated below.

The values of n can range from 1 to ∞ in integer steps.
The values of ℓ can range from 0 to $n - 1$ in integer steps.
The values of m_ℓ can range from $-\ell$ to ℓ in integer steps.

For example, if $n = 1$, only $\ell = 0$ and $m_\ell = 0$ are permitted. If $n = 2$, the value of ℓ may be 0 or 1; if $\ell = 0$, then $m_\ell = 0$, but if $\ell = 1$, then m_ℓ may be 1, 0, or -1.

TABLE 28.2
Three Quantum Numbers for the Hydrogen Atom

Quantum Number	Name	Allowed Values	Number of Allowed States
n	Principal quantum number	$1, 2, 3, \ldots$	Any number
ℓ	Orbital quantum number	$0, 1, 2, \ldots, n-1$	n
m_ℓ	Orbital magnetic quantum number	$-\ell, -\ell+1, \ldots,$ $0, \ldots, \ell-1, \ell$	$2\ell+1$

Table 28.2 summarizes the rules for determining the allowed values of ℓ and m_ℓ for a given value of n.

States that violate the rules given in Table 28.2 cannot exist. For instance, one state that cannot exist is the $2d$ state, which would have $n=2$ and $\ell=2$. This state is not allowed because the highest allowed value of ℓ is $n-1$, or 1 in this case. Thus, for $n=2$, $2s$ and $2p$ are allowed states but $2d$, $2f$, $\ldots$ are not. For $n=3$, the allowed states are $3s$, $3p$, and $3d$.

EXAMPLE 28.4   The $n=2$ Level of Hydrogen

Determine the number of states in the hydrogen atom corresponding to the principal quantum number $n=2$ and calculate the energies of these states.

Solution   For $n=2$, ℓ can have the values 0 and 1. For $\ell=0$, m_ℓ can only be 0; for $\ell=1$, m_ℓ can be -1, 0, or 1. Hence we have a state designated as the $2s$ state associated with the quantum numbers $n=2$, $\ell=0$, and $m_\ell=0$, and three states designated as $2p$ states, for which the quantum numbers are $n=2$, $\ell=1$, $m_\ell=-1$; $n=2$, $\ell=1$, $m_\ell=0$; and $n=2$, $\ell=1$, $m_\ell=1$.

Because all of these states have the same principal quantum number, $n=2$, they also have the same energy, which can be calculated using Equation 28.13, that is, $E_n=-13.6/n^2$. For $n=2$, this gives

$$E_2 = -\frac{13.6}{2^2}\ \text{eV} = \boxed{-3.4\ \text{eV}}$$

Exercise   How many possible states are there for the $n=3$ level of hydrogen? For the $n=4$ level?

Answers   9 states for $n=3$, and 16 states for $n=4$.

28.7

THE SPIN MAGNETIC QUANTUM NUMBER

Example 28.4 was presented to give you some practice in manipulating quantum numbers, but as we shall see in this section, there actually are *eight* states corresponding to $n=2$ for hydrogen, not four. This happens since it was later found that another quantum number, m_s, the **spin magnetic quantum number,** had to be introduced to explain the splitting of each level into two.

As pointed out in Section 28.4, the need for this new quantum number first came about because of an unusual feature in the spectra of certain gases, such as sodium vapor. Close examination of one of the prominent lines of sodium shows that it is, in fact, two very closely spaced lines. The wavelengths of these lines occur in the yellow region at 589.0 nm and 589.6 nm. In 1925, when this doublet was first noticed, atomic theory could not explain it. To resolve the dilemma, Samuel Goudsmidt and George Uhlenbeck, following a suggestion by the Austrian physicist Wolfgang Pauli, proposed that a fourth quantum number, called the spin quantum number, be introduced to describe any atomic level.

In order to describe the spin quantum number, it is convenient (but incorrect) to think of the electron as spinning on its axis as it orbits the nucleus, just as the Earth spins on its axis as it orbits the Sun. There are only two ways in which the electron can spin as it orbits the nucleus, as shown in Figure 28.11. If the direction of spin is as shown in Figure 28.11a, the electron is said to have "spin up." If the direction of spin is reversed, as in Figure 28.11b, the electron is said to have "spin down." The energy of the electron is slightly different for the two spin directions, and this energy difference accounts for the sodium doublet. The quantum numbers associated with electron spin are $m_s = \frac{1}{2}$ for the spin-up state and $m_s = -\frac{1}{2}$ for the spin-down state. As we shall see in Example 28.5, this new quantum number doubles the number of allowed states specified by the quantum numbers n, ℓ, and m_ℓ.

This classical description of electron spin is incorrect because quantum mechanics tells us that, since the electron cannot be precisely located in space, it cannot be considered to be spinning as pictured in Figure 28.11. In spite of this conceptual difficulty, all experimental evidence supports the fact that an electron does have some intrinsic property that can be described by the spin magnetic quantum number.

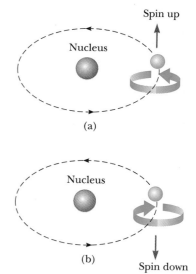

FIGURE 28.11
As an electron moves in its orbit about the nucleus, its spin can be either (a) up or (b) down.

EXAMPLE 28.5 Adding Electron Spin to Hydrogen

For a hydrogen atom, determine the quantum numbers associated with the possible states that correspond to the principal quantum number $n = 2$.

Solution With the addition of the spin quantum number, we have the following possibilities:

n	ℓ	m_ℓ	m_s	Subshell	Shell	Number of Electrons in Subshell
2	0	0	$\frac{1}{2}$	2s	L	2
2	0	0	$-\frac{1}{2}$			
2	1	1	$\frac{1}{2}$	2p	L	6
2	1	1	$-\frac{1}{2}$			
2	1	0	$\frac{1}{2}$			
2	1	0	$-\frac{1}{2}$			
2	1	−1	$\frac{1}{2}$			
2	1	−1	$-\frac{1}{2}$			

Exercise Show that for $n = 3$, there are 18 possible states. (This follows from the restrictions that the maximum number of electrons in the 3s state is 2, the maximum number in the 3p state is 6, and the maximum number in the 3d state is 10.)

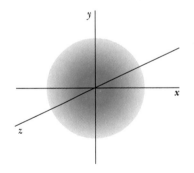

P_{1s}

$a_0 = 0.0529$ nm r

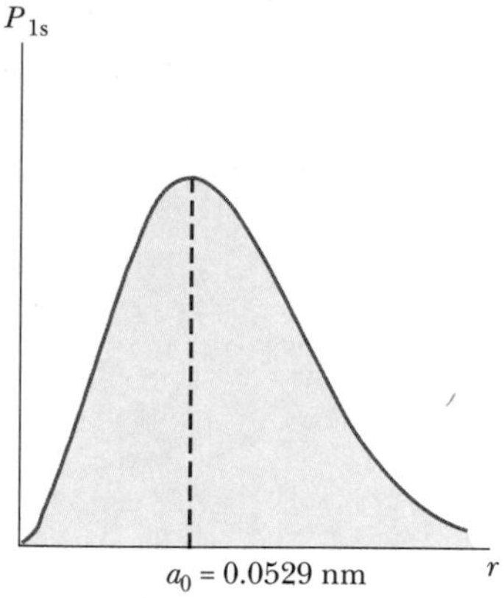

FIGURE 28.12
The probability of finding the electron versus distance from the nucleus for the hydrogen atom in the 1s (ground) state. Note that the probability has its maximum value when r equals the first Bohr radius, a_0.

FIGURE 28.13
The spherical electron cloud for the hydrogen atom in its 1s state.

28.8

ELECTRON CLOUDS

The solution of the wave equation for the wave function Ψ, discussed in Chapter 27, Section 27.10, yields a quantity dependent on the quantum numbers n, ℓ, and m_ℓ. Let us assume that we have found a value for Ψ and see what it may tell us about the hydrogen atom. Let us choose a value of $n = 1$ for the principal quantum number, which corresponds to the lowest energy state for hydrogen. For $n = 1$, the restrictions placed on the remaining quantum numbers are that $\ell = 0$ and $m_\ell = 0$. We can now substitute these values into our expression for Ψ.

The quantity Ψ^2 has great physical significance because it is proportional to *the probability of finding the electron at a given position*. Figure 28.12 gives the probability of finding the electron at various distances from the nucleus in the 1s state of hydrogen. Some useful and surprising information can be extracted from this curve. First, the curve peaks at a value of $r = 0.0529$ nm, the Bohr value of the radius of the first electron orbit in hydrogen. This means that there is a probability of finding the electron at this distance from the nucleus. However, as the curve indicates, there is also a probability of finding the electron at some other distance from the nucleus. In other words, the electron is not confined to a particular orbital distance from the nucleus, as assumed in the Bohr model. The electron may be found at various distances from the nucleus, but *the probability of finding it at a distance corresponding to the first Bohr orbit is a maximum*. Quantum mechanics also predicts that the wave function for the hydrogen atom in the ground state is spherically symmetric; hence the electron can be found in a spherical region surrounding the nucleus. This is in contrast to the Bohr theory, which confines the position of the electron to points in a plane. This result is often interpreted by viewing the electron as a cloud surrounding the nucleus. An attempt at picturing this cloud-like behavior is shown in Figure 28.13. The densest regions of the cloud represent those locations where the electron is most likely to be found.

If a similar analysis is carried out for the $n = 2$, $\ell = 0$ state of hydrogen, a peak of the probability curve is found at $4a_0$. Likewise, for the $n = 3$, $\ell = 0$ state, the curve peaks at $9a_0$. Thus, quantum mechanics predicts a most probable electron location that is in agreement with the location predicted by the Bohr theory.

28.9

THE EXCLUSION PRINCIPLE AND THE PERIODIC TABLE

Earlier we found that the state of an electron in an atom is specified by four quantum numbers: n, ℓ, m_ℓ, and m_s. For example, an electron in the ground state of hydrogen could have quantum numbers of $n = 1$, $\ell = 0$, $m_\ell = 0$, $m_s = \frac{1}{2}$. As it turns out, the state of an electron in any other atom may also be specified by this same set of quantum numbers. In fact, these four quantum numbers can be used to describe all the electronic states of an atom regardless of the number of electrons in its structure.

An obvious question that arises here is "How many electrons in an atom can have a particular set of quantum numbers?" This important question was an-

TABLE 28.3
The Number of Electrons in Filled Subshells and Shells

Shell	Subshell	Number of Electrons in Filled Subshell	Number of Electrons in Filled Shell
K ($n = 1$)	s ($\ell = 0$)	2	2
L ($n = 2$)	s ($\ell = 0$) p ($\ell = 1$)	$\left.\begin{array}{c} 2 \\ 6 \end{array}\right\}$	8
M ($n = 3$)	s ($\ell = 0$) p ($\ell = 1$) d ($\ell = 2$)	$\left.\begin{array}{c} 2 \\ 6 \\ 10 \end{array}\right\}$	18
N ($n = 4$)	s ($\ell = 0$) p ($\ell = 1$) d ($\ell = 2$) f ($\ell = 3$)	$\left.\begin{array}{c} 2 \\ 6 \\ 10 \\ 14 \end{array}\right\}$	32

swered by Pauli in 1925 in a powerful statement known as the **exclusion principle:**

> No two electrons in an atom can ever be in the same quantum state; that is, no two electrons in the same atom can have the same set of quantum numbers n, ℓ, m_ℓ, and m_s.

It is interesting to note that if this principle were not valid, every electron would end up in the lowest energy state of the atom and the chemical behavior of the elements would be grossly different. Nature as we know it would not exist! In reality, we can view the electronic structure of complex atoms as a succession of filled levels increasing in energy, where the outermost electrons are primarily responsible for the chemical properties of the element.

As a general rule, the order that electrons fill an atom's subshell is as follows. Once one subshell is filled, the next electron goes into the vacant subshell that is lowest in energy. One can understand this principle by recognizing that if the atom were not in the lowest energy state available to it, it would radiate energy until it reached this state. Recall that a subshell is filled when it contains $2(2\ell + 1)$ electrons. This rule is based on the analysis of quantum numbers to be described below. Following this rule, shells and subshells are filled according to the pattern given in Table 28.3.

The exclusion principle can be illustrated by an examination of the electronic arrangement in a few of the lighter atoms.

Hydrogen has only one electron, which, in its ground state, can be described by either of two sets of quantum numbers: $1, 0, 0, \frac{1}{2}$ or $1, 0, 0, -\frac{1}{2}$. The electronic configuration of this atom is often designated as $1s^1$. The notation $1s$ refers to a state for which $n = 1$ and $\ell = 0$, and the superscript indicates that one electron is present in this level.

Neutral *helium* has two electrons. In the ground state, the quantum numbers for these two electrons are $1, 0, 0, \frac{1}{2}$ and $1, 0, 0, -\frac{1}{2}$. No other possible combinations of quantum numbers exist for this level, and we say that the K shell is filled. The helium electronic configuration is designated as $1s^2$.

TABLE 28.4
Electronic Configuration of Some Elements

Z	Symbol	Ground-State Configuration	Ionization Energy (eV)	Z	Symbol	Ground-State Configuration	Ionization Energy (eV)
1	H	$1s^1$	13.595	19	K	$[\text{Ar}]\,4s^1$	4.339
2	He	$1s^2$	24.581	20	Ca	$4s^2$	6.111
				21	Sc	$3d4s^2$	6.54
3	Li	$[\text{He}]\,2s^1$	5.390	22	Ti	$3d^24s^2$	6.83
4	Be	$2s^2$	9.320	23	V	$3d^34s^2$	6.74
5	B	$2s^22p^1$	8.296	24	Cr	$3d^54s^1$	6.76
6	C	$2s^22p^2$	11.256	25	Mn	$3d^54s^2$	7.432
7	N	$2s^22p^3$	14.545	26	Fe	$3d^64s^2$	7.87
8	O	$2s^22p^4$	13.614	27	Co	$3d^74s^2$	7.86
9	F	$2s^22p^5$	17.418	28	Ni	$3d^84s^2$	7.633
10	Ne	$2s^22p^6$	21.559	29	Cu	$3d^{10}4s^1$	7.724
				30	Zn	$3d^{10}4s^2$	9.391
11	Na	$[\text{Ne}]\,3s^1$	5.138	31	Ga	$3d^{10}4s^24p^1$	6.00
12	Mg	$3s^2$	7.644	32	Ge	$3d^{10}4s^24p^2$	7.88
13	Al	$3s^23p^1$	5.984	33	As	$3d^{10}4s^24p^3$	9.81
14	Si	$3s^23p^2$	8.149	34	Se	$3d^{10}4s^24p^4$	9.75
15	P	$3s^23p^3$	10.484	35	Br	$3d^{10}4s^24p^5$	11.84
16	S	$3s^23p^4$	10.357	36	Kr	$3d^{10}4s^24p^6$	13.996
17	Cl	$3s^23p^5$	13.01				
18	Ar	$3s^23p^6$	15.755				

Note: The bracket notation is used as a shorthand method to avoid repetition in indicating inner-shell electrons. Thus, [He] represents $1s^2$, [Ne] represents $1s^22s^22p^6$, [Ar] represents $1s^22s^22p^63s^23p^6$, and so on.

Neutral *lithium* has three electrons. In the ground state, two of these are in the $1s$ subshell and the third is in the $2s$ subshell because it is lower in energy than the $2p$ subshell. Hence, the electronic configuration for lithium is $1s^22s^1$.

A list of electronic ground-state configurations for a number of atoms is provided in Table 28.4. In 1871 Dmitri Mendeleev (1834–1907), a Russian chemist, arranged the elements known at that time in a table according to their atomic weights and chemical similarities. The first table Mendeleev proposed contained many blank spaces, and he boldly stated that the gaps were there only because those elements had not yet been discovered. By noting the column in which these missing elements should be located, he was able to make rough predictions about their chemical properties. Within 20 years of this announcement, these elements were indeed discovered.

The elements in our current version of the periodic table are still arranged so that all those in a vertical column have similar chemical properties. For example, consider the elements in the last column: He (helium), Ne (neon), Ar (argon), Kr (krypton), Xe (xenon), and Rn (radon). The outstanding characteristic of these elements is that they do not normally take part in chemical reactions; that is, they do not join with other atoms to form molecules, and are therefore classified as inert. Because of this aloofness, they are referred to as the noble gases. We can partially understand their behavior by looking at the electronic configurations shown in Table 28.4. The element helium has the electronic configuration $1s^2$. In other words, one shell is filled. Additionally, the electrons in this filled shell are considerably separated in energy from the next available level, the $2s$ level.

Wolfgang Pauli and Niels Bohr watch a spinning top.
(Courtesy of AIP Niels Bohr Library, Margarethe Bohr Collection)

Wolfgang Pauli was an extremely talented Austrian theoretical physicist who made important contributions in many areas of modern physics. At the age of 21 he gained public recognition with a masterful review article on relativity that is still considered one of the finest and most comprehensive introductions to the subject. Other major contributions were the discovery of the exclusion principle, the explanation of the connection between particle spin and statistics, theories of relativistic quantum electrodynamics, the neutrino hypothesis, and the hypothesis of nuclear spin. Pauli was a forceful and colorful character, well known for his witty and often caustic remarks directed at those who presented new theories in a less than perfectly clear manner. Pauli exerted great influence on his students and colleagues by forcing them with his sharp criticism to a deeper and clearer understanding. Victor Weisskopf, one of Pauli's famous students, has aptly described him as ''the conscience of theoretical physics.'' Pauli's sharp sense of humor was nicely captured by Weisskopf in the following anecdote:

''In a few weeks, Pauli asked me to come to Zurich. I came to the big door of his office, I knocked, and no answer. I knocked again and no answer. After about five minutes he said, rather roughly, ''Who is it? Come in!'' I opened the door, and here was Pauli—it was a very big office—at the other side of the room, at his desk, writing and writing. He said, ''Who is this? First I must finish calculating.'' Again he let me wait for about five minutes and then: ''Who is that? ''I am Weisskopf.'' ''Uhh, Weisskopf, ja, you are my new assistant.'' Then he looked at me and said, ''Now, you see I wanted to take Bethe, but Bethe works now on the solid state. Solid state I don't like, although I started it. This is why I took you.'' Then I said, ''What can I do for you sir?'' and he said ''I shall give you right away a problem.'' He gave me a problem, some calculation, and then he said, ''Go and work.'' So I went, and after 10 days or so, he came and said, ''Well, show me what you have done.'' And I showed him. He looked at it and exclaimed: ''I should have taken Bethe!''*

*From *Physics in the Twentieth Century: Selected Essays, My Life as a Physicist,* Victor F. Weisskopf, The MIT Press, 1972, p. 10.

Wolfgang Pauli (1900–1958)

(SPL/Photo Researchers)

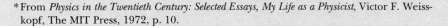

The electronic configuration for neon is $1s^2 2s^2 2p^6$. Again, the outer shell is filled and there is a large difference in energy between the $2p$ level and the $3s$ level. Argon has the configuration $1s^2 2s^2 2p^6 3s^2 3p^6$. Here, the $3p$ subshell is filled and there is a wide gap in energy between the $3p$ subshell and the $3d$ subshell. We could continue this procedure through all the noble gases, but the pattern remains the same. A noble gas is formed when either a shell or a subshell is filled and there is a large gap in energy before the next possible level is encountered.

The elements in the first column of the periodic table are called the alkali metals and are characterized by the fact that they are very chemically active. Referring to Table 28.4, we can understand why these elements interact so strongly with other elements. All of these alkali metals have a single outer electron in an s subshell. This electron is shielded from the nucleus by all the electrons in the inner shells. Thus, it is only loosely bound to the atom and can readily be accepted by other atoms to form molecules.

All the elements in the seventh column of the periodic table (called the halogens) are also very active chemically. Note that all these elements are lacking one electron in a subshell. Consequently, they readily accept electrons from other atoms to form molecules.

EXAMPLE 28.6   The Quantum Number for the 2p Subshell

List the quantum numbers for electrons in the 2p subshell.

Solution   For this subshell, $n = 2$ and $\ell = 1$. The magnetic quantum number can have the values -1, 0, 1, and the spin quantum number is always $+\frac{1}{2}$ or $-\frac{1}{2}$. Thus, the six possibilities are

n	ℓ	m_ℓ	m_s
2	1	-1	$-\frac{1}{2}$
2	1	-1	$\frac{1}{2}$
2	1	0	$-\frac{1}{2}$
2	1	0	$\frac{1}{2}$
2	1	1	$-\frac{1}{2}$
2	1	1	$\frac{1}{2}$

SPECIAL TOPIC

BUCKYBALLS

Diamond and graphite are well known as two forms of pure carbon, but now physicists have discovered a third form that goes under the strange name of **buckminsterfullerene,** or **buckyballs** for short. The description "ball" is appropriate for these molecules because they are the roundest molecule in existence. The simplest buckyball contains 60 carbon atoms bonded together in a molecular structure that looks much like a soccer ball (Fig. 28.14). The architecture that underlies such a surface is at the heart of the geodesic dome invented by visionary engineer R. Buckminster Fuller, hence the name for this new form of carbon. But the story doesn't stop with C_{60}. Another large molecule consisting of 70 carbon atoms and looking like a rugby ball has also been produced in abundance. Buckyballs were discovered by two astrophysicists, W. Kratschmer and D. Huffman, who were interested in studying the carbon gas that exists between stars. Because this carbon has coalesced and formed under myriad conditions, the two scientists were investigating as many methods as possible for vaporizing and then condensing it. One of their techniques consisted of evaporating graphite in a helium atmosphere. They found that the residue had some peculiar properties, one of which was its strong absorption of high-frequency ultraviolet light. They later found that chemistry literature had predicted that pure carbon-60 would have the properties they were finding. Carbon-60 had been produced in the past under exacting conditions, but only in samples of a few hundred molecules. They were producing samples containing billions of molecules. The technique they used was so simple that scientists around the world became involved in related investigations, and within a short time buckyballs became a fascinating item to study in the world of materials science. Modest amounts of even more exotic buckyballs have been produced as a result of these investigations. What has not yet been found is a use for buckyballs, but researchers are confident that they will discover a multitude of applications in the future. For example, scientists are trying to understand and exploit their superconducting properties and their ability to trap other atoms.

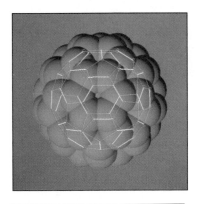

FIGURE 28.14
Structure of one of the recently discovered fullerene molecules, C_{60}, also known as the "buckyball." Another rendition is presented on page 896. *(Charles Winters)*

28.10

CHARACTERISTIC X-RAYS

X-rays are emitted when a metal target is bombarded with high-energy electrons. The x-ray spectrum typically consists of a broad continuous band and a series of sharp lines that are dependent on the type of metal used for the target, as shown in Figure 28.15. These discrete lines, called **characteristic x-rays,** were discovered in 1908, but their origin remained unexplained until the details of atomic structure were developed.

The first step in the production of characteristic x-rays occurs when a bombarding electron collides with an electron in an inner shell of a target atom with sufficient energy to remove the electron from the atom. The vacancy created in the shell is filled when an electron in a higher level drops down into the lower energy level containing the vacancy. The time it takes for this to happen is very short, less than 10^{-9} s. This transition is accompanied by the emission of a photon whose energy equals the difference in energy between the two levels. Typically, the energy of such transitions is greater than 1000 eV, and the emitted x-ray photons have wavelengths in the range of 0.01 nm to 1 nm.

Let us assume that the incoming electron has dislodged an atomic electron from the innermost shell, the K shell. If the vacancy is filled by an electron dropping from the next higher shell, the L shell, the photon emitted in the process is referred to as the K_α line on the curve of Figure 28.15. If the vacancy is filled by an electron dropping from the M shell, the line produced is called the K_β line.

Other characteristic x-ray lines are formed when electrons drop from upper levels to vacancies other than those in the K shell. For example, L lines are produced when vacancies in the L shell are filled by electrons dropping from higher shells. An L_α line is produced as an electron drops from the M shell to the L shell, and an L_β line is produced by a transition from the N shell to the L shell.

We can estimate the energy of the emitted x-rays as follows. Consider two electrons in the K shell of an atom whose atomic number is Z. Each electron partially shields the other from the charge of the nucleus, Ze, and so each is subject to an effective nuclear charge of $Z_{\text{eff}} = (Z - 1)e$. We can now use a modified form of Equation 28.12 to estimate the energy of either electron in the K shell (with $n = 1$):

$$E_{\text{K}} = -mZ_{\text{eff}}^2 \frac{k^2 e^4}{2\hbar^2} = -Z_{\text{eff}}^2 E_0$$

where E_0 is the ground-state energy. Substituting $Z_{\text{eff}} = Z - 1$ gives

$$E_{\text{K}} = -(Z - 1)^2 (13.6 \text{ eV}) \qquad \textbf{[28.22]}$$

As the following example shows, we can estimate the energy of an electron in an L or M shell in a similar fashion. Taking the energy difference between these two levels, we can then calculate the energy and wavelength of the emitted photon.

In 1914 Henry G. J. Moseley plotted the Z values for a number of elements versus $\sqrt{1/\lambda}$, where λ is the wavelength of the K_α line for each element. He found that such a plot produced a straight line, as in Figure 28.16. This is consistent with our rough calculations of the energy levels based on Equation 28.22. From this plot, Moseley was able to determine the Z values of other elements, providing a periodic chart in excellent agreement with the known chemical properties of the elements.

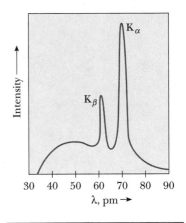

FIGURE 28.15
The x-ray spectrum of a metal target consists of broad continuous spectrum plus a number of sharp lines that are due to *characteristic x-rays.* The data shown were obtained when 35-keV electrons bombarded a molybdenum target. Note that 1 pm = 10^{-12} m = 0.001 nm.

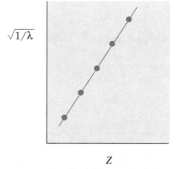

FIGURE 28.16
A Moseley plot. A straight line is obtained when $\sqrt{1/\lambda}$ is plotted versus Z for the K_α x-ray lines of a number of elements.

EXAMPLE 28.7 **Estimating the Energy of an X-Ray**

Estimate the energy of the characteristic x-ray emitted from a tungsten target when an electron drops from an M shell ($n = 3$ state) to a vacancy in the K shell ($n = 1$ state).

Solution The atomic number for tungsten is $Z = 74$. Using Equation 28.22, we see that the energy of the electron in the K shell state is approximately

$$E_K = -(74 - 1)^2(13.6 \text{ eV}) = -72\,500 \text{ eV}$$

The electron in the M shell ($n = 3$) is subject to an effective nuclear charge that depends on the number of electrons in the $n = 1$ and $n = 2$ states, which shield the nucleus. Because there are eight electrons in the $n = 2$ state and one electron in the $n = 1$ state, roughly nine electrons shield the nucleus, and so $Z_{eff} = Z - 9$. Hence, the energy of an electron in the M shell ($n = 3$), following Equation 28.22, is equal to

$$E_M = -Z_{eff}^2 E_3 = -(Z - 9)^2 \frac{E_0}{3^2} = -(74 - 9)^2 \frac{(13.6 \text{ eV})}{9} = -6380 \text{ eV}$$

where E_3 is the energy of an electron in the $n = 3$ level of the hydrogen atom. Therefore, the emitted x-ray has an energy equal to $E_M - E_K = -6380 \text{ eV} - (-72\,500 \text{ eV}) = 66\,100 \text{ eV}$. Note that this energy difference is also equal to hf, where $hf = hc/\lambda$, and where λ is the wavelength of the emitted x-ray.

Exercise Calculate the wavelength of the emitted x-ray for this transition.

Answer 0.0188 nm.

FIGURE 28.17
Energy level diagram of an atom with various allowed states. The lowest energy state, E_1, is the ground state. All others are excited states.

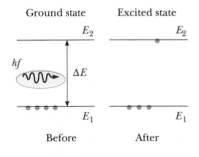

FIGURE 28.18
Diagram representing the process of *stimulated absorption* of a photon by an atom. The blue dots represent electrons in the various states. One electron is transferred from the ground state to the excited state when the atom absorbs one photon whose energy is $hf = E_2 - E_1$.

28.11

ATOMIC TRANSITIONS

We have seen that an atom will emit radiation only at certain frequencies that correspond to the energy separation between the various allowed states. Consider an atom with many allowed energy states, labeled $E_1, E_2, E_3, \ldots$, as in Figure 28.17. When light is incident on the atom, only those photons whose energy, hf, matches the energy separation, ΔE, between two levels can be absorbed by the atom. A schematic diagram representing this **stimulated absorption process** is shown in Figure 28.18. At ordinary temperatures, most of the atoms are in the ground state. If a vessel containing many atoms of a gaseous element is illuminated with a light beam containing all possible photon frequencies (that is, a continuous spectrum), only those photons of energies $E_2 - E_1$, $E_3 - E_1$, $E_4 - E_1$, and so on, can be absorbed. As a result of this absorption, some atoms are raised to various allowed higher energy levels called **excited states.**

Once an atom is in an excited state, there is a certain probability that it will jump back to a lower level by emitting a photon, as shown in Figure 28.19. This process is known as **spontaneous emission.** Typically, an atom will remain in an excited state for only about 10^{-8} s.

A third process which is important in lasers, **stimulated emission,** was predicted by Einstein in 1917. Suppose an atom is in the excited state E_2, as in Figure 28.20, and a photon with energy $hf = E_2 - E_1$ is incident on it. The incoming photon increases the probability that the excited electron will return to the ground state and thereby emit a second photon having the same energy hf. Note that two identical photons result from stimulated emission—the incident photon and the emitted photon. The emitted photon is exactly in phase with the incident photon. These photons can stimulate other atoms to emit photons in a chain of similar processes. The many photons produced in this fashion are the source of the intense, coherent light in a laser.

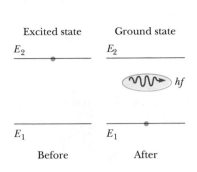

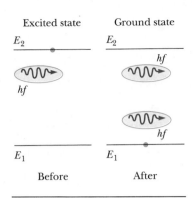

FIGURE 28.19
Diagram representing the process of *spontaneous emission* of a photon by an atom that is initially in the excited state E_2. When the electron falls to the ground state, the atom emits a photon whose energy is $hf = E_2 - E_1$.

FIGURE 28.20
Diagram representing the process of *stimulated emission* of a photon by an incoming photon of energy hf. Initially, the atom is in the excited state. The incoming photon stimulates the atom to emit a second photon of energy $hf = E_2 - E_1$.

28.12

LASERS AND HOLOGRAPHY

We have described how an incident photon can cause atomic transitions either upward (stimulated absorption) or downward (stimulated emission). The two processes are equally probable. When light is incident on a system of atoms, there is usually a net absorption of energy because, when the system is in thermal equilibrium, there are many more atoms in the ground state than in excited states. However, if the situation can be inverted so that there are more atoms in an excited state than in the ground state, a net emission of photons can result. Such a condition is called **population inversion**. This is the fundamental principle involved in the operation of a laser, an acronym for *l*ight *a*mplification by *s*timulated *e*mission of *r*adiation. The amplification corresponds to a buildup of photons in the system as the result of a chain reaction of events. The following three conditions must be satisfied in order to achieve laser action:

1. The system must be in a state of population inversion (that is, more atoms in an excited state than in the ground state).
2. The excited state of the system must be a *metastable state*, which means its lifetime must be long compared with the usually short lifetimes of excited states. When that is the case, stimulated emission will occur before spontaneous emission.
3. The emitted photons must be confined in the system long enough to allow them to stimulate further emission from other excited atoms. This is achieved by the use of reflecting mirrors at the ends of the system. One end is totally reflecting, while the other is slightly transparent to allow the laser beam to escape.

One device that exhibits stimulated emission of radiation is the helium-neon gas laser. Figure 28.21 is an energy level diagram for the neon atom in this

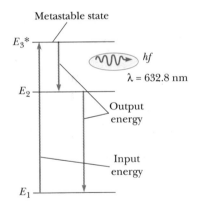

FIGURE 28.21
Energy level diagram for the neon atom, which emits photons at a wavelength of 632.8 nm through stimulated emission. The photon at this wavelength arises from the transition $E_3^* \rightarrow E_2$. This is the source of coherent light in the helium-neon gas laser.

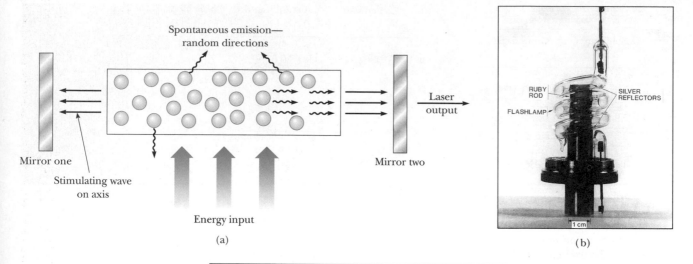

Spontaneous emission—
random directions

Mirror one

Stimulating wave
on axis

Energy input

(a)

Laser
output

Mirror two

RUBY
ROD

SILVER
REFLECTORS

FLASHLAMP

1 cm

(b)

FIGURE 28.22

(a) A schematic of a laser design. The tube contains atoms, which represent the active medium. An external source of energy (optical, electrical, etc.) is needed to "pump" the atoms to excited energy states. The parallel end mirrors provide the feedback of the stimulating wave. (b) Photograph of the first ruby laser showing the flash lamp surrounding the ruby rod. *(Courtesy of Hughes Aircraft Company)*

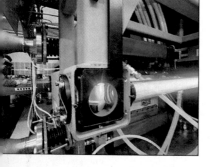

An engineer using a tunable dye laser to separate isotopes. *(Dan McCoy, Rainbow)*

system. The mixture of helium and neon is confined to a glass tube sealed at the ends by mirrors. An oscillator connected to the tube causes electrons to sweep through the tube, colliding with the atoms of the gas and raising them into excited states. Neon atoms are excited to state E_3 through this process and also as a result of collisions with excited helium atoms. Stimulated emission occurs as the neon atoms make a transition to state E_2 and neighboring excited atoms are stimulated. This results in the production of coherent light at a wavelength of 632.8 nm. Figure 28.22 summarizes the steps in the production of a laser beam.

Since the development of the first laser in 1960, laser technology has experienced tremendous growth. Lasers that cover wavelengths in the infrared, visible, and ultraviolet regions are now available. Applications include surgical "welding" of detached retinas, precision surveying and length measurement, a potential source for inducing nuclear fusion reactions, precision cutting of metals and other materials, and telephone communication along optical fibers. These and other applications are possible because of the unique characteristics of laser light. In addition to being highly monochromatic and coherent, laser light is also highly directional and can be sharply focused to produce regions of extremely intense light energy.

HOLOGRAPHY

One interesting application of the laser is holography, the production of three-dimensional images of objects. Figure 28.23a shows how a hologram is made.

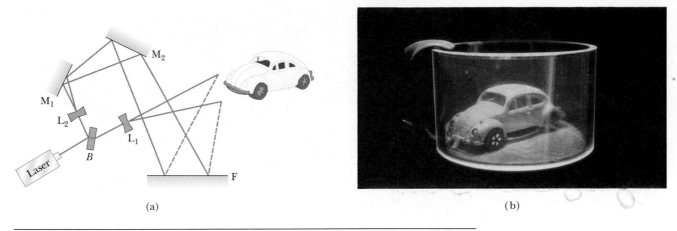

FIGURE 28.23

(a) Experimental arrangement for producing a hologram. (b) Photograph of a hologram made using a cylindrical film. Note the detail of the Volkswagen image. *(Courtesy of Central Scientific Company)*

Light from the laser is split into two parts by a half-silvered mirror at *B*. One part of the beam reflects off the object to be photographed and strikes an ordinary photographic film. The other half of the beam is diverged by lens L_2, reflects from mirrors M_1 and M_2, and finally strikes the film. The two beams overlap to form an extremely complicated interference pattern on the film, one that can be produced only if the phase relationship of the two waves is constant throughout the exposure of the film. This condition is met through the use of light from a laser because such light is coherent (all of the photons in the beam have the same phase). The hologram records not only the intensity of the light scattered from the object (as in a conventional photograph) but also the phase difference between the reference beam and the beam scattered from the object. Because of this phase difference, an interference pattern is formed that produces an image with full three-dimensional perspective.

A hologram is best viewed by allowing coherent light to pass through the developed film while looking back along the direction from which the beam comes. Figure 28.23b is a photograph of a hologram made using a cylindrical film.

SPECIAL TOPIC

COMPACT DISC PLAYERS

One of the most recent applications of the laser that has found its way into the life of college students is the compact disc (CD) player, a device that is rapidly replacing phonographs and tape players in ultra-high-fidelity sound reproduction. Figure 28.24 shows the important components of one of these devices. Light from a laser is sent

FIGURE 28.24
A compact disc player.

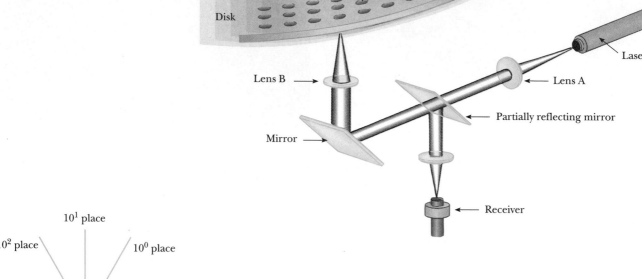

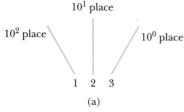

(a)

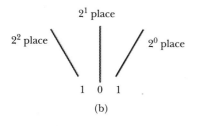

(b)

FIGURE 28.25
Weighting the digits in (a) the decimal system and (b) the binary system.

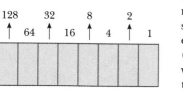

FIGURE 28.26
A byte consists of eight bits weighted as shown.

through lens A toward a partially reflecting mirror. The portion of the light that passes through this mirror is redirected by another mirror through lens B toward a laser disc. The light reflected off this disc retraces the original path until it reaches the partially reflecting mirror, where a portion of it is reflected into a receiver. The receiver then uses this reflected signal to drive a loudspeaker and to recreate the information stored on the laser disc.

In order to explain how the light retrieves information from the disc, let us first describe the binary number system. Consider an ordinary decimal number such as 123, where each digit has a place value associated with it. In this case, the number 123 contains 1 hundred, 2 tens, and 3 ones. Figure 28.25a shows an alternative way of expressing this same information. Note that in the figure we say that the number 3 occupies the 10^0 location, which is equivalent to saying that it occupies the one's place because $10^0 = 1$. Also, the number 2 occupies the 10^1 location, and the number 1 occupies the 10^2 location. This procedure can be extended upward as high as you care to count, where the weight of each position would be determined by some power of the number system base, which is 10 for decimal numbers.

Each bit position of a binary number also carries a particular weight that determines its magnitude, except that the place value of each bit is determined by some power of the base number 2. For example, Figure 28.25b shows a binary number 101 with its corresponding weight positions. The 1 on the right is in the $2^0 = 1$ location, the 0 is in the 2^1 location, and the 1 on the left is in the 2^2 position. Each digit in a binary number is called a bit, and a collection of eight bits is a byte. Thus, we can represent a byte with a box constructed as shown in Figure 28.26, indicating the weight of each bit. The box says that the weight of the lowest order bit (the rightmost one) is 1 ($=2^0$), and the weight of the highest order bit (the leftmost one) is 128 ($=2^7$). Thus, we see that the binary number 101 is the decimal number 5. As an exercise, convert the following binary numbers to decimal numbers: 00101101, 11011100, and 11111111. (The answers are 45, 220, and 255, respectively.)

Modern electronic devices such as computers use binary numbers in the form of a series of on and off signals. To understand this, consider the binary number 00000001, which is the binary representation of the decimal number 1. This binary

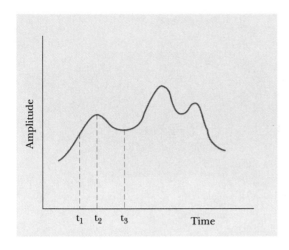

FIGURE 28.27
Sampling a sound signal whose amplitude varies with time.

number could be manipulated by an electronic circuit by turning a switch "off" for seven equal intervals of time, and then "on" for one interval.

The operation of a CD player is based on the procedure just described. The amplitude of a sound signal is sampled at regular intervals of time as in Figure 28.27, and the loudness of the sound at each sampling is then converted to a binary number by an electronic circuit. For example, suppose the amplitude of the sound signal in Figure 28.27 at t_1, t_2, and t_3 is decimal 4, 5, 4, respectively. The circuitry will translate this to binary 00000100, 00000101, and 00000100. This pattern of binary numbers is then recorded by placing a series of pits and smooth places on a laser disc (each pit is smaller than the dot over an "i" in this textbook). When laser light is reflected from one of the smooth places on the disc, it is reflected back to the detector in Figure 28.24, which interprets the presence of the reflection as an "on" signal, or as a binary 1. On the other hand, when the light strikes a pit on the rotating disc, it reflects in some random direction and does not return to the detector, which interprets the absence of a reflection as an "off," or a binary 0. Electrical equipment connected to the detector then drives a loudspeaker according to the size of the binary number it receives, thus reconstructing the original sound that was used to etch the disc.

Obviously, if a CD player is to reproduce sound faithfully, the laser beam must follow the spiral path of information perfectly. However, sometimes the laser beam can drift off track, and without a feedback procedure to let the player know this is happening, the fidelity of the music would be greatly reduced. Figure 28.28 shows how a diffraction grating is used to provide information to keep the beam on track. The central maximum of the diffraction pattern is used to read the information on the CD, and the two first order maxima are used for steering. The grating is designed so that the first order maxima fall on the smooth surface at each side of the information track. Both of these reflected beams have their own detectors, and since both are reflected from smooth, nonpitted surfaces, they should both have the same high intensity when they are detected. However, if the beam wanders off track, one of the beams will begin to strike pits on the information track and the amount of light reflected will diminish. This information can be used by suitable electronic circuits to guide the beam back to its desired location.

A new entrant in the high fidelity game is the Digital Audio Tape, DAT. The method of recording information is essentially the same as that used on compact discs in that the data is stored in the form of ones and zeros. The difference is that the medium is cassette tape. As noted in Chapter 20, information is recorded on cassette by

FIGURE 28.28
The laser beam in a CD player is able to follow the spiral track by using three beams produced with diffraction grating.

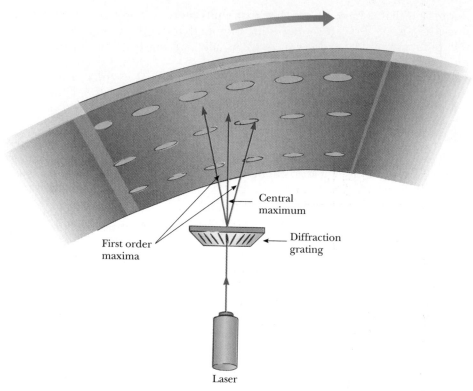

Central maximum

Diffraction grating

First order maxima

Laser

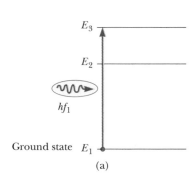

(a)

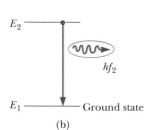

(b)

FIGURE 28.29
The process of fluorescence. (a) An atom absorbs a photon with energy hf_1 and ends up in an excited state, E_3. (b) The atom emits a photon of energy hf_2 when the electron moves from an intermediate state, E_2, back to the ground state.

magnetizing the tape, and the same process is used in DAT technology. Magnetizing a small portion of the tape in one direction is interpreted as the number one, while magnetizing it in the opposite direction is interpreted as a zero. The fidelity of CDs and the convenience of cassettes would seem to make DATs the perfect audio medium —except for the price.

28.13

FLUORESCENCE AND PHOSPHORESCENCE

When an atom absorbs a photon and ends up in an excited state, it can return to the ground state via some intermediate states, as shown in Figure 28.29. The photons emitted by the atom will have lower energy, and therefore lower frequency, than the absorbed photon. The process of converting high frequency radiation to lower frequency radiation by this means is called **fluorescence.**

The common fluorescent light, which makes use of this principle, works as follows. Electrons are produced in the tube as a filament at the end is heated to sufficiently high temperatures. The electrons are accelerated by an applied voltage, and this causes them to collide with atoms of mercury vapor present in the tube. As a result of the collisions, many mercury atoms are raised to excited states. As the excited atoms drop to their normal levels, some ultraviolet photons are emitted, and these strike a phosphor-coated screen. The screen absorbs these photons and emits visible light by means of fluorescence. Different phosphors on the screen emit light of different colors. "Cool white" fluorescent lights emit nearly all the visible colors and hence the light is very white. "Warm white" fluorescent lights have a phosphor that emits more red light and thereby produces a "warm" glow. It is interesting to note that the fluorescent lights above the meat counter in a grocery store are usually "warm white" to give the meat a redder color.

Two common fluorescent materials that you may have in your medicine cabinet are Murine eye drops and Pearl Drops toothpaste. If you use these products and then stand under a "black light," your eyes and teeth will glow with a beautiful yellow color. (A black light is simply a lamp that emits ultraviolet light along with some visible violet-blue light.)

Fluorescence analysis is often used to identify compounds. This is made possible by the fact that every compound has a "fingerprint" associated with the specific wavelength at which it fluoresces.

Another class of materials, called **phosphorescent** materials, continue to glow long after the illumination has been removed. An excited atom in a fluorescent material drops to its normal level in about 10^{-8} s, but an excited atom of a phosphorescent material may remain in an excited metastable state for periods ranging from a few seconds to several hours. Eventually, the atom will drop to its normal state and emit a visible photon. For this reason, phosphorescent materials emit light long after being placed in the dark. Paints made from these substances are often used to decorate the hands of watches and clocks, and to outline doors and stairways in large buildings so that these exits will be visible during power failures.

SUMMARY

The **Bohr model** of the atom is successful in describing the spectra of atomic hydrogen and hydrogen-like ions. One of the basic assumptions of the model is that the electron can exist only in certain orbits such that its angular momentum, mvr, is an integral multiple of $h/2\pi = \hbar$, where h is Planck's constant. Assuming circular orbits and a Coulomb force of attraction between electron and proton, the energies of the quantum states for hydrogen are

$$E_n = -\frac{mk^2e^4}{2\hbar^2}\left(\frac{1}{n^2}\right) \qquad n = 1, 2, 3, \ldots \qquad \text{[28.12]}$$

where k is the Coulomb constant, e is the charge on the electron, and n is an integer called a **quantum number.**

If the electron in the hydrogen atom jumps from an orbit whose quantum number is n_i to an orbit whose quantum number is n_f, it emits a photon of frequency f, given by

$$f = \frac{mk^2e^4}{4\pi\hbar^3}\left(\frac{1}{n_f^2} - \frac{1}{n_i^2}\right) \qquad \text{[28.14]}$$

Bohr's **correspondence principle** states that quantum mechanics is in agreement with classical physics when the quantum numbers for a system are very large.

One of the many great successes of quantum mechanics is that the quantum numbers n, ℓ, and m_ℓ associated with atomic structure arise directly from the mathematics of the theory. The quantum number n is called the **principal** quantum number, ℓ is the **orbital quantum number**, and m_ℓ is the **orbital magnetic quantum number**. In addition, a fourth quantum number, called the **spin magnetic quantum number**, m_s, is needed to explain certain features of atomic structure.

An understanding of the periodic table of the elements became possible when Pauli formulated the **exclusion principle,** which states that no two electrons in an atom can ever be in the same quantum state; that is, no two electrons in the same atom can have the same set of quantum numbers, n, ℓ, m_ℓ, and m_s.

Characteristic x-rays are produced when a bombarding electron collides with an electron in an inner shell of an atom with sufficient energy to remove the electron from the atom. The vacancy thus created is filled when an electron from a higher level drops down into the level containing the vacancy.

Lasers are monochromatic, coherent light sources that work on the principle of **stimulated emission** of radiation from a system of atoms.

ADDITIONAL READING

E. N. Andrade, *Rutherford and the Nature of the Atom*, New York, Doubleday Anchor, 1964.

M. W. Berns, "Laser Surgery," *Sci. American*, June 1991, p. 84.

R. F. Curl and R. E. Smalley, "Fullerenes," *Sci. American*, October 1991, p. 54.

G. Gamow, *Mr. Tomkins Explores the Atom*, New York, Cambridge University Press, 1945.

A. V. La Rocca, "Laser Applications in Manufacturing," *Sci. American*, March 1982, p. 94.

Lasers and Light, San Francisco, Freeman, 1969 (readings from *Scientific American*).

R. E. Latham, "Holography in the Science Classroom," *The Physics Teacher*, October 1986, p. 395.

E. N. Leith, "White-Light Holograms," *Sci. American*, October 1976, p. 80.

R. Moore, *Niels Bohr, The Man, His Science, and the World They Changed*, New York, Knopf, 1966.

P. Schewe, "Lasers," *The Physics Teacher*, November 1981, p. 534.

C. L. Strong, "How to Make Holograms," *Sci. American*, February 1967, p. 122.

J. A. Wheeler, "Niels Bohr, the Man," *Physics Today*, October 1985, p. 66.

CONCEPTUAL QUESTIONS

Example Suppose that the electron in the hydrogen atom obeyed classical mechanics rather than quantum mechanics. Why should such a hypothetical atom emit a continuous spectrum rather than the observed line spectrum?

Reasoning Classically, the electron can occupy any energy state. That is, all energies would be allowed. Therefore, if the electron obeyed classical mechanics, its spectrum, which originates from transitions between states, would be continuous rather than discrete.

Example If the exclusion principle were not valid, every electron would end up in the lowest energy state of the atom, and the chemical behavior of the elements would be grossly modified. Explain this statement.

Reasoning If it were not for the exclusion principle, there would be no restrictions on the number of electrons in an atom having the same set of quantum numbers. Thus, the electrons would all move to the lowest energy state, the ground state, which would grossly change the nature of matter as we know it.

Example If matter has a wave nature, why is this not observable in our daily experiences?

Reasoning The de Broglie wavelength of macroscopic objects such as a baseball moving with a typical speed such as 30 m/s is very small and impossible to measure. That is, $\lambda = h/mv$, is a very small number for macroscopic objects. We are not able to observe diffraction effects because the wavelength is much smaller than any aperture through which the object could pass.

Example Can the electron in the ground state of hydrogen absorb a photon of energy less than 13.6 eV? Can it absorb a photon of energy greater than 13.6 eV? Explain.
Reasoning In both cases, the answer is yes. Recall that the ionization energy of hydrogen is 13.6 eV. The electron can absorb a photon of energy less than 13.6 eV by making a transition to some intermediate state such as one with $n = 2$. It can absorb a photon of energy greater than 13.6 eV, but in doing so, the electron would be separated from the proton and have some residual kinetic energy.

Example List some ways in which quantum mechanics altered our view of the atom as pictured by the Bohr theory.
Reasoning It replaced the simple circular orbits in the Bohr theory with electron clouds. More important, quantum mechanics is consistent with Heisenberg's uncertainty principle, which tells us about the limits of accuracy in making measurements. In quantum mechanics, we talk about the probabilistic nature of the outcome of a measurement on a system, a concept which is incompatible with the Bohr theory. Finally, the Bohr theory of the atom contains only one quantum number, n, while quantum mechanics provides the basis for additional quantum numbers to explain the finer details of atomic structure.

Example In quantum mechanics, the angular momentum of an electron in an atom can be zero, whereas in classical mechanics, such a configuration is not meaningful. Does the quantum model introduce any philosophical difficulties in our view of the atom?
Reasoning In quantum mechanics, an electron in a state for which $L = 0$ simply means that the electron is in an orbital which has spherical symmetry. That is, the electron cloud has the same density at all points at a given distance r from the nucleus. It is incorrect to view an $L = 0$ electron as one which travels in a straight line through the nucleus.

Example When a hologram is produced, the system (including light source, object, beam splitter, and so on) must be held motionless within a quarter of a wavelength. Why?
Reasoning A hologram is an interference pattern between light scattered from the object and the reference beam. If anything moves by a distance comparable to the wavelength of the light, or more, the pattern will wash out. The effect is analogous to vibrating the slits in Young's double-slit experiment.

1. (a) Explain why the electrons in Rutherford's model of the atom cannot be at rest. (b) Discuss the interpretation of Rutherford's experiments if no large-angle deflection of the α particles had been seen.

2. The Bohr theory of the hydrogen atom is based on several assumptions. Discuss these assumptions and their significance. Do any of these assumptions contradict classical physics?

3. Does the light emitted by a neon sign constitute a continuous spectrum or only a few colors? Defend your answer.

4. Define the following terms: (a) energy level; (b) quantum jump; (c) ionization energy; (d) ground state.

5. The binding energy (in electron volts) of an electron in an atom is numerically equal to the minimum voltage (ionization potential) necessary to ionize the atom by electrical discharge. Explain this statement.

6. Must an atom first be ionized before it can emit light? Discuss.

7. Suppose an electron in the ground state of an atom has its first excited level at an energy of 3 eV above its ground-state value. What could happen to this electron if it were struck by an incoming electron having an energy of (a) 3 eV? (b) 4 eV?

8. Why aren't the relative intensities of the various spectral lines in a gas all the same?

9. What is excluded by the exclusion principle?

10. Why do lithium, potassium, and sodium exhibit similar chemical properties?

11. Which of the following elements would you expect to have properties most like those of silver: copper, cadmium, or palladium?

12. In what ways is the introduction of the quantum numbers via quantum mechanics more satisfying than their introduction through improvements on the Bohr theory?

13. Discuss why the term *electron clouds* is used to describe the electronic arrangement in the quantum mechanical view of the atom.

14. Why are the frequencies of characteristic x-rays dependent on the type of material used for the target?

15. In what sense do characteristic x-rays indicate that the electrons in the inner shells are tightly bound? Discuss.

16. Explain the difference between laser light and incandescent light.

17. Why is *stimulated emission* so important in the operation of a laser? (Interestingly, the concept of stimulated emission was first discussed by Albert Einstein 35 years before the first successful laser.)

18. Why does a laser usually emit only one particular color of light rather than several colors or perhaps a continuous spectrum?

19. Do all the emission lines in the Balmer series correspond to visible light? Explain.

PROBLEMS

Section 28.1 Early Models of the Atom

Section 28.2 Atomic Spectra

1. In a Rutherford scattering experiment, an α-particle (charge $= +2e$) with a kinetic energy of 5.0 MeV heads directly toward a gold nucleus (charge $= +79e$). Assuming the gold nucleus to be fixed in space, determine the distance of closest approach. (*Hint:* Use conservation of energy for the α-particle. Assume that the initial energy of the α-particle when far away is all kinetic and that at the point of closest approach it is all electrostatic potential energy.)

2. Use Equation 28.1 to calculate the wavelength of the first three lines in the Balmer series for hydrogen.

3. The "size" of the nucleus in Rutherford's model of the atom is about 1.0×10^{-15} m. (a) Determine the repulsive electrostatic force between two protons separated by this distance. (b) Determine (in MeV) the electrostatic potential energy of the pair of protons.

4. The "size" of the atom in Rutherford's model is about 1.0×10^{-10} m. (a) Determine the attractive electrostatic force between an electron and a proton separated by this distance. (b) Determine (in eV) the electrostatic potential energy of the atom.

5. Show that the Balmer series formula (Equation 28.1) can be written as

$$\lambda = \frac{364.5\,n^2}{n^2 - 4}\ \text{nm}$$

where $n = 3, 4, 5, \ldots$.

6. The "size" of the atom in Rutherford's model is about 1.0×10^{-10} m. (a) Determine the speed of an electron moving about the proton using the attractive electrostatic force between an electron and a proton separated by this distance. (b) Does this speed suggest that Einsteinian relativity must be considered when studying the atom? (c) Compute the de Broglie wavelength of the electron as it moves about the proton. (d) Does this wavelength suggest that wave effects, such as diffraction and interference, must be considered when studying the atom?

7. (a) Suppose the Rydberg constant in Balmer's formula were given by $R_H = 2.0 \times 10^7$ m^{-1}. What part of the electromagnetic spectrum would the Balmer series correspond to? (b) Repeat for

$$R_H = 0.50 \times 10^7\ \text{m}^{-1}.$$

Section 28.3 The Bohr Theory of Hydrogen

8. A hydrogen atom emits a photon of wavelength 657.7 nm. From what energy orbit to what lower energy orbit did the electron jump?

9. Show that the speed of the electron in the first (ground state) Bohr orbit of the hydrogen atom is given by $v = (1/137)c$.

10. Consider a large number of hydrogen atoms, with electrons all initially in the $n = 4$ state. (a) How many different wavelengths would be observed in the emission spectrum of these atoms? (b) What is the longest wavelength that could be observed? To which series does it belong?

11. Two hydrogen atoms collide head on and end up with zero kinetic energy. Each then emits a photon with a wavelength of 121.6 nm. At what speed were the atoms moving before the collision?

12. How much energy is required to ionize hydrogen when it is in (a) the ground state and (b) the state for which $n = 3$?

13. A hydrogen atom initially in its ground state ($n = 1$) absorbs a photon and ends up in the state for which $n = 3$. (a) What is the energy of the absorbed photon? (b) If the atom returns to the ground state, what photon energies could the atom emit?

14. Use Equation 28.10 to show that the radius of the ground state orbit of hydrogen is 0.529×10^{-10} m.

15. A particle of charge q and mass m, moving with a constant speed, v, perpendicular to a constant magnetic field, B, follows a circular path. If the angular momentum about the center of this circle is quantized so that $mvr = n\hbar$, show that the allowed radii for the particle are

$$r_n = \sqrt{\frac{n\hbar}{qB}}$$

for $n = 1, 2, 3, \ldots$.

16. (a) If an electron makes a transition from the $n = 4$ Bohr orbit to the $n = 2$ orbit, determine the wavelength of the photon created in the process. (b) Assuming that the atom was initially at rest, determine the recoil speed of the hydrogen atom when this photon is emitted.

17. Show that the speed of the electron in the nth Bohr orbit in hydrogen is given by

$$v_n = \frac{ke^2}{n\hbar}$$

18. A hydrogen atom is in its first excited state ($n = 2$). Using the Bohr theory of the atom, calcu-

☐ indicates problems that have full solutions available in the Student Solutions Manual and Study Guide.

late (a) the radius of the orbit, (b) the linear momentum of the electron, (c) the angular momentum of the electron, (d) the kinetic energy, (e) the potential energy, and (f) the total energy.

19. When a muon with charge $-e$ is captured by a proton, the resulting bound system forms a "muonic atom," which is the same as hydrogen except with a muon (of mass 207 times the mass of an electron) replacing the electron. For this "muonic atom" determine: (a) the Bohr radius, (b) the five lowest energy levels, (c) the first four wavelengths and the short wavelength limit in the Lyman series, and (d) the first three wavelengths and the short wavelength limit in the Balmer series.

20. Calculate the Coulomb force of attraction on the electron when it is in the ground state of the hydrogen atom.

21. In this problem you will estimate the lifetime of the hydrogen atom. An accelerating charge loses electromagnetic energy at a rate given by $P = -2kq^2a^2/(3c^3)$, where k is the Coulomb constant, q is the charge of the particle, a is its acceleration, and c is the speed of light in a vacuum. Assuming that the electron is one Bohr radius (0.0529 nm) from the center of the hydrogen atom, (a) determine its acceleration. (b) Show that P has units of energy per unit time and determine the rate of energy loss. (c) Calculate the energy of the electron and determine how long it will take for all of this energy to be converted into electromagnetic waves, assuming that the rate calculated in part (b) remains constant throughout the electron's motion.

22. What is the energy of the photon that, when absorbed by a hydrogen atom, could cause (a) an electronic transition from the $n = 3$ state to the $n = 5$ state and (b) an electronic transition from the $n = 5$ state to the $n = 7$ state?

23. Four possible transitions for a hydrogen atom are listed below:

(A) $n_i = 2; n_f = 5$ (B) $n_i = 5; n_f = 3$

(C) $n_i = 7; n_f = 4$ (D) $n_i = 4; n_f = 7$

(a) Which transition will emit the shortest wavelength photon? (b) For which transition will the atom gain the most energy? (c) For which transition(s) does the atom lose energy?

24. Analyze the Earth-Sun system following the Bohr model, where the gravitational force between Earth (mass m) and Sun (mass M) replaces the Coulomb force between the electron and proton (so that $F = GMm/r^2$ and $PE = -GMm/r$). Show that (a) the total energy of the Earth in an orbit

of radius r is given by $E = -GMm/(2r)$, (b) the radius of the nth orbit is given by $r_n = r_0 n^2$, where $r_0 = \hbar^2/(GMm^2) = 2.32 \times 10^{-138}$ m, and (c) the energy of the nth orbit is given by $E_n = -E_0/n^2$, where $E_0 = G^2M^2m^3/(2\hbar^2) = 1.71 \times 10^{182}$ J. (d) Using the Earth-Sun orbit radius of $r = 1.49 \times 10^{11}$ m, determine the value of the quantum number n. (e) Should you expect to observe quantum effects in the Earth-Sun system?

25. An electron is in the first Bohr orbit of hydrogen. Find (a) the speed of the electron, (b) the time required for the electron to circle the nucleus, and (c) the current in amperes corresponding to the motion of the electron.

26. (a) Calculate the angular momentum of the Moon due to its orbital motion about the Earth. In your calculation, use 3.84×10^8 m as the average Earth-Moon distance and 2.36×10^6 s as the period of the Moon in its orbit. (b) The Moon is in its lowest energy state while orbiting the Earth. Determine the corresponding quantum number. (c) By what fraction would the Earth-Moon radius have to be increased to increase the quantum number by 1?

27. Consider a hydrogen atom. (a) Calculate the frequency f of the $n = 2 \rightarrow n = 1$ transition and compare with the frequency f_{orb} of the electron orbital motion in the $n = 2$ state. (b) Make the same calculation for the $n = 10\,000 \rightarrow n = 9999$ transition. Comment on the results.

Section 28.4   Modification of the Bohr Theory

Section 28.5   De Broglie Waves and the Hydrogen Atom

28. Plot an energy level diagram like that in Figure 28.7 for singly ionized helium.

29. (a) Find the energy of the electron in the ground state of doubly ionized lithium, which has an atomic number $Z = 3$. (b) Find the radius of its ground-state orbit.

30. Construct an energy level diagram for doubly ionized lithium (Li^{2+}), for which $Z = 3$.

31. Using the concept of standing waves, de Broglie was able to derive Bohr's stationary orbit postulate by assuming that only those orbits that lead to standing waves, as in Figure 28.10a, are stable. Figure 28.10b shows the standing wave pattern for a vibrating string of length L fixed at its ends. If we imagine the wave pattern of the string to be that for a free particle of mass m confined to a one-dimensional box of length L, (a) show that the linear momentum of this particle is quantized with $mv = hn/(2L)$ and (b) show that the allowable en-

ergy levels for this particle are $E = E_0 n^2$, where $E_0 = h^2/(8mL^2)$.

32. Determine the wavelength of an electron in the third excited orbit of the hydrogen atom.

33. (a) Substitute numerical values into Equation 28.19 to find a value for the Rydberg constant for singly ionized helium, He^+. (b) Use the result of part (a) to find the wavelength associated with a transition from the $n = 2$ state to the $n = 1$ state of He^+. (c) Identify the region of the electromagnetic spectrum associated with this transition.

34. Using the Sommerfeld model, (a) identify all the possible subshells that can exist for the P shell, (b) find the maximum number of electrons that can occupy each subshell, and (c) find the maximum number of electrons that can occupy the P shell.

Section 28.6 Quantum Mechanics and the Hydrogen Atom

Section 28.7 The Spin Magnetic Quantum Number

35. The ρ-meson has a charge of $-e$, a spin quantum number of 1, and a mass 1507 times that of the electron. If the electrons in atoms were replaced by ρ-mesons, list the possible sets of quantum numbers for ρ-mesons in the $3d$ subshell.

36. List the possible sets of quantum numbers for electrons in the $3p$ subshell.

37. When the principal quantum number is $n = 4$, how many different values of (a) ℓ and (b) m_ℓ are possible?

Section 28.9 The Exclusion Principle and the Periodic Table

38. Zirconium ($Z = 40$) has two unpaired electrons in the d subshell. (a) What are all possible values of ℓ and s for each electron? (b) What are all possible values of n, m_ℓ, and m_s? (c) What is the electron configuration in zirconium?

39. (a) Write out the electronic configuration of the ground state for oxygen ($Z = 8$). (b) Write out the values for the set of quantum numbers n, ℓ, m_ℓ, and m_s for each of the electrons in oxygen.

40. How many different sets of quantum numbers are possible for an electron for which (a) $n = 1$, (b) $n = 2$, (c) $n = 3$, (d) $n = 4$, and (e) $n = 5$? Check your results to show that they agree with the general rule that the number of different sets of quantum numbers is equal to $2n^2$.

41. Suppose two electrons in the same system each have $\ell = 0$ and $n = 3$. (a) How many states would be possible if the exclusion principle were inoperative? (b) List the possible states, taking the exclusion principle into account.

Section 28.10 Characteristic X-Rays

42. The K-shell ionization energy of copper is 8979 eV. The L-shell ionization energy is 951 eV. Determine the wavelength of the K_α emission line of copper. What must the minimum voltage be on an x-ray tube with a copper target in order to see the K_α line?

43. When an electron drops from the M shell ($n = 3$) to a vacancy in the K shell ($n = 1$), the measured wavelength of the emitted x-ray is found to be 0.101 nm. Identify the element.

44. Use the method illustrated in Example 28.7 to calculate the wavelength of the x-ray emitted from a molybdenum target ($Z = 42$) when an electron undergoes a transition from the L shell ($n = 2$) to the K shell ($n = 1$).

45. The K series of the discrete spectrum of tungsten contains wavelengths of 0.0185 nm, 0.0209 nm, and 0.0215 nm. The K-shell ionization energy is 69.5 keV. Determine the ionization energies of the L, M, and N shells. Sketch the transitions.

ADDITIONAL PROBLEMS

46. The Lyman series for a (new!?) one-electron atom is observed in a distant galaxy. The wavelengths of the first four lines and the short-wavelength limit of this Lyman series are given by the energy-level diagram in Figure 28.30. Based on this information, calculate (a) the energies of the ground state

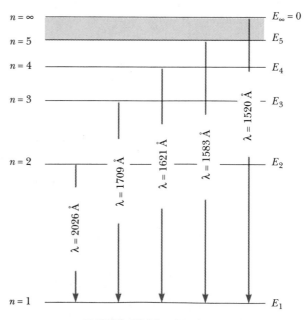

FIGURE 28.30 (Problem 46)

and first four excited states for this one-electron atom, and (b) the wavelengths of the first three lines and the short-wavelength limit in the Balmer series for this atom.

47. (a) Show that the wavelengths of the first four lines and the short wavelength limit of the Lyman series for the hydrogen atom are all exactly 60% of the wavelengths for the Lyman series in the one-electron atom described in Problem 46. (b) Based on this observation, explain why this atom could be hydrogen. (*Hint:* See Problem 63 in Chapter 26.)

48. (a) How much energy is required to cause an electron in hydrogen to move from the $n = 1$ state to the $n = 2$ state? (b) If the electrons gain this energy by collision with hydrogen atoms at a high temperature, find the minimum temperature of the heated hydrogen gas. The thermal energy of the heated atoms is given by $3k_B T/2$.

49. An electron is in the nth Bohr orbit of the hydrogen atom. (a) Show that the time it takes the electron to circle the nucleus once can be expressed as $T = \tau_0 n^3$, and determine the numerical value of τ_0. (b) On the average, an electron in the $n = 2$ orbit will exist in that orbit for about 1.0×10^{-8} s before it jumps down to the $n = 1$ (ground-state) orbit. How many revolutions about the nucleus are made by the electron before it jumps to the ground state? (c) If one revolution of the electron about the nucleus is defined as an "electron year" (analogous to an Earth year being one revolution of the Earth around the Sun), does the electron in the $n = 2$ orbit "live" very long? Explain. (d) How does the above calculation support the "electron cloud" concept?

50. A laser used in a holography experiment has an average output power of 5.0 mW. The laser beam is actually a series of pulses of electromagnetic radiation at a wavelength of 632.8 nm, each having a duration of 25 ms. Calculate (a) the energy (in joules) radiated with each pulse and (b) the number of photons in each pulse.

51. An electron in a hydrogen atom jumps from some initial Bohr orbit, n_i, to some final Bohr orbit, n_f, as in Figure 28.31. (a) If the photon emitted in the process is capable of ejecting a photoelectron from tungsten (work function = 4.58 eV), determine n_f. (b) If a minimum stopping potential of $V_0 = 7.51$ volts is required to prevent the photoelectron from hitting the anode, determine the value of n_i.

52. A pi meson (π^-) of charge $-e$ and mass 273 times greater than that of the electron is captured by a helium nucleus ($Z = +2$) as shown in Figure 28.32. (a) Draw an energy level diagram (in units

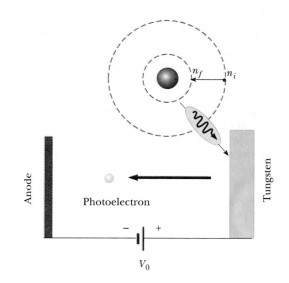

FIGURE 28.31 (Problem 51)

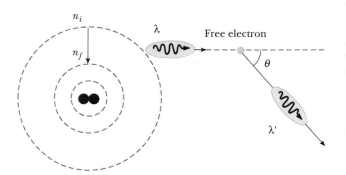

"Pi mesonic" He$^+$ atom
($Z = 2$, $m_\pi = 273 m_e$)

FIGURE 28.32 (Problem 52)

of eV) for this "Bohr-type" atom up to the first six energy levels. (b) When the pi meson makes a transition between two orbits, a photon is emitted that Compton scatters off a free electron initially at rest, producing a scattered photon of wavelength $\lambda' = 0.089\ 929\ 3$ nm at an angle of $\theta = 50.00°$, as shown on the right-hand side of Figure 28.32. Between which two orbits did the pi meson make a transition?

53. Mercury's ionization energy is 10.39 eV. The three longest wavelengths of the absorption spectrum of mercury are 253.7 nm, 185.0 nm, and 158.5 nm. (a) Construct an energy-level diagram for mercury. (b) Indicate all possible emission lines, starting from the highest energy level on the diagram.

(c) Disregarding recoil, determine the minimum speed an electron must have in order to make an inelastic collision with a mercury atom.

54. Suppose the ionization energy of an atom is 4.10 eV. In this same atom, we observe emission lines with wavelengths 310 nm, 400 nm, and 1377.8 nm. Use this information to construct the energy-level diagram with the least number of levels. Assume the higher energy levels are closer together.

55. An electron has a de Broglie wavelength equal to the diameter of the hydrogen atom. (a) What is the kinetic energy of the electron? (b) How does this energy compare with the ground-state energy of the hydrogen atom?

56. In order for an electron to be confined to a nucleus, its de Broglie wavelength would have to be less than 1.0×10^{-14} m. (a) What would be the kinetic energy of an electron confined to this region? (b) On the basis of this result, would you expect to find an electron in a nucleus? Explain.

57. Use Bohr's model of the hydrogen atom to show that when the atom makes a transition from the state n to the state $n - 1$, the frequency of the emitted light is given by

$$f = \frac{2\pi^2 mk^2 e^4}{h^3} \left(\frac{2n - 1}{(n - 1)^2 n^2} \right)$$

58. Calculate the classical frequency for the light emitted by an atom. To do so, note that the frequency of revolution is $v/2\pi r$, where r is the Bohr radius. Show that as n approaches infinity in the equation of Problem 57, the expression given there varies as $1/n^3$ and reduces to the classical frequency. (This is an example of the correspondence principle, which requires that the classical and quantum models agree for large values of n.)

59. A dimensionless number that often appears in atomic physics is the *fine-structure constant* α, given by

$$\alpha = \frac{ke^2}{\hbar c}$$

where k is the Coulomb constant. (a) Obtain a numerical value for $1/\alpha$. (b) In terms of α, what is the ratio of the Bohr radius, a_0, to the Compton wavelength, $\lambda = h/m_0 c$? (c) In terms of α, what is the ratio of the reciprocal of the Rydberg constant, $1/R$, to the Bohr radius?

ISAAC D. ABELLA   The University of Chicago

LASERS AND APPLICATIONS

Laser Principles

We encounter lasers in many applications in the home and workplace. The CD (compact disc) player in home audio systems incorporates a small laser disc reader. The laser printer in the office uses a laser beam to paint an image directly from the computer onto the xerographic surface, dispensing with the photographic lens of conventional copiers. The phone company uses laser communications devices on the fiber-optic network. The supermarket checkout counter is where most people see the laser beam in action. This enormous growth of laser applications was stimulated by many scientific and engineering advances that exploit some of the unique properties of laser light.

These properties derive from the distinctive way laser light is produced in contrast to the generation of ordinary light. Laser light originates from *energized* ("excited") atoms, ions, or molecules through a process of *stimulated emission* of radiation. The suitably prepared *active* laser medium is contained in an enclosure or *cavity* which organizes the normally random emission process into an intense, directional, monochromatic, and coherent wave. The end *mirrors* provide the essential optical feedback which selectively builds up the stimulating wave along the tube axis. However, in an ordinary sodium vapor street lamp, for example, the energized atoms *spontaneously* emit in random directions and at irregular times, over a broad spectrum, resulting in isotropic illumination of incoherent light. Laser light has a well-defined phase, permitting a wide variety of applications based on interference or wave modulation.

Currently operating laser systems utilize a variety of atomic gases, solids, or liquids as the working laser substance. These devices are designed to emit either continuous or pulsed monochromatic beams, and operate over a broad range of the optical spectrum (ultraviolet, visible, infrared) with output powers from milliwatts (10^{-3}) to megawatts (10^6). The particular application determines the choice of laser system, wavelength, power level, or other relevant variables, since no one laser has all the desirable properties.

For a laser to operate successfully, several *atomic physics* conditions must be satisfied, as noted in Section 28.12. The requirement for population inversion, that is, more atoms in a particular *excited state* than in a lower state, essentially means that energy must be supplied from outside the system. Otherwise, atoms would eventually radiate and develop an increasing probability for absorbing light. Finally, they would fall to the lowest energy state and stop emitting altogether.

Therefore, all laser systems must be connected to external energy sources, usually electrical, as required by conservation of energy, to maintain this nonthermal equilibrium situation. For example, we can energize atoms by electron impact in gaseous discharges (so called "electrical pumping"). We can also supply energy to lamps whose light populates excited states by photon absorption ("optical pumping") for those solids or liquids which do not conduct electric charge. These pumping mechanisms tend to have low efficiency (ratio of laser energy output to electric energy supplied), typically a few percent, with the balance discharged as heat into cooling water or circulating air.

Controlling the electrical input into the laser system allows a variable laser energy output, which may be important in many applications. Thus, the argon ion laser system can emit up to about 10 W in the green optical beam by adjustment of the elec-

933

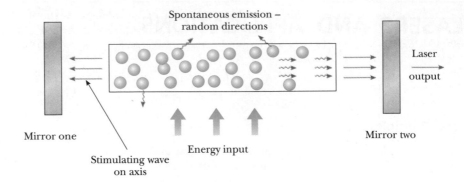

tric current in the argon gas, which in turn controls the degree of population inversion. Chemical lasers, on the other hand, operate without direct electrical input. Several highly reactive gases are mixed in the laser chamber, with the energy released in the ensuing reaction populating the excited levels in the molecule. In this case, the reactants need to be resupplied for the laser to operate for any length of time.

Some laser systems have fluid media, containing dissolved dye molecules. The dye lasers are usually pumped to excited levels by an external laser. The advantage of this arrangement is that dye lasers can be continuously "tuned" over a wide range of wavelengths, using prisms or gratings, whereas the pump source has a fixed wavelength. Color variability is important for those cases in which the laser is directed at materials whose absorption depends on wavelength. Thus, the laser can be tuned into exact coincidence with selected energy states. For example, blood does not absorb red light to any extent, which excludes red light use for most surgical applications on blood-rich tissue.

Recent developments in tunable solid state materials have permitted design of tunable lasers without the need for unstable dye molecules; most notable are sapphire crystals containing titanium ions. These materials are optically pumped by flashlamps or fixed wavelength lasers acting to populate the upper state directly.

A variety of laser systems are in general use today. They include the 1-mW helium-neon laser, usually appearing as a red beam at 632.8 nm (although yellow and green beams are available); the argon ion laser, which operates in the green or blue up to 10 W; the carbon dioxide gas laser, which emits in the infared at 10 mm and can produce several hundred watts; and the neodymium YAG laser, a powerful solid-state optically pumped system which emits at 1.06 mm either continuous or pulsed. The recently perfected diode junction laser emits in the near infrared and operates by passage of current through the semiconductor material. The recombination radiation is essentially direct conversion of electrical energy to laser light and is a very efficient process. The diodes can emit up to 5 W and can be used to energize other laser materials.

Applications

We shall describe a few applications that should serve to illustrate the wide variety of laser uses. Other applications are discussed as Special Topics in this chapter. First, there is the use of lasers in precision long-range distance measurement (range finding). It has become important, for astronomical and geophysical purposes, to measure as precisely as possible the distance from various points on the surface of the Earth to a point on the Moon's surface. To facilitate this, the Apollo astronauts set up a compact array, a 0.5-m square of reflector prisms on the Moon, which allows laser pulses directed from an Earth station to be retroreflected to the same station. Using the known speed of light and the measured round-trip travel time of a 1-ns pulse, one can determine the Earth-Moon distance, 380 000 km, to a precision of better than 10 cm.

Such information would be useful, for example, in making more reliable earthquake predictions and for learning more about the motions of the Earth-Moon system. This technique requires a high-power pulsed laser for its success, since a sufficient burst of photons must return to a collecting telescope on Earth and be detected. Variants of this method are also used to measure the distance to inaccessible points on the Earth as well as in military range finders.

Novel medical applications utilize the fact that the different laser wavelengths can be absorbed in specific biological tissues. A widespread eye condition, glaucoma, is manifested by a high fluid pressure in the eye, which can lead to destruction of the optic nerve. A simple laser operation (iridectomy) can "burn" open a tiny hole in a clogged membrane, relieving the destructive pressure. Along the same lines, a serious side effect of diabetes is neovascularization, the formation of weak blood vessels, which often leak blood into extremities. When this occurs in the eye, vision deteriorates (diabetic retinopathy) leading to blindness in diabetic patients. It is now possible to direct the green light from the argon ion laser through the clear eye lens and eye fluid, focus on the retina edges, and photo-coagulate the leaky vessels. These procedures have greatly reduced blindness in glaucoma and diabetes patients.

An argon laser passing through a cornea and lens during eye surgery. (© *Alexander Tsiaras, Science Source/Photo Researchers, Inc.*)

Laser surgery is now a practical reality. Infrared light at 10 mm from a carbon dioxide laser can cut through muscle tissue, primarily by heating and evaporating the water contained in cellular material. Laser power of about 100 W is required in this technique. The advantage of the "laser knife" over conventional methods is that laser radiation cuts and coagulates at the same time, leading to substantial reduction of blood loss. In addition, the technique virtually eliminates cell migration, which is very important in tumor removal. Furthermore, a laser beam can be trapped in fine glass-fiber light-guides (endoscopes) by means of total internal reflection (Section 22.9). The light fibers can be introduced through natural orifices, conducted around internal organs, and directed to specific interior body locations, eliminating the need for massive surgery. For example, bleeding in the gastrointestinal tract can be optically cauterized by fiber-optic endoscopes inserted through the mouth.

Finally, we describe an application to biological and medical research. It is often important to isolate and collect unusual cells for study and growth. A laser cell separator exploits the fact that specific cells can be tagged with fluorescent dyes. All cells are then dropped from a tiny charged nozzle and laser-scanned for the dye tag. If triggered by the correct light-emitting tag, a small voltage applied to parallel plates deflects the falling electrically charged cell into a collection beaker. This is an efficient method for extracting the proverbial needles from the haystack.

Problems

1. (a) In the case of the Earth-Moon laser range finder, what is the round-trip time for a laser pulse? (b) What precision in timing is required, that is, how small a time change needs to be detectable, to be able to measure the distance to an error of 10 cm? What effect does the Earth's atmosphere have?

2. A laser beam of wavelength $\lambda = 600$ nm is directed at the Moon from a laser tube 1 cm in diameter. Does the beam spread at all and what is the diameter of the "spot" on the Moon's surface? Does your answer demand a good strategy for successful lunar-array illumination and how would you do it?

3. Estimate how much chemical reagent is required to produce a chemical laser of 1-kW output, in moles/second. Pick a reasonable exothermic reaction rate in kcal/mole. Estimate how many kilograms each of hydrogen and chlorine gas would be needed to make an HCl laser operate for an hour at 1-kW output.

29 NUCLEAR PHYSICS

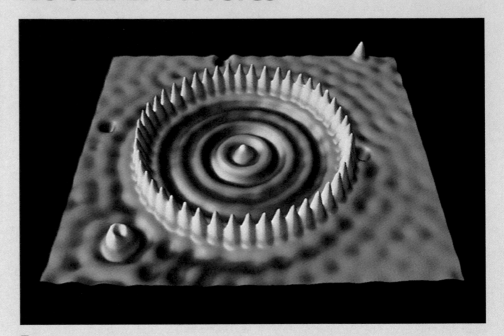

This photograph is a demonstration of a "quantum-corral" consisting of a ring of 48 iron atoms located on a copper surface. The diameter of the ring is 14.3 nm, and the photograph was obtained using a low-temperature scanning tunneling microscope. Corrals and other structures are able to confine surface-electron waves, and the study of such structures will play an important role in determining the future of small electronic devices. (IBM Research Corporation)

In 1896, the year that marks the birth of nuclear physics, Henri Becquerel (1852–1908) discovered radioactivity in uranium compounds. A great deal of activity followed this discovery as researchers attempted to understand and characterize the radiation that we now know to be emitted by radioactive nuclei. Pioneering work by Rutherford showed that the radiation was of three types, which he called alpha, beta, and gamma rays. These types are classified according to the nature of their electric charge and according to their ability to penetrate matter. Later experiments showed that alpha rays are helium nuclei, beta rays are electrons, and gamma rays are high-energy photons.

In 1911 Rutherford and his students Geiger and Marsden performed a number of important scattering experiments involving alpha particles. These experiments established that the nucleus of an atom can be regarded as essentially a point mass and point charge and that most of the atomic mass is contained in the nucleus. Furthermore, such studies demonstrated a wholly new type of force, the *nuclear force*, which is predominant at distances of less than about 10^{-14} m and zero at great distances.

Other milestones in the development of nuclear physics include:

1. The observation of nuclear reactions by Cockcroft and Walton in 1930
2. The discovery of the neutron by Chadwick in 1932
3. The discovery of artificial radioactivity by Joliot and Irene Curie in 1933
4. The discovery of nuclear fission by Hahn and Strassman in 1938
5. The development of the first controlled fission reactor by Fermi and his collaborators in 1942

In this chapter we discuss the properties and structure of the atomic nucleus. We start by describing the basic properties of nuclei and follow with a discussion of the phenomenon of radioactivity. Finally, we explore nuclear reactions and the various processes by which nuclei decay.

29.1

SOME PROPERTIES OF NUCLEI

All nuclei are composed of two types of particles: protons and neutrons. The only exception is the ordinary hydrogen nucleus, which is a single proton. In describing some of the properties of nuclei, such as their charge, mass, and radius, we make use of the following quantities:

1. The **atomic number,** Z, which equals the number of protons in the nucleus
2. The **neutron number,** N, which equals the number of neutrons in the nucleus
3. The **mass number,** A, which equals the number of nucleons in the nucleus (*Nucleon* is a generic term used to refer to either a proton or a neutron.)

The symbol we use to represent nuclei is $^A_Z X$, where X represents the chemical symbol for the element. For example, $^{27}_{13}Al$ has the mass number 27 and the atomic number 13; therefore, it contains 13 protons and 14 neutrons. When no confusion is likely to arise, we omit the subscript Z because the chemical symbol can always be used to determine Z.

The nuclei of all atoms of a particular element must contain the same number of protons but different numbers of neutrons. Nuclei that are related in this way are called **isotopes.**

> The isotopes of an element have the same Z value but different N and A values.

The natural abundances of isotopes can differ substantially. For example, $^{11}_6C$, $^{12}_6C$, $^{13}_6C$, and $^{14}_6C$, are four isotopes of carbon. The natural abundance of the $^{12}_6C$ isotope is about 98.9%, whereas that of the $^{13}_6C$ isotope is only about 1.1%. Some isotopes do not occur naturally but can be produced in the laboratory through nuclear reactions. Even the simplest element, hydrogen, has isotopes: ^1_1H, hydrogen; ^2_1H, deuterium; and ^3_1H, tritium.

CHARGE AND MASS

The proton carries a single positive charge, $+e$, the electron carries a single negative charge, $-e$, where $e = 1.6 \times 10^{-19}$ C, and the neutron is electrically neutral. Because the neutron has no charge, it is difficult to detect. The proton is about 1836 times as massive as the electron, and the masses of the proton and the neutron are almost equal (Table 29.1).

It is convenient to define, for atomic masses, the **unified mass unit,** u, in such a way that the mass of the isotope ^{12}C is exactly 12 u. That is, the mass of a nucleus (or atom) is measured relative to the mass of an atom of the neutral carbon-12

TABLE 29.1
Rest Mass of the Proton, Neutron, and Electron in Various Units

Particle	Mass		
	kg	u	MeV/c^2
Proton	1.6726×10^{-27}	1.007 276	938.28
Neutron	1.6750×10^{-27}	1.008 665	939.57
Electron	9.109×10^{-31}	5.486×10^{-4}	0.511

Ernest Rutherford (1871–1937), a physicist from New Zealand, was awarded the Nobel Prize in 1908 for discovering that atoms can be broken apart by alpha rays, and for studying radioactivity. "On consideration, I realized that this scattering backward must be the result of a single collision, and when I made calculations I saw that it was impossible to get anything of that order of magnitude unless you took a system in which the greater part of the mass of the atom was concentrated in a minute nucleus. It was then that I had the idea of an atom with a minute massive center carrying a charge." (*North Wind Picture Archives*)

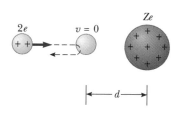

FIGURE 29.1
An alpha particle on a head-on collision course with a nucleus of charge *Ze*. Because of the Coulomb repulsion between the like charges, the alpha particle will stop instantaneously at a distance *d* from the nucleus.

isotope (the nucleus plus six electrons). Thus, the mass of ^{12}C is defined to be 12 u, where 1 u = $1.660\ 559 \times 10^{-27}$ kg. The proton and neutron each have a mass of about 1 u, and the electron has a mass that is only a small fraction of an atomic mass unit.

Because the rest energy of a particle is given by $E_0 = m_0 c^2$, it is often convenient to express the particle's mass in terms of its energy equivalent. In particular, it is important to know the energy equivalent of one atomic mass unit, 1u.

$$E_0 = m_0 c^2 = (1.67 \times 10^{-27}\ \text{kg})(3.00 \times 10^8\ \text{m/s})^2$$

$$= 1.50 \times 10^{-10}\ \text{J} = 9.39 \times 10^8\ \text{eV} = 939\ \text{MeV}$$

The rest energy of an electron is 0.511 MeV, and that of a neutron is 940 MeV.

The rest masses of the proton, neutron, and electron are given in Table 29.1. The masses and some other properties of selected isotopes are provided in Appendix B.

EXAMPLE 29.1   The Unified Mass Unit

Use Avogadro's number to show that 1 u = 1.66×10^{-27} kg.

Solution   We know that 12.0 g of ^{12}C contains Avogadro's number of atoms. That is, 1 mol of any substance contains 6.02×10^{23} atoms. Thus the mass of one carbon atom is

$$\text{Mass of one }^{12}\text{C atom} = \frac{12.0\ \text{g}}{6.02 \times 10^{23}\ \text{atoms}} = 1.99 \times 10^{-23}\ \text{g}$$

Since one atom of ^{12}C is defined to have a mass of 12.0 u, we find that

$$1\ \text{u} = \frac{1.99 \times 10^{-23}\ \text{g}}{12.0} = 1.66 \times 10^{-27}\ \text{kg}$$

THE SIZE OF NUCLEI

The size and structure of nuclei were first investigated in the scattering experiments of Rutherford, discussed in Chapter 28, Section 28.1. Using the principle of conservation of energy, Rutherford found an expression for how close an alpha particle moving directly toward the nucleus can come to the nucleus before being turned around by Coulomb repulsion.

In such a head-on collision, the kinetic energy of the incoming alpha particle must be converted completely to electrical potential energy when the particle stops at the point of closest approach and turns around (Fig. 29.1). If we equate the initial kinetic energy of the alpha particle to the maximum electrical potential energy of the system (alpha particle plus target nucleus), we have

$$\tfrac{1}{2}mv^2 = k_e \frac{q_1 q_2}{r} = k_e \frac{(2e)(Ze)}{d}$$

where *d* is the distance of closest approach. Solving for *d*, we get

$$d = \frac{4k_e Ze^2}{mv^2}$$

From this expression, Rutherford found that alpha particles approached to within 3.2×10^{-14} m of a nucleus when the foil was made of gold. Thus, the radius of the gold nucleus must be less than this value. For silver atoms, the distance of closest approach was 2×10^{-14} m. From these results, Rutherford concluded that the positive charge in an atom is concentrated in a small sphere, which he called the nucleus, whose radius is no greater than about 10^{-14} m. Because such small lengths are common in nuclear physics, a convenient unit of length is the *femtometer* (fm), sometimes called the **fermi,** defined as

$$1 \text{ fm} \equiv 10^{-15} \text{ m}$$

Since the time of Rutherford's scattering experiments, a multitude of other experiments have shown that most nuclei are approximately spherical and have an average radius given by

$$r = r_0 A^{1/3} \qquad \text{[29.1]}$$

where A is the mass number and r_0 is a constant equal to 1.2×10^{-15} m. Because the volume of a sphere is proportional to the cube of its radius, it follows from Equation 29.1 that the volume of a nucleus (assumed to be spherical) is directly proportional to A, the total number of nucleons. This suggests that *all nuclei have nearly the same density.* Nucleons combine to form a nucleus *as though* they were tightly packed spheres (Fig. 29.2).

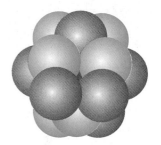

FIGURE 29.2
A nucleus can be visualized as a cluster of tightly packed spheres, where each sphere is a nucleon.

EXAMPLE 29.2 Nuclear Volume and Density

Find (a) an approximate expression for the mass of a nucleus of mass number A, (b) an expression for the volume of this nucleus in terms of the mass number, and (c) a numerical value for its density.

Solution   (a) The mass of the proton is approximately equal to that of the neutron. Thus, if the mass of one of these particles is m, the mass of the nucleus is approximately Am.

(b) Assuming the nucleus is spherical and using Equation 29.1, we find that the volume is

$$V = \tfrac{4}{3}\pi r^3 = \tfrac{4}{3}\pi r_0^3 A$$

(c) The nuclear density can be found as follows:

$$\rho_n = \frac{\text{mass}}{\text{volume}} = \frac{Am}{\tfrac{4}{3}\pi r_0^3 A} = \frac{3m}{4\pi r_0^3}$$

Taking $r_0 = 1.2 \times 10^{-15}$ m and $m = 1.67 \times 10^{-27}$ kg, we find that

$$\rho_n = \frac{3(1.67 \times 10^{-27} \text{ kg})}{4\pi(1.2 \times 10^{-15} \text{ m})^3} = 2.3 \times 10^{17} \text{ kg/m}^3$$

Note that the nuclear density is about 2.3×10^{14} times greater than the density of water $(1 \times 10^3 \text{ kg/m}^3)$!

NUCLEAR STABILITY

Given that the nucleus consists of a closely packed collection of protons and neutrons, you might be surprised that it can exist. The very large repulsive electrostatic forces between protons should cause the nucleus to fly apart. However, nuclei are stable because of the presence of another, short-range (about

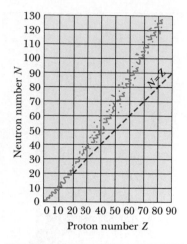

FIGURE 29.3

A plot of the neutron number N versus the proton number Z for the stable nuclei. The dashed straight line corresponding to the condition $N = Z$ is called the *line of stability*.

2 fm) force, the **nuclear force.** This is an attractive force that acts between all nuclear particles. The protons attract each other via the nuclear force, and at the same time they repel each other through the Coulomb force. The nuclear force also acts between pairs of neutrons and between neutrons and protons.

The nuclear force dominates the Coulomb repulsive force within the nucleus (at short ranges). If this were not the case, stable nuclei would not exist. Moreover, the strong nuclear force is nearly independent of charge. In other words, the nuclear forces associated with the proton-proton, proton-neutron, and neutron-neutron interactions are approximately the same, apart from the additional repulsive Coulomb force for the proton-proton interaction.

There are about 400 stable nuclei; hundreds of others have been observed but are unstable. A plot of N versus Z for a number of stable nuclei is given in Figure 29.3. Note that light nuclei are most stable if they contain equal numbers of protons and neutrons—that is, if $N = Z$—but heavy nuclei are more stable if $N > Z$. This can be partially understood by recognizing that, as the number of protons increases, the strength of the Coulomb force increases, which tends to break the nucleus apart. As a result, more neutrons are needed to keep the nucleus stable, since neutrons experience only the attractive nuclear forces. Eventually, when $Z = 83$, the repulsive forces between protons cannot be compensated by the addition of more neutrons. In effect, the additional neutrons ''dilute'' the nuclear charge density. Elements that contain more than 83 protons do not have stable nuclei.

29.2

BINDING ENERGY

The total mass of a nucleus is always less than the sum of the masses of its individual nucleons. According to the Einstein mass-energy relationship, if the mass difference, Δm, is multiplied by c^2, we obtain the binding energy of the nucleus. In other words, *the energy of the bound system (the nucleus) is less than the sum of the energies of the separated nucleons.* Therefore, in order to separate a nucleus into protons and neutrons, energy must be delivered to the system.

EXAMPLE 29.3   The Binding Energy of the Deuteron

The nucleus of the deuterium atom, called the deuteron, consists of a proton and a neutron. Calculate the deuteron's binding energy, given that its mass is 2.014 102 u.

Solution We know that the proton and neutron masses are

$$m_p = 1.007\ 825\ \text{u} \qquad m_n = 1.008\ 665\ \text{u}$$

Note that the masses used for the proton and deuteron in this example are actually those of the neutral atoms, as found in Appendix B. We are able to use atomic masses for these calculations since the electron masses cancel. Therefore,

$$m_p + m_n = 2.016\ 490\ \text{u}$$

Maria Goeppert-Mayer was born in Germany and received a Ph.D. in physics from Göttingen University. Her thesis work, which dealt with quantum mechanical effects in atoms, was encouraged by Paul Ehrenfest. She moved to the United States in 1930 after her husband received a professorship in chemistry at Johns Hopkins University. While at home raising two children, she wrote a book with her husband on statistical mechanics. Following the publication of the book, she was offered a lectureship in chemistry at Johns Hopkins University, but her presence as a woman at faculty functions was awkward.

In the late 1940s she and her husband received appointments at the University of Chicago, but her position was without pay because of a strict nepotism rule. While in Chicago, she worked with Enrico Fermi and Edward Teller. During her collaboration with Teller, she became interested in why certain elements in the periodic table were so abundant and stable. She eventually realized that the most stable elements were characterized by particular values of atomic and neutron numbers, which she called "magic numbers." She labored over a theoretical explanation of these numbers for about one year, and finally arrived at a solution during a conversation with Fermi. This resulted in a 1950 publication in which she described the "shell" model of the nucleus. (It is interesting to note that the paper confused many Russian scientists, who translated "shell" as "grenade.") As often happens in scientific research, a similar model was simultaneously developed by Hans Jensen, a German scientist. Maria Goeppert-Mayer and Hans Jensen were awarded the Nobel prize in physics in 1963 for their extraordinary work in understanding the structure of the nucleus. When Goeppert-Mayer heard that she had won the Nobel prize, she said "Oh, how wonderful, I've always wanted to meet a king."

Maria Goeppert-Mayer (1906–1972)

(Courtesy of Louise Barker/AIP Niels Bohr Library)

To calculate the mass difference, we subtract the deuteron mass from this value:

$$\Delta m = (m_p + m_n) - m_d$$

$$= 2.016\ 490\ \text{u} - 2.014\ 102\ \text{u} = 0.002\ 388\ \text{u}$$

Since 1 u corresponds to an equivalent energy of 931.50 MeV (that is, $1\ \text{u} \cdot c^2 = 931.50$ MeV), the mass difference corresponds to the binding energy

$$E_b = (0.002\ 388\ \text{u})(931.5\ \text{MeV/u}) = \boxed{2.224\ \text{MeV}}$$

This result tells us that in order to separate a deuteron into its constituent proton and a neutron, it is necessary to add 2.224 MeV of energy. One way of supplying the deuteron with this energy is by bombarding the nucleus with energetic particles.

If the binding energy of a nucleus were zero, the nucleus would separate into its constituent protons and neutrons without the addition of any energy; that is, it would spontaneously break apart.

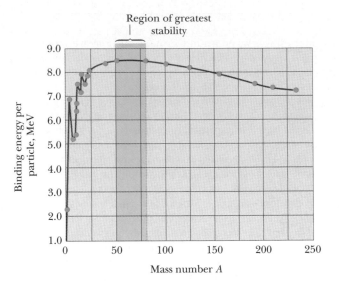

FIGURE 29.4

A plot of the binding energy per nucleon versus the mass number A for nuclei that are along the line of stability shown in Figure 29.3.

It is interesting to examine a plot of binding energy per nucleon, E_b/A, as a function of mass number for various stable nuclei (Fig. 29.4). Except for the lighter nuclei, the average binding energy per nucleon is about 8 MeV. Note that the curve peaks in the vicinity of $A = 60$. That is, nuclei with mass numbers greater or less than 60 are not as strongly bound as those near the middle of the periodic table. As we shall see later, this fact allows energy to be released in fission and fusion reactions. The curve is slowly varying for $A > 40$, which suggests that the nuclear force saturates. In other words, a particular nucleon can interact with only a limited number of other nucleons, which can be viewed as the "nearest neighbors" in the close-packed structure illustrated in Figure 29.2.

29.3
RADIOACTIVITY

In 1896 Becquerel accidentally discovered that uranium salt crystals emit an invisible radiation that can darken a photographic plate even if the plate is covered to exclude light. After several such observations under controlled conditions, he concluded that the radiation emitted by the crystals was of a new type, one that required no external stimulation. This spontaneous emission of radiation was soon called **radioactivity.** Subsequent experiments by other scientists showed that other substances were also radioactive.

The most significant investigations of this type were conducted by Marie and Pierre Curie. After several years of careful and laborious chemical separation processes on tons of pitchblende, a radioactive ore, the Curies reported the discovery of two previously unknown elements, both of which were radioactive. These were named polonium and radium. Subsequent experiments, including Rutherford's famous work on alpha-particle scattering, suggested that radioactivity was the result of the decay, or disintegration, of unstable nuclei.

Three types of radiation can be emitted by a radioactive substance: alpha (α) rays, where the emitted particles are ^{4}He nuclei; beta (β) rays, in which the emitted particles are either electrons or positrons; and gamma (γ) rays, in which the emitted "rays" are high-energy photons. A positron is a particle similar to the electron in all respects except that it has a charge of $+e$. (The positron is said to be the antiparticle of the electron.)

Marie Sklodowska Curie was born in Poland shortly after the unsuccessful Polish revolt against the Russians in 1863. Following her high school education, she worked diligently to help meet the educational expenses of her older brother and sister who had left for Paris. At the same time, she managed to save enough money for her own trip to Paris and entered the Sorbonne in 1891. Although she lived under very frugal conditions during this period (fainting once from hunger in the classroom), she graduated at the top of her class.

In 1895 she married the French chemist Pierre Curie, who was already known for the discovery of piezoelectricity. (A piezoelectric crystal exhibits a potential difference when it is under pressure.) Using piezoelectric materials to measure the activity of radioactive substances, she demonstrated the radioactive nature of the elements uranium and thorium. In 1898 she and her husband discovered a new radioactive element in uranium ore; they called it polonium, after Madame Curie's native land. By the end of 1898, the Curies had succeeded in isolating trace amounts of an even more radioactive element, which they named radium. In an effort to produce weighable quantities of radium, they embarked on a painstaking effort of isolating radium from pitchblende, an ore rich in uranium. After four years of purifying and repurifying tons of ore, and using their own life savings to finance their work, the Curies succeeded in preparing about 0.1 g of radium. In 1903 Marie and Pierre Curie were awarded the Nobel prize in physics, which they shared with A. H. Becquerel, for their studies of radioactive substances.

After her husband's death in a tragic accident in 1906, Madame Curie took over his professorship at the Sorbonne. Unfortunately, she experienced prejudice in the scientific community because she was a woman. For example, after being nominated to the French Academy of Sciences, she was refused membership, losing by one vote.

In 1911 she was awarded a second Nobel prize in chemistry for the discovery of radium and polonium. She spent the last few decades of her life supervising the Paris Institute of Radium. Madame Curie died of leukemia caused by years of exposure to radioactive substances.

Marie Curie
(1867–1934)

(FPG)

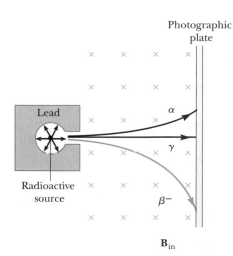

FIGURE 29.5
The radiation from a radioactive source, such as radium, can be separated into three components using a magnetic field to deflect the charged particles. The photographic plate at the right records the events.

The hands and numbers of this luminous watch contain minute amounts of radium salt. The radioactive decay of radium causes the watch to glow in the dark.
(© Richard Megna 1990, Fundamental Photographs)

It is possible to distinguish these three forms of radiation by using the scheme described in Figure 29.5. The radiation from a radioactive sample is directed into a region with a magnetic field, and the beam splits into three components, two bending in opposite directions and the third not changing

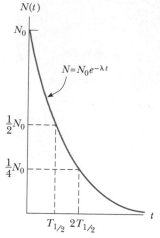

FIGURE 29.6
Plot of the exponential decay law for radioactive nuclei. The vertical axis represents the number of radioactive nuclei present at any time, t, and the horizontal axis is time. The parameter $T_{1/2}$ is the half-life of the sample.

Decay rate

Exponential decay

direction. From this simple observation it can be concluded that the radiation of the undeflected beam carries no charge (the gamma ray), the component deflected upward contains positively charged particles (alpha particles), and the component deflected downward contains negatively charged particles (electrons).

The three types of radiation have quite different penetrating powers. Alpha particles barely penetrate a sheet of paper, beta particles can penetrate a few millimeters of aluminum, and gamma rays can penetrate several centimeters of lead.

THE DECAY CONSTANT AND HALF-LIFE

If a radioactive sample contains N radioactive nuclei at some instant, it is found that the number of nuclei, ΔN, that decay in a small time interval Δt is proportional to N:

$$\Delta N = -\lambda N \Delta t \qquad \text{[29.2]}$$

where λ is a constant called the **decay constant**. The negative sign signifies that N decreases with time; that is, ΔN is negative. The value of λ for any isotope determines the rate at which that isotope will decay.

> The **decay rate**, or **activity**, R, of a sample is defined as the number of decays per second. From Equation 29.2, we see that the decay rate is

$$R = \left| \frac{\Delta N}{\Delta t} \right| = \lambda N \qquad \text{[29.3]}$$

Thus we see that isotopes with a large λ value decay at a rapid rate and those with a small λ value decay slowly.

A general decay curve for a radioactive sample is shown in Figure 29.6. It can be shown from Equation 29.3 (using calculus) that the number of nuclei present varies with time according to the expression

$$N = N_0 e^{-\lambda t} \qquad \text{[29.4]}$$

where N is the number of radioactive nuclei present at time t, N_0 is the number present at time $t = 0$, and $e = 2.718 \ldots$ is the base of the natural logarithms. Processes that obey Equation 29.4 are sometimes said to undergo exponential decay.[1]

Another parameter that is useful for characterizing radioactive decay is the **half-life**, $T_{1/2}$. The half-life of a radioactive substance is the time it takes half of a

---

[1] Other examples of exponential decays were discussed in Chapter 18 in connection with RC circuits, and in Chapter 20 in connection with RL circuits.

given number of radioactive nuclei to decay. Setting $N = N_0/2$ and $t = T_{1/2}$ in Equation 29.4 gives

$$\frac{N_0}{2} = N_0 e^{-\lambda T_{1/2}}$$

Writing this in the form $e^{\lambda T_{1/2}} = 2$ and taking the natural logarithm of both sides, we get

$$T_{1/2} = \frac{\ln 2}{\lambda} = \frac{0.693}{\lambda}$$ [29.5]   Half-life

This is a convenient expression relating the half-life to the decay constant. Note that after an elapsed time of one half-life, $N_0/2$ radioactive nuclei remain (by definition); after two half-lives, half of these will have decayed and $N_0/4$ radioactive nuclei will be left; after three half-lives, $N_0/8$ will be left; and so on.

The unit of activity is the **curie** (Ci), defined as

$$1 \text{ Ci} \equiv 3.7 \times 10^{10} \text{ decays/s}$$ [29.6]

This unit was selected as the original activity unit because it is the approximate activity of 1 g of radium. The SI unit of activity is the **becquerel** (Bq):

$$1 \text{ Bq} = 1 \text{ decay/s}$$ [29.7]

Therefore, $1 \text{ Ci} = 3.7 \times 10^{10}$ Bq. The most commonly used units of activity are the millicurie (10^{-3} Ci) and the microcurie (10^{-6} Ci).

EXAMPLE 29.4   The Activity of Radium

The half-life of the radioactive nucleus $^{226}_{86}\text{Ra}$ is 1.6×10^3 years. If a sample contains 3.0×10^{16} such nuclei, determine the activity at this time.

Solution   First, let us convert the half-life to seconds:

$$T_{1/2} = (1.6 \times 10^3 \text{ years})(3.16 \times 10^7 \text{ s/year}) = 5.0 \times 10^{10} \text{ s}$$

Now we can use this value in Equation 29.5 to get the decay constant:

$$\lambda = \frac{0.693}{T_{1/2}} = \frac{0.693}{5.0 \times 10^{10} \text{ s}} = 1.4 \times 10^{-11} \text{ s}^{-1}$$

We can calculate the activity of the sample at $t = 0$ using $R_0 = \lambda N_0$, where R_0 is the decay rate at $t = 0$ and N_0 is the number of radioactive nuclei present at $t = 0$:

$$R_0 = \lambda N_0 = (1.4 \times 10^{-11} \text{ s}^{-1})(3.0 \times 10^{16}) = 4.1 \times 10^5 \text{ decays/s}$$

Since $1 \text{ Ci} = 3.7 \times 10^{10}$ decays/s, the activity, or decay rate, at $t = 0$ is

$$R_0 = \boxed{11.1 \ \mu\text{Ci}}$$

EXAMPLE 29.5   The Activity of Radon Gas

Radon $^{222}_{86}\text{Rn}$ is a radioactive gas that can be trapped in the basement of homes, and its presence in high concentrations is a known health hazard. Radon has a half-life of 3.83 days. A gas sample contains 4.0×10^8 radon atoms initially.

(a) How many atoms will remain after twelve days have passed?

Reasoning First note, that 12 days corresponds to about 3.1 half-lives. In three half-lives, the number of radon atoms is reduced by a factor $2^3 = 8$. Therefore, the number of remaining radon atoms is approximately $4.0 \times 10^8/8 = 5.0 \times 10^7$.

Solution A more precise answer is obtained by first finding the decay constant from Equation 29.5:

$$\lambda = \frac{0.693}{T_{1/2}} = \frac{0.693}{3.83 \text{ days}} = 0.181 \text{ days}^{-1}$$

We now use Equation 29.4, taking $N_0 = 4.0 \times 10^8$ and the value of λ just found to obtain the number N remaining after twelve days:

$$N = N_0 e^{-\lambda t} = (4.0 \times 10^8 \text{ atoms})e^{-(0.181 \text{ days}^{-1})(12 \text{ days})} = \boxed{4.6 \times 10^7 \text{ atoms}}$$

This is very close to our original estimate of 5.0×10^7 atoms.

(b) What is the initial activity of the radon sample?

Solution First, we must express the decay constant in units of s^{-1}:

$$\lambda = \frac{0.693}{(3.83 \text{ days})(8.64 \times 10^4 \text{ s/day})} = 2.09 \times 10^{-6} \text{ s}^{-1}$$

From Equation 29.3 and the above value of λ, we find that the initial activity is

$$R = \lambda N_0 = (2.09 \times 10^{-6} \text{ s}^{-1})(4.0 \times 10^8) = 840 \text{ decays/s} = \boxed{840 \text{ Bq}}$$

Exercise Find the activity of the radon sample after 12 days have elapsed.

Answer 95 Bq.

29.4

THE DECAY PROCESSES

As stated in the previous section, a radioactive nucleus spontaneously decays via alpha, beta, and gamma decay. Let us discuss these in more detail.

ALPHA DECAY

If a nucleus emits an alpha particle ($^4_2$He), it loses two protons and two neutrons. Let us adopt the historical terminology of calling the nucleus before decay the **parent nucleus** and the nucleus remaining after decay the **daughter nucleus.** The nucleus $^{238}_{92}$U decays by alpha emission, and it is easy to visualize the process in symbolic form as follows:

$$^{238}_{92}\text{U} \longrightarrow \, ^{234}_{90}\text{Th} + \, ^4_2\text{He} \qquad\qquad \textbf{[29.8]}$$

This says that a parent nucleus, $^{238}_{92}$U, emits an alpha particle, $^4_2$He, and thereby changes to a daughter nucleus, $^{234}_{90}$Th. Note the following facts about this reaction: (1) the atomic number (number of protons) on the left is the same as on the right ($92 = 90 + 2$) since charge must be conserved, and (2) the mass number (protons plus neutrons) on the left is the same as on the right ($238 = 234 + 4$). The half-life for ^{238}U decay is 4.47×10^9 years.

When one element changes into another, as in alpha decay, the process is called **transmutation**. In order for alpha emission to occur, the mass of the parent must be greater than the combined mass of the daughter and the alpha particle. In the decay process, this excess mass is converted into energy of other forms and appears in the form of kinetic energy in the daughter nucleus and the alpha particle. Most of the kinetic energy is carried away by the alpha particle because it is much less massive than the daughter nucleus. This can be understood by first noting that a particle's kinetic energy and momentum, p, are related as follows:

$$KE = \frac{p^2}{2\,m}$$

Because momentum is conserved, the two particles in a decay must have equal, but oppositely directed, momenta. Thus, the lighter particle has more kinetic energy than the more massive particle.

EXAMPLE 29.6 The Alpha Decay of Radium

Radium, $^{226}_{88}\text{Ra}$, decays by alpha emission. What is the daughter element formed?

Solution The decay can be written symbolically as

$$^{226}_{88}\text{Ra} \longrightarrow X + ^{4}_{2}\text{He}$$

where X is the unknown daughter element. Requiring that the mass numbers and atomic numbers balance on the two sides of the arrow, we find that the daughter nucleus must have a mass number of 222 and an atomic number of 86:

$$^{226}_{88}\text{Ra} \longrightarrow ^{222}_{86}X + ^{4}_{2}\text{He}$$

The periodic table shows that the nucleus with an atomic number of 86 is radon, Rn. Thus, the process is

$$^{226}_{88}\text{Ra} \longrightarrow ^{222}_{86}\text{Rn} + ^{4}_{2}\text{He}$$

This decay is shown schematically in Figure 29.7

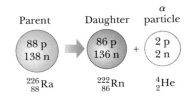

FIGURE 29.7
(Example 29.6) Alpha decay of the $^{226}_{88}\text{Ra}$ nucleus.

EXAMPLE 29.7 The Energy Liberated When Radium Decays

In Example 29.6, we showed that the $^{226}_{88}\text{Ra}$ nucleus undergoes alpha decay to $^{222}_{86}\text{Rn}$. Calculate the amount of energy liberated in this decay. Take the mass of $^{226}_{88}\text{Ra}$ to be 226.025 406 u, that of $^{222}_{86}\text{Rn}$ to be 222.017 574 u, and that of $^{4}_{2}\text{He}$ to be 4.002 603 u, as found in Appendix B.

Solution After decay, the mass of the daughter, m_d, plus the mass of the alpha particle, m_α, is

$$m_d + m_\alpha = 222.017\ 574\ \text{u} + 4.002\ 603\ \text{u} = 226.020\ 177\ \text{u}$$

Thus, calling the mass of the parent nucleus M_p, we find that the mass lost during decay is

$$\Delta m = M_p - (m_d + m_\alpha) = 226.025\ 406\ \text{u} - 226.020\ 177\ \text{u} = 0.005\ 229\ \text{u}$$

Using the relationship 1 u = 931.5 MeV, we find that the energy liberated is

$$E = (0.005\ 229\ \text{u})(931.50\ \text{MeV/u}) = \boxed{487\ \text{MeV}}$$

BETA DECAY

When a radioactive nucleus undergoes beta decay, *the daughter nucleus contains the same number of nucleons as the parent nucleus but the atomic number is increased by 1.* A typical beta decay event is

$$^{14}_{6}\text{C} \longrightarrow {}^{14}_{7}\text{N} + {}^{0}_{-1}\text{e} \qquad \text{[29.9]}$$

The superscripts and subscripts on the carbon and nitrogen nuclei follow our usual conventions, but those on the electron may need some explanation. The -1 indicates that the electron has a charge whose magnitude is equal to that of the proton but negative. The 0 used for the electron's mass number indicates that the mass of the electron is almost zero relative to that of carbon and nitrogen nuclei.

The emission of electrons from a *nucleus* is surprising because, in all our previous discussions, we stated that the nucleus is composed of protons and neutrons only. This apparent discrepancy can be explained by noting that the emitted electron is created in the nucleus by a process in which a neutron is transformed into a proton. This can be represented by the equation

$$^{1}_{0}\text{n} \longrightarrow {}^{1}_{1}\text{p} + {}^{0}_{-1}\text{e} \qquad \text{[29.10]}$$

Let us consider the energy of the system of Equation 29.9 before and after decay. As with alpha decay, energy must be conserved in beta decay. The following example illustrates how to calculate the amount of energy released in the beta decay of $^{14}_{6}\text{C}$.

EXAMPLE 29.8   The Beta Decay of $^{14}_{6}\text{C}$

Find the energy liberated in the beta decay of $^{14}_{6}\text{C}$ to $^{14}_{7}\text{N}$ as represented by Equation 29.9. Refer to Appendix B for the atomic masses.

Solution   We find from Appendix B that $^{14}_{6}\text{C}$ has a mass of 14.003 242 u and $^{14}_{7}\text{N}$ has a mass of 14.003 074 u. Here, the mass difference between the initial and final states is[2]

$$\Delta m = 14.003\ 242\ \text{u} - 14.003\ 074\ \text{u} = 0.000\ 168\ \text{u}$$

This corresponds to an energy release of

$$E = (0.000\ 168\ \text{u})(931.50\ \text{MeV/u}) = \boxed{0.156\ \text{MeV}}$$

From Example 29.8, we see that the energy released in the beta decay of ^{14}C is approximately 0.16 MeV. As with alpha decay, we expect the electron to carry away virtually all of this as kinetic energy because apparently it is the lightest particle produced in the decay. However, as Figure 29.8 shows, only a small number of electrons have this maximum kinetic energy, represented as $KE_{\max}$ on the graph; most of the electrons emitted have kinetic energies lower than this predicted value. If the daughter nucleus and the electron are not carrying away this liberated energy, then the energy conservation requirement leads to the

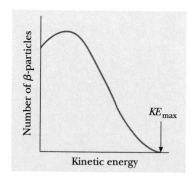

FIGURE 29.8
A typical beta-decay spectrum.

---

[2] In beta decay, we must keep track of the electrons involved in the process. In the initial state, the neutral ^{14}C atom has six electrons. In the final state, the ^{14}N atom also has six electrons. Thus, we must conclude that ^{14}N is not neutral. However, if we include the electron emitted in the decay process, we effectively have a mass equivalent to a neutral ^{14}N atom.

Enrico Fermi, an Italian-American physicist, received his doctorate from the University of Pisa in 1922 and did postdoctorate work in Germany under Max Born. After returning to Italy in 1924, he became a professor of physics at the University of Rome in 1926. He received the Nobel Prize for physics in 1938 for his work dealing with the production of transuranic radioactive elements by neutron bombardment.

Fermi first became interested in physics at the age of 14, after reading an old physics book in Latin. He had an excellent scholastic record and was able to recite Dante's *Divine Comedy* and much of Aristotle from memory. His great ability to solve problems in theoretical physics and his skill for simplifying very complex situations made him somewhat of an oracle. He was also a gifted experimentalist and teacher of physics. During one of his early lecture trips to the United States in the depths of the Depression, his car became disabled and he pulled into a nearby gas station. After Fermi repaired the car with ease, the station owner immediately offered him a job.

Fermi and his family immigrated to the United States, where he became a naturalized citizen in 1944. Once in America, Fermi accepted a position at Columbia University and later became a professor at the University of Chicago. After the Manhattan Project was established, Fermi was selected to design and build a structure (called an atomic pile) in which a self-sustained chain reaction might occur. The structure, built in the squash court of the University of Chicago, contained uranium in combination with graphite blocks to slow the neutrons to thermal velocities. Cadmium rods inserted in the "pile" were used to absorb neutrons and control the reaction rate. History was made at 3:45 P.M. on December 2, 1942 as the cadmium rods were slowly withdrawn, and a self-sustained chain reaction was observed. Fermi's earthshaking achievement of the world's first nuclear reactor marked the beginning of the atomic age.

Fermi died of cancer in 1954 at the age of 53. One year later, the one-hundredth element was discovered and named *fermium* in his honor.

**Enrico Fermi
(1901–1954)**

(National Accelerator Laboratory)

question, "What accounts for the missing energy?" As an additional complication, further analysis of beta decay shows that the principles of conservation of both angular momentum and linear momentum appear to be violated!

In 1930 Pauli proposed that a third particle must be present to carry away the "missing" energy and to conserve momentum. Enrico Fermi later named this particle the **neutrino** ("little neutral one") because it had to be electrically neutral and have little or no rest mass. Although it eluded detection for many years, the neutrino (ν) was finally detected experimentally in 1950. The properties of the neutrino are

1. It has zero electric charge.
2. It has a rest mass smaller than that of the electron, and in fact its rest mass may be zero (although recent experiments suggest that this may not be true).
3. It interacts very weakly with matter and is therefore very difficult to detect.

Properties of the neutrino

Thus, with the introduction of the neutrino, we are now able to represent the beta decay process of Equation 29.9 in its correct form:

$$^{14}_{6}\text{C} \longrightarrow {}^{14}_{7}\text{N} + {}_{-1}^{0}\text{e} + \bar{\nu} \qquad \text{[29.11]}$$

where the bar in the symbol $\bar{\nu}$ indicates an **antineutrino.** To explain what an antineutrino is, let us first consider the following decay.

$$^{12}_{7}N \longrightarrow {}^{12}_{6}C + {}^{0}_{1}e + \nu \qquad [29.12]$$

Here we see that when ^{12}N decays into ^{12}C, a particle is produced that is identical to the electron except that it has a positive charge of $+e$. This particle is called a positron. Because it is like the electron in all respects except charge, the positron is said to be **antiparticle** to the electron. We shall discuss antiparticles further in Chapter 30. For now, it suffices to say that *a neutrino is emitted in positron decay and an antineutrino is emitted in electron decay.*

GAMMA DECAY

Very often a nucleus that undergoes radioactive decay is left in an excited energy state. The nucleus can then undergo a second decay to a lower energy state, perhaps to the ground state, by emitting one or more photons. The process is very similar to the emission of light by an atom. An atom emits radiation to release some extra energy when an electron ''jumps'' from a state of high energy to a state of lower energy. Likewise, the nucleus uses essentially the same method to release any extra energy it may have following a decay or some other nuclear event. In nuclear de-excitation, the ''jumps'' that release energy are made by protons or neutrons in the nucleus as they move from a higher energy level to a lower level. The photons emitted in such a de-excitation process are called **gamma rays,** which have very high energy relative to the energy of visible light.

A nucleus may reach an excited state as the result of a violent collision with another particle. However, it is more common for a nucleus to be in an excited state as a result of alpha or beta decay. The following sequence of events represents a typical situation in which gamma decay occurs:

$$^{12}_{5}B \longrightarrow {}^{12}_{6}C^{*} + {}^{0}_{-1}e \qquad [29.13]$$

$$^{12}_{6}C^{*} \longrightarrow {}^{12}_{6}C + \gamma \qquad [29.14]$$

Equation 29.13 represents a beta decay in which ^{12}B decays to $^{12}C^{*}$ where the asterisk indicates that the carbon nucleus is left in an excited state following the decay. The excited carbon nucleus then decays to the ground state by emitting a gamma ray, as indicated by Equation 29.14. Note that gamma emission does not result in any change in either Z or A.

PRACTICAL USES OF RADIOACTIVITY

Carbon Dating.   The beta decay of ^{14}C given by Equation 29.11 is commonly used to date organic samples. Cosmic rays (high-energy particles from outer space) in the upper atmosphere cause nuclear reactions that create ^{14}C from ^{14}N. In fact, the ratio of ^{14}C to ^{12}C isotopic abundance in the carbon dioxide molecules of our atmosphere has a constant value of about 1.3×10^{-12} as determined by measuring carbon ratios in tree rings. All living organisms have the same ratio of ^{14}C to ^{12}C because they continuously exchange carbon dioxide with their surroundings. When an organism dies, however, it no longer absorbs ^{14}C from the atmosphere, and so the ratio of ^{14}C to ^{12}C decreases as the result of the beta decay of ^{14}C. It is therefore possible to determine the age of a material by measuring its activity per unit mass due to the decay of ^{14}C. Using carbon dating, samples of wood, charcoal, bone, and shell have been identified as having lived

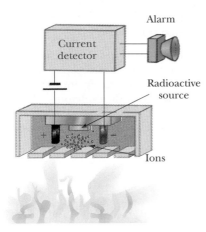

FIGURE 29.9
An ionization-type smoke detector. Smoke entering the chamber reduces the detected current, causing the alarm to sound.

from 1000 to 25 000 years ago. This knowledge has helped us to reconstruct the history of living organisms—including humans—during this time span.

A particularly interesting example is the dating of the Dead Sea Scrolls. This group of manuscripts was first discovered by a shepherd in 1947. Translation showed them to be religious documents, including most of the books of the Old Testament. Because of their historical and religious significance, scholars wanted to know their age. Carbon dating applied to fragments of the scrolls and to the material in which they were wrapped established that they were about 1950 years old.

Smoke Detectors. Smoke detectors are frequently used in homes and industry for fire protection. Most of the common ones are the ionization-type that use radioactive materials (see Fig. 29.9). A smoke detector consists of an ionization chamber, a sensitive current detector, and an alarm. A weak radioactive source ionizes the air in the chamber of the detector, which creates charged particles. A voltage is maintained between the plates inside the chamber, setting up a small but detectable current in the external circuit. As long as the current is maintained, the alarm is deactivated. However, if smoke drifts into the chamber, the ions become attached to the smoke particles. These heavier particles do not drift as readily as do the lighter ions, which causes a decrease in the detector current. The external circuit senses this decrease in current and sets off the alarm.

Radon Detecting. Radioactivity can also affect our daily lives in harmful ways. Soon after the discovery of radium by the Curies, it was found that the air in contact with radium compounds becomes radioactive. It was shown that this radioactivity came from the radium itself, and the product was therefore called "radium emanation." Rutherford and Soddy succeeded in condensing this "emanation," confirming that it is a real substance—the inert, gaseous element now called **radon,** Rn. We now know that the air in uranium mines is radioactive because of the presence of radon gas. The mines must therefore be well ventilated to help protect the miners. The fear of radon pollution has now moved from uranium mines into our own homes (see Example 29.5). Since certain types of rock, soil, brick, and concrete contain small quantities of radium, some of the resulting radon gas finds its way into our homes and other buildings. The most serious problems arise from leakage of radon from the ground into the structure. One practical remedy is to exhaust the air through a pipe just above the underlying soil or gravel directly to the outdoors by means of a small fan or blower.

EXAMPLE 29.9   Should We Report This to Homicide?

A 50.0-g sample of carbon is taken from the pelvis bone of a skeleton and is found to have a carbon-14 decay rate of 200.0 decays/min. It is known that carbon from a living organism has a decay rate of 15.0 decays/min·g and that ^{14}C has a half-life of 5730 y $= 3.01 \times 10^9$ min. Find the age of the skeleton.

Solution   Let us start with Equation 29.4

$$N = N_0 e^{-\lambda t}$$

and multiply both sides of λ to get

$$\lambda N = \lambda N_0 e^{-\lambda t}$$

But from Equation 29.3 we see that this is equivalent to

$$R = R_0 e^{-\lambda t}$$

where R is the present activity and R_0 was the activity when the skeleton was a part of a living organism. We are given that $R = 200.0$ decays/min, and we can find R_0 as

$$R_0 = \left(15.0 \frac{decays}{min \cdot g}\right)(50.0\ g) = 750 \frac{decays}{min}$$

The decay constant is found from Equation 29.5 as

$$\lambda = \frac{0.693}{T_{1/2}} = \frac{0.693}{3.01 \times 10^9\ min} = 2.30 \times 10^{-10}\ min^{-1}$$

Thus, we make the following substitutions:

$$R = R_0 e^{-\lambda t}$$

$$200.0 \frac{decays}{min} = \left(750 \frac{decays}{min}\right) e^{-(2.30 \times 10^{-10}\ min^{-1})\, t}$$

$$0.266 = e^{-(2.30 \times 10^{-10}\ min^{-1})\, t}$$

Now we take the natural log of both sides of the equation, to give

$$\ln(0.266) = -(2.30 \times 10^{-10}\ min^{-1})\, t$$

$$-1.32 = -(2.30 \times 10^{-10}\ min^{-1})\, t$$

$$t = 5.74 \times 10^9\ min = \boxed{10\ 900\ y}$$

S P E C I A L   T O P I C

CARBON-14 AND THE SHROUD OF TURIN

Since the Middle Ages, many people have marveled at a 14-foot-long, yellowing piece of linen found in Turin, Italy, purported to be the burial shroud of Jesus Christ. The cloth bears a remarkable, full-size likeness of a crucified body, with wounds on the head that could have been caused by a crown of thorns, and another in the side that could have been the cause of death. Skepticism over the authenticity of the shroud has existed since its first public showing in 1354; in fact, a French bishop declared it to be a fraud at the time. Because of its controversial nature, religious bodies have taken a neutral stance on its authenticity.

In 1978 the bishop of Turin allowed the cloth to be subjected to scientific analysis, but notably missing from these tests was a carbon-14 dating. The reason for this omission was that, at the time, carbon-dating techniques required a piece of cloth about the size of a handkerchief. In 1988 the process had been refined to the point that pieces as small as one square inch were sufficient, and permission was granted to allow the dating to proceed. Three labs were selected for the testing, and each was given four pieces of material. One of these was a piece of the shroud, while the other three were control pieces similar in appearance to the shroud.

The testing procedure consisted of burning the cloth to produce carbon dioxide, which was then converted chemically to graphite. The graphite sample was subjected to carbon-14 analysis, and in the end all three labs agreed amazingly well on the age of the shroud. The average of their results gave a date for the cloth of A.D. 1320 ± 60 years, with an absolute assurance that the cloth could not possibly be older than A.D. 1200. Carbon-14 dating has thus unraveled the most important mystery concerning the shroud, but others remain. For example, investigators have not yet been able to explain how the image was imprinted.

The Shroud of Turin as it appears in a photographic negative image. *(Santi Visali/The IMAGE Bank)*

29.5

NATURAL RADIOACTIVITY

Radioactive nuclei are generally classified into two groups: (1) unstable nuclei found in nature, which give rise to what is called **natural radioactivity**, and (2) nuclei produced in the laboratory through nuclear reactions, which exhibit **artificial radioactivity**.

Three series of naturally occurring radioactive nuclei exist (Table 29.2). Each series starts with a specific long-lived radioactive isotope whose half-life exceeds that of any of its descendants. The fourth series in Table 29.2 begins with ^{237}Np, a transuranic element (one having an atomic number greater than that of uranium) not found in nature. This element has a half-life of "only" 2.14×10^6 years.

The two uranium series are somewhat more complex than the ^{232}Th series (Fig. 29.10). Also, there are several naturally occurring radioactive isotopes, such as ^{14}C and ^{40}K, that are not part of either decay series.

Natural radioactivity constantly resupplies our environment with radioactive elements that would otherwise have disappeared long ago. For example, because the Solar System is about 5×10^9 years old, the supply of ^{226}Ra (whose half-life is only 1600 years) would have been depleted by radioactive decay long ago if it were not for the decay series that starts with ^{238}U, with a half-life of 4.47×10^9 years.

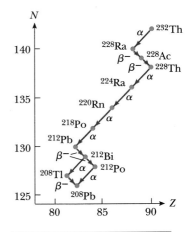

FIGURE 29.10
Decay series beginning with ^{232}Th.

TABLE 29.2
The Four Radioactive Series

Series	Starting Isotope	Half-life (years)	Stable End Product
Uranium	$^{238}_{92}$U	4.47×10^9	$^{206}_{82}$Pb
Actinium	$^{235}_{92}$U	7.04×10^8	$^{207}_{82}$Pb
Thorium	$^{232}_{90}$Th	1.41×10^{10}	$^{208}_{82}$Pb
Neptunium	$^{237}_{93}$Np	2.14×10^6	$^{209}_{83}$Bi

29.6

NUCLEAR REACTIONS

It is possible to change the structure of nuclei by bombarding them with energetic particles. Such changes are called **nuclear reactions.** Rutherford was the first to observe nuclear reactions, using naturally occurring radioactive sources for the bombarding particles. He found that protons were released when alpha particles were allowed to collide with nitrogen atoms. The process can be represented symbolically as

$$\ce{^4_2He} + \ce{^{14}_7N} \longrightarrow X + \ce{^1_1H} \qquad \text{[29.15]}$$

This equation says that an alpha particle (^4_2He) strikes a nitrogen nucleus and produces an unknown product nucleus (X) and a proton (^1_1H). Balancing atomic numbers and mass numbers, as we did for radioactive decay, enables us to conclude that the unknown is characterized as $^{17}_8\text{X}$. Since the element with atomic number 8 is oxygen, we see that the reaction is

$$^4_2\text{He} + {}^{14}_7\text{N} \longrightarrow {}^{17}_8\text{O} + {}^1_1\text{H} \qquad \text{[29.16]}$$

This nuclear reaction starts with two stable isotopes, helium and nitrogen, and produces two different stable isotopes, hydrogen and oxygen.

Since the time of Rutherford, thousands of nuclear reactions have been observed, particularly following the development of charged-particle accelerators in the 1930s. With today's advanced technology in particle accelerators and particle detectors, it is possible to achieve particle energies of at least 1000 GeV = 1 TeV. These high-energy particles are used to create new particles whose properties are helping to solve the mysteries of the nucleus.

EXAMPLE 29.10 The Discovery of the Neutron

A nuclear reaction of significant historical note occurred in 1932 when Chadwick, in England, bombarded a beryllium target with alpha particles. Analysis of the experiment indicated that the following reaction occurred:

$$^4_2\text{He} + {}^9_4\text{Be} \longrightarrow {}^{12}_6\text{C} + X$$

What is X in this reaction?

Solution Balancing mass numbers and atomic numbers, we see that the unknown particle must be represented as ^1_0X, that is, with a mass of 1 and zero charge. Hence, the particle X is the neutron, ^1_0n. This experiment was the first to provide positive proof of the existence of neutrons.

EXAMPLE 29.11 Synthetic Elements

(a) A beam of neutrons is directed at a target of $^{238}_{92}\text{U}$. The reaction products are a gamma ray and another isotope. What is the isotope? (b) This isotope is radioactive and emits a beta particle. Write the equation symbolizing this decay and identify the resulting isotope. (c) This isotope is also radioactive and decays by beta emission. What is the end product? (d) What is the significance of these reactions?

Solution
(a) Balancing input with output gives

$$^1_0\text{n} + {}^{238}_{92}\text{U} \longrightarrow {}^{239}_{92}\text{U} + \gamma$$

(b) The decay of ^{239}U by beta emission is

$$^{239}_{92}\text{U} \longrightarrow {}^{239}_{93}\text{Np} + {}^{0}_{-1}\text{e} + \bar{\nu}$$

(c) The decay of $^{239}_{93}$Np by beta emission gives

$$^{239}_{93}\text{Np} \longrightarrow {}^{239}_{94}\text{Pu} + {}^{0}_{-1}\text{e} + \bar{\nu}$$

(d) The interesting feature of these reactions is the fact that uranium is the element with the greatest number of protons, 92, that exists in nature in any appreciable amount. The reactions in parts (a), (b), and (c) do occur occasionally in nature; hence minute traces of neptunium and plutonium are present. In 1940, however, researchers bombarded uranium with neutrons to produce plutonium and neptunium by the steps given above. These two elements were thus the first elements made in the laboratory, and by bombarding them with neutrons and other particles, the list of synthetic elements has been extended to include those up to atomic number 108 (and possibly 109).

Q VALUES

We have just examined some nuclear reactions for which mass numbers and atomic numbers must be balanced in the equations. We shall now consider the energy involved in these reactions, since energy is another important quantity that must be conserved.

Let us illustrate this procedure by analyzing the following nuclear reaction:

$$^{2}_{1}\text{H} + {}^{14}_{7}\text{N} \longrightarrow {}^{12}_{6}\text{C} + {}^{4}_{2}\text{He} \qquad \textbf{[29.17]}$$

The total mass on the left side of the equation is the sum of the mass of $^{2}_{1}$H (2.014 102 u) and the mass of $^{14}_{7}$N (14.003 074 u), which equals 16.017 176 u. Similarly, the mass on the right side of the equation is the sum of the mass of $^{12}_{6}$C (12.000 000 u) plus the mass of $^{4}_{2}$He (4.002 603 u), for a total of 16.002 603 u. Thus, the total mass before the reaction is greater than the total mass after the reaction. The mass difference in this reaction is equal to 16.017 176 u − 16.002 603 u = 0.014 573 u. This "lost" mass is converted to the kinetic energy of the nuclei present after the reaction. In energy units, 0.014 573 u is equivalent to 13.567 MeV of kinetic energy carried away by the carbon and helium nuclei.

The energy required to balance the equation is called the *Q* value of the reaction. In Equation 29.17 the *Q* value is 13.567 MeV. Nuclear reactions in which there is a release of energy, that is, positive *Q* values, are said to be **exothermic reactions.**

The energy balance sheet is not complete, however. We must also consider the kinetic energy of the incident particle before the collision. As an example, let us assume that the deuteron in Equation 29.17 has a kinetic energy of 5 MeV. Adding this to our *Q* value, we find that the carbon and helium nuclei have a total kinetic energy of 18.567 MeV following the reaction.

Now consider the reaction

$$^{4}_{2}\text{He} + {}^{17}_{7}\text{N} \longrightarrow {}^{17}_{8}\text{O} + {}^{1}_{1}\text{H} \qquad \textbf{[29.18]}$$

Before the reaction, the total mass is the sum of the masses of the alpha particle and the nitrogen nucleus: 4.002 603 u + 14.003 074 u = 18.005 677 u. After the reaction, the total mass is the sum of the masses of the oxygen nucleus and the proton: 16.999 133 u + 1.007 825 u = 18.006 958 u. In this case, the total mass after the reaction is *greater* than the total mass before the reaction. The mass deficit is 0.001 281 u, equivalent to an energy deficit of 1.193 MeV. This deficit is expressed by the negative *Q* value of the reaction, −1.193 MeV. Reactions with negative Q values are called **endothermic reactions.** Such reactions will not take place unless the incoming particle has at least enough kinetic energy to overcome the energy deficit.

At first it might appear that the reaction in Equation 29.18 could take place if the incoming alpha particle had a kinetic energy of 1.193 MeV. In practice, however, the alpha particle must have more energy than this. If it had an energy of only 1.193 MeV, energy would be conserved but careful analysis would show that momentum was not. This can easily be understood by recognizing that the incoming alpha particle has some momentum before the reaction. However, if its kinetic energy were only 1.193 MeV, the products (oxygen and a proton) would be created with zero kinetic energy and, thus, zero momentum. It can be shown that, in order to conserve both energy and momentum, the incoming particle must have a minimum kinetic energy given by

$$KE_{min} = \left(1 + \frac{m}{M}\right) |Q| \qquad \text{[29.19]}$$

where m is the mass of the incident particle, M is the mass of the target, and the absolute value of the Q value is used. For the reaction given by Equation 29.19, we find

$$KE_{min} = \left(1 + \frac{4.002\ 603}{14.003\ 074}\right) |-1.193\ \text{MeV}| = 1.534\ \text{MeV}$$

This minimum value of the kinetic energy of the incoming particle is called the **threshold energy.** The nuclear reaction shown in Equation 29.19 will not occur if the incoming alpha particle has an energy of less than 1.534 MeV, but will occur if the kinetic energy is equal to or greater than 1.534 MeV.

SUMMARY

Nuclei are represented symbolically as $^A_Z X$, where X represents the chemical symbol for the element. The quantity A is the **mass number,** which equals the total number of nucleons (neutrons plus protons) in the nucleus. The quantity Z is the **atomic number,** which equals the number of protons in the nucleus. Nuclei that contain the same number of protons but different numbers of neutrons are called **isotopes.** In other words, isotopes have the same Z value but different A values.

Most nuclei are approximately spherical, with an average radius given by

$$r = r_0 A^{1/3} \qquad \text{[29.1]}$$

where A is the mass number and r_0 is a constant equal to 1.2×10^{-15} m.

The total mass of a nucleus is always less than the sum of the masses of its individual nucleons. This mass difference, Δm, multiplied by c^2 gives the **binding energy** of the nucleus.

The spontaneous emission of radiation by certain nuclei is called **radioactivity.** There are three processes by which a radioactive substance can decay: alpha (α) decay, in which the emitted particles are $^4_2$He nuclei; beta (β) decay, in which the emitted particles are electrons; and gamma (γ) decay, in which the emitted particles are high-energy photons.

The **decay rate**, or **activity**, R, of a sample is given by

$$R = \left| \frac{\Delta N}{\Delta t} \right| = \lambda N \qquad \text{[29.3]}$$

where N is the number of radioactive nuclei at some instant and λ is a constant for a given substance called the **decay constant.**

Nuclei in a radioactive substance decay in such a way that the number of nuclei present varies with time according to the expression

$$N = N_0 e^{-\lambda t} \qquad \text{[29.4]}$$

where N is the number of radioactive nuclei present at time t, N_0 is the number at time $t = 0$, and $e = 2.718 \ldots$.

The **half-life**, $T_{1/2}$, of a radioactive substance is the time required for half of a given number of radioactive nuclei to decay. The half-life is related to the decay constant as

$$T_{1/2} = \frac{0.693}{\lambda} \qquad \text{[29.5]}$$

If a nucleus decays by alpha emission, it loses two protons and two neutrons. A typical alpha decay is

$$^{238}_{92}\text{U} \longrightarrow \,^{234}_{90}\text{Th} + \,^{4}_{2}\text{He} \qquad \text{[29.8]}$$

Note that in this decay, as in all radioactive decay processes, the sum of the Z values on the left equals the sum of the Z values on the right; the same is true for the A values.

A typical beta decay is

$$^{14}_{6}\text{C} \longrightarrow \,^{14}_{7}\text{N} + \,^{0}_{-1}\text{e} + \bar{\nu}$$

When a nucleus decays by beta emission, a particle called an **antineutrino** ($\bar{\nu}$) is also emitted. In positron decay a **neutrino** (ν) is emitted. A neutrino has zero electric charge and a small rest mass (which may be zero) and interacts weakly with matter.

Nuclei are often in an excited state following radioactive decay, and release their extra energy by emitting a high-energy photon called a **gamma ray** (γ). A typical gamma ray emission is

$$^{12}_{6}\text{C*} \longrightarrow \,^{12}_{6}\text{C} + \gamma \qquad \text{[29.14]}$$

where the asterisk indicates that the carbon nucleus was in an excited state before gamma emission.

Nuclear reactions can occur when a bombarding particle strikes another nucleus. A typical nuclear reaction is

$$^{4}_{2}\text{He} + \,^{14}_{7}\text{N} \longrightarrow \,^{17}_{8}\text{O} + \,^{1}_{1}\text{H} \qquad \text{[29.16]}$$

In this reaction an alpha particle strikes a nitrogen nucleus, producing an oxygen nucleus and a proton. As in radioactive decay, atomic numbers and mass numbers balance on the two sides of the arrow.

Nuclear reactions in which energy is released are said to be **exothermic reactions** and are characterized by positive Q values. Reactions with negative Q values, called **endothermic reactions,** cannot occur unless the incoming particle has at least enough kinetic energy to overcome the energy deficit. In order to conserve both energy and momentum, the incoming particle must have a minimum kinetic energy, called the **threshold energy**, given by

$$KE_{\min} = \left(1 + \frac{m}{M}\right) |Q|$$ [29.19]

where m is the mass of the incident particle and M is the mass of the target atom.

ADDITIONAL READING

G. F. Bertsch, "Vibrations of the Atomic Nucleus," *Sci. American,* May 1983, p. 62.

H. Bethe, "What Holds the Nucleus Together?" *Sci. American,* September 1953, p. 201.

D. A. Bromley, "Nuclear Models," *Sci. American,* December 1978, p. 58.

B. N. Da Costa Andrade, *Rutherford and the Nature of the Atom.* Garden City, NY, Science Study Series, Doubleday, 1964.

S. A. Fetter and K. Tsipis, "Catastrophic Releases of Radioactivity," *Sci. American,* April 1981, p. 41.

W. Greiner and A. Sandulescu, "New Radioactivities," *Sci. American,* March 1990, p. 58.

O. Hahn, "The Discovery of Fission," *Sci. American,* February 1958, p. 76.

R. E. Hedges and J. Gowlett, "Radiocarbon Dating by Accelerator Mass Spectrometry," *Sci. American,* January 1986, p. 100.

R. E. M. Hedges, "Radioisotope Clocks in Archaeology," *Nature,* Vol. 281, No. 5725, 1979, p. 19.

M. Jacob and P. Landshoff, "The Inner Structure of the Proton," *Sci. American,* March 1980, p. 66.

J. G. Learned and D. Eichler, "A Deep-Sea Neutrino Telescope," *Sci. American,* February 1981, p. 138.

R. B. Marshak, "The Nuclear Force," *Sci. American,* March 1960, p. 98.

M. K. Moe and S. P. Rosen, "Double-Beta Decay," *Sci. American,* November 1989, p. 48.

P. R. Moran, R. J. Nickles, and J. A. Zagzebski, "The Physics of Medical Imaging," *Physics Today,* July 1983, p. 36.

I. I. Pykett, "NMR Imaging in Medicine," *Sci. American,* May 1982, p. 78.

M. M. Ter-Pogossian, M. E. Raichle, and B. E. Sobol, "Positron-Emission Tomography," *Sci. American,* October 1980, p. 170.

A. C. Upton, "The Biological Effects of Low-Level Ionizing Radiation," *Sci. American,* February 1982, p. 41.

S. Weinberg, "The Decay of the Proton," *Sci. American,* June 1981, p. 64.

V. F. Weisskopf and E. P. Rosenbaum, "A Model of the Nucleus," *Sci. American,* December 1955, p. 261.

E. Witten, "New Ideas About Neutrino Masses," *The Physics Teacher,* February 1983, p. 78.

CONCEPTUAL QUESTIONS

Example Why do heavier elements require more neutrons to maintain stability?

Reasoning As more protons are added to form heavier elements, the Coulomb repulsive force between protons tends to make the nucleus more unstable. Additional neutrons tend to dilute this effect by increasing the average separation between protons in the nucleus. From a different perspective, neutrons supply more "strong force" bonds between nucleons while not producing any repulsive Coulomb forces.

Example Suppose it could be shown that cosmic ray intensity was much greater 10 000 years ago. How would this affect the ages we assign to ancient samples of once-living matter?

Reasoning We would have to revise our age values upward for ancient materials. That is, we would conclude that the materials were older than we had thought, because the greater cosmic ray intensity would have left the samples with a larger percentage of carbon-14 when they died, and a longer time would have been necessary for it to decay to the percentage found at present.

Example Why is carbon dating unable to provide accurate estimates of very old materials?

Reasoning The amount of carbon-14 left in very old materials is extremely small, and detection cannot be accomplished with a high degree of accuracy.

Example What fraction of a radioactive sample has decayed after two half-lives have elapsed?

Reasoning After one half-life has elapsed, one half of the sample remains radioactive. After two half-lives have elapsed, one quarter of the sample remains radioactive; therefore three quarters of the sample has decayed.

Example If a nucleus such as ^{226}Ra that is initially at rest undergoes alpha decay, which has more kinetic energy after the decay, the alpha particle or the daughter nucleus?

Reasoning The alpha particle and the daughter nucleus must carry equal and opposite momenta because momentum is conserved. Since kinetic energy can be expressed as $p^2/2m$, the smaller-mass alpha particle carries more energy than the recoiling nucleus.

1. In Rutherford's experiment, assume that an alpha particle is headed directly toward the nucleus of an atom. Why does the alpha particle not make physical contact with the nucleus?

2. If the atom were indeed like the watermelon model, what results would Geiger and Marsden have obtained in their alpha-scattering experiment?

3. Estimate the mass of a pinhead if it were composed entirely of densely packed nuclear matter.

4. How many protons are there in the nucleus $^{222}_{86}$Rn? How many neutrons? How many electrons are there in the neutral atom?

5. Element X has several isotopes. What do these isotopes have in common? How do they differ?

6. Why do heavier elements require more neutrons in order to maintain stability?

7. Explain the main differences between alpha, beta, and gamma rays.

8. If film is kept in a box, alpha particles from a radioactive source outside the box cannot expose the film but beta particles can. Explain.

9. An alpha particle has twice the charge of a beta particle. Why does the former deflect less than the latter when passing between electrically charged plates, assuming they both have the same speed?

10. If no more people were born, the law of population growth would strongly resemble the radioactive decay law. Discuss this statement.

11. If a nucleus has a half-life of one year, does this mean that it will be completely decayed after two years? Explain.

12. "The more probable the decay, the shorter the half-life." Explain this statement.

13. Two samples of the same radioactive nuclide are prepared. Sample A has twice the initial activity of sample B. How does the half-life of A compare to the half-life of B? After each has passed through five half-lives, what is the ratio of their activities?

14. Use the analogy of a bullet and a rifle to explain why the recoiling nucleus carries off only a very small fraction of the disintegration energy in the α decay of a nucleus.

15. Explain why many heavy nuclei undergo alpha decay but do not spontaneously emit neutrons or protons.

16. Explain why neutron-rich isotopes are more likely to be β^- emitters, whereas neutron-deficient isotopes are more likely to be β^+ emitters.

17. Pick any beta decay process and show that the neutrino must have zero charge.

18. What is the difference between a neutrino and a photon?

19. If a nucleus captures a slow-moving neutron, the product is left in a highly excited state with an energy approximately 8.0 MeV above the ground state. Explain the source of the excitation energy.

20. Suppose a radioactive parent substance with a very long half-life has a daughter with a very short half-life. Describe what happens, starting with a freshly purified sample of the parent substance.

21. The radioactive nucleus $^{226}_{88}$Ra has a half-life of about 1.6×10^3 years. Since the Solar System is about 5 billion years old, how can you explain why we can still find this nucleus in nature?

22. A free neutron undergoes beta decay with a half-life of about 15 min. Can a free proton undergo a similar decay?

23. If two radioactive samples have the same activity measured in curies, will they necessarily create the same amount of damage in a medium? Explain.

24. Radiation can be used to sterilize such things as surgical equipment and packaged foods. Why do you suppose it works?

25. One method of treating cancer of the thyroid is to insert a small radioactive source directly into the tumor. The radiation emitted by the source can destroy cancerous cells. Very often, the radioactive isotope ^{133}I ($Z = 53$) is injected into the bloodstream in this treatment. Why is iodine used?

26. Many radiopharmaceuticals must be made immediately prior to use. Why?

27. Is it energetically possible for a ^{12}C ($Z = 6$) nucleus to spontaneously decay into three alpha particles? Explain.

PROBLEMS

Table 29.3 will be useful for many of these problems. A more complete list of atomic masses is given in Appendix B.

Section 29.1   Some Properties of Nuclei

1. Find the radius of a nucleus of (a) $^4_2$He and (b) $^{238}_{92}$U.
2. Find the atomic number of a nucleus with a radius of 4.36×10^{-15} m.
3. The compressed core of a star formed in the wake of a supernova explosion can consist of pure nuclear material and is called a pulsar, or neutron star. Calculate the mass of a pulsar whose volume is 10 cm^3.
4. Compare the nuclear radii of the following nuclides: $^2_1$H, $^{60}_{27}$Co, $^{197}_{79}$Au, $^{239}_{94}$Pu.
5. The unified mass unit u is exactly $\frac{1}{12}$ of the mass of an atom of ^{12}C. This unit was adopted in 1961. Before that date, there had been two different units in general use:

    u (physical scale) = 1/16 of the mass of ^{16}O

    u (chemical scale) = 1/16 of the average mass of oxygen, taking into account the relative isotopic abundance

    Calculate the percentage difference between the mass unit on the present scale and the older (a) physical and (b) chemical scales. Use values of atomic masses and percentage abundances of ^{16}O and ^{18}O from Appendix B, and use 16.999 131 u and 0.038, respectively, for the atomic mass and percentage abundance of ^{17}O.
6. Find the diameter of a sphere of nuclear matter that would have a mass equal to that of the Earth. Base your calculation on an Earth radius of 6.37×10^6 m and an average Earth density of 5.52×10^3 kg/m^3.
7. Consider the hydrogen atom to be a sphere of radius equal to the Bohr radius, 0.53×10^{-10} m, and calculate the approximate value of the ratio of the nuclear density to the atomic density.
8. What would be the gravitational force between two golf balls (each of 4.3-cm diameter), 1 m apart, if they were made of nuclear matter?
9. Use energy methods to calculate the distance of closest approach for a head-on collision between an alpha particle with an initial energy of 0.50 MeV and a gold nucleus (^{197}Au) at rest. Assume the gold nucleus remains at rest during the collision.
10. (a) Find the speed an alpha particle requires to come within 3.2×10^{-14} m of a gold nucleus. (b) Find the energy of the alpha particle in MeV.

TABLE 29.3
Some Atomic Masses

Element	Atomic Mass (u)
$^4_2$He	4.002 603
$^7_3$Li	7.016 004
$^9_4$Be	9.012 182
$^{10}_5$B	10.012 938
$^{12}_6$C	12.000 000
$^{13}_6$C	13.003 355
$^{14}_7$N	14.003 074
$^{15}_7$N	15.000 109
$^{15}_8$O	15.003 065
$^{18}_8$O	17.999 159
$^{18}_9$F	18.000 937
$^{20}_{10}$Ne	19.992 439
$^{23}_{11}$Na	22.989 770
$^{23}_{12}$Mg	22.994 127
$^{27}_{13}$Al	26.981 541
$^{30}_{15}$P	29.978 310
$^{40}_{20}$Ca	39.962 591
$^{43}_{20}$Ca	42.958 770
$^{56}_{26}$Fe	55.934 939
$^{64}_{30}$Zn	63.929 145
$^{64}_{29}$Cu	63.929 599
$^{93}_{41}$Nb	92.906 378
$^{197}_{79}$Au	196.966 560
$^{202}_{80}$Hg	201.970 632
$^{216}_{84}$Po	216.001 790
$^{220}_{86}$Rn	220.011 401
$^{238}_{92}$U	238.050 786

11. Certain stars are thought to collapse at the end of their lives, combining their protons and electrons to form a neutron star. Such a star could be thought of as a giant atomic nucleus. If a star with a mass equal to that of the Sun ($M = 1.99 \times 10^{30}$ kg) collapsed into neutrons ($m = 1.67 \times 10^{-27}$ kg), what would its radius be?
12. The nucleus of an iron atom has a radius equal to 4.60×10^{-15} m. What must be the minimum speed of an α particle if it is to reach the nucleus? Disregard the effect of the outer electrons.
13. An α particle ($Z = 2$, mass 6.64×10^{-27} kg) approaches to within 1.0×10^{-14} m of a carbon nu-

□ indicates problems that have full solutions available in the Student Solutions Manual and Study Guide.

cleus ($Z = 6$). What is the (a) maximum Coulomb force on the α particle, (b) acceleration of the α particle at this point, and (c) potential energy of the α particle at this point?

Section 29.2 Binding Energy

14. Calculate the total binding energy of $^{20}_{10}$Ne and $^{40}_{20}$Ca.
15. The energy required to construct a uniformly charged sphere of total charge Q and radius R is $E = 3kQ^2/5R$, where k is the Coulomb constant. Assume that a ^{40}Ca nucleus consists of 20 protons uniformly distributed in a spherical volume. (a) How much energy is required to counter the electrostatic repulsion given by the above equation? (*Hint:* First calculate the radius of a ^{40}Ca nucleus.) (b) Calculate the binding energy of ^{40}Ca and compare it to the result in part (a). (c) Explain why the result of part (b) is larger than that of part (a).
16. Calculate the binding energy per nucleon of $^{93}_{41}$Nb and $^{197}_{79}$Au.
17. The energy required to dismantle a uniform sphere of total mass M and radius R is given by $E = 3GM^2/5R$, where G is the gravitational constant. Assume that a ^{40}Ca calcium nucleus consists of 20 protons and 20 neutrons (each having a mass of 1.67×10^{-27} kg) uniformly distributed in a spherical volume. (a) How much energy is required to dismantle a ^{40}Ca nucleus? (*Hint:* First calculate the radius of a ^{40}Ca nucleus.) (b) Calculate the binding energy of ^{40}Ca and compare it to the result in part (a). (c) Based on the results of parts (a) and (b), why does the gravitational potential energy not account for the binding energy of the ^{40}Ca nucleus?
18. Compare the average binding energy per nucleon of $^{24}_{12}$Mg and $^{85}_{37}$Rb.
19. A pair of nuclei for which $Z_1 = N_2$ and $Z_2 = N_1$ are called *mirror isobars* (the atomic and neutron numbers are interchangeable). Binding energy measurements on such pairs can be used to obtain evidence of the charge independence of nuclear forces. Charge independence means that the proton-proton, proton-neutron, and neutron-neutron forces are approximately equal. Calculate the difference in binding energy for the two mirror nuclei, $^{15}_{8}$O and $^{15}_{7}$N.
20. Determine the difference of the binding energy of $^{3}_{1}$H and $^{3}_{2}$He.
21. Two isotopes having the same mass number are known as isobars. Calculate the difference in binding energy per nucleon for the isobars $^{23}_{11}$Na and $^{23}_{12}$Mg. How do you account for this difference?

22. Calculate the binding energy of the last neutron in the $^{43}_{20}$Ca nucleus. (*Hint:* You should compare the mass of $^{43}_{20}$Ca with the mass of $^{42}_{20}$Ca plus the mass of a neutron. The mass of $^{42}_{20}$Ca = 41.958 622 u.)

Section 29.3 Radioactivity

23. A radioactive sample contains 3.50 μg of pure ^{11}C, which has a half-life of 20.4 min. (a) Determine the number of nuclei present initially. What is the activity of the sample (b) initially and (c) after 8 h?
24. A 3.00-g sample of an unknown type of radioactive material has decreased after two days to the point that 2.52 g remain. (a) What is the half-life of this material? (b) Can you identify it by using the table of isotopes in Appendix B?
25. Radon gas has a half-life of 3.83 days. If a radon gas sample has a mass of 3.00 g, what mass will remain after two days have passed?
26. How long will it take for a sample of polonium with a half-life of 140 days to decay to one tenth of its original strength?
27. Suppose that you start with 1.00×10^{-3} g of a pure radioactive substance and 2.0 h later determine that only 0.25×10^{-3} g of the substance remains. What is the half-life of this substance?
28. The half-life of an isotope of phosphorus is 14 days. If a sample contains 3.0×10^{16} such nuclei, determine its activity.
29. A drug tagged with $^{99}_{43}$Tc (half-life = 6.05 h) is prepared for a patient. If the original activity of the sample was 1.1×10^4 Bq, what is its activity after it has sat on the shelf for 2 h?
30. The half-life of ^{131}I is 8.04 days. (a) Calculate the decay constant for this isotope. (b) Find the number of ^{131}I nuclei necessary to produce a sample with an activity of 0.50 μCi.
31. The ^{14}C content decreases after the death of a living system with a half-life of 5739 years. If an archaeologist finds an ancient firepit containing partially consumed firewood, and the ^{14}C content of the wood is only 12.5% that of an equal carbon sample from a present-day tree, what is the age of the ancient site?
32. A freshly prepared sample of a certain radioactive isotope has an activity of 10.0 mCi. After 4.0 h, the activity is 8.0 mCi. (a) Find the decay constant and half-life of the isotope. (b) How many atoms of the isotope were contained in the freshly prepared sample? (c) What is the sample's activity 30 h after it is prepared?
33. Tritium has a half-life of 12.33 years. What percentage of the nuclei in a tritium sample will decay in five years?

34. A radioactive sample contains 3.00×10^{-14} kg of $^{108}_{47}$Ag (half-life = 2.42 min) at some instant. What is the activity of this sample in millicuries?

35. A building has become accidentally contaminated with radioactivity. The longest-lived material in the building is strontium-90 ($^{90}_{38}$Sr, atomic mass 89.9077). If the building initially contained 5.0 kg of this substance and the safe level is less than 10.0 counts/min, how long will the building be unsafe?

Section 29.4 The Decay Processes

36. Complete the following radioactive decay formulas:

$$^{212}_{83}\text{Bi} \longrightarrow ? + {}^{4}_{2}\text{He}$$

$$^{95}_{36}\text{Kr} \longrightarrow ? + {}^{0}_{-1}\text{e}$$

$$? \longrightarrow {}^{4}_{2}\text{He} + {}^{140}_{58}\text{Ce}$$

37. Complete the following radioactive decay formulas:

$$^{12}_{5}\text{B} \longrightarrow ? + {}^{0}_{-1}\text{e}$$

$$^{234}_{90}\text{Th} \longrightarrow {}^{230}_{88}\text{Ra} + ?$$

$$? \longrightarrow {}^{14}_{7}\text{N} + {}^{0}_{-1}\text{e}$$

38. The mass of ^{56}Fe is 55.9349 u and the mass of ^{56}Co is 55.9399 u. Which isotope decays into the other, and by what process?

39. Find the energy released in the alpha decay of $^{238}_{92}$U. The following mass values will be useful:

$$M(^{238}_{92}\text{U}) = 238.050\ 786\ \text{u}$$

$$M(^{234}_{90}\text{Th}) = 234.043\ 583\ \text{u}$$

$$M(^{4}_{2}\text{He}) = 4.002\ 603\ \text{u}$$

40. An ^{3}H nucleus beta decays into ^{3}He by creating an electron and an antineutrino according to the reaction

$$^{3}_{1}\text{H} \longrightarrow {}^{3}_{2}\text{He} + {}^{0}_{-1}\text{e} + \bar{\nu}$$

Use Appendix B to determine the total energy released in this reaction.

41. Determine which of the following suggested decays can occur spontaneously:

(a) $^{40}_{20}\text{Ca} \longrightarrow {}^{0}_{1}\text{e} + {}^{40}_{19}\text{K}$;

(b) $^{144}_{60}\text{Nd} \longrightarrow {}^{4}_{2}\text{He} + {}^{140}_{58}\text{Ce}$

42. Figure 29.11 shows the steps by which $^{235}_{92}$U decays to $^{207}_{82}$Pb. Enter the correct isotope symbol in each square.

43. $^{66}_{28}$Ni (mass = 65.9291 u) decays by β^{-} emission to $^{66}_{29}$Cu (mass = 65.9289 u). (a) Write the complete decay formula for this process. (b) Find the maximum kinetic energy of the emerging β^{-} particles.

44. A wooden artifact is found in an ancient tomb. Its

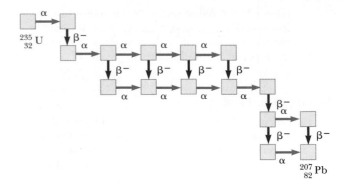

FIGURE 29.11 (Problem 42)

carbon-14 ($^{14}_{6}$C) activity is measured to be 60.0% of that in a fresh sample of wood from the same region. Assuming the same amount of carbon-14 was initially present in the wood from which the artifact was made, determine the age of the artifact.

45. A piece of charcoal used for cooking is found at the remains of an ancient campsite. A 1.00-kg sample of carbon from the wood has an activity of 2.00×10^{-3} decays per minute. Find the age of the charcoal.

Section 29.6 Nuclear Reactions

46. Natural gold has only one isotope, ^{197}Au ($Z = 79$). If gold is bombarded with slow neutrons, β^{-} particles are emitted. (a) Write the appropriate reaction equations. (b) Calculate the maximum energy of the emitted beta particles. The mass of ^{198}Hg ($Z = 80$) is 197.966 75 u.

47. The first nuclear reaction utilizing particle accelerators was performed by Cockcroft and Walton. Accelerated protons were used to bombard lithium nuclei, producing the following reaction:

$$^{1}_{1}\text{H} + {}^{7}_{3}\text{Li} \longrightarrow {}^{4}_{2}\text{He} + {}^{4}_{2}\text{He}$$

Since the masses of the particles involved in the reaction were well known, these results were used to obtain an early proof of the Einstein mass-energy relation. Calculate the Q value of the reaction.

48. Determine the Q values of the following reactions:

(a) $^{27}_{13}\text{Al} + \text{n} \longrightarrow {}^{24}_{11}\text{Na} + {}^{4}_{2}\text{He}$;

(b) $^{11}_{5}\text{B} + {}^{1}_{1}\text{H} \longrightarrow {}^{11}_{6}\text{C} + \text{n}$

49. (a) Determine the product of the reaction $^{7}_{3}\text{Li} + {}^{4}_{2}\text{He} \rightarrow ? + \text{n}$. (b) What is the Q value of the reaction?

50. Complete the following nuclear reactions:

$$? + {}^{14}_{7}\text{N} \longrightarrow {}^{1}_{1}\text{H} + {}^{17}_{8}\text{O}$$

$$^{7}_{3}\text{Li} + {}^{1}_{1}\text{H} \longrightarrow {}^{4}_{2}\text{He} + ?$$

51. Complete the following nuclear reactions:

$$^{27}_{13}\text{Al} + ^{4}_{2}\text{He} \longrightarrow ? + ^{30}_{15}\text{P}$$

$$^{1}_{0}\text{n} + ? \longrightarrow ^{4}_{2}\text{He} + ^{7}_{3}\text{Li}$$

52. The first known reaction in which the product nucleus was radioactive (achieved in 1934) was one in which $^{27}_{13}\text{Al}$ was bombarded with alpha particles. Produced in the reaction were a neutron and a product nucleus. (a) What was the product nucleus? (b) Find the Q value of the reaction.

53. When a $^{6}_{3}\text{Li}$ nucleus is struck by a proton, an alpha particle and a product nucleus are released. (a) What is the product nucleus? (b) Find the Q value of the reaction.

54. Find the energy released in the fission reaction

$$\text{n} + ^{235}_{92}\text{U} \longrightarrow ^{98}_{40}\text{Zr} + ^{135}_{52}\text{Te} + 3\text{n}$$

The atomic masses of the fission products are $^{98}_{40}\text{Zr}$, 97.9120 u, and $^{135}_{52}\text{Te}$, 134.9087 u.

55. A beam of 6.61 MeV protons is incident on a target of $^{27}_{13}\text{Al}$. Those that collide produce the reaction

$$\text{p} + ^{27}_{13}\text{Al} \longrightarrow ^{27}_{14}\text{Si} + \text{n}$$

($^{27}_{14}\text{Si}$ has a mass of 26.986 721 u.) Neglect any recoil of the product nucleus and determine the kinetic energy of the emerging neutrons.

56. (a) Suppose $^{10}_{5}\text{B}$ is struck by an alpha particle, releasing a proton and a product nucleus in the reaction. What is the product nucleus? (b) An alpha particle and a product nucleus are produced when $^{13}_{6}\text{C}$ is struck by a proton. What is the product nucleus?

57. Determine the Q value associated with the spontaneous fission of ^{236}U into ^{90}Rb and ^{143}Cs fragments, with masses of 89.914 811 u and 142.927 220 u, respectively. The masses of the other reaction products are given in Appendix B.

58. When ^{18}O is struck by a proton, ^{18}F and another particle are produced. (a) What is the other particle? (b) This reaction has a Q value of -2.453 MeV, and the atomic mass of ^{18}O is 17.999 160 u. What is the atomic mass of ^{18}F?

59. Find the threshold energy that the incident neutron must have to produce the reaction

$$^{1}_{0}\text{n} + ^{4}_{2}\text{He} \longrightarrow ^{2}_{1}\text{H} + ^{3}_{1}\text{H}$$

ADDITIONAL PROBLEMS

60. After determining that the Sun has existed for hundreds of millions of years, but before the discovery of nuclear physics, scientists could not explain why the Sun has continued to burn for such a long time. [If it used a non-nuclear burning process (e.g., coal), it would have burned up in 3000 years or so.] Assume that the Sun, whose mass is 1.99×10^{30} kg, consists entirely of hydrogen and that its total power output (or luminosity) is 3.9×10^{26} W. (a) If the energy generating mechanism of the Sun is the "burning" or transforming of hydrogen into helium via the reaction,

$$4(^{1}_{1}\text{H}) \longrightarrow ^{4}_{2}\text{He} + 2(^{0}_{1}\text{e}) + 2\nu + \gamma$$

calculate the energy (in joules) given off by this reaction. (b) Determine how many hydrogen atoms are available for burning. Take the mass of one hydrogen atom (proton) to be 1.67×10^{-27} kg. (c) Assuming that the total power output remains constant, how long will it be before all the hydrogen is converted into helium, and the Sun dies? (d) Why are your results larger than the accepted lifetime of about 10 billion years?

61. A sample of organic material is found to contain 18 g of carbon. The investigators believe the material to be 20 000 years old based on samples of pottery found at the site. If so, what is the expected activity of the organic material?

62. A 200.0-mCi sample of a radioactive isotope is purchased by a medical supply house. If the sample has a half-life of 14 days, how long will it keep before its activity is reduced to 20.0 mCi?

63. One method for producing neutrons for experimental use is to bombard $^{7}_{3}\text{Li}$ with protons. The neutrons are emitted according to the following reaction:

$$^{1}_{1}\text{H} + ^{7}_{3}\text{Li} \longrightarrow ^{7}_{4}\text{Be} + ^{1}_{0}\text{n}$$

What is the minimum kinetic energy the incident proton must have if this reaction is to occur?

64. Deuterons that have been accelerated are used to bombard other deuterium nuclei, resulting in the reaction

$$^{2}_{1}\text{H} + ^{2}_{1}\text{H} \longrightarrow ^{3}_{2}\text{He} + ^{1}_{0}\text{n}$$

Does this reaction require a threshold energy? If so, what is its value?

65. A medical laboratory stock solution is prepared with an initial activity due to ^{24}Na of 2.5 mCi/ml, and 10.0 ml of the stock solution is diluted at $t_0 = 0$ to a working solution whose total volume is 250 ml. After 48 h, a 5.0-ml sample of the working solution is monitored with a counter. What is the measured activity? (Note that 1 ml = 1 milliliter.)

66. A by-product of some fission reactors is the isotope $^{239}_{94}\text{Pu}$, which is an alpha emitter with a half-life of 24 000 years:

$$^{239}_{94}\text{Pu} \longrightarrow ^{235}_{92}\text{U} + \alpha$$

Consider a sample of 1.0 kg of pure $^{239}_{94}$Pu at $t = 0$. Calculate (a) the number of $^{239}_{94}$Pu nuclei present at $t = 0$ and (b) the initial activity of the sample. (c) How long does the sample have to be stored if a "safe" activity level is 0.10 Bq?

67. A fission reactor is hit by a nuclear weapon, causing 5.0×10^6 Ci of ^{90}Sr ($T_{1/2} = 28.7$ years) to evaporate into the air. The ^{90}Sr falls out over an area of 10^4 km^2. How long will it take the activity of the ^{90}Sr to reach the agriculturally "safe" level of 2.0 μCi/m^2?

68. Free neutrons have a characteristic half-life of 12 min. What fraction of a group of free neutrons at thermal energy (0.040 eV) will decay before traveling a distance of 10.0 km?

69. The theory of nuclear astrophysics is that all the heavy elements like uranium are formed in the interior of massive stars. These stars eventually explode, releasing the elements into space. If we assume that at the time of explosion there were equal amounts of ^{235}U and ^{238}U, how long ago were the elements that formed our Earth released, given that the present ^{235}U/^{238}U ratio is 0.0007? (The half-lives of ^{235}U and ^{238}U are 0.70×10^9 y and 4.47×10^9 y, respectively.)

70. A 25-g piece of charcoal is known to be about 25 000 years old. (a) Determine the number of decays per minute expected from this sample. (b) If the radioactive background in the counter without a sample is 20.0 counts/min and we assume 100% efficiency in counting, explain why 25 000 years is close to the limit of dating with this technique.

71. Many radioisotopes have important industrial, medical, and research applications. One of these is ^{60}Co, which has a half-life of 5.2 years and decays by the emission of a beta particle (energy 0.31 MeV) and two gamma photons (energies 1.17 MeV and 1.33 MeV). A scientist wishes to prepare a ^{60}Co sealed source that will have an activity of at least 10 Ci after 30 months of use. What is the minimum initial mass of ^{60}Co required?

72. During the manufacture of a steel engine component, radioactive iron (^{59}Fe) is included in the total mass of 0.20 kg. The component is placed in a test engine when the activity due to the isotope is 20.0 μCi. After a 1000-h test period, oil is removed from the engine and found to contain enough ^{59}Fe to produce 800 disintegrations/min per liter of oil. The total volume of oil in the engine is 6.5 liters. Calculate the total mass worn from the engine component per hour of operation. (The half-life for ^{59}Fe is 45.1 days.)

MEDICAL APPLICATIONS OF RADIATION

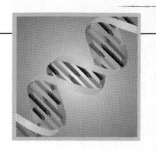

Radiation Damage in Matter

Radiation absorbed by matter can cause severe damage. The degree and type of damage depend upon several factors, including the type and energy of the radiation and the properties of the absorbing material. Radiation damage in biological organisms is primarily due to ionization effects in cells. The normal function of a cell may be disrupted when highly reactive ions or radicals are formed as the result of ionizing radiation. For example, hydrogen and hydroxyl radicals produced from water molecules can induce chemical reactions that may break bonds in proteins and other vital molecules. Large acute doses of radiation are especially dangerous because damage to a great number of molecules in a cell may cause death of the cell. Also, cells that do survive the radiation may become defective, which can lead to cancer.

In biological systems, it is common to separate radiation damage into two categories, somatic damage and genetic damage. **Somatic damage** is radiation damage to any cells except the reproductive cells. Such damage can lead to cancer at high radiation levels or seriously alter the characteristics of specific organisms. **Genetic damage** affects only reproductive cells. Damage to the genes in reproductive cells can lead to defective offspring. Clearly, we must be concerned about the effect of diagnostic treatments, such as x-rays and other forms of radiation exposure.

Several units are used to quantify radiation exposure and dose. The **roentgen** (R) is defined as *that amount of ionizing radiation that will produce 2.08×10^9 ion pairs in $1 \ cm^3$ of air under standard conditions.* Equivalently, the roentgen is *that amount of radiation that deposits $8.76 \times 10^{-3} \ J$ of energy into 1 kg of air.*

For most applications, the roentgen has been replaced by the **rad** (which is an acronym for r̲adiation a̲bsorbed d̲ose), defined as follows: *one rad is that amount of radiation that deposits $10^{-2} \ J$ of energy into 1 kg of absorbing material.*

Although the rad is a perfectly good physical unit, it is not the best unit for measuring the degree of biological damage produced by radiation. This is because the degree of biological damage depends not only on the dose but also on the type of radiation. For example, a given dose of alpha particles causes about ten times more biological damage than an equal dose of x-rays. The **RBE** (r̲elative b̲iological e̲ffectiveness) factor is defined as *the number of rad of x-radiation or gamma radiation that produces the same biological damage as 1 rad of the radiation being used.* The RBE factors for different types of radiation are given in Table 1. Note that the values are only approximate because they vary with particle energy and form of damage.

Finally, the **rem** (r̲oentgen e̲quivalent in m̲an) is defined as *the product of the dose in rad and the RBE factor:*

$$\text{Dose in rem} = \text{dose in rad} \times \text{RBE}$$

According to this definition, 1 rem of any two radiations will produce the same amount of biological damage. From Table 1, we see that a dose of 1 rad of fast neutrons represents an effective dose of 10 rem and that 1 rad of x-radiation is equivalent to a dose of 1 rem.

Low-level radiation from natural sources, such as cosmic rays and radioactive rocks and soil, delivers to each of us a dose of about 0.13 rem/year. The upper limit of radiation dose recommended by the U.S. government (apart from background radiation and exposure related to medical procedures) is 0.5 rem/year. Many occupa-

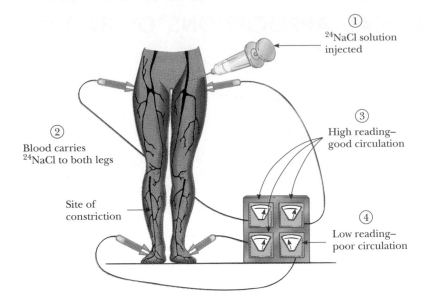

① ^{24}NaCl solution
injected

② Blood carries
^{24}NaCl to both legs

③ High reading–
good circulation

Site of
constriction

④ Low reading–
poor circulation

FIGURE 1

A tracer technique for determining the condition of the human circulatory system.

TABLE 1
RBE Factors for Several Types of Radiation

Radiation	RBE Factor
X-rays and gamma rays	1.0
Beta particles	1.0–1.7
Alpha particles	10–20
Slow neutrons	4–5
Fast neutrons and protons	10
Heavy ions	20

tions involve higher levels of radiation exposure, and for individuals in these occupations an upper limit of 5 rem/year has been set for whole-body exposure. Higher upper limits are permissible for certain parts of the body, such as the hands and forearms. An acute whole-body dose of 400 to 500 rem results in a mortality rate of about 50%. The most dangerous form of exposure is ingestion or inhalation of radioactive isotopes, especially those elements the body retains and concentrates, such as ^{90}Sr. In some cases, a dose of 1000 rem can result from ingesting 1 mCi of radioactive material.

Tracing

Radioactive particles can be used to trace chemicals participating in various reactions. One of the most valuable uses of radioactive tracers is in medicine. For example, ^{131}I is an artificially produced isotope of iodine (the natural, nonradioactive isotope is ^{127}I). Iodine, which is a necessary nutrient for our bodies, is obtained largely through the intake of iodized salt and seafood. The thyroid gland plays a major role in the distribution of iodine throughout the body. In order to evaluate the performance of the thyroid, the patient drinks a very small amount of radioactive sodium iodide. Two hours later, the amount of iodine in the thyroid gland is determined by measuring the radiation intensity at the neck area.

A second medical application is indicated in Figure 1. Here a salt containing radioactive sodium is injected into a vein in the leg. The time at which the the radioisotope arrives at another part of the body is detected with a radiation counter. The elapsed time is a good indication of the presence or absence of constrictions in the circulatory system.

The tracer technique is also useful in agricultural research. Suppose the best method of fertilizing a plant is to be determined. A certain material in the fertilizer, such as nitrogen, can be tagged with one of its radioactive isotopes. The fertilizer is then sprayed on one group of plants, sprinkled on the ground for a second group, and raked into the soil for a third. A Geiger counter is then used to track the nitrogen through the three types of plants.

Tracing techniques are as wide-ranging as human ingenuity can devise. Present applications range from checking the absorption of fluorine by teeth to checking contamination of food-processing equipment by cleansers to monitoring deterioration inside an automobile engine. In the latter case, a radioactive material is used in the manufacture of the pistons, and the oil is checked for radioactivity to determine the amount of wear on the pistons.

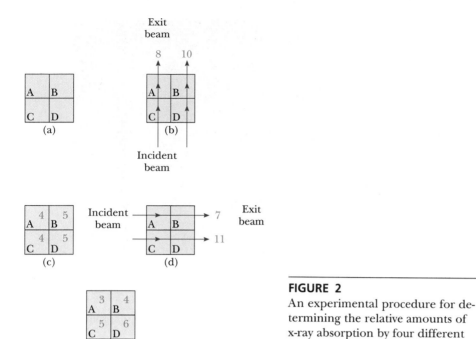

FIGURE 2
An experimental procedure for determining the relative amounts of x-ray absorption by four different compartments in a box.

Computed Axial Tomography (CAT Scans)

The normal x-ray of a human body has two primary disadvantages when used as a source of clinical diagnosis. First, it is difficult to distinguish between various types of tissue in the body because they all have similar x-ray absorption properties. Second, a conventional x-ray absorption picture is indicative of the average amount of absorption along a particular direction in the body, leading to somewhat obscured pictures. To overcome these problems, a device called a CAT scanner was developed in England in 1973; it is capable of producing pictures of much greater clarity and detail than were previously obtainable.

The operation of a CAT scanner can be understood by considering the following hypothetical experiment. Suppose a box consists of four compartments, labeled A, B, C, and D as in Figure 2a. Each compartment has a different amount of absorbing material from any other compartment. What set of experimental procedures will enable us to determine the relative amounts of material in each compartment? The following steps outline one method that will provide this information. First, a beam of x-rays is passed through compartments A and C, as Figure 2b. The intensity of the exiting radiation is reduced by absorption by some number that we assign as 8. (The number 8 could mean, for example, that the intensity of the exiting beam is reduced by eight-tenths of one percent from its initial value.) Since we do not know which of the compartments, A or C, was responsible for this reduction in intensity, half the loss is assigned to each compartment as in Figure 2c. Next, a beam of x-rays is passed through compartments B and D, as in Figure 2b. The reduction in intensity for this beam is 10, and again we assign half the loss to each compartment. We now redirect the x-ray source so that it sends one beam through compartments A and B and another through compartments C and D, as in Figure 2d, and again measure the absorption. Suppose the absorption through compartments A and B in this experiment is measured to be 7 units. On the basis of our first experiment, we would have guessed it would be 9 units, 4 by compartment A and 5 by compartment B. Thus, we have reduced the guessed absorption for each compartment by 1 unit so that the sum is 7 rather than 9, to give the numbers shown in Figure 2e. Likewise, when the beam is passed through compartments C and D as in Figure 2d, we may find the total absorp-

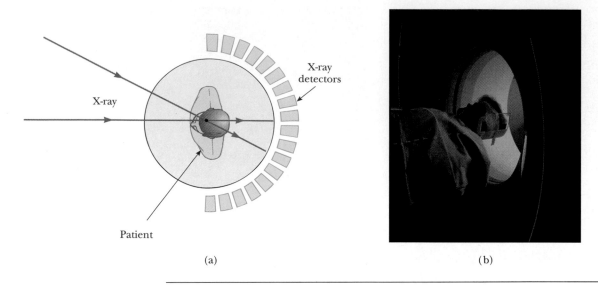

X-ray detectors

X-ray

Patient

(a)

(b)

FIGURE 3

(a) CAT scanner detector assembly. (b) Photograph of a patient undergoing a CAT scan in a hospital. (*Jay Freis/The Image Bank*)

tion to be 11 as compared to our first experiment of 9. In this case, we add 1 unit of absorption to each compartment to give a sum of 11 as in Figure 2e. This somewhat crude procedure could be improved by measuring the absorption along other paths. However, these simple measurements are sufficient to enable us to conclude that compartment D contains the most absorbing material and A the least. A visual representation of these results can be obtained by assigning to each compartment a shade of gray corresponding to the particular number associated with the absorption. In our example, compartment D would be very dark while compartment A would be very light.

The steps outlined above are representative of how a CAT scanner produces images of the human body. A thin slice of the body is subdivided into perhaps 10 000 compartments, rather than 4 compartments as in our simple example. The function of the CAT scanner is to determine the relative absorption in each of these 10 000 compartments and to display a picture of its calculations in various shades of gray. Note that CAT stands for **computed axial tomography.** The term axial is used because the slice of the body to be analyzed corresponds to a plane perpendicular to the head-to-toe axis. *Tomos* is the Greek word for slice and *graph* is the Greek word for picture. In a typical diagnosis, the patient is placed in the position shown in Figure 3 and a narrow beam of x-rays is sent through the plane of interest. The emerging x-rays are detected and measured by photomultiplier tubes behind the patient. The x-ray tube is then rotated a few degrees, and the intensity is recorded again. An extensive amount of information is obtained by rotating the beam through 180 degrees at intervals of about one degree per measurement, resulting in a set of numbers assigned to each of the 10 000 "compartments" in the slice. These numbers are then converted by the computer to a photograph in various shades of gray for this segment of the body.

A brain scan of a patient can now be made in about 2 s, while a full-body scan requires about 6 s. The final result is a picture containing much greater quantitative information and clarity than a conventional x-ray photograph. Since CAT scanners use x-rays, which are an ionizing form of radiation, the technique presents a health risk to the patient being diagnosed.

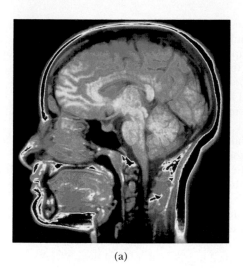

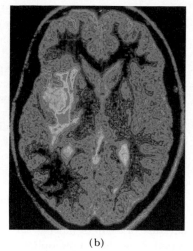

(a)   (b)

FIGURE 4
Computer-enhanced MRI images of (a) a normal human brain with the pituitary gland highlighted and (b) a human brain with a glioma tumor. *(Scott Camazine/Science Source)*

Magnetic Resonance Imaging (MRI)

At the heart of magnetic resonance imaging (MRI) is the fact that when a nucleus having a magnetic moment is placed in an external magnetic field, its moment will precess about the magnetic field with a frequency that is proportional to the field. For example, a proton, whose spin is 1/2, can occupy one of two energy states when placed in an external magnetic field. The lower energy state corresponds to the case where the spin is aligned with the field, while the higher energy state corresponds to the case where the spin is opposite the field. Transitions between these two states can be observed using a technique known as **nuclear magnetic resonance.** A dc magnetic field is applied to align the magnetic moments, while a second, weak oscillating magnetic field is applied perpendicular to the dc field. When the frequency of the oscillating field is adjusted to match the precessional frequency of the magnetic moments, the nuclei will "flip" between the two spin states. These transitions result in a net absorption of energy by the spin system, which can be detected electronically.

In MRI, image reconstruction is obtained using spatially varying magnetic fields and a procedure for encoding each point in the sample being imaged. Some MRI images taken on a human head are shown in Figure 4. In practice, a computer-controlled pulse sequencing technique is used to produce signals that are captured by a suitable processing device. This signal is then subjected to appropriate mathematical manipulations to provide data for the final image. The main advantage of MRI over other imaging techniques in medical diagnostics is that it causes minimal damage to cellular structures. Photons associated with the rf signals used in MRI have energies of only about 10^{-7} eV. Since molecular bond strengths are much larger (of the order of 1 eV), the rf photons cause little cellular damage. In comparison, x-rays or γ-rays have energies ranging from 10^4 to 10^6 eV and can cause considerable cellular damage.

Problems

1. A particular radioactive source produces 100 mrad of 2-MeV gamma rays per hour at a distance of 1.0 m. (a) How long could a person stand at this distance before accumulating an intolerable dose of 1 rem? (b) Assuming the gamma radiation is emitted isotropically, at what distance would a person receive a dose of 10 mrad/h from this source?

2. Assume that an x-ray technician takes an average of eight x-rays per day and receives a dose of 5 rem/year as a result. (a) Estimate the dose in rem per x-ray taken. (b) How does this result compare with the amount of low-level background radiation the technician is exposed to?

3. A person whose mass is 75 kg is exposed to a whole-body dose of 25 rad. How many joules of energy are deposited in the person's body?

4. In terms of biological damage, how many rad of heavy ions is equivalent to 100 rad of x-rays?

5. A 200-rad dose of radiation is administered to a patient in an effort to combat a cancerous growth. Assuming all of the energy deposited is absorbed by the growth, (a) calculate the amount of energy delivered. (b) Assuming the growth has a mass of 0.25 kg and a specific heat equal to that of water, calculate its temperature rise.

6. A "clever" technician decides to heat some water for his coffee with an x-ray machine. If the machine produces 10 rad/s, how long will it take to raise the temperature of a cup of water by 50°C?

7. A patient swallows a radiopharmaceutical tagged with phosphorus-32($^{32}_{15}$P), a β^- emitter with a half-life of 14.3 days. The average kinetic energy of the emitted electrons is 700 keV. If the initial activity of the sample is 1.31 MBq, determine: (a) the number of electrons emitted in a 10-day period, (b) the total energy deposited in the body during the 10 days, and (c) the absorbed dose if the electrons are completely absorbed in 100 g of tissue.

NUCLEAR ENERGY AND ELEMENTARY PARTICLES

30

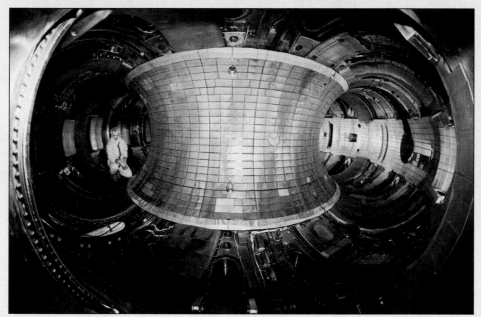

Interior view of the Tokamak Fusion Test Reactor (TFTR) vacuum vessel. The TFTR is located at the Princeton Plasma Physics Laboratory (PPPL), Princeton University, New Jersey. At the center, on the inner wall of the vessel, is the bumper limiter composed of graphite and graphite-composite tiles. Graphite tiles, which protect the vessel from neutral beams, can be seen to the right along the middle of the outer wall of the chamber. Arrays of graphite tiles are also clearly visible at the top and bottom of the vessel. Ion cyclotron radio-frequency launchers are seen at the left of the vacuum vessel near the shoulders of the kneeling person. (Courtesy of Princeton Plasma Physics Laboratory)

In this concluding chapter we discuss the two means by which energy can be derived from nuclear reactions. These two techniques are fission, in which a nucleus of large mass number splits, or fissions, into two smaller nuclei, and fusion, in which two light nuclei fuse to form a heavier nucleus. In either case, there is a release of large amounts of energy, which can be used destructively through bombs or constructively through the production of electric power.

We end our study of physics by examining the known subatomic particles and the fundamental interactions that govern their behavior. We also discuss the current theory of elementary particles, which states that all matter in nature is constructed from only two families of particles, quarks and leptons. Finally, we describe how clarifications of such models might help scientists understand the evolution of the Universe.

30.1

NUCLEAR FISSION

Nuclear fission occurs when a heavy nucleus, such as ^{235}U, splits, or fissions, into two smaller nuclei. In such a reaction, *the total rest mass of the products is less than the original rest mass of the heavy nucleus.*

Nuclear fission was first observed in 1939 by Otto Hahn and Fritz Strassman, following some basic studies by Fermi. After bombarding uranium ($Z = 92$) with

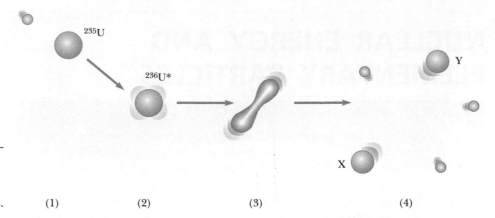

FIGURE 30.1

The stages involved in a nuclear fission event as described by the liquid-drop model of the nucleus.

(1)   (2)   (3)   (4)

neutrons, Hahn and Strassman discovered among the reaction products two medium-mass elements, barium and lanthanum. Shortly thereafter, Lisa Meitner and Otto Frisch explained what had happened. The uranium nucleus had split into two nearly equal fragments after absorbing a neutron. Such an occurrence was of considerable interest to physicists attempting to understand the nucleus, but it was to have even more far-reaching consequences. Measurements showed that about 200 MeV of energy is released in each fission event, and this fact was to affect the course of human history.

The fission of ^{235}U by slow (low energy) neutrons can be represented by the equation

$$ {}^{1}_{0}n + {}^{235}_{92}U \longrightarrow {}^{236}_{92}U^{*} \longrightarrow X + Y + \text{neutrons} \qquad \textbf{[30.1]} $$

where ^{236}U* is an intermediate state that lasts only for about 10^{-12} s before splitting into X and Y. The resulting nuclei, X and Y, are called **fission fragments.** There are many combinations of X and Y that satisfy the requirements of conservation of mass-energy and charge. In the fission of uranium, there are about 90 different daughter nuclei that can be formed. The process also results in the production of several (typically two or three) neutrons per fission event. On the average, 2.47 neutrons are released per event.

A typical reaction of this type is

$$ {}^{1}_{0}n + {}^{235}_{92}U \longrightarrow {}^{141}_{56}Ba + {}^{92}_{36}Kr + 3\,{}^{1}_{0}n \qquad \textbf{[30.2]} $$

The fission fragments, barium, and krypton, and the released neutrons have a great deal of kinetic energy following the fission event.

The breakup of the uranium nucleus can be compared to what happens to a drop of water when excess energy is added to it. All of the atoms in the drop have energy, but not enough to break up the drop. However, if enough energy is added to set the drop vibrating, it will undergo elongation and compression until the amplitude of vibration becomes large enough to cause the drop to break apart. In the uranium nucleus, a similar process occurs (Fig. 30.1). The sequence of events is

1. The ^{235}U nucleus captures a thermal (slow-moving) neutron.

2. This capture results in the formation of ^{236}U*, and the excess energy of this nucleus causes it to undergo violent oscillations.

3. The ^{236}U* nucleus becomes highly distorted, and the force of repulsion between protons in the two halves of the dumbbell shape tends to increase the distortion.

4. The nucleus splits into two fragments, emitting several neutrons in the process.

Let us estimate the disintegration energy, Q, released in a typical fission process. From Figure 29.4 we see that the binding energy per nucleon is about 7.6 MeV for heavy nuclei (those having a mass number of approximately 240) and about 8.5 MeV for nuclei of intermediate mass. This means that the nucleons in the fission fragments are more tightly bound and therefore have less mass than the nucleons in the original heavy nucleus. This decrease in mass per nucleon appears as released energy when fission occurs. The amount of energy released is (8.5—7.6) MeV per nucleon. Assuming a total of 240 nucleons, we find that the energy released per fission event is

$$Q = (240 \text{ nucleons}) \left(8.5 \frac{\text{MeV}}{\text{nucleon}} - 7.6 \frac{\text{MeV}}{\text{nucleon}} \right) = 220 \text{ MeV}$$

This is indeed a very large amount of energy relative to the amount released in chemical processes. For example, the energy released in the combustion of one molecule of the octane used in gasoline engines is about one hundred-millionth the energy released in a single fission event!

EXAMPLE 30.1   The Fission of Uranium

Two other possible ways by which ^{235}U can undergo fission when bombarded with a neutron are (1) by the release of ^{140}Xe and ^{94}Sr as fission fragments and (2) by the release of ^{132}Sn and ^{101}Mo as fission fragments. In each case, neutrons are also released. Find the number of neutrons released in each of these events.

Solution   By balancing mass numbers and atomic numbers, we find that these reactions can be written

$$^{1}_{0}n + {}^{235}_{92}U \longrightarrow {}^{140}_{54}Xe + {}^{94}_{38}Sr + 2\,{}^{1}_{0}n$$

$$^{1}_{0}n + {}^{235}_{92}U \longrightarrow {}^{132}_{50}Sn + {}^{101}_{42}Mo + 3\,{}^{1}_{0}n$$

Thus, two neutrons are released in the first event and three in the second.

EXAMPLE 30.2   The Energy Released in the Fission of ^{235}U

Calculate the total energy released if 1.00 kg of ^{235}U undergoes fission, taking the disintegration energy per event to be $Q = 208$ MeV (a more accurate value than the estimate given previously).

Solution   We need to know the number of nuclei in 1.00 kg of uranium. Since $A = 235$, the number of nuclei is

$$N = \left(\frac{6.02 \times 10^{23} \text{ nuclei/mol}}{235 \text{ g/mol}} \right) (1.00 \times 10^{3} \text{ g}) = 2.56 \times 10^{24} \text{ nuclei}$$

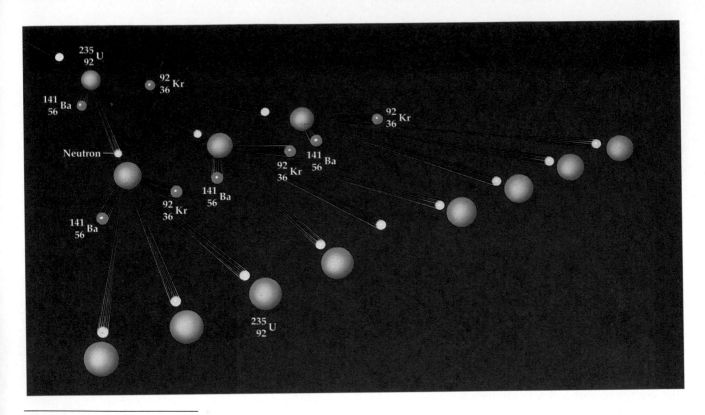

FIGURE 30.2
A nuclear chain reaction initiated by capture of a neutron. (Many pairs of different isotopes are produced, but only one pair is shown.)

Hence the disintegration energy

$$E = NQ = (2.56 \times 10^{24} \text{ nuclei}) \left(208 \, \frac{\text{MeV}}{\text{nucleus}} \right) = 5.32 \times 10^{26} \text{ MeV}$$

Since 1 MeV is equivalent to 4.45×10^{-20} kWh, $E = 2.37 \times 10^7$ kWh. This is enough energy to keep a 100-W lightbulb burning for about 30 000 years. Thus, 1.00 kg of ^{235}U is a relatively large amount of fissionable material.

30.2

NUCLEAR REACTORS

We have seen that when ^{235}U undergoes fission, an average of about 2.5 neutrons are emitted per event. These neutrons can in turn trigger other nuclei to undergo fission, with the possibility of a chain reaction (Fig. 30.2). Calculations show that if the chain reaction is not controlled (that is, if it does not proceed slowly), it could result in a violent explosion, with the release of an enormous amount of energy, even from only 1 g of ^{235}U. If the energy in 1 kg of ^{235}U were released, it would equal that released by the detonation of about 20 000 tons of TNT! This, of course, is the principle behind the first nuclear bomb, an uncontrolled fission reaction.

A nuclear reactor is a system designed to maintain what is called a **self-sustained chain reaction.** This important process was first achieved in 1942 by Fermi at the University of Chicago, with natural uranium as the fuel. Most reactors in operation today also use uranium as fuel. Natural uranium contains only about

0.7% of the ^{235}U isotope, with the remaining 99.3% being the ^{238}U isotope. This is important to the operation of a reactor because ^{238}U almost never undergoes fission. Instead, it tends to absorb neutrons, producing neptunium and plutonium. For this reason, reactor fuels must be artificially enriched so that they end up with a few percent of the ^{235}U isotope.

Earlier, we mentioned that an average of about 2.5 neutrons are emitted in each fission event of ^{235}U. In order to achieve a self-sustained chain reaction, one of these neutrons, on the average, must be captured by another ^{235}U nucleus and cause it to undergo fission. A useful parameter for describing the level of reactor operation is the **reproduction constant**, K, defined as *the average number of neutrons from each fission event that will cause another event.* As we have seen, K can have a maximum value of 2.5 in the fission of uranium. However, in practice K is less than this because of several factors, which we shall soon discuss.

A self-sustained chain reaction is achieved when $K = 1$. Under this condition, the reactor is said to be **critical**. When K is less than unity, the reactor is subcritical and the reaction dies out. When K is greater than unity the reactor is said to be supercritical, and a runaway reaction occurs. In a nuclear reactor used to furnish power to a utility company, it is necessary to maintain a K value close to unity.

The basic design of a nuclear reactor is shown in Figure 30.3. The fuel elements consist of enriched uranium. The function of the remaining parts of the reactor and some aspects of its design will now be described.

FIGURE 30.3
Cross-section of a reactor core surrounded by a radiation shield.

NEUTRON LEAKAGE

In any reactor, a fraction of the neutrons produced in fission will leak out of the core before inducing other fission events. If the fraction leaking out is too large,

Painting of the world's first reactor. Because of wartime secrecy, there are no photographs of the completed reactor. The reactor was composed of layers of graphite interspersed with uranium. A self-sustained chain reaction was first achieved on December 2, 1942. Word of the success was telephoned immediately to Washington with this message: "The Italian navigator has landed in the New World and found the natives very friendly." The historic event took place in an improvised laboratory in the racquet court under the west stands of the University of Chicago's Stagg Field, and the Italian navigator was Fermi. (*Courtesy of Chicago Historical Society*)

the reactor will not operate. The percentage lost is large if the reactor is very small because leakage is a function of the ratio of surface area to volume. Therefore, a critical requirement of reactor design is choosing the correct surface-area-to-volume ratio so that a sustained reaction can be achieved.

REGULATING NEUTRON ENERGIES

The neutrons released in fission events are very energetic, with kinetic energies of about 2 MeV. It is found that slow neutrons are far more likely than fast neutrons to produce fission events in ^{235}U. Furthermore, ^{238}U does not absorb slow neutrons. Therefore, in order for the chain reaction to continue, the neutrons must be slowed down. This is accomplished by surrounding the fuel with a **moderator** substance.

In order to understand how neutrons are slowed down, consider a collision between a light object and a very massive one. In such an event, the light object rebounds from the collision with most of its original kinetic energy. However, if the collision is between objects whose masses are nearly the same, the incoming projectile will transfer a large percentage of its kinetic energy to the target. In the first nuclear reactor ever constructed, Fermi placed bricks of graphite (carbon) between the fuel elements. Carbon nuclei are about 12 times more massive than neutrons, but after about 100 collisions with carbon nuclei, a neutron is slowed sufficiently to increase its likelihood of fission with ^{235}U. In this design the carbon is the moderator; most modern reactors use heavy water (D_2O) as the moderator.

NEUTRON CAPTURE

In the process of being slowed down, the neutrons may be captured by nuclei that do not undergo fission. The most common event of this type is neutron capture by ^{238}U. The probability of neutron capture by ^{238}U is very high when the neutrons have high kinetic energies and very low when they have low kinetic energies. Thus the slowing down of the neutrons by the moderator serves the dual purpose of making them available for reaction with ^{235}U and decreasing their chances of being captured by ^{238}U.

CONTROL OF POWER LEVEL

It is possible for a reactor to reach the critical stage ($K = 1$) after all the neutron losses described above are minimized. However, a method of control is needed to maintain a K value near unity. If K were to rise above this value, the heat produced in the runaway reaction would melt the reactor. To control the power level, control rods are inserted into the reactor core (see Fig. 30.3). These rods are made of materials such as cadmium that are very efficient in absorbing neutrons. By adjusting the number and position of these control rods in the reactor core, the K value can be varied and any power level within the design range of the reactor can be achieved.

A diagram of a pressurized-water reactor is shown in Figure 30.4. This type of reactor is commonly used in electric power plants in the United States. Fission events in the reactor core supply heat to the water contained in the primary (closed) system, which is maintained at high pressure to keep it from boiling.

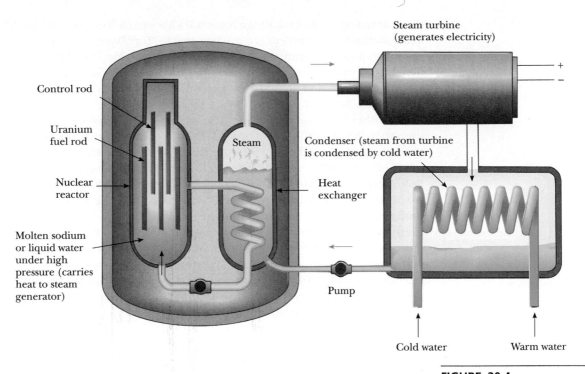

Steam turbine
(generates electricity)

Control rod

Uranium
fuel rod

Nuclear
reactor

Molten sodium
or liquid water
under high
pressure (carries
heat to steam
generator)

Steam

Heat
exchanger

Condenser (steam from turbine
is condensed by cold water)

Pump

Cold water   Warm water

FIGURE 30.4
Main components of a pressurized-water reactor.

This water also serves as the moderator. The hot water is pumped through a heat exchanger, and the heat is transferred to the water contained in the secondary system. There the hot water is converted to steam, which drives a turbine-generator to create electric power. Note that the water in the secondary system is isolated from the water in the primary system in order to prevent contamination of the secondary water and steam by radioactive nuclei from the reactor core.

REACTOR SAFETY[1]

The safety aspects of nuclear power reactors are often sensationalized by the media and misunderstood by the public. The 1979 near-disaster of Three Mile Island in Pennsylvania and the accident at the Chernobyl reactor in Ukraine rightfully focused attention on reactor safety. Yet the safety record in the United States is enviable. The records show no fatalities attributed to commercial nuclear power generation in the history of the United States nuclear industry.

Commercial reactors achieve safety through careful design and rigid operating procedures. Radiation exposure and the potential health risks associated with such exposure are controlled by three layers of containment. The fuel and radioactive fission products are contained inside the reactor vessel. Should this vessel rupture, the reactor building acts as a second containment structure to prevent radioactive material from contaminating the environment. Finally, the reactor facilities must be in a remote location to protect the general public from exposure should radiation escape the reactor building.

---

[1] The authors are grateful to Professor Gene Skluzacek of the University of Nebraska at Omaha for rewriting this section on reactor safety.

According to the Oak Ridge National Laboratory Review, "the health risk of living within 8 km (5 miles) of a nuclear reactor for 50 years is no greater than the risk of smoking 1.4 cigarettes, drinking 0.5 liters of wine, traveling 240 km by car, flying 9600 km by jet, or having one chest x-ray in a hospital. Each of these activities is estimated to increase a person's chances of dying in any given year by one in a million."

Another potential danger in nuclear reactor operations is the possibility that the water flow could be interrupted. Even if the nuclear fission chain reaction were stopped immediately, residual heat could build up in the reactor to the point of melting the fuel elements. The molten reactor core would melt to the bottom of the reactor vessel and conceivably melt its way into the ground below —the so-called "China syndrome." Although it might appear that this deep underground burial site would be an ideal safe haven for a radioactive blob, there would be danger of a steam explosion should the molten mass encounter water. This nonnuclear explosion could spread radioactive material to the areas surrounding the power plant. To prevent such an unlikely chain of events, nuclear reactors are designed with emergency core cooling systems requiring no power that automatically flood the reactor with water in the event of coolant loss. The emergency cooling water moderates heat build-up in the core, which in turn prevents the occurrence of melting.

A continuing concern in nuclear fission reactors is the safe disposal of radioactive material when the reactor core is replaced. This waste material contains long-lived, highly radioactive isotopes and must be stored over long periods of time in such a way that there is no chance of environmental contamination. At present, sealing radioactive wastes in waterproof containers and burying them in deep salt mines seems to be the most promising solution.

Transportation of reactor fuel and reactor wastes poses additional safety risks. However, the danger of theft during transport (say by a terrorist group) is greatly exaggerated. Furthermore, neither the waste nor the fuel of nuclear power reactors can be used to construct a nuclear bomb.

Accidents during transportation of nuclear fuel could expose the public to harmful levels of radiation. The Department of Energy requires stringent crash tests on all containers used to transport nuclear materials. Container manufacturers must demonstrate that their containers will not rupture even in high-speed collisions.

The safety issues associated with nuclear power reactors are complex and often emotional. All sources of energy have associated risks. In each case, one must weigh the risks against the benefits and the availability of the energy source.

30.3

NUCLEAR FUSION

Figure 29.4 shows that the binding energy for light nuclei (those having a mass number lower than 20) is much smaller than the binding energy for heavier nuclei. This suggests a possible process that is the reverse of fission. *When two light nuclei combine to form a heavier nucleus, the process is called nuclear fusion.* Because the mass of the final nucleus is less than the rest masses of the original nuclei, there is a loss of mass accompanied by a release of energy. It is important to recognize that although fusion power plants have not yet been developed, a great world-wide effort is under way to harness the energy from fusion reactions in the laboratory. Later we shall discuss the possibilities and advantages of this process for generating electric power.

FUSION IN THE SUN

All stars generate their energy through fusion processes. About 90% of the stars, including our own Sun, fuse hydrogen, while some older stars fuse helium or other heavier elements. Stars are born in regions of space containing vast clouds of dust and gas. Recent mathematical models of these clouds indicate that star formation is triggered by shock waves passing through a cloud. These shock waves are similar to sonic booms and are produced by events such as the explosion of a nearby star, called a supernova explosion. The shock wave compresses certain regions of the cloud, causing these regions to collapse under their own gravity. As the gas falls inward toward the center, the atoms gain speed, which causes the temperature of the gas to rise. Two conditions must be met before fusion reactions in the star can sustain its energy needs: (1) The temperature must be high enough (about 10^7 K for hydrogen) to allow the kinetic energy of the positively charged hydrogen nuclei to overcome their mutual Coulomb repulsion and collide, and (2) the density of nuclei must be high enough to ensure a high probability of collision.

When fusion reactions occur in a star, the energy liberated eventually becomes sufficient to prevent further collapse of the star under its own gravity. The star then continues to live out the remainder of its life under a balance between the inward force of gravity pulling it toward collapse and the outward force due to thermal effects and radiation pressure. The proton-proton cycle is a series of three nuclear reactions that are believed to be the stages in the liberation of energy in our Sun and other stars rich in hydrogen. An overall view of the proton-proton cycle is that four protons combine to form an alpha particle and two positrons, with the release of 25 MeV of energy in the process.

The three steps in the proton-proton cycle are

$$\begin{aligned} {}_1^1\text{H} + {}_1^1\text{H} &\longrightarrow {}_1^2\text{H} + {}_1^0\text{e} + \nu \\ {}_1^1\text{H} + {}_1^2\text{H} &\longrightarrow {}_2^3\text{He} + \gamma \end{aligned}$$

[30.3]

This second reaction is followed by either

$$ {}_1^1\text{H} + {}_2^3\text{H} \longrightarrow {}_2^4\text{H} + {}_1^0\text{e} + \nu $$

or

$$ {}_2^3\text{He} + {}_2^3\text{He} \longrightarrow {}_2^4\text{He} + {}_1^1\text{H} + {}_1^1\text{H} $$

The energy liberated is carried primarily by gamma rays, positrons, and neutrinos, as can be seen from the reactions. The gamma rays are soon absorbed by the dense gas, thus raising its temperature. The positrons combine with electrons to produce gamma rays, which in turn are also absorbed by the gas within a few centimeters. The neutrinos, however, almost never interact with matter; hence they escape from the star, carrying about 2% of the generated energy with them. These energy-liberating fusion reactions are called **thermonuclear fusion reactions.** The hydrogen (fusion) bomb, first exploded in 1952, is an example of an uncontrolled thermonuclear fusion reaction.

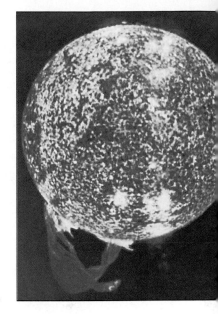

This photograph of the Sun, taken on December 19, 1973 during the third and final manned Skylab mission, shows one of the most spectacular solar flares ever recorded, spanning more than 588 000 km (365 000 mi) across the solar surface. The last picture, taken some 17 hours earlier, showed this feature as a large quiescent prominence on the eastern side of the Sun. The flare gives the distinct impression of a twisted sheet of gas in the process of unwinding itself. In this photograph the solar poles are distinguished by a relative absence of granulation and a much darker tone than the central portions of the disk. Several active regions are seen on the eastern side of the disk. The photograph was taken in the light of ionized helium by the extreme ultraviolet spectroheliograph instrument of the U.S. Naval Research Laboratory. *(NASA)*

FUSION REACTORS

The enormous amount of energy released in fusion reactions suggests the possibility of harnessing this energy for useful purposes here on Earth. A great deal of effort is currently under way to develop a sustained and controllable thermonuclear reactor—a fusion power reactor. Controlled fusion is often called the ultimate energy source because of the availability of its fuel source: water. For example, if deuterium were used as the fuel, 0.6 g of it could be extracted from 1 gal of water at a cost of about four cents. Such rates would make the fuel costs of even an inefficient reactor almost insignificant. An additional advantage of fusion reactors is that comparatively few radioactive by-products are formed. As noted in Equation 30.3, the end product of the fusion of hydrogen nuclei is safe, nonradioactive helium. Unfortunately, a thermonuclear reactor that can deliver a net power output over a reasonable time interval is not yet a reality, and many difficulties must be solved before a successful device is constructed.

We have seen that the Sun's energy is based, in part, on a set of reactions in which ordinary hydrogen is converted to helium. Unfortunately, the proton-proton interaction is not suitable for use in a fusion reactor because the event requires very high pressures and densities. The process works in the Sun only because of the extremely high density of protons in the Sun's interior. In fact, even at the densities and temperatures that exist at the center of the Sun, the average proton takes 14 billion years to react.

The fusion reactions that appear most promising in the construction of a fusion power reactor involve deuterium and tritium, which are isotopes of hydrogen. These reactions are

$$\begin{aligned}
{}^2_1\text{H} + {}^2_1\text{H} &\longrightarrow {}^3_2\text{He} + {}^1_0\text{n} & Q &= 3.27 \text{ MeV} \\
{}^2_1\text{H} + {}^2_1\text{H} &\longrightarrow {}^3_1\text{H} + {}^1_1\text{H} & Q &= 4.03 \text{ MeV} \\
{}^2_1\text{H} + {}^3_1\text{H} &\longrightarrow {}^4_2\text{He} + {}^1_0\text{n} & Q &= 17.59 \text{ MeV}
\end{aligned}$$

[30.4]

where the Q values refer to the amount of energy released per reaction. As noted earlier, deuterium is available in almost unlimited quantities from our lakes and oceans and is very inexpensive to extract. Tritium, however, is radioactive ($T_{1/2} = 12.3$ years) and undergoes beta decay to ${}^3\text{He}$. For this reason, tritium does not occur naturally to any great extent and must be artificially produced.

One of the major problems in obtaining energy from nuclear fusion is the fact that the Coulomb repulsion force between two charged nuclei must be overcome before they can fuse. The fundamental challenge is to give the two nuclei enough kinetic energy to overcome this repulsive force. This can be accomplished by heating the fuel to extremely high temperatures (about 10^8 K, far greater than the interior temperature of the Sun). As you might expect, such high temperatures are not easy to obtain in a laboratory or a power plant. At these high temperatures the atoms are ionized, and the system consists of a collection of electrons and nuclei, commonly referred to as a plasma.

In addition to the high temperature requirements, there are two other critical factors that determine whether or not a thermonuclear reactor will be successful: **plasma ion density,** n, and **plasma confinement time,** τ, the time the interacting ions are maintained at a temperature equal to or greater than that required for the reaction to proceed successfully. The density and confinement time must both be large enough to ensure that more fusion energy will be released than is required to heat the plasma.

Lawson's criterion states that a net power output in a fusion reactor is possible under the following conditions:

$$n\tau \geq 10^{14} \text{ s/cm}^3 \quad \text{Deuterium-tritium interaction}$$
$$n\tau \geq 10^{16} \text{ s/cm}^3 \quad \text{Deuterium-deuterium interaction}$$

[30.5] Lawson's criterion

The problem of plasma confinement time has yet to be solved. How can a plasma be confined at a temperature of 10^8 K for times on the order of 1 s? The basic plasma-confinement technique under investigation is discussed following Example 30.3.

EXAMPLE 30.3   The Deuterium-Deuterium Reaction

Find the energy released in the deuterium-deuterium reaction

$$^2_1\text{H} + {}^2_1\text{H} \longrightarrow {}^3_1\text{H} + {}^1_1\text{H}$$

Solution   The mass of the ^2_1H atom is 2.014 102 u. Thus, the total mass before the reaction is 4.028 204 u. After the reaction, the sum of the masses is equal to 3.016 049 u + 1.007 825 u = 4.023 874 u. Thus, the excess mass is 0.004 33 u. In energy units, this is equivalent to 4.03 MeV.

MAGNETIC FIELD CONFINEMENT

Most fusion experiments use magnetic field confinement to contain a plasma. One device, called a **tokamak,** has a doughnut-shaped geometry (a toroid), as shown in Figure 30.5a. This device, first developed in the USSR, uses a combina-

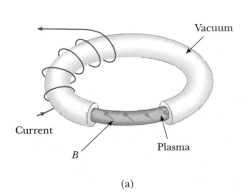

(a)

(b)

FIGURE 30.5
(a) Diagram of a tokamak used in the magnetic confinement scheme. The plasma is trapped within the spiraling magnetic field lines as shown. (b) Photograph of the Princeton Tokamak Fusion Test Reactor (TFTR), which uses magnetic field confinement. *(Courtesy of Princeton Plasma Physics Laboratory, Princeton, NJ)*

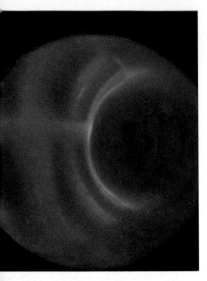

During TFTR operation, the plasma discharge is monitored using an optical system that views the interior of the vacuum vessel. In this view of a heated deuterium plasma, an ICRF antenna and a segment of an outboard limiter are visible in the left-side background. The bright radiation belt in the foreground, which encircles the toroidal bumper limiter, is known as a MARFE. This phenomenon occurs as the plasma density approaches the disruption limit for a given set of discharge conditions. *(Courtesy of Princeton Plasma Physics Laboratory)*

tion of two magnetic fields to confine the plasma inside the doughnut. A strong magnetic field is produced by the current in the windings, and a weaker magnetic field is produced by the current in the toroid. The resulting magnetic field lines are helical, as in Figure 30.5a. In this configuration, the field lines spiral around the plasma and prevent it from touching the walls of the vacuum chamber. If the plasma comes into contact with the walls, the temperature of the plasma is reduced, and impurities ejected from the walls ''poison'' the plasma and lead to large power losses.

In order for the plasma to reach ignition temperature, some form of auxiliary heating is necessary. A successful and efficient auxiliary heating technique that has been used recently is the injection of a beam of energetic neutral particles into the plasma.

Figure 30.5b is a photograph of the largest tokamak in the U.S. fusion program, the Tokamak Fusion Test Reactor (TFTR) at the Princeton Plasma Physics Laboratory. In December 1993 this reactor set a new record, with a fusion output power of 6.1 MW and an input power of 29.4 MW in a plasma lasting for 0.7 s. Researchers throughout the world continue their quest to reach the breakeven condition, where the power out equals the power in.

30.4
ELEMENTARY PARTICLES

The word ''atom'' is from the Greek word *atomos,* meaning ''indivisible.'' At one time atoms were thought to be the indivisible constituents of matter; that is, they were regarded as elementary particles. Discoveries in the early part of the 20th century revealed that the atom is not elementary, but has as its constituents protons, neutrons, and electrons. Until 1932 physicists viewed these three constituent particles as elementary because, with the exception of the free neutron, they are very stable. The theory soon fell apart, however, and beginning in 1945, many new particles were discovered in experiments involving high-energy collisions between known particles. These new particles are characteristically very unstable and have very short half-lives, ranging between 10^{-6} and 10^{-23} s. So far more than 300 of them have been catalogued.

Until the 1960s, physicists were bewildered by the large number and variety of subatomic particles being discovered. They wondered if the particles were like animals in a zoo, with no systematic relationship connecting them, or if a pattern was emerging that would provide a better understanding of the elaborate structure in the subnuclear world. In the last 25 years, physicists have made tremendous advances in our knowledge of the structure of matter by recognizing that all particles (with the exception of electrons, photons, and a few others) are made of smaller particles called quarks. Thus, protons and neutrons, for example, are not truly elementary but are systems of tightly bound quarks. The quark model has reduced the bewildering array of particles to a manageable number and has successfully predicted new quark combinations that were subsequently found in many experiments.

30.5
THE FUNDAMENTAL FORCES IN NATURE

The key to understanding the properties of elementary particles is to be able to describe the forces between them. All particles in nature are subject to four fundamental forces: strong, electromagnetic, weak, and gravitational.

TABLE 30.1
Particle Interactions

Interaction (Force)	Relative Strength	Range of Force	Mediating Field Particle
Strong	1	Short (≈ 1 fm)	Gluon
Electromagnetic	10^{-2}	Long ($\propto 1/r^2$)	Photon
Weak	10^{-13}	Short ($< 10^{-3}$ fm)	$W^{\pm}$ and Z bosons
Gravitational	10^{-38}	Long ($\propto 1/r^2$)	Graviton

The **strong force** is responsible for the binding of neutrons and protons into nuclei. This force represents the "glue" that holds the nucleons together and is the strongest of all the fundamental forces. It is very short-ranged and is negligible for separations greater than about 10^{-15} m (the approximate size of the nucleus). The **electromagnetic force,** which is about 10^{-2} times the strength of the strong force, is responsible for the binding of atoms and molecules. It is a long-range force that decreases in strength as the inverse square of the separation between interacting particles. The **weak force** is a short-range nuclear force that tends to produce instability in certain nuclei. It is responsible for most radioactive decay processes such as beta decay, and its strength is only about 10^{-13} times that of the strong force. (As we shall discuss later, scientists now believe that the weak and electromagnetic forces are two manifestations of a single force called the *electroweak* force). Finally, the **gravitational force** is a long-range force with a strength of only about 10^{-38} times that of the strong force. Although this familiar interaction is the force that holds the planets, stars, and galaxies together, its effect on elementary particles is negligible. Thus, the gravitational force is the weakest of all the fundamental forces.

Modern physics, often describes the interactions between particles in terms of the actions of field particles or quanta. In the case of the familiar electromagnetic interaction, the field particles are photons. In the language of modern physics, it can be said that the electromagnetic force is *mediated* by photons, which are the quanta of the electromagnetic field. Likewise, the strong force is mediated by field particles called *gluons,* the weak force is mediated by particles called the W and Z *bosons,* and the gravitational force is mediated by quanta of the gravitational field called *gravitons.* All of these field quanta have been detected except for the graviton, which may never be found directly because of the weakness of the gravitational field. These interactions, their ranges, and their relative strengths are summarized in Table 30.1.

The main tunnel at Fermilab, now housing both the older ring of conventional magnets (red and blue, above) and the new ring of superconducting magnets (yellow, below). In its present mode of operation as the Tevatron, the super-conducting ring receives injections of protons and antiprotons from the conventional ring. The two beams are accelerated in opposite directions around the ring to their final energy of 1 TeV; they then collide together at interaction points. The Tevatron is the most powerful accelerator in the world today. *(Courtesy of Fermi National Accelerator Laboratory)*

30.6
POSITRONS AND OTHER ANTIPARTICLES

In the 1920s, the theoretical physicist Paul Adrien Maurice Dirac (1902–1984) developed a version of quantum mechanics that incorporated special relativity. Dirac's theory successfully explained the origin of the electron's spin and its magnetic moment. But it had one major problem: its relativistic wave equation required solutions corresponding to negative energy states, and if negative energy states existed, we would expect an electron in a state of positive energy to make a rapid transition to one of these states, emitting a photon in the process.

Dirac circumvented this difficulty by postulating that all negative energy states are filled. The electrons that occupy the negative energy states are said to

FIGURE 30.6
(a) Bubble-chamber tracks of electron-positron pairs produced by 300-MeV gamma rays striking a lead sheet. *(Courtesy of Lawrence Berkeley Laboratory, University of California.)* (b) Sketch of the pertinent pair-production events. Note that the positrons deflect upward while the electrons deflect downward in an applied magnetic field directed into the diagram.

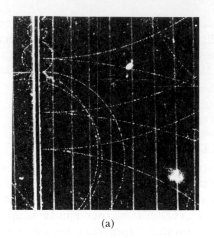

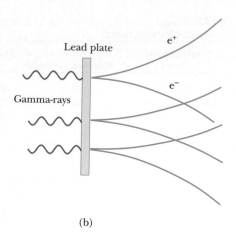

(a)

(b)

be in the ''Dirac sea'' and not directly observable because the Pauli exclusion principle does not allow them to react to external forces. However, if one of these negative energy states is vacant, leaving a hole in the sea of filled states, the hole can react to external forces and therefore be observable. The profound implication of this theory is that, *for every particle, there is an antiparticle.* The antiparticle has the same mass as the particle, but the opposite charge. For example, the electron's antiparticle (now called a *positron*) has a mass of 0.511 MeV and a positive charge of 1.6×10^{-19} C. As noted in Chapter 29, we designate an antiparticle with a bar over the symbol for the particle. Thus, $\bar{p}$ denotes the antiproton and $\bar{\nu}$ the antineutrino.

The positron was discovered by Carl Anderson in 1932 and in 1936 he was awarded the Nobel Prize for his achievement. Anderson discovered the positron while examining tracks created by electron-like particles of positive charge in a cloud chamber. (These early experiments used cosmic rays—mostly energetic protons passing through interstellar space—to initiate high-energy reactions on the order of several GeV.) In order to discriminate between positive and negative charges, the cloud chamber was placed in a magnetic field, causing moving charges to follow curved paths. Anderson noted that some of the electron-like tracks deflected in a direction corresponding to a positively charged particle.

Since Anderson's initial discovery, the positron has been observed in a number of experiments. Perhaps the most common process for producing positrons is **pair production**. In this process, a gamma ray with sufficiently high energy collides with a nucleus, creating an electron-positron pair. Since the total rest energy of the electron-positron pair is $2\,m_0\,c^2 = 1.02$ MeV (where m_0 is the rest mass of the electron), the gamma ray must have at least this much energy to create an electron-positron pair. Figure 30.6 shows tracks of electron-positron pairs created by 300-MeV gamma rays striking a lead sheet.

The reverse process can also occur. Under the proper conditions, an electron and positron can annihilate each other and produce two photons with a combined energy of at least 1.02 MeV:

$$e + \bar{e} \longrightarrow 2\gamma$$

Very rarely, a proton-antiproton pair can annihilate each other to produce two gamma ray photons.

Practically every known elementary particle has an antiparticle. Among the exceptions are the photon and the neutral pion (π^0). Following the construction of high-energy accelerators in the 1950s, many of these antiparticles were discovered. They included the antiproton ($\overline{p}$) discovered by Emilio Segrè and Owen Chamberlain in 1955 and the antineutron ($\overline{n}$) discovered shortly thereafter.

30.7

MESONS AND THE BEGINNING OF PARTICLE PHYSICS

Physicists in the mid-1930s had a fairly simple view of the structure of matter. The building blocks were the proton, the electron, and the neutron. Three other particles were known or postulated at the time: the photon, the neutrino, and the positron. These six particles were considered the fundamental constituents of matter. Although the accepted picture of the world was marvelously simple, no one was able to provide an answer to the following important question: Since the many protons in proximity in any nucleus should strongly repel each other due to their like charges, what is the nature of the force that holds the nucleus together? Scientists recognized that this mysterious force must be much stronger than anything encountered up to that time.

The first theory to explain the nature of the strong force was proposed in 1935 by the Japanese physicist Hideki Yukawa (1907–1981), an effort that later earned him the Nobel Prize. In order to understand Yukawa's theory, it is useful to first note that *two atoms can form a covalent chemical bond by the exchange of electrons*. Similarly, in the modern views of electromagnetic interactions, *charged particles interact by sharing a photon*. Yukawa used this same idea to explain the strong force by proposing a new particle that is shared by nucleons in the nucleus to produce the strong force. Furthermore, he established that the range of the force is inversely proportional to the mass of this particle, and predicted that the mass would be about 200 times the mass of the electron. Since the new particle would have a mass between that of the electron and the proton, it was called a **meson** (from the Greek *meso,* meaning ''middle'').

In an effort to substantiate Yukawa's predictions, physicists began looking for the meson by studying cosmic rays that enter the Earth's atmosphere. In 1937 Carl Anderson and his collaborators discovered a particle whose mass was 106 MeV/c^2, about 207 times the mass of the electron. However, subsequent experiments showed that the particle interacted very weakly with matter, and hence could not be the carrier of the strong force. This puzzling situation inspired several theoreticians to propose that there are two mesons with slightly different masses, an idea that was confirmed in 1947 with the discovery of the pi meson (π), or simply *pion*, by Cecil Frank Powell (1903–1969) and Guiseppe P. S. Occhialini (1907–   ). The lighter meson discovered earlier by Anderson, now called a *muon* (μ), has only weak and electromagnetic interaction and plays no role in the strong interaction.

The pion comes in three varieties, corresponding to three charge states: π^+, π^-, and π^0. The π^+ and π^- particles have masses of 139.6 MeV/c^2, while the π^0 has a mass of 135.0 MeV/c^2. Pions and muons are very unstable particles. For example, the π^-, which has a lifetime of about 2.6×10^{-8} s, decays into a muon and an antineutrino. The muon, with a lifetime of 2.2 μs, then decays into an electron, a neutrino, and an antineutrino. The sequence of decays is

$$\pi^- \longrightarrow \mu^- + \overline{\nu}$$
$$\mu^- \longrightarrow e + \nu + \overline{\nu}$$

[30.6]

Richard P. Feynman (1918–1988)

Richard Phillips Feynman was a brilliant theoretical physicist who together with Julian S. Schwinger and Shinchiro Tomonaga shared the 1965 Nobel Prize for Physics for their fundamental work in the principles of quantum electrodynamics. His many important contributions to physics include the invention of simple diagrams to represent particle interactions graphically, the theory of the weak interaction of subatomic particles, a reformulation of quantum mechanics, the theory of superfluid helium, and his contribution to physics education through the magnificent three-volume text, *The Feynman Lectures on Physics.*

Feynman did his undergraduate work at MIT and received his Ph.D. in 1942 from Princeton University, where he worked under John Archibald Wheeler. During World War II, he worked on the Manhattan Project at Princeton and then at Los Alamos, New Mexico. He joined the faculty at Cornell University in 1945 and was appointed professor of physics at the California Institute of Technology in 1950, where he remained for the rest of his career.

It is well known that Feynman had a passion for finding new and better ways to formulate each problem, or, as he would say, "turning it around." In the early part of his career, he was fascinated with electrodynamics and developed an intuitive view of quantum electrodynamics. Convinced that the electron could not interact with its own field, he said, "That was the beginning, and the idea seemed so obvious to me that I fell deeply in love with it . . . ." Often called the outstanding intuitionist of our age, he said in his Nobel acceptance speech, "Often, even in a physicist's sense, I did not have a demonstration of how to get all of these rules and equations, from conventional electrodynamics . . . I never really sat down, like Euclid did for the geometers of Greece, and made sure that you could get it all from a single set of axioms."

In 1986 Feynman was a member of the presidential commission to investigate the explosion of the space shuttle *Challenger.* In this capacity he performed a simple experiment for the commission members showing that one of the shuttle's O-ring seals was the likely cause of the disaster. After placing a seal in a

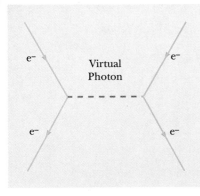

FIGURE 30.7
Feynman diagram representing a photon mediating the electromagnetic force between two electrons.

The interaction between two particles can be represented in a simple diagram called a *Feynman diagram,* developed by Richard P. Feynman (1918–1988). Figure 30.7 is a Feynman diagram for the electromagnetic interaction between two electrons. In this simple case, a photon is the field particle that mediates the electromagnetic force between the electrons. The photon transfers energy and momentum from one electron to the other in this interaction. Such a photon, called a *virtual photon,* can never be detected directly because it is absorbed by the second electron very shortly after being emitted by the first electron. The existence of a virtual photon would violate the law of conservation of energy, but because of the uncertainty principle and its very short lifetime, Δt, the photon's excess energy is less than the uncertainty in its energy, given by $\Delta E \approx \hbar/\Delta t$.

Now consider the pion exchange between a proton and a neutron via the strong force (Fig. 30.8). One can reason that the energy, ΔE, needed to create a pion of mass m_π is given by $\Delta E = m_\pi c^2$. Again, the existence of the pion violates conservation of energy only if this energy is surrendered in a time, Δt, the time it takes the pion to transfer from one nucleon to the other. From the uncertainty pitcher of ice water and squeezing it with a clamp, he demonstrated that the seal failed to spring back into shape once the clamp was removed.[1]

Feynman worked in physics with energy, vitality, and humor, a style reflective of his personality. The following quotes from some of his colleagues are characteristic of the great impact he made on the scientific community.[2]

"A brilliant, vital, and amusing neighbor, Feynman was a stimulating (if sometimes exasperating) partner in discussions of profound issues—we would exchange ideas and silly jokes in between bouts of mathematical calculation—we struck sparks off each other, and it was exhilarating." *Murray Gell-Mann*

"Reading Feynman is a joy and a delight, for in his papers, as in his talks, Feynman communicated very directly, as though the reader were watching him derive the results at the blackboard." *David Pines*

"He loved puzzles and games. In fact, he saw all the world as a sort of game, whose progress of "behavior" follows certain rules, some known, some unknown . . . . Find places or circumstances where the rules don't work, and invent new rules that do." *David L. Goodstein*

"Feynman was not a theorist's theorist, but a physicist's physicist and a teacher's teacher." *Valentine L. Telegdi*

Laurie M. Brown, one of his graduate students at Cornell, noted that Feynman, a playful showman, was "undervalued at first because of his rough manners, who in the end triumphs through native cleverness, psychological insight, common sense and the famous Feynman humor. . . . Whatever else Dick Feynman may have joked about, his love for physics approached reverence."

[1] Feynman's own account of this inquiry can be found in *Physics Today,* 4:26, February 1988.

[2] For more on Feynman's life and contributions, see the numerous articles in a special memorial issue of *Physics Today* 42, February 1989. For a personal account of Feynman, see his popular autobiographical books, *Surely You're Joking Mr. Feynman!,* New York, Bantam Books, 1985, and *What Do You Care What Other People Think?,* New York, W. W. Norton & Co., 1987.

principle, $\Delta E \, \Delta t \approx \hbar$, we get

$$\Delta t \approx \frac{\hbar}{\Delta E} = \frac{\hbar}{m_\pi c^2} \qquad \textbf{[30.7]}$$

Since the pion cannot travel faster than the speed of light, the maximum distance, d, it can travel in a time Δt is $c \, \Delta t$. Using Equation 30.7 and $d = c \, \Delta t$, we find this maximum distance to be

$$d \approx \frac{\hbar}{m_\pi c} \qquad \textbf{[30.8]}$$

The range of the strong force is about 1.5×10^{-15} m. Using this value for d in Equation 30.8, the rest energy of the pion is calculated to be

$$m_\pi c^2 \approx \frac{\hbar c}{d} = \frac{(1.05 \times 10^{-34} \, \text{J} \cdot \text{s})(3.00 \times 10^8 \, \text{m/s})}{1.5 \times 10^{-15} \, \text{m}}$$

$$= 2.1 \times 10^{-11} \, \text{J} \cong 130 \, \text{MeV}$$

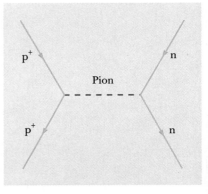

FIGURE 30.8

Feynman diagram representing a proton interacting with a neutron via the strong force. In this case, the pion mediates the strong force.

This corresponds to a mass of 130 MeV/c^2 (about 250 times the mass of the electron), which is in good agreement with the observed mass of the pion.

The concept we have just described is quite revolutionary. In effect, it says that a proton can change into a proton plus a pion, as long as it returns to its original state in a very short time. High-energy physicists often say that a nucleon undergoes "fluctuations" as it emits and absorbs pions. As we have seen, these fluctuations are a consequence of a combination of quantum mechanics (through the uncertainty principle) and special relativity (through Einstein's energy-mass relation $E = mc^2$).

This section has dealt with the particles that mediate the strong force, namely the pions, and the mediators of the electromagnetic force, the photons. The graviton, which is the mediator of the gravitational force, has yet to be observed. The W and Z particles that mediate the weak force were discovered in 1983 by Carlo Rubbia (1934–   ) and his associates. Rubbia and Simon van der Meer, both at CERN, shared the 1984 Nobel Prize in Physics for the discovery of the $W^{\pm}$ and Z^0 particles and the development of the proton-antiproton collider.

30.8

CLASSIFICATION OF PARTICLES

HADRONS

All particles other than photons can be classified into two broad categories, hadrons and leptons, according to their interactions. Particles that interact through the strong force are called *hadrons*. There are two classes of hadrons, known as *mesons* and *baryons,* distinguished by their masses and spins. All mesons are known to decay finally into electrons, positrons, neutrinos, and photons. The pion is the lightest of known mesons, with a mass of about 140 MeV/c^2 and a spin of 0. Another is the K meson, with a mass of about 500 MeV/c^2 and spin 0.

Baryons, have masses equal to or greater than the proton mass (the name *baryon* means "heavy" in Greek), and their spin is always a noninteger value (1/2 or 3/2). Protons and neutrons are baryons, as are many other particles. With the exception of the proton, all baryons decay in such a way that the end products include a proton. For example, the baryon called the Ξ hyperon first decays to a Λ^0 in about 10^{-10} s. The Λ^0 then decays to a proton and a π^- in about 3×10^{-10} s.

Today it is believed that hadrons are composed of quarks. (Later we shall have more to say about the quark model.) Some of the important properties of hadrons are listed in Table 30.2.

LEPTONS

Leptons (from the Greek *leptos* meaning "small" or "light") are a group of particles that participate in the weak interaction. All leptons have a spin of 1/2. Included in this group are electrons, muons, and neutrinos, which are less massive than the lightest hadron. Although hadrons have size and structure, leptons appear to be truly elementary, with no structure (that is, point-like).

TABLE 30.2
A Table of Some Particles and Their Properties

Category	Particle Name	Symbol	Anti-particle	Rest Mass (MeV/c^2)	B	L_e	L_μ	L_τ	S	Lifetime (s)	Principal Decay Modes[a]
Photon	Photon	γ	Self	0	0	0	0	0	0	Stable	
Leptons	Electron	e^-	e^+	0.511	0	+1	0	0	0	Stable	
	Neutrino (e)	ν_e	$\bar{\nu}_e$	0(?)	0	+1	0	0	0	Stable	
	Muon	μ^-	μ^+	105.7	0	0	+1	0	0	2.20×10^{-6}	$e^- \bar{\nu}_e \nu_\mu$
	Neutrino (μ)	ν_μ	$\bar{\nu}_\mu$	0(?)	0	0	+1	0	0	Stable	
	Tau	τ^-	τ^+	1784.	0	0	0	−1	0	$<4 \times 10^{-13}$	$\mu^- \bar{\nu}_\mu \nu_\tau$, $e^- \bar{\nu}_e \nu_\tau$, hadrons
	Neutrino (τ)	ν_τ	$\bar{\nu}_\tau$	0(?)	0	0	0	−1	0	Stable	
Hadrons											
Mesons	Pion	π^+	π^-	139.6	0	0	0	0	0	2.60×10^{-8}	$\mu^+ \nu_\mu$
		π^0	Self	135.0	0	0	0	0	0	0.83×10^{-16}	2γ
	Kaon	K^+	K^-	493.7	0	0	0	0	+1	1.24×10^{-8}	$\mu^+ \nu_\mu$, $\pi^+ \pi^0$
		K^0_S	$\overline{K^0_S}$	497.7	0	0	0	0	+1	0.89×10^{-10}	$\pi^+ \pi^-$, $2\pi^0$
		K^0_L	$\overline{K^0_L}$	497.7	0	0	0	0	+1	5.2×10^{-8}	$\pi^\pm e^\mp (\bar{\nu})_e$ $\pi^\pm \mu^\mp (\bar{\nu})_\mu$ $3\pi^0$
	Eta	η^0	Self	548.8	0	0	0	0	0	$<10^{-18}$	2γ, 3π
Baryons	Proton	p	$\bar{p}$	938.3	+1	0	0	0	0	Stable	
	Neutron	n	$\bar{n}$	939.6	+1	0	0	0	0	920	$pe^- \bar{\nu}_e$
	Lambda	Λ^0	$\overline{\Lambda^0}$	1115.6	+1	0	0	0	−1	2.6×10^{-10}	$p\pi^-$, $n\pi^0$
	Sigma	Σ^+	$\overline{\Sigma}^-$	1189.4	+1	0	0	0	−1	0.80×10^{-10}	$p\pi^0$, $n\pi^+$
		Σ^0	$\overline{\Sigma}^0$	1192.5	+1	0	0	0	−1	6×10^{-20}	$\Lambda^0 \gamma$
		Σ^-	$\overline{\Sigma}^+$	1197.3	+1	0	0	0	−1	1.5×10^{-10}	$n\pi^-$
	Xi	Ξ^0	$\overline{\Xi}^0$	1315	+1	0	0	0	−2	2.9×10^{-10}	$\Lambda^0 \pi^0$
		Ξ^-	Ξ^+	1321	+1	0	0	0	−2	1.64×10^{-10}	$\Lambda^0 \pi^-$
	Omega	Ω^-	Ω^+	1672	+1	0	0	0	−3	0.82×10^{-10}	$\Xi^0 \pi^0$, $\Lambda^0 K^-$

[a] A notation in this column such as $p\pi^-$, $n\pi^0$ means two possible decay modes. In this case, the two possible decays are $\Lambda^0 \rightarrow p + \pi^-$ or $\Lambda^0 \rightarrow n + \pi^0$.

Quite unlike hadrons, the number of known leptons is small. Currently, scientists believe there only are six leptons (each having an antiparticle)—the electron, the muon, the tau, and a neutrino associated with each:

$$\begin{pmatrix} e^- \\ \nu_e \end{pmatrix} \qquad \begin{pmatrix} \mu^- \\ \nu_\mu \end{pmatrix} \qquad \begin{pmatrix} \tau^- \\ \nu_\tau \end{pmatrix}$$

The tau lepton, discovered in 1975, has a mass about twice that of the proton. The neutrino associated with the tau has not yet been observed in the laboratory.

Although neutrinos are thought to be massless, there is a possibility that they have some small nonzero mass. As we shall see later, a firm knowledge of the neutrino's mass could have great significance in cosmological models and the future of the Universe.

30.9

CONSERVATION LAWS

A number of conservation laws are important in the study of elementary particles. Although the two described here have no theoretical foundation, they are supported by abundant empirical evidence.

BARYON NUMBER

The law of conservation of baryon number tells us that whenever a baryon is created in a reaction or decay, an antibaryon is also created. This can be quantified by assigning a baryon number: $B = +1$ for all baryons, $B = -1$ for all antibaryons, and $B = 0$ for all other particles. Thus, the **law of conservation of baryon number** states that whenever a nuclear reaction or decay occurs, the sum of the baryon numbers before the process must equal the sum of the baryon numbers after the process.

Conservation of baryon number

Note that if baryon number is absolutely conserved, the proton must be absolutely stable. If it were not for the law of conservation of baryon number, the proton could decay into a positron and a neutral pion. However, such a decay has never been observed. At present, we can only say that the proton has a half-life of at least 10^{31} years (the estimated age of the Universe is about 10^{10} years). In one recent version of a so-called grand unified theory, or GUT, physicists have predicted that the proton is actually unstable. According to this theory, the baryon number (sometimes called the baryonic charge) is not absolutely conserved, whereas electric charge is always conserved.

EXAMPLE 30.4   Checking Baryon Numbers

Determine whether or not each of the following reactions can occur based on the law of conservation of baryon number.

$$(1) \qquad p + n \longrightarrow p + p + n + \overline{p}$$

$$(2) \qquad p + n \longrightarrow p + p + \overline{p}$$

Solution   Recall that $B = +1$ for baryons and $B = -1$ for antibaryons. Hence the left side of (1) gives a total baryon number of $1 + 1 = 2$. The right side gives a total baryon number of $1 + 1 + 1 + (-1) = 2$. Thus the reaction can occur provided the incoming proton has sufficient energy.

The left side of (2) gives a total baryon number of $1 + 1 = 2$. However, the right side gives $1 + 1 + (-1) = 1$. Since baryon number is not conserved, the reaction cannot occur.

LEPTON NUMBER

Conservation of lepton number

There are three conservation laws involving lepton numbers, one for each variety of lepton. The **law of conservation of electron-lepton number** states that the sum of the electron-lepton numbers before a reaction or decay must equal the sum of the electron-lepton numbers after the reaction or decay. The electron and the electron neutrino are assigned a positive electron-lepton number, $L_e = +1$; the antileptons e^+ and $\bar{\nu}_e$ are assigned the electron-lepton number $L_e = -1$; and all other particles have $L_e = 0$. For example, consider the decay of the neutron

$$n \longrightarrow p^+ + e^- + \bar{\nu}_e$$

Neutron decay

Before the decay, the electron-lepton number is $L_e = 0$; after the decay it is $0 + 1 + (-1) = 0$. Thus, the electron-lepton number is conserved. It is important to recognize that the baryon number must also be conserved. This can easily be seen by noting that before the decay $B = +1$, whereas after the decay $B = +1 + 0 + 0 = +1$.

Similarly, when a decay involves muons, the muon-lepton number, L_μ, is conserved. The μ^- and the ν_μ are assigned $L_\mu = +1$, the antimuons μ^+ and $\bar{\nu}_\mu$ are assigned $L_\mu = -1$, and all other particles have $L_\mu = 0$. Finally, the tau-lepton number, L_τ, is conserved, and similar assignments can be made for the τ lepton and its neutrino.

EXAMPLE 30.5   Checking Lepton Numbers

Determine which of the following decay schemes can occur on the basis of conservation of lepton number.

$$(1) \qquad \mu^- \longrightarrow e^- + \bar{\nu}_e + \nu_\mu$$

$$(2) \qquad \pi^+ \longrightarrow \mu^+ + \nu_\mu + \nu_e$$

Solution   Since decay 1 involves both a muon and an electron, L_μ and L_e must both be conserved. Before the decay, $L_\mu = +1$ and $L_e = 0$. After the decay, $L_\mu = 0 + 0 + 1 = +1$, and $L_e = +1 - 1 + 0 = 0$. Thus, both numbers are conserved, and on this basis the decay mode is possible.

Before decay 2 occurs, $L_\mu = 0$ and $L_e = 0$. After the decay, $L_\mu = -1 + 1 + 0 = 0$, but $L_e = +1$. Thus, the decay is not possible because the electron-lepton number is not conserved.

Exercise   Determine whether the decay $\mu^- \rightarrow e^- + \bar{\nu}_e$ can occur.

Answer   No. The muon-lepton number is $+1$ before the decay and 0 after.

30.10
STRANGE PARTICLES AND STRANGENESS

Many particles discovered in the 1950s were produced by the nuclear interaction of pions with protons and neutrons in the atmosphere. A group of these particles, namely the K, Λ, and Σ, were found to exhibit unusual properties in their production and decay, and hence were called *strange particles*.

One unusual property of strange particles is that they are always produced in pairs. For example, when a pion collides with a proton, two neutral strange particles are produced with high probability (Fig. 30.9) following the reaction

$$\pi^- + p^+ \longrightarrow K^0 + \Lambda^0$$

On the other hand, the reaction $\pi^- + p^+ \rightarrow K^0 + n$ never occurred, even though no known conservation laws were violated and the energy of the pion was sufficient to initiate the reaction.

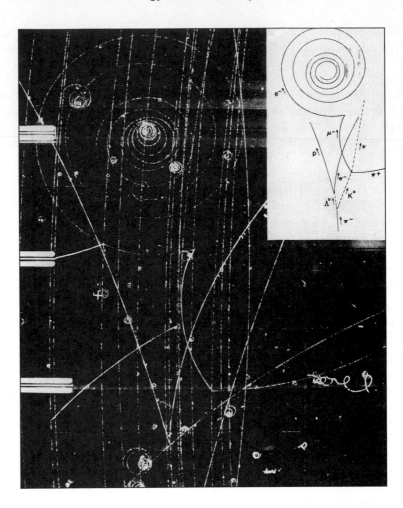

FIGURE 30.9
This bubble-chamber photograph shows many events, and the inset represents a drawing of identified tracks. The strange particles Λ^0 and K^0 are formed (at the bottom) as the π^- interacts with a proton according to $\pi^- + p \rightarrow \Lambda^0 + K^0$. (Note that the neutral particles leave no tracks, as indicated by the dashed lines.) The Λ^0 and K^0 then decay according to $\Lambda^0 \rightarrow \pi^- + p$ and $K^0 \rightarrow \pi + \mu^- + \nu_\mu$. *(Courtesy of Lawrence Berkeley Laboratory, University of California, Photographic Services.)*

The second peculiar feature of strange particles is that although they are produced by the strong interaction at a high rate, they do not decay into particles that interact via the strong force at a very high rate. Instead, they decay very slowly, which is characteristic of the weak interaction. Their half-lives are in the range 10^{-10} s to 10^{-8} s; most other particles that interact via the strong force have lifetimes on the order of 10^{-23} s.

To explain these unusual properties of strange particles, a law called conservation of strangeness was introduced, together with a new quantum number, S, called **strangeness**. The strangeness numbers for some particles are given in Table 30.2. The production of strange particles in pairs is explained by assigning $S = +1$ to one of the particles and $S = -1$ to the other. All nonstrange particles are assigned strangeness $S = 0$. The **law of conservation of strangeness** states that whenever a nuclear reaction or decay occurs, the sum of the strangeness numbers before the process must equal the sum of the strangeness numbers after the process.

One can explain the slow decay of strange particles by assuming that the strong and electromagnetic interactions obey the law of conservation of strangeness, whereas the weak interaction does not. Since the decay reaction involves the loss of one strange particle, it violates strangeness conservation and hence proceeds slowly via the weak interaction.

Conservation of strangeness number

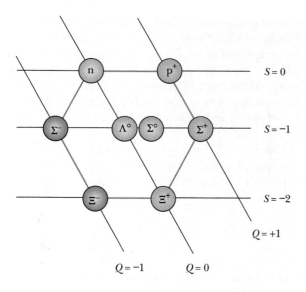

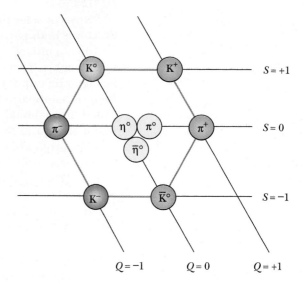

(a)

(b)

EXAMPLE 30.6   Is Strangeness Conserved?

(a) Determine whether the following reaction occurs on the basis of conservation of strangeness.

$$\pi^0 + n \longrightarrow K^+ + \Sigma^-$$

Solution   The initial state has a total strangeness of $S = 0 + 0 = 0$. Since the strangeness of the K^+ is $S = +1$, and the strangeness of the Σ^- is $S = -1$, the total strangeness of the final state is $+1 - 1 = 0$. Thus, strangeness is conserved and the reaction is allowed.

(b) Show that the following reaction does not conserve strangeness.

$$\pi^- + p \longrightarrow \pi^- + \Sigma^+$$

Solution   The initial state has strangeness $S = 0 + 0 = 0$, while the final state has strangeness $S = 0 + (-1) = -1$. Thus strangeness is not conserved.

Exercise   Show that the observed reaction $p^+ + \pi^- \rightarrow K^0 + \Lambda^0$ obeys the law of conservation of strangeness.

FIGURE 30.10
(a) The hexagonal Eightfold Way pattern for the eight spin-$\frac{1}{2}$ baryons. This strangeness versus charge plot uses a sloping axis for the charge number Q.
(b) The Eightfold Way pattern for the nine spin-zero mesons.

30.11

THE EIGHTFOLD WAY

As we have seen, quantities such as spin, baryon number, lepton number, and strangeness are labels we associate with particles. Many classification schemes that group particles into families based on such labels have been proposed. First, consider the first eight baryons listed in Table 30.2, all having a spin of one half. The family consists of the proton, the neutron, and six other particles. If we plot their strangeness versus their charge using a sloping coordinate system, as in Figure 30.10a, a fascinating pattern emerges. Six of the baryons form a hexagon, while the remaining two are at the hexagon's center.

Now consider the family of mesons listed in Table 30.2 with spins of zero. If we count both particles and antiparticles, there are nine such mesons. Figure 30.10b is a plot of strangeness versus charge for this family. Again, a fascinating hexagonal pattern emerges. In this case, the particles on the perimeter of the hexagon lie opposite their antiparticles, and the remaining three (which form their own antiparticles) are at its center. These and related symmetric patterns, called the **eightfold way,** were proposed independently in 1961 by Murray Gell-Mann and Yuval Ne'eman.

The groups of baryons and mesons can be displayed in many other symmetric patterns within the framework of the eightfold way. For example, the family of spin-3/2 baryons contains ten particles arranged in a pattern like the tenpins in a bowling alley. After the pattern was proposed, one of the particles was missing — it had yet to be discovered. Gell-Mann predicted that the missing particle, which he called the omega minus (Ω^-), should have a spin of 3/2, a charge of -1, a strangeness of -3, and a rest energy of about 1680 MeV. Shortly thereafter, in 1964, scientists at the Brookhaven National Laboratory found the missing particle through careful analyses of bubble chamber photographs, and confirmed all its predicted properties.

The patterns of the eightfold way in the field of particle physics have much in common with the periodic table. Whenever a vacancy (a missing particle or element) occurs in the organized patterns, experimentalists have a guide for their investigations. Furthermore, the existence of the eightfold-way patterns suggests that baryons and mesons have a more elemental substructure, to which we now turn.

30.12

QUARKS—FINALLY

As we have noted, leptons appear to be truly elementary particles because they have no measurable size or internal structure, are limited in number, and do not seem to break down into smaller units. Hadrons, on the other hand, are complex particles with size and structure. Furthermore, we know that hadrons decay into other hadrons and are many in number. Table 30.2 lists only those hadrons that are stable against hadronic decay; hundreds of others have been discovered. These facts strongly suggest that hadrons cannot be truly elementary but have some substructure.

THE ORIGINAL QUARK MODEL

In 1963 Gell-Mann and George Zweig independently proposed that hadrons have a more elementary substructure. According to their model, all hadrons are composite systems of two or three fundamental constitutents called **quarks.** Gell-Mann borrowed the word *quark* from the passage "Three quarks for Muster Mark" in James Joyce's book *Finnegans Wake*. In the original model, there were three types of quarks designated by the symbols *u, d,* and *s*. These were given the arbitrary names *up, down,* and *sideways* (or, now more commonly, *strange*).

An unusual property of quarks is that they have fractional electronic charges, as shown—along with other properties—in Table 30.3. Associated with each quark is an antiquark of opposite charge, a baryon number, and a

TABLE 30.3
Properties of Quarks and Antiquarks

Quarks								
Name	Symbol	Spin	Charge	Baryon Number	Strangeness	Charm	Bottomness	Topness
Up	u	$\frac{1}{2}$	$+\frac{2}{3}e$	$\frac{1}{3}$	0	0	0	0
Down	d	$\frac{1}{2}$	$-\frac{1}{3}e$	$\frac{1}{3}$	0	0	0	0
Strange	s	$\frac{1}{2}$	$-\frac{1}{3}e$	$\frac{1}{3}$	-1	0	0	0
Charmed	c	$\frac{1}{2}$	$+\frac{2}{3}e$	$\frac{1}{3}$	0	$+1$	0	0
Bottom	b	$\frac{1}{2}$	$-\frac{1}{3}e$	$\frac{1}{3}$	0	0	$+1$	0
Top (?)	t	$\frac{1}{2}$	$+\frac{2}{3}e$	$\frac{1}{3}$	0	0	0	$+1$

Antiquarks								
Name	Symbol	Spin	Charge	Baryon Number	Strangeness	Charm	Bottomness	Topness
Up	$\bar{u}$	$\frac{1}{2}$	$-\frac{2}{3}e$	$-\frac{1}{3}$	0	0	0	0
Down	$\bar{d}$	$\frac{1}{2}$	$+\frac{1}{3}e$	$-\frac{1}{3}$	0	0	0	0
Strange	$\bar{s}$	$\frac{1}{2}$	$+\frac{1}{3}e$	$-\frac{1}{3}$	$+1$	0	0	0
Charmed	$\bar{c}$	$\frac{1}{2}$	$-\frac{2}{3}e$	$-\frac{1}{3}$	0	-1	0	0
Bottom	$\bar{b}$	$\frac{1}{2}$	$+\frac{1}{3}e$	$-\frac{1}{3}$	0	0	-1	0
Top(?)	$\bar{t}$	$\frac{1}{2}$	$-\frac{2}{3}e$	$-\frac{1}{3}$	0	0	0	-1

strangeness. The compositions of all hadrons known when Gell-Mann and Zweig presented their models could be completely specified by three simple rules:

1. Mesons consist of one quark and one antiquark, giving them a baryon number of 0, as required.
2. Baryons consist of three quarks.
3. Antibaryons consist of three antiquarks.

Table 30.4 lists the quark compositions of several mesons and baryons. Note that just two of the quarks, u and d, are contained in all hadrons encountered in ordinary matter (protons and neutrons). The third quark, s, is needed only to construct strange particles with a strangeness of either $+1$ or -1. Figure 30.11 is a pictorial representation of the quark compositions of several particles.

CHARM AND OTHER RECENT DEVELOPMENTS

Although the original quark model was highly successful in classifying particles into families, there were some discrepancies between predictions of the model and certain experimental decay rates. Consequently, a fourth quark was proposed by several physicists in 1967. They argued that if there are four leptons (as was thought at the time), then there should also be four quarks because of an underlying symmetry in nature. The fourth quark, designated by c, was given a property called **charm**. A *charmed* quark would have the charge $+2e/3$, but its charm would distinguish it from the other three quarks. The new quark would have a charm of $C = +1$, its antiquark would have a charm of $C = -1$, and all other quarks would have $C = 0$, as indicated in Table 30.3. Charm, like strangeness, would be conserved in strong and electromagnetic interactions, but not in weak interactions.

TABLE 30.4
Quark Composition of Several Hadrons

Particle	Quark Composition
Mesons	
π^+	$u\bar{d}$
π^-	$\bar{u}d$
K^+	$u\bar{s}$
K^-	$\bar{u}s$
K^0	$d\bar{s}$
Baryons	
p	uud
n	udd
Λ^0	uds
Σ^+	uus
Σ^0	uds
Σ^-	dds
Ξ^0	uss
Ξ^-	dss
Ω^-	sss

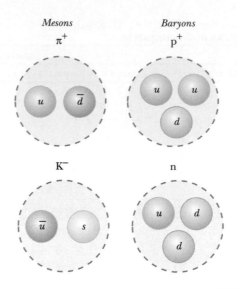

FIGURE 30.11
Quark compositions of several particles. Note that the mesons on the left contain two quarks, while the baryons on the right contain three quarks.

TABLE 30.5
The Fundamental Particles and Some of Their Properties

Particle	Rest Energy	Charge
Quarks		
u	360 MeV	$+\frac{2}{3}e$
d	360 MeV	$-\frac{1}{3}e$
c	1500 MeV	$+\frac{2}{3}e$
s	540 MeV	$-\frac{1}{3}e$
t	170 GeV	$+\frac{2}{3}e$
b	5 GeV	$-\frac{1}{3}e$
Leptons		
e^-	511 keV	$-e$
μ^-	107 MeV	$-e$
τ^-	1784 MeV	$-e$
ν_e	<30 eV	0
ν_μ	<0.5 MeV	0
ν_τ	<250 MeV	0

In 1974 a new heavy meson called the J/ψ particle (or simply ψ) was discovered independently by a group led by Burton Richter at the Stanford Linear Accelerator (SLAC) and another group led by Samuel Ting at the Brookhaven National Laboratory. Richter and Ting were awarded the Nobel Prize in 1976 for this work. The J/ψ particle did not fit into the three-quark model, but had the properties of a combination of a charmed quark and its antiquark ($c\bar{c}$). It was much heavier than the other known mesons ($\sim$3100 MeV/c^2) and its lifetime was much longer than those of other particles that decay via the strong force. In 1975 researchers at Stanford University reported strong evidence for the tau (τ) lepton, with a mass of 1784 MeV/c^2. Such discoveries led to more elaborate quark models and the proposal of two new quarks, named *top* (t) and *bottom* (b). (Some physicists prefer the whimsical names *truth* and *beauty*.) To distinguish these quarks from the old ones, quantum numbers called *topness* and *bottomness* were assigned to these new particles and are included in Table 30.3. In 1977 researchers at the Fermi National Laboratory, under the direction of Leon Lederman, reported the discovery of a very massive new meson, Υ, whose composition is considered to be $b\bar{b}$. Strong evidence for the existence of the top quark has been reported by researchers at Fermilab.

You are probably wondering whether or not such discoveries will ever end. How many "building blocks" of matter really exist? At the present, physicists believe that the fundamental particles in nature include six quarks and six leptons (together with their antiparticles). Some of the properties of these particles are given in Table 30.5.

In spite of many extensive experimental efforts, no isolated quark has ever been observed. Physicists now believe that quarks are permanently confined inside ordinary particles because of an exceptionally strong force that prevents them from escaping. This force, called the "color" force, increases with separation distance (similar to the force of a spring); the properties of this force are discussed in the next section. The great strength of the force between quarks has been described by one author as follows:[2]

> Quarks are slaves of their own color charge, . . . bound like prisoners of a chain gang. . . . Any locksmith can break the chain between two prisoners, but no locksmith is expert enough to break the gluon chains between quarks. Quarks remain slaves forever.

30.13

THE STANDARD MODEL

Shortly after the concept of quarks was proposed, scientists recognized that certain particles had quark compositions that were in violation of the Pauli exclusion principle. Because all quarks have spin 1/2, they are expected to follow the exclusion principle. One example of a particle that violates the exclusion principle is the Ω^- (*sss*) baryon, which contains three *s* quarks having parallel spins, giving it a total spin of 3/2. Other examples of baryons that have identical quarks with parallel spins are the Δ^{++} (*uuu*) and the Δ^- (*ddd*). To resolve this problem, it was suggested that quarks possess a property called **color.** This property is similar in many respects to electric charge except that it occurs in three varieties called red, green, and blue. Of course, the antiquarks have the colors antired, antigreen, and antiblue. In order to satisfy the exclusion principle, all three quarks in a baryon must have different colors. A meson consists of a quark of one color and an antiquark of the corresponding anticolor. The result is that baryons and mesons are always colorless (or white). Furthermore, the property of color increases the number of quarks by a factor of three.

Although the concept of color in the quark model was originally conceived to satisfy the exclusion principle, it also provided a better theory for explaining certain experimental results. For example, the modified theory correctly predicts the lifetime of the π^0 meson. The theory of how quarks interact with each other is called **quantum chromodynamics,** or QCD, to parallel quantum electrodynamics (the theory of interaction between electric charges). In QCD, the quark is said to carry a *color charge,* in analogy with electric charge. The force between quarks is often called the *color force.*

As stated earlier, the strong interaction between hadrons is mediated by massless particles called gluons (analogous to photons for the electromagnetic force). According to the theory, there are eight gluons, six of which have color charge. Because of their color charge, quarks can attract each other and form composite particles. When a quark emits or absorbs a gluon, its color changes. For example, a blue quark that emits a gluon may become a red quark, and the red quark that absorbs this gluon becomes a blue quark. The color force between quarks is analogous to the electric force between charges; like colors repel and opposite colors attract. Therefore, two red quarks repel each other, but a red quark will be attracted to an antired quark. The attraction between quarks of opposite color to form a meson ($q\bar{q}$) is indicated in Figure 30.12a. Differently colored quarks also attract each other, but with less intensity than opposite

[2] Harald Fritzsch, *Quarks, The Stuff of Matter,* London, Allen Lane, 1983.

An artist's version of a high-energy particle colliding with a nucleus. The quark structure of the nucleus is indicated by the small colored spheres inside the nucleus. *(Courtesy of Janie Martz/ CEBAF)*

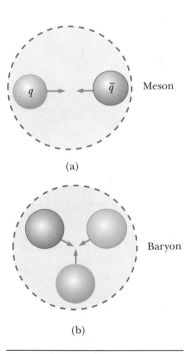

FIGURE 30.12
(a) A red quark is attracted to an antired quark. This forms a meson whose quark structure is ($q\bar{q}$). (b) Three different colored quarks attract each other to form a baryon.

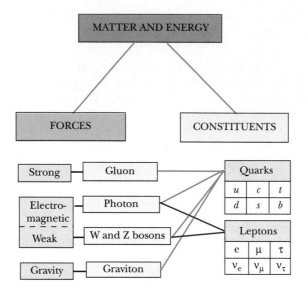

FIGURE 30.13
The Standard Model of particle physics.

colors of quark and antiquark. For example, a cluster of red, blue, and green quarks all attract each other to form baryons as indicated in Figure 30.12b. Thus, all baryons contain three quarks, each of which has a different color.

Recall that the weak force is believed to be mediated by the W^+, W^-, and Z^0 bosons (spin 1 particles). These particles are said to have *weak charge* just as a quark has color charge. Thus, each elementary particle can have mass, electric charge, color charge, and weak charge. Of course, one or more of these could be zero. Scientists now believe that the truly elementary particles are leptons and quarks, and the force mediators are the gluon, the photon, $W^\pm$, Z^0, and the graviton.

In 1979 Sheldon Glashow, Abdus Salam, and Steven Weinberg won a Nobel Prize for developing a theory that unified the electromagnetic and weak interactions. This **electroweak theory** postulates that the weak and electromagnetic interactions have the same strength at very high particle energies. Thus, the two interactions are viewed as two different manifestations of a single unifying electroweak interaction. The combination of the electroweak theory and QCD for the strong interaction is referred to as the *Standard Model.* Although the details of the Standard Model are complex, its essential ingredients can be summarized with the help of Figure 30.13. The strong force, mediated by gluons, holds quarks together to form composite particles such as protons, neutrons, and mesons. Leptons participate only in the electromagnetic and weak interactions.

However, the Standard Model does not answer all questions. A major question is why the photon has no mass while the W and Z bosons do. Because of this mass difference, the electromagnetic and weak forces are quite distinct at low energies, but become similar in nature at very high energies. This change in behavior from low to high energies, called *symmetry breaking,* leaves open the question of the origin of particle masses. In order to resolve this problem, a hypothetical particle called the *Higgs boson,* which provides a mechanism for breaking the electroweak symmetry, has been proposed. The Standard Model, including the Higgs mechanism, provides a logically consistent explanation of the massive nature of the W and Z bosons. Unfortunately, the Higgs boson has not yet been found, but physicists know that its mass should be less than 1 TeV (10^{12} eV).

A computer reconstruction of the magnetic field distribution in the cross-section of a proto-type SSC magnet. In the green central area, the field is uniform to better than 0.02%. Other colors denote deviations of 0.04%, 0.06%, and so on. High-quality field almost fills the whole beam chamber, ensuring that the proton beam will have excellent stability. *(Courtesy of Brookhaven National Laboratory)*

In order to determine whether the Higgs boson exists, two quarks of at least 1 TeV of energy must collide, but calculations show that this requires injecting 40 TeV of energy into the volume of a proton.

Scientists are convinced that because of the limited energy available in conventional accelerators using fixed targets, it is necessary to build colliding-beam accelerators called **colliders.** The concept of colliders is straightforward. Particles with equal masses and kinetic energies, traveling in opposite directions in an accelerator ring, collide head on to produce the required reaction and the formation of new particles. Because the total momentum of the interacting particles is zero, all of their kinetic energy is available for the reaction. The Large Electron-Positron Storage ring at CERN near Geneva, Switzerland and the Stanford Linear Collider in California collide both electrons and protons. The Super Proton Synchrotron at CERN accelerates protons and antiprotons to energies of 270 GeV, while the world's highest-energy proton accelerator, the Tevatron, at the Fermi National Lab in Illinois (see photo on p. 983) produces protons and antiprotons at almost 1000 GeV (or 1 TeV). The Superconducting Super Collider (SSC), which was being built in Texas, was an accelerator designed to produce 20-TeV protons in a ring 52 mi in circumference. After much debate in Congress, and an investment of almost 2 billion dollars, the SSC project was canceled by the U.S. Department of Energy in October, 1993. Figure 30.14 shows the evolution of the stages of matter that scientists have been able to investigate with various types of microscopes.

30.14

THE COSMIC CONNECTION

As we have seen, the world around us is dominated by protons, electrons, neutrons, and neutrinos. Other particles can be seen in cosmic rays. However, most of the new particles are produced using large, expensive machines that accelerate protons and electrons to energies in the GeV and TeV range. These energies are enormous when compared to the thermal energy in today's Universe. For example, $k_B T$ at the center of the Sun is only about 1 keV, but the temperature of the early Universe was high enough to reach energies of TeV and higher.

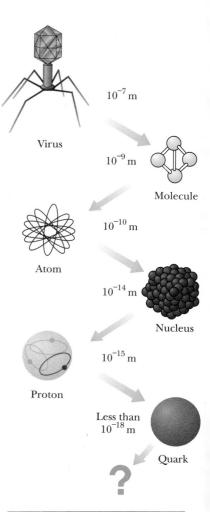

FIGURE 30.14

Observation of matter with microscopes reveals structures ranging in size from the smallest living thing, a virus, down to a quark, which has not yet been seen as an isolated particle.

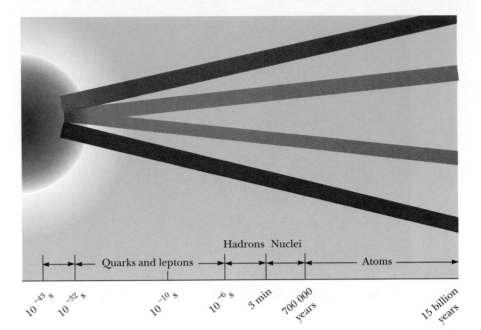

FIGURE 30.15

A brief history of the Universe from the Big Bang to the present. The fundamental forces became distinguishable during the first microsecond. Following this, all the quarks combined to form the strongly interacting particles. The leptons, however, remained separate and exist as individually observable particles to this day.

In this section we describe one of the most fascinating theories in all of science—the Big Bang theory of the creation of the Universe—and the experimental evidence that supports it. This theory of cosmology states that the Universe had a beginning, and that this beginning was so cataclysmic that it is impossible to look back beyond it. According to the theory, the Universe erupted from a point-like singularity about 15 to 20 billion years ago. The first few minutes after the Big Bang saw such extremes of energy that all four interactions of physics were unified and all matter melted down into an undifferentiated "quark soup."

The evolution of the four fundamental forces from the Big Bang to the present is shown in Figure 30.15. During the first 10^{-43} s (the ultra-hot epoch where $T \approx 10^{32}$ K), it is presumed that the strong, electroweak, and gravitational forces were joined to form a completely unified force. Between 10^{-43} s and 10^{-32} s following the Big Bang (the hot epoch where $T \approx 10^{29}$ K), gravity broke free of this unification while the strong and electroweak forces remained as one (they are described by a grand unification theory). This was a period when particle energies were so great ($>10^{16}$ GeV) that very massive particles as well as quarks, leptons, and their antiparticles existed. Then the Universe rapidly expanded and cooled during the warm epoch when the temperatures ranged from 10^{29} to 10^{15} K, the strong and electroweak forces parted company, and the grand unification scheme was broken. As the Universe continued to cool, the electroweak force split into the weak force and the electromagnetic force about 10^{-10} s after the Big Bang.

Until about 700 000 years after the Big Bang, the Universe was dominated by radiation; ions absorbed and re-emitted photons, thereby ensuring thermal equilibrium of radiation and matter. Energetic radiation also prevented matter from forming clumps or even single hydrogen atoms. By the time the Universe was about 700 000 years old, it had expanded and cooled to about 3000 K, and protons could bind to electrons to form hydrogen atoms. Since neutral atoms do

FIGURE 30.16
Robert W. Wilson *(left)* and Arno A. Penzias *(right)* with Bell Telephone Laboratories' horn-reflector antenna. *(AT&T Bell Laboratories)*

not appreciably scatter photons, the Universe suddenly became transparent to photons. Radiation no longer dominated the Universe and clumps of neutral matter steadily grew—first atoms, followed by molecules, gas clouds, stars, and finally galaxies.

OBSERVATION OF RADIATION FROM THE PRIMORDIAL FIREBALL

In 1965 Arno A. Penzias and Robert W. Wilson of Bell Labs made an amazing discovery while testing a sensitive microwave receiver. A pesky signal producing a faint background hiss was interfering with their satellite communications experiments. In spite of their valiant efforts, the signal remained. Ultimately it became clear that they were observing microwave background radiation (at a wavelength of 7.35 cm) representing the leftover glow from the Big Bang.

The microwave horn that served as their receiving antenna is shown in Figure 30.16. The intensity of the detected signal remained unchanged as the antenna was pointed in different directions. The fact that the radiation has equal strengths in all directions suggested that the entire Universe was the source of this radiation. Evicting a flock of pigeons from the 20-foot horn and cooling the microwave detector both failed to remove the "spurious" signal. Through a casual conversation, Penzias and Wilson discovered that a group at Princeton had predicted the residual radiation from the Big Bang and were planning an experiment to confirm the theory. The excitement in the scientific community was high when Penzias and Wilson announced that they had already observed an excess microwave background compatible with a 3-K blackbody source.

Because the measurements of Penzias and Wilson were taken at a single wavelength, they did not completely confirm the radiation as 3-K blackbody radiation. Subsequent experiments by other groups added intensity data at different wavelengths. The results confirm that the radiation is that of a black body at 2.9 K. This figure is perhaps the most clearcut evidence for the Big Bang theory. The 1978 Nobel prize in physics was awarded to Penzias and Wilson for their important discovery.

30.15

PROBLEMS AND PERSPECTIVES

While particle physicists have been exploring the realm of the very small, cosmologists have been exploring cosmic history back to the first microsecond of the Big Bang. Observation of the events that occur when two particles collide in an accelerator is essential in reconstructing the early moments in cosmic history. Perhaps the key to understanding the early Universe is first to understand the world of elementary particles. Cosmologists and particle physicists find that they have many common goals and are joining hands to attempt to study the physical world at its most fundamental level.

Our understanding of physics at short distances is far from complete. Particle physics is faced with many questions. Why is there so little antimatter in the Universe? Do neutrinos have a small rest mass, and if so, how do they contribute to the "dark matter" of the Universe? (Measurements on Supernova 1987A established an upper limit of 16 eV for the neutrino mass.) Is it possible to unify the strong and electroweak theories in a logical and consistent manner? Why do quarks and leptons form three similar but distinct families? Are muons the same as electrons (apart from their different masses), or do they have other subtle differences that have not been detected? Why are some particles charged and others neutral? Why do quarks carry a fractional charge? What determines the masses of the fundamental constituents? Can isolated quarks exist? The questions go on and on. Because of the rapid advances and new discoveries in the field of particle physics, by the time you read this book some of these questions will likely have been resolved while others may have emerged.

An important and obvious question that remains is whether leptons and quarks have a substructure. If they do, one could envision an infinite number of deeper structure levels. However, if leptons and quarks are indeed the ultimate constituents of matter, as physicists today tend to believe, we should be able to construct a final theory of the structure of matter as Einstein dreamed of doing. In the view of many physicists, the end of the road is in sight, but how long it will take to reach that goal is anyone's guess.

SUMMARY

In **nuclear fission** and **nuclear fusion,** the total rest mass of the products is always less than the original rest mass of the reactants. Nuclear fission occurs when a heavy nucleus splits, or fissions, into two smaller nuclei. In nuclear fusion, two light nuclei combine to form a heavier nucleus.

A **nuclear reactor** is a system designed to maintain a self-sustaining chain reaction. Nuclear reactors using controlled fission events are currently being used to generate electric power.

Controlled fusion events offer the hope of plentiful supplies of energy in the future. The nuclear fusion reactor is considered by many scientists to be the ultimate energy source because its fuel is water. **Lawson's criterion** states that a fusion reactor will provide a net output power if the product of the plasma ion density, n, and the plasma confinement time, τ, satisfies the following relationships:

$$n\tau \geqslant 10^{14} \text{ s/cm}^3 \qquad \text{Deuterium-tritium interaction}$$

$$n\tau \geqslant 10^{16} \text{ s/cm}^3 \qquad \text{Deuterium-deuterium interaction}$$

[30.5]

There are four fundamental forces in nature: **strong** (hadronic), **electromagnetic, weak,** and **gravitational.** The strong force is the force between nucleons that keeps the nucleus together. The weak force is responsible for beta decay. The electromagnetic and weak forces are now considered to be manifestations of a single force called the **electroweak** force.

Every fundamental interaction is said to be mediated by the exchange of field particles. The electromagnetic interaction is mediated by the photon; the weak interaction is mediated by the $W^\pm$ and Z^0 bosons; the gravitational interaction is mediated by gravitons; the strong interaction is mediated by gluons.

An antiparticle and a particle have the same mass but opposite charge, and other properties may also have opposite values, such as lepton number and baryon number. It is possible to produce particle-antiparticle pairs in nuclear reactions if the available energy is greater than $2mc^2$, where m is the rest mass of the particle (or antiparticle).

Particles other than photons are classified as hadrons or leptons. **Hadrons** interact primarily through the strong force. They have size and structure and hence are not elementary particles. There are two types of hadrons, *baryons* and *mesons*. Mesons have a baryon number of zero and have either zero or integer spin. Baryons, which generally are the most massive particles, have nonzero baryon numbers and spins of $1/2$ or $3/2$. The neutron and proton are examples of baryons.

Leptons have no structure or size and are considered truly elementary particles. Leptons interact only through the weak and electromagnetic forces. There are six leptons: the electron, e^-; the muon, μ^-; the tau, τ^-; and their associated neutrinos, ν_e, ν_μ, and ν_τ.

In all reactions and decays, quantities such as energy, linear momentum, angular momentum, electric charge, baryon number, and lepton number are strictly conserved. Certain particles have properties called **strangeness** and **charm.** These unusual properties are conserved only in those reactions and decays that occur via the strong force.

Recent theories postulate that all hadrons are composed of smaller units known as **quarks,** which have fractional electric charges and baryon numbers of $1/3$, and come in six "flavors": up, down, strange, charmed, top, and bottom. Each baryon contains three quarks, and each meson contains one quark and one antiquark.

According to the theory of **quantum chromodynamics,** quarks have a property called **color,** and the strong force between quarks is referred to as the **color force.**

Observation of background microwave radiation by Penzias and Wilson strongly confirmed that the Universe started with a Big Bang about 15 billion years ago. The background radiation is equivalent to that of a black body at a temperature of about 3 K.

ADDITIONAL READING

S. Aftergood, D. W. Hafemeister, O. F. Prilutsky, J. R. Primack, and S. N. Rodionov, "Nuclear Power in Space," *Sci. American,* June 1991, p. 42.

E. D. Bloom and G. J. Feldmann, "Quarkonium," *Sci. American,* May, 1982.

Frank Close, *The Cosmic Onion: Quarks and the Nature of the Universe,* The American Institute of Physics, 1986. A timely monograph on particle physics, including lively discussions of the Big Bang theory.

B. L. Cohen, "The Disposal of Radioactive Wastes from Fission Reactors," *Sci. American,* June 1977, p. 21.

R. W. Conn, "The Magnetic Fusion Reactors," *Sci. American,* October 1983, p. 176.

R. W. Conn, V. A. Chuyanov, N. Inoue, and D. R. Sweetman, "The International Thermonuclear Experimental Reactor," *Sci. American,* April 1992, p. 102.

L. Fermi, *Atoms in the Family: My Life with Enrico Fermi,* Chicago, University of Chicago Press, 1954.

Harald Fritzsch, *Quarks, The Stuff of Matter,* London, Allen and Lane, 1983. An excellent introductory overview of elementary particle physics.

H. P. Furth, "Progress Toward a Tokamak Fusion Reactor," *Sci. American,* August 1979, p. 50.

M. W. Golay and N. E. Todreas, "Advanced Light-Water Reactors," *Sci. American,* April 1990, p. 82.

W. Hafele, "Energy from Nuclear Power," *Sci. American,* September 1990, p. 136.

H. Harari, "The Structure of Quarks and Leptons," *Sci. American,* April, 1983.

Leon M. Lederman, "The Value of Fundamental Science," *Sci. American,* November, 1984.

R. K. Lester, "Rethinking Nuclear Power," *Sci. American,* March 1986, p. 31.

H. W. Lewis, "The Safety of Fission Reactors," *Sci. American,* March 1980, p. 3.

N. B. Mistry, R. A. Poling, and E. H. Thorndike, "Particles with Naked Beauty," *Sci. American,* July, 1983.

Chris Quigg, "Elementary Particles and Forces," *Sci. American,* April, 1985.

James S. Trefil, *From Atoms to Quarks,* New York, Scribner, 1980. This is an excellent introduction to the world of particle physics.

Steven Weinberg, *The Discovery of Elementary Particles,* New York, Scientific American Library, W. H. Freeman and Company, 1983. This book emphasizes the important discoveries, experiments, and intellectual exercises that reshaped physics in the 20th century.

Steven Weinberg, "The Decay of the Proton," *Sci. American,* June, 1981.

CONCEPTUAL QUESTIONS

Example   What factors make a fusion reaction difficult to achieve?

Reasoning   The two factors presenting the most technical difficulties are the requirements of a high plasma density and a high plasma temperature. These two conditions must occur simultaneously.

Example   Particles known as resonances have very short lifetimes, of the order of 10^{-23} s. Would you guess they are hadrons or leptons?

Reasoning   They are hadrons. Such particles decay into other strongly interacting particles such as p, n, and π with very short lifetimes. In fact, they decay so quickly that they cannot be detected directly. Decays which occur via the weak force have lifetimes of 10^{-13} s or longer; particles that decay via the electromagnetic force have times in the range of 10^{-16} s to 10^{-19} s.

Example   In the theory of quantum chromodynamics, quarks come in three colors. How would you justify the statement that "all baryons and mesons are colorless?"

Reasoning   Each flavor of quark can have three colors, designated as red, green, and blue. Antiquarks are colored antired, antigreen, and antiblue. Baryons consist of three quarks, each having a different color. Mesons consist of a quark of one color and an antiquark with a corresponding anticolor. Thus, baryons and mesons are colorless or white.

Example   Identify the particle decays in Table 30.2 that occur by the electromagnetic interaction. Justify your answer.

Reasoning   The decays of the neutral pion, eta, and neutral sigma occur by the electromagnetic interaction. These are the three shortest lifetimes in Table 30.2. All produce photons, which are the quanta of the electromagnetic force, and all conserve strangeness.

Example   When an electron and a positron meet at low speeds in free space, why are *two* 0.511-MeV gamma rays produced, rather than *one* gamma ray with an energy of 1.02 MeV?

Reasoning   Gamma rays are photons, and photons carry momentum. If only one were produced, momentum would not be conserved. The two gamma-ray photons that are produced travel off in opposite directions.

Example Why is a neutron stable inside the nucleus? (In free space, the neutron decays in 900 s.)

Reasoning A neutron inside a nucleus is stable because it is in a lower-energy state than a free neutron and lower in energy than it would be if it decayed into a proton (plus electron and antineutrino). The nuclear force gives it this lower energy by binding it inside the nucleus and by favoring pairing between neutrons and protons.

1. Why is water a better shield against neutrons than lead or steel?

2. Explain the function of a moderator in a fission reactor.

3. In a fission reactor, nuclear reactions produce heat to drive a turbine generator. How is this heat produced?

4. A breeder reactor produces more fissionable fuel than it consumes. Does this conflict with the law of conservation of energy? Discuss.

5. Discuss the advantages and disadvantages of fission reactors from the perspectives of safety, pollution, and resources. Make a comparison with power generated from the burning of fossil fuels.

6. Discuss the similarities and differences between fission and fusion.

7. Why would a fusion reactor produce less radioactive waste than a fission reactor?

8. Lawson's criterion states that the product of ion density and confinement time must exceed a certain number before a break-even fusion reaction can occur. Why should these two parameters determine the outcome?

9. Why is the temperature required for deuterium-tritium fusion lower than that needed for deuterium-deuterium fusion? Estimate the relative importance of coulombic repulsion and nuclear attraction in each case.

10. Name the four fundamental interactions and the particles that mediate each interaction.

11. "All mesons are hadrons, but not all hadrons are mesons." Is this true? Discuss.

12. Describe the properties of baryons and mesons and the important differences between them.

13. Discuss the differences between hadrons and leptons.

14. Is the term "heavy lepton" a contradiction? Discuss.

15. The W and Z bosons were first produced at CERN in 1983 (by having a beam of protons and a beam of antiprotons meet at high energy). Why was this an important discovery?

16. Discuss the following conservation laws: energy, linear momentum, angular momentum, electric charge, baryon number, lepton number, and strangeness. Are all of these laws based on fundamental properties of nature? Explain.

17. The family of K mesons all decay into final states that contain no protons or neutrons. What is the baryon number of the K mesons?

18. An antibaryon interacts with a meson. Can a baryon be produced in such an interaction? Explain.

19. Identify the particle decays listed in Table 30.2 that occur by the weak interaction. Justify your answers.

20. The Ξ^0 particle decays by the weak interaction according to the decay mode $\Xi^0 \rightarrow \Lambda^0 + \pi^0$. Would you expect this decay to be fast or slow? Explain.

21. Two protons in a nucleus interact via the strong interaction. Are they also subject to the weak interaction?

22. Discuss the quark model of hadrons, and describe the properties of quarks.

23. Discuss the essential features of the Standard Model of particle physics.

24. How many quarks are there in (a) a baryon, (b) an antibaryon, (c) a meson, and (d) an antimeson? How do you account for the fact that baryons have half-integral spins while mesons have spins of 0 or 1? (*Hint:* Quarks have spins of $\frac{1}{2}$.)

25. What is the quark composition of the Ξ^- particle? (See Table 30.4.)

26. The sky is illuminated with a uniform background of radiation corresponding to a temperature of 2.9 K. Discuss the spectrum of this radiation, thought to be a remnant of the Big Bang as the Universe began.

27. Explain why keeping the water in the reactor shown in Figure 30.4 at high pressure prevents it from boiling.

28. How could you experimentally verify that the lifetime of a proton is 10^{31} years if the Universe is only 10^{10} years old?

29. (a) Does Equation 30.8 predict the correct range for the electromagnetic force? Explain. (b) Based on Equation 30.8, what should the mass of a graviton be?

PROBLEMS

Section 30.1   Nuclear Fission

Section 30.2   Nuclear Reactors

1. Find the energy released in the following fission reaction:

$$^{1}_{0}n + ^{235}_{92}U \longrightarrow ^{144}_{56}Ba + ^{89}_{36}Kr + 3\,^{1}_{0}n$$

2. Another fission reaction similar to the one in Problem 1 leads to the formation of ^{141}Ba and ^{92}Kr when ^{235}U absorbs a neutron. Write down this reaction. How many neutrons are released?

3. Strontium-90 is a particularly dangerous fission product of ^{235}U because it is radioactive, and it re-

□ indicates problems that have full solutions available in the Student Solutions Manual and Study Guide.

places calcium in bones. What other direct fission products would accompany it in the neutron-induced fission of ^{235}U? (*Note:* This reaction may release two, three, or four free neutrons.)

4. Find the energy released in the following fission reaction:

$$^{1}_{0}\text{n} + ^{235}_{92}\text{U} \longrightarrow ^{88}_{38}\text{Sr} + ^{136}_{54}\text{Xe} + 12\,^{1}_{0}\text{n}$$

5. If the average energy released in a fission event is 208 MeV, find the total number of fission events required to keep a 100-W lightbulb burning for 1.0 h.

6. In order to minimize neutron leakage from a reactor, the ratio of the surface area to the volume must be as small as possible. Assume that a sphere and a cube both have the same volume. Find the surface-to-volume ratio for (a) the sphere and (b) the cube. (c) Which of these shapes would have the minimum leakage?

7. How many grams of ^{235}U must undergo fission to operate a 1000-MW power plant for one day if the conversion efficiency is 30.0%? (Assume 208 MeV released per fission event.)

8. An all-electric home uses approximately 2000 kWh of electric energy per month. How much ^{235}U would be required to provide this house with its energy needs for one year? (Assume 100% conversion efficiency and 208 MeV released per fission.)

9. It has been estimated that the Earth contains 1.0×10^9 tons of natural uranium that can be mined economically. Of this total, 0.70% is ^{235}U. If all the world's energy needs (7.0×10^{12} J/s) were supplied by ^{235}U fission, how long would this supply last?

10. The first atomic bomb released an energy equivalent to 20 kilotons of TNT. If 1 ton of TNT releases about 4.0×10^9 J, how much uranium was lost through fission in this bomb? (Assume 208 MeV released per fission.)

11. Suppose that the water exerts an average frictional drag of 1.0×10^5 N on a nuclear-powered ship. How far can the ship travel per kilogram of fuel if the fuel consists of enriched uranium containing 1.7% of the fissionable isotope ^{235}U, and the ship's engine has a efficiency of 20%? (Assume 208 MeV released per fission event.)

Section 30.3   Nuclear Fusion

12. Find the energy released in the fusion reaction

$$^{2}_{1}\text{H} + ^{2}_{1}\text{H} \longrightarrow ^{3}_{1}\text{H} + ^{1}_{1}\text{H}$$

13. Find the energy released in the fusion reaction

$$^{2}_{1}\text{H} + ^{3}_{1}\text{H} \longrightarrow ^{4}_{2}\text{He} + ^{1}_{0}\text{n}$$

14. If an all-electric home uses approximately 2000 kWh of electric energy per month, how many fusion events of the type described in Problem 13 would be required to keep this house running for one year?

15. Another series of nuclear reactions that can produce energy in the interior of stars is the cycle described below. This process is most efficient when the central temperature in a star is above 1.6×10^7 K. Because the temperature at the center of the Sun is only 1.5×10^7 K, the cycle below produces less than 10% of the Sun's energy. (a) A high-energy proton is absorbed by ^{12}C. Another nucleus, A, is produced in the reaction, along with a gamma ray. Identify nucleus A. (b) Nucleus A decays through positron emission to form nucleus B. Identify nucleus B. (c) Nucleus B absorbs a proton to produce nucleus C and a gamma ray. Identify nucleus C. (d) Nucleus C absorbs a proton to produce nucleus D and a gamma ray. Identify nucleus D. (e) Nucleus D decays through positron emission to produce nucleus E. Identify nucleus E. (f) Nucleus E absorbs a proton to produce nucleus F plus an alpha particle. What is nucleus F? (*Note:* If nucleus F is not ^{12}C, the nucleus you started with, you have made an error and should review the sequence of events.)

16. When a star has exhausted its hydrogen fuel, it may fuse other nuclear fuels. At temperatures above 1.0×10^8 K, helium fusion can occur. Write the equation for the processes described below. (a) Two alpha particles fuse to produce a nucleus A and a gamma ray. What is nucleus A? (b) Nucleus A absorbs an alpha particle to produce a nucleus B and a gamma ray. What is nucleus B? (c) Find the total energy released in the reactions given in (a) and (b).

17. Of all the hydrogen nuclei in the ocean, 0.0156% are deuterium. (a) How much deuterium could be obtained from 1.0 gal of ordinary tap water? (b) If all of this deuterium could be converted to energy through reactions such as the first reaction in Equation 30.4, how much energy of forms other than mass could be obtained from the 1.0 gal of water? (c) The energy released through the burning of 1.0 gal of gasoline is approximately 2.0×10^8 J. How many gallons of gasoline would have to be burned in order to produce the same amount of energy as was obtained from 1.0 gal of water?

18. To understand why containment of a plasma is necessary, consider the rate at which a plasma would be lost if it were not contained. (a) Estimate the rms speed of deuterons in a plasma at 1.0×10^8 K. (b) Estimate the time such a plasma

would remain in a cube 10 cm on an edge if no steps were taken to confine it.

19. The oceans have a volume of 317 million cubic miles and contain 1.32×10^{21} kg of water. Of all the hydrogen nuclei in this water, 0.0156% are deuterium. (a) If all of these deuterium nuclei were fused to helium via the first reaction in Equation 30.4, determine the total amount of energy that could be released. (b) Present world electric power consumption is about 7.00×10^{12} W. If consumption were 100 times greater, how many years would the energy supply calculated in (a) last?

Section 30.6 Positrons and Other Antiparticles

20. Two photons are produced when a proton and an antiproton annihilate each other. What is the minimum frequency and corresponding wavelength of each photon?

21. A photon with an energy of 2.09 GeV creates a proton-antiproton pair in which the proton has a kinetic energy of 95 MeV. What is the kinetic energy of the antiproton?

Section 30.7 Mesons and the Beginning of Particle Physics

22. One of the mediators of the weak interaction is the Z^0 boson, whose mass is 96 GeV/c^2. Use this information to find an approximate value for the range of the weak interaction.

23. Occasionally, high-energy muons will collide with electrons and produce two neutrinos according to the reaction $\mu^+ + e \rightarrow 2\nu$. What kind of neutrinos are these?

24. When a high-energy proton or pion traveling near the speed of light collides with a nucleus, it travels an average distance of 3.0×10^{-15} m before interacting. From this information, estimate the time for the strong interaction to occur.

25. If a π^0 at rest decays into two γ's, what is the energy of each of the γ's?

Section 30.9 Conservation Laws

Section 30.10 Strange Particles and Strangeness

26. Identify the unknown particle on the left side of the reaction

$$? + p \longrightarrow n + \mu^+$$

27. The neutral ρ meson decays by the strong interaction into two pions according to $\rho^0 \rightarrow \pi^+ + \pi^-$, with a half-life of about 10^{-23} s. The neutral K meson also decays into two pions according to

$K^0 \rightarrow \pi^+ + \pi^-$, but with a much longer half-life of about 10^{-10} s. How do you explain these observations?

28. Which of the following processes are allowed by the strong interaction, the electromagnetic interaction, the weak interaction, or no interaction at all?
   (a) $\pi^- + p \rightarrow 2\eta^0$ (d) $\Omega^- \rightarrow \Xi^- + \pi^0$
   (b) $K^- + n \rightarrow \Lambda^0 + \pi^-$ (e) $\eta^0 \rightarrow 2\gamma$
   (c) $K^- \rightarrow \pi^- + \pi^0$

29. For the following two reactions, the first may occur but the second cannot. Explain.

$$K^0 \longrightarrow \pi^+ + \pi^- \qquad \text{(can occur)}$$
$$\Lambda^0 \longrightarrow \pi^+ + \pi^- \qquad \text{(cannot occur)}$$

30. Determine whether or not strangeness is conserved in the following decays and reactions.
   (a) $\Lambda^0 \rightarrow p + \pi^-$
   (b) $\pi^- + p \rightarrow \Lambda^0 + K^0$
   (c) $\bar{p} + p \rightarrow \bar{\Lambda}^0 + \Lambda^0$
   (d) $\pi^- + p \rightarrow \pi^- + \Sigma^+$
   (e) $\Xi^- \rightarrow \Lambda^0 + \pi^-$
   (f) $\Xi^0 \rightarrow p + \pi^-$

31. Each of the following decays is forbidden. For each process, determine a conservation law that is violated.
   (a) $\mu^- \rightarrow e + \gamma$ (d) $p \rightarrow e^+ + \pi^0$
   (b) $n \rightarrow p + e + \nu_e$ (e) $\Xi^0 \rightarrow n + \pi^0$
   (c) $\Lambda^0 \rightarrow p + \pi^0$

32. Identify the conserved quantities in the following processes:
   (a) $\Xi^- \rightarrow \Lambda^0 + \mu^- + \nu_\mu$
   (b) $K^0 \rightarrow 2\pi^0$
   (c) $K^- + p \rightarrow \Sigma^0 + n$
   (d) $\Sigma^0 \rightarrow \Lambda^0 + \gamma$
   (e) $e^+ + e^- \rightarrow \mu^+ + \mu^-$
   (f) $\bar{p} + n \rightarrow \bar{\Lambda}^0 + \Sigma^-$

33. Fill in the missing particle. Assume that (a) occurs via the strong interaction while (b) and (c) involve the weak interaction.
   (a) $K^+ + p \rightarrow \underline{\quad} + p$
   (b) $\Omega^- \rightarrow \underline{\quad} + \pi^-$
   (c) $K^+ \rightarrow \underline{\quad} + \mu^+ + \nu_\mu$

34. Each of the following reactions is forbidden. Determine a conservation law that is violated for each reaction.
   (a) $p + \bar{p} \rightarrow \mu^+ + e$ (d) $p + p \rightarrow p + p + n$
   (b) $\pi^- + p \rightarrow p + \pi^+$ (e) $\gamma + p \rightarrow n + \pi^0$
   (c) $p + p \rightarrow p + \pi^+$

35. (a) Show that baryon number and charge are conserved in the following reactions of a pion with a proton.
   (1) $\pi^+ + p \longrightarrow K^+ + \Sigma^+$
   (2) $\pi^+ + p \longrightarrow \pi^+ + \Sigma^+$

(b) The first reaction is observed, but the second never occurs. Explain these observations.

36. The following reactions or decays involve one or more neutrinos. Supply the missing neutrinos.
    (a) $\pi^- \rightarrow \mu^- + ?$ (d) $? + n \rightarrow p + e$
    (b) $K^+ \rightarrow \mu^+ + ?$ (e) $? + n \rightarrow p + \mu^-$
    (c) $? + p \rightarrow n + e^+$ (f) $\mu^- \rightarrow e + ? + ?$

37. Determine the type of neutrino or antineutrino involved in each of the following processes:
    (a) $\pi^+ \rightarrow \pi^0 + e^+ + ?$
    (b) $? + p \rightarrow \mu^- + p + \pi^+$
    (c) $\Lambda^0 \rightarrow p + \mu^- + ?$
    (d) $\tau^+ \rightarrow \mu^+ + ? + ?$

38. Determine which of the reactions below can occur. For those that cannot occur, determine the conservation law (or laws) that each violates.
    (a) $p \rightarrow \pi^+ + \pi^0$ (d) $\pi^+ \rightarrow \mu^+ + \nu_\mu$
    (b) $p + p \rightarrow p + p + \pi^0$ (e) $n \rightarrow p + e + \bar{\nu}_e$
    (c) $p + p \rightarrow p + \pi^+$ (f) $\pi^+ \rightarrow \pi^+ + n$

39. Two protons approach each other with equal and opposite velocities. What is the minimum kinetic energy of each of the protons if they are to produce a π^+ meson at rest in the reaction,

$$p + p \longrightarrow p + n + \pi^+$$

40. A Σ^0 particle at rest decays according to

$$\Sigma^0 \longrightarrow \Lambda^0 + \gamma$$

Find the gamma-ray energy.

Section 30.12   Quarks—Finally

41. The quark compositions of the K^0 and Λ^0 particles are $d\bar{s}$ and uds, respectively. Show that the charge, baryon number, and strangeness of these particles equal the sums of these numbers for the quark constituents.

42. The quark composition of the proton is uud, while that of the neutron in udd. Show that the charge, baryon number, and strangeness of these particles equal the sums of these numbers for their quark constituents.

43. What is the electrical charge of the baryons with the quark compositions (a) $\bar{u}\bar{u}\bar{d}$ and (b) $\bar{u}\bar{d}\bar{d}$. What are these baryons called?

44. Describe each of the reactions in terms of their constituent quarks:
    (a) $\pi^- + p \rightarrow K^0 + \Lambda^0$
    (b) $\pi^+ + p \rightarrow K^+ + \Sigma^+$
    (c) $K^- + p \rightarrow K^+ + K^0 + \Omega^-$
    (d) $p + p \rightarrow K^0 + p + \pi^+ + ?$
    In the last reaction, identify the mystery particle.

45. Identify the particles corresponding to the following quark states: (a) uds; (b) $\bar{u}d$; (c) $\bar{s}d$; (d) ssd.

46. Neglect binding energies and estimate the mass of the u and d quarks from the mass of the proton and neutron.

Section 30.13   The Standard Model

47. Classical general relativity views the space-time manifold as a deterministic structure completely well-defined down to arbitrarily small distances. On the other hand, quantum general relativity forbids distances smaller than the Planck length given by $L = (\hbar G/c^3)^{1/2}$. (a) Calculate the value of L. The answer suggests that after the Big Bang (when all the known Universe was reduced to a singularity), nothing could be observed until that singularity grew larger than the Planck length, L. Since the size of the singularity grew at the speed of light, we can infer that during the time it took for light to travel the Planck length, no observations were possible. (b) Determine this time (known as the Planck time, T) and compare it to the ultra-hot epoch discussed in the text. (c) Does this suggest that we may never know what happened between the time $t = 0$ and the time $t = T$?

48. If baryon number is not conserved, then one possible mechanism by which a proton can decay is

$$p \longrightarrow e^+ + \gamma$$

(a) Show that this reaction violates conservation of baryon number. (b) Assuming that this reaction occurs, and that the proton is initially at rest, determine the energy and momentum of the positron and photon after the reaction. (*Hint:* Recall that energy and momentum must be conserved in the reaction.) (c) Determine the speed of the positron after the reaction.

Section 30.14   The Cosmic Connection

49. Treat the cosmic microwave background as blackbody radiation at 3.0 K. (a) Determine the wavelength in which this blackbody radiation has its maximum value. (b) In which part of the electromagnetic spectrum does it lie?

Section 30.15   Problems and Perspectives

50. Determine an upper limit, expressed in kilograms, to the mass of the neutrino.

ADDITIONAL PROBLEMS

51. Calculate the range of the force that might be produced by the virtual exchange of a proton.

52. A particle cannot generally be localized to dis-

tances smaller than its de Broglie wavelength. This means that a slow neutron appears to be larger than a fast neutron to a target particle because the slow neutron will probably be found over a larger volume of space. For a thermal neutron at room temperature (300 K) assume that its energy is given by $k_B T$ and find (a) its linear momentum and (b) the de Broglie wavelength. Compare this effective neutron size with both nuclear and atomic dimensions.

53. Calculate the mass of ^{235}U required to provide the total energy requirements of a nuclear submarine during a 100-day patrol, assuming a constant power demand of 100 000 kW and a conversion efficiency of 30%. (Assume that the average energy released per fission is 208 MeV.)

54. A 2.0-MeV neutron is emitted in a fission reactor. If it loses one half of its kinetic energy in each collision with a moderator atom, how many collisions must it undergo in order to achieve thermal energy (0.039 eV)?

55. A K^0 particle at rest decays into a π^+ and a π^-. What will be the speed of each of the pions? The mass of the K^0 is 497.7 MeV/c^2 and the mass of each pion is 139.6 MeV c^2.

56. (a) Show that about 1.0×10^{10} J would be released by the fusion of the deuterons in 1.0 gal of water. Note that 1 out of every 6500 hydrogen atoms is a deuteron. (b) The average energy consumption rate of a person living in the United States is about 1.0×10^4 J/s (an average power of 10 kW). At this rate, how long would the energy needs of one person be supplied by the fusion of the deuterons in 1.0 gal of water? Assume that the energy released per deuteron is 1.64 MeV.

"PARTICLES, PARTICLES, PARTICLES"

RONALD C. DAVIDSON
Director of the Princeton Plasma Physics Laboratory and Professor of
Astrophysical Sciences at Princeton University

RECENT FUSION ENERGY DEVELOPMENT—RESULTS FROM THE TOKAMAK TFTR*

Well before the middle of the next century, the world faces an energy deficit of extraordinary proportions. The total population is expected to double to about 10 billion people by the year 2040. With continued industrialization of Asia, Africa, and the Americas, world power consumption is projected to triple, to 30 trillion watts, over the same period of time. This assumes large gains in efficiency of energy use. At the present rate of consumption, however, the world's known oil supply will be depleted in about 60 years, and natural gas in about 100 years. While coal reserves could sustain some of the world's hungry appetite for energy for several centuries, the problems associated with mining and the high level of environmental pollution produced by coal-fired power plants would only aggravate an already precarious ecological balance. Energy is fundamental to an acceptable quality of life, and the requirements of the developing world are not to be denied. By any measure, the world *must* find new sources of energy in the coming decades—sources that will augment the inevitable increase in reliance on solar, renewables, and nuclear fission. Not to do so is to sentence the United States and the rest of the world to severe deprivation.

The search for alternative energy sources has led to a highly successful international research effort to develop fusion—the process of combining hydrogen nuclei that powers the Sun and the stars—as a practical energy source. This process is illustrated in Figure 1, where deuterium and tritium ions, in a high-temperature "plasma"

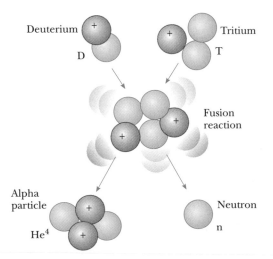

FIGURE 1
Schematic of the deuterium (D)–tritium (T) fusion reaction. Eighty percent (80%) of the energy released is in the 14-MeV neutrons.

*Figures in this essay are provided courtesy of the Princeton Plasma Physics Laboratory, which is supported by the U.S. Department of Energy.

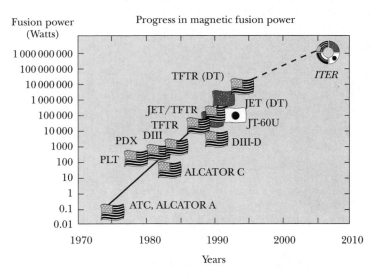

<table>
<tr></tr>
</table>

PLT Princeton Large Torus
PDX Princeton Divertor Experiment
JET Joint European Torus
JT-60 Japan
ITER International Thermonuclear Experimental Reactor
DIII & DIII-D General Atomics Tokamak Experiments
ATC & TFTR Princeton Plasma Physics Laboratory
ALCATOR A, C Massachusetts Institute of Technology

FIGURE 2

Progress in the fusion power achieved since construction authorization of the Toka-mak Fusion Test Reactor in 1976 has been a factor of 60 million.

confined by a strong magnetic field, undergo fusion reactions and produce both 3.5-MeV helium ions (called alpha particles) that are trapped in the plasma and ener-getic 14-MeV neutrons that carry away the heat from the reaction. Unlike fission, the reaction produces no long-lived radioactive by-products. Fusion offers the prospect of an environmentally attractive and secure long-term energy source, with a virtually un-limited fuel supply of deuterium obtained from ordinary water.

As illustrated in Figure 2, the power produced in laboratory fusion experiments has increased by a factor of more than 60 million since construction of the largest U.S. tokamak, the Tokamak Fusion Test Reactor (TFTR), was authorized at the Princeton Plasma Physics Laboratory in the mid-1970s by the U.S. Department of Energy. In fact, this rate of increase is even greater than the increase during the same period in com-puter memory density, that often-cited benchmark of technical progress. Scientific breakthroughs and world records in plasma performance have see-sawed among the large tokamaks in Europe, Japan, and the United States. The United States moved into the lead in December 1993, when 6.2 MW of fusion power was generated during the historic experiments on the TFTR tokamak at Princeton. With these experiments—which used a 50%-50% deuterium-tritium fuel mixture for the first time—the United States surpassed the European tokamak record of 1.7 MW set in 1991, and the ques-tions about scientific feasibility are being resolved.

FIGURE 3

Photograph of the Tokamak Fusion Test Reactor at the Princeton Plasma Physics Laboratory. The neutral beam injectors surrounding the tokamak device are capable of delivering up to 33 MW of auxiliary heating power to the plasma.

Figure 3 shows a photograph of the TFTR device, including the high-power neutral beam injectors around the periphery of the tokamak that are used to heat and fuel the plasma to ion temperatures approaching $T_i = 4 \times 10^8$ K and densities of $n_i = 0.6 \times 10^{14}$ cm^{-3}. Once the high-energy (100 keV) neutral tritium and deuterium atoms enter the plasma chamber, they are ionized and collisionally heat the background plasma. The high-temperature plasma in TFTR is confined and insulated from contact with the chamber wall by a strong toroidal magnetic field with strength $B = 50$ kG, which is sustained for about two seconds by external power supplies. Referring to Figure 30.5, the major radius of the donut-shaped plasma in TFTR is about 2.5 m, whereas the diameter of the plasma cross-section is about 1.6 m.

Properties of the plasma in TFTR are measured by an impressive array of noninvasive diagnostic instruments (more than 50 in total), ranging from detectors that sense microwave and light emission from the plasma, to detectors that measure energetic ions escaping from the plasma, and 14-MeV neutrons produced in the deuterium-tritium fusion process. Typical TFTR data on fusion-power production are illustrated in Figure 4, where the fusion power (in the neutrons *and* the alpha particles) is plotted versus time for three different experiments with varying concentrations of tritium and deuterium in the fuel mixture. In each case, the neutral beam injectors used for heating and fueling the plasma are "on" between $t = 0$ s and $t = 0.75$ s. Note from the figure that a 50% tritium–50% deuterium mixture is the optimum for maximizing the fusion power production at 6.2 MW. (Not shown in the figure are experiments where the maximum fusion power production is less than 6.2 MW, when the tri-

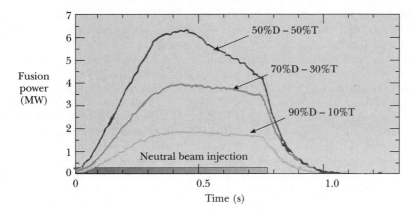

FIGURE 4

Plot of total fusion power versus time obtained in the historic experiments on the Tokamak Fusion Test Reactor in December 1993.

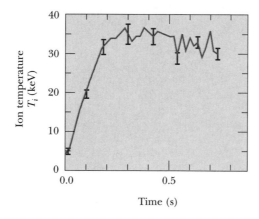

FIGURE 5

Plot of the peak ion temperature T_i versus time obtained in the highest fusion power experiment (6.2 MW) in Figure 4.

tium concentration exceeds 50%.) Also impressive are the high ion temperatures achieved in these experiments. This is illustrated in Figure 5, where the peak ion temperature T_i in the center of the plasma is plotted versus time for the same interval as the fusion power data shown in Figure 4. Note that T_i increases to about 35 keV = 3.5×10^4 eV $\times$ $(1.1 \times 10^4$ K/eV$) = 3.9 \times 10^8$ K in these experiments.

Also in the high-power experiments with deuterium-tritium plasmas there are preliminary indications that the plasma electrons are heated collisionally by the energetic 3.5-MeV alpha particles (^{4}He nuclei) created in the fusion process. In particular, the electrons experience a temperature increase due to alpha heating of about $\Delta T_e = 0.75$ keV $\cong 0.8 \times 10^7$ K above the baseline electron temperature of $T_{eo} = 9$ keV = 10^8 K obtained in otherwise similar deuterium plasmas. Heating of the background plasma by the alpha particles will be very important in future-generation power reactors and experimental facilities such as the International Thermonuclear Experimental Reactor (ITER), which is presently being designed. The ITER device will be a tokamak much larger in physical size than TFTR, and it is designed to produce 1000 MW of fusion power, about a factor of 100 larger than TFTR's ultimate design capability of 10 MW. In ITER, the energy content of the alpha particles will be so intense that the alpha particles will heat and sustain the plasma, allowing the auxiliary heating sources to be turned off once the reaction is initiated. Such a state of "sustained burn" of a deuterium-tritium plasma is referred to as "ignition."

Several scientific and technological challenges remain in the development of fusion as a practical energy source. These challenges include the demonstration of plasma ignition and controlled burn at high fusion power; the development of durable materials for the reactor-vessel wall that will withstand the large heat and particle fluxes for long periods of time (several years); and the demonstration of continuous tokamak operation at high plasma pressure. These technical issues are well understood, and the remaining national and international test facilities required for fusion energy development are in the design and planning phase. The scientific feasibility of fusion is being established on TFTR and the world's other large tokamaks, the Joint European Torus in Europe and the JT-60 Upgrade in Japan. What is required now is a national and international commitment to the next generation of facilities that will demonstrate the engineering and economic feasibility of fusion as a practical energy source.

For Further Study

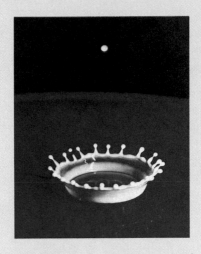

A "crown" produced by a drop of milk falling on a plate covered with a thin layer of milk. Surface tension affects the delicate shapes that change too quickly for any eye to discern. However, the high-speed camera is able to capture this event. (See For Further Study in Chapter 9.) (Harold E. Edgerton. Courtesy of Palm Press, Inc.)

The following pages contain material for further study on selected topics. The first mini-chapter can be used after completion of Chapter 7 on gravitational forces. This material is an extension of the study of gravitational potential energy to situations in which an object is too far away from the Earth for the free-fall acceleration to be considered constant. An example of the application of these ideas to astrophysics is then given. Finally, this first section ends with a set of problems based on this, and related, subject matter. All the problems have the answers given in brackets immediately following the problem.

The second mini-chapter includes material related to fluids both moving and stationary. This is a supplement to Chapter 9 in the textbook and should be of particular interest to students with a major in biology or the medical professions.

Finally the last mini-chapter discusses Gauss' law, a subject that can be covered following Chapter 15.

FOR FURTHER STUDY IN CHAPTER 7

GRAVITATIONAL POTENTIAL ENERGY

In Chapter 5 we introduced the concept of gravitational potential energy and found that the potential energy of an object could be calculated using the equation $PE = mgh$, where h is the height of the object above or below some reference level. This equation is actually valid only when the object is near the Earth's surface. For objects high above the Earth's surface, such as a satellite, an alternative expression must be used to compute the gravitational potential energy. The general expression for the gravitational potential energy for an object of mass m at a distance r *from the center of the Earth* can be shown (using integral calculus) to be

Gravitational potential energy

$$PE = -G\frac{M_E m}{r}$$ **[7a.1]**

where M_E is the mass of the Earth.

This equation assumes that the zero level for potential energy is at an infinite distance from the center of the Earth. This point is necessary because the gravitational force goes to zero when r is set equal to infinity.

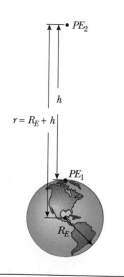

FIGURE 7a.1
(Example 7a.1)

EXAMPLE 7a.1   Does the Potential Energy Reduce to *mgh*?

(a) Find expressions for the gravitational potential energy of an object at the surface of the Earth and for the same object when at a height h above the surface of the Earth (Fig. 7a.1). (b) From the answers to part (a), show that the difference in potential energy between these two points reduces to the familiar expression $PE = mgh$ when h is small compared to the Earth's radius.

Solution   Equation 7a.1 gives the potential energy at the surface of the Earth as

$$PE_1 = -G\frac{M_E m}{R_E}$$

We can also use Equation 7a.1 to find the potential energy of the object at the height h above the surface of the Earth. Taking $r = R_E + h$, we find

$$PE_2 = -G\frac{M_E m}{(R_E + h)}$$

(b) The difference in potential energy between these two points is found as follows:

$$PE_2 - PE_1 = -G\frac{M_E m}{(R_E + h)} - \left(-G\frac{M_E m}{R_E}\right)$$

$$= -GM_E m\left[\frac{1}{(R_E + h)} - \frac{1}{R_E}\right]$$

After finding a common denominator and applying some algebra to the equation above, we find

$$PE_2 - PE_1 = \frac{GM_E mh}{R_E(R_E + h)}$$

When the height h is very small compared to R_E, the denominator in the expression above is approximately equal to R_E^2. Thus, we have

$$PE_2 - PE_1 \approx \frac{GM_E}{R_E^2} mh$$

Now note that the free-fall acceleration at the surface of the Earth is given by $g = GM_E/R_E^2$ (see Chapter 7). Thus,

$$PE_2 - PE_1 \approx mgh$$

ESCAPE SPEED

If an object is projected upward with a large enough speed it can escape the gravitational pull of the Earth and go soaring off into space. This particular speed is called the escape speed of an object from the Earth. We shall now proceed to derive an expression which will enable us to calculate the escape speed.

Suppose an object of mass m is projected vertically upward from the Earth's surface with an initial speed v_i as in Figure 7a.2. The initial mechanical energy (kinetic plus potential energy) is given by

$$KE_i + PE_i = \frac{1}{2} mv_i^2 - G\frac{M_E m}{R_E}$$

We neglect air resistance and assume that the initial speed is just large enough to allow the object to reach infinity with a speed of zero. We call this value of v_i the escape speed v_e. When the object is at an infinite distance from the Earth, its kinetic energy is zero, because $v_f = 0$, and the gravitational potential energy is also zero because our zero level of potential energy was selected at $r = \infty$. Hence, the total mechanical energy is zero, and the law of conservation of energy gives

$$\frac{1}{2} mv_e^2 - G\frac{M_E m}{R_E} = 0$$

and

$$v_e = \sqrt{\frac{2GM_E}{R_E}} \qquad \text{[7a.2]}$$

From this equation, one finds that the escape speed for Earth is about 11.2 km/s, which corresponds to about 25 000 mi/h (see Example 7a.2). Note that this expression for v_e is independent of the mass of the object projected from the Earth. For example, a spacecraft has the same escape speed as a molecule. A list of escape speeds for the planets and the Moon is given in Table 7a.1. The data presented in this table help in understanding why some planets have atmospheres while others do not. For example, on the very hot planet Mercury, gas molecules often reach speeds that are greater than the escape speed of 4.3 km/s. Consequently, any gases that might have been present on the surface of the

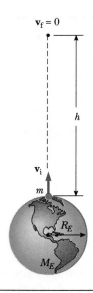

FIGURE 7a.2
An object of mass m projected upward from the Earth's surface with an initial speed v_i reaches a maximum altitude h (where $M_E \gg m$).

TABLE 7a.1
Escape Speeds for the Planets and the Moon

Planet	v_e(km/s)
Mercury	4.3
Venus	10.3
Earth	11.2
Moon	2.3
Mars	5.0
Jupiter	60
Saturn	36
Uranus	22
Neptune	24
Pluto	1.1

Escape speed

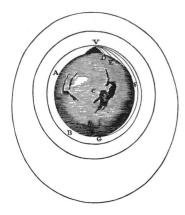

"The greater the velocity . . . with which [a stone] is projected, the farther it goes before it falls to the Earth. We may therefore suppose the velocity to be so increased, that it would describe an arc of 1, 2, 5, 10, 100, 1000 miles before it arrived at the Earth, till at last, exceeding the limits of the Earth, it should pass into space without touching.''—Newton, *System of the World*.

planet at its formation have long since wandered off into space. Likewise, in the atmosphere of our own Earth, hydrogen and helium molecules through collisions often gain speeds greater than 11.2 km/s, the escape speed on Earth. Therefore, these gases are not retained in the Earth's atmosphere. On the other hand, heavier gases such as oxygen and nitrogen in the Earth's atmosphere have average speeds less than 11.2 km/s, and do not escape.

EXAMPLE 7a.2   Escape Speed of a Rocket

Calculate the escape speed from the Earth for a 5000-kg spacecraft.

Solution   Using Equation 7a.2 with $M_E = 5.98 \times 10^{24}$ kg and $R_E = 6.38 \times 10^6$ m gives

$$v_e = \sqrt{\frac{2GM_E}{R_E}} = \sqrt{\frac{2(6.68 \times 10^{-11})(5.98 \times 10^{24})}{6.37 \times 10^6}} = \boxed{1.12 \times 10^4 \text{ m/s}}$$

This corresponds to about 25 000 mi/h or about 7 mi/s. Note that the mass of the spacecraft was not required for this calculation.

GRAVITY IN THE EXTREME, BLACK HOLES

The maternity wards of space are large clouds of gas and dust called nebulae. The mutual gravitational attraction between particles in a cloud pulls the particles together until eventually a huge collapsing spherical ball is formed. As the separation between the particles decreases, the gravitational force on each increases; the particles become part of a collapsing sphere, and fall together towards the center of the sphere at an increasing speed. Finally, their speed becomes great enough near the center of the collapsing sphere that fusion reactions begin to occur. One manifestation of these reactions is the continuous release of tremendous amounts of energy that prevents further gravitational collapse of the star. The end result is a star living out its active life in equilibrium between the inward pull of gravitational forces tending to collapse the star and the outward radiation and thermal forces tending to expand the star.

During the first stages of nuclear burning at the center of a star, hydrogen nuclei are fused together to produce helium nuclei. When the supply of hydrogen is exhausted, the fusion process ceases. At this point, the balance of forces is destroyed since no energy is being released from fusion processes to prevent the collapse by gravitational forces. As the star collapses, the temperature inside the star increases further, and other fusion reactions can occur involving elements heavier than hydrogen. The most spectacular deaths occur for stars that are originally very massive. Such a massive star may undergo a catastrophic explosion, hurling debris into space and leaving behind a central core. Such an exploding star is called a supernova. Supernovae are very rare events, and the only ones that have been observed since the invention of the telescope in 1609 have been in distant galaxies. One exception to this is the one found in 1987 in the large Magellanic Cloud, an irregular galaxy close to our Milky Way galaxy. In A.D. 1054, Chinese observers recorded a ''guest star'' in the constellation Taurus, in our galaxy, which was so bright that it was visible in daylight. In this same region of space, we now find a nebula, called the Crab nebula, which is believed to be the remnants thrown off by the 1054 supernova.

As mentioned above, when a star explodes as a supernova, most of the star's material is ejected, but a central core is left behind that continues to collapse. What will be the ultimate fate of the collapsing core? The answer to this intriguing question depends on the mass of the core. If the core's mass is greater than about 1.4 solar masses, the star may collapse due to gravity such that the electrons and protons combine to form neutrons. Such a star, in which all the remaining matter is in the form of neutrons, is called a neutron star. The contraction of a neutron star eventually ceases when the neutron gas is compressed to a radius of about 10 km (about 6 mi). A teaspoonful of this material on Earth would weigh about 5 billion tons!

An even more unusual death of a star may occur when the core has a mass greater than about 3 solar masses. No known forces in nature are strong enough to prevent the collapse of such a star. In fact, the collapse of the star may continue until the star becomes a mere point in space called a singularity. Such a singularity is commonly referred to as a **black hole**. *In effect, black holes are remains of stars that have collapsed under their own mass.*

Objects such as spaceships, atoms, and molecules can enter a black hole, but can never escape once in its grip. In fact, even light cannot escape the tremendous gravitational forces set up inside a black hole; hence, the origin of the terminology ''black hole.''

Using the concept of escape speed given by Equation 7a.2, we can obtain some understanding of how particles become trapped by black holes. (A proper derivation of the results below requires some ideas and equations from relativity.) Recall that the escape speed from a spherical body of mass M and radius R is given by

$$v_e = \sqrt{\frac{2GM}{R}}$$

If the escape speed exceeds the speed of light, $c = 3 \times 10^8$ m/s, light within the body will not be able to escape and the body will appear to be black. The critical radius for which this occurs is called the **Schwarzschild radius** after Karl Schwarzschild, who predicted the existence of black holes in 1916 (see Fig. 7a.3). Taking $v_e = c$ in the expression above, and solving for R, we find

$$R = \frac{2GM}{c^2} \qquad \text{[7a.3]}$$

Schwarzschild radius

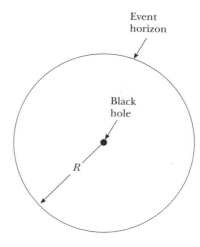

FIGURE 7a.3
The event horizon surrounding a black hole. The distance R equals the Schwarzschild radius.

For example, the value of R for a black hole whose mass is equal to that of the Sun is calculated to be 3.0 km (about 2 mi). The boundary of the region of radius R surrounding a black hole that permits entrance but prevents escape is called the **event horizon**. Any event occurring inside this horizon is invisible to an outside observer.

It is interesting to talk about exotic objects such as black holes, but do they really exist, and how does one go about detecting them? One way to find a black hole is to look for evidence of matter falling into it. As matter is pulled into a black hole, gravitational energy is converted into thermal energy, and the matter becomes hot enough to emit radiation in the form of x-rays. If these x-rays are emitted before the in-falling matter crosses the event horizon, they can be detected by astronomers. Although there are several objects in the sky that are suspected to contain black holes, the best known of these is Cygnus X-1, which is a strong source of x-rays in the constellation Cygnus. Astrophysicists believe that these objects consist of a black hole and a normal star orbiting about each other. X-rays are constantly emitted from the black hole as it "feeds on" the normal star.

PROBLEMS

1. A satellite of mass 200 kg is placed in Earth orbit at a height 200 km above the surface. (a) Assuming a circular orbit, how long does the satellite take to complete one orbit? (b) What is the satellite's speed? (c) What is the minimum energy necessary to place this satellite in orbit (assuming no air friction)? [(a) 88.5 min, (b) 7.99 km/s, (c) 6.43×10^9 J]

2. Sketch a graph of the gravitational potential energy, PE, versus r for a particle above the Earth's surface. Your graph should extend from the Earth's surface to infinity.

3. A satellite of the Earth has a mass of 100 kg and is at an altitude of 2.00×10^6 m. (a) What is the potential energy of the satellite at this location?

(b) What is the magnitude of the gravitational force on the satellite? [(a) -4.77×10^9 J, (b) 5.68×10^2 N]

4. Use the table of planetary data in Chapter 7 to find the escape speed from (a) the Moon, (b) Mercury, and (c) Jupiter. [(a) 2.38×10^3 m/s, (b) 4.18×10^3 m/s, (c) 6.02×10^4 m/s]

5. (a) Find the free-fall acceleration at the surface of a neutron star of mass 1.5 solar masses having a radius of 10.0 km. (b) Find the weight of a 0.120-kg baseball on this star. (c) Assume the equation $PE = mgh$ applies and calculate the energy that a 70.0-kg person would expend climbing a 1.00-cm-tall mountain on this star. [(a) 1.99×10^{12} m/s², (b) 2.38×10^{11} N, (c) 1.39×10^{12} J]

FOR FURTHER STUDY IN CHAPTER 9

SURFACE TENSION

If you look closely at a dewdrop sparkling in the morning sunlight, you will find that the drop is spherical. The drop takes this shape because of a property of liquid surfaces called **surface tension.** In order to understand the origin of surface tension, consider a molecule at point A in a container of water, as in Figure 9a.1. Although nearby molecules exert forces on this molecule, the net force on it is zero because it is completely surrounded by other molecules and hence is attracted equally in all directions. The molecule at B, however, is not attracted equally in all directions. Since there are no molecules above it to exert upward forces, the molecule is pulled toward the interior of the liquid. The contraction at the surface of the liquid ceases when the inward pull exerted on the surface molecules is balanced by the outward repulsive forces that arise from collisions with molecules in the interior of the liquid. *The net effect of this pull on all the surface molecules is to make the surface of the liquid contract and consequently to make the surface area of the liquid, as small as possible.* Drops of water take on a spherical shape because a sphere has the smallest surface area for a given volume.

If you place a sewing needle very carefully on the surface of a bowl of water, you will find that the needle floats even though the density of steel is about eight times that of water. This also can be explained by surface tension. A close examination of the needle shows that it actually rests in a depression in the liquid surface, as shown in Figure 9a.2. The water surface acts like an elastic membrane under tension. The weight of the needle produces a depression, thus increasing the surface area of the film. Molecular forces now act at all points along the depression in an attempt to restore the surface to its original horizontal position. The vertical components of these forces act to balance **w,** the weight of the needle.

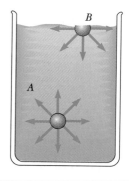

FIGURE 9a.1
The net force on a molecule at A is zero because such a molecule is completely surrounded by other molecules. The net force on a surface molecule at B is downward because it is not completely surrounded by other molecules.

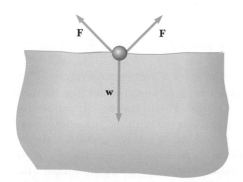

FIGURE 9a.2
End view of a needle resting on the surface of water. The components of surface tension balance the weight force.

1021

TABLE 9a.1
Surface Tensions for Various Liquids

Liquid	$T(°C)$	Surface Tension (N/m)
Ethyl alcohol	20	0.022
Mercury	20	0.465
Soapy water	20	0.025
Water	20	0.073
Water	100	0.059

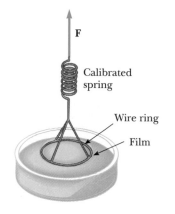

FIGURE 9a.3
An apparatus for measuring the surface tension of liquids. The force on the wire ring is measured just before it breaks free of the liquid.

The **surface tension,** γ, in a film of liquid is defined as the ratio of the magnitude of the surface tension force, **F**, to the length along which the force acts:

$$\gamma \equiv \frac{F}{L}$$ [9a.1]

The SI units of surface tension are newtons per meter, and values for a few representative materials are given in Table 9a.1.

The concept of surface tension can be thought of as the energy content of the fluid at its surface per unit surface area. To see that this is reasonable, we can manipulate the units of surface tension as

$$[\gamma] = \frac{N}{m} = \frac{N \cdot m}{m^2} = \frac{J}{m^2}$$

In general, *any equilibrium configuration of an object is one in which the energy is a minimum.* Consequently, a fluid will take on a shape such that its surface area is as small as possible. For a given volume, a spherical shape is the one that has the smallest surface area; therefore, a drop of water takes on a spherical shape.

An apparatus used to measure the surface tension of liquids is shown in Figure 9a.3. A circular wire with a circumference ℓ is lifted from a body of liquid. The surface film clings to the inside and outside edges of the wire, holding back the wire and causing the spring to stretch. If the spring is calibrated, one can measure the force required to overcome the surface tension of the liquid. In this case, the surface tension is given by

$$\gamma = \frac{F}{2\ell}$$

We must use 2ℓ for the length because the surface film exerts forces on the inside and outside of the ring.

The surface tension of liquids decreases with increasing temperature. This occurs because the faster moving molecules of a hot liquid are not bound together as strongly as are those in a cooler liquid. Furthermore, certain ingredients added to liquids decrease surface tension. For example, soap or detergent decreases the surface tension of water. This reduction in surface tension makes it easier for soapy water to penetrate the cracks and crevices of your clothes to clean them better than plain water. An effect similar to this occurs in the lungs. The

A razor blade floats on water because of surface tension. *(Courtesy of Henry Leap and Jim Lehman)*

surface tissue of the air sacs in the lungs contains a fluid that has a surface tension of about 0.050 N/m. A liquid with a surface tension this high would make it very difficult for the lungs to expand as one inhales. However, as the area of the lungs increases with inhalation, the body secretes into the tissue a substance that gradually reduces the surface tension of the liquid. At full expansion, the surface tension of the lung fluid can drop to as low as 0.005 N/m.

EXAMPLE 9a.1 Walking on Water

In this example, we shall illustrate how an insect is supported on the surface of water by surface tension. Let us assume that the insect's "foot" is spherical. When the insect steps onto the water with all six legs, a depression is formed in the water around each foot, as shown in Figure 9a.4. The surface tension of the water produces upward forces on the water which tend to restore the water surface to its normally flat shape. If the insect has a mass of 2.0×10^{-5} kg and if the radius of each foot is 1.5×10^{-4} m, find the angle θ.

Solution   From the definition of surface tension, we can find the net force F directed tangential to the depressed part of the water surface:

$$F = \gamma L$$

The length L along which this force acts is equal to the distance around the insect's foot, $2\pi r$. (It is assumed that the insect depresses the water surface such that the radius of the depression is equal to the radius of the foot.) Thus,

$$F = \gamma\, 2\pi r$$

and the net vertical force is

$$F_v = \gamma\, 2\pi r \cos \theta$$

Since the insect has six legs, this upward force must equal one sixth the weight of the insect, assuming its weight is equally distributed on all six feet. Thus,

$$\gamma\, 2\pi r \cos \theta = \frac{1}{6}\, w = \frac{1}{6}\, mg$$

(1) $$\cos \theta = \frac{mg}{12\pi r\gamma} = \frac{(2.0 \times 10^{-5}\text{ kg})(9.80\text{ m/s}^2)}{12\pi(1.5 \times 10^{-4}\text{ m})(0.073\text{ N/m})}$$

$$\theta = \boxed{62°}$$

Note that if the weight of the insect were great enough to make the right side of (1) greater than unity, a solution for θ would be impossible because the cosine can never be greater than unity. Under these conditions, the insect would sink.

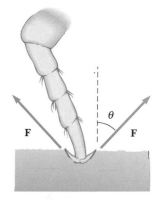

FIGURE 9a.4
(Example 9a.1)   One foot of an insect resting on the surface of water.

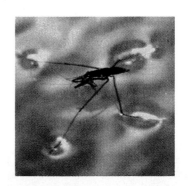

This water strider resting on the surface of a lake is able to remain on the surface, rather than sink, because an upward surface tension force acts on each leg, which balances the weight of the insect. The interesting pattern around the legs is due to the optical effect of the polarized light in the vicinity of the distorted water surface.

THE SURFACE OF LIQUIDS

If you have ever closely examined the surface of water in a glass container, you may have noticed that the surface of the liquid near the walls of the glass curves upward, as shown in Figure 9a.5a. However, if mercury is placed in a glass

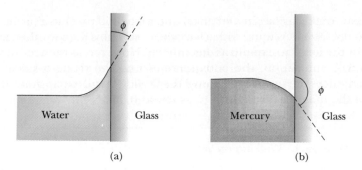

FIGURE 9a.5
A liquid in contact with a solid surface. (a) For water, the adhesive force is greater than the cohesive force. (b) For mercury, the adhesive force is less than the cohesive force.

container, the mercury surface curves downward, as in Figure 9a.5b. These surface effects can be explained by considering the forces between molecules. In particular, we must consider the forces that the molecules of the liquid exert on one another and the forces that the molecules of the glass surface exert on those of the liquid. In general terms, forces between like molecules, such as the forces between water molecules, are called *cohesive forces* and forces between unlike molecules, such as those of glass on water, are called *adhesive forces*.

Water tends to cling to the walls of the glass because the adhesive forces between the liquid molecules and the glass molecules are *greater* than the cohesive forces between the liquid molecules. In effect, the liquid molecules cling to the surface of the glass rather than fall back into the bulk of the liquid. When this condition prevails, the liquid is said to wet the glass surface. The surface of the mercury curves downward near the walls of the container because the cohesive forces between the mercury atoms are greater than the adhesive forces between mercury and glass. That is, a mercury atom near the surface is pulled more strongly toward other mercury atoms than toward the glass surface; hence mercury does not wet the glass surface.

The angle ϕ between the solid surface and a line drawn tangent to the liquid at the surface is called *the angle of contact* (Fig. 9a.6a and 9a.6b). Note that ϕ is less than 90° for any substance in which adhesive forces are stronger than cohesive forces and greater than 90° if cohesive forces predominate. For example, if a drop of water is placed on paraffin, the contact angle is approximately 107° (Fig. 9a.6a). If certain chemicals, called wetting agents or detergents, are added to the water, the contact angle becomes less than 90°, as shown in Figure 9a.6b. The addition of such substances to water is of value when one wants to ensure that

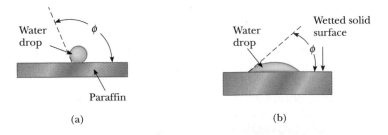

FIGURE 9a.6
(a) The contact angle between water and paraffin is about 107°. In this case, the cohesive force is greater than the adhesive force. (b) When a chemical called a wetting agent is added to the water, it wets the paraffin surface, and $\phi < 90°$. In this case, the adhesive force is greater than the cohesive force.

water makes intimate contact with a surface and penetrates it. For this reason, detergents are added to water to wash clothes or dishes. On the other hand, it is often necessary to keep water from making intimate contact with a surface, as in waterproofing clothing, where a situation somewhat the reverse of that shown in Figure 9a.6 is called for. The clothing is sprayed with a waterproofing agent, which changes ϕ from less than 90° to greater than 90°. Thus, the water beads up on the surface and does not easily penetrate the clothing.

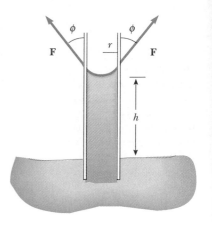

FIGURE 9a.7
A liquid rises in a narrow tube because of capillary action, a result of surface tension and adhesive forces.

CAPILLARY ACTION

Capillary tubes are tubes in which the diameter of the opening is very small. In fact, the word *capillary* means "hair-like." If such a tube is inserted into a fluid for which adhesive forces dominate over cohesive forces, the liquid will rise into the tube, as shown in Figure 9a.7. The rising of the liquid in the tube can be explained in terms of the shape of the surface of the liquid and in terms of the surface tension effects in the liquid. At the point of contact between liquid and solid, the upward force of surface tension is directed as shown in Figure 9a.7. From Equation 9a.1, the magnitude of this force is

$$F = \gamma L = \gamma (2\pi r)$$

We use $L = 2\pi r$ here because the liquid is in contact with the surface of the tube at all points around its circumference. The vertical component of this force due to surface tension is

$$F_v = \gamma(2\pi r)(\cos \phi) \qquad \text{[9a.2]}$$

In order for the liquid in the capillary tube to be in equilibrium, this upward force must be equal to the weight of the cylinder of water of height h inside the capillary tube. The weight of this water is

$$w = Mg = \rho Vg = \rho g \pi r^2 h \qquad \text{[9a.3]}$$

Equating F_v in Equation 9a.2 to w in Equation 9a.3, we have

$$\gamma(2\pi r)(\cos \phi) = \rho g \pi r^2 h$$

Thus, the height to which water is drawn into the tube is

$$h = \frac{2\gamma}{\rho g r} \cos \phi \qquad \text{[9a.4]}$$

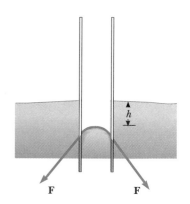

FIGURE 9a.8
When cohesive forces between molecules of the liquid exceed adhesive forces, the liquid level in the capillary tube is below the surface of the surrounding fluid.

If a capillary tube is inserted into a liquid in which cohesive forces dominate over adhesive forces, the level of the liquid in the capillary tube will be below the surface of the surrounding fluid, as shown in Figure 9a.8. An analysis similar to that done above would show that the distance h the surface is depressed is given by Equation 9a.4.

Capillary tubes are often used to draw small samples of blood from a needle prick in the skin. Capillary action must also be considered in the construction of concrete-block buildings because water seepage through capillary pores in the blocks or the mortar may cause damage to the inside of the building. To prevent this, the blocks are usually coated with a waterproofing agent either outside or inside the building. Water seepage through a wall is an undesirable effect of capillary action, but paper towels use capillary action in a useful manner to absorb spilled fluids.

EXAMPLE 9a.2　How High Does the Water Rise?

Find the height to which water would rise in a capillary tube with a radius equal to 5.0×10^{-5} m. Assume that the angle of contact between the water and the material of the tube is small enough to be considered zero.

Solution　The surface tension of water is 0.073 N/m. For a contact angle of $0°$, we have $\cos \phi = \cos 0° = 1$, so that Equation 9a.4 gives

$$h = \frac{2\gamma}{\rho g r} = \frac{2(0.073 \text{ N/m})}{(1.00 \times 10^3 \text{ kg/m}^3)(9.80 \text{ m/s}^2)(5.0 \times 10^{-5} \text{ m})} = \boxed{0.29 \text{ m}}$$

VISCOSITY

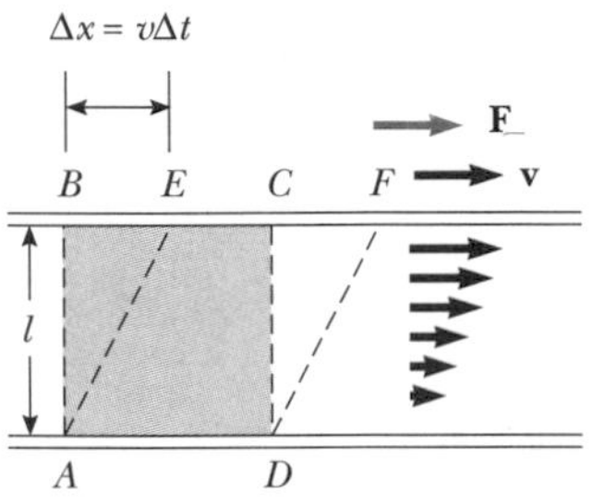

$\Delta x = v \Delta t$

FIGURE 9a.9

A layer of liquid between two solid surfaces in which the lower surface is fixed and the upper surface moves to the right with a velocity v.

It is considerably easier to pour water out of a container than to pour syrup. This is because syrup has a higher viscosity than water. In a general sense, *viscosity refers to the internal friction of a fluid.* It is very difficult for layers of a viscous fluid to slide past one another. Likewise, it is very difficult for one solid surface to slide past another if there is a highly viscous fluid, such as soft tar, between them.

To better understand the concept of viscosity, consider a liquid layer placed between two solid surfaces, as in Figure 9a.9. The lower surface is fixed in position, and the top surface moves to the right with a velocity v under the action of an external force F. Because of this motion, a portion of the liquid is distorted from its original shape, *ABCD*, at one instant to the shape *AEFD* a moment later. You will recognize that the liquid has undergone a constantly increasing shear strain. Ideal fluids that have no internal friction forces between adjacent layers could not have their shape distorted. However, in viscous fluids, there are cohesive forces between molecules in the various layers that can lead to strains which change with time. By definition, the shear stress on the liquid is

$$\text{Shear stress} \equiv \frac{F}{A}$$

where A is the area of the top plate. Furthermore, the shear strain is defined as

$$\text{Shear strain} \equiv \frac{\Delta x}{l}$$

The velocity of the fluid changes from zero at the lower plate to v at the upper. Thus, in a time Δt, the fluid at the upper plate moves a distance $\Delta x = v \Delta t$. Therefore,

$$\frac{\text{Shear strain}}{\Delta t} = \frac{\Delta x / l}{\Delta t} = \frac{v}{l}$$

This equation states that the rate of change of the shearing strain is v/l.

The **coefficient of viscosity,** η (lowercase Greek leter *eta*), for the fluid is defined as the ratio of the shearing stress to the rate of change of the shear strain:

$$\eta \equiv \frac{Fl}{Av} \qquad \text{[9a.5]}$$

Coefficient of viscosity

TABLE 9a.2
The Viscosities of Various Fluids

Fluid	$T(°C)$	Viscosity $\eta(N \cdot s/m^2)$
Water	20	1.0×10^{-3}
Water	100	0.3×10^{-3}
Whole blood	37	2.7×10^{-3}
Glycerine	20	1500×10^{-3}
10 wt motor oil	30	250×10^{-3}

The SI units of viscosity are $N \cdot s/m^2$. You should note that the units of viscosity in many reference sources are often expressed in $dyne \cdot s/cm^2$, called 1 **poise** in honor of the French scientist J. L. Poiseuille (1799–1869). The relationship between the SI unit of viscosity and the poise is

$$1 \text{ poise} = 10^{-1} \text{ N} \cdot \text{s/m}^2 \qquad \textbf{[9a.6]}$$

Small viscosities are often expressed in centipoise (cp), where $1 \text{ cp} = 10^{-2} \text{ poise}$. The coefficients of viscosity for some common substances are listed in Table 9a.2.

POISEUILLE'S LAW

Figure 9a.10 shows a section of a tube containing a fluid under a pressure P_1 at the left end and a pressure P_2 at the right. Because of this pressure difference, the fluid will flow through the tube. The rate of flow (volume per unit time) depends on the pressure difference $(P_1 - P_2)$, the dimensions of the tube, and the viscosity of the fluid. The result, known as **Poiseuille's law,** is

$$\text{Rate of flow} = \frac{\Delta V}{\Delta t} = \frac{(P_1 - P_2)(\pi R^4)}{8L\eta} \qquad \textbf{[9a.7]}$$

where R is the radius of the tube, L is its length, and η is the coefficient of viscosity. We shall not attempt to derive this equation here because the methods of integral calculus are required. However, you should note that the equation does agree with common sense. That is, it is reasonable that the rate of flow should increase if the pressure difference across the tube or the tube radius increases. Likewise, the flow rate should decrease if the viscosity of the fluid or the length of the tube increases. Thus, the presence of R and the pressure difference in the numerator of Equation 9a.7 and of L and η in the denominator makes sense.

From Poiseuille's law, we see that in order to maintain a constant flow rate, the pressure difference across the tube has to increase if the viscosity of the fluid increases. This is important when one considers the flow of blood through the circulatory system. The viscosity of blood increases as the number of red blood cells rises. Thus, blood with a high concentration of red blood cells requires greater pumping pressure from the heart to keep it circulating than does blood of lower red blood cell concentration.

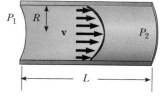

FIGURE 9a.10
Velocity profile of a fluid flowing through a uniform pipe of circular cross-section. The rate of flow is given by Poiseuille's law. Note that the fluid velocity is greatest at the middle of the pipe.

Poiseuille's law

Note that the flow rate varies as the radius of the tube raised to the fourth power. Consequently, if a constriction occurs in a vein or artery, the heart will have to work considerably harder in order to produce a higher pressure drop and hence to maintain the required flow rate.

EXAMPLE 9a.3   A Blood Transfusion

A patient receives a blood transfusion through a needle of radius 0.20 mm and length 2.0 cm. The density of blood is 1050 kg/m³. The bottle supplying the blood is 0.50 m above the patient's arm. What is the rate of flow through the needle?

Solution   The pressure differential between the level of the blood and the patient's arm is

$$P_1 - P_2 = \rho g h = (1050 \text{ kg/m}^3)(9.80 \text{ m/s}^2)(0.50 \text{ m}) = 5.15 \times 10^3 \text{ Pa}$$

Thus, the rate of flow, from Poiseuille's law, is

$$\frac{\Delta V}{\Delta t} = \frac{(P_1 - P_2)(\pi R^4)}{8L\eta}$$

$$= \frac{(5.15 \times 10^3 \text{ Pa})(\pi)(2.0 \times 10^{-4} \text{ m})^4}{8(2.0 \times 10^{-2} \text{ m})(2.7 \times 10^{-3} \text{ N·s/m}^2)} = \boxed{6.0 \times 10^{-8} \text{ m}^3/\text{s}}$$

Exercise   How long will it take to inject 1 pint (500 cm³) of blood into the patient?

Answer   140 min.

REYNOLDS NUMBER

As we mentioned earlier, at sufficiently high velocities, fluid flow changes from simple streamline flow to turbulent flow, that is, flow characterized by a highly irregular motion of the fluid. Experimentally it is found that the onset of turbulence in a tube is determined by a dimensionless factor called the **Reynolds number,** given by

Reynolds number

$$RN = \frac{\rho v d}{\eta} \qquad \text{[9a.8]}$$

where ρ is the density of the fluid, v is the average speed of the fluid along the direction of flow, d is the diameter of the tube, and η is the viscosity of the fluid. If RN is below about 2000, the flow of fluid through a tube is streamline; turbulence occurs if RN is above 3000. In the region between 2000 and 3000, the flow is unstable, meaning that the fluid can move in streamline flow but any small disturbance will cause its motion to change to turbulent flow.

EXAMPLE 9a.4   Turbulent Flow of Blood

Determine the speed at which blood flowing through an artery of diameter 0.20 cm would become turbulent. Assume that the density of blood is 1.05×10^3 kg/m³ and that its viscosity is 2.7×10^{-3} N·s/m².

Solution   At the onset of turbulence, the Reynolds number is 3000. Thus, the speed of the blood would have to be

$$v = \frac{\eta(RN)}{\rho d} = \frac{(2.7 \times 10^{-3}\,\text{N}\cdot\text{s/m}^2)(3000)}{(1.05 \times 10^3\,\text{kg/m}^3)(0.20 \times 10^{-2}\,\text{m})} = \boxed{3.9\ \text{m/s}}$$

TRANSPORT PHENOMENA

When a fluid flows through a tube, the basic mechanism that produces the flow is a difference in pressure across the ends of the tube. This pressure difference is responsible for the transport of a mass of fluid from one location to another. The fluid may also move from place to place because of a second mechanism, one that depends on a concentration difference between two points in the fluid, as opposed to a pressure difference. When the concentration (the number of molecules per unit volume) is higher at one location than at another, molecules will flow from the point where the concentration is high to the point where it is lower. The two fundamental processes involved in fluid transport resulting from concentration differences are called *diffusion* and *osmosis*. The following sections examine the nature and importance of these processes.

DIFFUSION

You can imagine what happens when someone wearing a strong shaving lotion or perfume strolls into a crowded room. All eyes turn to seek out the source of the delightful smell. The aroma spreads through the room by a process called diffusion.

> In a **diffusion** process, molecules move from a region where their concentration is high to a region where their concentration is lower.

That is, the molecules of the lotion or perfume move from the source (near the person's face), where there are many molecules per unit volume, throughout the room, to regions where the concentration of these molecules is lower. Although the example used here is one of diffusion in air, the process also occurs in liquids and, to a lesser extent, in solids. For example, if a drop of food coloring is placed in a glass of water, the coloring soon spreads throughout the liquid by diffusion. In either case, diffusion ceases when there is a uniform concentration at all locations in the fluid.

To understand why diffusion occurs, consider Figure 9a.11, which represents a container in which a high concentration of molecules has been introduced into the left side. For example, this could be accomplished by releasing a few drops of perfume into the left side of the container. The dashed line in Figure 9a.11 represents an imaginary barrier separating the region of high concentration from the region of lower concentration. Because the molecules are moving with high speeds in random directions, many of them will cross the imaginary barrier moving from left to right. Very few molecules of perfume will pass through this area moving from right to left simply because there are very few of them on the right side of the container at any instant. Thus, there will always be a *net* movement from the region where there are many molecules to the region where there are fewer molecules. For this reason, the concentration on

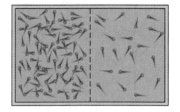

FIGURE 9a.11
When the concentration of gas molecules on the left side of the container exceeds the concentration on the right side, there will be a net motion (diffusion) of molecules from left to right.

TABLE 9a.3
Diffusion Coefficients for Various Substances at 20°C

Substance	$D(m^2/s)$
Oxygen through air	6.4×10^{-5}
Oxygen through tissue	1×10^{-11}
Oxygen through water	1×10^{-9}
Sucrose through water	5×10^{-10}
Hemoglobin through water	76×10^{-11}

the left side of the container will decrease in time and that on the right side will increase. There will be no *net* movement across the cross-sectional area once a concentration equilibrium has been reached. That is, when the concentration is the same on both sides, the number of molecules diffusing from right to left in a given time interval will equal the number moving from left to right in the same time interval.

The basic equation for diffusion is **Fick's law,** which in equation form is

Fick's law

$$\frac{\text{Diffusion}}{\text{rate}} = \frac{\text{mass}}{\text{time}} = \frac{\Delta M}{\Delta t} = DA \left(\frac{C_2 - C_1}{L} \right) \qquad \text{[9a.9]}$$

where D is a constant of proportionality. The left side of this equation is called the diffusion rate and is a measure of the mass being transported per unit time. This equation says that

> the rate of diffusion is proportional to the cross-sectional area A and to the change in concentration per unit distance, $(C_2 - C_1)/L$, which is called the concentration gradient.

The concentrations C_1 and C_2 are measured in kilograms per cubic meter. The proportionality constant D is called the **diffusion coefficient** and has units of square meters per second. Table 9a.3 lists diffusion coefficients for a few substances.

THE SIZE OF CELLS AND OSMOSIS

Diffusion through cell membranes is extremely vital in carrying oxygen to the cells of the body and in removing carbon dioxide and other waste products from them. Oxygen is required by the cells for those metabolic processes in which substances are either synthesized or broken down. In such metabolic processes, the cell uses up oxygen and produces carbon dioxide as a by-product. A fresh supply of oxygen diffuses from the blood, where its concentration is high, into the cell, where its concentration is low. Likewise, carbon dioxide diffuses from the cell into the blood, where it is in lower concentration. Water, ions, and other nutrients also pass into and out of cells by diffusion.

A common characteristic of cells in all plants and animals is their extremely small size. The adult human body contains literally trillions of cells. In order to understand why cells are so small, we must consider the relationship between the surface area of an object and its volume.

Let us consider a cube 2 cm on a side. The area of one of its faces is 2 cm $\times$ 2 cm = 4 cm^2, and because a cube has six sides, the total surface area is 24 cm^2. Its volume is 2 cm $\times$ 2 cm $\times$ 2 cm = 8 cm^3. Hence, the ratio of surface area to volume is 24/8 = 3. Now consider a larger cube, one measuring 3 cm on a side. Repeating the calculations gives us a surface area of 54 cm^2 and a volume of 27 cm^3. In this case, the ratio of surface area to volume is 54/27 = 2. Thus, we see that as the size of an object decreases, the ratio of its surface area to its volume increases. This, of course, says that a small cell has a larger surface-area-to-volume ratio than a large cell. But how does this pertain to the operation of a cell?

A cell can function properly only if it can (a) rapidly receive vital substances such as oxygen and (b) rapidly eliminate waste products. If such substances are to readily move into and out of cells, the cells should have a large surface area. However, if the volume of the cell is too large, it could take a considerable period of time for the nutrients to diffuse into the interior of the cell where they are needed. Under optimum conditions, the surface area of the cell should be large enough so that the exposed membrane area can exchange materials effectively while at the same time the volume should be small enough so that materials can reach or leave particular locations rapidly. To reach these optimum conditions, a small cell with its high surface-area-to-volume ratio is necessary.

As we have seen, the movement of material through cell membranes is necessary for the efficient functioning of cells. The diffusion of material through a membrane is partially determined by the size of the pores (holes) in the membrane wall. That is, small molecules, such as water, may pass through the pores easily while larger molecules, such as sugar, may pass through only with difficulty or not at all. A membrane that allows passage of some molecules but not others is called a selectively permeable membrane.

Osmosis is defined as the movement of water from a region where its concentration is high, across a selectively permeable membrane, into a region where its concentration is lower.

As in the case of diffusion, osmosis continues until the concentrations on the two sides of the membrane are equal. Osmosis is often described simply as the diffusion of water across a membrane.

To understand the effect of osmosis on living cells, let us consider a particular cell in the body that contains a sugar concentration of 1%. (That is, 1 g of sugar is dissolved in enough water to make 100 ml of solution.) Now assume that this cell is immersed in a 5% sugar solution (5 g of sugar dissolved in enough water to make 100 ml). In such a situation, water would diffuse from inside the cell, where its concentration is higher, across the cell wall membrane, to the outside solution, where the concentration of water is lower. This loss of water from the cell would cause it to shrink and perhaps become damaged through dehydration. If the concentrations were reversed, water would diffuse into the cell, causing it to swell and perhaps burst. It should be obvious from this description that normal osmotic relationships must be maintained in the body. If solutions are introduced into the body intravenously, care must be taken to ensure that these solutions do not disturb the osmotic balance of the body because such a disturbance could lead to cell damage. For example, if 9% saline solution surrounds a red blood cell, the cell will shrink. On the other hand, if the saline solution is about 1%, the cell will eventually burst.

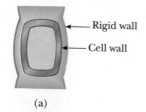

(a)

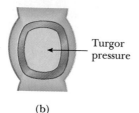

(b)

(c)

FIGURE 9a.12
(a) The structure of a plant cell. (b) As water accumulates in its interior, the cell expands under turgor pressure. (c) When water in the cell's interior is depleted, the walls of the cell collapse.

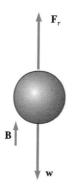

FIGURE 9a.13
A sphere falling through a viscous medium. The forces acting on the sphere are the resistive frictional force **F$_r$**, the buoyant force **B,** and the weight of the sphere **w.**

Under normal circumstances, the cells of our bodies are in an environment such that there is no net movement of water into or out of them. However, certain one-celled organisms, such as *protozoa,* do not enjoy this osmotic equilibrium. These organisms usually live in fresh water, which obviously has a higher concentration of water than the solution inside the cell. To prevent an inflow of water to the point of bursting, these organisms possess an organelle (a tiny organ) that acts as a pump and continually forces water out of the cell.

Most plant cells are contained within a rigid wall, as shown in Figure 9a.12a. If water accumulates in the cell, it expands (Fig. 9a.12b) and exerts a pressure, called **turgor pressure,** against the rigid wall. The rigidity of the wall prevents the cell from bursting. If water within the cell is depleted, the rigid wall collapses inward slightly, as in Figure 9a.12c. This causes the plant to wilt.

MOTION THROUGH A VISCOUS MEDIUM

When an object falls through air, its motion is impeded by the force of air resistance. In general, this force is dependent on the shape of the falling object and on its velocity. This viscous drag acts on all falling objects, but the exact details of the motion can be calculated only for a few cases in which the object has a simple shape, such as a sphere. In this section, we shall examine the motion of a tiny spherical object falling slowly through a viscous medium.

In 1845 a scientist named George Stokes found that the magnitude of the resistive force on a very small spherical object of radius r falling slowly through a fluid of viscosity η with speed v is given by

$$F_r = 6\pi\eta r v \qquad \text{[9a.10]}$$

This equation, called **Stokes's law,** has many important applications. For example, it describes the sedimentation of particulate matter in blood samples. It was used by Robert Millikan (1886–1953) to calculate the radius of charged oil droplets falling through air. From this, Millikan was ultimately able to determine the smallest known unit of electric charge. Millikan was awarded the Nobel prize in 1923 for this pioneering work on elemental charge.

As a sphere falls through a viscous medium, three forces act on it, as shown in Figure 9a.13: **F$_r$** is the force of frictional resistance, **B** is the buoyant force of the fluid, and **w** is the weight of the sphere, whose magnitude is given by

$$w = \rho g V = \rho g \left(\frac{4}{3}\pi r^3 \right)$$

where ρ is the density of the sphere and $\frac{4}{3}\pi r^3$ is its volume. According to Archimedes' principle, the magnitude of the buoyant force is equal to the weight of the fluid displaced by the sphere:

$$B = \rho_f g V = \rho_f g \left(\frac{4}{3}\pi r^3 \right)$$

where ρ_f is the density of the fluid.

At the instant the sphere begins to fall, the force of frictional resistance is zero because the speed of the sphere is zero. As it accelerates, the speed increases and so does **F$_r$**. Finally, at a speed called the **terminal speed** v_t, *the resultant force goes to zero.* This occurs when the net upward force balances the downward weight force. Hence, the sphere reaches terminal speed when

$$F_r + B = w$$

or

$$6\pi\eta r v_t + \rho_f g \left(\frac{4}{3} \pi r^3 \right) = \rho g \left(\frac{4}{3} \pi r^3 \right)$$

When this is solved for v_t, we get

$$v_t = \frac{2r^2 g}{9\eta} (\rho - \rho_f)$$ [9a.11]   Terminal speed

---

EXAMPLE 9a.5   A Falling Pearl

A pearl of density 2.0×10^3 kg/m^3 and radius 2.0 mm falls through a liquid shampoo of density 1.4×10^3 km/m^3 and viscosity 0.50×10^{-3} N·s/m^2. Find the terminal speed of the pearl.

Solution   Substituting the given values into Equation 9a.11, we have

$$v_t = \frac{2r^2 g}{9\eta} (\rho - \rho_f)$$

$$= \frac{2(2.0 \times 10^{-3} \text{ m})^2 (9.80 \text{ m/s}^2)}{9(0.50 \times 10^{-3} \text{ N·s/m}^2)} (2.0 \times 10^3 \text{ kg/m}^3 - 1.4 \times 10^3 \text{ kg/m}^3)$$

$$= 1.1 \times 10^{-2} \text{ m/s}$$

SEDIMENTATION AND CENTRIFUGATION

If an object is not spherical, we can still use the basic approach just described to determine its terminal speed. The only difference will be that we shall not be able to use Stokes's law for the resistive force. Instead, let us assume that the resistive force has a magnitude given by $F_r = kv$, where k is a coefficient of frictional resistance that must be determined experimentally. As we discussed above, the object reaches its terminal speed when the weight downward is balanced by the net upward force, or

$$w = B + F_r$$ [9a.12]

where B is the buoyant force, given by $B = \rho_f g V$.

We can use the fact that volume, V, of the displaced fluid is related to the density of the falling object, ρ, by $V = m/\rho$. Hence, we can express the buoyant force as

$$B = \frac{\rho_f}{\rho} mg$$

Let us substitute this expression for B and $F_r = kv_t$ into Equation 9a.12 (terminal speed condition):

$$mg = \frac{\rho_f}{\rho} mg + kv_t$$

or

$$v_t = \frac{mg}{k} \left(1 - \frac{\rho_f}{\rho} \right)$$ [9a.13]

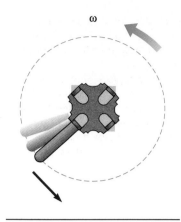

FIGURE 9a.14
Simplified diagram of a centrifuge (top view).

The terminal speed for particles in biological samples is usually quite small. For example, the terminal speed for blood cells falling through plasma is about 5 cm/h in the gravitational field of the Earth. The terminal speeds for the molecules that make up a cell are many orders of magnitude smaller than this because of their much smaller mass. The speed at which materials fall through a fluid is called the *sedimentation rate*. This number is often important in clinical analysis.

It is often desired to increase the sedimentation rate in a fluid. A common method used to accomplish this is to increase the effective acceleration g that appears in Equation 9a.13. A fluid containing various biological molecules is placed in a centrifuge and whirled at very high angular speeds (Fig. 9a.14). Under these conditions, the particles experience a large radial acceleration, $a_c = v^2/r = \omega^2 r$, which is much greater than the free-fall acceleration, and so we can replace g in Equation 9a.13 by $\omega^2 r$:

$$v_t = \frac{m\omega^2 r}{k}\left(1 - \frac{\rho_f}{\rho}\right)$$ **[9a.14]**

This equation indicates that those particles having the greatest mass will have the largest terminal speed. Therefore, the most massive particles will settle out on the bottom of a test tube first.

EXAMPLE 9a.6   The Spinning Test Tube

A centrifuge rotates at 50 000 rev/min, which corresponds to an angular frequency of 5240 rad/s (a typical speed). A test tube placed in this device has its top 5.0 cm from the axis of rotation and its bottom 13 cm from this axis. Find the effective value of g at the midpoint of the test tube, which corresponds to a distance 9.0 cm from the axis of rotation.

Solution   The acceleration experienced by the particles of the tube at a distance $r = 9.0$ cm from the axis of rotation is given by

$$a_c = \omega^2 r = \left(5240\,\frac{\text{rad}}{\text{s}}\right)^2 (9.0 \times 10^{-2}\,\text{m}) = \boxed{2.5 \times 10^6\,\text{m/s}^2}$$

Exercise   If the mass of the contents of the test tube is 15 g, find the centripetal force that the bottom of the tube must exert on the contents of the tube. Assume a centripetal acceleration equal to that found at the midpoint of the tube.

Answer 3.7×10^4 N, or about 8000 lb! (Because of such large forces, the base of the tube in a centrifuge must be rigidly supported to keep the glass from shattering.)

PROBLEMS

Surface Tension

 1. A vertical force of 1.61×10^{-2} N is required to lift a wire ring of radius 1.75 cm from the surface of a container of blood plasma. Calculate the surface tension of blood plasma from this information. [7.32×10^{-2} N/m]

2. Each of the six legs of an insect on a water surface makes a depression 0.25 cm in radius with a contact angle of 45°. Calculate the mass of the insect. [0.50 g]

FIGURE 9a.15 (Problem 4)

3. The surface tension of ethanol is 0.0227 N/m, and the surface tension of tissue fluid is 0.050 N/m. A force of 7.13×10^{-3} N is required to lift a 5.0-cm diameter wire ring vertically from the surface of ethanol. What should the diameter of a ring be so that this same force would lift it from the tissue fluid? [2.3 cm]

4. A square metal sheet 5.0 cm on a side and of negligible thickness is attached to a balance and inserted into a container of fluid. The contact angle is found to be zero, as shown in Figure 9a.15a and the balance to which the metal sheet is attached reads 0.40 N. A thin veneer of oil is then spread over the metal sheet and the contact angle becomes 180°, as shown in Figure 9a.15b. The balance now reads 0.39 N. What is the surface tension of the fluid? [5.0×10^{-2} N/m]

Capillary Action

5. A certain fluid has a density of 1080 kg/m³ and is observed to rise to a height of 2.1 cm in a 1.0-mm-diameter tube. The contact angle between the wall and the fluid is zero. Calculate the surface tension of the fluid. [5.6×10^{-2} N/m]

6. Whole blood has a surface tension of 0.058 N/m and a density of 1050 kg/m³. To what height can whole blood rise in a capillary blood vessel that has a radius of 2.0×10^{-6} m if the contact angle is zero? [5.6 m]

7. Use density and surface tension values to calculate the height to which water will rise in a capillary of diameter 1.00×10^{-4} m. Assume a contact angle of zero and a temperature of 20.0°C. [29.8 cm]

8. A staining solution used in a microbiology laboratory has a surface tension of 0.088 N/m and a

density 1.035 times the density of water. What must the diameter of a capillary tube be so that this solution will rise to a height of 5 cm? (Assume a zero contact angle.) [0.694 mm]

9. A capillary tube 1 mm in radius is immersed in a beaker of mercury. The mercury level inside the tube is found to be 0.536 cm *below* the level of the reservoir. Use the surface tension for mercury from Table 9.a1 and the density to determine the contact angle between mercury and glass. [140°]

Viscosity

Poiseuille's Law

10. The block of ice (temperature 0°C) shown in Figure 9a.16 is drawn over a level surface lubricated by a layer of water 0.10 mm thick. Determine the magnitude of force F needed to pull the block with a constant speed of 0.50 m/s. At 0°C, the viscosity of water is $\eta = 1.79 \times 10^{-3}$ N·s/m². [8.6 N]

11. A metal block is pulled over a horizontal surface that has been coated with a layer of lubricant 1.0 mm thick. The face of the block in contact with the surface has dimensions 0.40 m by 0.12 m. A force of 1.9 N is required to move the block at a constant speed of 0.50 m/s. Calculate the coefficient of viscosity of the lubricant. [7.9×10^{-2} N·s/m²]

12. A thin 1.5-mm coating of glycerine has been placed between two microscope slides of width 1.0 cm and length 4.0 cm. Find the force required to pull one of the microscope slides at a speed of 0.30 m/s relative to the other. [0.12 N]

13. A straight horizontal pipe with a diameter of 1.0 cm and a length of 50 m carries oil with a coefficient of viscosity of 0.12 Pa·s. At the output of the pipe, the flow rate is 8.6×10^{-5} m³/s and the pressure is 1.0 atm. Find the gauge pressure at the pipe input. [2.1×10^{6} Pa]

14. A hypodermic needle is 3.0 cm in length and 0.30 mm in diameter. What excess pressure is required along the needle so that the flow rate of water through it will be 1.0 g/s? (Use 1.0×10^{-3} Pa·s as the viscosity of water.) [1.5×10^{5} Pa]

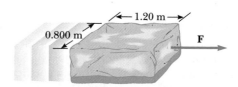

FIGURE 9a.16 (Problem 10)

15. A needle of radius 0.30 mm and length 3.0 cm is used to give a patient a blood transfusion. Assume the pressure differential across the needle is achieved by elevating the blood 1.0 m above the patient's arm. (a) What is the rate of flow of blood through the needle? (b) At this rate of flow, how long will it take to inject 1 pint (approximately 500 cm³) of blood into the patient? The density of blood is 1050 kg/m³, and its coefficient of viscosity is 4.0×10^{-3} N·s/m². [(a) 2.7×10^{-7} m³/s, (b) 31 min]

16. The water pressure in a horizontal pipe decreases 0.50 atm per 100 m when the flow rate in the pipe is 30.0 liter/min. Determine the radius of the pipe. (Assume a temperature of 20°C.) [7.1 mm]

17. What diameter needle should be used to inject a volume of 500 cm³ of a solution into a patient in 30 min? Assume that the needle length is 2.5 cm and that the solution is elevated 1.0 m. Furthermore, assume the viscosity and density of the solution are those of pure water. [0.41 mm]

18. The pulmonary artery, which connects the heart to the lungs, has an inner radius of 2.6 mm and is 8.4 cm long. If the pressure drop between the heart and lungs is 400 Pa, what is the average speed of blood in the pulmonary artery? [1.0 m/s]

19. A pipe carrying 20°C water has a diameter of 2.5 cm. Estimate the maximum flow speed if the flow is to be laminar. [8.0 cm/s]

20. What is the Reynolds number for the flow of liquid in the 1.2-m-diameter Alaska pipeline? The density of crude oil is 850 kg/m³, its speed is 3.01 m/s, and its viscosity is 0.30 Pa·s. Is the flow laminar or turbulent? [1.0×10^4, turbulent]

21. Determine the speed at which the flow of water through a 0.5-cm-diameter pipe will become turbulent ($RN \geq 3000$). [0.6 m/s]

22. Assume a value of 980 for the Reynolds number for blood in an artery and a viscosity of 4.0×10^{-3} N·s/m² for whole blood. If the density of whole blood is equal to 1.05×10^3 kg/m³, at what speed does blood flow through an artery 0.45 cm in diameter? [0.83 m/s]

23. The aorta in humans has a diameter of about 2.0 cm and, at certain times, the blood speed through it is about 55 cm/s. Is the blood flow turbulent? [2.9×10^3, unstable]

Transport Phenomena

24. Sucrose is allowed to diffuse along a 10-cm length of tubing filled with water. The tube is 6.0 cm² in cross-sectional area. The diffusion coefficient is 5.0×10^{-10} m²/s, and 8.0×10^{-14} kg is trans-ported along the tube in 15 s. What is the difference in the concentration levels of sucrose at the two ends of the tube? [1.8×10^{-3} kg/m³]

25. In a diffusion experiment, it is found that, in 60 s, 3.0×10^{-13} kg of sucrose will diffuse along a horizontal pipe of cross-sectional area 1.0 cm². If the diffusion coefficient is 5.0×10^{-10} m²/s, what is the concentration gradient (that is, the change in concentration per unit length along the path)? [0.10 kg/m⁴]

26. Glycerine in water diffuses along a horizontal column that has a cross-sectional area of 2.0 cm². The concentration gradient is 3.0×10^{-2} kg/m⁴, and the diffusion rate is found to be 5.7×10^{-15} kg/s. Determine the diffusion coefficient. [9.5×10^{-10} m²/s]

27. Use the data for sucrose given in Problem 24 to calculate how much sucrose will diffuse down a horizontal pipe of cross-sectional area 4.0 cm² in 10.0 s if the concentration gradient is 0.20 kg/m⁴. [4.0×10^{-13} kg]

Motion Through a Viscous Medium

28. Small spheres of diameter 1.00 mm fall through water with a terminal speed of 1.10 cm/s. Calculate the density of the spheres. [1.02×10^3 kg/m³]

29. A test tube 10.0 cm tall is filled with water that contains spherical particles of density 1.8 g/cm³ in suspension. If the radius of the particle is 2.0×10^{-4} cm, find the time it takes a particle near the top of the tube to reach the bottom. [1.4×10^4 s]

30. Calculate the viscous force on a spherical oil droplet that is 2.0×10^{-5} cm in diameter and falling with a speed of 0.40 mm/s in air. (Use $\eta = 1.8 \times 10^{-5}$ N·s/m² as the coefficient of viscosity of air.) [1.4×10^{-14} N]

31. The viscous force on an oil drop is measured to be 3.0×10^{-13} N when the drop is falling through air with a speed of 4.5×10^{-4} m/s. If the radius of the drop is equal to 2.5×10^{-6} m, what is the viscosity of air? [1.4×10^{-5} N·s/m²]

32. An oil drop ($\rho = 800$ kg/m³) is falling with a terminal velocity of 0.040 mm/s in air of density 1.29 kg/m³. Calculate the radius of the oil drop. (The coefficient of viscosity of air is $\eta = 1.8 \times 10^{-5}$ N·s/m².) [0.64 μm]

33. Spherical particles of a protein of density 1.8 g/cm³ are shaken up in a solution of water. The solution is allowed to stand for 1.0 h. If the depth of water in the tube is 5.0 cm, find the radius of the largest particles still in solution at the end of the hour. [2.82 μm]

FOR FURTHER STUDY IN CHAPTER 15

This section describes an elegant technique for calculating electric fields that was developed by Karl Friedrich Gauss (1777–1855). Even though the procedure developed by Gauss is applicable only to situations in which the charge distribution is highly symmetric, it serves as a guide for understanding more complicated problems. In order to develop the law, we must first understand the concept of electric flux.

ELECTRIC FLUX

Consider an electric field that is uniform in both magnitude and direction as in Figure 15a.1. The electric field lines penetrate a surface of area A, which is perpendicular to the field. The technique used for drawing a figure such as Figure 15a.1 is that the number of lines per unit area, Φ, is proportional to the magnitude of the electric field. Therefore,

$$\Phi = EA \qquad \text{[15a.1]}$$

where Φ is the electric flux, which has units of $\text{N} \cdot \text{m}^2/\text{C}$ in SI units. Thus, if the area shown in Figure 15a.1 is 1 m², the magnitude of the electric field is 14 N/C. (Count the lines in Fig. 15a.1.) If the surface under consideration is not perpendicular to the field, the expression for the electric flux is

$$\Phi = EA \cos \theta \qquad \text{[15a.2]}$$

Equation 15a.2 can be easily understood by considering Figure 15a.2, where the area A is at an angle θ with respect to the field. The number of lines that cross this area is equal to the number that cross the projected area A', which is perpendicular to the field. We see that the two areas are related by $A' = A \cos \theta$. From Equation 15a.2 we see that the flux through a surface of fixed area has the maximum value, EA, when the surface is perpendicular to the field (when $\theta = 0°$) and that the flux is zero when the surface is parallel to the field (when $\theta = 90°$).

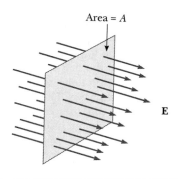

FIGURE 15a.1
Field lines of a uniform electric field penetrating a plane of area A perpendicular to the field. The electric flux, Φ, through this area is equal to EA.

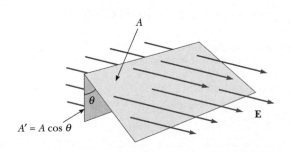

FIGURE 15a.2
Field lines for a uniform electric field through an area A that is at an angle θ to the field. Since the number of lines that go through the shaded area A' is the same as the number that go through A, we conclude that the flux through A' is equal to the flux through A and is given by $\Phi = EA \cos \theta$.

When the area is constructed such that a closed surface is formed, we shall adopt the convention that flux lines passing into the interior of the volume are negative and those passing out of the interior of the volume are positive.

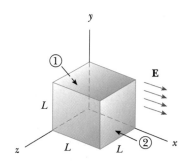

FIGURE 15a.3
(Example 15a.1)   A hypothetical surface in the shape of a cube in a uniform electric field parallel to the *x* axis. The net flux through the surface is zero when the net charge inside the cube is zero.

EXAMPLE 15a.1   Flux Through a Cube

Consider a uniform electric field oriented in the *x* direction. Find the net electric flux through the surface of a cube of edges *L* oriented as shown in Figure 15a.3.

Solution   The net flux can be evaluated by summing up the fluxes through each face of the cube. First, note that the flux through four of the faces is zero, since E is parallel to the area on these surfaces. These surfaces are those that are parallel to the *xy* and *xz* planes. For these surfaces, $\theta = 90°$, so $\Phi = EA \cos 90° = 0$. For surface 1 that lies in the *yz* plane in Figure 15a.3, the flux lines pass into the interior of the cube, and the flux is taken to be negative. We have

$$\Phi_1 = -EA = -EL^2$$

For surface 2 at the right face of the cube, the flux is positive and given by

$$\Phi_2 = EA = EL^2$$

The net flux through the surface of the cube is

$$\Phi_{net} = \Sigma EA = \Phi_1 + \Phi_2 = -EL^2 + EL^2 = 0$$

GAUSS' LAW

We shall now describe a general relation between the net electric flux through a closed surface (often called a *gaussian surface*) and the charge *enclosed* by the surface. In our discussion of electric field lines we saw that the number of field lines originating on a positive charge or terminating on a negative charge is proportional to the magnitude of the charge. That is, the net flux passing through a closed surface surrounding a charge Q is proportional to the magnitude of Q, or

$$\Phi_{net} = \Sigma EA \cos \theta \propto Q$$

In free space, the constant of proportionality is $1/\epsilon_0$ where ϵ_0 is called the permittivity of free space and has the value

$$\epsilon_0 = \frac{1}{4\pi k} = \frac{1}{4\pi(8.99 \times 10^9 \text{ N} \cdot \text{m}^2/\text{C}^2)} = 8.85 \times 10^{-12} \text{ C}^2/\text{N} \cdot \text{m}^2$$

Thus, the net electric flux through a closed surface is

Gauss' law

$$\Sigma EA \cos \theta = \frac{Q}{\epsilon_0}$$ [15a.3]

This result, known as **Gauss' law,** states that

> **The net electric flux through any closed gaussian surface is equal to the net charge inside the surface divided by ϵ_0.**

In principle, Gauss' law can always be used to calculate the electric field of a system of charges or a continuous distribution of charge. However, in practice, the technique is useful only in a limited number of situations where there is a high degree of symmetry.

To indicate the procedure for applying Gauss' law, consider the problem of calculating the electric field at a distance r from a positive point charge q. The first step is to construct a gaussian surface. To construct this surface, we must (1) consider the symmetry of the charge distribution, and (2) allow the gaussian surface to pass through the point at which we want to calculate the electric field. These two considerations lead us to the construction of a gaussian surface, which is a sphere with q at its center and having a radius of r as shown in Figure 15a.4.

Since there is only one surface to be considered, there is only one term in the sum on the left side of Gauss' law, and the left side of Equation 15a.3 can be expressed as $EA \cos \theta$. Furthermore, the electric field is perpendicular to the area A at all points, so $\cos \theta = \cos 0° = 1$ at all points on the gaussian surface. Thus, the left side of Equation 15a.3 reduces to

$$EA = E4\pi r^2$$

where $4\pi r^2$ is the surface area of the gaussian surface. Recall that the Q in Gauss' law is equal to the *net* charge enclosed by the gaussian surface, so the right side of Equation 15a.3 reduces to

$$\frac{Q}{\epsilon_0} = \frac{q}{\epsilon_0}$$

and we have

$$E4\pi r^2 = \frac{q}{\epsilon_0}$$

or

$$E = \frac{q}{4\pi r^2 \epsilon_0} = k\frac{q}{r^2},$$

which is the equation used to calculate the magnitude of an electric field set up by a point charge q used in Chapter 15.

The following examples should clarify the procedure for applying Gauss' law to some simple situations.

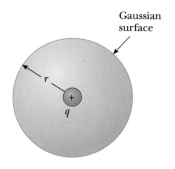

FIGURE 15a.4
A spherical surface of radius r surrounding a point charge q. When the charge is at the center of the sphere, the electric field is normal to the surface and constant in magnitude everywhere on the surface.

EXAMPLE 15a.2   The Electric Field of a Charged Thin Spherical Shell

A thin spherical shell of radius a has a total charge q distributed uniformly over its surface (Fig. 15a.5a). Find the electric field at points (a) outside and (b) inside the shell.

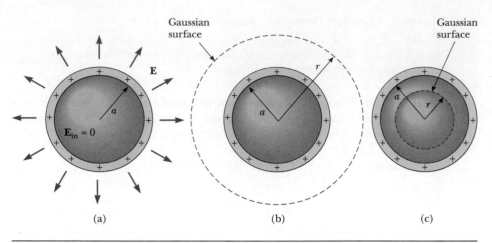

(a)   (b)   (c)

FIGURE 15a.5
(Example 15a.2)   (a) The electric field inside a uniformly charged spherical shell is *zero*. The field outside is the same as that of a point charge having a total charge Q located at the center of the shell. (b) The construction of a gaussian surface for calculating the electric field *outside* a spherical shell. (c) The construction of a gaussian surface for calculating the electric field *inside* a spherical shell.

Solution   (a) The calculation of the field outside the shell is identical to that carried out in the body of the text for a point charge. Since the charge distribution is spherically symmetric, we select a spherical gaussian surface of radius r, concentric with the shell, as in Figure 15a.5b. Following the same line of reasoning as that for a point charge, the left side of Gauss' law reduces to

$$\Sigma EA \cos \theta = E4\pi r^2$$

and the right side becomes

$$\frac{Q}{\epsilon_0} = \frac{q}{\epsilon_0}$$

Thus, we have

$$E4\pi r^2 = \frac{q}{\epsilon_0}$$

or

$$E = \frac{q}{4\pi r^2 \epsilon_0} = k\frac{q}{r^2}$$

Therefore, the field at a point outside the shell is equivalent to that of a point charge located at the center of the shell.

    (b) *The electric field inside the spherical shell is zero.* This follows from Gauss' law applied to a spherical gaussian surface of radius r placed inside the shell, as shown in Figure 15a.5c. Since the net charge inside the surface is zero, $Q = 0$, we see that $\mathbf{E} = 0$ in the region inside the shell.

EXAMPLE 15a.3 A Nonconducting Plane Sheet of Charge

Find the electric field due to a nonconducting infinite plane sheet of charge with uniform charge per unit area σ.

Solution The symmetry of the situation shows that the electric field must be perpendicular to the plane and that the direction of the field on one side of the plane must be opposite its direction on the other side, as shown in Figure 15a.6. It is convenient to choose for our gaussian surface a small cylinder whose axis is perpendicular to the plane and whose ends each have an area A. The left side of Gauss' law, $\Sigma EA \cos \theta$, now is the sum of three terms, for each of the three surfaces labeled (1), (2), and (3) in Figure 15a.6. Thus,

$$\Sigma EA \cos \theta = (EA \cos \theta)_1 + (EA \cos \theta)_2 + (EA \cos \theta)_3$$

But, we see that since **E** is parallel to the cylindrical surface, surface 1, $\cos \theta = \cos 90° = 0$, so $(EA \cos \theta)_1 = 0$.

Now consider surface (2). For this surface, we see that $\cos \theta = \cos 0° = 1$, and

$$(EA \cos \theta)_2 = EA$$

By the same reasoning, for surface (3),

$$(EA \cos \theta)_3 = EA$$

Therefore, the left side of Gauss' law becomes

$$(EA \cos \theta)_2 + (EA \cos \theta)_3 = 2EA$$

Since the total charge *inside* the gaussian surface is σA, the right side of Gauss' law becomes

$$\frac{Q}{\epsilon_0} = \frac{\sigma A}{\epsilon_0}$$

Applying Gauss' law to this situation gives

$$2EA = \frac{\sigma A}{\epsilon_0}$$

or,

$$E = \frac{\sigma}{2\epsilon_0}$$

Since the distance of the surfaces from the plane does not appear in our result, we conclude that $E = \sigma/\epsilon_0$ at *any* distance from the plane. That is, the electric field is uniform everywhere.

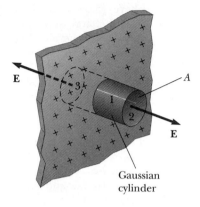

FIGURE 15a.6
(Example 15a.3) A cylindrical gaussian surface penetrating an infinite sheet of charge. The flux through each end of the gaussian surface is EA. There is no flux through the cylindrical surface.

PROBLEMS

1. A flat surface having an area of 3.2 m^2 is rotated in a uniform electric field of intensity $E = 6.2 \times 10^5$ N/C. (a) Determine the electric flux through this area when the electric field is perpendicular to the surface and (b) when the electric field is parallel to the surface. [(a) 1.98×10^6 N·m^2, (b) zero]

2. A 40-cm diameter loop is rotated in a uniform electric field until the position of maximum electric flux is found. The flux in this position is mea-

sured to be 5.2×10^5 N·m²/C. Calculate the electric field strength in this region. [4.1×10^6 N/C]

3. A point charge of $+5.00$ μC is located at the center of a sphere with a radius of 12.0 cm. Determine the electric flux through the surface of the sphere. [5.65×10^5 N·m²/C]

4. A point charge of magnitude q is located at the center of a spherical shell of radius a, which has a charge uniformly distributed on its surface of magnitude $-q$. Find the electric field (a) for all points outside the spherical shell and (b) for a point inside the shell a distance r from the center. [(a) zero, (b) $E = kq/r^2$]

5. Show that the electric field just outside the surface of a good conductor of any shape is given by $E = \sigma/\epsilon_0$, where σ is the charge per unit area on the conductor.

6. Use Gauss' law and the fact that the electric field inside any closed conductor in electrostatic equilibrium is zero to show that any excess charge placed on the conductor must reside on its surface.

7. An infinite plane conductor has charge spread out on its surface as shown in Figure 15a.7. Use Gauss'

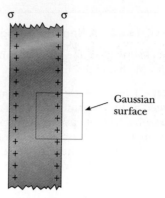

FIGURE 15a.7   (Problem 7)

law to show that the electric field at any point outside the conductor is given by $E = \sigma/\epsilon_0$, where σ is the charge per unit area on the conductor.
(*Hint:* Choose a gaussian surface in the shape of a cylinder with one end inside the conductor and one end outside the conductor.)

Appendix A

MATHEMATICAL REVIEW

A.1

SCIENTIFIC NOTATION

Many quantities that scientists deal with often have very large or very small values. For example, the speed of light is about 300 000 000 m/s and the ink required to make the dot over an i in this textbook has a mass of about 0.000 000 001 kg. Obviously, it is very cumbersome to read, write, and keep track of numbers such as these. We avoid this problem by using a method dealing with powers of the number 10:

$$10^0 = 1$$

$$10^1 = 10$$

$$10^2 = 10 \times 10 = 100$$

$$10^3 = 10 \times 10 \times 10 = 1000$$

$$10^4 = 10 \times 10 \times 10 \times 10 = 10\ 000$$

$$10^5 = 10 \times 10 \times 10 \times 10 \times 10 = 100\ 000$$

and so on. The number of zeros corresponds to the power to which 10 is raised, called the **exponent** of 10. For example, the speed of light, 300 000 000 m/s, can be expressed as 3×10^8 m/s.

For numbers less than one, we note the following:

$$10^{-1} = \frac{1}{10} = 0.1$$

$$10^{-2} = \frac{1}{10 \times 10} = 0.01$$

$$10^{-3} = \frac{1}{10 \times 10 \times 10} = 0.001$$

$$10^{-4} = \frac{1}{10 \times 10 \times 10 \times 10} = 0.0001$$

$$10^{-5} = \frac{1}{10 \times 10 \times 10 \times 10 \times 10} = 0.000\ 01$$

In these cases, the number of places the decimal point is to the left of the digit 1 equals the value of the (negative) exponent. Numbers that are expressed as some power of 10 multiplied by another number between 1 and 10 are said to be in **scientific notation**. For example, the scientific notation for 5 943 000 000 is 5.943×10^9 and that for 0.000 083 2 is 8.32×10^{-5}.

When numbers expressed in scientific notation are being multiplied, the following general rule is very useful:

$$10^n \times 10^m = 10^{n+m} \qquad \text{[A.1]}$$

where n and m can be *any* numbers (not necessarily integers). For example, $10^2 \times 10^5 = 10^7$. The rule also applies if one of the exponents is negative. For example, $10^3 \times 10^{-8} = 10^{-5}$.

When dividing numbers expressed in scientific notation, note that

$$\frac{10^n}{10^m} = 10^n \times 10^{-m} = 10^{n-m} \qquad \text{[A.2]}$$

Exercises

With help from the above rules, verify the answers to the following:

1. $86\ 400 = 8.64 \times 10^4$
2. $9\ 816\ 762.5 = 9.816\ 762\ 5 \times 10^6$
3. $0.000\ 000\ 039\ 8 = 3.98 \times 10^{-8}$
4. $(4.0 \times 10^8)(9.0 \times 10^9) = 3.6 \times 10^{18}$
5. $(3.0 \times 10^7)(6.0 \times 10^{-12}) = 1.8 \times 10^{-4}$
6. $\dfrac{75 \times 10^{-11}}{5.0 \times 10^{-3}} = 1.5 \times 10^{-7}$
7. $\dfrac{(3 \times 10^6)(8 \times 10^{-2})}{(2 \times 10^{17})(6 \times 10^5)} = 2 \times 10^{-18}$

A.2

ALGEBRA

A. SOME BASIC RULES

When algebraic operations are performed, the laws of arithmetic apply. Symbols such as x, y, and z are usually used to represent quantities that are not specified, what are called the **unknowns**.

First, consider the equation

$$8x = 32$$

If we wish to solve for x, we can divide (or multiply) each side of the equation by the same factor without destroying the equality. In this case, if we divide both sides by 8, we have

$$\frac{8x}{8} = \frac{32}{8}$$

$$x = 4$$

Next consider the equation

$$x + 2 = 8$$

In this type of expression, we can add or subtract the same quantity from each side. If we subtract 2 from each side, we get

$$x + 2 - 2 = 8 - 2$$

$$x = 6$$

In general, if $x + a = b$, then $x = b - a$.

Now consider the equation

$$\frac{x}{5} = 9$$

If we multiply each side by 5, we are left with x on the left by itself and 45 on the right:

$$\left(\frac{x}{5}\right)(5) = 9 \times 5$$

$$x = 45$$

In all cases, *whatever operation is performed on the left side of the equality must also be performed on the right side.*

The following rules for multiplying, dividing, adding, and subtracting fractions should be recalled, where a, b, and c are three numbers:

	Rule	Example
Multiplying	$\left(\dfrac{a}{b}\right)\left(\dfrac{c}{d}\right) = \dfrac{ac}{bd}$	$\left(\dfrac{2}{3}\right)\left(\dfrac{4}{5}\right) = \dfrac{8}{15}$
Dividing	$\dfrac{(a/b)}{(c/d)} = \dfrac{ad}{bc}$	$\dfrac{2/3}{4/5} = \dfrac{(2)(5)}{(4)(3)} = \dfrac{10}{12}$
Adding	$\dfrac{a}{b} \pm \dfrac{c}{d} = \dfrac{ad \pm bc}{bd}$	$\dfrac{2}{3} - \dfrac{4}{5} = \dfrac{(2)(5) - (4)(3)}{(3)(5)} = -\dfrac{2}{15}$

Exercises

In the following exercises, solve for x:

		Answers
1.	$a = \dfrac{1}{1 + x}$	$x = \dfrac{1 - a}{a}$
2.	$3x - 5 = 13$	$x = 6$
3.	$ax - 5 = bx + 2$	$x = \dfrac{7}{a - b}$
4.	$\dfrac{5}{2x + 6} = \dfrac{3}{4x + 8}$	$x = -\dfrac{11}{7}$

TABLE A.1
Rules of Exponents

$$x^0 = 1$$
$$x^1 = x$$
$$x^n x^m = x^{n+m}$$
$$x^n / x^m = x^{n-m}$$
$$x^{1/n} = \sqrt[n]{x}$$
$$(x^n)^m = x^{nm}$$

B. POWERS

When powers of a given quantity x are multiplied, the following rule applies:

$$x^n x^m = x^{n+m} \tag{A.3}$$

For example, $x^2 x^4 = x^{2+4} = x^6$.

When dividing the powers of a given quantity, note that

$$\frac{x^n}{x^m} = x^{n-m} \tag{A.4}$$

For example, $x^8 / x^2 = x^{8-2} = x^6$.

A power that is a fraction, such as $\frac{1}{3}$, corresponds to a root as follows:

$$x^{1/n} = \sqrt[n]{x} \tag{A.5}$$

For example, $4^{1/3} = \sqrt[3]{4} = 1.5874$. (A scientific calculator is useful for such calculations.)

Finally, any quantity x^n that is raised to the mth power is

$$(x^n)^m = x^{nm} \tag{A.6}$$

Table A.1 summarizes the rules of exponents.

Exercises

Verify the following:

1. $3^2 \times 3^3 = 243$
2. $x^5 x^{-8} = x^{-3}$
3. $x^{10} / x^{-5} = x^{15}$
4. $5^{1/3} = 1.709\,975$ (Use your calculator.)
5. $60^{1/4} = 2.783\,158$ (Use your calculator.)
6. $(x^4)^3 = x^{12}$

C. FACTORING

Some useful formulas for factoring an equation are

$$ax + ay + az = a(x + y + z) \qquad \text{common factor}$$

$$a^2 + 2ab + b^2 = (a + b)^2 \qquad \text{perfect square}$$

$$a^2 - b^2 = (a + b)(a - b) \qquad \text{differences of squares}$$

D. QUADRATIC EQUATIONS

The general form of a quadratic equation is

$$ax^2 + bx + c = 0 \tag{A.7}$$

where x is the unknown quantity and a, b, and c are numerical factors referred to as **coefficients** of the equation. This equation has two roots, given by

$$x = \frac{-b \pm \sqrt{b^2 - 4ac}}{2a} \tag{A.8}$$

If $b^2 \geq 4ac$, the roots will be real.

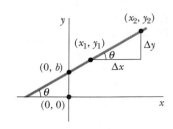

FIGURE A.1

EXAMPLE

The equation $x^2 + 5x + 4 = 0$ has the following roots corresponding to the two signs of the square-root term:

$$x = \frac{-5 \pm \sqrt{5^2 - (4)(1)(4)}}{2(1)} = \frac{-5 \pm \sqrt{9}}{2} = \frac{-5 \pm 3}{2}$$

that is,

$$x_+ = \frac{-5 + 3}{2} = \boxed{-1} \qquad x_- = \frac{-5 - 3}{2} = \boxed{-4}$$

where x_+ refers to the root corresponding to the positive sign and x_- refers to the root corresponding to the negative sign.

Exercises

Solve the following quadratic equations:

Answers

1. $x^2 + 2x - 3 = 0$	$x_+ = 1$	$x_- = -3$
2. $2x^2 - 5x + 2 = 0$	$x_+ = 2$	$x_- = 1/2$
3. $2x^2 - 4x - 9 = 0$	$x_+ = 1 + \sqrt{22}/2$	$x_- = 1 - \sqrt{22}/2$

E. LINEAR EQUATIONS

A linear equation has the general form

$$y = ax + \text{b} \qquad \text{[A.9]}$$

where a and b are constants. This equation is referred to as being linear because the graph of y versus x is a straight line, as shown in Figure A.1. The constant b, called the **intercept,** represents the value of y at which the straight line intersects the y axis. The constant a is equal to the **slope** of the straight line and is also equal to the tangent of the angle that the line makes with the x axis. If any two points on the straight line are specified by the coordinates (x_1, y_1) and (x_2, y_2), as in Figure A.1, then the *slope* of the straight line can be expressed

$$\text{Slope} = \frac{y_2 - y_1}{x_2 - x_1} = \frac{\Delta y}{\Delta x} \qquad \text{[A.10]}$$

Note that a and b can have either positive or negative values. If $a > 0$, the straight line has a *positive* slope, as in Figure A.1. If $a < 0$, the straight line has a *negative* slope. In Figure A.1, both a and b are positive. Three other possible situations are shown in Figure A.2: $a > 0$, $b < 0$; $a < 0$, $b > 0$; and $a < 0$, $b < 0$.

Exercises

1. Draw graphs of the following straight lines:
   (a) $y = 5x + 3$ (b) $y = -2x + 4$ (c) $y = -3x - 6$
2. Find the slopes of the straight lines described in Exercise 1.
   Answers: (a) 5 (b) -2 (c) -3
3. Find the slopes of the straight lines that pass through the following sets of points:
   (a) $(0, -4)$ and $(4, 2)$, (b) $(0, 0)$ and $(2, -5)$, and (c) $(-5, 2)$ and $(4, -2)$
   Answers: (a) $3/2$ (b) $-5/2$ (c) $-4/9$

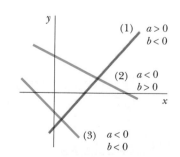

FIGURE A.2

F. SOLVING SIMULTANEOUS LINEAR EQUATIONS

Consider an equation such as $3x + 5y = 15$, which has two unknowns, x and y. Such an equation does not have a unique solution. That is, $(x = 0, y = 3)$, $(x = 5, y = 0)$, and $(x = 2, y = 9/5)$ are all solutions to this equation.

If a problem has two unknowns, a unique solution is possible only if we have *two* independent equations. In general, if a problem has n unknowns, its solution requires n independent equations. In order to solve two simultaneous equations involving two unknowns, x and y, we solve one of the equations for x in terms of y and substitute this expression into the other equation.

EXAMPLE

Solve the following two simultaneous equations:

$$(1)\quad 5x + y = -8 \qquad (2)\quad 2x - 2y = 4$$

Solution   From (2), $x = y + 2$. Substitution of this into (1) gives

$$5(y + 2) + y = -8$$
$$6y = -18$$
$$y = -3$$
$$x = y + 2 = \boxed{-1}$$

Alternate solution: Multiply each term in (1) by the factor 2 and add the result to (2):

$$10x + 2y = -16$$
$$\underline{2x - 2y = 4}$$
$$12x = -12$$
$$x = -1$$
$$y = x - 2 = \boxed{-3}$$

Two linear equations with two unknowns can also be solved by a graphical method. If the straight lines corresponding to the two equations are plotted in a conventional coordinate system, the intersection of the two lines represents the solution. For example, consider the two equations

$$x - y = 2$$
$$x - 2y = -1$$

These are plotted in Figure A.3. The intersection of the two lines has the coordinates $x = 5$, $y = 3$. This represents the solution to the equations. You should check this solution by the analytical technique discussed above.

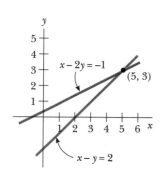

FIGURE A.3

Exercises

Solve the following pairs of simultaneous equations involving two unknowns:

Answers

1. $x + y = 8$ $x = 5$, $y = 3$
   $x - y = 2$
2. $98 - T = 10a$ $T = 65$, $a = 3.27$
   $T - 49 = 5a$
3. $6x + 2y = 6$ $x = 2$, $y = -3$
   $8x - 4y = 28$

G. LOGARITHMS

Suppose that a quantity x is expressed as a power of some quantity a:

$$x = a^y \qquad \text{[A.11]}$$

The number a is called the **base** number. The **logarithm** of x with respect to the base a is equal to the exponent to which the base must be raised in order to satisfy the expression $x = a^y$:

$$y = \log_a x \qquad \text{[A.12]}$$

Conversely, the **antilogarithm** of y is the number x:

$$x = \text{antilog}_a y \qquad \text{[A.13]}$$

In practice, the two bases most often used are base 10, called the *common* logarithm base, and base $e = 2.718 \ldots$, called the *natural* logarithm base. When common logarithms are used,

$$y = \log_{10} x \qquad (\text{or } x = 10^y) \qquad \text{[A.14]}$$

When natural logarithms are used,

$$y = \ln_e x \qquad (\text{or } x = e^y) \qquad \text{[A.15]}$$

For example, $\log_{10} 52 = 1.716$, so that $\text{antilog}_{10} 1.716 = 10^{1.716} = 52$. Likewise, $\ln_e 52 = 3.951$, so $\text{antiln}_e 3.951 = e^{3.951} = 52$.

In general, note that you can convert between base 10 and base e with the equality

$$\ln_e x = (2.302\ 585) \log_{10} x \qquad \text{[A.16]}$$

Finally, some useful properties of logarithms are

$$\log (ab) = \log a + \log b \qquad \ln e = 1$$

$$\log (a/b) = \log a - \log b \qquad \ln e^a = a$$

$$\log (a^n) = n \log a \qquad \ln \left(\frac{1}{a}\right) = -\ln a$$

A.3

GEOMETRY

Table A.2 gives the areas and volumes for several geometric shapes used throughout this text:

TABLE A.2
Useful Information for Geometry

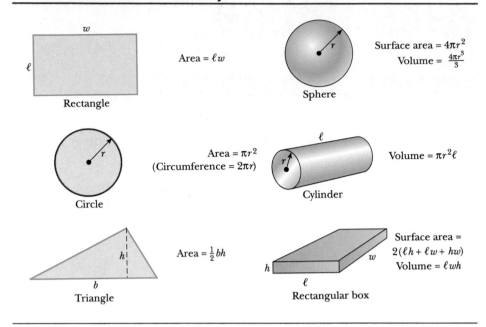

Rectangle — $\text{Area} = \ell w$

Sphere — $\text{Surface area} = 4\pi r^2$, $\text{Volume} = \frac{4\pi r^3}{3}$

Circle — $\text{Area} = \pi r^2$ ($\text{Circumference} = 2\pi r$)

Cylinder — $\text{Volume} = \pi r^2 \ell$

Triangle — $\text{Area} = \frac{1}{2} bh$

Rectangular box — $\text{Surface area} = 2(\ell h + \ell w + hw)$, $\text{Volume} = \ell wh$

A.4

TRIGONOMETRY

Some of the most basic facts concerning trigonometry are presented in Chapter 1, and we encourage you to study the material presented there if you are having trouble with this branch of mathematics. In addition to the discussion of Chapter 1, certain useful trig identities that can be of value to you follow.

$$\sin^2 \theta + \cos^2 \theta = 1$$

$$\sin \theta = \cos(90° - \theta)$$

$$\cos \theta = \sin(90° - \theta)$$

$$\sin 2\theta = 2 \sin \theta \cos \theta$$

$$\cos 2\theta = \cos^2 \theta - \sin^2 \theta$$

$$\sin(\theta \pm \phi) = \sin \theta \cos \phi \pm \cos \theta \sin \phi$$

$$\cos(\theta \pm \phi) = \cos \theta \cos \phi \mp \sin \theta \sin \phi$$

Appendix B

AN ABBREVIATED TABLE OF ISOTOPES

Atomic Number, Z	Element	Symbol	Mass Number, A	Atomic Mass[a]	Percent Abundance, or Decay Mode if Radioactive[b]	Half-life if Radioactive
0	(Neutron)	n	1	1.008 665	β^-	10.6 min
1	Hydrogen	H	1	1.007 825	99.985	
	Deuterium	D	2	2.014 102	0.015	
	Tritium	T	3	3.016 049	β^-	12.33 y
2	Helium	He	3	3.016 029	0.000 14	
			4	4.002 603	≈ 100	
3	Lithium	Li	6	6.015 123	7.5	
			7	7.016 005	92.5	
4	Beryllium	Be	7	7.016 930	EC, γ	53.3 days
			8	8.005 305	2α	6.7×10^{-17} s
			9	9.012 183	100	
5	Boron	B	10	10.012 938	19.8	
			11	11.009 305	80.2	
6	Carbon	C	11	11.011 433	β^+, EC	20.4 min
			12	12.000 000	98.89	
			13	13.003 355	1.11	
			14	14.003 242	β^-	5730 y
7	Nitrogen	N	13	13.005 739	β^+	9.96 min
			14	14.003 074	99.63	
			15	15.000 109	0.37	
8	Oxygen	O	15	15.003 065	β^+, EC	122 s
			16	15.994 915	99.76	
			18	17.999 159	0.204	
9	Fluorine	F	19	18.998 403	100	
10	Neon	Ne	20	19.992 439	90.51	
			22	21.991 384	9.22	
11	Sodium	Na	22	21.994 435	β^+, EC, γ	2.602 y
			23	22.989 770	100	
			24	23.990 964	β^-, γ	15.0 h

(Table continues)

Note: Data taken from *Chart of the Nuclides,* 12th ed., New York, General Electric, 1977, and from C. M. Lederer and V. S. Shirley (eds.), *Table of Isotopes,* 7th ed., New York, Wiley, 1978.
[a] Masses are those for the neutral atom, including the Z electrons, in u.
[b] The process EC stands for "electron capture."

Atomic Number, Z	Element	Symbol	Mass Number, A	Atomic Mass[a]	Percent Abundance, or Decay Mode if Radioactive[b]	Half-life if Radioactive
12	Magnesium	Mg	24	23.985 045	78.99	
13	Aluminum	Al	27	26.981 541	100	
14	Silicon	Si	28	27.976 928	92.23	
			31	30.975 364	β^-, γ	2.62 h
15	Phosphorus	P	31	30.973 763	100	
			32	31.973 908	β^-	14.28 days
16	Sulfur	S	32	31.972 072	95.0	
			35	34.969 033	β^-	87.4 days
17	Chlorine	Cl	35	34.968 853	75.77	
			37	36.965 903	24.23	
18	Argon	Ar	40	39.962 383	99.60	
19	Potassium	K	39	38.963 708	93.26	
			40	39.964 000	β^-, EC, γ, β^+	1.28×10^9 y
20	Calcium	Ca	40	39.962 591	96.94	
21	Scandium	Sc	45	44.955 914	100	
22	Titanium	Ti	48	47.947 947	73.7	
23	Vanadium	V	51	50.943 963	99.75	
24	Chromium	Cr	52	51.940 510	83.79	
25	Manganese	Mn	55	54.938 046	100	
26	Iron	Fe	56	55.934 939	91.8	
27	Cobalt	Co	59	58.933 198	100	
			60	59.933 820	β^-, γ	5.271 y
28	Nickel	Ni	58	57.935 347	68.3	
			60	59.930 789	26.1	
			64	63.927 968	0.91	
29	Copper	Cu	63	62.929 599	69.2	
			64	63.929 766	β^-, β^+	12.7 h
			65	64.927 792	30.8	
30	Zinc	Zn	64	63.929 145	48.6	
			66	65.926 035	27.9	
31	Gallium	Ga	69	68.925 581	60.1	
32	Germanium	Ge	72	71.922 080	27.4	
			74	73.921 179	36.5	
33	Arsenic	As	75	74.921 596	100	
34	Selenium	Se	80	79.916 521	49.8	
35	Bromine	Br	79	78.918 336	50.69	
36	Krypton	Kr	84	83.911 506	57.0	
			89	88.917 563	β^-	3.2 min
37	Rubidium	Rb	85	84.911 800	72.17	
38	Strontium	Sr	86	85.909 273	9.8	
			88	87.905 625	82.6	
			90	89.907 746	β^-	28.8 y
39	Yttrium	Y	89	88.905 856	100	
40	Zirconium	Zr	90	89.904 708	51.5	
41	Niobium	Nb	93	92.906 378	100	
42	Molybdenum	Mo	98	97.905 405	24.1	
43	Technetium	Tc	98	97.907 210	β^-, γ	4.2×10^6 y
44	Ruthenium	Ru	102	101.904 348	31.6	

(Table continues)

Atomic Number, Z	Element	Symbol	Mass Number, A	Atomic Mass[a]	Percent Abundance, or Decay Mode if Radioactive[b]	Half-life if Radioactive
45	Rhodium	Rh	103	102.905 50	100	
46	Palladium	Pd	106	105.903 48	27.3	
47	Silver	Ag	107	106.905 095	51.83	
			109	108.904 754	48.17	
48	Cadmium	Cd	114	113.903 361	28.7	
49	Indium	In	115	114.903 88	95.7; β^-	5.1×10^{14} y
50	Tin	Sn	120	119.902 199	32.4	
51	Antimony	Sb	121	120.903 824	57.3	
52	Tellurium	Te	130	129.906 23	34.5; β^-	2×10^{21} y
53	Iodine	I	127	126.904 477	100	
			131	130.906 118	β^-, γ	8.04 days
54	Xenon	Xe	132	131.904 15	26.9	
			136	135.907 22	8.9	
55	Cesium	Cs	133	132.905 43	100	
56	Barium	Ba	137	136.905 82	11.2	
			138	137.905 24	71.7	
			144	143.922 673	β^-	11.9 s
57	Lanthanum	La	139	138.906 36	99.911	
58	Cerium	Ce	140	139.905 44	88.5	
59	Praesodymium	Pr	141	140.907 66	100	
60	Neodymium	Nd	142	141.907 73	27.2	
			144	143.910 096	α, 23.8	2.1×10^{15} y
61	Promethium	Pm	145	144.912 75	EC, α, γ	17.7 y
62	Samarium	Sm	152	151.919 74	26.6	
63	Europium	Eu	153	152.921 24	52.1	
64	Gadolinium	Gd	158	157.924 11	24.8	
65	Terbium	Tb	159	158.925 35	100	
66	Dysprosium	Dy	164	163.929 18	28.1	
67	Holmium	Ho	165	164.930 33	100	
68	Erbium	Er	166	165.930 31	33.4	
69	Thulium	Tm	169	168.934 23	100	
70	Ytterbium	Yb	174	173.938 87	31.6	
71	Lutecium	Lu	175	174.940 79	97.39	
72	Hafnium	Hf	180	179.946 56	35.2	
73	Tantalum	Ta	181	180.948 01	99.988	
74	Tungsten	W	184	183.950 95	30.7	
75	Rhenium	Re	187	186.955 77	62.60, β^-	4×10^{10} y
76	Osmium	Os	191	190.960 94	β^-, γ	15.4 days
			192	191.961 49	41.0	
77	Iridium	Ir	191	190.960 60	37.3	
			193	192.962 94	62.7	
78	Platinum	Pt	195	194.964 79	33.8	
79	Gold	Au	197	196.966 56	100	
80	Mercury	Hg	202	201.970 63	29.8	
81	Thallium	Tl	205	204.974 41	70.5	
			210	209.990 069	β^-	1.3 min
82	Lead	Pb	204	203.973 044	β^-, 1.48	1.4×10^{17} y
			206	205.974 46	24.1	

(Table continues)

Atomic Number, Z	Element	Symbol	Mass Number, A	Atomic Mass[a]	Percent Abundance, or Decay Mode if Radioactive[b]	Half-life if Radioactive
			207	206.975 89	22.1	
			208	207.976 64	52.3	
			210	209.984 18	α, β^-, γ	22.3 y
			211	210.988 74	β^-, γ	36.1 min
			212	211.991 88	β^-, γ	10.64 h
			214	213.999 80	β^-, γ	26.8 min
83	Bismuth	Bi	209	208.980 39	100	
			211	210.987 26	α, β^-, γ	2.15 min
			214	213.998 702	β^-, α	19.7 min
84	Polonium	Po	210	209.982 86	α, γ	138.38 days
			214	213.995 19	α, γ	164 μs
85	Astatine	At	218	218.008 70	α, β^-	$\approx$2 s
86	Radon	Rn	222	222.017 574	α, γ	3.8235 days
87	Francium	Fr	223	223.019 734	α, β^-, γ	21.8 min
88	Radium	Ra	226	226.025 406	α, γ	1.60×10^3 y
			228	228.031 069	β^-	5.76 y
89	Actinium	Ac	227	227.027 751	α, β^-, γ	21.773 y
90	Thorium	Th	228	228.028 73	α, γ	1.9131 y
			232	232.038 054	100, α, γ	1.41×10^{10} y
91	Protactinium	Pa	231	231.035 881	α, γ	3.28×10^4 y
92	Uranium	U	232	232.037 14	α, γ	72 y
			233	233.039 629	α, γ	1.592×10^5 y
			235	235.043 925	0.72; α, γ	7.038×10^8 y
			236	236.045 563	α, γ	2.342×10^7 y
			238	238.050 786	99.275; α, γ	4.468×10^9 y
			239	239.054 291	β^-, γ	23.5 min
93	Neptunium	Np	239	239.052 932	β^-, γ	2.35 days
94	Plutonium	Pu	239	239.052 158	α, γ	2.41×10^4 y
95	Americium	Am	243	243.061 374	α, γ	7.37×10^3 y
96	Curium	Cm	245	245.065 487	α, γ	8.5×10^3 y
97	Berkelium	Bk	247	247.070 03	α, γ	1.4×10^3 y
98	Californium	Cf	249	249.074 849	α, γ	351 y
99	Einsteinium	Es	254	254.088 02	α, γ, β^-	276 days
100	Fermium	Fm	253	253.085 18	EC, α, γ	3.0 days
101	Mendelevium	Md	255	255.0911	EC, α	27 min
102	Nobelium	No	255	255.0933	EC, α	3.1 min
103	Lawrencium	Lr	257	257.0998	α	$\approx$35 s
104	Unnilquadium	Rf	261	261.1087	α	1.1 min
105	Unnilpentium	Ha	262	262.1138	α	0.7 min
106	Unnilhexium		263	263.1184	α	0.9 s
107	Unnilseptium		261		α	1–2 ms

Appendix C

SOME USEFUL TABLES

TABLE C.1
Mathematical Symbols Used in the Text and Their Meaning

Symbol	Meaning
$=$	is equal to
$\neq$	is not equal to
$\equiv$	is defined as
$\propto$	is proportional to
$>$	is greater than
$<$	is less than
$\gg$	is much greater than
$\ll$	is much less than
$\approx$	is approximately equal to
Δx	change in x or uncertainty in x
Σx_i	sum of all quantities x_i
$\mid x \mid$	absolute value of x (always a positive quantity)

TABLE C.2
Standard Abbreviations of Units

Abbreviation	Unit	Abbreviation	Unit
A	ampere	J	joule
Å	angstrom	K	kelvin
atm	atmosphere	kcal	kilocalorie
Btu	British thermal unit	kg	kilogram
C	coulomb	km	kilometer
°C	degree Celsius	kmol	kilomole
cal	calorie	lb	pound
cm	centimeter	m	meter
deg	degree (angle)	min	minute
eV	electron volt	N	newton
°F	degree Fahrenheit	rev	revolution
ft	foot	s	second
G	gauss	T	tesla
g	gram	u	atomic mass unit
H	henry	V	volt
h	hour	W	watt
hp	horsepower	Wb	weber
Hz	hertz	μm	micrometer
in.	inch	Ω	ohm

TABLE C.3
The Greek Alphabet

Alpha	A	α	Iota	I	ι	Rho	P	ρ
Beta	B	β	Kappa	K	κ	Sigma	Σ	σ
Gamma	Γ	γ	Lambda	Λ	λ	Tau	T	τ
Delta	Δ	δ	Mu	M	μ	Upsilon	Y	υ
Epsilon	E	ϵ	Nu	N	ν	Phi	Φ	ϕ
Zeta	Z	ζ	Xi	Ξ	ξ	Chi	X	χ
Eta	H	η	Omicron	O	o	Psi	Ψ	ψ
Theta	Θ	θ	Pi	Π	π	Omega	Ω	ω

TABLE C.4
Physical Data Often Used[a]

Average Earth-Moon distance	3.84×10^8 m
Average Earth-Sun distance	1.496×10^{11} m
Average radius of the Earth	6.37×10^6 m
Density of air (20°C and 1 atm)	1.20 kg/m^3
Density of water (20°C and 1 atm)	1.00×10^3 kg/m^3
Free-fall acceleration	9.80 m/s^2
Mass of the Earth	5.98×10^{24} kg
Mass of the Moon	7.36×10^{22} kg
Mass of the Sun	1.99×10^{30} kg
Standard atmospheric pressure	1.013×10^5 Pa

[a] These are the values of the constants as used in the text.

Appendix D

SI UNITS

TABLE D.1
SI Base Units

Base Quantity	SI Base Unit	
	Name	Symbol
Length	Meter	m
Mass	Kilogram	kg
Time	Second	s
Electric current	Ampere	A
Temperature	Kelvin	K
Amount of substance	Mole	mol
Luminous intensity	Candela	cd

TABLE D.2
Derived SI Units

Quantity	Name	Symbol	Expression in Terms of Base Units	Expression in Terms of Other SI Units
Plane angle	Radian	rad	m/m	
Frequency	Hertz	Hz	s^{-1}	
Force	Newton	N	$kg \cdot m/s^2$	J/m
Pressure	Pascal	Pa	$kg/m \cdot s^2$	N/m^2
Energy: work	Joule	J	$kg \cdot m^2/s^2$	$N \cdot m$
Power	Watt	W	$kg \cdot m^2/s^3$	J/s
Electric charge	Coulomb	C	$A \cdot s$	
Electric potential (emf)	Volt	V	$kg \cdot m^2/A \cdot s^3$	$W/A, J/C$
Capacitance	Farad	F	$A^2 \cdot s^4/kg \cdot m^2$	C/V
Electric resistance	Ohm	Ω	$kg \cdot m^2/A^2 \cdot s^3$	V/A
Magnetic flux	Weber	Wb	$kg \cdot m^2/A \cdot s^2$	$V \cdot s, T \cdot m^2$
Magnetic field intensity	Tesla	T	$kg/A \cdot s^2$	Wb/m^2
Inductance	Henry	H	$kg \cdot m^2/A^2 \cdot s^2$	Wb/A

ANSWERS TO ODD-NUMBERED PROBLEMS

Chapter 1

1. Based on units alone, the equation might be valid.
3. (a) MLT^{-2} (b) $kg \cdot m/s^2$
5. $h = \dfrac{4R^3}{3r^2}$, $[h] = \dfrac{L^1}{L^2} = L$
7. (a) 3.00×10^8 m/s (b) 2.9979×10^8 m/s
   (c) $2.997\,925 \times 10^8$ m/s
9. (a) 797 (b) 11 (c) 17.8
11. (a) 22 cm (b) 67.9 cm²
13. It will require about 47.5 years to count the money.
15. 3.16×10^9 y
17. 2.95×10^2 m³, 2.95×10^8 cm³
19. 35.7 m²
21. 2.9 cm
23. (a) 1 mi/h = 1.609 km/h (b) 88.5 km/h
    (c) 16.1 km/h
25. 6.71×10^8 mi
27. 10^{10} lb, 4.45×10^{10} N
29. 1.25×10^{10} lb, 4.17×10^7 head, (assuming 0.25 lb
    per burger and 300 lb net of meat per head of cattle)
31. 200 000 balls
33. 2.2 m
35. 8.1 cm
37. 8.60 m
39. (a) 1.5 m (b) 2.6 m
41. (a) 1 megaphone (b) 1 gigalo (c) 1 dekaration
    (d) 2 megacycles (e) 1 terapin (f) 1 dekadent
    (g) 2 kilomockingbirds (h) 1 microphone
    (i) 1 nanogoat
43. $V = \left(0.579 \dfrac{ft^3}{s} \right) t + \left(1.19 \times 10^{-9} \dfrac{ft^3}{s^2} \right) t^2$
45. (a) 127 y (b) 15 500 times
47. 6.97 m

Chapter 2

1. 12.2 mph
3. 2.4×10^4 s = 6.67 h
5. (a) 2.50 m/s (b) -2.27 m/s (c) 0
7. (a) 126 s (b) 12.6 m
9. (a) negative (b) positive (c) 0 (d) 0
11. (b) 41.03 m/s, 41.008 m/s, 41.002 m/s (c) 17 m/s
13. (a) 4.00 m/s (b) -0.50 m/s (c) -1.00 m/s (d) 0
15. (a) 2.34 min (b) 64.17 miles
17. 273 502 m/s² = 27 908 g
19. 8.60 m/s
21. (a) 5510 m (b) 20.8 m/s, 41.6 m/s, 20.8 m/s, 38.7 m/s
23. (a) 8.00 m/s² (b) 11.0 m/s²
25. -3.60 m/s²
27. (a) 2.32 m/s² (b) 14.4 s
29. (a) 12.5 s (b) -2.29 m/s² (c) 13.1 s
31. (a) 20 s (b) No, the minimum distance to stop =
    1000 m.
33. 1.50 m/s, 32.3 m
35. (a) 8.20 s (b) 134 m
37. (a) 2.88 s (b) 0.120 s (c) 40.7 m
39. (a) 308 m (b) 8.52 s (c) 16.5 s
41. 3.94 s
43. 941 m
45. (a) -21.1 m/s (b) 19.6 m (c) -18.1 m/s, 19.6 m
47. (a) 3.00 s (b) -24.5 m/s for each ball (c) 23.5 m
49. 4.16 m/s
51. 0.60 s
53. (a) 3.0 s (b) -15 m/s (c) -31 m/s, -35 m/s
55. (a) 6.5 s after sports car starts (b) 334 ft (c) v(sports
    car) = 103 ft/s, v(stock car) = 90 ft/s
57. (a) 5.0 s, 85 s (b) 200 ft/s (c) 18 500 ft from starting
    point (d) 10 s after starting to slow down (total trip
    time = 100 s)

Chapter 3

1. 1.31 km north and 2.81 km east
3. (a) 5.00 units, at an angle of 53° below the x axis.
   (b) 5.00 units, 53° above x axis.
5. 421 ft, 3.0° below horizontal
7. 83 m, 33° north of west

9. 103 ft horizontally, 86.8 ft vertically
11. 42.7 yd
13. 61.8 m at 14.0° from original line of travel, or 25.0 m at 36.9° from original line of travel
15. (a) 185 N at 77.8° from x axis (b) 185 N at 257.8°
17. (a) 35.1° or 54.9° (b) 42 m or 85 m, respectively
19. (a) clears by 0.85 m (b) $v_y = -13$ m/s falling
21. 29.4 m/s
23. 2.65 ft, 0.807 m
25. 7.5 min
27. (a) 22.3 s (b) 200 s
29. (a) (i) 1.2 m/s, (ii) 0 (b) 0.96 m (c) 0.50 m/s
31. (a) $v_{bs} = 10.1$ m/s at 8.53° east of north (b) 45.0 m
33. (a) 14.5° north of west (b) 194 m/s
35. 15.3 m
37. The swimmer that swims perpendicular to the stream's velocity returns first.
39. (a) 2.66 m/s (b) 0.64 m
41. 18 m on Moon, 7.9 m on Mars
43. 68.6 km/h
45. 14 m/s
47. 4.12 m (horizontal distance)
49. Answer given with problem.
51. less than 265 m or more than 3480 m from the western shore
53. Answer given with problem.
55. 10.4 m/s

Chapter 4

1. 7.4 min
3. 1.1×10^4 N upward
5. 13 s
7. (a) 12 N (b) 3.0 m/s²
9. 25 N
11. 310 N
13. (a) 789 N at 8.8° to right of forward direction (b) 0.266 m/s² in direction of resultant force
15. 1040 N rearward
17. 77.8 N in each wire
19. (a) $T_1 = T_2 = T_3 = Mg/2$, $T_4 = 3Mg/2$, and $T_5 = Mg$ (b) $F_A = Mg/2$
21. (a) 1.5 m (b) 1.4 m
23. 2.1 m/s² at 74° north of west.
25. $\mathbf{n} = \dfrac{m_2}{m_1 + m_2}\,\mathbf{F}$ when $\mathbf{F}$ is applied to m_1 and

   $\mathbf{n} = \dfrac{m_1}{m_1 + m_2}\,\mathbf{F}$ when $\mathbf{F}$ is applied to m_2
27. 7.90 m/s
29. 550 N
31. 100 N, 204 N
33. 4.43 m/s² up the incline, 53.7 N
35. 236 N (upper rope), 118 N (lower rope)
37. (a) 36.8 N (b) 2.45 m/s² (c) 1.23 m

39. $M_{max} = 0.612$ kg; answer would be the same on the Moon.
41. 3.16 s
43. (a) Slippage occurs between the bottom block and table first. (b) Must have $\mu > 0.400$ for slippage to occur between the two blocks first.
45. 72.0 N
47. (a) 0.366 m/s², No, the retarding force (friction plus tangential component of weight) exceeds the component of the applied force along the incline.
49. (a) 98.6 m (b) 16.4 m
51. 19.0 m
53. $\mu = \dfrac{F}{(3m_1 + m_2)g}$
55. (a) 2.13 s (b) 1.67 m
57. (a) -1.20 m/s² (b) $\mu_k = 0.122$ (c) 45.0 m
59. 104 N
61. 3320 N at 2.68° clockwise from the 3000 N force
63. $\mathbf{F} > 0$ for $t < 1.25$ s, $\mathbf{F} = 0$ for $t = 1.25$ s, $\mathbf{F} < 0$ for $t > 1.25$ s
65. 21.5 N
67. 7500 N, 50.0 m
69. $\mu_k = 0.813$
71. (a) 1.63 m/s² (b) 57.2 N for the string connecting the 5-kg and 4-kg block. $T = 24.5$ N for the string connecting the 4-kg and 3-kg block.
73. (a) 50.0 N (b) $\mu_s = 0.500$ (c) 25.0 N
75. (a) 1.78 m/s² (b) $\mu_k = 0.368$ (c) f = 9.37 N (d) 2.67 m/s
77. 0.69 m/s²
79. (b) 9.8 N, 0.58 m/s²
81. (b) 2-kg block: 5.7 m/s² to left
    3-kg block: 5.7 m/s² to right
    10-kg block: 5.7 m/s² downward
    (c) 17 N, 41 N
83. (a) 2.00 m/s² (b) 4.00 N, 6.00 N, and 8.00 N (all to right) (c) the force of m_2 on m_3 = 8 N, the force of m_1 on m_2 = 14 N
85. Answer given with problem.

Chapter 5

1. 1.50×10^7 J
3. 30.6 m
5. 1.6×10^3 J
7. 4.7×10^3 J
9. (a) 900 J (b) $\mu_k = 0.383$
11. 90.0 J
13. (a) 1.2 J (b) 5.0 m/s (c) 6.3 J
15.

v	KE (relativity)	KE (classical)
(a) 1.00×10^5 m/s	$5.000\,000\,4 \times 10^{10}$ J	5.00×10^{10} J
(b) 1.00×10^6 m/s	$5.000\,04 \times 10^{12}$ J	5.00×10^{12} J
(c) 1.00×10^7 m/s	$5.004\,17 \times 10^{14}$ J	5.00×10^{14} J
(d) 1.00×10^8 m/s	5.4594×10^{16} J	5.00×10^{16} J
(e) 1.00×10^9 m/s	undefined	5.00×10^{18} J

17. 160 m/s
19. (a) −170 J (b) 500 J (c) 150 J (d) 5.6 m/s
21. 2.0 m
23. (a) −19.6 J (b) 39.2 J (c) 0
25. 26.5 m/s
27. (a) −147 J (b) −147 J (c) −147 J, The results should be the same because gravitational forces are conservative.
29. (a) 544 N/m (b) 19.7 m/s
31. 5.1 m
33. (a) 10.9 m/s (b) 11.6 m/s
35. (a) 8.85 m/s (b) 54.1%
37. 77 m/s
39. (a) 2.29 m/s (b) 15.6 J
41. 1.5 m
43. 900 J
45. (a) 7.50×10^4 J (b) 33.5 hp (c) 44.7 hp
47. (a) 42 hp (b) 85 hp
49. 1.52×10^3 N
51. (a) 0.408 m/s (b) 2.45×10^3 J
53. (a) 7.92 hp (b) 14.9 hp
55. (a) 22.5 J (b) 6.71 m/s
57. (a) 2.29 m/s (b) 3.91 m/s (c) 4.50 m/s
59. (a) 3.13 m/s (b) 4.43 m/s (c) 1.00 m higher
61. (a) 9.90 m/s (b) 7.67 m/s
63. (a) 1.3 m (b) 0.46 m
65. (a) 9.90 m/s (b) 11.8 J
67. 9.80 m/s
69. (a) 4.4 m/s (b) 1.5×10^5 N
71. 4.89×10^5 J
73. (a) 0.883 m (b) 0.117 m to the right of point A
75. (a) mgh (b) mgh (c) $mgh + \frac{1}{2} mv_0^2$
77. 1.45 m
79. (a) 6.15 m/s (b) 9.87 m/s
81. (a) 0.225 J (b) 0.363 J (c) No, because the normal force and the force of friction change with the position of the object in the bowl.
83. Answer is given with the problem.

Chapter 6

1. (a) 8.35×10^{-21} kg·m/s (b) 4.50 kg·m/s (c) 750 kg·m/s (d) 1.78×10^{29} kg·m/s
3. (a) 31.0 m/s (b) the bullet, 337 J versus 74.2 J
5. 15.0 N
7. 260 N
9. (a) 13.5 N·s (b) 9000 N
11. 2.4 N
13. (a) 9.0 m/s (b) −15 m/s
15. 65 m/s
17. 120 m
19. (a) 1.15 m/s (b) 0.346 m/s in the direction opposite to the girl's velocity
21. (a) 9.0×10^{-24} m/s (b) The recoil velocity of the

Earth is essentially zero in comparison to the velocity of terrestrial objects.
23. $\mu_k = 0.410$
25. 3.00 m/s
27. 6.00 kg
29. −40.0 cm/s, 10.0 cm/s
31. (a) 6.0×10^5 m/s, 1.6×10^6 m/s (b) $KE_{\text{alpha after}} = 1.20 \times 10^{-15}$ J, $KE_{\text{proton after}} = 2.14 \times 10^{-15}$ J, $KE_{\text{alpha before}} = 3.34 \times 10^{-15}$ J, $KE_{\text{proton before}} = 0$
33. (a) 12.4 m/s at 14.9° north of east (b) 7.2%
35. (a) 9.90 m/s, −9.90 m/s; (b) − 16.5 m/s, + 3.3 m/s; (c) 13.9 m, 0.56 m; (d) After a second collision, the two objects return to their initial states and the process starts over. (e) 0.56 m
37. $v_{\text{orange}} = 4.00$ m/s, $v_{\text{yellow}} = 3.00$ m/s
39. 0.398%
41. 15 kg·m/s directed opposite to the initial velocity
43. (b) m_1 cannot be greater than m_2
45. 1.1×10^3 N
47. (a) 0, 1.50 m/s; (b) −1.00 m/s, 1.50 m/s; (c) 1.00 m/s, 1.50 m/s
49. 240 s (4 min)
51. 5.59 m/s
53. 0.980 m
55. (a) 8.0 N·s (b) 5.3 m/s (c) 3.3 m/s
57. (a) 1.1 m/s at 29.7° from +x direction (b) 0.32 or 32%
59. 1.9 m
61. 528 m/s
63. (a) 300 m/s (b) 3.75 m/s (c) 1.20 m
65. (a) 6.29 m/s (b) 6.16 m/s
67. (a) 0, 3.0 m/s (b) 0.21 m

Chapter 7

1. 0.52 rad, 0.79 rad, 1.1 rad, 3.1 rad, 4.7 rad, 6.3 rad
3. 2.2 m, 120 m, 770 m
5. 1.99×10^{-7} rad/s, 0.986 deg/day
7. 4.2×10^{-2} rad/s²
9. (a) 8.22×10^2 rad/s² (b) 4.21×10^3 rad
11. (a) 0.14 rad/s² (b) 8.4 rad/s
13. 50.0 rev
15. 119 rev
17. (a) 14.1 m/s (b) 200 m (c) 28.3 s
19. (a) 9.8 N (b) 9.8 N (c) 6.3 m/s
21. 36.5 rev
23. (a) The second sling gives the larger speed. (b) 1.5×10^3 m/s² (c) 1.3×10^3 m/s²
25. (a) 3.37×10^{-2} m/s² (b) 0
27. (a) 3.5×10^{-1} m/s² (b) 1.0 m/s (c) 3.5×10^{-1} m/s², 0.94 m/s², 1.0 m/s², at 20° with respect to the direction of a_r
29. (a) 1.58 m/s² (b) 455 N (c) 329 N (d) 397 N at 9.16° from the vertical
31. 12 m/s
33. (a) 2.49×10^4 N (b) 12.1 m/s

35. (a) 1.46×10^2 m (b) 5490 N
37. 4.08 km
39. 3.13 m/s
41. (a) 9.58×10^6 m (b) 5.55 h
43. (a) 4.23×10^7 m (b) 3.59×10^7 m (22 300 miles)
45. (a) 4.39×10^{20} N (b) 1.99×10^{20} N (c) 3.55×10^{22} N
47. (a) 1.63 m/s^2 (b) 1.68×10^3 m/s (c) 108 min
49. 9.00 m/s^2
51. 1.50×10^{11} m (from Sun)
53. (a) 2.51 m/s (b) 7.90 m/s^2 (c) 4.00 m/s
55. (a) 156° (b) 6.00 rad/s
57. (a) 126 rad/s (b) 2.51 m/s (c) 947 m/s^2 (d) 15.1 m
59. 8.3 s
61. (a) 8.42 N (b) 64.8° (c) 1.67 N
63. 9.0 rad/s

65. $w_{apparent} = w_{true} - \dfrac{mv^2}{r}$, so $w_{apparent} < w_{true}$ (b) 733 N

(equator), 735 N (poles)

67. (a) $v_0 = \sqrt{g\left(R - \dfrac{2h}{3}\right)}$ (b) $h' = \dfrac{R}{2} + \dfrac{2h}{3}$

69. $\mu = 0.131$
71. (b) 1.12×10^4 m/s

Chapter 8

1. 130 N
3. 170 N·m clockwise
5. 0.642 N·m counterclockwise
7. No, the resultant torque acting on the ladder is not zero.
9. 333 N, 567 N
11. (b) $T = 343$ N, $H = 171$ N, $V = 683$ N (c) 5.13 m
13. $F_{deltoid} = 724$ N, $F_s = 717$ N at 8.75° below line OA
15. $T = 157$ N, $R = 107$ N
17. $T = 1470$ N, $H = 1330$ N (to right), $V = 2580$ N (upward)
19. (a) 1300 N, 270 N (b) $\mu_s = 0.320$
21. 312 N
23. 67 N
25. 17.3 rad/s
27. $\tau_x = 150$ N·m, $\tau_y = 66$ N·m, $\tau_0 = 220$ N·m
29. 0.124 m (12.4 cm)
31. (a) 34 N (b) 33 cm
33. 177 N
35. 24 rad/s^2
37. (a) $mg - T = ma$ (b) $Tr = \frac{1}{2} Mr^2\alpha$
39. $\mu_k = 0.524$
41. 1.03×10^{-3} J
43. 0.150 kg·m^2
45. 0.63
47. Answer given in statement of problem.
49. (a) 1.37×10^8 J (b) 5.10 h
51. $KE_i = 2.6 \times 10^{29}$ J, $t = 1.8 \times 10^6$ y

53. (a) 7.08×10^{33} J·s (b) 2.66×10^{40} J·s
55. 6.73 rad/s
57. 7.5×10^{-11} s
59. 0.91 km/s
61. 0.235 J

63. (a) $\omega = \dfrac{I_1}{I_1 + I_2}\, \omega_0$ (b) $\dfrac{KE_f}{KE_i} = \dfrac{I_1}{I_1 + I_2}$

65. 36 rad/s
67. 11.2 N, 1.39 N, 7.23 N
69. $F_A = 6.59 \times 10^5$ N at 78.9° to the left of vertical, $F_B = 6.47 \times 10^5$ N horizontal to the right.
71. (a) 46.6 N (b) 0.233 kg·m^2 (c) 40.0 rad/s
73. $\mu_s = 0.268$
75. (a) 3.12 m/s^2 (b) $T_1 = 26.7$ N (c) $T_2 = 9.37$ N
77. (a) As the child walks to the right end of the boat, the boat moves left (toward pier). (b) 5.55 m (c) No, his maximum reach is 6.55 m.

Chapter 9

1. 1.8×10^6 Pa
3. 6.89 mm
5. 2.4×10^{-2} mm
7. Stress = 5.3×10^7 Pa; the arm should survive.
9. 2.1×10^7 Pa
11. 6.3×10^4 N (about 14 100 lb)
13. 2.0 cm
15. 1.9×10^4 N
17. 6.28 N
19. (a) 1.1×10^8 Pa (b) 2×10^6 N (4.4×10^5 lb)
21. 1.33 m
23. 12.6 cm
25. 2.31 lb
27. 1.05×10^5 Pa
29. 0.611 kg
31. 0.600 m
33. 1.65 m
35. 8.57×10^3 kg/m^3 (b) 714 kg/m^3
37. 15.9 cm
39. 60.8 m/s^2
41. 13 min
43. 31 kW
45. (a) 11.0 m/s (b) 2.64×10^4 Pa
47. 12.6 m/s
49. 1.5×10^5 N upward
51. 2.1×10^3 Pa
53. 9.00 cm
55. (a) 2.7 m/s (b) 2.3×10^4 Pa
57. 17.0 cm above bottom of tank
59. 1.03×10^4 Pa
61. 0.080 of the volume is exposed
63. 7.5×10^6 Pa
65. 0.780 mm

67. 4.14×10^3 m^3
69. (a) 1.16×10^5 Pa (b) 52.0 Pa
71. 6.3 m
73. 15 m
75. 1.71 cm
77. Answer is given in statement of problem.
79. (a) 16.0 m/s (b) 1.73×10^5 Pa

81. (b) $y_{\max} = \dfrac{P_{\text{atm}}}{\rho g}$

Chapter 10

1. (a) $-251°$C (b) 1.358 atm
3. (a) $37°$C, 310 K (b) $-20.6°$C, 252.6 K
5. (a) $T_{\text{TH}} = \frac{3}{2} T_{\text{C}} + 50$ (b) -360 TH
7. $57.8°$C, $-88.3°$C
11. 31 cm
13. 2.8 m
15. Answer is given with problem.
17. (a) $263.5°$C (b)$-262.2°$C
19. Answer is given with problem.
21. (a) 0.060 m (b) stress = 4.8×10^6 Pa; will not crumble
23. 1.1 L (0.29 gal)
25. (a) 3.0 mol (b) 1.8×10^{24} molecules
27. $V_f/V_i = 3/2$
29. 0.12 cm^3
31. (a) $630°$C (b)$930°$C
33. 16.0 cm^3
35. 7.1 m
37. 3.8 m
39. 0.131 kg/m^3
41. (a) 3740 J (b) 1930 m/s
43. 3.7×10^4 N
45. 1.8×10^4 Pa
47. 8.0 N
49. (a) 8.76×10^{-21} J/molecule (b) $v_{\text{He}} = 1620$ m/s, $v_{\text{Argon}} = 514$ m/s
51. 3.34×10^5 Pa
53. (a) 1.4×10^{-2} cm (b) 6.8×10^{-4} cm (c) 3.2×10^{-2} cm^3
55. (a) 491.67R, 671.67R (b) $T_f = T_R - 459.67$ (c) $T_K = \frac{5}{9} T_R$
57. Answer is given with problem.
59. 0.53 kg
61. $800°$C
63. 0.060 mm too short
65. 2.5 m

Chapter 11

1. $87°$C
3. $170°$C
5. 2.9×10^3 m
7. $47°$C
9. $10.1°$C
11. (a) 9.9×10^{-3} °C (b) The remaining energy is absorbed by the surface on which the block slides.
13. $23°$C
15. 190 g
17. 80 g
19. 470 pellets
21. 1.8×10^3 J/kg·°C (0.44 cal/g·°C)
23. 1.2×10^5 J
25. 4.9×10^4 J
27. 10 g
29. $11°$C
31. (a) all ice melts, $T_f = 40°$C (b) 8.0 g, $0°$C
33. (a) $0°$C, with 24 g of ice left over (b) $8.2°$C
35. (a) 220 J/s (b) 1.3×10^{-2} J/s (c) 4.5×10^{-2} J/s
37. 14 ft^2·F°·h/Btu
39. 39 m^3/day
41. 52 J/s (thermopane), 1.8×10^3 J/s (single pane)

43. $\dfrac{P_h}{P_c} = 1.4$

45. 7.2×10^{-2} J/s·m·°C
47. 9.0 cm
49. 2.4 min
51. $38°$C
53. 3.8×10^5 J (9.0×10^4 cal)
55. (a) System reaches $0°$C and 3.8×10^{-4} kg of ice is melted. (b) System reaches $100°$C and 1.6 g of water is vaporized.
57. $45°$C
59. (a) 2.0×10^3 W (b) $4.5°$C
61. 27 L
63. 20.02 cm
65. (b) 0.65 cal/g·°C
67. $110°$C

Chapter 12

1. 804 J
3. (c) More work is done in process (a) because of the higher pressure during the expansion part of the process.
5. (a) 610 J (b) 0 (c) -410 J (d) 0 (e) 200 J
7. 470 J
9. Answer is given in problem.
11. (a) -89.0 J (b) 720 J
13. 1.08×10^3 cal (4.5×10^3 J)
15. (a) 338 J (b) 4.52×10^3 J (c) 4.18×10^3 J
17. $W_{BC} = 0$, $W_{CA} < 0$, $W_{AB} > 0$, $Q_{BC} < 0$, $Q_{CA} < 0$, $Q_{AB} > 0$, $\Delta U_{AB} > 0$, $\Delta U_{BC} < 0$, $\Delta U_{CA} < 0$
19. (a) 26 J (b) 9.0×10^5 J (c) 9.0×10^5 J
21. -8.3×10^5 J
23. (a) $Q < 0$, $W = 0$, $\Delta U < 0$ (b) $Q > 0$, $W = 0$, $\Delta U > 0$

25. Max efficiency = 50%; claim is invalid.
27. (a) 667 J (b) 467 J
29. (a) 1.1×10^4 J (b) 0.53 s
31. Eff = 48.8%
33. 546°C
35. (a) 9.1 kW (b) 1.2×10^4 J
37. (a) Eff = 5.1% (b) 5.3×10^{12} J/h
39. $COP = 9.0$
41. 57.2 J/K
43. (a) -1.2×10^3 J/K (b) 1.2×10^3 J/K
45. (a)

Result	Possible Combinations	Total
all red	RRR	1
2R,1G	RRG,RGR,GRR	3
1R,2G	RGG,GRG,GGR	3
all green	GGG	1

(b)

Result	Possible Combinations	Total
all red	RRRRR	1
4R,1G	RRRRG,RRRGR,RRGRR,RGRRR, GRRRR	5
3R,2G	RRRGG,RRGRG,RGRRG,GRRRG, RRGGR,RGRGR,GRRGR,RGGRR, GRGRR,GGRRR	10
2R,3G	GGGRR,GGRGR,GRGGR, RGGGR,GGRRG,GRGRG, RGGRG,GRRGG,RGRGG,RRGGG	10
1R,4G	RGGGG,GRGGG,GGRGG, GGGRG,GGGGR	5
all green	GGGGG	1

47. 0.48°C
49. (a) 251 J (b) 314 J (c) 104 J (d) −104 J (e) 0 in both cases
51. (a) 0.95 J (b) 3.2×10^5 J (c) 3.2×10^5 J
53. 29.3 J
55. 78 W
57. 18°C
59. $\Delta S_h = -16.0$ J/K (b) $\Delta S_c = 26.7$ J/K (c) $\Delta S_{system} = 10.7$ J/K
61. (a) $W = 4 P_0 V_0$ (b) $Q = 4 P_0 V_0$ (c) 9.07×10^3 J
63. 8.4×10^6 J/K s

Chapter 13

1. (a) 8.0 s (b) No, the force is not of Hooke's law form.
3. (a) 110 N (b) The graph will be a straight line passing through the origin and with a slope of 1.0×10^3 N/m.
5. 9.97 cm

7. 2.68 m/s
9. (a) 100 J (b) 8.0 cm (c) 2.8 m/s (d) 90 m/s^2
11. (a) $PE = \dfrac{E}{4}$, $KE = \dfrac{3}{4} E$ (b) $x = \dfrac{A}{\sqrt{2}}$
13. 47.8 cm
15. 3.06 m/s
17. (a) 2.60 cm (b) $v = 0.866 v_{max}$
19. (a) 0.15 J (b) 0.78 m/s (c) 18 m/s^2
21. (a) 0 (b) −4.33 m/s (c) −5.00 m/s (d) 0 (e) 5.00 m/s
23. (a) You are observing the vertical motion of the boss given by $y = R \cos \omega t$, where R is the radius of the tire. (b) 0.63 s
25. (a) quadrupled (b) doubled (c) doubled (d) no change
27. 3.95 N/m
29. (a) 10.9 N toward left (b) 0.880 oscillations
31. $v = \pm A\omega \sin(\omega t)$, $a = -A\omega^2 \cos(\omega t)$
33. (a) 250 N/m (b) 0.28 s, 3.6 Hz, 22 rad/s (c) 0.31 J (d) 5.0 cm (e) $v_{max} = 1.1$ m/s, $a_{max} = 25$ m/s^2 (f) 0.92 cm
35. $F_{tangential} = -\left[(\rho_{air} - \rho_{He}) \dfrac{Vg}{L} \right] s = -ks$, $T = 1.40$ s
37. (a) slow (b) 9:47:17 A.M.
39. 21.9 m
41. (a) 1.14×10^{-8} s (b) 3.41 m
43. 8.2×10^{-2} m to 12 m
45. (a) 9.0 cm (b) 20 cm (c) 0.040 s (d) 500 cm/s
47. (a) 8.31 min (b) 1.28 s
49. 446 m
51. 80 N
53. (a) 30.0 N (b) 25.8 m/s
55. 13.5 N
57. (a) 0 (b) 0.30 m
59. 24.6 m/s
61. (a) slower (b) loses 5 s per hour
63. (a) 5.1×10^{-2} kg/m (b) 20 m/s
65. 0.990 m
67. (a) 19.8 m/s (b) 8.94 m
69. 3.14×10^6 Pa
71. (a) 100 m/s (b) 374 J
73. (a) −28.0 J (b) 0.446 m
75. (a) 588 N/m (b) 0.70 m/s
77. $\mu = 0.12$

Chapter 14

1. 32.2°C
3. 520 m
5. $\dfrac{\Delta \lambda}{\lambda} = 2.9\%$
7. 18.5 m
9. 20 km

11. (a) 5.0×10^{-17} W (b) 5.0×10^{-5} W
13. Answer is given in problem.
15. 5.00×10^4 m
17. (a) 7.96×10^{-2} W/m^2 (b) 109 dB (c) 2.82 m
19. Answer is given in problem.
21. 41 kHz
23. 32.1 m/s (72 mph) with the cyclist behind the car
25. (a) 537 Hz (b) 465 Hz
27. 599 Hz
29. (a) 59.7 s (b) 60.0 km (about 37 miles) away and 20.0 km high.
31. (a) 34.5 cm (b) constructive
33. (a) 0.431 m (b) 0.863 m
35. 0.0890 m, 0.304 m, 0.518 m, 0.732 m, 0.947 m, 1.16 m
37. (a) 79 N (b) 2.1×10^2 Hz
39. 1.00 cm toward the nut
41. (a) 3.5×10^2 Hz (b) 4.0×10^2 kg (fundamental mode)
43. 120 Hz
45. 0.352 Hz, 0.176 Hz, 0.117 Hz, etc.
47. $L_{open} = 65.9$ cm, $L_{closed} = 98.9$ cm
49. (a) open at both ends (b) 78.6 cm
51. (a) 0.550 m (b) 317 Hz
53. 328 m/s
55. 5.26 Hz
57. 29.7 cm
59. 1.76 cm
61. (a) 362 Hz (b) 287 Hz (c) 0.953 m and 1.20 m
63. (a) 0.555 m (b) 620 Hz
65. 3.01 dB
67. 2.82×10^8 m/s
69. 2.0×10^2 m/s
71. 32.9 m/s
73. 4.00 Hz
75. 1.20×10^3 Hz
77. 0.100 m/s

Chapter 15

1. (a) -80 μC (b) 4.8 μC
3. 1.1×10^{-8} N
5. 5.13×10^5 N
7. 5.71×10^{13} C
9. -1.2×10^{-8} N (negative x direction)
11. 2.25×10^{-9} N/m
13. 3.90×10^{-7} N at 11.3°
15. 91 N
17. 0.621 m
19. 1.45 m beyond the -3×10^{-9} C charge
21. 4.3×10^3 N/C at 91° CW from $+ x$ axis.
23. 2.94×10^{21} N/C
25. 5.54×10^{11} N/C (away from proton)
27. (a) 2.00×10^7 N/C to right (b) 40.0 N to left
29. 1.63×10^5 N/C

31. (a) 5.27×10^{13} m/s^2 (b) 5.27×10^5 m/s
33. (a) 6.13×10^{10} m/s^2 (b) 19.6 μs (c) 11.7 m
   (d) 1.20×10^{-15} J
35. 1.63×10^4 N/C
37. 756 N/C at 70.1° above $-x$ axis
39. 0
41. 1880 N/C at 4.4° below x axis
43. 1.82 m to left of -2.5×10^{-6} C charge
47. (a) $q_2 = 3q_1$ (b) $q_2 > 0$, $q_1 < 0$
51. (a) 0 (b) -5.00 μC inside, $+ 5.00$ μC outside
   (c) 5.00 μC outside, 0 inside (d) 5.00 μC outside, 0 inside
53. (a) 0 (b) 1.80×10^6 N/C (c) 1.13×10^5 N/C
55. (a) 5.3×10^{17} m/s^2 (b) 0.85 mm
57. No, the required charge would be $q = +2.25e$, which is not an integral multiple of e.
59. (a) 24 N/C in $+ x$ direction (b) 9.4 N/C at 64° above $-x$ axis
61. 5.25 μC
63. at $y = 9.8$ m
65. 4.4×10^5 N/C
67. (a) -4.4×10^{15} m/s^2 (b) 9.1×10^{-10} s (c) 1.8×10^{-3} m
69. (a) 0.307 s (b) Yes, neglecting gravity causes a 2.8% error.

Chapter 16

1. (a) 6.4×10^{-19} J (b) -6.4×10^{-19} J
3. (a) -6.0×10^{-4} J (b) -50 V
5. 4.3×10^6 J
7. 1.54×10^6 J
9. 3.20×10^{-19} C
11. (a) 1.52×10^5 m/s (b) 6.49×10^6 m/s
13. (a) 9.22×10^4 V (b) 1.69×10^8 V
15. 20 000 eV, 8.39×10^7 m/s
17. 7.26×10^6 m/s
19. (a) 3.83×10^4 V (b) -6.36×10^3 V
21. 8.10×10^{-7} J
23. -9.1 J
25. (a) 5.93×10^5 m/s (b) 1.38×10^4 m/s
27. 2.8×10^{-14} m
29. (a) 48.0 μC (b) 6.00 μC
31. 2.26×10^{-5} m^2
33. 5.3×10^{-12} C
35. (a) 90.4 V (b) 9.04×10^4 V/m
37. 17.0 μF
39. (a) $Q_{25} = 1.25 \times 10^3$ μC, $Q_{40} = 2.00 \times 10^3$ μC
   (b) $Q'_{25} = 290$ μC, $Q'_{40} = 460$ μC, $V = 11.5$ V.
41. (a) 18 μF (b) 1.8 μF
43. $Q_1 = 16$ μC, $Q_4 = 32$ μC, $Q_5 = 80$ μC, $Q_8 = 64$ μC
45. (a) 0, 0, 0, 65.0 μC (b) 20.0 μC, 20.0 μC, 15.0 μC, 30.0 μC

47. 30.0 μF
49. 83.7 μC
51. $Q_1 = \frac{10}{3}\ \mu$C, $Q_2 = \frac{20}{3}\ \mu$C
53. 121 V
55. 2.55×10^{-11} J
57. (a) 3.00×10^3 V/m (b) 42.5 nC (c) 5.31×10^{-12} C (or 5.31 pC)
59. Energy is multiplied by 3.4.
61. (a) 8.13 nF (b) 2.40 kV
63. 4.3 μF
65. 4.22×10^5 V
67. 6.25 μF
69. 1.8×10^{-12} F

71. $W = k\dfrac{q^2}{d}\,(2\sqrt{2} - 3)$

73. 8.00×10^3 V
75. (a) 400 μC on the 2.00 μF and 800 μF on the 4.00 μF (b) 200 V
77. (a) 7.54×10^{-9} s (b) $v_e = 1.33 \times 10^7$ m/s, $v_p = 7.23 \times 10^3$ m/s (c) 7.55 ns (d) 0.323 μs

Chapter 17

1. 45 C
3. (a) Answer is given in statement of problem. (b) 1.05 mA
5. 3.75×10^{14} electrons/s
7. 0.159 A
9. 5.90×10^{28} electrons/m^3
11. 27 y
13. (a) 3.0 A (b) 2.9 A
15. 1.7×10^{-4} m
17. 32 V, 0.16 V
19. 48 Ω
21. $\rho = \rho_{silver} = 1.59 \times 10^{-8}$ $\Omega \cdot$m
23. (a) 1.8 m (b) 0.28 mm
25. (a) 4.7×10^{-4} $\Omega \cdot$m (b) 30 Ω
27. (a) 2.82×10^8 A (b) 1.76×10^7 A
29. 10.8 Ω
31. 30.6°C
33. 1.08×10^{-3} (°C)$^{-1}$
35. (a) 1.1968 Ω (b) 8.00×10^{-4} (0.08%)
37. (a) 0.65 mV (b) 1.1 mV
39. (a) 140 Ω (b) 26 m (c) to accommodate the required length (d) 25 m
41. 50 MW
43. 1.75×10^{-8} m^2
45. 36.1% increase
47. 1.56 cm
49. 450 A
51. 13 cents
53. 23 cents

55. $I = 16$ A. Thus, use 20-A fuse.
57. (a) 19 A (b) 7.5×10^6 J (2.1 kWh) (c) 17 cents
59. 3.24×10^5 C
61. 48 kg/s
63. 1.1×10^3 m
65. 3.77×10^{28} electrons/m^3
67. 0.400 μA
69. (a) 9.4×10^{-7} Ω (b) 8.0×10^{-7} Ω
71. (a) 18 C (b) 3.6 A
73. (a) 9.3 m (b) 9.3×10^{-4} m
75. (a) Answer is given in statement of problem. (b) 1.420 Ω versus 1.418 Ω

Chapter 18

1. (a) 24 Ω (b) 1.0 A
3. (a) $\dfrac{24}{11}$ Ω (b) $I_4 = 6$ A, $I_8 = 3$ A, $I_{12} = 2$ A
5. There are 7 distinct values possible: (1) Use one alone, value = R; (2) Use two in series, value = $2R$; (3) three in series, value = $3R$; (4) two in parallel, value = $\frac{R}{2}$; (5) three in parallel, value = $\frac{R}{3}$; (6) two in series with one in parallel, value = $\frac{2}{3}R$; (7) two in parallel with one in series, value = $\frac{3}{2}R$.
7. (a) Connect two 50-Ω resistors in parallel, and then connect this combination in series with a 20-Ω resistor. (b) Connect two 50-Ω resistors in parallel; connect two 20-Ω resistors in parallel, and then connect these two combinations in series with each other.
9. 15.0 Ω
11. 9.83 Ω
13. (a) 5.13 Ω (b) 4.53 V
15. 0.429 A
17. (a) $I_3 = 1.82$ A, $I_6 = 1.09$ A, $I_9 = 0.727$ A (b) Same as (a).
19. 0.52 W
21. 2.40 V, 320 Ω
23. (a) 3.00 mA (b) -19.0 V (c) 4.50 V
25. Answer is given in statement of problem.
27. (a) 9.0 V wth b at a higher potential than a (b) 0.42 A flowing b to a.
29. 10.7 V
31. (a) 12 s (b) 120 μC
33. $O = 0.863 Q_f$ or 86.3%
35. (a) 5.0×10^5 Ω (b) 78 μC
37. (a) Toaster = 8.33 A, heater = 10.8 A, grill = 12.5 A. (b) $I_{total} = 31.6$ A so a 30-A circuit is insufficient.
39. (a) 6.25 A (b) 750 W
41. 16 A
43. Use three identical parallel combinations (each consisting of 3 resistors) connected in series.

45. (a) $R = 0.0999\ \Omega$, current in $R_1 = 50$ A; current in 100 Ω, R_2 and $R_3 = 0.045$ A. (b) $R = 1.09\ \Omega$, current in R_1 and $R_2 = 4.55$ A; current in 100 Ω and $R_3 = 0.045$ A. (c) $R = 9.99\ \Omega$, current in 100 $\Omega = 0.050$ A; current in R_1, R_2, and $R_3 = 0.450$ A.

47. (a) $I_1 = 1$ A, $I_2 = I_3 = 0.500$ A, $I_4 = 0.300$ A, $I_5 = 0.200$ A. (b) $V_{ac} = 6.00$ V, $V_{ce} = 1.20$ V, $V_{ed} = 1.80$ V, $V_{db} = 6.00$ V, $V_{fd} = 1.80$ V, $V_{cd} = 3.00$ V. (c) $P_{ac} = 6.00$ W, $P_{ce} = 0.600$ W, $P_{ed} = 0.540$ W, $P_{fd} = 0.360$ W, $P_{cd} = 1.50$ W, $P_{db} = 6.00$ W.

49. (a) 1.00×10^{-2} F (b) 0.414 C

51. $I_1 = 0$, $I_2 = I_3 = 0.500$ A

53. (a) $R_x = R_2 - \frac{1}{4} R_1$ (b) $R_x = 2.8\ \Omega$, antenna is inadequately grounded.

55. (a) 14 Ω (b) 56 W (c) 2.0 A

57. (a) 0.50 A (b) 3.0 W (c) 1.1 W

59. (a) 12.4 V (b) 9.65 V

61. 0.390 A, 1.50 V

63. 112 V, 0.200 Ω

Chapter 19

1. 2.6×10^{-11} N westward

3. (a) 8.7×10^{-14} N (b) 5.2×10^{13} m/s^2

5. **(a)** (a) to left, (b) into page, (c) out of page, (d) toward top of page, (e) into page, (f) out of page **(b)** If the charge is negative, the direction of all forces above are reversed 180°.

7. (a) into page (b) toward right (c) toward bottom of page.

9. 2.83×10^7 m/s

11. 2.1×10^{-2} T in $-y$ direction

13. 200 A eastward

15. (a) 1.3×10^{-3} N (b) 20 A

17. (a) to the left (b) into page (c) out of page (d) toward top of page (e) into page (f) out of page

19. 0.131 T downward

21. 0.245 A

23. (a) 9.0×10^{-3} N at 15° above horizontal in northward direction (b) 2.3×10^{-3} N horizontal and westward

25. 0.193 N · m

27. 0.167 Ω

29. $1.94 \times 10^3\ \Omega$

31. $R_1 = 3.33 \times 10^{-4}\ \Omega$, $R_2 = 2.998 \times 10^{-3}\ \Omega$, $R_3 = 2.997 \times 10^{-2}\ \Omega$

33. 1.39×10^{-2} T out of page

35. 8.65×10^{-8} s

37. (a) 0.0500 m (b) 8.78×10^6 m/s

39. $r = \sqrt{2}\, r_p$ for both deuterons and α particles.

41. 0.28 m

43. 3.20 cm

45. (a) 2.00×10^{-4} T (CCW) (b) 1.33×10^{-4} T (CW)

47. 500 A

49. (a) 4.00×10^{-5} T, into page (b) 5.00×10^{-6} T out of page (c) 1.67×10^{-6} T out of page

51. 82 A

53. (a) 2.00×10^{-4} N/m, (attracted) (b) 2.00×10^{-4} N/m (repelled)

55. 120 N · m

57. 2.26×10^{-4} N, torque = 0

59. No, $r = 3.1 \times 10^4$ m $< 1.3 \times 10^7$ m

61. 8.4×10^{-5} T

63. 2.12 cm

65. (a) 5.00×10^{-7} T (out of page) (b) 3.89×10^{-6} T in a plane parallel to the x-y plane and at 31.0° CCW from the $-y$ axis.

67. 0.331 T

69. (a) 30.0 A (b) 1.60×10^{-4} T (out of page)

71. 3.8×10^{-25} kg

73. 2.7×10^{-5} N to the left

Chapter 20

1. 3.1×10^{-4} T · m^2

3. 7.71×10^{-1} T · m^2

5. 2.96×10^{-5} T · m^2

7. zero

9. 0.67 V

11. 1.15×10^5 V

13. -6.28 V

15. 5.20×10^{-5} T

17. 34 mV

19. 0.88 C

21. (a) 1.88×10^{-7} T · m^2 (b) 6.28×10^{-8} V

23. 1.11×10^{-3} V

25. 2.00 mV, (the western end is positive)

27. (a) 100 μV (b) 20.0 μA

29. 2.87 mV

31. (a) left to right (b) right to left

33. (a) right to left (b) right to left (c) left to right (d) left to right

35. left to right

37. (a) 0.157 mV, end B will be at the higher potential (b) 5.89 mV, end A will be at the higher potential.

39. 79.6 Hz, 240 V

41. 0.13 V

43. (a) 180 V (b) 6.0 A (c) $\mathcal{E} = (180\ \text{V}) \sin (377\ \text{s}^{-1}\ t)$ (d) $i = (6.0\ \text{A}) \sin (377\ \text{s}^{-1}\ t)$

45. 76 V, 11 Ω

47. (a) 60 V (b) 57 V (c) 0.13 s

49. 256 turns

51. Answer is given in statement of problem.

53. Answer is given in statement of problem.

55. 12 mH

57. (a) 4.00 A (b) 2.52 A

59. 0.140 J
61. 1.9×10^{-3} T
63. 4.7 V
65. 9.7 V
67. negative, $V_a < V_b$
69. 4.7×10^{-3} T · m²
71. (a) Amplitude doubles, period unchanged
    (b) amplitude doubles, period cut in half
    (c) amplitude unchanged, period cut in half.
73. 440 μV
75. (a) 0.125 V (b) 0.02 A

Chapter 21

1. (a) 15 Ω (b) 830 W
3. $R_1 = R_2 = 80.7$ Ω, $R_3 = 121$ Ω, $I = 3.64$ A
5. (a) 106 V (b) 60.0 Hz (c) 0 (d) 3.00 A
7. 0.360 Ω
9. 0.224 A
11. 2.63 A
13. $L > 7.03$ H
15. (a) 12.5 A (b) 0.125 A
17. (a) 0.361 A (b) 18.1 V (c) 23.9 V (d) $-53.0°$
19. (a) 4.68 A (b) 35.3 V (c) 93.6 V (d) 20.7°
21. (a) 138 V (b) 104 V (c) 729 V (d) 640 V
23. 1.88 V
25. (a) 650 mA (b) 1.4 H
27. (a) 1.4×10^3 Ω (b) 98 mA (c) 51° (d) current lags
    voltage
29. (a) power factor = 0.492, 48.3 W (b) power factor = 0.404, 32.6 W
31. (a) $P = 100$ W, $\cos \phi = 0.634$; (b) $P = 156$ W,
    $\cos \phi = 0.790$
33. (a) $I_{10} = 0.50$ mA, $I_{100} = 6.3$ mA, $I_{1000} = 2.4$ mA,
    $I_{10\,000} = 0.23$ mA (b) $\phi_{10} = -89°$, $\phi_{100} = -80°$,
    $\phi_{1000} = 86°$, $\phi_{10\,000} = 89.6°$ (c) 210 Hz
35. 2.05 MHz, 146 m
37. 18.9 pF to 930 pF
39. (a) Answer given in statement of problem. (b) $Q = 33.3$
41. 0.18%
43. (a) 1.7×10^4 A (b) 10 A (c) 1.4×10^5 W/m, 5.0×10^{-2} W/m (d) 170 lines, 0.1 line (or one line)
45. 75.0 MHz
47. 1.0×10^{-3} J
49. 3.74×10^{26} W
51. (a) 141 V (b) 24.0 Ω (c) 28.8 Ω (d) 56.3° (e) $X_L = 24.0$ Ω, $X_C = 12.0$ Ω, or 36.0 Ω
53. 3.3×10^3 m²
55. 11.0 m
57. (a) 190 m to 560 m (b) 2.8 m to 3.4 m
59. The year will be 2674.
61. People 100 km away receive the news 8.74 ms before
    people across the room because radio waves travel
    faster than sound waves.

63. (a) 28.1 μF (b) 0
65. (a) 1.76×10^{13} m/s², 1.57×10^{-27} W
67. 0.429 A
69. 1.7 cents
71. (a) 2.7 cm, 27 turns (b) 9.4×10^{-5} H (c) 0.11 Ω
    (d) 180 A
73. (a) 2.9×10^4 W (b) $\dfrac{P_{loss}}{P} = 0.0058$ (c) Power loss in
    line = 72 × (delivered power).
75. 5.17 μH
77. (a) 0.552 N (b) 9.20×10^{-5} m/s² (c) 33.4 days
79. (a) 2.0×10^{-8} T (b) 600 kW
81. 4.47×10^{-9} J

Chapter 22

1. 2.3×10^8 m/s
3. (a) 2.1×10^4 rev/s, 4.2×10^4 rev/s (b) 2.1×10^2 rev/s, 4.2×10^2 rev/s
5. (b) 2.9997×10^8 m/s
7. (a) 1.94 m (b) 50.0° above horizontal (parallel to incident ray)
9. 26°
11. Answer is given in statement of problem.
13. $\theta_1 = 79°$, $\theta_2 = 22°$, $\theta_3 = 68°$, $\theta_4 = 29°$
15. 25.5°, 442 nm
17. 6.30 cm
19. First surface: $\theta_i = 30.0°$, $\theta_r = 19.5°$, second surface:
    $\theta_i = 19.5°$, $\theta_r = 30.0°$
21. 32°
23. (a) First surface: $\theta_i = 30°$, $\theta_r = 20°$, second surface:
    $\theta_i = 41°$, $\theta_r = 77°$ (b) first surface: θ (reflection) =
    30° = angle of incidence; second surface:
    θ (reflection) = 41° = angle of incidence
25. 23.1°
27. (a) 1.2 (b) 3.4 ns
29. 8.00°
31. 3.4 m
33. 245 nm
35. 4.61°
37. (a) 31.3° (b) 44.2° (c) 49.8°
39. 48.5°
41. 24.4°
43. $n \geq \sqrt{2}$
45. 1.08 cm $< h <$ 1.17 cm
47. $n = 1.38$
49. 2.09×10^{-11} s
51. (a) $\theta_2 = 53.1°$ (b) $\theta_1 \geq 38.7°$
53. $n = 1.93$
55. 40.0°
57. (a) $n = \sqrt{1 + \left(\dfrac{4t}{d}\right)^2}$ (b) 2.10 cm (c) violet-tinged

59. $\dfrac{n_w}{n_b} = 0.890$

61. Answer is given in statement of problem.

73. 32.0 cm beyond second surface (real image)

75. (a) $p = \dfrac{4f}{3}$ (b) $p = \dfrac{3f}{4}$ (c) $M_1 = -3.00$, $M_2 = 4.00$

Chapter 23

1. Answer is given in statement of problem.
3. 10.0 ft, 30.0 ft, 40.0 ft
5. Answer is given in statement of problem.
7. (a) Image is a small trapezoid. (b) 3.55 cm^2
9. (a) 2.22 cm (b) $M = 10.0$
11. (a) 15.0 cm (b) 60.0 cm
13. $M = +0.130$
15. 48.0 cm
17. concave, $R = 40.0$ cm
19. 10.0 cm in front of mirror
21. 30.0 cm in front of mirror
23. concave, with $f = 9.52$ cm, $M = -20.0$
25. (a) 1.50 m (b) 1.75 m
27. 52.5 cm
29. 9.00 cm inside the bowl
31. (a) 120 cm (b) -24.0 cm (c) -8.00 cm (d) -3.43 cm
33. 4.80 cm
35. 16.0 mm
37. (a) 40.0 cm past lens, real, inverted, $M = -1.00$
    (b) No image is formed. Rays are parallel. (c) 20.0 cm in front of lens, erect, virtual, $M = 2.00$.
39. 7.50 cm in front of lens.
41. (a) 13.3 cm in front of lens, virtual, erect, $M = \frac{1}{3}$
    (b) 10.0 cm in front of lens, virtual, erect, $M = \frac{1}{2}$
    (c) 6.67 cm in front of lens, virtual, erect, $M = 0.667$
43. (a) converging (b) $+50.0$ cm
45. 5.68 cm
47. -80.0 cm
49. $f/9$ beyond the focal point
51. either 36.2 cm or 13.8 cm from object, $M = -2.62$ for $p = 13.8$ cm, $M = -0.382$ for $p = 36.2$ cm
53. 30.0 cm in front of second lens, $M = -3.00$
55. 9.30 cm in front of second lens, $M = 0.370$
57. from 0.232 m to 187 m
59. (a) 13.3 cm (b) $M = -5.90$ (c) inverted
61. 25.3 cm to right of mirror, virtual, erect, $M = 8.05$
63. (a) 1.99 (b) 10.0 cm to left of lens, $M = -2.50$
    (c) inverted
65. (a) 33 cm beyond lens, $M = -0.67$, real, inverted
    (b) 60 cm beyond lens, $M = -2.0$, real, inverted
    (c) 20 cm in front of lens, $M = 2.0$, erect and virtual
67. (a) 10.0 cm beyond second lens (b) real, erect and same size as object, $M = 1.00$ (c) 20.0 cm beyond lens combination
69. $R_2 = -30.0$ cm
71. Final image is at center of upper mirror, real, erect, and same size as object.

Chapter 24

1. 1.25 m
3. (a) 2.62 mm (b) 2.62 mm
5. 11.3 m
7. 2.61 m
9. 34.0 cm
11. (a) 55.7 m (b) 124 m
13. 99.8 m
15. 148 m
17. 91.9 nm
19. 85.4 nm (Other possible thicknesses are odd integral multiples of this.)
21. 290 nm, 580 nm, 870 nm
23. 290 nm
25. 647 nm
27. 1.50
29. No, the wavelengths intensified are 276 nm, 138 nm, 92 nm. . . .
31. (a) 123 nm (b) 81.6 nm
33. 147 bands
35. $4.23\ \mu\mathrm{m} \leq t \leq 4.45\ \mu\mathrm{m}$
37. Anwer is given in statement of problem.
39. Answer is given in statement of problem.
41. 0.227 mm
43. (a) 1.09 m (b) 1.70 mm
45. Answer is given in statement of problem.
47. 31.2°
49. 48.4°
51. (a) $n = 1.11$ (b) 42°
53. 60.5°
55. 58.1°
57. (a) 6.00 m (b) 3.00 m
59. Answer is given in statement of problem.
61. 432 nm
63. 6.89 units
65. 0.350 mm
67. 313 nm
69. 0.149 mm

Chapter 25

1. 5.71 m
3. 177 m
5. 0.544 mm
7. 2.20 mm farther from film
9. (a) 33.3 cm (b) 3.0 diopters
11. (a) $+50.8$ to $+60.0$ diopters (b) -0.800 diopters

13. 17.0 diopters
15. 40 cm
17. Power of upper portion = -1.00 diopters, Power of lower portion = 2.50 diopters.
19. (a) $M = -1.22$ (b) $\frac{\theta}{\theta_o} = 6.08$
21. 4.07 cm
23. (a) $M = 2.00$ (b) $M = 1.00$
25. (a) $7.5 \times$ (b) 94.4 cm
27. (a) 0.0602 cm (b) $p = -2.00$ cm (c) 6.02 cm (d) $M = 4.00$
29. 0.810 μm
31. $f_e = +2.00$ cm, $f_o = +90.0$ cm
33. (a) 9.50 cm past the eyepiece (b) Probably not. When looking through a telescope, the eye is relaxed (focused at infinity) and would not properly focus on the bee. Also, the overall magnification of the bee is 0.1. Thus, the final image is 1/10 the size of the bee.
35. 157 miles
37. 0.772 m (30.4 in.)
39. 9.76 km
41. 38.3 cm
43. 0.400 μrad
45. (a) 2.29×10^{-4} rad (b) 43.7 m
47. (a) 7.75×10^{-3} cm (b) 1.18 mm (c) 174 m
49. 98 fringe shifts
51. 1310 fringe shifts
53. 40 fringe shifts
55. 6.56×10^3 lines/cm
57. (a) 478.7 nm, 647.6 nm, 696.6 nm (b) 20.5°, 28.3°, 30.7°
59. (a) 5.93° (b) 6.42°
61. 469 nm, 156 nm
63. No, a resolving power of 2.0×10^5 is needed and that of the grating is only 1.8×10^5.
65. (a) 102 cm (b) $M = 66.7$
67. 603 lines/mm
69. $V_1 G_1 R_1 V_2 G_2 V_3 R_2 V_4 G_3 V_5 G_4 R_3$
71. 660 nm
73. (a) 1.96 cm (b) $f/3.27$ (c) $f/9.8$
75. $n = 1.70$
77. 5.07 mm

Chapter 26

1. (a) $t_I = 1667$ s, $t_{II} = 2041$ s (b) $t'_I = 2500$ s, $t'_{II} = 2041$ s (c) $\Delta t = 85$ s
3. (a) 70 beats/min (b) 31 beats/min
5. 0.87c
7. (a) 1.38 y (b) 1.31 lightyear
9. Yes, with 18.8 m to spare
11. 229.4 m
13. 0.80c
15. (a) 20 m (b) 19 m (c) 0.31c

17. Ratio = 0.088.
19. 3600 s $+ 3.20 \times 10^{-9}$ s
21. (a) 2.7×10^{-24} kg·m/s (b) 1.6×10^{-22} kg·m/s (c) 5.6×10^{-22} kg·m/s
23. (a) 0.85c (b) 0.85c
25. 0.285c
27. 0.95c
29. 1.0×10^3 m
31. (a) 337 kg/m³ (b) 64.0 kg/m³
33. 0.985c
35. (a) 76 min (b) 63 min
37. 0.98c
39. 0.160 pulses/s
41. (a) 0.98c (b) 0.695c to the right, or 0.9992c to the left
43. (a) 939 MeV (b) 3010 MeV (c) 2070 MeV
45. 0.98c
47. (a) 4.44×10^9 kg (b) 1.43×10^{13} y
49. (a) Answer is given with problem (b) 0.942c
51. (a) 5.09×10^4 MeV/c (b) 0.9998c
53. (a) $L'_1 = 83.3$ m, $L'_2 = 62.5$ m; (b) $L_1 = 27.0$ m, $L_2 = 20.3$ m; (c) 6000 s (1.67 h); (d) $T_1 = 5.33 \times 10^3$ s (1.48 h); (e) $T_2 = 7.10 \times 10^3$ s (1.97 h); (f) casualties occur on rocket #1.
55. 0.37c
57. (a) 2.50 MeV/c (b) 4.60×10^3 MeV/c
59. 1.2×10^{13} m
61. (a) 7.05 μs (b) 1.12×10^4 muons
63. Answer is given in statement of problem.

Chapter 27

1. (a) 1000 nm (b) Peak is in the infrared region.
3. 5.7×10^3 photons
5. 500 nm
7. 0.62 nm
9. 7.73×10^3 K
11. 150 days
13. potassium ($\phi = 2.23$ eV)
15. (a) 1.4 eV (b) 3.4×10^{14} Hz
17. (a) lithium and iron (b) 1.8 eV and 0.24 eV, respectively
19. (a) given in problem (b) 3.9 eV (c) 8.8 eV
21. (a) 0.022c (b) 0.9992c
23. (a) 0.0830 nm (b) 0.0120 nm
25. 26 sec of arc
27. $d_1 = \frac{d_0}{\sqrt{2}}$, $d_2 = \frac{d_0}{\sqrt{5}}$
29. 0.124 nm
31. 0.150 keV
33. (a) $\theta = \phi = 43°$ (b) $E'_\gamma = 0.60$ MeV, $p'_\gamma = 0.60$ MeV/$c = 3.2 \times 10^{-22}$ kg · m/s; (c) $KE = 0.28$ MeV, $p_e = 0.60$ MeV/$c = 3.2 \times 10^{-22}$ kg · m/s
35. 1.8 keV, 1.8 keV/c
37. (a) 2.9×10^{-3} nm (b) 100°

39. (a) 1.88×10^4 MeV (b) 6.62×10^{-16} m
41. 1.98 MeV
43. 2.27×10^{23} Hz, 1.32×10^{-15} m
45. (a) $v \leq 1.1 \times 10^{-34}$ m/s (b) $t \geq 1.4 \times 10^{33}$ s (c) No. This time is over $10^5 \times$ (age of universe)

47. $\dfrac{c}{\sqrt{2}}$

49. 1.06×10^{-34} m
51. 150 V
53. 546 eV
55. 3.5×10^{-32} m
57. 23 m/s
59. $E_{\text{electron}} = 96.8E_0$ (relativistic), $E_{\text{proton}} = 1.001E_0$ (classical)
61. (a) Answer is given in statement of problem (b) 5.2 MeV
63. Answer is given in statement of problem.
65. 2.96×10^{20} photons
67. 2.00 eV
69. 1.35 eV

Chapter 28

1. 46 fm
3. (a) 230 N (b) 1.4 MeV
5. Answer is given in statement of problem.
7. (a) 200 nm $\leq \lambda \leq$ 360 nm (ultraviolet)
   (b) 800 nm $\leq \lambda \leq$ 1440 nm (infrared)
9. Answer is given in statement of problem.
11. 4.43×10^4 m/s
13. (a) 12.09 eV (b) 12.09 eV, 1.89 eV, 10.2 eV
15. Answer is given in statement of problem.
17. Answer is given in statement of problem.
19. (a) $a_0 = 2.56 \times 10^{-4}$ nm; (b) -2810 eV, -702.4 eV, -312.2 eV, -175.6 eV, and -112.4 eV; (c) 0.590 nm, 0.498 nm, 0.472 nm, 0.461 nm, and 0.442 nm (d) 3.19 nm, 2.36 nm, 2.11 nm, and 1.77 nm.
21. (a) 9.04×10^{22} m/s^2 (b) -2.90 eV/s (c) 4.69×10^{11} s
23. (a) Transition B (b) Transition A (c) Transitions B and C
25. (a) 2.19×10^6 m/s (b) 1.52×10^{-16} s (c) 1.05×10^{-3} A
27. (a) 2.46×10^{15} Hz, $f_{\text{orb}} = 8.31 \times 10^{14}$ Hz (b) 6.57×10^3 Hz, $f_{\text{orb}} = 6.65$ Hz, For large n, classical theory and quantum theory approach one another in their results.
29. (a) -122.4 eV (b) 1.76×10^{-11} m
31. Answers are given in statement of problem.
33. (a) 4.39×10^7 m^{-1} (b) 30.3 nm (c) deep ultraviolet region
35. The 3d subshell had $l = 2$ and $n = 3$. Also, we have $s = 1$. Therefore, we can have $n = 3$, $l = 2$, $m_l = -2$, $-1, 0, 1, 2$, $s = 1$, and $m_s = -1, 0, 1$, leading to the following table.

n	l	m_l	s	m_s
3	2	-2	1	-1
3	2	-2	1	0
3	2	-2	1	$+1$
3	2	-1	1	-1
3	2	-1	1	0
3	2	-1	1	$+1$
3	2	0	1	-1
3	2	0	1	0
3	2	0	1	$+1$
3	2	$+1$	1	-1
3	2	$+1$	1	0
3	2	$+1$	1	$+1$
3	2	$+2$	1	-1
3	2	$+2$	1	0
3	2	$+2$	1	$+1$

37. (a) 4 ($l = 3,2,1$, and 0) (b) 7 ($-3, -2, -1, 0, 1, 2, 3$)
39. (a) $1s^2 2s^2 2p^4$
    (b) The quantum numbers are

$1s^2$ state	$n = 1$	$l = 0$	$m_l = 0$ $m_s = \pm 1/2$
$2s^2$ state	$n = 2$	$l = 0$	$m_l = 0$ $m_s = \pm 1/2$

$2p^4$ state $n = 2$ $l = 1$ $\begin{cases} m_l = -1 \ m_s = \pm 1/2 \\ m_l = 0 \ m_s = \pm 1/2 \end{cases}$

41. (a) 4 (b) 2
43. Germanium, since $Z = 32$.
45. L shell: 11.8 keV; M shell: 10.1 keV; N shell: 2.39 keV
47. (a) Answer is given in statement of problem. (b) The observed spectrum is a hydrogen spectrum Doppler-shifted to longer wavelengths by a source moving away from Earth at $v = 0.47c$.
49. (a) $\tau_0 = 1.5 \times 10^{-16}$ s (b) 8.2×10^6 revolutions (c) Yes, for 8.2 million "electron years," (d) With that many revolutions, it appears that the electron (or at least some part of it) is present everywhere in the $n = 2$ orbit.
51. $n_f = 1$ (b) $n_i = 3$
53. (b) The wavelengths of the emission lines are: 158.5 nm, 185 nm, 253.7 nm, 422 nm, 682 nm, 1109 nm (c) 1.31×10^6 m/s
55. (a) 135 eV (b) This is about 10 times the ground state energy of hydrogen.
57. Answer is given in statement of problem.
59. (a) $\dfrac{1}{\alpha} = 137.036$ (b) $\dfrac{r_0}{\lambda} = \dfrac{1}{2\pi\alpha}$ (c) $\dfrac{1}{Rr_0} = \dfrac{4\pi}{\alpha}$

Chapter 29

1. (a) 1.90×10^{-15} m (b) 7.40×10^{-15} m
3. 2.30×10^{12} kg
5. (a) -0.0318% (b) -0.0018%
7. $\dfrac{\rho_n}{\rho_a} = 8.6 \times 10^{13}$

9. 4.6×10^{-13} m
11. 12.7 km
13. (a) 28 N (b) 4.2×10^{27} m/s^2 (c) 1.7 MeV
15. (a) 84.2 MeV (b) 342 MeV (c) In stable nuclei, E_b must exceed the energy required to overcome electrostatic repulsion.
17. (a) 2.72×10^{-28} eV (b) 342 MeV (c) nuclei are bound by nuclear forces, not gravitation.
19. 3.54 MeV
21. 0.211 MeV/nucleon, The difference is largely due to increased coulomb repulsion caused by the extra proton in $^{23}_{12}$Mg
23. (a) 1.92×10^{17} nuclei (b) 1.09×10^{14} decays/s (c) 8.96×10^6 decays/s
25. 2.09 g
27. 1.0 h
29. 8.8×10^3 Bq
31. 1.72×10^4 y
33. 24.5%
35. 1.65×10^3 y
37. $^{12}_{6}$C, $^{4}_{2}$He, $^{14}_{6}$C
39. 4.28 MeV
41. (a) No, $\Delta m < 0$ (b) Yes, $\Delta m > 0$
43. $^{66}_{28}$Ni $\rightarrow$ $^{66}_{29}$Cu + $_{-1}^{0}e$ + $\bar{\nu}$ (b) 187 keV
45. 1.66×10^4 y
47. 17.4 MeV
49. (a) $^{10}_{5}$B (b) -2.80 MeV
51. $^{1}_{0}$n, $^{10}_{5}$B
53. (a) $^{3}_{2}$He (b) 4.02 MeV
55. 0.989 MeV
57. 165.4 MeV
59. 22.0 MeV
61. 24 decays/min
63. 1.88 MeV
65. 5.5×10^{-2} mCi
67. 230 y
69. 5.9 billion years
71. 12 mg

Chapter 30

1. 174 MeV
3. $^{144}_{54}$Xe (2 neutrons released), $^{143}_{54}$Xe (3 neutrons released), $^{142}_{54}$Xe (4 neutrons released)
5. 1.1×10^{16} fissions
7. 3.38 kg
9. 2.5×10^3 y

11. 2.9×10^3 km (1.8×10^3 mi)
13. 17.6 MeV
15. (a) $^{13}_{7}$N (b) $^{13}_{6}$C (c) $^{14}_{7}$N (d) $^{15}_{8}$O (e) $^{15}_{7}$N (f) $^{12}_{6}$C
17. (a) 4.0×10^{22} nuclei (b) 1.0×10^{10} J (c) 52 gallons
19. (a) 3.6×10^{30} J (b) 160 million years
21. 118 MeV
23. $\bar{\nu}_\mu$, and ν_e.
25. 67.5 MeV
27. The first decay is via the strong interaction; the second via the weak interaction.
29. First can occur via the weak interaction; the second violates conservation of baryon number.
31. (a) Lepton number not conserved (b) Lepton number not conserved (c) charge not conserved (d) baryon number not conserved (e) strangeness violated by 2 units.
33. (a) K$^+$ (b) Ξ^0 (c) π^0
35. (a) Answer is given in statement of problem. (b) The second reaction does not conserve strangeness
37. (a) ν_e (b) ν_μ (c) $\bar{\nu}_\mu$ (d) ν_μ and $\bar{\nu}_\tau$
39. 70.5 MeV
41. (a)

	K^0	d	$\bar{s}$	Total
strangeness	1	0	1	1
baryon number	0	$\frac{1}{3}$	$-\frac{1}{3}$	0
charge	0	$-\frac{1}{3}e$	$\frac{1}{3}e$	0

(b)

	Λ^0	u	d	s	Total
strangeness	-1	0	0	-1	-1
baryon number	1	$\frac{1}{3}$	$\frac{1}{3}$	$\frac{1}{3}$	1
charge	0	$\frac{2}{3}e$	$-\frac{1}{3}e$	$-\frac{1}{3}e$	0

43. (a) $-e$ (antiproton) (b) 0, antineutron
45. (a) Λ^0 or Σ^0 (b) π^- (c) K^0 (d) Ξ^-
47. (a) 1.61×10^{-35} m (b) 5.38×10^{-44} s
49. (a) $\lambda = 0.97$ mm (b) microwave region
51. 0.21 fm
53. 34 kg
55. $0.828c$

Note: Page numbers in *italics* refer to illustrations; page numbers followed by "t" refer to tables; page numbers in **bold** refer to definitions; and "n" following a page number refers to footnotes.